FEATURES AND BENEFITS

Algebra 2 and Trigonometry

■ The lessons and exercises were written by an **experienced author team,** led by Dr. Mary P. Dolciani, to provide honors students with a rigorous, comprehensive course of study in second-year algebra and trigonometry. See pp. iii–x (Table of Contents).

■ Algebra concepts and skills are developed using these time-tested and effective varieties of **exercises:**

Oral Exercises check students' understanding of basic lesson ideas. See p. 298.

Written Exercises, graded A, B, and C, range from straightforward exercises reinforcing lesson concepts to thought-provoking exercises that challenge better students. See pp. 298–299.

Numerous **tests and reviews** check students' mastery of algebra skills. See pp. 332–336.

Problems give students a chance to apply the problem solving methods discussed in the lessons. See pp. 50–54, 272–273.

NEW optional **Computer Exercises** following selected sections enable students to use a computer to help them learn algebra. See p. 285.

NEW Mixed Reviews test students' ability to solve problems encountered out of context. See p. 436.

■ **NEW** student learning aids:

More **worked-out examples** in the lessons and exercise sets help students apply lesson concepts to solve exercises and problems. See pp. 343–344.

Reading Algebra helps students read mathematics more effectively. See p. 212.

Preparing for College Entrance Exams gives students opportunities to practice on test items like those on college entrance exams. See pp. 289, 570.

■ Realistic **Applications** enhance students' knowledge. See pp. 698–699.

■ Challenging **Contest Problems** motivate students to sharpen their logical reasoning skills. See p. 523.

■ **Programming in Pascal** sections enable students to explore algebra topics using the Pascal programming language, which is frequently recommended for advanced studies. See pp. 337–339.

■ **On the Calculator** features extend algebra lesson concepts. See p. 472.

■ **Teaching aids** include the following:

Comprehensive **Teacher's Edition** provides teacher's commentary followed by annotated pupil book pages with side-column notes. See Contents, p. T3.

Tests on duplicating masters provide tests on groups of sections, chapter tests, and cumulative tests.

Solution Key contains worked solutions to pupil book exercises.

Teacher's Edition

Algebra 2
and Trigonometry

Mary P. Dolciani

John A. Graham

Richard A. Swanson

Sidney Sharron

Editorial Adviser
Andrew M. Gleason

Teacher Consultants
Sonja Jernigan
Carmen Hinds Lee
Ann C. Seeboth

HOUGHTON MIFFLIN COMPANY · Boston

Atlanta   Dallas   Geneva, Ill.   Palo Alto   Princeton   Toronto

Authors

Mary P. Dolciani Former Professor of Mathematical Sciences, Hunter College of the City University of New York

John A. Graham Mathematics Teacher, Buckingham Browne and Nichols School, Cambridge, Massachusetts

Richard A. Swanson Supervisor of Mathematics, Liverpool Central Schools, Liverpool, New York

Sidney Sharron formerly Supervisor, Los Angeles Unified School District, Los Angeles, California

Editorial Adviser

Andrew M. Gleason Hollis Professor of Mathematics and Natural Philosophy, Harvard University, Cambridge, Massachusetts

Teacher Consultants

Sonja Jernigan Mathematics Teacher, Northview High School, Dothan, Alabama

Carmen Hinds Lee Mathematics Teacher, Theodore Roosevelt High School, San Antonio, Texas

Ann C. Seeboth Mathematics Teacher, Queen Anne School, Upper Marlboro, Maryland

Acknowledgment The authors wish to thank Dr. Larry D. Wiley of the Nightingale-Bamford School in New York City for contributing the material on programming in the Pascal computer language; and Hector Hirigoyen of the Dade County Public Schools in Miami, Florida, for contributing the material on discrete mathematics.

Printed in U.S.A.

ISBN: 0-395-43059-3

ABCDEFGHIJ-RM-96543210/898

Contents

Using the Teacher's Edition

This Teacher's Edition includes nearly full-sized textbook pages annotated with answers. Time-saving references, suggestions, and extra examples and exercises appear in the adjacent side-columns, where they will be most useful. The simulated Teacher's Edition pages below illustrate the material in the side columns.

For each section in the textbook, page references are given for the corresponding **Teaching Suggestions** and **Related Activities** in the Lesson Commentary at the front of the Teacher's Edition.

Supplementary Materials provide references to the Tests that accompany the textbook.

Key Ideas state the major concepts presented in the lesson.

Chalkboard Examples provide additional examples to use in presenting the lesson to your students.

The **Common Error** sections alert you to errors that students often make and suggest how you can help your students avoid these errors.

Reading Algebra sections suggest ways you may help your students read algebra more easily and with greater comprehension.

Teaching Suggestions
p. T83

Related Activities
p. T84

Supplementary Material
Test 17

Key Ideas
Use the distributive property and the laws of exponents to multiply polynomials.

Chalkboard Examples
Express each product as a polynomial in simple form.

1. $3x(x^2 - 4)^2$
$3x(x^4 - 8x^2 + 16)$
$3x^5 - 24x^3 + 48x$

2. $(3 + 2x)(5 + 7x)$
$15 + 21x + 10x + 14x^2$
$15 + 31x + 14x^2$

Common Errors
Stress the necessity of like bases in theorems such as $b^m b^n = b^{m+n}$. Students may want to multiply bases (especially if they are numerical). For example, show that
$$4^2 \cdot 2^3 \neq 8^5.$$
In general, then,
$$b^m \cdot a^n \neq (ba)^{m+n}.$$

Reading Algebra
Have the students read the first paragraph of Section 6-1 and give a verbal explanation of what b^n or b to the nth power means. In words, b to the nth power is b used as a factor n times.

6–2 Multiplying Polynomials   SIMULATED PAGE

You can find the product of two polynomials by using the familiar axioms of addition and multiplication and the first law of exponents. For example, to find the product of the *binomial* $3x - 2$ and the *trinomial* $5x^4 - x^3 + 4x$, you can proceed as follows:

$(3x - 2)(5x^4 - x^3 + 4x) = 3x(5x^4 - x^3 + 4x) - 2(5x^4 - x^3 + 4x)$
$= 15x^5 - 3x^4 + 12x^2 - 10x^4 + 2x^3 - 8x$
$= 15x^5 - 13x^4 + 2x^3 + 12x^2 - 8x$

You write the product as a polynomial in simple form (page 42).

> To obtain the product of two polynomials, multiply each term of one of the polynomials by each term of the other, and then add all the monomial products.

You are less likely to make errors in adding like terms, if you use a vertical arrangement in multiplying.

$$
\begin{array}{r}
5x^4 - \quad x^3 \qquad\quad + 4x \\
3x - 2 \\
\hline
15x^5 - \quad 3x^4 \qquad + 12x^2 \\
-10x^4 + 2x^3 \qquad\qquad - 8x \\
\hline
15x^5 - 13x^4 + 2x^3 + 12x^2 - 8x
\end{array}
$$

Three special cases of binomial products that are useful to know are given here:

$$(a + b)^2 = a^2 + 2ab + b^2$$
$$(a - b)^2 = a^2 - 2ab + b^2$$
$$(a + b)(a - b) = a^2 - b^2$$

EXAMPLE Express each product as a polynomial in simple form.
a. $(3x - 2)(x + 4)$ **b.** $(4 + m)(4 - m)$
c. $(w - 5)^2$ **d.** $(3x - 4y^2)^2$

SOLUTION **a.** $(3x - 2)(x + 4) = 3x^2 + 12x - 2x - 8$
$= 3x^2 + 10x - 8$
b. $(4 + m)(4 - m) = 4^2 - m^2 = 16 - m^2$
c. $(w - 5)^2 = w^2 - 2(5w) + 5^2$
$= w^2 - 10w + 25$
d. $(3x - 4y^2)^2 = (3x)^2 - 2(3x)(4y^2) + (4y^2)^2$
$= 9x^2 - 24xy^2 + 16y^4$

One of the features of the Teacher's Edition not illustrated is **Problem Solving Strategies** (see p. 41). **Permission-to-reproduce tests** and **reviews** and a number of other useful features are located in the front of the Teacher's Edition. See Contents, p. T3, for a complete listing.

Oral Exercises

SIMULATED PAGE

Express each product as a polynomial in simple form.

1. $(2a - b)(5a + 3b)$
2. $(8w + 3)(w - 5)$
3. $(2 - 3m)(5 - m)$
4. $(6 - r)(6 + r)$
5. $(3v + 1)(3v - 1)$
6. $(r + 7s)^2$
7. $(4p - 5q)^2$
8. $(h - k^3)^2$
9. $(x^2 + y)^2$

1. $10a^2 + ab - 3b^2$
2. $8w^2 - 37w - 15$
3. $10 - 17m + 3m^2$
4. $36 - r^2$
5. $9v^2 - 1$
6. $r^2 + 14rs + 49s^2$
7. $16p^2 - 40pq + 25q^2$
8. $h^2 - 2hk^3 + k^6$
9. $x^4 + 2x^2y + y^2$

Written Exercises

Write each product as a polynomial in simple form.

A   1. $(x + 3y)(x - 2y)$ $x^2 + xy - 6y^2$ 2. $(6a + 5)(a + 3)$ $6a^2 + 23a + 15$ 3. $(t + 4)^2$
4. $(3n - 2)^2$ $9n^2 - 12n + 4$ 5. $(5 + s)(5 - s)$ $25 - s^2$ 6. $(2b + 1)(2b - 1)$
7. $(x^3 - 2y)(x^3 + 2y)$ $x^6 - 4y^2$ 8. $(3c + 5d)^2$ 9. $(2u^2 - 5v)^2$
10. $9x^7(x^3 - 4x)$ $9x^{10} - 36x^8$ 11. $(a^2 - b)(a^2 - 2b)$ 12. $(3t - 7)^2$
13. $(n^5 - 8n^3)6n^2$ $6n^7 - 48n^5$ 14. $(2m - 3)(m - 6)$ 15. $(d^5 + e^3)^2$
16. $(4 - 7h^3)^2$ $16 - 56h^3 + 49h^6$ 17. $(r^n - 5)(r^n + 5)$ $r^{2n} - 25$ 18. $(k^n - 3p^m)^2$
19. $(2x - 3)(2x^2 + 3x - 5)$ $4x^3 - 19x + 15$ 20. $(a + 2b)(3a^2 - ab - 7b^2)$
21. $(u - 4)(u^2 + 4u + 16)$ $u^3 - 64$ 22. $(5 - t)(25 + 5t + t^2)$ $125 - t^3$

B   23. $(z + 3)^3$ $z^3 + 9z^2 + 27z + 27$ 24. $(v - 4w)^3$
25. $(x - y)(x + y)(x^2 + y^2)$ $x^4 - y^4$ 26. $(2c - 3d)^2(2c + 3d)^2$
27. $(p + q)(p - q)(p^4 + p^2q^2 + q^4)$ $p^6 - q^6$ 28. $(h - 2k)^4$
29. Write a formula for simplifying a product that has the form

$$(a - b)(a^2 + ab + b^2).$$ $a^3 - b^3$

Check the formula you wrote by applying it to the products in Exercises 21 and 22.
30. Write a formula for expressing $(a + b)^3$ as a polynomial in simple form. Check the formula you wrote by applying it to the product in Exercise 23. $a^3 + 3a^2b + 3ab^2 + b^3$

Write each product as a polynomial in simple form.

C   31. $(x^n + y^n)(x^{2n} + x^ny^n + y^{2n})$ 32. $(r^{4n} - s^{3m})^2$ $r^{8n} - 2r^{4n}s^{3m} + s^{6m}$
33. $(a^{p+1} - a^2b^q)(a^{p-1} + b^q)$ $a^{2p} - a^2b^{2q}$ 34. $(v^2 - vw + w^2)(v^2 + vw + w^2)$ $v^4 + v^2w^2 + w^4$
35. $(a - b)(a^4 + a^3b + a^2b^2 + ab^3 + b^4)$ $a^5 - b^5$
36. $(z - 1)(z^9 + z^8 + \ldots + z + 1)$ $z^{10} - 1$

37. Show that $(x^n - x^{-n})^2 + 4 = (x^n + x^{-n})^2$.
38. Use the fact that $(a - b)^2 \geq 0$ to prove that the average of the squares of two real numbers is always at least as large as their product.

Suggested Assignments

Minimum
   206/1–17 odd, 19–30
Extended Alg. with Trig.
   206/7–33 odd
Enriched Alg.
   206/13, 19–31 odd, 33–38
Enriched Alg. with Trig.
   206/13, 19–31 odd, 33–38

Additional A Exercises

Write each product as a polynomial in simple form.
1. $(1 - x)(1 + x)$ $1 - x^2$
2. $(1 - x)^2$ $1 - 2x + x^2$
3. $(2c^3 + 7d^2)(3c - 2d)$
   $6c^4 - 4c^3d - 14d^3 + 21cd^2$

Mixed Review

1. Simplify $(5a^6)(-6a^{-5})$.
   $-30a$
2. Solve.
   $5x - y + z = 5$
   $3x + y - z = 3$
   $x + 2y - z = 3$ $\{(1, 2, 2)\}$
3. If $f(x) = \dfrac{x}{3x - 1}$, find $f(1)$.
   $\dfrac{1}{2}$

Additional Answers
Written Exercises

3. $t^2 + 8t + 16$
6. $4b^2 - 1$

Quick Quiz

Factor completely.
1. $6x^2 + 7x + 2$
   $(2x + 1)(3x + 2)$
2. $r^6 - 27s^3$
   $(r^2 - 3s)(r^4 + 3r^2s + 9s^2)$
3. $x^2 - 15x + 36$
   $(x - 12)(x - 3)$

Suggested Assignments for minimum, extended algebra with trigonometry, enriched algebra, and enriched algebra with trigonometry courses are given with each lesson and also in the **Assignment Guide** at the front of the Teacher's Edition.

You can use the **Additional A Exercises** to check whether your students are ready to start the Written Exercises on their own.

Mixed Reviews provide practice in skills taught in previous lessons and help keep these skills alive.

Answers that cannot be annotated on the textbook page appear in the side-column.

For each Self-Test in the textbook, there is a corresponding **Quick Quiz.**

Teaching the Course

The Teacher's Edition and Solution Key have been designed to help you teach algebra. For each chapter in the textbook, the Teacher's Edition provides Lesson Commentary and slightly reduced reproductions of student pages and annotated answers. Side columns next to the student pages present additional teaching aids. The Teacher's Edition also includes extra tests and reviews, an assignment guide, term projects and special Reading Algebra, Problem Solving Strategies, and Error Analysis sections.

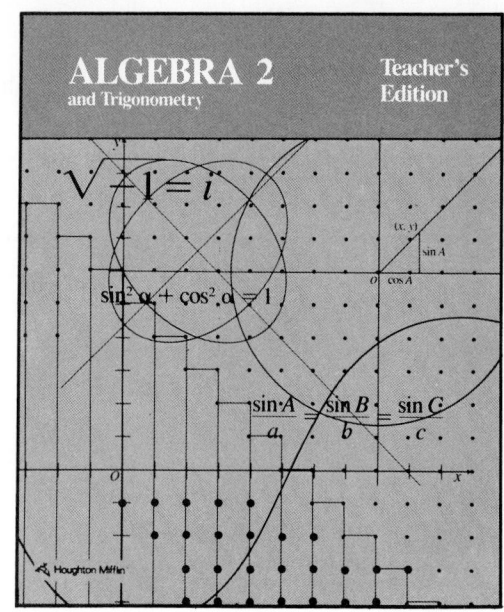

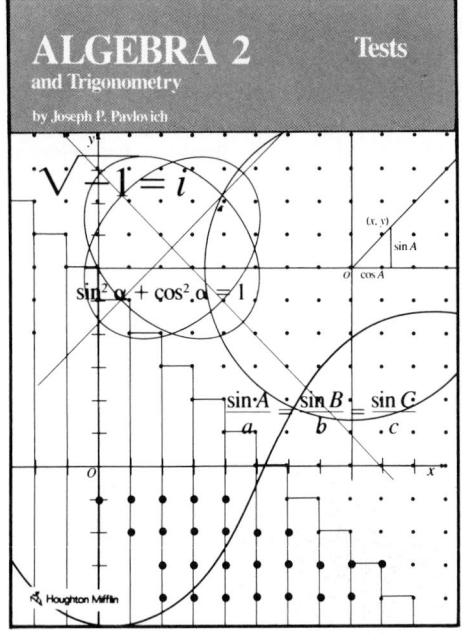

The optional Tests on duplicating masters contain tests on chapter subdivisions, chapters tests, and cumulative tests. The masters are keyed to the student textbook, and a separate Answer Key with answers annotated on reduced facsimiles of the tests is provided.

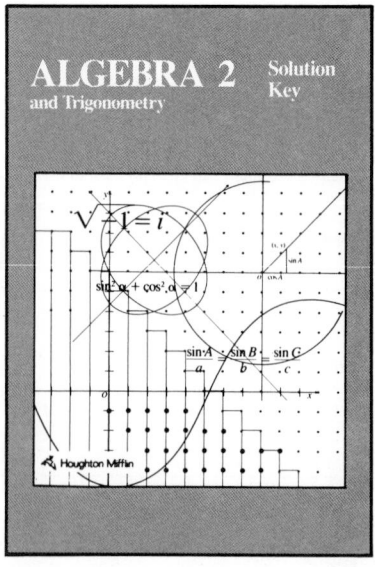

The Solution Key provides step-by-step solutions, including diagrams, for all written exercises in the student book.

Diagnostic Test

Chapters 1 and 2 of the student textbook review first-year algebra. You may wish to use this diagnostic test to determine what review material your students are able to omit. It is recommended that all students review Sections 1-3 through 1-5, 2-3, 2-4, and 2-6.

Classify each statement as true or false.

1. $7 \notin \{\text{positive integers}\}$

2. $\emptyset \subset \left\{\frac{1}{2}, \frac{1}{3}, \frac{1}{4}\right\}$ 1-1

3. If $A = \{1, 2, 3\}$ and $B = \{2, 3, 4, 5\}$, then $A \cap B = \{1, 2, 3, 4, 5\}$.

Solve each open sentence over $\{-2, -1, 0, 1, 2\}$.

4. $2n - 3 = 1$ **5.** $4 + c \neq 4$ 1-2

6. $z(z - 1) = 0$ **7.** $x^2 > 1$

Match each statement with the axiom or property that justifies the statement.

8. $3(n + 4) = 3n + 12$ **a.** Commutative axiom for addition   1-3

9. If $r = s$ and $s = t$, then $r = t$.   **b.** Symmetric property

10. $(6 + n) + 2 = 6 + (n + 2)$ **c.** Associative axiom for addition

11. If $2x + 1 = 7$, then $7 = 2x + 1$.   **d.** Distributive axiom

                e. Transitive property

Simplify.

12. $2r - (6 - r) + (-9)$ **13.** $5c[7 - 3(2 - 6d)]$ 1-6

14. Evaluate $(3 - r)[-2r + 3(1 + 2r)]$ for $r = -2$.

Simplify.

15. $\left(-\frac{2}{3}\right) \div \left(-\frac{1}{3}n + 5n\right)$ **16.** $4 \cdot 6 - 10 \div \frac{d}{2} + (-2)(5)$ 1-7

17. Evaluate $\left(\frac{1}{x} - 3\right)\left(\frac{x}{4} - 6\right)$ for $x = 3$.

Simplify.

18. $(3x^2 - xy + 5y^2) - (-x^2 + 2xy + 8y^2)$ 2-1

19. $(2c^3 - 6c^2 + c - 11) + (c^3 - 7c - 1)$

20. $[3y^2 - (6y - 2)] - [y^2 - (2y + 5)]$

Solve each equation over $\mathscr{R}$.

21. $y + 7 = -3$

22. $5d - (d - 3) + 9 = 7d$ 2-2

23. $\frac{2}{3}r = -10$

24. $\frac{-6a}{7} - \frac{4n}{7} = 8$

Solve each open sentence over $\mathscr{R}$ and graph its solution set.

25. $3x - 4 \leq 11$

26. $-5 < 2r - 1 \leq 5$ 2-5

27. $5 + 3y > 2$ or $-(5 + 3y) > 1$

Simplify.

28. $|5| - |-3|$

29. $|5 - 1| - 6|2 - 3|$ 2-7

Solve each open sentence.

30. $|n| \leq 3$

31. $|2y - 5| = 7$

Alternate Chapter Tests

(The starred problems may be used in place of the unstarred ones to individualize the tests or make them more challenging.)

Chapter Test 1

Let $A = \{0, 2, 4, 6, 8\}$, $B = \{1, 3, 5, 7, 9\}$, and $C = \{5, 6, 7, 8, 9\}$. List each set by roster.

1. $A \cup B$ **2.** $A \cup C$ **3.** $B \cap C$

4. List all the subsets of $\{1, 3, 5\}$.

Solve each sentence over $\{-2, -1, 0, 1, 2\}$. If the open sentence has no solution, so state.

5. $3y - 3 = 0$ **6.** $6 - z = -2$

7. $7a - 3a = 4a$ **8.** $\frac{x-1}{2}$ is an integer.

Graph each set of numbers on a number line.

9. $\{-3, -1, 0, 3, 5\}$ **10.** the set of all negative odd integers

State the axiom or property that justifies each statement. Assume each variable represents a real number.

11. $3(4 - b) = 12 - 3b$ **12.** $2x + 3y = 3y + 2x$

13. If $a = b$ and $b = 7$, then $a = 7$.

Simplify.

14. $(w - (-2)) + [-(-2w - 1)]$ **15.** $2[4 + (-3)h] + 5h$

16. $-8j[-(3k + 3m)]$ **17.** $-[-(4cd + 8d) + (-3c)(d + 5)]$

18. $-5f[g - (k - 7)]$ **19.** $10[12 - 3v + 2(4 - v)]$

20. $4 \cdot 7 \div 2 - 6 + 5 \cdot (-3)$ **21.** $(-18x)(-3y) \div \left[\frac{1}{2}(-6 - 3)\right]$

★**11.** $(4 \times 5) \times 2 = 4 \times (5 \times 2)$

★**15.** $\left[12\left(3 - \frac{1}{4}\right) - 15\left(\frac{2}{5} + \frac{1}{3}\right)\right]$

Chapter Test 2

Simplify.

1. $2(-2t^2 + 4t + 1) - (-3t^2 + 2t) + 2(5 - t)$
2. $[-2e^2 - (4e - 3f)] - [5f^2 - 2(4e + 3f)]$

Solve.

3. $y - 5(4 - y) = 20 - 2y$

4. $\frac{1}{2}(2x - 8) = 2(3x - 7)$

Solve for _n_.

5. $3ny = 5 + 7y$

6. $c - q = 2p^2n$

7. Find three consecutive integers such that the sum of the second and the third is four less than three times the first.

8. Each of the two congruent sides of an isosceles triangle is 7 cm longer than half the base. Find the length of each side of the triangle if its perimeter is 50 cm.

Solve each inequality and graph its solution set.

9. $3x - 4 > \frac{x}{2} + 1$

10. $3 - \frac{1}{3}(2x + 1) \le 2(6 - x)$

11. $-2 < \frac{1}{3}(2p - 2) < 4$

12. $5p - 4 < 6$ or $1 - 3p \le -8$

13. Two pieces of carpet, each 25 ft wide, are placed on the floor of a large room. One piece is 10 ft longer than the other. Find the least length the shorter piece must be if the total area covered by the carpets is at least 1750 ft^2.

14. Use an indirect proof to prove if $a \cdot c > b \cdot c$ then $a > b$ (a, b, and $c > 0$).

Solve each open sentence and graph its solution set.

15. $2|2x + 1| \le 10$

16. $\left|\frac{2 - t}{4}\right| \le 1$

★7. The difference of the squares of two consecutive even integers is 22 greater than the smaller of the two integers. What are the integers?

★15. $1 < |x - 4| < 5$

Chapter Test 3

If $f: x \rightarrow x^2 + 1$ and $g: x \rightarrow \left|\frac{x}{4}\right|$, find the value of each of the following.

1. $f(-3)$
2. $g(12)$
3. $[f \circ g](-8)$
4. $[g \circ f](2)$

Graph each relation over the domain $\{-1, 0, 1, 2, 3\}$. Is the relation a function?

5. $x + y = 3$
6. $|y| = |x + 2|$
7. $6y - 3x = 6$
8. $y = |2x| - x$

Graph each equation.

9. $y = [x] + 2$
10. $y = |x - 2| + 1$

Graph each inequality as a shaded region on a coordinate plane.

11. $y \geq |x| - 2$
12. $3x - 2y \geq 6$

Determine the slope, if any, of the line through each pair of points.

13. $(1, 3), (2, 5)$
14. $(-2, 5), (1, -4)$

Determine an equation in the form $y = mx + b$ of a line satisfying the given conditions.

15. passes through the points $(2, 5), (4, -1)$
16. passes through the point $(-3, -3)$ and is parallel to the line $3y - 2x = 3$
17. Solve for a if the points $\left(\frac{4}{3}, -5\right)$ and $(-4, a)$ are in a direct variation.
18. In a certain city, the amount of property tax paid varies directly with the assessed value of the property. If a family whose house is assessed at $80,000 pays $2000 in property tax during a certain year, what is the assessed value of a house whose owners paid $1850 in taxes during the same year?

★4. If $f(x) = 5 - x^2$ and $g(x) = |x + 2|$, find $f(g(-3))$ and $g(f(-3))$.
★9. $|x| = |y|$
★13. (b, b) and $(3b, -b)$
★18. A granola recipe that serves 8 calls for 750g of oatmeal and 150g of coconut. How many people can be served if a total of 675g of both oatmeal and coconut is used?

Chapter Test 4

Determine the apparent solution set of the system by graphing each system of equations. If the system has exactly one apparent solution, verify that the coordinates of the ordered pair satisfy all three equations.

1. $3x - 9y = 6$
$x - 3y = 2$
$3y - x = -5$

2. $-4x - 2y = 6$
$-2x = 4y$
$4x - 4y = -12$

Solve each system by either the linear-combination method or the substitution method, whichever seems simpler.

3. $4x - 4y = -12$
$-6x + 2y = -10$

4. $\dfrac{(x + 1)}{3} = \dfrac{(y - 2)}{5}$
$2y - 4x = 6$

Use Cramer's Rule to solve each system. If the system is inconsistent or has an infinite solution set, so state.

5. $3x - y = 2$
$2y = 5x$

6. $-\dfrac{2}{3}x + \dfrac{1}{4}y = -\dfrac{1}{2}$
$8x - 3y = 5$

7. $5x - 2y = -3$
$2x + 5y = 7$

8. $5y = \left(-\dfrac{10}{3}\right)x + 2$
$\dfrac{6}{5} = 2x + 3y$

9. With a given head wind, a plane can fly 5600 km in 8 h. Flying in the opposite direction with the same wind blowing, the plane can fly the same distance in 1 h less. Find the plane's air speed and the speed of the wind.

10. Pete earns $4.75/h and Beth earns $5.00/h. Together, they worked a total of 45 h and earned a total of $220.00. How many hours did each work?

Graph the solution set of each system in a coordinate plane.

11. $3x + y - 1 > 0$
$3x - 4y \geq -8$

12. $-2 < x < 9 - 3y$
$y \geq -5$

13. A baker, selling pumpernickel bread and rye bread, has shelf space for 120 loaves of bread and sells all of the bread during one day. Customers buy at least three times as many loaves of pumpernickel as loaves of rye. If the baker makes a profit of 9¢ on each loaf of rye bread and 8¢ on each loaf of pumpernickel bread, how many of each type of bread should the baker sell each day to earn the maximum profit?

★**5.** $bx - ay = 0$
$ax - by = a - b$

★**9.** Gormley Florists sells long-stemmed red roses at 1 for $2.00 or 4 for $6.00. In one day there were 46 roses sold and $76.00 collected. How many single long-stemmed red roses were sold?

★**11.** $|x - y| \leq 4$
$-4 \leq x \leq 4$

Chapter Test 5

1. Sketch the coordinate box of $(2, 4, -3)$.

2. Sketch the triangle in space whose vertices have the coordinates $(0, -3, 0)$, $(-3, 0, 0)$, and $(0, 0, 4)$.

3. Graph the part of the plane that has the given traces.
   xy-trace: $6x + 5y = 30$, $z = 0$
   yz-trace: $y + 3z = 6$, $x = 0$
   xz-trace: $2x + 5z = 10$, $y = 0$

4. Solve the system: $\begin{aligned} 2x + y - z &= 1 \\ 4x - 3y + 2z &= 8 \\ 3x + 2y + z &= 0 \end{aligned}$

Solve each system by using Cramer's Rule.

5. $\begin{aligned} x + 2y + 3z &= -7 \\ 2y - 5z &= 8 \\ 2x - 3y + z &= 3 \end{aligned}$

6. $\begin{aligned} 3x - 2y - z &= -5 \\ x + 3y + 4z &= 7 \\ 2x + 3y - 2z &= -6 \end{aligned}$

7. Points A, B, C, and D are the endpoints of legs of a cross-country course that is 10 km long. A run from B to D and back to A is 17 km, while a run from D to C, back to D, and then to B is 11 km. How long is each leg of the course?

8. Expand by minors of the second row and then evaluate:
$$\begin{vmatrix} 4 & -1 & 6 \\ 5 & -2 & -3 \\ 2 & 3 & 1 \end{vmatrix}$$

9. Evaluate: $\begin{vmatrix} 1 & 4 & 5 & -3 \\ -1 & 0 & 3 & -2 \\ 0 & 2 & 0 & 1 \\ -2 & 6 & 1 & 0 \end{vmatrix}$

★7. A pyramid has a base that is an isosceles triangle with perimeter 7 cm. The lateral edges of the pyramid are all equal. If the triangular face whose base is also the base of the isosceles triangle has perimeter 9 cm and a triangular face whose base is also one of the legs of the isosceles triangle has perimeter 11 cm, find the length of a lateral edge of the pyramid.

★8. Prove:

$$\text{For } A = \begin{bmatrix} a_1 & b_1 & c_1 & d_1 \\ 0 & b_2 & c_2 & d_2 \\ 0 & 0 & c_3 & d_3 \\ 0 & 0 & 0 & d_4 \end{bmatrix}, \; |A| = a_1 b_2 c_3 d_4$$

★9. Use the properties of determinants to prove:

$$\begin{vmatrix} a_1 & b_1 & c_1 \\ a_2 & b_2 & c_2 \\ ka_1 & kb_1 & kc_1 \end{vmatrix} = 0$$

Chapter Test 6

Write an equivalent expression using only positive exponents. Assume that no variable that appears in a denominator or with a negative exponent is zero.

1. $(2^{-1}cd^{-4})(8c^{-3}d^4)$

2. $\dfrac{(a^2b)^{-4}}{(abc)^{-3}}$

Write each product as a polynomial in simple form.

3. $(3x - 2)(4x^2 - 7x + 7)$

4. $[(4y + 3)(4y - 3)]^2$

Factor each polynomial completely.

5. $2x^2 - 10x + 12$

6. $27y^3 - 1$

Solve.

7. $3m^2 - 22m = 16$

8. $(5x + 3)(x - 4) = 11x$

9. A rectangle that is 5 cm longer than it is wide has an area of 84 cm². Find its width.

10. Solve $x^3 + 10x^2 + 25x \leq 0$ over $\mathcal{R}$ and graph its solution set.

11. Simplify $\dfrac{3x^2 - 12x - 63}{6x^2 - 294}$.

12. Express $\dfrac{y^2 + 3 - 5y^3 + 6y^4 + 6y}{2y - 1}$ as a sum by using division.

13. Simplify $\dfrac{x^4 - 16}{x + 2} \div \dfrac{x^2 + 4}{7}$.

14. Simplify $\dfrac{n^4 - 1}{n^2 + 1} + \dfrac{(n - 1)^2}{n^2 - 1}$.

15. How much of a 15% salt solution should be added to 600 g of a 6% salt solution to produce a 10% salt solution?

16. Solve $\dfrac{n}{n - 5} + \dfrac{17}{25 - n^2} = \dfrac{1}{n + 5}$.

★2. Rewrite $\dfrac{u^{-2} + v^{-2}}{u^{-2} - v^{-2}}$ using positive exponents, then simplify.

★16. Solve $\dfrac{-2k^2}{k^3 - 1} + \dfrac{1}{k - 1} = \dfrac{-1}{k^2 + k + 1}$.

Chapter Test 7

1. If y is directly proportional to x^2 and y is 12 when x is 1, determine the value of y when x is 3.

2. The distance that a ball will roll down an inclined plane is directly proportional to the square of the time it rolls. If a ball rolls 36 ft in 3 s, how far will it roll in 4 s?

3. Solve $x^2 - 1.44 = 0$ over $\mathcal{R}$.

4. Find the rational roots, if any, of $y^3 - 3y^2 + y - 3 = 0$.

5. Express 79,070 in standard notation.

6. Find a one-significant-digit estimate of $\dfrac{397 \times 0.00584}{8250 \times 0.973}$.

7. Express $\dfrac{5}{11}$ as a decimal.

8. Express $1.6\overline{81}$ as a fraction in lowest terms.

9. Find a rational number and an irrational number between $\dfrac{2}{3}$ and 0.68.

Simplify. Assume all radicals denote real numbers.

10. $\dfrac{6a^2}{\sqrt[4]{8a^2}}$

11. $\dfrac{3\sqrt[4]{144}}{5\sqrt[4]{9}}$

12. $\sqrt{\dfrac{(4x^2)^2}{48x^3}}$

13. $\sqrt{4a^2b + 4ab^2} \cdot \sqrt[6]{64a^9}$

14. $m\sqrt[4]{81m^4} + \sqrt{m^4a} + m\sqrt{64m^2a}$

15. $\dfrac{\sqrt{3} + 5}{\sqrt{3} - 2}$

16. Solve $y - 2 - \sqrt{4y - 3} = 0$ over $\mathcal{R}$.

17. Express $\dfrac{6i}{\sqrt{-36}} + i\sqrt{-12}$ in simplest form as a real number or pure imaginary number.

18. Express $\bar{a} + b$ as a complex number, given $a = 7 + 2i$ and $b = \dfrac{1}{5}i$.

19. Express $\dfrac{3 - 2i}{3 + 5i}$ in the form $a + bi$.

★15. Find the ratio of the surface area of a sphere to the lateral area of a cone that can be inscribed in the same cube as the sphere.
SA sphere $= 4\pi r^2$ LA cone $= \pi r\sqrt{r^2 + h^2}$

★16. Solve $\sqrt{2x + 6} - \sqrt{x - 1} = 2$ over $\mathcal{R}$.

Chapter Test 8

1. Give the first four terms of an arithmetic sequence in which $a_1 = -2$ and $a_{n+1} = a_n + 3$.

2. Specify the sequence 8, 4, 0, −4 . . . both explicitly and recursively.

3. Find the first term of an arithmetic sequence in which the fifth term is 63 and the eighth term is −105.

4. Find the sum of the series $5 + 17 + 29 \ldots + 101$.

5. Find the common difference in an arithmetic series in which $S_6 = 78$ and $a_1 = 3$.

6. Find the seventh term of the following geometric sequence:
   $-1216 + 608 - 304 + \ldots$.

7. A ski shop is having an end of season sale on ski boots. During the first week of the sale, the price of a certain pair of ski boots is $100. At the start of each new week the price is cut to $\frac{7}{10}$ of what it was the previous week. What will a pair of ski boots cost at the beginning of the fourth week?

8. Find the geometric mean of 6 and 24.

9. Find the value of $\sum_{n=1}^{5} 7\left(\frac{1}{2}\right)^{n-1}$

10. Give the general formula for $|L - a_n|$ for an infinite sequence with $a_n = \frac{n^2}{n^3}$ and $L = 0$.

11. Convert $0.\overline{450}$ to a fraction in lowest terms.

★3. Insert four arithmetic means between 11 and 31.

★7. If a car that was purchased originally for $5120 has a book value of $1620 at the end of 4 years, what was the annual rate of decrease in the value of the car?

★9. Find the sum of all the integral powers of 2 between $\frac{1}{4}$ and 128, inclusive.

★11. Convert $0.\overline{129}$ to a fraction in lowest terms.

Chapter Test 9

1. Solve $x^2 - 6x = -1$ over $\mathcal{C}$ by completing the square.

2. Solve $16v^2 = 1 - 20v$ over $\mathcal{C}$ by using the quadratic formula.

3. How many real or complex roots does the equation $x^2 + 5x + 2 = 0$ have? Tell whether any real roots are rational or irrational.

4. Give the sum and the product of the roots of $3x^2 + 18x + 2 = 0$.

5. Write a quadratic equation with integral coefficients whose solution set is $\{1 + \sqrt{2}, 1 - \sqrt{2}\}$.

6. Find a function in the form $y = a(x - h)^2 + k$ whose graph has vertex $(3, 7)$ and passes through the point $(2, 9)$.

7. Give the equation of the axis of symmetry and the coordinates of the vertex of the graph of $x^2 - 6x + 9 = y$. Sketch the graph.

8. **a.** Sketch the function related to $x^2 - 5x \leq 6$.
   b. Give the solution set of the inequality.

9. For $P(x) = x^3 + 3x^2 - 5x + 2$, use synthetic substitution to find
   (a) $P(-1)$ and **(b)** $P(2i)$.

10. For $P(x) = 3x^3 - 7x^2 - 20x + 3$, use synthetic division to divide $P(x)$ by $x - 3$. Express the result as an equation whose left side is $\dfrac{P(x)}{x - 3}$.

11. Solve $x^4 - 4x^3 + 7x^2 - 16x + 12 = 0$ over $\mathcal{C}$ given that $2i$ is a root.

12. Determine the possible number of positive roots and negative roots of $x^3 + x^2 - 3x - 3 = 0$. Summarize the results in a table.

13. Find all intervals of length one-half unit that contain a zero of $P(x) = x^3 - 3x^2 + 3$.

★5. Find a if $2 - i$ is a root of the equation $ax^2 - 4ix + 10 = 0$.

★6. Find a function of the form $y = a(x - h)^2 + k$ whose graph has vertex $(1, -3)$ and passes through the point $(3, 5)$.

★8. **a.** Sketch the function related to $3x^2 + 4x > 2$.
   b. Give the solution set of the inequality.

★10. Determine the value m so that $x + 2$ will be a factor of $3x^3 + 5x^2 + mx - 2 = 0$.

Chapter Test 10

1. Determine the coordinates of A if the midpoint of $\overline{AB}$ is the point $(3, 2)$ and B is the point $(5, 1)$.

2. Determine an equation of the line containing $(5, 3)$ and perpendicular to the line passing through $(-3, 1)$ and $(-2, 2)$.

3. Determine the center and radius of the circle with equation $x^2 - 4x + y^2 - 3y = 0$

4. Sketch the graph of $x = 3y^2 + 6y - 2$. Identify the vertex and the axis of symmetry.

5. Sketch the graph of $\dfrac{(x + 1)^2}{4} + \dfrac{y^2}{9} = 1$.

6. Sketch the graph of $4x^2 - 16y^2 = 64$. Label the equations of the asymptotes on the graph.

7. If x varies directly as y^2 and inversely as z, and $x = 2$ when $y = 3$ and $z = 18$, find the value of x when $y = 5$ and $z = 10$.

8. Solve over $\mathcal{R}$ by graphing:
$x^2 - y^2 = 9$
$2y - 3 = x$

9. Solve over $\mathcal{C}$ by substitution:
$4x^2 + y^2 = 16$
$y = 2x - 4$

10. Solve over $\mathcal{C}$.
$9x^2 - 4y^2 = 49$
$x^2 + y^2 = 17$

★3. Determine the equation of the circle whose diameter has endpoints $(3, -1)$ and $(7, -3)$.

★4. State, in the form $y = ax^2 + bx + c$, an equation for the parabola with focus $(3, -\frac{1}{2})$ and directrix $y = -\frac{3}{2}$.

★5. Find an equation of the ellipse symmetric to the origin whose major axis is on the x-axis and is twice as long as it minor axis, and such that $(8, 3)$ is on the ellipse.

★10. Solve over $\mathcal{C}$.
$x^2 + y^2 = 8$
$5x + y^2 = 14$

Chapter Test 11

Express in simplest radical form.

1. $\dfrac{\sqrt[12]{81}}{\sqrt[9]{27}}$

2. $\sqrt[10]{32} \cdot \sqrt[18]{64}$

3. Solve $\left(\dfrac{1}{8}\right)^{4x-9} = 64^{3x+1}$

4. Find the inverse function of $y = \dfrac{2}{5}x - 2$. Graph both the function and its inverse.

Solve for x.

5. $\log_{\frac{1}{5}} x = -4$

6. $\log_x \dfrac{1}{243} = -5$

7. $\log_7 5x = 2 \log_7 20 - (\log_7 4 + \log_7 10)$

8. $\log_2 x + \log_2 (x - 6) = 4$

9. $13x^7 = 143$

10. $\sqrt[5]{19x^3} = 220$

11. $3^{5x} = 62$

12. $14^{2x+1} = 15^{3x}$

13. A colony of bacteria consists of 2.8×10^5 bacteria at 2 PM. If the bacteria triples every 15 minutes, how many bacteria will there be at 3:30 PM?

14. Find the rate of interest on an account in which a $2500 investment has grown to $2790 in 1 year if the interest on the account is compounded continuously.

★3. Solve $9^{x^2-1} = \left(\dfrac{1}{27}\right)^x$.

★7. Solve $\log_{10} (x + 4) + \log_{10} (x + 1) = 1$.

★12. $5^{x+2} = 4^{x-1}$

★13. What amount of money, to the nearest dollar, compounded quarterly at 8% will grow to $6120 in 9 years?

Chapter Test 12

1. How many 3-letter code words can be formed from the letters in the word NUMBERS?

2. Evaluate $_9P_4$.

3. How many different permutations exist of all the letters in the word ADDITION?

4. Given a set with 9 elements. How many 4-element subsets are there?

5. How many different hands consisting of 4 jacks and 1 ace can be chosen from a standard 52-card bridge deck?

6. Find the third term of $(\frac{1}{3}x - 1)^4$.

7. Use Pascal's triangle to expand $(x - 2)^5$.

8. Three coins are tossed. List the sample space.

Three marbles are drawn at random from an urn containing 8 black, 7 white, and 5 red marbles.

9. What is the probability that none of them is red?

10. What is the probability that at least one of them is white?

11. A drawer contains 8 blue socks, 4 green socks, and 6 brown socks. If you choose 2 socks at random, what is the probability that they are both brown?

★5. How many different 5-digit numbers can be formed using the digits 1 through 9 without repetition, if each number must contain 3 even and 2 odd digits?

★11. If 3 letters are chosen at random from the 26 letters of the alphabet, what is the probability that they are all vowels (a, e, i, o, u) and are chosen in alphabetical order?

Chapter Test 13

1. Find the values of x and y for which the following statement is true:
$$[3x \quad 2y] - [2x \quad 6] = [7 \quad -14]$$

2. Solve for the matrix X:
$$X - \begin{bmatrix} -3 & 2 & 0 \\ 4 & -2 & 6 \end{bmatrix} = \begin{bmatrix} -6 & 3 & -2 \\ 0 & -5 & 1 \end{bmatrix}$$

3. Find the 2×2 matrix X that satisfies the following equation:
$$3X - 3\begin{bmatrix} -6 & 2 \\ 3 & -1 \end{bmatrix} = 4\begin{bmatrix} 3 & -3 \\ 0 & -3 \end{bmatrix}$$

4. Find the product: $\begin{bmatrix} 2 & 0 & 1 \\ 1 & 0 & -3 \end{bmatrix}\begin{bmatrix} -2 & 6 \\ 2 & 1 \\ 8 & 0 \end{bmatrix}$

5. For $A = \begin{bmatrix} 2 & -1 \\ 0 & 3 \end{bmatrix}$ and $B = \begin{bmatrix} 1 & -2 \\ 0 & 4 \end{bmatrix}$, compute $(2A)B$ and $A(2B)$ and state whether or not they are equal.

6. Solve for X: $\begin{bmatrix} 1 & -3 \\ -2 & 5 \end{bmatrix}X = \begin{bmatrix} -10 & -3 \\ 3 & 4 \end{bmatrix}$

7. Find a matrix equation of the translation of the plane that transforms the point $P(-3, 6)$ into the point $P'(-2, 1)$.

8. Use the matrix equation of Exercise 7 to find the image of $Q(-3, 0)$.

9. Find the coordinates of the image P' of the point $P(0, 2)$ under the linear transformation $\begin{bmatrix} x' \\ y' \end{bmatrix} = \begin{bmatrix} 6 & -2 \\ 0 & 1 \end{bmatrix}\begin{bmatrix} x \\ y \end{bmatrix}$.

10. Find the coordinates of the preimage P of the point $P'(3, -2)$ under the linear transformation $\begin{bmatrix} x' \\ y' \end{bmatrix} = \begin{bmatrix} 2 & 4 \\ 1 & 3 \end{bmatrix}\begin{bmatrix} x \\ y \end{bmatrix}$.

★4. Solve for x, y, and z: $\begin{bmatrix} -1 & 0 & 2 \\ 0 & -2 & 1 \\ 3 & 4 & 1 \end{bmatrix}\begin{bmatrix} x \\ y \\ z \end{bmatrix} = \begin{bmatrix} 6 \\ 5 \\ -1 \end{bmatrix}$

★5. Show that if $A = \begin{bmatrix} a & 0 \\ 0 & b \end{bmatrix}$ and $B = \begin{bmatrix} c & 0 \\ 0 & d \end{bmatrix}$, then $AB = BA$.

★6. Show that if $A = \begin{bmatrix} a & 1-a \\ 1+a & -a \end{bmatrix}$ for any real number a, then $A = A^{-1}$.

★10. Describe the mapping of the plane under the linear transformation
$$\begin{bmatrix} x' \\ y' \end{bmatrix} = \begin{bmatrix} 1 & -3 \\ -2 & 6 \end{bmatrix}\begin{bmatrix} x \\ y \end{bmatrix}.$$

Chapter Test 14

1. If a wheel travels 7.4 m as it makes 3.5 revolutions, find the diameter of the wheel to the nearest 0.01 m. Use $\pi \approx 3.14$.

2. Find the length of the arc on a circle with radius 15 cm that is intercepted by a central angle of $\frac{7\pi}{6}^{R}$. Use $\pi \approx 3.14$.

3. Express $135°$ as a radian measure using π.

4. Express $\frac{\pi}{3}^{R}$ as a degree measure.

5. If $\sin \alpha = \frac{5}{13}$ and $\cos \alpha < 0$ find $\cos \alpha$.

6. If $\frac{3\pi}{2} < x < 2\pi$, and $\cos x = \frac{1}{\sqrt{2}}$, find x.

7. Find all the values of $m(\alpha)$ for which the terminal side of α in standard position passes through $(-4, 4\sqrt{3})$.

8. Find a four-decimal place approximation for $\cos 2.4^{R}$.

9. Find the measure of α in degrees and minutes (to the nearest 10') for the first quadrant angle, such that $\cos \alpha = 0.9381$.

10. Find the amplitude of the function $y = 2 \sin x$ and sketch its graph over the interval $-\pi \le x \le \pi$.

11. Sketch the graph of $y = 2 \sin \frac{1}{2}x$ over one fundamental period.

12. Find the values of the other five trigonometric functions of α if $\sin \alpha = \frac{\sqrt{3}}{2}$ and $\cos \alpha < 0$.

13. If $\triangle ABC$ is a right triangle with $m(C) = 90°$, the length of $\overline{AC} = 8$, and the length of $\overline{AB} = 17$, find $m(B)$ to the nearest 10 minutes.

★5. Find $\sin \alpha$ if the terminal side of α in standard position intersects the unit circle in the point (u, v) in Quadrant I such that $u = 2\sqrt{2}v$.

★11. Find the amplitude and the fundamental period of the function
$$y = \pi \sin \frac{\pi}{2}x.$$

Chapter Test 15

1. Express $\sec^2 \alpha$ in terms of $\cos \alpha$.

2. Prove that $(1 + \cos \alpha) \dfrac{\cos \alpha - 1}{\cos^2 \alpha} = (\sin^2 \alpha)(-\sec^2 \alpha)$ and state any necessary restrictions.

3. Find $\sin (x - y)$, given that $\sin x = \dfrac{5}{13}$, $2\pi < x < \dfrac{5\pi}{2}$, and $\sin y = \dfrac{5}{13}$, $0 < y < \dfrac{\pi}{2}$.

4. Use a formula for the cosine of a sum or difference to evaluate $\cos \dfrac{11\pi}{12}$.

5. Find $\cos 2x$ if $\cos x = 0.85$.

6. Given $\tan x = -0.75$ and $\dfrac{\pi}{2} < x < \pi$, find $\tan \dfrac{x}{2}$.

7. Prove that $\dfrac{\cos 2x}{\tan x} + 2 \sin x \cos x = \cot x$ and state any necessary restrictions.

8. In $\triangle ABC$, if $b = 9$, $c = 17$, and $m(A) = 120°$, find a to the nearest tenth.

9. In $\triangle ABC$, if $m(A) = 30°$, $m(B) = 75°$, and $b = 16$, find a to the nearest tenth.

★5. Find $\sin 2\alpha$ if $\sin \alpha = \dfrac{3}{4}$ and $90° < m(\alpha) < 180°$.

★7. Prove that $\dfrac{1 - \cos 2\alpha}{\sin^2 \dfrac{\alpha}{2} \cos^2 \dfrac{\alpha}{2}} = 8$.

★8. Find the length of the base of an isosceles triangle to the nearest unit if each leg has length 14 cm and a median to a leg has length 9 cm.

T25

Chapter Test 16

1. Find the value of $\text{Sin}^{-1}\left(\cos \frac{5\pi}{3}\right)$.

2. Solve $\cot x \cos x = \sin x$ over $\mathcal{R}$.

3. Transform $x + y - 6 = 0$ into an equation in polar coordinates.

Let $z_1 = 4(\cos 225° + i \sin 225°)$ and $z_2 = 2(\cos 75° + i \sin 75°)$.

4. Express $z_1 z_2$ in the form $a + bi$.

5. Express $\frac{z_1}{z_2}$ in the form $a + bi$.

6. Express $(z_1)^3$ in the form $a + bi$.

7. Find the three cube roots of $27(\cos 90° + i \sin 90°)$ in the form $a + bi$.

8. For the vectors $\mathbf{u}$ and $\mathbf{v}$ such that $\|\mathbf{u}\| = 5$ with direction angle $-20°$ and $\|\mathbf{v}\| = 8$ with direction angle $85°$, find $\|\mathbf{u} + \mathbf{v}\|$ to the nearest tenth and the direction angle of $\mathbf{u} + \mathbf{v}$ to the nearest degree.

9. An airplane flies due east with an air speed of 400 km/h with a 50 km/h wind blowing with a bearing of 214°. What is the ground speed (to the nearest km/h) and the true course (to the nearest tenth of a degree) of the airplane?

★2. Solve for $0° \le m(\alpha) < 360°$: $\sin 2\alpha - \cos 2\alpha - 1 = 0$

★3. Sketch the graph of $r = \sin 2\theta$.

★6. Express the four fourth roots of $-8 + 8i\sqrt{3}$ in the form $a + bi$.

Chapter Test 17

1. Find the mean, median and mode of the following data. The number of days in a year that employees called in sick: 3, 12, 5, 0, 8, 1, 3, 7, 9

2. Find the variance and the standard deviation of the following data. Scores on a math test: 90, 82, 91, 75, 87, 85, 80, 90

3. Use the table on page 721 to find the value of $P(-1.5 < z < 1)$.

4. The mean age of the graduates from college is 24 with a standard deviation of 2.5 years. What percent of the graduates are over 29?

5. An appliance dealer claims that 95% of its dishwashers will last for eight years. Of 30 dishwashers sold eight years ago, 23 are still working. Use the 95% confidence interval to show that the dealer's claim is unlikely.

6. On a given day, 16 out of 150 students were late. Find a 95% confidence interval for the number of students that will be late on any given day.

7. Draw a scatter diagram for the following paired set of data points.

cost of car ($1000's)	5	7	8	10	15
life of car	6	4	5	7	8

8. For the set of data in Exercise 7, find the correlation coefficient, r.

★2. Lee and Phil scored 92 and 76 respectively, on a history test. Their z-scores were 1.4 and −1.8. What were the mean and the standard deviation for this test?

★3. Find the percent of the area under the standard normal curve that is less than $z = 0.4$.

Cumulative Tests

Test for Chapters 1–4

Simplify.

1. $-(-a + 2) + a$
2. $2(x + y) - [3x - (x + y)]$

State the axiom or property that justifies each statement.

3. $2(x - 5) = 2x - 10$
4. $a(bc) = (ab)c$
5. $-(x + y) = (-x) + (-y)$
6. $a + (-a) = 0$

Solve.

7. $7y + 4 - 2(y - 1) = 21$

8. $|5 - 2z| = 3$

9. Solve for x: $4x + 3a = 5(2b - x)$

10. Find three consecutive integers such that the sum of the first and three times the second is five less than the third.

Solve and graph the solution set.

11. $3(z - 3) + 3 > 4z - 3$

12. $-1 \leq 3 - 2a < 5$

13. $|4y + 2| < 6$

14. $\left|\dfrac{2x - 1}{5}\right| \geq 1$

If $f(x) = 2x^2$ and $g(x) = \dfrac{3}{x + 1}$, find the value of each of the following.

15. $f(-5)$

16. $[f \circ g](-2)$

17. $[g \circ f](-2)$

Graph each open sentence.

18. $y = |x + 3| + 2$

19. $4x - 2y < -8$

20. Determine the slope of the line through the points $(-2, 3)$ and $(3, -7)$.

21. Determine an equation in the form $y = mx + b$ of the line passing through $(-2, 5)$ and parallel to the line $2x + y = 6$.

22. If the property tax on a house assessed at $80,000 is $1800, what is the property tax on the house next door that is assessed at $90,000?

Solve each system.

23. $3x + y = 1$
$5x - 2y = 9$

24. $6x + 5y = 2$
$10x - 7y = 11$

25. $x - y = 4$
$-8 - 2y = -2x$

26. Graph the system: $x - y \geq -2$
$-3 < y < 3x + 1$

Test for Chapters 5–8

Solve each system.

1. $2x - y - z = 1$
$x + 2y + z = 0$
$3x - y - 2z = -1$

2. $x - 2y - z = 3$
$2x + y - 2z = 1$
$2x + 3y = 3$

3. A coin bank contains a total of 24 nickels, dimes, and quarters, with a total value of $2.90. If there are three times as many dimes as there are nickels and quarters combined, how many of each kind of coin are there?

Simplify.

4. $(a^3b^{-2}c^4)^2(ab^{-2}c^2)^{-3}$

5. $(x - 2)(x^2 + 2x - 3)$

6. $\dfrac{x}{x - 4} - \dfrac{x + 1}{x + 4} + \dfrac{x^2 + 8}{x^2 - 16}$

Solve.

7. $6a^2 + a = 15$

8. $x^2 - 8x \leq -15$

9. $\dfrac{x + 1}{x + 5} - \dfrac{2x + 1}{x - 2} = \dfrac{5 - x^2}{x^2 + 3x - 10}$

10. Factor completely: $x^4 - 2x^2 - 8$

11. Find the rational roots, if any, of $x^3 + 2x^2 - 3x - 10 = 0$.

Simplify.

12. $2\sqrt{25a^5}$

13. $\sqrt[3]{-\dfrac{1}{64}} + \sqrt[3]{\dfrac{1}{9}}$

14. $\dfrac{\sqrt{2}}{5 + 2\sqrt{5}}$

15. Solve $x - \sqrt{2x - 5} = 4$ over $\mathcal{R}$.

16. Express in the form $a + bi$: $\dfrac{4 - i}{3 - 2i}$

17. Express $0.2\overline{7}$ as a fraction in lowest terms.

Find the missing value, using the given values for an arithmetic sequence.

18. $a_{16} = -30$, $d = -3$, $a_1 = \underline{}$

19. $a_5 = 5$, $a_{15} = 30$, $a_{21} = \underline{}$

20. Find the common ratio of a geometric sequence in which $a_1 = 20$ and $a_4 = -\frac{5}{16}$.

21. Find the geometric mean of 9 and 25.

22. Find the sum S_6 for a geometric series in which $a_1 = 3$ and $r = 2$.

Evaluate.

23. $\displaystyle\sum_{a=1}^{7} (8a - 3)$

24. $\displaystyle\sum_{n=1}^{8} 16(-\tfrac{1}{2})^{n-1}$

25. $\displaystyle\sum_{n=1}^{\infty} \tfrac{2}{3}(-\tfrac{3}{4})^{n-1}$

Test for Chapters 9–12

1. Solve $x^2 - 8x = -6$ over $\mathcal{C}$ by completing the square.
2. Solve $3x^2 - 2x + 2 = 0$ over $\mathcal{C}$ by using the quadratic formula.
3. Write a quadratic equation with solution set $\{-1 + i\sqrt{2}, -1 - i\sqrt{2}\}$.
4. Give the equation of the axis of symmetry and the coordinates of the vertex of the graph of $y = x^2 + 4x + 3$. Sketch the graph.
5. Solve $x^4 - 2x^3 + 7x^2 - 10x + 10 = 0$ over $\mathcal{C}$ given that $1 + i$ is a root.
6. Determine an equation of the perpendicular bisector of the line segment with endpoints $(-3, 3)$ and $(-1, -1)$.
7. Write an equation of the form $x^2 + y^2 + ax + by + c = 0$ for the circle with center $(0, -4)$ and radius 5.
8. Sketch the graph of the equation $4x^2 + y^2 = 36$.
9. If z varies directly as x and inversely as y, and $z = \frac{3}{5}$ when $x = 2$ and $y = 3$, find z when $x = 10$ and $y = \frac{1}{3}$.

Solve over $\mathcal{C}$.

10. $x^2 - y^2 = 9$
    $-3 = x - 2y$

11. $y^2 - 6x^2 = 4$
    $y = 3x^2 - 2$

12. Find the inverse of the function $f(x) = x^3 + 4$.

Evaluate.

13. $125^{-\frac{2}{3}}$

14. $\sqrt[10]{32}\sqrt{8}$

15. $\log_2 \frac{1}{64}$

Solve.

16. $\left(\frac{1}{9}\right)^{x+2} = 27^{x-2}$

17. $\log_{3x} 16 = 2$

18. $\log_2 x + \log_2(x - 7) = 3$

Given that $\log 1.42 = 0.1523$, evaluate.

19. $\log 0.0142$

20. antilog 3.1523

21. How many different permutations exist of the letters in CAREER?
22. How many 5-element subsets does a set with 9 elements have?
23. How many 5-marble combinations drawn from an urn containing 9 white and 5 red marbles consist of 3 white and 2 red marbles?
24. If two dice are rolled, what is the probability that the sum is a 6?
25. Find the fourth term of $(a^2 - 2)^5$.

Test for Chapters 13–17

Chapters 14–16

1. Convert $150°$ to radian measure.

2. Convert $\dfrac{5\pi}{4}^{R}$ to degree measure.

3. If $\sin \alpha = \dfrac{\sqrt{7}}{4}$ and α is in Quadrant II, find the value of the other five trigonometric functions.

4. State the amplitude and the fundamental period p of the function $y = 3 \cos \dfrac{1}{2}x$. Sketch the graph over the interval $0 \le x \le p$.

5. Write $\dfrac{1 - \csc^2 \alpha}{1 - \sin^2 \alpha}$ in simplest form, using only the sine function and/or the cosine function. Note any restrictions.

If $\sin \alpha = \dfrac{3}{5}$ and $\sin \beta = \dfrac{8}{17}$, and α and β are both in Quadrant II, find each of the following.

6. $\sin(\alpha + \beta)$ **7.** $\cos 2\alpha$ **8.** $\tan \dfrac{\beta}{2}$

9. Find all the real values of y such that $\cos y = -\dfrac{1}{2}$.

10. Evaluate $\text{Sin}^{-1}(\cos 300°)$.

11. Solve $2 \sin^2 \alpha - 5 \sin \alpha - 3 = 0$ in the interval $0° \le m(\alpha) < 360°$.

12. Use DeMoivre's Theorem to express $(-1 - i)^3$ in the form $a + bi$.

13. Find the fourth roots of $-8 - 8i\sqrt{3}$ in the form $a + bi$.

14. Prove the identity $\dfrac{\sec^2 x}{1 + \cot^2 x} = \tan^2 x$. Note any restrictions.

For Exercises 15–21 use a calculator or a table. Unless otherwise requested, find angle measures to the nearest degree and lengths to the nearest unit.

15. Find $m(B)$ in $\triangle ABC$ if $m(C) = 90°$, $a = 14$, and $b = 13$.

16. Find $m(C)$ in $\triangle ABC$ if $a = 7$, $b = 9$, and $c = 13$.

17. In $\triangle ABC$, $b = 6$, $c = 11$, and $m(A) = 53°$. Find a to the nearest tenth of a unit and $m(B)$ and $m(C)$ to the nearest degree.

18. Solve $\triangle DEF$ if $d = 16$, $m(E) = 37°$, and $m(F) = 74°$. Give lengths of sides to the nearest tenth of a unit.

19. Two planes leave from the same point at the same time on flight paths 47° apart, and at speeds of 180 km/h and 210 km/h, respectively. How far apart are the planes after one hour?

20. Find a pair of polar coordinates for the point (8, 15) with $-180° < m(\alpha) \le 180°$.

21. The air speed of an airplane is 400 km/h and its heading is 100°. A 50 km/h wind is blowing with a bearing of 200°. Find the ground speed and the true course of the plane.

Chapter 13

22. Find the product: $\begin{bmatrix} -2 & 3 \\ 1 & -1 \end{bmatrix}\begin{bmatrix} -3 & 2 \\ -2 & 1 \end{bmatrix}$

23. Find the inverse of $\begin{bmatrix} 1 & 2 \\ 3 & 4 \end{bmatrix}$.

Solve for the matrix X.

24. $4X + \begin{bmatrix} 2 & -4 \\ 2 & -2 \end{bmatrix} = \begin{bmatrix} -3 & 0 \\ -1 & 3 \end{bmatrix}$

25. $\begin{bmatrix} 1 & 2 \\ 3 & 7 \end{bmatrix}X = \begin{bmatrix} 4 & 3 \\ -2 & -2 \end{bmatrix}$

26. Find a matrix equation of the translation of the plane that transforms the point $P(-2, 6)$ into the point $P'(3, -1)$.

27. Find the coordinates of the image of the point $P(2, -2)$ under the following transformation:
$$\begin{bmatrix} x' \\ y' \end{bmatrix} = \begin{bmatrix} 3 & -1 \\ 0 & 2 \end{bmatrix}\begin{bmatrix} x \\ y \end{bmatrix}$$

Chapter 17

28. Find the mean, median, and mode of the following data. New cars sold during February by the eight sales representatives at a car dealership:
$$6, 5, 7, 15, 10, 11, 9, 5$$

29. Find the variance and standard deviation of the following data:
$$30, 28, 34, 40, 29, 31$$

The table at the right shows the area under the standard normal curve from 0 to
x. Use it as needed.

30. What percent of the data in a normal distribution will be within 1.2 standard deviations of the mean?

31. The mean score on a certain test is 80 and the standard deviation is 8. If the scores are normally distributed, what percent of the scores are above 92?

32. Twenty-six pairs of socks of a production run of 1000 pairs were found to be defective. Find a 95% confidence interval for the number of pairs of defective socks that will result for every 1000 pairs manufactured.

33. Find the correlation coefficient for the following set of paired data points.

Hours spent studying	2	3	4	5	6
Test score	70	75	85	90	100

x	Area, A(x)
1.0	0.3413
1.1	0.3643
1.2	0.3849
1.3	0.4032
1.4	0.4192
1.5	0.4332
1.6	0.4452
1.7	0.4554
1.8	0.4641
1.9	0.4713
2.0	0.4772
2.1	0.4821
2.2	0.4861
2.3	0.4893
2.4	0.4918
2.5	0.4938

Topical Reviews

Review of Quadratics

Solve over $\mathcal{C}$ by factoring.

1. $64x^2 = 49$

2. $3x = 5x(x - 2)$

Solve over $\mathcal{C}$ by completing the square.

3. $x^2 - 12x = -27$

4. $\dfrac{x^2}{4} + \dfrac{x}{3} = 1$

Solve over $\mathcal{C}$ by using the quadratic formula.

5. $\dfrac{4}{x - 1} = \dfrac{5}{2x - 2} + \dfrac{3x}{4}$

6. $5x^2 + 8 = -12x$

Solve over $\mathcal{C}$ by the most appropriate method.

7. $(x - 2)^2 = 25$

8. $16x^2 - 8x + 1 = 0$

9. $8x^2 - 2x = 15$

10. $x^2 - 2x + 5 = 0$

11. $2x(x + 2) = 3(1 - x)$

12. $x^2 - x\sqrt{5} - 1 = 0$

13. One side of a rectangle is 7 cm longer than an adjacent side. If each diagonal of the rectangle has a length of 13 cm, find the lengths of the sides of the rectangle.

Write a quadratic equation having the given solution set.

14. $\{-4, 6\}$

15. $\{2 + \sqrt{5}, 2 - \sqrt{5}\}$

16. If one root of $2x^2 + kx - 24 = 0$ is -4, find the value of k.

Find the solution set over $\mathcal{C}$.

17. $x^2 + y^2 = 25$
$x + y = -7$

18. $y^2 - x^2 = 3$
$x^2 - 9y^2 = 5$

19. Solve $\sqrt{2x + 7} - \sqrt{x + 3} = 1$ over $\mathcal{R}$.

20. A large pipe can fill a tank in 3 h less time than a smaller pipe, and the two together can fill the tank in 2 h. How long would it take each pipe alone to fill the tank?

21. If the base of a square is increased 21 cm and its height is decreased 2 cm, the area of the rectangle formed will be three times the area of the square. Find the length of each side of the square.

Review of Equations and Inequalities

Solve over $\mathcal{R}$.

1. $5x^3 = 30x - 25x^2$

2. $|3y - 1| = 5$

3. $27^{x-1} = 3^{5x+4}$

4. $\dfrac{2x - 1}{x - 1} + \dfrac{x}{x + 1} = \dfrac{1}{x^2 - 1}$

5. $2 \cos x - \sqrt{3} = 0$

6. $\log_x 8 = -3$

7. $\sqrt{y^2} = -y$

8. $2\sqrt{a + 9} = a + 1$

9. $\sqrt{z + 1} = \sqrt{5 - z}$

10. $36y^4 - 13y^2 + 1 = 0$

11. $2x^3 + 3x^2 - 11x - 6 = 0$

12. $x^3 - 2x^2 + 1 = 0$

13. Solve $4 \sin \alpha - 4 \sin^2 \alpha - 1 = 0$ for $0° \le m(\alpha) < 360°$.

14. Solve $5x^2 + 4 = 3x$ over $\mathcal{C}$.

15. Solve $x^4 - 4x^3 + 6x^2 - 4x + 5 = 0$ over $\mathcal{C}$, given that $2 + i$ is one root.

Solve and graph the solution set.

16. $|2x - 1| < 3$

17. $|3 - x| > 2$

18. $x^2 - 5x \ge 0$

Find the solution set of each system.

19. $2x + 3y = -1$
    $x + 2y = 2$

20. $y = x^2 - 7x - 3$
    $2x - y = 3$

21. $x^2 + 2y^2 = 10$
    $3x^2 - y^2 = 9$

22. $3x - 2y + z = 13$
    $2x + 3y + 5z = 0$
    $5x + y = 13$

Graph.

23. $5x - y > 3$

24. $x^2 + y^2 \leq 25$

25. $2x + 3y \leq 6$
$2x - y < -2$

26. The length of a rectangle is 3 cm less than five times its width. If the area of the rectangle is 110 cm², find the dimensions of the rectangle.

27. The product of two integers is -12. The sum of the squares of the two integers is 25. Find the integers.

Review of Graphing

Graph.

1. $y = 4 - 3x$

2. $\frac{x}{2} - \frac{y}{3} = 1$

3. $x - 5 > 4y$

Graph the solution set of each system in a coordinate plane.

4. $x + 3y \leq 6$
$2x - y < 3$

5. $-3 < y < 2$
$y \geq -x - 3$

6. Sketch the triangle in space whose vertices have the coordinates $A(3, 0, 0)$, $B(0, -4, 0)$, $C(0, 0, 5)$.

Determine an equation in the form $y = mx + b$ of the line described.

7. passing through the points $(-2, 1)$ and $(-3, -2)$

8. passing through $(1, -3)$ and perpendicular to the line $x - 5y = 7$

For the given parabola, state (a) an equation of the axis of symmetry and (b) the coordinates of the vertex.

9. $y = -x^2 + 3$

10. $x = 2(y + 2)^2 - 1$

11. For the circle $x^2 + y^2 - 10y = 0$, find the following:
 a. the center of the circle
 b. the radius of the circle

12. Determine the coordinates of the foci of the ellipse whose equation is $9x^2 + y^2 = 225$.

13. State the equations for the asymptotes of the hyperbola whose equation is $\frac{x^2}{25} - \frac{y^2}{9} = 1$.

Sketch the graph.

14. $y = x^2 - 4x - 5$

15. $y = -\frac{1}{3}x^2 + 2x$

16. $x^2 + y^2 > 9$

17. $x^2 + y^2 + 6x - 3y = 9$

18. $\frac{(x + 1)^2}{4} + \frac{(y - 1)^2}{9} = 1$

19. $\frac{y^2}{16} - \frac{x^2}{25} = 1$

20. $4x^2 + 25y^2 = 100$

21. $4x^2 - y^2 = 36$

Solve each system over $\mathscr{R}$ by graphing.

22. $3x - y = 8$
$x + 4y = 7$

23. $4x^2 + 9y^2 = 72$
$x^2 - 2y^2 = 1$

24. Sketch the graph of $y = 2^x$.

Sketch the graph of each equation in the interval $0 \le x \le 2\pi$.

25. $y = 2 \cos x$

26. $y = \frac{1}{3} \sin 2x$

Review of Operations with Numbers

If $x = 6$, $y = -3$, and $z = -1$, find the value of each of the following.

1. $2x^2 + xy - y^2$

2. $(y^2 + z)^{\frac{1}{3}}$

3. $\cos (x + y)\pi$

4. $\log_6 x^2z^2$

5. $3^y + 3^z$

6. $(|xyz|)^{-\frac{1}{2}}$

7. $\dfrac{x^3 - y^3}{x + y}$

8. $1 + \left(\dfrac{y}{x}\right) + \left(\dfrac{y}{x}\right)^2 + \left(\dfrac{y}{x}\right)^3 + \ldots$

Simplify.

9. $\log_5 125$

10. $32^{\frac{3}{5}}$

11. $\frac{3}{5}\sqrt{50}$

12. $2\sqrt[3]{\dfrac{11}{9}}$

13. $\left(\dfrac{27}{64}\right)^{-\frac{1}{3}}$

14. $\dfrac{2 + \sqrt{7}}{5 - 2\sqrt{7}}$

15. $(\sqrt{5} - \sqrt{2})^2$

16. $4\sqrt{48} - 2\sqrt{75}$

17. $2\sqrt{12}(4\sqrt{3} - 3\sqrt{2})$

Simplify. Assume each radical represents a real number.

18. $\sqrt[5]{-32a^5c^{20}}$

19. $\sqrt[4]{(-3)^4r^8s^{10}}$

20. $\sqrt{49x^{10}}$

Express each of the following in the form $a + bi$. When possible, you may give your answer in the form a, bi, or 0.

21. $\sqrt{-12}$

22. $\sqrt{-20} + 2\sqrt{-45}$

23. $(3 + 4i) - (7 - i)$

24. $(4 - 7i)^2$

25. $\dfrac{10\sqrt{-40}}{\sqrt{-10}}$

26. $\dfrac{5 + 2i}{3 - i}$

Find the given values over $\mathcal{C}$ of $P(x) = 2x^5 - x^3 - 4x^2 - 3x - 10$.

27. $P(-2)$

28. $P(i)$

29. $P(2i)$

30. a. Express 16^3 as a power of 2.
    b. Express 27^n as a power of 3.

Simplify.

31. $\dfrac{4^{-1}}{4^{-1} + 4^{-2}}$

32. $\dfrac{x^2 - 2x + 4}{x^4 + 8x}$

33. $\dfrac{a^2 - 5ab + 4b^2}{a^2 - 3ab - 4b^2}$

34. Evaluate $\left(\dfrac{-1 + i\sqrt{3}}{2}\right)^3$.

35. Express $2(\log 63 - \log 7) + \frac{1}{2}\log 36$ as the logarithm of a single number.

36. Express $4(\cos 30° + i \sin 30°)$ in the form $a + bi$.

Evaluate.

37. $\sin 675°$

38. $\cot 225°$

39. $\text{Cos}^{-1}(-\frac{1}{2})$

40. Express $\tan 15°$ in radical form.

Answers

Diagnostic Test

1. False   **2.** True   **3.** False   **4.** {2}
5. {−2, −1, 1, 2}   **6.** {0, 1}   **7.** {−2, 2}
8. d   **9.** e   **10.** c   **11.** b   **12.** $3r - 15$
13. $5c + 90cd$ **14.** -25 **15.** $-\frac{1}{7n}$

16. $\frac{14d - 20}{d}$ **17.** 14   **18.** $4x^2 - 3xy - 3y^2$
19. $3c^3 - 6c^2 - 6c - 12$ **20.** $2y^2 - 4y + 7$
21. {−10}   **22.** {4}   **23.** {−15}   **24.** $\left\{-\frac{28}{5}\right\}$

25. $\{x: x \le 5\}$
26. $\{r: -2 < r \le 3\}$
27. $\{y: y > -1 \text{ or } y < -2\}$

28. 2   **29.** −2   **30.** $\{n: -3 \le n \le 3\}$ **31.** {−1, 6}

Alternate Chapter Tests

Chapter 1 Test

1. {0, 1, 2, 3, 4, 5, 6, 7, 8, 9}
2. {0, 2, 4, 5, 6, 7, 8, 9}   **3.** {5, 7, 9}
4. Ø, {1}, {3}, {5}, {1, 3}, {1, 5}, {3, 5}, {1, 3, 5}
5. {1}   **6.** no solution   **7.** {−2, −1, 0, 1, 2}
8. {−1, 1}

9.

10.

11. Distributive axiom
12. Commutative axiom for addition
13. Transitive property of equality
14. $3w + 3$ **15.** $8 - h$ **16.** $24jk + 24jm$
17. $7cd + 8d + 15c$ **18.** $-5fg + 5fk - 35f$
19. $200 - 50v$ **20.** -7 **21.** $-12xy$
★**11.** Associative axiom for multiplication
★**15.** 22

Chapter 2 Test

1. $-t^2 + 4t + 12$ **2.** $-2e^2 + 4e + 9f - 5f^2$
3. {5}   **4.** {2}   **5.** $n = \frac{5 + 7y}{3y}$ **6.** $n = \frac{c - q}{2p^2}$

7. 7, 8, 9   **8.** Each of the congruent sides has a
length of 16 cm, and the base has a length of 18 cm.
9. $\{x: x > 2\}$ **10.** $\{x: x \le 7\}$

11. $\{p: -2 < p < 7\}$ **12.** $\{p: p < 2\} \cup \{p: p \ge 3\}$

13. 30 ft
14. Case 1: Assume that $a = b$.   1. $a \cdot c = b \cdot c$
(Multiplication Property)   2. $a \cdot c = b \cdot c$ contradicts
the fact that $a \cdot c > b \cdot c$. (Hypothesis) ∴ the assumption that $a = b$ must be incorrect. Case 2: Assume that
$a < b$.   1. $a \cdot c < b \cdot c$ (Multiplication Property for
Order)   2. $a \cdot c < b \cdot c$ contradicts the fact that $a \cdot c$
$> b \cdot c$. (Hypothesis) ∴ the assumption that $a < b$ must
be incorrect. Hence, $a \cdot b > b \cdot c$. (Comparison Axiom)

15. $\{x: -3 \le x \le 2\}$

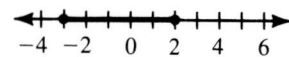

16. $\{t: -2 \le t \le 6\}$

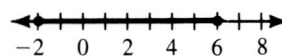

★**7.** 6 and 8   ★**15.** $\{x: -1 < x < 3\} \cup \{x: 5 < x < 9\}$

Chapter 3 Test

1. 10   **2.** 3   **3.** 5   **4.** $\dfrac{5}{4}$

5. yes

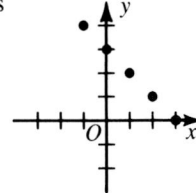

6. no

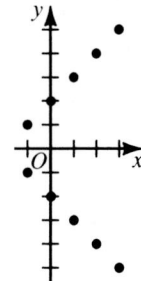

7. yes

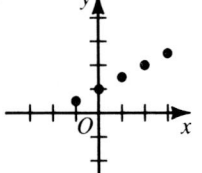

8. yes

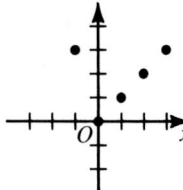

9.

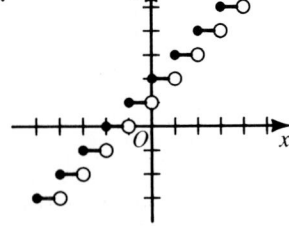

10.

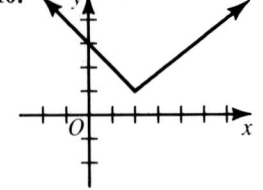

11.

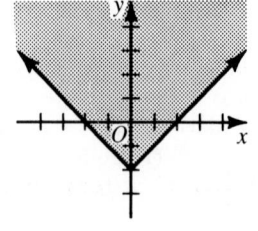

12.

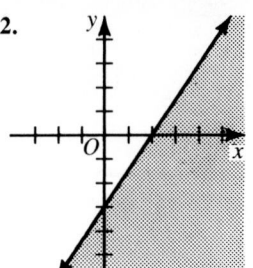

13. 2   **14.** −3

15. $y = -3x + 11$

16. $y = \dfrac{2}{3}x - 1$

17. 15

18. $74,000

★**4.** $f(g(-3)) = 4$; $g(f(-3)) = 2$

★**9.**

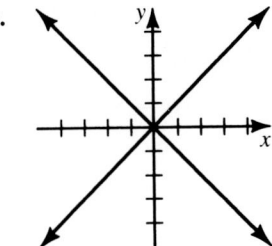

★**13.** −1

★**18.** 6

Chapter 4 Test

1. Ø is the solution set.

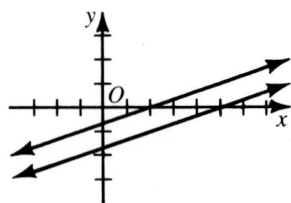

2. $\{(-2, 1)\}$ is the solution set.

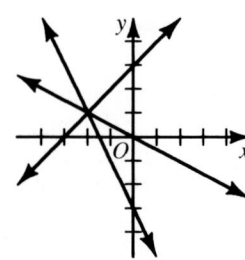

3. $\{(4, 7)\}$ **4.** $\{(2, 7)\}$ **5.** $\{(4, 10)\}$

6. Ø; the system is inconsistent.

7. $\left\{\left(-\dfrac{1}{29}, \dfrac{41}{29}\right)\right\}$ **8.** infinite solution set

9. The plane's air speed is 750 km/h, and the speed of the wind is 50 km/h.

10. Pete worked 20 hours, and Beth worked 25 hours.

11.

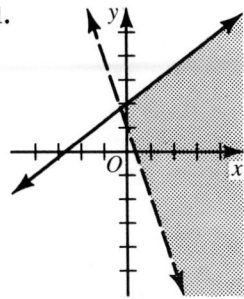

12.

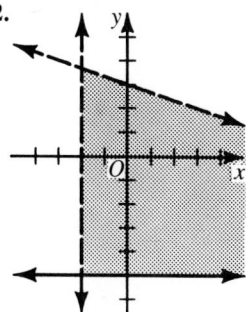

13. To earn the maximum profit, the baker should sell 30 loaves of rye bread and 90 loaves of pumpernickel.

★**5.** $\left\{\dfrac{a}{a+b}, \dfrac{b}{a+b}\right\}$ ★**9.** 14 single roses

★**11.**

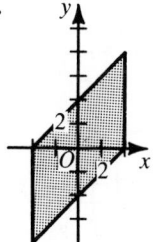

Chapter 5 Test

1.

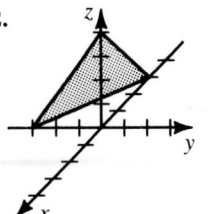

$(2, 4, -3)$

2.

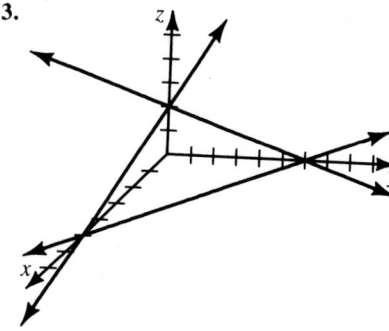

3.

4. $\left\{\left(\dfrac{31}{29}, -\dfrac{42}{29}, -\dfrac{9}{29}\right)\right\}$ **5.** $\{(1, -1, -2)\}$

6. $\{(-1, 0, 2)\}$ **7.** $AB = 3$ km, $BC = 5$ km, $CD = 2$ km   **8.** 153   **9.** 54

★**7.** 4 cm   ★**8.** $|A| = a_1 \begin{vmatrix} b_2 & c_2 & d_2 \\ 0 & c_3 & d_3 \\ 0 & 0 & d_4 \end{vmatrix} = a_1 b_2 \cdot \begin{vmatrix} c_3 & d_3 \\ 0 & d_4 \end{vmatrix}$

$= a_1 b_2 c_3 d_4$

★**9.** $\begin{vmatrix} a_1 & b_1 & c_1 \\ a_2 & b_2 & c_2 \\ ka_1 & kb_1 & kc_1 \end{vmatrix} = k \cdot \begin{vmatrix} a_1 & b_1 & c_1 \\ a_2 & b_2 & c_2 \\ a_1 & b_1 & c_1 \end{vmatrix} = k \cdot 0 = 0$

Chapter 6 Test

1. $\dfrac{4}{c^2}$ **2.** $\dfrac{c^3}{a^5 b}$ **3.** $12x^3 - 29x^2 + 35x - 14$

4. $256y^4 - 288y^2 + 81$ **5.** $2(x - 3)(x - 2)$

6. $(3y - 1)(9y^2 + 3y + 1)$ **7.** $\left\{-\dfrac{2}{3}, 8\right\}$ **8.** $\left\{-\dfrac{2}{5}, 6\right\}$

9. 7 cm   **10.** $\{x : x \le 0\}$

‹—+—+—+—+—+—+—›
−3 −2 −1 0 1 2 3

11. $\dfrac{x + 3}{2(x + 7)}$ or $\dfrac{x + 3}{2x + 14}$ **12.** $3y^3 - y^2 + 3 + \dfrac{6}{2y - 1}$ **13.** $7(x - 2)$, or $7x - 14$

14. $\dfrac{n^3 + n^2 - 2}{n + 1}$ **15.** 480 g   **16.** $\{-6, 2\}$

★**2.** $\dfrac{v^2 + u^2}{v^2 - u^2}$ ★**16.** $\{0, 2\}$

Chapter 7 Test

1. $y = 108$ **2.** $d = 64$ ft   **3.** $\{\pm1. 2\}$ **4.** $\{3\}$

5. 7.907×10^4 **6.** 3×10^{-4} **7.** $0.\overline{45}$

8. $1\frac{15}{22}$ **9.** $0.67; 0.676776777 \ldots$

10. $3|a|\sqrt[4]{2a^2}$ **11.** $1\frac{1}{5}$ **12.** $\frac{\sqrt{|3x|}}{3}$

13. $4a^2\sqrt{ab} + b^2$ **14.** $3m^2(1 + 3\sqrt{a})$ **15.** $-13 - 7\sqrt{3}$ **16.** $\{7\}$ **17.** $1 - 2\sqrt{3}$ **18.** $7 - \frac{9}{5}i$

19. $\frac{1}{16} + \frac{21}{16}i$

★15. $\frac{4\sqrt{5}}{5}$ **★16.** $\{5\}$

Chapter 8 Test

1. $-2, 1, 4, 7$ **2.** explicitly, $a_n = 12 - 4n$: recursively, $a_{n+1} = a_n - 4$ **3.** 287   **4.** 477   **5.** 4

6. 19   **7.** \$34.30   **8.** 12   **9.** $13\frac{9}{16}$ **10.** $|L - a_n| = \frac{1}{n}$ **11.** $\frac{50}{111}$

★3. 15, 19, 23, 27   **★7.** 25%   **★9.** $\frac{1023}{4}$

★11. $\frac{43}{333}$

Chapter 9

1. $\{3 \pm 2\sqrt{2}\}$ **2.** $\left\{\frac{-5 \pm \sqrt{29}}{8}\right\}$ **3.** 2 real irrational roots   **4.** $-6, \frac{2}{3}$ **5.** $x^2 - 2x - 1 = 0$

6. $y = 2(x - 3)^2 + 7$

7. $x = 3; (3, 0)$

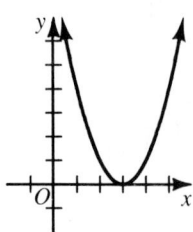

8. a.

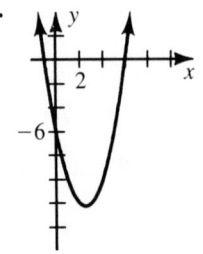

b. $\{x: -1 \le x \le 6$ **9. a.** 9   **b.** $-10 - 18i$

10. $3x^2 + 2x - 14 - \frac{39}{x - 3}$ **11.** $\{2i, -2i, 1, 3\}$

12. one positive, two or zero negative roots

+	−	imag.
1	2	0
1	0	2

13. $-1.0 < r_1 < -0.5; 1.0 < r_2 < 1.5; 2.5 < r_3 < 3.0$

★5. -2 **★6.** $y = 2(x - 1)^2 - 3$

★8. a.

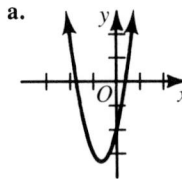

b. $\left\{x: x < \dfrac{-2 - \sqrt{10}}{3}\right\}$ or $\left\{x: x > \dfrac{-2 + \sqrt{10}}{3}\right\}$

★10. -3

Chapter 10 Test

1. $(1, 3)$ **2.** $y = -x + 8$ **3.** $\left(2, \frac{3}{2}\right); \frac{5}{2}$

4.

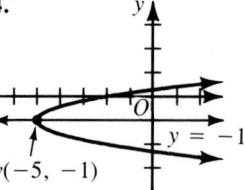

5.

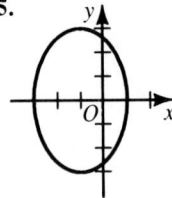

6.

7. 10

8.

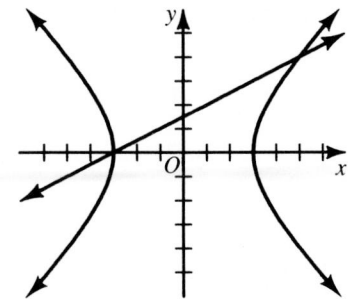

9. $\{(0, -4), (2, 0)\}$ **10.** $\{(\pm 3, \pm 2\sqrt{2})\}$

★**3.** $(x - 5)^2 + (y + 2)^2 = 5$

★**4.** $y = \frac{1}{2}x^2 - 3x + \frac{7}{2}$ ★**5.** $\frac{x^2}{100} + \frac{y^2}{25} = 1$

★**10.** $\{(2, \pm 2), (3, \pm i)\}$

Chapter 11 Test

1. 1 **2.** $\sqrt[6]{32}$ **3.** $\left\{\frac{7}{10}\right\}$

4. $y = \frac{5}{2}x + 5$

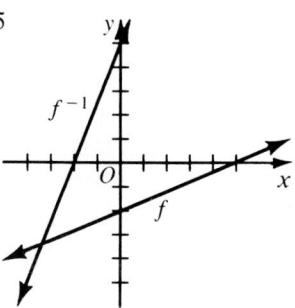

5. $\{625\}$ **6.** $\{3\}$ **7.** $\{2\}$ **8.** $\{8\}$ **9.** $\{1.41\}$
10. $\{3000\}$ **11.** $\{0.751\}$ **12.** $\{0.927\}$
13. 2.04×10^8 **14.** 11%
★**3.** $\left\{\frac{1}{2}, -2\right\}$ ★**7.** $\{1\}$ ★**12.** $\{-20.6\}$
★**13.** 3000

Chapter 12 Test

1. 343 **2.** 3024 **3.** 10,080 **4.** 126 **5.** 4
6. $\frac{2}{3}x^2$ **7.** $x^5 - 10x^4 + 40x^3 - 80x^2 + 80x - 32$
8. {(H, H, H), (H, H, T), (H, T, H), (H, T, T),
(T, H, H), (T, H, T), (T, T, H), (T, T, T)} **9.** $\frac{91}{228}$

10. $\frac{427}{570}$ **11.** $\frac{5}{51}$

★**5.** 4800 ★**11.** $\frac{1}{1560}$

Chapter 13 Test

1. $x = 7; y = -4$ **2.** $\begin{bmatrix} -9 & 5 & -2 \\ 4 & -7 & 7 \end{bmatrix}$

3. $\begin{bmatrix} -2 & -2 \\ 3 & -5 \end{bmatrix}$ **4.** $\begin{bmatrix} 4 & 12 \\ -26 & 6 \end{bmatrix}$

5. $(2A)B = A(2B) = \begin{bmatrix} 4 & -16 \\ 0 & 24 \end{bmatrix}$ **6.** $X = \begin{bmatrix} 41 & 3 \\ 17 & 2 \end{bmatrix}$

7. $\begin{bmatrix} x' \\ y' \end{bmatrix} = \begin{bmatrix} x \\ y \end{bmatrix} + \begin{bmatrix} 1 \\ -5 \end{bmatrix}$ **8.** $Q'(-2, -5)$

9. $P'(-4, 2)$ **10.** $\begin{bmatrix} \frac{17}{2} \\ -\frac{7}{2} \end{bmatrix}$

★**4.** $x = 0, y = -1, z = 3$

★**5.** $AB = \begin{bmatrix} ac & 0 \\ 0 & bd \end{bmatrix} = BA$

★**6.** $A^{-1} = \frac{1}{\det A}\begin{bmatrix} -a & a - 1 \\ -a - 1 & a \end{bmatrix} =$

$-1\begin{bmatrix} -a & a - 1 \\ -a - 1 & a \end{bmatrix} = A$, or $AA =$

$\begin{bmatrix} a^2 + 1 - a^2 & a - a^2 - a + a^2 \\ a + a^2 - a - a^2 & 1 - a^2 + a^2 \end{bmatrix} = \begin{bmatrix} 1 & 0 \\ 0 & 1 \end{bmatrix} = I$

★**10.** The plane is mapped onto the line $y' = -2x'$.

Chapter 14 Test

1. 0.67 m **2.** 55.0 cm **3.** $\frac{3\pi}{4}^R$ **4.** $60°$

5. $-\frac{12}{13}$ **6.** $\frac{7\pi}{4}^R$ **7.** $120° + k \cdot 360°$, or

$\frac{2\pi}{3}^R + 2k\pi^R$ **8.** -0.7374 **9.** $20°20'$

10. amplitude: 2

11.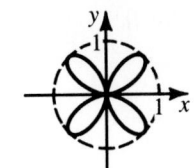

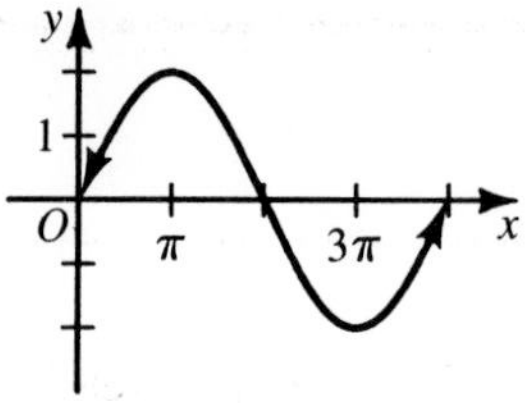

12. $\cos \alpha = -\dfrac{1}{2}$;

$\tan \alpha = -\sqrt{3}$;

$\sec \alpha = -2$;

$\csc \alpha = \dfrac{2}{\sqrt{3}}$;

$\cot \alpha = -\dfrac{1}{\sqrt{3}}$

13. $m(B) = 28°0'$

★**5.** $\dfrac{1}{3}$ ★**11.** amplitude: π; period: 4

Chapter 15 Test

1. $\sec^2 \alpha = \dfrac{1}{\cos^2 \alpha}$, $\cos \alpha \neq 0$ **2.** $(1 + \cos \alpha) \cdot$

$\dfrac{\cos \alpha - 1}{\cos^2 \alpha} = \dfrac{\cos^2 \alpha - 1}{\cos^2 \alpha} = \dfrac{-(1 - \cos^2 \alpha)}{\cos^2 \alpha} =$

$(\sin^2 \alpha)\left(\dfrac{-1}{\cos^2 \alpha}\right) = (\sin^2 \alpha)(-\sec^2 \alpha)$, $\cos \alpha \neq 0$

3. 0   **4.** $\dfrac{-\sqrt{2} - \sqrt{6}}{4}$ **5.** 0.445   **6.** 3

7. $\dfrac{\cos 2x}{\tan x} + 2 \sin x \cos x =$

$\dfrac{\cos^2 x - \sin^2 x}{\dfrac{\sin x}{\cos x}} + \dfrac{2 \sin x \cos x \left(\dfrac{\sin x}{\cos x}\right)}{\dfrac{\sin x}{\cos x}} =$

$\dfrac{\cos^2 x - \sin^2 x + 2 \sin^2 x}{\dfrac{\sin x}{\cos x}} = \dfrac{\cos^2 x + \sin^2 x}{\dfrac{\sin x}{\cos x}} =$

$\dfrac{1}{\tan x} = \cot x$, $\cos x \neq 0$, $\sin x \neq 0$

8. 22.9   **9.** 8.3

★**5.** $-\dfrac{3\sqrt{7}}{8}$ ★**7.** $\dfrac{1 - \cos 2\alpha}{\sin^2 \dfrac{\alpha}{2} \cos^2 \dfrac{\alpha}{2}} =$

$\dfrac{1 - (2\cos^2 \alpha - 1)}{\left(\dfrac{1 - \cos \alpha}{2}\right)\left(\dfrac{1 + \cos \alpha}{2}\right)} = \dfrac{2(1 - \cos^2 \alpha)}{\dfrac{1 - \cos^2 \alpha}{4}} = 8$

★**8.** 8 cm

Chapter 16 Test

1. $\dfrac{\pi}{6}$ **2.** $\left\{\dfrac{\pi}{4} + k\dfrac{\pi}{2}\right\}$ **3.** $r(\cos \theta + \sin \theta) = 6$

4. $4 - 4i\sqrt{3}$ **5.** $-\sqrt{3} + i$ **6.** $32\sqrt{2} - 32i\sqrt{2}$

7. $\dfrac{3}{2}\sqrt{3} + \dfrac{3}{2}i$, $-\dfrac{3}{2}\sqrt{3} + \dfrac{3}{2}i$, $-3i$

8. $\|\mathbf{u} + \mathbf{v}\| = 8.3$; $\theta = 49°$

9. ground speed: 374 km/h; true course: 96.3°

★**2.** $\{45°, 90°, 225°, 270°\}$

★**3.**

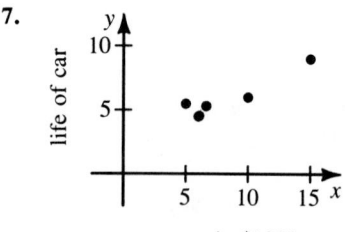

★**6.** $\sqrt{3} + i$, $-\sqrt{3} - i$, $-1 + \sqrt{3}i$, $1 - \sqrt{3}i$

Chapter 17 Test

1. mean: 5.3; median: 5; mode: 3

2. $\sigma^2 = 28$; $\sigma \approx 5.29$ **3.** 0.7745 or 77.45%

4. 2.28%   **5.** $0.6871 < p < 0.8462$

6. between 8 and 24 students

7.

[scatter plot: y-axis labeled "life of car" with marks at 5 and 10; x-axis labeled "cost in $1000" with marks at 5, 10, 15]

8. $r \approx 0.75$

★**2.** mean: 85; $\sigma \approx 5$ ★**3.** 65.54%

Cumulative Tests

Test for Chapters 1–4

1. $2a - 2$ **2.** $3y$ **3.** Distributive axiom
4. Commutative axiom for multiplication
5. Property of the opposite of a sum
6. Axiom of additive inverses **7.** $\{3\}$ **8.** $\{1, 4\}$
9. $x = \dfrac{10b - 3a}{9}$ **10.** $-2, -1, 0$
11. $\{z: z < -3\}$

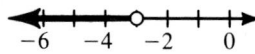

12. $\{a: -1 < a \le 2\}$

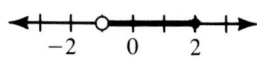

13. $\{y: -2 < y < 1\}$

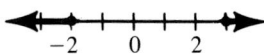

14. $\{x: x \le -2 \text{ or } x \ge 3\}$

15. 50 **16.** 18 **17.** $\dfrac{1}{3}$

18. 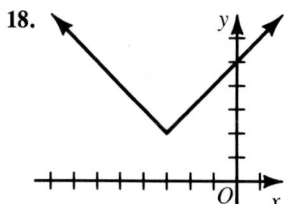 **19.**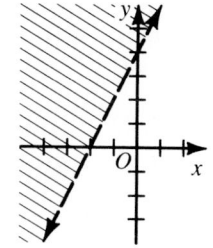

20. -2 **21.** $y = -2x + 1$ **22.** \$2025
23. $\{(1, -2)\}$ **24.** $\{(\frac{3}{4}, -\frac{1}{2})\}$
25. $\{(x, y): x - y = 4\}$

26.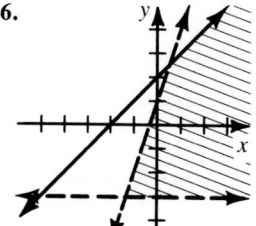

Test for Chapters 5–8

1. $\{(1, -2, 3)\}$ **2.** $\{(3, -1, 2)\}$ **3.** 2 nickels,
18 dimes, 4 quarters **4.** $a^3b^2c^2$ **5.** $x^3 - 7x + 6$
6. $\dfrac{x + 3}{x - 4}$ **7.** $\left\{\dfrac{3}{2}, -\dfrac{5}{3}\right\}$ **8.** $\{x: 3 \le x \le 5\}$
9. $\{-1\}$ **10.** $(x + 2)(x - 2)(x^2 + 2)$ **11.** 2
12. $10a^2\sqrt{a}$ **13.** $-\dfrac{1}{4} + \dfrac{1}{3}\sqrt[3]{3}$ **14.** $\dfrac{5\sqrt{2} - 2\sqrt{10}}{5}$
15. $\{7\}$ **16.** $\dfrac{14}{13} + \dfrac{5}{13}i$ **17.** $\dfrac{5}{18}$ **18.** 15
19. 45 **20.** $-\dfrac{1}{4}$ **21.** 15 **22.** 189 **23.** 203
24. $10\dfrac{5}{8}$ **25.** $\dfrac{8}{21}$

Test for Chapters 9–12

1. $\{4 - \sqrt{10}, 4 + \sqrt{10}\}$ **2.** $\left\{\dfrac{1}{3} + \dfrac{i}{3}\sqrt{5}, \dfrac{1}{3} - \dfrac{i}{3}\sqrt{5}\right\}$
3. $x^2 + 2x + 3 = 0$
4. $x = -2; (-2, -1)$ **5.** $\{1 + i, 1 - i, i\sqrt{5}, -i\sqrt{5}\}$
6. $x - 2y = -4$
7. $x^2 + y^2 + 8y - 9 = 0$

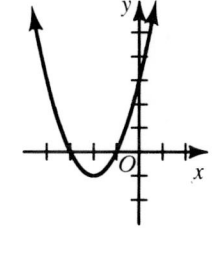

8.

9. 27   **10.** {(−3, 0), (5, 4)}   **11.** {(0, −2), ($\sqrt{2}$, 4), (−$\sqrt{2}$, 4)}   **12.** $f^{-1}(x) = \sqrt[3]{x - 4}$ **13.** $\frac{1}{25}$

14. 4   **15.** −6   **16.** $\left\{\frac{2}{5}\right\}$ **17.** $\left\{\frac{4}{3}\right\}$ **18.** {8}

19. 0.1523 − 2   **20.** 1420   **21.** 180   **22.** 126

23. 840   **24.** $\frac{5}{36}$ **25.** $-80a^4$

Test for Chapters 13–17

1. $\frac{5\pi}{6}^{\text{R}}$ **2.** 225°   **3.** $\cos \alpha = -\frac{3}{4}$, $\tan \alpha = -\frac{\sqrt{7}}{3}$, $\cot \alpha = -\frac{3}{\sqrt{7}}$, $\sec \alpha = -\frac{4}{3}$, $\csc \alpha = \frac{4}{\sqrt{7}}$

4. amplitude: 3; period: 4π

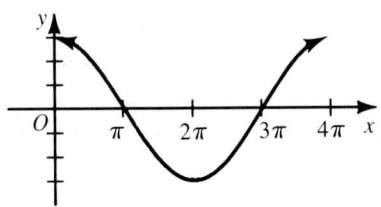

5. $-\frac{1}{\sin^2 \alpha}$, $\sin \alpha \neq 0$, $\cos \alpha \neq 0$ **6.** $-\frac{77}{85}$ **7.** $\frac{7}{25}$

8. 4   **9.** $\frac{2\pi}{3} + 2k\pi$ and $\frac{4\pi}{3} + 2k\pi$, $k \in J$

10. 30°   **11.** {210°, 330°}   **12.** 2 − 2i

13. $1 + i\sqrt{3}$, $-\sqrt{3} + i$, $-1 - i\sqrt{3}$, $\sqrt{3} - i$

14. $\frac{\sec^2 x}{1 + \cot^2 x} = \frac{\sec^2 x}{\csc^2 x} = \frac{1}{\cos^2 x} \div \frac{1}{\sin^2 x} = \frac{\sin^2 x}{\cos^2 x} = \tan^2 x$, $\sin x \neq 0$, $\cos x \neq 0$ **15.** 43°

16. 108°   **17.** $a = 8.8$, $m(B) = 33°$, $m(C) = 94°$

18. $m(D) = 69°$, $e = 10.3$, $f = 16.5$ **19.** 158 km

20. (17, 62°)   **21.** 394 km/h; 107°   **22.** $\begin{bmatrix} 0 & -1 \\ -1 & 1 \end{bmatrix}$

23. $\begin{bmatrix} -2 & 1 \\ \frac{3}{2} & -\frac{1}{2} \end{bmatrix}$ **24.** $\left\{\begin{bmatrix} -\frac{5}{4} & 1 \\ -\frac{3}{4} & \frac{5}{4} \end{bmatrix}\right\}$ **25.** $\left\{\begin{bmatrix} 32 & 25 \\ -14 & -11 \end{bmatrix}\right\}$

26. $\begin{bmatrix} x' \\ y' \end{bmatrix} = \begin{bmatrix} x \\ y \end{bmatrix} + \begin{bmatrix} 5 \\ -7 \end{bmatrix}$ **27.** $P'(8, -4)$

28. 8.5; 8; 5   **29.** 16.33; 4.04   **30.** 76.98%

31. about 6.68%   **32.** $16 < p < 36$ **33.** 0.99

Topical Reviews

Review of Quadratics

1. $\left\{\frac{7}{8}, -\frac{7}{8}\right\}$ **2.** $\left\{0, \frac{13}{5}\right\}$ **3.** {9, 3}

4. $\left\{\frac{-2 - 2\sqrt{10}}{3}, \frac{-2 + 2\sqrt{10}}{3}\right\}$ **5.** {−1, 2}

6. $\left\{\frac{-6 - 2i}{5}, \frac{-6 + 2i}{5}\right\}$ **7.** {−3, 7}   **8.** $\left\{\frac{1}{4}\right\}$

9. $\left\{-\frac{5}{4}, \frac{3}{2}\right\}$ **10.** {1 + 2i, 1 − 2i}

11. $\left\{\frac{-7 - \sqrt{73}}{4}, \frac{-7 + \sqrt{73}}{4}\right\}$

12. $\left\{\frac{\sqrt{5} - 3}{2}, \frac{\sqrt{5} + 3}{2}\right\}$ **13.** 5 cm and 12 cm

14. $x^2 - 2x - 24 = 0$ **15.** $x^2 - 4x - 1 = 0$

16. 2   **17.** {(−4, −3), (−3, −4)}

18. {(2i, i), (2i, −i), (−2i, i), (−2i, −i)}   **19.** {−3, 1}

20. 3 h and 6 h   **21.** 6 cm

Review of Equations and Inequalities

1. {−6, 0, 1}   **2.** $\left\{-\frac{4}{3}, 2\right\}$ **3.** $\left\{-\frac{7}{2}\right\}$

4. $\left\{-\frac{\sqrt{6}}{3}, \frac{\sqrt{6}}{3}\right\}$ **5.** $\left\{\frac{\pi}{6} + 2k\pi, \frac{11\pi}{6} + 2k\pi\right\}$

6. $\left\{\frac{1}{2}\right\}$ **7.** {y: y ≤ 0}   **8.** {7}   **9.** {1}

10. $\left\{\frac{1}{3}, -\frac{1}{3}, \frac{1}{2}, -\frac{1}{2}\right\}$ **11.** $\left\{-3, -\frac{1}{2}, 2\right\}$

12. $\left\{1, \frac{1 + \sqrt{5}}{2}, \frac{1 - \sqrt{5}}{2}\right\}$ **13.** {30°, 150°}

14. $\left\{\frac{3 - i\sqrt{71}}{10}, \frac{3 + i\sqrt{71}}{10}\right\}$ **15.** {i, −i, 2 + i, 2 − i}

16. {x: −1 < x < 2}   ←|—|—○—|—|—○—|—|→
                                 −2   0   2

17. {x: x < 1 or 5 < x}   ←|—○—|—|—|—○—|—|→
                                 0   2   4   6

18. $\{x: x \le 0 \text{ or } 5 \le x\}$

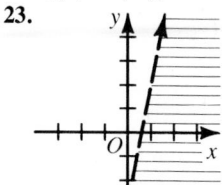

19. $\{(-8, 5)\}$ **20.** $\{(0, -3), (9, 15)\}$ **21.** $\{(2, \sqrt{3}), (2, -\sqrt{3}), (-2, \sqrt{3}), (-2, -\sqrt{3})\}$

22. $\{(3, -2, 0)\}$

23.

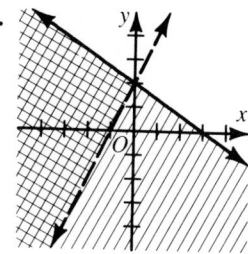

24.

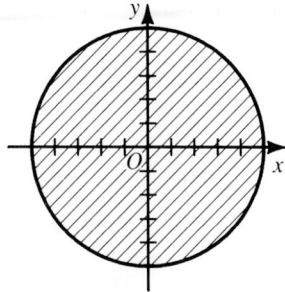

5.

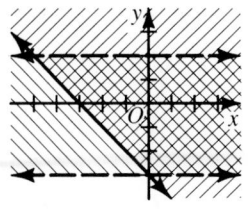

6.

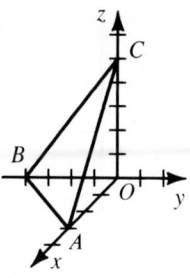

7. $y = 3x + 7$ **8.** $y = -5x + 2$ **9. a.** $x = 0$
b. $(0, 3)$ **10. a.** $y = -2$ **b.** $(-1, -2)$
11. a. $(0, 5)$ **b.** 5 **12.** $(0, 10\sqrt{2}), (0, -10\sqrt{2})$
13. $y = \frac{3}{5}x$ and $y = -\frac{3}{5}x$

25.

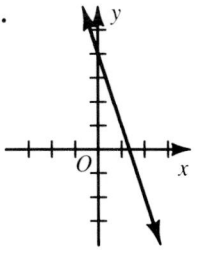

26. 5 cm and 22 cm
27. -4 and 3 or -3 and 4

14.

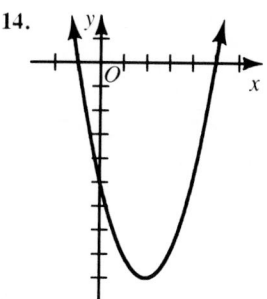

15.

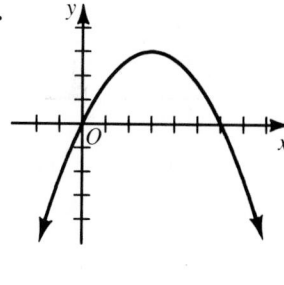

Review of Graphing

1.

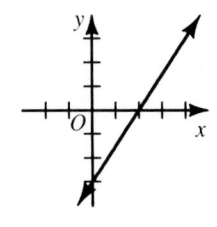

2.

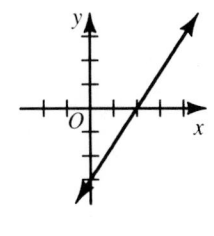

16.

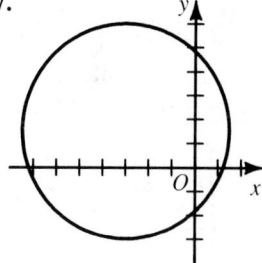

17.

18.

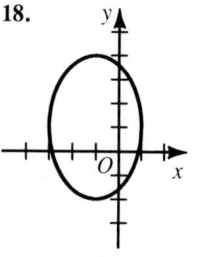

19.

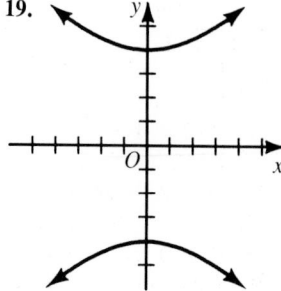

3.

4.

T49

20.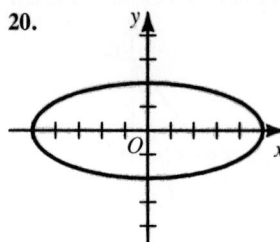

21.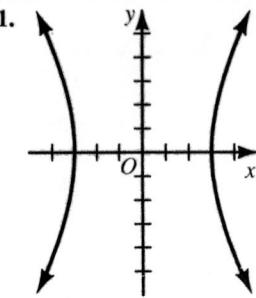

22. $\{(3, 1)\}$

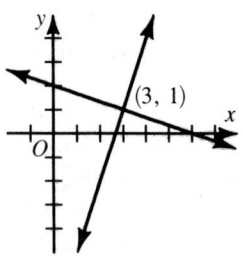

(3, 1)

23. $\{(3, 2), (3, -2),$
$(-3, 2), (-3, -2)\}$

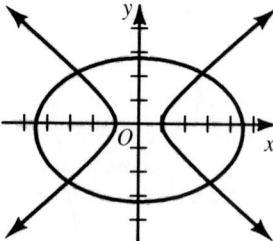

24.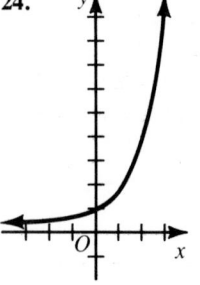

25. y

26.

2π

___

Review of Operations with Numbers

1. 45   **2.** 2   **3.** -1 **4.** 2   **5.** $\dfrac{10}{27}$ **6.** $\dfrac{\sqrt{2}}{6}$

7. 81   **8.** $\dfrac{2}{3}$ **9.** 3   **10.** 8   **11.** $3\sqrt{2}$

12. $\dfrac{2}{3}\sqrt[3]{33}$ **13.** $\dfrac{4}{3}$ **14.** $-8 - 3\sqrt{7}$

15. $7 - 2\sqrt{10}$ **16.** $6\sqrt{3}$ **17.** $48 - 12\sqrt{6}$

18. $-2ac^4$ **19.** $3r^2s^2\sqrt[4]{|s|^2}$ **20.** $7|x|^5$ **21.** $2i\sqrt{3}$

22. $8i\sqrt{5}$ **23.** $-4 + 5i$ **24.** $-33 - 56i$ **25.** 20

26. $\dfrac{13}{10} + \dfrac{11}{10}i$ **27.** -76 **28.** -6 **29.** $6 + 66i$

30. a. 2^{12} **b.** 3^{3n} **31.** $\dfrac{4}{5}$ **32.** $\dfrac{1}{x^2 + 2x}$

33. $\dfrac{a - b}{a + b}$ **34.** 1   **35.** $\log 486$ **36.** $2\sqrt{3} + 2i$

37. $-\dfrac{1}{\sqrt{2}}$ **38.** 1   **39.** $\dfrac{2\pi}{3}$, or $120°$

40. $\dfrac{1}{2 + \sqrt{3}}$, or $2 - \sqrt{3}$

Reading Algebra

Suggestions for helping your students learn how to read an algebra textbook

The question is sometimes asked, "Why do we need to teach reading in an algebra class?" A study of reading algebra reveals that it is integrally tied to the learning of algebra. Difficulties that students encounter in algebra are often the result of difficulties in reading. To improve reading algebra is to improve algebra. To improve reading is to increase the chance of making the student a more independent learner since reading is a learning-to-learn skill. Once a student learns to read proficiently, doors are opened that were previously closed.

The ideas of algebra are expressed in compact form, with extensive use of mathematical symbols, and are applied in the solution of a wide variety of problems. Exposition and examples are illustrated with diagrams that must be read and understood along with the text. Following through the solution of an equation often requires up-and-down eye movements as well as the familiar left-to-right motion, together with pauses to consider how successive steps in the solution are related. For all these reasons, even students who can read other material proficiently may have trouble in reading algebra. Throughout the course, students will profit from the teacher's specific suggestions for improving their reading skills.

Many familiar words—for example, *power, variation, imaginary*—take on specialized meanings in algebra. These meanings need to be pointed out to students, and the importance of learning definitions and using mathematical terms correctly needs to be emphasized.

Diagrams are of great value in almost all branches of mathematics. Their importance in algebra should not be overlooked. Students need to realize that diagrams are essential parts of the discussion, the examples, and often the exercises, and that the diagrams must be read along with the words of the text. Identifying and interpreting the information provided by diagrams, graphs, and charts is a skill that has to be learned. Without adequate preparation, students are likely to become frustrated as they attempt to cope with material presented in this way.

Clearly, successful reading calls for practice and patience on the part of both teacher and students. To help the student, this textbook has been organized in a way that is easy to follow, with important material highlighted. Sections titled *Reading Algebra* contain hints that are designed to help the student use the textbook more effectively. Additional suggestions of ways for the teacher to aid students in improving reading are included in appropriate sections of the Lesson Commentary. The following six major reading objectives are emphasized.

Objective 1

Reading and Communicating Orally

This first objective is to ensure that students have a basic understanding of the relationship between what is spoken and what is written. A mathematics textbook contains many symbols, diagrams, charts, and graphs that are not commonly used in other books. It is often difficult to express their content in words. Further, the interrelationship between symbols is often complex, and the method of verbalizing may be unclear. It is not uncommon for a set of symbols to be verbalized in several ways within one class lesson. For example, a^2 may be called the square of a, the second power of a, a squared, or a to the second power. Practice is necessary if the relationship between the spoken and the written is to be understood.

Math teachers need to stress the use of oral

language by having students verbalize in sentences, summarize, repeat, say another way, read aloud, work together, and perform other tasks that enhance both their own understanding and their ability to communicate the meaning of algebraic expressions.

Objective 2

Reading Silently

This second objective is to provide students with the necessary skills to be able to read silently, recognizing that one's purpose determines the speed and type of reading. When students read silently it is often beneficial if they do so with a purpose—to find the main idea, to look for a specific detail, to summarize, to answer a question, to study an example.

Different purposes demand different types of reading. For previewing or reviewing a lesson, skimming (rapid reading) is used. By contrast, slow reading is used when the student is trying to learn the main ideas and important details of a lesson. For slow reading, it is beneficial for teachers to provide questions in advance that will be discussed upon completion of the reading. When beginning a section, students may be directed first to skim the page to find new ideas and symbols, then to read carefully, and finally to answer questions or to work exercises.

Objective 3

Using Symbols

The third objective is to assist students in recognizing symbols and in associating symbols with words and diagrams. As was suggested earlier, the verbalization of symbols creates a reading problem for many students. The word forms of symbols often are not obvious to students. Once learned, they must be reinforced to ensure retention.

Other difficulties introduced by the use of symbols are their conciseness and their association with concepts not actually stated. The following example demonstrates these difficulties.

$$7(x + 10) - 4x = (-6 + x)(-5)$$
$$7x + 70 - 4x = 30 - 5x$$
$$3x + 70 = 30 - 5x$$
$$8x = -40$$
$$x = -5$$

One way to recognize the conciseness of the symbolism is to try to write out in words the ideas that appear in symbolic form. While this compactness demands slower reading and greater concentration than reading English sentences, it has the advantage of showing the structure of a complex reasoning process. The reader is expected to mentally supply the property that justifies replacing each equation by the equivalent equation that follows it.

Students will benefit from the teacher's explicit demonstration of the appropriate choice and ordering of symbols in the solution of equations and word problems. When several different symbolic forms can be used to represent the same idea, comparisons of these forms will help students recognize the flexibility of algebraic symbolism.

Objective 4

Using Mathematical Words

This fourth objective is to stress the mathematical meaning of words. A mathematics page contains words of everyday language (and, when, the), words that are unique to mathematics (monomial, abscissa, secant), and familiar words with specialized meanings in mathematics (prime, variable, complex). The most commonly used words are primarily *structure* words, whereas most of the *content* words are mathematical. This implies that the meaning of the content words needs to be taught in the mathematics classroom.

You can help students find clues for words within the words themselves and from content clues on the page, such as charts, figures, and symbols. Call attention to prefixes such as *bi-* and *quadr-* that students may know already and that they can use in figuring out the meaning of new words. Point out also that certain important words—*equivalent*, for example—are used without change of meaning in combination with many other words. You should be aware that students may have been exposed to several different names for the same concept—distributive rule, distributive axiom, distributive property, for instance—and that they may therefore be confused.

Understanding definitions of terms is of primary importance. Use of the glossary and the index should be encouraged. It is not out of place to have a spelling quiz from time to time.

Objective 5

Reading Word Problems

This fifth objective is to give students confidence in attacking word problems independently and to help them develop their reasoning power in a variety of situations.

If teachers or students were asked how reading skills are used in algebra, they would probably say, "In word problems." Few mathematics teachers would question the importance of careful reading as the first step in solving a problem, but one needs to do more than say, "Read the problem carefully." How does the student "read carefully"? More specific help is needed.

One learns to read carefully by having some specific questions to ask oneself while reading, by taking advantage of all available resources that aid in comprehension, and by thinking about the problem in a well-organized way. The teacher can help students by working with them through the solution of a variety of problems, following the plan of attack suggested in the textbook on page 50. You may need to point out that many problems that *look* very different are basically alike and can be solved by the same method.

Objective 6

Reading Charts, Graphs, and Diagrams

This sixth objective is to make sure that students can relate the reading of charts, graphs, and diagrams to the rest of the exposition and exercises. Tables and graphs play an integral role in the development of many lessons in algebra. Geometric diagrams are used in problems throughout the textbook, and the discussion of a number of important concepts is illustrated by diagrams. Many students fail to use these visual aids to their advantage. Through oral reading and questioning you can assist students in relating these aids to the words and symbols on the page.

Many students may need a good deal of help in understanding how a table or chart is organized and in relating a table to the graphic representation of the data. Construction of a chart on the chalkboard, with a discussion of the procedure, will help to make the structure clearer to students. The best way to check students' understanding is to involve them in constructing and explaining charts and diagrams.

Problem Solving Strategies

Some strategies that your students can use to become better problem solvers

A problem solving strategy is simply a plan or technique for solving a problem. There are a number of well-known problem solving strategies that relate specifically to algebra. For example, applying the quadratic formula is a strategy for solving quadratic equations. One of the goals of an algebra course is to familiarize students with these standard techniques and to give them enough practice with these techniques so they can use them confidently and successfully to solve problems.

Algebra, however, provides an excellent opportunity to teach not only the standard algorithms but also more general problem solving strategies that are useful in other branches of mathematics and in other subject areas as well as in this course. Since these general strategies provide an *approach* to solving a problem rather than a specific method of solution, they are useful for attacking a problem when the method of solution is not obvious.

Students will learn problem solving skills as they have successful experiences in nonthreatening situations. The teacher should endeavor to expose them to a variety of problems that appeal to their interests and that they can solve with a reasonable amount of effort. Number problems, coin problems, probability problems, and problems involving sequences and series, to name but a few, often lend themselves to many different approaches.

Just to understand a problem thoroughly requires careful reading, knowing the vocabulary, being able to recognize necessary, irrelevant, and contradictory data, identifying special conditions and restraints, and visualizing the situation described. Even if the statement of a problem is understood, a correct solution is not guaranteed. The method presented on page 50 of the textbook is a basic strategy that students can use throughout the course. Some extensions and refinements of the method are suggested in the Reading Algebra feature on page 107.

The following list suggests some questions that students may ask themselves as they search for a plan of attack on an unfamiliar problem.

- Is it similar to a problem I have seen before?
- Can I write and solve a simpler problem of the same type?
- Is there a recognizable pattern in the data?
- Can I draw a diagram to represent the given conditions?
- Can the data be organized in a table, chart, or graph?
- Is there a theorem or formula that I can apply?
- Is trial and error a suitable approach?
- Can I devise a simulation of the situation or an experiment that will lead to a solution?
- Can I frame and test a hypothesis?
- Can I work the problem backward?
- Can I reason deductively?
- Is it possible that the problem has no solution?

Once a strategy has been devised, the solution should be attempted. If a strategy does not work, it should be examined to see whether it should be rejected or modified and tried again. Feedback is essential.

Discussion of problems, strategies, and results should be an ongoing activity in almost every mathematics class. By taking students' special interests into account and building on them, the teacher can do much to create an atmosphere of enthusiasm for problem solving.

In the side column at the beginning of most chapters (see, for example, page 41), there is a short list of problem solving strategies related to topics presented in the chapter, together with references to places in the chapter where these strategies might profitably be used and discussed.

Error Analysis

Anticipating common errors and helping students avoid them

Since mathematics builds on previously learned symbols, concepts, and skills, error patterns that are left uncorrected will impede students' progress. Of course, there are many different reasons for errors, but certain types of errors are more common than others. If you are aware of these common errors, you can help students avoid them and you can be better prepared to help students overcome them if they do occur. Throughout the book, in the side columns next to the textbook pages, common errors have been identified and suggestions for avoiding them have been provided. (See, for example, "Common Errors" on pages 163, 187, and 344.) The errors that are discussed in the side columns are fairly specific. However, many of them can be grouped into one or more of the following categories.

Errors in Reading and Translating

(See, for example, pp. 143, 149, and 512.)

Students often have difficulty in translating English phrases and sentences into mathematical expressions and sentences. For example, students may translate the expression "five less than a number" as $5 - n$. Not reading word problems carefully, with concentration on their meaning, is another frequent cause of difficulty. Students may make mistakes because they do not fully understand the meanings of mathematical terms—for example, "maximum value of a function" or "angle of elevation." Words that have a different meaning in mathematics than in everyday speech, such as *or*, may cause confusion.

Failure to Understand Symbols

(See, for example, pp. 2, 86, and 295.)

Students often do not fully understand the mean-

ings of mathematical symbols. As a result, they may make the following errors:

$$\frac{6}{0} = 0 \qquad \frac{0}{0} = 1 \qquad -x^2 = (-x)^2 \qquad b^n = nb$$

$$2^{-2} = -4 \qquad (a + b)^2 = a^2 + b^2 \qquad -2 > x > 3$$

$$0.\overline{5} = \frac{1}{2} \qquad f(4) = 4f \qquad 2\sqrt{3} = \sqrt{6} \qquad \sqrt{x^2} = x$$

Misunderstanding of Properties

(See, for example, pp. 14, 57, 201, and 215.)

Recurring errors often stem from students' misapplying properties in the ways shown below.

Addition property of equality: $x + 2 = 6$
$$x = 8$$

Multiplication property of equality: $\frac{x}{2} + \frac{x}{3} = 5$
$$3x + 2x = 5$$

Division property of equality: $3x(x + 2) = 0$
$$x + 2 = 0$$

Multiplication property of order: $2x > -8$
$$x < -4$$

Distributive axiom: $-4(x + 1) = 9$
$$-4x + 1 = 9$$

Theorem, p. 213: $(x - 3)(x + 2) = 14$
$$x - 3 = 14 \quad \text{or} \quad x + 2 = 14$$

Errors in Using Standard Forms

(See, for example, pp. 344 and 458.)

Although students may have memorized the Pythagorean theorem or the quadratic formula, they may not understand the importance of using the standard form when they apply these formulas.

Consequently, they often try to work with an expression or an equation without first putting it into standard form. This means that they may substitute incorrect values in the formula $a^2 + b^2 = c^2$ or try to solve an equation by the quadratic formula before transforming it so that one side is 0. Other errors that students are liable to make are illustrated below.

$$2x + 3y = 6 \qquad y = 1 - 3x$$
$$\text{slope} = 2 \qquad \text{slope} = 1$$
$$y\text{-intercept} = -3$$

$$x^2 + 1 - 2x = 0$$
$$a = 1, b = 1, c = -2$$

Use of Incorrect Formulas

(See, for example, pp. 100, 397, 404, and 509.)

Many errors are the result of students' using formulas that are incorrect. Some of the more common "impostors" are shown below.

$$p = l + w \qquad C = \pi r^2 \qquad \text{slope} = \frac{x_2 - x_1}{y_2 - y_1}$$

$$x = -b \pm \frac{\sqrt{b^2 - 4ac}}{2a} \qquad \text{discriminant} = \sqrt{b^2 - 4ac}$$

Errors in Simplifying

(See, for example, pp. 42, 221, 227, and 268.)

In addition to some of the reasons already given, students may make errors in simplifying expressions because of incorrect assumptions such as $\sqrt{a^2 - b^2} = a^2 - b^2$. They may also make errors because they do not take notice of grouping symbols such as the fraction bar or because they do

not follow the prescribed order of operations or because they add unlike terms. Students sometimes confuse the rules of exponents—multiplying exponents when they are multiplying and dividing exponents when they are dividing. Students may forget to (or not realize that they must) change the sign of every term of a polynomial that they are subtracting. Simplifying fractions seems to be particularly troublesome. Thinking that $\frac{n}{n} = 0$ can lead to errors such as $6x^2 + 8x + 2 = 2(3x^2 + 4x)$. Errors

such as $\frac{n + 3}{n} = 3$ and $\frac{\overset{3}{\cancel{6}}y(x + 3)}{\overset{}{\cancel{2}}y^2(x + 1)} = \frac{3(x + 3)}{2(x + 1)}$ are

common.

Errors in Checking

(See, for example, pp. 172, 209, 236, and 274.)

Checking can help students develop self-confidence and alert them to errors. However, a check that is incorrectly performed is not useful. Students often fail to realize that it is not only helpful but necessary to check the roots of fractional and radical equations. The following checking errors may occur: Students may occasionally substitute a value for a variable such as 8 for x, get a true statement such as $4 = 4$, and conclude that the solution is 4. Students may not realize that they must check their answers with the *words* of word problems or that they must check their solutions to systems of linear equations in *all* the *original* equations. Some students may think that they should always discard negative solutions. Students may not think of checking their answers when the method involves considering whether an answer is reasonable or multiplying to check factoring.

Assignment Guide

The following Assignment Guide may be of help in planning the year's work. Four courses are outlined in order to allow flexibility in response to your curriculum requirements and your students' abilities:

(1) an average course covering only the topics in algebra,
(2) an average course covering algebra and trigonometry,
(3) an extended algebra course,
(4) an extended course covering algebra and trigonometry.

Because students' interests and backgrounds differ widely from class to class, the optional features are not listed in the Assignment Guide. You will want to choose those features which best suit your individual classes. If you have access to a computer that accepts Pascal, you may wish to allow some time for your students to do the Programming in Pascal features or the Computer Exercises. Please see the note on page xi regarding these features.

As each section in the text includes more exercises than students would normally be expected to complete, these assignments include only a portion of the exercises.

All the assignments refer to the Written Exercises, with the letter "P" indicating word problems. The letter "S" indicates the spiraled portion of the assignment, which reviews earlier work. The letter "R" indicates a review built into the text.

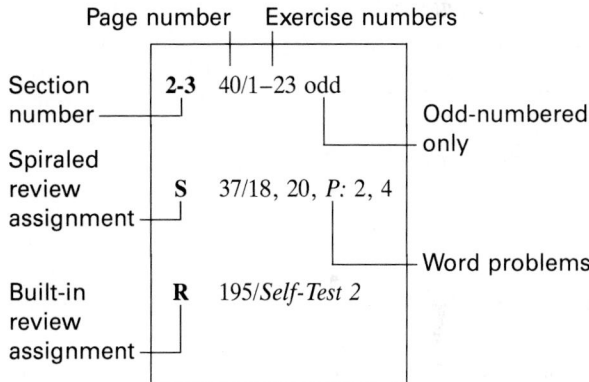

Approximate Time Schedule

Chapter	1	2	3	4	5	6	7	8	9	10	11	12	13	14	15	16	17	Total
Average Algebra Course	7	8	12	12	11	19	17	15	22	18	19	—	—	—	—	—	—	160
Extended Algebra and Trigonometry Course	6	6	9	8	8	17	13	14	14	15	15	—	—	14	12	9	—	160
Enriched Algebra Course	4	5	8	7	7	14	13	12	14	12	14	16	13	14	—	—	7	160
Enriched Algebra and Trigonometry Course	4	5	8	7	7	13	12	11	14	12	14	14	—	13	11	8	7	160

Trimester   Semester   Trimester

Day	Minimal Course		Extended Course Covering Trigonometry		Enriched Course in Algebra		Enriched Course in Algebra and Trigonometry	
1	**1-1**	2/1-7	**1-1**	2/1-9 odd	**1-1**	2/1-9 odd	**1-1**	2/1-9 odd
	1-2	6/1-10 odd, 11-19	**1-2**	6/1-19 odd	**1-2**	6/1-17 odd, 18-22	**1-2**	6/1-17 odd, 18-22
					1-3	9/1-25 odd	**1-3**	9/1-25 odd
					R	11/*Self-Test 1*	**R**	11/*Self-Test 1*
2	**1-3**	9/1-16 odd, 17-22	**1-3**	9/1-19 odd	**1-4**	15/7-39 odd	**1-4**	15/7-39 odd
	R	11/*Self-Test 1*	**R**	11/*Self-Test 1*	**1-5**	20/15-41 odd	**1-5**	20/15-41 odd
					1-6	23/9-33 odd	**1-6**	23/9-33 odd
					1-7	26/5-35 odd	**1-7**	26/5-35 odd
					R	29/*Self-Test 2*	**R**	29/*Self-Test 2*
3	**1-4**	15/1-26	**1-4**	15/1-29 odd	*Prepare for Chapter Test*		*Prepare for Chapter Test*	
	S	9/1-16 even	**S**	9/20-22				
			1-5	20/7-27 odd				
4	**1-5**	20/1-25	**1-6**	23/1-29 odd	*Administer Chapter Test*		*Administer Chapter Test*	
	1-6	23/1-24	**S**	6/4, 6, 8				
	S	6/1-10 even	**1-7**	27/1-27 odd				
			R	29/*Self-Test 2*				
5	**1-7**	26/1-22	*Prepare for Chapter Test*		**2-1**	44/19-33 odd	**2-1**	44/19-33 odd
	R	29/*Self-Test 2*			**2-2**	47/23-45 odd	**2-2**	47/23-45 odd
					2-3	53/*P*: 11-21 odd	**2-3**	53/*P*: 11-21 odd
					R	55/*Self-Test 1*	**R**	55/*Self-Test 1*
6	*Prepare for Chapter Test*		*Administer Chapter Test*		**2-4**	59/5, 7, 15-33 odd	**2-4**	59/5, 7, 15-33 odd
					2-5	62/7-27 odd	**2-5**	62/7-27 odd
						63/*P*: 3-11 odd		63/*P*: 3-11 odd
7	*Administer Chapter Test*		**2-1**	44/5, 9, 13-29 odd	**2-6**	67/3, 4, 5-17 odd	**2-6**	67/3, 4, 5-17 odd
			2-2	47/3-39 odd	**2-7**	71/5, 13-37 odd	**2-7**	71/5, 13-37 odd
					R	72/*Self-Test 2*	**R**	72/*Self-Test 2*
8	**2-1**	44/1-26	**2-3**	52/*P*: 7-19 odd	*Prepare for Chapter Test*		*Prepare for Chapter Test*	
	2-2	47/1-9, 11-23 odd, 24-28	**R**	55/*Self-Test 1*				
			S	47/12-20 even				
			2-4	59/5-25 odd				
9	**2-3**	52/*P*: 1-13	**2-5**	62/5-19 odd, 22-24	*Administer Chapter Test*		*Administer Chapter Test*	
	R	55/*Self-Test 1*		63/*P*: 1-8	**R**	76/*Mixed Review*	**R**	76/*Mixed Review*
	S	47/10-22 even						
10	**2-4**	59/1-20	**2-6**	67/1-5, 7-15 odd, 16	**3-1**	86/1-19 odd, 20-33	**3-1**	86/1-19 odd, 20-33
			S	62/20, 21				
			2-7	71/5-35 odd				
			R	72/*Self-Test 2*				

Day	Minimal Course	Extended Course Covering Trigonometry	Enriched Course in Algebra	Enriched Course in Algebra and Trigonometry
11	**2-5** 62/1-7, 9, 11, 13-18 63/*P*: 1-6 **S** 59/25, 26	*Prepare for Chapter Test*	**3-2** 90/3-27 odd **R** 91/*Self-Test 1*	**3-2** 90/3-27 odd **R** 91/*Self-Test 1*
12	**2-6** 67/1-4 **S** 62/8, 10, 12	*Administer Chapter Test* **R** 76/*Mixed Review*	**3-3** 94/5-17 odd, 18-28 **3-4** 97/5-19 odd, 20-24 **R** 98/*Self-Test 2*	**3-3** 94/5-17 odd, 18-28 **3-4** 97/5-19 odd, 20-24 **R** 98/*Self-Test 2*
13	**2-7** 71/1-18 **R** 72/*Self-Test 2*	**3-1** 86/1-23 odd, 24-30 **3-2** 90/1-21 odd **R** 91/*Self-Test 1* **S** 86/2-8 even	**3-5** 102/5-23 odd, 24-26	**3-5** 102/5-23 odd, 24-26
14	*Prepare for Chapter Test*	**3-3** 94/1-11 odd, 12-24	**3-6** 105/7-33 odd, 35-40 **R** 106/*Self-Test 3* **S** 87/14-18 even	**3-6** 105/7-33 odd, 35-40 **R** 106/*Self-Test 3* **S** 87/14-18 even
15	*Administer Chapter Test* **R** 76/*Mixed Review*	**3-4** 97/1-13 odd, 15-21 **R** 98/*Self-Test 2* **S** 90/10-18 even	**3-7** 110/5, 9, 11, 15-31 odd 112/*P*: 3-9 **R** 113/*Self-Test 4*	**3-7** 110/5, 9, 11, 15-31 odd 112/*P*: 3-9 **R** 113/*Self-Test 4*
16	**3-1** 86/1-22 odd, 23-30	**3-5** 102/1-23 odd	*Prepare for Chapter Test*	*Prepare for Chapter Test*
17	**3-2** 90/1-7, 9-17 odd, 18, 19	**3-6** 105/5-39 odd **R** 106/*Self-Test 3* **S** 87/14-20 even	*Administer Chapter Test*	*Administer Chapter Test*
18	**R** 91/*Self-Test 1* **S** 86/2-12 even	**3-7** 110/1-29 odd	**4-1** 126/9-27 odd **4-2** 131/7-35 odd, 36-39	**4-1** 126/9-27 odd **4-2** 131/7-35 odd, 36-39
19	**3-3** 94/1-16	**3-7** 112/*P*: 1-5 **R** 113/*Self-Test 4*	**4-3** 135/7-27 odd **S** 54/22 64/8, 10, 12	**4-3** 135/7-27 odd **S** 54/22 64/8, 10, 12
20	**3-4** 97/1-18	*Prepare for Chapter Test*	**4-4** 142/7-25 odd **R** 144/*Self-Test 1*	**4-4** 142/7-25 odd **R** 144/*Self-Test 1*
21	**R** 98/*Self-Test 2* **S** 90/8-16 even	*Administer Chapter Test*	**4-5** 146/7-35 odd **S** 71/18-24 even 97/12-18 even	**4-5** 146/7-35 odd **S** 71/18-24 even 97/12-18 even
22	**3-5** 102/1-20	**4-1** 126/9-23 odd, 14, 24-28	**4-6** 150/5-19 odd **R** 152/*Self-Test 2*	**4-6** 150/5-19 odd **R** 152/*Self-Test 2*

Day	Minimal Course		Extended Course Covering Trigonometry		Enriched Course in Algebra		Enriched Course in Algebra and Trigonometry	
23	**3-6**	105/1-28	**4-2**	131/1-31 odd, 32	*Prepare for Chapter Test*		*Prepare for Chapter Test*	
	R	106/*Self-Test 3*						
	S	81/14-22 even						
24	**3-7**	110/1-26	**4-3**	135/1-21 odd	*Administer Chapter Test*		*Administer Chapter Test*	
			S	53/*P*: 14, 16, 18	**R**	155/*Cumulative Review*	**R**	155/*Cumulative Review*
				64/*P*: 9-12				
25	**3-7**	112/*P*: 1-6	**4-4**	142/1-21 odd	**5-1**	164/3-21 odd, 22-26	**5-1**	164/3-21 odd, 22-26
	R	113/*Self-Test 4*	**R**	144/*Self-Test 1*				
26	*Prepare for Chapter Test*		**4-5**	146/1-31 odd	**5-2**	170/19-29 odd	**5-2**	170/19-29 odd
			S	71/20, 22, 24	**S**	131/22-34 even	**S**	131/22-34 even
				97/22-24				
27	*Administer Chapter Test*		**4-6**	150/5-17 odd	**5-3**	174/7-19 odd, 20-26	**5-3**	174/7-19 odd, 20-26
			R	152/*Self-Test 2*	**R**	176/*Self-Test 1*	**R**	176/*Self-Test 1*
					S	135/20-26 even	**S**	135/20-26 even
28	**4-1**	126/1-26	*Prepare for Chapter Test*		**5-4**	179/1-15 odd	**5-4**	179/1-15 odd
					5-5	181/*P*: 1-13	**5-5**	181/*P*: 1-13
29	**4-2**	131/2-18 even,	*Administer Chapter Test*		**5-6**	187/3-11 odd	**5-6**	187/3-11 odd
		26-32 even	**R**	155/*Cumulative Review*	**R**	189/*Self-Test 2*	**R**	189/*Self-Test 2*
30	**4-3**	135/1-20	**5-1**	164/3-21 odd, 22-26	*Prepare for Chapter Test*		*Prepare for Chapter Test*	
31	**S**	53/*P*: 14-19	**5-2**	170/5-27 odd	*Administer Chapter Test*		*Administer Chapter Test*	
		63/*P*: 7-10	**S**	131/16-30 even				
32	**4-4**	142/1-22	**5-3**	174/5-23 odd	**6-1**	203/13-33 odd	**6-1**	203/13-33 odd
			R	176/*Self-Test 1*				
			S	135/12-22 even				
33	**R**	144/*Self-Test 1*	**5-4**	179/1-13 odd	**6-2**	206/13, 19-31 odd, 33-38	**6-2**	206/13, 19-31 odd, 33-38
	S	62/19-28						
		94/17, 18						
34	**4-5**	146/1-30	**5-5**	181/*P*: 1-11	**6-3**	210/1-25	**6-3**	210/1-25
					R	211/*Self-Test 1*	**R**	211/*Self-Test 1*
35	**S**	71/19-24	**5-6**	187/1-9 odd	**6-4**	216/5, 7, 17-25 odd	**6-4**	216/5, 7, 17-25 odd
		97/19-24	**R**	189/*Self-Test 2*		216/*P*: 1-24		216/*P*: 1-24

Day	Minimal Course		Extended Course Covering Trigonometry		Enriched Course in Algebra		Enriched Course in Algebra and Trigonometry	
36	**4-6**	150/1-13	*Prepare for Chapter Test*		**6-5** **R** **S**	219/19-23 odd 220/*Self-Test 2* 203/18-32 even	**6-5** **R** **S**	219/19-23 odd 220/*Self-Test 2* 203/18-32 even
37	**4-6** **R**	151/14-17 152/*Self-Test 2*	*Administer Chapter Test*		**6-6**	222/7-25 odd, 27-30	**6-6**	222/7-25 odd, 27-30
38	*Prepare for Chapter Test*		**6-1**	203/9-17 odd, 18-27	**6-7** **S**	225/3, 13, 15, 17-26 210/26-37	**6-7** **S**	225/3, 13, 15, 17-26 210/26-37
39	*Administer Chapter Test*		**6-2**	206/7-33 odd	**6-8**	228/9-17 odd, 18-22	**6-8**	228/9-17 odd, 18-22
40	**5-1**	164/1-23 odd	**6-3** **R**	210/1-16, 17-27 odd 211/*Self-Test 1*	**6-9**	230/7-19 odd	**6-9**	230/7-19 odd
41	**5-2**	170/1-26	**6-4**	216/1-24	**6-10** **S**	232/*P*: 1-12 219/19-23 odd 230/20-24	**6-10** **S**	232/*P*: 1-12 219/19-23 odd 230/20-24
42	**S**	131/25-31 odd 164/2-24 even	**6-4**	216/*P*: 1-12	**6-11**	235/3-23 odd	**6-11** **R**	235/3-23 odd 236/*P*: 7-13 238/*Self-Test 3*
43	**5-3**	174/1-22	**6-5**	219/1-12, 13-17 odd	**6-11** **R**	236/*P*: 1-13 238/*Self-Test 3*	*Prepare for Chapter Test*	
44	**R** **S**	176/*Self-Test 1* 135/21, 22, 26, 27	**R** **S**	220/*Self-Test 2* 203/2-16 even	*Prepare for Chapter Test*		*Administer Chapter Test*	
45	**5-4**	179/1-14	**6-6**	222/5-29 odd	*Administer Chapter Test*		**7-1**	249/5-21 odd 250/*P*: 1-9
46	**S**	142/2-16 even	**6-7** **S**	225/1-23 odd 210/18-28 even, 19-35	**7-1**	249/5-21 odd 250/*P*: 1-9	**7-2**	253/7-25 odd, 26-30
47	**5-5**	181/*P*: 1-9	**6-8** **S**	228/1-21 odd 203/16-24 even	**7-2**	253/7-25 odd, 26-30	**7-3** **R**	255/9-27 odd, 28-30 256/*Self-Test 1*
48	**5-6** **R**	187/1-9 189/*Self-Test 2*	**6-9**	230/1-23 odd	**7-3** **R**	255/9-27 odd, 28-30 256/*Self-Test 1*	**7-4** **S**	260/3, 7, 11-21 odd, 24-26 222/2-12 even
49	*Prepare for Chapter Test*		**6-10**	232/*P*: 1-12	**7-4** **S**	260/3, 7, 11-21 odd, 24-26 222/2-12 even	**7-5** **R**	264/7, 11, 13-25 odd, 27-29 265/*Self-Test 2*

Day	Minimal Course		Extended Course Covering Trigonometry		Enriched Course in Algebra		Enriched Course in Algebra and Trigonometry	
50	*Administer Chapter Test*		S	219/14-18 even 230/2-22 even	**7-5**	264/7, 11, 13-25 odd, 27-29	**7-6**	269/7-37 odd
					R	265/*Self-Test 2*		
51	**6-1**	203/1-23 odd	**6-11**	235/3-23 odd	**7-6**	269/7-37 odd	**7-7**	271/17-41 odd 272/*P*: 1-10
							S	206/6-18 even
52	**6-2**	206/1-17 odd, 19-30	**6-11**	236/*P*: 1-13	**7-7**	271/5-41 odd 272/*P*: 1-9	**7-8**	275/9, 15-29 odd
			R	238/*Self-Test 3*	**S**	206/6-18 even	**R**	275/*Self-Test 3*
							S	269/6-26 even
53	**6-3**	210/1-16, 17-27 odd	*Prepare for Chapter Test*		**7-8**	275/9-33 odd	**7-9**	278/1-35 odd
	R	211/*Self-Test 1*			**R**	275/*Self-Test 3*	**7-10**	282/9-25 odd, 26-28
54	**6-4**	216/1-23	*Administer Chapter Test*		**7-9**	278/1-33 odd	**7-11**	284/5-25 odd
							R	285/*Self-Test 4*
55	**6-4**	216/*P*: 1-13	**7-1**	249/5-21 odd 250/*P*: 1-9	**7-10**	282/9-25 odd, 26-28	*Prepare for Chapter Test*	
56	**6-5**	219/1-12, 13-17 odd	**7-2**	253/3-27 odd	**7-11**	284/5-25 odd	*Administer Chapter Test*	
					R	285/*Self-Test 4*		
57	**R**	220/*Self-Test 2*	**7-3**	255/1-27 odd	*Prepare for Chapter Test*		**8-1**	295/9-31 odd
	S	201/2-14 even	**R**	256/*Self-Test 1*				
58	**6-6**	222/2-12 even, 13-24	**7-4**	260/1-23 odd	*Administer Chapter Test*		**8-2**	298/17-43 odd 299/*P*: 3-8
			S	222/2-12 even				
59	**6-7**	225/1-22	**7-5**	264/5-27 odd	**8-1**	295/9-31 odd	**8-3**	303/7-37 odd 304/*P*: 5-12
			R	265/*Self-Test 2*				
60	**S**	210/18-28 even, 29-35	**7-6**	269/5-33 odd	**8-2**	298/17-43 odd 299/*P*: 3-8	**R**	305/*Self-Test 1*
							S	298/18-32 even
61	**6-8**	228/1-22	**7-7**	271/5-37 odd 272/*P*: 1-9	**8-3**	303/7-37 odd 304/*P*: 5-12	**8-4**	309/5, 9, 15-31 odd 311/*P*: 5-11
	S	203/16-24 even	**S**	206/2-18 even				
62	**6-9**	230/1-23 odd	**7-8**	275/9-33 odd	**R**	305/*Self-Test 1*	**8-5**	314/5, 11, 17-31 odd, 32-36
			R	275/*Self-Test 3*	**S**	298/18-32 even		

Day	Minimal Course		Extended Course Covering Trigonometry		Enriched Course in Algebra		Enriched Course in Algebra and Trigonometry	
63	**6-10**	232/*P*: 1-12	**7-9**	278/1-33 odd	**8-4**	309/5, 9, 15-31 odd 310/*P*: 1-11	**8-6** **R**	317/15-33 odd 319/*P*: 4-8 320/*Self-Test 2*
64	**S**	219/14-18 even 230/2-22 even	**7-10**	282/5-27 odd	**8-5**	314/5, 11, 17-31 odd, 32-36	**8-7** **S**	323/1-20 264/20-28 even
65	**6-11**	235/1-20	**7-11** **R**	284/1-21 odd 285/*Self-Test 4*	**8-6** **R**	317/17-27 odd 318/*P*: 1-7 320/*Self-Test 2*	**8-8** **R**	327/7-21 odd 328/*P*: 5-10 330/*Self-Test 3*
66	**6-11** **R**	236/*P*: 1-12 238/*Self-Test 3*	*Prepare for Chapter Test*		**8-7** **S**	323/1-19 264/20-28 even	*Prepare for Chapter Test*	
67	*Prepare for Chapter Test*		*Administer Chapter Test*		**8-8** **R**	327/7-33 odd 328/*P*: 1-6 330/*Self-Test 3*	*Administer Chapter Test*	
68	*Administer Chapter Test*		**8-1**	295/3-29 odd	*Prepare for Chapter Test*		**9-1**	345/9, 19, 29-45 odd
69	**R**	241/*Mixed Review*	**8-2**	298/13-41 odd 299/*P*: 1-8	*Administer Chapter Test*		**9-1**	346/*P*: 3-10
70	**7-1**	249/1-20	**8-3**	303/5-33 odd 303/*P*: 1-8	**R**	334/*Cumulative Review*	**9-2**	349/7-13 odd, 14-20
71	**7-1**	250/*P*: 1-5	**R** **S**	305/*Self-Test 1* 298/14-32 even	**9-1**	345/9, 19, 29-45 odd	**9-3** **R** **S**	352/7-15 odd, 16-20 353/*Self-Test 1* 349/1-12 even
72	**7-2**	253/1-28	**8-4**	309/1-25 odd 310/*P*: 1-11	**9-1**	346/*P*: 3-10	**9-4**	356/7-27 even
73	**7-3**	255/1-17 odd, 18-25	**8-5**	314/1-29 odd	**9-2**	349/7-13 odd, 14-20	**9-5**	360/1-16 361/*P*: 1-7
74	**R** **S**	256/*Self-Test 1* 222/1-11 odd	**8-6**	317/5-27 odd	**9-3** **R** **S**	352/7-15 odd, 16-20 353/*Self-Test 1* 349/1-12 even	**9-6** **R** **S**	364/1-11 364/*Self-Test 2* 255/2-16 even
75	**7-4**	260/1-24	**8-6** **R**	318/*P*: 1-7 320/*Self-Test 2*	**9-4**	356/7-27 odd	**9-7**	367/11-33 odd
76	**7-5** **R**	264/1-18, 20-28 even 265/*Self-Test 2*	**8-7** **S**	323/1-12 264/18-28 even	**9-5**	360/1-16 361/*P*: 1-7	**9-8**	371/3, 7, 11-33 odd

Day	Minimal Course	Extended Course Covering Trigonometry	Enriched Course in Algebra	Enriched Course in Algebra and Trigonometry
77	**7-6** 269/2-34 even	**8-8** 327/1-31 odd	**9-6** 364/1-11 **R** 364/*Self-Test 2* **S** 255/2-16 even	**9-9** 375/5-13 odd, 14-19 **R** 376/*Self-Test 3* **S** 367/2-22 even
78	**7-7** 271/1-28	**8-8** 328/*P*: 1-6 **R** 330/*Self-Test 3*	**9-7** 367/11-33 odd	**9-10** 379/1-16
79	**7-7** 272/*P*: 1-9	*Prepare for* *Chapter Test*	**9-8** 371/3, 7, 11-33 odd	**9-11** 382/1-18 **R** 383/*Self-Test 4*
80	**7-8** 275/1-34 **S** 269/1-33 odd	*Administer* *Chapter Test*	**9-9** 375/5-13 odd, 14-19 **R** 376/*Self-Test 3* **S** 367/2-22 even	*Prepare for* *Chapter Test*
81	**R** 275/*Self-Test 3* **S** 206/2-18 even	**R** 334/*Cumulative* *Review*	**9-10** 379/1-16	*Administer* *Chapter Test*
82	**7-9** 278/1-33	**9-1** 345/7-43 odd	**9-11** 382/1-18 **R** 383/*Self-Test 4*	**10-1** 396/15, 17, 19-29
83	**7-10** 282/1-24	**9-1** 346/*P*: 1-9	*Prepare for* *Chapter Test*	**10-2** 399/9, 11-18 **R** 400/*Self-Test 1* **S** 345/2-10 even
84	**7-11** 284/1-21 **R** 285/*Self-Test 4*	**9-2** 349/1-11 odd, 12-18	*Administer* *Chapter Test*	**10-3** 402/5, 15, 19-37 odd
85	*Prepare for* *Chapter Test*	**9-3** 352/1-13 odd, 15-18 **R** 353/*Self-Test 1* **S** 349/2-14 even	**10-1** 396/15, 17, 19-29	**10-4** 407/11-31 odd
86	*Administer* *Chapter Test*	**9-4** 356/5-21 odd	**10-2** 399/9, 11-18 **R** 400/*Self-Test 1* **S** 345/2-10 even	**10-5** 411/11-35 odd
87	**8-1** 295/1-27	**9-5** 360/7-19 odd 361/*P*: 1-7	**10-3** 402/5, 15, 19-37 odd	**10-6** 416/13-33 odd
88	**8-2** 298/2-36 even, 37-42 299/*P*: 1-8	**9-6** 364/1-10 **R** 364/*Self-Test 2* **S** 255/2-16 even	**10-4** 407/11-31 odd	**10-7** 420/5-21 odd 421/*P*: 1-7 **R** 422/*Self-Test 2*
89	**8-3** 303/1-34 303/*P*: 1-8	**9-7** 367/7-27 odd	**10-5** 411/11-35 odd	**10-8** 424/1-15 odd **S** 345/36-44 even 131/2-14 even
90	**R** 305/*Self-Test 1* **S** 298/1-35 odd	**9-8** 371/1-31 odd	**10-6** 416/13-33 odd	**10-9** 427/1-13 odd 427/*P*: 1-11

Day	Minimal Course	Extended Course Covering Trigonometry	Enriched Course in Algebra	Enriched Course in Algebra and Trigonometry
91	**8-4** 309/1-26	**9-9** 375/1-19 **R** 376/*Self-Test 3* **S** 367/2-22 even	**10-7** 420/5-21 odd 421/*P*: 1-7 **R** 422/*Self-Test 2*	**10-10** 431/1-13 odd, 14-16 432/*P*: 3-7 **R** 433/*Self-Test 3*
92	**8-4** 310/*P*: 1-11	**9-10** 379/1-16	**10-8** 424/1-15 odd **S** 345/36-44 even 131/2-14 even	*Prepare for Chapter Test*
93	**8-5** 314/1-30	**9-11** 382/1-16 **R** 383/*Self-Test 4*	**10-9** 427/1-13 odd 427/*P*: 1-11	*Administer Chapter Test*
94	**8-6** 317/1-28	*Prepare for Chapter Test*	**10-10** 431/1-13 odd, 14-16 432/*P*: 3-7 **R** 433/*Self-Test 3*	**11-1** 443/7-41 odd, 42-48
95	**8-6** 318/*P*: 1-7 **R** 320/*Self-Test 2*	*Administer Chapter Test*	*Prepare for Chapter Test*	**11-2** 446/7-17 odd, 18-22
96	**8-7** 323/1-14 **S** 264/21-27 odd	**10-1** 396/7-27 odd	*Administer Chapter Test*	**R** 447/*Self-Test 1* **S** 111/8-18 even
97	**8-8** 327/1-32	**10-2** 399/1-15 odd **R** 400/*Self-Test 1* **S** 345/2-10 even	**11-1** 443/7-41 odd, 42-48	**11-3** 450/5-11 odd, 13-18
98	**8-8** 328/*P*: 1-6 **R** 330/*Self-Test 3*	**10-3** 402/3-29 odd **S** 356/1-22 even	**11-2** 446/7-17 odd, 18-22	**11-4** 453/9-37 odd
99	*Prepare for Chapter Test*	**10-4** 407/1-27 odd	**R** 447/*Self-Test 1* **S** 111/8-18 even	**11-5** 456/1-35 odd
100	*Administer Chapter Test*	**10-5** 411/7-31 odd	**11-3** 450/5-11 odd, 13-18	**R** 456/*Self-Test 2* **S** 453/14-24 even
101	**R** 334/*Cumulative Review*	**10-6** 416/5-27 odd	**11-4** 453/9-37 odd	**11-6** 460/2-4, 7-31 odd
102	**9-1** 345/1-9 odd, 11-28	**10-7** 420/1-19 odd	**11-5** 456/1-35 odd	**11-7** 463/7-31 odd
103	**9-1** 346/29-41 odd	**10-7** 421/*P*: 1-7 **R** 422/*Self-Test 2*	**R** 456/*Self-Test 2* **S** 453/14-24 even	**11-8** 466/*P*: 1-10
104	**9-1** 346/*P*: 1-9	**10-8** 424/1-12 **S** 345/30-44 even 131/2-14 even	**11-6** 460/2-4, 7-31 odd	**11-9** 471/3, 7, 11-29 odd

Day	Minimal Course	Extended Course Covering Trigonometry	Enriched Course in Algebra	Enriched Course in Algebra and Trigonometry
105	**9-2** 349/2-16 even, 17, 18	**10-9** 427/1-14	**11-7** 463/7-31 odd	**R** 473/*Self-Test 3*
106	**9-3** 352/1-18	**10-9** 427/*P:* 1-9	**11-8** 466/*P:* 1-10	*Prepare for Chapter Test*
107	**R** 353/*Self-Test 1* **S** 349/1-15 odd	**10-10** 431/1-16	**11-9** 471/3, 7, 11-29 odd	*Administer Chapter Test*
108	**9-4** 356/1-12	**10-10** 432/*P:* 1-7 **R** 433/*Self-Test 3*	**R** 473/*Self-Test 3*	**12-1** 485/1-13 odd, 14
109	**9-4** 356/13-22	*Prepare for Chapter Test*	*Prepare for Chapter Test*	**12-2** 489/1-13 odd, 15-25
110	**9-5** 360/1-16	*Administer Chapter Test*	*Administer Chapter Test*	**12-3** 491/5-15 odd **R** 491/*Self-Test 1*
111	**9-5** 361/*P:* 1-7	**11-1** 443/1-43 odd	**12-1** 485/1-13 odd, 14	**12-4** 494/5-17 odd, 18-20
112	**9-6** 364/1-5	**11-2** 446/1-19 odd	**12-2** 489/1-13 odd, 15-25	**12-5** 496/3-15 odd, 16-22 **R** 497/*Self-Test 2* **S** 206/24, 28, 30
113	**9-6** 364/6-10 **R** 364/*Self-Test 2*	**R** 447/*Self-Test 1* **S** 111/8-18 even	**12-3** 491/5-15 odd	**12-6** 500/1-25 odd
114	**9-7** 367/2-24 even	**11-3** 450/1-9 odd, 10-16	**R** 491/*Self-Test 1* **S** 489/2-12 even	**12-7** 504/1-15 odd, 16-19
115	**9-7** 367/23, 25-28	**11-4** 453/1-33 odd	**12-4** 494/1-11 odd, 12-20	**R** 504/*Self-Test 3* **S** 485/2-12 even
116	**9-8** 371/1-8	**11-5** 456/1-15 odd	**12-5** 496/3-15 odd, 16-22	**12-8** 506/1-15 odd, 16 **S** 494/6, 8, 10
117	**9-8** 372/9-28	**11-5** 456/17-33	**R** 497/*Self-Test 2* **S** 206/24, 28, 30	**12-9** 510/9-15 odd
118	**9-9** 375/1-15	**R** 456/*Self-Test 2* **S** 453/14-24 even	**12-6** 500/1-25 odd	**12-10** 513/1-9 odd
119	**R** 376/*Self-Test 3* **S** 367/1-23 odd	**11-6** 460/1, 2, 3-27 odd	**12-7** 504/1-15 odd, 16-19	**12-11** 517/1-15 odd **R** 518/*Self-Test 4*
120	**9-10** 379/1-14	**11-7** 463/1-15 odd	**R** 504/*Self-Test 3* **S** 485/2-12 even	*Prepare for Chapter Test*

Day	Minimal Course	Extended Course Covering Trigonometry	Enriched Course in Algebra	Enriched Course in Algebra and Trigonometry
121	**9-11** 382/1-16 **R** 383/*Self-Test 4*	**11-7** 463/17-25 odd	**12-8** 506/1-15 odd, 16 **S** 494/2-10 even	*Administer Chapter Test*
122	*Prepare for Chapter Test*	**11-8** 466/*P*: 1-9	**12-9** 510/9-15 odd	**14-1** 578/3, 7, 11, 13-25 odd
123	*Administer Chapter Test*	**11-9** 471/3-27 odd **R** 473/*Self-Test 3*	**12-10** 513/1-9 odd	**14-2** 582/7-31 odd, 32-34
124	**10-1** 396/1-22	*Prepare for Chapter Test*	**12-11** 517/1-16 **R** 518/*Self-Test 4*	**R** 584/*Self-Test 1* **S** 578/6-12 even
125	**10-2** 399/1-14	*Administer Chapter Test*	*Prepare for Chapter Test*	**14-3** 588/5-25 odd, 24
126	**R** 400/*Self-Test 1* **S** 345/2-10 even	**14-1** 578/3-23 odd, 4, 6, 22	*Administer Chapter Test*	**14-4** 591/1-17 odd, 19-23
127	**10-3** 402/1-24	**14-2** 582/3-33 odd	**13-1** 531/3-17 odd, 18, 19	**14-5** 594/9-47 odd
128	**10-3** 403/25-30 **S** 356/1-22 even	**R** 584/*Self-Test 1* **S** 578/10-20 even	**13-2** 534/3-17 odd	**14-6** 598/1-19 odd
129	**10-4** 407/1-24	**14-3** 588/1-23 odd	**13-3** 537/7-25 odd, 26-28	**14-7** 600/3-19 odd
130	**10-5** 411/1-24	**14-4** 591/1-13, 15-21	**13-4** 541/7-21 odd, 22-30	**R** 601/*Self-Test 2* **S** 588/6-22 even
131	**10-6** 416/1-24	**14-5** 594/7-33 odd	**13-5** 545/5-17 odd, 19-27	**14-8** 605/1-51 odd
132	**10-7** 420/1-20	**14-6** 598/1-13 odd, 15-19	**R** 546/*Self-Test 1* **S** 531/2-16 even 541/8-20 even	**14-9** 610/1-19 odd 610/*P*: 1-6 **R** 612/*Self-Test 3*
133	**10-7** 421/*P*: 1-7 **R** 422/*Self-Test 2*	**14-7** 600/1-17 odd	**13-6** 551/1-27 odd	*Prepare for Chapter Test*
134	**10-8** 424/1-12	**R** 601/*Self-Test 2* **S** 588/2-22 even	**R** 552/*Self-Test 2* **S** 534/2-12 even	*Administer Chapter Test*
135	**S** 345/30-44 even 131/1-11 odd	**14-8** 605/1-47 odd	**13-7** 556/1-27 odd, 28, 29	**15-1** 621/7-35 odd
136	**10-9** 427/1-14	**14-9** 610/1-19 odd	**13-8** 562/1-29 odd, 30	**15-2** 623/1-9, 11-19 odd, 21-28 **R** 624/*Self-Test 1*

Day	Minimal Course	Extended Course Covering Trigonometry	Enriched Course in Algebra	Enriched Course in Algebra and Trigonometry
137	**10-9** 427/*P*: 1-9 **S** 131/13-17 odd	**14-9** 610/*P*: 1-6 **R** 612/*Self-Test 3*	**R** 563/*Self-Test 3*	**15-3** 630/1-35 odd, 36 **S** 621/10-20 even
138	**10-10** 431/1-15	*Prepare for Chapter Test*	*Prepare for Chapter Test*	**15-4** 635/3-59 odd
139	**10-10** 432/*P*: 1-7 **R** 433/*Self-Test 3*	*Administer Chapter Test*	*Administer Chapter Test*	**15-5** 640/5, 11, 13-33 odd, 35-38 **S** 623/10-20 even
140	*Prepare for Chapter Test*	**15-1** 621/9-31 odd	**14-1** 578/3, 7, 11, 13-25 odd	**15-6** 643/7-25 odd, 27-32 **R** 644/*Self-Test 2*
141	*Administer Chapter Test*	**15-2** 623/1-9, 11-19 odd, 21-26 **R** 624/*Self-Test 1*	**14-2** 582/7-31 odd, 32-34	**15-7** 646/5-17 odd, 14, 16
142	**11-1** 443/1-24	**15-3** 630/1-35 odd **S** 621/10-20 even	**R** 584/*Self-Test 1* **S** 578/6-12 even	**15-7** 647/*P*: 3-10
143	**11-1** 444/25-38	**15-4** 635/1-55 odd	**14-3** 588/5-25 odd, 24	**15-8** 651/5-23 odd 652/*P*: 2-8 **R** 653/*Self-Test 3*
144	**11-2** 446/1-20	**15-5** 640/1-37 odd **S** 623/10-20 even	**14-4** 591/1-19 odd, 20-23	*Prepare for Chapter Test*
145	**R** 447/*Self-Test 1*	**15-6** 643/1-23 odd, 25-29 **R** 644/*Self-Test 2*	**14-5** 594/7-43 odd	*Administer Chapter Test*
146	**11-3** 450/1-15	**15-7** 646/1-17 odd	**14-6** 598/1-19 odd	**16-1** 666/5-51 odd **S** 643/10-20 even
147	**11-4** 453/1-19 odd	**15-7** 647/*P*: 1-9	**14-7** 600/3-19 odd	**16-2** 669/9-33 odd, 35-38 **R** 670/*Self-Test 1* **S** 600/4-12 even
148	**11-4** 453/21-34	**15-8** 651/1-17 odd, 18-22	**R** 601/*Self-Test 2* **S** 588/2-22 even	**16-3** 675/5-33 odd, 35-37
149	**11-5** 456/1-20	**15-8** 652/*P*: 1-6 **R** 653/*Self-Test 3*	**14-8** 605/1-51 odd	**16-4** 678/5-21 odd, 23-26
150	**11-5** 456/21-34	*Prepare for Chapter Test*	**14-9** 610/1-19 odd	**16-5** 683/5-23 odd, 24-28 **R** 684/*Self-Test 2*

Day	Minimal Course	Extended Course Covering Trigonometry	Enriched Course in Algebra	Enriched Course in Algebra and Trigonometry
151	**R** 456/*Self-Test 2* **S** 453/8-20 even	*Administer* *Chapter Test*	**14-9** 610/*P*: 1-6 **R** 612/*Self-Test 3*	**16-6** 690/1-15 odd, 22 **16-7** 693/4-10 **R** 694/*Self-Test 3*
152	**11-6** 460/1-24	**16-1** 666/1-47 odd **S** 643/10-20 even	*Prepare for* *Chapter Test*	*Prepare for* *Chapter Test*
153	**11-7** 463/1-10	**16-2** 669/1-33 odd **R** 670/*Self-Test 1* **S** 600/4-12 even	*Administer* *Chapter Test*	*Administer* *Chapter Test*
154	**11-7** 463/11-24	**16-3** 675/1-33 odd	**17-1** 708/3-15 odd, 16, 21	**17-1** 708/3-15 odd, 16, 21
155	**11-8** 466/*P*: 1-8	**16-4** 678/1-23 odd	**17-2** 714/1-13 odd, 14-17 **R** 716/*Self-Test 1* **S** 708/1, 2	**17-2** 714/1-13 odd, 14-17 **R** 716/*Self-Test 1* **S** 708/1, 2
156	**11-9** 471/1-20	**16-5** 683/1-23 odd **R** 684/*Self-Test 2*	**17-3** 724/1-10 **S** 714/2, 4, 6	**17-3** 724/1-10 **S** 714/2, 4, 6
157	**11-9** 471/21-27	**16-6** 690/1-19	**17-4** 730/1-13 **R** 731/*Self-Test 2*	**17-4** 730/1-13 **R** 731/*Self-Test 2*
158	**R** 473/*Self-Test 3*	**16-7** 693/*P*: 1-9 **R** 694/*Self-Test 3*	**17-5** 738/1-17 odd **R** 739/*Self-Test 3*	**17-5** 738/1-17 odd **R** 739/*Self-Test 3*
159	*Prepare for* *Chapter Test*	*Prepare for* *Chapter Test*	*Prepare for* *Chapter Test*	*Prepare for* *Chapter Test*
160	*Administer* *Chapter Test*	*Administer* *Chapter Test*	*Administer* *Chapter Test*	*Administer* *Chapter Test*

Term Projects

Following is a list of topics suitable for written term papers or class presentations by students. A brief description of each topic aimed at generating student interest and a short list of suggested references designed to give the student a starting point are provided.

Semester I

Curve Fitting

Graphing equations is a familiar task. How about trying the reverse—finding an equation which describes as closely as possible the graph of a set of data?

Baker, Justine C., and Oakley, Cletus O. "Least Squares and the 3:40-Minute Mile." *Mathematics Teacher* 70: 322–324.

Batschelet, Edward. *Introduction to Mathematics for Life Scientists*. 3d ed. New York: Springer Verlag, 1979.

Kastner, Bernice. *Applications of Secondary School Mathematics*. Reston, VA: National Council of Teachers of Mathematics, 1978.

Mosteller, Frederick, et al. eds. *Statistics by Example*. Reading, MA: Addison-Wesley Publishing Company, 1976.

Smith, C. A. *Biomathematics, the Principles of Mathematics for Students of Biological and General Science*. 4th ed. Algebra, Geometry, Calculus, Volume I. New York: Hefner, 1966.

Farey Sequences

Farey Sequences consist of the irreducible fractions $\frac{a}{b}$ arranged in order of increasing magnitude, where $0 \leq a \leq b$. The sequences possess some interesting properties, the proofs of which can be very challenging.

Alladi, K. A. "A Farey Sequence of Fibonacci Numbers." *Fibonacci Quarterly* 13: 1–10; Feb., 1975.

Enrichment Mathematics for High School. Twenty-eighth Yearbook of the National Council of Teachers of Mathematics. Washington, D.C.: The Council, 1963.*

Honsberger, Ross. *Ingenuity in Mathematics*. Washington, D.C.: Mathematical Association of America, 1975. Chapter 5.

Kriewall, Thomas E. "McKay's Theorem and Farey Fractions." *Mathematics Teacher* 68: 28–31.

Fibonacci Numbers

In 1202 the mathematician Fibonacci wrote about a problem concerning the breeding of rabbits. The pattern of population growth under the given constraints formed a sequence now known as a *Fibonacci Sequence,* famous for its applications in many fields of mathematics and nature.

Brown, Stephen I. "From the Golden Rectangle and Fibonacci to Pedagogy and Problem Posing." *Mathematics Teacher* 69: 180–188.

Dalton, Leroy C., and Snyder, Henry D., eds. *Topics for Mathematics Clubs*. 2d ed. Reston, VA: National Council of Teachers of Mathematics, 1983.

Gardner, Martin. "The Multiple Fascinations of the Fibonacci Sequence." *Scientific American*, Vol. 220, No. 3, pp. 116–120; March, 1969.

---

*The current address of the National Council of Teachers of Mathematics is 1906 Association Drive, Reston, VA 22091.

The Golden Ratio

If a point divides a segment in such a way that the ratio of the shorter segment to the longer is equal to the ratio of the longer segment to the entire segment, then each ratio formed is known as the *Golden Ratio* and the resulting proportion as the *Divine Proportion*. How is this ratio used in architecture and art, and what applications exist in nature? How can this ratio be constructed? How is it related to a Fibonacci Sequence? How does the Golden Ratio apply to the theory of polyhedra?

Bergamini, David, et al. *Mathematics* ("Life Science Library"). New York: Time-Life Books, 1963.

Huntley, H. E. *The Divine Proportion: A Study in Mathematical Beauty.* New York: Dover Publications, 1970.

Insights into Modern Mathematics. Twenty-third Yearbook of the National Council of Teachers of Mathematics. Washington, D.C.: The Council, 1957.

Thompson, D'Arcy. *On Growth and Form.* New York: Cambridge University Press, 1952.

Verno, C. Ralph. "The Golden Section and Conic Sections." *Mathematics Teacher* 67: 361–363.

The Infinite Set of Primes

Euclid's proof of the infinitude of the primes demonstrates a method of generating a new prime number given any set of primes.

Barnett, I. A. *Some Ideas about Number Theory.* Washington, D.C.: National Council of Teachers of Mathematics, 1961.

Historical Topics for the Mathematics Classroom. Thirty-first Yearbook of the National Council of Teachers of Mathematics. Washington, D.C.: The Council, 1969.

Pomerance, Carl. "The Search for Prime Numbers." *Scientific American,* Vol. 247, pp. 136-147; Dec., 1982.

Tietze, Heinrich. *Famous Problems of Mathematics.* Translation of 2d German Edition (1959) by Beatrice Kevitt Hofstader and Horace Komm. Baltimore, MD: Graylock Press, 1965.

Linear Programming

Linear programming is a method developed to help solve economic and other problems. Solutions to simple problems often can be found by the graphing method introduced in Section 4-6, but more complicated problems may necessitate the use of matrices and/or computers.

Garvin, G. *Introduction to Linear Programming.* New York: McGraw-Hill Book Company, 1960.

Goodman, Adolph W., and Ratti, J. S. *Finite Mathematics with Applications.* New York: Macmillan Company, 1971.

Kastner, Bernice. *Applications of Secondary School Mathematics.* Reston, VA: National Council of Teachers of Mathematics, 1978.

Mersenne Numbers

Numbers of the form $2^n - 1$ are called *Mersenne numbers* and are of special interest when prime. An unproven conjecture is that there is an infinite number of Mersenne primes.

Barnett, I. A. "Mersenne Numbers," in *Historical Topics for the Mathematics Classroom.* Thirty-first Yearbook of the National Council of Teachers of Mathematics. Washington, D.C.: The Council, 1969.

Bezuszka, Stanley J. "Even Perfect Numbers—An Update." *Mathematics Teacher* 74: 460–463.

Cohen, David B. "Another Way of Finding Prime Numbers." *Mathematics Teacher* 69: 398–400.

Feigelstock, Shalom. "Mersenne Primes and Group Theory." *Mathematics Magazine* 49: 198–199.

Gardner, Martin. "Mathematical Games." *Scientific American*, Vol. 241, No. 5, pp. 20–34; Nov., 1979.

Non-Euclidean Geometry

Many attempts have been made to prove Euclid's fifth postulate, all unsuccessful. A system of geometry that is constructed without the use of the Parallel Postulate is known as a non-Euclidean geometry. What is absolute geometry? hyperbolic geometry? elliptic geometry?

Courant, Richard, and Robbins, Herbert. *What Is Mathematics?* New York: Oxford University Press, 1978.

Insights into Modern Mathematics. Twenty-third Yearbook of the National Council of Teachers of Mathematics. Washington, D.C.: The Council, 1957.

Mathematics in the Modern World: Readings from "Scientific American." San Francisco: W. H. Freeman and Co., 1968.

Perfect, Deficient, and Abundant Numbers

If the sum of the proper divisors of a number is equal to the number, then that number is called a *perfect number*. A number is called *deficient* or *abundant* depending on whether the sum of its proper divisors is, respectively, less than or greater than the number. Is there a finite number of perfect numbers? Does an odd perfect number exist? These are some of the unanswered questions about perfect numbers.

Bezuszka, Stanley J. "Even Perfect Numbers—An Update." *Mathematics Teacher* 74: 460–463.

Francis, Richard L. "A Note on Perfect Numbers." *Mathematics Teacher* 68: 606–607.

Historical Topics for the Mathematics Classroom. The Thirty-first Yearbook of the National Council of Teachers of Mathematics. Washington, D.C.: The Council, 1969.

Honsberger, Ross. *Mathematical Gems.* Washington, D.C.: The Mathematical Association of America, 1973.

Shoemaker, Richard W. *Perfect Numbers.* Reston, VA: National Council of Teachers of Mathematics, 1973.

Transfinite Numbers

What are cardinal and ordinal numbers? How can one infinite set be larger than another infinite set? Are there different kinds of infinities?

Eves, Howard. *An Introduction to the History of Mathematics.* 5th ed. Philadelphia: Saunders College Publishing, 1983.

Gamow, George. *One, Two, Three, . . . Infinity.* New York: Viking Press, 1948.

Historical Topics for the Mathematics Classroom. Thirty-first Yearbook of the National Council of Teachers of Mathematics. Washington, D.C.: The Council, 1969.

Newman, James R. *The World of Mathematics, Volume III.* New York: Simon & Schuster, 1956.

Semester II

Algebraic and Transcendental Numbers

Any number that is a root of a polynomial equation with rational coefficients is called an *algebraic number*. Any number that is not algebraic is called a *transcendental number*. It is known that e and π, for example, are transcendental numbers, but many other numbers, such as π^{π}, have not yet been classified.

Eves, Howard. *An Introduction to the History of Mathematics.* 5th ed. Philadephia: Saunders College Publishing, 1983.

Historical Topics for the Mathematics Classroom. The Thirty-first Yearbook of the National Council of Teachers of Mathematics. Washington, D.C.: The Council, 1969.

Niven, Ivan. *Numbers: Rational and Irrational.* Washington, D.C.: Mathematical Association of America, 1975.

Group Theory

A group is a simple but very important mathematical system consisting of a set of elements and one operation, such as multiplication. For a set to be a group, what postulates must it satisfy? What is an Abelian group? What are isomorphic groups? cyclic groups? How are the theory of braids and group theory related?

Adler, Irving. *The New Mathematics.* New York: John Day Company, 1972.

Dalton, LeRoy C., and Snyder, Henry D., eds. *Topics for Mathematics Clubs.* 2d ed. Reston, VA: National Council of Teachers of Mathematics, 1983.

Enrichment Mathematics for High School. Twenty-eighth Yearbook of the National Council of Teachers of Mathematics. Washington, DC: The Council, 1963.

Gardner, Martin. *Martin Gardner's New Mathematical Diversions from "Scientific American."* Chicago, IL: University of Chicago Press, 1984.

Lichtenberg, Donovan R. "A Group Whose Elements Are Functions." *Mathematics Teacher* 74: 521–523.

Investigating Beyond the Third Dimension

We perceive our world as one of three dimensions. Mathematicians of vision, however, have ventured beyond these limits, conceptualizing "spaces" of four and even more dimensions. How can we use algebra to extend our knowledge of the first three dimensions to higher dimensions? Can you build a model of a tesseract, the fourth-dimensional equivalent of a cube?

Abbott, Edwin. *Flatland.* Many editions.

Burger, Dionys. *Sphereland.* (trans. by Cornelie J. Rheinboldt) Scranton, PA: Apollo Editions. (Distributed by Harper and Row Publishers, Inc., NY)

Henry, Boyd. "The Fourth Dimension and Beyond . . . with a Surprise Ending!" *Mathematics Teacher* 67: 274–279.

Hess, Adrien L. *Four-Dimensional Geometry.* Reston, VA: National Council of Teachers of Mathematics, 1977.

Manning, Henry. *The Fourth Dimension Simply Explained.* New York: Dover Publications, 1960.

Marr, Richard F. *4-Dimensional Geometry.* Boston: Houghton Mifflin Company, 1970.

Sommerville, D. M. Y. *An Introduction to Geometry of N Dimensions.* New York: Dover Publications, 1958.

Knot Theory

What does it mean for two knots to be the same? How can you tell if knots you have tied are the same or not? Knot theory, a branch of three-dimensional topology, answers some of these questions.

Enrichment Mathematics for High School. Twenty-eighth Yearbook of the National Council of Teachers of Mathematics. Washington, D.C.: The Council, 1963.

Gardner, Martin. *The Unexpected Hanging and Other Mathematical Diversions.* New York: Simon & Schuster, 1972.

Neuwirth, L. P. *Knot Groups.* Princeton, NJ: Princeton University Press, 1964.

Steinhaus, H. *Mathematical Snapshots.* New York: Oxford University Press, 1969.

Mathematical Induction

Mathematical induction is closely allied to the creative side of mathematics. Mathematicians often make intelligent "guesses" from a few examples, formulating hypotheses inductively, and then use deduction to prove these hypotheses.

Enrichment Mathematics for High School. Twenty-eighth Yearbook of the National Council of Teachers of Mathematics. Washington, D.C.: The Council, 1963.

Sitomer, Harry. "Motivating Deduction." *Mathematics Teacher* 63: 661–664.

Wiscamb, Margaret. "A Geometric Introduction to Mathematical Induction." *Mathematics Teacher* 63: 402–404.

Models of Growth

Mathematical models can be used to study patterns of population growth. Some problems concerning population growth can be solved using logarithms, some necessitate the use of a computer.

Enrichment Mathematics for High School. The Twenty-eighth Yearbook of the National Council of Teachers of Mathematics. Washington, D.C.: The Council, 1963.

Kastner, Bernice. *Applications of Secondary School Mathematics.* Reston, VA: National Council of Teachers of Mathematics, 1978.

Lotka, Alfred J. *Elements of Mathematical Biology.* New York: Dover Publications, 1957.

Pielou, Evelyn C. *An Introduction to Mathematical Ecology.* New York: Wiley-Interscience, 1969.

Schiffer, Max M. *Applied Mathematics in the High School.* Studies in Mathematics, Vol. 10. Stanford, CA: School Mathematics Study Group, 1963.

Music and Mathematics

The ancient Greeks thought of music as a branch of mathematics. Many surprising relationships exist between the two fields.

Benade, Arthur H. *Horns, Strings, and Harmony.* Science Study Series. Garden City, NY: Doubleday, 1960.

Haak, Sheila. "Using the Monochord: A Classroom Demonstration on the Mathematics of Musical Scales." *Mathematics Teacher* 75: 238–244.

Kastner, Bernice. *Applications of Secondary School Mathematics.* Reston, VA: National Council of Teachers of Mathematics, 1978.

Maor, Eli. "What Is There So Mathematical about Music?" *Mathematics Teacher* 72: 414–422.

O'Shea, Thomas. "Geometric Transformations and Musical Composition." *Mathematics Teacher* 72: 523–528.

Networks

The city of Königsberg, Germany, is famous for a problem posed about its seven bridges. Euler's solution to this problem was the beginning of the theory of networks, a specialized branch of topology.

Arnold, B. H. *Intuitive Concepts in Elementary Topology.* Englewood Cliffs, NJ: Prentice Hall, Inc., 1962.

Johnson, Donovan, and Glenn, William. *Topology, the Rubber Sheet Geometry.* New York: McGraw-Hill Book Co., 1960.

Reeves, Charles A. "Network Theory." *Mathematics Teacher* 67: 175–178.

Pascal's Triangle

There are many relationships involving Pascal's Triangle. Those that you studied in Chapter 12, probability and binomial expansion, are just samples.

Bidwell, James K. "Pascal's Triangle Revisited." *Mathematics Teacher* 66: 448–452.

Dalton, Leroy C., and Snyder, Henry D., eds. *Topics for Mathematics Clubs.* 2d ed. Reston, VA: National Council of Teachers of Mathematics, 1983.

Gardner, Martin. *Mathematical Carnival.* New York: Alfred A. Knopf, 1975.

Jansson, Lars C. "Spaces, Functions, Polygons, and Pascal's Triangle." *Mathematics Teacher* 66: 71–77.

Quaternions

In 1843, William Rowan Hamilton generalized from a complex number, $a + bi$, representing a vector in a plane, to a number of the form

$$a + bi + cj + dk$$

(a, b, c, d real; $i^2 = j^2 = k^2 = -1 = ijk$) representing a vector in space. He called numbers of this form *quaternions;* one of their most amazing properties is that they do not obey the commutative axiom for multiplication. The real numbers, complex numbers, and quaternions are all related through 2×2 matrices.

Enrichment Mathematics for High School. Twenty-eighth Yearbook of the National Council of Teachers of Mathematics. Washington, D.C.: The Council, 1963.

Historical Topics for the Mathematics Classroom. Thirty-first Yearbook of the National Council of Teachers of Mathematics. Washington, D.C.: The Council, 1969.

Mathematics in the Modern World: Readings from "Scientific American." San Francisco: W. H. Freeman and Co., 1968.

Some Special Numbers

The constants 0, 1, *i*, π, and *e* have many important and unique properties. What are these numbers? How are they related to each other? How did they develop in the history of mathematics?

Bell, E. T. *Men of Mathematics.* New York: Simon & Schuster, 1937.

Gardner, Martin. "Mathematical Games." *Scientific American*, Vol. 241, No. 2, pp. 18–24; Aug., 1979.

Green, D. R. "The Historical Development of Complex Numbers." *The Mathematical Gazette*, Vol. 60, No. 412, pp. 99–107; June, 1976.

Historical Topics for the Mathematics Classroom. Thirty-first Yearbook of the National Council of Teachers of Mathematics. Washington, D.C.: The Council, 1969.

Kasner, Edward, and Newman, James. *Mathematics and the Imagination.* New York: Simon & Schuster, 1940.

Mathematics in the Modern World: Readings from "Scientific American." San Francisco: W. H. Freeman and Co., 1968.

Unsolved Problems in Mathematics

There are many problems in mathematics that can be stated simply but have never been proved. Among these are Fermat's last theorem and Goldbach's conjecture.

Bell, E. T. *The Last Problem.* New York: Simon & Schuster, 1961.

Kolata, Gina. "Number Theory Problem Is Solved." *Science,* Vol. 221, pp. 349–350; July 22, 1983.

Newman, James. *The World of Mathematics.* New York: Simon & Schuster, 1956.

Shanks, Daniel. *Solved and Unsolved Problems in Number Theory.* 3d ed. New York: Chelsea Publishing, 1985.

Lesson Commentary

1 Review of Essentials

This chapter reviews the basic concepts of set theory which are used throughout the book. The field axioms for the real numbers are presented and used to prove some basic algebraic theorems. Because this is essentially a review, you may want to omit some sections, but it is recommended that Sections 1-3 through 1-5 be covered by all students.

1-1 (pages 1–3)

Teaching Suggestions

This section (one of the ones that may be omitted) reviews some of the basic concepts of set theory that students need in order to be able to read later sections of the book. Emphasize understanding (and memorization) of the meaning of the symbols in the chart on page 2. Make sure students understand the difference between the symbols "$\in$" and "$\subset$." You can use the following example to illustrate this:

Correct $2 \in \{2, 4, 6\}$

Correct $\{2\} \subset \{2, 4, 6\}$

Note that the symbol $=$ is used for both equality of sets and equality of elements.

1-2 (pages 3–6)

Teaching Suggestions

Here again, understanding of the terms introduced is essential for comprehension of later material. "Root" may be a word unfamiliar to many students. Make sure they understand that "root" in the sense of an equation should not be confused with the concept of "square root," "cube root," etc., although for better students you might point out that historically there is a connection, illustrated by the example: "The *square root* of 2 is a *root* of the equation $x^2 - 2 = 0$."

1-3 (pages 6–10)

Teaching Suggestions

Make sure your students understand the logical necessity of axioms in mathematics. You can relate this logical framework to their prior experience in geometry (where the axioms may have been called "postulates"), but in any case, they should realize that finding the solution of even the simplest problem depends (ultimately) on the axioms. Even more important, although the axioms may seem intuitively obvious and even unnecessary, they enable us, as in geometry, to deduce much less obvious facts that will hold in *any* system that conforms to the original axioms. As a diversion, you might ask students why 0 has no multiplicative inverse. One answer is that if 0 had a multiplicative inverse we could carry out the following "proof" that $1 = 2$:

$0 = 0$ (Reflexive prop. of $=$)

$0 \cdot 1 = 0 \cdot 2$ (Mult. prop. of 0 and subs. prin.)

$\frac{1}{0}(0 \cdot 1) = \frac{1}{0}(0 \cdot 2)$ (Mult. prop. of $=$)

$(\frac{1}{0} \cdot 0)1 = (\frac{1}{0} \cdot 0)2$ (Assoc. ax. for mult.)

$1 \cdot 1 = 1 \cdot 2$ (Ax. of mult. inv.)

$1 = 2$ (Iden. ax. for mult.)

Faced with a choice of allowing $1 = 2$ or *not* allowing division by 0, we choose the latter.

Emphasize that the Substitution Principle allows substitution in expressions involving sums or products only. "If $a = b$, and $\frac{1}{a} = \frac{1}{a}$, then $\frac{1}{a} = \frac{1}{b}$" is not a correct application of this principle.

Related Activities

1. Define new symbols $\oplus$ and $\otimes$ as follows:

    $$a \oplus b = a + b + 1 \text{ and } a \otimes b = 2ab.$$

    Which of the axioms on page 7 are true for these operations?
    Closure, Commutative, and Associative
2. Suppose the only numbers that existed were fractions whose denominators (in lowest terms) were *not* divisible by 3. Would the set of all such numbers (with the usual addition and multiplication rules) satisfy the axioms on page 7? No multiplicative inverse

1-4 (pages 12–17)

Teaching Suggestions

First note that this section should be covered by all students and can be treated in two ways.

(1) It can be used primarily as a technical review with only the *results* of the theorems stressed and with no emphasis on formal proof. In this case, only exercises from among Written Exercises 1–26 should be assigned.
(2) Formal proof and logical sequence can be stressed. This approach should be used only with better students. Even then, you should not spend an inordinate amount of time on this section at the expense of later material. In this case, Written Exercises 25–39 may be assigned.

1-5 (pages 17–21)

Teaching Suggestions

This section can be treated in the same manner as the preceding section. As a check that students appreciate the concepts of this section you might ask them to distinguish, *in their own words*, between the *meanings* of the two (equal) numbers $\frac{1}{-a}$ and $-\frac{1}{a}$, and to explain the significance of the fact, demonstrated on page 20, that $\frac{1}{-a} = -\frac{1}{a}$.

Related Activities

Suppose the integers 1 through 12 were arranged as on a clock, and addition and multiplication were defined as usual except that if an answer came out greater than 12, it would be reinterpreted as if time were involved. For example, $6 + 8 = 2$, since 6 o'clock "plus" eight hours is 2 o'clock, and similarly $5 \times 4 = 8$. Does the Cancellation Property for Multiplication hold in such a system? Can you find three numbers a, b, c, such that $ac = bc$, but $a \neq b$?   No: $3 \times 6 = 5 \times 6 = 6$, but $3 \neq 5$.

1-6 (pages 22–24)

Teaching Suggestions

Explain that subtraction is not really a separate operation but is rather a more convenient way of writing the addition of the additive inverse of a number. Of more importance to all students is the geometric interpretation of subtraction, which is essential in analytic applications.

1-7 (pages 24–28)

Teaching Suggestions

As with subtraction, you should explain that division is a convenient way of denoting multiplication by the multiplicative inverse of a number. This is a good time to review the rules for combining fractions, since students must be thoroughly familiar with them if they are to succeed in more complicated algebraic manipulations in the future.

Related Activities

Another "proof" that $1 = 2$:
    Let $a = 1$ and $b = 2$. Then $2a = b \longrightarrow$
$4a^2 = 2ab \longrightarrow 4a^2 - b^2 = 2ab - b^2 \longrightarrow$
$(2a + b)(2a - b) = b(2a - b) \longrightarrow$
$2a + b = b \longrightarrow \frac{2a + b}{b} = \frac{b}{b} \longrightarrow 2 = 1.$
Find the fallacy in the proof.
When you divide both sides by $2a - b$, you are dividing by 0.

2 Review of Essentials

This chapter continues the review of the basic concepts of first-year algebra. Operations with polynomials, solving linear equations in one variable, the order properties of the real numbers, and absolute value are some of the topics presented. As with Chapter 1, you may want to omit some sections of this chapter. Sections 2-3, 2-4, and 2-6, however, should be covered by all students.

2-1 (pages 41–45)

Teaching Suggestions

This section introduces many technical terms, at least some of which most students should already recognize. Emphasize that it is important to understand the language introduced here because it will be used throughout the book, but memorization of the formal definition of each new term is not recommended. You should not insist that students work out each problem according to the axioms, as is done on page 43 for two examples. The B and C problems review the rules governing polynomials but involve enough juggling of signs to maintain interest.

2-2 (pages 45–48)

Teaching Suggestions

This section, like the preceding one, is much more important in application than in theory. Better students may be able to omit it. You can use this section to verify, however, that your students no longer rely on guesswork to solve an equation such as $2x - 3 = 15$. The "operational" properties on page 46 should be regarded as tools with which to solve equations rather than abstractions to memorize.

2-3 (pages 50–54)

Teaching Suggestions

Word problems sometimes seem to cause students extraordinary difficulty. You can lessen the psychological impact of this section by emphasizing the five-step method on page 50. Students should choose a variable, x, for example, to represent one of the quantities to be found. The other quantities should depend on x in some simple manner. The assignments must be as specific as possible. Encourage the students to write "Let $x =$ the rate of the boat" rather than "Let $x =$ boat." This will help them avoid writing equations in which one side represents a distance, for example, and the other side represents a duration of time.

The equation to be solved will involve x but, in all likelihood, will not have x as one side (otherwise, there would be no need of algebra). The equation must have two sides, at least one of which must involve x, that *express the same quantity in different ways*. The other side will often be a given number. Students who have difficulty should be urged to set up the equations for some problems without solving as practice in translating the words into symbols.

2-4 (pages 56–60)

Teaching Suggestions

As with some of the sections in the preceding chapter, it is possible to approach this section from a purely technical point of view, or from a more theoretical one. For both approaches, however, it is important for students to be able to apply the order properties on page 57, by means of the transformational rules on page 58, in solving inequalities.

To challenge students, you may use a more theoretical approach, including formal proof (and have students try the C exercises in the Written Exercises). The graphing of inequalities, begun on page 57, suggests that this might be a good time to give your students a deeper feeling for the set of real numbers than (perhaps) they already have. In particular, many will probably cling to the notion that the set graphed on page 57 has a largest element, even if, as some may maintain, "We can't ever know what it is." You can convince them that this is not so by showing that for any number n less than 3, the number $\frac{n+3}{2}$ is always larger than n but still less than 3. If the number 2.999 . . . is suggested by someone, make sure it is understood that this is just another decimal representation of the number 3 (as can be seen from the fact that if $x = 2.99 \ldots$, then $10x = 29.99 \ldots$.

Thus, $10x = 29.99 \ldots$

$$\underline{-x = 2.99 \ldots}$$
$$9x = 27.00 \qquad \text{and } x = 3).$$

2-5 (pages 60–64)

Teaching Suggestions

From a technical standpoint, the first part of this section can be regarded as a simple extension of the previous one, for, as students will quickly discover, the same rules that govern the symbols $>$ and $<$ apply unchanged to the symbols $\geq$ and $\leq$. In order to make clear the contrast between disjunction and conjunction, you might use Venn diagrams (with which students should be familiar) like those below:

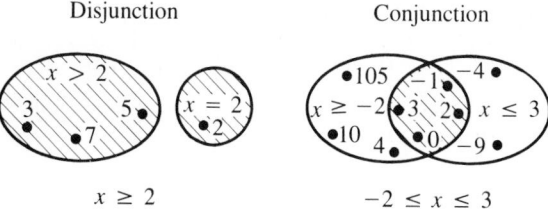

Put these diagrams on the board, name some other numbers, and ask your students in what region they belong. Mention some fractions and irrational numbers. Note that the notions of disjunction and conjunction will be used in the discussion of absolute value in Section 2-7.

2-6 (pages 65–69)

Teaching Suggestions

Students should be familiar with the notion of indirect proof from geometry, but if they are unsure about the logic of this method, you might use a simple example (unrelated to inequalities) to demonstrate this, such as the following:

Theorem: There is no real number x satisfying the equation $x = x + 1$.

Proof: Assume $a = a + 1$ for some real number a. Subtracting a from both members, we get $0 = 1$, an obvious contradiction.

Explain that in an indirect proof, the contradiction shows that "something is wrong" in the proof. If all the steps are carried out correctly, then the only thing that could be wrong is the assumption.

Related Activities

Suppose there were a number i that had the property: $i^2 = -1$. Show by means of the indirect method that i could be neither positive, negative, nor 0 (that is, that the statements $i < 0$, $i > 0$, and $i = 0$ are all false).

Hypothesis: $i^2 = -1$

Case I:
Assume $0 < i$.
1. $0 \cdot i < i \cdot i$ 1. Mult. prop. of order
2. $0 < i \cdot i$ 2. Mult. prop. of zero
3. $0 < -1$ 3. Substitution principle

(Case II and Case III are on the next page.)

Case II:

Assume $0 = i$.

1. $0 \cdot i = i \cdot i$ 1. Mult. prop. of eq.
2. $0 = i \cdot i$ 2. Mult. prop. of zero
3. $0 = -1$ 3. Substitution principle

Case III:

Assume $i < 0$.

1. $i \cdot i > 0 \cdot i$ 1. Mult. prop. of order
2. $i \cdot i > 0$ 2. Mult. prop. of zero
3. $-1 > 0$ 3. Substitution principle

Since the last statement in the proof of each case contradicts a known fact (i.e., $0 > -1$, $0 \neq -1$, $-1 < 0$), the assumption in each case must be false. Therefore, i is neither positive, negative, nor zero.

2-7 (pages 69–72)

Teaching Suggestions

Many students have a tendency to think of the absolute value of a number intuitively as "the positive of the number." Although this vague notion may serve at times, you should emphasize that in algebraic operations like those of this section, the precise algebraic definition on page 69 is essential.

    Examples 2 and 3 illustrate two typical kinds of problems dealing with absolute value. It is important to stress the logical distinction embodied in the words "or" and "and." You can stress this by first calling attention to the two separate inequalities that appear as the final step in each case; then choose specific numbers in the solution set of each and show that in Example 2 each number satisfies *both* inequalities, while in Example 3 each number satisfies only *one* of the inequalities. Warn students strongly not to write the solution set of Example 3 as $\{z: 4 \leq z \leq -6\}$.

Related Activities

1. Show that for all real numbers a and b, $|a + b| \leq |a| + |b|$. Divide the problem into four cases: (1) a, b both positive; (2) a negative, b positive; (3) a positive, b negative; (4) a, b both negative.

Case 1: $a > 0$, $b > 0$

1. $a > 0$, $b > 0$ (hypothesis)
2. $|a| = a$, $|b| = b$ (Def. abs. value)
3. $a + b > 0$ (Closure ax. for $\mathscr{R}_+$)
4. $|a + b| = a + b$ (Def. abs. value)
5. $|a + b| = |a| + |b|$ (Subst. princ.)
   Case 2: $a < 0$, $b > 0$

   Consider the Case (i) where $|a| > |b|$ and (ii) where $|a| < |b|$.

1. $a < 0$, $b > 0$ (Hypothesis)
2. $|a| = -a$, $|b| = b$ (Def. abs. value)
3. $|a| > |b|$ (Hypothesis (i))
4. $|a| - |b| > 0$ (Add. prop. of order)
5. $-a - b > 0$ (Subst. princ.)
6. $a + b < 0$ (Theorem, p. 65)
7. $|a + b| = -a - b$ (Def. abs. value)
8. $-b < 0$ (Theorem, p. 65)
9. $-b < b$ (Trans. prop. of order)
10. $-a - b < -a + b$ (Add. prop. of order)
11. $|a + b| < |a| + |b|$ (Subst. princ.)

Proof of Case 2 (ii) is similar to proof of Case 2 (i). Proof of Case 3 is similar to proof of Case 2. Proof of Case 4 is similar to proof of Case 1.

2. Define the absolute value of a point in the Cartesian plane by

$$|(x, y)| = \sqrt{x^2 + y^2}.$$

(Note that this definition is consistent with the definition of the absolute value of a real number in that it also measures distance from the origin.) Define the addition of two points by

$$(x_1, y_1) + (x_2, y_2) = (x_1 + x_2, y_1 + y_2).$$

Then show that for any two points (a, b) and (c, d), the points $(0, 0)$, (a, b), (c, d), and $(a + c, b + d)$ are the vertices of a parallelogram, and explain geometrically the significance of the inequality that corresponds to the one in the previous activity, namely

$$|(a, b) + (c, d)| \leq |(a, b)| + |(c, d)|.$$

(1) Show that the slopes of opposite sides of the parallelogram are equal.

(2) The inequality above is another expression for the triangle inequality.

3 Linear Functions and Relations

This chapter presents the idea of a function, concentrating on linear functions. Equations and inequalities are discussed; the various standard forms of a linear equation are presented and their geometric significance noted.

3-1 (pages 83–87)

Teaching Suggestions

Your students may already be familiar with the notion of a function, but their concept may be less rigorous than that demanded here. The heuristic device of a "black box" into which a number from the domain is fed and out of which a number from the range is produced, is useful for emphasizing that function values (coming out of the box) must be unique, but the same value may come out for two different values in the domain (inputs). Several concrete examples of functions may help reinforce this idea:

(1) The function that assigns to each Fahrenheit temperature the equivalent Celsius temperature (this function is one-to-one: each x has a different y-value);

(2) the function that assigns to each mass the postage necessary to mail a package of that mass first class (this function is not one-to-one: many different x's will have the same y-value).

Note that many functions are not defined by a simple formula (see (2) above). You might mention an example like the following,

$$f(x) = \begin{cases} x + 1 \text{ if } x \geq 0 \\ 2 \text{ if } x < 0, \end{cases}$$

to show that *one function* may require two different formulas (although for each x there is still only one y-value). In this example, you might point out that if the second inequality were altered to "$\leq$," the relation would no longer define a function.

3-2 (pages 88–91)

Teaching Suggestions

This section provides two geometric interpretations of the notion of a function. Note that the Cartesian-coordinate method is obviously less cumbersome and is used for most applications of the function concept in this book. For this method *only*, emphasize the criterion for a function in the box on page 88.

If there is time, this would be a good opportunity to show your students graphs of some unfamiliar functions. For example,

(1) $y = \begin{cases} x + 1 \text{ if } x \geq 0 \\ 2 \text{ if } x < 0 \end{cases}$

(2) $y = $ the amount paid for x hours at a parking lot that charges 75¢ for the first hour or part of an hour and 25¢ for each additional hour or part of an hour

Related Activities

Graph each of the following.

1. $|x + y| + |x - y| = 4$

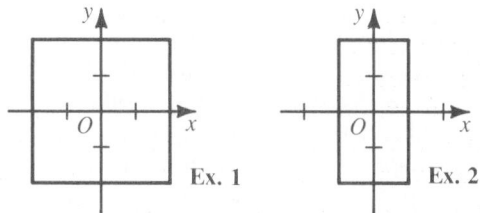
Ex. 1   Ex. 2

2. $|2x + y| + |2x - y| = 4$
Make a conjecture about the graph of any equation of the form

$$|ax + by| + |ax - by| = c, \quad a, b, c > 0.$$

The graph is a rectangle with vertices at $\left(\pm \dfrac{c}{2a}, \pm \dfrac{c}{2b} \right)$.

Teaching Suggestions

Although most students will probably be familiar with this material, some may need to review it. Remind students that it is a good idea to solve an equation for y (if possible) before graphing. You can sustain interest by discussing the greatest integer function introduced at the end of the section. Written Exercises 23 and 24 provide graphing practice related to this concept. Remind students to use both positive and negative values for the variable, especially when they are graphing equations involving absolute value, or the greatest integer function.

Teaching Suggestions

This material is straightforward and should present no problems, except that occasionally students may have difficulty determining which half-plane is the graph of the inequality. You may want to present an alternative method of doing this. After the equation associated with the inequality has been graphed, choose any point in one of the half-planes and test its coordinates in the inequality. If this point satisfies the inequality, then this half-plane is its graph. If not, the other half-plane is its graph.

Make sure students understand why the boundary line is dashed in the graph of a strict inequality. If they have difficulty, point out that the dashed line is analogous to the open circle used to graph one-dimensional inequalities in Section 2-4. In particular, you can take an inequality such as $y > 2x - 1$ and demonstrate that for a fixed value of x, say $x = 3$, the graph is a vertical "half-line." In this case it would be the half-line consisting of all points on the line $x = 3$ whose y-coordinates are greater than 5, since $y > 2(3) - 1$. Note that the point $(3, 5)$ is part of the dashed line that marks the boundary of the whole graph, which consists of all the vertical half-lines above the graph of $y = 2x - 1$.

Teaching Suggestions

Two technical points should be emphasized.

1) The y-coordinates are always in the numerator, the x-coordinates always in the denominator.
2) It does not matter which of the two points is called P and which Q, but in computing the slope, consistency must be maintained in subtracting each coordinate of P from the corresponding coordinate of Q.

Try to get across the idea that slope is the numerical embodiment of a geometric notion. Give students an intuitive idea of this notion by drawing lines (all through the origin, for example) of various slopes, some negative, and giving approximate values of the different slopes. You could use statements like, "A steep uphill (from left to right) line has a large positive slope," which, although somewhat imprecise, will help students to determine the reasonableness of answers later on.

Teaching Suggestions

Although the students are already familiar with the concept of slope, the idea of substituting for x and y to find the constant b in the slope-intercept form can be confusing, since x and y generally represent unknown quantities. Emphasize that since m and b are constants for a particular line, their values as determined by two pairs of values (x, y) will be the same for *any* pair (x, y) on that same line. Although the slope-intercept form (which you should insist your students memorize) can be used to solve a problem in which the slope of a line and a point on the line are given, you should urge students to use the point-slope form in such a situation. Since the point-slope form need only be simplified to produce the equation of the line, no confusion in substituting in values and solving for constants can arise.

Introduce the *intercept form* of the equation of a line (for nonvertical, nonhorizontal lines only):

$$\frac{x}{a} + \frac{y}{b} = 1.$$

Have students find a and b in terms of A and B in the standard form $Ax + By = C$. Find the x- and y-intercepts of the line in terms of a and b, and explain why this form is called intercept form.

$$a = \frac{C}{A}, \ b = \frac{C}{B}; \ x\text{-intercept} = a, \ y\text{-intercept} = b$$

3-7 (pages 108–113)

Teaching Suggestions

You can relate the definition of direct variation, $y = mx$, to the property on page 109,

$$\frac{y_1}{x_1} = \frac{y_2}{x_2},$$

by extending the proportion as follows:

$$\frac{y_1}{x_1} = \frac{y_2}{x_2} = m.$$

Therefore, $y_1 = mx_1$ and $y_2 = mx_2$.

There are many examples of direct variation from the sciences and social sciences. Here are a few:

1) Hooke's Law: The displacement of a spring varies directly with the force on the spring.

2) Gay-Lussac's Law: The pressure of a gas varies directly with (Kelvin) temperature (if the volume is held constant).

3) Distance traveled at a constant speed varies directly with time.

4) The amount paid in sales tax varies directly with the purchase price of an item.

Note that every direct variation is a linear function, but not every linear function is a direct variation.

Related Activities

For points (x, y) in the plane, define

$$t(x, y) = (tx, ty) \text{ and}$$
$$(x_1, y_1) + (x_2, y_2) = (x_1 + x_2, y_1 + y_2).$$

With these definitions, show that if (x_1, y_1) and (x_2, y_2) are any two points on a line in the plane, then the point

$$t(x_1, y_1) + (1 - t)(x_2, y_2)$$

is also a point on the same line.

(x_1, y_1), (x_2, y_2), and $t(x_1, y_1) + (1 - t)(x_2, y_2)$ lie on the same line if and only if the slopes of the lines determined by two pairs of these points are equal.

$$t(x_1, y_1) + (1 - t)(x_2, y_2)$$
$$= (tx_1, ty_1) + (x_2 - tx_2, y_2 - ty_2)$$
$$= (tx_1 + x_2 - tx_2, ty_1 + y_2 - ty_2)$$

$$\frac{(ty_1 + y_2 - ty_2) - y_2}{(tx_1 + x_2 - tx_2) - x_2} = \frac{t(y_1 - y_2)}{t(x_1 - x_2)} = \frac{y_1 - y_2}{x_1 - x_2}$$

---

4 Systems of Linear Equations or Inequalities

This chapter discusses graphing, methods of solving, and applications of systems of linear equations or inequalities. Determinants are introduced as a method for solving linear systems, and linear programming, a major application, is treated in the final section.

4-1 (pages 123–127)

Teaching Suggestions

At the risk of boring a few students, you may do well to remind the class, at this point, of the relationship between a graph and its equation: namely, that the two coordinates of any point on the graph, when substituted into the equation, satisfy it—that is, make it true. It will then be easier for students to understand that the coordinates of the common point(s) of two graphs satisfy *both* equations. Of course, several concrete examples done in complete detail will help reinforce the basic concept. (See the Chalkboard Examples in the side column on page 124.)

4-2 *(pages 128–133)*

Teaching Suggestions

Students should have little or no trouble applying the methods of this section, but they may have difficulty understanding the underlying theoretical concepts. You can clarify these by showing graphically the steps involved in the solution of the Example on page 129, as follows:

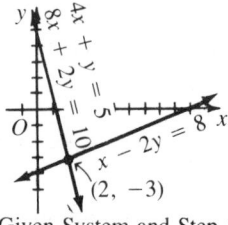

Given System and Step 1

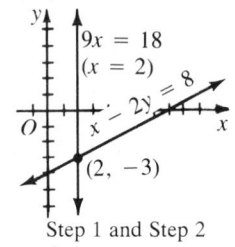

Step 1 and Step 2

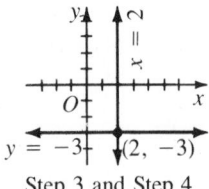

Step 3 and Step 4

Note that the graphs of the equations produced by linear combinations of the original pair of equations are different from those of the original equations, but the new graphs always contain the original point of intersection. (This was proved in Exercise 28 of Written Exercises 4-1.) The transforma-

tions simply make this point of intersection immediately readable from the equations.

Related Activities

Prove that the graphs of the equations

$$y = m_1 x + b_1,$$
$$y = m_2 x + b_2,$$
$$y = m_3 x + b_3,$$

are concurrent (have a common point) if and only if the points (b_1, m_1), (b_2, m_2), and (b_3, m_3) are collinear. (*Hint*: The equations have a common solution if the x-coordinates of the solutions of each pair of equations are equal.)
Given: The graphs of the equations are concurrent.
Prove: The points are collinear.
Show that the slope from (b_1, m_1) to (b_2, m_2) equals the slope from (b_1, m_1) to (b_3, m_3).
To prove the converse, show that there exists an x_1 such that $m_1 x_1 + b_1 = m_2 x_1 + b_2 = m_3 x_1 + b_3$.
$\left(\text{Let } x_1 = \dfrac{b_2 - b_1}{m_1 - m_2}.\right)$

4-3 *(pages 133–136)*

Teaching Suggestions

The justification of the determinant formulas (*4*) is given on page 134. After you have explained this derivation, you might have students check the two formulas (*3*) by substituting back for x and y in the two original equations (*1*) and (*2*). In addition to convincing them of the validity of solving systems of equations by means of Cramer's Rule, this is a good algebraic exercise. Insist that students memorize formulas (*4*).

Although students usually have few problems with determinants, motivation may be a problem. You should point out that although determinants may seem a laborious method of solving systems of two linear equations in two variables, the generalization of the concepts introduced here provides an efficient means of solving systems of more than two linear equations in more than two variables.

Related Activities

You can find the area of triangle ABC shown below as follows:

$$\text{Area of } \triangle ABC = \frac{1}{2}\left(\begin{vmatrix} a & b \\ c & d \end{vmatrix} + \begin{vmatrix} c & d \\ e & f \end{vmatrix} + \begin{vmatrix} e & f \\ a & b \end{vmatrix}\right)$$

Prove this formula.

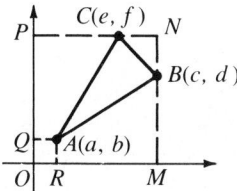

Area $\triangle ABC$ = area rectangle $MNPO$ – area rectangle $ORAQ$ – area trapezoid $PQAC$ – area trapezoid $RABM$ – area $\triangle BCN$

4-4 (pages 138–144)

Teaching Suggestions

As with the previous section on word problems (2-3), you should encourage a systematic approach. Insist that your students *write down* (in words) what quantities the variables represent before they set up the equations. As before, each equation should contain two sides that are *different expressions for the same quantity*. Using Oral Exercises 1 and 2 as examples, point out that the equation for Exercise 1 involves the *numbers* of cucumbers and heads of lettuce, while that for Exercise 2 involves the *costs* of the vegetables. Students should check that both sides of an equation have the same units.

Because students find setting up equations one of the most difficult aspects of solving word problems, they should be advised to read the problem slowly and carefully several times.

4-5 (pages 145–147)

Teaching Suggestions

As a matter of practical convenience when graphing the intersection of two inequalities, a good plan is to shade the graph of one inequality with lines having a slope of about $+1$ and to shade the other with lines having a slope of about -1. Then the intersection will show clearly as being shaded with "diamonds" or cross-hatching. A review of Section 3-4 may be needed here. Emphasize that when two (or more) inequalities are graphed, the system is to be regarded as a conjunction (connected by "and").

When graphing a system of linear inequalities, it is helpful to students to rewrite each inequality in the form y (inequality sign) $mx + b$, if possible, before graphing. Students should realize that $x + 2 < y$ is equivalent to $y > x + 2$ and $x - y < 4$ is equivalent to $y > x - 4$.

4-6 (pages 147–151)

Teaching Suggestions

Students may find the word problems of this section extremely challenging because there is so much information to correlate. Emphasize the importance of writing down

(1) the quantities that the variables represent
(2) *all* the restrictions on the variables (inequalities)
(3) the function, of the form $ax + by$, to be maximized or minimized.

Note that the function $ax + by$ is a function of two variables, a notion that will be unfamiliar to most students; therefore, the domain of the function is the xy-plane (or a subset thereof). If we were to graph such a function, its range would be represented on a third coordinate axis (perpendicular to the plane).

Related Activities

Write a linear programming problem about a situation such as manufacturing sneakers or truck farming. Do enough research so the numbers chosen are reasonable representations of real-life situations.

5 Graphs in Space; Determinants

In this chapter students study the Cartesian coordinate system in three dimensions. Systems of three linear equations in three variables are solved by means of determinants, and applications are made to problem solving.

5-1 (pages 161–164)

Teaching Suggestions

Although many students have some familiarity with three-dimensional space from geometry, most will encounter some difficulty in drawing a two-dimensional diagram of a three-dimensional coordinate system. Some teaching aids you might want to try are the following:

1) Have each student sketch the lower left-hand corner of the front wall of the room, indicating carefully the lines where each pair of walls meets. (This will represent the first octant.) You might even want to tack thin pieces of cardboard along these lines, marked off in large units.

2) Have each student construct a model of a three-dimensional coordinate system, out of pipe cleaners, for example, and then sketch it. Be sure units are marked off on the model, and caution students to take account of the "foreshortening" of the x-axis described on page 162.

3) Bring a rectangular carton (preferably without printing on the outside) to class and have students sketch this; then have them draw as dotted lines the edges of the box that are hidden from view. When the drawing is complete, have them label one vertex of the box (for example, the lower left-hand rear corner) as the origin and give hypothetical coordinates of the diagonally opposite vertex. (For example, if the lower left-hand rear vertex is taken as the origin, then the diagonally opposite vertex will have three positive coordinates.) Note that the orientation of the diagram on page 161 is used throughout this book (and in almost all other math and science books). If students are confused about this orientation, call attention to the so-called "right-hand rule," by which, if the fingers of the right hand point from the positive x-axis to the positive y-axis, then the thumb points in the direction of the positive z-axis.

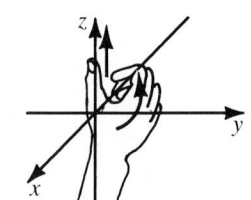

4) Draw a picture of a three-dimensional coordinate system on the chalkboard and note the two-dimensional characteristics of the drawing.

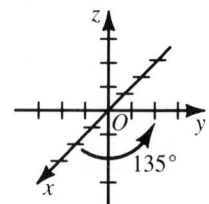

a. Draw a right angle to make the positive z-axis and positive y-axis.

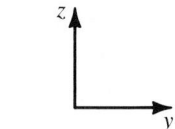

b. Extend the z-axis as a dashed line below the y-axis, and extend the y-axis as a dashed line to the left of the z-axis.

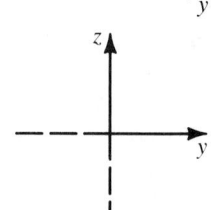

c. Now draw the x-axis to cut the 90° angle between these axes in half, as a dashed line between the solid y- and z- axes, and as a solid line between the dashed y- and z- axes.

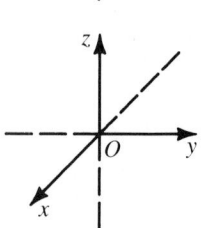

Related Activities

Construct a three-dimensional ticktacktoe board, consisting of three levels, each with nine squares. A winning position is three X's or O's along any line in space.

Teaching Suggestions

Visualizing the graphs of linear equations in space may be difficult for many students, so you might plan on a good deal of practice with this material.

First note that any equation involving x, y, and z (and possibly higher powers of these, such as x^2, xy^3, z^4, etc.) defines a surface in space, not necessarily a flat surface. If the equation is of the general form given at the top of page 165, then its graph is a plane, i.e., a flat surface. The situation is analogous to the two-dimensional case, where a single equation defines a "curve" (which may be straight), while a linear equation necessarily defines a straight line.

To continue this analogy, in a plane any two points determine a line, and in graphing we often choose these to be the x- and y-intercepts. Similarly, in space any three (noncollinear) points determine a plane, and (if possible) we choose these to be the x-, y-, and z-intercepts.

At this point you might perhaps discuss in detail Example 1 on page 166. Note that if A, B, and C in the general form are positive, then the three intercepts are positive. In any case, the traces of the graph in the three coordinate planes are easily found by joining the intercepts to form a triangle.

Two further items should be noted:

1) The triangle usually pictured as the graph of a linear equation in three variables is, of course, only a part of the graph, the whole of which extends infinitely in two dimensions. You can use the carton suggested under Teaching Suggestions in Section 5-1 with an isosceles triangle cut out of cardboard to aid students in visualizing the graph of a linear equation in space. Place the triangle obliquely in one corner to form a tetrahedron.

2) An equation such as $x + 2y = 5$ has among its solutions (in space) the points $(1, 2, 0)$, $(1, 2, 3)$, $(1, 2, -7)$, and $(1, 2, 17)$. This situation is similar to that of the equation $x = 2$ in the plane, which has among its solutions $(2, 3)$, $(2, 0)$, and $(2, 358)$.

This makes sense logically, since in each case no restriction is placed on one of the coordinates, and therefore that coordinate can have any real value.

Teaching Suggestions

Technically, solving systems of equations in three variables should present few difficulties for students. Most of the solutions follow the general pattern of the Example on page 173, but many systems contain at least one equation in which one of the variables is missing. In this case, that equation can be used to substitute for one of the variables in the other two equations, thereby reducing immediately to two equations in two variables. Students are apt to complain that, although not difficult, this material seems tedious. This might be a good time to emphasize the importance of finding the *simplest method* of solution first (by a quick inspection of the system) before plunging into the algebra.

From a theoretical standpoint, if a system has a unique solution, then the planes defined by its equations have exactly one point in common. The planes defined by the original system, however, are usually not all vertical or horizontal as in Figure 7, page 171. By analogy with the two-dimensional case, however, you can explain that when the solution is found, the original "slanted" planes are transformed into vertical and horizontal ones with the same common point as had the planes defined by the original system.

Related Activities

Any two nonparallel planes in space intersect in a line. Try drawing such a situation analytically. For example, use the planes defined by the equations

$$x + 2y + 3z = 6 \quad \text{and}$$
$$-2x + y \qquad = 0.$$

The second equation defines a vertical plane, whose trace in the xy-plane is the line $y = 2x$.

The intercepts of the plane defined by the first equation are $(6, 0, 0)$, $(0, 3, 0)$, and $(0, 0, 2)$. We get the situation pictured at the right.

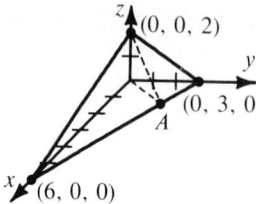

To find the coordinates of point A, the intersection in the xy-plane, take the traces of the two planes in the xy-plane:

$$y = 2x \qquad \text{and}$$
$$x + 2y = 6$$

and solve them simultaneously. We get $\left(\frac{6}{5}, \frac{12}{5}\right)$ for point A. Try these other examples:

1. $2x + 3y \qquad = 6$
$\quad x \qquad + 2z = 2 \quad \left(2, \frac{2}{3}\right)$

2. $x + 3y \qquad = 6$
$\quad\quad 2y + z = 4 \qquad (0, 2)$

5-4 (pages 176–179)

Teaching Suggestions

It is obviously a good idea to relate the material of this section to that of Section 4-3, *Determinants*. The technique itself is straightforward, and will certainly be appreciated after the cumbersome method of solving systems introduced in the previous section. For a proof of Cramer's Rule you must refer students to an analogy of the kind of derivation for two variables done on page 134. (A proof, easily generalizable, is suggested in the Related Activities on this page. Because this sort of derivation for three variables is quite involved, you may not want to assign it as an exercise.)

Note, for future reference, that the diagonal method does *not* work for determinants with more than three rows and three columns.

Related Activities

Prove Cramer's Rule for three equations in three variables using the following plan:

1) Let
$$a_1 x + b_1 y + c_1 z = d_1$$
$$a_2 x + b_2 y + c_2 z = d_2$$
$$a_3 x + b_3 y + c_3 z = d_3$$

be the system of equations and let

$$D = \begin{vmatrix} a_1 & b_1 & c_1 \\ a_2 & b_2 & c_2 \\ a_3 & b_3 & c_3 \end{vmatrix}$$

2) Multiply the first equation of the system by A_1 = the minor of a_1; the second equation by $-A_2 = -(\text{the minor of } a_2)$; the third equation by $A_3 =$ the minor of a_3.

3) Add the three equations you obtain in (2) and show that

a) the coefficient of $x = D$

b) the coefficient of $y = 0$

c) the coefficient of $z = 0$

d) the right-hand member $= D_x$

4) Use the results of (3) to show that $x = \dfrac{D_x}{D}$.

5) Use the same method to show $y = \dfrac{D_y}{D}$ and $z = \dfrac{D_z}{D}$.

5-5 (pages 180–183)

Teaching Suggestions

As in previous sections dealing with word problems, emphasize the importance of following the five step plan. They must state what quantities are represented by the variables (even more important now with three variables) and set up three equations, in each of which two expressions for the same quantity are equated.

5-6 (pages 183–189)

Teaching Suggestions

Three points should be emphasized here:
1) Regardless of which row or column you choose to expand by, expansion of a determinant by minors yields the same answer.

2) In view of (1), students should take a moment to inspect each determinant to see which row or column will make calculation by minors easiest (i.e., which row or column contains the most zeros).

3) If no row or column has many zeros, it may be possible to use Property 5 on page 186 to produce an equivalent determinant with more zeros.

Some students may have problems visualizing the expansion of a 4×4 determinant. To help the students prevent errors in expanding a determinant by minors, you may want to present the illustration below. To expand the determinant

$$\begin{vmatrix} 4 & -1 & 3 & 5 \\ 2 & 0 & 1 & 6 \\ -3 & 2 & -4 & 0 \\ 7 & 1 & -5 & 8 \end{vmatrix}$$

over the second column first take the first element in the second column and block out its row and column.

$$\begin{vmatrix} 4 & -1 & 3 & 5 \\ 2 & 0 & 1 & 6 \\ -3 & 2 & -4 & 0 \\ 7 & 1 & -5 & 8 \end{vmatrix}$$

The remaining numbers form the 3×3 minor of -1.

$$\begin{vmatrix} 2 & 1 & 6 \\ -3 & -4 & 0 \\ 7 & -5 & 8 \end{vmatrix}$$

The illustrations below continue the expansion along the second column.

$$\begin{vmatrix} 4 & -1 & 3 & 5 \\ 2 & 0 & 1 & 6 \\ -3 & 2 & -4 & 0 \\ 7 & 1 & -5 & 8 \end{vmatrix} \quad \text{minor:} \begin{vmatrix} 4 & 3 & 5 \\ -3 & -4 & 0 \\ 7 & -5 & 8 \end{vmatrix}$$

$$\begin{vmatrix} 4 & -1 & 3 & 5 \\ 2 & 0 & 1 & 6 \\ -3 & 2 & -4 & 0 \\ 7 & 1 & -5 & 8 \end{vmatrix} \quad \text{minor:} \begin{vmatrix} 4 & 3 & 5 \\ 2 & 1 & 6 \\ 7 & -5 & 8 \end{vmatrix}$$

$$\begin{vmatrix} 4 & -1 & 3 & 5 \\ 2 & 0 & 1 & 6 \\ -3 & 2 & -4 & 0 \\ 7 & 1 & -5 & 8 \end{vmatrix} \quad \text{minor:} \begin{vmatrix} 4 & 3 & 5 \\ 2 & 1 & 6 \\ -3 & -4 & 0 \end{vmatrix}$$

This may help the students visualize the expansion by minors.

6 Polynomials and Rational Expressions

Beginning with the laws of exponents, this chapter reviews the basic operations with polynomials and introduces factoring as a method for solving both polynomial equations and inequalities. The field of rational expressions is presented as being analogous to the field of rational numbers.

6-1 (pages 199–203)

Teaching Suggestions

Try to get students to treat powers of coefficients as they treat powers of variables: that is, leave them unmultiplied-out and in exponential form (at least until all cancellation has been carried out). A technical shortcut that will work if it is applied correctly is the following: a power of any quantity that is a *factor* of the numerator (or denominator) of a rational expression can be moved to the denominator (or numerator) if the sign of the exponent is reversed. If you mention this, however, be sure to caution against reversing the sign of a *term* of the numerator (or denominator) and then switching it to the denominator (or numerator). (See, for example, Written Exercise 20 on page 203.)

6-2 (pages 205–206)

Teaching Suggestions

Students should not only memorize the special formulas on page 205 but also have them at their fingertips. They should not confuse $a^2 - b^2$ with $(a - b)^2$.

Related Activities

Under what condition(s) will $(a + b)^2 = a^2 + b^2$? Have students multiply the left side to get:

$$a^2 + 2ab + b^2 = a^2 + b^2$$

Subtract a^2 and b^2 from both sides of the equation.

$$\therefore 2ab = 0 \text{ and}$$
$$ab = 0$$

Although this theorem is applied and proved later in the chapter, students should be able to make the conjectures that either $a = 0$ or $b = 0$.

6-3 (pages 207–211)

Teaching Suggestions

Much of this material may have been covered in Algebra I, but some of it, in particular, formulas IV and V, page 209, may be unfamiliar. The chief difficulty that students are likely to run into here is uncertainty about which factoring formula or technique to apply (or to apply *first*) to a given problem. You can help them by making two points: (1) it is important to check for monomial factors first; (2) formulas I–V should be so familiar to the student that they suggest themselves immediately when a problem fitting the pattern arises. For clarity, you might suggest that students include an extra step (in brackets below) to show what quantities are "playing the roles" of a and b in the problem; for example, in applying formula V:

$$8x^6 - 125 = [(2x^2)^3 - (5)^3]$$
$$= (2x^2 - 5)((2x^2)^2 + 2x^2 \cdot 5 + 5^2)$$
$$= (2x^2 - 5)(4x^4 + 10x^2 + 25)$$

6-4 (pages 213–217)

Teaching Suggestions

For the word problems, emphasize that all the problems of this section should be set up using only *one* variable. Make sure you discuss Example 2 on pages 214–215 in detail, to illustrate that a theoretically correct answer may not check in the original physical situation, owing to implicit conditions in the problem. Note that many word problems contain the implicit condition $x > 0$.

6-5 (pages 218–220)

Teaching Suggestions

The logic of the solution of quadratic inequalities may present problems for students. You can help them by insisting that all work be written out carefully and all connectives ("or" or "and") be included. A diagram like the one below may be helpful:

6-6 (pages 221–223)

Teaching Suggestions

The point to stress here is that in order to cancel a quantity from the numerator and denominator of a rational expression, that quantity must be a *factor* of both the numerator and the denominator. However, although students at this stage should know what a factor is, it might be worthwhile to remind them in plain language that the factors of an expression are any quantities which, when multiplied, give the whole expression.

6-7 (pages 223–226)

Teaching Suggestions

It is important for students to understand the connection between division of integers and division of polynomials with real coefficients. In each case there is a division algorithm, which states, roughly, that if $b \neq 0$, then for any A, there exist Q and R, with "R less than B," such that

$$\frac{A}{B} = Q + \frac{R}{B}$$
$$A = BQ + R.$$

In the case of integers, A, B, R, and Q are all integers, and "R less than B" means $0 \leq R < |B|$; in the case of polynomials, these letters, of course, stand for polynomials, and "R less than B" means that the *degree* of R is less than the *degree* of B.

From a technical standpoint, the method of carrying out the division algorithm for polynomials (as is illustrated on page 224) is quite similar to the method for integers. Emphasize that terms of the dividend must be arranged in order of decreasing powers of the variable and that missing terms must be accounted for (see Example 2).

6-8 (pages 226–228)

Teaching Suggestions

This material should present few technical difficulties if students remember to factor the expressions in the numerators and denominators and, if possible, reduce by cancellation, before carrying out the indicated products or quotients. You may wish to advise students to leave all answers in factored form (combined into a single rational expression in

lowest terms). Caution against performing too many operations in any one step (that is, factoring, inverting, cancelling).

6-9 (pages 228–231)

Teaching Suggestions

In this section, it is essential for students to copy original expressions accurately and set up their work carefully on paper. Tell the students to use the worked-out examples in the text as models for simplifying expressions step by step. Example 2 is especially worthy of study in class, since it demonstrates again how to deal with complex fractions.

Related Activities

1. *Fact:* If a and b are relatively prime integers (that is, they have no common factor except 1), then there are integers m and n (possibly negative) such that $ma + nb = 1$. Use this fact to prove that if a and b are relatively prime, then there are integers c and d such that $\frac{k}{ab} = \frac{c}{a} + \frac{d}{b}$, for any integer k. (This fact is clearly not true unless a a. b are relatively prime, as can be seen if you let $a = 2$, $b = 4$, and $k = 3$.)

$$\frac{k}{ab} = \frac{k \cdot 1}{ab} = \frac{(ma + nb)k}{ab} = \frac{mk}{b} + \frac{nk}{a}$$

2. Prove: If A and B are polynomials of the form $x - a$ and $x - b$, respectively, then there are real numbers M and N such that $MA + NB = 1$. Use this to prove a result for polynomials similar to the one above for integers.

$$M(x - a) + N(x - b) = 1 \longleftrightarrow$$

$$M + N = 0 \text{ and } -aM - bN = 1 \longleftrightarrow$$

$$N = \frac{1}{a - b} \text{ and } M = -\frac{1}{a - b}$$

Thus, if A and B are polynomials of the form $x - a$ and $x - b$, respectively, then there are real numbers C and D such that $\frac{K}{AB} = \frac{C}{A} + \frac{D}{B}$ for any real number K.

6-10 (pages 231–233)

Teaching Suggestions

As before, you should advise students to write down what quantities the variables in the solution of a word problem represent. A general formula here may be of assistance to students. Although they should be familiar with the formula

$$\text{distance} = \text{rate} \cdot \text{time},$$

they may not realize that in many situations where motion is not involved, a generalized version of this formula may be applicable:

$$\text{total amount} = (\text{amount/item})(\text{no. of items}).$$

Some examples of this are the following:

$$\text{total km} = (\text{no. of km/L})(\text{no. of liters});$$
$$\text{total pay} = (\text{hourly wage})(\text{no. of h worked}).$$

Note that in Example 1 the first factor of each term on the right side of the equation represents a *rate*.

6-11 (pages 234–237)

Teaching Suggestions

As the text indicates, it is important for students to check each apparent solution by substituting it back into the original equation. The reason that extraneous roots can appear in the solution sets of problems like those of this section can be shown by the following example.

$$x = 3$$
$$x(x - 2) = 3(x - 2)$$
$$x^2 - 2x = 3x - 6$$
$$x^2 - 5x + 6 = 0$$
$$(x - 3)(x - 2) = 0$$

Hence {2, 3} is the apparent solution set, but it is obvious that the solution set of the original equation is {3}. Note that if both sides of an equation are multiplied by a quantity containing a variable, then the equation is true when that quantity is 0.

Suggest that students begin solving fractional equations by listing the values of the variable for which any fractions are meaningless, that is, values that make the denominator equal to zero. Substituting apparent solutions in the original equation may also be helpful in eliminating extraneous roots.

Related Activities

In the diagram below, R_1 and R_2 are resistances connected in parallel in an electrical circuit with voltage source E.

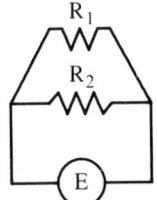

Kirchoff's laws for circuits such as the one illustrated state that if I_1 and I_2 are the currents (in amperes) through R_1 and R_2 (in ohms), respectively, I the current through the voltage source, and R the total resistance of the circuit, then these relationships hold:

(1) $I_1 + I_2 = I$, and
(2) $I_1 R_1 = I_2 R_2 = IR$

Use these laws to prove that

$$\frac{1}{R_1} + \frac{1}{R_2} = \frac{1}{R}$$

(1) and (2) $\longrightarrow$
$I_1 R_1 = (I_1 + I_2)R$ and
$I_2 R_2 = (I_1 + I_2)R \longrightarrow$
$I_1 R_1 R_2 = (I_1 + I_2)RR_2$ and
$I_2 R_1 R_2 = (I_1 + I_2)RR_1 \longrightarrow$
$(I_1 + I_2)(R_1 R_2) = (I_1 + I_2)(RR_2 + RR_1) \longrightarrow$
$R_1 R_2 = RR_2 + RR_1 \longrightarrow$
$\frac{1}{R} = \frac{1}{R_1} + \frac{1}{R_2}$

7 Radicals and Irrational Numbers

A discussion in the first section of this chapter of the general power function leads directly to a discussion of roots of a real number and roots of polynomial equations. Properties of and operations with radicals are studied. The basic operations with complex numbers are developed.

7-1 (pages 247–251)

Teaching Suggestions

Several points should be made about functions of the form $f(x) = ax^n$. First, note the similarity between the graphs of all functions of this form in which n is even, typified by the graph of $y = x^2$, as well as the similarity between the graphs of all those in which n is odd, typified by the graph of $y = x^3$. (See page 247.) Second, note the effect of different values of a for the same n. (See Written Exercises 1, 3, 4.) Finally, note the effect of different values of n for the same value of a particularly when $0 < x < 1$ and when $x > 1$.

To relate this material to your students' prior experience with equations, it would be a good idea to point out that although we know that the equation $x^2 = 9$, for example, has two roots (by factoring the equivalent equation $x^2 - 9 = 0$), we can now demonstrate this fact graphically by showing that the graphs of $y = x^2$ and $y = 9$ intersect exactly twice: at $(3, 9)$ and at $(-3, 9)$. Similarly, we can show graphically that an equation like $x^3 = 8$ has only one real root. This discussion will be made more general in the next section.

7-2 (pages 251–253)

Teaching Suggestions

Although the graphs on page 251 are those of specific functions, Figures 3 and 5 could well be taken as approximate graphs of any equation of the form $y = x^n$ where n is even, and Figure 4 as the graph of such an equation in which n is odd. Make sure your students see the relationship between the graphs and the chart at the top of the next page.

A problem almost invariably arises at this point about the meaning of the symbol $\sqrt{n}$. Does it mean, students ask, the positive or negative square root? The answer is that it means the positive square root *only*. Mathematicians have agreed to let the symbol mean only the positive square root because by so doing, they avoid ambiguity. If one wants to refer to the positive *or* negative square root, one can always write $\pm\sqrt{n}$. None of the foregoing, however, alters the fact that the equation $x^2 = 3$, for example, has two roots: $\sqrt{3}$ and $-\sqrt{3}$. Make sure students understand this.

7-3 (pages 254–256)

Teaching Suggestions

The theorem on page 254 may be somewhat mystifying to students, but you can remove some of the mystery by referring to a concrete example:

$$6x^2 - x - 35 = 0$$
$$(2x - 5)(3x + 7) = 0 \quad \text{and hence}$$
$$x = \frac{5}{2} \text{ or } x = -\frac{7}{3}.$$

Note that both denominators are factors of the leading coefficient, 6, and both numerators are factors of the constant term, -35, and it should be obvious from the method why this is the case. You might add that essentially the same idea of factoring extends to higher degree polynomial equations: namely, that $\frac{5}{2}$, for example, is a root of the equation $f(x) = 0$ if and only if $2x - 5$ is a factor of the polynomial $f(x)$. This, however, is not a proof of the theorem (for a partial proof, see Written Exercise 29), but only a heuristic argument.

If your students have not seen a proof that $\sqrt{2}$ is irrational, you might want to show them one now.

Assume $\sqrt{2} = \frac{a}{b}$, a, b both integers. Then $2 = \frac{a^2}{b^2}$ and $2b^2 = a^2$. The last equation implies that 2 is a prime factor of a^2, and hence of a, and therefore it occurs an even number of times in the prime factorization of a^2. On the other hand, 2 obviously occurs an odd number of times in the prime factorization of $2b^2$, a contradiction. Hence $\sqrt{2}$ is irrational.

Make certain that students realize a legitimate question is involved here: How do we know that there is no fraction (even one with hundreds of digits in numerator and denominator) whose square is 2?

Related Activities

Show that if $2^x = 3$, then x cannot be a rational number. (*Hint*: Assume $x = \frac{a}{b}$, a, b, integers, and show that this leads to a contradiction.)

$2^{\frac{a}{b}} = 3 \rightarrow 2^a = 3^b$. This implies 2 is a factor of 3^b—contradiction.

7-4 (pages 257–261)

Teaching Suggestions

Standard notation is, of course, used by virtually all scientists who deal with numbers, especially with numbers larger than 1000 or smaller than $\frac{1}{1000}$. In fact, many calculators have a display that can be converted (or converts automatically) to standard notation.

From a technical standpoint, see to it that students keep straight the rules for moving the decimal point when converting into or out of standard notation. One way to do this is to visualize the zeros in the power of 10 in the standard-notation representation, rather than try to memorize the rules blindly. All students should know that a positive exponent denotes a large (positive) number. A check that their answers are reasonable

should guarantee that the decimal point is not moved the wrong way. Checking to see if answers are reasonable is also the rationale behind the material on estimation, which, you should emphasize, is important in any numerical work.

7-5 (pages 261–265)

Teaching Suggestions

Make sure your students understand which fractions can be represented by terminating decimals. If they are uncertain, you could lead them to the correct necessary and sufficient condition—namely, that in lowest terms the denominator of the fraction has no prime factors other than 2 and 5 (a proof of which is called for in Written Exercise 32)—by suggesting they experiment with some examples. They should know that fractions with 2, 4, 5, and 8 as denominators can be represented by terminating decimals, but have them try fractions with 16, 40, 50, and perhaps 75, as denominators.

It is important to make a careful distinction between a repeating decimal (with a finite "block" of digits that repeats literally) and a decimal with some sort of pattern, like the one given on page 263. Note that, as has been shown, only those decimals with literally repeating blocks of digits (and terminating ones) represent rational numbers.

From their mathematical experience so far, students may have the idea that there are, in some sense, "more" rational numbers than irrationals. In fact, just the opposite can be shown, and you can give your students some feeling of why this is true by emphasizing the fact that *any* nonterminating, nonrepeating decimal represents an irrational number, not just expressions like $\sqrt{2}$, $\sqrt[3]{5}$, and so on.

Related Activities

If you have learned about systems of numeration based on numbers other than 10, try representing some common fractions as "decimals" in, for example, base 6. What are the terminating "decimals"? $\frac{1}{2}$, $\frac{1}{3}$, $\frac{2}{3}$, $\frac{1}{6}$, $\frac{5}{6}$, $\frac{1}{9}$, $\frac{2}{9}$, ... Can you make a generalization as to which fractions are represented by terminating "decimals" in base n?

those fractions whose denominators have no prime factors other than prime factors of n

7-6 (pages 267–269)

Teaching Suggestions

You should point out that the theorem on page 268 assures us that successive roots and powers of a given number can be computed in any order. In view of this fact, it is best to use the simplest method when faced with a specific example. When given the choice of finding the nth root of the mth power of a number or the mth power of the nth root, the latter is almost always easier, since large numbers are avoided.

Point out, also, that in the case of rationalizing a denominator, it is always easier to simplify the denominator first, then rationalize. In the following example, we could multiply numerator and denominator at once by $\sqrt{48x^3}$, but an easier method is the one shown below:

$$\frac{4x^2}{\sqrt{48x^3}} = \frac{4x^2}{4x\sqrt{3x}} \cdot \frac{\sqrt{3x}}{\sqrt{3x}} = \frac{x\sqrt{3x}}{3x} = \frac{\sqrt{3x}}{3}$$

7-7 (pages 270–273)

Teaching Suggestions

It is important for students to realize that in simplifying a sum or difference, each term should be simplified first, so that any common factors can be spotted.

In rationalizing denominators with two terms, one or both of which contain square roots, note that the formula $(a + b)(a - b) = a^2 - b^2$ is used in every case.

Finally, note that in factoring $x^2 - 16$, for example, over the set of polynomials with real coefficients, after we have rewritten the expression as $(x + 4)(x - 4)$, we cannot further factor $(x - 4)$ into $(\sqrt{x} - 2)(\sqrt{x} + 2)$, since $\sqrt{x} \pm 2$ is *not* a polynomial (although $x - \sqrt{2}$ is).

7-8 (pages 273–275)

Teaching Suggestions

The solution of some of the problems in this section is involved, and you must encourage patience and perseverance. In problems that require squaring both sides a second time, suggest that students cancel common factors on both sides before squaring, since this will ease the burden slightly.

7-9 (pages 276–279)

Teaching Suggestions

In working with complex numbers, you should emphasize that most of the same rules apply that apply to real numbers. An important exception, which is a frequent cause of error, is that $\sqrt{a} \cdot \sqrt{b} \neq \sqrt{ab}$, if a and b are negative. To avoid applying the rule incorrectly, make it a point always to change $\sqrt{a}\,(a < 0)$ to a real multiple of i before performing any other operations. Then, correct use of the relationship $i^2 = -1$ should prevent error.

7-10 (pages 279–282)

Teaching Suggestions

One underlying objective of this section and the following is to demonstrate that complex numbers satisfy all the field axioms (those given in Section 1-4; but not the order axioms in Section 2-4). It is important to express all answers in the form $a + bi$, where a and b are real numbers, both from the standpoint of satisfying the axioms of closure and from a practical standpoint; it is the numbers a and b that represent real physical quantities. Note that any term containing a power of i higher than the first power should be simplified according to the rule $i^2 = -1$.

Related Activities

Prove that if z_1 and z_2 are nonreal numbers and $z_1 + z_2$ and $z_1 z_2$ are both real, then z_1 and z_2 are

complex conjugates. (*Hint*: Start by letting $z_1 = a + bi$, $z_2 = c + di$, with $b \neq 0$, $d \neq 0$.)
Since $z_1 + z_2 \in \mathcal{R}$, $b + d = 0$, so $d = -b$. Since $z_1 z_2 \in \mathcal{R}$, $ad + bc = 0$, and hence, $a(-b) + bc = 0$, and $bc = ab$. $\therefore c = a$ (since $b \neq 0$), and we get $z_2 = a - bi$.

7-11 (pages 282–285)

Teaching Suggestions

Aside from the important algebraic techniques of multiplying and dividing complex numbers taught here, an important underlying theoretical by-product is that the set of complex numbers is closed under multiplication and that multiplicative inverses exist for all numbers except 0. (The general case for inverses is proved in Written Exercise 22.) In order to clarify this result, you should explain carefully that a number of the form $\dfrac{1}{c + di}$ is not in $a + bi$ form, and therefore is not obviously equal to any complex number. To demonstrate that it is complex, we must rationalize the denominator, as in Example 2. (Emphasize that complex numbers must be of the form $a + bi$ with a and b real.)

Note that both the product and the sum of a complex number and its conjugate are real numbers. In fact, this property characterizes the conjugate of a nonreal number.

8   Sequences and Series

The chief properties of arithmetic and geometric sequences and series are presented. In particular, formulas are developed for the general term of each type of sequence and for the sum of the first n terms of each type of series. A discussion of infinite geometric series is preceded by a definition of the limit of an infinite sequence and the introduction of the Completeness Axiom.

last example above could *not* be denoted simply as $1, 2, 4, \ldots$, because there is more than one sequence that begins in the same way:

$$1, 2, 4, 8, 16, \ldots, \text{ and}$$
$$1, 2, 4, \tfrac{15}{2}, 13, \ldots ; \quad a_n = \frac{n^3 + 5n + 6}{12}$$

Point out to the students that when a sequence is defined it is assumed to be infinite unless otherwise specified.

8-1 (pages 293–296)

Teaching Suggestions

Because this is probably the students' initial exposure to sequences, it is important to give a few examples of non-arithmetic sequences to put things in perspective:

$$3, 6, 12, 24, \ldots$$
$$1, 4, 9, 16, 25 \ldots$$
$$1, 2, 4, 7, 11, 16, \ldots$$

These examples may raise a question: how many terms must be written out before the dots? The answer is, enough so that there can be no doubt about the rule for generating the sequence. The

---

Related Activities

Let $f(x)$ be any function of the form $f(x) = ax^2 + bx + c$, and let
$$a_1 = f(1) - f(0)$$
$$a_2 = f(2) - f(1)$$
$$\vdots$$
$$a_n = f(n) - f(n - 1).$$
Show that $a_1, a_2, a_3, \ldots a_n$ is an arithmetic sequence.
$$a_n = an^2 + bn + c - [a(n - 1)^2 + b(n - 1) + c]$$
$$= 2an + (b - a)$$

$$a_{n-1} = a(n-1)^2 + b(n-1) + c - [a(n-2)^2 + b(n-2) + c]$$
$$= 2an + (b - 3a)$$
$$a_n - a_{n-1} = 2an + (b-a) - [2an + (b - 3a)]$$
$$= 2a$$

8-2 (pages 296–300)

Teaching Suggestions

Although a rigorous proof of the result in the box on page 296 would involve the principle of mathematical induction, students will nevertheless have no trouble believing it, in view of the justification given.

Note that the notion of arithmetic means is a generalization of the notion of the average of two numbers. There are many applications of this idea in daily life, for example, subdividing a beam at equally spaced points in order to attach supports to it.

8-3 (pages 300–305)

Teaching Suggestions

It is important that students understand the difference between a sequence and a series, and the connection between the two: namely, given a sequence, we can construct a series by choosing the first n terms of the sequence and writing them as a formal sum. Among other things, this means that although a sequence may have infinitely many terms, we do not (yet) know what it means to have an *infinite series*.

The summation notation introduces no new mathematical concepts; it is simply a shorthand symbol. When students have any doubt about its meaning or use, they should immediately revert to expanded form (with dots, if necessary) and refer to the axioms for the real numbers. For example, is it true that $\sum_{i=1}^{n} ca_i = c\sum_{i=1}^{n} a_i$? The left-hand side, expanded, is $ca_1 + ca_2 + \cdots + ca_n$; the right-hand side is $c(a_1 + a_2 + \cdots + a_n)$. The distributive

axiom tells us that these are equal. On the other hand,

$$\sum_{i=1}^{n} a_i \sum_{i=1}^{n} b_i \neq \sum_{i=1}^{n} a_i b_i,$$

and it might be a good test of your students' comprehension of the summation notation to have them explain why this is the case.

Note that it is not strictly necessary to know both formulas on page 302, since the first formula combined with the formula for a term of an arithmetic sequence gives the second formula, as Example 3 shows. But it will often be convenient to know both formulas (see Oral Exercises 9–20).

You may want to stress the fact that an infinite sequence can have an infinite number of series associated with it. That is, you have a different series depending upon the number of terms being summed. The series S_1 sums the first term of the sequence, the series S_2 sums the first two terms of the sequence, the series S_3 sums the first three terms of the sequence, and so on.

Related Activities

Demonstrate to your students the formation of Pascal's Triangle, below.

The rule for generating this array is to make the borders of the triangle 1's, and make any element not along a border equal to the sum of the two elements closest to it in the row above. Ask your students to look for sequences in the triangle. For example, they may see that if they take sums along one diagonal, then each sum is the respective element in the diagonal below. If you sum along d_0 you get $1 = 1$, $1 + 2 = 3$, $1 + 2 + 3 = 6$ and so on. Note that the sums 1, 3, 6, etc. make the diagonal d_1. Encourage your students to discover more patterns in the triangle.

Teaching Suggestions

Many examples of geometric sequences in daily life can be adduced to motivate students here:
(1) a ball bouncing in such a way that the height of each bounce is a fixed fraction of the height of the preceding bounce (an essentially accurate physical case);
(2) any quantity, such as cost of living, population, etc., that increases by a fixed percent each year;
(3) the number of radioactive atoms in a substance at equal time intervals (the time interval in which half of the nuclei of a given mass undergo radioactive decay is called the half-life).

As before, the formula in the box on page 306 would require mathematical induction for a rigorous proof, but students should have little trouble believing it. From a technical standpoint, the numbers involved in geometric sequences (and later, series) will be more cumbersome than those in arithmetic series and sequences, but you can assist students by urging that they leave all powers of numbers in exponential form until all possible cancellation has been carried out.

Teaching Suggestions

The geometric mean of two numbers can be regarded as an "average with respect to multiplication." A concrete interpretation is that it is the length of the side of a square of area equal to the area of a rectangle whose sides have the given numbers as lengths (if these are positive). Remind students that the altitude to the hypotenuse of a right triangle is the geometric mean of the two segments into which it divides the hypotenuse. Two other examples of the use of geometric means are the following:
(1) the value of a house at yearly intervals between two sales of the house, if it is assumed that the house appreciates by a fixed percentage of its value each year;
(2) the amount of bacteria in a culture at regular intervals between two readings of an instrument that measures this amount (if the culture increases by a fixed percentage over equal time intervals).

Teaching Suggestions

Note that, as with arithmetic series, the two formulas for the sum of a geometric series should both be learned for the sake of convenience, one being more practical to employ in a given situation than the other.

Unfortunately, the formula for the sum of a geometric series often allows for little cancellation, so make sure students do not cancel incorrectly —for example, cancel r in the denominator with r^n in the numerator.

Teaching Suggestions

Conceptually, this section is liable to prove difficult for students. To minimize this difficulty, you should first make sure they fully understand the somewhat informal (but logically correct) definition of the limit of an infinite sequence on page 322. Emphasize here that if L is the limit, then by choosing n sufficiently large we can make the error $|L - a_n|$ smaller than any positive number (but note that 0 is *not* a positive number). This means that if a small positive number (like $\frac{1}{1000}$) is given, then by going "far enough out" in the sequence, we can find a term within that distance of L, and every term thereafter will also be within that distance of L.

Next, note that the Completeness Axiom ensures that certain types of sequences converge but gives no hint as to their limits. It is an abstract property, but one that is crucial to operations involving irrational numbers. For example, consider the sequence we generate when trying to approximate the square root of 2:

$$1, 1.4, 1.41, 1.414, \ldots$$

This sequence is bounded and nondecreasing and converges to $\sqrt{2}$. But how do we know there is such a number as $\sqrt{2}$? Because the Completeness Axiom assures us there is. Indeed, in higher mathematics, real numbers are often identified with sequences like the one above. Intuitively, the Completeness Axiom assures us that the real numbers form a "continuum" and do not have any "holes."

You may want to emphasize to the students the fact that a sequence can be bounded, but *not* convergent. If a sequence is bounded, but neither nondecreasing or nonincreasing (as the example in the text p. 322) the sequence is not "settling down to one point" as a convergent sequence is.

From a practical standpoint, you might emphasize two facts for students to keep in mind:

(1) If $a, b > 0$ and $a > b$, then $\frac{1}{a} < \frac{1}{b}$.

(2) If $0 < |a| < 1$, then
$|a| > |a^2| > |a^3| > \ldots$ and in fact,
$\lim\limits_{n \to \infty} a^n = 0$.

Related Activities

Present an interesting sequence of numbers to your students. The sequence 1, 1, 2, 3, 5, 8, 13, . . . was encountered by Fibonacci, a mathematician of the middle ages, when he posed the following problem:

A pair of adult rabbits is placed in an enclosed area. How many rabbits will be produced in a year if each month this pair produces another pair of rabbits who in turn become productive at two months of age?

Assuming none of the rabbits dies, the sequence generated by the numbers of rabbits produced

each month is the Fibonacci sequence. You may want to demonstrate this in class by showing that in the first month the adult pair produces one pair, in the second month the adult pair produces one pair, in the third month the adult pair and the young pair each produce a new pair, etc.

Ask your students to write a recursive rule and/or equation to define the Fibonacci sequence. If the nth term is f_{k+2}, $f_1 = 1$, $f_2 = 2$, and $f_k + f_{k+1} = f_{k+2}$.

8-8 *(pages 324–330)*

Teaching Suggestions

From your students' standpoint, the notion of a sequence of partial sums (each of which is the sum of a finite series), whose limit is defined to be the sum of the infinite series, may seem conceptually complex. You can perhaps clarify things somewhat with another interpretation of the example on page 324:

$$1 + \frac{1}{2} + \frac{1}{4} + \frac{1}{8} + \frac{1}{16} + \cdots$$

$$S_1 = 1$$
$$S_2 = \frac{3}{2}$$
$$S_3 = \frac{7}{4}$$
$$S_4 = \frac{15}{8}$$

If students are asked to guess the limit from the foregoing, they will undoubtedly guess 2, and you can point out that the basis for their guess must have been the observation of "where the partial sums are heading," that is, their limit. Note carefully that the definition of the limit of an infinite series (as the limit of the sequence of partial sums) is just that—a definition; and as such, it cannot be "checked" by actually adding up all the terms. (If we could do the latter, we wouldn't need the definition.)

9 Polynomial Functions

This chapter begins by examining quadratic equations and functions. Quadratic equations are solved and methods are developed for determining the nature of the roots of quadratic equations and for relating their roots and coefficients. Quadratic functions and inequalities are then graphed and discussed. In the next part of the chapter techniques and theorems for dealing with polynomial equations and functions, including synthetic substitution (division), the Remainder and Factor Theorems, and the Fundamental Theorem of Algebra are presented. The chapter concludes by locating and estimating real roots of polynomial equations using graphing and linear interpolation.

9-1 (pages 341–347)

Teaching Suggestions

The concept of completing the square involves finding the third term of a perfect square trinomial. Students will understand this better if you first review the squares of binomials, such as $(x + 5)^2 = (x + 5)(x + 5) = x^2 + 10x + 25$, with emphasis on how the middle term is found. A useful class activity is to ask students to find the missing third term of perfect square trinomials. For example, find the missing term:

$$x^2 + 10x + \text{?}$$

Emphasize that the missing term can be found by squaring one-half the coefficient of x.

The quadratic formula is one of the most important formulas your students will study this year. Its derivation should be stressed and the formula itself memorized. When using the formula to solve quadratic equations, remind students that the values of a, b, and c can only be determined when the quadratic equation is in standard form: $ax^2 + bx + c = 0$. It is also important to note that b or c can equal zero, but a cannot. If $a = 0$, there would not be an x^2 term and hence you would not have a quadratic equation.

9-2 (pages 347–350)

Teaching Suggestions

It is important to stress that the theorem on page 348 holds only for equations with *real* coefficients. Emphasize these two key points:

1. A quadratic equation with real coefficients cannot have one real and one non-real root.
2. If a quadratic equation with real coefficients has a double root, the root must be real.

Also mention that it is possible to determine whether the real roots of a quadratic equation with *rational* coefficients are rational or irrational. The roots will be rational if the discriminant is a perfect square and irrational if it is not.

Related Activities

Show that if x_1 and x_2 are the roots of $x^2 + bx + c = 0$, then $(x_1 - x_2)^2 =$ the discriminant.

Let $x_1 = \dfrac{-b + \sqrt{b^2 - 4c}}{2}$ and $x_2 = \dfrac{-b - \sqrt{b^2 - 4c}}{2}$

Then $x_1 - x_2 = \dfrac{-b + \sqrt{b^2 - 4c}}{2} - \dfrac{-b - \sqrt{b^2 - 4c}}{2}$

$x_1 - x_2 = \dfrac{-b + \sqrt{b^2 - 4c} + b + \sqrt{b^2 - 4c}}{2}$

$x_1 - x_2 = \dfrac{2\sqrt{b^2 - 4c}}{2}$

Then $(x_1 - x_2)^2 = b^2 - 4c$, the discriminant.

9-3 (pages 350–353)

Teaching Suggestions

The relationship between the roots and coefficients of a quadratic equation can be discovered by asking your students to do the following:

1. Find the roots of the two quadratic equations given on the next page.
2. Write the sum of these roots.
3. Write the product of these roots.

4. Compare these sums and products to the co-efficients of the original quadratic equations.

quadratic equation:	$x^2 - 6x + 5 = 0$	$x^2 + 4x - 21 = 0$
roots:	1, 5	3, −7
sum of the roots:	6	−4
product of the roots:	5	−21

You can now point out the fact that the sum of the roots is always $-\dfrac{b}{a}$ and the product of the roots is always $\dfrac{c}{a}$.

One important point to make in this section is that the roots of a quadratic do not determine a unique equation. In Example 2 on page 351 the roots $\frac{1}{2} + i\sqrt{5}$ and $\frac{1}{2} - i\sqrt{5}$ were used to find the equation $4x^2 - 4x + 21 = 0$. But these are also the roots of $8x^2 - 8x + 42 = 0$, and in fact of any equation of the form $4ax^2 - 4ax + 21a = 0$.

9-4 (pages 354–357)

Teaching Suggestions

Some interesting class activities to reinforce this lesson involve the overhead projector. Graph a function such as $f(x) = 2x^2$ on a transparent overlay and ask your students to describe how to shift the graph as h and k vary. For example, the graph of $f(x) = 2x^2 - 3$ would shift the graph 3 units down. As the students describe the shift(s), actually move the parabola to the appropriate position on the coordinate axes of the overhead projector. Conversely, you could move the parabola to different locations, and ask your students to give the new equation.

Spend some time considering the graph of $f(x) = ax^2$ and the effect a has on the parabola: if $a > 0$, the parabola opens upward; if $a < 0$, the parabola opens downward; the $|a|$ determines how "wide" or "narrow" the parabola is.

Emphasize that in the general form $f(x) = a(x - h)^2 + k$, a alone determines the shape of the parabola while h and k locate its position.

9-5 (pages 357–362)

Teaching Suggestions

A good deal of algebraic technique is demanded in this section. Students should be able to do the derivation on pages 358–359, namely, rewriting an equation of the form $y = ax^2 + bx + c$ in the form $y = a(x - h)^2 + k$. As you proceed with the derivation, it may be helpful to do it on the chalkboard next to Example 1, pointing out the corresponding steps. Explain that the derivation includes the use of completing the square in a situation other than solving quadratic equations.

As you can see below, the equation of the axis of symmetry also furnishes the x-coordinate of the vertex point.

equation of the axis of symmetry	vertex
$x = -\dfrac{b}{2a}$	$\left(-\dfrac{b}{2a},\ -\dfrac{b^2 - 4ac}{4a}\right)$

Therefore, students need only memorize the equation of the axis of symmetry. To find the y-coordinate of the vertex, they can simply substitute the x-coordinate, $-\dfrac{b}{2a}$, back into the original equation.

Related Activities

Suppose the graphs of $y = mx + k$ and $y = ax^2 + bx + c$ intersect only once (i.e., the line $y = mx + k$ is tangent to the parabola $y = ax^2 + bx + c$).
1. Solve for x under this assumption, by solving the two equations simultaneously and assuming that the discriminant of the resulting equation is 0 (giving only one double root).

$$ax^2 + bx + c = mx + k$$
$$ax^2 + (b - m)x + (c - k) = 0$$
$$x = \frac{-(b - m) \pm \sqrt{(b - m)^2 - 4a(c - k)}}{2a}$$
$$= \frac{-b + m}{2a},\ \text{since } D = 0$$

2. Use (1) to show that the assumption implies that if (x, y) is the point of intersection, then m must equal $2ax + b$. (This fact can be found using calculus as well.) This gives a formula for the slope of a line tangent to a parabola at a point (x, y).

$$2ax = -b + m$$
$$m = 2ax + b$$

9-6 *(pages 362–364)*

Teaching Suggestions

As mentioned at the top of page 363, the example in this section can be solved without having to test a value of x from each of the three regions of the number line. Reconsider the example: $-x^2 + 2x + 9 > 0$. After determining the roots and placing them on a number line (the x-axis), graph the parabola $y = -x^2 + 2x + 9$ (below). The parabola opens down because the value of a is negative.

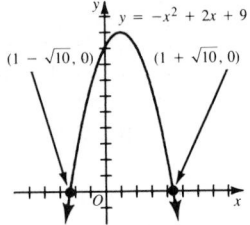

Where is $-x^2 + 2x + 9 > 0$? Remembering that greater than zero (> 0) is *above* the x-axis, the part of the parabola which is above the x-axis lies *between* $1 - \sqrt{10}$ and $1 + \sqrt{10}$. Thus, the solution set is $\{x: 1 - \sqrt{10} < x < 1 + \sqrt{10}\}$.

9-7 *(pages 365–368)*

Teaching Suggestions

Synthetic substitution (division) is an important process that will be used extensively in this and the next four sections. It may be helpful for your students to see an actual example of the process before you consider the general justification given on page 366. Given the polynomial function $P(x) = 5x^3 - 12x^2 - 20x + 1$, ask your students to find

$P(7)$ by direct substitution. Answer: $P(7) = 988$. Then ask them to again consider the same polynomial function and do the following:

1. Multiply the coefficient of x^3 by 7. 35
2. Add to that the coefficient of x^2. 23
3. Multiply this result by 7. 161
4. Add to that the coefficient of x. 141
5. Multiply this result by 7. 987
6. Add to that the constant term. $988 = P(7)$

Stress the saving in time and effort using the latter procedure.

When you explain the general justification of synthetic substitution, you may want to point out that the process simply makes use of the Distributive Property several times. As pointed out in the text, students can verify this by simplifying Step 6 on page 366.

Explain that this procedure, synthetic substitution, can be accomplished more simply if the following format is used.

$$
\begin{array}{r|rrrr}
7 & 5 & -12 & -20 & 1 \\
  &   & 35 & 161 & 987 \\
\hline
  & 5 & 23 & 141 & 988 = P(7)
\end{array}
$$

Remind your students to put in zeros for any missing terms when using this method.

9-8 *(pages 368–372)*

Teaching Suggestions

Students should learn the Remainder Theorem and the Factor Theorem thoroughly since they are used extensively for the remainder of the chapter. It may prove helpful to label the key statement in the Remainder Theorem on page 369 as follows:

$$P(x) = (x - r)\, Q(x) + P(r)$$

dividend   divisor   quotient   remainder

When Example 1 on page 369 is done, you may want to compare the two methods on the next page to show how synthetic division eliminates repetitious rewriting of the variable as well as some of the coefficients.

$$
\begin{array}{r}
1x^2 + 0x + 4 \quad \text{R } 0 \\
x - 3 \,)\, \overline{1x^3 - 3x^2 + 4x - 12} \\
\underline{1x^3 - 3x^2} \\
0x^2 + 4x \\
\underline{0x^2 + 0x} \\
4x - 12 \\
\underline{4x - 12} \\
0
\end{array}
\qquad
\begin{array}{r|rrrr}
3 & 1 & -3 & 4 & -12 \\
  &   & 3 & 0 & 12 \\
\hline
  & 1 & 0 & 4 & 0
\end{array}
$$

When finding the roots of a polynomial equation, as in Example 3 on pages 370–371, students should be encouraged to use the compact form of synthetic division illustrated. Notice that the table contains two separate synthetic divisions. This method is best utilized when a number of divisions must be done using the same dividend.

9-9 (pages 374–375)

Teaching Suggestions

The problem of finding the roots of a polynomial equation occurs frequently in mathematics and science. When searching for these roots, your students can use the two theorems in this section along with the Factor Theorem. Key points are:

1. Every polynomial equation of degree n has exactly n roots (first theorem on page 374).
2. These roots may be real and/or complex.
3. Some roots may be double roots.
4. Imaginary roots always occur in conjugate pairs (second theorem on page 374).

Note that this second theorem holds only for polynomial equations with *real* coefficients.

Remind students that once it is determined that $x^2 + 4x + 5$ (bottom of page 374) is a factor of $P(x)$, the other factor is found by dividing $P(x)$ by $x^2 + 4x + 5$ by standard methods of polynomial division. Synthetic division cannot be used because the divisor is a quadratic and not a linear factor.

9-10 (pages 376–379)

Teaching Suggestions

Remind students that Descartes' Rule of Signs and the theorem on bounds in this section do not find the roots of a polynomial equation but merely give information about them. Example 1 on page 377 should be carefully reviewed so that students understand how the information is being used to narrow down the possible combinations of roots of the polynomial equation. In this case, there is only one positive root and either one or three negative roots. The number of imaginary roots in each case is determined by the fact that the sum of the numbers of positive, negative, and imaginary roots must equal 4. (We know that a fourth degree equation has exactly four roots.) The chart at the end of the example lists the possible root combinations.

9-11 (pages 380–382)

Teaching Suggestions

There is no easy way to determine a real root of a polynomial equation if the root is not rational and the degree of the equation is 3 or more. This section will provide students with a method for estimating these real roots.

The roots of a polynomial equation occur where the graph of the polynomial function crosses the x-axis. As can be seen from the graph in Example 1, roots occur between -1 and 0, between 1 and 2, and between 2 and 3. Students should understand the correlation between using a graph to determine between which two integers a root of the polynomial lies and using the observed sign changes in the $P(x)$ column of the chart to determine between which two integers a root of the polynomial lies. Remind students to set up their synthetic division charts from least x to greatest x so that the sign changes can be easily found.

Linear interpolation enables students to estimate the real roots with a greater degree of accuracy. Emphasize that the results obtained by linear interpolation are not exact because, as can be seen from Figure 5 on page 381, the line and the curve do not necessarily cross the x-axis at exactly the same point.

10 Quadratic Relations and Systems

This chapter presents some of the fundamentals of analytic geometry. After the basic formulas for finding the length and midpoint of a line segment are developed, they are used to derive the general formulas for the conic sections circle, parabola, ellipse, and hyperbola. Graphing aids that minimize the need for tables of values are introduced, and in the latter part of the chapter, both graphic and algebraic methods for solving linear-quadratic systems and quadratic-quadratic systems are presented.

know the slope-intercept form of an equation ($y = mx + b$). A brief review of this material will aid in understanding the relationship shown in Figure 3.

The theorem on page 398 can be rephrased in the following form:

Two nonvertical lines are perpendicular if and only if their slopes are negative reciprocals of each other.

Emphasize that the theorem does not apply to vertical and horizontal lines. Ask students to find the slope of a vertical line (no slope) and a horizontal line (zero slope). They should see that there is no defined product for these slopes.

10-1 (pages 393–397)

Teaching Suggestions

The distance formula is an important formula from analytic geometry. Students should be aware that it will be used to derive the equations of the circle, parabola, ellipse, and hyperbola.

You may want to start the lesson by plotting two points in the coordinate plane and asking students for ways to find the distance between the points. Stress the relationship between distance and length and the use of the Pythagorean Theorem. Once this is accomplished, the derivation of the formula will be more easily understood. Point out that it does not matter which point is labeled (x_1, y_1) and which is labeled (x_2, y_2).

Once the students realize that the midpoint of a segment is the point one-half the distance from the endpoints of the segment, the midpoint formula follows easily from the distance formula. Both the distance and the midpoint formulas should be memorized.

10-3 (pages 401–403)

Teaching Suggestions

The circle is the first of four quadratic relations to be considered in this chapter. Students should be aware that the equation of a circle is not a function. Point out the similarities between the distance formula and the equation of a circle. This may help students to remember both more easily.

Example 1 should be explained completely, but it may be a good idea to review the technique of completing the square before you do the example. Also remind students of the need to divide each side of the equation by the coefficients of x^2 and y^2 before attempting to complete the square. (This is necessary in Written Exercises 15 and 16.) Students should be able to graph circles without using a table of values.

10-2 (pages 397–399)

Teaching Suggestions

Before beginning this lesson, make sure students recall the concept of the slope of a line and still

10-4 (pages 404–408)

Teaching Suggestions

The parabola is the second quadratic relation to be examined in this chapter. The parabola occurs in a wide variety of situations in science such as the paths of projectiles and the shapes of headlight reflectors in automobiles.

Much of this material was covered in Section 9-4 where it was shown that the graph of a quadratic function is a parabola. Emphasize here that not all parabolas are functions. You may want to remind students that a parabola with an x^2 term opens up or down (and is a function) while a parabola with a y^2 term opens right or left (and is not a function).

Related Activities

Take a piece of ordinary notebook paper and mark a point P on it near the bottom and about halfway across the page. Fold the paper so that a point on the bottom of the page coincides with point P, as shown. Do this many times. The curve suggested by the fold lines should be a parabola with focus at P and directrix the line coinciding with the bottom of the page.

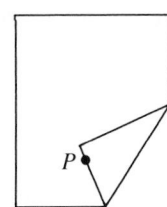

What happens to the curve as the distance between P and the bottom of the page increases? decreases?

Curve gets "wider."
Curve gets "narrower."

If you do this using waxed paper, the resulting curve can be mounted on colored paper for an interesting display.

10-5 (pages 408–412)

Teaching Suggestions

The ellipse is the third quadratic relation to be studied in this chapter. Like parabolas, ellipses have important scientific applications. One example is the elliptical orbits of planets.

It may be useful to draw an ellipse for the students using string, a pencil, and two thumb tacks. Remind students that since the definition of the ellipse involves distance, the distance formula is important in determining the equation of an ellipse. Stress that the relationship between the

x-intercepts $((a, 0),$ and $(-a, 0))$, the y-intercepts $((0, b),$ and $(0, -b))$, and the foci $((c, 0),$ and $(-c, 0))$ is $b^2 = a^2 - c^2$. Students should know that the major axis is always the longer of the two axes and the vertices and foci are located on the major axis, and that the sum of the distances to any point on the ellipse from the foci is equal to the length of the major axis $(2a)$.

Point out to the students that the fact that ellipses are symmetric allows one to graph an ellipse (whose center is the origin) with only a small table of values from the first quadrant.

Related Activities

Another definition of an ellipse is that it is the set of points whose distances from a fixed point and a fixed line are in a constant (positive) ratio less than 1. Show that the set of points P such that

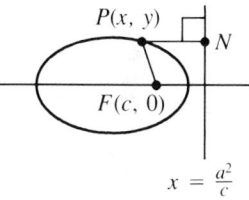

$\dfrac{PF}{PN} = \dfrac{c}{a}$ (where PN is the shortest distance from P to the line $x = \dfrac{a^2}{c}$) is the ellipse $\dfrac{x^2}{a^2} + \dfrac{y^2}{b^2} = 1,$
where $b^2 = a^2 - c^2$.

$$\frac{PF}{PN} = \frac{\sqrt{(x - c)^2 + y^2}}{\left|\dfrac{a^2}{c} - x\right|} = \frac{c}{a}$$

$$(x - c)^2 + y^2 = \frac{c^2}{a^2}\left(\frac{a^2}{c} - x\right)^2$$

$$a^2(x^2 - 2cx + c^2 + y^2) = c^2\left(\frac{a^4}{c^2} - \frac{2a^2x}{c} + x^2\right)$$

$$a^2x^2 - 2a^2cx + a^2c^2 + a^2y^2 = a^4 - 2a^2cx + c^2x^2$$

$$(a^2 - c^2)x^2 + a^2y^2 = a^2(a^2 - c^2)$$

$$\frac{x^2}{a^2} + \frac{y^2}{a^2 - c^2} = 1$$

10-6 (pages 412–417)

Teaching Suggestions

The equation of a hyperbola is very similar to that of an ellipse. To help the students distinguish between the equations of hyperbolas and ellipses, you should note that an addition sign is used for the ellipse while a subtraction sign is used for the hyperbola. In discussing hyperbolas, you may have to dispel the notion that a hyperbola is simply "two parabolas laid end to end." The fact that hyperbolas have asymptotes provides a simple way of distinguishing, geometrically at least, between the two curves. A hyperbola gets closer and closer to its asymptotes, whereas a parabola eventually "curves away" from any line that does not intersect it. See the diagram below.

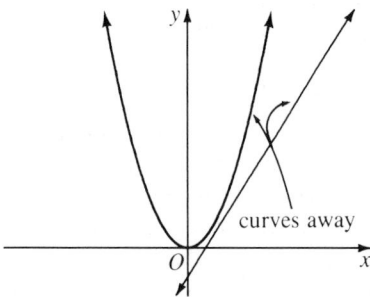

In order to graph a hyperbola, students should first find the vertices and asymptotes. You should mention that asymptotes are *not* part of the hyperbola but merely serve as aids in graphing. Once the vertices and asymptotes have been found the graph can be completed using only a table of first-quadrant values.

The four quadratic relations discussed in sections 10-3 through 10-6 are sometimes referred to as the *conic sections*. Direct the students' attention to Figure 15 which shows how each is generated using a right circular cone and a plane. Your school may have a three-dimensional model that will illustrate this.

Related Activities

If a shaded incandescent bulb is considered as a point-source of light, what curve is described by the shadow of the shade on a wall if the opening of the shade is a perfect circle?
A hyperbola

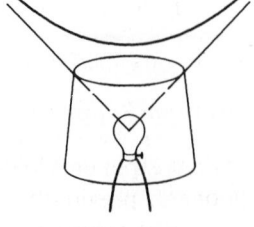

10-7 (pages 417–422)

Teaching Suggestions

Encourage students to think of practical examples in which one quantity decreases as another related quantity increases. Graphs of inverse variations of this type ($xy = k$) are hyperbolas. Note that the asymptotes of this hyperbola are the x- and y- axis.

It may be useful to make a chart listing the four types of variation (direct, inverse, joint, and combined) along with a description of each and the general equations for each type.

10-8 (pages 423–424)

Teaching Suggestions

Before beginning this section, it may be wise to review the standard forms of the equations of the circle, parabola, ellipse, hyperbola, and straight line. It is essential for students to know which figure they are graphing before they begin their work.

Stress that a graphical solution to a system of equations is a set of ordered pairs corresponding to the points of intersection of the graphs of the two equations. A useful activity may be to ask your students to sketch all the possible graphical solution situations for each system. For example, the diagrams below and on the next page show how a system consisting of a parabola and an ellipse can have 0, 1, 2, 3, or 4 solutions.

Emphasize that graphical solutions lack the accuracy of algebraic methods, but are helpful in approximating the solutions to a system. The solutions, therefore, may not check exactly.

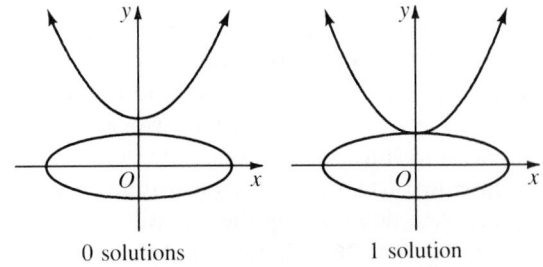

0 solutions   1 solution

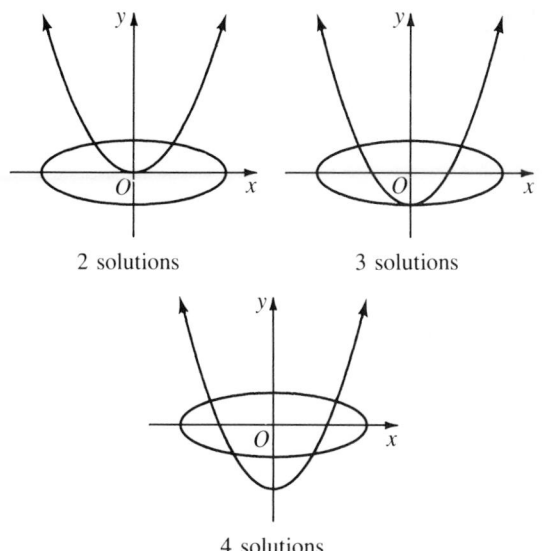

2 solutions   3 solutions

4 solutions

10-9   (pages 424–428)

Teaching Suggestions

A system consisting of one linear and one quadratic equation can be most easily solved by substitution. The substitution method is preferable to the graphical method for two reasons. First, as mentioned in Section 10-8, graphing can produce solutions that lack accuracy, and second, graphing yields only real number solutions. The substitution method yields any complex solutions as well.

To avoid the possibility of introducing extraneous roots, encourage students to substitute into the linear equation to find the second coordinate after the first coordinate has been found.

10-10   (pages 430–432)

Teaching Suggestions

Remind students that quadratic-quadratic systems of equations can have from zero to four solutions, and any complex solutions will appear in conjugate pairs. Encourage students to consider both the substitution method and the linear-combination method for each system and then have them select the method that will provide the simplest solution. You may want to mention that the substitution method in this case may produce a fourth-degree equation.

11   *Exponents and Logarithms*

This chapter begins with a discussion of rational exponents and then extends the concept to real number exponents. After the general idea of an inverse of a function is explored, the logarithmic function is defined as the inverse of the exponential function. The laws of exponents then provide rules for operations with logarithms and these rules are used in numerical computations with common logarithms.

Students will learn how to solve exponential equations and see their application to the area of exponential growth and decay. Finally, an introduction to natural logarithms provides students with a sound basis for future study.

11-1   (pages 441–444)

Teaching Suggestions

Impress upon your students the fact that this section is important because it provides meaning to rational exponents while preserving the laws of exponents previously studied.

The key concept in this lesson is the relationship between exponential form and radical form. This relationship can be illustrated using the following informal argument:

Since $3^{\frac{1}{2}} \cdot 3^{\frac{1}{2}} = 3^1$ or 3
and $\sqrt{3} \cdot \sqrt{3} = 3$
then $3^{\frac{1}{2}}$ must be equal to $\sqrt{3}$.

You may want to extend the definition on page 442 to: $b^{\frac{p}{r}} = (\sqrt[r]{b})^p = \sqrt[r]{b^p}$. This extension permits the base b to be raised to the power p either *before* or *after* the root r is taken. Sometimes one definition is more useful than the other in simplifying expressions. Finally, note that the definition holds only for positive numbers b. This must be insisted upon because, although real odd roots of negative numbers exist, real even ones do not, and we want to restrict our attention to real numbers for the present.

11-2 (pages 444–446)

Teaching Suggestions

This lesson extends the definitions of powers to include irrational numbers as exponents. When considering the value of $2^{\sqrt{3}}$, for example, remind students that an irrational number such as $\sqrt{3}$ cannot be expressed as a terminating or repeating decimal. You may want to encourage the use of a scientific calculator to find approximate values such as:

$$2^{\sqrt{3}} \approx 2^{1.7320508 \cdots} \approx 3.321997 \ldots$$

Ask students to list some common characteristics of the graphs of all exponential functions of the form $y = b^x$. Here are some examples:

1. The graphs lie entirely above the x-axis.
2. The x-axis is an asymptote.
3. The graphs all pass through the point $(0, 1)$.

Another interesting fact about the graphs of exponential functions is that the graph of $y = \left(\frac{1}{b}\right)^x$ is always symmetric to the graph of $y = b^x$ relative to the y-axis.

Finally, an equation containing a variable exponent, such as the example on page 446, is called an exponential equation. Mention that an equation such as this can sometimes be solved by writing both sides in the same base and equating the exponents.

11-3 (pages 447–450)

Teaching Suggestions

Inverse functions often present conceptual problems for students. Before beginning the lesson, you may want to consider a function f and its inverse function f^{-1} using sets of ordered pairs as follows:

If $f = \{(2, 1), (-3, 5), (4, -2)\}$
then $f^{-1} = \{(1, 2), (5, -3), (-2, 4)\}$.

The inverse function f^{-1} is obtained by switching the x- and y-coordinates of each ordered pair. When a function is defined by an equation, as in Example 2 on page 449, it is then easy for students to see why the inverse is obtained by interchanging x and y and then solving for y. Stress that not all functions have inverse functions.

11-4 (pages 451–453)

Teaching Suggestions

In section 11-3 the inverse of the function $y = 3x - 2$ was found by interchanging x and y and then solving for y. To find the inverse of the exponential function $y = b^x$, it is easy enough to interchange x and y and get $x = b^y$. The problem now is that it is not algebraically possible to solve this equation for y. Although this is difficult for students to understand, it is the reason logarithmic notation has been adopted to write the inverse of an exponential function.

In order to continue the study of logarithms in this chapter, students should have a thorough understanding of the relationship between logarithmic form and exponential form. The following may be helpful:

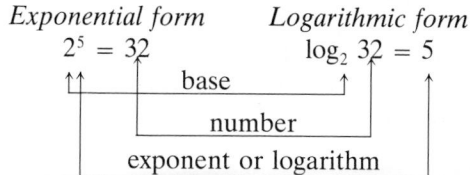

Exponential form   Logarithmic form

$2^5 = 32$ $\log_2 32 = 5$

base

number

exponent or logarithm

To this end, you should spend considerable time having your students convert equations from exponential form to logarithmic form and vice versa.

11-5 (pages 454–456)

Teaching Suggestions

Logarithms are exponents. Once your students understand this, the Laws of Logarithms on page 454 will be more comprehensible. You may find it helpful to restate those laws as follows:

1. To find the logarithm of a product, add the logarithms of the factors.
2. To find the logarithm of a quotient, subtract the logarithm of the divisor from the logarithm of the dividend.
3. To find the logarithm of the power of a number, find the product of the power and the logarithm of the number.

Related Activities

Prove that for m, n, a, and b all positive real numbers and $a \neq 1$ and $b \neq 1$,

$$\frac{\log_a m}{\log_a n} = \frac{\log_b m}{\log_b n}.$$

(*Hint*: Let the left-hand side equal x, the right-hand side equal y, and show that $x = y$.)

$$x = \frac{\log_a m}{\log_a n}$$

$$x \log_a n = \log_a m$$

$$\log_a n^x = \log_a m$$

$$n^x = m$$

$$y = \frac{\log_b m}{\log_b n}$$

$$y \log_b n = \log_b m$$

$$\log_b n^y = \log_b m$$

$$n^y = m$$

$$\therefore n^x = n^y$$

$$\therefore x = y$$

11-6 (pages 458–461)

Teaching Suggestions

You may want to start the lesson by conducting a brief review of scientific notation, since this is the basis for determining the characteristic of a common logarithm. Stress that a common logarithm is the sum of an integer (called the characteristic), and a nonnegative number less than 1 (called the mantissa). Remind students that, since logarithms are exponents and common logarithms have a base of 10, then when the number is greater than 1, its exponent (when in scientific notation) is also the characteristic of the logarithm. In the example below the exponent 4 is the characteristic of the log.

scientific notation: $57{,}900 = 5.79 \times 10^4$

logarithm of 57,900: $\log 57{,}900 = \log 5.79 + \log 10^4$

$$= 0.7627 + 4$$

$$= 4.7627$$

Finding logarithms of numbers less than 1, however, causes confusion for students because the characteristic is negative. The student cannot just take the characteristic and place it before the mantissa to get the logarithm, but must either subtract carefully or keep the logarithm in the form of the sum of the mantissa and the characteristic. The example below illustrates the need to carefully follow the steps of writing the number in scientific notation, writing and then adding the logarithms of the product.

$$\log 0.00579 = \log (5.79 \times 10^{-3})$$

$$= \log 5.79 + \log 10^{-3}$$

$$= 0.7627 - 3 = -2.2373$$

Emphasize that this is *not* the same as -3.7627.

11-7 (pages 461–463)

Teaching Suggestions

The most common mistake that occurs in this section is for students to confuse $\dfrac{\log a}{\log b}$ (an expression which is almost always used in the solution to the kinds of problems considered here) and $\log \frac{a}{b}$ (which students have seen quite frequently in the previous two sections). Since some students may attempt to replace $\frac{\log a}{\log b}$ by the expression $\log a - \log b$, assure them that the notation $\frac{\log a}{\log b}$ means just what it says: that $\log a$ is to be divided by $\log b$. This division of $\log a$ by $\log b$ (see Examples 1 and 2 on page 462) can be time consuming. You may want to permit students to use calculators to save time.

When solving exponential equations, encourage students to check their work to determine if the solution is reasonable. Example 1 on page 462 has a solution of 0.925.

$$5^{3x} = 87$$
$$5^{3(0.925)} = 87$$
$$5^{2.775} = 87$$

This solution can be checked as follows:

$$5^2 = 25, \quad 5^{2.775} = 87, \quad 5^3 = 125$$

Since 87 is between 25 and 125, and since 2.775 is between 2 and 3, then the solution is reasonable.

11-8 (pages 464–467)

Teaching Suggestions

There are many situations in chemistry, biology, economics, and other areas that involve exponential growth and decay. Applications in these areas will interest most students. In particular, they may find the contrast between simple and compound interest quite intriguing, as well as the method of carbon dating found in the Application at the end.

Related Activities

Have students compare the interest earned on a given principal invested for one year at 8% if the interest is:

a. simple
b. compounded semiannually

c. compounded quarterly
d. compounded monthly
e. compounded weekly
f. compounded daily

11-9 (pages 468–471)

Teaching Suggestions

The number e is so useful in advanced mathematical applications that logarithms to the base e are computed and found in table form in many mathematical texts. Also, most scientific calculators have a natural logarithm key marked $\ln x$.

You may want to point out some facts regarding the graphs of exponential functions. All exponential functions of the form $y = b^x$ (Section 11-2) pass through the point $(0, 1)$. Particularly interesting is the fact that the graph of $y = e^x$ is the only graph of an exponential function whose tangent at $(0, 1)$ makes an angle of exactly 45° with the y-axis. That is, the tangent line has the equation $y = x + 1$. (See the sketch below.) This means that when $|x|$ is very small, $e^x \approx x + 1$.

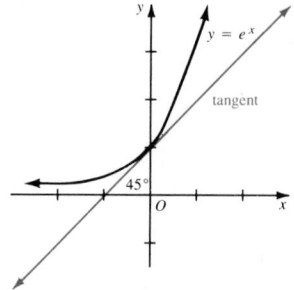

Related Activities

This section introduces the irrational number $e = 2.718 \ldots$ as an important base for logarithms. Although it may seem unlikely that such a number should be chosen as a base, it makes *finding* logarithms (that is, making up a table) relatively easy. In fact, for $0 < x \le 2$,

$$\ln x = (x - 1) - \frac{(x - 1)^2}{2} + \frac{(x - 1)^3}{3} - \cdots$$

(the right-hand side is an infinite series.) Use a

calculator or computer and the series on the preceding page to compute an approximation of $\ln \frac{4}{5}$ and $\ln \frac{5}{4}$ to five terms each. Show that, using these approximations,

$$\ln \frac{4}{5} + \ln \frac{5}{4} \approx \ln 1 = 0.$$

Compute $\ln \frac{4}{5}$ and $\ln \frac{5}{4}$ to *seven terms* each. Does $\ln \frac{4}{5} + \ln \frac{5}{4}$ get closer to zero? Why?

12 *Permutations, Combinations, and Probability*

This chapter begins with the fundamental counting principles in preparation for development of permutation and combination formulas. The Binomial Theorem is presented as a means of expanding a binomial and for finding a specific term. Once these patterns of expansion are clear, the use of Pascal's Triangle is introduced. Presentation of sample spaces leads to the final sections on probability, including mutually exclusive events and conditional probability.

diagram below illustrates the Cartesian product.

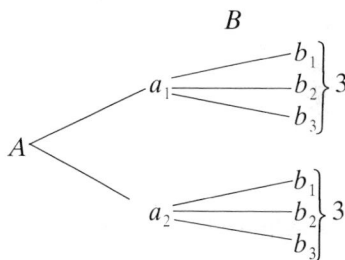

Therefore the Cartesian product of A and B produces 6 elements.

12-1 *(pages 483–486)*

Teaching Suggestions

Emphasize the fact that $A \cup B$ contains unique elements, without repetitions.

Be sure to carefully review the concept of Cartesian product here. An analogy the students will be familiar with is the multiplying of polynomials. For example, a binomial (2 terms) times a trinomial (3 terms) produces 6 terms before simplification. Tell students that just as they must multiply each term in a polynomial by each term in the other polynomial, they must match each element in one set with every element in the other set when they are finding the Cartesian product.

The "box" method as illustrated in Example 2, page 484 is useful in determining the solution to most counting problems. Another useful method for illustrating situations where the sets have a small number of elements is a "tree" diagram. For example, for $A = \{a_1, a_2\}$ and $B = \{b_1, b_2, b_3\}$ the

12-2 *(pages 486–489)*

Teaching Suggestions

The "box" method used with the counting principle is also useful in illustrating linear permutation problems. Physical illustrations will also help to impress upon students the meaning of a permutation. Use actual books on a shelf or students sitting in a row to demonstrate linear permutations, and a set of keys on a ring or students seated in a circle to show circular permutations. Stress the idea that the order makes a difference in these situations.

Some students may question the last factor $(n - r + 1)$ in the permutation formula on page 487. Explain that 1 has been subtracted from n for each box after the first one. Since there are r boxes, n has been diminished by one, $r - 1$ times, thus making the final box entry $n - (r - 1)$, or $n - r + 1$.

12-3 (pages 490–491)

Teaching Suggestions

Have students find the number of permutations of the letters of a short word such as ALL (6) using the formula from Section 12.2. Ask students to find the number of distinguishable permutations (3) and the relationship between the number of letter repetitions and the total number of permutations. After drawing conclusions, have students test conjectures on other words such as ROOM (12), INTEREST (10,080), and PERSEVERE (7560). Have a successful student present the necessary reasoning for those who have not already discovered it for themselves.

12-4 (pages 492–495)

Teaching Suggestions

Stress the difference between permutations and combinations. Permutations are possible arrangements of elements while combinations are groups of elements without regard to order.

Relate combinations to previously-learned permutations by finding the permutations of three digits chosen from the digits 1, 2, 3, 4, 5.

```
123 234 135 124  .  .  .
132 243 153  .
213 324  .  .
231  .  .  .
     .  .  .
        .  .  .
```

Since a combination is a group of elements without regard to order; each column is all the different permutations of a single combination. Since the number of permutations is the same in each column, to derive the number of combinations simply divide the total number of permutations by the number of permutations in (any) one column. In this example then, the number of combinations of 3 elements chosen from five is $_5C_3 = \frac{_5P_3}{_3P_3}$. More generally,

$$_nC_r = \frac{_nP_r}{_rP_r}$$
$$= \frac{n(n-1)(n-2)\ldots(n-r+1)}{r!}$$
$$= \frac{n!}{r!(n-r)!}$$

The last form is obtained by multiplying the numerator and denominator by $(n-r)!$.

12-5 (pages 496–497)

Teaching Suggestions

This section extends the students' ability to solve counting problems. Stress that the counting principle is used only when two events both take place, one event having r possible outcomes and the other having s possible outcomes.

12-6 (pages 498–500)

Teaching Suggestions

When expanding a binomial initially, emphasize the fact that coefficients within a binomial term do not affect the coefficient calculated as described in suggestion 3 on page 498. For example,

$$(2x - 3y)^4 = 1(2x)^4 + 4(2x)^3(-3y) + 6(2x)^2(-3y)^2 + \ldots$$

Students should be encouraged to write out terms in their unsimplified forms as above, so as not to confuse coefficients. Point out to your students that the expansion of a binomial *difference* will produce terms with alternating signs.

Related Activities

Write the first 3 terms of the expansion of $(a + b)^{\frac{1}{2}}$ according to the Binomial Theorem. Then use these terms to approximate

    a. $\sqrt{1.1} = (1 + 0.1)^{\frac{1}{2}}$
    b. $\sqrt{17} = (16 + 1)^{\frac{1}{2}}$

Compare your results with answers computed on a calculator.

Teaching Suggestions

After presenting several expansions of $(a + b)^n$ starting with $n = 0$, list corresponding rows of Pascal's Triangle and have students try to write successive rows. Be sure to point out the symmetry of each row.

Some students may wonder why the recursive pattern of the triangle proceeds as it does. The following demonstration can be used to explain. When we multiply $(a + b)^4$ by $(a + b)$, we get

$$a^4 + 4a^3b^2 + 6a^2b^2 + \ldots$$
$$\underline{a + b}$$
$$a^5 + 4a^4b + 6a^3b^2 + \ldots$$
$$\underline{\qquad 1a^4b + 4a^3b^2 + \ldots}$$
$$a^5 + 5a^4b + 10a^3b^2 + \ldots$$

It should be clear that each coefficient of the expansion of $(a + b)^5$ is the sum of two adjacent coefficients of $(a + b)^4$.

Related Activities

Count 1 1 as the first row of Pascal's Triangle. Find the sum of the squares of the numbers in a row of the triangle, and find this sum in the triangle. Find the sum of the squares in each of several more rows, and find each sum in the triangle. In general, where is the sum of the squares of the numbers in a row located? Express the location in terms of the row number of the numbers being squared.

If n is the row number, the sum of the squares of the numbers of the row is the middle entry in row number $2n$.

Teaching Suggestions

The concept of a sample space is introduced in this section in preparation for the topic of probability. It will be helpful to review Cartesian product before going on to determining the sample space represented by a lattice, in Example 2. Explain to the students that the problems of this section assume ideal conditions (truly random experiments) which means that dice are not weighted, coins are balanced, cards are not worn or marked, etc. Be sure students understand the idea of the sample space in this section since it will be used to develop the formulas in subsequent lessons.

Related Activities

A famous problem in probability is the following: If a needle (or a toothpick, matchstick, and so on) is dropped on a piece of paper that is marked off with parallel lines, the distance between the lines being exactly the length of the needle, what is the probability that the needle will fall so that it is crossing one of the lines? The answer (the derivation of which is beyond the scope of this book but is not too hard a problem in calculus) is $\frac{2}{\pi}$. By performing this experiment a large number of times, determine an approximate value of π.

Teaching Suggestions

A priori probability, that which is based on the inherent mathematical properties of an experiment, is presented here and should be contrasted with *a posteriori* or *empirical* probability, in which predictions of outcomes are based upon observation or sampling. (A common example of the latter is weather-forecasting.) Stress the fact that neither type provides any certainties about actual future outcomes, but they do suggest what may happen in an experiment. The more times an experiment is repeated, the closer the results will get to the mathematical probability.

This topic provides an opportunity for many classroom activities. Tossing a coin, spinning a spinner, drawing colored marbles from a bag, drawing cards from a deck, or rolling dice are experiments which will help students grasp the meaning of mathematical probability symbols. Do some experiments as a class and have students do others individually or in small groups.

Teaching Suggestions

Encourage students to use Venn diagrams to visualize the relationships between events. If the intersection of the sets representing the two events is empty, then they are mutually exclusive.

Use Figure 3 on page 511 to show the meaning of $P(A \cup B) = P(A) + P(B) - P(A \cap B)$. Since 2 is in both A and B, the probability of its occurring is counted twice in the total probability $P(A) + P(B)$. The probability $P(A \cap B)$ must therefore be subtracted to obtain the correct probability. Mutually exclusive events have an empty intersection. Thus, when A and B are mutually exclusive events, no subtraction is necessary. That is, $P(A \cup B) = P(A) + P(B)$.

Point out that addition of individual probabilities is also used for more than two mutually exclusive events, as in the example on page 512.

Related Activities

With the World Series partly played, the Bears must win two games and the Coyotes must win four games to win the Series. If the odds for either team to win any game are equal, find the odds favoring the Bears to win the Series.
Winning combinations and their probabilities are $BB(\frac{1}{4})$, $BCB(\frac{1}{8})$, $CBB(\frac{1}{8})$, $BCCB(\frac{1}{16})$, $CBCB(\frac{1}{16})$, $CCBB(\frac{1}{16})$, $BCCCB(\frac{1}{32})$, $CBCCB(\frac{1}{32})$, $CCBCB(\frac{1}{32})$, $CCCBB(\frac{1}{32})$. Since these are mutually exclusive events, find the sum of the probabilities to find the probability that the Bears win the Series. $\frac{13}{16}$; the odds, therefore, are 13 to 3.

Teaching Suggestions

Be sure students understand the notation $P(B|A)$ to mean "the probability of event B occurring given event A has already occurred."

Confusion often arises among students in distinguishing between mutually exclusive events and independent events. Remind them that if events A and B are mutually exclusive, $A \cap B = \emptyset$, so $P(A \cap B) = 0$, whereas, if A and B are independent events, $P(A \cap B) = P(A) \cdot P(B)$.

13 Matrices

The algebra of matrices is developed in this chapter. The operations of addition, multiplication, and scalar multiplication are defined, with emphasis on similarities and differences between this system and the algebra of real numbers. The last three sections present practical applications of matrices to solving linear systems of equations and to transformation geometry.

Teaching Suggestions

Since statistical data is usually entered into a computer in the form of a matrix, most students will understand an analogy between data and matrix like the following:

Central High School Enrollment

	Fresh.	Soph.	Jr.	Sr.
1960	215	208	230	210
1970	190	200	185	180
1980	160	155	170	175

Matrix Representation

$$\begin{bmatrix} 215 & 208 & 230 & 210 \\ 190 & 200 & 185 & 180 \\ 160 & 155 & 170 & 175 \end{bmatrix}$$

Be sure the students understand what is meant by the dimensions of a matrix. Particularly emphasize the importance of the *order* of the dimension statement, that is, rows by columns. Point out that a 2 × 3 matrix is *not* the same as a 3 × 2 matrix, since the first has two rows and three columns, and the second has three rows and two columns. Also, make it clear that equality of two matrices requires the matrices to have the same dimensions and equal corresponding entries.

13-2 (pages 532–534)

Teaching Suggestions

The five addition properties of matrices stated on page 532 should indicate to students that an algebra of matrices can be established to correspond closely to the algebra of real numbers. Point out that subtraction of matrices is defined in terms of addition.

New in this section for some students may be the use of subscript notation. If you assign Written Exercises 11–16 be sure it is clear to the student that the notation is used to differentiate between the elements of the matrix.

13-3 (pages 535–538)

Teaching Suggestions

Although scalar multiplication has no corresponding property in the set of real numbers, the students should grasp the concept easily. Scalar multiplication must be emphasized since it is used to solve a variety of problems. In addition to introducing this new concept, this section can be used for further practice in using subscript notation.

13-4 (pages 538–542)

Teaching Suggestions

To perform matrix multiplication, the number of columns of the first matrix must be equal to the number of rows of the second. If this does not hold, then the product of the two matrices is not defined. But if, for example, the dimensions of one matrix are $m \times n$ and those of a second are $n \times p$,

the product matrix does exist and has dimensions $m \times p$.

Use an example like the following to illustrate:

$$\underset{(3 \times 1)}{\begin{bmatrix} 3 \\ 2 \\ -1 \end{bmatrix}} \underset{(1 \times 3)}{\begin{bmatrix} 1 & -5 & 2 \end{bmatrix}} = \underset{(3 \times 3)}{\begin{bmatrix} 3 & -15 & 6 \\ 2 & -10 & 4 \\ -1 & 5 & -2 \end{bmatrix}}$$

Again, it is important to emphasize correct notation, since whether or not a product is defined depends upon the dimensions of the two matrices multiplied together.

13-5 (pages 543–546)

Teaching Suggestions

Emphasize the differences between matrix multiplication and real number multiplication. Among other differences, matrix multiplication is not commutative, i.e., $AB \neq BA$. You can contrast matrix multiplication with that of real numbers by showing that although $(A + B) \cdot (A + B) = A^2 + 2AB + B^2$ for real numbers A and B, the same expression using matrices A and B will result in $(A + B) \cdot (A + B) = A^2 + AB + BA + B^2$.

Point out the existence of an identity matrix for multiplication of square matrices. Its importance will be obvious in the next lesson where matrices are used to solve systems of equations.

13-6 (pages 547–552)

Teaching Suggestions

First note that throughout the remainder of the chapter, the notation $\det \begin{bmatrix} a_1 & b_1 \\ a_2 & b_2 \end{bmatrix}$ is used to denote the determinant that was denoted by $\begin{vmatrix} a_1 & b_1 \\ a_2 & b_2 \end{vmatrix}$ in Chapters 4 and 5. The determinant, as will be seen from the derivation, is of crucial importance here, as it *determines* whether or not a given matrix has an inverse.

The interpretation of the inverse matrix given

on page 548 should be emphasized. The system
$$a_1x + b_1y = c_1$$
$$a_2x + b_2y = c_2$$
or
$$\begin{bmatrix} a_1 & b_1 \\ a_2 & b_2 \end{bmatrix}\begin{bmatrix} x \\ y \end{bmatrix} = \begin{bmatrix} c_1 \\ c_2 \end{bmatrix} \text{ or simply } AX = B$$
can be compared to the simple equation $ax = b$. To solve $ax = b$, we would multiply both sides by the multiplicative inverse of a. Theoretically, the solution of the system $AX = B$ is identical to this, except that, if det $A = 0$, the multiplicative inverse of A does not exist. Left-multiplying both sides by A^{-1}, if it exists, we get

$$A^{-1}AX = A^{-1}B$$
$$IX = A^{-1}B$$
$$X = A^{-1}B.$$

Students may wonder why an equation of the form $ax = b$ "always has a solution," while the corresponding type of matrix equation $AX = B$ may not. The answer is that the equation $ax = b$ has *no* solution if $a = 0$ and $b \neq 0$, or has infinitely many solutions if $a = b = 0$. The case where $a = 0$ is analogous to the case where the matrix A is singular.

Related Activities

Show that a nonsingular matrix A has a *unique* inverse. That is, show that if B and C are both inverses of A, then $B = C$.

$BA = I$; $CA = I$
$BA = CA$
$BAB = CAB$
$BI = CI$, since B is an inverse of A
$B = C$

13-7 *(pages 553–556)*

Teaching Suggestions

This lesson can be very effectively presented using an overhead projector. (Have a coordinate system drawn on an opaque lower sheet and a set of prepared overlays to illustrate various linear trans-

lations.) Emphasize the following two facts: (1) every point is translated the same distance in the same direction, thereby allowing the generalization in matrix form $\begin{bmatrix} x' \\ y' \end{bmatrix} = \begin{bmatrix} x \\ y \end{bmatrix} + \begin{bmatrix} h \\ k \end{bmatrix}$; (2) any translation in the plane can be interpreted as one horizontal motion h followed by one vertical motion k (or the reverse).

13-8 *(pages 558–563)*

Teaching Suggestions

The main point to be made in this section is that an equation of the form $X' = AX$ is a linear transformation of the plane onto itself (assuming det $A \neq 0$). Some examples like the following may help students get a better understanding of this idea:

1. Using $A = \begin{bmatrix} 2 & 0 \\ 0 & 2 \end{bmatrix}$ produces a transformation which maps a point (a, b) into the point $(2a, 2b)$. Such a mapping "expands" the plane along lines radiating from the origin.

2. Using $A = \begin{bmatrix} 0 & -1 \\ 1 & 0 \end{bmatrix}$ produces a transformation which maps a point (a, b) into the point $(-b, a)$. The diagram below shows that this mapping rotates the plane 90° counterclockwise. (Remind students that nonvertical perpendicular lines have slopes which are negative reciprocals of each other.)

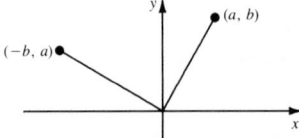

3. Finally, note that if $AX = X'$ and $BX = X'$ are two linear transformations, and (x', y') is the image of (x, y) under B and (x'', y'') is the image of (x', y') under A, then (x'', y'') is the image of (x, y) under AB, the product matrix. (See Written Exercise 30, where this is proved.)

Related Activities

1. Show that if complex numbers $x + yi$ are identified with points (x, y) in the plane, then the transformation defined by multiplying each point by a *fixed* complex number $a + bi$ is a linear transformation.

$$(a + bi)(x + yi) = (ax - by) + (bx + ay)i;$$
$$\begin{bmatrix} a & -b \\ b & a \end{bmatrix}\begin{bmatrix} x \\ y \end{bmatrix} = \begin{bmatrix} ax - by \\ bx + ay \end{bmatrix}$$ defines the transformation.

2. A linear transformation is a rotation of the plane about the origin if (1) each point in the plane and its image are the same distance from the origin and (2) the angles with vertices at the origin and whose sides contain a point and its image are congruent. Show that the linear transformation

$$\begin{bmatrix} x' \\ y' \end{bmatrix} = \begin{bmatrix} a & -b \\ b & a \end{bmatrix}\begin{bmatrix} x \\ y \end{bmatrix}$$

is a rotation of the plane about the origin if $a^2 + b^2 = 1$. Use the following steps:

(1) Show that (x', y') and (x, y) are the same distance from the origin. (This is a necessary condition for (x', y') and (x, y) to lie on the same circle with center at the origin.)

(2) Show that the measure of the angle with vertex at the origin and whose sides contain $A(x, y)$ and $B(x', y')$ is constant. To do this, show that any chord, $\overline{RS}$, on the unit

circle intercepted by $\overline{OA}$ and $\overline{OB}$ is of constant length, i.e., its length is independent of x and y. (Remember, if central angles intercept chords of equal length, the angles are congruent.)

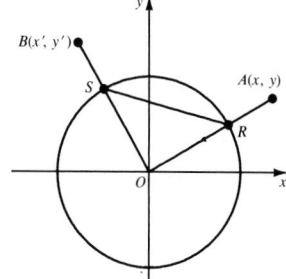

(1) $OB = \sqrt{(ax - by)^2 + (bx + ay)^2}$
$\qquad = \sqrt{(a^2 + b^2)x^2 + (a^2 + b^2)y^2}$
$\qquad = \sqrt{(a^2 + b^2)(x^2 + y^2)} = \sqrt{x^2 + y^2}$
$\qquad = OA$

(2) $RS = \sqrt{\dfrac{(x' - x)^2}{x^2 + y^2} + \dfrac{(y' - y)^2}{x^2 + y^2}}$

$\qquad = \sqrt{\dfrac{((a - 1)x - by)^2 + ((a - 1)y + bx)^2}{x^2 + y^2}}$

$\qquad = \sqrt{\dfrac{(a - 1)^2(x^2 + y^2) + b^2(x^2 + y^2)}{x^2 + y^2}}$

$\qquad = \sqrt{(a - 1)^2 + b^2}$

---

14 Trigonometric and Circular Functions

This chapter begins with a discussion of directed angles and the measurement of these angles using both degree and radian measure. The sine and cosine functions are then defined and examined with the distinction made between trigonometric functions (with sets of angles as domains), and circular functions (with sets of real numbers as domains). Students will learn the values of the sine and cosine of special angles and also how to use the fundamental period or reference angles to find the sine and cosine of an angle of any size. A detailed discussion of the graphs of the sine and cosine functions follows. The last part of the chapter expands the students' knowledge of trigonometry to include the tangent, cotangent, secant, and cosecant functions and the applications of all six trigonometric functions to solving right triangles.

14–1 (pages 575–578)

Teaching Suggestions

The terminology introduced in this section will be used throughout the chapter, so it is important that the students understand the terms and their usage. Although most students are familiar with the concept of an angle from geometry, stress that a directed angle is determined by the rotation of a ray.

It may be helpful to draw several different angles on the chalkboard, showing directed angles, coterminal angles, and angles in standard position. Remind students to mark the arrows that indicate the direction of the rotation of the angle they are sketching.

14–2 (pages 579–583)

Teaching Suggestions

Although students undoubtedly will be more familiar with degree measure than with radian measure, point out to the students that the radian is more useful than the degree because radian measure is a real number and therefore can be used with other functions of real numbers.

Students should be able to convert between degree measure and radian measure with ease. You may have to emphasize the relationship between the two types of measure by repeating that since the circumference of a unit circle is 2π and one complete revolution is 360°, we have:

$$360° = 2\pi^R$$

or

$$180° = \pi^R$$

Some conversions occur so frequently that presenting the table below to your students may be useful.

degree	0	30	45	60	90	180	270	360
radian	0	$\frac{\pi}{6}$	$\frac{\pi}{4}$	$\frac{\pi}{3}$	$\frac{\pi}{2}$	π	$\frac{3\pi}{2}$	2π

Related Activities

1. Write a formula to describe in radians all the positive angles that are coterminal with $\frac{\pi}{2}$.

   $k \cdot 2\pi + \frac{\pi}{2}$, where k is an integer ≥ 0.

2. Write a formula to describe in radians all the negative angles that are coterminal with π.
   $k \cdot 2\pi - \pi$, where k is an integer ≤ 0.

3. Write a formula that converts an angle measure of x degrees and y minutes to a radian measure.
   $\frac{\pi x}{180} + \frac{\pi y}{10800}$

14–3 (pages 584–588)

Teaching Suggestions

Your students' prior exposure to trigonometry probably encompassed only acute angles (right triangle trigonometry). This section will present a more complete discussion of trigonometric functions. Remind students that the definition of similar triangles ensures that the sine and cosine functions have the same value for a particular angle, no matter what the value of r.

Since the terminal ray of an angle can lie in any quadrant, it is important to be able to determine the signs of trigonometric functions in particular quadrants. Point out that r is always positive and u and v vary according to the quadrant in which the terminal ray lies. Students can rely on their understanding of the coordinate plane if they get confused about the signs of u and v. They know, for example, that if a point lies in Quadrant II, then its first coordinate (u) is negative and its second coordinate (v) is positive.

Students should be encouraged to use a systematic approach when doing exercises that require finding sines and cosines of angles. They should:

1. Sketch the angle.
2. Find u, v, and r using the relationship $u^2 + v^2 = r^2$.
3. Find the sine and/or the cosine.

14–4 (pages 589–592)

Teaching Suggestions

In this section, students will learn to make use of their knowledge of the relationships between the lengths of the sides of 30°–60°–90° triangles and 45°–45°–90° triangles. Although the trigonometric values of angles which are multiples of 30° and 45°, and quadrantal angles appear in the table on page 590, encourage students to memorize the two special right triangles and *not* the table. They can use the triangles to derive trigonometric values whenever any of the related angles come up. Remind students that when they derive values of the quadrantal angles, r is always equal to positive 1, and the values of u and v are always equal to -1, 0, or 1, depending upon the position of the angle (see Figure 19).

14–5 (pages 592–595)

Teaching Suggestions

Since most tables only give the trigonometric values of angles from 0° to 90°, it is important for students to be able to "reduce" any angle to an angle that can be found in the table.

Emphasize the fact that the reference angle θ of a given angle α is the angle formed by the terminal side of α and the x-axis. Explain to students that they will first use the reference angle to determine the absolute value of the desired trigonometric function, and then determine the correct sign for the value by looking at the position of the angle. The sketch below should help students remember the position of the reference angle R and the signs in each quadrant. Remind students again, that the radius r is always positive.

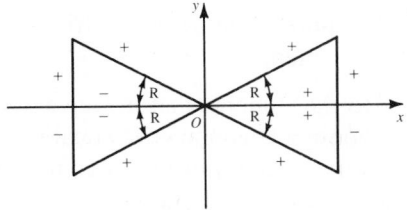

Be sure students understand that the reference angle for all angles, even those whose measure is less than 0° or greater than 360°, is determined by the angle formed by the terminal ray and the x-axis.

14–6 (pages 595–599)

Teaching Suggestions

Before beginning the graphs, make sure students are aware that only radian measure will be used here. Explain that the use of radian measure allows the system of units on the x-axis to be related to the units on the y-axis. In this way, the circular functions of this section will fit in with other real-valued functions already studied: polynomial, logarithmic, and exponential functions.

The graphs of the functions $y = \sin x$ and $y = \cos x$ should be made carefully using the table of values on page 590. Students should then study these graphs, noting the period (2π) and amplitude (1) of each. You may want to have students sketch the graphs of various sine and cosine functions (other than those found in the Oral Exercises) on the chalkboard.

14–7 (pages 599–601)

Teaching Suggestions

Students will need to memorize the graphs of $y = \sin x$ and $y = \cos x$. Graphs of the form $y = A \sin Bx$ and $y = A \cos Bx$ can then be drawn using the shapes of the graphs of $y = \sin x$ and $y = \cos x$ respectively as guides.

Explain to students that they need only find a few key points to draw the graph of a function $y = A \sin Bx$ or $y = A \cos Bx$, where A and B are not 1. Instruct them to first find the amplitude and period of the graph, and then use these values to mark the minimum and maximum points and intercepts of the graph. Once these points are known, the student needs only to fill in the curve, whose general shape is known. Consider the Example on page 600. Once students find the amplitude (2) and period (4π), the 5 key points which will help determine the graph are indicated below.

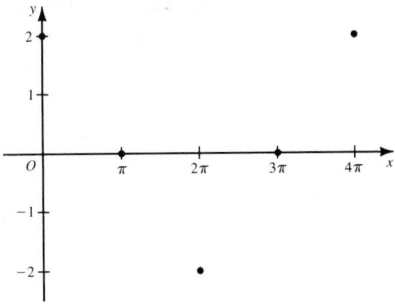

To help demonstrate the effects of changing the values of A and B on graphs of the form $y = A \sin Bx$ and $y = A \cos Bx$, you may want to graph some examples on the same set of axes or on sets of axes with the same units.

14–8 (pages 602–606)

Teaching Suggestions

There is a considerable amount of information to be memorized in this section. The six trigonometric functions, their definitions, notations, and reciprocal relationships are summarized below.

Reciprocal Functions

$$\sin \alpha = \frac{v}{r} \longleftrightarrow \csc \alpha = \frac{1}{\sin \alpha} = \frac{r}{v}$$

$$\cos \alpha = \frac{u}{r} \longleftrightarrow \sec \alpha = \frac{1}{\cos \alpha} = \frac{r}{u}$$

$$\tan \alpha = \frac{\sin \alpha}{\cos \alpha} = \frac{v}{u} \longleftrightarrow \cot \alpha = \frac{\cos \alpha}{\sin \alpha} = \frac{u}{v}$$

Figure 31, which gives the sign of each function in each quadrant, is also very important. To help your students remember these signs, the sentence, "**A**ll **s**tudents **t**ake **c**alculus," may be helpful. It tells which trigonometric functions are positive in each quadrant. Although the sentence only indicates the signs for sine, cosine, and tangent, it applies to their reciprocal functions as well.
All: **All** functions are positive in Quad. I.
students: Only **s**in (csc) is positive in Quad. II.
take: Only **t**an (cot) is positive in Quad. III.
calculus: Only **c**os (sec) is positive in Quad. IV.

Related Activities

Have students graph different pairs of reciprocal functions on the same coordinate axes. Then have them use the graphs to show how the range of each function is related to the range of its reciprocal function.

Both student and teacher should be aware that answers will vary slightly depending upon whether a calculator or a table is used to find trigonometric values.

14–9 (pages 607–612)

Teaching Suggestions

If students do not already know the definitions of the six trigonometric functions (inside the box on page 607), they should be memorized. The acronym soh-cah-toa may help.

sine is	cosine is	tangent is
opposite over	**a**djacent over	**o**pposite over
hypotenuse	**h**ypotenuse	**a**djacent

Remind students that the word soh-cah-toa and the definitions of the reciprocal functions provide them with the definitions of cotangent $\left(\frac{a}{o}\right)$, secant $\left(\frac{h}{a}\right)$, and cosecant $\left(\frac{h}{o}\right)$ as well.

When solving practical problems, students should be encouraged to use a systematic approach:

1. Draw a diagram.
2. Label all given information on the diagram.
3. Locate the right triangle.
4. Place the unknown (variable) on the diagram.
5. Select the appropriate trigonometric function to solve the problem.

Emphasize to the students that in step 5 above, they must choose a trigonometric equation in which they know two out of the three necessary quantities.

Related Activities

A man is walking on level ground toward a building 120 m tall. He stops and sights the top of the building at an angle of elevation having measure 23°.

a. If his line of sight is 2 m above ground how much farther (to the nearest meter) must he walk to reach the building?

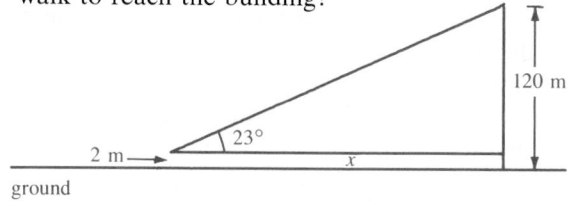

$$\tan 23° = \frac{\text{height of building}}{\text{distance to building}}$$

$$0.4245 = \frac{120 - 2}{x}; \ x \approx 278 \text{ m}$$

b. If, at the same distance from the building, the man's line of sight is at ground level, what is the measure of angle of elevation? Give the answer correct to the nearest minute, or in decimal degrees correct to the nearest tenth.

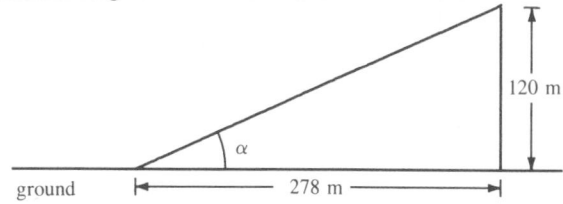

$$\tan \alpha = \frac{\text{height of building}}{\text{distance to building}}$$

$$\tan \alpha = \frac{120}{278}$$

$$\tan \alpha \approx 0.4317$$

$$\alpha \approx 23° 21'$$

$$\alpha \approx 23.3°$$

---

15 Trigonometric Identities and Formulas

This chapter begins by presenting the fundamental trigonometric identities and the methods of proving identities. It continues by deriving the sum and difference formulas, the reduction formulas, and the double-angle and half-angle relationships. To conclude the chapter, the law of cosines and the law of sines are developed and used in solving general triangles.

15–1 (pages 619–622)

Teaching Suggestions

In this section we establish the eight basic identities (which students should memorize) that will be used to prove other identities later on. Identities (1), (7), and (8), sometimes known as the Pythagorean Identities, are really theorems, while (2)–(5), as students already know, are definitions. Arrange the identities in the following way to facilitate easy memorization.

Pythagorean Identities

$$\sin^2 \alpha + \cos^2 \alpha = 1$$
$$\tan^2 \alpha + 1 = \sec^2 \alpha$$
$$1 + \cot^2 \alpha = \csc^2 \alpha$$

Reciprocal Identities

$$\csc \alpha = \frac{1}{\sin \alpha}$$

$$\sec \alpha = \frac{1}{\cos \alpha}$$

$$\tan \alpha = \frac{1}{\cot \alpha}$$

Quotient Identities

$$\tan \alpha = \frac{\sin \alpha}{\cos \alpha}$$

$$\cot \alpha = \frac{\cos \alpha}{\sin \alpha}$$

Students need practice in transforming expressions involving functions before beginning to prove identities in the next section. Emphasize that all work should be written out in equation form in a way that makes sense mathematically. Be sure to observe that most problems in this section, as well as most identities in subsequent sections, can be attacked in more than one way. Emphasize, too, that, as stated on page 620, all the results of this and subsequent sections hold equally well for trigonometric *and* circular functions.

15–2 (pages 622–624)

Teaching Suggestions

Point out that there are two approaches to proving identities: (1) transform one side of the equation until the other side is produced; (2) make substitutions on both sides of the equation until the left and right sides are identical, or until they are the two sides of a *known* identity. In using the first approach, it is usually best to choose the more complicated side, so it can be simplified.

Students should be advised to make substitutions in terms of sine and cosine if no other substitutions are immediately obvious.

15–3 (pages 625–631)

Teaching Suggestions

Before presenting the derivation of the cosine of a difference, students should review the distance formula, $P_1P_2 = \sqrt{(x_2 - x_1)^2 + (y_2 - y_1)^2}$ (page 394). Emphasize the idea that a derivation is produced by starting with a known or accepted statement and proceeding to the desired result, justifying each intervening step. Explain to the students that once they know the sum and difference formulas they will be able to derive the many other reduction, double-angle, and half-angle formulas by variable substitutions. Encourage students to derive formulas whenever they have any doubt about a formula. If they do this instead of referring to the text, they will gain valuable experience in deriving formulas, as well as gain confidence in being able to work things out for themselves.

15–4 (pages 631–637)

Teaching Suggestions

Students will find the derivation of the sine formulas to be straightforward from the previously-obtained reduction formulas (page 629). However, the derivation of the tangent of a sum or difference often causes consternation among some students when the terms of the right side are divided by the product $\cos x_1 \cos x_2$ and replaced by functions in terms of $\tan x_1$ and $\tan x_2$. Reassure students that although they might not have thought of this step on their own this time, after more experience with derivations, techniques like these will occur to them more readily.

15–5 (pages 637–641)

Teaching Suggestions

Double-angle formulas tend to cause few problems with students. Half-angle formulas, however, are more difficult to use because students must decide on the positive or negative form of the root. Drill students on the signs of the functions of angles in various quadrants. Ask also for the half-angle of an angle, the quadrant in which the half-angle lies, and the sign of a function of the half-angle. These drills will help students feel more comfortable with making the decisions necessary to use half-angle formulas.

Related Activities

In the diagram at the right $\overline{BD}$ is a diameter of semicircle O and $\overline{AC}$ $\perp \overline{BD}$. Find an expression for CD in terms of the radius r of the semicircle and angle θ only. (*Hint*: Consider angle AOC.)

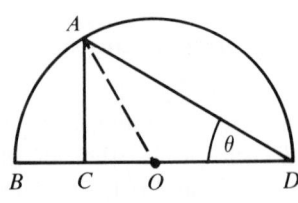

$CD = r + r \cos 2\theta = r(1 + \cos 2\theta) =$
$r(1 + \cos^2 \theta - \sin^2 \theta) = 2r \cos^2 \theta$

15–6 (pages 642–643)

Teaching Suggestions

The summary of identities on page 642 should be studied carefully. Point out that notation such as
$$\sin (x_1 \pm x_2) = \sin x_1 \cos x_2 \pm \cos x_1 \sin x_2$$
will help students remember the sum and difference formulas.

Problems involving terms like $\sin 3x$ and $\sin 4x$ (Written Exercises 17–20) can be difficult for many students. Give hints such as $\sin 3x = \sin (2x + x)$ and $\sin 4x = \sin (2x + 2x)$ to help them successfully apply the angle formulas.

Related Activities

Use the law of cosines to find x in the diagram at the right.

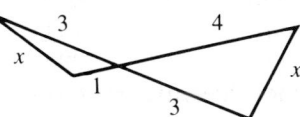

$$\frac{3^2 + 1^2 - x^2}{2 \cdot 3 \cdot 1} = \frac{4^2 + 3^2 - x^2}{2 \cdot 3 \cdot 4}$$
$$\frac{10 - x^2}{6} = \frac{25 - x^2}{24}$$
$$90 = 18x^2$$
$$x^2 = 5$$
$$x = \sqrt{5}$$

15–7 (pages 644–647)

Teaching Suggestions

It is essential that students understand how the coordinates of A in Figure 4 are obtained before you step through the derivation of the first form of the law of cosines. Once this is clear, students should be able to follow the straightforward use of the distance formula and simplification to obtain the first form of the law of cosines.

Also be sure it is clear to students that the triangles in Figures 4–6 are all the same triangle, turned so that a different angle is in standard position in each diagram. Students should understand that each angle is placed in standard position so that the coordinates of the vertex not on the x-axis can be referred to in terms of sine and cosine, so that by using the distance formula the length of the side opposite the angle in standard position can be found. Have students use Figures 5 and 6 to derive the other forms of the law of cosines. This will help ensure their understanding.

You may want to point out that the case where C is a right angle, which yields $c^2 = a^2 + b^2$ from the law of cosines, does not *prove* the Pythagorean theorem. Remind students that the Pythagorean theorem was used to establish the distance formula, which was used to derive the law of cosines. In other words, the law of cosines is a generalization of the Pythagorean theorem.

15–8 (pages 648–653)

Teaching Suggestions

It is essential that students understand how the area formula for triangles is converted from $A = \frac{1}{2}bh$ to $A = \frac{1}{2}ab \sin C$. The base of the triangle in Figure 7 is of length a, and since the coordinates of B are $(b \cos C, b \sin C)$, the height of the triangle is $b \sin C$. Substituting these values respectively into b for base and h for height in the formula $A = \frac{1}{2}bh$, you get $A = \frac{1}{2}ab \sin C$. Be sure you use Figure 7 carefully to demonstrate this to your students.

After deriving the law of sines, you can categorize the various data combinations for oblique triangles in terms of the formulas used to solve them.

Use the *law of cosines* when you have:

> three sides, or
> two sides and one included angle

Use the *law of sines* when you have:

> two angles and one side (included or non-included), or
> two sides and one non-included angle

Be sure to illustrate how the given information for the last case above can result in no triangle, one triangle, or two triangles (Example 3). Cover this example in detail as it is important for students to know how to handle the ambiguous case.

You may want to point out that the case where C is a right angle, which yields the $\sin A = \dfrac{a}{c}$ relationship from the law of sines, does not *prove* the relationship $\sin A = \dfrac{a}{c}$. This relationship was used to express the area of a triangle as Area = $\frac{1}{2}ab \sin C$, which *precedes* the derivation of the law of sines.

Related Activities

Ask students to make up word problems like those at the end of this section. Encourage them to be creative in applying the formulas learned. Students can exchange problems for solution, or work together to solve problems.

16 Inverses: Polar Coordinates; Vectors

Inverses of both trigonometric and circular functions are introduced at the beginning of this chapter, and are used in solving trigonometric equations. Converting between Cartesian and polar coordinates is then presented, followed by operations on complex numbers in polar form (De Moivre's Theorem). In the last two sections of the chapter vectors are examined. Sums and dot products of vectors are discussed along with the use of vectors to solve problems.

16-1 *(pages 661–667)*

Teaching Suggestions

It may be best to start this lesson by reviewing the concept of the inverse of a function. For example, the inverse of the function $y = x^2$ is obtained by interchanging x and y: $x = y^2$. Graphing both the function and its inverse might help students understand that the inverse is not an inverse function (vertical line test) unless its range is restricted to $y > 0$ for example.

You may want your students to graph each trigonometric function and its inverse by first making a table of values.

There are several important points to stress in this section.
1. The inverses of the trigonometric functions are not themselves functions.

2. The inverses of the trigonometric functions can be made functions by restricting their ranges.
3. The inverse trigonometric functions obtained by restricting the range are denoted with a capital letter.

Related Activities

Prove that $\text{Sin}^{-1}(\sin x) = x$ and $\text{Cos}^{-1}(\cos x) = x$ are not identities by finding values of x for which the statements are false.
For example, for $x = 2\pi$, $\text{Sin}^{-1}(\sin x) = 0 \neq 2\pi$ and $\text{Cos}^{-1}(\cos x) = 0 \neq 2\pi$.

16-2 *(pages 667–670)*

Teaching Suggestions

It might be informative to mention that equations are sometimes classified into two categories: conditional equations and identities. A conditional equation is true for only certain values of the variable; an identity is true for all values. Simple examples such as $2x + 3x = 10$ and $2x + 3x = 5x$ may help illustrate the difference.

If students have difficulty understanding why a trigonometric equation has an infinite number of solutions (when the solution set is not empty), the graphs of the circular functions may be helpful. For example, to find the general solution over $\mathscr{R}$ of sin $x = 1$, consider the graph of the sine curve on page 596. By looking at the graph, students will see that x can equal $\frac{\pi}{2}, \frac{5\pi}{2}$, etc. and $-\frac{3\pi}{2}, -\frac{7\pi}{2}$, etc.

When solving the trigonometric equations in this section, several methods will prove useful. Factoring is frequently used, especially when quadratic equations are involved (Examples 1 and 3). Transforming an equation so that it contains only one trigonometric function is also a valuable technique. This can be done by substituting trigonometric identities into the equation (Example 2).

16-3 (pages 672–676)

Teaching Suggestions

In many applications, using polar coordinates is more convenient than using Cartesian coordinates. The following example may help illustrate that to students: If a tour guide, atop the Empire State Building, wishes to help a person locate the Statue of Liberty, the guide might say that the statue is 4 miles away in a direction 40° west of south.

It is important for students to be able to change from one system to the other. Note that the conversion formulas at the bottom of page 673 can be easily derived from the following figure.

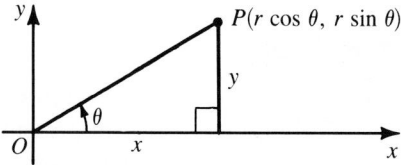

Since $\cos \theta = \frac{x}{r}$, then $x = r \cos \theta$ and

since $\sin \theta = \frac{y}{r}$, then $y = r \sin \theta$.

Similarly, the formulas for converting from Cartesian to polar coordinates (page 674) can be derived once students see that $r^2 = x^2 + y^2$ (Pythagorean Theorem).

If polar coordinate paper is available, it will greatly assist students in graphing equations such as those in Written Exercises 24–38.

Related Activities

Write an equation of the parabola with focus at the origin and directrix the line $y = -k$ in polar coordinates.

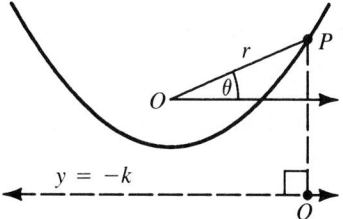

By definition. $PO = PQ$. Using the distance formula, we get

$$\sqrt{(r \cos \theta - 0)^2 + (r \sin \theta - 0)^2} =$$
$$\sqrt{(r \cos \theta - r \cos \theta)^2 + (r \sin \theta + k)^2} \longrightarrow$$
$$\sqrt{r^2(\cos^2 \theta + \sin^2 \theta)} = \sqrt{(r \sin \theta + k)^2}$$
$$|r| = |r \sin \theta + k|.$$

So $r = \dfrac{k}{1 - \sin \theta}$ or $r = \dfrac{-k}{1 + \sin \theta}$.

16-4 (pages 676–680)

Teaching Suggestions

When graphing in the complex plane, students should notice many similarities to graphing in the Cartesian system. In the Cartesian plane, both the x- and y-axes are real number lines. Stress that in the complex plane, although the x-axis is a real number line, the y-axis is an imaginary axis.

For practice, you may want to have students plot some complex numbers in the complex plane and determine the absolute value and amplitude of each. Point out that the absolute value, $|z|$, of a complex number is equal to its distance from the origin. The amplitude, θ, is the angle, in standard position, with the least positive value.

After deriving the theorem on page 677, go through Example 2 carefully, making sure students understand how to multiply and divide complex numbers in polar form. This theorem is important since it is the basis for DeMoivre's Theorem in the next section.

16-5 (pages 680–684)

Teaching Suggestions

There is no substantial saving in time when adding, subtracting, multiplying, or dividing complex numbers in polar form. But when finding a power or a root of a complex number, the savings in time can be sizable. You may want to compare the solving of Example 1 on page 681, which uses polar form, to the solving when complex numbers are used. Express $(1 + i)^6$ as a product of six factors and multiply. Note the additional time and amount of computation needed.

Be sure to discuss Example 3 in detail. To help simplify the computation of nth roots, you may want to point out that once one root is found, the others are equally spaced every $\frac{360}{n}$ degrees around a circle of radius r.

Related Activities

Find a formula for sin $4x$ and cos $4x$ in terms of sin x and cos x by the following method:
(a) Let $z = \cos x + i \sin x$. Find z^4 by De Moivre's Theorem.
(b) Find z^4 using the Binomial Theorem.
(c) Use these to find sin $4x$ and cos $4x$ in terms of sin x and cos x.
(a) $z^4 = \cos 4x + i \sin 4x$
(b) $z^4 = \cos^4 x + 4 \cos^3 x(i \sin x) +$
$\qquad 6 \cos^2 x(i \sin x)^2 + 4 \cos x(i \sin x)^3 +$
$\qquad (i \sin x)^4$
$\qquad = (\cos^4 x - 6 \cos^2 x \sin^2 x + \sin^4 x) +$
$\qquad\qquad i(4 \cos^3 x \sin x - 4 \cos x \sin^3 x)$
(c) Equating the real and imaginary parts of (a) and (b), we get the formulas.

16-6 (pages 686–691)

Teaching Suggestions

The concept of a vector may be new to your students. Vectors are extremely useful in many branches of physics where quantities such as force, velocity, and acceleration possess both magnitude and direction. It might be helpful to have students practice drawing vectors before considering the resultant of two vectors.

Students may understand vector addition better, from a physical point of view, if they think of two forces acting on an object. Students should be encouraged to represent the two vectors graphically even if the solution is to be done algebraically (Example 1). The graphical representation will help students visualize the problem and also act as a check of the algebraic solution.

Remind students that although the sum of two vectors is another vector, the dot product of two vectors is a real number.

Related Activities

If the norm and direction angle of a vector are given, tell how the vector could be written using horizontal and vertical components ($a\mathbf{i} + b\mathbf{j}$ form). Rewrite $\mathbf{u}$: 17, 118° in $a\mathbf{i} + b\mathbf{j}$ form. Round a and b to the nearest integer.
Use the polar coordinates to transform the norm and direction angle of the vector to Cartesian coordinates. $\mathbf{u} = -8\mathbf{i} + 15\mathbf{j}$

16-7 (pages 692–694)

Teaching Suggestions

In this section students will see how vectors can help solve practical problems. Before presenting the examples, you may want to go over carefully the terms associated with navigation and the terms associated with work and energy. One important point is that, unlike previous angle measurement, bearing is measured clockwise from due north (positive y-axis).

Encourage students to make a careful sketch for each problem. An alternate graphical representation of the three vectors shown in Example 1 on page 692 is shown on the following page. The vector $\mathbf{v}$ is the diagonal of a parallelogram having $\mathbf{u}$ and $\mathbf{w}$ as adjacent sides.

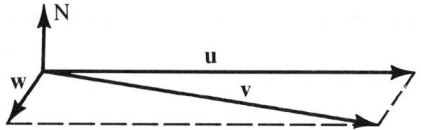

If the drawing is made to scale using a ruler and protractor, then approximate solutions can be obtained by measuring the magnitude and direction of **v**.

17 Statistics

The chapter presents some of the basic concepts from the field of statistics, including measures of central tendency and ways of measuring variation from the mean. Normal distribution, statistical inference, and correlation of data complete the chapter.

17-1 (pages 703–710)

Teaching Suggestions

Because students often confuse mean, median, and mode, it is necessary to emphasize the differences among the three. You may want to use the set {1, 1, 5, 6, 7}; the mean is 4; the median is 5; the mode is 1. Point out that the mean is not always a member of the set but that the mode (if any) is always a member of the set. The median is either a member or an average of two members.

You may want to spend some time explaining the summation notation (Σ). The initial value of i is given under the symbol (Greek "sigma") and the value of i is incremented by 1 until the final value, n, is reached. Thus the symbol $\sum_{i=1}^{n} x_i$ represents the sum $x_1 + x_2 + x_3 + \ldots + x_n$. This expression is used in the formula for the mean (page 706). The more complicated formula on page 707 is used to find the mean in Example 4a. Having students read the formulas may help them become more familiar with the notation.

Related Activities

If the "extreme" values, the highest and the lowest, were removed from the fuel economy ratings

data list on page 703, what would be the effect on the mean, the median, and the mode?

Mean (with extremes) $= \dfrac{2062}{93}$

≈ 22.17

Mean (without extremes) $= \dfrac{2006}{91}$

≈ 22.04

The mean decreases by 0.13.
The median stays at 21 and the mode stays at 19.

17-2 (pages 710–715)

Teaching Suggestions

Sets that have the same mean, median, and mode can have other characteristics that are very different. For example the data in one set may vary greatly from the mean. The measure of the amount that the data points in a set vary, or deviate, from the mean of the set is called the standard deviation. An important extension of standard deviation is the z-score, which tells the *number* of standard deviations that a data point is away from the mean. Z-scores are used to compare data points from different sets of data. You may want to point out that these concepts are used widely to interpret data from standardized tests such as SAT and achievement tests.

The symbol, σ, for variance and standard deviation is the Greek letter "sigma." Students may find the new notation a little overwhelming so careful explanation of the concepts is important. Use Example 1 to illustrate how to calculate range, variance (formula 1a), and standard deviation (formula 2a).

17-3 (pages 717–725)

Teaching Suggestions

Students may be interested in conducting an experiment that will increase their understanding of the standard normal curve. Have them gather data from their peers about a common activity or characteristic (e.g. the number of hours per week spent doing homework). One suggestion is to have the students make three graphs: the first by plotting only a small amount of the data, the second by plotting half of the data, and the third by plotting all of the data gathered. Notice that the graph with the most data plotted is the one that most closely resembles a standard normal curve.

The properties of the standard normal curve should be emphasized, particularly the area under the curve. Using the method of area of rectangles, illustrate how to approximate the area under the normal curve. This method is preliminary to the study of the calculus.

17-4 (pages 726–731)

Teaching Suggestions

Considering the importance which American politics places on the results of opinion polls, students should realize that pollsters must be well versed in the field of statistics so that the results of their samplings can be interpreted correctly. Some students may wish to do further study to determine which careers require knowledge of statistics.

Emphasize the importance of calculating correctly the 95% confidence interval. Illustrate carefully the procedure for hypothesis testing (Example 1 on page 728) and for estimation (Example 2 on page 729). Use Written Exercises 1–2 to allow students to check their understanding of hypothesis testing and Written Exercises 3–7 to test their understanding of estimation.

Related Activities

Choose a page of text from a nontechnical book.

Find the following for each of the first three sentences:

a. the number of words in each sentence.
b. the average (mean) length of a word in each sentence.
c. the variance of the word length in each sentence.

What inferences can you make about each of the above for the fourth sentence? Now find these statistics for the fourth sentence and compare with your inferences.

Answers will vary.

17-5 (pages 732–739)

Teaching Suggestions

Constructing a scatter diagram should be relatively easy for students since they have had much practice with graphing. However, they should be made aware of the importance of the scatter diagram as a visual representation of the correlation between two variables. You may want to do this exercise with the students. Give one item: a scatter diagram, the correlation (positive, negative, or none), or an approximate value of r. Ask students to describe the other two. That should help students understand the relationship between the three ways of showing correlation.

Related Activities

Using the data from Related Activity 17-4, what inferences can you make about the following:

a. the average number of words in a sentence.
b. the average length of a word in the book.
c. the variance of the word length in the book.
d. the correlation between word length and sentence length.

How would you check your inferences about variance of word length and about the correlation between word length and sentence length?

Check by randomly choosing a sentence.

Algebra 2
and Trigonometry

Mary P. Dolciani
John A. Graham
Richard A. Swanson
Sidney Sharron

Editorial Adviser
Andrew M. Gleason

Teacher Consultants
Sonja Jernigan
Carmen Hinds Lee
Ann C. Seeboth

HOUGHTON MIFFLIN COMPANY · Boston

Atlanta   Dallas   Geneva, Ill.   Palo Alto   Princeton   Toronto

Authors

Mary P. Dolciani Former Professor of Mathematical Sciences, Hunter College of the City University of New York

John A. Graham Mathematics Teacher, Buckingham Browne and Nichols School, Cambridge, Massachusetts

Richard A. Swanson Supervisor of Mathematics, Liverpool Central Schools, Liverpool, New York

Sidney Sharron formerly Supervisor, Los Angeles Unified School District, Los Angeles, California

Editorial Adviser

Andrew M. Gleason Hollis Professor of Mathematics and Natural Philosophy, Harvard University, Cambridge, Massachusetts

Teacher Consultants

Sonja Jernigan Mathematics Teacher, Northview High School, Dothan, Alabama

Carmen Hinds Lee Mathematics Teacher, Theodore Roosevelt High School, San Antonio, Texas

Ann C. Seeboth Mathematics Teacher, Queen Anne School, Upper Marlboro, Maryland

Acknowledgment The authors wish to thank Dr. Larry D. Wiley of the Nightingale-Bamford School in New York City for contributing the material on programming in the Pascal computer language; and Hector Hirigoyen of the Dade County Public Schools in Miami, Florida, for contributing the material on discrete mathematics.

Contents

Chapter 6 *Polynomials and Rational Expressions* *Lesson 6*   **199**

Chapter 7 *Radicals and Irrational Numbers* *Lessa 7*   **247**

Chapter 8 *Sequences and Series* *Lesson 2* 293

Chapter 9 *Polynomial Functions* *Lesson 8* 341

Chapter 10 *Quadratic Relations and Systems* *Lesson 9* 393

Solving Quadratic Systems

Chapter 11   *Exponents and Logarithms*   *Lesson 5*   **441**

Chapter 12   *Permutations, Combinations, and*   *Lesson 3*
Probability   **483**

Chapter 13 *Matrices* Lesson 1   *527*

Chapter 14 *Trigonometric and Circular Functions* Lesson 2 & 6   *575*

Chapter 15   *Trigonometric Identities*   *Lesson 7*   **619**

Identities

Functions: Sums and Differences

Solving General Triangles

Chapter 16   *Inverses; Polar Coordinates; Vectors*   *Lesson 8*   **661**

Inverse Functions

Polar Coordinates

Vectors

Chapter 17   *Statistics*   *Lesson 4*   **703**

Organizing and Measuring Statistical Data

Applying the Normal Distribution

Examining Relationships between Sets of Data
17-5 Correlation *732*

Features Computer Exercises *710, 715, 739*, Biography: Leonardo Torres Quevedo *732*, Preparing for College Entrance Exams *743*, Contest Problems *744*, Programming in Pascal *744*

Reviews and Tests Self-Tests *716, 731, 739*, Chapter Summary *740*, Chapter Review *741*, Chapter Test *742*

Using a Computer with This Course

There are two types of optional computer material in this book: *Computer Exercises* and *Programming in Pascal* features. The Computer Exercises are designed for students who have some familiarity with programming. Students are usually asked to write one or more programs related to the lesson just presented. These Exercises may be done using BASIC or any other convenient computer programming language.

The optional Programming in Pascal features are designed for students who are familiar with the rudiments of Pascal. These features usually apply the computer to a topic presented in the chapter. Sometimes an aspect of the Pascal language is presented.

An appendix that presents the major features of Pascal may be found beginning on page 772. This appendix may be used as review by students who are familiar with Pascal, or it may be used by teachers to introduce Pascal to students who are familiar with other programming languages.

Symbols

READING ALGEBRA

An algebra textbook requires a different type of reading than a novel or short story. You will need to read every paragraph with great care and concentration. Keep a paper and pencil handy for doing calculations and drawing sketches. It is also a good practice to keep an algebra notebook in which you summarize important ideas.

Vocabulary

Important vocabulary words are printed in red when they are first explained. At the beginning of each Self-Test, there is a list of all such words that have appeared in that part of the chapter. You can also find these words referenced in the Glossary and Index at the back of the book. The Glossary will give you a definition of the word, and the Index will give page references for more information.

Symbols

The language of algebra is primarily symbolic. Thus, in order to understand algebra, you must be able to read symbols. For example, $\sqrt[n]{a^3}$ is read "the nth root of a cubed." A list of symbols that appear in this book faces this page. If you do not recall what a symbol means, check this list.

Diagrams

Diagrams are often used in this book to illustrate a concept that is difficult to express in words only. Study these diagrams carefully when you read the text that accompanies them.

Displayed Material

Throughout this book, red boxes are used to display axioms, theorems, summaries, and other important material. These boxes will help you identify key concepts as you read and when you study for a test.

Examples and Solutions

Each section of this book contains one or more clearly labeled examples with detailed solutions. Study these carefully, since they will help you in doing many of the exercises and problems that follow.

Reading Aids

Throughout the book, the sections titled *Reading Algebra* contain hints that are designed to help you use your book more effectively.

The architecture of the National Aquarium in Baltimore, shown in the photo, displays a dramatic use of geometric shapes. A knowledge of mathematics was essential in the design and engineering of this complex structure.

Chapter 1

Review of Essentials

Basic Properties

OBJECTIVES for Sections 1-1 through 1-3:
1. To use set notation.
2. To solve simple open sentences with a given domain.
3. To graph sets of numbers on the number line.
4. To apply the basic properties of real numbers.

1–1 Sets and Symbols

From previous mathematics courses you recall that a set is a well-defined collection of objects and that each member of a set is called an element. One way to specify a set is to list the names of its members within braces. For example, $\{0, -1, 2\}$ is read "the set whose members are 0, -1, and 2." A set can be specified by roster (or list), by rule (or description), or by graph. It is sometimes convenient to describe a set using set builder notation. For example, {all real numbers greater than -3} becomes $\{x: x > -3$ and x is a real number} when set builder notation is used. This can be read "the set of all x such that x is greater than -3, and x is a real number."

Some familiar set symbols are reviewed in the table on page 2 where A and B each denote a set.

Review of Essentials   **1**

Teaching Suggestions
p. T76

Key Ideas
Describe sets and use set notation.

Complete each sentence with $\in$, $\subset$, $=$ or the negation of one of these symbols.

1. 6 _?_ {2, 4, 6} $\in$
2. {6} _?_ {2, 4, 6} $\subset$
3. 0 ___ {2, 4, 6} $\notin$
4. Ø ___ {2, 4, 6} $\subset$

Common Errors

Students frequently confuse the null set (or empty set) notation (Ø) with the number zero (0) since they are similar in appearance. Repeat that the empty set notation has the slash through it (Ø) and that zero is represented as usual (0). Also point out to students that if it is a solution set of an open sentence that is being referred to, then it must be a *set* (even if it is the empty set) and *not a number*. They would therefore be reminded in such a case that Ø is "the empty set" and not "zero." If zero is the solution to the open sentence it must appear in braces, as an element of the solution set: {0}.

Suggested Assignments

Minimum
   2/1–7
Extended Alg. with Trig.
   2/1–9 odd
Enriched Alg.
   2/1–9 odd
Enriched Alg. and Trig.
   2/1–9

Symbolism	How to Read	Meaning
{ }	the set whose members are	a collection or set
$\in$	is a member of, or is an element of	is in the collection or set
$A \subset B$	A is a subset of B	Every member of A is a member of B. For every set A, A is a subset of A.
Ø	the null set, or the empty set	The set that contains no elements. The empty set is considered a subset of every set.
$A = B$	A is equal to B	A and B contain exactly the same elements.
/	is not	Used with $\in$, $=$, or $\subset$ to show negation.
$A \cap B$	A intersect B	The set of elements belonging to both A and B.
$A \cup B$	A union B	The set of elements in at least one of the given sets.

Oral Exercises

Classify each statement as true or false.

1. $0.5 \in \{\frac{1}{2}, \frac{1}{3}, \frac{1}{4}, \frac{1}{5}\}$ true
3. $\{0, -2\} \subset$ {nonpositive integers}   true
5. $-5 \subset \{5, -5\}$ false
7. $4 \notin \{1, 2^2, 3^2, 4^2\}$ false
9. Ø $\subset$ {negative integers}   true

2. $8 \notin \{2, 4, 6\}$ true   false
4. $\{1, 3, 5\} \not\subset$ {odd integers}
6. {Ø} = Ø   false
8. $\{\frac{2}{5}, \frac{3}{5}, \frac{4}{5}\} = \{0.2, 0.4, 0.8\}$
10. $\{1, 3\} = \{3, 1\}$ true   false

Written Exercises

Let $A = \{5, 10, 15, 20, 25\}$, $B = \{1, 9, 11, 13, 15\}$, and $C = \{4, 5, 6, 7, 8, 10\}$.
Specify each set by roster.

A **1.** $A \cup B$ **2.** $A \cup C$ **3.** $A \cap B$ **4.** $B \cup C$ **5.** $A \cap C$ **6.** $B \cap C$

1. {1, 5, 9, 10, 11, 13, 15, 20, 25}   2. {4, 5, 6, 7, 8, 10, 15, 20, 25}   3. {15}
4. {1, 4, 5, 6, 7, 8, 9, 10, 11, 13, 15}   5. {5, 10}   6. Ø

B **7.** If $X \subset Y$, describe $X \cap Y$ and $X \cup Y$. Give an example to illustrate your answer. $X \cap Y = X$, $X \cup Y = Y$; Example: $X = \{1, 3\}$, $Y = \{1, 3, 5, 7\}$

C **8.** List all the subsets of $\{a, b, c, d\}$. $\emptyset$, $\{a\}$, $\{b\}$, $\{c\}$, $\{d\}$, $\{a, b\}$, $\{a, c\}$ $\{a, d\}$ $\{b, c\}$, $\{b, d\}$, $\{c, d\}$, $\{a, b, c\}$, $\{a, b, d\}$, $\{a, c, d\}$, $\{b, c, d\}$, $\{a, b, c, d\}$

9. If a set contains n elements, what is the total number of subsets that the set has? 2^n

Computer Exercises   For students with computer experience

The optional Computer Exercises that appear throughout this book can be worked using Pascal or any other programming language. Students using Pascal may find the appendix on page 772 summarizing the fundamental concepts and terminology of Pascal useful.

1. Write a program that will divide an interval on a number line into a given number of equal segments. The program should input the endpoints of the interval and the number of equal segments. The output should give the coordinates of each division point.

Run the program in Exercise 1 dividing each interval into the given number of equal segments.

2. $2 \leq x \leq 4$; 5   **3.** $3 \leq x \leq 10$; 11   **4.** $-1 \leq x \leq 5$; 8
2. 2, 2.4, 2.8, 3.2, 3.6, 4   **3.** 3, 3.6, 4.3, 4.9, 5.5, 6.2, 6.8, 7.5, 8.1, 8.7, 9.4, 10
4. -1, -0.25, 0.5, 1.25, 2, 2.75, 3.5, 4.25, 5

1–2 Open Sentences and Graphs

In mathematics a group of symbols such as

$$7 + 9, \qquad \frac{4 + x}{3}, \qquad 5(y - 7) + 2, \qquad \text{or} \qquad (4 + 1)^2$$

is called an expression. An expression is a number, a *variable*, or a sum, difference, product, or quotient that contains one or more variables. A variable is a symbol, usually a letter, that represents any of the members of a specified set. This set is called the replacement set or domain of the variable, and its members are called the values of the variable. A variable with only one value is called a constant. Sometimes numerals are also called constants.

A group of symbols that states a relationship between two mathematical expressions is called a mathematical sentence. The following are examples of mathematical sentences that contain numerical expressions.

$$\frac{3}{4} \neq \frac{2}{5} \qquad 1 + 2 = 17 \qquad 14 - 8 = 6$$

These mathematical sentences can easily be classified as true or false.

Review of Essentials **3**

3

1. Solve $y + 2 \neq 4$ over $\{1, 2, 3, 4\}$.
$1 + 2 \neq 4$ True
$2 + 2 \neq 4$ False
$3 + 2 \neq 4$ True
$4 + 2 \neq 4$ True
∴ the solutions are 1, 3, and 4, and the solution set is $\{1, 3, 4\}$.

2. Solve $5x - 1 = x$ over $\{0, 0.5, 1.0\}$.
The open sentence is false for every value of the variable, and the solution set is ∅.

3. Solve $1 - 2x = 3$ over $\{-1, 0, 1, 2\}$.
$1 + 2 = 3$ True
$1 - 0 = 3$ False
$1 - 2 = 3$ False
$1 - 4 = 3$ False
∴ the solution is -1 and the solution set is $\{-1\}$.

Common Errors

Students may forget to indicate the continuation of a roster with three ellipses. Make sure they understand that it is necessary to include these in the notation of a set to indicate that the set roster continues on indefinitely, according to the pattern made clear by the numbers actually listed. Without this notation the only elements of the set are those actually listed.

Point out to the students that they should use ellipses (dots) in their written work and arrows on their graphs to indicate the continuation of a roster. Be sure, also, that it is clear to them that the continuation of the set can go on in *either direction* or *both directions*.

A mathematical sentence that contains one or more variables is called an open sentence. The truth or falsity of an open sentence usually cannot be determined without knowing the value that the variable represents. Each of the following is an example of an open sentence.

$$x - 7 > 4 \qquad m + p = 2m + 5 \qquad h < 93$$

The values of the variable that make an open sentence true are called the solutions, or roots, of the open sentence. The solution set is the set of *all* solutions that make the open sentence true. All members of the solution set are said to satisfy the open sentence. To solve an open sentence over a given domain means to find the solution set using this domain.

EXAMPLE 1   Solve $2x - 3 \neq 7$ over $\{2, 3, 4, 5\}$.

SOLUTION   Replace x in the open sentence with each value in the domain.

$2 \cdot 2 - 3 \neq 7$ True $2 \cdot 3 - 3 \neq 7$ True
$2 \cdot 4 - 3 \neq 7$ True $2 \cdot 5 - 3 \neq 7$ False

∴ (read "therefore") the solutions are 2, 3, and 4, and the solution set is $\{2, 3, 4\}$.

EXAMPLE 2   Solve $y = y + 1$ over $\{0, 1, 2, 3, 4, 5\}$.

SOLUTION   Since the open sentence is false for every value of the variable, it has no solutions, and the solution set is ∅.

Sometimes it is useful to graph a set such as the solution set of an open sentence on a number line. The graph below pictures the set $\{-2, \frac{3}{2}, \sqrt{5}\}$.

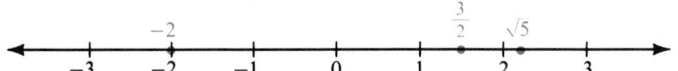

The point representing a number is called the graph of the number while the number is called the coordinate of the point.

A basic assumption of mathematics is that for each real number there corresponds a point on the number line, and, conversely, for each point on the number line there corresponds a real number. Thus, there is a *one-to-one correspondence* between the points on any number line and the set of real numbers, $\mathcal{R}$. Any number that is either a positive number, a negative number, or zero is a real number. Some important subsets of $\mathcal{R}$ are shown below and on the next page.

1. $N = \{$the natural numbers$\} = \{$the positive integers$\} = \{1, 2, 3, \ldots\}$

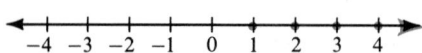

2. W = {the whole numbers} = {0, 1, 2, 3, . . .}

3. J = {the integers} = {. . . , $-3, -2, -1, 0, 1, 2, 3,$ . . .}

4. {the even integers} = {. . . , $-4, -2, 0, 2, 4,$ . . .}

5. {the odd integers} = {. . . , $-3, -1, 1, 3,$ . . .}

Notice that in the examples above each set is specified by an *incomplete roster*, or list. The three dots indicate that the pattern shown in the list continues in one or both directions without end. The heavy arrowheads on the accompanying diagrams indicate that the graph of the set similarly continues without end.

Oral Exercises

Classify each open sentence as true or false for the given replacement set of the variable.

false

1. $x + 3 = 10$; {$-7, 7$} false   **2.** $2y + 3 \neq 6$; {2, -1} true   **3.** $a - 3 = 2a$; {3, 0}

4. $8 - 3b \neq 5$; {1, -2} false   **5.** $c + 4c = 5c$; {6, -8} true   **6.** $9d - d = 7d$; {0, 6}

false

Specify the solution set of each open sentence.

7. $8r = 3r + 5r$
{all real numbers}

8. $3p + 1 = 3p + 4$ Ø

9. $2x = 7x - 5x$
{all real numbers}

Describe each graph. Answers may vary.

EXAMPLE

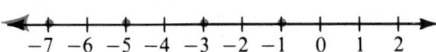

SOLUTION   The set of negative odd integers

10.

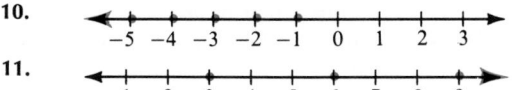

The set of negative integers

11.

The set of positive integers
that are multiples of 3

12.

The set of real numbers n such
that $n + \frac{1}{2}$ is a positive integer

Review of Essentials   **5**

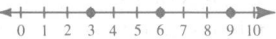

11.

12.
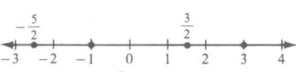

13.

14.

15.

16.

17.

18.

19.

20.

21.

22.

Mixed Review

Simplify.

1. $3r - 4(2r - 4)$ $-5r + 16$

2. $-((s)(t^2)(x^3))$ $-st^2x^3$

3. $4n - (2n - 1) - 5(3n - 8)$
   $-13n + 41$

4. $3(4 - 16) + 2.5 - (-6 + 2)$
   -29.5

Classify each statement as true or false.

5. $-3 \subset \{-4, -3, 6\}$ False

6. $14 \in \{$positive even integers$\}$ True

Written Exercises

Solve each open sentence over $\{-2, -1, 1, 2, 3\}$.

A   **1.** $x - 3 = 0$ {3}
   3. $2b - 1 = 7$ $\varnothing$
   5. $3c + 2 \neq 3c$ {-2, -1, 1, 2, 3}
   7. $\dfrac{n+1}{2}$ is an integer. {-1, 1, 3}

   2. $5 + y \neq 7$ {-2, -1, 1, 3}
   4. $5k - 2k = 3k$ {-2, -1, 1, 2, 3}
   6. $(w - 2)(w + 2) = 0$ {-2, 2}
   8. $\dfrac{n-1}{3}$ is an integer. {-2, 1}

   9. $2n + 1$ is an odd integer. {-2, -1, 1, 2, 3}   **10.** $2n$ is not an even integer. $\varnothing$

Graph each set of numbers on a number line.

11. $\left\{3, -1, 0, \dfrac{4}{3}\right\}$

12. $\left\{-\dfrac{5}{2}, -1, \dfrac{3}{2}, 3\right\}$

13. the set of all negative even integers
14. the set of all integers that are multiples of 4

Solve each open sentence over the set of positive integers and graph the solution set.

B   **15.** $x^2 = 9$ {3}
   17. $y^2 < 3$ {1}

   16. $25 - r^2 = 9$ {4}
   18. $z^2 > 5$ {z > 2}

   19. $2d$ is a prime number.
      {1}

   20. $\dfrac{p}{2}$ is an integer.
      {p: p is a positive even integer}

C   **21.** $n^2 + 1$ is a multiple of 4.
      $\varnothing$

   22. $\dfrac{m+1}{3}$ is a natural number.
      {m: m = 3n - 1 and n is a positive integer}

1–3 Axioms for the Real Numbers

There are two basic operations used in working with real numbers, *addition* and *multiplication*. Each of these operations is called a binary operation because it assigns any two real numbers to a third real number. The addition operation, $+$, assigns any two real numbers a and b to another real number, $a + b$, called the sum of the two real numbers.

Multiplication assigns to any two real numbers a and b their product denoted by $a \times b$, $a \cdot b$, $a(b)$, $(a)(b)$, or simply ab. In the sum $a + b$, a and b are called terms; in the product ab, a and b are called factors.

The properties of these operations in $\mathcal{R}$ all stem from a few basic statements, called axioms or postulates, that are assumed to be true. These familiar assumptions are listed on the next page.

Notice that parentheses, ( ), are used in some of the axioms to indicate an order of operations. For example, $(a + b) + c$ represents the

6   *Chapter 1*

result of first adding a and b, and then adding c to the sum. In some cases, such as in the distributive law where $(ab) + (ac)$ occurs, the parentheses usually are omitted. Thus, $(ab) + (ac) = ab + ac$.

Teaching Suggestions
p. T76

Related Activities p. T77

Axioms of Addition and Multiplication in $\mathcal{R}$

Let a, b, and c denote any real numbers (a, b, and $c \in \mathcal{R}$).

1. $a + b$ is a unique real number.   Closure Axiom for Addition

2. $(a + b) + c = a + (b + c)$ Associative Axiom for Addition

3. $a + b = b + a$ Commutative Axiom for Addition

4. There exists an element $0 \in \mathcal{R}$ such that for each $a \in \mathcal{R}$, $0 + a = a$ and $a + 0 = a$.   Identity Axiom for Addition

5. There exists an element $-a \in \mathcal{R}$ for each $a \in \mathcal{R}$, such that $a + (-a) = 0$ and $(-a) + a = 0$.   Axiom of Additive Inverses

6. ab is a unique real number.   Closure Axiom for Multiplication

7. $(ab)c = a(bc)$ Associative Axiom for Multiplication

8. $ab = ba$ Commutative Axiom for Multiplication

9. There exists an element $1 \in \mathcal{R}$, $1 \neq 0$, such that for each $a \in \mathcal{R}$, $a \cdot 1 = a$ and $1 \cdot a = a$.   Identity Axiom for Multiplication

10. There exists an element $\frac{1}{a} \in \mathcal{R}$ for each nonzero $a \in \mathcal{R}$ such that $\frac{1}{a} \cdot a = 1$ and $a \cdot \frac{1}{a} = 1$.   Axiom of Multiplicative Inverses

11. $a(b + c) = ab + ac$ and $(b + c)a = ba + ca$ Distributive Axiom

---

Supplementary Material
Test 1

Key Ideas
Use addition and multiplication axioms.

Chalkboard Examples
State the axiom that justifies each statement.
1. $(2 + 3) + 4 = 2 + (3 + 4)$
Associative axiom for addition
2. $6 + (-6) = 0$ and $(-6) + 6 = 0$
Axiom of additive inverses
3. $3(x + 2) = 3x + 6$
Distributive axiom
4. If $a = 4$ and $4 = c$ then $a = c$.   Transitive property

Complete each open sen-
tence so that it is true for all
real values of the variable.
State the axiom or property
that justifies each statement.

1. $(a + b) + c = \underline{\ ?\ } + (b + c)$
   a; Associative axiom for
   addition

2. $(4p)\left(\dfrac{1}{?}\right) = 1$ $4p, p \neq 0$
   Axiom of multiplicative in-
   verses

3. If $h = 8$, then $8 = \underline{\ ?\ }$ h
   Symmetric property

4. $2 + \underline{\ ?\ } = 0$ -2
   Axiom of additive inverses

5. $7(x + y) = \underline{\ ?\ }x + \underline{\ ?\ }y$
   7, 7
   Distributive axiom

6. If $a = b$ and $b = c$,
   then $\underline{\ ?\ } = \underline{\ ?\ }$. $a = c$
   Transitive property

The uniqueness of sums or products as shown in the axioms has this implication.

Substitution Principle

Since $a + b$ and ab are unique, changing the numeral by which a number is named in an expression involving sums or products does not change the value of the expression.

For example, since

$$8 + 2 = 10 \text{ and } 10 - 3 = 7,$$

you know that

$$(8 + 2) - 3 = 10 - 3 = 7.$$

In $\mathcal{R}$, 0 and 1 are called the **identity elements** for addition and multiplication, respectively. The expressions $-a$ and $\frac{1}{a}$ are called the **additive inverse** and the **multiplicative inverse** (or **reciprocal**) of a, respectively. Note also that you should always read $-a$ as "the additive inverse of a" or "the opposite of a." The expression "negative a" should not be used since $-a$ may represent either a negative number, a positive number, or zero.

The way you use the symbol $=$ in sentences is consistent with the following assumptions.

Axioms of Equality

Let a, b, and c be any elements of $\mathcal{R}$.

$a = a$	Reflexive Property
If $a = b$, then $b = a$.	Symmetric Property
If $a = b$ and $b = c$, then $a = c$.	Transitive Property

How do you add or multiply three or more real numbers? If a, b, c, d, . . . are real numbers, we define

$$a + b + c \text{ to be } (a + b) + c,$$
$$abc \text{ to be } (ab)c,$$
$$a + b + c + d \text{ to be } (a + b + c) + d,$$

and so on.

The commutative and associative properties allow you to add (or multiply) in any order and in any convenient groups of two, and still

8   *Chapter 1*

obtain the same result. We assume that the basic number facts are known and can be used. Thus, for example,

$$17 + 12 + 3 + 18 = (17 + 3) + (12 + 18)$$
$$= 20 + 30$$
$$= 50.$$

Oral Exercises

State the axiom or property that justifies each statement. Assume that each variable represents a real number.

1. If $x = 5$ and $5 = y$, then $x = y$. Trans. prop. of $=$

2. $7 + (5 + x) = (7 + 5) + x$ Assoc. ax. for add.

3. $(1)(-9) = -9$ Ident. ax. for mult.

4. $2(a + 6) = 2a + 12$ Dist. ax.

5. $8 \cdot \frac{1}{8} = 1$ Ax. of mult. inv.

6. $x + 5$ is a real number. Clos. ax. for add.

7. $3 + (-3) = 0$ Ax. of add. inv.

8. $y + 7 = 7 + y$ Comm. ax. for add.

9. If $z = 4$, then $4 = z$. Symm. prop. of $=$

10. $3(2a) = 6a$ Subst.

Written Exercises

Complete each open sentence so that it is true for all real values of the variable. State the axiom or property that justifies each statement.

A 1. $3 + (6 + x) = (6 + x) + \underline{\ ?\ }$ 3; Comm. ax. for add.

2. $(2a)b = 2(\underline{\ ?\ })$ ab; Assoc. ax. for mult.

3. If $y = 7$, then $7 = \underline{\ ?\ }$. y; Symm. prop. of $=$

4. $3(5 + n) = 15 + \underline{\ ?\ }$ 3n; Dist. ax.

5. $4 + \underline{\ ?\ } = 0$ -4; Ax. of add. inv.

6. $\underline{\ ?\ } + (c + d) = c + d$ 0; Iden. ax. for add.

7. $(5 + m) + (-2) = (5 + m) + \underline{\ ?\ }$ (-2); Reflex. prop. of $=$

8. $(2r)(\underline{\ ?\ }) = 1$ $\frac{1}{2r}$; Ax. of mult. inv.

9. $a(8 + c) = (\underline{\ ?\ })a$ $8 + c$; Comm. ax. for mult.

10. $(ef + 3)(\underline{\ ?\ }) = ef + 3$ 1; Iden. ax. for mult.

11. If $p = q$ and $q = r$, then $\underline{\ ?\ } = \underline{\ ?\ }$. p, r; Trans. prop. of $=$

12. If x is a real number, then so is $(-6)(\underline{\ ?\ })$. x; Clos. ax. for mult.

Using the commutative and associative axioms, determine the value of each expression.

13. $16 + (-8) + 5 + (-11) + 9$ 11

14. $-12 + 29 + (-5) + (-7) + 17$ 22

15. $(-4)(3)(5)(-2)$ 120

16. $(-3)(7)(-5)(-8)$ -840

Review of Essentials **9**

Suggested Assignments
Minimum
  9/1–16 odd, 17–22
R 11/*Self-Test 1*
Extended Alg. with Trig.
  9/1–19 odd
R 11/*Self-Test 1*
Enriched Alg.
  9/1–25 odd
R 11/*Self-Test 1*
Enriched Alg. with Trig.
  9/1–25 odd
R 11/*Self-Test 1*

Additional Answers
Written Exercises

18. yes; yes; nth row is identical to the nth column.

21. $(1 \oplus 2) \oplus 3 =$
$3 \oplus 3 = 0$; $1 \oplus (2 \oplus 3) =$
$1 \oplus 1 = 0$; $\therefore (1 \oplus 2) \oplus 3 =$
$1 \oplus (2 \oplus 3)$

22. $(1 \star 2) \star 3 = 2 \star 3 = 3$;
$1 \star (2 \star 3) = 1 \star 3 = 3$;
$\therefore (1 \star 2) \star 3 = 1 \star (2 \star 3)$

23. $2 \star (1 \oplus 3) = 2 \star 2 = 1$;
$2 \star 1 \oplus 2 \star 3 = 2 \oplus 2 \star 3 = 2 \oplus 3 = 1$;
$\therefore 2 \star (1 \oplus 3) = 2 \star 1 \oplus 2 \star 3$

24. $3 \star (1 \oplus 2) = 3 \star 3 = 0$;
$3 \star 1 \oplus 3 \star 2 = 3 \oplus 3 \star 2 = 3 \oplus 3 = 0$;
$\therefore 3 \star (1 \oplus 2) = 3 \star 1 \oplus 3 \star 2$

25. $(x + a)(x + b) =$
$(x + a)x + (x + a)b =$
$x \cdot x + a \cdot x + x \cdot b + a \cdot b$
(Dist. ax.)
$= x^2 + ax + bx + ab$
(Comm. ax. for mult.)
$= x^2 + (a + b)x + ab$
(Dist. ax.)

9

Quick Quiz

1. Use set notation to represent the sentence: "The set consisting of only 1 is not a subset of the set whose members are even positive integers less than 6 and greater than 0."
$\{1\} \not\subset \{2, 4\}$

Solve each open sentence over $\{1, 2, 3, 4\}$.

2. $3t + 3 = 9$ $\{2\}$

3. $4x - 12 = 0$ $\{3\}$

4. Graph $\left\{-1, 0, \frac{1}{2}, 1\frac{1}{2}\right\}$ on a number line.

5. Graph the solution set of $3x - 1 = 2$ over $\{-1, 0, 1\}$ on a number line.

State the axiom that justifies each statement. Assume that each variable represents a real number.

6. $(c + 6) + 2 = c + (6 + 2)$.
Associative axiom for addition

7. $(3t)\left(\dfrac{1}{3t}\right) = 1$
Axiom of multiplicative inverses

8. $(3 \cdot 4)(2) = (3)(4 \cdot 2)$
Associative axiom for multiplication

9. $a(b + 5) = ab + a \cdot 5$
Distributive axiom

10. Solve over $\mathcal{R}$. $\dfrac{1}{5}(-y) = 15$

$\{-75\}$

The following tables define the operations of $\oplus$ and $\star$ on a set of four elements: 0, 1, 2, and 3. (For example, $1 \oplus 3 = 2$, because 2 is in the box to the right of 1 and below 3 in the $\oplus$ table.) Use these tables in Exercises 17–24.

$\oplus$	0	1	2	3
0	0	1	2	3
1	1	0	3	2
2	2	3	0	1
3	3	2	1	0

$\star$	0	1	2	3
0	0	0	0	0
1	0	1	2	3
2	0	2	1	3
3	0	3	3	0

B **17.** Is the set closed with respect to $\oplus$? with respect to $\star$? yes; yes

18. Is $\oplus$ commutative? Is $\star$ commutative? What pattern in the tables tells you this?

19. What is the identity element for $\oplus$? Does every element in $\oplus$ have an inverse?

20. What is the identity element for $\star$? Does every element in $\star$ have an inverse?

21. Show that $\oplus$ is associative in the case: 19. 0; yes 20. 1; no
$$(1 \oplus 2) \oplus 3 = 1 \oplus (2 \oplus 3)$$

22. Show that $\star$ is associative in the case:
$$(1 \star 2) \star 3 = 1 \star (2 \star 3)$$

C **23.** Show that the distributive axiom holds in the case:
$$2 \star (1 \oplus 3) = (2 \star 1) \oplus (2 \star 3)$$

24. Show that the distributive axiom holds in the case:
$$3 \star (1 \oplus 2) = (3 \star 1) \oplus (3 \star 2)$$

25. Use the axioms for the real numbers to show that
$$(x + a)(x + b) = x^2 + (a + b)x + ab.$$

[*Hint*: Start by applying the distributive axiom:
$$(x + a)(x + b) = (x + a)x + (x + a)b.]$$

Computer Exercises   For students with computer experience

1. Write a program that will find the additive inverse and the multiplicative inverse (if it exists) of a given real number. If no multiplicative inverse exists, the output should state this.

Run the program in Exercise 1 for each number.

2. -5 5, -0.2 **3.** 0.625 $-0.625, 1.6$ **4.** 0 0, none **5.** 7 $-7, 0.\overline{142857}$ **6.** -0.83333 $0.8\overline{3}, -1.2$ **7.** $-3\frac{1}{3}$ $3\frac{1}{3}, -0.3$

10 *Chapter 1*

Self-Test 1

VOCABULARY   set (p. 1)
member or element of a set
    (p. 1)
subset (p. 2)
empty set or null set (p. 2)
expression (p. 3)
variable (p. 3)
replacement set, or domain, of
    a variable (p. 3)
values of a variable (p. 3)
constant (p. 3)
mathematical sentence (p. 3)
open sentence (p. 4)
solutions, or roots, of an open
    sentence (p. 4)
solution set (p. 4)
graph of a number (p. 4)
number line (p. 4)
coordinate of a point (p. 4)
real number (p. 4)

binary operation (p. 6)
sum (p. 6)
product (p. 6)
term (p. 6)
factor (p. 6)
axioms, or postulates (p. 6)
identity element for addition
    (p. 8)
identity element for
    multiplication (p. 8)
additive inverse (p. 8)
multiplicative inverse (p. 8)

1. Use set notation to represent the sentence: "Three is a member of   *Obj. 1, p. 1*
the set of all y such that y is less than 5 and y is a real number."
$3 \in \{y: y < 5 \text{ and } y \in \mathcal{R}\}$

Solve each open sentence over $\{-3, 0, 1\}$.

2. $5m + 3 = 3$ {0}   *Obj. 2, p. 1*
3. $3x - 7 = 1$ ∅
4. Graph $\{-2\frac{1}{2}, -1, 0, 3\}$ on a number line.   *Obj. 3, p. 1*
5. Graph the solution set of $4y - 1 = 3$ over $\{-1, 0, 1\}$.   {1}

State the axiom that justifies each statement.

6. $7 + 0 = 7$ Iden. ax. for add.   *Obj. 4, p. 1*
7. If $x \in \mathcal{R}$, then $(x + 6) + 4 = x + (6 + 4)$. Assoc. ax. for add.
8. If $y \in \mathcal{R}$, then $3(y - 2) = 3y - 6$. Dist. ax.
9. If $m \in \mathcal{R}$, then $\left(\frac{1}{3} \cdot 3\right)m = 1m$. Ax. of mult. inv.

10. Solve over $\mathcal{R}$: $\frac{1}{4}(-x) = 12$ {−48}

Check your answers with those at the back of the book.

Review of Essentials   **11**

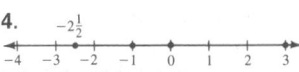

Mixed Review

Simplify.

1. $(p - 3)^2$ $p^2 - 6p + 9$
2. $(2u)^3(3u)(u)$ $24u^5$
3. $g(2gh - 1) + 5g(2gh)$
     $12g^2h - g$
4. If $2p - 1 = p + 6$ then
    $p = \underline{\ ?\ }$.   7
5. Graph the set $\{-3, 3, 4\}$.
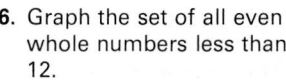
6. Graph the set of all even
    whole numbers less than
    12.

Properties of Operations in $\mathcal{R}$

OBJECTIVES for Sections 1-4 through 1-7:
1. *To understand the meaning and methods of direct proof.*
2. *To understand important theorems that give properties of real numbers.*
3. *To apply these theorems in simplifying expressions for sums, products, differences, and quotients.*

Teaching Suggestions
p. T77

Key Ideas

Use a hypothesis and a conclusion to give a direct proof for addition theorems. Apply theorems in simplifying expressions.

Chalkboard Examples

Simplify.
1. $c + [-(c + 5)]$

    $c + [-(c + 5)]$
    $= c + (-c) + (-5)$
    $= 0 + -5$
    $= -5$

2. $(-2b) + (-2b)$

    $(-2b) + (-2b)$
    $= -(2b + 2b)$
    $= -(4b)$
    $= -4b$

1–4 Theorems and Proof: Addition

The basic properties of $\mathcal{R}$ stated in Section 1-3 imply other properties of $\mathcal{R}$. These implications are stated as theorems. A theorem consists of two parts, a hypothesis (or premise) and a conclusion. The hypothesis states what is assumed to be true, and the conclusion states something which logically follows from the assumptions. To give a direct proof of a theorem, you start with its hypothesis and by a logical chain of steps arrive at its conclusion. Here is an example of a direct proof.

> *Theorem.* For all real numbers b and c,
> $$(b + c) + (-c) = b.$$

PROOF

First note that the hypothesis is that b and c denote real numbers. Reasoning from this assumption, you have:

Statements	*Reasons*
1. b and c are real numbers.	Hypothesis
2. $b + c$ is a real number.	Closure axiom for addition
3. $-c$ is a real number.	Axiom of additive inverses
4. $(b + c) + (-c)$ is a real number.	Closure axiom for addition
5. $(b + c) + (-c) =$ $b + [c + (-c)]$	Associative axiom for addition
6. $c + (-c) = 0$	Axiom of additive inverses
7. $(b + c) + (-c) = b + 0$	Substitution principle
8. $b + 0 = b$	Identity axiom for addition
9. $(b + c) + (-c) = b$	Transitive property of equality (or Substitution principle)

12 *Chapter 1*

Observe that each step in the proof is guaranteed either by hypothesis or by an axiom. Frequently, simple steps involving closure, substitution, and other basic properties of equality are omitted. For example, the preceding proof might be replaced by the following:

PROOF

Statements	Reasons
1. b and c are real numbers.	Hypothesis
2. $-c$ is a real number.	Axiom of additive inverses
3. $(b + c) + (-c) =$ $b + [c + (-c)]$	Associative axiom for addition
4. $\qquad\qquad = b + 0$	Axiom of additive inverses
5. $\qquad\qquad = b$	Identity axiom for addition
6. $(b + c) + (-c) = b$	Transitive property of equality

Theorems that have been proved can then be used to help prove other theorems. For example, the following theorem uses the theorem proved on page 12 as a reason.

Theorem.   For all real numbers a, b, and c,

$$\text{if } a + c = b + c, \text{ then } a = b.$$

PROOF

Statements	Reasons
1. a, b, and c are real numbers and $a + c = b + c$.	Hypothesis
2. $(a + c) + (-c) =$ $(b + c) + (-c)$	Substitution principle
3. $(b + c) + (-c) = b$	Theorem proved on page 12
4. $(a + c) + (-c) = b$	Transitive property of equality
5. $(a + c) + (-c) = a$	Theorem proved on page 12 with a in place of b
6. $a = b$	Substitution principle

From one theorem you can sometimes quickly deduce a closely related theorem, called a **corollary**. Because addition in $\mathcal{R}$ is a commutative operation, you can easily prove the corollary on page 14 which is closely related to the preceding theorem (Exercise 27, page 16).

> *Corollary.* For all real numbers a, b, and c,
>
> $$\text{if } c + a = c + b, \text{ then } a = b.$$

The theorem on page 13 and its corollary can be restated in the following combined form.

Cancellation Property of Addition

For all real numbers a, b, and c, if

$$a + c = b + c \qquad \text{or} \qquad c + a = c + b,$$

then $a = b$.

The next theorem is useful in computing sums..

Property of the Opposite of a Sum

For all real numbers a and b,

$$-(a + b) = (-a) + (-b).$$

That is, the opposite of a sum of real numbers is the sum of the opposites of the numbers.

PROOF

Plan: Show that $(a + b) + [(-a) + (-b)] = 0$. Then use the axiom of additive inverses $((a + b) + [-(a + b)] = 0)$ and the cancellation law to obtain the desired result.

Statements	Reasons
1. a and b are real numbers.	Hypothesis
2. $(a + b) + [(-a) + (-b)]$ $\quad = [a + (-a)] + [b + (-b)]$	Commutative and associative axioms for addition
3. $\quad = 0 + 0$	Axiom of additive inverses
4. $\quad = 0$	Identity axiom for addition
5. $(a + b) + [(-a) + (-b)] = 0$	Transitive property of equality
6. $(a + b) + [-(a + b)] = 0$	Axiom of additive inverses
7. $(a + b) + [-(a + b)]$ $\quad = (a + b) + [(-a) + (-b)]$	Transitive property of equality
8. $-(a + b) = (-a) + (-b)$	Cancellation property of addition

14 *Chapter 1*

The property of the opposite of a sum is useful in simplifying expressions.

EXAMPLE 1 Simplify $x + [-(x + 3)]$.

SOLUTION $x + [-(x + 3)] = x + (-x) + (-3)$
$= 0 + (-3) = -3$

EXAMPLE 2 Simplify $(-y) + (-y)$.

SOLUTION $(-y) + (-y) = -(y + y)$
$= -(2y) = -2y$

The following property, which you are asked to prove in Exercise 26 on page 16, is also useful in simplifying expressions.

Cancellation Property of Additive Inverses

For all real numbers a, $-(-a) = a$.

Oral Exercises

4. Dist. prop.
6. Canc. prop. of add.

State the axiom or theorem that justifies each statement. Assume each variable represents a real number.

Opp. of sum prop.

1. If $3 + x = 7$, then $x = 4$. Canc. prop. of add.   **2.** $-(2 + x) = -2 + (-x)$
3. $(5 + (-3)) + 3 = 5$ Theorem, p. 12   **4.** $-3(-7 + x) = 21 + (-3)x$
5. $-(-9 + y) = 9 + (-y)$ Opp. of sum prop.   **6.** If $r + (-12) = 1$, then $r = 13$.
7. $(5 + (-2)) + 0 = 3$ Iden. ax. of add.   **8.** $-(-2x) = 2x$

Canc. prop. of add. invs.

Written Exercises

Replace each __?__ with a variable or numeral so that a true statement results. Assume each variable represents a real number.

A **1.** $-(z + 3) = -z + $ __?__ . -3
3. $-(2 + (-y)) = -2 + $ __?__ y
5. If $5 + (-x) = 2$, then $x = $ __?__ . 3
7. $(n + (-4)) + $ __?__ $= n$ 4

2. If $x + 8 = -7$, then $x = $ __?__ . -15
4. $-6 + (a + 6) = a + $ __?__ 0
6. $-2(c + (-4)) = (-2)c + $ __?__ 8
8. If $p + (-9) = -9$, then $p = $ __?__ 0

Simplify.

9. $3 + (z + (-3))$ z
11. $7(-4 + c) + 12$ $-16 + 7c$

10. $-(b + (-10)) + (-10)$ $-b$
12. $-(-k + 4) + k$ $-4 + 2k$

Review of Essentials **15**

Suggested Assignments
Minimum
15/1–26
S 9/1–16 even
Extended Alg. with Trig.
15/1–29 odd
S 9/20–22
Enriched Alg.
15/7–39 odd
Enriched Alg. with Trig.
15/7–39 odd

Additional A Exercises

Replace each __?__ with a variable or a numeral so that a true statement results. Assume each variable represents a real number.

1. $-(w + 1) = -w + $ __?__ .
-1
2. If $c + (-2) = 0$ then
$c = $ __?__ . 2
3. $-7(d + (-2)) = -7d + $
__?__ 14
4. If $2x + (-4) = -4$ then
$x = $ __?__ . 0
5. $(b + (-5)) + $ __?__ $= b$ 5
6. If $6 + (-z) = 2$ then $z = $
__?__ . 4

28. 1. $a + (-c) = b + (-c)$
Hypothesis
2. $-c$ is a real number
Ax. of add. inv.
3. $a = b$
Canc. prop. of add.

30. 1. $a + b = 0$
Hypothesis
2. $0 = a + (-a)$
Ax. of add. inv.
3. $a + b = a + (-a)$
Trans. prop. of =
4. $b = -a$
Canc. prop. of add.

32. 1. $a = b$ and $c = d$
Hypothesis
2. $a + c = a + c$
Ref. prop. of =
3. $a + c = b + c$
Subs. principle
4. $a + c = b + d$
Subs. principle

34. 1. $x + 3 = 7$
Hypothesis
2. $= 7 + 0$
Iden. ax. for add.
3. $= 7 + [(-3) + 3]$
Ax. of add. inv.
4. $= [7 + (-3)] + 3$
Assoc. ax. for add.
5. $= 4 + 3$
Subs. prin.
6. $x + 3 = 4 + 3$
Trans. prop. of =
7. $x = 4$
Canc. prop. of add.

36. 1. $x + 9 = -1$
Hypothesis
2. $= -1 + 0$
Iden. ax. for add.
3. $= -1 + [(-9) + 9]$
Ax. of add. inv.
4. $= [-1 + (-9)] + 9$
Assoc. ax. for add.
5. $= (-10) + 9$
Subs. prin.
6. $x + 9 = (-10) + 9$
Trans. prop. of =
7. $x = -10$
Canc. prop. of add.

Simplify.

13. $8x + (-9) + (-5)x$ $3x + (-9)$

14. $(m + (-2)) + [-(6 + m)]$ -8

15. $-(-q) + (-[q + (-2)])$ 2

16. $8(-5 + r) + (-7)r$ $-40 + r$

17. $-[(-w) + (-11)] + [w + (-14)]$ $2w + (-3)$

18. $-[-(n + 15)] + [(-4) + (-n)]$ 11

Solve each equation over $\mathcal{R}$.

19. $y + (-4) = -27$ $\{-23\}$

20. $13 + p = -5$ $\{-18\}$

21. $-16 + (-x) = 8$ $\{-24\}$

22. $(-h) + 21 = 14$ $\{7\}$

23. $2x + (-13) = 7 + x$ $\{20\}$

24. $-(-t + 6) = 2t + (-9)$ $\{3\}$

State the axiom or theorem that justifies each step in the following proofs.

B   25. Prove: For all real numbers b and c, $[b + (-c)] + c = b$.

   1. b and c are real numbers. Hypothesis

   2. $-c$ is a real number. Axiom of additive inverses

   3. $[b + (-c)] + c = b + [(-c) + c]$ Associative axiom for addition

   4. $= b + 0$ Axiom of additive inverses

   5. $= b$ Identity axiom for addition

   6. $[b + (-c)] + c = b$ Transitive property of equality

26. Prove: For all real numbers a, $-(-a) = a$. (Cancellation property of additive inverses)

   1. a is a real number. Hypothesis

   2. $-a$ and $-(-a)$ are real numbers. Axiom of additive inverses

   3. $-(-a) = -(-a) + 0$ Identity axiom for addition

   4. $= -(-a) + [(-a) + a]$ Axiom of additive inverses

   5. $= [-(-a) + (-a)] + a$ Associative axiom for addition

   6. $= 0 + a$ Axiom of additive inverses

   7. $= a$ Identity axiom for addition

   8. $-(-a) = a$ Transitive property of equality

Prove each theorem. Assume that each variable represents a real number. You may use the results of a previous exercise in any proof.

27. If $c + a = c + b$, then $a = b$.

28. If $a + (-c) = b + (-c)$, then $a = b$.

29. If $b + a = a$, then $b = 0$. (Uniqueness of additive identity)

30. If $a + b = 0$, then $b = -a$. (Uniqueness of additive inverse)

31. $-[(-a) + (-b)] = a + b$ (*Hint:* Use the result of Exercise 26.)

32. If $a = b$ and $c = d$, then $a + c = b + d$.

33. If $a + c = b + d$, and $c = d$, then $a = b$.

34. If $x + 3 = 7$, then $x = 4$.

16   *Chapter 1*

35. If $x + (-6) = -1$, then $x = 5$.

36. If $x + 9 = -1$, then $x = -10$.

C **37.** If $x + b = a$, then $x = a + (-b)$.

38. If $x = a + (-b)$, then $x + b = a$.

39. If $2x + b = x$, then $x = -b$.

1–5 Properties of Products

Multiplication in $\mathcal{R}$ has properties similar to those of addition in $\mathcal{R}$. Compare the three theorems stated below with those for addition on pages 12–14. (See Exercises 24–27 on page 21 for proofs.)

Theorem. For all real numbers b and all nonzero real numbers c,

$$(bc)\frac{1}{c} = b.$$

Cancellation Property of Multiplication

For all real numbers a and b and all nonzero real numbers c, if

$$ac = bc \qquad \text{or} \qquad ca = cb,$$

then $a = b$.

Property of the Reciprocal of a Product

For all nonzero real numbers a and b,

$$\frac{1}{ab} = \frac{1}{a} \cdot \frac{1}{b}.$$

That is, the reciprocal of a product of nonzero real numbers is the product of the reciprocals of the numbers.

EXAMPLE 1 Simplify. **a.** $36 \cdot \frac{1}{4}$ **b.** $9 \cdot \frac{1}{36}$

SOLUTION **a.** $36 \cdot \frac{1}{4} = (9 \cdot 4) \cdot \left(\frac{1}{4}\right)$

$$= 9 \cdot \left(4 \cdot \frac{1}{4}\right) = 9 \cdot 1 = 9$$

b. $9 \cdot \frac{1}{36} = 9 \cdot \left(\frac{1}{9 \cdot 4}\right) = 9 \cdot \left(\frac{1}{9} \cdot \frac{1}{4}\right)$

$$= \left(9 \cdot \frac{1}{9}\right) \cdot \frac{1}{4} = 1 \cdot \frac{1}{4} = \frac{1}{4}$$

Review of Essentials **17**

---

38. 1. $x = a + (-b)$
   Hypothesis
2. $x + b = x + b$
   Refl. prop. of =
3. $x + b = [a + (-b)] + b$
   Subs. prin.
4. $= a + [(-b) + b]$
   Assoc. ax. for add.
5. $= a + 0$
   Ax. of add. inv.
6. $= a$
   Iden. ax. for add.
7. $x + b = a$
   Trans. prop. of =

---

1. a. $15 \cdot \dfrac{1}{5}$

$$= (3 \cdot 5) \cdot \left(\dfrac{1}{5}\right)$$

$$= 3 \cdot \left(5 \cdot \dfrac{1}{5}\right)$$

$$= 3 \cdot 1 = 3$$

b. $2 \cdot \dfrac{1}{12}$

$$= 2 \cdot \left(\dfrac{1}{2} \cdot \dfrac{1}{6}\right)$$

$$= \left(2 \cdot \dfrac{1}{2}\right) \cdot \dfrac{1}{6}$$

$$= 1 \cdot \dfrac{1}{6} = \dfrac{1}{6}$$

2. $6y(-2)$

$$= (-2)(6y)$$

$$= (-1)(2)(6y)$$

$$= (-1)(12y) = -12y$$

3. $(-5)(-9v)$

$$= (-1 \cdot 5)(-1 \cdot 9v)$$

$$= (-1)(-1)(5 \cdot 9)v$$

$$= (1 \cdot 45)v = 45v$$

4. $-4(s + (-2))$

$$= (-4)s + (-4)(-2)$$

$$= (-1)(4)s +$$
$$(-1)(-1)(4 \cdot 2)$$

$$= (-1)(4s) + (1)(8)$$

$$= -4s + 8$$

Suggested Assignments

Minimum
  20/1–25
Extended Alg. with Trig.
  20/7–27 odd
Enriched Alg.
  20/15–41 odd
Enriched Alg. with Trig.
  20/15–41 odd

Some real numbers, such as the identity element 1, have special properties with respect to multiplication. Two other such numbers are 0 and −1.

Multiplicative Property of Zero

For all real numbers a,

$$a \cdot 0 = 0 \qquad \text{and} \qquad 0 \cdot a = 0.$$

PROOF

Statements	Reasons
1. a is a real number.	Hypothesis
2. $0 + 0 = 0$	Identity axiom for addition
3. $a \cdot (0 + 0) = a \cdot 0$	Substitution principle
4. $a \cdot 0 + a \cdot 0 = a \cdot 0$	Distributive axiom
5. $a \cdot 0 = a \cdot 0 + 0$	Identity axiom for addition
6. $a \cdot 0 + a \cdot 0 = a \cdot 0 + 0$	Transitive property of equality
7. $a \cdot 0 = 0$	Cancellation property of addition
8. $0 \cdot a = 0$	Commutative axiom for multiplication

The next theorem leads to the familiar rules for multiplying additive inverses.

Multiplicative Property of −1

For all real numbers a,

$$a(-1) = -a \qquad \text{and} \qquad (-1)a = -a.$$

PROOF

Statements	Reasons
1. $(-1)a = (-1)a + 0$	Identity axiom for addition
2. $\quad = (-1)a + [a + (-a)]$	Axiom of additive inverses
3. $\quad = [(-1)a + a] + (-a)$	Associative axiom for addition
4. $\quad = [(-1)a + 1 \cdot a] + (-a)$	Identity axiom for multiplication
5. $\quad = (-1 + 1)a + (-a)$	Distributive axiom
6. $\quad = 0 \cdot a + (-a)$	Axiom of additive inverses

18 Chapter 1

7. $= 0 + (-a)$ Multiplicative property of zero
8. $= -a$ Identity axiom for addition
9. $(-1)a = -a$ Transitive property of equality

The second part of the theorem, $a(-1) = -a$, is proved in a similar way.

The following examples show how the preceding theorem can be used to simplify expressions for products involving negative numbers.

EXAMPLE 2 $5x(-3) = (-3)(5x)$
$\qquad\qquad\quad = (-1)(3)(5x)$
$\qquad\qquad\quad = (-1)(15x) = -15x$

EXAMPLE 3 $(-2)(-7y) = (-1 \cdot 2)(-1 \cdot 7)y$
$\qquad\qquad\qquad\quad = (-1)(-1)(2 \cdot 7)y$
$\qquad\qquad\qquad\quad = (1 \cdot 14)y = 14y$

EXAMPLE 4 $-3(z + (-2)) = (-3)z + (-3)(-2)$
$\qquad\qquad\qquad\qquad = (-1)(3)z + (-1)(-1)(3 \cdot 2)$
$\qquad\qquad\qquad\qquad = (-1)(3z) + (1)(6)$
$\qquad\qquad\qquad\qquad = -3z + 6$

These examples suggest how to deduce the following corollary of the multiplicative property of -1. (See Exercises 31–33 on page 21.)

Properties of Opposites in Products

For all real numbers a and b,

$$(-a)b = -ab, \qquad a(-b) = -ab, \qquad (-a)(-b) = ab.$$

Can you explain why the following statements are true?

A product of several nonzero real numbers of which an even number are negative is a positive number.
A product of several nonzero real numbers of which an odd number are negative is a negative number.

The fact that $(-1)(-1) = 1$ means that the reciprocal of -1 is -1; that is, $-1 = \dfrac{1}{-1}$. You can use this fact to show that for every nonzero real number a, *the reciprocal of the opposite of a is the opposite of the reciprocal of a.*

Review of Essentials   **19**

Additional A Exercises
Simplify.
1. $(2)(-3)(5)$ -30
2. $(-3)(-4)(2)$ 24
3. $(-2c)(-d)(7)$ $14cd$
4. $-3m[(-8) + 6t]$
   $24m - 18mt$
5. $-12\left[\left(-\dfrac{1}{4}\right) + \dfrac{1}{3}\right]$ -1
6. $-2q[-(2s + 3r)]$
   $4qs + 6qr$

Additional Answers
Written Exercises

26. 1. $ca = cb, c \ne 0$
       Hypothesis
    2. $ca = ac$
       $cb = bc$
       Comm. ax. for mult.
    3. $ac = bc$
       Trans. prop. of $=$
    4. $a = b$
       Theorem, Exercise 25

28. 1. $ab = a, a \ne 0$
       Hypothesis
    2. $a = a \cdot 1$
       Iden. ax. for mult.
    3. $ab = a \cdot 1$
       Trans. prop. of $=$
    4. $b = 1$
       Canc. prop. of mult.

30. 1. $a \ne 0$
       Hypothesis
    2. $\dfrac{1}{a} \cdot a = 1$
       Ax. of mult. inv.
    3. $a = \dfrac{1}{\dfrac{1}{a}}$
       Exercise 29
    4. $\dfrac{1}{\dfrac{1}{a}} = a$
       Symm. prop. of $=$

(continued)

Thus, we have the following:

$$\frac{1}{-a} = \frac{1}{(-1)a} = \frac{1}{-1} \cdot \frac{1}{a} = -1 \cdot \frac{1}{a} = -\frac{1}{a}$$

Additional Answers
Written Exercises
(continued)

32. 1. $a(-b) = a[b(-1)]$
   Mult. prop. of -1
   2. $= (ab)(-1)$
   Assoc. ax. of mult.
   3. $= -ab$
   Mult. prop. of -1
   4. $a(-b) = -ab$
   Trans. prop. of $=$

34. 1. $a = b$
   $a \neq 0, b \neq 0$
   Hypothesis
   2. $a \cdot \frac{1}{a} = 1, b \cdot \frac{1}{b} = 1$
   Ax. of mult. inv.
   3. $a \cdot \frac{1}{a} = b \cdot \frac{1}{b}$
   Trans. prop. of $=$
   4. $a \cdot \frac{1}{a} = a \cdot \frac{1}{b}$
   Subs. prin.
   5. $\frac{1}{a} = \frac{1}{b}$
   Canc. prop. of mult.

36. 1. $xb = a, b \neq 0$
   Hypothesis
   2. $xb\left(\frac{1}{b}\right) = a\left(\frac{1}{b}\right)$
   Subs. prin.
   3. $x\left[b\left(\frac{1}{b}\right)\right] = a\left(\frac{1}{b}\right)$
   Assoc. prop. of mult.
   4. $x \cdot 1 = a\left(\frac{1}{b}\right)$
   Ax. of mult. inv.
   5. $x = a\left(\frac{1}{b}\right)$
   Iden. ax. of mult.

Oral Exercises

State the axiom, theorem, or property of multiplication that justifies each sentence.

1. $\frac{1}{2} \cdot \frac{1}{x} = \frac{1}{2x}$ Recip. of a prod. prop.

2. $3[-(2c)] = -(6c)$ Opp. in prod. prop.

3. $(-5)(-k) = 5k$ Opp. in prod. prop.

4. Mult. prop. of zero

4. If $y = -3$, then $(y + 3)(y - 5) = 0$.

5. If $4x = 28$, then $x = 7$. Canc. prop. of mult.

6. $(-n)(9m) = -(9nm)$
   Opp. in prod. prop., Comm. ax. for mult.

7. $8a \cdot \frac{1}{8} = a$ Theorem, p. 17

8. If $r \neq 0$ and $ar = br$, then $a = b$.
   Canc. prop. of mult.

9. $(-1)(c + d) = -(c + d)$
   Mult. prop. of -1

10. If $z \neq -2$, then $(z + 2) \cdot \frac{1}{z + 2} = 1$.
   Ax. of mult. inv.

Written Exercises

Simplify.

A

1. $(-5)(7)(-12)$ 420

2. $\left(-\frac{1}{6}\right)(-22)(-15)$ -55

3. $-5(x)(-y)$ 5xy $\frac{19}{6}$

4. $(-3a)(-b)(8)$ 24ab

5. $-\frac{1}{3}[(-5) + 23]$ -6

6. $-6\left[\left(-\frac{1}{4}\right) + \left(-\frac{5}{18}\right)\right]$

7. $-5p[(-6) + 8q]$
   $30p + (-40)pq$

8. $-[-(r + 9s)](-2)$
   $-2r + (-18)s$

9. $-4c[-(a + b)]$
   4ac + 4bc

10. $(2x)\left[(-9) + \frac{1}{2}(-[y + 5])\right]$ $-23x + (-xy)$

11. $6h[(-3) + 11k](-2)$ 36h + (−132)hk

12. $\frac{1}{2}\left[(-8w) + \left(-\frac{1}{3}\right)(12w + (-9))\right]$ $-6w + \frac{3}{2}$

13. $-\frac{1}{4}[(-7c)(2d + (-6)) + (-10cd)]$
   6 cd + $(-\frac{21}{2})$c

14. $-6a\left[(-5)\left(\frac{1}{2} + (-2b)\right) + (-10b)\right]$ 15a

15. $-[4x + (-3y)] + (-5 + x)(-2y)$
   $-4x + 13y + (-2)xy$

16. $-[-(7mn + 5m) + (-3n)(m + 2)]$
   10mn + 5m + 6n

17. $\frac{1}{3}[6ab + 9a] + (-5a + 20b)(-2)$
   2ab + 13a + (−40)b

Without computing, state whether the value of each expression is positive, negative, or zero.

18. $(-5.9)(8)(6)(-6.2)$ positive

19. $(21)(-7)(-8)(5)$ positive

20. $(-5)[8 + (-8)](-9)$ zero

21. $\left(\frac{-2}{3} + \frac{7}{-6}\right)(-4)(-1)$ negative

22. $\left(7 + \frac{3}{2}\right)(-3)(-11)$ positive

23. $(-8)(-10)(0)(-6)(2)$ zero

20 *Chapter 1*

State the axiom or theorem that justifies each step of the following proofs.

B **24.** Prove: For all real numbers b and all nonzero real numbers c, $bc\left(\frac{1}{c}\right) = b$.

    1. $bc\left(\frac{1}{c}\right) = b\left(c \cdot \frac{1}{c}\right)$ Associative axiom for multiplication

    2. $\quad\quad = b \cdot 1$ Axiom of multiplicative inverses

    3. $\quad\quad = b$ Identity axiom for multiplication

    4. $bc\left(\frac{1}{c}\right) = b$ Transitive property of equality

25. Prove: For all real numbers a, b, c, $c \neq 0$, if $ac = bc$, then $a = b$.

    1. $ac = bc$ Hypothesis

    2. $\frac{1}{c}$ is a real number. Axiom of multiplicative inverses

    3. $(ac)\left(\frac{1}{c}\right) = (bc)\left(\frac{1}{c}\right)$ Substitution principle

    4. $a\left(c \cdot \frac{1}{c}\right) = b\left(c \cdot \frac{1}{c}\right)$ Associative axiom for multiplication

    5. $a \cdot 1 = b \cdot 1$ Axiom of multiplicative inverses

    6. $\quad a = b$ Identity axiom for multiplication

Prove the following theorems. Assume that each variable represents a real number. You may use the results of a previous exercise in any proof.

26. If $ca = cb$ and $c \neq 0$, then $a = b$.

27. If $a \neq 0$, $b \neq 0$, then $\frac{1}{ab} = \frac{1}{a} \cdot \frac{1}{b}$.

28. If $ab = a$, and $a \neq 0$, then $b = 1$.

    (Uniqueness of multiplicative identity)

29. If $ab = 1$, then $b = \frac{1}{a}$.

    (Uniqueness of multiplicative inverse)

30. If $a \neq 0$, then $\frac{1}{\frac{1}{a}} = a$.

31. $(-a)b = -ab$

32. $a(-b) = -ab$

33. $(-a)(-b) = ab$

34. If $a = b$ and $a \neq 0$, $b \neq 0$, then $\frac{1}{a} = \frac{1}{b}$.

35. If $ab = 0$ and $b \neq 0$, then $a = 0$.

C **36.** If $xb = a$ and $b \neq 0$, then $x = a \cdot \frac{1}{b}$.

37. If $x = a \cdot \frac{1}{b}$ and $b \neq 0$, then $bx = a$.

38. If $ax + b = c$ and $a \neq 0$, then $x = \frac{1}{a}[c + (-b)]$.

39. If $\frac{1}{x} = a\left(\frac{1}{b}\right)$ and a, b, and $x \neq 0$, then $x = b\left(\frac{1}{a}\right)$.

40. If $\frac{1}{x} = ab$ and a, b, and $x \neq 0$, then $x = \frac{1}{ab}$.

41. If $a\left(\frac{1}{x}\right) = b$ and a, b, and $x \neq 0$, then $x = a\left(\frac{1}{b}\right)$.

38. 1. $ax + b = c$, $a \neq 0$
    Hypothesis
2. $ax + b = c + 0$
    Iden. ax. of add.
3. $\quad\quad = c + [(-b) + b]$
    Ax. of add. inv.
4. $\quad\quad = [c + (-b)] + b$
    Assoc. ax. of add.
5. $ax + b = [c + (-b)] + b$
    Trans. prop. of $=$
6. $ax = c + (-b)$
    Canc. prop. of add.
7. $xa = c + (-b)$
    Comm. prop. for mult.
8. $x = [c + (-b)]\frac{1}{a}$
    Exercise 36
9. $x = \frac{1}{a}[c + (-b)]$
    Comm. ax. for mult.

40. 1. $\frac{1}{x} = ab$
    Hypothesis
2. $\frac{1}{\frac{1}{x}} = \frac{1}{ab}$
    Exercise 34
3. $x = \frac{1}{\frac{1}{x}}$
    Exercise 30
4. $x = \frac{1}{ab}$
    Trans. prop. of $=$

Simplify.

1. $2c + (4 - c) - 24$ $c - 20$

2. $-9 + (-3.2 - 8)$ -20.2

3. $15 - [-(6 - 7)] - 2(5 - 4)$ 12

4. Graph the set of all negative odd integers

$-6\ -5\ -4\ -3\ -2\ -1\ \ 0\ \ 1\ \ 2\ \ 3\ \ 4$

Solve each open sentence.

5. $3(g - 4) = 9$ $\{7\}$

6. $k - 4 = -k + 16$ $\{10\}$

Teaching Suggestions
p. T77

Key Ideas

Define the relationship between addition and subtraction. Prove and use subtraction theorems.

Chalkboard Examples

Simplify.

1. $-13 - (-4)$
$\quad = -13 + 4 = -9$

2. $-4 - (-13)$
$\quad = -4 + 13 = 9$

Computer Exercises   For students with computer experience

1. Write a program that will tell whether the product of two real numbers is positive, negative, or zero *without actually multiplying* the numbers.

Run the program in Exercise 1 for each pair of numbers.

2. $5, -3$ negative

3. $4, 0$ zero

4. $10, 3$ positive

5. $-7, -5$ positive

6. $-6, 7$ negative

7. $0, -9$ zero

8. Modify the program in Exercise 1 to tell whether the product of three real numbers is positive, negative, or zero *without actually multiplying* the numbers.

Run the program in Exercise 8 for each set of numbers.

9. $3, 4, 7$ positive

10. $-2, -5, -11$ negative

11. $-6, 12, 17$ negative

12. $9, -16, -3$ positive

13. $-1, 0, 15$ zero

14. $-10, -4, 0$ zero

1–6 Properties of Differences

Two other operations are defined in terms of the basic operations of addition and multiplication in $\mathcal{R}$. In this section, we shall consider the first of these. The **difference** between a and b, $a - b$, is defined in terms of addition.

Definition of Subtraction

For all real numbers a and b,

$$a - b = a + (-b).$$

From this definition we have the rule: To subtract b from a, add the opposite of b to a. For example,

$$-4 - (-5) = -4 + 5 = 1$$

and
$$7 - 15 = 7 + (-15) = -8.$$

Since
$$\begin{aligned} b + (a - b) &= (a - b) + b \\ &= [a + (-b)] + b \\ &= a + [(-b) + b] \\ &= a + 0 \\ &= a \end{aligned}$$

22   *Chapter 1*

you can see that $a - b$ is the number which when added to b produces a.

Because $\mathscr{R}$ is closed under addition, and every real number has an additive inverse, it follows from the definition of subtraction that $\mathscr{R}$ is also closed under subtraction. If you notice that

$$8 - 5 = 3 \quad \text{whereas} \quad 5 - 8 = -3$$

you can see that subtraction in $\mathscr{R}$ is not commutative. The fact that

$$(6 - 2) - 9 = 4 - 9 = -5$$

but that

$$6 - (2 - 9) = 6 - (-7) = 13$$

demonstrates that subtraction in $\mathscr{R}$ is not associative.

Knowing the relationship between addition and subtraction, you can prove several theorems about subtraction. For example, Exercise 19 on page 24 outlines a proof of the fact that *multiplication is distributive with respect to subtraction*. Thus, for each real number m,

$$3(m - 4) = 3 \cdot m - 3 \cdot 4 = 3m - 12.$$

Oral Exercises

Simplify each expression, using subtraction whenever possible.

1. $(-a) + b + (-c)$ b − a − c
2. $x + [-(y + z)]$ x − y − z
3. $(-p)[(-q) + r]$ pq − pr
4. $m[(-k) + (-n)]$ −mk − mn
5. $c - [(-d) + (-e)]$ c + d + e
6. $[-(-w) + (-z)](t)$ wt − zt

Is the statement true for all real numbers? If not, give an example that shows this. Let a = 3, b = 2, c = 1

7. $3 - (2 - 1) = 2; (3 - 2) - 1 = 0$
7. $a - (b - c) = (a - b) - c$
8. $3 - 2 = 1; 2 - 3 = -1$
8. $a - b = b - a$
9. $a - (b + c) = (a - b) - c$ true
10. $a - (b - c) = (a - b) + c$ true

Written Exercises

Simplify.

A

1. $-2(x - y)$ −2x + 2y
2. $(a - b + c)(-5)$ 5b − 5a − 5c
3. $2m(p - (q + 3))$ 2mp − 2mq − 6m
4. $-7e[d - (f - 5)]$ 7ef − 7ed − 35e
5. $[r - (5s - 4t)](-12)$ 60s − 12r − 48t
6. $-6n(k - 9)(-11)$ 66nk − 594n
7. $[b - a(6 + c)](-3)$ 18a + 3ac − 3b
8. $4m[9 - 5(3 - 7n)]$ 140mn − 24m
9. $-[7(3 - w) - 6(8 - z)]$ 27 + 7w − 6z
10. $-3x[4(-y) + 7(3 - y)] - 5xy$
11. $-4(f + g) - 3[5 - (2f + 6g)]$ 2f + 14g − 15
10. $28xy - 63x$
12. $-2[-3p - 6q(p - 5)]$ 6p + 12qp − 60q

Review of Essentials   **23**

---

3. $-12 - (6 - 8)$
$= -12 - (-2)$
$= -12 + 2$
$= -10$

4. $(-12 - 6) - 8$
$= -18 - 8 = -26$

5. $2m(n - 7)$
$2m(n - 7) = 2mn - 14m$

Suggested Assignments

Minimum
  23/1–24
S 6/1–10 even
Extended Alg. with Trig.
  23/1–29 odd
S 6/4, 6, 8
Enriched Alg.
  23/9–33 odd
Enriched Alg. with Trig.
  23/9–33 odd

Additional A Exercises

Simplify.

1. $-4(p - q)$
$-4p + 4q$

2. $(a - b - c)(-3)$
$-3a + 3b + 3c$

3. $-4f[g - (j - 6)]$
$-4fg + 4fj - 24f$

4. $3a[b - (c + 2)]$
$3ab - 3ac - 6a$

5. $[e - (3d - 5f)](-15)$
$-15e + 45d - 75f$

6. $-8m(p - 7)(-12)$
$96mp - 672$

23

19. 1. Hypothesis
2. Def. of subt.
3. Subs. prin.
4. Dist. ax.
5. Prop. of opp. in prod.
6. Def. of subt.
7. Trans. prop. of =

20. 1. Hypothesis
2. Ax. of add. inv.
3. Def. of subt.
4. Canc. prop. of add.
5. Subs. prin.

22. 1. $a, b, c, \in \mathcal{R}$
Hypothesis
2. $-a(b + c) = -[a(b + c)]$
Prop. of opp. in prod.
3. $= -(ab + ac)$
Dist. ax.
4. $= (-ab) + (-ac)$
Prop. of opp. of sum
5. $= -ab - ac$
Def. of subt.
6. $-a(b + c) = -ab - ac$
Trans. prop. of =

(continued on p. 30)

Mixed Review

Simplify.

1. $((m)^3(m))^2$
m^8

2. $(p + 3)(p - 1)$
$p^2 + 2p - 3$

3. $5 + [-5(2 - 4)] - 2(-10)$
35

4. Graph the odd integers greater than or equal to 3.

Solve each open sentence.

5. $2w + 4w + 1 = 4w + 13$
$\{6\}$

6. $2v - 4 = 2(v + 1)$
$\varnothing$

Evaluate each expression for the given value of the variable.

13. $-3 - 5(x - 2); x = 1$ 2
14. $-2(7 - k) - (3)(-5); k = 4$ 9
15. $-2y[3 - (6 - y)]; y = -2$ -20
16. $[4 - (9 + z)](-3z); z = -3$ -18
17. $(n + 7)[-5 - 6(4 + n)]; n = -8$ -19
18. $(8 - r)[-6r - 4(7 - r)]; r = 2$
-192

Justify each step in the proofs of the following theorems. Assume that each variable represents a real number.

B 19. $a(b - c) = ab - ac$ 20. $a - (-b) = a + b$

PROOF	PROOF
1. $a, b,$ and c are real numbers.	1. a and b are real numbers.
2. $b - c = b + (-c)$	2. $-b$ is a real number.
3. $a(b - c) = a[b + (-c)]$	3. $a - (-b) = a + [-(-b)]$
4. $= ab + a(-c)$	4. $-(-b) = b$
5. $= ab + (-ac)$	5. $a - (-b) = a + b$
6. $= ab - ac$	
7. $a(b - c) = ab - ac$	

Prove each theorem. Assume that each variable represents a real number. You may use the results of a previous exercise in any proof.

21. $(a - b) + b = a$
22. $-a(b + c) = -ab - ac$
23. $-a(b - c) = ac - ab$
24. If $a - c = b - c$, then $a = b$.
25. If $c - a = c - b$, then $a = b$.
26. If $a = b$, then $a - c = b - c$.
27. If $a = b$, then $c - a = c - b$.
28. If $a - b = 0$, then $a = b$.
29. If $x - b = a$, then $x = a + b$.
30. If $x = a + b$, then $x - b = a$.

C 31. $(a - b)(c + d) = ac - bc + ad - bd$
32. $(a - b)(c - d) = ac - bc - ad + bd$
33. $a[b - (c + d)] = ab - ac - ad$
34. $a[b - (c - d)] = ab - ac + ad$

1–7 Properties of Quotients

Division is defined in terms of multiplication. The **quotient** of a and b, that is $a \div b$, or $\frac{a}{b}$, where $b \neq 0$, is defined as follows.

Definition of Division

For all real numbers a and all nonzero real numbers b,

$$\frac{a}{b} = a \cdot \frac{1}{b}.$$

---

24   Chapter 1

For example,

$$-\frac{32}{8} = (-32)\left(\frac{1}{8}\right) = (-4 \cdot 8)\left(\frac{1}{8}\right) = -4\left(8 \cdot \frac{1}{8}\right) = -4(1) = -4$$

and

$$15 \div \left(-\frac{1}{3}\right) = 15 \cdot (-3) = -45.$$

Since

$$(a \div b) \cdot b = \left(a \cdot \frac{1}{b}\right) \cdot b$$
$$= a \cdot \left(\frac{1}{b} \cdot b\right)$$
$$= a \cdot 1 = a,$$

you can see that $a \div b$ is the number which when multiplied by b produces a.

Notice that *division by 0* is *not defined*. This is because:

1. If $a \neq 0$, $\frac{a}{0} = c$ would imply that $0 \cdot c = a$, but $0 \cdot c = 0$ for every real number c (page 18).

2. If $a = 0$, $\frac{a}{0} = c$ would not be unique, since $0 \cdot c = 0$ for every real number c.

Because $\mathcal{R}$ is closed under multiplication and $a \div b = a \cdot \frac{1}{b}$, the set $\mathcal{R}$ is closed with respect to division, excluding division by zero.

Knowing the relationship between multiplication and division can help to prove some of the properties of division. For example, Exercise 21 on page 27 outlines a proof that division is distributive over addition from the right. The theorem could have been written $(a + b) \div c = (a \div c) + (b \div c)$. Since division is not commutative, $(a + b) \div c \neq c \div (a + b)$; thus division is not distributive over addition from the left. Exercises 22–35 on pages 27–28 prove additional properties of division.

You will notice that grouping symbols are frequently used when more than one operation is used in an expression. What is the value of an expression like $13 - 2 \div 18 \cdot 2 \div 4$ that has no grouping symbols to show the order in which operations are to be performed? To avoid ambiguity, mathematicians have agreed on the following steps to simplify expressions when grouping symbols have been omitted.

Order of Operations

1. Simplify powers.
2. Then simplify products and quotients in order from left to right.
3. Then simplify sums and differences in order from left to right.

Teaching Suggestions
p. T77

Related Activities
p. T77

Supplementary Material
Test 2

Key Ideas
Define the relationship between multiplication and division.
Prove and use properties of division.

Chalkboard Examples
Simplify.

1. $9 - 14 + 3 \cdot 10 \div 6$
$9 - 14 + 3 \cdot 10 \div 6$
$= 9 - 14 + 30 \div 6$
$= 9 - 14 + 5$
$= -5 + 5 = 0$

2. $\frac{1}{2}[-8 + 3(8 \div 4)]$

$\frac{1}{2}[-8 + 3(8 \div 4)]$

$= \frac{1}{2}[-8 + 3(2)]$

$= \frac{1}{2}[-8 + 6]$

$= \frac{1}{2}(-2) = -1$

3. $2d + 6d \div 3d + 2 \cdot 4$
$2d + 6d \div 3d + 2 \cdot 4$
$= 2d + 2 + 8 = 2d + 10$

When grouping symbols have been used, apply these rules within each grouping symbol beginning with the innermost grouping symbol, and work out to the entire expression.

EXAMPLE 1   Simplify $13 - 2 + 18 \cdot 2 \div 4$.

SOLUTION $13 - 2 + 18 \cdot 2 \div 4 = 13 - 2 + 36 \div 4$
$$= 13 - 2 + 9 = 11 + 9 = 20$$

EXAMPLE 2   Simplify $2\left[-5 - \frac{1}{3}(17 + 4)\right]$.

SOLUTION $2\left[-5 - \frac{1}{3}(17 + 4)\right] = 2\left[-5 - \frac{1}{3}(21)\right]$
$$= 2(-5 - 7) = 2(-12)$$
$$= -24$$

EXAMPLE 3   Simplify $9x + 6x \div 2x - 8 \cdot 3$.

SOLUTION $9x + 6x \div 2x - 8 \cdot 3 = 9x + 3 - 24 = 9x - 21$

Note that in Example 3 grouping symbols were not used for $6x$ and $2x$, but that each was considered as a single quantity.

Oral Exercises

a. **State the value of each expression as printed.**
b. **State where to put grouping symbols so that the given expression has the value in red.**

1. $7 + 3 \cdot 5$; 50 22

2. $24 \div 2 \cdot (3 + 4)$; 84 40

3. $(5 \cdot 2 + 16) \div 2$; 13 18

4. $36 \div (3 \cdot 4) + 5$; 8   53

5. $18 \div \left(\frac{2}{3} + \frac{4}{3}\right)$ 9 $\frac{85}{3}$

6. $6 + \left(1 - \frac{1}{5}\right) \div \frac{4}{5}$; 7 $\frac{27}{4}$

7. $3 \cdot (2 + x)$ 6 + 3x   6 + x

8. $(y - 12) \div 4$; $\frac{y - 12}{4}$ y - 3

Written Exercises

Simplify.

A   1. $(-25 - 14) \div 13$ -3

2. $-42 \div (-15 + 8)$ 6

3. $8x + (-24) \div 4$ 8x - 6

4. $(-10)(9y) \div (-7 + 1)$ 15y

5. $\left(\frac{1}{3}k - 5k\right) \div \frac{7}{3}$ -2k

6. $\left(-\frac{4}{5}\right) \div \left(-\frac{2}{5}c + 6c\right)$ $-\frac{1}{7c}$

7. $\left(-\frac{1}{2}m + 8 \cdot 2m\right) \div \left(\frac{3}{4} - 1\right)$ -62m

8. $-5p + \left(\frac{1}{2} \div \frac{1}{12}\right)p - 9p$ -8p

26   *Chapter 1*

Simplify.

9. $5 \cdot 7 - 16 \div \frac{n}{4} + (-3)(2)$ $29 - \frac{64}{n}$

10. $(-20a)(-9b) \div \left[\frac{1}{2}(-2 - 11)\right]$ $-\frac{360ab}{13}$

11. $\left(-\frac{1}{5}\right)\left(33r - \frac{1}{3}r\right) \div \left(-\frac{1}{4}\right)$ $\frac{392r}{15}$

12. $-\frac{4}{7} \div \left[\frac{1}{2}\left(-\frac{v}{4} + \frac{v}{3}\right)\right]$ $-\frac{96}{7v}$

In Exercises 13–20:
a. **Evaluate both sides of each equation when the variable has the value in red.**
b. **State whether the equation is true for this value.**

18. $3 \neq -3$; false
20. $3 = 3$; true

13. $\frac{9 + x}{2} - 5 = \frac{x - 7}{5}$; -3 $-2 = -2$; true

14. $\frac{y + 3}{7} = 1 - \frac{y + 10}{7}$;   11 $2 \neq -2$; false

15. $a(a - 5) + \frac{a - 10}{a - 1} = -9a$; -2 $18 = 18$; true

16. $\frac{a - 6}{5} - \frac{1 - a}{5} = \frac{1}{a + 1}$;   4 $\frac{1}{5} = \frac{1}{5}$; true

17. $\left(\frac{1}{b} - 2\right)\left(\frac{b}{5} - 9\right) = 3b + 2$;   3 $14 \neq 11$; false

18. $\frac{2w + 1}{-3} = \left(-\frac{2}{w} - 4\right)\left(-\frac{w}{6}\right)$; -5

19. $z\left(\frac{1}{6} - z\right) + \frac{z - 2}{3} = \frac{1}{3}(z - 1)$; 2 $-\frac{11}{3} \neq \frac{1}{3}$; false

20. $\left(\frac{2}{d} - 3\right)\left(\frac{d}{5} - 2\right) = 11 - 2d$;   4

Justify each step in the proofs of the following theorems. Assume that each variable represents a real number.

B   21. $\frac{a + b}{c} = \frac{a}{c} + \frac{b}{c}$ $(c \neq 0)$

22. $\frac{-a}{b} = -\frac{a}{b}$ $(b \neq 0)$

PROOF

1. a, b, and c are real numbers;   Hypothesis
   $c \neq 0$.

2. $\frac{a + b}{c} = (a + b) \cdot \frac{1}{c}$ Def. of div.

3. $\phantom{\frac{a+b}{c}} = a \cdot \frac{1}{c} + b \cdot \frac{1}{c}$ Dist. ax.

4. $\phantom{\frac{a+b}{c}} = \frac{a}{c} + \frac{b}{c}$ Def. of div.

5. $\frac{a + b}{c} = \frac{a}{c} + \frac{b}{c}$ Trans. prop. of $=$

PROOF

1. a and b are real numbers;   Hypothesis
   $b \neq 0$.

2. $\frac{-a}{b} = -a\left(\frac{1}{b}\right)$ Def. of div.

3. $\phantom{\frac{-a}{b}} = -\left(a \cdot \frac{1}{b}\right)$ Prop. of opp. in prod.

4. $\phantom{\frac{-a}{b}} = -\frac{a}{b}$ Def. of div.

5. $\frac{-a}{b} = -\frac{a}{b}$ Trans. prop. of $=$

Prove each of the following theorems. Assume that each variable represents a real number. You may use the result of a previous exercise in any proof.

23. $\frac{a}{a} = 1$ $(a \neq 0)$

24. $\frac{-a}{a} = -1$ $(a \neq 0)$

25. $(ab) \div b = a$ $(b \neq 0)$

26. If $ax = b$ and $a \neq 0$, then $x = \frac{b}{a}$.

27. If $x = \frac{b}{a}$ and $a \neq 0$, then $ax = b$.

28. $\frac{ac}{bc} = \frac{a}{b}$ $(b, c \neq 0)$

Review of Essentials   **27**

Prove each of the following theorems. Assume each variable denotes a real number.

1. $-[c - d] = d - c$

   1. c and d are real numbers.   Hypothesis
   2. $-[c - d] = -[c + (-d)]$ Def. of subt.
   3. $-[c + (-d)] = -(c) + [-(-d)]$ Prop. of opp. of a sum
   4. $-(c) + [-(-d)] = -c + d$ Canc. prop. of add. inv.
   5. $-c + d = d + (-c)$ Comm. ax. for add.
   6. $d + (-c) = d - c$ Def. of subt.
   7. $-[c - d] = d - c$ Trans. prop. of $=$

2. $\frac{a}{b} \cdot \frac{b}{a} = 1$, $a \neq 0$, $b \neq 0$

   1. a and b are real numbers and $a \neq 0$, $b \neq 0$.   Hypothesis
   2. $\frac{a}{b} \cdot \frac{b}{a} = \left(a \cdot \frac{1}{b}\right) \cdot \left(b \cdot \frac{1}{a}\right)$ Definition of division
   3. $\left(a \cdot \frac{1}{b}\right) \cdot \left(b \cdot \frac{1}{a}\right) = \left(a \cdot \frac{1}{b}\right) \cdot \left(\frac{1}{a} \cdot b\right)$ Comm. ax. for $\times$
   4. $\left(a \cdot \frac{1}{b}\right) \cdot \left(\frac{1}{a} \cdot b\right) = \left(a \cdot \frac{1}{a}\right) \cdot \left(\frac{1}{b} \cdot b\right)$ Comm. and assoc. ax. for $\times$.
   5. $\left(a \cdot \frac{1}{a}\right) \cdot \left(\frac{1}{b} \cdot b\right) = 1 \cdot 1 = 1$ Ax. of mult. inv. and iden. ax. for mult.
   6. $\frac{a}{b} \cdot \frac{b}{a} = 1$ Subst.

(continued)

(continued)

Simplify

3. $-[2(4 - 1) + (7 - 9)] - 3$
  -7

4. $(-5y - (-3))(-4 + 7)$
  $-15y + 9$

5. $3x(-2) \div 6 - 7 \quad -x - 7$

6. $-(y + 4) - (y + 2)$
  $-2y - 6$

Reading Algebra

Students will read the expression $12 \div 4 + 7 \cdot 8 - 11$ as "twelve divided by four plus seven times eight minus eleven." Point out that this reading is correct, but it does not convey the order in which the operations are to be performed. Suggest that placing parentheses around the numbers that are to be multiplied and/or divided will serve as a reminder that these operations are to be performed first. Students should think of this expression as first, divide 12 by 4; second, multiply 7 by 8; then add and subtract as indicated.

Mixed Review

Simplify.

1. $(h)^3(n)^2(n) \quad h^3n^3$

2. $2(n - 1)^2 \quad 2n^2 - 4n + 2$

3. $2 - [-6(2 - 14)] + 2(-10) \quad -90$

4. $\left(\dfrac{1}{b}\right)\left(\dfrac{1}{b}\right)\left(\dfrac{1}{b}\right) \quad \dfrac{1}{b^3}$

5. Graph the set $\{\ldots -6, -4, -2, 0\}$

6. Solve $2.5z = 10 - 2.5z$. $\{2\}$

Prove each of the following theorems. Assume that each variable represents a real number. You may use the result of a previous exercise in any proof.

29. $\dfrac{a}{b} \cdot \dfrac{c}{d} = \dfrac{ac}{bd} \quad (b, d \neq 0)$

30. $a + \dfrac{c}{b} = \dfrac{ab + c}{b} \quad (b \neq 0)$ (See Exercises 21 and 25.)

C **31.** $\dfrac{1}{a} + \dfrac{1}{b} = \dfrac{b + a}{ab} \quad (a, b \neq 0)$ (See Exercise 28.)

32. $\dfrac{a}{b} + \dfrac{c}{d} = \dfrac{ad + bc}{bd} \quad (b, d \neq 0)$

33. $\dfrac{a}{b} \cdot \dfrac{b}{a} = 1 \quad (a, b \neq 0)$

34. $\dfrac{1}{\frac{a}{b}} = \dfrac{b}{a} \quad (a, b \neq 0)$ (See Exercise 33.)

35. $\dfrac{\frac{a}{b}}{\frac{c}{d}} = \dfrac{ad}{bc} \quad (b, c, d \neq 0)$ (See Exercise 34.)

ON THE CALCULATOR

To determine if a calculator can follow the order of operations used in simplifying expressions, enter the expression below exactly as it is written:

$$18 + 6 \div 2$$

The correct answer is 21. If the calculator displays 21, it operated algebraically and performed division before addition. If the calculator displays 12, it performed the operations in the exact order in which the keys were pressed. This calculator can perform accurately but you must enter the data in the correct order. You can obtain the correct answer by following the order of operations. In the expression above divide 6 by 2 and then add 18.

Exercises

Use a calculator to simplify each expression.

1. $35 - 4.1 \times 0.05$ 34.795

2. $0.25 + 2.5 \div 0.5$ 5.25

3. $20.02 + 0.2 - 0.199 \times 10$ 18.23

4. $0.027 \times 0.52 - 2 + 0.101$

5. $141 \div 3 + 16 \times 110 + 3$ 1810

6. $-18 \times 4 - 16 \div (-4) + 7$

4. -1.88496 6. -61

28   *Chapter 1*

Self-Test 2

VOCABULARY
theorem (p. 12)
hypothesis or premise (p. 12)
conclusion (p. 12)
direct proof (p. 12)
corollary (p. 13)
difference (p. 22)
quotient (p. 24)

Prove each of the following theorems. Assume each variable denotes a real number.

1. $-[(-a) + (-b)] = a + b$ *Obj. 1, p. 12*

2. $b \cdot \dfrac{1}{ab} = \dfrac{1}{a}, \ a \neq 0, \ b \neq 0$ *Obj. 2, p. 12*

Simplify.

3. $-[3(9 - 7) + (8 - 11)] - 8 \ -11$ 4. $(-6n - (-5))(-2 + 6)$ *Obj. 3, p. 12*

5. $2b(-3) \div 6 + 9 \ 9 - b$ 6. $-(x + 2) - (-3 - x) \ 1$

                                                                          4. $20 - 24n$

Check your answers with those at the back of the book.

Chapter Summary

1. Some familiar set symbols include { } (indicates a set), ∈ (indicates a member of a set), ⊂ (indicates a subset), Ø (indicates the empty set), ∪ (indicates union), and ∩ (indicates intersection).

2. A *variable* is a symbol that represents any of the members of a specified set called the *domain* of the variable. A sentence that contains one or more variables is called an *open sentence*. The values of the variable that make an open sentence true are called the *solutions* of the open sentence. The *solution set* is the set of all solutions that make the open sentence true.

3. *Axioms*, or *postulates*, are statements that are assumed to be true. These statements are used to prove *theorems*. The "if-clause" of a theorem is called the *hypothesis* and the "then-clause" of a theorem is called the *conclusion*. By reasoning from the hypothesis to the conclusion, you can give a *direct proof* of a theorem. Every step in the proof of a theorem can be justified by an axiom, a definition, a given fact, or a theorem that has been previously proved.

4. The axioms for real numbers (page 7) determine the properties of the system of real numbers.

5. Equality among real numbers is *reflexive*, *symmetric*, and *transitive*.

Review of Essentials **29**

Chapter Review

Write the letter of the correct answer.

In Exercises 1–4, indicate whether the statement is (a) true or (b) false.

1-1

1. $-2 \subset \{-3, -2, -1, 0\}$ b

2. $\{5, 10, 15\} = \{15, 5, 10\}$ a

3. $\emptyset \subset \{\text{even integers}\}$ a

4. $0 \notin \{\text{nonpositive integers}\}$ b

Solve each open sentence over $\{-5, -3, -1, 1, 3, 5\}$.

1-2

5. $2t = 10$
   a. $\{-3\}$ **b.** $\{-5\}$ **c.** $\{-5, 5\}$ **d.** $\{5\}$

6. $5x = 2x + 3x$
   a. $\{-1\}$ **b.** $\{1, 3, 5\}$ **c.** $\emptyset$ **d.** $\{-5, -3, -1, 1, 3, 5\}$

7. Which of the following is the graph of the set of all nonnegative integers?

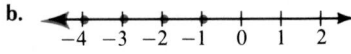

Choose the axiom that justifies the given statement.

1-3

8. $2 + (3 + 4) = (3 + 4) + 2$
   a. Associative axiom for addition
   c. Closure axiom for addition
   b. Distributive axiom
   d. Commutative axiom for addition

9. $(7)\left(\frac{1}{7}\right) = 1$
   a. Distributive axiom
   c. Axiom of multiplicative inverses
   b. Identity axiom for multiplication
   d. Commutative axiom for multiplication

Solve.

1-4

10. $4x + (-2) = 5 + 3x$
   a. $\{0\}$ **b.** $\{1\}$ **c.** $\{7\}$ **d.** $\{-2\}$

11. $-(-m + 3) = 2m + (-6)$
   a. $\{3\}$ **b.** $\emptyset$ **c.** $\{2\}$ **d.** $\{9\}$

Simplify.

1-5

12. $-2(z + (-3))$
   a. $2z - 6$ **b.** $2z + 6$ **c.** $-2z + 6$ **d.** $-2z - 6$

13. $(-4)(-2)(-1)(3)$
   a. -12 **b.** -24 **c.** 21 **d.** 24

30   *Chapter 1*

14. Evaluate $(-5 + r)(-6r + 2(1 + r))$ for $r = -1$. \hfill *1-6*
a. -60 b. 0 (c.) -36 d. 24

Simplify.

15. $3x(y - (z + 2))$
a. $3xy - z - 2$ (b.) $3xy - 3xz - 6x$
c. $3xy - 3xz + 2$ d. $3xy - 3xz - 2$

16. $16 - 12 \div 4 + 2 \cdot (-1)$ \hfill *1-7*
(a.) 11 b. -11 c. -14 d. 14

17. $4x + 8x \div 2x - 6x \cdot (-2) + 3$
(a.) $16x + 7$ b. $4x + 7$
c. $20x + 3$ d. $-4x + 3$

Chapter Test

Let $P = \{0, 1, 2, 3\}$, $Q = \{\text{positive integers}\}$, and $R = \{2, 3, 4, 5, 6\}$. List each set by roster.

                                         {nonnegative integers}

1. $P \cap Q$ {1, 2, 3}   **2.** $P \cup Q$ **3.** $P \cap R$ {2, 3} \hfill *1-1*

4. List all the subsets of P. {0}, {1}, {2}, {3}, {0, 1}, {0, 2}, {0, 3}, {1, 2}, {1, 3}, {2, 3}, {0, 1, 2}, {0, 1, 3}, {0, 2, 3}, {1, 2, 3}, {0, 1, 2, 3}, $\emptyset$

Solve each sentence over $\{-1, 0, 1, 5, 10\}$. If the open sentence has no solution, so state.

5. $2z + 1 = 1$ {0}   **6.** $6 - m = -1$ no solution \hfill *1-2*

7. $8q - 2q = 6q$ {−1, 0, 1, 5, 10}   **8.** $\dfrac{n + 2}{3}$ is an integer. {1, 10}

Graph each set of numbers on a number line.

9. $\{-4, -2, 0, 2, 4\}$ **10.** the set of all positive odd integers

State the axiom or property that justifies each statement. Assume each variable represents a real number.

11. If $m = n$ and $n = p$, then $m = p$. Transitive property of equality \hfill *1-3*

12. $(xy)z = x(yz)$ Associative axiom for multiplication

13. $6(x + 2) = 6x + 12$ Distributive axiom

Simplify.

                            2x − 7

14. $(x + (-3)) + [-(4 + (-x))]$ **15.** $6[5 + (-2)x] + 5x$ 30 − 7x \hfill *1-4*

16. $-7d[-(r + 2s)]$ 7dr + 14ds **17.** $-[-(3ab + 6b) + (-2a)(b + 4)]$ \hfill *1-5*

18. $-4m[k - (h - 8)]$ **19.** $10[7 - 2(3 - 2c) - 8c]$ 10 − 40c \hfill *1-6*

               −4mk + 4mh − 32m

20. $3 \cdot 8 \div 2 - 5 + 4 \cdot (-2)$ −1 **21.** $(-12\dot{m})(-3n) \div \left[\frac{1}{2}(-4 - 5)\right]$ −8mn \hfill *1-7*

17. 5ab + 6b + 8a

Review of Essentials   **31**

---

32. 1. $a, b, c, d \in \mathcal{R}$
     Hypothesis
2. $(a - b)(c - d) = [a + (-b)][c + (-d)]$
     Def. of subt.
3. $= [a + (-b)]c + [a + (-b)](-d)$
     Dist. ax.
4. $= ac + (-b)c + a(-d) + (-b)(-d)$
     Dist. ax.
5. $= ac - bc - ad + bd$
     Prop. of opp. in prod.
6. $(a - b)(c - d) = ac - bc - ad + bd$
     Trans. prop. of =

34. 1. $a, b, c, d \in \mathcal{R}$
     Hypothesis
2. $a[b - (c - d)] = ab - [a(c - d)]$
     Exercise 19
3. $= ab - (ac - ad)$
     Exercise 19
4. $= ab - [ac + (-ad)]$
     Def. of subt.
5. $= ab - ac - (-ad)$
     Prop. of opp. of sum
6. $= ab - ac + ad$
     Prop. of opp. in prod.
7. $a[b - (c - d)] = ab - ac + ad$
     Trans. prop. of =

Additional Answers
Chapter Test

9.
      −4 −2 0 2 4

10.
      1 3 5 7 9

APPLICATION

Ohm's Law

A flashlight contains one or more dry cells, a bulb, and a switch, all connected together by materials that conduct electricity. A flashlight circuit can be shown as in Figure A. When the switch is closed, electrons can travel from one end of the cell to the other by way of the circuit. Energy-producing chemical reactions cannot take place in the cell unless electrons are transferred via the circuit, since the chemicals that react are kept apart in the cell. The voltage (V) of a cell is measured in volts (V) and is a measure of the energy the cell gives to the electrons that travel, or flow, in the circuit. The rate of electron flow, or current (I), is measured in amperes (A). The amount of current depends on the voltage and on the resistance (R) in the circuit. Resistance is measured in ohms (Ω).

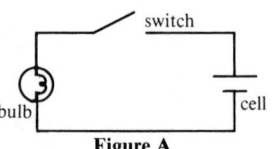

Figure A

The current, voltage, and resistance in circuits similar to the flashlight circuit are related by a formula first stated by Georg Simon Ohm, a German physicist who lived from 1787 to 1854. According to Ohm's law, the voltage is equal to the product of the current and the resistance.

$$V = IR$$

When two of the quantities are known, the third quantity can be determined. For example, the current in a flashlight circuit that has a voltage of 3.0 V and a resistance of 30 Ω is:

$$I = \frac{V}{R} = \frac{3.0 \text{ V}}{30 \ \Omega} = 0.10 \text{ A}.$$

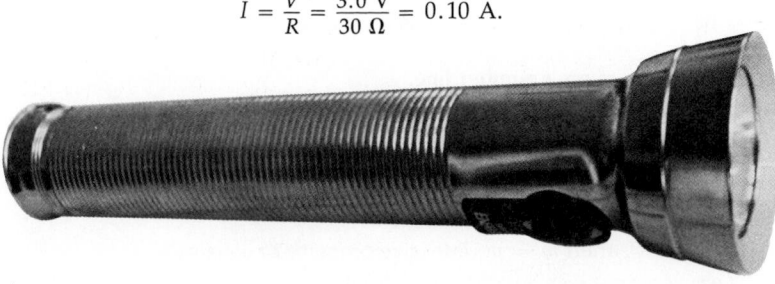

Exercises

1. What would be the effect on the current in a flashlight circuit if (a) the voltage is halved, (b) the resistance is doubled? a. halved   b. halved
2. What is the resistance of the bulb in a flashlight circuit if a voltage of 6.0 V produces a current of 0.15 A in the circuit? 40 Ω
3. Dry cells weaken with age. What is the effective voltage provided by a "6-V" cell if it produces a current of 0.15 A in a flashlight circuit having a resistance of 36 Ω? 5.4 V

32 Chapter 1

PREPARING FOR
COLLEGE ENTRANCE EXAMS

Strategy for Success: The method that is used in scoring a multiple-choice exam determines whether or not it is worthwhile to guess an answer. If it *is* worthwhile, use your knowledge of algebra to eliminate several choices. For example, if a certain answer must be a positive integer, you can eliminate fraction and decimal answers, and any answer that is a negative number or zero.

Decide which is the best of the choices given and write the corresponding letter on your answer sheet.

1. Let $A = \{1, 3, 5, 7\}$ and $B = \{0, 1, 2, 3, \ldots\}$. Which of the following must be true. D
   I. $A \subset B$ II. $A \cup B = B$ III. $A \cap B = B$
   (A) I only   **(B)** II only   **(C)** III only
   (D) I and II only   **(E)** I, II, and III

2. Which of the following statements describes the graph below? B

   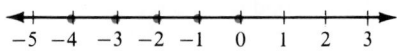

   I. Negative integers greater than -5
   II. Integers between -5 and 1
   III. Non-positive integers greater than -5
   (A) I and II only   **(B)** II and III only   **(C)** I and II only
   (D) I, II, and III   **(E)** None

3. What is the value of the expression $-8(7 + 12 - 8)$? D
   (A) 104   **(B)** 88   **(C)** -216 **(D)** -88 **(E)** 216

4. Simplify $4x - (7 + x) - (-2x)$. A
   (A) $5x - 7$ **(B)** $7x - 7$ **(C)** $5x + 7$ **(D)** $x - 7$ **(E)** $2x - 7$

5. Evaluate the expression $m(3 - 2m)$ if $m = -1$. C
   (A) -1 **(B)** 4   **(C)** -5 **(D)** 5   **(E)** -4

6. Which of the following must be true for all real numbers? D
   I. $-(-a) = a$ II. $a - b = b - a$ III. $a(b - c) = ab - ac$
   (A) I only   **(B)** II only   **(C)** III only
   (D) I and III only   **(E)** II and III only

7. Which of the following illustrates an Axiom of Equality? Assume x, y, and z are real numbers. C
   I. If $x = y$, then $y = x$.   II. $x + (y + z) = (x + y) + z$
   III. If $x = y$ and $y = z$, then $x = z$.
   (A) I and II only   **(B)** II and III only   **(C)** I and III only   **(D)** I, II, and III   **(E)** None

8. If $x = a + 3$ and $y = a - 2$, then $x - y$ equals which of the following? A
   (A) 5   **(B)** 1   **(C)** $2a$ **(D)** $2a + 1$ **(E)** -1

Review of Essentials   **33**

---

28. 1. $a, b, c \in \mathcal{R}$
    $b, c \neq 0$
    Hypothesis

    2. $\dfrac{ac}{bc} = ac \cdot \dfrac{1}{bc}$
    Def. of div.

    3. $= ac\left(\dfrac{1}{b} \cdot \dfrac{1}{c}\right)$
    Prop. of recip. of a prod.

    4. $= \left[a\left(c \cdot \dfrac{1}{c}\right)\right]\dfrac{1}{b}$
    Comm. and assoc. ax. for mult.

    5. $= (a \cdot 1)\dfrac{1}{b}$
    Ax. of mult. inv.

    6. $= a \cdot \dfrac{1}{b}$
    Iden. ax. of mult.

    7. $= \dfrac{a}{b}$
    Def. of div.

    8. $\dfrac{ac}{bc} = \dfrac{a}{b}$
    Trans. prop. of $=$

30. 1. $a, b, c \in \mathcal{R}$
    $b \neq 0$
    Hypothesis

    2. $a + \dfrac{c}{b} = \dfrac{ab}{b} + \dfrac{c}{b}$
    Exercise 25

    3. $= \dfrac{ab + c}{b}$
    Exercise 21

    4. $a + \dfrac{c}{b} = \dfrac{ab + c}{b}$
    Trans. prop. of $=$

(continued)

33

32. 1. $a, b, c, d \in \mathcal{R}$
   $b, d \neq 0$
   Hypothesis

   2. $\dfrac{a}{b} = \dfrac{ad}{bd}, \dfrac{c}{d} = \dfrac{cb}{db}$
   Exercise 28

   3. $\dfrac{cb}{db} = \dfrac{bc}{bd}$
   Comm. ax. for mult.

   4. $\dfrac{a}{b} + \dfrac{c}{d} = \dfrac{ad}{bd} + \dfrac{bc}{bd}$
   Subs. prin.

   5. $= \dfrac{ad + bc}{bd}$
   Exercise 21

   6. $\dfrac{a}{b} + \dfrac{c}{d} = \dfrac{ad + bc}{bd}$
   Trans. prop. of $=$

34. 1. $a, b \in \mathcal{R}$
   $a, b \neq 0$
   Hypothesis

   2. $\dfrac{1}{\frac{a}{b}} = \dfrac{1}{a\left(\frac{1}{b}\right)}$
   Def. of div.

   3. $= \dfrac{1}{a} \cdot \dfrac{1}{\frac{1}{b}}$
   Prop. of recip. of prod.

   4. $= \dfrac{1}{a} \cdot b$
   Exercise 30, page 21

   5. $= b \cdot \dfrac{1}{a}$
   Comm. ax. for mult.

   6. $= \dfrac{b}{a}$
   Def. of div.

   7. $\dfrac{1}{\frac{a}{b}} = \dfrac{b}{a}$
   Trans. prop. of $=$

READING ALGEBRA   *Reading Speed*

Textbooks require a different kind of reading than novels or newspapers. When you read a history textbook, you pay particular attention to dates, names of people, places, and other facts as well as to chapter headings, sub-headings and topic sentences.

Your algebra book also requires special reading techniques. When you begin a new chapter in the book you will first want to know its main idea. The title, the main subdivisions, and the objectives listed under each subdivision will tell you not only what you will be reading about, but also what new skills and techniques you will learn in the chapter. If there are particular kinds of problems to be solved using the ideas of the section, they will be mentioned as well. Objectives are listed for several sections of a chapter at a time. At the end of those sections, a self-test based on the objectives will help you determine whether or not you have adequately mastered the objectives. Rereading the objectives before you begin to work exercises and problems from a particular section will help you to focus your attention on the skills and techniques which are emphasized in the exercises. Another good time to reread the objectives is when you are reviewing for a test or quiz, or when you encounter difficulty with a set of exercises.

You have undoubtedly already realized that the rate at which you read your mathematics textbook must be considerably slower than the rate you would use for a book written without symbols. The purpose of symbols, of course, is to provide a brief, concise way of expressing a quantity or an idea. For example, $f: x \rightarrow 3x$ puts a large amount of information into a small amount of space. If you skim by the symbols as quickly as you would skim by words, you will probably not grasp the full meaning of what you are reading. Reading aloud is an excellent way to slow down your reading and to ensure that you appreciate the meaning of the various symbols which are used in mathematical language.

If there are diagrams or figures in the text, you should sketch a copy of them on a piece of scrap paper. You should also study the worked-out examples in the text one step at a time. Then cover the example with a piece of paper and see if you can work out the solution yourself. As all students have noticed at one time or another, it is much easier to watch an example being worked out either by a teacher or in a textbook than to do it oneself.

When you have carefully read a section of the text, keeping in mind as you read, what the objectives are, when you can duplicate the diagrams in the section, and can do the worked-out examples correctly by yourself, then you will be prepared to start working out additional exercises at the end of the section.

34   *Chapter 1*

Careers

Optometry

Optometrists examine patients eyes and test their vision. Although optometrists are not medical doctors, they can prescribe eyeglasses or contact lenses, eye exercises, and treatments other than drugs and surgery. They often supply eyeglasses and fit and repair frames.

Many functions of the human eye can be described using mathematical relationships and principles. The optometrist needs to understand these relationships and principles before studying optics, the science of light and vision.

Training for this profession is specialized, requiring two years of preoptometry training, courses in mathematics, physics, biology and physics, and four years of optometry school.

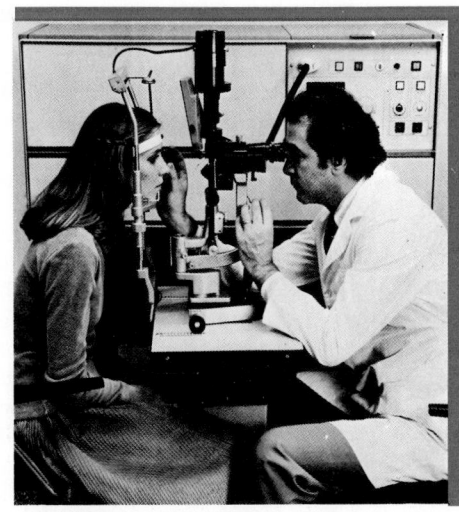

EXAMPLE Optometrists fit their patients with lenses to correct their vision. The diagram below shows a convex or converging lens. Lenses of this type are used to correct farsightedness.

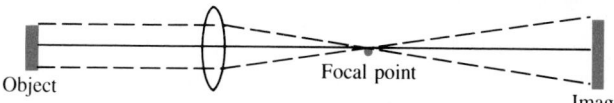

Object   Focal point   Image

The focal length of the lens, or the distance from its center to the point where the light rays converge, is 20 cm. An object is placed 30 cm from the center of the lens. How far from the lens will the image be formed?

SOLUTION From physics we have the following relationship:

$$\frac{1}{\text{distance to object}} + \frac{1}{\text{distance to image}} = \frac{1}{\text{focal length}}$$

Substituting the values for this lens:

$$\frac{1}{30} + \frac{1}{D_i} = \frac{1}{20}$$

$$\frac{1}{D_i} = \frac{1}{20} - \frac{1}{30} = \frac{1}{60}$$

$$D_i = 60 \text{ cm}$$

∴ the image will be formed 60 cm from the lens.

PROGRAMMING IN PASCAL

Pascal is one of a few widely available computer languages that offers a built-in data structure (a ready-made method of organizing information) called SET and various operations for manipulating sets. Most versions of Pascal restrict the size and kinds of sets that are allowed so that the computer can perform operations on the sets efficiently. Usually, sets must be rather small in size (for example, with no more than 512 elements), sets must consist of elements taken from a fairly small *ordinal* (ordered) *base type*, and must not contain negative integers. For example, the base type might be a subset of the positive integers. Or you may declare *charset* for which the base type is all or part of the collection of characters on a computer keyboard.

The Pascal notations for the operations on and relations between sets differs from those used in mathematics. Some of the symbols used are shown below. Compare these with the mathematical symbols in the beginning of this book.

Set Symbol	Meaning
[. . .]	the set whose members are . . .
IN	is an element of
<=	is a subset of
[]	the empty set
=	equals
NOT	is not
*	intersection
+	union
−	difference
>=	is a superset of
<>	is not equal to

An example of a type declaration useful in creating set variables is:

```
TYPE
    charset = SET OF char;
    nums = SET OF 1..100;
```

The identifier *charset* represents the set of all characters that appear on the keyboard. The identifier *nums* represents the set of all integers from 1 to 100 inclusive. With those identifiers, variables that are to be used in a program can be defined:

```
VAR
    ok_replies : charset;
    set_a, set_b, set_c : nums;
```

The variable *ok_replies* can now represent any subset of *charset* and the variables *set_a*, *set_b*, *set_c* can now represent any subset of numbers 1 to 100 inclusive. For example, in the main program *set_a* can be assigned to represent a set having elements 1, 2, and 3:

```
set_a := [1, 2, 3]
```

1.b. english_vowels := ['a', 'e', 'i', 'o', 'u'];

1.c. punctuation_marks : charset;

Exercises **1.d.** punctuation_marks := ['.', ',', ';', ':', '?', '!'];

1. Write a Pascal statement to do each of the following:
   a. declare a *charset* variable called *english_vowels* english_vowels : charset;
   b. assign an appropriate value to *english_vowels*
   c. declare a *charset* variable called *punctuation_marks*
   d. assign an appropriate value to *punctuation_marks*
   e. declare a set variable called *digits* digits : charset;
   f. assign an appropriate value to *digits* digits := ['0'..'9'];

Exercises 2 and 3 refer to the following program.

```
PROGRAM sets 2 (OUTPUT);

TYPE
    nums = SET OF 1 . . 25;

VAR
    set_a, set_b, set_c : nums;
    i : integer;

( ************************************************************)
BEGIN (* main *)
    set_a := [5, 10, 15, 20, 25]; (* initialize the set variables *)
    set_b := [1, 7, 11, 13, 15];
    set_c := [4 . . 8, 10];
    write('A = {');
    FOR i := 1 TO 25 DO (* loop to display set *)
        IF i IN set_a
            THEN write (i:3);
    writeln('  }');
    writeln;
END.
```

2. Insert code that will display the contents of *set_b* and *set_c*. Run the program to check whether or not your code is correct. In order for the computer to display the union of *set_a* and *set_b*, add the following lines to the program:

```
write('A ∪ B = {');
FOR i := 1 to 25 DO
  IF i IN set_a + set_b
    THEN write(i:3);
writeln('  }');
writeln;
```

3. Insert code that will display the contents of the intersection of *set_b* and *set_c*.

You may have noticed the lack of generality of the program *sets2*. It is possible for *set_a*, *set_b*, and *set_c* to be user-defined. That is, each time the program is run, the user decides what the elements of each set should be. To do this, the following program makes use of the powerful modular construct in Pascal called the PROCEDURE.

Exercises 4 and 5 refer to the following program.

```
PROGRAM sets 3 (INPUT, OUTPUT);

TYPE
  nums = SET of 1 . . 100;

VAR
  set_a, set_b : nums;
  i : integer ;


(***********************************************************)
PROCEDURE get_elements (VAR temp : nums);

VAR
  j, x, last : integer;

BEGIN
  writeln;
  writeln('How many elements are in the set: ');
  readln(last);
  FOR j := 1 TO last DO
    BEGIN
      REPEAT
        write('Enter element number , j:2, ': ');
        readln(x);
      UNTIL x IN [1 . . 25];
        temp := temp + [x];
    END;
END;
```

38 *Chapter 1*

```
(*********************************************************)
PROCEDURE display (temp : nums);
VAR
   j : integer;
BEGIN
   write ('{');
   FOR j := 1 TO 25 DO
       IF j IN temp
           THEN write (j:3);
   writeln('  }');
END;


(*********************************************************)
BEGIN (* main *)
   set_a := [ ]; (* initialize the set variables *)
   set_b := [ ];
   get_elements(set_a); (* call procedure *)
   get_elements(set_b);
   writeln;
   writeln;
   write('A = ');
   display(set_a); (* call procedure *)
END.
```

4. Add lines to the above program that will display the contents of set *B*.

5. Insert lines in the above program so the user can define a set *C*. (Remember to declare a *set_c* and initialize it before any calls to a procedure are made.)

6. Research Problem
   Modify the program so the union and intersection of sets *A* and *B*, sets *A* and *C*, and sets *B* and *C* are found and displayed.

The photo shows an untitled Alexander Calder sculpture that hangs in the East Wing of the National Gallery of Art in Washington, D.C. A mathematical sense of precision was required to balance the free-floating elements that make up the sculpture.

40 *Chapter 2*

Chapter 2

Review of Essentials

Problem Solving Strategies

Word Problem Plan
In Section 2–3 students are introduced to the five-step plan by which word problems are solved. Students employ the five-step plan and inequalities to solve word problems in Section 2-5.
Drawing a Diagram
Through the use of number lines students learn the meaning of absolute value and its relation to distance in Section 2-7.

Solving and Applying Equations

OBJECTIVES for Sections 2-1 through 2-3:
1. *To simplify expressions for sums and differences of polynomials.*
2. *To solve first-degree equations in one variable.*
3. *To use linear equations to solve problems.*

2–1 Sums and Differences of Polynomials

A **monomial** is a numeral, a variable, or an indicated product of a numeral and one or more variables. Some examples of monomials are

$$-4, \; m^3, \; 6x^2, \; 32n^4p.$$

A monomial in a single variable x is an expression of the form

$$ax^n,$$

where $a \in \mathscr{R}$, and n denotes a positive integer. The number denoted by a is called the **numerical coefficient** (or simply the **coefficient**) of the monomial.

Teaching Suggestions
p. T78

Key Ideas

Define like and unlike terms. Add and subtract polynomials.

Review of Essentials   **41**

Simplify each expression.

1. $(7p^2 - 4p) +$
$(3p^2 + 2p - 5)$
$= (7 + 3)p^2 + (-4 + 2)p +$
$(0 + (-5))$
$= 10p^2 + (-2)p + (-5)$
$= 10p^2 - 2p - 5$

2. $(7p^2 - 4p) -$
$(3p^2 + 2p - 5)$
$= (7 - 3)p^2 + (-4 - 2)p +$
$(0 - (-5))$
$= 4p^2 + (-6)p + 5$
$= 4p^2 - 6p + 5$

3. $(2y + z) - (4y - 2z) -$
$(y + z)$
$= (2 - 4 - 1)y +$
$(1 + 2 - 1)z$
$= -3y + 2z$

Common Errors

Students may think that coefficients play a role in determining similar terms. Stress that two terms are similar (like terms) only if the variables and the exponents of the variables used in the terms are the same. The coefficients may be different.

When they are subtracting polynomials, students often forget to change the sign of every term in the polynomial that is being subtracted.

The symbol x^n represents a power of x, where x is called the base and n the exponent. In general, the nth power of x denotes the product of n factors, each of which is x. For example:

$$x^1 = x$$
$$x^2 = x \cdot x \text{ (read ''}x\text{-squared'' or ''the square of }x\text{'')}$$
$$x^3 = x \cdot x \cdot x \text{ (read ''}x\text{-cubed'' or ''the cube of }x\text{'')}$$
$$x^4 = x \cdot x \cdot x \cdot x \text{ (read ''}x\text{-fourth'' or ''}x\text{ to the fourth'')}$$

In the monomial ax^n ($a \neq 0$), n is called the degree of the monomial. Thus, for the monomial $-x^4$, the coefficient is -1 and the degree is 4. Monomials such as -9 and 5 are called constant monomials or simply constants and are assigned *degree zero*. The constant monomial 0 has *no degree*. Notice that *degree zero* and *no degree* are not the same.

A monomial that contains more than one variable, such as

$$4x^2y^3,$$

has as its degree the sum of the exponents of the variables. Thus, the degree of $4x^2y^3$ is $2 + 3$, or 5, and its coefficient is 4.

A monomial or a sum of monomials, such as

$$5x^3 + 0x^2 + (-2x) + (-5)$$

is called a polynomial. The monomials in the expression are called the terms of the polynomial and the coefficients of the terms are called the coefficients of the polynomial. Thus, the terms of the preceding polynomial are $5x^3$, $0x^2$, $-2x$, and -5, and the coefficients are 5, 0, -2, and -5. The usual way to write the polynomial is

$$5x^3 - 2x - 5,$$

where the term with 0 coefficient is omitted and the connecting $+$ signs are taken as understood.

Two monomials are said to be like or similar if they are exactly the same or if they differ only in numerical coefficients. Thus,

$$5x^5, \; -3x^5, \text{ and } x^5$$

are *like* monomials, while

$$5x^2, \; 5x^4, \text{ and } 5x^5y$$

are *unlike*. A polynomial is said to be in simple form when no two of its terms are like terms. For example,

$$2x^3 - 5x + 7$$

is in simple form, but

$$2x^3 - 3x - 2x + 7$$

is not. The terms of a simplified polynomial are usually written in order of decreasing degree of one of the variables from left to right.

A polynomial that has two terms is called a **binomial** and a polynomial that has three terms is called a **trinomial**. Thus, $4x^2 + 3$ is a binomial and $2x^3 - 5x + 7$ is a trinomial.

The **degree** of a polynomial in simple form is defined to be the greatest of the degrees of its terms. Thus, the polynomial $2x^3 - 5x + 7$ is of degree 3. A polynomial of the form $ax^2 + bx + c$ where $a \neq 0$ and a, b, $c \in \mathcal{R}$ is called a **quadratic polynomial**.

Given any two polynomials such as

$$4x^2 - 3x \text{ and } x^2 + 2x - 1,$$

the expression

$$(4x^2 - 3x) + (x^2 + 2x - 1)$$

is called the **sum** of the polynomials, and the expression

$$(4x^2 - 3x) - (x^2 + 2x - 1)$$

is called their **difference**. To replace the sum or difference by polynomials in simple form, you use the following rules.

Rules for Adding and Subtracting Polynomials

1. To add polynomials, add the coefficients of similar terms in the polynomials.
2. To subtract one polynomial from another, subtract the coefficient of each term in the one polynomial from the coefficient of the similar term in the other polynomial.

Using these rules, you can simplify an expression. For example,

$$
\begin{aligned}
(4x^2 - 3x) + (x^2 + 2x - 1) &= (4 + 1)x^2 + (-3 + 2)x + (0 + (-1)) \\
&= 5x^2 + (-1)x + (-1) \\
&= 5x^2 - x - 1
\end{aligned}
$$

and

$$
\begin{aligned}
(4x^2 - 3x) - (x^2 + 2x - 1) &= (4 - 1)x^2 + (-3 - 2)x + (0 - (-1)) \\
&= 3x^2 + (-5)x + (1) \\
&= 3x^2 - 5x + 1.
\end{aligned}
$$

Because it can be proved, by using properties of the real numbers, that the equation

$$(4x^2 - 3x) + (x^2 + 2x - 1) = 5x^2 - x - 1$$

is a true statement for *every* numerical replacement of the variable, the two sides of the equation (the expressions related by the = symbol) are called **equivalent expressions**. Whenever you replace a given polynomial by an equivalent polynomial in simple form, you say that you have **simplified** the given polynomial.

Additional A Exercises

Add each pair of polynomials.

1. $4p^2 + 2p + 6$
   $\underline{3p^2 + p + 7}$
   $7p^2 + 3p + 13$

2. $4p^3 - 2p^2q - 6q^3$
   $\underline{-6p^3 + 8pq^2 + 5q^3}$
   $-2p^3 - 2p^2q + 8pq^2 - q^3$

Subtract the lower polynomial from the one above it.

3. $3s^2 + 2s - 8$
   $\underline{4s^2 - s + 2}$
   $-s^2 + 3s - 10$

4. $2a^3 - 3a^2b + 7b^3$
   $\underline{-4a^3 + 3ab^2 - 4b^3}$
   $6a^3 - 3a^2b - 3ab^2 + 11b^3$

Simplify.

5. $(4t + v) - (t - 2v)$
   $3t + 3v$

6. $(5m - 6) + (-4m + 3)$
   $m - 3$

Suggested Assignments

Minimum
   44/1–26
Extended Alg. with Trig.
   44/5, 9, 13–29 odd
Enriched Alg.
   44/19–33 odd
Enriched Alg. with Trig.
   44/19–33 odd

Oral Exercises

State the coefficients and the degree of each polynomial.

1. $-3x + 5$ $-3, 5; 1$ **2.** $a^3 - 3a^2 - a + 2$ $1, -3, -1, 2; 3$

3. $-4k^5 + k^3 - 6k^2 - k + \frac{1}{2}$ **4.** $-\frac{1}{3}x^2y^3 + 2xy^2 - 7xy + 3$
   $-4, 0, 1, -6, -1, \frac{1}{2}; 5$ $-\frac{1}{3}, 2, -7, 3; 5$

Simplify.

5. $6x + 11x$ $17x$ **6.** $-5x^3 - 7x^3$ $-12x^3$ **7.** $x^2 - 3x^2 + 12x^2$ $10x^2$ **8.** $-x^4 - 7x^4 + x^4$
                                                                                      $-7x^4$

Each of the letters A, B, C, and D stands for the specified polynomial. State the indicated sum or difference in simple form.

$A: x^2 - 5$ $B: 7x - 3$ $C: -6x^2 + 4x$ $D: x^2 - x - 2$

9. $A + B$ $x^2 + 7x - 8$ **10.** $A + C$ $-5x^2 + 4x - 5$ **11.** $C + D$ $-5x^2 + 3x - 2$ **12.** $A + D$

13. $D - B$ $x^2 - 8x + 1$ **14.** $C - A$ $-7x^2 + 4x + 5$ **15.** $B - C$ $6x^2 + 3x - 3$ **16.** $D - C$

12. $2x^2 - x - 7$

16. $7x^2 - 5x - 2$

Written Exercises   1–30   Show work!

Add each pair of polynomials.

A   **1.** $x^3 + 5x^2 - 7x + 3$ **2.** $7y^3 - 6y^2 + 3y - 9$
           $\underline{4x^3 - 2x^2 + 3x - 11}$ $\underline{-8y^3 + y^2 + 4}$
           $5x^3 + 3x^2 - 4x - 8$ $-y^3 - 5y^2 + 3y - 5$

       3. $-z^4 - 7z^3 + 3z - 1$ **4.** $3n^4 - n^2 - 2n + 3$
           $\underline{z^4 - 6z^2 - 2z - 10}$ $\underline{-5n^4 + 6n^3 - 5n^2 + n - 3}$
               $-7z^3 - 6z^2 + z - 11$ $-2n^4 + 6n^3 - 6n^2 - n$

       5. $5a^3 - 4a^2b - ab^2$ **6.** $-6p^3 + 9pq^2 - 7q^3$
           $\underline{-3a^3 + 8a^2b - ab^2 - b^3}$ $\underline{-5p^3 - 13p^2q + 12q^3}$
           $2a^3 + 4a^2b - 2ab^2 - b^3$ $-11p^3 - 13p^2q + 9pq^2 + 5q^3$

7–12. Subtract the lower polynomial from the one above it in Exercises 1–6.

Simplify.   **15.** $-4a + 5b$ **16.** $6cd + 2d^2$
                   17. $-9p^2 - 3$ **18.** $q^2 - 4q - 4$

13. $(3x - 4) + (-2x + 1)$ $x - 3$ **14.** $(8 - 9y) - (-7 + 3y)$ $15 - 12y$

15. $(5a - b) - (7a + 4b) - (2a - 10b)$ **16.** $(6cd + 8d^2) - (5d^2 - cd) - (cd + d^2)$

17. $(p^2 - 7p + 5) - (10p^2 - 7p + 8)$ **18.** $(2q^2 - q - 15) - (q^2 + 3q - 11)$

19. $(r^3 - 3r^2) + (3r^3 - 5r - 12) - (-r^3 - 8r^2 + 4r - 9)$ $5r^3 + 5r^2 - 9r - 3$

20. $(-v^3 + 6v^2 - v + 13) - (3v^3 - v + 17) - (-4v^3 + v^2 - 4)$ $5v^2$

21. $(w^4 - 6w^3 - w^2 + 5) - (w^4 - 2w^3 - w - 3) - (-4w^3 - w + 8)$ $-w^2 + 2w$

22. $(-7k^4 + k^3 - 3) - (2k^4 - 8k^2 - 14) - (-k^4 + 2k^3 + 8k^2 - 1)$ $-8k^4 - k^3 + 12$

23. $(9x^2 - 7xy + y^2) - (-3x^2 - 4xy - y^2) - (5x^2 - xy + 2y^2)$ $7x^2 - 2xy$

24. $(-p^3 - p^2q + 2pq^2) - (6p^3 + 5p^2q - 3pq^2) - (-p^3 - p^2q)$ $-6p^3 - 5p^2q + 5pq^2$

B   **25.** $[-3c^2 - (7c - 4d)] - [c^2 - (5c + 10d)]$ $-4c^2 - 2c + 14d$

     26. $[5m^2 - (8mn + 11n^2)] - [6m^2 - (-8mn - 9n^2)]$ $-m^2 - 16mn - 20n^2$

$^{-}$**7.** $-3x^3 + 7x^2 - 10x + 14$

8. $15y^3 - 7y^2 + 3y - 13$

$-$**9.** $-2z^4 - 7z^3 + 6z^2 + 5z + 9$

10. $8n^4 - 6n^3 + 4n^2 - 3n + 6$

$-$**11.** $8a^3 - 12a^2b + b^3$

12. $-p^3 + 13p^2q + 9pq^2 - 19q^3$

32. 1. $-(x^2 - 2) + x^2 =$
         $-[x^2 + (-2)] + x^2$
       Def. of subt.

     2. $= (-x^2 + [-(-2)]) + x^2$
       Prop. of opp. of sum

     3. $= (-x^2 + 2) + x^2$
       Canc. prop. of add. inv.

     4. $= (-x^2 + x^2) + 2$
       Comm. and assoc. ax. for add.

     5. $= 0 + 2$
       Ax. of add. inv.

     6. $= 2$
       Iden. ax. for add.

     7. $-(x^2 - 2) + x^2 = 2$
       Trans. prop. of $=$

34. 1. $(x^2 - 8) - (x^2 - 8) =$
         $(x^2 - 8) + [-(x^2 - 8)]$
       Def. of subt.

     2. $= 0$
       Ax. of add. inv.

     3. $(x^2 - 8) - (x^2 - 8) = 0$
       Trans. prop. of $=$

27. $2[-3x - 7(4 - x)] - 8[x - (2x - 5)]$ 16x − 96

28. $-6[y - 2(9 - 3y)] + 4[-y - 8(y + 5)]$ −78y − 52

29. What polynomial must be added to $5x^4 - 3x^3 + 7x^2 - x + 1$ to obtain the polynomial $x - 4$? −5x⁴ + 3x³ − 7x² + 2x − 5

30. What polynomial must be subtracted from $x + 3$ in order to obtain $9x^5 - 3x^3 + 5x - 2$? −9x⁵ + 3x³ − 4x + 5

Prove that each of the following is a pair of equivalent expressions by using the properties of real numbers.

EXAMPLE $5x^2 - 5x^2$; 0

SOLUTION

1.	$5x^2 - 5x^2 = 5x^2 + (-5x^2)$	Definition of subtraction
2.	$= 5x^2 + (-5)x^2$	Property of opposites in products
3.	$= [5 + (-5)]x^2$	Distributive axiom
4.	$= 0 \cdot x^2$	Axiom of additive inverses
5.	$= 0$	Multiplicative property of zero

C **31.** $x^2 - (x^2 + 1)$; -1

32. $-(x^2 - 2) + x^2$; 2

33. $3x^2 - x - (3x^2 + x)$; $-2x$

34. $(x^2 - 8) - (x^2 - 8)$; 0

2–2 Transforming Equations

The equations below have the same solution set over $\mathcal{R}$, namely, {2}.

$$3x - 2(x + 3) = 2x - 8 \quad \text{and} \quad x = 2$$

Equations that have the same solution set over a given set are called **equivalent equations** over that set.

To solve an equation, either you identify the solution(s) by inspection or transform it into an equivalent equation whose solution is evident by inspection. The properties of real numbers guarantee that the following transformations of a given equation always produce an equivalent equation.

Transformations Producing an Equivalent Equation

1. Substituting for either side of the given equation an expression equivalent to it.
2. Adding to or subtracting from each side of the given equation the same polynomial in any variable(s) appearing in the equation.
3. Multiplying or dividing each side of the given equation by the same nonzero number.

Review of Essentials **45**

Mixed Review

Simplify.

1. $13 - (-4) + 6(-5)$ −13

2. $(3x^2)(-y^3)$ −3x²y³

3. $3cd(cd) + 9c(cd^3) + 5c^2d^2$ 9c²d³ + 8c²d²

4. $3x - (-5x) + 4(-2x)$ 0

5. $(n^4)(n^5)$ n⁹

Teaching Suggestions
p. T78

Key Ideas

Perform a sequence of transformations to solve a linear equation in one variable.

Common Errors

Make sure that in solving equations of the form $ax = b$, students *divide* both sides by a (rather than subtract a from both sides), and that in solving those of the form $\frac{1}{a}x = b$, they *multiply* both sides by a (rather than add). Even good students may have trouble solving equations like those given in Written Exercises 27 and 28. Make sure all your students know how to do these.

45

Chalkboard Examples

Solve each equation over $\mathcal{R}$.

1. $2v - 5(v - 2) = 4 - v$
   $2v - 5v + 10 = 4 - v$
   $-3v + 10 = 4 - v$
   $-3v + 10 + v = 4 - v + v$
   $-2v + 10 = 4$
   $-2v + 10 - 10 = 4 - 10$
   $-2v = -6$
   $-2v \div (-2) = -6 \div (-2)$
   $v = 3$
   Check the solution.
   $2v - 5(v - 2) = 4 - v$
   $2(3) - 5(3 - 2) \overset{?}{=} 4 - 3$
   $6 - (5)(1) \overset{?}{=} 4 - 3$
   $6 - 5 \overset{?}{=} 4 - 3$
   $1 = 1 \quad \checkmark \quad \{3\}$

2. $12p - (6p - 2) = -10$
   $12p - 6p + 2 = -10$
   $6p + 2 = -10$
   $6p + 2 - 2 = -10 - 2$
   $6p = -12$
   $6p \div 6 = -12 \div 6$
   $p = -2$
   Check the solution.
   $12p - (6p - 2) = -10$
   $12(-2) - (6(-2) - 2) \overset{?}{=} -10$
   $-24 - (-12 - 2) \overset{?}{=} -10$
   $-24 - (-14) \overset{?}{=} -10$
   $-24 + 14 \overset{?}{=} -10$
   $-10 = -10 \quad \checkmark \quad \{-2\}$

Additional A Exercises

Name the transformation used to produce the second equation from the first.

1. $3 - 2n = 17; \; -2n = 14$
   Subtract 3 from each side.

2. $3(2 - r) = 12; \; 6 - 3r = 12$
   Distributive axiom

Solve each equation.

3. $2u + 4 = -16 \quad \{-10\}$

4. $6 - (1 + 4p) = 25 \quad \{-5\}$

5. $3x - 18 = 9x - 30 \quad \{2\}$

6. $\frac{2}{3}t = 10 \quad \{15\}$

EXAMPLE   Solve $2y - 5(y - 3) = 7 - y$ over $\mathcal{R}$.

SOLUTION
1. Copy the equation.
2. Use the distributive axiom to help simplify the left side.
3. Add y to each side.

4. Subtract 15 from each side.

5. Divide each side by -2 (or multiply each side by $-\frac{1}{2}$).

$$2y - 5(y - 3) = 7 - y$$
$$2y - 5y + 15 = 7 - y$$
$$-3y + 15 = 7 - y$$
$$-3y + 15 + y = 7 - y + y$$
$$-2y + 15 = 7$$
$$-2y + 15 - 15 = 7 - 15$$
$$-2y = -8$$
$$-2y \div (-2) = -8 \div (-2)$$
$$y = 4$$

Because errors may occur in transforming equations, you should always check each solution in the original equation.

$$2y - 5(y - 3) = 7 - y$$
$$2(4) - 5(4 - 3) \overset{?}{=} 7 - 4$$
$$8 - (5)(1) \overset{?}{=} 3$$
$$8 - 5 \overset{?}{=} 3$$
$$3 = 3 \quad \checkmark$$

∴ the solution set is $\{4\}$.

Note that, in the preceding example, the domain of the variable was specified as $\mathcal{R}$. From here on, in this book, *unless otherwise stated, all open sentences are to be solved over* $\mathcal{R}$.

The "operational" properties of equality upon which the transformations used to solve equations are based are contained in the following theorem that is proved in Exercises 41–44 on page 48.

For all real numbers a, b, and c, if $a = b$, then:	
$a + c = b + c, \; c + a = c + b$	Addition Property
$ac = bc, \; ca = cb$	Multiplication Property
$a - c = b - c$	Subtraction Property
$\dfrac{a}{c} = \dfrac{b}{c}$, provided $c \neq 0$	Division Property

Oral Exercises

In Exercises 1–9:
a. State the transformation(s) you would use to solve each equation.
b. State the new equation produced by each transformation.

1. $x + 9 = 17$

2. $8 - y = 5$

3. $3c - 4 = 8$

4. $\frac{m}{5} + 2 = 6$ **5.** $\frac{3r}{4} = -9$ **6.** $\frac{1}{2}s + 5 = 1$

7. $7 - 2x = -5$ **8.** $-5 + \frac{1}{2}p = 4$ **9.** $3q = q + 15$

Solve each equation for the variable shown in red. Assume that each variable represents a real number and no denominator is 0.

10. $\frac{x}{a} = b$ $x = ab$ **11.** $\frac{e}{y} = f$ $y = \frac{e}{f}$ **12.** $c - z = d$
$z = c - d$

13. $\frac{r}{v} + s = 0$ $v = -\frac{r}{s}$ **14.** $\frac{1}{aw} = b$ $w = \frac{1}{ab}$ **15.** $gx + h = fx$
$x = \frac{h}{f-g}$

Written Exercises
A = Add   S = Subtract
M = Multiply   D = Divide

State the transformations used to produce the second equation from the first and the third equation from the second.

A **1.** $-7 + 5x = -22$; $5x = -15$; $x = -3$ A 7; D 5

2. $a - 3 = -4a + 17$; $5a - 3 = 17$; $5a = 20$ A 4a; A 3

~~1-40~~
~~Show w/c~~

3. $\frac{1}{3}y - 2 = 9$; $\frac{1}{3}y = 11$; $y = 33$ A 2; M 3

4. $7 - \frac{z}{4} = -2$; $-\frac{z}{4} = -9$; $z = 36$ S 7; M -4

5. $5r - 2(r - 7) = -4$; $5r - 2r + 14 = -4$; $3r + 14 = -4$ Dist. ax; Simplify

6. $2k + 14 = -10 - 4k$; $6k + 14 = -10$; $6k = -24$ A 4k; S 14

7. $-\frac{1}{2}p - 5 = \frac{1}{2}p + 8$; $-p - 5 = 8$; $-p = 13$ S $\frac{1}{2}p$; A 5

8. $-4 = 11 - 3(n + 2)$; $-4 = 11 - 3n - 6$; $-4 = 5 - 3n$ Dist. ax; Simplify

Solve each equation.

9. $6u + 3 = -45$ {−8}   **10.** $2 = 27 - 5n$ {5}

11. $8 - (1 + 3p) = -8$ {5}   **12.** $5x - (x - 6) = -6$ {−3}

13. $7y - 2(y - 13) = 26$ {0}   **14.** $4v - 21 = 19 - v$ {8}

15. $c - 3(c - 17) = 11$ {20}   **16.** $40 - 5(d + 12) = 2d$ {$-\frac{20}{7}$}

17. $\frac{3}{5}r = 42$ {70}   **18.** $-\frac{7w}{6} = 14$ {−12}

19. $\frac{2}{9}z + 7 = 1$ {−27}   **20.** $5 - \frac{3}{7}e = 26$ {−49}

21. $-3[7 - 3(x + 6)] + x = 9$ {$-\frac{12}{5}$}   **22.** $5[12 - 3(2 - y) - 2y] = 2(1 - y)$ {−4}

23. $-\frac{1}{4}a + \frac{7}{4}a = -21$ {−14}   **24.** $\frac{-5t}{9} - \frac{2t}{9} = 14$ {−18}

25. $\frac{1}{3}(8 - m) + \frac{5}{3}(2m - 4) = 14$ {6}   **26.** $\frac{3}{4}(2b - 5) - \frac{5}{4}(3b + 9) = -6$ {−4}

27. $\frac{6}{z} + 7 = 5$ {−3}   **28.** $9 - \frac{12}{w} = -3$ {1}

Review of Essentials   **47**

1. a. Subtract 9 from each side.
b. $x = 8$

2. a. Add y to each side; subtract 5 from each side.
b. $8 = 5 + y$, $3 = y$

3. a. Add 4 to each side; divide each side by 3.
b. $3c = 12$, $c = 4$

4. a. Subtract 2 from each side; multiply each side by 5.
b. $\frac{m}{5} = 4$, $m = 20$

5. a. Multiply each side by 4; divide each side by 3.
b. $3r = -36$, $r = -12$

6. a. Subtract 5 from each side; multiply each side by 2.
b. $\frac{1}{2}s = -4$, $s = -8$

7. a. Subtract 7 from each side; divide each side by −2.
b. $-2x = -12$, $x = 6$

8. a. Add 5 to each side; multiply each side by 2.
b. $\frac{1}{2}p = 9$, $p = 18$

9. a. Subtract q from each side; divide each side by 2.
b. $2q = 15$, $q = \frac{15}{2}$

Suggested Assignments

Minimum
  47/1–9, 11–23 odd, 24–28
Extended Alg. with Trig.
  47/3–39 odd
Enriched Alg.
  47/23–45 odd
Enriched Alg. with Trig.
  47/23–45 odd

**Additional Answers
Written Exercises**

42. 1. $a = b; a, b, c \in \mathcal{R}$
    Hypothesis
 2. $a - c = a - c$
    Refl. prop. of =
 3. $a - c = b - c$
    Subs. prin.

44. 1. $a = b; a, b, c \in \mathcal{R}$,
    $c \neq 0$
    Hypothesis
 2. $\frac{a}{c} = \frac{a}{c}$
    Refl. prop. of =
 3. $\frac{a}{c} = \frac{b}{c}$
    Subs. prin.

Reading Algebra

When they are reading equations such as $2v - 5(v - 2) = 4 - v$ students should also think about the meaning of the equation. Point out that the reading of this equation, "Two times v minus five times the quantity $v - 2$ equals four minus v," although exact, does not convey the meaning of the equation which can be interpreted as follows: "Subtract two from a number, then multiply by five. Subtract this answer from the double of the number. The result is the original number subtracted from four. What is the original number?" (3) This type of activity will be beneficial to students when they are representing the conditions in a problem and setting up an equation to solve the problem.

Solve each equation for the variable shown in red. Assume that each variable represents a real number and no denominator is 0.

B 29. $3ar - b = 6b$ $r = \frac{7b}{3a}$

30. $5c - \frac{1}{2}dn = 3c$ $n = \frac{4c}{d}$

31. $4kx - 7h = kx$ $x = \frac{7h}{3k}$

32. $ay - by = c$ $y = \frac{c}{a - b}$

33. $5rv - sv = 2r$ $v = \frac{2r}{5r - s}$

34. $2dw - 3 = ew$ $w = \frac{3}{2d - e}$

35. $v = -gt + p$ $t = \frac{p - v}{g}$

36. $A = \frac{1}{2}h(a + x)$ $x = \frac{2A - ha}{h}$

37. $A = \pi rh + 2\pi r^2$ $h = \frac{A - 2\pi r^2}{\pi r}$

38. $9p = kp - 7n$ $p = \frac{7n}{k - 9}$

39. $\frac{a}{x} = \frac{b}{c}$ $x = \frac{ac}{b}$

40. $\frac{d}{v} + c = 5d$ $v = \frac{d}{5d - c}$

Prove each of the following theorems.

C 41. For all real numbers a, b, and c, if $a = b$, then $a + c = b + c$.

42. For all real numbers a, b, and c, if $a = b$, then $a - c = b - c$.

43. For all real numbers a, b, and c, if $a = b$, then $ac = bc$.

44. For all real numbers a and b and nonzero real numbers c,
$$\text{if } a = b, \text{ then } \frac{a}{c} = \frac{b}{c}.$$

45. For all real numbers b and c and nonzero real numbers a,
$$\text{if } ax + b = c, \text{ then } x = \frac{c - b}{a}.$$

Computer Exercises   For students with computer experience

1. Write a program that will solve any equation of the form
$$ax + b = cx + d$$
when the values of a, b, c, and d are input. If the equation has no solution or is satisfied by all real numbers, the output should state this.

Run the program in Exercise 1 for each equation.

2. $3x - 7 = 8x + 19$ -5.2
3. $4x + 3 = 4x - 15$ no solution
4. $5x - 3 = x + 7$ 2.5
5. $11x + 2 = 16x + 2$ 0
6. $-25x + 9 = -25x + 9$ all real numbers
7. $13x + 2 = 42x - 27$ 1

48   *Chapter 2*

Careers

Actuarial Science

Actuaries collect and analyze statistical data relating to insurance. Actuaries must apply their knowledge of mathematics and statistics, as well as principles of business and finance. Some of the typical problems actuaries solve include determining mortality, accident, sickness, disability, and retirement rates, calculating premiums, and determining the amount of money an insurance company needs to cover payment of its benefits.

Most actuaries are employed by private insurance companies. They may also serve as consultants for large corporations, where they advise on insurance and pension plans for the employees.

EXAMPLE In calculating the premium for a one-year term life insurance policy, an actuarial worker has the following information. Out of a given number of people alive at a given age at the beginning of a year, a certain number will be likely to die during that year. Under the terms of the policy, the beneficiary is to receive $1000 in the event of the death of the policyholder. The problem is to determine the amount of the premium to be charged in order to cover payments to the beneficiaries.

SOLUTION The premium for persons of a given age is calculated by the following formula:

$$\text{Premium} = \frac{\left(\begin{array}{c}\text{amt. pd. to}\\\text{ea. beneficiary}\end{array}\right) \cdot \frac{100}{103} \cdot \left(\begin{array}{c}\text{number likely to}\\\text{die during year}\end{array}\right)}{\text{number living at beginning of year}}$$

The factor $\frac{100}{103}$ takes into account the interest the collected premiums will earn before any benefits are paid. For age 35, probability tables show that of 9374 people alive at the beginning of a year, 24 will die during the year.

$$\text{Premium} = \frac{1000 \cdot \frac{100}{103} \cdot 24}{9374}$$

$$\approx \$2.49 \qquad \begin{array}{l}\text{(for \$1000 life insurance}\\\text{for 1 year, at age 35)}\end{array}$$

Review of Essentials **49**

Supplementary Material

Test 4

Key Ideas

Solve word problems involving linear equations in one variable.

Chalkboard Examples

1. The manager of a soccer team bought 4 pieces of artificial turf to cover the entire 100 m² area of the soccer field. Three of the pieces each cover 20 m² less in area than the fourth piece covers. How much area did each piece of artificial turf cover?

Let x = the area covered by the big piece. Then $x - 20$ = the area covered by each small piece.

100 m² = the total area covered.

$$x + 3(x - 20) = 100$$
$$x + 3x - 60 = 100$$
$$4x = 160$$
$$x = 40$$
$$x - 20 = 20$$

∴ the large piece covers 40 m² and each small piece covers 20 m².

2. Karen and Andrew start out from towns 164 km apart and drive toward each other. Andrew travels at a rate of 40 km/h and Karen travels at a rate of 48 km/h. If Karen begins the journey 15 min after Andrew does, how long after Andrew leaves will they meet?

2–3 Solving Problems Using Linear Equations

No simple mathematical procedure, or algorithm, will solve every kind of practical problem. However, the steps listed and illustrated in the following examples provide a general method applicable to many real-life situations.

EXAMPLE 1   A landscaper has determined that together 1 small bag of lawn seed and 3 large bags will cover 330 m² of ground. If the large bag covers 50 m² more than the small bag, what is the area covered by each size bag?

SOLUTION   1. Read the problem carefully several times and decide what number or numbers are asked for. The problem asks for the area that each size bag will cover.

2. Choose a variable to represent one of the numbers asked for or described in the problem. A sketch or a chart may be helpful. Let s represent the number of square meters covered by one small bag. Then $s + 50$ is the number of square meters covered by one large bag.

3. Write an open sentence representing the relationship(s) stated or implied in the problem.

Area covered by 1 small bag	added to	area covered by 3 large bags	is	total area covered.
s	$+$	$3(s + 50)$	$=$	330

4. Solve the open sentence.

$$s + 3(s + 50) = 330$$
$$s + 3s + 150 = 330$$
$$4s + 150 = 330$$
$$4s = 180$$
$$s = 45$$
$$s + 50 = 95$$

5. Check your results with the requirements stated in the problem. One small bag and three large bags will cover 330 m²:

$$45 + 3(95) \stackrel{?}{=} 330$$
$$45 + 285 \stackrel{?}{=} 330$$
$$330 = 330 \ \checkmark$$

∴ one small bag will cover 45 m² and one large bag will cover 95 m².

Formulas from the social and physical sciences are often useful in solving practical problems. Example 2 uses the distance formula; Example 3 uses the formula for the area of a trapezoid.

EXAMPLE 2   From the point where a tanker springs a leak 100 km offshore, the edge of an oil spill approaches the shore at 2 km/h. A boat traveling at 16 km/h leaves the shore 30 min after the accident to try to contain the spill. How long after the accident will the boat reach the edge of the spill?

SOLUTION   1. The problem asks for the amount of time after the accident that it will take for the boat to reach the edge of the spill.

2. Let t = the number of hours after the accident that the boat reaches the spill. This must be in *hours* in order to be consistent with the given rates. Thus, $t - \frac{1}{2}$ = the time of the boat's trip (30 min = $\frac{1}{2}$ h).

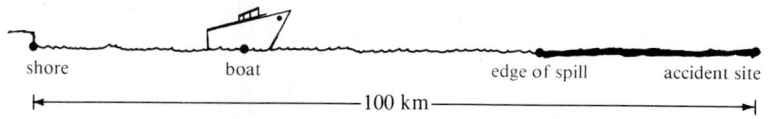

shore   boat   edge of spill   accident site

———————————100 km———————————

Using the relationship distance = rate × time, or $d = rt$, for motion at a constant rate, the following table can be made:

	r	t	d
spill	2	t	$2t$
boat	16	$t - \frac{1}{2}$	$16(t - \frac{1}{2})$

3. $\underbrace{\text{Distance covered}}_{2t}$ $\underbrace{\text{added}}_{+}$ $\underbrace{\text{distance covered}}_{16\left(t - \frac{1}{2}\right)}$ $\underbrace{\text{is}}_{=}$ $\underbrace{\text{total distance}}_{100}$
   by spill   to   by boat   ↓   offshore.

4–5. Completing the solution and checking the result is left to you. The boat reaches the spill 6 h after the accident happened.

EXAMPLE 3   The length of one base of a trapezoid is 6 cm greater than the length of the other base. The height of the trapezoid is 11 cm and its area is 165 cm². What are the lengths of the bases?

SOLUTION   1. The problem asks for the lengths of the bases of the trapezoid.

2. Let x = the length of the shorter base. Then $x + 6$ represents the length of the longer base.

x

11

———$x + 6$———

3. Recall that the formula for the area of a trapezoid is $A = \frac{1}{2}h(b_1 + b_2)$.

$$\underbrace{\text{Area}}_{\frac{1}{2}(11)(x + x + 6)} \quad \underset{=}{\text{is}} \quad \underbrace{165 \text{ cm}^2.}_{165}$$

4–5. Completing the solution and checking are left to you. You should find that the lengths of the bases are 12 cm and 18 cm.

Let t = the time in hours Andrew travels and $t - \frac{1}{4}$ = the time in hours Karen travels. Then using the relationship distance = rate × time, $40t$ = the distance Andrew travels, and

$48\left(t - \frac{1}{4}\right)$ = the distance Karen travels.

164 km = the total distance.

$40t + 48\left(t - \frac{1}{4}\right) = 164$

$40t + 48t - 12 = 164$

$88t = 176$

$t = 2$

∴ they will meet 2 h after Andrew leaves.

3. The length of one base of a trapezoid is three times the length the other base. The height of the trapezoid is 12 cm and its area is 120 cm². What are the lengths of the bases?

Let x = the length of the shorter base. Then $3x$ = the length of the longer base.

Using the formula for the area of a trapezoid:

$A = \frac{1}{2}h(b_1 + b_2)$

$120 = \frac{1}{2}(12)(x + 3x)$

$120 = 6(x + 3x)$

$120 = 6(4x)$

$120 = 24x$

$x = 5$

$3x = 15$

∴ the lengths of the bases are 5 cm and 15 cm.

Suggested Assignments

Minimum
52/*P*: 1–13
R 55/*Self-Test 1*
S 47/10–22 even
Extended Alg. with Trig.
52/*P*: 7–19 odd
R 55/*Self-Test 1*
S 47/12–20 even
Enriched Alg.
53/*P*: 11–21 odd
R 55/*Self-Test 1*
Enriched Alg. with Trig.
53/*P*: 11–21 odd
R 55/*Self-Test 1*

Additional A Exercises

1. The sum of three consecutive integers is 72. Find the integers.   23, 24, 25

2. Find three consecutive integers whose sum is 16 more than two times the middle integer.   15, 16, 17

3. Find the measures of the three angles of a triangle such that the measure of the second angle is twice the measure of the first angle and the measure of the third angle is three times the measure of the first.   30°, 60°, 90°

4. The length of a rectangle is 4 less than twice the width. Find the dimensions of the rectangle if its perimeter is 58 cm.
   18 cm by 11 cm

Oral Exercises

Use the given assignment of the variable to set up an equation that represents the given conditions. You need not solve the equation.

1. Twice the sum of two consecutive integers is 246. Let $n =$ the smaller integer. $2(n + n + 1) = 246$

2. Each of the two congruent sides of an isosceles triangle is 10 cm shorter than its base, and the perimeter of the triangle is 205 cm. Let $x =$ the length of the base. $x + 2(x - 10) = 205$

3. The measure of one of a pair of supplementary angles is 5 more than 6 times the measure of the second angle. Let $y =$ the measure of the second angle. $6y + 5 + y = 180$

4. Leaving on a vacation trip, a family traveled at a constant speed for $3\frac{1}{2}$ h. Returning by the same route, they traveled at a speed 5 mi/h less than their original speed and took 4 h. Let $r =$ their original speed. $(3\frac{1}{2})r = 4(r - 5)$

Problems 1-22

A 1. The sum of three consecutive integers is 159. Find the integers. 52, 53, 54

2. A copy service charges a fixed price for the first copy and 6¢ less for all subsequent copies. Laura Kennedy paid $1.42 for 17 copies of a one-page report. What was the charge for the first copy? $0.14

3. Find two consecutive integers such that the sum of the first and twice the second is 203. 67, 68

4. Find three consecutive integers whose sum is 11 less than 5 times the smallest of the three. 7, 8, 9

5. Find three consecutive *odd* integers whose sum is 13 more than twice the largest of the three. 15, 17, 19

6. Each of the two congruent sides of an isosceles triangle is 7 cm shorter than twice the base. Find the lengths of the three sides if the perimeter of the triangle is 106 cm. 24 cm, 41 cm, 41 cm

7. The measures of the three angles of a triangle are integers such that, if written in increasing order, each is 6 greater than the preceding one. What is the measure of each angle? 54°, 60°, 66°

Triangle ABC is isosceles with $AC = BC$; $\overline{BD}$ is the median to side $\overline{AC}$.

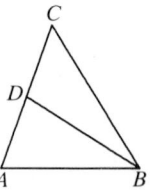

8. If DB is 2 cm greater than DC and the perimeter of $\triangle BDC$ is 70 cm, find BC. 34 cm

9. Suppose DB is 2 cm greater than DC, and AB is 1 cm greater than DC. Find BC if the sum of the lengths of all the segments in the diagram is 45 cm. 14 cm

52   *Chapter 2*

10. The trapezoid in the diagram has a height of 12. The length of the longer base is 17. Find the value of x if the area of the trapezoid is 168. 3

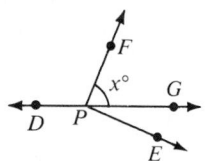

11. In the diagram $\angle EPF$ is a right angle and the measure of $\angle DPF$ is 5 times the measure of $\angle EPG$. Find the value of x. 67.5

12. The fuel efficiency rating for a certain car is 28 mi/gal on the highway and 20 mi/gal in the city. During one week the car traveled 220 mi and used 9 gal of fuel. How much fuel was used for city driving? How many miles were driven on the highway? city; 4 gal; highway; 5 gal

13. George Germain wants to buy 5 lb of ground beef. Extra lean ground beef costs \$3.10/lb, while regular ground beef costs \$2.20/lb. How much extra lean ground beef can George buy if he has \$12.35 to spend on ground beef? 1.5 lb

B 14. Traveling by highways, Carrie Brattle's speed on her way to work is 60 km/h. If she uses local streets to drive the same distance, the trip takes 30 min longer and her speed is 36 km/h. How long does it take her to drive to work by highways? 45 min

15. On Tuesday it took Julia Ward 2 h to drive 40 km and then make 3 round trips to a construction site. On Wednesday she drove 17 km and made 2 round trips to the site, all in 1 h. If all her driving was done at the same constant speed, what is the round-trip distance to the construction site? 6 km

16. Harry Gilmore rode his bike the length of a new bike trail and then rode an extra 8 km, all in 5 h. The next day it took him 3 h to ride just the new bike trail. His speed on the second day was $1\frac{1}{2}$ times the speed of the first day. How long is the bike trail? 72 km

   17. 4 min

17. Each of two computers can process bills at 160/min. A newer model can process bills at 200/min. After the first two computers have worked $1\frac{1}{2}$ min, they are joined by the third computer. How long will it take, from the time the first two computers are turned on, to process 1780 bills?

18. The hypotenuse of a right triangle is 2 cm longer than one leg. The other leg has length 8 cm. Find the lengths of the three sides of the triangle. 8 cm, 15 cm, 17 cm

19. A cylinder with height 15 has a *total* surface area that is twice the area of a circle whose radius is 6 more than the radius of the cylinder. What is the radius of the cylinder? 12

Review of Essentials **53**

Mixed Review

Simplify.

1. $4x - (3x - 4) - 7(3x - 1)$
$-20x + 11$

2. $(n)(q^2)(n^3)$ n^4q^2

3. $\left(\dfrac{1}{c}\right)\left(\dfrac{1}{d}\right)$ $\dfrac{1}{cd}$

Write a mathematical expression for the given phrase.

4. Twice a number plus three less than four times another number.
$2a + (4b - 3)$

5. The product of thirteen and two minus the quantity 16 times *y*.
$(13 \cdot 2) - (16y)$

Complete.

6. If $\dfrac{1}{2}z + 12 = 6 - z$, then
$z = ?$ -4

---

20. In baseball, a player's batting average equals the number of hits divided by the number of times at bat. Fernando Lopez has 10 more times at bat than hits. Paul Montague's times at bat exceed his hits by 15. Montague has 3 more hits than Lopez, and both players have the same batting average. Find their common batting average. 0.375

21. At the opening of the stock exchange Mel sold part of his 30 shares of the Heartbreak Gold Mine for a total of $165. Later in the day the price of a share doubled, and he sold the rest of his shares for $120. How many shares did he sell at each price? 22 shares at $7.50; 8 shares at $15

22. What speed must a motorist average on a return trip from a city 100 km from home in order to have an average speed of 32 km/h for the round trip, if the average speed going out was 20 km/h? (*Hint*: Average speed = total distance ÷ total time.) 80 km/h

Computer Exercises   For students with computer experience

1. Write a program that will compute a customer's monthly electric bill when the number of kW·h for a certain month is input. Use the following rate schedule.

Basic service charge per month: $7.42

kW·h use per month	Cost per kW·h
first 350 kW·h	5.876¢
next 400 kW·h	8.252¢
over 750 kW·h	9.064¢

Run the program in Exercise 1 for each monthly use.

2. 528 kW·h $42.67   **3.** 275 kW·h $23.58   **4.** 812 kW·h $66.61

5. In baseball, a player's batting average is given by

$$\text{batting average} = \frac{\text{number of hits}}{\text{number of times at bat}}.$$

Suppose a player already has *h* hits in *t* times at bat. Write a program that will tell how many hits the player will need in *x* *more* times at bat in order to have a batting average of at least *A*. (Input *h*, *t*, *x*, and *A*.) If an average *A* is impossible to achieve in *x* more times at bat, the output should state this.

Run the program in Exercise 5 for each set of data.

6. $h = 12$, $t = 41$, $x = 20$, $A = .310$ 7
7. $h = 69$, $t = 285$, $x = 23$, $A = .300$ impossible to achieve

54   *Chapter 2*

Self-Test 1

monomial (p. 41)
numerical coefficient (p. 41)
power (p. 42)
base (p. 42)
exponent (p. 42)
nth power of x (p. 42)
degree of a monomial (p. 42)
constant (p. 42)
polynomial (p. 42)
terms of a polynomial (p. 42)
coefficients of a polynomial
  (p. 42)
like monomials (p. 42)
similar monomials (p. 42)
simple form of a polynomial
  (p. 42)

binomial (p. 43)
trinomial (p. 43)
degree of a polynomial (p. 43)
quadratic polynomial (p. 43)
equivalent expressions (p. 43)
simplifying a polynomial
  (p. 43)
equivalent equations (p. 45)
algorithm (p. 50)

Simplify each expression. Assume that each variable denotes a real number.

1. $(-4y^2 + y + 5) + (5y^2 - 3y - 7)$ $y^2 - 2y - 2$ *Obj. 1, p. 41*
2. $2(x^2 + 6) - 8(x^2 - 3)$ $-6x^2 + 36$
3. $(a^3 + 4) - (a^2 + 5a - 2) - (2a^3 + 4a^2 - 7)$ $-a^3 - 5a^2 - 5a + 13$
4. $(3xy - 5) - (2xy + x + 2y) + (4xy + 3x - 11)$ $5xy + 2x - 2y - 16$

Solve over $\mathscr{R}$.

5. $3(y + 2) - 8 = 3y - (4y - 2)$ $\{1\}$ *Obj. 2, p. 41*
6. $\frac{3}{4}x - 9 = 3$ $\{16\}$
7. Solve for a: $4ax + y = 2y - 3$ $a = \frac{y-3}{4x}$

8. An estate that is valued at $10,000 is to be divided among one adult   *Obj. 3, p. 41*
   and three children. Each person receives $100 and then the remain-
   der of the money is divided such that the adult receives three times
   as much as each child. What is the total amount of money each
   person receives? Each child receives $1700, the adult receives $4900.

9. One side of a larger square is 8 cm longer than a side of a smaller
   square. Find the dimensions of both squares if the sum of their
   perimeters is 128 cm. 12 cm and 20 cm

10. Find three consecutive even integers such that the sum of the first
    and the third is eight less than three times the second. 6, 8, 10

Check your answers with those at the back of the book.

Review of Essentials   **55**

Teaching Suggestions
p. T78

Key Ideas

Solve linear inequalities.
Graph solution sets of linear
inequalities.

Chalkboard Examples

Solve each inequality and
graph its solution set.

1. 5x − 4 < 3x + 2
   5x − 4 + 4 < 3x + 2 + 4
   5x < 3x + 6
   5x − 3x < 3x + 6 − 3x

   2x < 6

   $\dfrac{2x}{2} < \dfrac{6}{2}$

   x < 3
   ∴ the solution set is
   {x: x < 3}.

   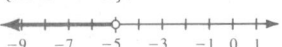
   −2 −1 0 1 2 3 4 5 6 7 8

2. 4x − 13 > 7x + 2
   4x − 13 + 13 > 7x + 2 +
   13
   4x > 7x + 15
   4x − 7x > 7x + 15 − 7x
   −3x > 15

   $\dfrac{-3x}{-3} < \dfrac{15}{-3}$

   x < −5
   ∴ the solution set is
   {x: x < −5}.

   ◄─┼──┼──◊──┼──┼──┼──┼──►
   −9   −7   −5   −3   −1 0 1

Order in the Set of Real Numbers

OBJECTIVES for Sections 2-4 through 2-7:
1. *To solve linear inequalities.*
2. *To apply linear inequalities to practical problems.*
3. *To give simple direct proofs.*
4. *To solve inequalities involving absolute value.*

2–4 Properties of Order

The symbol < is read "is less than." It is used to show the relative order of two real numbers. You say that 3 < 5 (read "3 is less than 5"), because there is a positive number, 2, such that 3 + 2 = 5. The statement 3 < 5 can be written equivalently as 5 > 3 (read "5 is greater than 3"). In general, we have the following definition:

If a and b are real numbers, then

$$a < b \qquad (\text{or } b > a)$$

if and only if there is a positive real number c such that

$$a + c = b.$$

Note the phrase "if and only if," which condenses two statements into one. In this case it means:

If there is a positive real number c such that $a + c = b$, then $a < b$;

and

If $a < b$, then there is a positive real number c such that $a + c = b$.

These statements are called converses of each other, as are any two "If . . . , then . . ." statements each of which can be obtained from the other by interchanging hypothesis and conclusion.

To compare real numbers, we make the following assumption:

Comparison Axiom

If a and b are real numbers, then one and only one of the following statements is true:

$$a > b, \qquad a = b, \qquad a < b.$$

---

56 *Chapter 2*

The set of *positive real numbers* is denoted by the symbol $\mathcal{R}_+$. Another assumption we make is that the sum of two positive real numbers is a positive real number and the product of two positive real numbers is a positive real number.

Closure Axiom for $\mathcal{R}_+$

If a and $b \in \mathcal{R}_+$, then

$$a + b \in \mathcal{R}_+ \quad \text{and} \quad ab \in \mathcal{R}_+;$$

that is, $\mathcal{R}_+$ is closed under addition and multiplication.

Using the definition of "less than" and the closure axiom for $\mathcal{R}_+$, we can prove the following three theorems about order in $\mathcal{R}$.

Transitive Property of Order

If a, b, and c are real numbers, and if $a < b$ and $b < c$, then $a < c$.

Addition Property of Order

If a, b, and c are real numbers, and if $a < b$, then $a + c < b + c$.

Multiplication Property of Order

Let a, b, and $c \in \mathcal{R}$.

1. If $a < b$ and c is positive, then $ac < bc$.
2. If $a < b$ and c is negative, then $ac > bc$.

A proof of the transitive property of order is outlined in Oral Exercise 10 on page 58. A proof of the addition property of order is outlined in Exercise 25 on page 59. The proofs of the two parts of the multiplication property of order are left for you as Exercises 29 and 30 on page 60. Of course, the transitive, addition, and multiplication properties of order are also true when $<$ is replaced by $>$, and $>$ is replaced by $<$, throughout.

Graphically, given two different real numbers, the graph of the smaller lies to the *left* of the graph of the greater on a number line with positive direction to the right. Thus, if $x \in \mathcal{R}$, the graph of the solution set of $x < 3$ appears below. Note that the endpoint is depicted by an open dot, which indicates that the graph of that point is not in the set.

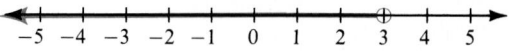

Review of Essentials   **57**

Common Errors

Students may forget to reverse the direction of the inequality sign after they have multiplied or divided each side of the inequality by a negative number. Explain that the properties of order correspond to those of equality *with this one exception.*

Additional A Exercises

State whether each statement is true for all real numbers. If it is true, give the transformation or property that justifies the statement. If it is false, give a value of each variable that demonstrates this.

1. If $r > s$, then $r + 7 > s + 7$.   True by Transformation 2.

2. If $-c > -4$, then $c < 4$. True by Transformation 4.

3. If $x > y$, then $-3x > -3y$. False; for example, $x = 5$ and $y = 2$, $x = {}^-2$ and $y = -5$, $x = 3$ and $y = -3$.

Solve each inequality and graph its solution set.

4. $x + 4 > 6$ $\{x: x > 2\}$

5. $4z - 3 < z + 9$ $\{z: z < 4\}$

6. $-2t + 7 < -1$ $\{t: t > 4\}$

57

The addition and multiplication properties of order imply that the following transformations on inequalities produce **equivalent inequalities** over $\mathcal{R}$, that is, inequalities having the same solution set over $\mathcal{R}$.

Transformations Producing an Equivalent Inequality

1. Substituting for either side of the inequality an expression equivalent to that side.
2. Adding to or subtracting from each side of the inequality the same polynomial in any variable(s) appearing in the inequality.
3. Multiplying or dividing each side by the same *positive* number.
4. Multiplying or dividing each side by the same *negative* number and reversing the direction of the inequality symbol.

EXAMPLE Solve $2x - 8 > 7x + 2$ and graph the solution set.

SOLUTION

1. Copy the inequality. $\qquad\qquad$ $2x - 8 > 7x + 2$
2. Add 8 to each side and simplify. $\qquad$ $2x - 8 + 8 > 7x + 2 + 8$
   $\qquad\qquad\qquad\qquad\qquad\qquad$ $2x > 7x + 10$
3. Subtract $7x$ from each side and simplify. $\quad$ $2x - 7x > 7x + 10 - 7x$
   $\qquad\qquad\qquad\qquad\qquad\qquad$ $-5x > 10$
4. Divide each side by -5 and reverse the direction of the inequality symbol. $\quad$ $\dfrac{-5x}{-5} < \dfrac{10}{-5}$
   $\qquad\qquad\qquad\qquad\qquad\qquad$ $x < -2$

∴ the solution set is $\{x: x < -2\}$.

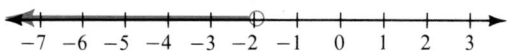

Oral Exercises

In Exercises 1–9:
a. State the solution of each inequality.
b. State the transformation that justifies the step needed to arrive at the solution.

1. $x + 7 < -2$
2. $z - 5 > 3$
3. $-8 > y + 9$
4. $5a < -30$
5. $-4d < -36$
6. $15 < -3c$
7. $-\frac{1}{4}b > 8$
8. $\frac{2}{3}k < -12$
9. $-\frac{1}{5}n > -15$

10. Supply reasons for the following proof of the transitive property of order that is stated on page 57.
    1. a, b, and c are real numbers, $a < b$, and $b < c$. $\qquad$ Hypothesis
    2. For some positive numbers p and q, $a + p = b$ and $b + q = c$. $\quad$ Def. of $<$

58 Chapter 2

3. $(a + p) + q = c$ Subs. prin.

4. $a + (p + q) = c$ Assoc. ax. for add.

5. $p + q$ is a positive number. Clos. ax. for $\mathcal{R}_+$.

6. $a < c$ Def. of $<$

Written Exercises

State whether each statement is true for all real numbers. If it is true, give the transformation or property that justifies the statement. If it is false, give a value of each variable that demonstrates this.

A

1. If $a < b$, then $-2a < -2b$.

2. If $x - 7 < 5$, then $x < 12$.

3. If $-\frac{1}{2}r > 5$, then $r < -10$.

4. If $p > q - 3$ and $q - 3 > 6$, then $p > 6$.

5. If $a < b$ and $c < d$, then $ac < bd$.

6. If $\frac{k}{-4} > 3$, then $k > -12$.

7. If $a + b > 0$, then $a > -b$.

8. If $5n < -35$, then $n > -7$.

Solve each inequality and graph its solution set.

9. $y + 11 < -2$

10. $3 - x > 5$

11. $-2a + 1 < 13$

12. $5 - 7k < -23$

13. $-\frac{1}{4}c - 9 > -1$

14. $4 + \frac{2}{3}p < -8$

15. $\frac{r}{6} - 5 > 7 + \frac{r}{2}$

16. $-\frac{1}{3}(n - 5) > \frac{11}{3}$

17. $6(1 - 2h) + 7h < -9$

18. $\frac{1}{2}(12 - 5m) - \frac{7}{2}m < 18$

B

19. $10 - \frac{1}{3}(2d + 4) < 2(d + \frac{1}{3})$

20. $-\frac{1}{4}(3z - 8) > -9 + \frac{1}{3}(5z + 4)$

21. $\frac{2(4 - 5v)}{3} - 1 > -\frac{3(5v - 6)}{2} + 1$

22. $\frac{3(-5 + 2w) - 1}{4} < \frac{7w - 1}{5} - 3$

23. $-\frac{1}{3}(2(4 - x)) + 5 < \frac{1}{4}(3(1 - 3x)) - 13$

24. $\frac{5(2 - 3x)}{4} - 1 < 4 - \frac{5(1 - x)}{2}$

Supply reasons for the following proofs.

25. Prove: If $a < b$, then
     $a + c < b + c$.

   1. $a < b$

   2. For some positive number p, $a + p = b$.

   3. $(a + p) + c = b + c$

   4. $(a + c) + p = b + c$

   5. $a + c < b + c$

26. Prove: If $a < b$, then
     $c - a > c - b$.

   1. $a < b$

   2. For some positive number q, $a + q = b$.

   3. $a = b - q$

   4. $c - a = c - (b - q)$

   5. $ = (c - b) + q$

   6. $c - a > c - b$

Review of Essentials   **59**

(continued on p. 78)

Simplify.
1. $3(x - 2)$ $3x - 6$
2. $4rs^2 - (2rs^2 - 4s^2) + 7s(3r - s)$
   $2rs^2 - 3s^2 + 21rs$

Graph each set.
3. $\{-7, \ldots, -3, -2, -1\}$

4. $\{-2, 0, 2, 4, 6, 8\}$

List by roster.
5. All the even integers between -6 and 3 inclusive.
   $\{-6, -4, -2, 0, 2\}$
6. All the integers greater than 5.
   $\{6, 7, 8, 9, \ldots\}$

Teaching Suggestions
p. T79

Key Ideas

Solve compound sentences. Graph solution sets of compound sentences. Solve word problems involving compound sentences.

Common Errors

Students are often confused by a set such as $\{x: x > 6$ or $x < 2\}$, because they conceive of this set as a union of two sets (true), which suggests "and" (false). Remind students that "$x > 6$ or $x < 2$" is a *condition* on x, and no real number is both less than 2 *and* greater than 6.

C 27. Prove: For all real numbers a, if $a > 0$, then $-a < 0$.
28. Prove: For all real numbers a, if $a < 0$, then $-a > 0$.
29. Prove that for all real numbers a, b, and c, if $a < b$ and $c > 0$, then $ac < bc$.
30. Prove that for all real numbers a, b, and c, if $a < b$ and $c < 0$, then $ac > bc$.
31. Prove: For all real numbers a, b, and c, if $a + c > b + c$, then $a > b$.
32. Is it true that for all real numbers a and b, if $a > b$, then $a^2 > b^2$? If so, prove it. If not, give an example that demonstrates this.
33. Prove: For all real numbers a, b, c, and d, if $a > b$ and $c > d$, then $a + c > b + d$. (*Hint:* Show that $a + c > b + c$ and $b + c > b + d$.)
34. Prove: For all *positive* real numbers a, b, c, and d, if $a > b$ and $c > d$, then $ac > bd$. (*Hint:* Show that $ac > bc$ and $bc > bd$.)

2–5 Compound Sentences

Is it true that "$3 < 5$ or $3 = 5$"? The answer is "yes" because $3 < 5$. Of course, $3 = 5$ is *not* true, but the compound sentence "$3 < 5$ or $3 = 5$" is true because one part of it is true. A sentence such as

$$3 < 5 \quad \text{or} \quad 3 = 5$$

that is formed by joining two sentences with the word **or** is called a disjunction of sentences. For a disjunction to be true, *at least one* of the joined sentences must be true. Disjunctions such as "$3 < 5$ or $3 = 5$" are ordinarily written $3 \le 5$ and read "3 is less than or equal to 5." Similarly, $5 \ge 3$ is read "5 is greater than or equal to 3."

The graph of the solution set of the open sentence $x \le 2$ over $\mathcal{R}$ is shown below, where the solid dot at the endpoint of the graph indicates that the point is in the set.

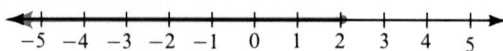

A compound sentence such as

$$2 < 3 \quad \text{and} \quad 3 < 5$$

that is formed by joining two sentences with the word and is called a conjunction of sentences, and is true if and only if *both* sentences are true. For example, the conjunction $2 < 3$ and $3 < 5$ is true, while the conjunction $2 < 3$ and $5 < 3$ is false, because $5 < 3$ is false.

Conjunctions of the form "$a < b$ and $b < c$" are ordinarily written

$$a < b < c$$

and read "a is less than b and b is less than c."

60   Chapter 2

Conjunctions and disjunctions frequently are combined in compound sentences. For example,

$$3 \leq x \leq 7$$

represents "3 is less than or equal to x, and x is less than **or** equal to 7."

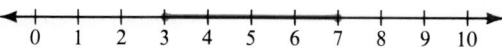

This sentence is true provided *both* disjunctions are true. Its graph over $\mathcal{R}$ shows both endpoints as solid dots to denote that the endpoints are in the set.

The transitive, addition, and multiplication properties of order given on page 57 also hold with $<$ replaced by $\leq$ and $>$ replaced by $\geq$. Similarly, the transformations shown on page 58 will produce equivalent sentences when used in sentences containing $\leq$ or $\geq$.

EXAMPLE   Solve $-(3x + 1) + 2 \leq 4 - x$ and graph its solution set.

SOLUTION
$$-(3x + 1) + 2 \leq 4 - x$$
$$-3x - 1 + 2 \leq 4 - x$$
$$-3x + 1 \leq 4 - x$$
$$-2x + 1 \leq 4$$
$$-2x \leq 3$$
$$x \geq -\tfrac{3}{2}$$

$\therefore$ the solution set is $\{x: x \geq -\tfrac{3}{2}\}$.

It is often useful to think of conjunctions and disjunctions of sentences in terms of set operations on the number line. For example:

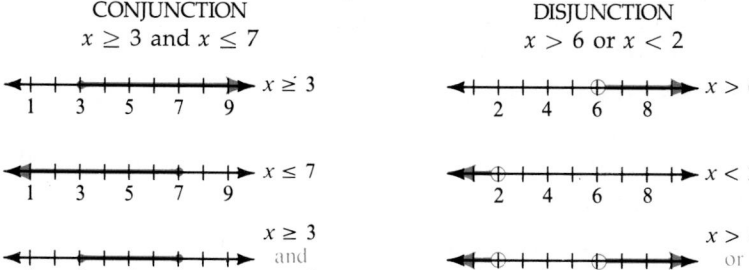

CONJUNCTION
$x \geq 3$ and $x \leq 7$

DISJUNCTION
$x > 6$ or $x < 2$

Using set notation, $x \geq 3$ and $x \leq 7$ can be written

$$\{x: x \geq 3\} \cap \{x: x \leq 7\}$$

and $x > 6$ or $x < 2$ can be written

$$\{x: x > 6\} \cup \{x: x < 2\}.$$

Review of Essentials   **61**

Additional A Exercises
(continued)

5. $-1 < x + 3 < 2$
$\{x: x > -4\} \cap \{x: x < -1\}$

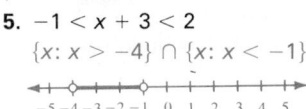

-5 -4 -3 -2 -1 0 1 2 3 4 5

6. $-7 < 2x + 5 < 1$
$\{x: x > -6\} \cap \{x: x < -2\}$

-6 -4 -2 0 2

Suggested Assignments

Minimum
    62/1–7, 9, 11, 13–18
    63/*P*: 1–6
S 59/25, 26
Extended Alg. with Trig.
    62/5–19 odd, 22–24
    63/*P*: 1–8
Enriched Alg.
    62/7–27 odd
    63/*P*: 3–11 odd
Enriched Alg. with Trig.
    62/7–27 odd
    63/*P*: 3–11 odd

Additional Answers
Written Exercises

1. $\{x: x \le -2\}$

-4 -2 0 2

2. $\{r: r \ge -7\}$

-9 -7 -5 -3 -1

3. $\{a: a \le 9\}$

3 5 7 9

4. $\{y: y \le 9\}$

3 5 7 9

5. $\left\{g: g \le \dfrac{4}{3}\right\}$

0 1 2

6. $\{p: p \le 6\}$

0 2 4 6 8

Oral Exercises

State a simple or compound open sentence describing each graph.

1.

0 1 2 3 4 5 6

$x > 3$

2.

-4 -3 -2 -1 0 1 2

$x \le -1$

3.

-2 -1 0 1 2 3 4

$x < 1$

4.

1 2 3 4 5 6 7

$x \ge 5$

5.

-2 -1 0 1 2 3 4

$x \le 0$ or $x \ge 2$

6.

-4 -3 -2 -1 0 1 2 3

$-3 < x < 2$

7.

-2 -1 0 1 2 3 4

$-1 < x \le 4$

8.

0 1 2 3 4 5 6

$x \le 1$ or $x > 5$

Describe in words the graph of each sentence.

9. $x > 2$ and $x < 5$
the points between the points
representing 2 and 5
not including the end points

10. $x > 3$ or $x < -1$
the points at the right of the
point representing 3, not including
the point at 3, and the points at
the left of the point representing
-1 not including the point at -1

Written Exercises

Solve each open sentence and graph its solution set.

A
1. $3x + 7 \le 1$
3. $a + 3 \ge 6(a - 7)$
5. $\frac{1}{2}(8 - 3g) \ge 2$
7. $-1 < k - 5 < 0$
9. $-8 < 2(n + 1) < 2$
11. $-4 \le \frac{1}{3}(v - 7) \le 1$
13. $7(u - 3) - 1 \ge u - 4$

2. $5 - 2r \le 19$
4. $-4(3 - y) \le 3(17 - y)$
6. $5 - \frac{1}{3}p \ge \frac{1}{6}p + 2$
8. $-3 < 2d + 1 < 7$
10. $1 < 3(m - 5) + 1 < 13$
12. $6 \le 9 - \frac{1}{2}t \le 10$
14. $8 - 2(w + 5) \ge 6 - 4w$

B
15. $3[(5 - z) - 2] \le \frac{1}{2}(9 - 3z)$

16. $-4(2 - 3w) \ge \frac{2}{3}[5 - (2 - 15w)]$

17. $\dfrac{2 - 3x}{4} - \dfrac{5 - x}{3} \le \dfrac{7}{6} - x$

18. $\dfrac{4 - y}{5} - \dfrac{7 + 3y}{2} \le -2 - \dfrac{1 + 11y}{10}$

19. $3b - 1 < -7$ or $2b + 5 > 13$

20. $-6c > 0$ and $8 - 7c < -20$

21. $4 + \dfrac{t}{3} < 3$ and $\dfrac{9 + 4t}{3} > -5$

22. $7 - \dfrac{4d}{5} < \dfrac{3}{5}$ or $1 - \dfrac{d}{2} > 4$

23. $-2(7 - 3h) < -8 + 2h$ or $-7h + 22 < 9 - 3(h + 5)$

24. $\dfrac{1}{2}(2 - 5r) < 6 - r$ and $r > 7 - \dfrac{1}{4}(r + 8)$

C
25. $-11 < 4v - 3 < 5$ or $v - 5 < 7 - 2v \le 3$
26. $20 \ge 8 - 3x \ge -7$ and $-7 < 6x - 1 < 41$
27. $7 \le 5 - 2y \le 13$ and $1 \le 9 - 4y \le 13$
28. $-2 < 4 - 3n \le 7$ or $17 > 5n + 12 > 7$

62   *Chapter 2*

Problems

A 1. Marie Melmotte has invited 16 adults and 15 children to a family party. She has a total of 125 oz of meat, of which the adults will have one size portion and the children another size that will be 2 oz smaller. What is the maximum weight of one adult portion? 5 oz

2. The height of a trapezoid is 15 cm and one base is 17 cm shorter than the other. What is the greatest length the shorter base can have if the area must be less than 165 cm²? 2.5 cm

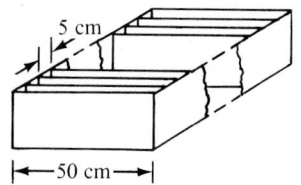

3. A drawer organizer has a front and a back each 50 cm long and dividers of the same length spaced 5 cm apart. What is the maximum number of dividers there can be if the *total* length of the front, back, sides, and dividers cannot exceed 770 cm? 11 dividers

4. Studs (vertical beams) for the wall of a house are $1\frac{1}{2}$ in. thick, and there must be no more than 14 in. between the sides of two adjacent studs. If there are two studs at each end of a wall that is $116\frac{1}{2}$ in. long, what is the smallest number of additional studs needed to support the wall? 7 studs

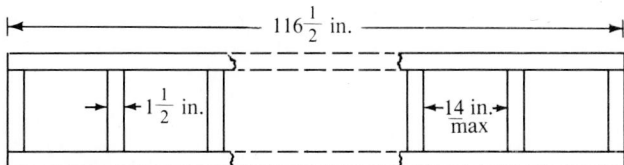

B 5. A cylindrical container including a top and bottom is made from no more than 108π cm² of sheet metal. If the radius of the top is 3 cm, what is the maximum volume that the container can hold? 135π cm³

6. Jose wants to swim part of the way and run part of the way to a friend's house. The distance from Jose's house to his friend's house along the straight shore of a lake is 2 km. If he runs at 18 km/h and swims at 2 km/h, what is the greatest distance he can swim and still arrive in less than 20 minutes? 0.5 km

7. The local train heading for Westford at 60 km/h and the express train heading for Eastford at 80 km/h pass each other at Centerville while heading in opposite directions along a straight roadbed. Westford and Eastford are 195 km apart, and if the local train makes one additional 15 min stop after passing Centerville the express completes its run before the local does. What is the maximum distance to the nearest kilometer between Westford and Centerville? 75 km

7. $\{k: 4 < k < 5\}$

8. $\{d: -2 < d < 3\}$

9. $\{n: -5 < n < 0\}$

10. $\{m: 5 < m < 9\}$

11. $\{v: -5 \le v \le 10\}$

12. $\{t: -2 \le t \le 6\}$

13. $\{u: u \ge 3\}$

14. $\{w: w \ge 4\}$

15. $\{z: z \ge 3\}$

16. $\{w: w \ge 5\}$

17. $\{x: x \le 4\}$

18. $\{y: y \ge -1\}$

19. $\{b: b < -2\} \cup \{b: b > 4\}$

20. $\varnothing$

21. $\{t: -6 < t < -3\}$

22. $\{d: d < -6\} \cup \{d: d > 8\}$

(continued)

23. $\left\{h: h < \frac{3}{2}\right\} \cup \{h: h > 7\}$

24. $\{r: r > 4\}$

25. $\{v: -2 < v < 4\}$

26. $\{x: -1 < x \le 5\}$

27. $\{-1\}$

28. $\{n: -1 \le n < 2\}$

Mixed Review

1. Simplify $2x - (3 - 4x) - (2 + 4x)$. $\quad 2x - 5$

2. Complete: If $4m - 4 = 2m + 6$, then $m = \underline{?}$ $\quad 5$

3. Solve $y - 5(2 - y) = 7 + 3y$ over $\mathscr{R}$. $\quad \left\{\frac{17}{3}\right\}$

4. Solve $6ab - 5c = 2ab + 7$ for b. $\quad b = \dfrac{5c + 7}{4a}$

5. Solve $6x + 3 < 8x - 7$ and graph its solution set. $\quad \{x: x > 5\}$

6. Each of the two congruent sides of an isosceles triangle is 7 cm less than twice the base. Find the length of each side of the triangle if its perimeter is 36 cm. Each congruent side is 13 cm long, and the base is 10 cm long.

---

8. The Crawleys' hallway has 300 ft² of wall space. They received an estimate of \$1.40/ft² to paint a wall and 60¢/ft² to wallpaper a wall. With \$324 to spend, what is the maximum area they can have painted, if the rest has to be papered? 180 ft²

9. In order to fill a water tank three pipes producing 12 L/min, 16 L/min, and 20 L/min of water are turned on at 30 min intervals. If they are turned on in the order given, how long after the first pipe is turned on will a 1200 L tank be filled? 60 min

10. Max runs the first 4 km of her training session at 8 km/h. At least how many kilometers must she run at 15 km/h in order to average 12 km/h for the whole session? 10 km

C 11. A tennis ball can in the shape of a cylinder with a flat top and bottom of the same radius as the tennis balls is designed so that the space inside the can that is *not* occupied by the balls has volume at most equal to the volume of one ball. What is the largest number of balls the can will contain? 2 tennis balls

12. Chris needs 50 m of fencing to enclose two separate square gardens whose areas must differ by no more than 100 m². What is the greatest length that a side of the larger square garden can have? 10.25 m

---

Emilie du Châtelet
1706–1749

As a child, Emilie de Breteuil amazed the French author and philosopher Voltaire when she divided nine figures by another nine figures entirely in her head. Later, as the Marquise du Châtelet, she dazzled France and much of Europe with her achievements in science and mathematics.

Emilie du Châtelet's most significant contribution to science was in her role as expounder of the principles of the great English physicist, Sir Isaac Newton. The last years of her life were spent translating Newton's *Principia* into French; her version is still the only one in that language. Earlier she had produced *Institutions de Physique*, begun as a textbook for her son, which is a comprehensive study of developments in the physical sciences during the seventeenth century. Châtelet's intellectual power was also displayed in an essay on the nature of fire, in which she introduced points later confirmed by research.

64 *Chapter 2*

2–6 Additional Properties of Order

The results of Exercises 27 and 28 on page 60 can be stated in the following theorem about the order of opposites and 0.

> *Theorem.*   For all real numbers a,
>
> $$\text{if } a > 0, \text{ then } -a < 0;$$
> $$\text{if } a < 0, \text{ then } -a > 0.$$

Another useful inequality is stated below.

> *Theorem.*   If a is a nonzero real number, then $a^2 > 0$.

Chalkboard Examples
Write an indirect proof for this theorem:
For all real numbers a and b, if $a > 0$ and $b < 0$, then $ab < 0$.
First assume the opposite of the conclusion; that is, assume $ab \geq 0$. Separate this into two different cases; in the first assume $ab = 0$, in the second assume $ab > 0$.

PROOF

If $a > 0$, then $a \cdot a > a \cdot 0$, from which $a^2 > 0$. On the other hand, if $a < 0$, then $-a > 0$, and $(-a)(-a) > (-a)(0)$, or $a^2 > 0$.

A corollary of the preceding theorem is that $1 > 0$, because for $a = 1$, $1^2 > 0$ and $1^2 = 1$. Thus, by the first theorem in this section, $-1 < 0$.

We can use the fact that $1 > 0$ to prove the following property of reciprocals.

Case 1: Assume that $ab = 0$.

1. $ab \cdot \dfrac{1}{b} = 0 \cdot \dfrac{1}{b}$

   Mult. prop. of equality

2. $a\left(b \cdot \dfrac{1}{b}\right) = 0 \cdot \dfrac{1}{b}$

   Assoc. axiom for mult.

3. $a \cdot 1 = 0 \cdot \dfrac{1}{b}$

   Ax. of mult. inverses

4. $a = 0 \cdot \dfrac{1}{b}$

   Iden. ax. for ×

5. $a = 0$

   Mult. prop. of zero

6. $a = 0$ contradicts the fact that $a > 0$.

   Hypothesis
   ∴ the assumption that $ab = 0$ must be incorrect.

(continued)

> *Theorem.*   For all nonzero real numbers a,
>
> $$\text{if } a > 0, \text{ then } \tfrac{1}{a} > 0;$$
> $$\text{if } a < 0, \text{ then } \tfrac{1}{a} < 0.$$

To prove this theorem, we shall use a method of reasoning called an **indirect proof** In an indirect proof, begin by assuming that the conclusion of a theorem is false, even though the hypothesis is accepted as true. Then show that a sequence of logically correct steps leads to a contradiction of an accepted fact, such as the hypothesis, an axiom, or a previously proved theorem. Because the assumption that the conclusion of the theorem is false leads to a contradiction, you know that the conclusion cannot be false, and thus that the theorem must be true.

Case 2: Assume that $ab > 0$.

1. $ab \cdot \dfrac{1}{b} < 0 \cdot \dfrac{1}{b}$

   Mult. prop. of order

2. $a\left(b \cdot \dfrac{1}{b}\right) < 0 \cdot \dfrac{1}{b}$

   Assoc. axiom for mult.

3. $a \cdot 1 < 0 \cdot \dfrac{1}{b}$

   Ax. of mult. inverses

4. $a < 0 \cdot \dfrac{1}{b}$

   Iden. Ax for $\times$

5. $a < 0$

   Mult. property of zero

6. $a < 0$ contradicts the fact
   that $a > 0$.
   Hypothesis
   ∴ the assumption that
   $ab > 0$ must be incorrect.

   Hence, $ab < 0$.
   Comparison axiom

Common Errors

Students may assume that
the hypothesis instead of the
conclusion is false when they
are writing indirect proofs.
Have them identify the hypothesis and conclusion for
several theorems, then have
them negate each conclu-
sion. Stress that the negation
of "$a > b$" is not "$a < b$" but
"$a \leq b$."

As an example of an indirect proof, let us prove the first part of the preceding theorem.

PROOF

Suppose that a is a real number such that $a > 0$. To show that $\dfrac{1}{a} > 0$, we shall show that assuming $\dfrac{1}{a}$ is **not** greater than 0 (in symbols, $\dfrac{1}{a} \not> 0$) leads to a contradiction.

If $\dfrac{1}{a} \not> 0$, then by the comparison axiom of inequality there are two cases to consider: (1) $\dfrac{1}{a} = 0$, and (2) $\dfrac{1}{a} < 0$.

Case 1: Assume that $\dfrac{1}{a} = 0$.

1. $\dfrac{1}{a} \cdot a = 0 \cdot a$ Multiplication property of equality

2. $\dfrac{1}{a} \cdot a = 0$ Multiplicative property of zero

3. $1 = 0$ Axiom of multiplicative inverses

Case 2: Assume that $\dfrac{1}{a} < 0$.

1. $\dfrac{1}{a} < 0$ and $a > 0$ Hypothesis

2. $\dfrac{1}{a} \cdot a < 0 \cdot a$ Multiplication property of order

3. $\dfrac{1}{a} \cdot a < 0$ Multiplicative property of zero

4. $1 < 0$ Axiom of multiplicative inverses

In each case, the last step contains a statement that contradicts the fact that $1 > 0$, which was deduced on page 65. Therefore, the assumption that $\dfrac{1}{a} \not> 0$ leads to contradictions and must be incorrect. Hence, $\dfrac{1}{a} > 0$.

The proof of the second part of the theorem is left as Exercises 1 and 2 on page 67.

To Write an Indirect Proof of a Theorem

1. Assume that the conclusion of the theorem is false.
2. Reason from this assumption until you obtain a statement contradicting the hypothesis, an axiom, or a previously proved theorem.
3. Point out that the assumption must be incorrect, so that the conclusion of the theorem must be true.

Oral Exercises

State the assumption with which you would begin an indirect proof of each theorem. Assume that each variable represents a real number.

1. If $a \neq b$, then $a + c \neq b + c$. $a + c = b + c$
2. If $a^2 > 0$, then $a \neq 0$. $a = 0$
3. If $a \geq b$, then $-a \leq -b$. $-a > -b$
4. If $ac \geq bc$ and $c > 0$, then $a \geq b$. $a < b$
5. If $a > 0$ and $ab > 0$, then $b > 0$. $b \leq 0$
6. If $ac \neq bc$, then $a \neq b$. $a = b$
7. If the bisectors of $\angle A$ and $\angle B$ in $\triangle ABC$ are congruent, then $\triangle ABC$ is isosceles. $\triangle ABC$ is not isosceles
8. In $\triangle PQR$, if $m\angle P > m\angle Q$, then $QR > PR$. $QR \leq PR$
9. If a or b is even, then ab is even. ab is not even
10. If $a = b + 3$, then $a \neq b$. $a = b$
11. If $a \neq b$, then $-a \neq -b$. $-a = -b$
12. If $a > 1$, then $\frac{1}{a} < 1$. $\frac{1}{a} \geq 1$
13. If $a < 0$, then $-a > 0$. $-a \leq 0$
14. If $c \neq 0$ and $a < b$, then $\frac{a}{c} < \frac{b}{c}$. $\frac{a}{c} \geq \frac{b}{c}$

Written Exercises

Justify each step in the indirect proofs of the following theorems. For the step designated "contradiction" tell what fact is contradicted. Assume that each variable represents a real number.

Theorem I. If $a < 0$, then $\frac{1}{a} < 0$.

PROOF

A 1. Case 1: Assume that $\frac{1}{a} = 0$.

   1. $\frac{1}{a} \cdot a = 0 \cdot a$ Mult. prop. of =

   2. $\frac{1}{a} \cdot a = 0$ Mult. prop. of 0

   3. $1 = 0$ Ax. of mult. inv.

   4. Contradiction $1 > 0$

2. Case 2: Assume that $\frac{1}{a} > 0$.

   1. $\frac{1}{a} > 0$ and $a < 0$ Hyp.

   2. $\frac{1}{a} \cdot a < 0 \cdot a$ Mult. prop. of order

   3. $\frac{1}{a} \cdot a < 0$ Mult. prop. of 0

   4. $1 < 0$ Ax. of mult. inv.

   5. Contradiction $1 > 0$

Suggested Assignments

Minimum
  67/1–4
S 62/8, 10, 12
Extended Alg. with Trig.
  67/1–5, 7–15 odd, 16
S 62/20, 21
Enriched Alg.
  67/3, 4, 5–17 odd
Enriched Alg. with Trig.
  67/3, 4, 5–17 odd

Additional A Exercises

Justify each step in the indirect proofs of the following theorems. For the step designated "contradiction," tell what fact is contradicted. Assume that each variable represents a real number.

1. Theorem: If $b = 0$, then $a + b = a$.

   Case 1: Assume that $a + b > a$.

   1. $a + b + (-a) > a + (-a)$
   Addition property for order
   2. $b + a + (-a) > a + (-a)$
   Comm. axiom for addition
   3. $b + [a + (-a)] > a + (-a)$
   Assoc. axiom for addition
   4. $b + 0 > 0$
   Axiom of additive inverses
   5. $b > 0$
   Identity axiom for addition
   6. Contradiction $b = 0$

(continued)

Theorem II. If $a > b > 0$, then $\dfrac{a}{b} > 1$.

PROOF

3. Case 1: Assume that $\dfrac{a}{b} = 1$.

1. $\dfrac{a}{b} \cdot b = 1 \cdot b$ Mult. prop. of $=$

2. $\dfrac{a}{b} \cdot b = b$ Iden. ax. for mult.

3. $\left(a \cdot \dfrac{1}{b}\right)b = b$ Def. of div.

4. $a\left(\dfrac{1}{b} \cdot b\right) = b$ Assoc. ax. for mult.

5. $a \cdot 1 = b$ Ax. of mult. inv.

6. $a = b$ Iden. ax. for mult.

7. Contradiction Hypothesis:
$a > b$

4. Case 2: Assume that $\dfrac{a}{b} < 1$.

1. $\dfrac{a}{b} < 1$ and $a > b > 0$ Hypothesis

2. $\dfrac{a}{b} \cdot b < 1 \cdot b$ Mult. prop. of order

3. $\dfrac{a}{b} \cdot b < b$ Iden. ax. for mult.

4. $\left(a \cdot \dfrac{1}{b}\right)b < b$ Def. of div.

5. $a\left(\dfrac{1}{b} \cdot b\right) < b$ Assoc. ax. for mult.

6. $a \cdot 1 < b$ Ax. of mult. inv.

7. $a < b$ Iden. ax. for mult.

8. Contradiction Hypothesis:
$a > b$

In Exercises 5–18 prove each theorem using either a direct or an indirect proof. Assume that each variable represents a real number.

B   **5.** If $a > 1$, then $\dfrac{1}{a} < 1$.

6. If $a > 0$ and $\dfrac{1}{a} < 1$, then $a > 1$.

7. If $a > b$, then $\dfrac{a + b}{2} < a$. (The number $\dfrac{a + b}{2}$ is called the *arithmetic mean* or *average* of a and b.)

8. If $a > b$, then $\dfrac{a + b}{2} > b$.

C   **9.** If $0 < a < b$, then $\dfrac{1}{a} > \dfrac{1}{b}$.

10. If $a > b > 0$, then $a^2 > b^2$.

11. If $a > 0$, $b > 0$, and $a^2 > b^2$, then $a > b$.

12. If $a < b < 0$, then $\dfrac{1}{a} > \dfrac{1}{b}$.

13. If $a < 0$, $b < 0$, and $a^2 > b^2$, then $a < b$.

14. If $a < 0$, $b < 0$, and $a > b$, then $a^2 < b^2$.

15. If a, b, c, and d are all *positive* with $ab < cd$ and $b > d$, then $a < c$. (*Hint*: First show that $ab > ad$.)

16. If a, b, c, and d are all *negative* with $ab < cd$ and $b < d$, then $a > c$. (*Hint*: First show that $ab > ad$.)

17. If $a > b > 0$, then $\dfrac{2ab}{a + b} > b$. (The number $\dfrac{2ab}{a + b}$ is called the *harmonic mean* of a and b.)

18. If $a > b > 0$, then $\dfrac{2ab}{a + b} < a$.

2–7 Absolute Value and Order

For every nonzero real number a, the **absolute value** of a, denoted $|a|$, is the positive number of the pair a and the opposite of a. For example, $|-3|$ is the positive number of the pair -3 and 3, so $|-3| = 3$. Similarly, $|7| = 7$ and $|-8| = 8$. The absolute value of 0, denoted $|0|$, is defined to be 0. Formally, we make this definition.

> If a is a real number, then
> $$|a| = \begin{cases} a, & \text{if } a \geq 0, \\ -a, & \text{if } a < 0. \end{cases}$$

EXAMPLE 1   Solve $|x - 4| = 7$.

SOLUTION   By definition, you have the disjunction

$$
\begin{array}{lll}
(x - 4) = 7 & \text{or} & -(x - 4) = 7, \\
x - 4 = 7 & \text{or} & x - 4 = -7, \\
x = 11 & \text{or} & x = -3.
\end{array}
$$

∴ the solution set is $\{-3, 11\}$.

On a number line $|a|$ or $|-a|$ measures the distance between the origin and the graph of a or $-a$ shown in the diagram below.

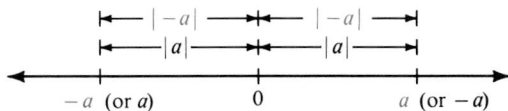

Graphically, an expression of the form

$$|a - b| \text{ or } |b - a|$$

represents the nondirected distance between the graph of a and the graph of b as shown below.

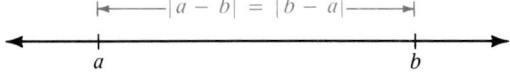

Review of Essentials **69**

Teaching Suggestions
p. T80

Related Activities p. T80

Supplementary Material
Test 5

Key Ideas
Solve open sentences involving absolute value.

Chalkboard Examples
1. Solve $|n - 2| = 3$.
$(n - 2) = 3$ or $-(n - 2) = 3$
$n - 2 = 3$ or $n - 2 = -3$
$n = 5$ or $n = -1$
∴ the sol. set is $\{-1, 5\}$.

Solve each open sentence and graph its solution set.

2. $|n - 1| \leq 2$
$(n - 1) \leq 2$ and $-(n - 1) \leq 2$
$n - 1 \leq 2$ and $n - 1 \geq -2$
$n \leq 3$ and $n \geq -1$
∴ the solution set is
$\{n: n \geq -1\} \cap \{n: n \leq 3\}$
or, $\{n: -1 \leq n \leq 3\}$.

3. $|q + 4| \geq 1$
$(q + 4) \geq 1$ or $-(q + 4) \geq 1$
$q + 4 \geq 1$ or $q + 4 \leq -1$
$q \geq -3$ or $q \leq -5$
∴ the solution set is
$\{q: q \leq -5\} \cup \{q: q \geq -3\}$.

Additional A Exercises

Simplify.

1. $|9| - |-4|$ 5

2. $|-7 - (-3)|$ 4

3. $|8| \cdot |-2| - |-4|$ 12

Solve each open sentence and graph its solution set.

4. $|p| \leq 3$
   $\{p: p \geq -3\} \cap \{p: p \leq 3\}$
   or, $\{p: -3 \leq p \leq 3\}$

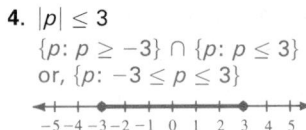

5. $|a - 4.5| \leq 2.5$
   $\{a: a \geq 2\} \cap \{a: a \leq 7\}$ or,
   $\{a: 2 \leq a \leq 7\}$

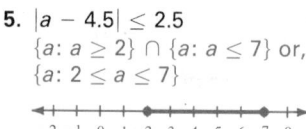

6. $|y + 4| > 2$
   $\{y: y < -6\} \cup \{y: y > -2\}$

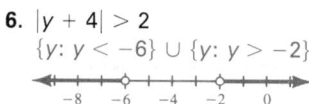

EXAMPLE 2   Solve $|x - 6| \leq 4$ and graph its solution set.

SOLUTION   By inspection, $|x - 6|$ represents the distance between the graph of 6 and the graph of x. The sentence asserts that this distance is 4 or less. Since 2 and 10 are 4 units from 6, you see at once that $|x - 6| \leq 4$ is equivalent to $2 \leq x \leq 10$. Thus, by inspection, the solution set is $\{x: 2 \leq x \leq 10\}$. More formally,

$$|x - 6| \leq 4$$

is equivalent to the compound sentence

$$
\begin{array}{lcl}
(x - 6) \leq 4 & \text{and} & -(x - 6) \leq 4, \\
x - 6 \leq 4 & \text{and} & x - 6 \geq -4, \\
x \leq 10 & \text{and} & x \geq 2.
\end{array}
$$

∴ the solution set is $\{x: x \leq 10\} \cap \{x: x \geq 2\}$ or, $\{x: 2 \leq x \leq 10\}$ whose graph is shown below.

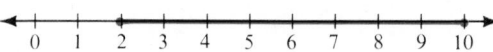

Example 2 suggests that sentences of the form

$$|\text{expression}| \leq b$$

are equivalent to conjunctions. The next example suggests that sentences of the form

$$|\text{expression}| \geq b$$

are equivalent to disjunctions.

EXAMPLE 3   Solve $|z + 1| \geq 5$ and graph its solution set.

SOLUTION   Write $|z + 1| \geq 5$ as $|z - (-1)| \geq 5$. Then, by inspection, the distance between the graph of z and the graph of -1 must be 5 or greater. Since -6 and 4 are each 5 units from -1 the solution set must be

$$\{z: z \leq -6\} \cup \{z: z \geq 4\}.$$

More formally,

$$|z + 1| \geq 5$$

is equivalent to

$$
\begin{array}{lcl}
(z + 1) \geq 5 & \text{or} & -(z + 1) \geq 5, \\
z + 1 \geq 5 & \text{or} & z + 1 \leq -5, \\
z \geq 4 & \text{or} & z \leq -6.
\end{array}
$$

∴ the solution set is $\{z: z \geq 4\} \cup \{z: z \leq -6\}$ whose graph is shown below.

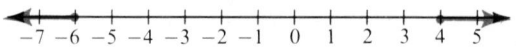

Oral Exercises

State the value of each expression.

1. $|-7|$ 7 **2.** $|-5| + |5|$ 10 **3.** $|-9 + 9|$ 0 **4.** $|4| - |-4|$ $\overset{0}{}$

5. $|7 - 10|$ 3 **6.** $|4| - |9|$ -5 **7.** $|6||-2|$ 12 **8.** $|0|$ 0

State a compound sentence equivalent to each sentence. You need not solve.

EXAMPLE $|x| = 3$ SOLUTION $x = 3$ or $x = -3$

9. $|x + 2| = 8$ $x + 2 = 8$ or $-(x + 2) = 8$ **10.** $|z - 4| = 3$ $z - 4 = 3$ or $-(z - 4) = 3$

11. $|y - 3| \le 5$ $y - 3 \le 5$ and $-(y - 3) \le 5$ **12.** $|r + 2| \ge 7$ $r + 2 \ge 7$ or $-(r + 2) \ge 7$

13. $|n + 1| > 9$ $n + 1 > 9$ or $-(n + 1) > 9$ **14.** $|k - 6| < 2$ $k - 6 < 2$ and $-(k - 6) < 2$

Express as a single inequality using absolute value.

15. $-6 < x < 6$ $|x| < 6$

16. $x > 10$ or $x < -10$ $|x| > 10$

17. $x + 3 \ge 2$ or $x + 3 \le -2$
$|x + 3| \ge 2$

18. $-5 < x - 1 < 5$ $|x - 1| < 5$

Classify each statement as true or false.

19. The inequalities $|a - 3| < 2$ and $|3 - a| < 2$ are equivalent. true

20. For all $a \in \mathscr{R}$, $|a| + |-a| = 0$. false

21. For all $a \in \mathscr{R}$, $|a|$ is positive. false

22. The solution set for $|x| = -2$ is $\emptyset$. true

Written Exercises

Simplify.

A

1. $|7| - |-2|$ 5 **2.** $|-5 - (-8)|$ 3 **3.** $|-6| + |6|$ 12

4. $|3| - |-3|$ 0 **5.** $|-4| \cdot |3| - |-5|$ 7 **6.** $|6 - 4| - |4 - 6|$
$\overset{0}{}$

Solve each open sentence and graph its solution set.

7. $|r| \ge 4$ **8.** $|t| < \frac{5}{2}$ **9.** $|x - 7| \le 1$

10. $|y + 1| > 3$ **11.** $|c - 6| \ge 2$ **12.** $|d + 3| < 4$

13. $|2k - 3| = 1$ **14.** $|7 - 3p| = 5$ **15.** $|2n + 9| \ge 7$

16. $|\frac{1}{2}r + 1| < 4$ **17.** $\frac{1}{3}|q - 4| \le 2$ **18.** $3|w - 6| < 12$

B **19.** $\left|2m - \frac{1}{2}\right| \le \frac{5}{2}$ **20.** $-2\left|x + \frac{1}{2}\right| < -6$ **21.** $\left|\frac{2u + 7}{3}\right| \ge 3$

22. $\left|\frac{3 - 2v}{5}\right| < 1$ **23.** $\left|\frac{2}{3}a - 3\right| > 1$ **24.** $\left|\frac{7 - 3c}{2}\right| \le 4$

Review of Essentials **71**

---

9. $\{x: 6 \le x \le 8\}$

10. $\{y: y > 2\} \cup \{y: y < -4\}$

11. $\{c: c \ge 8\} \cup \{c: c \le 4\}$

12. $\{d: -7 < d < 1\}$

13. $\{1, 2\}$

14. $\left\{\frac{2}{3}, 4\right\}$

15. $\{n: n \ge -1\} \cup \{n: n \le -8\}$

16. $\{r: -10 < r < 6\}$

17. $\{q: -2 \le q \le 10\}$

18. $\{w: 2 < w < 10\}$

19. $\left\{m: -1 \le m \le \frac{3}{2}\right\}$

20. $\left\{x: x > \frac{5}{2}\right\} \cup \left\{x: x < -\frac{7}{2}\right\}$

21. $\{u: u \ge 1\} \cup \{u: u \le -8\}$

22. $\{v: -1 < v < 4\}$

23. $\{a: a > 6\} \cup \{a: a < 3\}$

24. $\left\{c: -\frac{1}{3} \le c \le 5\right\}$

State whether each statement is true for *all* real numbers *a*, *b*, and c (excluding 0 denominators). If the statement is false, give a value for each variable that shows this.

25. $|a + b| = |a| + |b|$ false   **26.** $|a| \cdot |b| = |ab|$ true
                             $a = -1, b = 2$
27. $\dfrac{|a|}{|b|} = \left|\dfrac{a}{b}\right|$ true   **28.** If $|a| = |b|$, then $a = b$. false;
                                        $a = -1, b = 1$
29. $|a| \ge a$ true   **30.** $|a - b| = |a| - |b|$ false; $a = -1, b = 2$
31. If $|a - b| = 0$, then   **32.** $|a| = |-a|$ true
     $a = b$. true
33. $|a + b|^2 = (|a| + |b|)^2$ false   **34.** $|a - b| \ge |a| - |b|$ true
                      $a = -1, b = 2$

Prove each theorem for all real numbers *a* and *b*, using the formal definition of $|a|$.

C   **35.** $|a|^2 = a^2$
      36. $ab \le |ab|$ (*Hint*: Consider several cases.)
      37. $|a + b|^2 \le (|a| + |b|)^2$ (*Hint*: Use the results of Exercises 35 and 36.)
      38. Use the theorem proved in Exercise 11 on page 68 to prove:

$$|a + b| \le |a| + |b|$$

(*Hint*: Use the result of Exercise 37.)

Computer Exercises   For students with computer experience

1. Write a program that will solve an inequality of the form

$$|x + a| < b \quad \text{or} \quad |x + a| > b, (b > 0).$$

Input the values of a and b as well as the inequality sign.

Run the program in Exercise 1 for each inequality.

2. $|x - 5| < 7$ $-2 < x < 12$ **3.** $|x + 4| > 3$ $x < -7$ or $x > -1$
4. $|x + 2| > 15$ **5.** $|x - 28| < 9$
     $x < -17$ or $x > 13$ $19 < x < 37$

■ Self-Test 2

VOCABULARY   is less than (p. 56)   compound sentence (p. 60)
                     order (p. 56)   disjunction (p. 60)
                     is greater than (p. 56)   conjunction (p. 60)
                     converse (p. 56)   indirect proof (p. 65)
                     equivalent inequalities (p. 58)   absolute value (p. 69)

Solve each inequality over $\mathcal{R}$ and graph its solution set.

1. $5a - 3 > 17$ *Obj. 1, p. 56*

2. $1 - 2x < 3$

3. $3(y + 14) \geq y - (4 + 2y)$

4. $0 < 2(4b - 2) \leq 8$

5. The charge for the first 3 min of a certain long-distance telephone call   *Obj. 2, p. 56*
   is \$2.05. The charge for each additional minute (or fraction of a
   minute) is \$.35. What is the greatest number of minutes Beth can talk
   on the telephone if she has \$4.00 to spend on this long distance call?   8 min.

6. Use an indirect proof to show that if $a < 0$ and $ab > 0$, then $b < 0$.   *Obj. 3, p. 56*

Solve each open sentence over $\mathcal{R}$ and graph its solution set.

7. $|x - 5| \leq 7$ *Obj. 4, p. 56*

8. $|3y + 2| > 4$

Check your answers with those at the back of the book.

Chapter Summary

1. One polynomial is added to or subtracted from another by adding or
   subtracting coefficients of like terms. The result of this procedure is a
   simpler polynomial *equivalent* to the original sum or difference.

2. Transformations that produce an equivalent equation are:
   1. Substituting for either side of the given equation an expression
      equivalent to it.
   2. Adding to or subtracting from each side of the given equation the
      same polynomial in any variable(s) appearing in the equation.
   3. Multiplying or dividing each side by the same nonzero number.

3. To solve a word problem:
   1. Read the problem carefully several times and decide what
      number or numbers are asked for.
   2. Choose a variable to represent one of the numbers asked for or
      described in the problem. A sketch or a chart may be helpful.
   3. Write an open sentence representing the relationship(s) stated or
      implied in the problem.
   4. Solve the open sentence.
   5. Check your results with the requirements stated in the problem.

4. The real numbers are ordered. See the *comparison axiom* on page 56
   and the *transitive, addition,* and *multiplication properties of order* on
   page 57.

Review of Essentials   **73**

73

---

1. $\{a: a > 4\}$

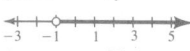

2. $\{x: x > -1\}$

3. $\left\{y: y \geq -\dfrac{23}{2}\right\}$

4. $\left\{b: \dfrac{1}{2} < b \leq \dfrac{3}{2}\right\}$

6. Case 1: Assume $b = 0$
   1. $a < 0$, $ab > 0$, $b = 0$
      Hypothesis
   2. $ab = ab$
      Refl. prop. of $=$
   3. $ab = a \cdot 0$
      Subs. prin.
   4. $ab = 0$
      Mult. prop. of 0
   5. Contradiction
      Hypothesis: $ab > 0$
   Case 2: Assume $b > 0$
   1. $a < 0$, $ab > 0$, $b > 0$
      Hypothesis
   2. $a \cdot b < 0 \cdot b$
      Mult. prop. of order
   3. $ab < 0$
      Mult. prop. of 0
   4. Contradiction
      Hypothesis: $ab > 0$
   Therefore, $b < 0$

7. $\{x: -2 \leq x \leq 12\}$

8. $\left\{y: y > \dfrac{2}{3}\right\} \cup \{y: y < -2\}$

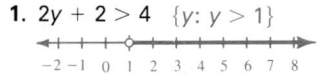
Solve each inequality over $\mathcal{R}$ and graph its solution set.

1. $2y + 2 > 4$ $\{y: y > 1\}$

$\xleftarrow{\hspace{0.3cm}}\overset{\hspace{0.2cm}-2\ -1\ \ 0\ \ 1\ \ 2\ \ 3\ \ 4\ \ 5\ \ 6\ \ 7\ \ 8}{\longmapsto\!\!\!\!\!\circ\!\!-\!\!\!+\!\!\!+\!\!\!+\!\!\!+\!\!\!+\!\!\!+\!\!\!+\!\!\!+}\xrightarrow{\hspace{0.3cm}}$

2. $6 - 3t > -3$ $\{t: t < 3\}$

$\xleftarrow{\hspace{0.3cm}}\overset{\hspace{0.2cm}-2\ -1\ \ 0\ \ 1\ \ 2\ \ 3\ \ 4\ \ 5\ \ 6\ \ 7\ \ 8}{\longmapsto\!\!-\!\!\!+\!\!\!+\!\!\!+\!\!\!+\!\!\!\circ\!\!\!+\!\!\!+\!\!\!+\!\!\!+\!\!\!+}\xrightarrow{\hspace{0.3cm}}$

3. $4(c + 12) \le c - (6 + 3c)$

$\{c: c \le -9\}$

$\xleftarrow{\hspace{0.3cm}}\overset{\hspace{0.2cm}-11\ \ \ -9\ \ \ -7\ \ \ -5\ \ \ -3}{\longmapsto\!\!\!\bullet\!\!-\!\!\!+\!\!\!+\!\!\!+\!\!\!+\!\!\!+\!\!\!+\!\!\!+}\xrightarrow{\hspace{0.3cm}}$

4. $-3 < 2x - 7 \le 1$ $\{x: 2 < x \le 4\}$

$\xleftarrow{\hspace{0.3cm}}\overset{\hspace{0.2cm}-2\ -1\ \ 0\ \ 1\ \ 2\ \ 3\ \ 4\ \ 5\ \ 6\ \ 7\ \ 8}{\longmapsto\!\!-\!\!\!+\!\!\!+\!\!\!+\!\!\!\circ\!\!-\!\!\!\bullet\!\!\!+\!\!\!+\!\!\!+\!\!\!+}\xrightarrow{\hspace{0.3cm}}$

5. The two congruent sides of an isosceles triangle are each 2 cm less than three times the base. If the perimeter of the triangle is to be at most 66 cm, what is the maximum length of the base?
10 cm

6. Use an indirect proof to show that if $ab < 0$ and $a < 0$ then $b > 0$.

Case 1: Assume that $b = 0$.

1. $ab = a \cdot 0$ Mult. prop. of equality
2. $ab = 0$ Mult. prop. of zero
3. $ab = 0$ contradicts the fact that $ab < 0$.
   Hypothesis ∴ the assumption that $b = 0$ must be incorrect.

Case 2: Assume that $b < 0$.

1. a is negative.
   Hypothesis
2. $ab > a \cdot 0$ Mult. prop. of order
3. $ab > 0$ Mult. prop. of zero

5. Transformations that produce an equivalent inequality are:
   1. Substituting for either side of the inequality an expression equivalent to that side.
   2. Adding to or subtracting from each side of the inequality the same polynomial in any variable(s) appearing in the inequality.
   3. Multiplying or dividing each side by the same *positive* number.
   4. Multiplying or dividing each side by the same *negative* number and reversing the direction of the inequality symbol.

6. Compound sentences involving the word "or" are *disjunctions*, and are true whenever at least one of the simple sentences involved is true.

7. Compound sentences involving the word "and" are *conjunctions*, and are true whenever both of the simple sentences involved are true.

8. To write an *indirect proof*, start with the assumption that the conclusion of a theorem is false and reason from this to a contradiction of the hypothesis, an axiom, or a previously proved theorem.

9. The *absolute value* of a number is defined by

$$|a| = \begin{cases} a, & \text{if } a \ge 0, \\ -a, & \text{if } a < 0. \end{cases}$$

Geometrically, $|a|$ measures the distance between the origin and the graph of a or $-a$.

Chapter Review

Write the letter of the correct answer.

Simplify.

1. $(2p^2 + 7p - 4) - (p^2 - 8p + 6)$ 2-1
   a. $p^2 - p - 10$ **b.** $p^2 + 15p + 2$
   c. $p^2 + 15p - 10$ **d.** $p^2 - p + 2$

2. $[-3x^2 - (8x + 2)] - 2[x^2 - 3(x - 4)]$
   a. $-x^2 - 11x + 14$ **b.** $-5x^2 - 11x + 22$
   c. $-5x^2 - 2x - 26$ **d.** $-4x^2 - 14x - 26$

Solve.

3. $10 - 2(2 - f) = 2(f + 4)$ 2-2
   a. $\{-\frac{4}{5}\}$ **b.** $\varnothing$ **c.** $\{2\}$ **d.** $\{4\}$

4. $5 + \dfrac{18}{x} = 8$
   a. $\{0\}$ **b.** $\{6\}$ **c.** $\{18\}$ **d.** $\{3\}$

74 Chapter 2

Exercises 5 and 6 refer to the following situation. When the Linn family drives to visit their relatives, they usually travel at a constant speed of 75 km/h. During snowstorms, it takes 1 h longer to travel the same route at a constant speed of 60 km/h. Let x = the time traveling during snowstorms.

5. Which equation represents the given conditions?

   a. $75x = 60(x + 1)$ **b.** $75x + 1 = 60x$

   c. $75x = 60x + 1$ **d.** $75(x - 1) = 60x$

6. What is the time traveling during snowstorms?

   a. 2.5 h   **b.** 5 h   **c.** 4 h   **d.** 1 h

Solve.

7. $5 + \frac{1}{4}m < -3$

   a. $\{m: m < 2\}$ **b.** $\{m: m < -2\}$ **c.** $\{m: m > 32\}$ **d.** $\{m: m < -32\}$

8. $4 - \frac{1}{5}(2x + 6) > 2(3 - \frac{2}{5}x)$

   a. $\{x: x > 2\}$ **b.** $\{x: x > 8\}$ **c.** $\{x: x < 8\}$ **d.** $\{x: x > \frac{3}{5}\}$

9. Which of the following is equivalent to $x - 8 \geq 2$?

   a. $x - 8 = 2$ and $x - 8 > 2$ **b.** $x - 8 = 2$ or $x - 8 > 2$

   c. $x = 2$ and $x - 8 > 2$ **d.** $x = 2$ or $x - 8 < 2$

10. Solve the conjunction $3e + 1 < 7$ and $-\frac{2}{9}e + \frac{1}{3} < 1$ for e.

   a. $\{e: e < 2 \text{ or } e > -3\}$ **b.** $\{e: e > 2 \text{ or } e < -3\}$

   c. $\{e: -3 < e < 2\}$ **d.** $\{e: e > 2 \text{ and } e < -3\}$

In Exercises 11 and 12, state the assumption with which you would begin an indirect proof of each statement.

11. If $a \geq b$, then $a + c \geq b + c$.

   a. $a + c \geq b + c$ **b.** $a \leq b$

   c. $a \geq b$ **d.** $a + c < b + c$

12. If $a \neq b$, then $ac \neq bc$ $(c \neq 0)$.

   a. $a = b$ **b.** $ac = bc$ **c.** $a \neq b$ **d.** $ac \neq bc$

Solve.

13. $|5 - 2p| = 7$

   a. $\{-1\}$ **b.** $\{-1, 1\}$ **c.** $\{-1, -6\}$ **d.** $\{-1, 6\}$

14. $|4x + 6| \geq 2$

   a. $\{x: -2 \leq x \leq 1\}$ **b.** $\{x: x \leq 1 \text{ or } x \geq 2\}$

   c. $\{x: x \leq -2 \text{ or } x \geq -1\}$ **d.** $\{x: -1 \leq x \leq 1\}$

2-3

2-4

2-5

2-6

2-7

---

4. $ab > 0$ contradicts the fact that $ab < 0$.
   Hypothesis
   ∴ the assumption that $b < 0$ must be incorrect.
   Hence, $b > 0$.
   Comparison axiom

Solve each open sentence over $\mathscr{R}$ and graph its solution set.

7. $|p - 3| \leq 2$
   $\{p: p \geq 1\} \cap \{p: p \leq 5\}$,
   or $\{p: 1 \leq p \leq 5\}$

   −2 −1 0 1 2 3 4 5 6 7 8

8. $|3w + 6| > 6$
   $\{w: w < -4\} \cup$
   $\{w: w > 0\}$

   −6 −4 −2   0   2

Reading Algebra

Give your students practice reading open sentences with absolute values. For example, $2|x - 3| > 5(x + 4)$ should be read "Two times the absolute value of the quantity x minus three is greater than five times the quantity x plus four."

Mixed Review

Simplify.

1. $2cp^2 + c(2 - c) - p(2c + 1)$
   $2cp^2 - c^2 - 2cp + 2c - p$

2. $\left(\frac{1}{m}\right)\left(\frac{1}{n}\right)$ $\frac{1}{mn}$

3. $12 - |-(2 + 23)| - 25$ -38

Solve.

4. $2(t - 4) = 4t - 12$ $\{2\}$

5. $s - 4 > -3s + 16$
   $\{s: s > 5\}$

6. Evaluate $4v + 3 - 2v - 8$ for $v = -6$. -17

Additional Answers
Chapter Test

9. $\{x: x < -10\}$

10. $\{x: x > 2\}$

11. $\{n: -1 < n < 5\}$

12. $\{q: q < 2\} \cup \{q: q \geq 5\}$

14. Case 1: Assume $a = b$
  1. $a = b$, $a + c > b + c$
    Hypothesis
  2. $a + c = b + c$
    Add. prop. of $=$
  3. Contradiction of
    Hypothesis:
    $a + c > b + c$
  Case 2: Assume $a < b$
  1. $a < b$, $a + c > b + c$
    Hypothesis
  2. $a + c < b + c$
    Add. prop. of order
  3. Contradiction
    Hypothesis: $a + c >$
    $b + c$
  Therefore, $a > b$.

15. $\{x: -2 \leq x \leq 1\}$

16. $\{v: -2 \leq v \leq 8\}$

Chapter Test

Simplify.

1. $3(-5x^2 + 2x + 1) - (-16x^2 + 5x) + 4(6 - x)$ $x^2 - 3x + 27$ 2-1

2. $[-4d^2 - (3e - 5d)] - [9d^2 - 3(2d - 4e)]$ $-13d^2 - 15e + 11d$

Solve.

3. $x - 9(3 - x) = 5 + 2x$ $\{4\}$ **4.** $\frac{1}{2}(3x + 7) - \frac{2}{3}(1 - 5x) = -2$ $\{-1\}$ 2-2

Solve for a.

5. $2ax = 3 + 5a$ $a = \frac{3}{2x - 5}$ **6.** $\frac{b}{a} - f = 3m$ $a = \frac{b}{3m + f}$

7. Find three consecutive integers such that the sum of the first and   2-3
twice the third is one more than three times the second.   Any three consecutive integers

8. Each of the two congruent sides of an isosceles triangle is 4 cm
longer than half the base. Find the length of each side of the triangle
if its perimeter is 36 cm.   14 cm, 11 cm, 11 cm

Solve each inequality and graph its solution set.

9. $\frac{x}{5} - 2 > \frac{x}{2} + 1$ **10.** $5 - \frac{1}{4}(2x + 4) < -3(1 - x)$ 2-4

11. $-3 < \frac{1}{3}(4n - 5) < 5$ **12.** $6q - 8 < 4$ or $1 - 2q \leq -9$ 2-5

13. Amy wants to fence in a garden whose length is 2 m more than twice
its width. What is the greatest width her garden can be, if she has at
most 70 m of fencing?   11 m

14. Use an indirect proof to prove if $a + c > b + c$, then $a > b$ (a, b, and   2-6
$c \neq 0$).

Solve each open sentence and graph its solution set.

15. $4|2x + 1| \leq 12$ **16.** $\left|\frac{3 - v}{5}\right| \leq 1$ 2-7

Mixed Review

Simplify.

1. $(-1)(6)(0)(-3)(-4)$ 0   **2.** $15 - 36 \div 9 \cdot 2 + 3$ 10

3. $10z[3w - 2(4 - w)]$ **4.** $(3x^2 - 5x + 1) - (2x^2 + 6x - 4) - (-6x^2 - 2)$
     $50zw - 80z$ $7x^2 + 11x + 7$

Solve.

5. $3 + (-x) = 7$ $\{-4\}$ **6.** $11 - 7(d + 3) = 4$ $\{-2\}$

76   Chapter 2

7. $-[z + (-2)] = 5z + (-4)$ $\{1\}$ 8. $5 - [10 - 3(2a + 1) + 6a] = 2(4 - a)$ $\{5\}$

Solve for b.

9. $4ab - 5b = 2a$ $b = \frac{2a}{4a - 5}$ 10. $\frac{m}{b} = \frac{t}{m}$ $b = \frac{m^2}{t}$

Evaluate for $r = -2$.

11. $-4 - 3(2 - r)$ -16 12. $(5 + r)[-2r - 6(r + 1)]$ 30

Solve each open sentence and graph its solution set.

13. $-2 < a - 4 \leq 3$ 14. $-2b + 3 < -5$ or $b + 1 > 4b + 4$

15. $|2t + 3| \leq 5$ 16. $\frac{x}{3} + 4 > -1 - \frac{x}{2}$

17. Use an indirect proof to show that if $a > b$, then $a - b > 0$.
18. Find three consecutive even integers whose sum is ten less than four times the smallest integer. 16, 18, 20
19. Francesca traveled for 5 h in her car at a constant speed of 90 km/h and then traveled in a bus at a constant speed for 3 h. If the total distance she traveled was 795 km, what was the speed of the bus? 115 km/h
20. The student government can sell a total of 400 tickets to a basketball game. An adult ticket costs $3 and a student ticket costs $2. What is the minimum number of adult tickets that must be sold if the student government wants to collect at least $1100? 300 adult tickets

Contest Problems

The following are mathematical problems with a strong emphasis on logical reasoning. They are similar to problems that you might encounter in a mathematical contest or competition.

1. Starting at point A, a man rows 1 mi upstream (against the current) to point B where his hat falls overboard. The man, however, does not notice the loss of the hat and continues rowing upstream. After 10 min he realizes that he has lost his hat and immediately turns and rows downstream. Exactly at point A he overtakes his hat, which has been carried downstream by the current. If the man rows at a constant rate relative to the water, what is the rate of the current? 3 mi/h
2. The distance from Rollins to Plainville is 2 mi. If Carol drives the first mile at a constant speed of 30 mi/h, at what constant speed must she drive the second mile so that her average speed for the whole trip is 60 mi/h?

Additional Answers
Mixed Review

13. $\{a: 2 < a \leq 7\}$

14. $\{b: b > 4\} \cup \{b: b < -1\}$

15. $\{t: -4 \leq t \leq 1\}$

16. $\{x: x > -6\}$

Additional Answers
Contest Problems

2. If Carol drives the first mile at 30 mi/h, her time will be $\frac{1}{30}$ h or 2 min. But if she wishes to average 60 mi/h for 2 mi, her total time must be 2 min. Hence, she cannot possibly have an average speed of 60 mi/h for 2 mi.

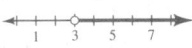
19. $\{d: d > 3\}$

20. $\{z: z < 4\}$

21. $\{v: v > 2\}$

22. $\{w: w < 8\}$

23. $\{x: x < -5\}$

24. $\{x: x > 0\}$

25. 1. Hypothesis
2. Def. of $<$
3. Add. prop. of $=$
4. Comm. and assoc. ax. for add.
5. Def. of $<$

26. 1. Hypothesis
2. Def. of $<$
3. Subt. prop. of $=$
4. Subs. prin.
5. Dist. ax. and assoc. ax. for add.
6. Def. of $>$

28. 1. $a < 0,\ a \in \mathscr{R}$
Hypothesis
2. $a + (-a) < 0 + (-a)$
Add. prop. of order
3. $0 < 0 + (-a)$
Ax. of add. inv.
4. $0 < -a$
Iden. ax. for add.
5. $-a > 0$
Inequal. rewritten

PROGRAMMING IN PASCAL

Now that you have studied compound sentences, you are prepared to understand how sets can be used to simplify and clarify various compound sentences used to control the flow of execution of a program. Suppose that the following base type and variables have been declared.

```
TYPE
    ok_nums = 1..100;

VAR
    x, y : integer;
    set_s, set_t : SET OF ok_nums;
```

Once *set_s* and *set_t* have been assigned appropriate values, then the following are equivalent.

Compound sentence	Pascal statement
$3 < x \le 10$	x IN [4..10]
$3 \le y \le 5$ or $8 \le y \le 15$	y IN [3..5] + [8..15]
$x = 3$ or $x = 5$ or $x = 7$	x IN [3, 5, 7]
$y - 10 \le x \le y + 10$	x IN [y−10 .. y+10]
$x \in S$ and $y \in S$	[x, y] <= set_s
$x \in S$ or $x \in T$	x IN set_s + set_t
$y \in S$ and $y \in T$	y IN set_s * set_t
$x \in S$ and $x \notin T$	x IN set_s − set_t
$x \le 10$ and $x \le 20$ or $x = 30$	x IN [10..20, 30]

There is an absolute value function in Pascal denoted by *abs*(x). This function takes its real or integer argument and returns its absolute value. For example:

$$abs(-5.3) = 5.3 \qquad abs(0) = 0 \qquad abs(3.41459) = 3.41459$$

The absolute value function can be used to write programs to solve simple absolute value equations or inequalities.

Exercises

1. Suppose that *s* and *t* have been declared as integer variables and that *set_u* and *set_v* have been assigned appropriate sets of integers. Translate each of the following compound sentences into an equivalent Pascal statement.

   a. $2 \leq s < 18$ s IN [2..17]

   b. $4 < t \leq 7$ or $10 \leq t < 15$ t IN [5..7] + [10..14]

   c. $s \in U$ and $s \in V$ s IN set_u * set_v

   d. $t \in U$ and $s \in U$ [t,s] <= set_u

   e. $s \in U$ or $s \in V$ s IN set_u + set_v

   f. $s = 2$ or $s = 4$ or $s = 6$ s IN [2,4,6]

   g. $s \notin U$ and $s \in V$ s IN set_v − set_u

   h. $s - 3 < t \leq s + 2$ t IN [s − 2 .. s + 2]

2. **a.** Write a Pascal program *abs_val* that prompts the user to enter a number, and displays the absolute value of that number.

   b. Modify the program you wrote in part (a) to execute repeatedly until the user enters 0, at which point the program ends after correctly processing the 0. An algorithm for such a program, (an outline of the steps needed to solve the problem) is shown below. Note that a REPEAT . . . UNTIL loop is used in the main program.

   1. Print out brief explanation of program to user.
   2. Prompt user for entry of number.
   3. REPEAT ⌈Read number
      loop ⎨abs(number)
           ⌊Print result
      UNTIL value of number is 0.

   c. Explain what happens if the user enters something other than a number when your program requests the entry of a number.

   d. Describe how you could modify the program so that any data entered is checked for validity. In cases where incorrect entries are found, a message should be printed out to the user and a new entry read and checked.

   e. Make the modifications you specified in part (d) and run your program to test the data check.

30. 1. $a < b, c < 0$;
       $a, b, c \in \mathcal{R}$
       Hypothesis
    2. For some positive
       number p, $a + p = b$
       Def. of $<$
    3. $(a + p)(-c) = b(-c)$
       Mult. prop. of =
    4. $a(-c) + p(-c) = b(-c)$
       Dist. ax.
    5. $-ac + p(-c) = -bc$
       Prop. of opp. in prod.
    6. $ac + (-ac) + p(-c) = ac + (-bc)$
       Add. prop. of =
    7. $p(-c) = ac + (-bc)$
       Ax. of add. inv.
    8. $p(-c) + bc = ac + (-bc) + bc$
       Add. prop. of =
    9. $p(-c) + bc = ac$
       Ax. of add. inv.
    10. $-c > 0$
        Exercise 28
    11. $p(-c)$ is a positive number
        Clos. ax. for $\mathcal{R}_+$
    12. $bc < ac$
        Def. of $<$
    13. $ac > bc$
        Ineq. rewritten

32. No; Example: $a = -2$, $b = -3$.

34. 1. $a > b, c > d$;
       $a, b, c, d \in \mathcal{R}_+$
       Hypothesis
    2. $ac > bc$
       $bc > bd$
       Mult. prop. of order
    3. $ac > bd$
       Trans. prop. of order

Additional Answers
Written Exercises (p. 68)

6. 1. $a > 0$ and $\frac{1}{a} < 1$

Hypothesis

2. $\frac{1}{a} \cdot a < 1 \cdot a$

Mult. prop. of order

3. $1 < 1 \cdot a$

Ax. of mult. inv.

4. $1 < a$

Iden. ax. for mult.

5. $a > 1$

Inequal. rewritten

8. 1. $a > b$

Hypothesis

2. $a + b > b + b$

Add. prop. of $=$

3. $a + b > 2b$

Dist. ax.

4. $(a + b)\frac{1}{2} > 2b\left(\frac{1}{2}\right)$

Mult. prop. of order

5. $\frac{a + b}{2} > 2b\left(\frac{1}{2}\right)$

Def. of div.

6. $\frac{a + b}{2} > b\left(2 \cdot \frac{1}{2}\right)$

Comm. and assoc. ax. for mult.

7. $\frac{a + b}{2} > b \cdot 1$

Ax. of mult. inv.

8. $\frac{a + b}{2} > b$

Iden. ax. for mult.

10. 1. $a > b > 0$

Hypothesis

2. $a \cdot a > a \cdot b$
$a \cdot b > b \cdot b$

Mult. prop. of order

3. $a \cdot a > b \cdot b$

Trans. prop. of order

4. $a \cdot a = a^2$
$b \cdot b = b^2$

Def. of x^2

5. $a^2 > b^2$

Subs. prin.

12. 1. $a < b < 0$

Hypothesis

The following program is used in Exercises 3 and 4.

```
(* A program to solve absolute value inequalities of the form *)
(* |ax − b| >= c for integer values of a, b, and c. *)
(* Solution is subject to rounding errors. *)

PROGRAM inequalities (INPUT,OUTPUT);
VAR
    a,b,c, : integer;
    left_end_point, right_end_point : real;

( *****************************************************)
PROCEDURE instruct;
BEGIN
    (* student supplies code for instructions to the user here *)
END; (* Instruct *)

( *****************************************************)
PROCEDURE get_coefficients;
BEGIN
    writeln('Press <SPACEBAR> after entering');
    writeln('each of a, b, and c.');
    write('Enter a, b, c : ');
    readln(a, b, c);
END; (* Get Coefficients *)

( *****************************************************)
PROCEDURE solve_and_show_solution;
BEGIN
    writeln;
    writeln('The solution of |', a, 'x − ', b, '| >= ', c, ' is ');
    IF c <= 0
      THEN writeln('the set, R, of all real numbers.')
      ELSE IF a = 0
              THEN IF abs(b) >= c
                      THEN writeln('the set, R, of all real numbers.')
                      ELSE writeln('the empty set.')
              ELSE
                BEGIN
                    left_end_point := b/a − c/abs(a);
                    right_end_point := b/a + c/abs(a);
                    write('x <= ', left_end_point:3:3);
                    write(' OR ');
                    writeln('x >= ', right_end_point:3:3);
                END
END;   (* solve_and_show_solution *)
```

```
(  ********************************************************)
   BEGIN (* Main Program *)
      (* student supplies code here *)
   END.
```

3. Carefully study the program beginning on page 80, which is called *inequalities*. Insert the code for the procedure *instruct* as indicated in the comments. Note that the procedure should be a series of *writeln* statements to explain the form of the inequality to be solved and any restrictions on the coefficients. Write code in the main program that will call the procedures in an appropriate order. Run the completed version of *inequalities* and comment briefly on the form of the output.

4. By making a slight change to the procedure *solve_and_show_solution*, do each of the following.

   a. Modify the program to solve an inequality of the form $|ax - b| > c$.

   b. Modify the program to solve an inequality of the form $|ax - b| \leq c$.

5. Write a Pascal program *linear_equation*, complete with appropriate comments and instructions, to solve any linear equation of the form $ax + b = c$, for a, b, and c real numbers such that a is not 0.

2. $\frac{1}{a} < 0$ and $\frac{1}{b} < 0$
   Theorem, page 65

3. $\frac{1}{a} \cdot \frac{1}{b} > 0$
   A product of two negative numbers is a positive number.

4. $a\left(\frac{1}{a} \cdot \frac{1}{b}\right) < b\left(\frac{1}{a} \cdot \frac{1}{b}\right)$
   Mult. prop. of order

5. $\left(a \cdot \frac{1}{a}\right)\left(\frac{1}{b}\right) <$
   $\left(b \cdot \frac{1}{b}\right)\left(\frac{1}{a}\right)$
   Comm. and assoc. ax. for mult.

6. $1 \cdot \frac{1}{b} < 1 \cdot \frac{1}{a}$
   Ax. of mult. inv.

7. $\frac{1}{b} < \frac{1}{a}$
   Iden. ax. for mult.

8. $\frac{1}{a} > \frac{1}{b}$
   Inequal. rewritten

14. 1. $a < 0, b < 0, a > b$
    Hypothesis
    2. $a \cdot a < b \cdot a$
    $b \cdot a < b \cdot b$
    Mult. prop. of order
    3. $a \cdot a < b \cdot b$
    Trans. prop. of order
    4. $a \cdot a = a^2; b \cdot b = b^2$
    Def. of x^2
    5. $a^2 < b^2$
    Subs. prin.

16. 1. $a < 0, b < 0, c < 0,$
    $d < 0, ab < cd,$
    $b < d$
    Hypothesis
    2. $ab > ad$
    Mult. prop. of order
    3. $ad < ab$
    Inequal. rewritten
    4. $ad < cd$
    Trans. prop. of order
    5. $\frac{1}{d} < 0$
    Theorem, page 65

(continued)

6. $ad\left(\dfrac{1}{d}\right) > cd\left(\dfrac{1}{d}\right)$

   Mult. prop. of order

7. $a\left(d \cdot \dfrac{1}{d}\right) > c\left(d \cdot \dfrac{1}{d}\right)$

   Assoc. prop. for mult.

8. $a \cdot 1 > c \cdot 1$
   Ax. of mult. inv.

9. $a > c$
   Iden. ax. for mult.

18.   1. $a > b > 0$
      Hypothesis

   2. $a + b > b + b$
      Add. prop. of order

   3. $a + b > 2b$
      Dist. ax.

   4. $a(a + b) > a(2b)$
      Mult. prop. of order

   5. $a + b > 0$
      Clos. ax. for $\mathcal{R}_+$

   6. $\dfrac{1}{a + b} > 0$

      Theorem, page 65

   7. $[a(a + b)]\left(\dfrac{1}{a + b}\right) >$

      $2ab\left(\dfrac{1}{a + b}\right)$

      Mult. prop. of order

   8. $a\left[(a + b)\left(\dfrac{1}{a + b}\right)\right] >$

      $2ab\left(\dfrac{1}{a + b}\right)$

      Assoc. ax. for mult.

   9. $a \cdot 1 > 2ab\left(\dfrac{1}{a + b}\right)$

      Ax. of mult. inv.

   10. $a > 2ab\left(\dfrac{1}{a + b}\right)$

       Iden. ax. for mult.

   11. $a > \dfrac{2ab}{a + b}$

       Def. of div.

   12. $\dfrac{2ab}{a + b} < a$

       Inequal. rewritten

The steep slope of the raceway shown in the photograph challenges the skate boarder's athletic skills. See pages 98-101 for a mathematical discussion of slope.

82   *Chapter 3*

Chapter 3

Linear Functions and Relations

Problem Solving Strategies

Generalizing from Specific
The notion of slope in general is presented by way of a concrete example on page 98. Specific examples of constant rate of change show the usefulness of direct variation in Section 3–7.
Drawing a Diagram
Diagrams are used in Section 3–2 to illustrate the concept of function and in Section 3–3 to show graphs of linear equations.

Specifying Functions and Relations

OBJECTIVES for Sections 3-1 and 3-2:
1. To determine when a relation is a function.
2. To determine the range of a function or a relation with a given domain.
3. To picture a function or relation by graphing.

3–1 Relations and Functions

Teaching Suggestions
p. T81

Key Ideas
Use an open sentence to define a relation.
Determine whether or not a relation is a function.
Evaluate functions.

In mathematics any set of *ordered pairs*, such as

$$\{(1, -1), \quad (1, 1), \quad (4, 4), \quad (4, -4), \quad (0, 0)\}$$

is called a **relation,** and we say that the first and second *components* of each pair are **related**. The set of all first components is called the **domain** of the relation, and the set of all second components is called the **range** of the relation. In the preceding example, the domain is $\{1, 4, 0\}$, and the range is $\{-1, 1, 4, -4, 0\}$. Often there is an open sentence, or **rule,** that defines a given relation. In the example, if we let x represent an element of the domain and y represent an element of the range, the relation is defined by

$$x = |y|.$$

Linear Functions and Relations **83**

1. State the domain of the function f defined by

$$f: x \rightarrow \frac{1}{x + 1}.$$

The expression $\frac{1}{x + 1}$ produces a unique real number for each value of x except -1. Thus the domain is $\{x: x \neq -1, x \in \mathcal{R}\}$.

2. State a rule that defines the function.
$\{(1, 1), (2, 8), (-3, -27),$
$(5, 125), (3, 27), (-1, -1)\}$.
a. Use an open sentence.
b. Use arrow notation.
a. $y = x^3$ **b.** $f: x \rightarrow x^3$

3. If $f(x) = 2x - 3$, determine the value of each of the following: **a.** $f(5)$ **b.** $f(-1)$
 c. $f(0)$
a. $f(5) = 2 \cdot 5 - 3 = 7$
b. $f(-1) = 2(-1) - 3 = -5$
c. $f(0) = 2 \cdot 0 - 3 = -3$

4. If $f(x) = 5x - 1$ and $g(x) = x^3$, determine the value of $[f \circ g](2)$ and $[g \circ f](2)$.
$[f \circ g](2) = f(g(2))$
$\qquad = f(2^3)$
$\qquad = f(8)$
$\qquad = 5(8) - 1$
$\qquad = 39$
$[g \circ f](2) = g(f(2))$
$\qquad = g(5(2) - 1)$
$\qquad = g(9)$
$\qquad = 9^3$
$\qquad = 729$

If a relation is such that *no* first component appears in *more than one* ordered pair, then the relation is called a function. In the preceding relation the two ordered pairs, (4, 4) and (4, −4) have the same first component. Thus, the relation is *not* a function. On the other hand, the relation

$$\{(1, 2), (2, 4), (-3, -6), (4, 8), (0, 0)\}$$

is a function. An open sentence that defines this function is

$$y = 2x.$$

Some other open sentences that define relations are

$$y = x^2, \quad y = 3x + 1, \quad y > x, \quad y \neq x + 1, \quad \text{and} \quad x^2 + y^2 = 25.$$

Notice that if the domain is $\mathcal{R}$, only the first two open sentences define functions. Each of the other open sentences can be satisfied by ordered pairs of real numbers that have the same first component and different second components. For example, (0, 5) and (0, −5) are in the set of ordered pairs defined by $x^2 + y^2 = 25$.

Since a function has the property that exactly one second component is related to each first component, an alternative definition of a function is the following. A *function* is a rule that associates with each element of one set exactly one element of another set.

Functions are often denoted by letters, such as f, g, and h. If the function defined by the rule $y = 2x$ is called f, the following "arrow notation" can also be used to define the function:

$$f: x \rightarrow 2x$$

This is read "f is the function that associates with a number x the number $2x$."

Another common way of defining a function is to give an equation or formula that relates an element in the domain with a unique element in the range. For example $f: x \rightarrow 2x$ can be thought of as the function $y = 2x$ and we say that $y = 2x$ defines y as a function of x.

Of course, in defining a function, you should specify its domain. Often, however, we will simply state the rule of a function either as an open sentence or with arrow notation. In that case *the domain will be assumed to be the set of all real numbers for which the rule produces real numbers*.

EXAMPLE 1   State the domain of the function f defined by

$$f: x \rightarrow \frac{1}{x(x - 1)}.$$

SOLUTION   The expression $\frac{1}{x(x - 1)}$ produces a unique real number for each value of x except 0 and 1. Thus the domain is $\{x: x \neq 0, x \neq 1, x \in \mathcal{R}\}$.

84   *Chapter 3*

EXAMPLE 2   State a rule that defines the function

$$\{(1, 1), (2, 4), (-3, 9), (5, 25), (3, 9), (-1, 1), (-2, 4)\}.$$

   a. Use an open sentence.   **b.** Use arrow notation.

SOLUTION   **a.** $y = x^2$ **b.** $f: x \longrightarrow x^2$

   Note that in the function defined in Example 2, the same *second* component, 1, appears in more than one ordered pair. The relation does, however, associate only one y-value with each x-value and is therefore a function.

   The $f(x)$ notation can also be used to define a function. If f is a function, the symbol $f(x)$, read "f of x," is used to denote the y-value (element of the range) that the function f associates with x. Thus $f(3)$ denotes "the y-value when x is 3." Note that $f(3)$ does not mean the product of f and 3.

EXAMPLE 3   If $f(x) = 3x - 5$, determine the value of each of the following.
   a. $f(8)$ **b.** $f(-2)$ **c.** $f(0)$

SOLUTION   **a.** $f(8) = 3 \cdot 8 - 5 = 19$
           b. $f(-2) = 3 \cdot (-2) - 5 = -11$
           c. $f(0) = 3 \cdot 0 - 5 = -5$

   Two functions can often be combined to form a new function. Applying one function after another is called the **composition** of functions. The **composite** of f with g is written $f \circ g$ and is defined as follows:

$$[f \circ g](x) = f(g(x))$$

EXAMPLE 4   If $f(x) = 2x + 1$ and $g(x) = x^2$, determine the value of $[f \circ g](3)$ and $[g \circ f](3)$.

SOLUTION $[f \circ g](3) = f(g(3))$
            $= f(3^2)$
            $= f(9)$
            $= 2(9) + 1$
            $= 19$
        $[g \circ f](3) = g(f(3))$
            $= g(2(3) + 1)$
            $= g(7)$
            $= 7^2$
            $= 49$

   Note that $[f \circ g](x)$ is not the same as $[g \circ f](x)$. Thus composition of functions is not commutative.

Reading Algebra

Students should read the function notation $f(x)$ as "f of x," not "f times x." Similarly, the arrow notation should be read correctly. For example, $f: x \rightarrow x - 2$ should be read as "f is the function that associates with a number x the range value $x - 2$."

Additional A Exercises

If $f: x \rightarrow 3x^2 - 5x + 2$ and $g: x \rightarrow \dfrac{|x - 6|}{3}$, determine the value of each of the following.

1. $f(2)$ 4

2. $f(-3)$ 44

3. $g(3)$ 1

4. $g(-6)$ 4

State an open sentence that defines each relation and tell whether or not the relation is a function.

5. $\{(8, 4), (6, 3), (1, .5),$
   $(-2, -1), (0, 0)\}$

   $y = \dfrac{x}{2}$ or $f: x \rightarrow \dfrac{x}{2}$

   The relation is a function.

6. $\{(1, 4), (2, 7), (3, 10),$
   $(-1, -2), (-2, -5), (0, 1)\}$
   $y = 3x + 1$ or
   $f: x \rightarrow 3x + 1$
   The relation is a function.

Linear Functions and Relations   **85**

Oral Exercises

State an open sentence that defines each relation and tell whether or not the relation is a function.

EXAMPLE $\quad$ {(1, 0), (3, 2), (10, 9), (0, −1), (−5, −6)}

SOLUTION $\quad y = x − 1$
$\qquad\qquad$ The relation is a function.

1. {(3, 5), (7, 9), (−2, 0), (1, 3), (−6, −4)} $y = x + 2$; a function
2. {(−1, 1), (2, −2), (3, −3), (−5, 5), (0, 0)} $y = −x$; a function
3. {(9, 3), (1, 1), (9, −3), (16, 4), (16, −4), (1, −1)} $x = y^2$; not a function
4. $\left\{\left(3, \frac{1}{3}\right), \left(-2, -\frac{1}{2}\right), \left(5, \frac{1}{5}\right), (1, 1), \left(4, \frac{1}{4}\right)\right\}$ $y = \frac{1}{x}$, a function
5. {(8, 1), (8, 2), (8, −3), (8, −6), (8, 0)} $x = 8$; not a function
6. {(3, −1), (6, −1), (0, −1), (4, −1), (−2, −1)} $y = −1$; a function

State the value of $f(3)$ and $f(−2)$ for each function.

7. $f(x) = 4x + 1$ 13; −7
8. $f(x) = 5 − x^2$ −4; 1
9. $f(x) = \dfrac{|x − 1|}{2}$ 1; $\frac{3}{2}$
10. $f(x) = (3x)^2$ 81; 36
11. $f: x \longrightarrow 2x^3$ 54; −16
12. $f: x \longrightarrow \dfrac{1}{x − 5}$ $-\frac{1}{2}$; $-\frac{1}{7}$

State the domain and range of each relation.

13. $y = \dfrac{1}{x − 3}$ {x: x ≠ 3, x ∈ ℛ}; {y: y ≠ 0, y ∈ ℛ}
14. $y = x^2 + 7$ {x: x ∈ ℛ} {y: y ≥ 7, y ∈ ℛ}
15. $y^2 = x$ {x: x ≥ 0, x ∈ ℛ}; {y: y ∈ ℛ}
16. $x^2 + y^2 = 1$ {x: |x| ≤ 1, x ∈ ℛ} {y: |y| ≤ 1, y ∈ ℛ}

Show that each relation is *not* a function by giving two distinct ordered pairs with the same first component for each. Examples are given.

17. $y > x + 1$ (1, 3), (1, 4)
18. $y \neq x + 3$ (2, 6), (2, −3)
19. $x^2 + y^2 = 36$ (0, 6), (0, −6)
20. $y^2 = x$ (4, 2), (4, −2)

Written Exercises

If $f: x \longrightarrow 2x^2 + 8x − 5$ and $g: x \longrightarrow \dfrac{|x − 3|}{2}$, determine the value of each of the following.

A $\quad$ 1. $f(3)$ 37
$\qquad$ 2. $f(−1)$ −11
$\qquad$ 3. $f(−2)$ −13
$\qquad$ 4. $f(4)$ 59

5. $g(-5)$ 4

6. $g(9)$ 3

7. $g(12)$ $\frac{9}{2}$

8. $g(-4)$ $\frac{7}{2}$

9. $f(g(9))$ 37

10. $g(f(4))$ 28

11. $[f \circ g](5)$ 5

12. $[g \circ f](1)$ 1

State an open sentence that defines each relation and tell whether or not the relation is a function.

13. $\left\{(6, 3), (-10, -5), (2, 1), \left(3, \frac{3}{2}\right)\right\}$ $y = \frac{x}{2}$; a function

14. $\{(6, 24), (-2, -8), (5, 20), (1, 4)\}$ $y = 4x$; a function

15. $\left\{\left(\frac{1}{2}, 2\right), \left(\frac{1}{3}, 3\right), \left(\frac{1}{3}, -3\right), \left(\frac{1}{4}, 4\right), \left(\frac{1}{4}, -4\right)\right\}$ $x = \frac{1}{|y|}$; not a function

16. $\{(3, 7), (5, 11), (100, 201), (-7, -13), (0, 1)\}$ $y = 2x + 1$; a function

17. $\left\{\left(3, \frac{1}{4}\right), \left(5, \frac{1}{6}\right), \left(10, \frac{1}{11}\right), \left(-8, -\frac{1}{7}\right), \left(2, \frac{1}{3}\right)\right\}$ $y = \frac{1}{x+1}$; a function

18. $\{(7, 1), (7, -2), (7, 3), (7, 6)\}$ $x = 7$; not a function

B **19.** $\{(3, -9), (4, -16), (-4, -16), (5, -25), (-10, -100), (1, -1)\}$ $y = -x^2$; a function

20. $\{(2, 3), (-2, 3), (2, -3), (1, -4), (1, 4), (-1, -4), (0, 5), (0, -5)\}$ $|x| + |y| = 5$; not a function

21. $\{(3, 2), (30, 11), (60, 21), (21, 8), (-15, -4)\}$ $y = \frac{1}{3}x + 1$; a function

22. $\{(4, 17), (5, 26), (10, 101), (-10, 101), (-7, 50), (7, 50)\}$ $y = x^2 + 1$; a function

If $f(x) = 2x - 5$ and $g(x) = \dfrac{1}{x^2}$, determine the value of each of the following.

23. $[f \circ g](2)$ $-4\frac{1}{2}$

24. $[f \circ g]\left(\frac{1}{2}\right)$ 3

25. $[f \circ g](3)$ $-4\frac{7}{9}$

26. $[g \circ f](2)$ 1

27. $[g \circ f](-1)$ $\frac{1}{49}$

28. $[g \circ f]\left(-\frac{1}{2}\right)$ $\frac{1}{36}$

29. $[f \circ g]\left(-\frac{5}{2}\right)$ $-4\frac{17}{25}$

30. $[g \circ f]\left(\frac{5}{2}\right)$ Not defined

C **31.** If f is a function such that $f(x + 2) = f(x) + f(2)$ for all $x \in \mathcal{R}$, show each of the following.
a. $f(0) = 0$
b. $f(-2) = -f(2)$

32. If f is a function such that $f(-x) = -f(x)$, show that $f(0) = 0$.

33. Suppose f is a function such that
    i. $f(0) = 0$ and
    ii. $f(x + 2) = f(x) + 5$ for all real numbers x.
a. Determine the value of $f(2)$, $f(4)$, and $f(24)$. 5, 10, 60
b. Use an open sentence to define a function on the even integers with properties **i** and **ii**. $2y = 5x$

Linear Functions and Relations **87**

87

Key Ideas

Picture and graph functions and relations.

Chalkboard Examples

1. Graph the function $f(x) = |x| - 3$ over the domain $\{-2, -1, 0, 1, 2, 3\}$.
Make a table of values for x and y and graph the ordered pairs.

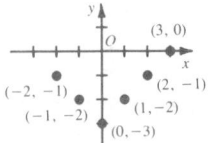

2. Graph the function $f(x) = |x - 1| + 1$ over the domain $\{-2, -1, 0, 1, 2, 3\}$.

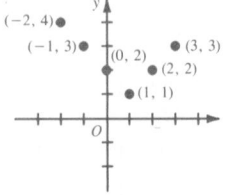

Common Errors

The confusion about whether it is duplicate x-values or duplicate y-values that are permissible in a function leads to confusion about whether it is a vertical line or a horizontal line that should be used to test a graph of a relation to see if it is the graph of a function. Since they are

3–2 Graphing Functions and Relations

It is often useful in mathematics to picture relations and functions so that the effects of the defining rule can be seen quickly and easily. One way to do this is to use a **mapping diagram** such as the one below, which pictures the relation

$$\{(1, -1), (1, 2), (4, 2), (5, 3), (-2, -3)\}.$$

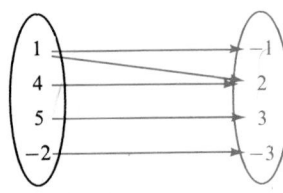

Such a diagram shows how the domain, at the left, is "mapped onto" the range, at the right.

A more convenient way to picture a relation or a function is by means of a **plane rectangular,** or **Cartesian coordinate, system,** in which each point corresponds to exactly one ordered pair of real numbers. In such a coordinate system, a point is identified with an ordered pair. The horizontal and vertical position of the point is determined by the first and second component of the ordered pair respectively. The first component of the ordered pair is called the x-coordinate, or **abscissa,** and the second component is called the y-coordinate, or **ordinate.** The graph in Figure 1 pictures the relation shown in the mapping diagram. Since this relation has two distinct ordered pairs whose first component is 1, it is not a function. The graph clearly shows that the points

$$(1, 2) \text{ and } (1, -1)$$

have the same abscissa and different ordinates. The dashed line through $(1, 2)$ and $(1, -1)$ suggests a simple test to determine whether or not a given graph represents a function.

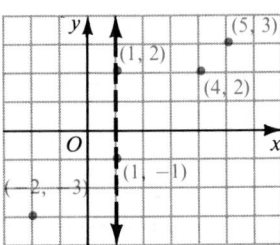

Figure 1

> A relation is a function if and only if no vertical line intersects the graph more than once.

EXAMPLE   Graph the function $f(x) = |x + 1| - 2$ over the domain $\{-3, -2, -1, 0, 1, 2, 3\}$.

88   *Chapter 3*

SOLUTION Make a table of values for x and y and then graph these ordered pairs.

x	$\lvert x + 1 \rvert - 2 = y$
-3	$\lvert -3 + 1 \rvert - 2 = 0$
-2	$\lvert -2 + 1 \rvert - 2 = -1$
-1	$\lvert -1 + 1 \rvert - 2 = -2$
0	$\lvert 0 + 1 \rvert - 2 = -1$
1	$\lvert 1 + 1 \rvert - 2 = 0$
2	$\lvert 2 + 1 \rvert - 2 = 1$
3	$\lvert 3 + 1 \rvert - 2 = 2$

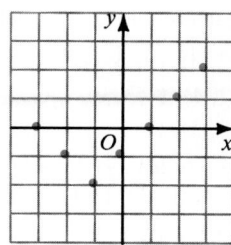

Oral Exercises

State the ordered pairs in the relation pictured by the mapping diagram or graph. Tell whether or not the relation is a function.

1. 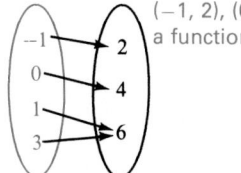 $(-1, 2)$, $(0, 4)$, $(1, 6)$, $(3, 6)$; a function

2. 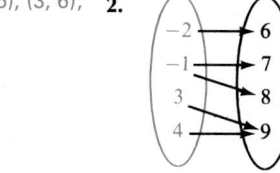 $(-2, 6)$, $(-1, 7)$, $(-1, 8)$, $(3, 9)$, $(4, 9)$; not a function

3. 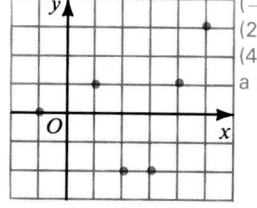 $(-1, 0)$, $(1, 1)$, $(2, -2)$, $(3, -2)$, $(4, 1)$, $(5, 3)$; a function

4. 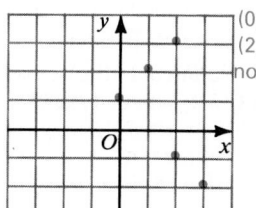 $(0, 1)$, $(1, 2)$, $(2, 3)$, $(2, -1)$, $(3, -2)$; not a function

5. $(-2, 7)$, $(0, 1)$, $(3, -8)$ **6.** $(-5, 0)$, $(0, 5)$, $(3, 4)$ **7.** $(2, 1)$, $(2, -1)$, $(6, 3)$ **8.** $(-1, -2)$, $(1, -2)$, $(3, 0)$

Name three distinct ordered pairs in the relation whose rule is given. Answers will vary.

5. $y = 1 - 3x$ **6.** $x^2 + y^2 = 25$ **7.** $2\lvert y \rvert = x$ **8.** $y = \lvert x \rvert - 3$

Name three distinct ordered pairs in the relation whose graph is shown. Answers will vary.
Give a defining rule for the relation.

9. $(2, 3)$, $(2, 1)$, $(2, -1)$; $x = 2$ **10.** $(-2, -3)$, $(0, -3)$, $(2, -3)$; $y = -3$

9. 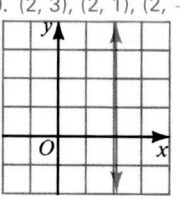 **10.** 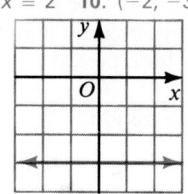 **11.** 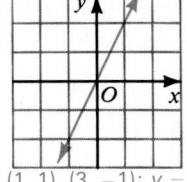 **12.**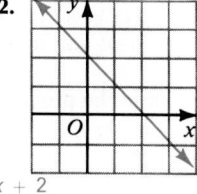

11. $(-1, -2)$, $(0, 0)$, $(1, 2)$; $y = 2x$ **12.** $(-1, 3)$, $(1, 1)$, $(3, -1)$; $y = -x + 2$

Linear Functions and Relations **89**

---

testing for repeated x-values with different y-values, the line should be vertical (where the x-value remains constant).

Additional A Exercises

Graph each relation. Tell whether or not the relation is a function.
1. $\{(4, 2), (-2, -2), (4, -2), (-2, 4)\}$ not a function

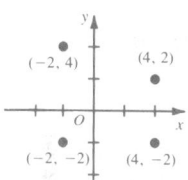

2. $\{(0, 0), (1, -1), (2, -2), (-1, 1)\}$ function

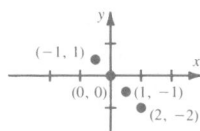

Graph the relation defined by each of the following rules over the domain $\{-2, -1, 0, 1, 2\}$.
3. $y = -2x$

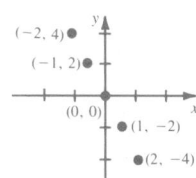

4. $y = \lvert x \rvert + 2$

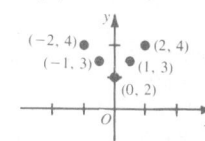

Suggested Assignments

Minimum
  First day
    90/1–7, 9–17 odd, 18, 19
  Second day
    R 91/Self–Test 1
    S 86/2–12 even
Extended Alg. with Trig.
  90/1–21 odd
  R 91/Self–Test 1
  S 86/2–8 even
Enriched Alg.
  90/3–27 odd
  R 91/Self–Test 1
Enriched Alg. with Trig.
  90/3–27 odd
  R 91/Self–Test 1

Additional Answers
Written Exercises
(See p. 114.)

Quick Quiz

State an open sentence that defines each relation and tell whether or not the relation is a function.

1. {(2, 5), (1, 4), (3, 6), (4, 7)}
  $y = x + 3$ function
2. {(1, 1), (2, 4), (−1, 1), (−2, 4)}
  $y = x^2$ function

If $f(x) = -x^2 + 1$ and $h(x) = 4x + 3$, find the value of each of the following.

3. $f(2)$ −3   4. $h(4)$ 19
5. $f(h(2))$ −120
6. $[h \circ f](1)$ 3

Graph the relation over the domain {−4, −2, 0, 2, 4}.
7. $y = |2x|$

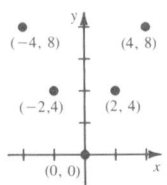

Written Exercises

Graph each relation. Tell whether or not the relation is a function.

A 1. {(−2, 3), (0, 5), (1, −2), (3, 4), (2, −3)} a function
    2. {(−1, 4), (−3, 0), (−1, −2), (3, 1), (4, 0)} not a function
    3. {(−3, 1), (1, 0), (2, 1), (−1, 0), (5, −2)} a function
    4. {(−2, −2), (−1, −1), (0, 0), (1, −1), (2, −2)} a function
    5. {(1, 4), (0, 1), (−1, 0), (0, −1), (1, −4)} not a function
    6. $\left\{(-2, 2), \left(-1, \frac{1}{2}\right), (0, 0), \left(1, \frac{1}{2}\right), (2, 2)\right\}$ a function

Graph the relation defined by each of the following rules over the domain {−2, −1, 0, 1, 2, 3}.

7. $y = 2x - 3$ 8. $y = \frac{1}{2}x + 4$
9. $5 - x = 2y$ 10. $x + y = 3$
11. $y = |x| - 2$ 12. $y = 5 - |x|$
13. $y = 2|1 - x|$ 14. $y = \frac{|x - 3|}{2}$
15. $2|x| = |y|$ 16. $|y| = |x - 1|$

B 17. $|x| + |y| = 3$ 18. $y = 4 - x^2$
   19. $y = x^2 - 3x + 2$ 20. $y = 7 - 2|x|$
   21. $xy = -6$ 22. $|x| \cdot |y| = 6$

Graph each relation over the domain {−3, −2, −1, 0, 1, 2, 3}. Restrict the range to {−3, −2, −1, 0, 1, 2, 3}.

C 23. $|y| < 3|x|$ 24. $|x - y| < 2$ 25. $x^2 + y^2 \leq 4$
   26. $x^2 - y^2 > 1$ 27. $x^2 + 2y^2 < 10$ 28. $|x + y| + |x - y| = 6$

Computer Exercises For students with computer experience

1. A parking lot charges $1.25 for the first hour (or part of an hour) and 45¢ for each additional hour (or part of an hour) that a car remains parked in the lot. If a car is parked for more than 12 h, however, the full-day rate of $6.50 is charged for each day (or part of a day) that the car remains parked in the lot. Write a program that finds the cost of parking a car in this lot for a given time period *in hours*.

Run the program in Exercise 1 for each time period.

2. 7 h   3. $9\frac{1}{4}$ h   4. $13\frac{1}{2}$ h   5. $90\frac{1}{5}$ h   6. 110 h
$3.95 $5.30 $6.50 $26.00 $32.50

90 *Chapter 3*

7. Modify the program in Exercise 1 so that the user has a choice of entering the time of the car's stay in the parking lot in days or hours. Remember that if the time is entered in hours, the total time may exceed 12 h; if the time is entered in days, the total time may be less than 12 h.

Run the program in Exercise 7 for each time period.

8. $7\frac{3}{4}$ h   **9.** $9\frac{1}{2}$ d   **10.** 56 h   **11.** $\frac{1}{2}$ d   **12.** 0.32 d
$4.40 $65.00 $19.50 $6.20 $4.40

Self-Test 1

VOCABULARY   relation (p. 83)
                     domain (p. 83)
                     range (p. 83)
                     rule (p. 83)
                     function (p. 84)
                     formula (p. 84)
                     composition of functions
                        (p. 85)
                     mapping diagram (p. 88)
                     coordinate system (p. 88)
                     abscissa, or x-coordinate
                        (p. 88)
                     ordinate, or y-coordinate
                        (p. 88)

State an open sentence that defines each relation and tell whether or not the relation is a function.

1. {(1, 2), (−3, −6), (2, 4), (0, 0)} $y = 2x$; a function   *Obj. 1, p. 83*
2. {(1, 7), (2, 7), (−1, 7), (0, 7)} $y = 7$; a function

If $f(x) = -x^2 + 3$ and $h(x) = 2x + 7$, find the value of each of the following.

3. $f(3)$ −6   **4.** $h(3)$ 13   **5.** $f(h(2))$ −118   **6.** $[h \circ f](1)$ 11 *Obj. 2, p. 83*

Graph each relation over the domain {−4, −2, 0, 2, 6}. Is the relation a function?

7. $y = |x|$ a function   **8.** $y = \frac{1}{2}x + 3$ a function   *Obj. 3, p. 83*

Check your answers with those at the back of the book.

Linear Functions and Relations   **91**

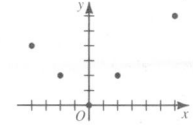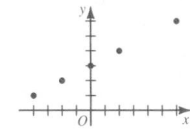

Mixed Review

1. Simplify $(-16a)(-3b) \div \left[\frac{1}{3}(-6 - 2)\right]$.   −18ab

2. Simplify $[-3c^2 - (4f - 5c)] - [8f^2 - 4(3f - 2c)]$.
$-3c^2 + 8f - 3c - 8f^2$

3. Solve $|2x - 1| \le 3$ and graph its solution set.
$\{x: -1 \le x \le 2\}$

4. Solve $y - 7(5 - y) = 5 + 4y$.   {10}

5. If $f: x \to 3x^2 - 5x + 2$ and $g: x \to \dfrac{|x - 4|}{3}$, determine the value of $[f \circ g](7)$ and $[g \circ f](3)$.
$[f \circ g](7) = 0$
$[g \circ f](3) = 3\frac{1}{3}$

Graphs of Linear Equations and Inequalities

OBJECTIVES for Sections 3-3 and 3-4:
1. *To graph a linear equation.*
2. *To use the graph of the associated linear equation to graph a linear inequality.*

Teaching Suggestions
p. T82

Key Ideas

Graph linear equations in two variables.

Chalkboard Examples

Graph each equation.

1. $2x - 3y = 9$

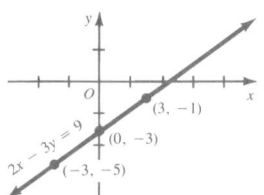

2. $y = -4$

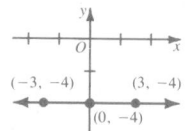

3. $x = 2$

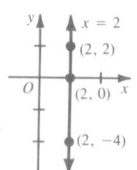

3–3 The Graph of a Linear Equation

A *solution* of an open sentence is any ordered pair (x, y) of real numbers that satisfies the equation. The *solution set* of the equation consists of all its solutions.

We have assumed unless stated otherwise that when a function is defined by an open sentence, the domain is all the x-values in $\mathcal{R}$ that produce y-values in $\mathcal{R}$. When we graph a function such as $3x + y = 6$ we obviously cannot graph all the ordered pairs that satisfy this equation. If we graph a few ordered pairs, however, we may determine a pattern that suggests the nature of the entire graph. It is easy to check that the ordered pairs $(-1, 9)$, $(0, 6)$, $(2, 0)$, and $(3, -3)$ satisfy the open sentence $3x + y = 6$ and belong to the function. These points, graphed in Figure 2, all appear to lie on a straight line. This suggests that all the other points on this line correspond to ordered pairs that also satisfy the given open sentence. In fact, this line consists of *all* the points and *only* those points whose coordinates satisfy the given equation. We call this line the graph of the equation. In general:

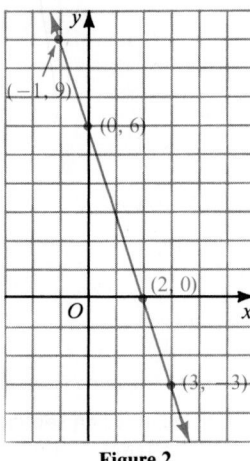

Figure 2

> The graph of an equation of the form $Ax + By = C$, where A, B, $C \in \mathcal{R}$ and A and B are not both zero, is a straight line. Conversely, every straight line in the plane is the graph of a *linear equation in two variables*, that is, an equation of the form $Ax + By = C$, where A and B are not both zero.

One convenient way to graph a linear equation is to transform the given equation into an equivalent equation with y alone on one side. Then choose a value for x, substitute it in the equation, and find the value for y. Although two ordered pairs would be enough to determine a line, at least three ordered pairs should be graphed as a check.

92   *Chapter 3*

EXAMPLE 1 Graph $3x - 4y = 8$.

SOLUTION Transform the equation into an equivalent equation with y alone on one side:

$$3x - 4y = 8$$
$$-4y = -3x + 8$$
$$y = \frac{3}{4}x - 2$$

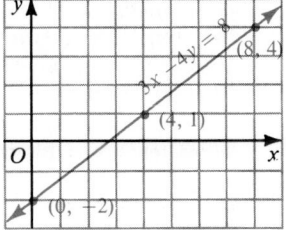

It is usually simpler to graph points whose coordinates are integers. Thus, use x-values such as 0, 4, and 8. These yield the ordered pairs

$$(0, -2), (4, 1), \text{ and } (8, 4).$$

Graph these points and draw the line through them. This line is the graph of $3x - 4y = 8$. The fact that the three points all lie on the line is a check that this is the correct graph.

The relation defined by

$$Ax + By = C$$

is a function as long as $B \neq 0$. This is true even if $A = 0$, as the next example shows.

EXAMPLE 2 Graph $y = 3$.

SOLUTION The points $(-1, 3)$, $(2, 3)$, and $(5, 3)$ satisfy the given equation and are therefore on the graph. Indeed, any point of the form $(x, 3)$ satisfies this equation. Since no vertical line intersects the horizontal line representing $y = 3$ in more than one point, the graph represents a function.

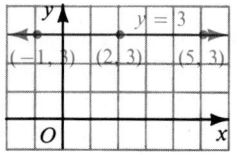

If $B = 0$ and $A \neq 0$, then the graph of $Ax + By = C$ will be on a vertical line. Such a relation is *not* a function, as the next example shows.

EXAMPLE 3 Graph $x = -\frac{5}{2}$.

SOLUTION The graph consists of all points of the form $\left(-\frac{5}{2}, y\right)$. Some examples are $\left(-\frac{5}{2}, -1\right)$, $\left(-\frac{5}{2}, 0\right)$, and $\left(-\frac{5}{2}, 3\right)$. The graph is the vertical line shown. Since this vertical line intersects *itself* in infinitely many points, the graph is not that of a function.

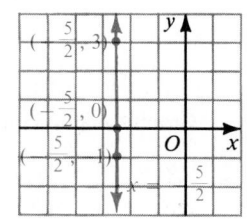

Linear Functions and Relations **93**

Graph each equation.

1. $y = 4$

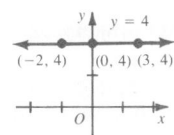

2. $x = 0$

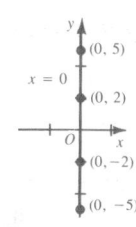

3. $x + y = 0$

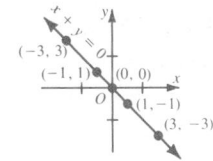

4. $\frac{1}{3}x - \frac{1}{2}y = 1$

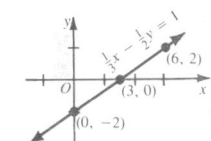

5. $y = -|x|$

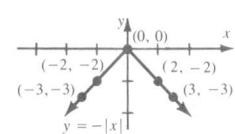

Suggested Assignments

Minimum
94/1–16
Extended Alg. with Trig.
94/1–11 odd, 12–24
Enriched Alg.
94/5–17 odd, 18–28
Enriched Alg. with Trig.
94/5–17 odd, 18–28

Additional Answers
Written Exercises

2.

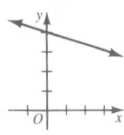

4.

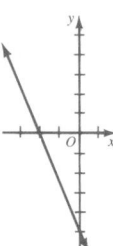

6.

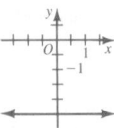

8.

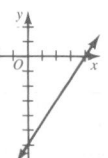

10.

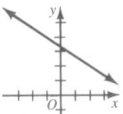

(continued on p. 118)

A function that is closely related to linear functions is the *greatest integer function* denoted by $f(x) = [x]$. The **greatest integer function** is defined as the greatest integer less than or equal to x. Thus $[4] = 4$, $[4\frac{1}{2}] = 4$, and $[-4\frac{1}{2}] = -5$. A partial graph of $y = [x]$ is shown at the right and illustrates why this function defines one kind of *step function*.

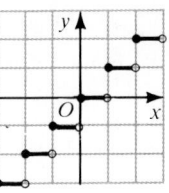

Oral Exercises

Tell whether or not the graph of each equation is a straight line.

1. $x = -3$ yes   **2.** $y = -\frac{2}{5}$ yes   **3.** $y = 7 - 5x$ yes   **4.** $-3x = 8y + 9$ yes

5. $\frac{1}{7} \cdot \frac{x + 2y}{3} = 5$ yes   **6.** $\frac{3}{x} - \frac{2}{y} = 6$ no   **7.** $y = x^2$ no   **8.** $xy = 5$ no

9. $y = 2|x|$ no   **10.** $y = |x| + 4$ no   **11.** $y = [x] + 1$ no   **12.** $y = [2x]$ no

State which ordered pair(s) are on the graph of the given equation.

13. $y - 2x = 7$; (9, 1), (−1, 5), (3, −5), (0, 7) (−1, 5), (0, 7)

14. $3y + x = 12$; (−3, 3), (3, −3), (3, 3), (−3, −3) (3, 3)

15. $3y - 4x = 6$; (6, 3), (0, 2), (−9, 10), (9, 14) (0, 2), (9, 14)

16. $5x + 2y = -10$; (2, 0), (−4, 5), (−5, 0), (6, −10) (−4, 5)

For each equation give the coordinates $(x, 0)$ and $(0, y)$ of the points, if any, where the graph intersects each axis.

17. (6, 0), (0, 3)   **18.** (0, −2.5)   **19.** (1.5, 0)   **20.** (4, 0), (0, −6)

17. $\frac{1}{2}x + y = 3$ **18.** $2y = -5$ **19.** $4x = 6$ **20.** $3x - 2y = 12$

State the value of each expression.

21. $[0]$ 0   **22.** $[3.99]$ 3   **23.** $[-1]$ −1   **24.** $[-1.3]$ −2

State a value of the missing coordinate that makes the ordered pair satisfy the equation.

25. $(8, \underline{\quad?\quad})$; $4x = 10 - 3y$ $-\frac{22}{3}$ **26.** $(\underline{\quad?\quad}, -7)$; $-2x + 5 = 3y$ 13

27. $(\underline{\quad?\quad}, -2)$; $3x - 5y = -8$ −6   **28.** $(6, \underline{\quad?\quad})$; $12 - 7y = 4x$ $-\frac{12}{7}$

29. $(\underline{\quad?\quad}, 10)$; $x = -4$ −4   **30.** $(\underline{\quad?\quad}, -8)$; $y = -8$ Any real number

Written Exercises

Graph each equation.

A   **1–8.** Use the equations in Oral Exercises 13–20.

9. $\frac{1}{3}x - \frac{1}{4}y = 2$ **10.** $y - 4 = -\frac{2}{3}(x + 1)$

94   *Chapter 3*

11. $\dfrac{x-5}{2} = \dfrac{y-1}{3}$ **12.** $\dfrac{x}{5} + \dfrac{y}{6} = 1$

13. $y = |x|$ **14.** $y = |x-3|$

B **15.** $y = |x+1| - 2$ **16.** $y = |x| - x$

17. $y = [2x]$ **18.** $y = [x] - 3$

Determine the value of k so that the given point will be on the graph of the given equation.

19. $kx + 3y = -5$; $(4, 9)$ −8   **20.** $-4x + 6k = 3y$; $(3, -10)$ −3

21. $2y - kx = k + 4$; $(-7, 5)$ −1   **22.** $ky + (k-2)x = 8$; $(-3, 7)$ $\frac{1}{2}$

23. A parking lot charges 75¢ for the first hour (or part of an hour) and 50¢ for each additional hour (or part of an hour). A function is defined for parking a car in this lot, such that the first component is the amount of time (in hours) that a car is in the lot and the second component is the corresponding cost (in cents). Graph this function over a domain of 5 h.

24. The cost of sending a letter is a function of its weight. Assume that first-class domestic postage is 22¢ for the first ounce or fraction of an ounce and 18¢ for each additional ounce up to and including 12 oz. Graph this function over a domain of 6 oz.

Graph each equation.

C **25.** $|x| + |y| = 5$ **26.** $|x| - |y| = 1$

27. $y = 2|x| - 3$ **28.** $|x+y| + |x-y| = 8$

3–4 The Graph of a Linear Inequality

When the line whose equation is $y = x$ is graphed, the coordinate plane is separated into two regions called **open half-planes**. The line $y = x$ is called the *boundary* of each half-plane. Open half-planes do not include the boundary.

If you start at any point of the line, say $(2, 2)$, and move vertically upward, the y-coordinate *increases* while the x-coordinate remains the same. Thus, the upper open half-plane is the graph of $y > x$. Every point located above $y = x$ and no other points in the plane have coordinates that satisfy the linear inequality $y > x$. This open half-plane is shown in the pink-shaded region in Figure 3.

By similar reasoning you can see that the graph of the inequality

$$y < x$$

is the shaded open half-plane below $y = x$ in Figure 3.

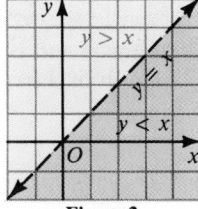

Figure 3

Linear Functions and Relations   **95**

Mixed Review

1. Solve $12 - 8(c + 3) = 12$ over $\mathscr{R}$. $\{-3\}$

2. Evaluate $(6 + g)[-3g - 7(g + 4)]$ for $g = -3$.   6

If $f: x \to x^2 - 5$ and

$h: x \to \left|\dfrac{x}{4}\right|$, find the value

of each of the following.

3. $f(-3)$ 4
4. $h(-8)$ 2
5. $[f \circ h](-16)$ 11
6. $[h \circ f](-5)$ 5

Teaching Suggestions
p. T82

Supplementary Material
Test 7

Key Ideas
Graph linear inequalities in two variables.

Chalkboard Examples
Graph each inequality.
1. $2x - 3y > 6$

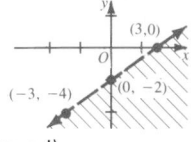

(continued)

2. $4x \geq 3$

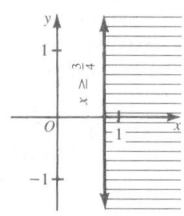

3. $y \leq |x|$

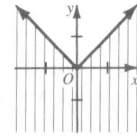

Additional A Exercises

Graph each inequality as a shaded region on a coordinate plane.

1. $-3y < 5$

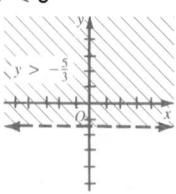

2. $x - \frac{1}{2}y \leq 1$

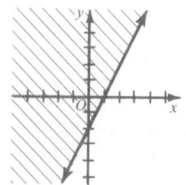

3. $x + 2y \geq 8$

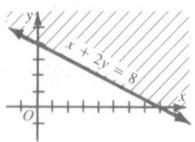

4. $y - x > 0$

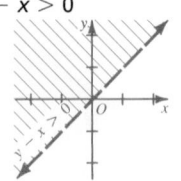

In Figure 4 the shaded, closed half-plane ("closed" means the boundary line is included) is the graph of the inequality

$$y \geq x.$$

The graph consists of all those points, but no others, with coordinates that satisfy either $y > x$ or $y = x$. The fact that the boundary line, $y = x$, is shown solid rather than dashed indicates that the half-plane is closed rather than open.

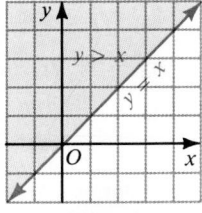

Figure 4

EXAMPLE 1 Graph $3x - 2y < 6$.

SOLUTION
1. Using transformations express y in terms of x.

$$3x - 2y < 6$$
$$-2y < 6 - 3x$$
$$y > \frac{3}{2}x - 3$$

2. Draw the graph of $y = \frac{3}{2}x - 3$ as a dashed line. Three points on the graph are $(0, -3)$, $(2, 0)$, and $(4, 3)$.
3. The graph of the inequality is shown by the shaded open half-plane above the dashed line.

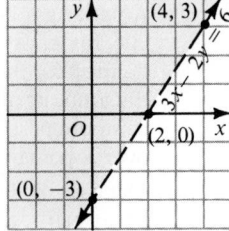

EXAMPLE 2 Graph $2x \leq 3$.

SOLUTION The boundary line is the graph of $2x = 3$ or $x = \frac{3}{2}$. The shaded closed half-plane is the graph of $2x \leq 3$.

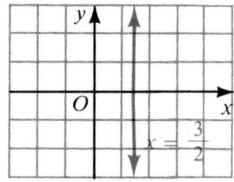

In general, the graph of a *linear inequality in two variables*, such as $Ax + By < C$ is either an open or closed half-plane. It is open for the signs $<$ and $>$, and closed for the signs $\leq$ and $\geq$. The boundary is the graph of the *associated linear equation*, $Ax + By = C$.

EXAMPLE 3 Graph $y \geq |x|$.

SOLUTION The graph of $y = |x|$ divides the plane into two regions. The shaded region including the solid lines of the graph of $y = |x|$ is the graph of $y \geq |x|$.

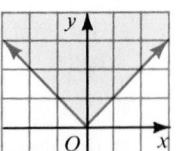

96 Chapter 3

Oral Exercises

State an inequality for each graph.

1.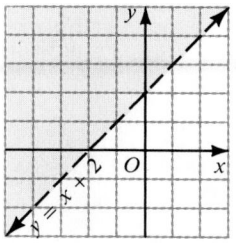

$y > x + 2$

2.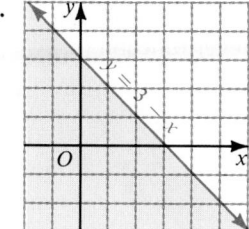

$y \leq 3 - x$

3.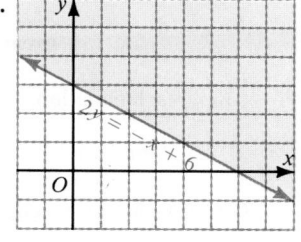

$2y \geq -x + 6$

4.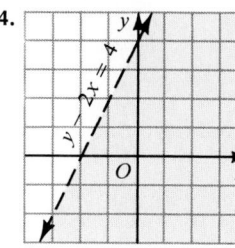

$y - 2x < 4$

5.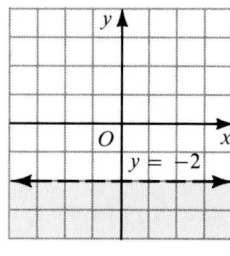

$y < -2$

6.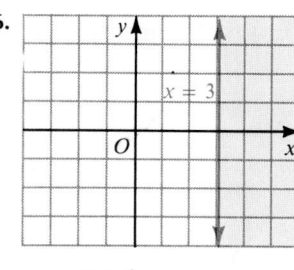

$x \geq 3$

State an equivalent inequality in which y is expressed in terms of x.

7. $3x - y > 2$ $\quad y < 3x - 2$ **8.** $-5x + 4y \leq 12$ $\quad y \leq \frac{5}{4}x + 3$ **9.** $-2y < 7$ $\quad y > -\frac{7}{2}$

10. $-x + \frac{1}{3}y \leq 2$ $\quad y \leq 3x + 6$ **11.** $-3y + 2x < 9$ $\quad y > \frac{2}{3}x - 3$ **12.** $\frac{x}{2} - 2y > 4$ $\quad y < \frac{x}{4} - 2$

Written Exercises

Graph each inequality as a shaded region on a coordinate plane.

A **1–6.** Use the inequalities in Oral Exercises 7–12.

7. $x + y > 0$

8. $4x \leq -3y$

9. $-2x + 5y \geq 10$

10. $-3y - 6 < 2x$

11. $2y > 5$

12. $7 - 2x \leq 4$

13. $x - 6 \geq 3y$

14. $8 - x < -4y$

15. $-\frac{y}{3} + \frac{x}{2} > 1$

16. $2x - \frac{y}{3} \leq 2$

B **17.** $y \leq |x|$

18. $y < |x| + 1$

19. $y > |x - 2|$

20. $y \geq |x + 3|$

21. $y + 3 < |x| - 1$

22. $y - 2 < |x + 4|$

C **23.** $y < |x| + x$

24. $x + y \geq |x|$

Linear Functions and Relations   **97**

Suggested Assignments

Minimum
  First day
  97/1–18
  Second day
R 98/*Self-Test 2*
S 90/8–16 even
Extended Alg. with Trig.
  97/1–13 odd, 15–21
R 98/*Self–Test 2*
S 90/10–18 even
Enriched Alg.
  97/5–19 odd, 20–24
R 98/*Self–Test 2*
Enriched Alg. with Trig.
  97/5–19 odd, 20–24
R 98/*Self–Test 2*

Additional Answers
Written Exercises
(See p. 119.)

Mixed Review

If $f(x) = x^2 - 2$ and $g(x) = \dfrac{1}{x + 3}$, find the value of each of the following.

1. $f(0)$ $\quad -2$

2. $f(-6)$ $\quad 34$

3. $g(0)$ $\quad \dfrac{1}{3}$

4. $[g \circ f](-1)$ $\quad \dfrac{1}{2}$

Tell whether or not the relation is a function.

5. $\{(-1, -5), (2, 4), (1, 1), (0, -2)\}$
The relation is a function.

6. $\{(-3, 6), (-1, 5), (3, 6), (-3, 2)\}$
The relation is not a function.

Quick Quiz

1. Graph $3x + 2y = 8$.

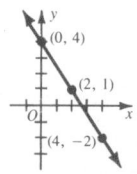

2. Graph $y - 3 > -4$.

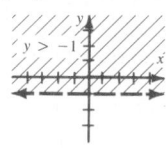

Additional Answers
Self-Test 2

1.

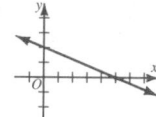

2.   **3.**

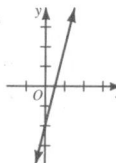

4.   **5.**

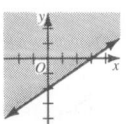

6.

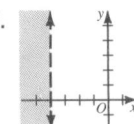

■ Self-Test 2

VOCABULARY   graph of an equation (p. 92)   open half-plane (p. 95)
                              greatest integer function (p. 94)   closed half-plane (p. 96)

Graph each equation.

 1. $2x + 5y = 10$ **2.** $x = -2$ **3.** $\frac{1}{2}y - 2x = -1$ *Obj. 1, p. 92*

Graph each linear inequality.

 4. $2x - 3y \leq 6$ **5.** $y - 4 > -7$ **6.** $9x < -36$ *Obj. 2, p. 92*

Check your answers with those at the back of the book.

Lines and Their Equations

OBJECTIVES for Sections 3-5 and 3-6:
1. To determine the slope of a line through two given points.
2. To determine an equation of a line through two given points.
3. To determine an equation of a line when its slope and the coordinates of
*   one of its points are given.*
4. To determine an equation of a line when its slope and y-intercept are given.

3–5 The Slope of a Line

Figure 5 shows a hill that is rising steadily at a rate of 50 m
of vertical "rise" for each 100 m of horizontal "run." The
steepness, or *grade*, of such an incline is defined to be the
ratio of rise to run, in this case $\frac{50}{100}$ or 50%.

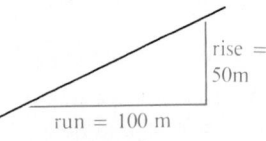

Figure 5

   This hill can be represented mathematically by the line

$$y = \tfrac{1}{2}x$$

shown in Figure 6. By observing the coordinates of P and
Q, you can readily confirm that this line rises at a rate of 1
vertical unit for each 2 horizontal units. Hence its steepness, or **slope**, is

$$\frac{\text{rise}}{\text{run}} = \frac{1}{2}.$$

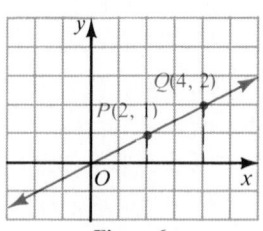

Figure 6

98   *Chapter 3*

In general, as in Figure 7, use *subscript notation* to name any two points $P(x_1, y_1)$ (read "x sub one, y sub one") and $Q(x_2, y_2)$ on a nonvertical line and form this ratio:

$$m = \frac{\text{rise}}{\text{run}} = \frac{\text{ordinate of } Q - \text{ordinate of } P}{\text{abscissa of } Q - \text{abscissa of } P} = \frac{y_2 - y_1}{x_2 - x_1}$$

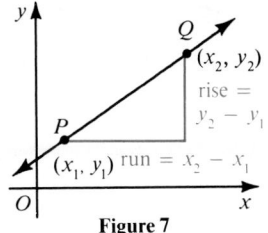

Figure 7

The proof of the theorem stated later in this section shows that this ratio does not depend upon the particular points chosen to be P and Q; therefore, the ratio of rise to run may be taken as the definition of **slope**. Thus in Figure 6, using the points P and Q, and O and P, you have

$$m = \frac{2-1}{4-2} = \frac{1}{2}, \quad \text{or} \quad m = \frac{1-0}{2-0} = \frac{1}{2}.$$

EXAMPLE 1 Determine the slope of each line and graph both lines on the same plane.

a. $y = \frac{1}{2}x + 1$ **b.** $-3x + 6y = -12$

SOLUTION Find two pairs of coordinates for each line, and then use the slope formula.

a. $(0, 1)$ and $(4, 3)$ satisfy $y = \frac{1}{2}x + 1$.

$$\therefore m = \frac{3-1}{4-0} = \frac{2}{4} = \frac{1}{2}.$$

b. $(0, -2)$ and $\left(3, -\frac{1}{2}\right)$ satisfy

$-3x + 6y = -12.$

$$\therefore m = \frac{-\frac{1}{2} - (-2)}{3 - 0} = \frac{\frac{3}{2}}{3} = \frac{1}{2}.$$

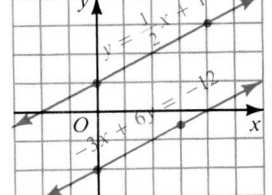

Notice that if you transform the equation $-3x + 6y = -12$ in Example 1 to the form $y = \frac{1}{2}x - 2$ and then compare the equations of the lines in Figure 6 and Example 1,

$$y = \frac{1}{2}x,$$

$$y = \frac{1}{2}x + 1$$

$$y = \frac{1}{2}x - 2$$

you see that *the coefficient of x is equal to the slope* in each case. This illustrates the theorem on page 100.

Linear Functions and Relations **99**

Key Ideas

Determine the slope of a line.

Chalkboard Examples

1. Determine the slope of each line and graph.
   a. $x - y = 3$
   b. $2y = 2x + 4$

   a. $m = \dfrac{-1 - (-3)}{2 - 0} = 1$

   b. $m = \dfrac{0 - 2}{-2 - 0} = 1$

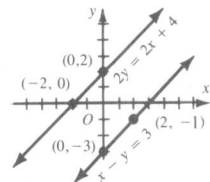

2. Determine the slope m of $2x + 3y = 9$ and graph.

   $m = -\dfrac{A}{B} = -\dfrac{2}{3}$

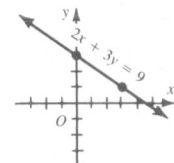

99

Common Errors

Be sure to distinguish carefully between horizontal lines, which have zero slope, and vertical lines, which have no slope. Again, vertical lines do not represent functions. Emphasize to the students the idea that in computing a slope the difference of the *y*-coordinates is always in the numerator of the fraction, and the difference of the *x*-coordinates is in the denominator of the fraction.

Additional A Exercises

Determine the slope of the line through each pair of points.

1. $(-2, 5), (-2, 9)$ undefined

2. $(-4, 0), (-1, -2)$ $-\dfrac{2}{3}$

3. $(-6, 3), (9, 3)$ 0

4. $(8, -2), (-2, 8)$ -1

Determine the slope of each line and graph the line.

5. $\dfrac{1}{2}x - 3y = 1$

The slope is $\dfrac{1}{6}$.

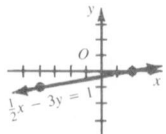

6. $4x + 3y = 12$

The slope is $-\dfrac{4}{3}$.

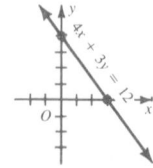

Theorem.   For all A, B, and $C \in \mathcal{R}$ and $B \neq 0$, the slope m of the line $Ax + By = C$ is $-\dfrac{A}{B}$.

PROOF

Since $B \neq 0$, the equation $Ax + By = C$ is equivalent to $y = -\dfrac{A}{B}x + \dfrac{C}{B}$.

Let $P(x_1, y_1)$ and $Q(x_2, y_2)$ be any two points on the graph of the equation. Since the coordinates of both P and Q must satisfy this equation, we can write the following:

$$y_2 = -\frac{A}{B}x_2 + \frac{C}{B} \tag{1}$$

$$y_1 = -\frac{A}{B}x_1 + \frac{C}{B} \tag{2}$$

Subtracting equation (2) from equation (1), we obtain

$$y_2 - y_1 = -\frac{A}{B}(x_2 - x_1) \tag{3}$$

Notice from (3) that $x_2 - x_1 \neq 0$, since otherwise we would have $x_2 - x_1 = 0$, $y_2 - y_1 = 0$, and P and Q would be the same point. Therefore, dividing (3) by $x_2 - x_1$, we get

$$\frac{y_2 - y_1}{x_2 - x_1} = m = -\frac{A}{B}.$$

EXAMPLE 2   Determine the slope m of $3x + 4y = 8$ and graph the line.

SOLUTION   From the theorem just proved

$$m = -\frac{A}{B} = -\frac{3}{4}.$$

Two pairs of coordinates that satisfy the given equation are $(0, 2)$ and $(4, -1)$. The graph is shown at the right.

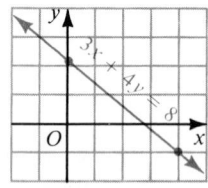

The line graphed in Example 2 has a *negative slope*. In general, on the coordinate plane a line with negative slope falls from left to right; a line with positive slope rises from left to right.

The slope of a horizontal line, such as the graph of $y = 1$ in Figure 8, is 0, because the numerator in the slope formula, $m = \dfrac{y_2 - y_1}{x_2 - x_1}$, is 0 for any two points on the line.

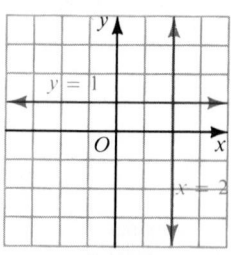

Figure 8

100   *Chapter 3*

On a vertical line, such as $x = 2$ in Figure 8, every point has the same x-coordinate. Hence the denominator $x_2 - x_1$ in the formula for m would always be 0 and, accordingly, *slope is not defined for a vertical line*. Note that a slope of 0 is not the same as an undefined slope.

Lines that have equal slopes or lines with no slope are *parallel*. For example, Figure 9 shows two distinct lines, $y - 2x = -1$ and $y - 2x = 3$, that are parallel because they both have slope 2.

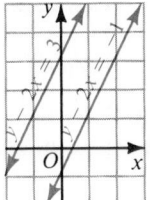

Figure 9

Oral Exercises

Give the slope, if any, of the line determined by the two given points.

1. (0, 10) and (5, 0) -2

2. (3, 5) and (1, 9) -2

3. (2, -1) and (6, 1) $\frac{1}{2}$

4. (-2, -5) and (1, 4) 3

5. (-3, 5) and $\left(-4, \frac{7}{2}\right)$ $\frac{3}{2}$

6. (9, 8) and (9, -7) not defined

7. (4, 2) and (-2, -1) $\frac{1}{2}$

8. (3, 7) and (-8, 7) 0

9. How are the lines determined by the points in Exercises 1 and 2 related? parallel

10. Describe the line in Exercise 6. vertical

11. Describe the line in Exercise 8. horizontal

Give the slope of each line, if any. If the line is vertical or horizontal, so state.

12. $y = -\frac{1}{2}x + 5$ $-\frac{1}{2}$

13. $2x + 3 = -7$ vertical

14. $-5x = -3y$ $\frac{5}{3}$

15. $x - y = 2$ 1

16. $4y = -10$ 0, horizontal

17. $3x + 2y = 9$ $-\frac{3}{2}$

18. $-x + 4y = 2$ $\frac{1}{4}$

19. $-8x + 3 = 2y$ -4

Determine the rise and the run between two convenient points on each line and state the slope of the line.

20. 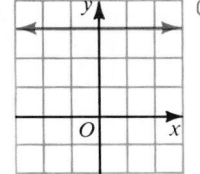 0

21. 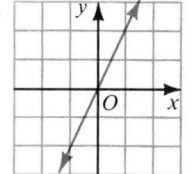 2

22. 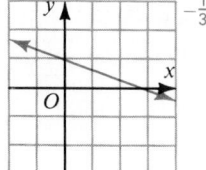 $-\frac{1}{3}$

23. 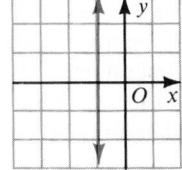 not defined

Linear Functions and Relations **101**

Reading Algebra

Have students give the meaning of the slope formula, $m = \frac{y_2 - y_1}{x_2 - x_1}$, as well as read it. Point out that the reading of this formula, "m equals the quantity y sub two minus y sub one divided by the quantity x sub two minus x sub one," is exact but does not convey the meaning. Have them read the formula instead as "the change in (or the difference of) the y-coordinates divided by the change in (or the difference of) the x-coordinates." The slope can also be stated as "rise" (vertical change) over "run" (horizontal change) or as "the ratio of rise to run."

Suggested Assignments

Minimum
  102/1–20
Extended Alg. with Trig.
  102/1–23 odd
Enriched Alg.
  102/5–23 odd, 24–26
Enriched Alg. with Trig.
  102/5–23 odd, 24–26

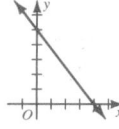
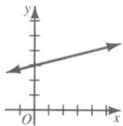
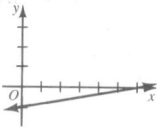
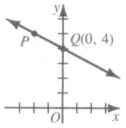
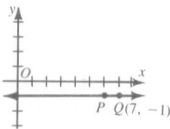

Mixed Review

If $f: x \rightarrow 3x^2 + 7x - 6$ and

$g: x \rightarrow \left| \dfrac{x - 2}{4} \right|$, find the

value of each of the following.

1. $g(-10)$ 3
2. $[f \circ g](-6)$ 20
3. Determine the value of k so that $(2, 3)$ will be on the graph of $kx - 3y = -5$. $k = 2$
4. Graph $y > |x| - 1$.

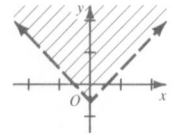

Written Exercises

Determine the slope of the line through each pair of points.

A   1. $(-2, 5), (-3, 2)$ 3
2. $(4, -1), (2, 7)$ -4
3. $(6, 0), (-2, 4)$ $-\frac{1}{2}$
4. $(7, 2), (-5, 2)$ 0
5. $(-6, 2), (-6, -9)$ undefined
6. $\left(\frac{1}{2}, 9\right), \left(\frac{5}{2}, -1\right)$ -5

Determine the slope of each line and then graph the line.

7. $4x - 3y = 24$ $\frac{4}{3}$
8. $5x + 4y = 20$ $-\frac{5}{4}$
9. $3x = 5y + 15$ $\frac{3}{5}$
10. $-x + 4y = 12$ $\frac{1}{4}$
11. $\frac{1}{2}x + \frac{1}{3}y = 1$ $-\frac{3}{2}$
12. $\frac{x}{3} - 2y = 2$ $\frac{1}{6}$

Use the given point P and the given slope m to determine the coordinates of a second point Q. Then draw the line through P and Q.

EXAMPLE $P(-1, 2)$; $m = \dfrac{3}{2}$

SOLUTION   Since the slope is $\dfrac{3}{2}$, use the rise of 3 and the run of 2 to determine the coordinates of Q:

$$Q = (-1 + 2, 2 + 3) = (1, 5)$$

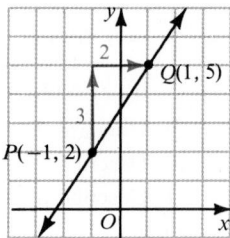

13. $P(2, -1)$; $m = \dfrac{2}{3}$ Q (5,1)
14. $P(-2, 5)$; $m = \dfrac{-1}{2}$ Q (0, 4)
15. $P(-1, 3)$; $m = -4$ Q (0, -1)
16. $P(-3, 7)$; $m = \dfrac{-5}{2}$ Q (-1, 2)
17. $P(1, -2)$; $m = \dfrac{4}{3}$ Q (4, 2)
18. $P(6, -1)$; $m = 0$ For example Q (7, -1)

Determine the value of a that makes the slope of the line through the two given points equal to the given value of m.

B   19. $(4, -3)$ and $(2, a)$; $m = \dfrac{1}{4}$ $-\frac{7}{2}$
20. $(a, 1)$ and $(-2, -4)$; $m = \dfrac{3}{2}$ $\frac{4}{3}$
21. $(3a, 1)$ and $(a, -4)$; $m = -\dfrac{1}{3}$ $-\frac{15}{2}$
22. $(a + 3, 5)$ and $(1, a - 2)$; $m = 4$ $-\frac{1}{5}$
23. $(a, 5)$ and $(-a, a)$; $m = -2$ $-\frac{5}{3}$
24. $(-3a, -a)$ and $(4, 2a)$; $m = \dfrac{1}{2}$ $\frac{4}{3}$

C   25. For three distinct numbers x_1, x_2, and x_3, give a necessary and sufficient condition for the three points (x_1, y_1), (x_2, y_2), and (x_3, y_3) to lie on one straight line. $\dfrac{y_2 - y_1}{x_2 - x_1} = \dfrac{y_3 - y_2}{x_3 - x_2} = \dfrac{y_3 - y_1}{x_3 - x_1}$
26. Find the value of c if the line through $(7, 2c - 5)$ and $(3, -2)$ is parallel to the line through $(-6, c + 4)$ and $(-9, 5)$. $\frac{5}{2}$

102   *Chapter 3*

3–6 Finding an Equation of a Line

Since two points determine a unique line, you can use the slope formula,

$$m = \frac{y_2 - y_1}{x_2 - x_1},$$

to determine an equation of a line through two given points, P and Q, with coordinates (x_1, y_1) and (x_2, y_2), if $x_2 \neq x_1$.

EXAMPLE 1   Determine an equation of the line containing $P(1, 2)$ and $Q(3, -4)$.

SOLUTION   The general equation of a nonvertical line, $Ax + By = C$, can be written equivalently in the form $y = -\frac{A}{B}x + \frac{C}{B}$. Since $-\frac{A}{B} = m$, the above equation can be simplified to $y = mx + b$, where b is the constant term, $\frac{C}{B}$. First, calculate $m = -3$ from the slope formula. Then substitute the coordinates of *either* P or Q in the equation $y = mx + b$ to find b. In this case, substituting $(1, 2)$ in $y = -3x + b$ yields $b = 5$. Now use the known values for m and b to write an equation of the desired line:

$$y = -3x + 5$$

Check: Do $(1, 2)$ and $(3, -4)$ both satisfy $y = -3x + 5$?

$$2 = -3(1) + 5 \quad \text{and} \quad -4 = -3(3) + 5 \ \checkmark$$

$\therefore$ an equation of the line is $y = -3x + 5$.

From Example 1 you can see that a line ℓ is uniquely determined by its slope m and the coordinates of just one point $P(x_1, y_1)$ on ℓ. That is, any point Q other than P lies on ℓ if and only if the coordinates of Q together with those of P satisfy the slope formula for the given value of m.

Theorem.   Given a line ℓ with slope m and a point $P(x_1, y_1)$ on ℓ. Then for P and any other point $Q(x, y)$:

1. If Q is on ℓ, then $\frac{y - y_1}{x - x_1} = m$.

2. If $\frac{y - y_1}{x - x_1} = m$, then Q is on ℓ.

If the equation for slope m in the theorem is rewritten as

$$y - y_1 = m(x - x_1),$$

we have the **point-slope form** of an equation of a line.

Linear Functions and Relations   **103**

Key Ideas

Find an equation of a line given:

1. two points on the line,

2. its slope and one point on the line, and

3. its slope and its y-intercept.

Chalkboard Examples

1. Determine an equation of the line containing $R(2, 3)$ and $S(-4, -9)$.

$$m = \frac{y_2 - y_1}{x_2 - x_1} = \frac{-9 - 3}{-4 - 2} = 2$$

An equation of the line is $y = 2x - 1$.

2. Determine an equation of the line with slope 3 and containing $Q(-2, -7)$.
$$y - y_1 = m(x - x_1)$$
$$y + 7 = 3(x + 2)$$
$$y + 7 = 3x + 6$$
$$y = 3x - 1$$
Check: Does $(-2, -7)$ satisfy $y = 3x - 1$?
$$-7 = 3(-2) - 1 \ \checkmark$$
An equation of the line is $y = 3x - 1$.

3. Determine an equation of the line with y-intercept 4 and x-intercept -3.
$$m = \frac{0 - 4}{-3 - 0} = \frac{4}{3}$$
$$y = \frac{4}{3}x + 4$$

If the given point $P(x_1, y_1)$ of the line ℓ lies on the y-axis as in Figure 10, so that P is the point where ℓ intersects the y-axis, then $x_1 = 0$ and the ordinate y_1 of P is called the y-intercept of ℓ. (In Figure 10 the y-intercept of ℓ is 2.) Using the point-slope form of ℓ, we have

$$y - y_1 = m(x - 0)$$
$$y = mx + y_1.$$

Thus an equation of a line with slope m and y-intercept b is

$$y = mx + b.$$

This equation is called the slope-intercept form of an equation of a line.

The abscissa of the point in which a line ℓ intersects the x-axis is called the x-intercept of the line. The x-intercept of the line in Figure 10 is -2.

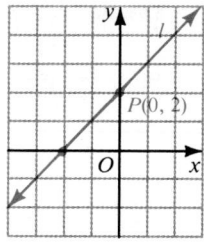

Figure 10

EXAMPLE 2   Determine an equation of the line with y-intercept -2 and x-intercept 6.

SOLUTION   To find m use the intercepts $(0, -2)$ and $(6, 0)$:

$$m = \frac{0 - (-2)}{6 - 0} = \frac{1}{3}$$

Then, use the slope-intercept form to obtain $y = \frac{1}{3}x - 2$.

Oral Exercises

State an equation of the line with the given slope m and y-intercept b.

1. $m = -5$; $b = 2$ $y = -5x + 2$ 2. $m = 3$; $b = -7$ $y = 3x - 7$ 3. $m = \frac{1}{2}$; $b = 0$ $y = \frac{1}{2}x$

4. $m = 0$; $b = -6$ $y = -6$ 5. $m = -\frac{1}{3}$; $b = \frac{3}{4}$ $y = -\frac{1}{3}x + \frac{3}{4}$ 6. $m = 0$; $b = 0$ $y = 0$

State an equation in the form $y = mx + b$ of the line passing through the given point and having the given slope m.

7. $(0, 8)$; $m = 3$ $y = 3x + 8$ 8. $(0, -5)$; $m = -4$ $y = -4x - 5$ 9. $(1, 4)$; $m = 9$ $y = 9x - 5$

10. $(-2, 7)$; $m = \frac{1}{2}$ $y = \frac{1}{2}x + 8$ 11. $(-3, -6)$; $m = 2$ $y = 2x$ 12. $(6, -8)$; $m = -\frac{2}{3}$ $y = -\frac{2}{3}x - 4$

State the slope and y-intercept of each line.

13. $y - 2x = -7$ $2, -7$ 14. $5x = y + 6$ $5, -6$ 15. $2y = 8x - 3$ $4, -\frac{3}{2}$

16. $-4y = 3x - 12$ $-\frac{3}{4}, 3$ 17. $4x + 5y = -20$ $-\frac{4}{5}, -4$ 18. $8x - 2y = 5$ $4, -\frac{5}{2}$

19. What is the y-intercept of the line $y - y_1 = m(x - x_1)$ in terms of x_1, y_1, and m?

20. How are the lines $y - 3x = -2$ and $y = 3x + 7$ related? How can you deduce this from their equations? They are parallel; both have $m = 3$.

Written Exercises

Determine an equation in the form $y = mx + b$ of the line passing through the given point P with the given slope m.

A **1.** $P(7, 1)$; $m = -1$
$y = -x + 8$

2. $P(-2, 5)$; $m = 3$
$y = 3x + 11$

3. $P(6, -2)$; $m = \frac{1}{2}$
$y = \frac{1}{2}x - 5$

4. $P(-8, 4)$; $m = 0$ $\;y = 4$

5. $P(9, 7)$; $m = \frac{4}{3}$ $\;y = \frac{4}{3}x - 5$

6. $P(5, 0)$; $m = -2$
$y = -2x + 10$

7. $P(3, 1)$; $m = \frac{3}{2}$ $\;y = \frac{3}{2}x - \frac{7}{2}$

8. $P(4, -1)$; $m = -\frac{2}{3}$
$y = -\frac{2}{3}x + \frac{5}{3}$

9. $P(0, 0)$; $m = \frac{3}{4}$ $\;y = \frac{3}{4}x$

10. $P(-8, -9)$; $m = 2$
$y = 2x + 7$

11. $P(15, 4)$; $m = -\frac{2}{5}$
$y = -\frac{2}{5}x + 10$

12. $P(-6, -10)$; $m = -\frac{1}{3}$
$y = -\frac{1}{3}x - 12$

Determine an equation in the form $y = mx + b$ of the line containing the two given points. **18.** $y = \frac{1}{2}x - 10$ **21.** $y = \frac{1}{2}x - \frac{13}{6}$

13. $(4, 3)$, $(7, 9)$ $y = 2x - 5$

14. $(5, -2)$, $(8, -2)$ $y = -2$

15. $(-1, 4)$, $(1, 2)$
$y = -x + 3$

16. $(6, -3)$, $(8, -9)$
$y = -3x + 15$

17. $(6, 5)$, $(-4, 0)$ $y = \frac{1}{2}x + 2$

18. $(0, -10)$, $(6, -7)$

19. $(-1, 5)$, $(1, -3)$
$y = -4x + 1$

20. $\left(\frac{1}{2}, 7\right)$, $\left(-\frac{1}{2}, 4\right)$
$y = 3x + \frac{11}{2}$

21. $\left(7, \frac{4}{3}\right)$, $\left(5, \frac{1}{3}\right)$

22. $\left(-2, -\frac{5}{2}\right)$, $\left(1, \frac{13}{2}\right)$
$y = 3x + \frac{1}{2}$

23. $\left(1, \frac{5}{3}\right)$, $(-3, -1)$
$y = \frac{2}{3}x + 1$

24. $(-4, 3)$, $\left(-1, -\frac{9}{2}\right)$
$y = -\frac{5}{2}x - 7$

Determine an equation in the form $y = mx + b$ of the line through the given point and satisfying the given condition.

B **25.** $(1, -7)$; parallel to the line $2x + y = 5$ $\;y = -2x - 5$

26. $(3, 10)$; parallel to the line $3y - x = 4$ $\;y = \frac{1}{3}x + 9$

27. $(-2, 9)$; parallel to the line through $(2, 8)$ and $(-1, 11)$ $\;y = -x + 7$

28. $(-3, -12)$; parallel to the line through $(6, 0)$ and $(4, -5)$ $\;y = \frac{5}{2}x - \frac{9}{2}$

29. $(6, -13)$; parallel to the line through $(7, 4)$ and $(-3, 9)$ $\;y = -\frac{1}{2}x - 10$

30. $(-5, 2)$; parallel to the line through $(-4, 1)$ and $(1, 3)$ $\;y = \frac{2}{5}x + 4$

Find the value of k so that the two given lines will be parallel.

31. $3x - ky = 7$ $\;-\frac{15}{2}$
$2x + 5y = -1$

32. $-5y + kx = 4$ $\;\frac{45}{2}$
$2y - 9x = 11$

33. $4x + ky = -2$
$-3x + (k - 1)y = 5\frac{4}{7}$

34. $(k + 2)x - 5y = 8$ $\;-\frac{3}{4}$
$kx + 3y = 9$

C **35.** Determine the value of k so that the line through the points $(-2, 3)$ and $(6, k)$ has y-intercept 4. $\;7$

36. Determine the value of k so that the line through the points $(0, 4)$ and $(k - 2, 6)$ has x-intercept k. $\;\frac{4}{3}$

37. A line through $(2, 1)$ has y-intercept b. Express the x-intercept, k, of the line in terms of b. $\;k = \frac{2b}{b - 1}$

38. A line through $(-1, 3)$ has x-intercept a. Express the y-intercept, k, of the line in terms of a. $\;\frac{3a}{a + 1}$

Linear Functions and Relations **105**

Suggested Assignments

Minimum
   105/1–28
R 106/Self-Test 3
S 81/14–22 even
Extended Alg. with Trig.
   105/5–39 odd
R 106/Self-Test 3
S 87/14–20 even
Enriched Alg.
   105/7–33 odd, 35–40
R 106/Self-Test 3
S 87/14–18 even
Enriched Alg. with Trig.
   105/7–33 odd, 35–40
R 106/Self-Test 3
S 87/14–18 even

Additional A Exercises

Determine an equation in the form $y = mx + b$ of the line passing through the given point P with the given slope m.

1. $P(0, 0)$; $m = 0$ $\;y = 0$

2. $P(-4, -3)$; $m = \frac{1}{2}$

$y = \frac{1}{2}x - 1$

3. $P(3, 7)$; $m = -\frac{2}{3}$

$y = -\frac{2}{3}x + 9$

Determine an equation in the form $y = mx + b$ of the line containing the two given points.

4. $(0, 5)$, $(5, 0)$ $\;y = -x + 5$

5. $(3, -7)$, $(-5, -7)$ $\;y = -7$

6. $(8, -3)$, $(-6, 4)$

$y = -\frac{1}{2}x + 1$

39. Determine the value of k so that the points $(k - 1, 2)$, $(k, 4)$, and $(3, 11)$ all lie on one line. $-\dfrac{1}{2}$

40. Show that a line with equal nonzero x- and y-intercepts and containing the point (c, d) has equation $x + y = c + d$.

Computer Exercises   For students with computer experience

1. Write a program that will find the slope-intercept form of the equation of a line when two distinct points on the line are given.

Run the program in Exercise 1 for each pair of points.

2. $(-3, 12)$, $(2, 29)$ $y = 3.4x + 22.2$ **3.** $(5, -18)$, $(-3, 37)$ $y = -6.875x + 16.375$

4. $(6, 19)$, $(-15, -14)$ $y = 1.57143x + 9.57143$ **5.** $(-7, 42)$, $(-39, 16)$ $y = 0.8125x + 47.6875$

6. Modify the program in Exercise 1 so that two points whose first coordinates are equal can be input. The output should give the equation and state that the line is vertical. Include a safeguard against the same point being entered twice.

Run the program in Exercise 6 for each pair of points.

7. $(5, -8)$, $(5, 43)$ $x = 5$; vertical   **8.** $(-8, 2)$, $(-8, -15)$ $x = -8$; vertical

9. $(-7, 2)$, $(-7, 19)$ $x = -7$; vertical   **10.** $(0, 17)$, $(0, -22)$ $x = 0$; vertical

 Self-Test 3

VOCABULARY   slope (p. 98)   slope-intercept form (p. 104)
point-slope form (p. 103) x-intercept (p. 104)
y-intercept (p. 104)

1. Determine the slope of the line through the points $P(-3, 1)$ and   *Obj. 1, p. 98*
$Q(2, 5)$. $\dfrac{4}{5}$

2. Determine an equation of the line passing through the points $P(2, 0)$ *Obj. 2, p. 98*
and $Q(0, -4)$. $y = 2x - 4$

3. Determine an equation of the line passing through the points

$P\left(\dfrac{1}{4}, 6\right)$ and $Q\left(-\dfrac{7}{4}, 3\right)$. $y = \dfrac{3}{2}x + \dfrac{45}{8}$

4. Determine an equation of the line having slope 4 and passing   *Obj. 3, p. 98*
through the point $(1, 2)$. $y = 4x - 2$

5. Determine an equation of the line having slope -2 and y-intercept 3.   *Obj. 4, p. 98*

Check your answers with those at the back of the book. $y = -2x + 3$

READING ALGEBRA   Problem Solving Strategies

Problem solving is not an automatic process. Accurate reading is an important part of solving problems. Another important step in problem solving is finding a solution method. There are often many ways to solve a problem. Sometimes you may need to reread and rethink to understand the problem. You may need to use several different strategies and approaches on the same problem or a combination of strategies. Do not be discouraged if your first approach does not work. In problem solving there are often false starts and reversals. If one strategy fails, try another. Learn what a strategy did not show you as well as what the strategy did show you.

Some of the strategies you can try are:

1. Reword or restate the problem in your own words.
2. Substitute similar but simpler quantities in the problem. Try using a straightforward example and develop an analogy or pattern.
3. Make a reasonable guess and check if it works in the problem. Is your guess possible? Is it reasonable? Refine your guess and check the results again.
4. Try to change a complex problem to a series of simpler problems.
5. Illustrate the relationships in the problem by a diagram, picture, or chart.

The last step in solving a problem is to check that your arithmetic is correct and that your answer is reasonable.

Exercises

1. The perimeter of a rectangle is 98 m. Can the dimensions of this rectangle be consecutive integers? Can the dimensions be consecutive even integers? Can the dimensions be consecutive odd integers? Explain your answers. yes; no; no
2. Tom is three years older than his sister. The product of their ages is 460. Make a reasonable guess of Tom's age and check your guess to see if it is correct. If your guess is not correct, refine it and check it again. Explain how you chose your second guess. Tom, 23 yr; sister, 20 yr
3. Measure the thickness of 50 pages of this book. Use this information to estimate the total number of pages in the book. Check your answer with the actual number of pages.
4. What are the last three digits of 15^8? Explain your strategy. 625
5. On the first Wednesday of April, Trudy announced that her birthday was 200 days away. On what day of the week is Trudy's birthday this year? Explain your strategy. Sunday

Mixed Review

1. State an open sentence that defines the relation
$\left\{(8, -4), (-12, 6), (4, -2), \left(5, -\frac{5}{2}\right)\right\}$ and tell whether or not the relation is a function.
$y = -\frac{1}{2}x$
The relation is a function.
2. If $f(x) = 3x - 7$ and $g(x) = -\frac{1}{x^2}$, determine the value of $[f \circ g](-3)$. $-7\frac{1}{3}$
3. Determine the value of k so that the point $(-2, 3)$ will be on the graph of $kx + 5y = -7$. $k = 11$
4. Determine the slope of the line passing through the points $(-6, 3)$ and $(-2, -1)$. $m = -1$
5. Determine the slope of the line whose equation is $3x - 4y = 20$. $m = \frac{3}{4}$
6. Determine the value of a that makes the slope of the line through the points $(2, -3)$ and $(4, a)$ equal to $\frac{1}{2}$. $a = -2$

Applications of Linear Relations

OBJECTIVE for Section 3-7:
1. To solve problems using direct variation.

Supplementary Material

Test 8

Key Ideas

Use direct variation to solve problems.

Chalkboard Examples

1. **a.** Find a rule that defines a linear function F if

   $F(3) = 5$ and $m = \dfrac{2}{3}$.

   b. Is (12, 11) on the graph of F?

   a. $F(x) = mx + b$

   $5 = \dfrac{2}{3}(3) + b$

   $b = 3$

   Hence a rule for F is

   $y = \dfrac{2}{3}x + 3$

   b. The pair (12, 11) is on the graph of F because

   $11 = \dfrac{2}{3}(12) + 3.$

3–7 Direct Variation

A function f in which the rule for pairing is given by a linear equation of the form

$$y = mx + b \quad (m, b \in \mathcal{R})$$

is called a linear function and can be written in functional notation as

$$f: x \longrightarrow mx + b.$$

Generally, the domain of a linear function is $\mathcal{R}$, and its graph is a nonvertical line.

EXAMPLE 1   **a.** Find a rule that defines a linear function G if $G(8) = 7$ and $m = \dfrac{3}{4}$.

   b. Is $\left(10, \dfrac{17}{2}\right)$ on the graph of G?

SOLUTION   **a.** Since $m = \dfrac{3}{4}$ and (8, 7) is a pair determined by G, we can substitute these values in $G(x) = mx + b$:

$$7 = \dfrac{3}{4} \cdot 8 + b \qquad \text{or} \qquad b = 7 - 6 = 1$$

   Hence a rule for G is $y = \dfrac{3}{4}x + 1$.

   b. The pair $\left(10, \dfrac{17}{2}\right)$ is on the graph of G because $\dfrac{17}{2} = \dfrac{3}{4} \cdot 10 + 1.$

When a linear function is written in the form $y = mx + b$ and $m = 0$, then $y = b$ for all $x \in \mathcal{R}$. In that case f is called a constant function, and its graph is the horizontal line through (0, b).

If $b = 0$ and $m \neq 0$, we have $y = mx$. Then the function f is called a direct variation, and we say that y varies directly as x, or y is directly proportional to x, and that m is the constant of variation or the constant of proportionality. When the domain of a direct variation f is $\mathcal{R}$, the ordered pair (0, 0) is always on the graph of f, and hence this graph is a line passing through the origin and having slope m.

Figure 11 shows the graph of the direct variation specified by

$$y = 2x \quad (x \in \mathcal{R}).$$

108   *Chapter 3*

For any two points (x_1, y_1) and (x_2, y_2) on the graph of this function, we have

$$y_1 = 2x_1 \quad \text{and} \quad y_2 = 2x_2.$$

If $x_1, x_2 \neq 0$, you can transform these equations to

$$\frac{y_1}{x_1} = 2 \quad \text{and} \quad \frac{y_2}{x_2} = 2.$$

Therefore $\frac{y_1}{x_1} = \frac{y_2}{x_2}$.

In general:

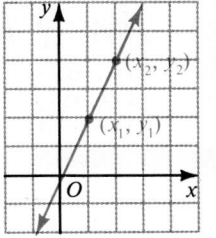

Figure 11

If a linear function f is a direct variation, then for any two ordered pairs (x_1, y_1) and (x_2, y_2) determined by f, with $x_1, x_2 \neq 0$,

$$\frac{y_1}{x_1} = \frac{y_2}{x_2}.$$

Such an equality of ratios, sometimes written as

$$y_1:x_1 = y_2:x_2,$$

is called a **proportion**. The terms y_1 and x_2 are called the **extremes**, and x_1 and y_2 the **means**, of the proportion. Since $\frac{y_1}{x_1} = \frac{y_2}{x_2}$ can be transformed to the equivalent equation $y_1 x_2 = x_1 y_2$, you can see that:

In any proportion the product of the means equals the product of the extremes.

EXAMPLE 2   When a bicycle is being pedaled in a certain gear, it travels 16 m for every 3 pedal revolutions. How many revolutions would be needed to travel 600 m?

SOLUTION   Let $x =$ the number of pedal revolutions. Distance traveled varies directly as the number of pedal revolutions. Two ordered pairs determined by the variation are $(3, 16)$ and $(x, 600)$. Therefore we have

$$\frac{16}{3} = \frac{600}{x}$$
$$16x = 1800$$
$$x = 112.5.$$

It would take 112.5 revolutions to go 600 m.

2. When a bicycle is being pedaled in a certain gear, it travels 10 m for every 4 pedal revolutions. How many revolutions would be needed to travel 500 m?

Let $x =$ the number of pedal revolutions. Two ordered pairs determined by the relation are $(4, 10)$ and $(x, 500)$. Therefore, we have

$$\frac{10}{4} = \frac{500}{x}$$
$$10x = 2000$$
$$x = 200$$

It would take 200 revolutions to go 500 m.

Suggested Assignments

Minimum
First day
110/1–26
Second day
112/P: 1–6
R 113/Self-Test 4
Extended Alg. with Trig.
First day
110/1–29 odd
Second day
112/P: 1–5
R 113/Self-Test 4
Enriched Alg.
110/5, 9, 11, 15–31 odd
112/P: 3–9
R 113/Self-Test 4
Enriched Alg. with Trig.
110/5, 9, 11, 15–31 odd
112/P: 3–9
R 113/Self-Test 4

Additional Answers
Oral Exercises

3. For $g: x, -\dfrac{1}{3}$; for $y = \dfrac{7}{2}x$,

$\dfrac{7}{2}$; for $y = \dfrac{4x}{5}, \dfrac{4}{5}$

Additional A Exercises

If y varies directly as x, determine the x- or y-values for each pair of points.

1. $(6, y), (10, 9)$ $y = 5.4$

2. $(7, 5), (x, 2)$ $x = \dfrac{14}{5}$

3. $(x, 7), (9, 12)$ $x = \dfrac{21}{4}$

Determine a rule for the linear function f, given the specified values.

4. $f(2) = 5; f(3) = 8$
   $y = 3x - 1$

5. $f(-2) = 6; f(1) = 12$
   $y = 2x + 10$

6. $f(-3) = -18; f(4) = 17$
   $y = 5x - 3$

Oral Exercises

Exercises 1–3 refer to the functions below.

$f: x \longrightarrow 3x^2$ $g: x \longrightarrow -\dfrac{x}{3}$ $h(x) = 7 - 2x$

$y = \dfrac{7}{2}x$ $y = 6$ $y = \dfrac{2}{x}$

$y = 3x - 8$ $y = \dfrac{4x}{5}$ $y = \dfrac{2 - 9x}{11}$

1. State all the linear functions in the list.

2. State all the direct variations in the list.

3. State the constant of variation for each direct variation stated in Exercise 2.

1. $g, h, y = \dfrac{7}{2}x, y = 6, y = 3x - 8,$ $y = \dfrac{4x}{5}, y = \dfrac{2 - 9x}{11}$

2. $g, y = \dfrac{7}{2}x, y = \dfrac{4x}{5}$

For each set of ordered pairs, examine the ratio $\dfrac{y}{x}$ and tell whether or not the pairs are in direct variation.

4. $\{(-1, 0), (2, 3), (5, 6)\}$ no
5. $\{(3, 1), (6, 2), (-3, -1)\}$ yes
6. $\{(-3, 5), (6, -10), (9, -15)\}$ yes
7. $\{(1, 2), (-2, -4), (8, 4)\}$ no

State an equation that expresses one of the variables mentioned as a linear function of the other. Tell whether the equation defines a direct variation.

8. The speed of an object dropped from the top of a building increases by 9.8 m/s every second. Relate the speed v to time t. $v = 9.8t$; a direct variation

9. In Exercise 8, suppose the object is *thrown* downward at 12 m/s (with the same increase of speed as before). Relate the speed v to time t. $v = 12 + 9.8t$; not a direct variation

10. To find the radius of a circle, multiply the circumference by $\dfrac{1}{2\pi}$. Relate the radius r to the circumference C. $r = \frac{1}{2\pi}C$; a direct variation

11. A taxicab charges $.90, plus $1.20 for each mile traveled. Relate the cost for a ride C (in dollars) to the distance x (in miles). $C = 0.90 + 1.20x$; not a direct variation

12. Eggs cost $1.35 a dozen. Relate the cost C (in dollars) to the number of eggs purchased N. $C = 1.35(\frac{N}{12})$; a direct variation

13. At 0°C a quantity of ice requires 670 J of heat to melt. For each additional 8 J, the temperature of the water produced rises by 1°C. Relate the temperature of the water T to the heat applied h (in joules). (Assume $h > 670$.) $T = \frac{h - 670}{8}$; not a direct variation

14. Sound can travel 1070 m in 4 s. Relate the distance traveled d to time t. $d = 267.5t$; a direct variation

Written Exercises

If y varies directly as x, determine the x- or y-values for each pair of points.

A   1. $(8, 5), (4, y)$ $\frac{5}{2}$ 2. $(16, 9), (x, 3)$ $\frac{16}{3}$ 3. $(6, y), (9, 14)$ $\frac{28}{3}$

110   Chapter 3

4. $(x, 12)$, $(5, 40)$ $\frac{3}{2}$ **5.** $(8, y)$, $(10, 7)$ $\frac{28}{5}$ **6.** $(6, 8)$, $(x, 6)$ $\frac{9}{2}$

Determine a rule for the linear function f, given the specified values.

7. $f(-1) = 4$; $f(1) = -2$ $y = -3x + 1$

8. $f(2) = -1$; $f(5) = 11$ $y = 4x - 9$

9. $f(2) = 9$; $f(-3) = -1$ $y = 2x + 5$

10. $f(-9) = 7$; $f(-3) = 3$

11. $f(6) = 1$; $f(14) = -3$ $y = -\frac{1}{2}x + 4$

12. $f(-6) = 9$; $f(8) = -12$

13. $f\left(\frac{1}{2}\right) = 5$; $f\left(\frac{3}{2}\right) = 9$ $y = 4x + 3$

14. $f\left(-\frac{1}{3}\right) = 8$; $f\left(\frac{5}{3}\right) = 2$ $y = -3x + 7$

Determine the value of k if the graph of the given linear function contains the given point.

15. $y = -\frac{2k}{3}x - 5$; $(12, 3)$ -1

16. $y = \frac{5x}{2k}$; $(8, 6)$ $\frac{10}{3}$

17. $y = 3kx - 4$; $(6, 17)$ $\frac{7}{6}$

18. $y = -\frac{kx}{2} + 1$; $(20, 6)$ $-\frac{1}{2}$

19. $y = \frac{2kx}{3}$; $(9, 5)$ $\frac{5}{6}$

20. $y = \frac{4kx + 1}{3}$; $(5, 17)$ $\frac{5}{2}$

21. A relation is defined by $y = \frac{3kx}{4}$. If $y = 30$ when $x = 8$, find y when $x = 6$. $\frac{45}{2}$

22. A relation is defined by $y = 6kx - 5$. If $y = 11$ when $x = 4$, find y when $x = 7$. 23

Find the missing value if each set of ordered pairs is determined by a linear function.

B   **23.** $\{(-3, 5), (4, -9), (7, \underline{\ ?\ })\}$ -15

24. $\{(4, -3), (12, 1), (\underline{\ ?\ }, 7)\}$ 24

25. $\{(6, 5), (-9, -5), (1, \underline{\ ?\ })\}$ $\frac{5}{3}$

26. $\{(8, 9), (-4, -9), (\underline{\ ?\ }, -2)\}$ $\frac{2}{3}$

For the proportion $\dfrac{y_1}{x_1} = \dfrac{y_2}{x_2}$ $(x_1, y_1, x_2, y_2 \neq 0)$, prove each of the following properties.

27. $\dfrac{x_1}{y_1} = \dfrac{x_2}{y_2}$

28. $\dfrac{y_1}{y_2} = \dfrac{x_1}{x_2}$

29. If $x_1 \neq x_2$, then $\dfrac{y_1 - y_2}{x_1 - x_2} = \dfrac{y_2}{x_2}$. (*Hint:* Let $y_1 = mx_1$.)

C   **30.** If the function g is a direct variation, prove that

$$g(r + s) = g(r) + g(s)$$

for all $r, s \in \mathcal{R}$.

31. If g is a linear function defined by

$$g(x) = mx + b, \text{ and } g(r + s) = g(r) + g(s)$$

for some $r, s \in \mathcal{R}$, prove that b must be 0.

Linear Functions and Relations   **111**

29. 1. $\dfrac{y_1}{x_1} = \dfrac{y_2}{x_2}$
$(x_1, x_2, y_1, y_2 \neq 0,$
$x_1 \neq x_2)$
Hypothesis

2. $y_1 = mx_1$
Def. of dir. variation

3. $\dfrac{mx_1}{x_1} = \dfrac{y_1}{x_1}$
Div. prop. of =

4. $\dfrac{mx_1}{x_1} = \dfrac{y_2}{x_2}$
Subst.

5. $m = \dfrac{y_2}{x_2}$
Ex. 25, p. 27

6. $m = \dfrac{y_1 - y_2}{x_1 - x_2}$
Def. of slope

7. $\dfrac{y_1 - y_2}{x_1 - x_2} = \dfrac{y_2}{x_2}$
Subst.

30. 1. $g(x) = mx$
Hypothesis (Def. of direct variation)

2. $g(r + s) = m(r + s)$
Subst.

3. $g(r + s) = mr + ms$
Dist. ax.

4. $g(r) = mr, g(s) = ms$
Subst.

5. $g(r + s) = g(r) + g(s)$
Subst.

31. 1. $g(x) = mx + b$
$g(r + s) = g(r) + g(s)$
Hypothesis

2. $g(r + s) = m(r + s) + b$
Subs.

3. $= mr + ms + b$
Dist. ax.

4. $g(r) = mr + b$
$g(s) = ms + b$
Subs.

5. $mr + ms + b = mr + b + ms + b$
Subs.

A 1. A convention organizer has calculated that it would cost $500 for refreshments for 240 people at the opening session. If 300 people actually attend the opening session, what is the total cost for the refreshments? $625

2. Because of refraction, an object 4 cm below the surface of a lake appears to be 3 cm below the surface. If the apparent depth varies directly with the actual depth, what is the actual depth of a rock that appears to be 5 m below the surface? $6\frac{2}{3}$ m

3. The voltage in the circuits of a transformer varies directly with the number of windings of wire around the iron core. If 250 windings produce a voltage of 10,000 V, what voltage is produced by 3 windings? 120 V

4. If 5 kg of coal can be burned to produce 36 kW·h of electricity, how many kg of coal would it take to produce 90 kW·h of electricity? 12.5 kg

5. The temperature of a laboratory sample is 8°C at 7:00 A.M. and −2°C at 11:00 A.M. If the temperature is a linear function of the time, what will the temperature be at 2:00 P.M.? −9.5°C

6. A repair service charges a fixed fee for a house call plus a fixed hourly rate. If a 2-hour call costs $57 and a $4\frac{1}{2}$-hour call costs $97, what is the fixed fee for a house call? $25

B 7. At 2:00 A.M. snow started falling at a constant rate. At 10:30 A.M. Harry Coldfoot began clearing his driveway at a constant rate. At 11:00 A.M. when it stopped snowing the average depth of the snow in Harry's driveway was 2 in. By 11:24 the driveway was clear. At what rate did the snow fall in in./h? $\frac{1}{2}$ in./h

8. Fahrenheit temperature (F) is a linear function of Celsius temperature (C). Water freezes at 0°C, or 32°F, and boils at 100°C, or 212°F. What Fahrenheit temperature is equivalent to 35°C? 95°F

9. In the seventeenth century the Danish astronomer Ole Roemer deduced that light takes longer to reach Earth from Jupiter when Earth is at E_2 than when Earth is at E_1. The modern figure for this time difference is 1000 s. If the approximate distance from Jupiter (J) to E_1 and E_2 is 600 Gm and 900 Gm respectively, how long does it take light from Jupiter to reach Earth at E_1? (A *gigameter* is defined as follows: 1 Gm = 1,000,000,000 m.) 2000 s

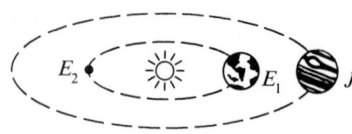

10. Anna Taylor had $3500 invested, part in a regular savings account paying interest at 5% per year and part in a high-yield account. In one year she made $235 in interest. The following year she transferred all her money to the high-yield account and made $280 interest. How much had been in the regular savings account? $1100

Self-Test 4

VOCABULARY linear function (p. 108)
constant function (p. 108)
direct variation (p. 108)
constant of proportionality (p. 108)

proportion (p. 109)
extremes (p. 109)
means (p. 109)

1. If y varies directly as x, and $y = 7$ when $x = 4$, find y when $x = 5$. $\frac{35}{4}$ *Obj. 1, p. 108*
2. Distance measured on a map varies directly with the actual distance. If 1 cm represents 15 km, how many kilometers are represented by 7 cm? 105 km

Check your answers with those at the back of the book.

Chapter Summary

1. A *relation* is any set of ordered pairs. A *function* is a relation in which each first component is paired with exactly one second component. Functions can be defined using an open sentence, a formula or equation, arrow notation, or $f(x)$ notation.
2. Functions and relations can be pictured as ordered pairs on a *coordinate system* where the first component is called the *x-coordinate*, or *abscissa*, and the second component is called the *y-coordinate*, or *ordinate*.
3. A *solution* of an *open sentence in two variables* is any ordered pair for which the sentence is true. The *solution set* of the sentence is the set of all its solutions.
4. A *linear equation in two variables* is an equation of the form

$$Ax + By = C$$

where A, B, $C \in \mathcal{R}$ and A and B are not both zero. The graph of a linear equation is a straight line.
5. The *graph of a linear inequality* is an *open* or *closed half-plane*. The *boundary* of the half-plane is the graph of the associated linear equation.

Linear Functions and Relations **113**

6. $0 = b$
   Canc. prop. of add.
7. $b = 0$
   Symm. prop. of $=$

Quick Quiz

1. If y varies directly as x, and $y = 6$ when $x = 5$, find y when $x = 8$. $\frac{48}{5}$

2. The distance between two points measured on a map varies directly with the actual distance. If 1 cm represents 12 km, how many kilometers are represented by 5 cm? 60 km

Mixed Review

Determine an equation in the form $y = mx + b$ for each of the following.

1. $P(4, -1)$; $m = \frac{2}{3}$

   $y = \frac{2}{3}x - \frac{11}{3}$

2. $P(3, 2)$; $m = -2$
   $y = -2x + 8$

3. $(0, 0)$; $m = \frac{2}{5}$

   $y = \frac{2}{5}x$

4. $(3, 4)$, $(9, 7)$ $y = \frac{1}{2}x + \frac{5}{2}$

5. $(4, -1)$, $(2, 1)$ $y = -x + 3$

6. $\left(\frac{1}{2}, 5\right)$, $\left(-\frac{1}{2}, 6\right)$

   $y = -x + \frac{11}{2}$

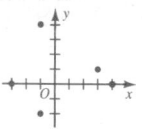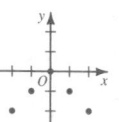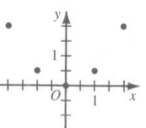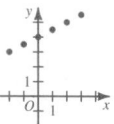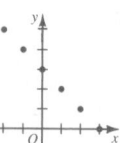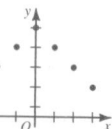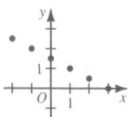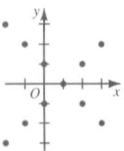
(continued on p. 116)

114

---

6. The *slope* of a nonvertical line is the ratio of *rise* to *run*. A horizontal line has *zero* slope, and a vertical line has *no* slope. The slope of the line $Ax + By = C$ is $-\frac{A}{B}$ where $A, B, C \in \mathcal{R}$ and $B \neq 0$.

7. The equation of a line can be determined using two points on the line, the slope of the line and one point, or the slope and y-intercept. The *slope-intercept* form of a line is $y = mx + b$ where m is the slope and b is the y-intercept. The *point-slope* form of a line is $y - y_1 = m(x - x_1)$ where m is the slope and (x_1, y_1) is a point on the line.

8. A *linear function* is a function in which the rule for pairing is given by the linear equation of the form $y = mx + b$ with $m, b \in \mathcal{R}$. If $b = 0$, then the linear function is a *direct variation*.

9. An equality of *ratios* is called a *proportion*; the product of the *means* equals the product of the *extremes*.

Chapter Review

Write the letter of the correct answer.

1. If $f(x) = x^2 - 1$ and $g(x) = \dfrac{1}{x + 2}$, find $[f \circ g](-3)$.   *3-1*

   a. $\dfrac{1}{10}$ **(b.)** 0   **c.** $-\dfrac{1}{8}$ **d.** -2

2. Which rule defines the relation $\left\{\left(1, \frac{1}{2}\right), \left(2, \frac{1}{4}\right), \left(3, \frac{1}{6}\right), \left(4, \frac{1}{8}\right)\right\}$?

   a. $y = \frac{1}{2}x$ **b.** $y = 2x$ **c.** $y = \dfrac{1}{x^2}$ **(d.)** $y = \dfrac{1}{2x}$

Classify each statement as (a) true or (b) false.

3. The relation $\{(2, 1), (3, -2), (4, 1), (6, 5)\}$ is a function. a   *3-2*

4. The relation $\{(5, -3), (5, -1), (3, 2), (4, 5)\}$ is not a function. a

5. The relation $|y| = |2x|$ where $x \in \{-1, 0, 1, 2\}$ is a function. b

6. Simplify $[-2.2]$.   *3-3*

   a. -2.2 **b.** 2.2 **c.** -2 **(d.)** -3

7. Which ordered pair is *not* on the graph of $5y - 2x = 3$?

   a. $\left(0, \frac{3}{5}\right)$ **b.** $(-4, -1)$ **(c.)** $\left(-\frac{2}{3}, 0\right)$ **d.** $(1, 1)$

8. If $(-2, p)$ is a point on the line $3x + 4y = 10$, find p.

   a. 8.5   **b.** 1   **(c.)** 4   **d.** 11.5

114   *Chapter 3*

9. Which point is *not* on the graph of the inequality $2y - 5x > 8$? 3-4

    a. $(-2, 1)$ **b.** $(-4, -3)$ **c.** $(0, 5)$ **(d)** $\left(\frac{3}{10}, \frac{5}{4}\right)$

10. Which point is *not* on the graph of the inequality $y \leq |x + 1|$?

    a. $(-3, 2)$ **(b)** $(-6, 7)$ **c.** $(4, 0)$ **d.** $(-10, 8)$

11. For what value of a does a line with slope $\frac{1}{4}$ pass through points 3-5
    $(-3, a)$ and $(5, 3)$?

    a. $\frac{1}{2}$ **b.** 0 **c.** 3 **(d)** 1

12. Determine an equation in the form $y = mx + b$ of a line containing 3-6
    the points $(3, 1)$ and $(-6, 7)$.

    a. $y = \frac{2}{3}x + 3$ **(b)** $y = \frac{-2}{3}x + 3$

    c. $y = \frac{-3}{2}x + 3$ **d.** $y = \frac{-2}{3}x$

13. Determine an equation in the form $y = mx + b$ of a line passing
    through the point $(-4, 2)$ and parallel to the line $y = \frac{3}{4}x - 2$.

    a. $y = \frac{-3}{4}x + 2$ **b.** $y = \frac{3}{4}x - 2$

    c. $y = \frac{3}{4}x - 5$ **(d)** $y = \frac{3}{4}x + 5$

14. Solve the proportion $5 : -2 = h : 4$ for h. 3-7

    a. $\frac{-8}{5}$ **(b)** -10 **c.** -20 **d.** 10

15. If y varies directly as x and $y = 18$ when $x = 12$, find x when $y = 27$.

    a. 6 **b.** 9 **(c)** 18 **d.** 36

Chapter Test

If $f: x \longrightarrow x^2 + 4$ and $g: x \longrightarrow \left|\dfrac{x}{3}\right|$, find the value of each of the following. 3-1

1. $f(-4)$ 20   **2.** $g(9)$ 3   **3.** $[f \circ g](-6)$ 8   **4.** $[g \circ f](2)$ $\frac{8}{3}$

Graph each relation over the domain $\{-1, 0, 1, 2, 3\}$. Is the relation a function?

5. $x + y = 4$ yes   **6.** $|y| = |x - 3|$ no 3-2

7. $4y - 2x = 4$ yes   **8.** $y = |x| - x$ yes

Linear Functions and Relations   **115**

Additional Answers
Chapter Test

5.

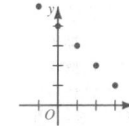

6.

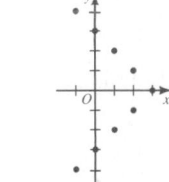

7.

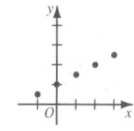

8.

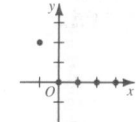

(continued)

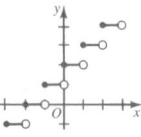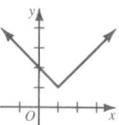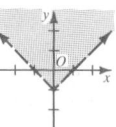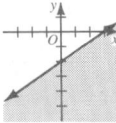
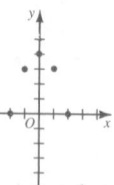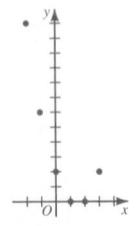
Graph each equation.

9. $y = [x] + 2$ 10. $y = |x - 1| + 1$ *3-3*

Graph each inequality as a shaded region on a coordinate plane.

11. $y > |x| - 1$ 12. $2x - 3y \geq 6$ *3-4*

Determine the slope of the line through each pair of points.

13. $(-1, 6)$, $(3, -2)$ -2 14. $(5, 2)$, $(1, 7)$ $-\frac{5}{4}$ *3-5*

Determine an equation in the form $y = mx + b$ of a line satisfying the given condition(s).

15. through the points $(5, -1)$ and $(3, 0)$ $y = -\frac{1}{2}x + \frac{3}{2}$ *3-6*
16. through the point $(-5, -5)$ and parallel to the line $5y - 2x = 5$ $y = \frac{2}{5}x - 3$

17. Solve for a if the points $\left(\frac{3}{2}, -4\right)$ and $(-6, a)$ are in a direct variation.   16   *3-7*

18. In a certain city, the amount of property tax paid varies directly with the assessed value of the property. If a family whose house is assessed at $60,000 pays $1500 in property tax during a certain year, what is the assessed value of a house whose owners paid $2125 in taxes during the same year? $85,000

APPLICATION

Paper Chromatography

Does the ink in your pen contain one colored substance or a mixture of them? One way to find out is to analyze the ink using a technique called *paper chromatography*.

In a typical procedure, a drop of ink is placed on a paper strip and allowed to dry. The paper strip is suspended, as shown in Figure A, with one end just under a liquid, or *solvent*, that will dissolve the ink. As the solvent moves up the strip (just as water moves up a paper towel that has been dipped in it), it dissolves the ink and carries the ink up the strip. If the ink is a mixture, each component of the mixture will move a different distance up the strip. The distance depends on how the tendency of the component to stay dissolved in the solvent compares to its attraction to the paper.

When the solvent has nearly reached the top, the strip is removed and allowed to dry. If the ink contains just one colored substance, only one spot of color is seen on the strip. If the ink is a mixture, two or more spots of color are seen on the strip. The dried strip is called a *chromatogram*.

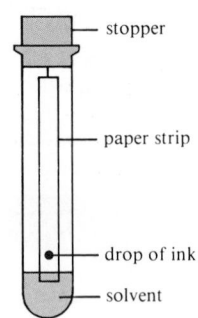

stopper

paper strip

drop of ink

solvent

Fig. A

The chromatogram in Figure B was produced by a drop of brown ink. It shows that the ink contains four different colored substances.

As the name implies, paper chromatography was first used to separate mixtures of colored substances. However, some of its most important applications (such as the separation of mixtures of amino acids, the building blocks of proteins) involve noncolored substances. The different spots on the chromatogram show up clearly when it is sprayed with a certain chemical.

Components of a mixture may be identified by their R_f, or *ratio front*, values. For a given component:

$$R_f = \frac{\text{distance moved by the component}}{\text{distance moved by the solvent front}}$$

The solvent front is the leading edge of the solvent as it moves up the paper. A component that moves 9.80 cm up a paper strip while the solvent front moves 12.6 cm has

$$R_f = \frac{9.80}{12.6} = 0.778.$$

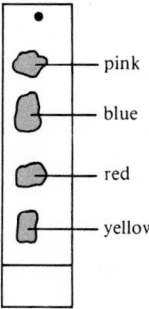

pink

blue

red

yellow

Fig. B

Exercises

1. In a certain chromatography procedure, substances A, B, and C have R_f values of 0.920, 0.742, and 0.540, respectively. When the procedure is carried out using a mixture of A, B, and C, which substance will move the farthest up the strip? substance A

2. In the Exercise 1 procedure, how far does substance C move up the strip if the solvent front moves a distance of 14.2 cm? 7.7 cm

Linear Functions and Relations **117**

20.
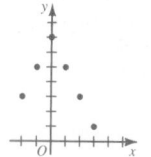

21.
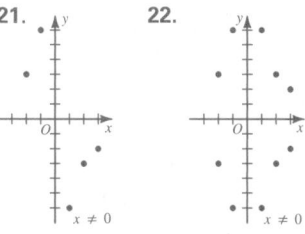
22.

$x \neq 0$ $x \neq 0$

23.
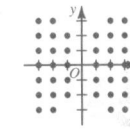

24.
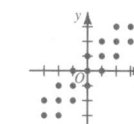

25.

26.
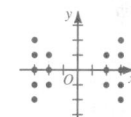

27.
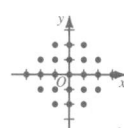

28.
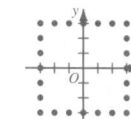

12.

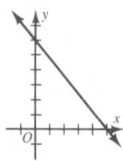

14.

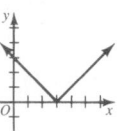

16.

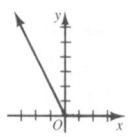

18.

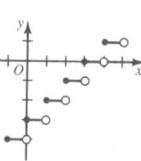

24.

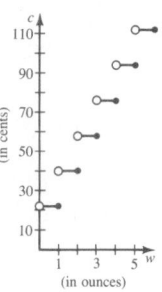

26.

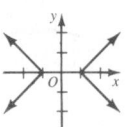

28.

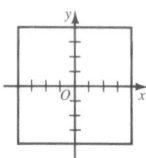

PREPARING FOR
COLLEGE ENTRANCE EXAMS

Strategy for Success: You receive just as much credit for correctly answering easy questions as you do for correctly answering difficult ones. Answer all the questions that seem easy to you first, and if time permits, return to the questions that seem difficult.

Decide which is the best of the choices given and write the corresponding letter on your answer sheet.

1. Which of the following must be true for $A = \{n: n^2 - (n - 1)^2, n$ is a nonpositive integer$\}$ and $B = \{$negative odd integers$\}$? D
   I. $A \subset B$ II. $A = B$ III. $A \in B$
   (A) I only   **(B)** II only   **(C)** III only
   (D) I and II only   **(E)** I, II, and III

2. Which of the following is true if $2x + b = x$? B
   (A) $x = b$ **(B)** $x = -b$ **(C)** $2x = x + b$ **(D)** $x = 2b$ **(E)** $3x = b$

3. Simplify $-3[4x - 6(x - 3)] - 5[-7x - 2(3 + 2x)]$. A
   (A) $61x - 24$ **(B)** $61x + 24$ **(C)** $49x - 24$
   (D) $-49x + 24$ **(E)** $85x - 24$

4. Which of the following is equivalent to $a(x + y) - 4a(x - y) = 2x - 7y$? C
   (A) $a = \dfrac{2x + 7y}{3x + 5y}$ **(B)** $a = \dfrac{2x - 7y}{3x - 5y}$ **(C)** $a = \dfrac{2x - 7y}{5y - 3x}$
   (D) $a = \dfrac{2x + 7y}{5y - 3x}$ **(E)** $a = \dfrac{7y - 2x}{5y + 3x}$

5. Which of the following is the solution set for $2|x - 3| > 8$? E
   (A) $\{x: x > 7\}$ **(B)** $\{x: x < -1\}$ **(C)** $\{x: -1 < x < 7\}$
   (D) $\{x: -7 < x < 1\}$ **(E)** $\{x: x > 7\} \cup \{x: x < -1\}$

6. What is an equation of the line with x-intercept 3 and y-intercept -4? E
   (A) $y = -\frac{4}{3}x - 4$ **(B)** $y = \frac{4}{3}x + 4$ **(C)** $y = \frac{3}{4}x - 4$
   (D) $y = -\frac{3}{4}x - 4$ **(E)** None of these

7. What is the value of k, so that the graph of $4x + ky = 8$ has the same slope as the line passing through the points $(6, -2)$ and $(4, -1)$? E
   (A) -8 **(B)** $\frac{1}{2}$ **(C)** $-\frac{1}{2}$ **(D)** 2   **(E)** 8

8. If $\dfrac{y_2}{y_1} = \dfrac{x_2}{x_1}$ $(x_1, x_2, y_1, y_2 \neq 0)$, which of the following must be true? E
   I. $\dfrac{x_1}{y_1} = \dfrac{x_2}{y_2}$ II. $\dfrac{y_2}{x_2} = \dfrac{y_1}{x_1}$ III. $\dfrac{y_1}{y_2} = \dfrac{x_1}{x_2}$
   (A) I only   **(B)** II only   **(C)** III only
   (D) I and II only   **(E)** I, II, and III

118   *Chapter 3*

PROGRAMMING IN PASCAL

The programming language of Pascal contains several pre-defined functions that are offered to a user. The identifier *trunc* (for truncate) represents a function that discards the decimal part, if any, of its argument. For example:

$$trunc(3.14154) = 3 \qquad trunc(5) = 5 \qquad trunc(-1.414) = -1$$

The function *round* will round its argument to the nearest integer. For example:

$$round(1.7) = 2 \qquad round(-2.3) = -2 \qquad round(6.8) = 7$$

In addition to the pre-defined functions, Pascal allows the user to define their own functions. The following is an example of a user-defined function that could be used in finding the corresponding y-value for any x in the domain of a linear relation.

```
FUNCTION y (m, x, b : real) : real;
BEGIN
  y := m * x + b;
END;
```

Most user-defined functions in Pascal are more involved than the above example. The following user-defined function is used to round to a real number to the nearest tens.

```
FUNCTION round_to_tens(num : real) : real;
BEGIN
  num := num/10;
  num := round(num);
  round_to_tens: = num*10;
END;
```

Linear Functions and Relations **119**

Additional Answers
Written Exercises(p. 97)

2.
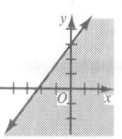

4.
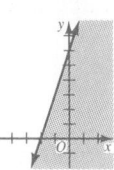

6.
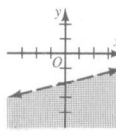

8.
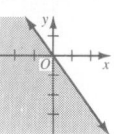

10.
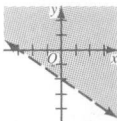

12.
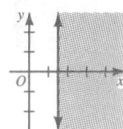

14.
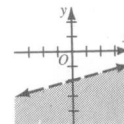

(continued)

16.

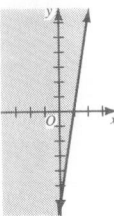

18.

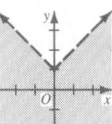

20.

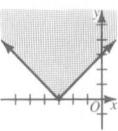

22.

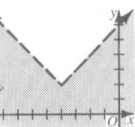

24.

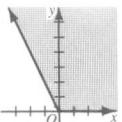

Exercises

1. Study round_to_tens and then write a Pascal function which rounds real numbers to the nearest tenth. Use the heading

    FUNCTION round_to_tenths(num : real) : real;

2. Recall the greatest integer function, [x], on page 94. Write a Pascal function with heading

    FUNCTION greatest_integer(x : real) : integer;

    which computes the greatest integer value for x a real number. Explain the restrictions, if any, imposed on the values of x by your computer. (*Hint*: On most systems this function gives peculiar results if x is too large or too small.)

As an example of an application of Pascal to linear functions, consider the following program:

```
PROGRAM find_slope (INPUT, OUTPUT) ;

VAR
   x1, y1, x2, y2 : real ;


(***********************************************************)
FUNCTION slope (x1, y1, x2, y2 : real ) : real;
BEGIN
   slope := (y1 − y2)/(x1 − x2)
END


(***********************************************************)
BEGIN (* main *)
   writeln('This program finds the slope of');
   writeln('the line between two distinct points');
   writeln;
   writeln('Enter the coordinates of two points.');
   write('Place a space between each of the values: ');
   readln(x1, y1, x2, y2);
   writeln;
   writeln('The slope is: ', slope(x1, y1, x2, y2):3:3);
END.
```

3. Run the program to find the slope of the line between the following pairs of points.
    a. (3, 5), (7, 21) 4.000
    b. (1, 4), (−3, −9) 3.250
    c. (−3, 0), (−4, −5) 5.000
    d. (0, 7), (−6, 11) −0.667
    e. (−5, −3), (−2, −4) −0.333

120   *Chapter 3*

f. What happens when the computer calculates the slope of the line between $(4, -2)$ and $(4, 7)$?

4. Modify the program so the function to find the slope will not be called when the points are on a line parallel to the y-axis. Include a line that tells the user that the points are on a line parallel to the x-axis.

5. Modify the program so it will check the coordinates of the points to make sure they are distinct. If the points are not distinct, the program should point this out to the user, then prompt the user for another pair of coordinates.

6. Write a Pascal program, that satisfies conditions similar to those of *find_slope* but displays the *point-slope* form of an equation of the line passing through the two points (if the slope exists) or which displays only an equation of the form $x = a$ (if the line is vertical).

3.f. The program is interrupted and an error message is printed.

The kayaker in the photo is using the current to help propel the kayak down the river. The speed of the kayak depends on the speed at which the kayak is being paddled and the speed of the current. Motion problems using current and boat speeds are discussed in Section 4-4.

122 *Chapter 4*

Chapter 4

Systems of Linear Equations or Inequalities

Problem Solving Strategies

Word Problem Plan
The five-step plan is used in Section 4–4 to solve problems involving an equation in two variables.
Drawing a Diagram
Graphs are used extensively in Section 4–6 to illustrate the concept of linear programming.

Systems of Equations in Two Variables

OBJECTIVES for Sections 4-1 through 4-4:
1. *To identify the apparent solution set of a system of linear equations in two variables from the graphs of the equations.*
2. *To solve a system of two linear equations in two variables.*
3. *To use determinants to solve a system of two linear equations in two variables.*
4. *To solve problems by using systems of linear equations.*

4–1 Graphing a System of Equations

A conjunction of linear equations such as $x + 3y = 5$ and $2x - y = 3$ is called a **system of linear equations** and can be pictured by one or more lines in one coordinate plane.

Systems of Linear Equations or Inequalities   **123**

Teaching Suggestions
p. T83

Key Ideas

Graph and identify the solution set of a system of linear equations in two variables.

Figures 1, 2, and 3 illustrate the three possibilities that can occur for a system of two equations.

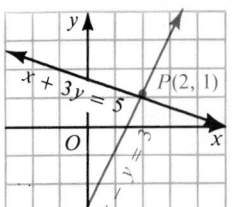
Figure 1

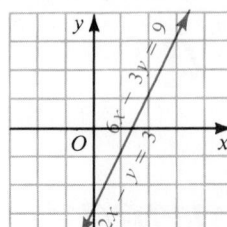
Figure 2

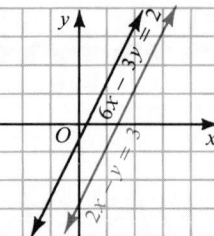
Figure 3

To determine the solution of a conjunction, it is necessary to find all the ordered pairs that satisfy *every* equation in the system. The set of all the ordered pairs of real numbers that satisfy every equation in the system is called the **solution set.**

In Figure 1 the two lines

$$2x - y = 3$$
$$x + 3y = 5$$

intersect in exactly one point P that appears to have coordinates (2, 1). It is easy to verify that the apparent solution, (2, 1), satisfies both equations.

$$2x - y = 3$$
$$2 \cdot 2 - 1 = 3 \checkmark$$

$$x + 3y = 5$$
$$2 + 3 \cdot 1 = 5 \checkmark$$

Note that the graphs of the two lines have *different slopes.*

In Figure 2 the two lines

$$2x - y = 3$$
$$6x - 3y = 9$$

coincide. Thus the solution set of the system consists of the infinite set of points on the line that is the graph of either equation. Note that the graphs of the two lines have the *same slope* and the *same y-intercept.*

In Figure 3 the two lines

$$2x - y = 3$$
$$6x - 3y = 2$$

are parallel and the solution set is the empty set, Ø. Note that the graphs of the two lines have the *same slope* and *different y-intercepts.*

Observe that the number of solutions for a system of two linear equations is related to the slopes of their graphs. Figure 1 and Figure 2 are examples of a system of equations that have at least one solution. These systems are said to be **consistent.** Figure 3 is an example of a system of equations that has no solution and is said to be **inconsistent.**

124 *Chapter 4*

Chalkboard Examples

Graph each system of linear equations and determine whether or not it is consistent.

1. $3x + 2y = 2$
$\quad\quad 4y = -6x + 4$

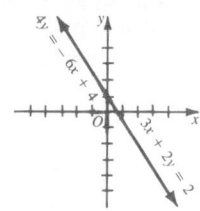

The system is consistent.

2. $\quad\quad y = -4$
$\quad y + 4 = \dfrac{x}{2}$
$\quad\quad 2y = -6x - 8$

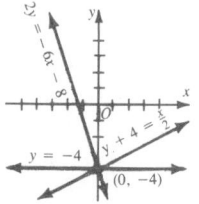

The system is consistent.

Common Errors

Most errors arise when students substitute values into the equation to be graphed. Remind students to be careful when determining the coordinates of points on the graph and also to check their graphs for reasonableness. They can do this by first doing a quick mental calculation of the slope using the fact that the slope of the graph of $Ax + By = C$ is $-\dfrac{A}{B}$.

Next, they should compare the slope of the graph on their coordinate system with the value of $-\dfrac{A}{B}$ to see if it is

124

When a system has an infinite solution set, as shown in Figure 2, the equations are said to be **dependent**.

The following theorem summarizes all the possibilities for a system of two linear equations, as illustrated by Figures 1, 2, and 3. (For a proof of this theorem see Exercises 35–39, pages 132 and 133.)

Theorem. A system of two linear equations in two variables in the same plane has:

1. exactly *one solution* if the two graphs have *different slopes*, or if just one of them has no slope;

2. an *infinite set of solutions* if both graphs have the *same slope and the same y-intercept*, or both have no slope and the same *x*-intercept;

3. *no solution* if both graphs have the *same slope but different y-intercepts*, or else no slope and different *x*-intercepts.

This theorem does not apply to systems of more than two equations. However, we can often determine the consistency or inconsistency of such a system by graphing it.

EXAMPLE   Graph the following system of linear equations and determine whether or not it is consistent.

$$x + y = 1$$
$$x + 2y = 4$$
$$x - y = -5$$

SOLUTION   Find the *x*- and *y*-intercepts for each line. Then draw the three lines through their pairs of intercepts and check to see that the apparent solution of the system, $(-2, 3)$, satisfies all three equations.

$$x + y = 1$$
$$-2 + 3 \overset{?}{=} 1$$
$$1 = 1 \checkmark$$

$$x + 2y = 4$$
$$-2 + 2 \cdot 3 \overset{?}{=} 4$$
$$4 = 4 \checkmark$$

$$x - y = -5$$
$$-2 - 3 \overset{?}{=} -5$$
$$-5 = -5 \checkmark$$

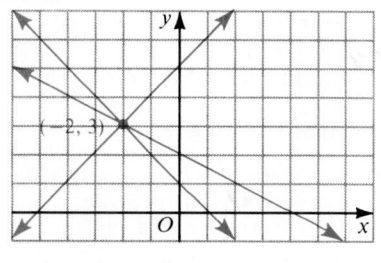

$\therefore \{(-2, 3)\}$ is the solution set of the system of equations, and the system is consistent.

Systems of Linear Equations or Inequalities **125**

Determine the apparent solution set by graphing each system of equations.

1. $y = 3$
    $y - 3 = -x$

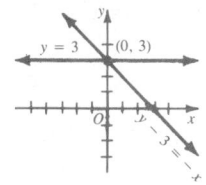

$\therefore \{(0, 3)\}$ is the solution set.

2. $y = -4$
    $x = 5$

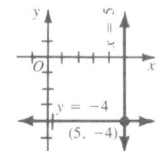

$\therefore \{(5, -4)\}$ is the solution set.

3. $y + 5 = 0$
    $y + 2x = -5$

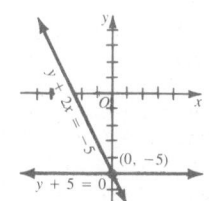

$\therefore \{(0, -5)\}$ is the solution set.

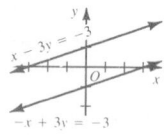
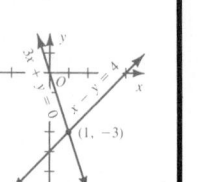
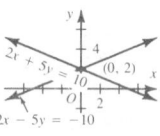
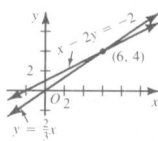
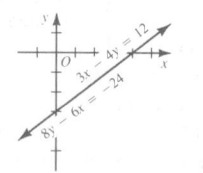
Oral Exercises

a. State the slope and y-intercept of each equation in the given system.
b. Tell whether the system is consistent or inconsistent. If it is consistent, tell whether it has one or infinitely many solutions.

1. $x + y = 5$ a. $-1, 5$ and
   $x - 2y = -1$ $\frac{1}{2}, \frac{1}{2}$
   b. consistent; one

2. $x - 3y = -3$ a. $\frac{1}{3}, 1$ and
   $-x + 3y = -3$ $\frac{1}{3}, -1$
   b. inconsistent

3. $x + 3y = 6$
   $\dfrac{x}{3} + y = 2$

a. Both $-\frac{1}{3}, 2$
b. consistent, many

4. $x - y = 4$ a. $1, -4$ and $-3, 0$
   $3x + y = 0$ b. consistent; one

5. $x + 2y = 2$
   $3x + 6y = 12$ a. $-\frac{1}{2}, 1$ and $-\frac{1}{2}, 2$
   b. inconsistent

6. $2x - 5y = -10$
   $2x + 5y = 10$
   a. $\frac{2}{5}, 2$ and $-\frac{2}{5}, 2$
   b. consistent, one

State the apparent solution set of each system.

7. $x + y = 3$
   $x + y = 10$ $\emptyset$

8. $x = -2y$
   $2x - y = -5$ $\{(-2, 1)\}$

9. $x + y = 3$
   $x - 2y = -6$
   $3x - y = -3$ $\{(0, 3)\}$

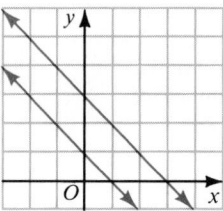

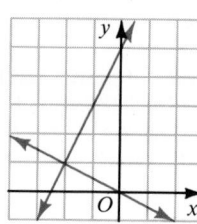

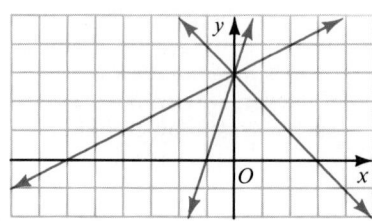

Written Exercises

Determine the apparent solution set by graphing each system of equations. Show that each system satisfies one of the three possibilities listed in the theorem on page 125 by finding the slope and (if necessary) the y-intercept of each graph.

A **1–6.** Use the systems in Oral Exercises 1–6.

7. $3x + y = 7$
   $x + 2y = -1$

8. $y = \dfrac{2}{3}x$
   $x - 2y = -2$

9. $x - 5y = -5$
   $2x - y = 8$

10. $3x - 4y = 12$
   $8y - 6x = -24$

11. $\dfrac{1}{2}x + 3y = 3$
   $-x - 6y = 6$

12. $2x - 3y = 6$
   $-2x + 3y = 3$

13. $3x + 2y = 5$
   $\dfrac{1}{2}x + y = \dfrac{7}{2}$

14. $\dfrac{3}{2}x - y = 3$
   $x + 2y = -6$

15. $-3x + 4y = 9$
   $x - \dfrac{4}{3}y = -3$

16. $6x - 5y = 9$
   $-x + 3y = -8$

17. $5x - 2y = 10$
   $\dfrac{5}{2}x - y = -3$

18. $4x - 6y = 6$
   $-x + \dfrac{3}{2}y = -\dfrac{3}{2}$

Determine the apparent solution set of the system by graphing each system of equations. If the system has exactly one apparent solution, verify that the coordinates of the ordered pair satisfy all three equations.

B **19.** $x + y = 4$
$x - y = -4$
$4x - y = -4$

20. $x = -2y$
$3x + y = 5$
$x - 4y = 6$

21. $x - 2y = 3$
$-x + 2y = 6$
$x + y = 0$

22. $x + 3y = 6$
$2x - y = -2$
$x - 4y = 6$

23. $x - y = 3$
$x - 4y = -3$
$x + y = 7$

24. $x + 2y = 6$
$x + 2y = 2$
$3x + 2y = 10$

25. Graph the equations $kx + y = 4$ for the values $k = -1, \frac{1}{2}, 1, 2$ on one set of axes. What is the apparent solution set of this system?

26. Graph the equations $x + ky = -3$ for the values $k = -6, 0, 3, 6$ on one set of axes. What is the apparent solution set of this system?

C **27.** Choose r and s so that $rs \neq 0$. Graph the following three equations to determine their apparent solution set. Verify that the apparent solution set is actually the solution set.

$$x + 2y = 5$$
$$x - y = -1$$
$$r(x + 2y) + s(x - y) = r(5) + s(-1)$$

28. Prove that any solution to the first two equations in Exercise 27 is a solution to the third.

Computer Exercises For students with computer experience

1. Write a program that will determine whether a given linear system has a unique solution, no solution, or infinitely many solutions. Let the system be in the form

$$ax + by = c$$
$$dx + ey = f$$

with a, b, c, d, e, and f input by the user.

Run the program in Exercise 1 for each system.

2. $3x - 5y = 2$ infinitely many
$-12x + 20y = -8$ solutions

3. $2x - 7y = 38$ unique solution
$6x + 21y = 15$

4. $12x + 18y = -5$ no solution
$-2x - 3y = 17$

5. $6x - 5y = 11$ unique solution
$-7x + 2y = 9$

Systems of Linear Equations or Inequalities **127**

12. Ø
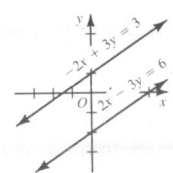

14. $\{(0, -3)\}$
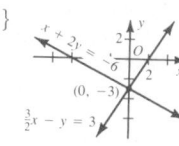

16. $\{(-1, -3)\}$
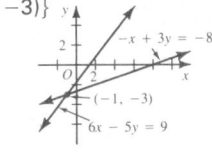

18. $\{(x, y): 4x - 6y = 6\}$
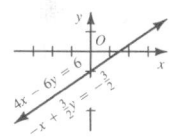

(continued on p. 155)

Mixed Review

1. State the slope and the y-intercept of the graph of $2x + 3y = 15$.

The slope is $-\frac{2}{3}$, and the y-intercept is 5.

2. State an equation in the form $y = mx + b$ of the line passing through the point $(-2, 6)$ and having a slope of -3.
$y = -3x + 0$ or $y = -3x$

3. Simplify $(y + (-5)) + [-(2 + (-y))]$. $2y - 7$

4. State the axiom or property that justifies the statement $7(a - 5) = 7a - 35$.
Distributive axiom

5. Solve $x - 3(9 - x) = 7 + 6x$. $\{-17\}$

Key Ideas

Solve systems of linear
equations in two variables
by the linear-combination
method and by the substitution method.

Chalkboard Examples

1. Solve the system
   $5x + 3y = 7$
   $3x - 5y = -23$
   by the linear-combination
   method.
   $5x + 3y = 7$
   $3x - 5y = -23$
   $5(5x + 3y = 7)$
   $3(3x - 5y = -23)$
   $25x + 15y = 35$
   $9x - 15y = -69$
   $34x = -34$
   $x = -1$
   $5(-1) + 3y = 7$
   $3y = 12$
   $y = 4$
   ∴ the solution set is
   $\{(-1, 4)\}$.

2. Solve the system
   $x + y = 11$
   $3x - y = 5$
   by the substitution method.
   $x + y = 11$ or
   $y = 11 - x$
   $3x - y = 5$
   $3x - (11 - x) = 5$
   $3x - 11 + x = 5$
   $4x = 16$
   $x = 4$
   $y = 11 - 4$
   $y = 7$
   ∴ the solution set is
   $\{(4, 7)\}$.

4–2 Solving a System of Equations

Your aim in solving a system of two linear equations in two variables is
the same as that in solving a linear equation in one variable. That is, you
want to *transform the system into an equivalent one.* In this case you want
equivalent equations of the form

$$x = a$$
$$y = b,$$

which gives the solution explicitly.

Consider the system: $2x + 3y = 4$ (1)
$5x - 2y = 6$ (2)

Using the properties of equality, you can *multiply* both sides of Equation
(1) by the same nonzero number, say, 2, and likewise multiply Equation
(2) by 3 and thereby obtain the *equivalent system:*

$$4x + 6y = 8 \qquad 2 \times (1) = (1')$$
$$15x - 6y = 18 \qquad 3 \times (2) = (2')$$

Equations (1) and (1') are equivalent, and so are (2) and (2'). The *sum*
of Equations (1') and (2'),

$$19x = 26, \qquad ((1') + (2'))$$

obtained by adding the left sides and right sides of both equations, is
called a *linear combination* of the two equations (1) and (2).

In general, when you multiply both sides of an equation by the same
nonzero constant, and add the resulting expressions to the corresponding sides of another equation, you obtain a **linear combination** of the
two equations.

The transformations used to solve systems of linear equations are
summarized below.

---

*Transformations That Produce an Equivalent System
of Linear Equations*

1. Replace any equation of the system with an equivalent equation in the same variable.

2. Replace any equation of the system with a linear combination
   of itself and another equation of the system.

3. Substitute for one variable in any equation either
   a. its actual value, or
   b. an equivalent expression for that variable obtained from
      another equation in the system.

---

128   *Chapter 4*

EXAMPLE   Solve the system:

$$4x + y = 5 \quad (1)$$
$$x - 2y = 8 \quad (2)$$

SOLUTION 1   *Linear-combination method*

1. The variable y can be eliminated from the system as follows: Replace Equation (1) with $2 \times$ Equation (1). (Transformation 1). Replace (1′) with the linear combination (1′) + (2). (Transformation 2). The purpose of obtaining Equation ((1′) + 2) was to eliminate one variable and enable you to solve for the other variable.

$$8x + 2y = 10 \quad (1')$$
$$x - 2y = 8 \quad (2)$$
$$9x = 18 \quad (1'')$$

2. Replace (1″) with an equation of the form $x = a$. (Transformation 1). The original system has been transformed to:

$$x = 2 \quad (1''')$$
$$x - 2y = 8 \quad (2)$$

$$x = 2$$
$$x - 2y = 8.$$

3. Substitute the value 2 for x in Equation (2). (Transformation 3).

$$x = 2 \quad (1''')$$
$$2 - 2y = 8 \quad (2')$$

4. Replace Equation (2′) with an equation of the form $y = b$. (Transformation 1). The system has now been transformed to an equivalent system that gives the solution explicitly.

$$x = 2$$
$$y = -3$$

5. Check by substituting $x = 2$ and $y = -3$ in both Equation (1) and Equation (2).

$$4x + y = 5 \quad (1)$$
$$4 \cdot 2 + (-3) = 5 \checkmark$$

$$x - 2y = 8 \quad (2)$$
$$2 - 2(-3) = 8 \checkmark$$

∴ the solution set is $\{(2, -3)\}$.

Note that we could just as easily have eliminated x instead of y in Step 1 by multiplying Equation (2) by -4 and adding the resulting equation to Equation (1).

Additional A Exercises

Solve each system by the linear-combination method.

1. $x + 4y = 6$
   $3x + 8y = 20$ $\left\{\left(8, -\frac{1}{2}\right)\right\}$

2. $5x = 11 - 3y$
   $9x + 2y = -4$ $\{(-2, 7)\}$

Solve each system by the substitution method.

3. $2x + 3y = 1$
   $x - 2y = 4$ $\{(2, -1)\}$

4. $y = 5 - x$
   $2x - y = -5$ $\{(0, 5)\}$

5. $x = -2y$
   $5x - 3y = 13$ $\{(2, -1)\}$

Suggested Assignments

Minimum
    131/2–18 even, 26–32 even
Extended Alg. with Trig.
    131/1–31 odd, 32
Enriched Alg.
    131/7–35 odd, 36–39
Enriched Alg. with Trig.
    131/7–35 odd, 36–39

Additional Answers
Oral Exercises

13. Add $-5x$ to each side of (2); subst. for y in (1).

15. Add $-8x$ to each side of (1); mult. each side by -1; subst. for y in (2).

17. Add $-4y$ to each side of (1); mult. each side by 2; subst. for x in (2).

Additional Answers
Written Exercises

1. $\{(3, 5)\}$ **2.** $\{(4, -1)\}$
3. $\{(-2, -6)\}$ **4.** $\{(-3, 0)\}$
5. $\{(7, 3)\}$ **6.** $\{(-4, 7)\}$
7. $\{(2, -5)\}$ **8.** $\{(0, 1)\}$
9. $\left\{\left(4, \frac{1}{2}\right)\right\}$ **10.** $\left\{\left(3, -\frac{1}{3}\right)\right\}$
11. $\{(6, 4)\}$ **12.** $\left\{\left(-9, \frac{1}{2}\right)\right\}$
13. $\{(8, 7)\}$ **14.** $\left\{\left(-2, -\frac{1}{3}\right)\right\}$
15. $\left\{\left(\frac{3}{2}, 5\right)\right\}$ **16.** $\left\{\left(\frac{1}{2}, -\frac{5}{2}\right)\right\}$
17. $\{(-2, 3)\}$ **18.** $\{(13, 12)\}$

35. 1. $\left.\begin{array}{l} y = m_1 x + b_1 \\ y = m_2 x + b_2 \end{array}\right\} m_1 \neq m_2$
    $b_1, b_2, m_1, m_2 \in \mathcal{R}$
    Hypothesis
  2. $0 = m_2 x - m_1 x + (b_2 - b_1); \; y = m_1 x + b_1$
    Trans. 2
  3. $-(b_2 - b_1) = m_2 x - m_1 x; \; y = m_1 x + b_1$
    Add. prop. of $=$

SOLUTION 2   *Substitution method*

1. Restate Equation (1) so that it gives an expression for y in terms of x. (Transformation 1).

    $y = 5 - 4x$ (1′)
    $x - 2y = 8$ (2)

2. Substitute the expression for y in Equation (1′) into Equation (2). (Transformation 3). This gives an expression in x that can be solved for x. (Equation (2‴)).

    $x - 2(5 - 4x) = 8$ (2′)
    $9x = 18$ (2″)
    $x = 2$ (2‴)

3. Solve for y in Equation (1′) by substituting the value for x found in Equation (2‴). (Transformation 3).

    $y = 5 - 4 \cdot 2$ (1″)
    $y = -3$

4. The original system has been transformed to an equivalent system that gives the solution explicitly.

    $x = 2$
    $y = -3$

5. Check the answer in the same way it was done in Step 5 of Solution 1.

    ∴ the solution set is $\{(2, -3)\}$.

Figure 4 shows that the original system was finally transformed into a system of equations whose graphs are a pair of lines parallel to the two axes and having the same point of intersection (that is, the same solution) as the graphs of the original system. In general, the graph of an equation formed from a given pair of equations by using one or more of the transformations listed on page 128 is a line through the point of intersection of the graphs of the given pair if these lines intersect. (If the graphs of the given pair of equations are parallel lines, the graph of the new equation is parallel to them; if they are coincident, the graph of the new equation is also coincident with them.)

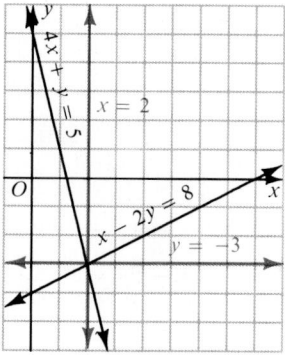

Figure 4

Oral Exercises

State a value by which both sides of one of the equations in each system can be multiplied in order to eliminate the variable in red.
Call the first equation (1) and the second (2).

1. $4x - 3y = -3$ Multiply (2) by -4.
    $x + 2y = 13$

2. $6x + 2y = 22$ Multiply (2) by -2.
    $3x - 5y = 17$

3. $7x - 2y = -2$ Multiply (1) by -3.
    $5x - 6y = 26$

4. $-3x + 10y = 9$
    $2x - 5y = -6$
    Multiply (2) by 2.

5. $2x - \frac{2}{3}y = 12$ Multiply (1) by 2.　**6.** $\frac{1}{2}x + 2y = 12$ Multiply (1) by -8

$-3x + \frac{4}{3}y = -17$　　　　　　　$4x + 5y = 19$

State how to form a linear combination of the two equations to eliminate the variable in red. Answers may vary.

7. Multiply (1) by 4 and (2) by -3, add.

EXAMPLE　　$5x - 2y = 1$
　　　　　　$7x + 3y = 13$

9. Multiply (1) by 5 and (2) by -3, add.

11. Multiply (1) by 5 and (2) by 2, add.

SOLUTION　Multiply the first equation by 3, the second by 2, and add the equations.

7. $3x - 2y = 16$
　　 $4x - 3y = 23$

8. $8x - 5y = -5$ Multiply (1) by 2 and (2) by 5,
　　 $3x + 2y = 2$　　add.

9. $3x + 4y = 14$
　　 $5x - 6y = 17$

10. $5x + 12y = 11$ Multiply (1) by 3 and (2) by 4,
　　　 $2x - 9y = 9$　 add.

11. $-\frac{2}{3}x + 3y = 8$

$\frac{5}{3}x - 4y = -6$

12. $\frac{1}{3}x + 8y = 1$

$-2x - 6y = 15$ Multiply (1) by 6 and add.

State how to transform one of the equations so that one variable is expressed in terms of the other. Then state how to use the transformed equation to solve the system by the substitution method. Answers may vary.

13. $6x - 5y = 13$
　　 $5x + y = 47$

14. $x - 15y = 3$　　Add $15y$ to each side of (1);
　　 $7x - 9y = -11$ subst. for x in (2).

15. $8x - y = 7$
　　 $6x - 3y = -6$

16. $3x - 5y = 14$　　Add $-7y$ to each side of (2);
　　 $-x + 7y = -18$ mult. each side by -1; subst.
　　　　　　　　　　 for x in (1).

17. $\frac{1}{2}x + 4y = 11$

$3x + 5y = 9$

18. $3x - 2y = 15$
　　 $5x - 6y = -7$
Add $-3x$ to each side of (1); divide each side by -2; subst. for y in (2).

Written Exercises

A　**1–12.** Solve the systems of equations in Oral Exercises 1–12 by the linear combination method. Check your solutions.

　　13–18. Solve the systems of equations in Oral Exercises 13–18 by the substitution method. Check your solutions.

Solve each system by either the linear combination method or the substitution method, whichever seems simpler.

19. $-5x + y = 20$ $\{(-6, -10)\}$　**20.** $12x - 7y = 16$ $\left\{\left(\frac{3}{4}, -1\right)\right\}$　**21.** $x = \frac{5y + 4}{2}$ $\{(7, 2)\}$
　　 $7x - 3y = -12$　　　　　　　 $8x - 6y = 12$　　　　　　　　　　$4x - 9y = 10$

Systems of Linear Equations or Inequalities　**131**

4. $b_1 - b_2 = m_2x - m_1x$
　 $y = m_1x + b_1$
　 Prop. of the opp. of a sum

5. $b_1 - b_2 = (m_2 - m_1)x$
　 $y = m_1x + b_1$
　 Dist. ax.

6. $x = \dfrac{b_1 - b_2}{m_2 - m_1}$
　 $y = m_1x + b_1$
　 Div. prop., symm. prop. of $=$

7. $x = \dfrac{b_1 - b_2}{m_2 - m_1}$
　 $y = m_1\left(\dfrac{b_1 - b_2}{m_2 - m_1}\right) + b_1$
　 Subst.

8. $x = \dfrac{b_1 - b_2}{m_2 - m_1}$
　 $y = m_1\left(\dfrac{b_1 - b_2}{m_2 - m_1}\right) +$
　 $b_1\left(\dfrac{m_2 - m_1}{m_2 - m_1}\right)$
　 Subst., Iden. ax. for $\times$

9. $x = \dfrac{b_1 - b_2}{m_2 - m_1}$
　 $y = [m_1(b_1 - b_2) +$
　 $b_1(m_2 - m_1)] \div$
　 $(m_2 - m_1)$
　 Addition of fractions

10. $x = \dfrac{b_1 - b_2}{m_2 - m_1}$
　 $y = [m_1b_1 - m_1b_2 +$
　 $b_1m_2 - b_1m_1] \div$
　 $(m_2 - m_1)$
　 Dist. ax.

11. $x = \dfrac{b_1 - b_2}{m_2 - m_1}$
　 $y = \dfrac{m_2b_1 - m_1b_2}{m_2 - m_1}$
　 Subst., $a - a = 0$, Comm. ax of $\times$

(continued)

Mixed Review

1. Simplify $4 \cdot 8 \div 2 - 6 + 3 \cdot (-4)$. -2

2. If $f(x) = 3x - 2$ and $g(x) = x^2$, determine the value of $[f \circ g](4)$ and $[g \circ f](4)$. 46; 100

3. Find three consecutive integers such that the sum of the first and twice the second is 4 less than three times the third. 16, 17, 18

Determine the apparent solution set by graphing each system of equations and state whether or not the system is consistent.

4. $x + y = 2$
$y = x$

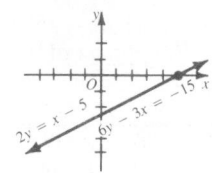

$\{(1, 1)\}$; consistent

5. $2y = x - 5$
$6y - 3x = -15$

$\{(x, y): 2y = x - 5\}$; consistent

Solve each system by either the linear combination method or the substitution method, whichever seems simpler.

22. $\begin{aligned} 2y &= 8 - 3x \\ 5x + 6y &= 40 \end{aligned}$ $\{(-4, 10)\}$

23. $\begin{aligned} \frac{1}{2}x - 5y &= 32 \\ \frac{3}{2}x - 7y &= 45 \end{aligned}$ $\left\{\left(\frac{1}{4}, -\frac{51}{8}\right)\right\}$

$\{(1, -3)\}$

24. $\begin{aligned} 4x - \frac{1}{3}y &= 5 \\ 9x + \frac{4}{3}y &= 5 \end{aligned}$

B 25. $\begin{aligned} \frac{x-1}{2} &= \frac{y-2}{8} \\ 12x - 3y &= 6 \end{aligned}$
$\{(x, y): 4x - y = 2\}$

26. $\begin{aligned} 3\left(\frac{x+y}{2}\right) &= y - 3 \\ 5\left(\frac{x-y}{2}\right) &= 7x \end{aligned}$ $\{(-5, 9)\}$

27. $\begin{aligned} \frac{x}{3} - \frac{y}{2} &= y + 2 \\ \frac{x}{2} - \frac{y}{5} &= x - 3 \end{aligned}$
$\{(6, 0)\}$

28. $\begin{aligned} x &= \frac{7x - 5y}{3} \\ y &= 9(x - y) - 20 \end{aligned}$
$\{(20, 16)\}$

29. $\begin{aligned} \frac{1}{3}(x - 1) + \frac{1}{2}(y + 1) &= 0 \\ \frac{2}{3}(x - 1) + \frac{5}{2}(y + 1) &= 12 \end{aligned}$
$\{(-11, 7)\}$

30. $\begin{aligned} \frac{x+7}{5} - 2y &= -3 \\ 2x + \frac{y+5}{4} &= 18 \end{aligned}$
$\{(8, 3)\}$

In Exercises 31 and 32, solve first for $\frac{1}{x}$ and $\frac{1}{y}$; then determine the values of x and y.

31. $\begin{aligned} \frac{3}{x} + \frac{1}{y} &= 1\frac{1}{1} \\ \frac{2}{x} + \frac{5}{y} &= 18 \end{aligned}$ $\frac{1}{x} = -1, \frac{1}{y} = 4; \left\{\left(-1, \frac{1}{4}\right)\right\}$

32. $\begin{aligned} \frac{5}{x} - \frac{2}{y} &= 5\frac{1}{x} = 3, \frac{1}{y} = 5; \\ \frac{4}{x} - \frac{3}{y} &= -3 \end{aligned}$ $\left\{\left(\frac{1}{3}, \frac{1}{5}\right)\right\}$

In Exercises 33 and 34, determine values of a and b so that $f(x) = ax^2 + bx$.

C 33. $f(-1) = 5$ and $f(3) = 21$ $a = 3, b = -2$

34. $f(3) = -15$ and $f(2) = -9$

35. Prove that for all real numbers $m_1, m_2, b_1,$ and b_2 such that $m_1 \neq m_2$, $a = -\frac{1}{2}, b = -\frac{7}{2}$ the following two systems of equations are equivalent.

$$\begin{aligned} y &= m_1x + b_1 \\ y &= m_2x + b_2 \end{aligned} \quad \text{and} \quad \begin{aligned} x &= \frac{b_1 - b_2}{m_2 - m_1} \\ y &= \frac{m_2b_1 - m_1b_2}{m_2 - m_1} \end{aligned}$$

36. Prove that for all real numbers $m, b_1,$ and b_2, the solution set of each of the following two systems is an infinite set or the empty set according as $b_1 = b_2$ or $b_1 \neq b_2$.

$$\begin{aligned} y &= mx + b_1 \\ y &= mx + b_2 \end{aligned} \quad \text{and} \quad \begin{aligned} x &= b_1 \\ x &= b_2 \end{aligned}$$

37. Prove that for all real numbers $m, b_1,$ and b_2, the solution set of the following system is $\{(b_2, mb_2 + b_1)\}$.

$$\begin{aligned} y &= mx + b_1 \\ x &= b_2 \end{aligned}$$

132 *Chapter 4*

38. Use the results of Exercises 35–37 to prove the theorem stated on page 125.

39. Use the results of Exercises 35–37 to prove that two lines are parallel if and only if each line has no slope and different x-intercepts or both lines have the same slope and different y-intercepts.

4–3 Determinants

The general formulas for the solution of a system of two linear equations in two variables can be found by solving the general system

$$a_1x + b_1y = c_1 \quad (1)$$
$$a_2x + b_2y = c_2 \quad (2)$$

where $a_1, b_1, c_1, a_2, b_2,$ and $c_2 \in \mathcal{R}$. Multiplying Equation (1) by b_2 and Equation (2) by $-b_1$, you get:

$$a_1b_2x + b_1b_2y = c_1b_2$$
$$-a_2b_1x - b_1b_2y = -c_2b_1.$$

Adding these equations gives $(a_1b_2 - a_2b_1)x = c_1b_2 - c_2b_1.$

Similarly, multiplying Equation (1) by $-a_2$ and Equation (2) by a_1, and adding, you get $(a_1b_2 - a_2b_1)y = a_1c_2 - a_2c_1.$

Therefore, if $a_1b_2 - a_2b_1 \neq 0$, you have

$$x = \frac{c_1b_2 - c_2b_1}{a_1b_2 - a_2b_1} \quad \text{and} \quad y = \frac{a_1c_2 - a_2c_1}{a_1b_2 - a_2b_1}. \quad (3)$$

You can check that the values for x and y given by the formulas (3) do in fact satisfy Equations (1) and (2).

EXAMPLE 1 Use formulas (3) to solve the system: $x - 3y = 6$
$2x + 7y = -1$

SOLUTION By inspection, $a_1 = 1, b_1 = -3, c_1 = 6$ and $a_2 = 2, b_2 = 7, c_2 = -1.$ Substituting in formulas (3):

$$x = \frac{6 \cdot 7 - (-1)(-3)}{1 \cdot 7 - 2(-3)} \qquad y = \frac{1(-1) - 2 \cdot 6}{1 \cdot 7 - 2(-3)}$$

$$= \frac{39}{13} = 3 \qquad\qquad = \frac{-13}{13} = -1$$

Checking that $(3, -1)$ is a solution is left to you.

$\therefore$ the solution set is $\{(3, -1)\}.$

There is a convenient way to denote the numerators and the denominators in formulas (3) for x and y that is shown on the next page.

Systems of Linear Equations or Inequalities   **133**

Key Ideas

Use determinants to solve systems of two linear equations in two variables.

Chalkboard Examples

1. Use formulas (3) (on page 133) to solve the system:
$3x + 4y = 7$
$2x - y = 12$

$$x = \frac{c_1b_2 - c_2b_1}{a_1b_2 - a_2b_1}$$

$$x = \frac{7(-1) - 12(4)}{3(-1) - 2(4)} = 5$$

$$y = \frac{a_1c_2 - a_2c_1}{a_1b_2 - a_2b_1}$$

$$y = \frac{3(12) - 2(7)}{3(-1) - 2(4)} = -2$$

$\therefore$ the solution set is $\{(5, -2)\}.$

2. Use Cramer's Rule to solve the system:
$4x - 3y = 5$
$-7x + 9y = -11$

$$D = \begin{vmatrix} 4 & -3 \\ -7 & 9 \end{vmatrix} = 36 - 21$$
$$= 15$$

$$x = \frac{D_x}{D} = \frac{\begin{vmatrix} 5 & -3 \\ -11 & 9 \end{vmatrix}}{15}$$

$$= \frac{45 - 33}{15} = \frac{12}{15} = \frac{4}{5}$$

$$y = \frac{D_y}{D} = \frac{\begin{vmatrix} 4 & 5 \\ -7 & -11 \end{vmatrix}}{15}$$

$$= \frac{-44 + 35}{15} = \frac{-9}{15} = \frac{-3}{5}$$

$\therefore$ the sol. set is $\left\{\left(\frac{4}{5}, -\frac{3}{5}\right)\right\}.$

Evaluate each determinant.

1. $\begin{vmatrix} 4 & 0 \\ 7 & -5 \end{vmatrix}$ -20

2. $\begin{vmatrix} k & 4 \\ 3 & k \end{vmatrix}$ $k^2 - 12$

3. $\begin{vmatrix} k & k+3 \\ 2k & k-4 \end{vmatrix}$ $-k^2 - 10k$

Use Cramer's Rule to solve each system. If the system is inconsistent or has an infinite solution set, so state.

4. $x + y = 3$
$2x - y = -9$ $\{(-2, 5)\}$

5. $3x + 7y = 22$
$2x - 8y = 2$ $\{(5, 1)\}$

6. $2x + 3y = 6$
$3x + 5y = 15$ $\{(-15, 12)\}$

Reading Algebra

Stress the meaning of determinant notation rather than the reading of the notation. Emphasize that the square array of numerals set off with vertical bars (not absolute value signs) is another way of writing "$a_1b_2 - a_2b_1$," where a_1, b_1, a_2, b_2 are the coefficients of x and y and are called entries or elements of the determinant.

For any $a_1, b_1, a_2, b_2 \in \mathcal{R}$, the **determinant** $D = \begin{vmatrix} a_1 & b_1 \\ a_2 & b_2 \end{vmatrix}$ has the value $a_1b_2 - a_2b_1$.

Notice that the square array of numerals, set off with vertical bars (not absolute-value signs!), is just another way of writing "$a_1b_2 - a_2b_1$." In the array, a_1, b_1, a_2, b_2 are called the **entries** (or **elements**) of the determinant.

A convenient way to remember how to evaluate the determinant is to take the difference of products indicated below.

$$\begin{vmatrix} a_1 & b_1 \\ a_2 & b_2 \end{vmatrix} = a_1b_2 - a_2b_1$$

From formulas (3) and the definition of D, you can see that, for $D \neq 0$,

$$x = \frac{D_x}{D} \quad \text{and} \quad y = \frac{D_y}{D} \quad (4)$$

where $D = \begin{vmatrix} a_1 & b_1 \\ a_2 & b_2 \end{vmatrix}$, $D_x = \begin{vmatrix} c_1 & b_1 \\ c_2 & b_2 \end{vmatrix}$, and $D_y = \begin{vmatrix} a_1 & c_1 \\ a_2 & c_2 \end{vmatrix}$.

Notice that the entries a_1, b_1, a_2, b_2 of D are just the coefficients of x and y in Equations (1) and (2); D is called the **determinant of coefficients**. To obtain the entries for D_x, replace the x-coefficients a_1, a_2 in D with the constants c_1, c_2. Similarly, to obtain the entries for D_y, replace the y-coefficients b_1, b_2 in D with the constants c_1, c_2.

The solution of a linear system in determinant form (4) is called **Cramer's Rule**. If $D = 0$, then either the system is inconsistent or the system has an infinite solution set.

EXAMPLE 2   Use Cramer's Rule to solve the system: $3x - 4y = 7$
$5x - 8y = 9$

SOLUTION   By inspection $D = \begin{vmatrix} 3 & -4 \\ 5 & -8 \end{vmatrix} = -24 - (-20) = -4.$

Then

$$x = \frac{D_x}{D} = \frac{\begin{vmatrix} 7 & -4 \\ 9 & -8 \end{vmatrix}}{-4} = \frac{-56 - (-36)}{-4} = \frac{-20}{-4} = 5,$$

and

$$y = \frac{D_y}{D} = \frac{\begin{vmatrix} 3 & 7 \\ 5 & 9 \end{vmatrix}}{-4} = \frac{27 - 35}{-4} = \frac{-8}{-4} = 2.$$

Checking that $(5, 2)$ is a solution is left to you.

$\therefore$ the solution set is $\{(5, 2)\}$.

134   *Chapter 4*

Oral Exercises

Evaluate each determinant.

1. $\begin{vmatrix} 4 & 5 \\ 3 & 7 \end{vmatrix}$ 13   2. $\begin{vmatrix} 3 & 5 \\ -2 & 7 \end{vmatrix}$ 31   3. $\begin{vmatrix} 0 & 8 \\ -1 & 6 \end{vmatrix}$ 8   4. $\begin{vmatrix} a & b \\ c & d \end{vmatrix}$ ad − bc

5. $\begin{vmatrix} x+y & a \\ x-y & b \end{vmatrix}$ 6. $\begin{vmatrix} 5 & 2 \\ 10 & 4 \end{vmatrix}$ 0   7. $\begin{vmatrix} y & 3 \\ ny & 3n \end{vmatrix}$ 0   8. $\begin{vmatrix} c+3 & 4 \\ 2c+6 & 8 \end{vmatrix}$ 0

bx + by − ax + ay

9. By looking at Exercises 6–8, what can you say about the value of any

determinant of the form $\begin{vmatrix} a & b \\ ka & kb \end{vmatrix}$? 0

10. How are the determinants $\begin{vmatrix} a & b \\ c & d \end{vmatrix}$ and $\begin{vmatrix} c & d \\ a & b \end{vmatrix}$ related?

Their values are additive inverses.

State the value of the determinants D, D_x, and D_y for each system of $D = -2$
equations. $\quad D_x = -8$

$D = 1 \qquad\qquad D = -1 \qquad\qquad D_y = -12$

11. $3x + 7y = -2$ $D_x = -38$ 12. $3x + 4y = 1$ $D_x = -3$ 13. $5x - 3y = 2$
$\quad 2x + 5y = 4$ $D_y = 16$ $\quad -5x - 7y = -1$ $D_y = 2$ $\quad -9x + 5y = -6$

14. $7x - 5y = 0$ $D = -3$ 15. $-3x + 11y = 1$ $D = -8$ 16. $4x - 3y = 2$
$\quad 5x - 4y = -3$ $D_x = -15$ $\quad -2x + 10y = 2$ $D_x = -12$ $\quad 9x - 5y = 1$
$\qquad\qquad D_y = -21$ $\qquad\qquad D_y = -4$ $\quad D = 7$
$\qquad\qquad\qquad\qquad\qquad\qquad\qquad\qquad\qquad\qquad\qquad D_x = -7$
$\qquad\qquad\qquad\qquad\qquad\qquad\qquad\qquad\qquad\qquad\qquad D_y = -14$

Written Exercises

Evaluate each determinant.

A 1. $\begin{vmatrix} 3 & -2 \\ 5 & 4 \end{vmatrix}$ 22   2. $\begin{vmatrix} 8 & -2 \\ -3 & 7 \end{vmatrix}$ 50   3. $\begin{vmatrix} 9 & 8 \\ -6 & 0 \end{vmatrix}$ 48   4. $\begin{vmatrix} -8 & -6 \\ 4 & 3 \end{vmatrix}$ 0

5. $\begin{vmatrix} 7 & k \\ k & 9 \end{vmatrix}$ 63 − k² 6. $\begin{vmatrix} k-3 & 4 \\ 5 & k-3 \end{vmatrix}$ 7. $\begin{vmatrix} k & k-5 \\ 2k & 2k-10 \end{vmatrix}$ 0   8. $\begin{vmatrix} k & -3k \\ 2k-1 & 3-6k \end{vmatrix}$

k² − 6k − 11   0

9–14. Use Cramer's Rule to solve the systems in Oral Exercises 11–16.

9. $\{(-38, 16)\}$ 10. $\{(3, -2)\}$ 11. $\{(4, 6)\}$ 12. $\{(5, 7)\}$

Use Cramer's Rule to solve each system. If the system is inconsistent or has
an infinite solution set, so state.   13. $\left\{\left(\frac{3}{2}, \frac{1}{2}\right)\right\}$ 16 $\{(-1, -2)\}$ 14 $(5, 7)$

15. $3x - 4y = 6$ $\{(2, 0)\}$ 16. $4x + 5y = 3$ $\{(-8, 7)\}$
$\quad 4x - 7y = 8$ $\quad 5x + 6y = 2$

B 17. $-7x + 4y = -5$ $\left\{\left(\frac{1}{3}, -\frac{2}{3}\right)\right\}$ 18. $2x - 4y = -9$ $\left\{\left(\frac{1}{2}, \frac{5}{2}\right)\right\}$
$\qquad 5x - 2y = 3$ $\qquad 3x + 5y = 14$

19. $11x + 6y = -12$ $\left\{\left(-2, \frac{5}{3}\right)\right\}$ 20. $9x - 4y = -1$ $\{(-5, -11)\}$
$\quad -5x - 3y = 5$ $\quad -8x + 3y = 7$

21. $2x + 5y = 3$ $\left\{\left(-\frac{1}{6}, \frac{2}{3}\right)\right\}$ 22. $7x - 6y = 4$ $\left\{\left(\frac{16}{13}, \frac{10}{13}\right)\right\}$
$\quad 8x - 7y = -6$ $\quad 4x - 9y = -2$

Systems of Linear Equations or Inequalities **135**

Suggested Assignments

Minimum
First day
135/1–20
Second day
S 53/P: 14–19
63/P: 7–10
Extended Alg. with Trig.
135/1–21 odd
S 53/P: 14, 16, 18
64/P: 9–12
Enriched Alg.
135/7–27 odd
S 54/P: 22
64/P: 8, 10, 12
Enriched Alg. with Trig.
135/7–27 odd
S 54/P: 22
64/P: 8, 10, 12

Additional Answers
Written Exercises

23. The area of the rectangle is $(a + c)(b + d)$. The area of each of the triangles with sides $\overline{AB}$ and $\overline{CD}$ is $\frac{1}{2}b(a + c)$; of those with sides $\overline{BC}$ and $\overline{AD}$ is $\frac{1}{2}c(b + d)$. The area of the parallelogram is $(a + c)(b + d) -$

$2 \cdot \frac{1}{2}b(a + c) -$

$2 \cdot \frac{1}{2}c(b + d) = ad -$

$bc = \begin{vmatrix} a & b \\ c & d \end{vmatrix}$.

24. If the graphs are parallel or coincident, then the slopes are the same,

$-\dfrac{a_1}{b_1} = -\dfrac{a_2}{b_2}$, or they are undefined, so that $b_1 = b_2 = 0$. If $-\dfrac{a_1}{b_1} = -\dfrac{a_2}{b_2}$,

$-a_1b_2 = -a_2b_1$ (means and extremes) and $a_1b_2 - a_2b_1 = 0$. If $b_1 = b_2 = 0$, then $a_1b_2 - a_2b_1 = a_1 \cdot 0 - a_2 \cdot 0 = 0$. In any case, $\begin{vmatrix} a_1 & b_1 \\ a_2 & b_2 \end{vmatrix} = a_1b_2 - a_2b_1 = 0$.

25. By hyp., $\begin{vmatrix} a_1 & b_1 \\ a_2 & b_2 \end{vmatrix} = a_1b_2 - a_2b_1 = 0$. If $b_1 \neq 0$ and $b_2 \neq 0$, then $\dfrac{a_1}{b_1} - \dfrac{a_2}{b_2} = 0$, and $-\dfrac{a_1}{b_1} = -\dfrac{a_2}{b_2}$, so that the slopes are the same. If $b_1 = 0$, then $a_1b_2 - a_2b_1 = 0$ means that $a_1b_2 = 0$, and $b_2 = 0$ (a_1 and b_1 can't

C **23.** Use the diagram to show that the area of parallelogram $ABCD$ is $\begin{vmatrix} a & b \\ c & d \end{vmatrix}$. (*Hint:* Find the area of the rectangle and subtract the areas of the four triangles.)

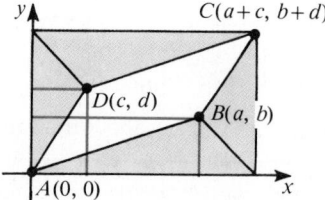

24. Prove that if the graphs of the system of equations

$$a_1x + b_1y = c_1$$
$$a_2x + b_2y = c_2$$

are parallel or coincident, then $\begin{vmatrix} a_1 & b_1 \\ a_2 & b_2 \end{vmatrix} = 0$.

25. Prove that for the system of equations in Exercise 24, if $\begin{vmatrix} a_1 & b_1 \\ a_2 & b_2 \end{vmatrix} = 0$, then the graphs of the system are parallel or coincident.

26. Use Cramer's Rule to prove that if one of the equations of the system of equations in Exercise 24 is multiplied by a nonzero constant k, the resulting system has the same solution as the original system.

27. Use Cramer's Rule to prove that if the first equation of the system of equations in Exercise 24 is replaced by the *sum* of the two equations, the resulting system has the same solution as the original system.

Computer Exercises   For students with computer experience

1. Write a program that will determine the unique solution (if there is one) of a given linear system. Let the system be in the form

$$ax + by = c$$
$$dx + ey = f$$

with a, b, c, d, e, and f input by the user. If the solution of the system is not unique, the output should state this.

Run the program in Exercise 1 for each system.

2. $7x - 19y = 28$
$15x + 11y = 53$ $\{(3.6, -0.1)\}$

4. $52x - 29y = -17$
$-12x + 33y = -75$ $\{(-2, -3)\}$

3. $-8x + 13y = 76$
$17x - 9y = -32$ $\{(1.8, 7.0)\}$

5. $-48x + 51y = 44$
$16x - 17y = -18$ no solution

136   *Chapter 4*

Careers

Environmental Protection

Concern over protection of the environment has increased in recent years with the realization that pollution has immediate and long-term effects on human health and the environment. This new concern has led to more careers in environmental protection. This field includes determining the effects of pollutants on health and the environment, setting standards for acceptable levels of pollutants, and developing ways to conform to them.

One particularly serious form of pollution is acid rain, which is caused by industrial use of high-sulfur coal for fuel. In the burning process, the sulfur in the coal changes to sulfur oxide, a compound that mixes with water in the air to form sulfuric acid, which then falls to the ground with rain. The sulfuric acid then causes damage to waterways and can corrode sculptures that are exposed to the elements.

One way to combat acid rain is to burn low-sulfur coal. There is however, a limited supply of this type of coal so that other methods of reducing the amount of sulfur oxide released in fuel-burning have been developed. One method is a process called flue gas desulfurization. A more effective method is a combined process of cleaning the coal and desulfurization.

EXAMPLE   A certain company wants to burn 50,000 t of high-sulfur coal and has allotted $180,000 to physically clean this coal. Flue gas desulfurization costs $5.82/t and the combined process costs $2.76/t. The company plans to use both processes. How many tons of coal should be cleaned using the combined process?

SOLUTION   Let x = the number of tons of coal cleaned using the combined process. Then $50000 - x$ = the number of tons of coal cleaned using flue gas desulfurization.
Then

$$2.76x + 5.82(50000 - x) = 180000$$
$$2.76x + 291000 - 5.82x = 180000$$
$$-3.06x = -111000$$
$$x = 36{,}274.51$$

$\therefore$ 36,275 tons of coal should be cleaned using the combined process.

Systems of Linear Equations or Inequalities   **137**

---

both be zero, or there is no line). Similarly, if $b_2 = 0$, then $b_1 = 0$. $\therefore$ the slopes are the same, or both lines have no slope and the lines are either parallel or coincident. (Theorem p. 125, and Ex. 39, p. 133)

26. Mult. the system by k to yield the following determinants.

$$D = ka_1b_2 - ka_2b_1$$
$$= k(a_1b_2 - a_2b_1)$$
$$D_x = kc_1b_2 - kc_2b_1$$
$$= k(c_1b_2 - c_2b_1)$$
$$D_y = ka_1c_2 - ka_2c_1$$
$$= k(a_1c_2 - a_2c_1)$$

Using Cramer's Rule and simplifying, the sol. set is

$$\left\{ \left(\frac{c_1b_2 - c_2b_1}{a_1b_2 - a_2b_1}, \frac{a_1c_2 - a_2c_1}{a_1b_2 - a_2b_1} \right) \right\}$$

which is the sol. set of the original system.

Mixed Review

Determine the slope of the line through each pair of points.

1. $(3, -7)$, $(3, 5)$ undefined
2. $(4, -1)$, $(6, 9)$ 5
3. $(7, -3)$, $(-1, -3)$ 0

Solve each system by either the linear combination method or the substitution method, whichever seems simpler.

4. $3x + y = 2$
   $x - 3y = 14$ $\{(2, -4)\}$
5. $6x + 2y = 22$
   $6x - 10y = 34$ $\{(4, -1)\}$
6. $12x - 7y = 16$
   $4x - 3y = 6$
   $\left\{ \left(\frac{3}{4}, -1 \right) \right\}$

Supplementary Material

Test 10

Key Ideas

Use systems of linear equations in two variables to solve word problems.

Chalkboard Examples

1. During a certain time period, one wall of a house containing 20 ft² of glass and 80 ft² of plaster lost 1360 Btu of heat. Another wall containing 50 ft² of glass and 110 ft² of plaster lost 2680 Btu of heat during the same time period. How much heat was lost per square foot of glass? How much heat was lost per square foot of plaster?

The problem asks for the heat lost per square foot of glass and heat lost per square foot of plaster.

Let x = the heat lost per square foot of glass and y = the heat lost per square foot of plaster.

total heat lost for first wall: $20x + 80y$

total heat lost for second wall: $50x + 110y$

Solve the system:
$20x + 80y = 1360$
$50x + 110y = 2680$
The solution is (36, 8).
∴ the heat lost through glass is 36 Btu/ft² and the heat lost through plaster is 8 Btu/ft².

4–4 Using Two Variables to Solve Problems

The following examples illustrate how systems of equations can be used to solve practical problems.

EXAMPLE 1   A heat loss survey showed that during a certain time period one wall of a house containing 30 ft² of glass and 90 ft² of plaster lost 1800 Btu of heat. Another wall containing 40 ft² of glass and 100 ft² of plaster lost 2240 Btu during the same time period. How much heat was lost per square foot of glass? How much heat was lost per square foot of plaster?

SOLUTION   1. The problem asks for heat lost per square foot of glass and heat lost per square foot of plaster in a certain time period.

2. Let x be the heat lost per square foot of glass and y be the heat lost per square foot of plaster. The facts are listed in the chart below.

first wall	Area in square feet	×	heat lost per square foot	=	total heat lost.
glass	30		x		$30x$
plaster	90		y		$90y$

total heat lost for first wall: $30x + 90y$

second wall	Area in square feet	×	heat lost per square foot	=	total heat lost.
glass	40		x		$40x$
plaster	100		y		$100y$

total heat lost for second wall: $40x + 100y$

3. Heat lost by first wall   was   1800 Btu.
$$30x + 90y = 1800$$

Heat lost by second wall   was   2240 Btu.
$$40x + 100y = 2240$$

4. Solve the system:
$$30x + 90y = 1800$$
$$40x + 100y = 2240$$

Showing that the solution of this system is (36, 8) and checking this result in the words of the problem (Step 5) is left to you.

∴ the heat lost through glass is 36 Btu/ft² and the heat lost through plaster is 8 Btu/ft².

In a system that represents a given practical situation it often happens that the sides of one equation refer to different quantities from those referred to by the sides of the other equation.

EXAMPLE 2   A coin bank contains 30 coins, all dimes and quarters, worth $5.70. How many dimes and how many quarters are in the bank?

SOLUTION   1. The problem asks for the number of dimes and the number of quarters.

2. Let d be the number of dimes and q be the number of quarters.

3.
Number of dimes	added to	number of quarters	is	30.
d	+	q	=	30

Value of dimes	added to	value of quarters	is	$5.70.
$10d$	+	$25q$	=	570

Notice that the first equation refers to the *number* of coins and the second equation refers to the *value* of the coins.

4. Solve the system: $d + q = 30$
$$10d + 25q = 570$$

Showing that the solution of this system is (12, 18) and checking this result in the problem (Step 5) is left to you.

∴ there are 12 dimes and 18 quarters in the bank.

To solve motion problems about airplanes it is necessary to know the meaning of the following phrases:

Tail wind: a wind blowing in the same direction as the one in which the airplane is heading.
Head wind: a wind blowing in the direction opposite to the one in which the airplane is heading.
Wind speed: the speed of the wind.
Air speed: the speed of the airplane in still air.
Ground speed: the speed of the airplane relative to the ground.

With a tail wind, an airplane's ground speed is the sum of its air speed and the wind speed. With a head wind, the ground speed is the difference between the air speed and the wind speed.

Systems of Linear Equations or Inequalities   **139**

2. A hardware store orders a shipment of two types of hammers for $168. One type of hammer costs $3, the other type of hammer costs $5. If the store ordered 40 hammers in all, how many of each type were ordered?
Let x = the number of $3 hammers and let y = the number of $5 hammers.
Solve the system:
$x + y = 40$
$3x + 5y = 168$
The solution set is
$\{(16, 24)\}$.
∴ there are sixteen $3 hammers and twenty-four $5 hammers.

3. With a given head wind, a certain airplane can travel 3500 km in 7 h. Flying in the opposite direction with the same wind blowing, the airplane can fly the distance in 2 hours less. Find the airplane's air speed and the wind speed.
Let x = the air speed of the airplane in km/h and y = the wind speed in km/h. Then $x - y$ is the ground speed with a head wind, $x + y$ is the ground speed with a tail wind. The distance with a head wind is $7(x - y)$ and the distance with a tailwind is $5(x + y)$.
Solve the system:
$7(x - y) = 3500$
$5(x + y) = 3500$
$x - y = 500$
$x + y = 700$
$2x = 1200$
$x = 600$
$y = 100$
∴ the airplane's speed is 600 km/h and the wind speed is 100 km/h.

Additional A Exercises

1. A grocer sells pecans at $1.20/lb and peanuts at $.70/lb. If he sells 50 lb of nuts for a total of $50, how many pounds of pecans and how many pounds of peanuts did he sell?   Pecans, 30 lb; peanuts, 20 lb.

2. A boat travels 16 km downstream in two hours. It then travels the same distance upstream in 8 hours. Find the speed of the boat in still water and the speed of the current.   Boat speed, 5 km/h; current speed, 3 km/h.

3. The perimeter of an isosceles triangle is 50 cm. If the length of each congruent side is 10 cm shorter than three times the length of the base, find the lengths of the three sides.   The length of the base is 10 cm and the length of each of the congruent sides is 20 cm.

4. The measure of one of the angles of a triangle is 40° less than three times the measure of another angle. The measure of the third angle is 80°. Find the measure of the other two angles.   65° and 35°

5. The sum of the angle measures of a quadrilateral is 360°. In quadrilateral $ABCD$, angles A and C are congruent, and angles B and D are congruent. The measure of angle D is 40° less than the measure of angle A. What is the measure of each angle?
$m(\angle A) = m(\angle C) = 110°$;
$m(\angle B) = m(\angle D) = 70°$.

EXAMPLE 3 With a given head wind, a certain airplane can travel 3600 km in 9 h. Flying in the opposite direction with the same wind blowing, the airplane can fly the same distance in 1 h less. Find the airplane's air speed and the wind speed.

SOLUTION 1. The problem asks for the speed of the airplane in still air and for the speed of the wind.

2. Let x = the air speed of the airplane in kilometers per hour (km/h);
y = the wind speed in kilometers per hour.
The facts of the problem are listed below in the chart. (Recall the use of the relationship $d = rt$ on page 51.)

	Ground speed (km/h) r	Time (h) t	Distance (km) $rt = d$
With a head wind	$x - y$	9	$9(x - y)$
With a tail wind	$x + y$	8	$8(x + y)$

3. Distance with a head wind   is   3600 km.
$$9(x - y) = 3600$$
Distance with a tail wind   is   3600 km.
$$8(x + y) = 3600$$

4. Solve the system: $9(x - y) = 3600$
$8(x + y) = 3600$

$$x - y = 400$$
$$x + y = 450$$

$$2x = 850$$
$$x = 425$$

Completing Step 4 and checking the results (Step 5) are left to you.

∴ the airplane's speed is 425 km/h and the wind speed is 25 km/h.

Oral Exercises

In Exercises 1–5 let c be the number of cucumbers and h be the number of heads of lettuce bought. Translate each sentence into an equation involving c and h.

1. The total number of cucumbers and heads of lettuce bought is 12. $c + h = 12$

2. If cucumbers cost 25¢ each and heads of lettuce cost 75¢ each, the total cost of these vegetables is $7.50. $0.25c + 0.75h = 7.50$

3. The number of heads of lettuce is 6 fewer than 5 times the number of cucumbers. $h + 6 = 5c$

4. If cucumbers cost 30¢ each and heads of lettuce cost 70¢ each, the total cost of these vegetables is $7.20. $0.30c + 0.70h = 7.20$

5. If twice as many cucumbers and three times as many heads of lettuce are bought, there is a total of 33 vegetables. $2c + 3h = 33$

In Exercises 6–11 let b be the speed (in kilometers per hour) of a boat in still water and c be the speed (in kilometers per hour) of the current in a river. Translate each sentence into an equation involving b and c.

6. The speed of the boat going upstream was 14 km/h. $b - c = 14$

7. The speed of the boat going downstream was 16 km/h. $b + c = 16$

8. The boat traveled 6 km upstream in 30 min. $\frac{1}{2}(b - c) = 6$

9. The boat traveled 6 km downstream in 20 min. $\frac{1}{3}(b + c) = 6$

10. With a current 2 km/h faster, the boat could travel 5 km downstream in 15 min. $\frac{1}{4}(b + c + 2) = 5$

11. If the speed of the current were twice as great, a trip of 6 km upstream would take 40 min. $\frac{2}{3}(b - 2c) = 6$

Exercises 12–16 refer to an isosceles triangle that has a base of length b and congruent sides of length a. Translate each sentence into an equation involving a and b.

12. The perimeter of the triangle is 140 cm. $2a + b = 140$

13. The length of each congruent side is 20 cm shorter than twice the length of the base. $a + 20 = 2b$

14. If two of the described isosceles triangles are joined as shown, to form a parallelogram with one diagonal, the sum of the lengths of all the line segments is 228 cm. $3a + 2b = 228$

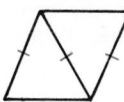   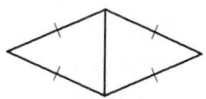

Ex. 14   **Ex. 15**

15. If two of the described isosceles triangles are joined as shown, so that their bases coincide, the sum of the lengths of all the line segments is 244 cm. $4a + b = 244$

16. Trapezoid $ABCD$ is formed from the described isosceles triangle by joining the midpoints of the two congruent sides, as shown. The perimeter of trapezoid $ABCD$ is 106 cm. (Recall that the length of a segment joining the midpoints of two sides of a triangle is one half the length of the third side.) $2 \cdot \frac{1}{2}a + b + \frac{1}{2}b = 106$

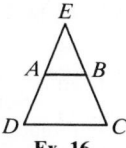

Ex. 16

6. With a given tail wind, a certain plane can fly 2880 km in 4 h. Flying in the opposite direction with the same wind blowing, the plane can fly the same distance in 4.5 h. What is the plane's air speed? 680 km/h

Systems of Linear Equations or Inequalities   **141**

Suggested Assignments

Minimum
  First day
    142/1–22
  Second day
R 144/*Self-Test 1*
S 62/19–28
    94/17, 18
Extended Alg. with Trig.
    142/1–21 odd
R 144/*Self-Test 1*
Enriched Alg.
    142/7–25 odd
R 144/*Self-Test 1*
Enriched Alg. with Trig.
    142/7–25 odd
R 144/*Self-Test 1*

Problems

A **1.** Use the relationships stated in Oral Exercises 1 and 2 to determine the number of cucumbers and the number of heads of lettuce that were bought. 3 cucumbers, 9 heads of lettuce

  2. Use the relationships stated in Oral Exercises 8 and 9 to determine the speed of the boat in still water and the speed of the current.

  3. Use the relationships stated in Oral Exercises 12 and 13 to determine the lengths of the 3 sides of the described isosceles triangle.

  4. Use the relationships stated in Oral Exercises 15 and 16 to determine the lengths of the 3 sides of the described isosceles triangle.

  5. The measure of one of the angles of a triangle is 26° less than twice the measure of another angle. The measure of the third angle is 35°. Find the measure of the other two angles. 57°, 88°

  6. The sum of the angle measures of a pentagon is 540°. In pentagon $ABCDE$, angles A, B, and C are congruent, and angles D and E are congruent. The measure of angle D is 30° greater than the measure of angle A. What is the measure of each angle?

  7. In a certain electric circuit the sum of the currents I_1 and I_2 is 4 A. The product of the resistance of resistor R_1 and the current I_1 equals the product of the resistor R_2 and the current I_2. If $R_1 = 80\ \Omega$ and $R_2 = 48\ \Omega$, find I_1 and I_2. (Current is measured in amps [A] and resistance is measured in ohms [Ω].) $I_1 = 1.5A$, $I_2 = 2.5A$

  8. With a given tail wind, a certain plane can fly 1440 km in 2.5 h. Flying in the opposite direction with the same wind blowing, the plane can fly the same distance in 3 h. What is the plane's air speed?

  9. On Family Night, tickets to a ball game cost $4.50 for the first person and $2.50 for every accompanying member of the same family. In one section 480 ticket holders paid $1550. How many $4.50 tickets were sold for that section? 175

  10. A boat took 2 h to travel a certain distance upstream against a 3 km/h current. Returning downstream with the same current, the boat traveled an additional 2 km in 1 h 20 min. What was the boat's speed? How far did the boat go upstream?

  11. In the diagram $HG = GE$ and $GF = FE$. The perimeters of trapezoid $DEGH$ and parallelogram $DEFH$ are 39 cm and 50 cm, respectively. Find the lengths of $\overline{HG}$, $\overline{GF}$, and $\overline{DE}$.
$HG = 7$ cm; $GF = 9$ cm; $DE = 16$ cm

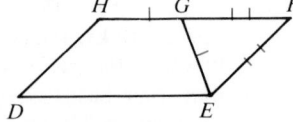

  12. Cecilia Holt made 75% of her free throws before the last basketball game of the season. In the last game she shot 4 free throws and made 1 of them, lowering her percentage to 65%. How many free throws had she attempted before the last game, and how many of those had she made? 16 attempted, 12 made

B 13. Vehicles passing through a certain toll booth are charged 25¢ for each axle. During a certain time period $72 was collected from 108 vehicles. If only passenger cars with 2 axles and trucks with 5 axles passed through this toll booth, how many vehicles were trucks? 24

14. A rectangular table can be set against one wall of a room, as shown, so that it is 5 ft from the two side walls and 8 ft from the fourth wall. If the perimeter of the room is 58 ft and the combined length of the three exposed sides of the table is $17\frac{1}{2}$ ft, what are the dimensions of the table? $4\frac{1}{2}$ ft by $6\frac{1}{2}$ ft

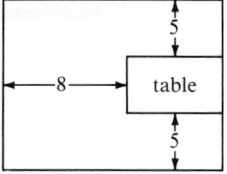

15. Two workers are at point P, part of the way across a 1.2 km railroad bridge when they hear a train approaching. They both run at 12 km/h toward opposite ends of the bridge. The train passes one worker as he gets to the near end of the bridge and the other worker 1 min later as she gets to the other end of the bridge. How far from each end of the bridge is point P? 0.5 km, 0.7 km

16. Working together, three large pipes and two small pipes can fill a 1000 L tank in 2 h. The tank can also be filled when the two small pipes run together for 4 h 10 min and then the three large pipes are opened and all run for an additional 50 min. What is the rate (in L/h) of each large pipe and each small pipe? 120 L/h; 70 L/h

Find values of A and B so that the graph of the given equation will contain the given points.

17. $Ax + By = 3$; (8, 3), (13, 7) $A = \frac{12}{17}, B = -\frac{15}{17}$
18. $y = Ax^2 + Bx$; (4, 28), (3, 12) $A = 3, B = -5$
19. $y = x^2 + Ax + B$; (1, 9), (−3, 25) $A = -2, B = 10$
20. $y = Ax^2 - 3x + B$; (−3, 6), (−4, 2) $A = -1, B = 6$

21. The height h (in meters) above the ground of an object propelled upward with an initial velocity v_0 (in meters per second) from an initial height h_0 (in meters) is given by the equation,

$$h = -5t^2 + v_0t + h_0,$$

where t represents the time (in seconds). Find the initial velocity v_0 and height h_0 for an object that is 30 m from the ground after 1 s and that hits the ground after 3 s. $v_0 = 5$ m/s; $h_0 = 30$ m

22. The equation $y = bx + ax^2$ describes the trajectory of a baseball thrown from a point considered to be the origin of a coordinate system. If the ball goes through the point (20, 8) and lands at the point (100, 0), find the value of a and b. $a = -\frac{1}{200}; b = \frac{1}{2}$

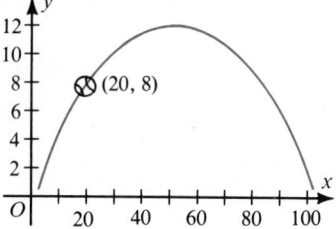

Common Errors
Students sometimes confuse the units of the quantities represented in equations. To help them avoid this, point out that products and quotients represent quantities other than those of the factors (or those of the divisor and dividend). For example, a rate times a time represents a *distance*; a distance divided by a rate represents a *time*. Also, in coin problems, students need to be careful to distinguish between the *number* of coins and the *value* of the coins. Coins must be represented in the same kind of units. For example either all cents as 25 cents or all dollars as $0.25.

Mixed Review
Evaluate each determinant.

1. $\begin{vmatrix} 4 & -3 \\ 6 & 8 \end{vmatrix}$ 50

2. $\begin{vmatrix} 6 & 0 \\ -9 & 3 \end{vmatrix}$ 18

3. $\begin{vmatrix} k & k-3 \\ 3k & 3k-5 \end{vmatrix}$ $4k$

Use Cramer's Rule to solve each system. If the system is inconsistent or has an infinite solution set, so state.

4. $2x - 4y = -8$
   $6x - 12y = -36$
   inconsistent; Ø

5. $2x + 5y = 13$
   $3x - 2y = 10$ {(4, 1)}

6. $3x - 5y = 10$
   $9x - 15y = 10$
   inconsistent; Ø

1. Graph the following system of equations and determine the apparent solution set.

$$x + y = -3$$
$$4x - 3y = 16$$

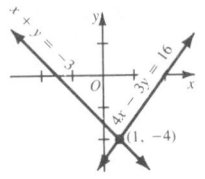

$$\{(1, -4)\}$$

Determine how many solutions each system has. Use the slope and y-intercept of the graphs if necessary.

2. $x + y = 5$
 $4x - 2y = 6$ one solution

3. $5 + y = x$
 $x - y = 4$ no solutions

4. $6y = 2x + 12$
 $x - 3y = -6$
 an infinite set of solutions

5. Solve by using the linear-combination method:
 $3x + 4y = 7$
 $4x + 3y = 7$ $\{(1, 1)\}$

6. Solve by the substitution method:
 $8x + 4y = 28$
 $3x - y = 3$ $\{(2, 3)\}$

7. Use determinants to solve the system:
 $3x + y = 6$
 $2x + y = 4$ $\{(2, 0)\}$

8. Paul spent $2.80 to buy 3 lb of tomatoes and 4 lb of oranges. He would have spent $3.40 if he had bought 4 lb of tomatoes and 2 lb of oranges. What is the cost per pound of each kind of fruit?   Tomatoes, $0.80/lb; oranges, $0.10/lb.

C 23. A car's fuel efficiency rating for city driving is a different constant from its fuel efficiency rating for highway driving. A certain car traveled 120 mi in the city and 280 mi on the highway using one tankful of 16 gal of fuel. On the next tankful of fuel, the car traveled 220 mi in the city and 140 mi on the highway. What is the fuel efficiency rating for this car on the highway? 28 miles per gallon

24. After leaving the boathouse, a rower rowing upstream passes a log 2 km upstream from the boathouse. The rower rows upstream for one more hour and then rows back to the boathouse, arriving at the same time as the log. How fast was the current flowing? 1 km/h

25. A coin bank contains 27 coins, all nickels, dimes, or quarters. The combined value of the nickels and dimes is 90¢; the combined value of the dimes and quarters is $4.00. How many coins of each denomination are there in the bank? 8 nickels, 5 dimes, 14 quarters

Self-Test 1

VOCABULARY
system of linear equations (p. 123)
solution set of a system of equations (p. 124)
consistent equations (p. 124)
inconsistent equations (p. 124)
dependent solutions (p. 125)
linear combination of equations (p. 128)

determinant (p. 134)
entry or element of a determinant (p. 134)
determinant of coefficients (p. 134)
Cramer's Rule (p. 134)

1. Graph the following system of equations and determine the apparent solution set.   *Obj. 1, p. 123*

$$x + y = -2$$
$$3x - 2y = 9$$

Determine how many solutions each system has. Use the slope and y-intercept of the graphs if necessary.

2. $2x + y = 4$
 $3x - 2y = -4$
 1 solution

3. $x - y = 3$
 $y - x = 4$
 no solution; same slope;
 y intercepts, -3 and 4

4. $4y - 2x = 8$
 $x - 2y = -4$
 infinitely many solutions
 same slope, same intercept
 Obj. 2, p. 123

5. Solve by using the linear-combination method.

$$3x + 2y = 5$$
$$2x + 3y = 7$$ $\left\{\left(\frac{1}{5}, \frac{11}{5}\right)\right\}$

6. Solve by the substitution method.

$$4x + 2y = 14$$
$$3x - y = 3$$ $\{(2, 3)\}$

7. Use determinants to solve the system. *Obj. 3, p. 123*

$$\begin{aligned} 2x + y &= 8 \\ 3x + 4y &= 17 \end{aligned} \quad \{(3, 2)\}$$

8. Nancy spent \$3.45 to buy 3 lb of oranges and 2 lb of apples. If she had bought 2 lb of oranges and 4 lb of apples she would have spent \$3.90. What is the cost per pound of each kind of fruit? *Obj. 4, p. 123*
oranges: \$0.75; apples: \$0.60

Check your answers with those at the back of the book.

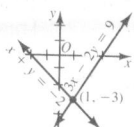

Systems of Inequalities in Two Variables

OBJECTIVES for Sections 4-5 and 4-6:
1. To graph the solution set of a system of linear inequalities in two variables.
2. To use systems of linear inequalities to solve linear programming problems in two variables.

4–5 Graphing a System of Linear Inequalities

How would you describe the graph of the solution set of the following of linear inequalities?

$$\begin{aligned} x + y &< 9 \\ y &\geq 2x \end{aligned}$$

Figure 5

In Figure 5, gray shading is used to show (part of) the open half-plane that consists of the points lying below the line with equation $x + y = 9$. This is the graph of the inequality

$$x + y < 9$$

(recall Section 3-4). Red shading in Figure 5 shows the closed half-plane that is the graph of

$$y \geq 2x.$$

The region in Figure 5 where the two colors overlap is the *intersection* of the two half-planes; it consists of the points whose coordinates satisfy *both* inequalities in the given system. Note that P itself is not in the region.

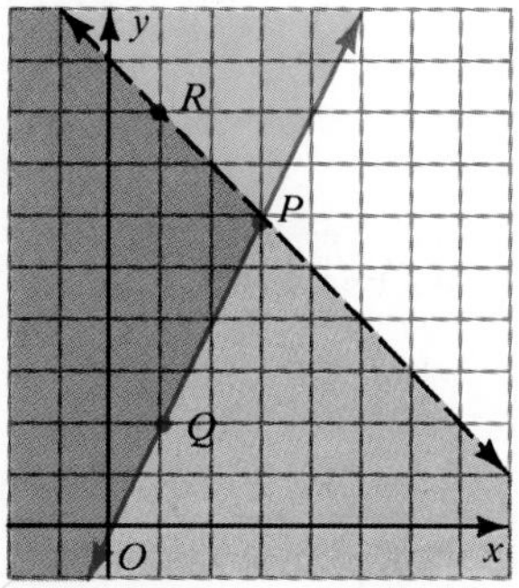

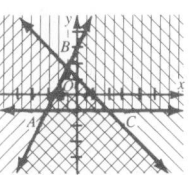

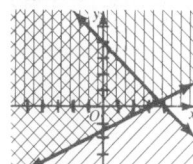

Additional Answers Self-Test 1

1. $\{(1, -3)\}$

Teaching Suggestions
p. T85

Key Ideas

Graph a system of two linear inequalities in two variables.

Chalkboard Examples

Graph the solution set of each system.

1. $-1 \leq y \leq 1 - x$
$3 + 2x \geq y$

The graph is $\triangle ABC$ and its interior.

2. $0 \leq y \leq 4 - x$
$x - 2y \leq 3$

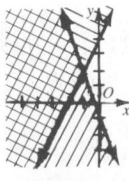
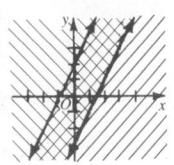
To show the graph of the solution set of a system of inequalities, you take these steps:

1. Graph each inequality in the system.
2. Show by heavier shading the region that consists of those points that belong to all of the graphs drawn in Step 1.

EXAMPLE   Graph the solution set of the system: $0 \leq y \leq 8 - 2x$
                                    $x - 2y \geq -1$

SOLUTION   The given system is equivalent to the following system of *three* inequalities whose graphs are shown.

$y \geq 0$ (red shading)
$y \leq 8 - 2x$ (gray shading)
$x - 2y \geq -1$ (diagonal hatching)

The graph is the region common to all of the graphs of the system, that is $\triangle ABC$ and its interior.

Oral Exercises

1. $y > -2x + 4$; $y \geq x + 1$
2. $y \leq -2$; $y > -2x + 4$
3. $y \geq -2$; $y \geq x + 1$
4. $y < -2x + 4$; $y \leq x + 1$

The lines whose equations are $y = -2$, $y = -2x + 4$, and $y = x + 1$ separate the plane into seven nonintersecting numbered regions as shown. State a system of equations whose graph is the given region.

1. 3   2. 7
3. $3 \cup 4$ 4. $1 \cup 6$
5. 5   6. $2 \cup 7$
7. 1   8. 2
9. 6   10. 4

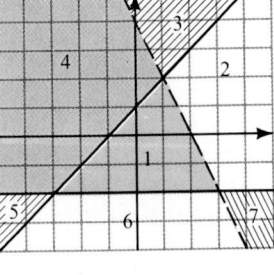

5. $y \leq -2$; $y \geq x + 1$
6. $y > -2x + 4$; $y \leq x + 1$
7. $y \geq -2$; $y < -2x + 4$;
   $y \leq x + 1$
8. $y \geq -2$; $y > -2x + 4$;
   $y \leq x + 1$
9. $y \leq -2$; $y < -2x + 4$;
   $y \leq x + 1$
10. $y \geq -2$; $y < -2x + 4$;
   $y \geq x + 1$

Written Exercises

Graph the solution set of each system in a coordinate plane.

A **1.** $x < 2$ **2.** $y > x$ **3.** $x \leq -y$
      $y > -1$ $y \leq 3$ $2y < x$

  4. $y < x + 3$ **5.** $y > 2 - x$ **6.** $y \geq -2x + 6$
      $y > 3x + 3$ $y > x - 2$ $y \leq 2x - 2$

146   *Chapter 4*

7. $y \le 1 - x$
$3y < x - 5$

8. $2y < x - 1$
$3y < x$

9. $-2y > x - 8$
$4y > x - 12$

*Additional Answers
Written Exercises*
(See p. 157.)

10. $2y \le 3x - 6$
$y > 2x - 5$

11. $x + y < 3$
$2x + y > 1$

12. $3x - y < -6$
$4x + y < -1$

13. $3x - 4y < 6$
$2x - y < -1$

14. $x + 3y \ge 7$
$2x - 3y \ge 5$

15. $5x + 2y > 13$
$4x + y > 11$

16. $-2 < x < 3$

17. $-1 \le y \le 2$

18. $x < y < x + 3$

B 19. $x - 4 < 2y < x + 2$

20. $2x + 1 \ge y \ge 2x - 3$

21. $2 < x < 4$
$2y > x - 6$

22. $-1 < y < 3$
$y + 2x < -1$

23. $y < -2x + 2$
$0 < y < x + 2$

24. $-1 < y < x + 1$
$2x + y < 7$

25. $y \le x$
$x + y < 4$
$x - 2y < 1$

26. $2x + y < 2$
$x - y > -2$
$x + y > -2$

27. $x + y \ge 3$
$x - y \le 3$
$x + 5y \ge 15$

28. $2x + y < 6$
$x + 2y < 12$
$4x - y > 12$

29. $x \ge 0$
$y \ge 0$
$x + 3y \le 9$
$2x + y \le 8$

30. $x > -2$
$y > 0$
$2y < x + 4$
$y < -3x + 9$

C 31. $y \ge |x|$
$3y < x + 8$

32. $|x + y| \ge 6$
$|x - y| \le 6$

33. $|x| + |y| \le 6$

34. $y > |x - 2|$
$y < x$
$y < 4 - x$

35. $|x| - |y| \le 3$
$|x| - 3 \ge y$

36. $|x + y| + |x - y| \le 6$

Mixed Review

Graph each linear inequality.

1. $y > -3$

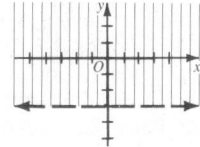

2. $x + 4 \le y$

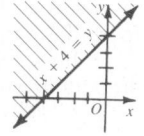

Solve each system by either the linear combination method or the substitution method, whichever seems simpler.

3. $4x - 3y = 17$
$3x + 2y = 0$ $\{(2, -3)\}$

4. $x = \dfrac{5y + 4}{7}$
$3x - 8y = 31$ $\{(-3, -5)\}$

4–6 Linear Programming ~skip~

The following situation illustrates a type of problem in economics that involves a system of inequalities in its solution.

A manufacturer of ice skates can produce as many as 60 pairs of hockey skates and as many as 45 pairs of figure skates per day. It takes 3 h of labor to produce a pair of hockey skates and 4 h of labor to produce a pair of figure skates. The company has up to 240 h of labor available for ice skate production each day. If the profit is $12 on each pair of hockey skates and $18 on each pair of figure skates, find the number of pairs of each kind of skate the firm should produce to gain the maximum profit each day.

To solve this problem, let x represent the number of pairs of hockey skates and y represent the number of pairs of figure skates that are produced each day. Then

$$12x = \text{the daily profit on hockey skates,}$$
$$18y = \text{the daily profit on figure skates,}$$
$$12x + 18y = \text{the daily profit on both kinds of skates.}$$

Systems of Linear Equations or Inequalities **147**

Teaching Suggestions
p. T85

Related Activities p. T85

Supplementary Material
Test 11

Key Ideas
Solve linear programming problems.

Chalkboard Examples

Space food *A* contains 3 calories per gram; space food *B* contains 4 calories per gram. An astronaut's food bar can contain no more than 30 g of space food *A* and no more than 20 g of space food *B*, and can have a maximum of 110 calories. If space food *A* contains 10 units of protein per gram and space food *B* contains 20 units of protein per gram, find the number of grams of each kind of space food that a food bar contains to produce the maximum amount of protein.

Let x = the number of grams of space food *A* and y = the number of grams of space food *B*.

Then $10x$ = the number of units of protein in space food *A*;

$20y$ = the number of units of protein in space food *B*;

$10x + 20y$ = the number of units of protein in both kinds of space food.

The following system of inequalities represents the constraints.

$x \geq 0$
$y \geq 0$
$x \leq 30$
$y \leq 20$
$3x + 4y \leq 110$

The shaded portion of the graph is the solution of the system.

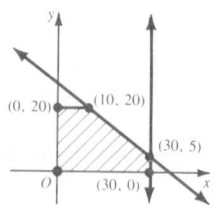

The values of x and y must satisfy certain conditions represented by the following system of inequalities. These conditions are called **constraints.**

$x \geq 0$	The firm must produce a nonnegative number of
$y \geq 0$	pairs of each kind of skate.
$x \leq 60$	At most 60 pairs of hockey skates can be produced each day.
$y \leq 45$	At most 45 pairs of figure skates can be produced each day.
$3x + 4y \leq 240$	Up to 240 h of labor is available for production each day.

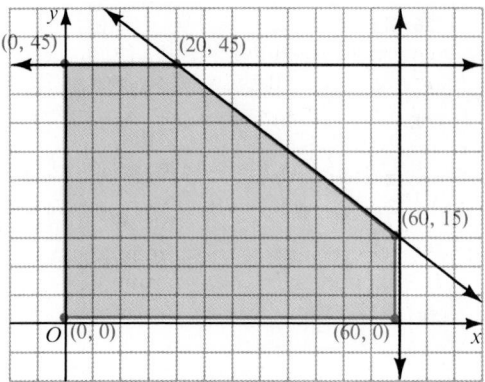

Figure 6

The shaded portion of Figure 6 is the solution set of the preceding system of constraints. Notice that the graph of this solution set is the intersection of a finite number of closed half-planes (actually, five closed half-planes). The graph is called the **feasibility region,** and the points of the graph where the lines that form the boundary intersect are called **corner points** of the feasibility region. We can then find the maximum value of $12x + 18y$ over the feasibility region by using the following theorem that we accept without proof.

Theorem. If a and b are any real numbers, and if the linear expression $ax + by$ has a maximum (greatest) value over a feasibility region that is the intersection of a finite number of closed half-planes and that has corner points, then the maximum occurs for the coordinates of some corner point.

Similarly, if $ax + by$ has a minimum (least) value over the region, then the minimum occurs for the coordinates of some corner point.

---

148 *Chapter 4*

Therefore, to **maximize** (or **minimize**) $12x + 18y$, that is, to find its greatest (or least) value, evaluate the expression at the five corner points, whose coordinates are found by solving simultaneously the equations of the boundary lines determining those points:

Corner Point	$12x + 18y$
$(0, 0)$	$12 \cdot 0 + 18 \cdot 0 = 0$
$(60, 0)$	$12 \cdot 60 + 18 \cdot 0 = 720$
$(60, 15)$	$12 \cdot 60 + 18 \cdot 15 = 990$
$(20, 45)$	$12 \cdot 20 + 18 \cdot 45 = 1050$
$(0, 45)$	$12 \cdot 0 + 18 \cdot 45 = 810$

Thus, over the feasibility region, the maximum value of $12x + 18y$ is 1050 and this occurs at the point $(20, 45)$. The minimum value of $12x + 18y$ is 0 and occurs at the origin, that is, when no ice skates are produced. Therefore, to maximize the monthly profit, the firm should produce 20 pairs of hockey skates and 45 pairs of figure skates and thus obtain a profit of $1050 each day.

The process illustrated in this example is called **linear programming** because it furnishes a means of finding maximum and minimum values of a linear expression over a feasibility region determined by linear inequalities.

Oral Exercises

State the value of each expression for the coordinates of the given point. Refer to the graph shown at the right.

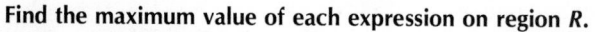

1. $x + y$; A 4
2. $x + y$; B 7
3. $x + y$; D 6
4. $2x + y$; B 10
5. $2x + y$; A 4
6. $2x + y$; D 12
7. $4x + 3y$; B 24
8. $4x + 3y$; C 26
9. $3x - y$; C 13

Ex. 1–13

Find the maximum value of each expression on region R.

10. $3x + 4y$ 25 at B
11. $3x + 2y$ 19 at C
12. $6x + 5y$ 40 at C
13. $2y - x$ 8 at A

Find the minimum value of each expression on region S shown at the right.

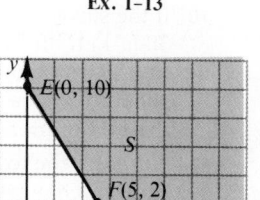

14. $x + y$ 7 at F and G
15. $2x + y$ 10 at E
16. $5x + 4y$ 33 at F
17. $2x + 3y$ 14 at G

Exs. 14–17

Systems of Linear Equations or Inequalities **149**

Evaluate $10x + 20y$ at the five corner points to maximize the amount of protein.

∴ the maximum amount is 500 units, and this occurs when the bar contains 10 g of space food A and 20 g of space food B.

Common Errors

Simply keeping track of all the conditions of the problem can be a formidable task in the exercises presented here. Emphasize that it is necessary to translate each given condition, separately, into an inequality.

Suggested Assignments

Minimum
First day
150/1–17
Second day
151/14–17
R 152/*Self-Test 2*
Extended Alg. with Trig.
150/5–17 odd
R 152/*Self-Test 2*
Enriched Alg.
150/5–19 odd
R 152/*Self-Test 2*
Enriched Alg. with Trig.
150/5–19 odd
R 152/*Self-Test 2*

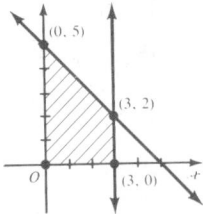
Written Exercises

In Exercises 1–6:
a. Graph the solution set of the system of inequalities.
b. Find the coordinates of the corner points of the graph.
c. Find the value of the linear expression printed in red at each of the corner points.
d. State the maximum and minimum values (if any) of the given linear expression under the given constraints.

A

1. $0 \le y \le 6 - x$
   $0 \le x \le 4$
   $x + 3y$

2. $2 \le x \le 5$
   $0 \le y \le x + 1$
   $3x - y$

3. $1 \le y \le 5$
   $x \ge 3$
   $2x + y \le 13$
   $2y - x$

4. $1 \le x \le 7$
   $2 \le y \le 6$
   $x + 2y \le 15$
   $5x - 2y$

5. $x \ge 0$
   $y \ge 0$
   $x + y \ge 9$
   $x + 2y \ge 14$
   $3x + 4y$

6. $0 \le x \le 8$
   $x + y \le 9$
   $x + 4y \le 24$
   $3x + y$

Exercises 7–12 refer to the following situation. A wood stove and an oil furnace are both used to heat a house. On any given day the stove can be used for at most 12 h and the furnace can be used for at most 8 h. It costs $1.00/h to operate the stove and $1.50/h to operate the furnace. At most $18.00 can be spent to operate both for one day.

7. Let x represent the number of hours the stove operates in one day and y represent the number of hours the furnace operates in one day. Give five inequalities that express the given conditions.

8. Graph the system of inequalities in Exercise 7.

9. Give the coordinates of the corner points of the graph in Exercise 8.

10. If the stove produces heat at a rate of 3000 kJ/h and the furnace produces heat at a rate of 5000 kJ/h, how long should each be operated to produce the maximum amount of heat?

11. Suppose the cost of operating the furnace changes to $2.00/h. Graph the resulting system of inequalities.

12. Answer Exercise 10 using $2.00/h as the operating cost of the furnace.

Exercises 13–16 refer to the following situation.
A cereal producer wants to make a breakfast cereal from a mixture of oats and wheat so that it contains at least 88 g of protein and at least 36 mg of iron. The following table gives the amounts of protein and iron per kilogram for each grain.

	protein	iron
wheat	80 g	40 mg
oats	100 g	30 mg

13. Graph the system of inequalities that expresses the given conditions.

B 14. If the company wants to minimize the costs, how many kilograms of each grain should the company use if the cost of wheat is 30¢ per kilogram and the cost of oats is 40¢ per kilogram? wheat: 1.1 kg, no oats

15. Answer question 14 if the cost of wheat is 50¢ per kilogram and the cost of oats is 40¢ per kilogram. wheat: 0.6 kg; oats: 0.4 kg

16. Answer question 14 if the cost of wheat is 70¢ per kilogram and the cost of oats is 50¢ per kilogram. no wheat; oats 1.2 kg

In Exercises 17 and 18, assume that a, b, and c are greater than 0.

17. Find the maximum of $ax + by$ over the rectangular region with vertices $(0, 0)$ $(0, c)$, $(1, c)$, and $(1, 0)$. a + bc

C 18. Find the minimum of $ax + by$ over the rectangular region with vertices $(-1, c)$, $(-1, -c)$, $(c, -c)$, and (c, c). −a − bc

19. Graph the solution set R of the system of inequalities $x \geq 0$, $y \geq 0$, $x + 2y \leq 8$, and $3x + 2y \leq 12$. On the same coordinate plane, graph the line with equation $x + y = p$ for $p = -2$, 0, 3, 5, and 8. Explain why the minimum value of $x + y$ over R is 0 and the maximum is 5.

Computer Exercises For students with computer experience

1. Write a program that will test 5 given corner points of a feasibility region for maximum and minimum values of a linear expression $ax + by$. The output should give the coordinates of the corner point that produces the maximum value and the coordinates of the corner point that produces the minimum value.

In Exercises 2 and 3 use the points (1, 26), (3, 22), (5, 18), (9, 16), and (2, 25) as corner points of the feasibility region. max., (1, 26); min., (5, 18)

2. Run the program in Exercise 1 for the linear expression $2.6x + 3.2y$.

3. Run the program in Exercise 1 for the linear expression
$$-4.1x + 5.2y.$$ max., (1, 26); min., (9, 16)

 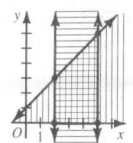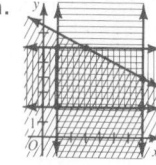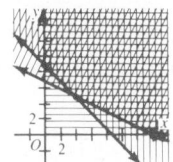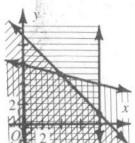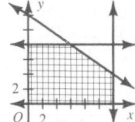

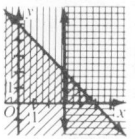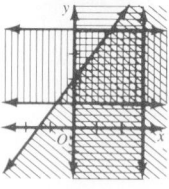

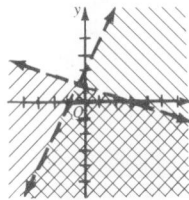
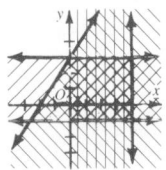
Self-Test 2

VOCABULARY   constraints (p. 148)   maximize (p. 149)
             feasibility region (p. 148)   minimize (p. 149)
             corner points (p. 148)   linear programming (p. 149)

Graph the solution set of each system.

1. $2x + 4y < 7$ 2. $y \ge 0$ 3. $0 \le x \le 3$ *Obj. 1, p. 123*
   $x - 3y > 2$ $3 \le x \le 5 - y$ $1 \le y \le 4$
                                                   $3y - 4x \le 6$

Determine the maximum and minimum values of each expression over the   *Obj. 2, p. 123*
feasibility region shown at the right.

4. $x + 2y$ max: 15 at (5, 5); min: 4 at (2, 1)   5. $3x + y$ max: 21 at (6, 3); min: 7 at (2, 1)

6. Find the maximum daily profit that can be realized producing x shirts and y slacks if the maximum number of shirts and slacks that can be produced in one day is 200 and 100, respectively. Each shirt requires 5 h of labor and each pair of slacks requires 11 h of labor; there are 1540 h of labor available each day. The profit on each shirt is \$3 and on each pair of slacks is \$5.
   \$845 from 200 shirts and 49 pairs of slacks

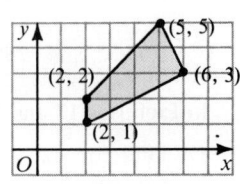

Exs. 4–5

Check your answers with those at the back of the book.

Chapter Summary

1. For a *system of two linear equations* in two variables, the graphs of the equations intersect in a single point, or the graphs coincide, or the graphs are parallel. Correspondingly, the system has a single solution, an infinite number of solutions, or no solution. A system that has at least one solution is *consistent*; otherwise, the system is *inconsistent*. A system that has an infinite solution set is *dependent*.

2. A system of linear equations in two variables can be solved by making transformations that yield equivalent systems, using the *linear-combination* method or the *substitution method*. Transformations that produce an equivalent system of linear equations are the following:

   1. Replace any equation of the system with an equivalent equation in the same variables.
   2. Replace any equation of the system with the sum of that equation and an equation obtained by multiplying both sides of another equation of the system by a real number.

152   *Chapter 4*

3. In any equation substitute for one variable (a) its value, if known, or (b) an equivalent expression for that variable obtained from another equation of the system.

3. *Determinants* can be used to solve a system of two linear equations in two variables by applying *Cramer's Rule*.

4. Systems of linear equations can be applied to solve problems by translating relationships into a system of equations.

5. The *solution set* of a system of linear inequalities in two variables is the intersection of open or closed half-planes representing the inequalities.

6. If a linear expression has a *maximum* or *minimum value* over a *feasibility region* which is the intersection of a finite number of closed half-planes and which has *corner points*, then that value occurs for the coordinates of some corner point.

Chapter Review

Write the letter of the correct answer.

Tell whether each system of equations has (a) one solution, (b) an infinite set of solutions, or (c) no solution.

1. $3x - 4y = 2$ c.
   $12y = 9x$

2. $-2x + 5y = 1$ b.
   $4x - 10y = -2$ 4-1

3. Solve the system by the linear combination method.   4-2

$$2x + 4y = 6$$
$$3x - 2y = 9$$

a. $\{(-3, 3)\}$ b. $\{(3, 0)\}$ c. $\{(0, 3)\}$ d. $\{(1, -3)\}$

4. Solve the system by the substitution method.

$$5x + 4y = -1$$
$$2x + y = 2$$

a. $\left\{\left(0, -\frac{1}{4}\right)\right\}$ b. $\{(1, 0)\}$ c. $\{(3, -4)\}$ d. $\{(3, 4)\}$

5. Evaluate the determinant $\begin{vmatrix} 3 & 1 \\ -5 & -2 \end{vmatrix}$.   4-3

a. -11 b. 1 c. -1 d. 11

6. Use Cramer's Rule to solve the system.

$$3x + y = -5$$
$$-5x - 2y = 8$$

a. $\{(-18, -1)\}$ b. $\{(-1, -2)\}$ c. $\{(-2, 1)\}$ d. $\{(-4, 6)\}$

Systems of Linear Equations or Inequalities **153**

Determine the maximum and minimum values of each expression over the feasibility region shown below.

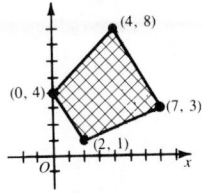

3. $2x + y$
   max: 17
   min: 4

4. $x + 4y$
   max: 36
   min: 6

Mixed Review

Evaluate each expression for $a = -6$ and $b = -3$.

1. $|a - b|$ 3
2. $|b - a|$ 3
3. $|3a - 4b|$ 6
4. $a^2 + b^2$ 45
5. $b^2 - a^2$ -27
6. $5b^2 - 3a^2$ -63

Exercises 7 and 8 refer to a situation where a boat traveled 24 km upstream in 3 h. Let s = rate of the boat in still water and c = rate of the current.

7. Which equation expresses this relationship correctly? *4-4*

    a. $24(s - c) = 3$ **b.** $3(s + c) = 24$

    c. $3(s - c) = 24$ **d.** $24(s + c) = 3$

8. Using the additional information that the boat then traveled 30 km downstream in 2 h, find the rate of the boat in still water and the rate of the current.

    a. $s = 11.5$ km/h, $c = 3.5$ km/h

    b. $s = 11.5$ km/h, $c = 2.5$ km/h

    c. $s = 27$ km/h, $c = 3$ km/h

    d. $s = 27$ km/h, $c = 2.5$ km/h

Graph the following system of inequalities. Use the graph to answer Exercises 9 and 10.

$$-2x \le y < 4$$
$$y \ge 3x - 8$$

9. Which of the following points lies on the graph of the system? *4-5*

    a. $(4, 4)$ **b.** $(2, 6)$ **c.** $(0, 0)$ **d.** $(-3, -1)$

10. Which of the following points does *not* lie on the graph of the system?

    a. $(0, 3)$ **b.** $(2, -2)$ **c.** $(1, -1)$ **d.** $(-2, 4)$

In Exercises 11 and 12, use the following system of constraints.

$$3 \le x \le 3y - 6 \qquad y \le 8 \qquad y \le -x + 14$$

11. Find the minimum value of $2x - y$. *4-6*

    a. 3   **b.** -2 **c.** -5 **d.** 4

12. Find the maximum value of $4x + 3y$.

    a. 48   **b.** 52   **c.** 51   **d.** 17

 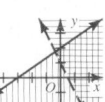
Chapter Test

Determine the apparent solution set of the system by graphing each system of equations. If the system has exactly one apparent solution, verify that the coordinates of the ordered pair satisfy all three equations.

1. $2x + y = -3$ 2. $4x - 6y = 10$ *4-1*

              $x = -2y$ $2x - 3y = 5$

     $-x + 3y = 5$ $3y - 2x = -4$

154   *Chapter 4*

Solve each system by either the linear combination method or the substitution method, whichever seems simpler.

4-2

3. $2x - 2y = -6$
$-3x + y = -5$ $\{(4, 7)\}$

4. $\dfrac{x - 1}{2} = \dfrac{y + 5}{3}$ $\{(3, -2)\}$
$3x + 4y = 1$

Use Cramer's Rule to solve each system. If the system is inconsistent or has an infinite solution set, so state.

5. $4x - y = 1$ $\left\{\left(\dfrac{3}{4}, 2\right)\right\}$
$3y = 8x$

6. $6y = -\dfrac{8}{3}x + 1$ infinite solution set
$\dfrac{3}{2} = 4x + 9y$

4-3

7. $\dfrac{1}{5}x + \dfrac{2}{3}y = 1$ $\{(-5, 3)\}$
$x + 3y = 4$

8. $-\dfrac{3}{4}x + \dfrac{1}{6}y = -\dfrac{1}{3}$ inconsistent
$9x - 2y = 6$

9. With a given head wind, a plane can fly 3000 km in 6 h. Flying in the opposite direction with the same wind blowing, the plane can fly the same distance in 1 h less. Find the plane's air speed and the speed of the wind. 550 km/h, 50 km/h

4-4

10. Ron earns $4.50/h and Jim earns $5.25/h. Together, they worked a total of 44 h and earned a total of $205.50. How many hours did each work? Ron 34 h; Jim 10 h

Graph the solution set of each system in a coordinate plane.

4-5

11. $6x + 3y - 3 > 0$
$2x - 3y \geq -6$

12. $-1 < x < 8 - 2y$
$y \geq -4$

13. A vender, selling peaches and apples, has space for 180 pieces of fruit and sells all of the fruit during one day. Customers buy at least twice as many apples as peaches. If the vender makes a profit of 8¢ on each peach and 10¢ on each apple, how many of each type of fruit should the vender sell each day to earn the maximum profit?
0 peaches, 180 apples

4-6

Cumulative Review

Chapter 1

Simplify.

1. $-2(-12)(5)$ 120

2. $-3(x - 5)$ $-3x + 15$

3. $-8\left[\left(-\dfrac{1}{4}\right) + \left(-\dfrac{3}{8}\right)\right]$ 5

4. $-3x[(-2) + 6y]$ $6x - 18xy$

5. $14z + 8z \div 4z - 3z \cdot (-5) + 6$ $29z + 8$

6. $-[-(3xy + 2x) + (-5y)(x + 6)]$
$8xy + 2x + 30y$

Systems of Linear Equations or Inequalities **155**

 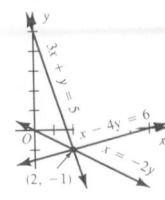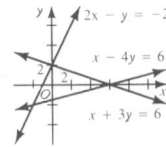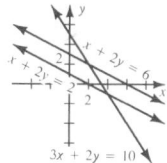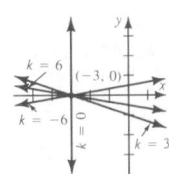

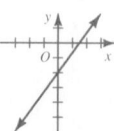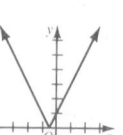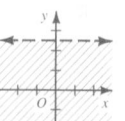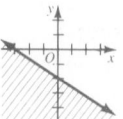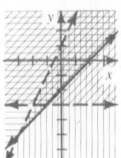
Solve each open sentence over $\{-4, -2. 0, 2, 4\}$. If the open sentence has no solution, so state.

7. $3x \div 4 = 5x$ $\{0\}$

8. $y^2 = 4$ $\{-2, 2\}$

No solution
9. $2p + 1 < -7$

10. $(x + 1)(x - 2) = 0$ $\{2\}$

11. $6m - 2m = 4m$ $\{-4, -2, 0, 2, 4\}$

12. $n^2 < 4$ $\{0\}$

13. Prove if $\frac{a}{c} = \frac{b}{c}$, then $a = b$ $(c \neq 0)$.

Graph each set of numbers on a number line.

14. $\{x: x \geq -1\}$

15. the set of all integers that are multiples of 6

Solve.

16. $-2(x - 1) = 3x - 8$ $\{2\}$

17. $5 + (-t) = 9$ $\{-4\}$

Chapter 2

Solve.

18. $\frac{x}{8} + 5 = 1$ $\{-32\}$

19. $9 - \frac{5}{12}x = 4$ $\{12\}$

20. $6(x - 2) + 4x = 0$ $\{\frac{6}{5}\}$

21. $3 - [8x - 2(3 - 4x)] + 16x = 3x$ $\{3\}$

Solve for b.

22. $x = \frac{1}{4}y(a + b)$ $b = \frac{4x - ay}{y}, y \neq 0$

23. $ab - cb = a - c$ $b = 1, a \neq c$

Solve each open sentence and graph its solution set.

24. $6 + \frac{2}{5}m < 4$ $\{m: m < -5\}$

25. $-\frac{3}{4}x > -6$ and $5 + 2x > 9$ $\{x: 2 < x < 8\}$

26. $|2 - 4r| < 10$ $\{r: -2 < r < 3\}$

27. $6|t + 4| \geq 12$ $\{t: t \geq -2\} \cup \{t: t \leq -6\}$

28. One base of a trapezoid is 4 cm longer than the other base. If the area of the trapezoid is 40 cm² and the height is 5 cm, what is the length of each base? 6 cm, 10 cm

29. Lisa has quiz grades of 83, 76, 92, and 85. What is the lowest grade she can receive on the next quiz and still maintain an average of at least 85? 89

Chapter 3

If $f(x) = 3x + 1$ and $g(x) = x^2$, find the value of each of the following.

30. $f(-5)$ -14

31. $g\left(\frac{4}{3}\right)$ $\frac{16}{9}$

32. $[f \circ g](-2)$ 13

33. $[g \circ f](-2)$ 25

State an open sentence that defines each relation and tell whether or not the relation is a function.

34. $\{(3, -1), (5, 1), (8, 4), (12, 8), (20, 16)\}$ $y = x - 4$; a function

156 *Chapter 4*

35. $\{(4, 25), (4, -2), (4, -6)\}$ $x = 4$, not a function

Graph each equation.

36. $6x - 4y = 8$ **37.** $y = |2x + 1|$

Graph each inequality as a shaded region on a coordinate plane.

38. $2y < 5$ **39.** $6 + 2x \leq -3y$

Determine the value of k so that the given point will be on the graph of the given equation.

40. $-4x + ky = 6$; $(-2, -1)$ $k = 2$ **41.** $kx + 5y = k - 3$; $(5, -3)$ $k = 3$

Determine an equation in the form $y = mx + b$ of a line satisfying the given condition(s).

42. through the points $(-2, 1)$ and $(3, 11)$ $y = 2x + 5$

43. through the point $(1, -2)$ and parallel to the line $3x + y = 7$ $y = -3x + 1$

44. If 1 gal of paint will cover 600 ft², how many gallons of paint are needed to cover 4000 ft²? $6\frac{2}{3}$ gal

Chapter 4

Tell whether each of the following systems has no solution, one solution, or infinitely many solutions. If there is one solution, determine what that solution is.

45. $2x + 5y = 1$ one solution;
$\quad\quad y = -\frac{3}{2}x - 2$ $\{(-2, 1)\}$

46. $\quad 15x - 3y = 12$ no solution
$\quad\quad -120x + 24y = 96$

47. $\quad\quad y = \frac{5}{3}x + 4$
$\quad 5x - 3y = -4$ no solution

48. $\quad 9x + 4y = 11$ one solution;
$\quad -12x + 10y = -4$ $\left\{\left(\frac{21}{23}, \frac{16}{23}\right)\right\}$

Graph the solution set of each system in a coordinate plane.

49. $4x + 5y - 8 \geq 2$
$\quad -3x + 2y < 6$

50. $\quad x - y \leq 2$
$\quad -3 < y < 2x + 1$

51. An auditorium has a seating capacity of 500. For a certain performance, an adult ticket costs $4.50 and a student ticket costs $1.50. A total of $1650 was collected in ticket sales one evening. If the performance was sold out, how many adult tickets and how many student tickets were sold that evening? Adult tickets: 300; student tickets: 200

52. The junior class at a certain school decides to sell shirts and hats with the school name. The class treasury has at most $875 that can be used to purchase the shirts and hats from a manufacturer who charges $3.50 for each shirt and $1.25 for each hat. The class has decided to purchase at least 150 more hats than shirts. The manufacturer, however, can provide at most 420 hats. If the class plans to sell each shirt for $5 and each hat for $2, how many shirts and how many hats should they purchase to earn a maximum profit?
100 shirts; 420 hats

Systems of Linear Equations or Inequalities **157**

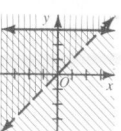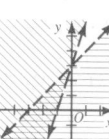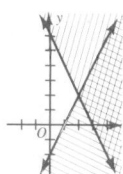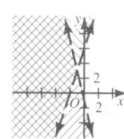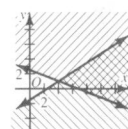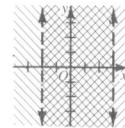

18.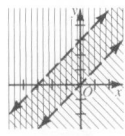

20.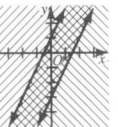

22.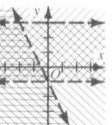

24.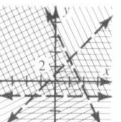

26.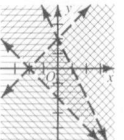

28.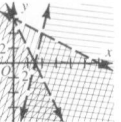

30.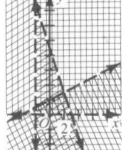

Contest Problems

1. When Jim ran up a certain "up" escalator, he counted 10 steps and the ride took 20 s. Running down the same escalator, he counted 50 steps and the ride took 30 s. How many steps of the escalator were visible at one time? 26 steps

2. A fruit dealer sells apples according to the following system:
   1. The cost of any number of apples is a whole number of dollars.
   2. The cost of 2 apples is $0, that is, they are free.
   3. The cost of 3 apples is at least $1.
   4. The cost of x apples, where $x = n + m$, is the cost of either
      a. the cost of n apples + the cost of m apples, or
      b. the cost of n apples + the cost of m apples + 1.
   5. The cost of 9999 apples is $3333.

   What is the cost of 1982 apples? $660

PROGRAMMING IN PASCAL

In this section, you will develop a Pascal program that uses Cramer's Rule to solve systems of two linear equations in two variables. Use notation consistent with the text whenever possible in working the following exercises.

Exercises

1. To use Cramer's Rule to solve the following system

$$a_1x + b_1y = c_1$$
$$a_2x_2 + b_2y = c_2$$

where $a_1, b_1, c_1, a_2, b_2, c_2$ are integers, the following determinants must be calculated.

$$D = \begin{vmatrix} a_1 & b_1 \\ a_2 & b_2 \end{vmatrix} \qquad D_x = \begin{vmatrix} c_1 & b_1 \\ c_2 & b_2 \end{vmatrix} \qquad D_y = \begin{vmatrix} a_1 & c_1 \\ a_2 & c_2 \end{vmatrix}$$

Write a Pascal function with the heading

        FUNCTION det(s, t, u, v : integer) : integer

that can be used to calculate a determinant whenever the call $det(a, b, c, d)$ is made in the main program or in a procedure that follows the main program.

2. Write a Pascal procedure *get_values* that prompts the user to enter the coefficients of each linear equation. The coefficients should be identified as *a1, b1, c1, a2, b2,* and *c2*. (*Hint*: These identifiers should be declared as variable parameters.)

3. Write a Pascal program *cramer2* to solve and display the answers of a system of linear equations in 2 variables. If the determinant of the system is 0, then the computer should display "No solution or infinitely many solutions."

The following identifiers and their types should be declared in the program heading:

```
VAR
    a1, b1, c1, a2, b2, c2 : integer
    x , y : real
    d , dx , dy : integer
```

The program should call the procedure *get_values* and then successively call the function *det* to get values for *d, dx,* and *dy*. Recall from Cramer's Rule that if D ≠ 0

$$x = \frac{D_x}{D} \text{ and } y = \frac{D_y}{D}.$$

4. Modify the program so the user can repeatedly solve a system of linear equations. When the user wants to terminate the session, a reply of "N" or "n" to the question "solve another system (Y/N)?" would end the loop.

5. Modify the main program so it determines the appropriate answer "No solution." or "Infinitely many solutions." when *D* = 0.

32.

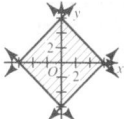

34.

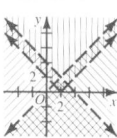

36.

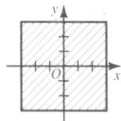

***Additional Answers
Written Exercises***
(continued from p. 151)

13.

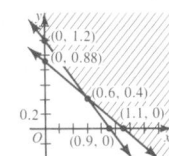

19.

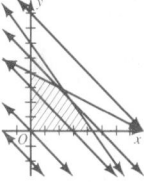

The restoration of the Statue of Liberty was undertaken to commemorate the centennial of the statue's dedication. The scaffolding seen in the photo suggests three-dimensional space. Graphs in space are studied in this chapter.

Problem Solving Strategies

Drawing a Diagram
Diagrams in Sections 5–1, 5–2, and 5–3 illustrate graphs in space.
Word Problem Plan
The five–step plan is used in Section 5–5 to solve problems involving equations in three variables.

Chapter 5

Graphs in Space; Determinants

Systems of Equations in Three Variables

OBJECTIVES for Sections 5-1 through 5-3:
1. *To draw graphs of ordered triples.*
2. *To determine the x-, y-, and z-intercepts of a given plane and the traces of the plane in the coordinate planes.*
3. *To draw the space-graph of a given linear equation in three variables.*
4. *To solve a system of three linear equations in three variables by transforming it into a simple equivalent system.*

5–1 Coordinates in Space

Just as a coordinate system in the two-dimensional plane establishes a one-to-one correspondence between the set of points in the plane and the set of ordered pairs of real numbers, so does a rectangular coordinate system in three-dimensional space establish a one-to-one correspondence between the set of points in space and the set of *ordered triples* of real numbers.

To set up a rectangular coordinate system in space, draw three mutually perpendicular number lines, or *axes*, passing through a common point O, the *origin*, of each. The *axes* are usually labeled x, y, and z, as in Figure 1, with an arrowhead on each to indicate the positive direction. The

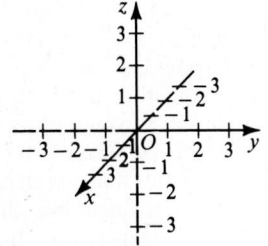

Figure 1

Graphs in Space; Determinants **161**

Key Ideas

Introduce the rectangular co-ordinate system in three-dimensional space.

Chalkboard Examples

State the coordinates of P and Q.

1.

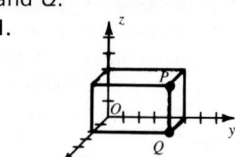

$P(2, 5, 3)$, $Q(2, 5, 0)$

2.

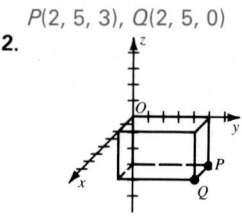

$P(0, 5, -3)$, $Q(2, 5, -3)$

Additional A Exercises

Sketch a diagram of a co-ordinate system in space. In the diagram show the co-ordinate box of the given point. Show the hidden edges as dashed segments and be sure to indicate units along the coordinate axes.

1. $(-2, 4, 3)$

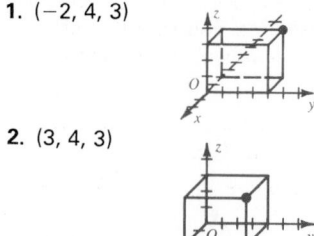

2. $(3, 4, 3)$

same scale ordinarily is used on each axis. However in order to give spatial perspective to the figure, the angle between the positive x- and y-axes is drawn as a 135° angle instead of a 90° angle. When equal units are used on all three axes, the units of length on the y- and z-axes are drawn the same, while the unit on the x-axis is *drawn* as if it were two thirds of the unit on each of the other axes; this "fore-shortening" helps give the appearance of depth to the drawing. Also, the negative portion of each axis is often shown as a dashed line.

The coordinate axes determine three coordinate planes as shown in Figure 2, each passing through the origin and each containing two of the axes:

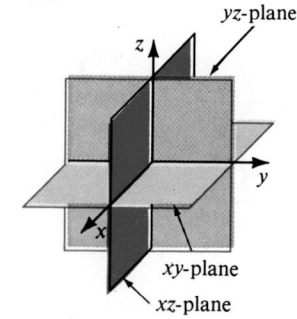

Figure 2

1. the xy-plane, which contains the x- and y-axes and is perpendicular to the z-axis;

2. the yz-plane, which contains the y- and z-axes and is perpendicular to the x-axis;

3. the xz-plane, which contains the x- and z-axes and is perpendicular to the y-axis.

The coordinate planes separate space into eight regions, called octants, each bounded by the positive part or the negative part of each of the three axes. Each octant is designated by a succession of three signs, either plus or minus, according as the octant is bounded by the positive or negative part of the x-axis, the y-axis, and the z-axis. Therefore the $(-, -, +)$-octant, read "the minus minus plus octant," is bounded in part by the negative part of the x-axis, the negative part of the y-axis, and the positive part of the z-axis. The $(+, +, +)$-octant is also called the *first octant*; the other octants are not numbered.

To assign an ordered triple of numbers, or coordinates, to a point such as P in Figure 3, draw three planes through P, the first (represented by $ABCP$) perpendicular to the x-axis, the second $(EDCP)$ perpendicular to the y-axis, and the third $(AFEP)$ perpendicular to the z-axis. The numbers paired with the points in which these planes intersect the respective axes are, in order, the x-coordinate, the y-coordinate, and the z-coordinate of P. For example, in Figure 3, P has coordinates $(3, 4, 5)$.

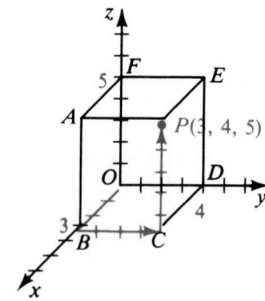

Figure 3

Together with the coordinate planes, the three planes drawn through P form a *rectangular parallelepiped*, or *box*, which we will call the coordinate box of P.

Notice that a path can be traced to point P by moving on the coordinate box choosing edges parallel to each axis in succession. This suggests how to locate a point whose coordinates are given. For example, Figure 4 on the next page shows the plotting of the point $R(2, -4, -1)$.

162 *Chapter 5*

1. From O move 2 units in the positive x-direction along the x-axis.

2. Then move -4 units (4 units in the negative direction) parallel to the y-axis.

3. Then move -1 unit parallel to the z-axis.

Point $S(-5, 1, -2)$ has also been plotted in Figure 4.

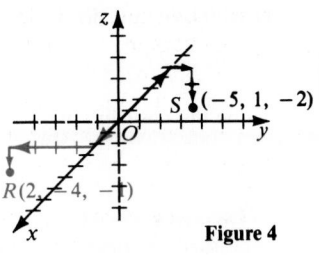

Figure 4

EXAMPLE State the coordinates of Q, R, S and T.

SOLUTION Q:$(-2, 0, 0)$ R:$(0, 3, 0)$
 S:$(-2, 3, 0)$ T:$(-2, 3, 4)$

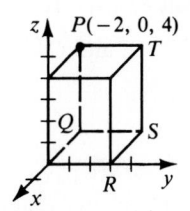

For better visualization, it is helpful in drawing space figures to show "hidden edges" by dashed segments and visible edges by darkened segments, as illustrated in the example for the coordinate box of $P(-2, 0, 4)$.

Oral Exercises

In Exercises 1–24 state the coordinates of the given vertex of a coordinate box shown below.

1. A (4, 0, 0)   2. B (4, 3, 0)
3. C (0, 3, 0)   4. D (0, 3, 2)
5. E (0, 0, 2)   6. F (4, 0, 2)
7. G (0, 0, -1) 8. H (0, 4, -1)
9. I (0, 4, 0)   10. J (-3, 4, 0)
11. K (-3, 0, 0) 12. L (-3, 0, -1)
13. M (3, 0, 0)   14. N (3, 0, -3)
15. Q (0, 0, -3) 16. R (0, 4, -3)
17. S (0, 4, 0)   18. T (3, 4, 0)
19. U (-3, 0, 4) 20. V (0, 0, 4)
21. W (-3, 0, 0) 22. X
23. Y (0, -4, 0) 24. Z
22. (-3, -4, 0)
24. (0, -4, 4)

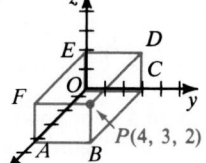

Ex. 1–6

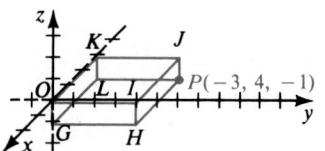

Ex. 7–12

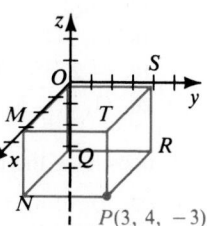

Ex. 13–18

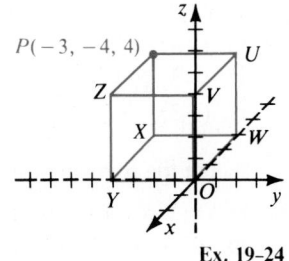

Ex. 19–24

In Exercises 25–32 name the coordinate plane that contains the two given points.

25. A and F xz 26. E and C yz 27. S and T xy 28. S and R yz
29. I and H yz 30. K and J xy 31. U and Z none   32. W and Y xy

Graphs in Space; Determinants **163**

Mixed Review

3. $(-2, -3, 4)$

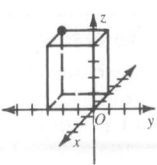

1. Solve $4t + 10 = 3[-(t - 10) + 5t - 12]$. {2}
2. Find the solution set of $3x - (9x - 5) \le 15 - 4(7 - 3x)$. $\{x: x \ge 1\}$
3. Solve $|3x + 9| = 6$ $\{-5, -1\}$
4. Find the value of $x + x^x + (x)(x^x)$ if $x = 2$. 14

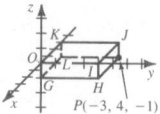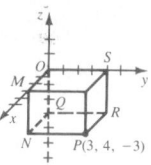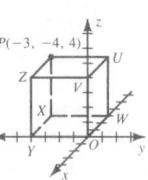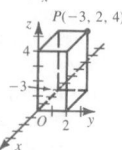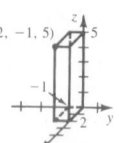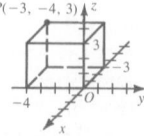
In which octant does the given point lie?

33. (6, 3, 7) first   **34.** (−14, 2, 5) (−, +, +)   **35.** (3, −6, −1) (+, −, −)

Name the coordinate plane or axis containing all points of the given form.

36. (0, a, b) yz - plane   **37.** (a, b, 0) xy - plane   **38.** (0, 0, a) z - axis

In Exercises 39–42, name a fourth point that is in the same plane (not necessarily a coordinate plane) as the three given points in the coordinate boxes on page 163.

39. P, D, and C B   **40.** M, N, and Q O   **41.** J, K, and H G   **42.** O, P, and Z W

Written Exercises

A 1–4. Copy the diagrams shown in the Oral Exercises on page 163, showing the hidden edges as dashed segments.

Sketch a diagram of a coordinate system in space. In the diagram show the coordinate box of the given point. Show the hidden edges as dashed segments and be sure to indicate units along the coordinate axes.

5. (3, 5, 2)   **6.** (−3, 2, 4)
7. (2, −1, 5)   **8.** (−3, −4, 3)
9. (5, 6, −2)   **10.** (4, −2, 3)
11. (−4, −2, −6)   **12.** (−3, 5, −2)

In Exercises 13–20 sketch a diagram of a coordinate system in space and locate the given point using a diagram like Figure 4 on page 163. Give the coordinates of each point at which the arrow changes direction.

13. (4, −2, 3)   **14.** (3, 2, −3)
15. (−4, 6, 2)   **16.** (5, −1, −2)
17. (−5, 2, 4)   **18.** (−2, 5, −4)
19. (−3, −5, −6)   **20.** (1, 4, −6)

Sketch the triangle in space whose vertices have the given coordinates.

B 21. (4, 0, 0), (0, −5, 0), (0, 0, 3)   **22.** (0, 0, 3), (−3, 0, 0), (0, 6, 0)
23. (0, 0, 0), (4, 0, −4), (0, −3, 5)   **24.** (0, 0, 0), (−5, 0, 4), (−3, −4, 2)

On a coordinate system in space, graph each of the following sets of points. In each set connect the four points that all lie in one plane.

C 25. {(0, 0, 2), (0, 0, 0), (0, 3, 0), (−4, 0, 0), (−6, 3, 0)}
26. {(0, −2, 0), (0, 5, 0), (−3, 0, 0), (0, −2, 4), (−3, 0, 4)}

164 *Chapter 5*

5–2 Graphs of Linear Equations in Three Variables

An equation such as

$$2x + 3y + 4z = 9$$

is called a *linear equation in three variables*. In general, any equation of the form

$$Ax + By + Cz = D,$$

where A, B, C, and D are real constants such that A, B, and C are *not all* 0, is a linear equation in the variables x, y, and z. In our work, *we will assume that the replacement set of each variable is $\mathcal{R}$.*

An equation in one or two variables, such as

$$2y = 3 \qquad \text{or} \qquad 4x - 3z = 5,$$

can be regarded as an equation in three variables for which one or two of the coefficients of the three variables are zero:

$$0x + 2y + 0z = 3 \qquad \text{or} \qquad 4x + 0y - 3z = 5.$$

The ordered triple

$$(1, 5, -2)$$

is called a solution of the linear equation

$$2x + 3y + 4z = 9$$

because the assertion

$$2 \cdot 1 + 3 \cdot 5 + 4 \cdot (-2) = 9$$

is a true statement. In general, a *solution* of an open sentence in three variables is an ordered triple of values of the variables for which the open sentence is true. Such an ordered triple of numbers is said to *satisfy* the sentence. The set of *all* ordered triples of real numbers that are solutions of the open sentence is the *solution set* of the sentence over $\mathcal{R}$. To denote the solution set of the equation

$$2x + 3y + 4z = 9$$

we can use the notation

$$\{(x, y, z): 2x + 3y + 4z = 9\}.$$

In Section 5-1 you saw that an ordered triple of real numbers gives the coordinates of a point in space. The set consisting of those points and only those points whose coordinates satisfy a given open sentence in three variables is the *graph* of the sentence. We accept the theorem on page 166 without proof.

Graphs in Space; Determinants **165**

Teaching Suggestions
p. T87

Key Ideas

Graph a linear equation in three variables.

Chalkboard Examples

a. Find the x-, y-, and z-intercepts of each equation.
b. Give three systems of equations whose solution sets are the traces of the graph in each of the three coordinate planes.
c. Sketch the traces and (part of) the graph of the given equation.

1. $3x + y + 2z = 6$
   a. Let $y = 0$, $z = 0$ to find x-intercept.
   $3x + (0) + 2(0) = 6$
   $3x = 6$
   $x = 2$
   ∴ the plane cuts the x-axis at $A(2, 0, 0)$.

   Let $x = 0$, $z = 0$ to find y-intercept.
   $3(0) + y + 2(0) = 6$
   $y = 6$
   ∴ the plane cuts the y-axis at $B(0, 6, 0)$.

   Let $x = 0$, $y = 0$ to find z-intercept.
   $3(0) + (0) + 2z = 6$
   $z = 3$
   ∴ the plane cuts the z-axis at $C(0, 0, 3)$.

   b. Trace in the xy-plane:
   $z = 0$
   $3x + y = 6$
   Trace in the yz-plane:
   $x = 0$
   $y + 2z = 6$
   Trace in the xz-plane:
   $y = 0$
   $3x + 2z = 6$

(continued)

c.

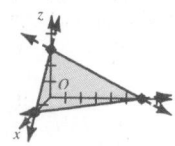

2. $4x + 3z = 12$

  a. Let $y = 0$, $z = 0$

    $4x + 3(0) = 12$

           $x = 3$

    $\therefore$ the x-intercept is 3.

    Let $x = 0$, $y = 0$

    $4(0) + 3z = 12$

           $z = 4$

    $\therefore$ the z-intercept is 4.

    Let $x = 0$, $z = 0$

    $4(0) + 3(0) = 12$

          $0 = 12$

    This is false for all

    values of y.

    $\therefore$ there is no

    y-intercept.

  b. Trace in xy-plane:

    $z = 0$

    $x = 3$

    Trace in yz-plane:

    $x = 0$

    $z = 4$

    Trace in xz-plane:

         $y = 0$

    $4x + 3z = 12$

c.

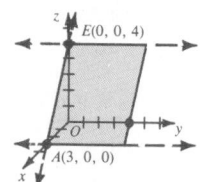

> *Theorem.* In space, the graph of a linear equation in three variables is a plane. Conversely, every plane is the graph of some linear equation in three variables, called an **equation of the plane.**

For example, $x = 0$ for all points in the yz-plane and for no other points. Accordingly, the linear equation

$$x = 0$$

is an equation of the yz-plane. Similarly,

$$y = 0 \quad \text{and} \quad z = 0$$

are equations of the xz-plane and the xy-plane, respectively.

Since three noncollinear points determine a plane, you can use the preceding theorem to graph a linear equation in three variables by finding three noncollinear points whose coordinates satisfy the equation. When possible, it often is easiest to choose the points where the plane cuts the coordinate axes.

Just as we can draw only a part of the line that is the graph of a linear equation in two variables, we can draw only a *part of* the plane that is the graph of a linear equation in three variables. Usually we will draw only the part of the graph that can be shown in the first octant.

EXAMPLE 1 Sketch (part of) the graph of the equation $2x + 3y + 6z = 12$.

SOLUTION 1. To find the coordinates of the point where the graph cuts the x-axis, replace y and z with 0 in the equation and solve for x:

$$2x + 3y + 6z = 12$$
$$2x + 3 \cdot 0 + 6 \cdot 0 = 12$$
$$2x = 12$$
$$x = 6$$

$\therefore$ the plane cuts the x-axis at the point $A(6, 0, 0)$.

2. Next, replace x and z with 0 in the given equation and solve for y:

$$2x + 3y + 6z = 12$$
$$2 \cdot 0 + 3y + 6 \cdot 0 = 12$$
$$3y = 12$$
$$y = 4$$

$\therefore$ the plane cuts the y-axis at the point $B(0, 4, 0)$.

3. Now replace x and y with 0 in the given equation and solve for z:

$$2x + 3y + 6z = 12$$
$$2 \cdot 0 + 3 \cdot 0 + 6z = 12$$
$$6z = 12$$
$$z = 2$$

$\therefore$ the plane cuts the z-axis at the point $C(0, 0, 2)$.

166   *Chapter 5*

4. Draw a sketch showing the three points where the plane cuts the three axes by drawing the line segments connecting the three points by pairs, and shade the space triangle as shown. This triangle shows a portion of the graph of $2x + 3y + 6z = 12$ that lies in the first octant.

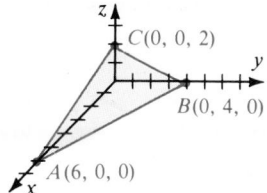

The x-coordinate, 6, of the point where the plane of Example 1 cuts the x-axis is called the x-intercept of the plane. Similarly, the y-intercept of the plane is 4 and the z-intercept is 2. In general, if a plane intersects the x-axis in a *single point*, then the x-coordinate of that point is called the x-intercept of the plane. The y-intercept and z-intercept are defined similarly.

In the graph of Example 1, the points A and B lie in the xy-plane and also in the plane that is the graph of

$$2x + 3y + 6z = 12.$$

Therefore the line $\overleftrightarrow{AB}$ is the line of intersection of the xy-plane and the graph.

A line in which a plane intersects a coordinate plane is called the trace of the given plane in that coordinate plane. Thus, the trace of the graph of Example 1 in the xy-plane is $\overleftrightarrow{AB}$, in the yz-plane is $\overleftrightarrow{BC}$, and in the xz-plane is $\overleftrightarrow{AC}$.

Since $\overleftrightarrow{AB}$ lies in the xy-plane, the coordinates of its points must satisfy the equation $z = 0$. But $\overleftrightarrow{AB}$ also lies in the graph of $2x + 3y + 6z = 12$, so the coordinates of its points must satisfy this equation, too. Since there are no other points on both of these planes, you can conclude that $\overleftrightarrow{AB}$ is the solution set of the system of equations:

$$2x + 3y + 6z = 12$$
$$z = 0$$

If you replace z with "0" in the first equation of the system, you obtain the equivalent system

$$2x + 3y = 12$$
$$z = 0$$

whose solution set is $\overleftrightarrow{AB}$.

Similarly, the trace $\overleftrightarrow{BC}$ in the yz-plane is the solution set of the system

$$3y + 6z = 12 \text{ (or } y + 2z = 4\text{)}$$
$$x = 0$$

and, the trace $\overleftrightarrow{AC}$ in the xz-plane is the solution set of the system

$$2x + 6z = 12 \text{ (or } x + 3z = 6\text{)}$$
$$y = 0$$

Graphs in Space; Determinants **167**

Additional A Exercises

For each equation, sketch part of the graph of the equation. Give three systems of equations whose solution sets are the traces of the graph in each of the coordinate planes. If the equation has no trace in a coordinate plane, so state.

1. $2x - y + 3z = 6$

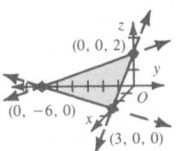

Trace in yz-plane:
$x = 0$
$3z - y = 6$
Trace in xz-plane:
$y = 0$
$2x + 3z = 6$
Trace in xy-plane:
$z = 0$
$2x - y = 6$

2. $3x + 4y = 12$

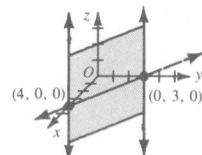

Trace in yz-plane:
$x = 0$
$y = 3$
Trace in xz-plane:
$y = 0$
$x = 4$
Trace in xy-plane:
$z = 0$
$3x + 4y = 12$

(continued)

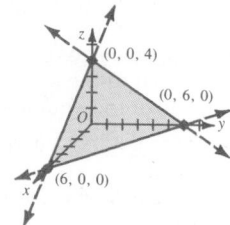
EXAMPLE 2   **a.** Find the x-, y-, and z-intercepts of the graph of
$$5x + 2z = 10.$$

  b. Give a system of equations whose solution set is the trace of the graph in each of the three coordinate planes.

  c. Sketch each trace and shade a portion of the graph in the first octant.

SOLUTION   **a.** $5x + 2 \cdot 0 = 10$; $x = 2$. The x-intercept is 2.
$5 \cdot 0 + 2 \cdot 0 = 10$; $0 = 10$. This equation is false for every value of y. There is no y-intercept, and the graph is parallel to the y-axis.
$5 \cdot 0 + 2z = 10$; $z = 5$. The z-intercept is 5.

  b. 1. Trace in the xy-plane ($\overleftrightarrow{AB}$ in the diagram):
$$z = 0$$
$$x = 2$$

    2. Trace in the yz-plane ($\overleftrightarrow{ED}$ in the diagram):
$$x = 0$$
$$z = 5$$

    3. Trace in the xz-plane ($\overleftrightarrow{AE}$ in the diagram):
$$y = 0$$
$$5x + 2z = 10$$

  c. The space-graph of $5x + 2z = 10$ in the first octant and its traces are shown above.

EXAMPLE 3   Sketch part of the graph of $3x + 2z = 0$ in the first octant.

SOLUTION   For every point on the y-axis, $x = 0$ and $z = 0$. Substitute 0 for x and for z in the given equation to obtain the statement
$$3 \cdot 0 + 2 \cdot 0 = 0, \quad \text{or} \quad 0 = 0,$$
which is true for every value of y. Thus, the coordinates of every point on the y-axis satisfy the equation $3x + 2z = 0$, so that the graph of the equation *contains* the y-axis. Substitute 2 for x and let y have the value 0, to obtain
$$3 \cdot 2 + 2z = 0, \quad \text{or} \quad z = -3.$$

Thus, the point with coordinates $(2, 0, -3)$ is on the graph. Sketch a part of the plane containing the y-axis and the point with coordinates $(2, 0, -3)$ as shown in red at the right.

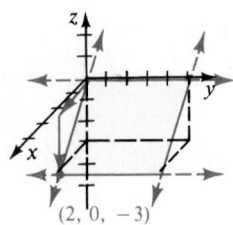

Examples 2 and 3 illustrate the facts on the following page.

168   *Chapter 5*

If the coefficient of a variable in an equation of a plane is zero, then:

1. The plane is parallel to the axis of that variable if the constant term is not zero;

2. The plane contains the axis of that variable if the constant term is zero.

You can apply the preceding facts in particular to graph a linear equation in three variables when the coefficients of *two* of the variables are zero. For example, in $y = 2.5$ both the coefficient of x and the coefficient of z are zero, but the constant term is not zero. Therefore, the graph is parallel to the x-axis and also to the z-axis; that is, the graph is parallel to the xz-plane. You can now sketch the graph (Figure 6) of the equation when you note that the ordered triple $(0, 2.5, 0)$ satisfies fhe equation.

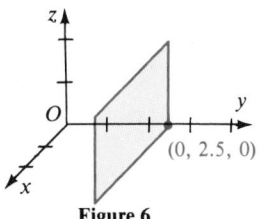

Figure 6

Oral Exercises

Exercises 1–8 refer to the diagram that shows part of the graph of the equation $3x - 2y + 4z = 12$. Name each requested part of the graph.

1. the x-intercept 4
2. the y-intercept -6
3. the z-intercept 3
4. the trace in the xy-plane
5. the trace in the yz-plane $\overleftrightarrow{BC}$
4. $\overleftrightarrow{AB}$
6. the trace in the xz-plane $\overleftrightarrow{AC}$
7. the part of the graph in the $(+, -, +)$-octant $\triangle ABC$
8. the part of the graph in the $(+, +, +)$-octant
   the plane bounded by $\overrightarrow{CD}$, $\overrightarrow{CA}$, and $\overrightarrow{AE}$

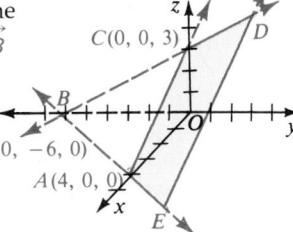

In Exercises 9–26:
a. Name the x-, y-, and z-intercepts.
b. Determine whether or not the graph is parallel to or contains any of the coordinate axes, and if so which one(s).
c. Determine whether or not the graph is parallel to or coincides with one of the coordinate planes, and if so which one.

9. $x + 2y + 3z = 6$
10. $3x + 2y = 6$
11. $2x - 3y + z = 6$
12. $x = -4$
13. $4x + 3y = 0$
14. $6x - 3y - 2z = 18$
15. $5x + 2z = 10$
16. $-4x + 2y - z = 8$
17. $y = 0$
18. $5x - 3y + 3z = 15$
19. $3x - y = 0$
20. $z = 2$
21. $-10x + 5y - 4z = 20$
22. $6x - 5y + 15z = 30$
23. $2y = -3$
24. $3y + 5z = -15$
25. $-8x - 3y - 4z = 24$
26. $4z = 0$

Graphs in Space; Determinants **169**

15. a. $(2, 0, 0)$, no y-intercept, $(0, 0, 5)$
    b. parallel to the y-axis
    c. none
16. a. $(-2, 0, 0)$, $(0, 4, 0)$, $(0, 0, -8)$
    b. none   c. none
17. a. the x-axis, $(0, 0, 0)$, the z-axis
    b. contains the x- and z-axes
    c. coincides with the xz-plane
18. a. $(3, 0, 0)$, $(0, -5, 0)$, $(0, 0, 5)$
    b. none   c. none
19. a. $(0, 0, 0)$, $(0, 0, 0)$, the z-axis
    b. contains the z-axis
    c. none
20. a. no x-intercept, no y-intercept, $(0, 0, 2)$
    b. parallel to the x- and y-axes
    c. parallel to the xy-plane
21. a. $(-2, 0, 0)$, $(0, 4, 0)$, $(0, 0, -5)$
    b. none   c. none
22. a. $(5, 0, 0)$, $(0, -6, 0)$, $(0, 0, 2)$
    b. none   c. none
23. a. no x-intercept, $\left(0, -\dfrac{3}{2}, 0\right)$, no z-intercept
    b. parallel to the x- and z-axes
    c. parallel to the xz-plane
24. a. no x-intercept, $(0, -5, 0)$, $(0, 0, -3)$
    b. parallel to the x-axis
    c. none
25. a. $(-3, 0, 0)$, $(0, -8, 0)$, $(0, 0, -6)$
    b. none   c. none
26. a. the x-axis, the y-axis, $(0, 0, 0)$
    b. contains the x- and y-axes
    c. coincides with the xy-plane

Suggested Assignments

Minimum
First day
170/1–26
Second day
S 131/25–31 odd
164/2–24 even
Extended Alg. with Trig.
170/5–27 odd
S 131/16–30 even
Enriched Alg.
170/19–29 odd
S 131/22–34 even
Enriched Alg. with Trig.
170/19–29 odd
S 131/22–34 even

Additional Answers
Written Exercises

2. b. $z = 0$; $3x + 2y = 6$
$x = 0$; $y = 3$
$y = 0$; $x = 2$

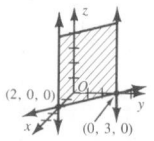

(2, 0, 0)   (0, 3, 0)

3. b. $z = 0$; $2x - 3y = 6$
$x = 0$; $-3y + z = 6$
$y = 0$; $2x + z = 6$

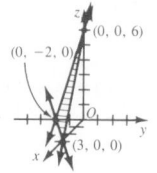

(0, 0, 6)
(0, -2, 0)
(3, 0, 0)

(continued on p. 192)

Mixed Review

1. If $\dfrac{x + y}{a - b} = \dfrac{2}{3}$, find the value

of $\dfrac{9x + 9y}{10a - 10b}$. $\dfrac{3}{5}$

2. Solve over $\mathcal{R}$.
$2p - (5p - 7) = 19$ $\{-4\}$

For each of the following equations of a plane state a system of equations whose solution set is the trace of the plane in the given coordinate plane.

$z = 0$, $4x + 2y = 12$ $y = 0$, $3x = 24$

27. $4x + 2y - 3z = 12$; xy-plane   **28.** $3x + 4y = 24$; xz-plane

29. $3y + 5z = 0$; yz-plane   **30.** $5x + 5y - 2z = 10$; xy-plane

31. $9x - 2y + 6z = 18$; xz-plane   **32.** $7x - z = 14$; yz-plane

33. What can you say about two planes that are the graphs of equations of the form

$$ax + by + cz = r \quad \text{and} \quad ax + by + cz = s \ (r \neq s)?$$ They are parallel.

29. $x = 0$, $3y + 5z = 0$ **30.** $z = 0$, $x + y = 2$

31. $y = 0$, $9x + 6z = 18$ **32.** $x = 0$, $z = -14$

Written Exercises

A **1–18.** For each of the equations in Oral Exercises 9–26:

    a. Sketch part of the graph of the equation.

    b. Give three systems of equations whose solution sets are the traces of the graph in each of the coordinate planes. If the equation has no trace in a coordinate plane, so state.

Graph part of the plane that has the given traces and that satisfies any additional condition given.

B **19.** xy-trace: $3x + 4y = 12$, $z = 0$;
yz-trace: $5y + 3z = 15$, $x = 0$;
xz-trace: $5x + 4z = 20$, $y = 0$

20. xy-trace: $6x + y = 6$; $z = 0$;
parallel to z-axis

21. xy-trace: $5x - y = 5$, $z = 0$;
yz-trace: $3y - 5z = -15$, $x = 0$;
xz-trace: $3x + z = 3$, $y = 0$

22. xz-trace: $5x + z = 10$;
parallel to y-axis

23. yz-trace: $z = 4$, $x = 0$;
parallel to xy-plane

24. xy-trace: $x + 2y = 8$, $z = 0$;
yz-trace: $y - 4z = 4$, $x = 0$;
xz-trace: $x - 8z = 8$, $y = 0$

25. xy-trace: $5x + 6y = 30$, $z = 0$;
yz-trace: $2y - 5z = 10$, $x = 0$;
xz-trace: $x - 3z = 6$, $y = 0$

26. xy-trace: $7x + 2y = 14$, $z = 0$;
yz-trace: $y = 7$, $x = 0$;
xz-trace: $x = 2$, $y = 0$

C **27.** Write a system of equations   **27.** $4x + y = 4$
for the xy-trace of the plane $z = 0$
with
yz-trace: $5y - 4z = 20$, $x = 0$ **28.** $x = -6$
xz-trace: $5x - z = 5$, $y = 0$ $y = 0$

28. Write a system of equations
for the xz-trace of the plane
with
xy-trace: $-3x + 2y = 18$
yz-trace: $y = 9$, $x = 0$

Give a single equation for each plane that satisfies the given conditions.

29. xy-trace: $2x - 3y = 6$, $z = 0$
yz-trace: $-2y + z = 4$, $x = 0$
$4x - 6y + 3z = 12$

30. yz-trace: $4y - 3z = 12$, $x = 0$
xz-trace: $2x + 5z = -20$, $y = 0$
$-6x + 20y - 15z = 60$

170   Chapter 5

The Bernoulli Family

17th–18th centuries

Bernoulli is one of the most famous names in the history of science and learning. Perhaps the most outstanding Bernoullis were the brothers Jacques (1654–1705) and Jean (1667–1748), rivals and critics of each other's work; their nephew Nicholas (1687–1759); and Jean's two sons, Nicholas II (1695–1726) and Daniel (1700–1782).

Jacques Bernoulli's most original and influential work, *Ars conjectandi* (*Art of conjecturing*), deals with probability, the theory of combinations, exponential series, and the expectation of profit to be made from certain games. Jean Bernoulli created a sensation in Paris scientific circles with his "golden theorem"—the determination of the radius of curvature of a curve.

Jean's sons, Daniel and Nicholas II, traveled throughout Europe and worked with another great Swiss mathematician, Leonhard Euler. Daniel's best-known work, *Hydrodynamica*, deals with the forces exerted by fluids.

Teaching Suggestions
p. T87

Related Activities p. T87

Supplementary Material
Test 14

Key Ideas

Find the solution set of systems of linear equations in three variables.

Chalkboard Examples

Determine the solution set.

$$
\begin{array}{ll}
x + 2y + z = 5 & (1) \\
2x - y + z = 4 & (2) \\
3x + y + 4z = 1 & (3)
\end{array}
$$

$$
\begin{array}{ll}
x + 2y + z = 5 & (1) \\
4x - 2y + 2z = 8 & (2) \times 2 \\
\hline
5x \quad + 3z = 13 & (4)
\end{array}
$$

$$
\begin{array}{ll}
2x - y + z = 4 & (2) \\
3x + y + 4z = 1 & (3) \\
\hline
5x \quad + 5z = 5 & (5)
\end{array}
$$

$$
\begin{array}{ll}
5x + 3z = 13 & (4) \\
5x + 5z = 5 & (5) \\
\hline
\quad -2z = 8 \\
\quad\quad z = -4 \\
\quad\quad x = 5
\end{array}
$$

Replace x with 5; z with -4 in any of the original equations,
$2(5) - y + (-4) = 4$
Solve for y.
$y = 2$
∴ the solution set is $\{(5, 2, -4)\}$.

5–3 Systems of Linear Equations in Three Variables

A *solution* of a system of linear equations in three variables over $\mathcal{R}$ is an ordered triple of real numbers that satisfies all equations of the system. The *solution set* of the system is the set of all its solutions.

The geometric interpretation of the solution of a system of two equations in two variables (Section 4-1) can be extended to systems of three equations in three variables.

A system such as

$$
\begin{array}{l}
x = 5 \\
y = 4 \\
z = 6
\end{array}
$$

can readily be solved, because you can see by inspection that its one and only solution is (5, 4, 6). Thus, the solution set is $\{(5, 4, 6)\}$. Figure 7 pictures this fact by showing $P(5, 4, 6)$ as the single point on the graphs of all three of the equations.

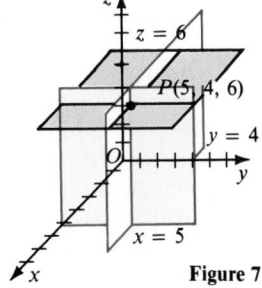

Figure 7

Graphs in Space; Determinants   **171**

Read aloud the definition of *consistent* as it applies to a system of three equations. Have the students describe in their own words how the three planes representing this system can be placed in order to represent a consistent system. Encourage the students to think of physical examples. For example, "Three planes are said to represent a consistent system if they intersect in only one point, or if they intersect in only one line (like a paddle wheel), or if two overlap and intersect a third in a single line, or if they all overlap."

Errors may occur when students solve systems in three variables because solving requires involved computations. Insist that students *organize* their written work carefully on the page, and step carefully through the computations. It should be clear with each step what method of solving is being used.

Note that the substitution method can be combined with the linear-combination method to solve systems in which one of the equations lacks a variable. Students sometimes make errors during the transformation of a system. To be sure to catch these errors, remind the students that when they check their work they should substitute their solutions back into the original system and not into the system after it has been transformed.

As Figure 7 suggests, two nonparallel planes intersect in a line, and three such planes *can* intersect in a single point. In that case, a system of three linear equations in three variables represented by the three planes is *consistent* and the solution is unique.

But such a system is also consistent if the graphs of the equations in such a system consist of one of the following:

1. three different planes intersecting in a single line (Figure 8),
2. two coincident planes intersecting a third plane in a line (Figure 9),
3. three coincident planes (Figure 10).

In each of these three cases the system of equations has an infinite set of solutions.

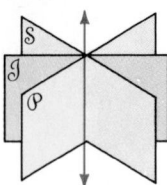 

      Figure 8   **Figure 9**   **Figure 10**

The solution set is the empty set $\emptyset$, and the system is *inconsistent*, if the graphs of the equations consist of:

1. three parallel planes (Figure 11),
2. two coincident planes parallel to a third plane (Figure 12),
3. three planes intersecting in three parallel lines (Figure 13),
4. two parallel planes intersecting a third plane in two parallel lines (Figure 14).

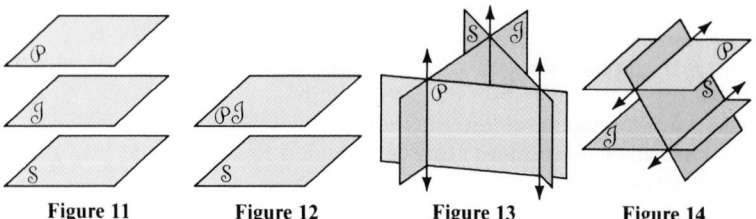

  Figure 11   **Figure 12**   **Figure 13**   **Figure 14**

Because it often is difficult to obtain accurate information about coordinates of points of intersection of space figures from flat drawings, you should not ordinarily attempt to solve systems of linear equations in three variables by graphing the equations. Instead, to solve such a system you transform it into an equivalent system which can be solved by inspection.

The transformations used in solving systems of linear equations in two variables (page 128) are also applicable in solving three-variable

172   *Chapter 5*

systems. In applying transformations, you can use the linear-combination method, the substitution method, or a combination of these. If the transformations yield a *false* statement such as $0 = 2$, then the system is inconsistent. If they do not yield a false statement but do yield a true statement such as $0 = 0$, then the solution set is an infinite set.

EXAMPLE   Determine the solution set of the system:

$$3x + 2y + z = 8 \quad (1)$$
$$2x - y + 2z = -7 \quad (2)$$
$$5x + 3y + 4z = 3 \quad (3)$$

SOLUTION

1. To obtain an equation in which the coefficient of y is 0, multiply each side of Equation (2) by 2 and add the resulting equation to Equation (1).

$$\begin{array}{rl} 3x + 2y + z = 8 & (1) \\ 4x - 2y + 4z = -14 & (2) \times 2 \\ \hline 7x + 5z = -6 & (4) \end{array}$$

2. To obtain a second equation in which the coefficient of y is 0, multiply both sides of Equation (2) by 3 and add the resulting equation to Equation (3).

$$\begin{array}{rl} 6x - 3y + 6z = -21 & (2) \times 3 \\ 5x + 3y + 4z = 3 & (3) \\ \hline 11x + 10z = -18 & (5) \end{array}$$

3. The equations obtained in Steps 1 and 2 involve only x and z. Use the method of transforming two equations in two variables to replace Equations (4) and (5) with an equivalent system that states the solution explicitly.

$$x = 2$$
$$z = -4$$

4. Replace x with 2 and z with -4 in Equation (2) and solve for y.

$$2x - y + 2z = -7 \quad (2)$$
$$2 \cdot 2 - y + 2(-4) = -7$$
$$y = 3$$

5. Thus the given system is equivalent to the system whose solution is $(2, 3, -4)$

$$x = 2$$
$$y = 3$$
$$z = -4$$

6. Checking in the given system:

$$3 \cdot 2 + 2 \cdot 3 + (-4) = 8 \quad \checkmark$$
$$2 \cdot 2 - 3 + 2(-4) = -7 \quad \checkmark$$
$$5 \cdot 2 + 3 \cdot 3 + 4(-4) = 3 \quad \checkmark$$

∴ solution set is $\{(2, 3, -4)\}$.

Additional A Exercises

Solve.

1. $3x + y = 5$
$x - z = 2$
$y + 2z = -1$
$\{(2, -1, 0)\}$

2. $2x - y - z = 1$
$x + 2y + z = 0$
$3x - y - 2z = -1$
$\{(1, -2, 3)\}$

3. $2x + y + z = 0$
$x - 2y + z = 2$
$3x + y + 2z = 2$
$\{(-2, 0, 4)\}$

4. $x + y + 2z = 1$
$-x - y + 3z = 4$
$2x - 3y - 2z = 11$
$\{(2, -3, 1)\}$

5. $3x - 2y + z = 9$
$x + 4y = -2$
$ - y + 4z = 5$
$\{(2, -1, 1)\}$

Suggested Assignments

Minimum
First day
174/1–22
Second day
R 176/*Self-Test 1*
S 135/21, 22, 26, 27
Extended Alg. with Trig.
174/5–23 odd
R 176/*Self-Test 1*
S 135/12–22 even
Enriched Alg.
174/7–19 odd, 22–26
R 176/*Self-Test 1*
S 135/20–26 even
Enriched Alg. with Trig.
174/7–19 odd, 20–26
R 176/*Self-Test 1*
S 135/20–26 even

Additional Answers
Oral Exercises

1. Add the first equation to two times the second equation; subtract the third equation from the first equation.

2. Add the three equations; subtract twice the second equation from the third equation.

3. Add six times the first equation to the second equation; add two times the third equation to the second equation.

4. Subtract the second equation from two times the first equation; subtract two times the third equation from five times the first equation.

5. Substitute $5 + z - x$ for y in the first and third equations.

6. Substitute $5 + 3y - z$ for x in the third equation.

7. Substitute $-\frac{3}{2}x - \frac{7}{2}$ for z in the first equation.

8. Substitute $2z - 3$ for x in the first and third equations.

Oral Exercises

Describe how you would use the linear-combination method to transform each system into two new equations in which the variable in red has been eliminated. Answers may vary.

1. x: $\quad 2x - y - z = 2$
$\quad\quad -x + 3y + z = -1$
$\quad\quad 2x - 4y - 3z = -5$

2. z: $\quad 2x + y + 6z = -8$
$\quad\quad x - 3y - 2z = 4$
$\quad\quad -3x + y - 4z = 2$

3. y: $\quad x - y + z = 3$
$\quad\quad 3x + 6y - 2z = 2$
$\quad\quad x - 3y + 3z = 9$

4. x: $\quad 2x + 3y + z = 11$
$\quad\quad 4x - y + 2z = 1$
$\quad\quad 5x + 4y - z = 3$

Describe how you would use the substitution method to transform each system into two equations in which the variable in red has been eliminated.

5. y: $\quad x - 2y = 3$
$\quad\quad x + y - z = 5$
$\quad\quad 3x + 4z = 1$

6. x: $\quad 2y + z = 1$
$\quad\quad x - 3y + z = 5$
$\quad\quad 4x - 3z = -5$

7. z: $x - 3y + z = 4$
$\quad\quad -3x - 2z = 7$
$\quad\quad 4x - 5y = -2$

8. x: $3x - y - 3z = 4$
$\quad\quad -x + 2z = 3$
$\quad\quad x + 5y - z = -4$

In Exercises 9–12:
a. **State whether each system of equations is consistent or inconsistent.**
b. **If the system is consistent, state whether there is a unique solution or an infinite set of solutions. Give reasons for your statements.**
c. **Describe the relationships between the graphs of the equations using Figures 7–14 on pages 171 and 172 as models.**

9. $x + 2y + z = 3$
$x + 2y + z = 6$
$\quad\quad\quad z = 0$

10. $x + y = 4$
$x - y = 2$
$\quad\quad z = 5$

11. $x + y = 0$
$x - y = 4$
$\quad\quad x = 2$

12. $\quad x + y - z = 2$
$2x + 2y - 2z = 7$
$3x + 3y - 3z = 9$

9. a. inconsistent
c. Figure 14

10. a. consistent
b. unique
c. Figure 7

11. a. consistent
b. infinite
c. Figure 8

a. inconsistent
c. Figure 11

Written Exercises
1. $\{(3, -1, 5)\}$ 2. $\{(4, 2, -3)\}$ 3. $\{(0, 2, 5)\}$ 4. $\{(-1, 3, 4)\}$

A   **1–8. Solve the systems in Oral Exercises 1–8.**   5. $\{(3, 0, -2)\}$ 6. $\left\{\left(\frac{7}{13}, -\frac{9}{13}, \frac{31}{13}\right)\right\}$
7. $\{(-3, -2, 1)\}$ 8. $\{(5, -1, 4)\}$ 9. $\{(7, -6, -9)\}$

In Exercises 9–18 each system has a single solution. Find the solution.

10. $\{(-1, 3, 8)\}$ 11. $\{(4, -7, 0)\}$

9. $2x - 5y + 4z = 8$
$\quad 2x + y + z = -1$
$\quad x - 3y + 2z = 7$

10. $5x - 3y + z = -6$
$\quad x + 3y - z = 0$
$\quad 4x + y + z = 7$

11. $\quad 3x + y = 5$
$\quad x + y + z = -3$
$\quad -y + 4z = 7$

12. $\quad 2x - z = -2$
$\quad 3x - y = 9$
$4x - 5y + 2z = 14$
$\{(5, 6, 12)\}$

13. $2x + 3y - z = 4$
$\quad x + 4y + 3z = 3$
$\quad 2y + z = 7$
$\left\{\left(-\frac{68}{9}, \frac{47}{9}, -\frac{31}{9}\right)\right\}$

14. $2x + 7y + z = 3$
$\quad -x + y - 3z = 7$
$\quad x + y + 2z = -3$
$\{(-10, 3, 2)\}$

174   *Chapter 5*

174

15. $x - y + 2z = 7$ $\left\{\left(\frac{1}{2}, -\frac{3}{2}, \frac{5}{2}\right)\right\}$
    $-2x + 4y + 4z = 3$
    $-x - 3y + 2z = 9$

16. $x - 3y + z = 1$ $\left\{\left(\frac{2}{3}, \frac{1}{3}, \frac{4}{3}\right)\right\}$
    $2x + 2y - 3z = -2$
    $5x + 2y = 4$

B   **17.** $2x - \frac{1}{3}y + z = 3$ $\{(8, 12, -9)\}$
    $-x + \frac{1}{6}y - \frac{2}{3}z = 0$
    $\frac{1}{4}x - \frac{1}{2}y - \frac{1}{3}z = -1$

18. $\frac{2}{3}x - y - z = 8$ $\{(18, -6, 10)\}$
    $\frac{1}{6}x + 2y + \frac{3}{2}z = 6$
    $-\frac{1}{2}x - \frac{1}{3}y + \frac{1}{2}z = -2$

In Exercises 19–22 tell whether each system is consistent or inconsistent. If the system is consistent, give the solution set.

EXAMPLE $x + y = 3$
         $x - y = 5$
         $x = 4$

SOLUTION   The system is consistent.
         The solution set
         is $\{(4, -1, z): z \in \mathcal{R}\}$.

19. $2x + y = 0$ consistent
    $x - 2y = 0$ $\{(0, 0, z): z \in \mathcal{R}\}$
    $y = 0$

20. $z = 0$ inconsistent
    $x - z = 0$
    $x = -2$

21. $x + y - z = 3$ consistent
    $x + y + z = 3$ $\{(x, y, 0): x + y = 3\}$
    $2x + 2y = 6$

22. $x + y = 4$ consistent
    $y + z = 4$ $\{(2, 2, 2)\}$
    $x + z = 4$

C   **23–26.** Sketch the graphs of the systems in Exercises 19–22.

3. parallel to yz-plane, $(-3, 0, 0)$
5. parallel to z-axis, $(20, 0, 0)$, $(0, -35, 0)$
7. $(2.8571, 0, 0)$, $(0, -2.5, 0)$, $(0, 0, 4)$

Computer Exercises   For students with computer experience

1. Write a program that will find the x-, y-, and z-intercepts of the plane defined by an equation of the form $ax + by + cz = d$. The input should give the values of a, b, c, and d. If the plane defined by the equation contains, or is parallel to, one of the coordinate axes or planes, the output should state this.

Run the program in Exercise 1 for the plane defined by each equation.

2. $z = 0$ contains xy-plane   **3.** $x = -3$ **4.** $9x - 12z = 0$ contains y-axis

5. $-7x + 4y = -140$ **6.** $-3x + 2y - 5z = 30$ **7.** $7x - 8y + 5z = 20$
                                 $(-10, 0, 0)$, $(0, 15, 0)$, $(0, 0, -6)$

8. Given the xy-trace in the form $ax + by = c$; $z = 0$; $a, b \neq 0$; and the yz-trace in the form $dy + ez = f$; $x = 0$; $d, e \neq 0$; write a program that will find the three intercepts of the plane and the equation of the xz-trace when the values of a, b, c, d, e, and f are input. Note that if the x- and z-intercepts are found, the xz-trace is the *line* in the xz-plane with these intercepts.

Run the program in Exercise 8 for each pair of traces.

9. $-6x + 15y = -30$; $z = 0$ **10.** $x - 2y = 7$; $z = 0$
    $3y - 4z = -6$; $x = 0$ $24y + 5z = -28$; $x = 0$ no solution
  xy-trace: $1.5x + 5z = 7.5$, $y = 0$; $(5, 0, 0)$, $(0, -2, 0)$, $(0, 0, 1.5)$

Graphs in Space; Determinants   **175**

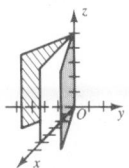
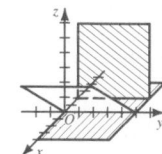
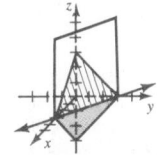
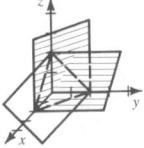
Mixed Review

1. Find the solution set of
    $|x + 7| < 2$.
    $\{x: -9 < x < -5\}$

2. Simplify.
    $\left(-\frac{1}{4}\right)\left(\frac{1}{-3}\right)\left(\frac{1}{-5}\right)(90)$ $-\frac{3}{2}$

3. True or false?
    -2 satisfies the equation
    $y(y - 5) + \dfrac{y - 10}{y - 1} = -9y$.
    T

4. Solve: $3x + 2y = 8$
             $2x - y = 3$
    $\{(2, 1)\}$

5. Solve by substitution.
    $3x + y = 1$
    $5x - 2y = 9$ $\{(1, -2)\}$

Quick Quiz

1. Sketch the coordinate box of (3, 4, 3).

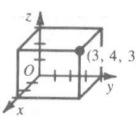

2. Determine the x-, y-, and z-intercepts and sketch the graph of $2x - 3y + 2z = 6$.

(3, 0, 0)
(0, −2, 0)
(0, 0, 3)

3. Give three systems of equations whose solution sets are traces in the three coordinate planes of the plane defined by $x - 8y - 2z = 8$.

$x = 0;\ 8y + 2z = -8$
$y = 0;\ x - 2z = 8$
$z = 0;\ x - 8y = 8$

4. Solve the system:
$5a - 5b + 2c = 13$
$2a - 4b + 3c = 8$
$3a + 2b - 4c = 2$
$\{(4, 3, 4)\}$

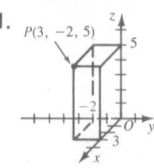
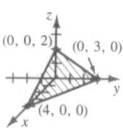

Self-Test 1

VOCABULARY

coordinate plane (p. 162)
xy-, yz-, xz-plane (p. 162)
octants (p. 162)
x-, y-, z-coordinate (p. 162)
coordinate box (p. 162)

linear equation in three variables (p. 165)
equation of a plane (p. 166)
x-, y-, z-intercept (p. 167)
trace of a plane (p. 167)

1. Sketch a diagram of a coordinate system in space and in the diagram show the coordinate box of (3, −2, 5).   *Obj. 1, p. 161*

2. Determine the x-, y-, and z-intercepts and sketch the graph of $3x + 4y + 6z = 12$.   *Obj. 2, p. 161*

3. Give three systems of equations whose solution sets are the traces in the three coordinate planes of the plane defined by $4x + 13y - 7z = 24$.

3. $4x + 13y = 24;\ z = 0$
$13y - 7z = 24;\ x = 0$
$4x - 7z = 24;\ y = 0$

4. Sketch part of the plane $3x + 8y + 12z = 24$ that lies in the first octant.   *Obj. 3, p. 161*

5. Solve the system: $3x - y - 2z = -1\ \{(2, 1, 3)\}$ *Obj. 4, p. 161*
$x + y + z = 6$
$2x + 3y - z = 4$

Check your answers with those at the back of the book.

Determinants

OBJECTIVES for Sections 5-4 through 5-6:
1. *To use determinants to solve a system of three linear equations in three variables.*
2. *To use three variables to solve problems.*
3. *To use properties of determinants to simplify the expansion of a determinant by minors.*

5–4 Third-order Determinants

The determinants introduced in Section 4-3, such as $\begin{vmatrix} a_1 & b_1 \\ a_2 & b_2 \end{vmatrix}$, have two (horizontal) rows, namely $a_1\quad b_1$ and $a_2\quad b_2$, and two (vertical) columns,

176   *Chapter 5*

namely $\dfrac{a_1}{a_2}$ and $\dfrac{b_1}{b_2}$; such a determinant is called a *second-order determinant*, or a determinant of **order 2**.

You can use *third-order determinants* in the solution of three linear equations in three variables over $\mathcal{R}$.

For any $a_1, b_1, c_1, a_2, b_2, c_2, a_3, b_3, c_3 \in \mathcal{R}$, the determinant

$$D = \begin{vmatrix} a_1 & b_1 & c_1 \\ a_2 & b_2 & c_2 \\ a_3 & b_3 & c_3 \end{vmatrix}$$

has the value

$$a_1b_2c_3 + a_2b_3c_1 + a_3b_1c_2 - a_1b_3c_2 - a_2b_1c_3 - a_3b_2c_1.$$

A convenient way to remember how to evaluate a *third-order* determinant is to copy the 3×3 (read "three by three") array and repeat the first two columns after the third column as shown at the right. Compute the product of the entries along each diagonal arrow as shown below. Add the products found from the descending arrows to the opposites of the products found from the ascending arrows. (This does not work for higher-order determinants.)

$$\begin{array}{ccc|cc} a_1 & b_1 & c_1 & a_1 & b_1 \\ a_2 & b_2 & c_2 & a_2 & b_2 \\ a_3 & b_3 & c_3 & a_3 & b_3 \end{array}$$

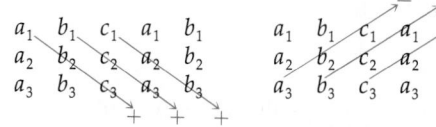

$$a_1b_2c_3 + b_1c_2a_3 + c_1a_2b_3 - a_3b_2c_1 - b_3c_2a_1 - c_3a_2b_1$$
$$= a_1b_2c_3 + a_2b_3c_1 + a_3b_1c_2 - a_1b_3c_2 - a_2b_1c_3 - a_3b_2c_1$$

EXAMPLE 1   Evaluate the determinant:
$$D = \begin{vmatrix} 4 & 2 & -1 \\ 3 & 5 & -3 \\ -2 & 0 & 1 \end{vmatrix}$$

SOLUTION

$$\begin{array}{ccc|cc} 4 & 2 & -1 & 4 & 2 \\ 3 & 5 & -3 & 3 & 5 \\ -2 & 0 & 1 & -2 & 0 \end{array}$$

$$D = 20 + 12 + 0 - 10 - 0 - 6 = 16$$

$$\therefore D = 16$$

Graphs in Space; Determinants   **177**

Teaching Suggestions
p. T88

Related Activities
p. T88

Key Ideas

Evaluate third-order determinants.
Use Cramer's Rule to solve systems of three linear equations in three variables.

Chalkboard Examples

1. Evaluate the determinant:

$$D = \begin{vmatrix} 3 & 1 & 2 \\ 4 & 3 & 0 \\ -1 & 3 & -4 \end{vmatrix}$$

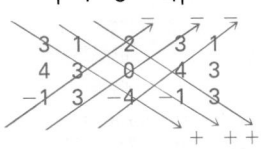

$$D = -36 + 0 + 24 - (-6) - 0 - (-16) = 10$$

2. Use Cramer's Rule to solve the system:
$$2x + y - z = 3$$
$$4x - y + 4z = 0$$
$$-3y + 2z = 6$$

$$D = \begin{vmatrix} 2 & 1 & -1 \\ 4 & -1 & 4 \\ 0 & -3 & 2 \end{vmatrix}$$
$$= -4 + 0 + 12 - 0 - (-24) - 8 = 24$$

$$D_x = \begin{vmatrix} 3 & 1 & -1 \\ 0 & -1 & 4 \\ 6 & -3 & 2 \end{vmatrix}$$
$$= -6 + 24 + 0 - 6 - (-36) - 0 = 48$$

(continued)

$$D_y = \begin{vmatrix} 2 & 3 & -1 \\ 4 & 0 & 4 \\ 0 & 6 & 2 \end{vmatrix}$$

$$= 0 + 0 - 24 - 0 -$$
$$48 - 24 = -96$$

$$D_z = \begin{vmatrix} 2 & 1 & 3 \\ 4 & -1 & 0 \\ 0 & -3 & 6 \end{vmatrix}$$

$$= -12 + 0 + (-36)$$
$$- 0 - 0 - 24 = -72$$

$$x = \frac{D_x}{D} = \frac{48}{24} = 2$$

$$y = \frac{D_y}{D} = \frac{-96}{24} = -4$$

$$z = \frac{D_z}{D} = \frac{-72}{24} = -3$$

$$\{(2, -4, -3)\}$$

Additional A Exercises

Evaluate each determinant.

1. $\begin{vmatrix} 2 & 1 & 1 \\ 2 & 1 & 4 \\ 1 & 5 & 3 \end{vmatrix}$ -27

2. $\begin{vmatrix} 3 & -1 & 4 \\ 0 & 2 & -5 \\ 2 & 0 & 2 \end{vmatrix}$ 6

3. $\begin{vmatrix} 4 & 0 & 5 \\ -3 & 5 & 0 \\ 0 & -2 & 1 \end{vmatrix}$ 50

Use Cramer's Rule to solve
each system.

4. $2x + y - 2z = 3$
$\quad x - y + z = 0$
$\quad x - 2y + z = -1$
$\quad \{(1, 1, 0)\}$

5. $x - 3y + z = 5$
$\quad\quad\quad y - z = 2$
$\quad x - y \quad\quad = 6$
$\quad \{(5, -1, -3)\}$

6. $3x + y + 2z = 4$
$\quad -x + 2y + z = 3$
$\quad\quad\quad y + z = 1$
$\quad \{(2, 4, -3)\}$

You could use transformations, as in Section 5-3, to solve the system

$$a_1x + b_1y + c_1z = d_1$$
$$a_2x + b_2y + c_2z = d_2$$
$$a_3x + b_3y + c_3z = d_3$$

over $\mathcal{R}$, and let

$$D = \begin{vmatrix} a_1 & b_1 & c_1 \\ a_2 & b_2 & c_2 \\ a_3 & b_3 & c_3 \end{vmatrix}, \quad D_x = \begin{vmatrix} d_1 & b_1 & c_1 \\ d_2 & b_2 & c_2 \\ d_3 & b_3 & c_3 \end{vmatrix},$$

$$D_y = \begin{vmatrix} a_1 & d_1 & c_1 \\ a_2 & d_2 & c_2 \\ a_3 & d_3 & c_3 \end{vmatrix}, \quad D_z = \begin{vmatrix} a_1 & b_1 & d_1 \\ a_2 & b_2 & d_2 \\ a_3 & b_3 & d_3 \end{vmatrix},$$

to find that, if $D \neq 0$, the system has the unique solution

$$x = \frac{D_x}{D}, \quad y = \frac{D_y}{D}, \quad z = \frac{D_z}{D}.$$

These equations are called **Cramer's Rule** for systems of three linear equations in three variables. The determinant D is the **determinant of coefficients**. If $D = 0$, then either the system is inconsistent or the system has an infinite solution set.

EXAMPLE 2 Use Cramer's Rule to solve the system:

$$3x + 2y - z = 5$$
$$4x + y - 2z = 8$$
$$-3y + z = 9$$

SOLUTION

$$D = \begin{vmatrix} 3 & 2 & -1 \\ 4 & 1 & -2 \\ 0 & -3 & 1 \end{vmatrix} \begin{matrix} 3 & 2 \\ 4 & 1 \\ 0 & -3 \end{matrix} = 3 + 0 + 12 - 0 - 18 - 8$$

$$= -11$$

$$D_x = \begin{vmatrix} 5 & 2 & -1 \\ 8 & 1 & -2 \\ 9 & -3 & 1 \end{vmatrix} \begin{matrix} 5 & 2 \\ 8 & 1 \\ 9 & -3 \end{matrix} = 5 + (-36) + 24 - (-9) - 30 - 16$$

$$= -44$$

$$D_y = \begin{vmatrix} 3 & 5 & -1 \\ 4 & 8 & -2 \\ 0 & 9 & 1 \end{vmatrix} \begin{matrix} 3 & 5 \\ 4 & 8 \\ 0 & 9 \end{matrix} = 24 + 0 + (-36) - 0 - (-54) - 20$$

$$= 22$$

$$D_z = \begin{vmatrix} 3 & 2 & 5 \\ 4 & 1 & 8 \\ 0 & -3 & 9 \end{vmatrix} \begin{matrix} 3 & 2 \\ 4 & 1 \\ 0 & -3 \end{matrix} = 27 + 0 + (-60) - 0 - (-72) - 72$$

$$= -33$$

Thus by substitution

$$x = \frac{D_x}{D} = \frac{-44}{-11} = 4, \qquad y = \frac{D_y}{D} = \frac{22}{-11} = -2, \qquad z = \frac{D_z}{D} = \frac{-33}{-11} = 3.$$

The check is left to you.

$\therefore$ the solution set is $\{(4, -2, 3)\}$.

Written Exercises

Evaluate each determinant.

A **1.** $\begin{vmatrix} 5 & 3 & -1 \\ 2 & 1 & 2 \\ -2 & 0 & 1 \end{vmatrix}$ -15 **2.** $\begin{vmatrix} 2 & -1 & 3 \\ 1 & -2 & 0 \\ 1 & 3 & 4 \end{vmatrix}$ 3 **3.** $\begin{vmatrix} 5 & 3 & 2 \\ -2 & 1 & -3 \\ 3 & 4 & 2 \end{vmatrix}$ 33 **4.** $\begin{vmatrix} 5 & -3 & 1 \\ 0 & 2 & 4 \\ 6 & 0 & 3 \end{vmatrix}$
-54

Use Cramer's Rule to solve each system.

5. $\begin{aligned} x - 3y - z &= 1 \\ 2x + y + z &= 3 \\ x + 2y + 2z &= 0 \end{aligned}$ $\{(2, 1, -2)\}$

6. $\begin{aligned} x - y + 2z &= 1 \\ -x + 3y + z &= 5 \\ -2x + y - z &= 4 \end{aligned}$ $\{(-3, 0, 2)\}$

7. $\begin{aligned} 2x + z &= 1 \\ 3y + 2z &= -2 \\ -3x + y - z &= -3 \end{aligned}$ $\{(-2, -4, 5)\}$

8. $\begin{aligned} 2x - 4y - z &= 5 \\ 3x - 2z &= -1 \\ x + y - z &= -3 \end{aligned}$ $\{(3, -1, 5)\}$

9. $\begin{aligned} x - 3y &= 1 \\ 2y - z &= 3 \\ -x + 4y + z &= -1 \end{aligned}$ $\{(4, 1, -1)\}$

10. $\begin{aligned} 2x - 3z &= 3 \\ x - 2y &= -4 \\ 3y - 2z &= 9 \end{aligned}$ $\{(6, 5, 3)\}$

In Exercises 11–14, evaluate D. If $D = 0$, state whether the system is inconsistent or has an infinite solution set. If the system has a single solution, find the solution.

$D = 0$; infinite solution set

B **11.** $\begin{aligned} x + y &= 0 \\ x - y &= 0 \\ x - 2y &= 0 \end{aligned}$ $D = 0$; infinite solution set

12. $\begin{aligned} x + y - z &= 0 \\ 2x + 4y - z &= 0 \\ -x - 5y - z &= 0 \end{aligned}$

13. $\begin{aligned} y + z &= 4 \\ y - z &= -4 \\ x &= 3 \end{aligned}$ $D = -2;\ \{(3, 0, 4)\}$

14. $\begin{aligned} x - 2y + z &= 3 \\ 2x - 4y + 2z &= 9 \\ x &= 4 \end{aligned}$ $D = 0$, inconsistent

C **15.** Prove that for all real numbers a, b, c, d, e, f, and k,

$$\begin{vmatrix} a & b & c \\ ka & kb & kc \\ d & e & f \end{vmatrix} = 0.$$

16. Prove that for all real numbers a, b, c, d, e, and f,

$$\begin{vmatrix} a & b & c \\ d & e & f \\ a + d & b + e & c + f \end{vmatrix} = 0.$$

Suggested Assignments

Minimum
First day
179/1–14
Second day
142/2–16 even
Extended Alg. with Trig.
179/1–13 odd
Enriched Alg.
179/1–15 odd
Enriched Alg. with Trig.
179/1–15 odd

Additional Answers
Written Exercises

15. $\begin{vmatrix} a & b & c \\ ka & kb & kc \\ d & e & f \end{vmatrix} =$

$kabf + kbcd + kace$
$- kbcd - kace - kabf = 0$

16. $\begin{vmatrix} a & b & c \\ d & e & f \\ a + d & b + e & c + f \end{vmatrix} =$

$ae(c + f) + bf(a + d) +$
$cd(b + e) - ce(a + d) -$
$af(b + e) - bd(c + f) =$
$aec + aef + bfa + bfd +$
$cdb + cde - cea - ced -$
$afb - afe - bcd - bdf =$
0

Mixed Review

1. Solve for k: $\begin{vmatrix} 4k & 3 \\ k & 2 \end{vmatrix} = -15$

$\{-3\}$

2. A rectangle with a perimeter of 24 cm is twice as long as it is wide. Find its dimensions.
4 cm × 8 cm

3. Find an equation of the line with y-intercept 0 and having the same slope as the line through $(-2, -5)$ and $(0, 2)$. $y = \frac{7}{2}x$

4. If $f(x) = 3x$ and $g(x) = x^2 - 1$, find $g(f(2))$. 35

Chalkboard Examples

1. In a coin bank there are three times as many dimes as there are nickels and quarters combined. The total value of the 24 coins (all nickels, dimes, and quarters) is $2.90. How many of each kind of coin are there?

1. The problem asks for the number of each kind of coin.

2. Let x = no. of nickels
y = no. of dimes
z = no. of quarters

3. There are 24 coins.
$x + y + z = 24$
There are three times as many dimes as there are nickels and quarters combined.
$y = 3(x + z)$
The total value of the coins is $2.90.
$5x + 10y + 25z = 290$

4. Transform the following equations and solve.
$x + y + z = 24$
$-3x + y - 3z = 0$
$5x + 10y + 25z = 290$
$z = 4$
$x = 2$
$y = 18$
∴ there are 2 nickels, 18 dimes, and 4 quarters.

5. Check.

5–5 Solving Problems with Three Variables

Systems of linear equations in three variables sometimes represent the conditions of practical problems.

EXAMPLE Soap comes in packages of one, two, or four bars, costing 40¢, 60¢, and 90¢ per package, respectively. In 1 h a store sold 50 bars of soap in 22 packages and took in $13.60. How many packages of each size were sold?

SOLUTION 1. The problem asks for the number of packages of each size.

2. Let x be the number of forty-cent packages sold, let y be the number of sixty-cent packages sold, and let z be the number of ninety-cent packages sold.

3. The total number of packages sold is 22.
$$x + y + z \quad = \quad 22$$

The total number of bars of soap is 50.
$$x + 2y + 4z \quad = \quad 50$$

The total value of the packages is $13.60.
$$40x + 60y + 90z \quad = \quad 1360$$

4. Transforming the equations
$$x + y + z = 22$$
$$x + 2y + 4z = 50$$
$$40x + 60y + 90z = 1360$$

into an equivalent system, or solving by Cramer's Rule is left to you.

5. The check to show that $x = 10$, $y = 4$, and $z = 8$ is also left to you.

∴ there were ten forty-cent packages, four sixty-cent packages, and eight ninety-cent packages sold.

Oral Exercises

In Oral Exercises 1–6, let n represent the number of nickels, d represent the number of dimes, and q represent the number of quarters in a certain parking meter. Translate each sentence into an equation in n, d, and q.

1. There is a total of 38 coins. $n + d + q = 38$

2. The total value of the coins is $3.70. $5n + 10d + 25q = 370$

3. There are twice as many dimes as quarters. $d = 2q$

4. The total number of coins is 69. $n + d + q = 69$

5. There are 3 more nickels than dimes and quarters combined. $n = d + q + 3$

6. The coins have a total value of $7.80. $5n + 10d + 25q = 780$

Problems

A 1. Use the relationships stated in Oral Exercises 1–3 to find the number of nickels, dimes, and quarters in the parking meter.

2. Use the relationships stated in Oral Exercises 4–6 to find the number of nickels, dimes, and quarters in the parking meter.

3. A rectangular box has a base that has a perimeter of 30 cm, a side A that has a perimeter of 24 cm, and a side B that has a perimeter of 26 cm. Find the dimensions of the box. 8 cm by 7 cm by 5 cm

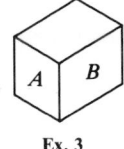
Ex. 3

4. The mass of a single sheet of 3 types of paper needs to be determined using only a 20-g weight, a 25-g weight, and a balance. It is found that 1 sheet of airmail paper and 3 sheets of typing paper balance with 25 g. One sheet of typing paper and 2 sheets of construction paper also balance with 25 g. Three sheets of paper, one of each type, balance with 20 g. What is the mass of each type of paper? airmail, 4 g; typing, 7 g; construction, 9 g

5. A breakfast consisting of a glass of milk, a serving of shredded wheat, and an egg supplies 17 g of protein. Three eggs and a glass of milk supply 29 g of protein, while a serving of shredded wheat and two glasses of milk supply 18 g of protein. How much protein is supplied by one glass of milk, one egg, and one serving of shredded wheat if each is consumed by itself. milk, 8 g; shredded wheat, 2 g; egg, 7 g

6. A combination of tiles in the shapes shown will be used for a mosaic. The shapes chosen contain a total of 40 tiles, 72 diagonals, and 153 sides. How many triangles, squares, and pentagons will be used? 13 triangles, 21 squares, 6 pentagons

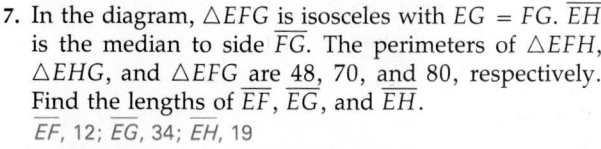
Ex. 6

7. In the diagram, $\triangle EFG$ is isosceles with $EG = FG$. $\overline{EH}$ is the median to side $\overline{FG}$. The perimeters of $\triangle EFH$, $\triangle EHG$, and $\triangle EFG$ are 48, 70, and 80, respectively. Find the lengths of $\overline{EF}$, $\overline{EG}$, and $\overline{EH}$.
$\overline{EF}$, 12; $\overline{EG}$, 34; $\overline{EH}$, 19

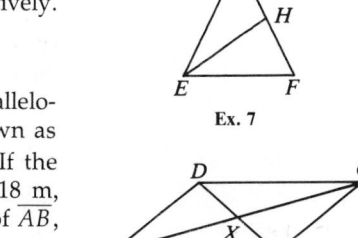
Ex. 7

8. A bridge support module is in the shape of a parallelogram. It has three sections of equal length shown as $\overline{AD}$, $\overline{DB}$, and $\overline{BC}$ in the diagram at the right. If the perimeters of $\triangle ABD$, $\triangle ABC$, and $\triangle ABX$ are 18 m, 20 m, and 12 m, respectively, find the lengths of $\overline{AB}$, $\overline{AD}$, and $\overline{AC}$.
$\overline{AB}$, 4 m; $\overline{AD}$, 7 m; $\overline{AC}$, 9 m

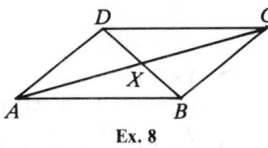
Ex. 8

B 9. Dick Ross rode his bike uphill for $\frac{3}{4}$ h and on level ground for $\frac{1}{2}$ h and traveled a distance of 22 km. On the trip home he traveled the same distance and spent $\frac{1}{2}$ h going downhill and $\frac{1}{2}$ h on level ground. If his speed on level ground is the average of his uphill and downhill speeds, what are these three speeds?
uphill, 16 km/h; downhill, 24 km/h; level, 20 km/h

Graphs in Space; Determinants **181**

Suggested Assignments

Minimum
   181/*P*: 1–9
Extended Alg. with Trig.
   181/*P*: 1–11
Enriched Alg.
   181/*P*: 1–13
Enriched Alg. with Trig.
   181/*P*: 1–13

Additional Answers
Problems

1. 20 nickels, 12 dimes, 6 quarters

2. 36 nickels, 15 dimes, 18 quarters

11. 2-packs, $0.84; 4-packs, $1.44; 5-packs, $2.40

12. single adult, 16 tickets; adult and child, 14 tickets; two adults and two children, 24 tickets

13. 30 km/h; 80 km/h

Mixed Review

1. State the coordinates of the *x*-intercept for $2x - y + 3z = 6$. (3, 0, 0)

2. Solve:
$2y - 5x = 10$
$10x - 4y = 6$ $\varnothing$

3. Simplify:
$\dfrac{|14 - 3| - |7 - 16|}{3|(-2) + 1|}$ $\dfrac{2}{3}$

4. Solve mentally:
$7a - (a + 11) = 7$ {3}

5. If $f(x) = -3x^2 - 2x + 5$, then find $f(-2)$. -3

6. Find the slope of the line passing through the points (a, a) and $(2a, -a)$. -2

1. The width of a rectangular box is twice the height. Twice the length of the box exceeds the sum of the width and the height by 5. Given that the sum of the length, width, and height is 16 cm, find the length, width, and height of the box. $l = 7$, $w = 6$, $h = 3$

2. If the sum of the digits of a certain three-digit number is doubled, the result is 16 more than 3 times the tens digit. The units digit is one more than twice the hundreds digit, and the hundreds digit is 11 decreased by twice the tens digit. Find the three-digit number. 347

3. The sum of three numbers is 42. The third number is twice the sum of the first two and is 4 more than three times their difference. What are the numbers? 3, 11, 28

4. Ron, Hank, and Louise have a total of 52 tropical fish. If Louise had 6 more fish she would have as many as the boys have put together. Hank has three more fish than Ron. How many fish does each of the three have? Ron, 13; Hank, 16; Louise, 23

5. The proceeds from the school car wash in one, five, and ten dollar bills come to a total of $355. If there are 120 bills altogether, and twice as many one dollar bills as five and ten dollar bills combined, how many bills of each denomination were there? 80 ones, 25 fives, 15 tens

10. The height above the ground, h, that an object reaches when it is thrown upward is given by the equation $h = -kt^2 + v_0 t + y_0$ where v_0 is the initial velocity in meters per second, y_0 is the initial height in meters and t is the time in seconds. If a ball reaches a height of 90 m in 1 s, 80 m in 2 s and hits the ground in 4 s, what was the initial velocity at which the ball was thrown. 20 m/s

C 11. A company produces 30 ballpoint pens in 1 min. If half this output is packaged in two-packs and the other half of this output is packaged in four-packs the company makes a profit of $11.70. If half this output is packaged in two-packs and the other half of this output is packaged in five-packs the company makes a profit of $13.50. If this output is packaged equally in two-packs, four-packs, and five-packs, the profit is $12.60. What is the company's profit for two-packs, four-packs, and five-packs?

12. A certain airline sells three types of tickets for a certain route: a single adult ticket costs $40, a ticket for one adult traveling with one child costs $60, and a ticket for two adults traveling with two children costs $100. The airline collected $3880 in fares from 140 passengers, 78 of whom were adults. How many tickets of each type were sold?

13. It is 48 km from A to B and 72 km from B to C. A train has to complete the trip from A to C in 3 h. When the train travels at one speed from A to B and then a higher speed from B to C, it can stop at B. When the whole trip is made at the higher speed the train can stop 3 times as long at B. When the train travels the higher speed from A to B and the lower speed from B to C, no stops can be made. The higher speed is $2\frac{2}{3}$ times as fast as the lower speed. What are the two speeds?

ON THE CALCULATOR

You can solve systems of three linear equations in three variables by using Cramer's Rule and a calculator. If the calculator has a memory and operates algebraically (or has parentheses), then the solution of the system is made easier.

Use the calculator to determine the value of D for the system and store it in the memory. Then evaluate D_x, D_y, and D_z and divide each by D to determine the values of x, y, and z respectively.

Exercises

Use a calculator to evaluate each determinant.

1. $\begin{vmatrix} 3 & -2 & 1.4 \\ 2.1 & 5 & 4.2 \\ -4 & 6.1 & 1 \end{vmatrix}$ 21.874

2. $\begin{vmatrix} 15 & -12 & 6 \\ -6 & 3 & 12 \\ 2 & -1 & -4 \end{vmatrix}$ 0

Use a calculator and Cramer's Rule to solve each system.

3. $5x - 8y - 10z = 87$
$1.4x + 5y + 4z = -3$
$8x - 3.2y - 20z = 86$
$\{(12, -5, 1.3)\}$

4. $\frac{2}{5}x + 3y - 4z = 17$

$5x + \frac{1}{8}y + 7z = 6$

$\frac{1}{2}x - y - \frac{3}{4}z = -2$

$\left\{ \left(\frac{5}{2}, 4, -1 \right) \right\}$

5–6 Properties of Determinants

In the defining equations $\begin{vmatrix} a_1 & b_1 \\ a_2 & b_2 \end{vmatrix} = a_1 b_2 - a_2 b_1$ and

$$\begin{vmatrix} a_1 & b_1 & c_1 \\ a_2 & b_2 & c_2 \\ a_3 & b_3 & c_3 \end{vmatrix} = a_1 b_2 c_3 + a_2 b_3 c_1 + a_3 b_1 c_2 - a_1 b_3 c_2 - a_2 b_1 c_3 - a_3 b_2 c_1$$

the sums of products in the right-hand sides are called *expansions* of the determinants. Notice that the first of these expansions shows all possible arrangements of the subscripts 1 and 2 of the letters a and b, and that the second expansion shows all possible arrangements of the subscripts 1, 2, and 3 of the letters a, b, and c.

The fact that a determinant of order 2 or 3 is a sum of terms involving all possible arrangements of the subscripts of the elements suggests a similar definition for determinants of order n when $n > 3$. A determinant that has n rows and n columns could be defined as a sum of terms involving all possible arrangements of the subscripts 1, 2, 3, . . . , n. However, since there are no simple arrow diagrams to help us make the rather lengthy computations (120 terms, for instance, in the sum for a determinant of order 5, and 720 for one of order 6), we give an alternative (but equivalent) definition of a determinant of higher order in terms of determinants of next lower order by using *minors*. The **minor** of an element in a determinant is the determinant resulting from the deletion of the row and column containing the element.

Graphs in Space; Determinants **183**

Teaching Suggestions
p. T88

Supplementary Material
Test 15

Key Ideas

Use minors to expand third-order determinants.
Extend to the expansion of fourth-order determinants by using minors.
Introduce and use properties of determinants.

1. Expand by minors of the second row and evaluate.

$$\begin{vmatrix} 5 & -1 & -2 \\ 3 & 6 & -7 \\ 2 & -3 & 4 \end{vmatrix}$$

$$\begin{vmatrix} 5 & -1 & -2 \\ ③ & ⑥ & ⑦ \\ 2 & -3 & 4 \end{vmatrix}$$

$$= -3\begin{vmatrix} -1 & -2 \\ -3 & 4 \end{vmatrix} + 6\begin{vmatrix} 5 & -2 \\ 2 & 4 \end{vmatrix} -$$

$$(-7)\begin{vmatrix} 5 & -1 \\ 2 & -3 \end{vmatrix}$$

$$= (-3)(-10) + 6(24) +$$
$$7(-13) = 83$$

2. Expand by minors and evaluate.

$$\begin{vmatrix} 4 & -1 & 1 & 2 \\ 3 & -1 & 0 & 2 \\ 0 & 4 & 0 & 1 \\ 3 & 1 & 1 & 2 \end{vmatrix}$$

Expand by minors of the third column since two elements are zero.

$$[3(8-1) + 3(-1-8)] -$$
$$[4(-1-8) - 3(-1-8)]$$
$$= -6 - (-9)$$
$$= 3$$

For example, in the determinant

$$\begin{vmatrix} 4 & 3 & -9 \\ 2 & 5 & -2 \\ 7 & 8 & 0 \end{vmatrix}$$

the minor of 4 is $\begin{vmatrix} 4 & 3 & -9 \\ 2 & 5 & -2 \\ 7 & 8 & 0 \end{vmatrix}$, or $\begin{vmatrix} 5 & -2 \\ 8 & 0 \end{vmatrix}$.

Similarly, the minor of 2 is $\begin{vmatrix} 3 & -9 \\ 8 & 0 \end{vmatrix}$ and the minor of 5 is $\begin{vmatrix} 4 & -9 \\ 7 & 0 \end{vmatrix}$.

If you rewrite the right-hand side of the expression defining a third-order determinant on page 177 as

$$a_1b_2c_3 - a_1b_3c_2 - (a_2b_1c_3 + a_2b_3c_1) + (a_3b_1c_2 - a_3b_2c_1),$$

you can factor it to obtain

$$a_1(b_2c_3 - b_3c_2) - a_2(b_1c_3 - b_3c_1) + a_3(b_1c_2 - b_2c_1).$$

It follows that if you let A_1, A_2, and A_3 represent the minors of a_1, a_2, and a_3, respectively, you can then write

$$\begin{vmatrix} a_1 & b_1 & c_1 \\ a_2 & b_2 & c_2 \\ a_3 & b_3 & c_3 \end{vmatrix} = a_1A_1 - a_2A_2 + a_3A_3.$$

The right-hand side of this equation is called the *expansion* of the determinant by minors of the elements in the first column.

By suitably arranging the terms in the definition of a third-order determinant, you can show that such a determinant can be expanded by minors about any row or any column as follows:

1. Choose a row or column and form the product of each element in the row or column with its minor.
2. Use the product obtained or its negative according as the sum of the number of the row and number of the column containing the element is even or odd.
3. The sum of the resulting numbers is the value of the determinant.

EXAMPLE 1   Expand by minors of the second row and evaluate. $\begin{vmatrix} -4 & 1 & 6 \\ 7 & 2 & -5 \\ 5 & 3 & -1 \end{vmatrix}$

SOLUTION   The elements of the second row are 7, 2, and −5. The element 7 is in the second row, first column and since $2 + 1 = 3$ which is odd, use the negative of its product with its minor. Similarly, use the product of 2 and

184   *Chapter 5*

its minor and the negative of the product of -5 and its minor. Thus

$$\begin{vmatrix} -4 & 1 & 6 \\ 7 & 2 & -5 \\ 5 & 3 & -1 \end{vmatrix} = -7 \begin{vmatrix} 1 & 6 \\ 3 & -1 \end{vmatrix} + 2 \begin{vmatrix} -4 & 6 \\ 5 & -1 \end{vmatrix} - (-5) \begin{vmatrix} -4 & 1 \\ 5 & 3 \end{vmatrix}$$

$$= (-7)(-19) + 2(-26) - (-5)(-17) = -4.$$

We can extend this idea to fourth-order (or higher-order) determinants. For example,

$$\begin{vmatrix} a_1 & b_1 & c_1 & d_1 \\ a_2 & b_2 & c_2 & d_2 \\ a_3 & b_3 & c_3 & d_3 \\ a_4 & b_4 & c_4 & d_4 \end{vmatrix} = a_1 A_1 - a_2 A_2 + a_3 A_3 - a_4 A_4,$$

where

$$A_1 = \begin{vmatrix} b_2 & c_2 & d_2 \\ b_3 & c_3 & d_3 \\ b_4 & c_4 & d_4 \end{vmatrix}, \quad A_2 = \begin{vmatrix} b_1 & c_1 & d_1 \\ b_3 & c_3 & d_3 \\ b_4 & c_4 & d_4 \end{vmatrix}, \quad \text{etc.}$$

Cramer's Rule can be extended to find solutions of four linear equations in four variables. Thus, the solution of

$$a_1 x + b_1 y + c_1 z + d_1 w = e_1$$
$$a_2 x + b_2 y + c_2 z + d_2 w = e_2$$
$$a_3 x + b_3 y + c_3 z + d_3 w = e_3$$
$$a_4 x + b_4 y + c_4 z + d_4 w = e_4$$

is $\quad x = \dfrac{D_x}{D}, \quad y = \dfrac{D_y}{D}, \quad z = \dfrac{D_z}{D}, \quad w = \dfrac{D_w}{D}, \quad$ *provided* $D \neq 0$ where

$$D = \begin{vmatrix} a_1 & b_1 & c_1 & d_1 \\ a_2 & b_2 & c_2 & d_2 \\ a_3 & b_3 & c_3 & d_3 \\ a_4 & b_4 & c_4 & d_4 \end{vmatrix}, \quad D_x = \begin{vmatrix} e_1 & b_1 & c_1 & d_1 \\ e_2 & b_2 & c_2 & d_2 \\ e_3 & b_3 & c_3 & d_3 \\ e_4 & b_4 & c_4 & d_4 \end{vmatrix}, \quad D_y = \begin{vmatrix} a_1 & e_1 & c_1 & d_1 \\ a_2 & e_2 & c_2 & d_2 \\ a_3 & e_3 & c_3 & d_3 \\ a_4 & e_4 & c_4 & d_4 \end{vmatrix}, \quad \text{etc.}$$

EXAMPLE 2   Expand by minors and evaluate. $\begin{vmatrix} 3 & 1 & -1 & 2 \\ 2 & 4 & 0 & 3 \\ 1 & -1 & 0 & -1 \\ 0 & 1 & 1 & 2 \end{vmatrix}$

SOLUTION   Expand by minors of the third column since two elements are 0 and their minors do not have to be evaluated.

$$-1 \begin{vmatrix} 2 & 4 & 3 \\ 1 & -1 & -1 \\ 0 & 1 & 2 \end{vmatrix} - 0 \begin{vmatrix} 3 & 1 & 2 \\ 1 & -1 & -1 \\ 0 & 1 & 2 \end{vmatrix} + 0 \begin{vmatrix} 3 & 1 & 2 \\ 2 & 4 & 3 \\ 0 & 1 & 2 \end{vmatrix} - 1 \begin{vmatrix} 3 & 1 & 2 \\ 2 & 4 & 3 \\ 1 & -1 & -1 \end{vmatrix}$$

$$= -[2(-2 + 1) - 1(8 - 3)] -$$
$$[3(-4 + 3) - 2(-1 + 2) + 1(3 - 8)]$$

$$= -(-7) - (-10) = 17$$

Graphs in Space; Determinants   **185**

3. Evaluate. $\begin{vmatrix} 2 & -1 & -6 \\ 3 & 4 & 2 \\ 5 & -2 & 3 \end{vmatrix}$

Use the properties of determinants to obtain zeros in the second column. Multiply row 1 by 4, then add it to row 2 to obtain a "new" row 2, with 0 in the second column.

$$\begin{vmatrix} 2 & -1 & -6 \\ 11 & 0 & -22 \\ 5 & -2 & 3 \end{vmatrix}$$

Now multiply row 1 by -2 and add it to row 3 to obtain another 0 in the second column third row.

$$\begin{vmatrix} 2 & -1 & -6 \\ 11 & 0 & -22 \\ 1 & 0 & 15 \end{vmatrix}$$

You now have a determinant which is easy to evaluate using minors of the second column.

$$\begin{vmatrix} 2 & -1 & -6 \\ 11 & 0 & -22 \\ 1 & 0 & 15 \end{vmatrix} =$$

$$-(-1) \begin{vmatrix} 11 & -22 \\ 1 & 15 \end{vmatrix} +$$

$$0 \begin{vmatrix} 2 & -6 \\ 1 & 15 \end{vmatrix} - 0 \begin{vmatrix} 2 & -6 \\ 11 & -22 \end{vmatrix} =$$

$$187 + 0 - 0 = 187$$

1. Column 3

$$\begin{vmatrix} 3 & 0 & 1 \\ 1 & 2 & -1 \\ -2 & 6 & 2 \end{vmatrix} \quad 40$$

2. Row 3

$$\begin{vmatrix} 3 & 5 & 29 \\ -1 & 2 & 35 \\ 0 & 0 & 4 \end{vmatrix} \quad 44$$

3. Column 1

$$\begin{vmatrix} 1 & 3 & 6 \\ -2 & 2 & -4 \\ -1 & 5 & 2 \end{vmatrix} \quad 0$$

Evaluate.

4.
$$\begin{vmatrix} 3 & 2 & 4 & 2 \\ 0 & 0 & -1 & 1 \\ -2 & 1 & -3 & 2 \\ 1 & 3 & 2 & 0 \end{vmatrix} \quad -21$$

5.
$$\begin{vmatrix} -1 & 3 & 2 & 1 \\ 1 & 0 & -1 & -2 \\ 3 & 0 & 1 & 1 \\ 2 & 4 & 0 & 6 \end{vmatrix} \quad -122$$

6.
$$\begin{vmatrix} 4 & 3 & -2 & 1 \\ 2 & -2 & 0 & 4 \\ 0 & 7 & 5 & 4 \\ -1 & 0 & 0 & 2 \end{vmatrix} \quad -338$$

Determinants have some properties that are useful in simplifying their expansion by minors. The properties are presented here without proof. While third-order determinants are used in illustrating them, the properties are valid for determinants of any order.

Property 1: If each element in any row (or each element in any column) is 0, then the determinant is equal to 0.

$$\begin{vmatrix} 2 & 1 & 4 \\ 0 & 0 & 0 \\ 1 & 3 & -1 \end{vmatrix} = -0\begin{vmatrix} 1 & 4 \\ 3 & -1 \end{vmatrix} + 0\begin{vmatrix} 2 & 4 \\ 1 & -1 \end{vmatrix} - 0\begin{vmatrix} 2 & 1 \\ 1 & 3 \end{vmatrix}$$

$$= 0 + 0 + 0 = 0$$

Property 2: If any two rows (or any two columns) of a determinant are interchanged, the resulting determinant is the negative of the original determinant.

$$\begin{vmatrix} 1 & 2 & -3 \\ 2 & 1 & 4 \\ 3 & 1 & 2 \end{vmatrix} = 1\begin{vmatrix} 1 & 4 \\ 1 & 2 \end{vmatrix} - 2\begin{vmatrix} 2 & -3 \\ 1 & 2 \end{vmatrix} + 3\begin{vmatrix} 2 & -3 \\ 1 & 4 \end{vmatrix} = 17$$

$$\begin{vmatrix} 2 & 1 & -3 \\ 1 & 2 & 4 \\ 1 & 3 & 2 \end{vmatrix} = -1\begin{vmatrix} 1 & 4 \\ 1 & 2 \end{vmatrix} + 2\begin{vmatrix} 2 & -3 \\ 1 & 2 \end{vmatrix} - 3\begin{vmatrix} 2 & -3 \\ 1 & 4 \end{vmatrix} = -17$$

Property 3: If two rows (or two columns) of a determinant have corresponding elements that are equal, the determinant is equal to 0.

$$\begin{vmatrix} 1 & 3 & -2 \\ 3 & 2 & 4 \\ 1 & 3 & -2 \end{vmatrix} = -3\begin{vmatrix} 3 & -2 \\ 3 & -2 \end{vmatrix} + 2\begin{vmatrix} 1 & -2 \\ 1 & -2 \end{vmatrix} - 4\begin{vmatrix} 1 & 3 \\ 1 & 3 \end{vmatrix}$$

$$= -3 \cdot 0 + 2 \cdot 0 - 4 \cdot 0 = 0$$

Property 4: If each element in one row (or one column) of a determinant is multiplied by a real number k, the determinant is multiplied by k.

$$\begin{vmatrix} 1 & 3 & -1 \\ 2 & 1 & 4 \\ 3(2) & 3(-1) & 3(5) \end{vmatrix} = 3\begin{vmatrix} 1 & 3 & -1 \\ 2 & 1 & 4 \\ 2 & -1 & 5 \end{vmatrix}$$

Verify this by expanding both determinants by minors of the elements in the third row.

Property 5: If each element of one row (or one column) is multiplied by a real number k and the resulting products are added to the corresponding elements of another row (or another column), then the resulting determinant is equal to the original determinant.

$$\begin{vmatrix} 1 & 3 & -2 \\ 0 & 4 & 3 \\ 1 & -2 & 5 \end{vmatrix} = \begin{vmatrix} 1 & 3 & -2 \\ 0 & 4 & 3 \\ 1+3(1) & -2+3(3) & 5+3(-2) \end{vmatrix} = \begin{vmatrix} 1 & 3 & -2 \\ 0 & 4 & 3 \\ 4 & 7 & -1 \end{vmatrix}$$

EXAMPLE 3 Evaluate. $\begin{vmatrix} 3 & 1 & 2 \\ 5 & -4 & 3 \\ 2 & -2 & -1 \end{vmatrix}$

SOLUTION Use the properties of determinants to obtain zeros in the second column. Multiply the first row by 4 and add to the second row.

$$\begin{vmatrix} 3 & 1 & 2 \\ 5+4(3) & -4+4(1) & 3+4(2) \\ 2 & -2 & -1 \end{vmatrix} = \begin{vmatrix} 3 & 1 & 2 \\ 17 & 0 & 11 \\ 2 & -2 & -1 \end{vmatrix}$$

Multiply the first row by 2 and add to the third row.

$$\begin{vmatrix} 3 & 1 & 2 \\ 17 & 0 & 11 \\ 2+2(3) & -2+2(1) & -1+2(2) \end{vmatrix} = \begin{vmatrix} 3 & 1 & 2 \\ 17 & 0 & 11 \\ 8 & 0 & 3 \end{vmatrix}$$

Expand the last determinant by minors of the second column.

$$\begin{vmatrix} 3 & 1 & 2 \\ 17 & 0 & 11 \\ 8 & 0 & 3 \end{vmatrix} = -1\begin{vmatrix} 17 & 11 \\ 8 & 3 \end{vmatrix} + 0\begin{vmatrix} 3 & 2 \\ 8 & 3 \end{vmatrix} - 0\begin{vmatrix} 3 & 2 \\ 17 & 11 \end{vmatrix}$$

$$= 37 + 0 - 0$$
$$= 37$$

Oral Exercises

Give the requested element or determinant using this expansion by minors. You need not evaluate the determinants.

$$\begin{vmatrix} 3 & -2 & 0 & 1 \\ 5 & 7 & 6 & -4 \\ -1 & -3 & 8 & 2 \\ 9 & 5 & 1 & 4 \end{vmatrix} = a_1\begin{vmatrix} 5 & 7 & -4 \\ -1 & -3 & 2 \\ 9 & 5 & 4 \end{vmatrix} - a_2\begin{vmatrix} 3 & -2 & 1 \\ -1 & -3 & 2 \\ 9 & 5 & 4 \end{vmatrix} + 8A_3 - 1A_4$$

1. a_1 0   **2.** a_2 6   **3.** (the determinant) A_3 **4.** A_4

$\begin{vmatrix} 3 & -2 & 1 \\ 5 & 7 & -4 \\ 9 & 5 & 4 \end{vmatrix}$ $\begin{vmatrix} 3 & -2 & 1 \\ 5 & 7 & -4 \\ -1 & -3 & 2 \end{vmatrix}$

Written Exercises

Expand each determinant by minors of the given row or column and then evaluate.

A

1. column 3

$\begin{vmatrix} -3 & 2 & -4 \\ 1 & 5 & 0 \\ -2 & -1 & 6 \end{vmatrix}$ −138

2. row 3

$\begin{vmatrix} 5 & 2 & 34 \\ -1 & 3 & 22 \\ 0 & 0 & 4 \end{vmatrix}$ 68

3. column 1

$\begin{vmatrix} -1 & 5 & 2 \\ 4 & 1 & 6 \\ 1 & -2 & 3 \end{vmatrix}$ −63

Suggested Assignments

Minimum
  187/1–9
R 189/*Self-Test 2*
Extended Alg. with Trig.
  187/1–9 odd
R 189/*Self-Test 2*
Enriched Alg.
  187/3–11 odd
R 189/*Self-Test 2*
Enriched Alg. with Trig.
  187/3–11 odd
R 189/*Self-Test 2*

Common Errors

Errors due to hasty computation without caution are probably the most common in this section. To help avoid this, stress the importance of careful, step-by-step work. In particular, remind your students to be careful when they are determining whether to use the product of an element and its minor or the negative of the product. One way to remember to take the *negative* of the product when the sum of the row and column numbers is odd is to think "negative one times itself an *odd* number of times is *negative*" but "negative one times itself an *even* number of times is *positive.*"

Mixed Review

1. Simplify $(2.6)(5)[4 + (-4)]$.
  0

2. Solve $\frac{2}{3}n - 6 = 2(n - 11)$.
  {12}

3. Solve $5 - 3x \le 17$.
  {$x: x \ge -4$}

4. If $g: x \to -2x^2 + 3x + 5$, find $g(-1)$.   0

Additional Answers
Written Exercises

9. a. $\begin{vmatrix} a_1 & 0 & 0 & 0 \\ 0 & b_2 & 0 & 0 \\ 0 & 0 & c_3 & 0 \\ 0 & 0 & 0 & d_4 \end{vmatrix} =$

$a_1 \begin{vmatrix} b_2 & 0 & 0 \\ 0 & c_3 & 0 \\ 0 & 0 & d_4 \end{vmatrix} =$

$a_1 \left(b_2 \begin{vmatrix} c_3 & 0 \\ 0 & d_4 \end{vmatrix} \right) =$

$a_1[b_2(c_3 d_4 - 0)] =$

$a_1(b_2 c_3 d_4) = a_1 b_2 c_3 d_4$

9. b. $\begin{vmatrix} a_1 & b_1 & 0 & 0 \\ a_2 & b_2 & 0 & 0 \\ 0 & 0 & c_3 & d_3 \\ 0 & 0 & c_4 & d_4 \end{vmatrix} =$

$a_1 \begin{vmatrix} b_2 & 0 & 0 \\ 0 & c_3 & d_3 \\ 0 & c_4 & d_4 \end{vmatrix} -$

$a_2 \begin{vmatrix} b_1 & 0 & 0 \\ 0 & c_3 & d_3 \\ 0 & c_4 & d_4 \end{vmatrix} =$

$a_1 \left(b_2 \begin{vmatrix} c_3 & d_3 \\ c_4 & d_4 \end{vmatrix} \right) -$

$a_2 \left(b_1 \begin{vmatrix} c_3 & d_3 \\ c_4 & d_4 \end{vmatrix} \right) =$

$(a_1 b_2 - a_2 b_1) \begin{vmatrix} c_3 & d_3 \\ c_4 & d_4 \end{vmatrix} =$

$\begin{vmatrix} a_1 & b_1 \\ a_2 & b_2 \end{vmatrix} \cdot \begin{vmatrix} c_3 & d_3 \\ c_4 & d_4 \end{vmatrix}$

10. Assume a 3×3 determinant has two rows alike. By Property 5,

$\begin{vmatrix} a_1 & b_1 & c_1 \\ a_1 & b_1 & c_1 \\ a_3 & b_3 & c_3 \end{vmatrix} =$

$\begin{vmatrix} a_1 & b_1 & c_1 \\ 0 & 0 & 0 \\ a_3 & b_3 & c_3 \end{vmatrix} = 0$

(by Property 1)

188 *Chapter 5*

Evaluate. Use properties of determinants to simplify the expansion.

4. $\begin{vmatrix} -1 & 3 & 2 & 4 \\ 0 & 0 & 0 & -1 \\ 5 & -1 & 2 & 9 \\ -2 & 3 & 1 & 8 \end{vmatrix}$ -6

5. $\begin{vmatrix} 1 & 3 & -2 & -1 \\ 0 & 2 & 4 & 1 \\ 0 & -5 & 0 & 3 \\ -1 & -3 & -1 & 1 \end{vmatrix}$ 33

6. $\begin{vmatrix} 3 & 4 & -2 & 1 \\ 5 & 1 & 0 & -1 \\ 0 & -3 & 2 & 3 \\ 4 & -1 & 0 & -5 \end{vmatrix}$
-66

Use Cramer's Rule to solve the system.

B **7.** $x - 2y - w = 0$ $w = 3$
$2x - y + z = 1$ $x = 1$
$-x + z + w = 0$ $y = -1$
$3y - z + w = 2$ $z = -2$

8. $-x + y - w = 5$ $w = -1$
$2y - z + w = 0$ $x = -2$
$x + z + w = 0$ $y = 2$
$-3x + y - 2z - w = 3$ $z = 3$

9. Use expansion by minors to prove each of the following.

a. $\begin{vmatrix} a_1 & 0 & 0 & 0 \\ 0 & b_2 & 0 & 0 \\ 0 & 0 & c_3 & 0 \\ 0 & 0 & 0 & d_4 \end{vmatrix} = a_1 b_2 c_3 d_4$

b. $\begin{vmatrix} a_1 & b_1 & 0 & 0 \\ a_2 & b_2 & 0 & 0 \\ 0 & 0 & c_3 & d_3 \\ 0 & 0 & c_4 & d_4 \end{vmatrix} = \begin{vmatrix} a_1 & b_1 \\ a_2 & b_2 \end{vmatrix} \cdot \begin{vmatrix} c_3 & d_3 \\ c_4 & d_4 \end{vmatrix}$

In Exercises 10–12 refer to properties 1–5 on page 186.

C **10.** Use Properties 1 and 5 to prove Property 3 for 3×3 determinants.

11. Use Property 2 *only* to prove Property 3 for 4×4 determinants. (*Hint:* First prove Property 3 for 3×3 determinants.)

12. Prove that, for any real numbers r and t,

$$\begin{vmatrix} a_1 & b_1 & c_1 \\ a_2 & b_2 & c_2 \\ ra_1 + ta_2 & rb_1 + tb_2 & rc_1 + tc_2 \end{vmatrix} = 0.$$

Computer Exercises For students with computer experience

1. Write a program that uses Cramer's Rule to solve a system of equations in three variables of the form

$$a_1 x + b_1 y + c_1 z = d_1$$
$$a_2 x + b_2 y + c_2 z = d_2$$
$$a_3 x + b_3 y + c_3 z = d_3.$$

Input values for each a, b, c, and d. If the system does not have a unique solution, the output should state this.

Run the program of Exercise 1 for each system.

2. $7x + 8y - 3z = 3$
$-5x - 9y + 7z = 32$
$4x - 2y - 11z = -25$ $\{(6, -3, 5)\}$

3. $8x - 7y + 10z = 69$
$-12x + 9y - z = -10$
$6x - y - 5z = -23$
$\{(2.5, 3, 7)\}$

Run the program of Exercise 1 for each system.

4. $6x - 3y - 8z = 7$
    $4x + 10y - 7z = -12$
    $-2x + 13y + z = -19$ infinite solution set

5. $3x - 5y + 7z = 8$
    $-2x + 7y - 8z = -14$
    $9x - 15y + 21z = -7$
    inconsistent

6. Write a program that will evaluate a 4×4 determinant using expansion by minors of the fourth row.

Run the program in Exercise 6 for each determinant.

7. $\begin{vmatrix} 3 & -2 & 5 & 4 \\ -7 & 0 & 6 & -11 \\ -9 & 1 & -7 & 8 \\ 4 & -1 & 9 & 13 \end{vmatrix}$ 3718

8. $\begin{vmatrix} 5 & -6 & -9 & 4 \\ -3 & 8 & 7 & -3 \\ -1 & 10 & 5 & -2 \\ 2 & 1 & 0 & 10 \end{vmatrix}$ 0

9. Run the program for Written Exercises 4–6 on page 188.
   $-6;\ 33;\ -66$

▮ Self-Test 2

VOCABULARY   rows of a determinant (p. 176)
columns of a determinant (p. 176)
order of a determinant (p. 177)
Cramer's Rule (p. 178)
determinant of coefficients (p. 178)
minor (p. 183)

1. Use determinants to solve the system: $x + y + z = 4$ *Obj. 1, p. 176*
   $2x + 3y = 3$
   $4y - z = -6$ $\{(3, -1, 2)\}$

2. A citrus grower purchased 5 orange trees, 3 grapefruit trees,   *Obj. 2, p. 176*
   and 2 lemon trees for $59. The following week, at the same
   price, the grower bought 7 orange trees, 12 grapefruit trees, orange, $6
   and 1 lemon tree for $130. During the following week, the grapefruit, $7
   grower purchased 4 orange trees and 1 grapefruit tree at the lemon, $4
   original prices, and paid $31. Find the cost of each type of
   citrus tree.

3. Evaluate. $\begin{vmatrix} -1 & -3 & 4 & -1 \\ 0 & 1 & 0 & -2 \\ 3 & 1 & 2 & 6 \\ 2 & 1 & 0 & 5 \end{vmatrix}$ -6 *Obj. 3, p. 176*

Check your answers with those at the back of the book.

Graphs in Space; Determinants **189**

---

11. Assume a 4×4 determinant D has two rows alike. Property 2 says that if the two rows are interchanged, the resulting determinant is $-D$. Yet clearly the resulting determinant is D. Thus $D = -D$, and $D = 0$.

12. By Property 5 the given determinant equals

$$\begin{vmatrix} a_1 & b_1 & c_1 \\ a_2 & b_2 & c_2 \\ ta_2 & tb_2 & tc_2 \end{vmatrix} =$$

$$\begin{vmatrix} a_1 & b_1 & c_1 \\ a_2 & b_2 & c_2 \\ 0 & 0 & 0 \end{vmatrix} = 0$$

by Property 1

Quick Quiz

1. Use determinants to solve the system:
   $2x + y = 3$
   $3x + y - z = 2$
   $2y - z = -5$
   $\{(2, -1, 3)\}$

2. The three angles of triangle ABC have the properties that the degree measure of $\angle C$ is three times that of $\angle B$, and the sum of the degree measures of $\angle A$ and the supplement of $\angle C$ is twice the degree measure of $\angle B$. Find the degree measures of the three angles.
   $m(\angle A) = 20°$, $m(\angle B) = 40°$, $m(\angle C) = 120°$

3. Evaluate.
   $\begin{vmatrix} 3 & -2 & 2 & 0 \\ 1 & -2 & 1 & 0 \\ 4 & 3 & 6 & 1 \\ -1 & 1 & 2 & 2 \end{vmatrix}$ -27

(continued from p. 164)

10.

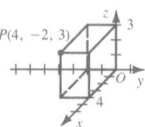

12.

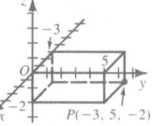

14.

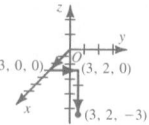

15.

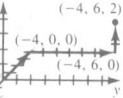

16.

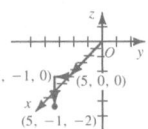

17.

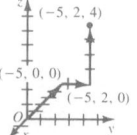

18.

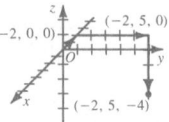

20.

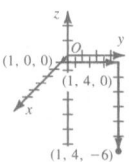

Chapter Summary

1. A rectangular coordinate system in three-dimensional space assigns an ordered triple of numbers to each point. The three *coordinate axes* determine three *coordinate planes*, and the coordinate planes separate space into eight *octants*.

2. The graph of a linear equation in three variables is a plane. The x-, y-, and z-*intercepts* of a plane are the x-, y-, and z-coordinates, respectively, of the points where the plane intersects the x-, y-, and z-axes. A line in which a plane cuts a coordinate plane is called the *trace* of the given plane in that coordinate plane.

3. The transformations used in solving systems of linear equations in two variables are also applicable in solving three-variable systems. See page 128.

4. Third-order determinants can be used to solve a system of three linear equations in three variables by a method called *Cramer's Rule*.

5. A determinant can be *expanded by minors* of the elements in any row or column.

6. Cramer's Rule can be extended to the solution of n linear equations in n variables.

7. You can use properties of determinants (page 186) in simplifying their expansion by minors.

Chapter Review

Write the letter of the correct answer.

1. In which octant does the point $(-13, 4, 9)$ lie?   5-1
   a. +, +, +   **b.** +, +, −   **c.** −, +, +   **d.** −, −, +

2. Which point lies on the yz-plane?
   a. $(1, 2, 0)$ **b.** $(2, 1, 0)$ **c.** $(1, 0, 2)$ **d.** $(0, 1, 2)$

3. Find the z-intercept of the graph of $3x + 2y + 7z = 140$.   5-2
   a. 20   **b.** 2   **c.** 70   **d.** no intercept

4. Which trace of the plane with equation $5x - 2y + 4z = 23$ is defined by the following equations?

$$5x - 2y = 23$$
$$z = 0$$

   a. xy-trace   **b.** xz-trace   **c.** yz-trace   **d.** none

190   *Chapter 5*

5. Determine the solution set of the following system of equations.

$$x + z = 9$$
$$3x + 2y - z = -15$$
$$5x + 3y = -7$$

a. $\{(5, -5, 4)\}$ **b.** $\{(4, -4, 5)\}$ **c.** $\{(-2, 1, 11)\}$ **d.** $\{(6, -2, -3)\}$

5-3

Exercises 6 and 7 refer to the following system of equations.

$$3x + 2y + z = 6$$
$$x + y + z = -6$$
$$2y - z = 7$$

6. Which determinant is D_y?

a. $\begin{vmatrix} 3 & 2 & 1 \\ 1 & 1 & 1 \\ 0 & 2 & -1 \end{vmatrix}$ **b.** $\begin{vmatrix} 3 & 6 & 1 \\ 1 & -6 & 1 \\ 0 & 7 & -1 \end{vmatrix}$ **c.** $\begin{vmatrix} 6 & 2 & 1 \\ -6 & 1 & 1 \\ 7 & 2 & -1 \end{vmatrix}$ **d.** $\begin{vmatrix} 3 & 2 & 6 \\ 1 & 1 & -6 \\ 0 & 2 & 7 \end{vmatrix}$

5-4

7. Evaluate D_z.

a. 7 **b.** -11 **c.** 2 **d.** 55

8. The sum of three numbers is 37. The third number is three less than the sum of the first and the second, and the second number is two more than twice the first. What are the numbers?

a. 6, 14, 17 **b.** 2, 16, 19 **c.** 2, 15, 20 **d.** 5, 11, 20

5-5

Exercises 9 and 10 refer to the following determinant. $\begin{vmatrix} 1 & 0 & 2 \\ -1 & 0 & 0 \\ -3 & 3 & -2 \end{vmatrix}$

9. What is the minor of 2?

a. $\begin{vmatrix} 0 & 0 \\ 3 & -2 \end{vmatrix}$ **b.** $\begin{vmatrix} 1 & 0 \\ -3 & 3 \end{vmatrix}$ **c.** $\begin{vmatrix} -1 & 0 \\ -3 & 3 \end{vmatrix}$ **d.** $\begin{vmatrix} -1 & 0 \\ -3 & -2 \end{vmatrix}$

5-6

10. Evaluate the determinant.

a. 8 **b.** -6 **c.** 6 **d.** -8

Chapter Test

1. Sketch the coordinate box of $(-1, 4, -5)$.

5-1

2. Sketch the triangle in space whose vertices have the coordinates $(3, 0, 0)$, $(0, 4, 0)$, and $(0, 0, 8)$.

3. Graph part of the plane that has the given traces.

$$xy\text{-trace: } 2x + y = -4, z = 0$$
$$yz\text{-trace: } -5y + 4z = 20, x = 0$$
$$xz\text{-trace: } 5x - 2z = -10, y = 0$$

5-2

22.
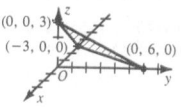

23.
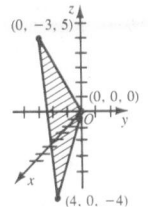

24.
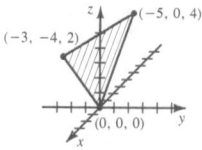

26.
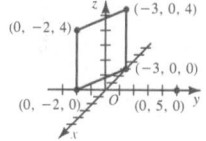

Additional Answers
Chapter Test

1.
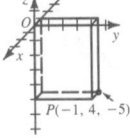

2.
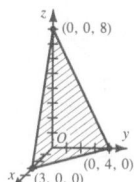

3.
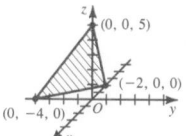

Additional Answers
Written Exercises
(continued from p. 170)

4. b. $z = 0; x = -4$
$y = 0; x = -4$
no trace in the yz-plane

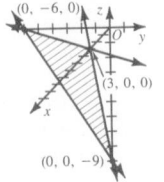

6. b. $z = 0; 2x - y = 6$
$y = 0; 3x - z = 9$
$x = 0; 3y + 2z = -18$

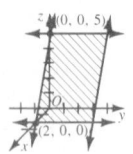

7. b. $z = 0; x = 2$
$y = 0; 5x + 2z = 10$
$x = 0; z = 5$

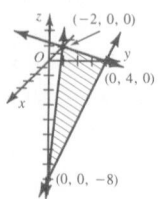

8. b. $z = 0; -4x + 2y = 8$
$y = 0; 4x + z = -8$
$x = 0; 2y - z = 8$

9. b. $z = 0; y = 0$
$x = 0; y = 0$
coincides with the xz-plane

4. Solve the system:
$$2x + 3y - z = 1 \quad \{(-1, 2, 3)\}$$
$$x - 4y + 4z = 3$$
$$-3x + y - 2z = -1$$
5-3

Solve each system using Cramer's Rule. If $D = 0$, state whether the system is inconsistent or has an infinite solution set.

5. $2x - 3y - z = -1 \quad \{(-3, -2, 1)\}$
$x + 2y + 3z = -4$
$3x - 5y + 2z = 3$

6. $3x - 2y + z = 2$
$-x + 3y - z = -1$
$6x - 4y + 2z = 5$
inconsistent
5-4

7. The perimeter of a certain triangle is 26 cm. Twice the sum of the two smaller sides is 4 cm greater than twice the largest side. Four times the smallest side is 1 cm less than the sum of the other two sides. Find the length of each side. 5 cm, 9 cm, 12 cm
5-5

Evaluate each determinant. Use the properties of determinants to simplify the expansion if necessary.

8. $\begin{vmatrix} -1 & 2 & 3 & 1 \\ 0 & 3 & 4 & 5 \\ 1 & 0 & 0 & -2 \\ 4 & 1 & -3 & 2 \end{vmatrix}$ 48

9. $\begin{vmatrix} 2 & -4 & 6 & 8 \\ -2 & 1 & 3 & 0 \\ 3 & 2 & -1 & 1 \\ 1 & -2 & 3 & 4 \end{vmatrix}$ 0
5-6

APPLICATION

Capacitors and Capacitance

In today's microelectronic world, a silicon chip the size of a postage stamp may contain thousands of circuit components. One of the most

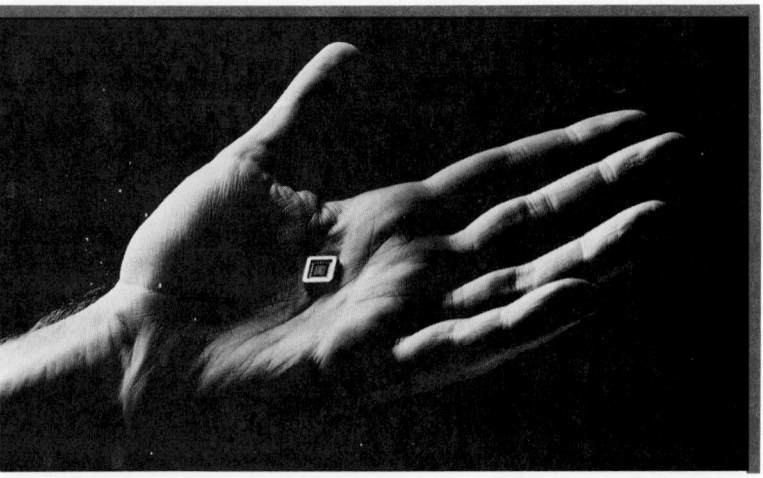

important of these components is the *capacitor*, a device that can store electric energy. In order to explain how a capacitor does this, it is helpful to consider the simple DC (direct current) circuit as shown in the diagram below. The circuit contains a capacitor made of two parallel metal plates separated by an air gap. A dry cell cannot supply enough energy to make electrons flow across an air gap, since air is a very poor conductor of electricity. In this circuit, current flows for only an instant when the switch is closed. Electrons flow from the negative side of the cell to one of the capacitor plates; at the same time, electrons flow away from the other plate to the positive side of the cell. No current passes

across the air gap. As a result, one plate of the capacitor is left with a negative charge, and the other has a positive charge. Thus electric charge is "stored" in the capacitor. If the capacitor is discharged, a brief pulse of current is produced.

The amount of charge that can be stored in a given capacitor is called its *capacitance*, and is denoted C. The capacitance depends on the voltage in the circuit and on the characteristics of the capacitor. The capacitance of a parallel-plate capacitor like the one shown is directly proportional both to the area of the plates and to the reciprocal of the distance between them. The capacitance also depends on the nature of the material, or *dielectric*, between the plates. If glass, rather than air, were the dielectric, the capacitance could be up to 10 times as great. In general, the charge denoted Q on a capacitor at a given voltage denoted V is given by the relationship $Q = CV$ where the charge is measured in coulombs (C) and voltage is measured in volts (V).

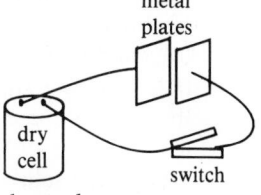

metal
plates

dry
cell

switch

One of the many practical uses of capacitors is in tuning circuits. You may have seen the variable capacitors used in radios made a decade or more ago. When turning the dial to tune in different stations, you could see one set of metal plates moving relative to another (stationary) set of plates. In that case air was the dielectric. The change in effective plate size produced a corresponding change in capacitance. Capacitors are also used in the oscillator circuits that produce radio waves.

Exercises

1. A given capacitor stores a charge of 2×10^{-6} C at 20 V. What charge is stored by a capacitor having twice the capacitance at 25 volts? 5×10^{-6} C
2. What is the effect on the capacitance of a parallel-plate capacitor if **(a)** the area of the plates is doubled, and **(b)** the distance between the plates is halved? **2. a.** The capacitance is doubled.
   2. b. The capacitance is doubled.

Graphs in Space; Determinants **193**

10. b. $z = 0$; $5x - 3y = 15$
$y = 0$; $5x + 3z = 15$
$x = 0$; $-y + z = 5$

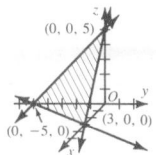
(0, 0, 5)
(3, 0, 0)
(0, -5, 0)

12. b. no trace in the xy-plane
$y = 0$; $z = 2$
$x = 0$; $z = 2$

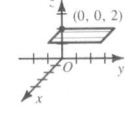
(0, 0, 2)

13. b. $z = 0$; $-2x + y = 4$
$y = 0$; $5x + 2z = -10$
$x = 0$; $5y - 4z = 20$

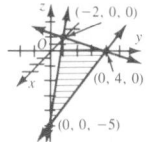
(-2, 0, 0)
(0, 4, 0)
(0, 0, -5)

14. b. $z = 0$; $6x - 5y = 30$
$y = 0$; $2x + 5z = 10$
$x = 0$; $-y + 3z = 6$

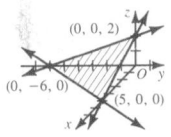
(0, 0, 2)
(0, -6, 0)
(5, 0, 0)

15. b. $z = 0$; $y = -\dfrac{3}{2}$

no trace in the xy-plane

$x = 0$; $y = -\dfrac{3}{2}$

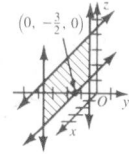
$(0, -\frac{3}{2}, 0)$

(continued)

16. b. $z = 0$; $y = -5$
$y = 0$; $z = -3$
$x = 0$; $3y + 5z = -15$

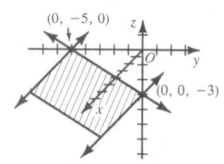

17. b. $z = 0$; $8x + 3y = -24$
$y = 0$; $2x + z = -6$
$x = 0$; $3y + 4z = -24$

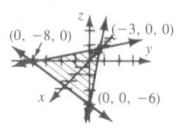

18. b. $x = 0$; $z = 0$
$y = 0$; $z = 0$
coincides with the xy-plane

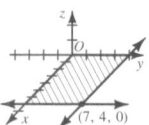

20.

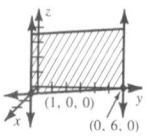

22.

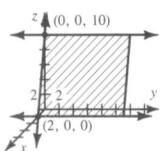

23.

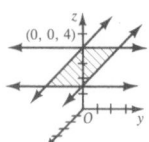

Strategy for Success: Often the answer to a question can be found by writing an equation or inequality and solving it. When a complete solution is time consuming, you may find the fastest way to answer the question is to test the suggested answers in your equation or inequality.

Decide which is the best of the choices given and write the corresponding letter on your answer sheet.

1. If $5x - y = 4$ and $4y - 20x = -3$ is a system of equations which of the following must be true for the system? E
   I. The system is consistent.
   II. The system is inconsistent.
   III. The system has no solution.
   (A) I only   **(B)** II only   **(C)** III only
   (D) I and II only   **(E)** II and III only

2. If $7x - 3y = 1$ and $-4x + 2y = -2$ is a system of equations what is the value of the determinant D_y? C
   (A) -2 **(B)** -4 **(C)** -10 **(D)** 2 **(E)** -5

3. What is the solution set for the following system: A
   $$y = mx + b_1$$
   $$x = b_2$$
   (A) $\{(b_2, mb_2 + b_1)\}$ **(B)** $\{(0, b_2)\}$ **(C)** $\emptyset$ **(D)** $\left\{\left(\frac{mx}{b_1}, b_2\right)\right\}$ **(E)** $\left\{\left(b_2, \frac{mx}{b_1}\right)\right\}$

4. If $r = \frac{3ka}{2}$ and $r = 15$ when $a = 8$, what is the value of r when $a = -0.5$? E
   (A) $\frac{2}{25}$ **(B)** $\frac{25}{2}$ **(C)** 0 **(D)** -12 **(E)** $-\frac{15}{16}$

5. Which of the following sentences is equivalent to $|x - 2| > 6$? A
   (A) $x - 2 > 6$ or $2 - x > 6$ **(B)** $x - 2 > 4$ or $-x - 2 > 6$
   (C) $x - 2 = 6$ or $2 - x > 6$ **(D)** $x - 2 > 6$ or $x + 2 < -6$
   (E) $x - 2 > 6$ and $2 - x > 6$

6. What is an equation for a plane that contains the point (r, s, t), where $t \neq 0$, and the plane is parallel to the xy-plane? E
   (A) $z = s$ **(B)** $y = s$ **(C)** $r = x$ **(D)** $z = r$ **(E)** $z = t$

7. For what value of x does the line with slope $-\frac{2}{3}$ pass through the points $(6, -3)$ and $(3, x)$? B
   (A) 1 **(B)** -1 **(C)** 2 **(D)** 4 **(E)** -2

8. A wire of uniform diameter and composition weighs 18 lb. When the wire was cut, one piece weighed 12 lb and was 36 yd long. What was the length of the wire before it was cut? B
   (A) 48 yd   **(B)** 54 yd   **(C)** 90 yd   **(D)** 24 yd   **(E)** 12 yd

194 *Chapter 5*

PROGRAMMING IN PASCAL

24.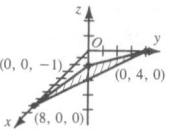

25.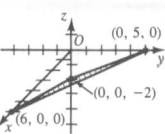

26.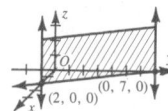

In Chapter 4, you should have written a program, *cramer2*, similar to the following:

```
PROGRAM cramer2(INPUT, OUTPUT);

(* This program solves a system of two linear equations in two variables *)
(* of the form            a1*x + b1*y = c1                               *)
(*                         a2*x + b2*y = c2                               *)
(* where a1, b1, c1, a2, b2, and c2 are integers, by Cramer's method.    *)
VAR
  a1, b1, c1, a2, b2, c2, d, dx, dy : integer;
  x, y : real;


( ***********************************************************)
PROCEDURE get_values
BEGIN
  writeln('Enter the coefficients and constant of the first equation,');
  write('place a space between each of the values:');
  readln(a1, b1, c1);
  writeln('Enter the coefficients and constant of the second equation,');
  write('place a space between each of the values: ');
  readln(a2, b2, c2);
END;   (* get_values *)


( ***********************************************************)
FUNCTION det(s, t, u, v : integer) : integer;
BEGIN
  det := s*v − t*u;
END;   (* det *)


( ***********************************************************)
PROCEDURE solve_system;
BEGIN
  d := det(a1,b1,a2,b2);
  IF d <> 0
    THEN
      BEGIN
        dx := det(c1, b1, c2, b2);
        dy := det(a1, c1, a2, c2);
        x := dx/d;
        y := dy/d;
      END
END;   (* solve_system *)
```

(Program continues on next page.)

Graphs in Space; Determinants   **195**

```
( ************************************************************)
PROCEDURE display_answer;
BEGIN
  IF d = 0
    THEN
      writeln('No solution or infinitely many solutions.')
    ELSE writeln('Solution is (',X,' , ',Y,').');
END;   (* display_answer *)


( ************************************************************)
BEGIN   (* cramer2 *)
  get_values;
  solve_system;
  display_answer
END.
```

You can extend this program to another program, cramer3, to solve systems of three linear equations in three variables and to still another program, cramer4, to solve systems of four linear equations in four variables.

Exercises

1. **a.** Use the expansion formula in Section 5-4, page 177 to write a Pascal function, *det*, to compute the determinants needed to solve a system of three linear equations in three variables. Use. the notation of the text. Check your function by solving a few determinants without the computer.

   b. With *cramer2* as a model write a Pascal program, *cramer3*, that solves systems of three linear equations in three variables. Comment on the limitations, if any, of your program.

2. Use expansion by minors and the function *det* from Exercise 1(a), and modify the procedure *solve_system* to solve a system of four linear equations in four variables.

3. Write a Pascal program, *cramer4*, that follows the general structure of *cramer2* to solve and display the result for a system of equations such as those in Exercises 7 and 8 of Section 5-6 on page 188. Solve these exercises using your *cramer4*. Compare the solutions you obtain by using Cramer's Rule to the solutions obtained by using the program *cramer4*.

4. We say that a determinant is *upper-triangular* if all of the elements below the main diagonal are zero. The *main diagonal* runs from the upper left corner to the lower right corner. The following is an example of an upper-triangular determinant:

$$\begin{vmatrix} 1 & -2 & -3 \\ 0 & 7 & 14 \\ 0 & 0 & -7 \end{vmatrix}$$

196   *Chapter 5*

We can use the properties of determinants in Section 5-6 to show that the determinant shown below is equivalent to the upper-triangular form shown on page 196.

$$\begin{vmatrix} 1 & -2 & -3 \\ 3 & 1 & 5 \\ 2 & 6 & 7 \end{vmatrix}$$

Multiply the first row by -3 and add the result to the second row.

$$\begin{vmatrix} 1 & -2 & -3 \\ 3 & 1 & 5 \\ 2 & 6 & 7 \end{vmatrix} = \begin{vmatrix} 1 & -2 & -3 \\ 0 & 7 & 14 \\ 2 & 6 & 7 \end{vmatrix}$$

Multiply the first row by -2 and add the result to the third row.

$$\begin{vmatrix} 1 & -2 & -3 \\ 0 & 7 & 14 \\ 2 & 6 & 7 \end{vmatrix} = \begin{vmatrix} 1 & -2 & -3 \\ 0 & 7 & 14 \\ 0 & 10 & 13 \end{vmatrix}$$

Multiply the second row by $-\dfrac{10}{7}$ and add the result to the third row. You obtain the desired upper-triangular determinant.

$$\begin{vmatrix} 1 & -2 & -3 \\ 0 & 7 & 14 \\ 0 & 10 & 14 \end{vmatrix} = \begin{vmatrix} 1 & -2 & -3 \\ 0 & 7 & 14 \\ 0 & 0 & -7 \end{vmatrix}$$

a. Evaluate both the original determinant and the upper-triangular determinant to verify their equivalence.

b. Explain why an upper-triangular determinant is equal to the product of the elements along its main diagonal.

5. Research Problem
Write a Pascal procedure to transform a 3×3 determinant into upper-triangular form when possible.

The pelican is one of the world's largest living birds. This brown pelican in the Galapagos Islands is soaring above the water in search of food. The force of the pelican's dive into the water stuns the fish, which are then scooped up in the pelican's pouch.

Chapter 6

Polynomials and Rational Expressions

Problem Solving Strategies

Word Problem Plan
The five-step plan is used in Section 6-4 to solve problems involving polynomial equations, in Section 6-10 to solve problems using polynomials with rational coefficients, and in Section 6-11 to solve problems involving fractional equations.
Trial and Error
Educated trial and error is used in factoring in Section 6-3.

Polynomials and Their Factors

OBJECTIVES for Sections 6-1 through 6-3:
1. *To apply the laws of exponents to simplify products and quotients of monomials.*
2. *To write products of polynomials in simple form.*
3. *To write a polynomial in factored form.*

6–1 Laws of Exponents

Teaching Suggestions
p. T89

Key Ideas

Prove the laws of exponents. Use the laws to simplify products and quotients of monomials.

You will recall from page 42 that b^n, the nth power of b, where n is a positive integer, denotes a product of n equal factors. That is:

$$n \text{ factors}$$
$$b^n = \overbrace{(b \times b \times \ldots \times b)}$$

Each factor is b, the *base*, and the number of such factors is n, the *exponent*. The laws for working with positive-integral exponents are summarized in the theorem on the next page.

Polynomials and Rational Expressions   **199**

Theorem.   If a and $b \in \mathscr{R}$, and m and n are positive integers, then:

1. $b^m b^n = b^{m+n}$ **2.** $(b^m)^n = b^{mn}$

3. $(ab)^m = a^m b^m$ **4.**   If $m > n$ and $b \neq 0$, $\dfrac{b^m}{b^n} = b^{m-n}$.

5.   If $m < n$ and $b \neq 0$, $\dfrac{b^m}{b^n} = \dfrac{1}{b^{n-m}}$.

6.   If $b \neq 0$, $\left(\dfrac{a}{b}\right)^m = \dfrac{a^m}{b^m}$.

7.   If $b \notin \{-1, 0, 1\}$, then $b^m = b^n$ if and only if $m = n$.

Knowing the definition of a power and the properties of real numbers, you can see why each of the laws stated in the preceding theorem holds.

PROOF OF LAW 1

Statements	*Reasons*
m factors *n* factors	
1. $b^m b^n = \overbrace{(b \times \ldots \times b)}\overbrace{(b \times \ldots \times b)}$	Definition of a power
$(m + n)$ factors	
2. $= \overbrace{(b \times \ldots \times b)}$	Associative axiom of multiplication
3. $= b^{m+n}$	Definition of a power
4. $b^m b^n = b^{m+n}$	Transitive property of equality

Laws 1–3 together with the properties of multiplication enable you to simplify a *product* of two or more monomials.

EXAMPLE 1   Simplify.

       a. $(3x^4y^5)(-7x^3y)$ **b.** $(-3p^5s^2)^4$ **c.** $(-4m^3q)^3(-5m^2q)^4$

SOLUTION   **a.** $(3x^4y^5)(-7x^3y) = 3(-7)(x^4x^3)(y^5y)$
$$= -21x^7y^6$$

       b. $(-3p^5s^2)^4 = (-3)^4(p^5)^4(s^2)^4 = 81p^{20}s^8$

       c. $(-4m^3q)^3(-5m^2q)^4 = (-4)^3(-5)^4(m^3)^3(m^2)^4q^3q^4$
$$= -40{,}000\, m^{17}q^7$$

You can deduce Laws 4 and 5 from Law 1 with the help of a corollary of the following *basic property of quotients*, which you may recall from an earlier algebra course.

200   *Chapter 6*

Common Errors

Stress the necessity of like bases in theorems such as $b^m b^n = b^{m+n}$. Students may want to multiply bases (especially if they are numerical). Emphasize that, for example,
$$4^2 \cdot 2^3 \neq 8^5.$$
In general, then,
$$b^m \cdot a^n \neq (ba)^{m+n}.$$

$$\boxed{\begin{array}{c} \textit{Basic Property of Quotients} \\ \text{For all } r, s, t, \text{ and } u \in \mathcal{R}, \text{ and } t, u \neq 0, \\[2mm] \dfrac{rs}{tu} = \dfrac{r}{t} \cdot \dfrac{s}{u}. \end{array}}$$

This theorem says that a quotient of products can be written as a product of quotients. If you let (1) $t = 1$, or (2) $r = 1$, you obtain the following corollary.

$$\boxed{\begin{array}{l} \textit{Corollary.} \quad \text{For all } r, s, t, \text{ and } u \in \mathcal{R}, \text{ and } t, u \neq 0: \\[3mm] \textbf{1. } \dfrac{rs}{u} = r \cdot \dfrac{s}{u} \qquad\qquad\qquad \textbf{2. } \dfrac{s}{tu} = \dfrac{1}{t} \cdot \dfrac{s}{u}. \end{array}}$$

PROOF OF LAW 4

First, note that if $m > n$, then $m - n$ is positive (Definition of ">").

Statements	Reasons
1. $\dfrac{b^m}{b^n} = \dfrac{b^{(m-n)+n}}{b^n}$	$(m - n) + n = m$
2. $= \dfrac{b^{m-n} \cdot b^n}{b^n}$	Law 1 for positive exponents
3. $= b^{m-n} \cdot \dfrac{b^n}{b^n}$	Corollary above
4. $= b^{m-n} \cdot 1$	$\dfrac{b^n}{b^n} = 1$
5. $= b^{m-n}$	Identity axiom for multiplication
6. $\dfrac{b^m}{b^n} = b^{m-n}$	Transitive property of equality

EXAMPLE 2   Simplify each quotient. Assume that no variable equals 0.

    a. $\dfrac{-15r^5 s^4}{3r^8 s^3}$ **b.** $\left(\dfrac{-2x^3}{x^2 z^3}\right)^5$

SOLUTION   **a.** $\dfrac{-15r^5 s^4}{3r^8 s^3} = \left(-\dfrac{15}{3}\right)\left(\dfrac{1}{r^{8-5}}\right)(s^{4-3}) = -\dfrac{5s}{r^3}$

          b. $\left(\dfrac{-2x^3}{x^2 z^3}\right)^5 = \left(\dfrac{-2x^{3-2}}{z^3}\right)^5$

                 $= \left(\dfrac{-2x}{z^3}\right)^5 = \dfrac{(-2)^5(x^5)}{(z^3)^5} = \dfrac{-32x^5}{z^{15}}$

Polynomials and Rational Expressions   **201**

Thus far we have used only *positive* integers as exponents. The laws for exponents will hold for *any* integral exponent with the addition of these definitions:

For all nonzero $b \in \mathcal{R}$ and all positive integers n,

$$b^0 = 1 \quad \text{and} \quad b^{-n} = \frac{1}{b^n}.$$

EXAMPLE 3 Show that Law 1 holds for $b^{-m}b^{-n}$ when $m = 7$, $n = 3$.

SOLUTION Show that $b^{-7}b^{-3} = b^{-7+-3} = b^{-10}$:

  1. $b^{-7}b^{-3} = \dfrac{1}{b^7} \cdot \dfrac{1}{b^3}$ Definition of b^{-n}
  2. $\qquad = \dfrac{1}{b^7 \cdot b^3}$ Basic property of quotients
  3. $\qquad = \dfrac{1}{b^{10}}$ Law 1
  4. $\qquad = b^{-10}$ Definition of b^{-n}
  5. $b^{-7}b^{-3} = b^{-10}$ Transitive property of equality

Oral Exercises

State an equivalent numeral without exponents.

1. $2^9 \cdot 2^{-6}$ 8

2. $3 \cdot 3^3$ 81

3. $\dfrac{2}{2^{-4}}$ 32

4. $3^{-6} \cdot 3^4$ $\frac{1}{9}$

5. $\dfrac{5^6}{5^8}$ $\frac{1}{25}$

6. $(2 + 5)^2$ 49

7. $(5 - 3)^4$ 16

8. $\left(\dfrac{2}{3}\right)^{-4}$ $\frac{81}{16}$

9. $\left(\dfrac{17}{23}\right)^0$ 1

State an equivalent expression using only positive exponents. Assume that no variable equals 0.

10. $x^5 \cdot x^3$ x^8

11. $\dfrac{a^8}{a^{-7}}$ a^{15}

12. $(b^3)^7$ b^{21}

13. y^0 1

14. $(x + y)^{-1}$ $\frac{1}{x+y}$

15. $2x^{-4}$ $\frac{2}{x^4}$

16. $\dfrac{3}{c^{-2}}$ $3c^2$

17. $p^{-1} + q^{-1}$ $\frac{1}{p} + \frac{1}{q}$

18. $(5c^3d)(-2d^4)$

19. $(3r^{-2})(5r^{-4})$ $\frac{15}{r^6}$

20. $(-3t^4)(-4t^{-4})$ 12

21. $(-5w^4)^3$

18. $-10c^3d^5$

$-125w^{12}$

Written Exercises

Write an equivalent expression using only positive exponents. Assume that no variable that appears in a denominator or with a negative exponent is zero.

A

1. $(-7u^2)(5u^8)$ $-35u^{10}$
2. $(-8v^3)(-6v^{-9})$ $\frac{48}{v^6}$
3. $(2x^4y^{-7})(3x^{-1}y^5)$ $\frac{6x^3}{y^2}$
4. $(9z^6w^8)(-5z^{-6}w^{-11})$ $-\frac{45}{w^3}$
5. $(5^{-3}a^{-2}b^{-1})(-25a^6b^{-4})$ $-\frac{a^4}{5b^5}$
6. $(-3p^4q^{-3})^2(4p^{-5}q^7)$ $36p^3q$
7. $\left(\frac{3}{4}d^3e^{-5}\right)^{-2}$ $\frac{16e^{10}}{9d^6}$
8. $\left(\frac{1}{2}a^2b^3\right)^4(a^{-5}b^{-10})$ $\frac{a^3b^2}{16}$
9. $(6x^8y^{-9})(x^{-2}y^{-3})^3$ $\frac{6x^2}{y^{18}}$
10. $(-v^7w^{-8})^3(-v^{-9}w^6)^4$ $-\frac{1}{v^{15}}$

11. $(u^6)^{-4}(u^3)^8$ 1
12. $(ab)^{-7}(a^3b^{-2})^4$ $\frac{a^5}{b^{15}}$
13. $\frac{c^4d^{-5}}{c^3d}$ $\frac{c}{d^6}$
14. $\left(\frac{f^5g^{-3}}{g^{-5}}\right)^2$ $f^{10}g^4$
15. $\frac{(2r^3s^{-3})^5}{4r^7s^{-10}}$ $\frac{8r^8}{s^5}$
16. $\frac{25x^{-5}y^3}{(5x^2y)^{-3}}$ $3125xy^6$
17. $\left(\frac{3u^{-4}v^3}{2u^{-5}v}\right)^{-2}$ $\frac{4}{9u^2v^4}$
18. $\left(-\frac{a^{-2}b^{-1}}{3a^{-3}b^3}\right)^{-4}$ $\frac{81b^{16}}{a^4}$

B

19. $\frac{(x+y)^{-2}}{(x+y)^{-1}}$ $\frac{1}{x+y}$
20. $\frac{a^{-1}-b^{-1}}{a^{-1}+b^{-1}}$ $\frac{b-a}{b+a}$
21. $(z-w)(w^{-1}-z^{-1})$ $\frac{w^2-2wz+z^2}{wz}$
22. $(d^{-1}+e^{-1})(d+e)^{-1}$ $\frac{1}{de}$
23. $\frac{a^{m+1}}{a^{m-1}}$ a^2
24. $(b^{2-p}b^{2+p})^2$ b^8
25. $2(4^k+4^k)4^{k+1}$
26. $5(5^{e+1})^{e-1}5^{e^2}$

27. Deduce Law 5 from Law 1.
28. Deduce Law 6 from Law 3.

29. Prove that for any positive integer n and any real number $b \neq 0$,
$$\frac{1}{b^{-n}} = b^n.$$

30. Prove that for any positive integer n and any nonzero real numbers a and b, $(ab)^{-n} = a^{-n}b^{-n}$.

C

31. Show that if Law 1 is extended to the case in which m or n equals 0, then we must define b^0 to be 1.

32. Show that if Law 1 is extended to the case in which m or n is negative, then we must define b^{-n} to be $\frac{1}{b^n}$. (*Hint:* Let $m = -n$ in Law 1.)

33. Prove the Basic Property of Quotients (page 201).
$$\left[\text{Hint: } \frac{rs}{tu} = rs \cdot \frac{1}{tu} = (r \cdot s) \cdot \left(\frac{1}{t} \cdot \frac{1}{u}\right).\right]$$

Polynomials and Rational Expressions **203**

Suggested Assignments

Minimum
   203/1–23 odd
Extended Alg. with Trig.
   203/9–17 odd, 18–27
Enriched Alg.
   203/13–33 odd
Enriched Alg. with Trig.
   203/13–33 odd

Mixed Review

1. Solve.
   $5x - y + z = 5$
   $3x + y - z = 3$
   $x + 2y - z = 3$ $\{(1, 2, 2)\}$

2. True or false?
   A system of equations is said to be inconsistent if the solution set is the empty set Ø. T

3. The relation $\{(-2, 1), (-1, 3), (0, 5)\}$ is a function. Would this be a true statement if the ordered pair $(1, 7)$ were added to the set? yes

4. Simplify.
   $[2 + 2(-8)] \div (-14 + 7)$.
   2

5. Write an equation of the line having slope 5 and y-intercept -7.
   $y = 5x - 7$

Careers

Business Management

Linear programming is used in a great variety of industrial and business operations. Its techniques are not only applied to production, but to purchasing and distribution, to job assignments, to budgeting, and even to advertising.

It is important to note that often the relationships among the variables of the problem are only probable. For example, in the ice skate example of Section 4-6, profits on each pair of skates are assumed to be constant while in fact such profits may vary depending on the sales volume. Linear programming is an approximation to a real-world situation. However, it is a sufficiently accurate mathematical model to be of great use to the world of business.

EXAMPLE An advertising manager of a company considers the alternatives in advertising various products in two weekly magazines. A half-page advertisement costs $300 in magazine A and $250 in magazine B per weekly issue. A recent advertising survey indicates that for every issue, 6000 readers will see the advertisement in magazine A and 5000 will see it in B. In addition, 200 readers of A and 300 of B per week will usually complete attached questionnaire cards for additional information. In order to profit from the advertising campaign, it was determined that at least 59,000 readers should be reached and at least 2900 request cards for additional information must be received. How many weekly advertisements should be placed with each magazine in order to minimize the cost for these advertisements?

SOLUTION Let x be the number of weekly advertisements in magazine A and y the number of weekly advertisements in magazine B. Graph the following equations:

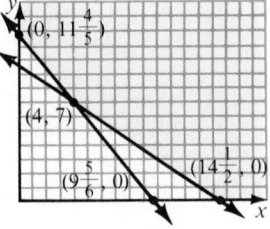

$$\begin{aligned} \text{Readers\}} \quad & 6000x + 5000y \geq 59{,}000 \\ \text{Cards\}} \quad & 200x + 300y \geq 2900 \\ & x \geq 0 \quad y \geq 0 \end{aligned}$$

The feasibility region is shown at the right.
Determining the minimum value of $300x + 250y$, you find that 4 ads should be placed in magazine A, and 7 in magazine B.

204 *Chapter 6*

6–2 Multiplying Polynomials

You can find the product of two polynomials by using the familiar axioms of addition and multiplication and the first law of exponents. For example, to find the product of the *binomial* $3x - 2$ and the *trinomial* $5x^4 - x^3 + 4x$, you can proceed as follows:

$$(3x - 2)(5x^4 - x^3 + 4x) = 3x(5x^4 - x^3 + 4x) - 2(5x^4 - x^3 + 4x)$$
$$= 15x^5 - 3x^4 + 12x^2 - 10x^4 + 2x^3 - 8x$$
$$= 15x^5 - 13x^4 + 2x^3 + 12x^2 - 8x$$

You write the product as a polynomial in simple form (page 42).

> To obtain the product of two polynomials, multiply each term of one of the polynomials by each term of the other, and then add all the monomial products.

You are less likely to make errors in adding like terms, if you use a vertical arrangement in multiplying.

$$
\begin{array}{r}
5x^4 - x^3 + 4x \\
3x - 2 \\
\hline
15x^5 - 3x^4 + 12x^2 \\
- 10x^4 + 2x^3 - 8x \\
\hline
15x^5 - 13x^4 + 2x^3 + 12x^2 - 8x
\end{array}
$$

Three special cases of binomial products that are useful to know are given here:

$$(a + b)^2 = a^2 + 2ab + b^2$$
$$(a - b)^2 = a^2 - 2ab + b^2$$
$$(a + b)(a - b) = a^2 - b^2$$

EXAMPLE   Express each product as a polynomial in simple form.

    a. $(3x - 2)(x + 4)$ **b.** $(4 + m)(4 - m)$

    c. $(w - 5)^2$ **d.** $(3x - 4y^2)^2$

SOLUTION   **a.** $(3x - 2)(x + 4) = 3x^2 + 12x - 2x - 8$
$$= 3x^2 + 10x - 8$$

    b. $(4 + m)(4 - m) = 4^2 - m^2 = 16 - m^2$

    c. $(w - 5)^2 = w^2 - 2(5w) + 5^2$
$$= w^2 - 10w + 25$$

    d. $(3x - 4y^2)^2 = (3x)^2 - 2(3x)(4y^2) + (4y^2)^2$
$$= 9x^2 - 24xy^2 + 16y^4$$

Teaching Suggestions
p. T90

Related Activities p. T90

Key Ideas

Use the distributive property and the laws of exponents to multiply polynomials.

Chalkboard Examples

Express each product as a polynomial in simple form.

1. $3x(x^2 - 4)^2$
   $3x(x^4 - 8x^2 + 16)$
   $3x^5 - 24x^3 + 48x$
2. $(3 + 2x)(5 + 7x)$
   $15 + 21x + 10x + 14x^2$
   $15 + 31x + + 14x^2$
3. $(3 + y)(3 - y)$
   $(3^2) - (y)^2$
   $9 - y^2$
4. $(2 - b)^2$
   $2^2 - 2(2 \cdot b) + b^2$
   $4 - 4b + b^2$

Additional A Exercises

Write each product as a polynomial in simple form.

1. $(1 - x)(1 + x)$ $1 - x^2$
2. $(1 - x)^2$ $1 - 2x + x^2$
3. $(2c^3 + 7d^2)(3c - 2d)$
   $6c^4 - 4c^3d - 14d^3 + 21cd^2$
4. $(3x + y)(x + y^2 + 3)$
   $3x^2 + 3xy^2 + 9x + xy + y^3 + 3y$
5. $(m^n + w)(m^n - w)$
   $m^{2n} - w^2$

Suggested Assignments

Minimum
206/1–17 odd, 19–30
Extended Alg. with Trig.
206/7–33 odd
Enriched Alg.
206/13, 19–31 odd, 33–38
Enriched Alg. with Trig.
206/13, 19–31 odd, 33–38

Additional Answers
Written Exercises

3. $t^2 + 8t + 16$

6. $4b^2 - 1$

8. $9c^2 + 30cd + 25d^2$

9. $4u^4 - 20u^2v + 25v^2$

11. $a^4 - 3a^2b + 2b^2$

12. $9t^2 - 42t + 49$

14. $2m^2 - 15m + 18$

15. $d^{10} + 2d^5e^3 + e^6$

18. $k^{2n} - 6k^np^m + 9p^{2m}$

20. $3a^3 + 5a^2b - 9ab^2 - 14b^3$

24. $v^3 - 12v^2w + 48vw^2 - 64w^3$

(continued on p. 212)

Mixed Review

1. Simplify $(5a^6)(-6a^{-5})$.
$-30a$

2. Evaluate $\begin{vmatrix} 2 & 1 & 1 \\ 2 & 1 & 4 \\ 1 & 5 & 3 \end{vmatrix}$. -27

3. If $f(x) = \dfrac{x}{3x - 1}$, find $f(1)$.
$\dfrac{1}{2}$

4. True or false? If $a > 0$ then $-a < 0$. T

5. Simplify $(28)\left(2\dfrac{2}{3}\right)(0)(5)$.
0

206

Oral Exercises

Express each product as a polynomial in simple form.

1. $(2a - b)(5a + 3b)$
2. $(8w + 3)(w - 5)$
3. $(2 - 3m)(5 - m)$
4. $(6 - r)(6 + r)$
5. $(3v + 1)(3v - 1)$
6. $(r + 7s)^2$
7. $(4p - 5q)^2$
8. $(h - k^3)^2$
9. $(x^2 + y)^2$

1. $10a^2 + ab - 3b^2$
2. $8w^2 - 37w - 15$
3. $10 - 17m + 3m^2$
4. $36 - r^2$
5. $9v^2 - 1$
6. $r^2 + 14rs + 49s^2$
7. $16p^2 - 40pq + 25q^2$
8. $h^2 - 2hk^3 + k^6$
9. $x^4 + 2x^2y + y^2$

Written Exercises

Write each product as a polynomial in simple form.

A

1. $(x + 3y)(x - 2y)$ $x^2 + xy - 6y^2$
2. $(6a + 5)(a + 3)$ $6a^2 + 23a + 15$
3. $(t + 4)^2$
4. $(3n - 2)^2$ $9n^2 - 12n + 4$
5. $(5 + s)(5 - s)$ $25 - s^2$
6. $(2b + 1)(2b - 1)$
7. $(x^3 - 2y)(x^3 + 2y)$ $x^6 - 4y^2$
8. $(3c + 5d)^2$
9. $(2u^2 - 5v)^2$
10. $9x^7(x^3 - 4x)$ $9x^{10} - 36x^8$
11. $(a^2 - b)(a^2 - 2b)$
12. $(3t - 7)^2$
13. $(n^5 - 8n^3)6n^2$ $6n^7 - 48n^5$
14. $(2m - 3)(m - 6)$
15. $(d^5 + e^3)^2$
16. $(4 - 7h^3)^2$ $16 - 56h^3 + 49h^6$
17. $(r^n - 5)(r^n + 5)$ $r^{20} - 25$
18. $(k^n - 3p^m)^2$
19. $(2x - 3)(2x^2 + 3x - 5)$ $4x^3 - 19x + 15$
20. $(a + 2b)(3a^2 - ab - 7b^2)$
21. $(u - 4)(u^2 + 4u + 16)$ $u^3 - 64$
22. $(5 - t)(25 + 5t + t^2)$ $125 - t^3$

B

23. $(z + 3)^3$ $z^3 + 9z^2 + 27z + 27$
24. $(v - 4w)^3$
25. $(x - y)(x + y)(x^2 + y^2)$ $x^4 - y^4$
26. $(2c - 3d)^2(2c + 3d)^2$
27. $(p + q)(p - q)(p^4 + p^2q^2 + q^4)$ $p^6 - q^6$
28. $(h - 2k)^4$

29. Write a formula for simplifying a product that has the form
$$(a - b)(a^2 + ab + b^2).\ a^3 - b^3$$
Check the formula you wrote by applying it to the products in Exercises 21 and 22.

30. Write a formula for expressing $(a + b)^3$ as a polynomial in simple form. Check the formula you wrote by applying it to the product in Exercise 23. $a^3 + 3a^2b + 3ab^2 + b^3$

Write each product as a polynomial in simple form.

C

31. $(x^n + y^n)(x^{2n} + x^ny^n + y^{2n})$
32. $(r^{4n} - s^{3m})^2$ $r^{8n} - 2r^{4n}s^{3m} + s^{6m}$
33. $(a^{p+1} - a^2b^q)(a^{p-1} + b^q)$ $a^{2p} - a^2b^{2q}$
34. $(v^2 - vw + w^2)(v^2 + vw + w^2)$ $v^4 + v^2w^2 + w^4$
35. $(a - b)(a^4 + a^3b + a^2b^2 + ab^3 + b^4)$ $a^5 - b^5$
36. $(z - 1)(z^9 + z^8 + \ldots + z + 1)$ $z^{10} - 1$
37. Show that $(x^n - x^{-n})^2 + 4 = (x^n + x^{-n})^2$.
38. Use the fact that $(a - b)^2 \geq 0$ to prove that the average of the squares of two real numbers is always at least as large as their product.

206   *Chapter 6*

6–3 Factoring a Polynomial

Teaching Suggestions
p. T90

Recall from Section 1-3 that when two or more numbers are multiplied, each of the numbers is called a *factor.* When a number is expressed as the product of two or more members of a given set, the number is **factored** over that set. The set from which the factors are chosen is called the **factor set.**

If a, b, and c are integers, and c is a factor of both a and b, then c is a common factor of a and b. The greatest integer that is a factor of each of the given integers is called their **greatest common factor,** or **GCF.** For example, 15 is the GCF of 30 and 45.

If the GCF of any two integers is 1, then the integers are said to be **relatively prime.** For example, 10 and 21 are relatively prime.

In Section 6-2 you saw that:

I. $a^2 - b^2 = (a - b)(a + b)$ Differences of squares
II. $a^2 - 2ab + b^2 = (a - b)(a - b) = (a - b)^2$ Trinomial square
III. $a^2 + 2ab + b^2 = (a + b)(a + b) = (a + b)^2$ Trinomial square

A polynomial, such as any of those in I–III, that can be expressed as a product of two or more *polynomials of lower positive degree* is said to be **reducible.** Each of the polynomials that is multiplied is called a *factor* of the given polynomial. To **factor a polynomial** over a designated set, express it as a product of polynomials whose coefficients are all in the designated set.

Unless stated otherwise, we assume that *all polynomials with integral coefficients are to be factored over the integers.*

EXAMPLE 1 Factor the polynomial $6xy^2 + 15x^3y - 3xy$.

SOLUTION Notice that $3xy$ is the monomial of greatest coefficient and degree that is a factor of each term in the given polynomial. Then by the distributive law,

$$6xy^2 + 15x^3y - 3xy = 3xy(2y + 5x^2 - 1).$$

In Example 1, $3xy$ is called the **greatest monomial factor** of the given polynomial because it is the monomial with the *greatest* numerical coefficient and the *greatest* degree that is a factor of each term of the polynomial. The other factor, $2y + 5x^2 - 1$, cannot be reduced to a product of factors of lower positive degree, and is hence **irreducible.** Moreover, its greatest monomial factor is 1.

A polynomial is said to be **factored completely** when it has been expressed as the product of a constant and one or more irreducible polynomials or powers of irreducible polynomials each of which has 1 as its greatest monomial factor. In Example 1 the constant is 3, and the irreducible polynomials are x, y, and $2y + 5x^2 - 1$.

Key Ideas
Express a polynomial in factored form.

Factor each polynomial completely.

1. $9x^2 - 30xy + 25y^2$
$(3x)^2 - 30xy + (5y)^2$
Trinomial square
$(3x - 5y)^2$

2. $18p^3 - 32pq^2$
$2p(9p^2 - 16q^2)$
$2p((3p)^2 - (4q)^2)$
Difference of squares
$2p(3p - 4q)(3p + 4q)$

3. $y^{12} - 8z^3$
$(y^4)^3 - (2z)^3$
Difference of cubes
$(y^4 - 2z)(y^8 + 2y^4z + 4z^2)$

EXAMPLE 2   Factor each polynomial completely.

       a. $5c^3 - 10c^2d + 5cd^2$

       b. $x^4 - y^4$

SOLUTION   **a.** $5c^3 - 10c^2d + 5cd^2 = 5c(c^2 - 2cd + d^2) = 5c(c - d)^2$

       b. $x^4 - y^4 = (x^2 - y^2)(x^2 + y^2) = (x + y)(x - y)(x^2 + y^2)$

       Since $x^2 + y^2$ cannot be reduced to a product of factors of lower positive degree, $x^2 + y^2$ is irreducible and no further factorization is possible.

   Recall (p. 43) that a quadratic polynomial is a polynomial of the form

$$ax^2 + bx + c,$$

where $a \neq 0$, and a, b, and $c \in \mathcal{R}$. We call the second degree term, or ax^2, the **quadratic term.** The first degree term, or bx, is called the **linear term,** and the numerical term, or c, is called the **constant term.**

   Every reducible *quadratic trinomial* of the form $ax^2 + bx + c$ has two binomial factors, of the form $Ax + B$ and $Cx + D$. Since

$$(Ax + B)(Cx + D) = ACx^2 + (AD + BC)x + BD,$$

the problem of factoring such a trinomial is to find values of the coefficients A, B, C, and D such that

$$AC = a, \quad\quad AD + BC = b, \quad\quad \text{and} \quad\quad BD = c.$$

EXAMPLE 3   Factor $10x^2 + 11x + 3$ completely.

SOLUTION   The coefficients can be analyzed as follows: Since the coefficient of x^2, which is a, or AC, is positive, we know that A and C must both be of the same sign, and likewise since the constant term c, or BD, is positive, B and D must be of the same sign. (Two numbers are of the "same sign" if both are positive or both are negative; they are of "opposite signs" if one is positive and the other negative.) Since the coefficient of x, which is b, or $AD + BC$, is also positive, A and C cannot have the opposite sign of D and B or else AD and BC would both be negative. Hence A, B, C, and D must all be of the same sign, which we can take to be positive. Since $AC = 10$ and $BC = 3$, the possible factorizations are:

A	B	C	D	AC	AD + BC	BD
$\downarrow$	$\downarrow$	$\downarrow$	$\downarrow$	$\downarrow$		$\downarrow$
$(10x + 1)(x + 3)$				$= 10x^2 +$	$31x$	$+\ 3$
$(10x + 3)(x + 1)$				$= 10x^2 +$	$13x$	$+\ 3$
$(5x + 1)(2x + 3)$				$= 10x^2 +$	$17x$	$+\ 3$
$(5x + 3)(2x + 1)$				$= 10x^2 +$	$11x$	$+\ 3$

   The combination of factors shown in red makes $AD + BC = b$, or 11.

     $\therefore$ in factored form, $10x^2 + 11x + 3 = (5x + 3)(2x + 1)$.

   Notice that in Example 3, if we had chosen A, B, C, and D as all

208   *Chapter 6*

negative instead of all positive, the result would have been equivalent. That is,

$$[-5x + (-3)][-2x + (-1)] = 10x^2 + 11x + 3.$$

EXAMPLE 4 Factor $6y^2 - 7y - 5$ completely.

SOLUTION The process of finding the possible factors can be shortened by writing only the coefficients for each factor.

6		−5		−7
$A \times C$		$B \times D$		$AD + BC$
3	2	1	−5	$-15 + 2 = -13$
3	2	−1	5	$15 - 2 = 13$
3	2	5	−1	$-3 + 10 = 7$
3	2	−5	1	$3 - 10 = -7$

Hence the factors are $(3y - 5)$ and $(2y + 1)$ and

$$6y^2 - 7y - 5 = (3y - 5)(2y + 1).$$

Of course, as soon as you find the correct factors to make

$$AC = a, \ AD + BC = b, \text{ and } BD = c,$$

there is no point in going through the other possibilities.

EXAMPLE 5 Factor $x^2 - 5x + 2$ completely.

SOLUTION Use the values of the coefficients to try to find the correct factors.

1		2		−5
$A \times C$		$B \times D$		$AD + BC$
1	1	2	1	3
1	1	−2	−1	−3

Since there are no other different factorizations possible, $x^2 - 5x + 2$ is irreducible.

Two other factor patterns that are useful to know besides the three at the beginning of this section are:

IV. $a^3 + b^3 = (a + b)(a^2 - ab + b^2)$ Sum of cubes
V. $a^3 - b^3 = (a - b)(a^2 + ab + b^2)$ Differences of cubes

EXAMPLE 6 Factor $8x^3 + 125$ completely.

SOLUTION Use formula IV with $a = 2x$ and $b = 5$.

$$\therefore 8x^3 + 125 = (2x + 5)(4x^2 - 10x + 25).$$

Polynomials and Rational Expressions   **209**

Suggested Assignments

Minimum
   210/1–16, 17–27 odd
   R 211/Self-Test 1
Extended Alg. with Trig.
   210/1–16, 17–27 odd
   R 211/Self-Test 1
Enriched Alg.
   210/1–25
   R 211/Self-Test 1
Enriched Alg. with Trig.
   210/1–25
   R 211/Self-Test 1

Additional Answers
Written Exercises

2. $(5a - 3b)(a - 2b)$
14. $(2g^2 - 5)(4g^4 + 10g^2 + 25)$
16. $(2p^5 - 3q^4)(2p^5 + 3q^4)$
20. $3c(c - 5d)^2$
22. $(t - 3)(t + 3)(t - 1)(t + 1)$
24. $(1 + 4a^6)(1 - 2a^3)(1 + 2a^3)$
26. $(d - 10e - 1)(d + 10e - 1)$
28. $[(r + 1) - s][(r + 1)^2 + s(r + 1) + s^2]$
32. $(7x^{2y} + 3)(x^{2y} - 2)$
34. $d(d^k + 2)(d^k - 2) \cdot (d^k + 1)(d^k - 1)$
36. $(x - y)(x^2 + xy + y^2) \cdot (x + y)(x^2 - xy + y^2);$
    $(x - y)(x + y) \cdot (x^4 + x^2y^2 + y^4);$
    $\therefore (x^2 + xy + y^2) \cdot (x^2 - xy + y^2) = x^4 + x^2y^2 + y^4$

37. Suppose m and n are not relatively prime. Then there are nonzero integers d, m_1, and n_1, such that $m = m_1 d$ and $n = n_1 d$, $(d \neq 1)$. Thus by subst. $am - bn = am_1 d - bn_1 d = (am_1 - bn_1)d$ $= 1$, and $am_1 - bn_1 = \frac{1}{d}$.
    But $am_1 - bn_1$ is an integer. $\therefore m$ and n are relatively prime.

Oral Exercises

State the greatest monomial factor of each polynomial.

1. $42xy^2 - 30x^2y$ $6xy$
2. $24a^2 + 16b^2$ 8
3. $c^4d^3 - c^6d^2$ c^4d^2
4. $35p^3q + 20pq^2$ $5pq$
5. $14r^2s^2 - 21rs^2t$ $7rs^2$
6. $x^3y + x^4y^2 - x^2y^3$ x^2y

Complete the factorization of each polynomial by filling in each blank with a binomial.

7. $8k^3 + 1 = (\underline{\ ?\ })(4k^2 - 2k + 1)$ $2k + 1$
8. $r^6 - s^3 = (\underline{\ ?\ })(r^4 + r^2s + s^2)$ $r^2 - s$
9. $x^{16} - y^2 = (\underline{\ ?\ })(x^8 + y)$ $x^8 - y$
10. $3w^2 - 2w - 8 = (\underline{\ ?\ })(w - 2)$ $3w + 4$

Factor each polynomial completely.

11. $p^3 - q^3$
12. $y^3 - 8$
13. $8x^3 - 1$
14. $r^3 + s^3$
15. $x^3 + 27$
16. $64n^3 + 1$
11. $(p - q)(p^2 + pq + q^2)$
12. $(y - 2)(y^2 + 2y + 4)$
13. $(2x - 1)(4x^2 + 2x + 1)$
14. $(r + s)(s^2 - rs + s^2)$
15. $(x + 3)(x^2 - 3x + 9)$
16. $(4n + 1)(16n^2 - 4n + 1)$

Written Exercises

Factor each polynomial completely. Write "irreducible" for any that cannot be factored over the set of polynomials with integral coefficients.

A
1. $25x^2 + 30x + 9$ $(5x + 3)^2$
2. $5a^2 - 13ab + 6b^2$
3. $49 - 36k^2$ $(7 - 6k)(7 + 6k)$
4. $25z^2 + 16$ irreducible
5. $1 - 18r + 81r^2$ $(1 - 9r)^2$
6. $h^3 + 125$ $(h + 5)(h^2 - 5h + 25)$
7. $27n^3 - 64m^3$ $(3n - 4m)(9n^2 + 12mn + 16m^2)$
8. $4c^2 - 28cd + 49d^2$ $(2c - 7d)^2$
9. $3u^2 + uv - 10v^2$ $(3u - 5v)(u + 2v)$
10. $6t^2 - 11t - 10$ $(3t + 2)(2t - 5)$
11. $27w^3 + 64z^3$ $(3w + 4z)(9w^2 - 12wz + 16z^2)$
12. $p^4 - 121q^2$ $(p^2 - 11q)(p^2 + 11q)$
13. $x^2 + 4x + 1$ irreducible
14. $8g^6 - 125$
15. $3a^2b^2 + ab - 24$ $(3ab - 8)(ab + 3)$
16. $4p^{10} - 9q^8$

B
17. $64xy^2 - 9x^3$ $x(8y - 3x)(8y + 3x)$
18. $80r^4 - 45s^2$ $5(4r^2 - 3s)(4r^2 + 3s)$
19. $81w^4 - 16$ $(3w - 2)(3w + 2)(9w^2 + 4)$
20. $3c^3 - 30c^2d + 75cd^2$
21. $z^6 - 14z^3 + 49$ $(z^3 - 7)^2$
22. $t^4 - 10t^2 + 9$
23. $50x^3y^2 + 40x^2y^3 + 8xy^4$ $2xy^2(5x + 2y)^2$
24. $1 - 16a^{12}$
25. $8b^4c + 27bc^4$ $bc(2b + 3c)(4b^2 - 6bc + 9c^2)$
26. $(d^2 - 2d + 1) - 100e^2$
27. $(m - 1)^2 - (n + 1)^2$ $(m + n)(m - n - 2)$
28. $(r + 1)^3 - s^3$

C
29. $p^{3n} - q^{3n}$ $(p^n - q^n)(p^{2n} + p^nq^n + q^{2n})$
30. $25k^{2m} - 10k^m + 1$ $(5k^m - 1)^2$
31. $16v^{6n} - 49u^{4n}$ $(4v^{3n} - 7u^{2n})(4v^{3n} + 7u^{2n})$
32. $7x^{4y} - 11x^{2y} - 6$
33. $c^{4n} - 13c^{2n} + 36$ $(c^n + 3)(c^n - 3)(c^n + 2)(c^n - 2)$
34. $d^{4k+1} - 5d^{2k+1} + 4d$

35. Factor $x^4 + 4y^4$. (*Hint:* Write the given polynomial as $x^4 + 4x^2y^2 + 4y^4 - 4x^2y^2$.) $(x^2 - 2xy + 2y^2)(x^2 + 2xy + 2y^2)$

36. Factor $x^6 - y^6$ both as a difference of squares and as a difference of cubes. From these two factorizations determine that a product of two special trinomials equals another special trinomial.

37. Show that if $a, b, m,$ and n are integers such that $am - bn = 1$, then m and n are relatively prime.

Computer Exercises For students with computer experience

1. Write a program that will factor a binomial of the form $ax^n + by^m$ as the sum of two cubes over the integers, if possible. (Note that on many computers, exponentiation is not exact. Instead of testing $\sqrt[3]{a}$ and $\sqrt[3]{b}$ to see if they are integers, you may have to test if the cube roots are close to an integer within a certain tolerance, and round off before displaying them.)

Run the program in Exercise 1 for each binomial.

2. $343x^{12} + 729y^3$ **3.** $1331x^6 + 2744y^{15}$ **4.** $576x^9 + 216y^{18}$ **5.** $x^{21} + 1728y^{39}$
2. $(7x^4 + 9y)(49x^8 - 63x^4y + 81y^2)$ **3.** $(11x^2 + 14y^5)(121x^4 - 154x^2y^5 + 196y^{10})$
4. irreducible   **5.** $(x^7 + 12y^{13})(x^{14} - 12x^7y^{13} + 144y^{26})$

▌Self-Test 1

VOCABULARY   factored over a set (p. 207)   greatest monomial factor
              factor set (p. 207)   (p. 207)
              greatest common factor, or   irreducible (p. 207)
                GCF (p. 207)   factor a polynomial completely
              relatively prime (p. 207)   (p. 207)
              reducible (p. 207)   quadratic term (p.208)
              factor a polynomial (p. 207)   linear term (p. 208)
                                                  constant term (p. 208)

Write an equivalent expression using only positive exponents. Assume that no variable equals 0.

1. $(\frac{2}{3}x^2y^3z)^2(x^{-3}y^2z^{-4})$ $\frac{4xy^8}{9z^2}$ **2.** $\frac{12a^{-3}b^{-2}c^3}{36a^{-2}b^5c^{-1}}$ $\frac{c^4}{3ab^7}$ *Obj. 1, p. 199*

3. Write $(3x - 2)^4$ as a polynomial in simple form.   *Obj. 2, p. 199*
   $81x^4 - 216x^3 + 216x^2 - 96x + 16$

Factor completely.

4. $8x^2 - 2x - 3$ $(4x - 3)(2x + 1)$ **5.** $8g^3 - h^3$ $(2g - h)(4g^2 + 2gh + h^2)$ *Obj. 3, p. 199*
6. $x^2 - 20x + 64$ $(x - 16)(x - 4)$ **7.** $z^2 - 4z + 1$ irreducible

Check your answers with those at the back of the book.

Polynomials and Rational Expressions   **211**

---

1. Compute 4^5.   1024

2. Find the slope of a line if its equation is $3y + 2x = $

3. $-\frac{2}{3}$

3. Simplify $-15(41 - 21)$.
   -300

4. State the formula for finding the area of a trapezoid. $A = \frac{1}{2}h(b_1 + b_2)$

5. True or false? If $a > 1$; then $\frac{1}{a} > 1$.   F

Write an equivalent expression using only positive exponents. Assume that no variable equals 0.

1. $(6x^3y^{-4})(\frac{1}{2}x^{-5}y^7)$ $\frac{3y^3}{x^2}$

2. $\frac{6a^{-2}b^{-5}c^{-1}}{18a^{-3}b^2c^3}$ $\frac{a}{3b^7c^4}$

3. Write $(5x + 7)^3$ as a polynomial in simple form.
   $125x^3 + 525x^2 + 735x + 343$

Factor completely.

4. $6x^2 + 7x + 2$
   $(3x + 2)(2x + 1)$

5. $r^6 - 27s^3$
   $(r^2 - 3s)(r^4 + 3r^2s + 9s^2)$

6. $x^2 - 15x + 36$
   $(x - 3)(x - 12)$

7. $x^2 + 3x + 1$ prime

26. $16c^4 - 72c^2d^2 + 81d^4$

28. $h^4 - 8h^3k + 24h^2k^2 - 32hk^3 + 16k^4$

31. $x^{3n} + 2x^{2n}y^n + 2x^ny^{2n} + y^{3n}$

37. $(x^n - x^{-n})^2 = x^{2n} - 2 + x^{-2n}$
$(x^n + x^{-n})^2 = x^{2n} + 2 + x^{-2n}$
Thus $(x^n - x^{-n})^2 + 4 = (x^n + x^{-n})^2$

38. Let a and b be any real numbers. Then the average of their squares is $\dfrac{a^2 + b^2}{2}$, and their product is ab. Since $(a - b)^2 \geq 0$, $a^2 - 2ab + b^2 \geq 0$ and $a^2 + b^2 \geq 2ab$. Thus $\dfrac{a^2 + b^2}{2} \geq ab$.

READING ALGEBRA Vocabulary

A good mathematics vocabulary is essential to reading algebra. The vocabulary used in mathematics contains many words that are the same as those used in everyday conversation. Some words, however, such as *root* and *imaginary* have different meanings when they are used in a mathematical context.

For example, the word imaginary came into use in the sixteenth and seventeenth centuries to describe the number $\sqrt{-1}$. The choice of this word reflects the uneasiness that mathematicians felt with these "nonreal" numbers. Today these numbers are used in map making, quantum mechanics, and electrical AC circuits.

A rational number is defined as a number that can be expressed as a ratio of two integers. (See page 221.) Thus the definition of rational numbers is based on the word *ratio*.

To use these mathematical words successfully you should know their precise mathematical meaning. Important mathematical terms are printed in red in the text and a vocabulary list appears at the beginning of each Self-Test. The Glossary, beginning on page 801, gives the mathematical definition of words and the Index beginning on page 807 gives you page references for additional information.

Exercises

1. Skim through Chapter 5 and list the vocabulary words that have a specialized mathematical meaning.

Explain the difference between the everyday usage and the mathematical usage of the following words.

2. average	**3.** complex
4. event	**5.** coordinate
6. irrational	**7.** polar
8. radical	**9.** real
10. root	**11.** translation
12. degree	**13.** index
14. mean	**15.** base
16. characteristic	**17.** factor
18. mapping	**19.** minor
20. odd	**21.** plane
22. power	**23.** rational
24. term	**25.** identity

Applications of Factoring

OBJECTIVES for Sections 6-4 and 6-5:
1. *To solve polynomial equations by factoring.*
2. *To solve problems using factorable polynomial equations.*
3. *To solve polynomial inequalities by factoring.*

Teaching Suggestions
p. T90

Key Ideas

Use factoring to solve polynomial equations.

6–4 Solving Equations by Factoring

The quadratic polynomial in the equation

$$3x^2 - 5x + 2 = 0$$

can be factored as

$$(3x - 2)(x - 1).$$

To find the solution set of the equation use the following theorem, which states that *a product of real numbers is zero if and only if at least one of the factors is zero.*

Theorem. For all a and $b \in \mathcal{R}$, $ab = 0$ if and only if

$$a = 0 \text{ or } b = 0.$$

PROOF

I. The "if" part says that if either a or b is zero, then $ab = 0$. This follows directly from the multiplication property of zero, that is:

$$a \cdot 0 = 0 \cdot a = 0$$

II. The "only if" part says that if $ab = 0$, then at least one of a and b is zero. The reasoning goes as follows: Suppose $b \neq 0$. We want to show that a must then be zero.

Statements	*Reasons*
1. $ab = 0$	Hypothesis
2. $ab\left(\frac{1}{b}\right) = 0 \cdot \frac{1}{b}$	Multiplication property of equality
3. $a\left(b \cdot \frac{1}{b}\right) = 0$	Associative axiom and multiplication property of zero
4. $a \cdot 1 = 0$	Axiom of multiplicative inverses
5. $a = 0$	Identity axiom for multiplication

Polynomials and Rational Expressions **213**

Solve by factoring.

1. $y^2 - 8y = -12$
$y^2 - 8y + 12 = 0$
$(y - 2)(y - 6) = 0$
$y - 2 = 0$ or $y - 6 = 0$
$y = 2$ or $y = 6$
$\{2, 6\}$

2. $x^3 - x^2 = 6x$
$x^3 - x^2 - 6x = 0$
$x(x^2 - x - 6) = 0$
$x(x - 3)(x + 2) = 0$
$x = 0$ or $x - 3 = 0$ or
$x + 2 = 0$
$x = 0$ or $x = 3$ or $x = -2$
$\{0, 3, -2\}$

3. The surface area of a cardboard box with rectangular sides and a square base is 2800 cm². If the box has a height of 25 cm, what is the volume of the box?
Let x = the length of the side of the base.
x^2 = the area of the base
$25x$ = the area of each side of the box.
The surface area of the box is:
$2x^2 + 4(25x) = 2800$
$2x^2 + 100x - 2800 = 0$
$x^2 + 50x - 1400 = 0$
$(x + 70)(x - 20) = 0$
$x + 70 = 0$ or $x - 20 = 0$
$x = -70$ or $x = 20$
Since the length must be positive -70 is not a reasonable solution and therefore 20 is the only possible solution.
∴ the volume of the box is $20 \cdot 20 \cdot 25 = 10{,}000$ cm³.

From the preceding theorem we know that the equation at the beginning of the section,

$$3x^2 - 5x + 2 = 0, \text{ or } (3x - 2)(x - 1) = 0,$$

is equivalent to the statement that either $3x - 2 = 0$, or $x - 1 = 0$. Solving these two linear equations, we obtain

$$x = \tfrac{2}{3} \quad \text{or} \quad x = 1.$$

Checking, you find that each of these values satisfies the original equation. Hence the solution set is $\{\tfrac{2}{3}\} \cup \{1\} = \{\tfrac{2}{3}, 1\}$.

Sometimes a polynomial has two or more factors that are identical. These factors yield a **double** or **multiple root**.

EXAMPLE 1 Solve $y^3 - 3y^2 = 10y$.

SOLUTION

1. Transform the equation into an equivalent equation that has 0 on one side.

$$y^3 - 3y^2 - 10y = 0$$

2. Factor completely.

$$y(y^2 - 3y - 10) = 0$$
$$y(y - 5)(y + 2) = 0$$

3. Solve the compound sentence:

$$y = 0 \quad \text{or} \quad y - 5 = 0 \quad \text{or} \quad y + 2 = 0$$
$$y = 0 \quad \text{or} \quad y = 5 \quad \text{or} \quad y = -2$$

4. Check each solution in the *original* equation, $y^3 - 3y^2 = 10y$.

$$0 - 0 = 0 \qquad 5^3 - 3 \cdot 5^2 = 10 \cdot 5 \qquad (-2)^3 - 3(-2)^2 = 10(-2)$$
$$0 = 0 \checkmark \qquad 125 - 75 = 50 \checkmark \qquad -8 - 12 = -20 \checkmark$$

∴ the solution set is $\{0, 5, -2\}$.

EXAMPLE 2 A manufacturer uses 1400 cm² of cardboard to make a box with rectangular sides. If the box has a square base and a height of 30 cm, what is the volume of the box?

SOLUTION

1. The problem asks for the volume of the box. To find the volume, we need to know only the length of one side of the base, since the base is a square and the height is given.

2. Let x = the length of a side of the base. Then the area of the top of the box is x^2 as is the area of the bottom of the box. The area of each side is $30x$.

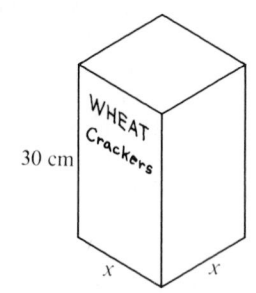

30 cm

WHEAT Crackers

x x

3. The surface area of the box is 1400 cm².

$$2(x^2) + 4(30x) = 1400$$

4. Solve the equation.

$$2(x^2) + 4(30x) = 1400$$
$$2x^2 + 120x = 1400$$
$$2x^2 + 120x - 1400 = 0$$
$$x^2 + 60x - 700 = 0$$
$$(x + 70)(x - 10) = 0$$

$$x + 70 = 0 \quad \text{or} \quad x - 10 = 0$$
$$x = -70 \quad \text{or} \quad x = 10$$

Since the length of the base must be positive, a solution of -70 is impossible. Hence 10 is the only possible solution.

5. The surface area of the box is $2 \cdot 100 + 4 \cdot 300 = 1400.\checkmark$

∴ the volume of the box = $lwh = 10 \cdot 10 \cdot 30 = 3000$ cm³.

Oral Exercises

State the solution set of each equation.

1. $(x - 4)(3x - 1) = 0$ $\{4, \frac{1}{3}\}$

2. $(a + 5)(2a - 3) = 0$ $\{-5, \frac{3}{2}\}$

3. $(2y + 1)(y - 7) = 0$ $\{-\frac{1}{2}, 7\}$

4. $(b + 5)b = 0$ $\{-5, 0\}$

5. $c^2 - 3c = 0$ $\{0, 3\}$

6. $z^2 - 16 = 0$ $\{4, -4\}$

A rectangular swimming pool is surrounded by a cement walk of uniform width. Let x represent this width. For each of the following conditions, set up an equation involving x.

7. The pool measures 6 m by 10 m, and the total area of the pool and walk is 96 m². $(6 + 2x)(10 + 2x) = 96$

8. The pool measures 8 m by 14 m, and the area of the walk alone is 104 m². $(8 + 2x)(14 + 2x) - 8 \cdot 14 = 104$

9. The outer dimensions of the walk are 16 m by 10 m, and the area of the pool is 112 m². $(16 - 2x)(10 - 2x) = 112$

10. The outer dimensions of the walk are 16 m by 22 m and the area of the walk alone is 192 m². $16 \cdot 22 - (16 - 2x)(22 - 2x) = 192$

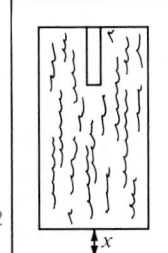

11. Explain why the following "solution" is not correct.

$$x^2 - 2x - 3 = 1$$
$$(x - 3)(x + 1) = 1$$
$$x - 3 = 1 \quad \text{or} \quad x + 1 = 1$$
$$x = 4 \quad \text{or} \quad x = 0$$

If the product of two factors is 1, then one of the factors does not necessarily equal 1.

Polynomials and Rational Expressions **215**

Suggested Assignments

Minimum
  216/1–23
  216/P: 1–13
Extended Alg. with Trig.
  216/1–24
  216/P: 1–12
Enriched Alg.
  216/5, 7, 17–25 odd
  216/P: 1–24
Enriched Alg. with Trig.
  216/5, 7, 17–25 odd
  216/P: 1–24

Additional Answers
Written Exercises

26. If $abc = 0$, then by the assoc. prop. $(ab)c = 0$. But $(ab)c = 0$ if and only if $ab = 0$ or $c = 0$ and $ab = 0$ if and only if $a = 0$ or $b = 0$. ∴ if $abc = 0$, $a = 0$, $b = 0$, or $c = 0$.

Additional A Exercises

Solve.

1. $(x - 3)\left(x + \dfrac{1}{2}\right) = 0$

$\left\{3, -\dfrac{1}{2}\right\}$

2. $5a^2 = a$ $\left\{0, \dfrac{1}{5}\right\}$

3. $4x^2 - 16 = 0$ $\{2, -2\}$

4. $x^2 - 4x = 12$ $\{-2, 6\}$

5. $9x^2 - 6x + 1 = 0$ $\left\{\dfrac{1}{3}\right\}$

Written Exercises

Solve.

A **1.** $x^2 - 3x - 28 = 0$ $\{7, -4\}$ **2.** $a^2 - 144 = 0$ $\{12, -12\}$

3. $50 - 2x^2 = 0$ $\{5, -5\}$ **4.** $x^2 - 14x + 49 = 0$ $\{7\}$

5. $4y^2 + 12y + 9 = 0$ $\left\{-\dfrac{3}{2}\right\}$ **6.** $2x^2 + x - 10 = 0$ $\left\{-\dfrac{5}{2}, 2\right\}$

7. $320 - 5b^2 = 0$ $\{8, -8\}$ **8.** $2r^2 - 5r = 0$ $\left\{0, \dfrac{5}{2}\right\}$

9. $20x^2 = 45x$ $\left\{0, \dfrac{9}{4}\right\}$ **10.** $15 - 4m - 4m^2 = 0$ $\left\{-\dfrac{5}{2}, \dfrac{3}{2}\right\}$

11. $x^2 - 13x = 48$ $\{16, -3\}$ **12.** $3n^2 - 10n + 8 = 0$ $\left\{\dfrac{4}{3}, 2\right\}$

13. $6x^2 = x + 12$ $\left\{-\dfrac{4}{3}, \dfrac{3}{2}\right\}$ **14.** $5 = 8d^2 - 6d$ $\left\{-\dfrac{1}{2}, \dfrac{5}{4}\right\}$

15. $4w^2 = 11w - 6$ $\left\{\dfrac{3}{4}, 2\right\}$ **16.** $(x - 4)(3x + 5) = 4x$ $\left\{-\dfrac{4}{3}, 5\right\}$

B **17.** $(5x - 6)^2 = 5x$ $\left\{\dfrac{9}{5}, \dfrac{4}{5}\right\}$

18. $(x - 8)(x - 7) - 2x(x + 3) = 8 + x$ $\{-24, 2\}$

19. $(2x + 3)(x + 1) - 3(x + 2)(x + 1) = -3(x + 5)$ $\{-4, 3\}$

20. $z^3 - 4z^2 - 21z = 0$ $\{0, 7, -3\}$

21. $3a^3 - 17a^2 + 10a = 0$ $\left\{0, \dfrac{2}{3}, 5\right\}$

22. $x^4 - 17x^2 + 16 = 0$ $\{-1, 1, 4, -4\}$

23. $x^4 - 21x^2 - 100 = 0$ $\{5, -5\}$

C **24.** $x^2(x^2 - 9) - 25(x^2 - 9) = 0$ $\{3, -3, 5, -5\}$

25. $3x^5 - x^4 - (3x - 1) = 0$ $\left\{\dfrac{1}{3}, 1, -1\right\}$

26. Prove that if $abc = 0$, then $a = 0$, $b = 0$, or $c = 0$. (*Hint*: Use the associative property.)

Problems

A **1.** The perimeter of a rectangle is 50 cm and its area is 144 cm². Find the dimensions of the rectangle. 16 cm by 9 cm

2. The area of a rectangle is 480 cm², and its length is 4 cm more than 3 times its width. Find the dimensions of the rectangle. 40 cm by 12 cm

3. Use the conditions of Oral Exercise 7 on page 215 to find the width of the walk. 1 m

4. Use the conditions of Oral Exercise 10 on page 215 to find the width of the walk. 3 m

5. The longer leg of a right triangle has length 1 cm less than twice the shorter leg. The hypotenuse has length 1 cm greater than the shorter leg. Find the lengths of the three sides of the triangle. 2.5 cm, 2 cm, 1.5 cm

6. One side of a rectangle is 0.8 cm longer than an adjacent side. If each diagonal of the rectangle has length of 4 cm, find the dimensions of the rectangle. 3.2 cm by 2.4 cm

216   Chapter 6

7. A layout artist wants to have 270 cm² of graphic art on a page that measures 20 cm by 28 cm. If the bottom border of the page must be 3 times as deep as the borders on the other three sides, how wide should each border be?
top and each side: 2.5 cm; bottom: 7.5 cm

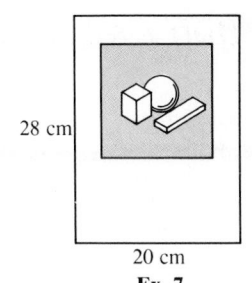

28 cm

20 cm
Ex. 7

8. A brick walk of uniform width is to be built around three sides of a garden that measures 12 m by 16 m. If there are enough bricks to cover 88 m², how wide can the walk be? 2 m

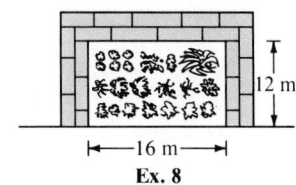

12 m

|←——16 m——→|
Ex. 8

9. Henry leaves Tankerville Square at 10:00 A.M. walking east at 2 km/h. At 10:30 A.M. Marlene leaves from the same place and walks south at 4 km/h. At what time are they 5 km apart? 11:30 A.M.

B 10. The daily number of riders on a certain bus route is 360, each of whom pay a 50¢ fare. The bus company has determined that for every fare increase of 5¢, there will be 20 fewer riders. What fare should the company charge in order to collect $196 per day from this route? 70¢

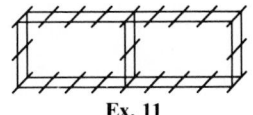

Ex. 11

11. A rectangular pen divided as shown, has a total area of 240 m² and is made from 84 m of fencing. What are the outside dimensions of the pen?
30 m by 8 m, or 20 m by 12 m

12. The foundation of a building is in the shape of two adjacent squares, as shown. The perimeter of the foundation is 80 m and its area is 325 m². Find the length of a side of each square.
17 m and 6 m or 15 m and 10 m

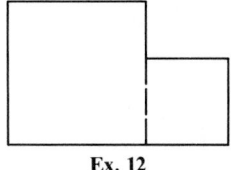

Ex. 12

13. A frying pan is in the shape of a truncated cone. The radius of the base of the pan is 10 cm and the depth of the pan is 3 cm. What should the radius of the top of the pan be so that the pan has a volume of 364π cm?

$$\left[\text{Volume of a truncated cone} = \frac{1}{3}\pi h(r_1^2 + r_1 r_2 + r_2^2).\right]$$ 12 cm

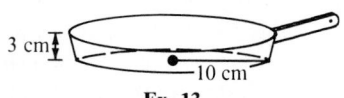

3 cm
10 cm
Ex. 13

C 14. A hollow metal ball has a wall that is 3 cm thick. If the total volume of metal used is 684π cm³, what is the radius of the ball? 9 cm

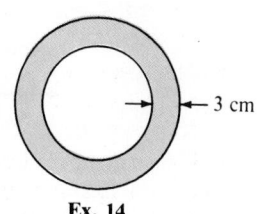

3 cm
Ex. 14

1. Multiply $(5x^2 y^{-3})(7xy^5)$.
$35x^3 y^2$

2. Multiply $(a + 2)(a^2 + 2a + 4)$.
$a^3 + 4a^2 + 8a + 8$

3. State the y-intercept for the line with equation $2x - 3y = 6$. -2

4. Solve for a.

$a : 3 = 5 : 6$ $\frac{5}{2}$

5. True or false? A function is a relation. T

Key Ideas

Use factoring to solve polynomial inequalities.

Chalkboard Examples

Solve over $\mathcal{R}$ and graph the solution set.

1. $x^2 - 2x - 15 > 0$
   $(x - 5)(x + 3) > 0$

   The solution set is
   $\{x: x > 5 \text{ or } x < -3\}$.

   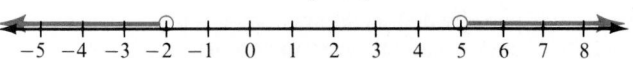

2. $y^2 - 2y < 3$
   $(y - 3)(y + 1) < 0$
   The solution set is
   $\{y: -1 < y < 3\}$.

Additional A Exercises

Solve each inequality over $\mathcal{R}$
and graph its solution set.

1. $(x - 2)(x + 3) \geq 0$
   $\{x: x \geq 2 \text{ or } x \leq -3\}$

2. $(y - 3)(y - 7) \leq 0$
   $\{y: 3 \leq y \leq 7\}$

3. $x^2 - 6x \geq 0$
   $\{x: x \geq 6 \text{ or } x \leq 0\}$

4. $(x - 2)^2 > 0$ $\{x: x \neq 2\}$

6–5 Solving Inequalities by Factoring

You know that for a and $b \in \mathcal{R}$, if $ab > 0$, then a and b are of the same sign, while if $ab < 0$, then a and b are of opposite signs. You can use these facts to solve an inequality in which one side consists of a reducible quadratic polynomial and the other side is 0.

EXAMPLE 1 Solve $x^2 - 3x > 10$ over $\mathcal{R}$ and graph the solution set.

SOLUTION
$$x^2 - 3x > 10$$
$$x^2 - 3x - 10 > 0$$
$$(x - 5)(x + 2) > 0$$

The inequality is satisfied if and only if $x - 5$ and $x + 2$ both have the same sign.

both factors positive	or	*both factors negative*
$x - 5 > 0$ and $x + 2 > 0$		$x - 5 < 0$ and $x + 2 < 0$
$x > 5$ and $\quad x > -2$		$x < 5$ and $\quad x < -2$
This compound sentence is equivalent to the sentence $x > 5$.		This compound sentence is equivalent to the sentence $x < -2$.

$\therefore$ the solution set of the inequality is $\{x: x > 5 \text{ or } x < -2\}$.

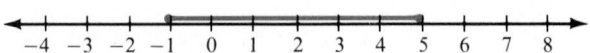

EXAMPLE 2 Solve $y^2 - 4y \leq 5$ over $\mathcal{R}$ and graph the solution set.

SOLUTION
$$y^2 - 4y \leq 5$$
$$y^2 - 4y - 5 \leq 0$$
$$(y - 5)(y + 1) \leq 0$$

The equation is satisfied if $y = 5$ or $y = -1$. The inequality is satisfied if and only if $y - 5$ and $y + 1$ have opposite signs.

$y - 5 > 0$ and $y + 1 < 0$	or	$y - 5 < 0$ and $y + 1 > 0$
$y > 5$ and $\quad y < -1$		$y < 5$ and $\quad y > -1$
There are no such real numbers y.		This compound sentence is equivalent to the sentence $-1 < y < 5$.

$\therefore$ the solution set of the inequality is the $\{y: -1 \leq y \leq 5\}$.

A visual method of solving inequalities is the use of a sign graph shown in Example 3. This method uses the fact that a product of n nonzero factors will be negative if and only if there is an odd number of negative factors.

EXAMPLE 3   Solve $x^3 < 9x$ over $\mathscr{R}$ and graph the solution set.

SOLUTION

$$x^3 < 9x$$
$$x^3 - 9x < 0$$
$$x(x^2 - 9) < 0$$
$$x(x + 3)(x - 3) < 0$$

To draw the sign graph, first draw four number lines, one for each factor and one for the product. Then label the number line for each factor to show where the factor is positive and where it is negative.

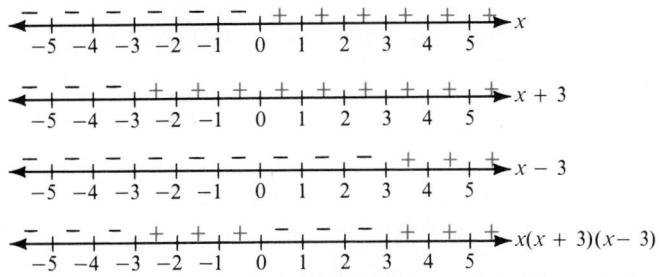

The product $x(x + 3)(x - 3)$ is negative when either exactly one factor is negative, or when exactly three factors are negative. The last number line shows that the solution set is $\{x: x < -3 \text{ or } 0 < x < 3\}$, whose graph is shown below.

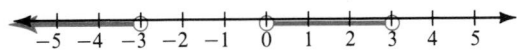

Oral Exercises

Describe the solution set of each inequality.

1. $x(x - 5) > 0$
2. $x(x + 2) > 0$
3. $x(x - 4) \leq 0$
4. $x(x + 3) < 0$
5. $(x - 1)(x - 6) > 0$
6. $(x + 2)(x - 3) < 0$
7. $(x - 3) > 0$
8. $2(x - 3)(x - 7) < 0$
9. $x^2(x - 2) \geq 0$

1. $\{x: x > 5\} \cup \{x: x < 0\}$
2. $\{x: x > 0\} \cup \{x: x < -2\}$
3. $\{x: 0 \leq x \leq 4\}$
4. $\{x: -3 < x < 0\}$
5. $\{x: x > 6\} \cup \{x: x < 1\}$
6. $\{x: -2 < x < 3\}$
7. $\{x: x > 3\}$
8. $\{x: 3 < x < 7\}$
9. $\{x: x = 0\} \cup \{x: x \geq 2\}$

Written Exercises

Solve each inequality over $\mathscr{R}$ and graph its solution set.

A
1. $(x + 3)(x - 4) \geq 0$
2. $(y - 5)(y - 2) \leq 0$
3. $a^2 - 3a - 4 < 0$
4. $x^2 - 9x + 14 < 0$
5. $x^2 - 7x \geq 0$
6. $5b^2 - 180 > 0$
7. $75 \geq 3c^2$
8. $3x^2 + 5x - 2 < 0$
9. $x^2 - 5x < 6$
10. $x^2 - 12x + 32 \geq -3$
11. $(k - 3)^2 > 0$
12. $r^2 + 14r + 49 \leq 0$

B
13. $y^2(y - 3) \geq 0$
14. $5x^3 - 15x^2 < 0$
15. $4x^3 < x^2$

Polynomials and Rational Expressions   **219**

Reading Algebra

Have students read the opening sentence of section 6-5. Then ask them to restate the sentence in their own words. You may get answers such as: "If the product of two real numbers is a positive number, then the product is made up of either two positive numbers or two negative numbers."

Common Errors

Students often try to solve quadratic inequalities without first getting 0 on one side. For example, they erroneously give the solution of $x^2 > 9$ as $x > 3$. Emphasize that 0 is the unique number with the property that if $ab > 0$, then $a > 0$ and $b > 0$ or $a < 0$ and $b < 0$.

Suggested Assignments

Minimum
  219/1–12, 13–17 odd
R 220/*Self-Test 2*
S 203/2–14 even
Extended Alg. with Trig.
  219/1–12, 13–17 odd
R 220/*Self-Test 2*
S 203/2–16 even
Enriched Alg.
  219/19–23 odd
R 220/*Self-Test 2*
S 203/18–32 even
Enriched Alg. with Trig.
  219/19–23 odd
R 220/*Self-Test 2*
S 203/18–32 even

Additional Answers
Written Exercises
(See p. 244.)

Mixed Review

1. Factor completely.
   $a(a + 2) - 3(a + 2)$
   $(a + 2)(a - 3)$

2. Find the x-intercept of the plane with equation $3x + 4y + 2z = 12$.   4

3. Solve over $\mathcal{R}$.
   $7a - (a + 11) = 7$ $\{3\}$

4. If y varies directly as x and $y = 15$ when $x = 3$, what does y equal when $x = 15$?   75

Additional Answers
Self-Test 2

4. [number line with $-\frac{3}{5}$ marked, open circles at $-\frac{3}{5}$ and 1, from -1 to 1]

5. [number line with open circles at -2 and 1, from -2 to 1]

Quick Quiz

Solve by factoring.

1. $2x^2 + 18x = 0$ $\{0, -9\}$

2. $5x^2 + 11x - 12 = 0$
   $\left\{\frac{4}{5}, -3\right\}$

3. A rectangle has a length that is 2 m longer than 4 times its width. If its area is 72 m², what is its perimeter?   44 m

Solve each inequality and graph its solution set.

4. $x^2 - 4x < 21$
   $\{x: -3 < x < 7\}$

   [number line from -3 to 7, open circles at -3 and 7]

5. $2b^2 - 4b > 0$
   $\{b: b < 0 \text{ or } b > 2\}$

   [number line from -4 to 6, open circles at 0 and 2]

Solve each inequality over $\mathcal{R}$ and graph its solution set.

16. $t(t - 6)^2 \geq 0$

17. $(z - 3)(z^2 + 4) > 0$

18. $18y^3 - 2y \leq 0$

C   19. $x^3 + 6x^2 + 9x \leq 0$

20. $4x^3 + 25x > 20x^2$

21. $x^4 - 16 > 0$

22. $4x^3 - 25x > 0$

23. $a^4 \geq 10a^2 - 9$

24. $x^{4n} + 3x^{2n} - 4 < 0$

Self-Test 2

VOCABULARY   double or multiple root (p. 214)
              sign graph (p. 218)

Solve by factoring.

1. $3x^2 - 15x = 0$ $\{0, 5\}$ 2. $3x^2 - 11x - 4 = 0$ $\{-\frac{1}{3}, 4\}$ *Obj. 1, p. 213*

3. A rectangle has a length that is 4 cm longer than twice its width.   *Obj. 2, p. 213*
   If its area is 30 cm², what is its perimeter? 26 cm

Solve each inequality and graph its solution set.

4. $5x^2 - 2x < 3$ 5. $3x^2 + 6x > 0$ *Obj. 3, p. 213*
   $\{x: -\frac{3}{5} < x < 1\}$ $\{x: x > 0 \text{ or } x < -2\}$

Check your answers with those at the back of the book.

Benjamin Banneker

1731–1806

From childhood, Benjamin Banneker was a natural mathematician, delighting in number puzzles and challenging problems.

Learning from borrowed books, he taught himself astronomy and was able to chart projections for solar and lunar eclipses. Out of this work came Banneker's most notable achievement—tables of astronomical positions for the year. These tables were eventually published in almanacs and became so popular that Banneker retired from tobacco farming to devote full time to astronomy.

In 1753 he built a striking clock almost entirely of hand-carved wooden parts. This remarkable timepiece ran for over twenty years.

In 1791 Banneker was asked to be a member of the party commissioned to survey the area designated by Congress to be the site of Washington, D.C. Banneker kept records and an astronomical field clock for this historic project.

Rational Algebraic Expressions

OBJECTIVES for Sections 6-6 through 6-11:
1. *To simplify a rational expression by factoring its numerator and denominator.*
2. *To find the quotient and remainder when one polynomial is divided by another.*
3. *To express a product or quotient of rational expressions as a rational expression in lowest terms.*
4. *To transform a sum or difference of rational expressions into an equivalent rational expression in lowest terms.*
5. *To solve problems involving equations with rational coefficients and problems with fractional equations.*

Teaching Suggestions
p. T90

Key Ideas
Reduce a rational algebraic expression to lowest terms.

6–6 Simplifying Rational Expressions

Just as any number that is the quotient of two integers is called a **rational number**, so the quotient of two polynomials is called a **rational expression** or, more fully, a **rational algebraic expression**. Every simplified rational expression in one variable defines a *rational function*. A function that is defined by a quotient of two polyomials and is written in simplest form is called a **rational function**. In no case can the divisor be zero.

EXAMPLE 1   For what values of x is the following rational function not defined?

$$f(x) = \frac{x^2 - 3x + 2}{x^3 - 4x}$$

SOLUTION   Because division by zero is undefined, the function is not defined when the denominator is zero.

$$f(x) = \frac{x^2 - 3x + 2}{x^3 - 4x} = \frac{(x-2)(x-1)}{x(x+2)(x-2)}$$

The denominator is zero when $x = 0, 2,$ or -2

$\therefore$ the function is undefined at $x = 0, 2,$ and -2.

The following theorem enables you to **reduce a fraction to lowest terms**, that is, to express it as an equivalent fraction whose numerator and denominator have no common factors except 1 and -1.

> *Theorem.*   For all r, s, and $t \in \mathcal{R}$, s and $t \neq 0$,
>
> $$\frac{r}{s} = \frac{r \div t}{s \div t}, \quad \text{and} \quad \frac{r}{s} = \frac{r \cdot t}{s \cdot t}.$$

Common Errors

A common mistake is to cancel the 5's, for example, in the expression $\frac{x+5}{5}$. If students make this sort of mistake, one way to correct it is to ask what they think the answer is if the 5's are canceled. Some may say "x," some "$x + 1$." You can show that both answers are wrong by substituting a value for x, say $x = 2$, and showing that the value of the supposedly simplified expression does not equal the value of the original expression. Emphasize that "cancelling" means dividing the numerator and denominator by the same quantity.

Another common mistake is illustrated by the following incorrect method:

$$\frac{x+y}{(x+y)(x+y)} = x + y$$

Remind students that

$$\frac{x+y}{(x+y)(x+y)} = \frac{1(x+y)}{(x+y)(x+y)}$$

so that when "everything cancels" in the numerator,

$$\frac{1(x+y)}{(x+y)(x+y)} = \frac{1}{(x+y)}$$

the 1 is left, and it is necessary to write it in (unlike the similar case for the denominator for obvious reasons).

Simplify.

1. $\dfrac{a^2 - 2ab - 3b^2}{a^2 - 4ab + 3b^2}$

   $\dfrac{(a - 3b)(a + b)}{(a - 3b)(a - b)} = \dfrac{a + b}{a - b}$

   $(a \neq b, 3b)$

2. $(x^2 - 9)(x - 3)^{-2}$

   $\dfrac{x^2 - 9}{(x - 3)^2} = \dfrac{(x - 3)(x + 3)}{(x - 3)(x - 3)} =$

   $\dfrac{x + 3}{x - 3}$ $(x \neq 3)$

3. $\dfrac{r^3 - 8}{r^2 - 4}$

   $\dfrac{(r - 2)(r^2 + 2r + 4)}{(r - 2)(r + 2)} =$

   $\dfrac{r^2 + 2r + 4}{r + 2}$ $(r \neq 2, -2)$

Minimum
  222/2–10 even, 12–24
Extended Alg. with Trig.
  222/5–29 odd
Enriched Alg.
  222/7–25 odd, 27–30
Enriched Alg. with Trig.
  222/7–25 odd, 27–30

Simplify.

1. $6(3x - 15)^{-1}$ $\dfrac{2}{x - 5}$

   $(x \neq 5)$

2. $(r - 4)^2(4 - r)^{-3}$ $\dfrac{1}{4 - r}$

   $(r \neq 4)$

3. $(k - 2)^3(2 - k)^{-1}$

   $-k^2 + 4k - 4$ $(k \neq 2)$

4. $\dfrac{m^3 + m}{(m + 1)^2}$ $\dfrac{m^3 + m}{m^2 + 2m + 1}$

   $(m \neq -1)$

5. $\dfrac{27y^3 - 1}{3y^2 + 5y - 2}$ $\dfrac{9y^2 + 3y + 1}{y + 2}$

   $\left(y \neq \dfrac{1}{3}, -2 \right)$

For example, $\dfrac{56}{42} = \dfrac{56 \div 14}{42 \div 14} = \dfrac{4}{3}$, and $\dfrac{\frac{2}{3}}{\frac{5}{6}} = \dfrac{\frac{2}{3} \cdot 6}{\frac{5}{6} \cdot 6} = \dfrac{4}{5}$.

Likewise you can simplify a rational expression by factoring the numerator and denominator completely and then dividing both by their greatest common factor. The rational expression is said to be **simplified**, or in **lowest terms**.

EXAMPLE 2 Simplify $\dfrac{y^5 - 3y^4 - 4y^3}{y^3 - 6y^2 + 8y}$.

SOLUTION $\dfrac{y^5 - 3y^4 - 4y^3}{y^3 - 6y^2 + 8y} = \dfrac{y^3(y^2 - 3y - 4)}{y(y^2 - 6y + 8)} = \dfrac{y \cdot y^2(y + 1)(y - 4)}{y(y - 2)(y - 4)}$

Divide numerator and denominator by the product $y(y - 4)$ of all their common irreducible factors (that is, their greatest common factor).

$$\dfrac{y^2(y + 1)}{(y - 2)} \quad (y \neq 0, 2, 4)$$

EXAMPLE 3 Simplify $(15 - 5a)^{-1}(27 - a^3)$.

SOLUTION $(15 - 5a)^{-1}(27 - a^3) = \dfrac{27 - a^3}{15 - 5a} = \dfrac{(3 - a)(9 + 3a + a^2)}{5(3 - a)}$

$= \dfrac{a^2 + 3a + 9}{5}$ $(a \neq 3)$

Hereafter in this book it will be assumed, usually without comment, that the replacement sets of the variables in a fraction include no numbers for which the denominator is zero.

Oral Exercises

Simplify.

1. $\dfrac{3x - y}{y - 3x}$ -1

2. $\dfrac{(x - 5)^3}{5 - x}$ $-(x - 5)^2$

3. $\dfrac{r + 2}{r - 3}$

4. $\dfrac{x^2 - 9}{(x + 3)^2}$ $\dfrac{x - 3}{x + 3}$

5. $(x^2 + 4x - 5)(x + 5)^{-1}$ $x - 1$

6. $\dfrac{a^2 - 3a - 10}{a}$

   $\dfrac{(r - 3)(r + 2)}{(r - 3)^2}$

7. $\dfrac{(c + d)^2}{(c + d)^{-3}}$ $(c + d)^5$

8. $\dfrac{3x^5 - 4x^3}{x^4}$ $\dfrac{3x^2 - 4}{x}$

6. $(a^3 - 3a^2 - 10a)a^{-2}$

9. $\dfrac{x^2 - 10x + 25}{x^2 - 25}$ $\dfrac{x - 5}{x + 5}$

Written Exercises

Simplify.

A 1. $8x(4x - 28)^{-1}$ $\dfrac{2x}{x - 7}$

2. $(6y^2)^{-1}(2y^2 - 10y)$ $\dfrac{y - 5}{3y}$

3. $(4a)^{-2}(6a^3 - 14a^2)$ $\dfrac{3a - 7}{8}$

4. $(27b^2 - 18b)(45b)^{-1}$ $\frac{3b-2}{5}$

6. $(6r^2 - 15r)^{-1}(5 - 2r)$ $-\frac{1}{3r}$

8. $(h - 3)^3(3 - h)^{-1}$ $-(h-3)^2$

10. $(x^4y^2 - 4x^3y)(x^2y^2 - 4xy)^{-1}$ x^2

5. $(z^5 - 8z^4)(3z - 24)^{-1}$ $\frac{z^4}{3}$

7. $(9k^3 + 12k^2)(-6k - 8)^{-1}$ $-\frac{3}{2}$

9. $(p - 7)^2(7 - p)^{-3}$ $\frac{1}{7-p}$

11. $(16h - 8k)(12h^2k - 6hk^2)^{-1}$

12. $(9c^2 - d^2)(3c - d)^{-2}$ $\frac{3c+d}{3c-d}$

13. $\frac{n^3 - 8}{3n^2 - 12}$ $\frac{n^2 + 2n + 4}{3(n+2)}$

14. $\frac{d^4 + d^2}{d^2(d+1)^2}$

15. $\frac{c^2 - 10c + 25}{c^3 - 25c}$ $\frac{c-5}{c(c+5)}$

16. $\frac{a^2 - 6ab + 8b^2}{a^2 - 3ab + 2b^2}$ $\frac{a-4b}{a-b}$

17. $\frac{64u^3 + 1}{4u^2 + 5u + 1}$

18. $\frac{5m^3 - 8m^2 + 3m}{25m^3 - 9m}$ $\frac{m-1}{5m+3}$

19. $\frac{r^4 - r^2 - 12}{r^4 + 3r^2}$ $\frac{(r+2)(r-2)}{r^2}$

20. $\frac{4x^2 - 12xy - 7y^2}{4x^2 - 49y^2}$

B **21.** $\frac{a^3 + b^3}{(a + b)^3}$ $\frac{a^2 - ab + b^2}{a^2 + 2ab + b^2}$

22. $\frac{p^4 - q^4}{(p + q)^2(q - p)}$ $-\frac{p^2 + q^2}{p + q}$

23. $\frac{(3x^2 - 48)(x^3 - 3x^2 - 28x)}{3(x + 4)^2(x^3 - 11x^2 + 28x)}$ 1

24. $\frac{(c^3 - d^3)(c^3 - cd^2)}{c^3 + c^2d + cd^2}$

25. $\frac{k^4 - 26k^2 + 25}{k^2 - 6k + 5}$ $(k + 5)(k + 1)$

26. $\frac{(54 - 2b^3)(b^2 + 9)}{2b^4 - 162}$

27. $\frac{r^6 - 8}{(4 - r^2)(4 + 2r + r^2)}$ $\frac{r^6 - 8}{(4 - r^2)(4 + 2r + r^2)}$

28. $\frac{(c^2 + 4)(c - 2)^2}{c^4 - 16}$ $\frac{c-2}{c+2}$

C **29.** $\frac{x^4 + 64y^4}{x^2 + 4xy + 8y^2}$ $x^2 - 4xy + 8y^2$

(*Hint:* Add $16x^2y^2$ to, and then subtract the same quantity from, the numerator.)

30. Prove that if $\frac{a + c}{b + c} = \frac{a}{b}$ for some nonzero real number c, then $a = b$.

6–7 Dividing One Polynomial by Another

The following theorem enables you to replace a rational number named by an improper fraction with an equivalent mixed numeral that names the sum of an integer and a proper fraction.

> **Theorem.** For all a, b, and $c \in \mathcal{R}$, $c \neq 0$, $\frac{a + b}{c} = \frac{a}{c} + \frac{b}{c}$.

Thus, $\quad \dfrac{29}{8} = \dfrac{24 + 5}{8} = \dfrac{3 \cdot 8 + 5}{8} = \dfrac{3 \cdot 8}{8} + \dfrac{5}{8} = 3 + \dfrac{5}{8}$, or $3\dfrac{5}{8}$.

Polynomials and Rational Expressions **223**

Additional Answers
Written Exercises

7. $-\frac{3k^2}{2}$ **11.** $\frac{4}{3hk}$

14. $\frac{d^2 + 1}{(d + 1)^2}$

17. $\frac{16u^2 - 4u + 1}{u + 1}$

20. $\frac{2x + y}{2x + 7y}$

24. $(c - d)^2(c + d)$

26. $-\frac{9 + 3b + b^2}{b + 3}$

(continued on p. 239)

Mixed Review

1. Simplify.
$$(36)\left[\frac{1}{4} + \left(-\frac{1}{6}\right)\right]\left[-\frac{1}{2} + 7\frac{1}{3}\right](0)$$
0

2. Solve for x.
$$\begin{vmatrix} -2 & x \\ 3 & 5 \end{vmatrix} = 8 \quad \{-6\}$$

3. Solve over $\mathcal{R}$.
$$\frac{1}{2}(y + 3) - \frac{3}{2}(y - 1) =$$
$-2 \quad \{5\}$

4. True or false?
$$\frac{1}{a}(a + b) = \frac{a}{b} \quad \text{F}$$

5. True or false?
If $|a| = 5$ and $|b| = 3$, then $|a - b| = 2$ F

Teaching Suggestions
p. T91

Key Ideas

Rewrite a rational algebraic expression as a sum.

223

1. Express $\dfrac{x^3 - 5x^2 + 7x - 2}{x - 2}$ as a sum by using division.

$$
\begin{array}{r}
x^2 - 3x + 1 \\
x - 2 \overline{)\, x^3 - 5x^2 + 7x - 2} \\
\underline{x^3 - 2x^2} \\
-3x^2 + 7x \\
\underline{-3x^2 + 6x} \\
x - 2 \\
\underline{x - 2} \\
0
\end{array}
$$

$\therefore \dfrac{x^3 - 5x^2 + 7x - 2}{x - 2} =$

$x^2 - 3x + 1$

2. Divide $4x^3 + 5x + 1$ by $2x + 1$.

$$
\begin{array}{r}
2x^2 - x + 3 \\
2x + 1 \overline{)\, 4x^3 + 0x^2 + 5x + 1} \\
\underline{4x^3 + 2x^2} \\
-2x^2 + 5x \\
\underline{-2x^2 - x} \\
6x + 1 \\
\underline{6x + 3} \\
-2
\end{array}
$$

$\therefore \dfrac{4x^3 + 5x + 1}{2x + 1} =$

$2x^2 - x + 3 + \dfrac{-2}{2x+1}$

3. Express

$\dfrac{9y^4 + 27y^3 - y^2 + 12}{3y^3}$ as a

sum.   Divide each term in the dividend by the monomial divisor.

$\dfrac{9y^4 + 27y^3 - y^2 + 12}{3y^3} =$

$3y + 9 - \dfrac{1}{3y} + \dfrac{4}{y^3}$

Likewise, using the theorem on page 223 along with the division algorithm, you obtain

$$
\frac{4233}{28} = \frac{151 \cdot 28 + 5}{28} = \frac{151 \cdot 28}{28} + \frac{5}{28} = 151 + \frac{5}{28}, \text{ or } 151\frac{5}{28}.
$$

It is often useful to transform a rational expression, by similar means, into the *sum of a polynomial and another rational expression.* The process, called **division**, consists of successively subtracting a monomial multiple of the divisor from the dividend until you finally obtain *either the remainder zero or a polynomial of lower degree than that of the divisor.*

$$
\begin{array}{r}
151 \\
28 \overline{)\, 4233} \\
\underline{28} \quad \text{subtract } 1 \times 28 \\
143 \\
\underline{140} \quad \text{subtract } 5 \times 28 \\
33 \\
\underline{28} \text{ subtract } 1 \times 28 \\
5
\end{array}
$$

EXAMPLE 1   Express $\dfrac{-22x - 7x^2 + 6x^3 - 3}{2x - 5}$ as a sum by using division.

SOLUTION   Before dividing, arrange the terms of both dividend and divisor in order of decreasing degree.

$$
\begin{array}{r}
3x^2 + 4x - 1 \\
2x - 5 \overline{)\, 6x^3 - 7x^2 - 22x - 3} \\
\underline{6x^3 - 15x^2} \longleftarrow \quad \text{subtract } 3x^2\,(2x - 5) \\
8x^2 - 22x \\
\underline{8x^2 - 20x} \longleftarrow \quad \text{subtract } 4x\,(2x - 5) \\
-2x - 3 \\
\underline{-2x + 5} \longleftarrow \text{subtract } -1\,(2x - 5) \\
- 8
\end{array}
$$

$\therefore \dfrac{-22x - 7x^2 + 6x^3 - 3}{2x - 5} = 3x^2 + 4x - 1 + \dfrac{-8}{2x - 5}.$

The following example illustrates the division process for polynomials involving two variables. In this case you first arrange the terms in order of decreasing degree in *one* of the variables.

EXAMPLE 2   Divide $9a^3 + 8ab^2 + 8b^3$ by $3a^2 - 2ab + 4b^2$.

SOLUTION   As given, the terms are in order of decreasing degree in the variable a. Note that the dividend has no second-degree term in a. When dividing, insert any such "missing" term with a 0 as its coefficient.

$$
\begin{array}{r}
3a + 2b \\
3a^2 - 2ab + 4b^2 \overline{)\, 9a^3 + 0a^2b + 8ab^2 + 8b^3} \\
\underline{9a^3 - 6a^2b + 12ab^2} \\
6a^2b - 4ab^2 + 8b^3 \\
\underline{6a^2b - 4ab^2 + 8b^3} \\
0
\end{array}
$$

$\therefore \dfrac{9a^3 + 8ab^2 + 8b^3}{3a^2 - 2ab + 4b^2} = 3a + 2b.$

EXAMPLE 3   Express $\dfrac{6n^4 + 18n^3 - n^2 + 15}{3n^3}$ as a sum by using division.

SOLUTION   Divide each term in the dividend by the monomial divisor.

$$\frac{6n^4 + 18n^3 - n^2 + 15}{3n^3} = 2n + 6 - \frac{1}{3n} + \frac{5}{n^3}$$

Suggested Assignments

Minimum
  225/1–22
S 210/18–28 even, 29–35
Extended Alg. with Trig.
  225/1–23 odd
S 210/18–28 even, 29–35
Enriched Alg.
  225/3, 13, 15, 17–26
S 210/26–37
Enriched Alg. with Trig.
  225/3, 13, 15, 17–26
S 210/26–37

Oral Exercises

Express each quotient as a sum by using division.

1. $\dfrac{8x^3 - 12x}{4x^2}$ $2x - \dfrac{3}{x}$

2. $\dfrac{5ab^2 + a^3b}{a^2b}$ $\dfrac{5b}{a} + a$

3. $\dfrac{6x^8 - 12x^6 + 14x}{2x^4}$ $3x^4 - 6x^2 + \dfrac{7}{x^3}$

Fill in the missing expressions for each division.

4.
$$\begin{array}{r} ?\ \ 4x \\ x - 3\,\overline{)\,4x^2 +\ \ \ \ x - 1} \\ 4x^2 - 12x \\ \hline ?\ \ 13x \end{array}$$

5.
$$\begin{array}{r} 4x^2 \\ 2x - 1\,\overline{)\,8x^3 - 9x^2 + x - 3} \\ ?\ \ 8x^3 - 4x^2 \\ \hline ?\ \ {-5x^2} \end{array}$$

6.
$$\begin{array}{r} ?\ \ {-2x^2} \\ 3x + 5\,\overline{)\,{-6x^3}\ \ \ \ \ \ \ \ \ - x + 1} \\ {-6x^3} - 10x^2 \\ \hline ?\ \ 10x^2 \end{array}$$

7.
$$\begin{array}{r} x \\ x^2 + 2x - 1\,\overline{)\,x^3 - 5x^2 - 4x - 3} \\ ?\ \ x^3 + 2x^2 - x \\ \hline ? \\ {-7x^2} - 3x \end{array}$$

Written Exercises

Express each rational expression as a sum by using division.

A

1. $\dfrac{14x^5 + 35x^3 - 21x^2}{7x^2}$ $2x^3 + 5x - 3$

2. $\dfrac{3c^2d^2 - 8cd - 18}{6cd}$ $\dfrac{cd}{2} - \dfrac{4}{3} - \dfrac{3}{cd}$

3. $\dfrac{ab^3 + a^2b^2 - a^3b}{-ab}$ $-b^2 - ab + a^2$

4. $\dfrac{45u^2v + 20uv^2 + v^3}{5uv}$ $9u + 4v + \dfrac{v^2}{5u}$

5. $\dfrac{x^2 - 6x + 3}{x - 4}$ $x - 2 + \dfrac{-5}{x - 4}$

6. $\dfrac{-10r^2 + 31r - 24}{2r - 3}$ $-5r + 8$

7. $\dfrac{6y^2 - 17y - 45}{3y + 5}$ $2y - 9$

8. $\dfrac{12z^2 - 108}{2z + 6}$ $6z - 18$

9. $\dfrac{r^3 + r^2 - r + 15}{r + 3}$ $r^2 - 2r + 5$

10. $\dfrac{3u^3 - 7u^2 - 22u + 8}{u - 4}$ $3u^2 + 5u - 2$

11. $\dfrac{-6v^3 - 10v^2 + 13v - 5}{3v - 1}$ $-2v^2 - 4v + 3 + \dfrac{-2}{3v - 1}$

12. $\dfrac{10w^3 + 8w^2 - 29w + 6}{5w - 6}$ $2w^2 + 4w - 1$

13. $\dfrac{2a^3 + a^2 - 17a + 14}{2a + 7}$ $a^2 - 3a + 2$

14. $\dfrac{64t^3 - 25}{4t - 3}$ $16t^2 + 12t + 9 + \dfrac{2}{4t - 3}$

15. $\dfrac{27n^3 + 8}{3n + 2}$ $9n^2 - 6n + 4$

16. $\dfrac{18x^3 - 15x^2 + 12x - 10}{6x - 5}$ $3x^2 + 2$

Additional A Exercises

Express each rational expression as a sum by using division.

1. $\dfrac{15y^4 - 25y^3 + 10y^2}{5y^2}$

  $3y^2 - 5y + 2$

2. $\dfrac{27r^2s + 15rs^2 + r}{3rs}$

  $9r + 5s + \dfrac{1}{3s}$

3. $\dfrac{18t^2 + 3t - 10}{3t - 2}$ $6t + 5$

4. $\dfrac{x^2 + 5x - 66}{x - 5}$

  $x + 10 + \dfrac{-16}{x - 5}$

5. $\dfrac{27t^3 - 8}{3t - 2}$ $9t^2 + 6t + 4$

6. $\dfrac{18 + 2c^3 - 5c^2}{2c + 3}$

  $c^2 - 4c + 6$

Polynomials and Rational Expressions   **225**

1. Solve $2x - 3 = 13$ over $\mathcal{R}$.

$\{8\}$

Simplify.

2. $\dfrac{|13 - 2| - |-1|}{|17 - 12|}$ 2

3. $\left[8\left(3 - \dfrac{1}{2}\right) - 12\left(\dfrac{2}{3} + \dfrac{3}{4}\right)\right]7$

21

4. Find the solution set over $\mathcal{R}$.

$3x - 6y = 18$
$5x + 3y = 17$ $\{(4, -1)\}$

Express each rational expression as a sum by using division.

17. $\dfrac{12y^3 + 5y^2 - 18y + 4}{4y - 1}$ $3y^2 + 2y - 4$

18. $\dfrac{4z^3 - 19z + 15}{2z + 5}$ $2z^2 - 5z + 3$

B **19.** $\dfrac{32b^5 + 9}{2b + 1}$ $16b^4 - 8b^3 + 4b^2 - 2b + 1 + \dfrac{8}{2b + 1}$

20. $\dfrac{5m^4 - 13m^3 - 19m^2 + 9}{5m - 3}$ $m^3 - 2m^2 - 5m - 3$

21. $\dfrac{2w^4 - 2w^3 - 13w^2 + 23w - 10}{w^2 - 3w + 2}$ $2w^2 + 4w - 5$

22. $\dfrac{4x^4 - 5x^3 - x + 2}{x^2 - 2x + 1}$ $4x^2 + 3x + 2$

23. $\dfrac{2v^4 + v^3 - 17v^2 - 5v}{v^2 - 6}$ $2v^2 + v - 5 + \dfrac{v - 30}{v^2 - 6}$

24. $\dfrac{8a^3 + 27}{4a^2 - 6a + 9}$ $2a + 3$

C **25.** By trying examples such as $\dfrac{x^5 - 32}{x - 2}$, make a conjecture about the pattern of the quotient of any division of the form $\dfrac{a^n - b^n}{a - b}$, in which n is a positive integer. $a^{n-1} + a^{n-2}b + a^{n-3}b^2 + \ldots + a^2b^{n-3} + ab^{n-2} + b^{n-1}$

26. Repeat the process mentioned in Exercise 25 to make a conjecture about the quotient of any division of the form $\dfrac{a^n + b^n}{a + b}$, in which n is a positive *odd* integer. $a^{n-1} - a^{n-2}b + a^{n-3}b^2 - \ldots + a^2b^{n-3} - ab^{n-2} + b^{n-1}$

6–8 Multiplying and Dividing Rational Expressions

You can multiply two rational expressions by using the same rule as that for multiplying rational numbers.

> **Theorem.**   For all r, s, t, and $u \in \mathcal{R}$, t and $u \neq 0$,
>
> $$\frac{r}{t} \cdot \frac{s}{u} = \frac{rs}{tu}.$$

EXAMPLE 1   Simplify $\dfrac{p^2 - 4p - 5}{p^3} \cdot \dfrac{p^2 + 3p}{2p - 10}$.

SOLUTION $\dfrac{p^2 - 4p - 5}{p^3} \cdot \dfrac{p^2 + 3p}{2p - 10} = \dfrac{(p - 5)(p + 1)}{p^3} \cdot \dfrac{p(p + 3)}{2(p - 5)}$

$= \dfrac{p(p - 5)(p + 1)(p + 3)}{p(p - 5)2p^2}$

$= \dfrac{p^3 + 4p + 3}{2p^2}$

Key Ideas

Multiply rational expressions.
Divide rational expressions.

Chalkboard Examples

Simplify.

1. $\dfrac{p^2 - 3p - 4}{p^3} \cdot \dfrac{p^2 + 2p}{2p - 8}$

$= \dfrac{(p - 4)(p + 1)}{p^3} \cdot \dfrac{p(p + 2)}{2(p - 4)}$

$= \dfrac{p^2 + 3p + 2}{2p^2}$

2. $\dfrac{v^2 + 4v + 4}{v^3 - 9v} \div \dfrac{v^2 - 4}{v^2 + 2v - 15}$

$= \dfrac{v^2 + 7v + 10}{v^3 + v^2 - 6v}$

3. $(9x - x^{-1}) \div (3x^{-1} + x^{-2})$

$= \dfrac{9x - x^{-1}}{3x^{-1} + x^{-2}}$

$= \dfrac{9x^2 - 1}{x} \cdot \dfrac{x^2}{3x + 1}$

$= (3x - 1)x$, or $3x^2 - x$

The definition of division (page 24) and the fact that the reciprocal of $\frac{s}{u}$ is $\frac{u}{s}$ if $s \neq 0$ and $u \neq 0$ lead to the following result.

> **Theorem.** For all r, s, t, and $u \in \mathcal{R}$, s, t, and $u \neq 0$,
>
> $$\frac{r}{t} \div \frac{s}{u} = \frac{r}{t} \cdot \frac{u}{s} = \frac{ru}{ts}.$$

EXAMPLE 2 Simplify $\dfrac{v^2 + 6v + 9}{v^3 - 25v} \div \dfrac{v^2 - 9}{v^2 - 3v - 10}$.

SOLUTION

$$\frac{v^2 + 6v + 9}{v^3 - 25v} \div \frac{v^2 - 9}{v^2 - 3v - 10} = \frac{v^2 + 6v + 9}{v^3 - 25v} \cdot \frac{v^2 - 3v - 10}{v^2 - 9}$$

$$= \frac{(v + 3)(v + 3)}{v(v + 5)(v - 5)} \cdot \frac{(v - 5)(v + 2)}{(v - 3)(v + 3)}$$

$$= \frac{(v + 3)(v + 2)}{v(v + 5)(v - 3)}$$

$$= \frac{v^2 + 5v + 6}{v^3 + 2v^2 - 15v}$$

EXAMPLE 3 Simplify $(4x - x^{-1}) \div \left(x + \dfrac{1}{2}\right)$.

SOLUTION

$$\frac{4x - x^{-1}}{x + \frac{1}{2}} = \frac{4x - \frac{1}{x}}{x + \frac{1}{2}} = \frac{\frac{4x^2 - 1}{x}}{\frac{2x + 1}{2}} = \frac{4x^2 - 1}{x} \cdot \frac{2}{2x + 1}$$

$$= \frac{2(2x + 1)(2x - 1)}{x(2x + 1)}$$

$$= \frac{2(2x - 1)}{x}, \text{ or } \frac{4x - 2}{x}$$

Oral Exercises

Simplify.

1. $\dfrac{5a^3}{6} \cdot \dfrac{12}{a^3}$ 10

2. $\dfrac{9x^2}{2} \cdot \dfrac{8}{x^2}$ 36

3. $\dfrac{7b^2}{5} \div \dfrac{7b}{10}$ 2b

4. $\dfrac{8y^2}{2} \div \dfrac{y^2}{5}$ 20

5. $\dfrac{3x - 21}{4} \cdot \dfrac{8}{x - 7}$ 6

6. $\dfrac{3}{x - 4} \cdot \dfrac{x^2 - 16}{9}$ $\frac{x + 4}{3}$

7. $\dfrac{(a + 6)^2}{9} \div \dfrac{a + 6}{27}$ 3(a + 6)

8. $\dfrac{7pq^2}{3^2} \div \dfrac{pq}{9}$ 7q

9. $\dfrac{(x - 3)^2}{6} \cdot \dfrac{42}{x^2 - 9}$ $\frac{7(x - 3)}{x + 3}$

Polynomials and Rational Expressions **227**

Additional A Exercises

Simplify.

1. $\dfrac{3p - 6}{p} \cdot \dfrac{p^2 + 2p}{p^2 - 4}$ 3

2. $\dfrac{3}{b^2 - 3b} \cdot \dfrac{b^2 - b - 6}{3b + 6}$ $\frac{1}{b}$

3. $\dfrac{d^2 + 4d}{d^2 + 3d - 4} \div \dfrac{d^3}{2d - 2}$ $\frac{2}{d^2}$

4. $\dfrac{x^3 - 8}{x^3 - 4x} \cdot \dfrac{x^3 + 2x^2}{x^2 + 2x + 4}$ x

5. $\dfrac{n^4 - 1}{n^2 + 1} \div \dfrac{(n - 1)^2}{n^2 - 1}$
   $(n + 1)^2$, or $n^2 + 2n + 1$

Reading Algebra

Read the first theorem in this section aloud, saying "r divided by t times s divided by u equals r times s divided by t times u." Now ask students to rewrite the statement in their own words. This should help students conceptualize the theorem rather than simply see a formula.

Common Errors

In exercises such as Written Exercises 15, 17, 18, 20, and 22 of this section, students are often tempted to invert the fractions (separately) in the second set of parentheses and change the sign to multiplication. Emphasize that when you invert a term, you are not simply "turning everything over"; you are taking "one divided by" the term. Therefore the inverse of $\left(\dfrac{x}{5} - \dfrac{3}{x}\right)$ is $\dfrac{1}{\left(\dfrac{x}{5} - \dfrac{3}{x}\right)} \neq \dfrac{5}{x} - \dfrac{x}{3}$.

The fractions must be combined inside the parentheses first in order for the inversion rule to be applicable.

Suggested Assignments

Minimum
228/1–22
S 203/16–24 even
Extended Alg. with Trig.
228/1–21 odd
S 203/16–24 even
Enriched Alg.
228/9–17 odd, 18–22
Enriched Alg. with Trig.
228/9–17 odd, 18–22

___

Mixed Review

1. After two tests, Maria's average in math was 78. After three tests, it was 82. If the teacher ignored Maria's lowest score (one of the first two tests), her average would be 86. What were Maria's test scores?   74, 82, 90

2. What is the slope of the line containing the points $(-1, 4)$ and $(3, -8)$? -3

3. Solve $\dfrac{3x}{2} - \dfrac{5x}{2} = 10$. $\{-10\}$

4. Determine k so that the given ordered pair will be a solution.
   $3x - ky = 17;\ (3, 2)$ -4

___

___

___

228

Written Exercises

Simplify.

A 1. $\dfrac{5c - 15}{c} \cdot \dfrac{c^2 + 3c}{c^2 - 9}$ 5

2. $\dfrac{7b^2 - 42b}{49b^3} \cdot \dfrac{b}{(b - 6)^2}$ $\frac{1}{7b(b - 6)}$

3. $\dfrac{16x^2 - 1}{(4x + 1)^2} \cdot \dfrac{3x^2}{12x^2 - 3x}$ $\frac{x}{4x + 1}$

4. $\dfrac{5}{p^2 - 5p} \cdot \dfrac{p^2 + 3p - 10}{5p - 10}$ $\frac{p + 5}{p(p - 5)}$

5. $\dfrac{7v - 63}{12} \div \dfrac{v^2 - 81}{6}$ $\frac{7}{2(v + 9)}$

6. $\dfrac{y^2 + 5y}{y^2 + 6y + 5} \div \dfrac{y^3}{3y + 3}$ $\frac{3}{y^2}$

7. $\dfrac{3a - 3b}{a^2 + 2ab + b^2} \div \dfrac{a^2 - 3ab + 2b^2}{a^2 - ab - 2b^2}$ $\frac{3}{a + b}$

8. $\dfrac{t^3 - t^2 - 2t}{t^2 - 3t + 2} \div \dfrac{t^2 + 3t + 2}{t^2 + 2t}$ $\frac{t^2}{t - 1}$

9. $\dfrac{u^3 - 3u^2 - 4u}{u^2 - 8u + 16} \cdot \dfrac{u^2 - 16}{3u^2 + 3u}$ $\frac{u + 4}{3}$

10. $\dfrac{(r^2 - s^2)^2}{r^2 - rs} \cdot \dfrac{r^2 + s^2}{r^4 - s^4}$ $\frac{r + s}{r}$

B 11. $\dfrac{x^4 - 81}{x^2 + 9} \div \dfrac{(x - 3)^3}{x^3 - 6x^2 + 9x}$ $x(x + 3)$

12. $\dfrac{6m^4 - 6n^4}{m^3 + n^3} \div \dfrac{3m^2 + 3n^2}{m^2 - mn + n^2}$ $2(m - n)$

13. $\dfrac{(a - b)^2}{5a^2 - 3ab - 2b^2} \div \dfrac{a^3 - b^3}{4b^2 - 25a^2}$ $\frac{2b - 5a}{a^2 + ab + b^2}$

14. $\dfrac{c^3(c - d)^3}{c^3 - d^3} \div \dfrac{c^3 - 2c^2d + cd^2}{c^2 + cd + d^2}$ c^2

15. $x\left(x - \dfrac{y^2}{x}\right) \div \left(\dfrac{1}{y} - \dfrac{1}{x}\right)$ $xy(x + y)$

16. $\left(\dfrac{p^2}{q} - q\right)\left(\dfrac{p}{q} - 1\right)^{-2}$ $\frac{q(p + q)}{p - q}$

17. $\left(2s + \dfrac{r}{3}\right) \div \left(\dfrac{36s}{r} - \dfrac{r}{s}\right)$ $\frac{rs}{3(6s - r)}$

18. $\left(\dfrac{x}{3} - \dfrac{2x - 3}{x}\right) \div \left(\dfrac{x}{3} - \dfrac{3}{x}\right)$ $\frac{x - 3}{x + 3}$

19. $\left(k + \dfrac{1}{k + 2}\right)\left(1 - \dfrac{2}{k^2 + k}\right)$ $\frac{k^2 - 1}{k}$

20. $\left(\dfrac{x^3}{y^3} - 1\right)\left(\dfrac{x}{y} - 1\right)^{-1} \div \left(\dfrac{x^2}{y^2} - 1\right)$

C 21. $\dfrac{a^3 - b^3}{a^4 + a^2b^2 + b^4} \cdot \dfrac{a^3 + b^3}{a^2 - b^2}$ 1

22. $\left(c - 2d + \dfrac{3d^2}{c + d}\right) \div \left(\dfrac{1}{c^3} + \dfrac{1}{d^3}\right)$

20. $\frac{x^2 + xy + y^2}{x^2 - y^2}$

22. $\frac{c^3 d^3}{(c + d)^2}$

___

6–9 Adding and Subtracting Rational Expressions

Two rational numbers having the same denominator can be added or subtracted in accordance with the following theorem (see Exercise 21 on page 27).

Theorem. For all a, b, and $c \in \mathcal{R}$, $c \neq 0$, $\dfrac{a}{c} + \dfrac{b}{c} = \dfrac{a + b}{c}$ and $\dfrac{a}{c} - \dfrac{b}{c} = \dfrac{a - b}{c}$.

The same rule applies in the case of rational expressions. For example,

$$\frac{3x^2}{x^2 - 1} - \frac{x + 2}{x^2 - 1} = \frac{3x^2 - x - 2}{x^2 - 1}$$

$$= \frac{(3x + 2)(x - 1)}{(x + 1)(x - 1)}$$

$$= \frac{3x + 2}{x + 1}.$$

If the denominators differ, then you must find a common denominator before adding or subtracting. Just as with fractions, it is simplest to use the **least common denominator** (LCD), that is, the polynomial of least degree and least positive constant factor that has each denominator as a factor.

EXAMPLE 1 Simplify $\dfrac{a^2}{a - 2} + \dfrac{5}{3} - \dfrac{7a - 2}{3a - 6}$ and state any restrictions on the variable.

SOLUTION To find the LCD, first factor the denominators completely.

$$\frac{a^2}{a - 2} + \frac{5}{3} - \frac{7a - 2}{3(a - 2)}$$

∴ the LCD is $3(a - 2)$.

Next replace each rational expression with an equivalent one having the LCD as denominator, and then simplify.

$$\frac{3a^2}{3(a - 2)} + \frac{5(a - 2)}{3(a - 2)} - \frac{7a - 2}{3(a - 2)} = \frac{3a^2 + 5a - 10 - 7a + 2}{3(a - 2)}$$

$$= \frac{3a^2 - 2a - 8}{3(a - 2)}$$

$$= \frac{(3a + 4)(a - 2)}{3(a - 2)}$$

$$= \frac{3a + 4}{3} \quad (a \neq 2)$$

EXAMPLE 2 Simplify $\dfrac{y^{-2} - x^{-2}}{y^{-1} + x^{-1}}$.

SOLUTION

$$\frac{y^{-2} - x^{-2}}{y^{-1} + x^{-1}} = \frac{\dfrac{1}{y^2} - \dfrac{1}{x^2}}{\dfrac{1}{y} + \dfrac{1}{x}} = \frac{\dfrac{x^2}{x^2y^2} - \dfrac{y^2}{x^2y^2}}{\dfrac{x}{xy} + \dfrac{y}{xy}}$$

$$= \frac{\dfrac{x^2 - y^2}{x^2y^2}}{\dfrac{x + y}{xy}} = \frac{x^2 - y^2}{x^2y^2} \cdot \frac{xy}{x + y}$$

$$= \frac{(x + y)(x - y)xy}{xyxy(x + y)} = \frac{x - y}{xy}$$

Suggested Assignments

Minimum
    230/1–23 odd
Extended Alg. with Trig.
    230/1–23 odd
Enriched Alg.
    230/7–19 odd
Enriched Alg. with Trig.
    230/7–19 odd

Additional A Exercises

Simplify.

1. $\dfrac{3x}{x-2} - \dfrac{6}{x-2}$ 3

2. $\dfrac{2r}{r+3} + \dfrac{12}{2r+6}$ 2

3. $3 + \dfrac{k^2}{k-5} + \dfrac{4k}{2k-10}$

    $\dfrac{k^2 + 5k - 15}{k-5}$

4. $\dfrac{c^2 d^{-1} - d}{cd^{-1} - 1}$ $c + d$

5. $\dfrac{4a^2}{a^2 - b^2} + \dfrac{a+b}{a-b} - \dfrac{a-b}{a+b}$

    $\dfrac{4a}{a-b}$

Mixed Review

1. If $f(x) = -3x^2 - 2x + 5$, find $f(-2)$. -3

2. Write an equation of the line passing through the point $P(0, 3)$ with slope $m = -2$. $y = -2x + 3$

3. A piece of wire 47 cm in length is cut so that the two pieces can be formed into a square and an equilateral triangle. The sum of the lengths of a side of the square and a side of the triangle is 15 cm. How long is a side of the square?   2 cm

4. Simplify $\left(-\dfrac{2}{3}\right)[(-8 + 3)(5)$
   $- (-7)]$. 12

Oral Exercises

State the least common denominator of the terms of each expression.

1. $\dfrac{3}{2a} + \dfrac{5}{4b}$ $4ab$

2. $\dfrac{1}{xy^2} + \dfrac{8}{x^2}$ $x^2 y^2$

3. $\dfrac{4}{x-3} - \dfrac{7}{x+3}$ $(x+3)(x-3)$

4. $\dfrac{x}{x-3} - \dfrac{4}{5(x-3)}$ $5(x-3)$

5. $\dfrac{3}{xy^2} + \dfrac{2}{x^2 y} - \dfrac{1}{x^2 y^2}$ $x^2 y^2$

6. $\dfrac{3}{a^2 - b^2} + \dfrac{2}{a-b}$ $a^2 - b^2$

Written Exercises

14. $\dfrac{-7x}{(x-2)(x+3)(x-4)}$

Simplify.

A

1. $\dfrac{5x}{x-4} - \dfrac{20}{x-4}$ 5

2. $\dfrac{3}{8b^2} - \dfrac{5}{12b}$ $\dfrac{9-10b}{24b^2}$

3. $\dfrac{3}{c^2 d} - \dfrac{9}{cd^2} + \dfrac{6}{cd}$ $\dfrac{3(d-3c+2cd)}{c^2 d^2}$

4. $\dfrac{4}{r+5} + \dfrac{3r+7}{2r+10}$ $\dfrac{3}{2}$

5. $\dfrac{2y}{(y-3)^2} + \dfrac{y}{y-3} - \dfrac{5y-9}{(y-3)^2}$ 1

6. $\dfrac{t-5}{4t^2} - \dfrac{2}{8t} - \dfrac{7}{6t^3}$ $\dfrac{-15t-14}{12t^3}$

7. $\dfrac{a}{a+6} - \dfrac{30-7a}{a^2-36}$ $\dfrac{a-5}{a-6}$

8. $\dfrac{x}{x+3y} - \dfrac{6xy}{x^2-9y^2} + \dfrac{3y}{x-3y}$ $\dfrac{x-3y}{x+3y}$

9. $\dfrac{3}{x+2} - \dfrac{3}{2-x} - \dfrac{3x^2}{4-x^2}$ $\dfrac{3x}{x-2}$

10. $\dfrac{b}{b+2} - \dfrac{b+1}{b+3}$ $\dfrac{-2}{(b+2)(b+3)}$

11. $\dfrac{x+y}{(x-5y)^2} - \dfrac{x-y}{x^2-25y^2}$ $\dfrac{12xy}{(x+5y)(x-5y)^2}$

12. $\dfrac{2c}{c^2-d^2} - \dfrac{c+d}{c^2-2cd+d^2}$ $\dfrac{c^2-4cd-d^2}{(c-d)^2(c+d)}$

13. $\dfrac{x-3}{x^2-x-6} - \dfrac{x+3}{x^2-5x+6}$ $\dfrac{-10x}{(x-2)(x+2)(x-3)}$

14. $\dfrac{x-8}{x^2-6x+8} - \dfrac{x+6}{x^2+x-6}$

15. $\dfrac{36p^2}{9p^2-q^2} + \dfrac{3p+q}{3p-q} - \dfrac{3p-q}{3p+q}$ $\dfrac{12p}{3p-q}$

16. $\dfrac{16c^2+d^2}{(4c-d)^3} - \dfrac{4c+d}{(4c-d)^2}$ $\dfrac{2d^2}{(4c-d)^3}$

17. $\dfrac{a^2 b^{-1} - b}{ab^{-1} + 1}$ $a - b$

18. $\left(\dfrac{r}{s} - \dfrac{s}{r}\right) \div (r^{-1} - s^{-1})$ $-r - s$

B

19. $\left(x - \dfrac{4x+3}{x+2}\right) \div \left(x - \dfrac{6x+3}{x+2}\right)$ $\dfrac{(x-3)(x+1)}{x^2-4x-3}$

20. $\left(\dfrac{2x}{x-2} - \dfrac{x}{x-1}\right) \div \left(\dfrac{3x}{x-3} - \dfrac{2x}{x-2}\right)$ **20.** $\dfrac{x-3}{x-1}$

21. $\dfrac{(a+b)(a^{-1} - b^{-1})}{(a-b)(a^{-1} + b^{-1})}$ -1

22. $\left(y + \dfrac{2y-2}{y-3}\right)\left(y - \dfrac{2y+2}{y+3}\right) \div \left(y^2 + \dfrac{4y^2+4}{y^2-9}\right)$ 1

23. $\left(u - v + \dfrac{2v^2}{u+v}\right)\left(\dfrac{1}{u^2} - \dfrac{1}{v^2}\right) \div \left(u^3 - \dfrac{u^3 v + v^4}{u+v}\right)$ $\dfrac{-1}{u^2 v^2}$

230

C **24.** Prove the following theorem: Any rational expression of the form

$\dfrac{Ax + B}{(x - 1)(x - 2)}$, where A and B are integers, can be written equiva-

lently in the form $\dfrac{p}{x - 1} + \dfrac{q}{x - 2}$ where p and q are integers. Prove

the theorem by solving the equation $\dfrac{Ax + B}{(x - 1)(x - 2)} = \dfrac{p}{x - 1} + \dfrac{q}{x - 2}$

for p and q in terms of A and B.

6–10 Using Polynomials with Rational Coefficients

The mathematical description of practical situations often involves equations whose sides are polynomials with rational coefficients.

EXAMPLE 1 An optical scanner can correct a batch of standardized tests in 15 min. A newer model will do the same job in 12 min. How many minutes would it take both machines to complete the job working together?

SOLUTION 1. The problem asks for the number of minutes required for the machines to process a batch of tests working together.

2. Let x represent the number of minutes for the machines to process the tests together. Then

$\dfrac{1}{15}$ = rate of the first machine $\left(\dfrac{1}{15} \text{ of the job per minute}\right)$,

$\dfrac{1}{12}$ = rate of the second machine $\left(\dfrac{1}{12} \text{ of the job per minute}\right)$,

1 = total work done together (one whole job) in x min.

3. Total work is part done by the first plus part done by the second.

$$1 \quad = \quad \dfrac{1}{15}x \quad + \quad \dfrac{1}{12}x$$

4. $\quad 60 \cdot 1 = 60 \cdot \dfrac{1}{15}x + 60 \cdot \dfrac{1}{12}x$

$\quad\quad 60 = 4x + 5x$

5. Completing the solution and checking the result is left to you. You should find that, working together, the machines would require $6\frac{2}{3}$ min, or 6 min 40 s.

A percent is equivalent to a fraction whose denominator is 100. For example, $41\% = \frac{41}{100}$, or 0.41, and $165\% = \frac{165}{100}$, or 1.65. When you multiply a number called the **base** (b), by a **percent** (r), the product is called the **percentage** (p). The formula $p = br$ is a basic tool in solving many problems in science and business.

Polynomials and Rational Expressions **231**

Teaching Suggestions
p. T92

Key Ideas

Use polynomials with rational coefficients to solve practical problems.

Chalkboard Examples

1. One computer system can prepare the weekly sales summary of a company in 10 hours. A faster system can do the job in 6 hours. How many hours would it take both systems to prepare the summary working together?

Let x = number of hours required for the systems to prepare the sales summary together.

$\dfrac{1}{10}$ = rate of first system (one-tenth of the job in 1 hour).

$\dfrac{1}{6}$ = rate of the second system (one-sixth of the job in 1 hour).

$1 = \dfrac{1}{10}x + \dfrac{1}{6}x$

$30 = 3x + 5x$, $3\dfrac{3}{4} = x$

(continued)

∴ working together, the systems can complete the job in 3 h 45 min.

2. At most how many grams of a 42% alcohol solution can be added to 20 grams of a 60% alcohol solution, if the resulting solution is to contain at least 50% alcohol?

Let y = number of grams of 42% solution to be added.

$20 + y$ = the number of grams in the resulting solution.

$0.50(20 + y)$ = 50% of the resulting solution.

$0.42y$ = the amount of alcohol in the added solution.

$0.60(20)$ = the amount of alcohol in the original solution.

$0.60(20) + 0.42y \geq 0.50(20 + y)$, $y \leq 25$

∴ at most, 25 grams of 42% solution can be added.

Minimum
  232/*P*: 1–12
S 219/14–18 even
S 230/2–22 even
Extended Alg. with Trig.
  232/*P*: 1–12
S 219/14–18 even
S 230/2–22 even
Enriched Alg.
  232/*P*: 1–12
S 219/19–23 odd
S 230/20–24
Enriched Alg. with Trig.
  232/*P*: 1–14
S 219/19–23 odd
S 230/20–24

EXAMPLE 2   At least how many liters of a 75% antifreeze solution must be added to a car radiator that already contains 3.6 L of a 50% antifreeze solution, if the resulting solution is to contain at least 60% antifreeze?

SOLUTION   1. The problem asks for the minimum number of liters of 75% solution to be added.

2. Let y represent the number of liters of a 75% solution to be added. Then

$3.6 + y$ = the number of liters in the resultant solution,

$0.60(3.6 + y)$ = 60% of the resultant solution,

$0.75y$ = the amount of antifreeze in the added solution,

$0.50(3.6)$ = the amount of antifreeze in the original solution.

3.
Antifreeze in original solution	plus	antifreeze in added solution	is at least	60% of resultant solution.
0.50(3.6)	+	0.75y	≥	0.60(3.6 + y)

4. Multiply each side of the inequality by 100.

$$50(3.6) + 75y \geq 60(3.6 + y)$$

5. Completing the solution and checking the result is left to you. You should find that at least 2.4 L of 75% solution must be added.

∴ at least 2.4 L of solution must be added.

Problems

A 1. One pump can fill a water storage tank in 30 min. Another pump can fill the tank in 45 min. How long would it take both pumps working together to fill the tank? 18 min

2. A pipe takes 90 min to empty the storage tank in Exercise 1. How long would it take for the two pumps in Exercise 1 to fill the tank if the drain pipe is open at the same time? 22.5 min

3. A stamping machine can produce a metal refrigerator part every 18 s. A newer model can produce the same part in 12 s. How long would it take the two machines to produce 50 parts working simultaneously? 6 min

4. Three casting machines can each fill a mold in 10 s. A more efficient model can fill the same mold in 6 s. How long would it take all four machines to fill 70 molds? 2.5 min

5. At 9:00 Mary Thorne set out from her house for a picnic, traveling at 15 km/h on her bike. Her sister left 15 min later and traveled at a speed of 30 km/h. If both sisters arrived at the picnic area at the same time, how far did each travel? 7.5 km

6. Isabel Boncassen ran the first leg of a 12 km road race at a speed of 9 km/h and the remainder of the race at 10 km/h. If her time for the whole race was $1\frac{1}{4}$ h, how long was the first leg of the course? 4.5 km

7. When Frank Gresham drove 150 mi in the city, the fuel efficiency rating for his car was 18 mi/gal. For highway driving, the fuel efficiency rating is 24 mi/gal. How far will Frank have to drive on the highway for his car to have a fuel efficiency rating of 20 mi/gal? 100 mi

8. Repeat Exercise 7 when an extra 0.5 gal of fuel is used for stopping and starting in city driving. 160 mi

9. How much pure hydrogen peroxide must be added to 240 g of a 45% hydrogen peroxide solution to produce a solution that is 70% hydrogen peroxide? 200 g

10. How much of an 80% nitric acid solution must be added to 175 g of a solution that is 12% nitric acid to produce a 60% nitric acid solution? 420 g

B 11. Working alone two metal crushers can each process one truckload of aluminum cans in 25 min. After the two machines have been working for 6 min they are joined by a third machine that by itself could process one truckload in 20 min. How long after the first two machines start working will one truckload of cans be processed? 10 min

12. A mechanical sorter can process a bag of mail in 18 min. After the sorter has been working for a time, it breaks down. The rest of the mail it was sorting is divided equally between two older machines, each of which would take 60 min to complete the job working alone. The mail is finished being sorted 20 min after the first machine started working. How long did the first machine work before breaking down? 15 min

C 13. A lean-to greenhouse extending 2 m from a wall, as shown, is to have one end and four fifths of its square slanting side made from transparent plastic. What is the tallest it can be if enough plastic is available to cover 6.5 m²? 1.5 m

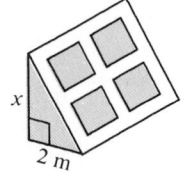

x

$2\ m$

14. Lydia Bennett traveled at a speed of 40 km/h for the first 30 km of her vacation trip. How far would she have to travel at a speed of 60 km/h to have an average speed of 54 km/h for the whole trip? 105 km

Polynomials and Rational Expressions **233**

Supplementary Material

Test 18

Key Ideas

Solve a fractional equation.

Chalkboard Examples

1. Solve $\dfrac{a}{2} - \dfrac{9 - 2a}{a - 7} = \dfrac{5}{a - 7}$.

$$\dfrac{a}{2} - \dfrac{9 - 2a}{a - 7} = \dfrac{5}{a - 7}$$

$$a(a - 7) - 2(9 - 2a) = 5 \cdot 2$$

$$a^2 - 7a - 18 + 4a = 10$$

$$a^2 - 3a - 28 = 0$$

$$(a - 7)(a + 4) = 0$$

$$a = 7, a = -4$$

Check: $a = 7$

$$\dfrac{7}{2} - \dfrac{9 - 14}{0} = \dfrac{5}{0}$$

zero appears in denominator.

∴ 7 is not a solution.

Check: $a = -4$

$$\dfrac{-4}{2} - \dfrac{9 + 8}{-4 - 7} = \dfrac{5}{-4 - 7}$$

$$-2 + \dfrac{17}{11} = \dfrac{-5}{11}$$

$$\dfrac{-5}{11} = \dfrac{-5}{11}$$

∴ −4 is a solution.

$\{-4\}$

6–11 Fractional Equations

An equation involving one or more rational expressions in which a variable appears in the denominator is called a fractional equation.
To solve the fractional equation

$$1 + \dfrac{30}{a^2 - 9} - \dfrac{5}{a - 3} = 0, \quad (1)$$

You can begin by multiplying both sides of the equation by their LCD, $(a + 3)(a - 3)$, or $a^2 - 9$:

$$(a^2 - 9)\left(1 + \dfrac{30}{a^2 - 9} - \dfrac{5}{a - 3}\right) = (a^2 - 9) \cdot 0$$

Then you have:

$$a^2 - 9 + 30 - 5(a + 3) = 0$$
$$a^2 + 21 - 5a - 15 = 0$$
$$a^2 - 5a + 6 = 0 \quad (2)$$
$$(a - 3)(a - 2) = 0$$
$$a = 3, a = 2$$

Checking each solution of Equation (2) in the original Equation (1), we have:

$$1 + \dfrac{30}{3^2 - 9} - \dfrac{5}{3 - 3} \overset{?}{=} 0$$

$$1 + \dfrac{30}{0} - \dfrac{5}{0} \overset{?}{=} 0$$

$$1 + \dfrac{30}{2^2 - 9} - \dfrac{5}{2 - 3} \overset{?}{=} 0$$

$$1 + (-6) - (-5) \overset{?}{=} 0$$

$$0 \overset{?}{=} 0 \checkmark$$

Since $a = 3$ produces zero divisors, 3 is not an allowable root of Equation (1). Hence, although the solution set of Equation (2) is {3, 2}, the solution set of Equation (1) is simply {2}.

Thus you can see that when you transform a fractional equation by multiplying both sides by their LCD, the resulting equation is not necessarily equivalent to the original one. The solution set of the transformed equation will, however, include all the roots of the original equation. *Always check the roots back in the original equation* to see which ones are allowable.

EXAMPLE   In one year Dave increased his running speed by 60 m/min. At the end of that year it took him 3 min less time to run a 3600 m course than it took him at the beginning of the year. What was Dave's original time to run the course?

SOLUTION   1. The problem asks for Dave's original time to run the course.
    2. Let t = his original time, then $t - 3$ = his new time.
    Let r = his original speed, then $r + 60$ = his new speed.

234   *Chapter 6*

3. Recall that rate $\times$ time = distance.
   Thus $r \cdot t = 3600$ and $(r + 60)(t - 3) = 3600$.

   original speed: $r = \dfrac{3600}{t}$

   new speed: $r + 60 = \dfrac{3600}{t - 3}$.

   $\underbrace{\text{Original speed}}_{\dfrac{3600}{t}} + \underbrace{\text{60 m/min}}_{60} \text{ is } \underbrace{\text{new speed}}_{\dfrac{3600}{t-3}}.$

4. Solve for t: $3600(t - 3) + 60t(t - 3) = 3600t$
$$3600t - 10{,}800 + 60t^2 - 180t = 3600t$$
$$60t^2 - 180t - 10{,}800 = 0$$
$$t^2 - 3t - 180 = 0$$
$$(t - 15)(t + 12) = 0$$
$$t = 15 \quad \text{or} \quad t = -12$$

   Since time for running must be positive, a solution of -12 is impossible. Hence, 15 is the only possible solution.

5. Checking the solution is left to you.

   $\therefore$ Dave's original time to run the course was 15 min.

Oral Exercises

State the least common denominator of the terms of each equation.

1. $\dfrac{3}{4x} + \dfrac{1}{6x} = 1$ $12x$

2. $\dfrac{9}{10b} - \dfrac{2}{5} = \dfrac{1}{2b^2}$ $10b^2$

3. $\dfrac{5}{2y} = \dfrac{4}{3y} - 2$ $6y$

4. $\dfrac{4}{3} = \dfrac{4c + 6}{5c - 3}$ $3(5c - 3)$

5. $\dfrac{3}{z - 2} - \dfrac{5z}{z + 2} = 0$ $(z - 2)(z + 2)$

6. $\dfrac{5}{x - 3} - \dfrac{2x}{x + 3} - \dfrac{9}{x^2 - 9} = 0$ $(x - 3)(x + 3)$

Written Exercises

A **1–6.** Solve the equations in Oral Exercises 1–6.

1. $\{\frac{11}{12}\}$ 2. $\{\frac{5}{4}, 1\}$ 3. $\{-\frac{7}{12}\}$ 4. $\{\frac{15}{4}\}$ 5. $\{-\frac{2}{5}, 3\}$ 6. $\{-\frac{1}{2}, 6\}$

Solve.

7. $\dfrac{17}{x^2 - 25} - \dfrac{x}{x - 5} = \dfrac{-1}{x + 5}$ $\{-6, 2\}$

8. $\dfrac{p^2 - 2}{3p - 6} - \dfrac{p}{3} = \dfrac{2}{p - 2}$ $\{4\}$

9. $\dfrac{y - 2}{2y + 1} + \dfrac{y - 2}{y - 4} = 2$ $\{7, -2\}$

10. $\dfrac{n}{2n - 1} - \dfrac{1}{4n^2 - 1} = \dfrac{4}{2n + 1}$ $\{3\}$

Polynomials and Rational Expressions **235**

2. A plane is to seed a field measuring 1 km wide and 1.2 km long. If the ratio of the time it takes to fly the length of the field with a 10 km/h tailwind to the time required to fly back against that same wind is 8 : 9, what is the speed of the plane in still air?

Let r = speed of the plane in still air.

$r + 10$ = speed of the plane with tailwind.

$r - 10$ = speed of the plane against tailwind.

t = time to fly the length with the wind.

t' = time to fly the length against the wind.

Since $rt = d$
$$t(r + 10) = 1.2$$
and $t'(r - 10) = 1.2$
$$\therefore t(r + 10) = t'(r - 10)$$
$$\dfrac{t}{t'} = \dfrac{r - 10}{r + 10} \text{ and } \dfrac{t}{t'} = \dfrac{8}{9}$$
$$\dfrac{8}{9} = \dfrac{r - 10}{r + 10}$$
$$8(r + 10) = 9(r - 10)$$
$$r = 170$$
$\therefore$ the speed of the plane in still air is 170 km/h.

Suggested Assignments

Minimum
First day
   235/1–16
Second day
   236/P: 1–12
Third day
   236/17–20
R 238/*Self-Test 3*
Extended Alg. with Trig.
   235/3–23 odd
   236/P: 1–13
R 238/*Self-Test 3*
Enriched Alg.
   235/3–23 odd
   236/P: 1–13
R 238/*Self-Test 3*
Enriched Alg. with Trig.
   235/3–23 odd
   236/P: 7–13
R 238/*Self-Test 3*

Common Errors

When multiplying both sides of an equation by the same expression in order to eliminate denominators, students often forget to multiply one side by the expression, especially if that side is a constant or lacks a denominator. Occasionally weaker students try to cancel the numerator of a rational expression on one side of an equation with a denominator on the other. In general, remind your students that when working with equations, they should keep in mind what transformation they are using at each step and remember to apply it to *both* sides.

Another error that frequently occurs is that after transforming an equation and solving for its roots, students assume all the roots are allowable. In fact, some of these roots may be roots of the transformed equation only. Remind the students that they must remember to check the roots by substituting them back into the original equation to see which ones are allowable.

Solve.

11. $\dfrac{4}{a^2 - 1} + \dfrac{a - 2}{a - 1} = \dfrac{a - 3}{a + 1}$ {$\frac{1}{3}$}

12. $\dfrac{k^2 + 9}{k^2 - 3k + 2} - \dfrac{6}{k - 1} = \dfrac{2k}{k - 2}$ {$-7, 3$}

B 13. $\dfrac{n + 4}{n + 3} - \dfrac{n - 3}{n - 1} = \dfrac{3 - n^2}{n^2 + 2n - 3}$ {$-1, -2$}

14. $\dfrac{2v + 4}{2v - 1} - 2 = \dfrac{17 - v}{2v^2 + 5v - 3}$ {$-\frac{1}{2}, 1$}

15. $\dfrac{2y - 5}{y - 3} = 3 + \dfrac{3}{2y^2 - 7y + 3}$ {$\frac{7}{2}, 1$}

16. $\dfrac{5}{x - 2} - \dfrac{4x + 1}{(x - 2)^2} + \dfrac{3x + 2}{(x - 2)^3} = 0$

17. $\dfrac{r + 3}{r^2 - 2r} - \dfrac{r + 4}{r^2 + 2r} = \dfrac{r - 10}{r^2 - 4}$ {$14, -1$}

18. $\dfrac{p - 2}{p^2 - 1} - \dfrac{3}{p^2 + 4p + 3} = \dfrac{2p - 1}{p^2 + 2p - 3}$

19. $\dfrac{2k^2 - 1}{k^3 + 1} + \dfrac{k + 2}{k^2 - k + 1} = \dfrac{7}{k + 1}$ {$1, \frac{3}{2}$}

20. $\dfrac{4}{a - 3} - \dfrac{a + 2}{a^2 + 3a + 9} = \dfrac{2a^2 + 6}{a^3 - 27}$

C 21. $\dfrac{2c^2 - 5}{c^2 - 4} - 1 = \dfrac{c^2 + 7}{c^2 + 1}$ {$3, -3$}

22. $\dfrac{3d - 2}{d + 2} - \dfrac{3d}{d + 1} = \dfrac{1}{d - 3}$ {$-\frac{1}{3}, 2$}

23. $\left(\dfrac{x - 1}{x}\right)^2 - \left(\dfrac{x - 1}{x}\right) - 6 = 0$ {$\frac{1}{3}, -\frac{1}{2}$}

24. $\left(\dfrac{y^2 - 36}{y}\right)^2 = 25$ {$9, -9, 4, -4$}

16. {4, 6}
18. {−2}
20. {−4, −9}

Problems

A 1. Boyle's Law, applied to the amount of gas used in a given experiment at a constant temperature, states that $pV = 240$ where p is pressure measured in kilopascals and V is volume measured in cubic centimeters. If a decrease in pressure of 1 kPa produces an increase in volume of 20 cm³, what was the original pressure on the gas? 4 kPa

2. Every camera lens has a characteristic measurement f, called the focal length, such that when an object is in focus, its distance D_0 from the lens and the distance D_i from the lens to the film satisfy the equation

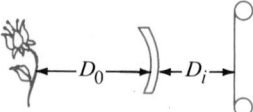

$$\frac{1}{D_0} + \frac{1}{D_i} = \frac{1}{f}.$$

If $D_0 = 60$ cm and D_i is 3 cm greater than the focal length, what is the focal length of the lens? 12 cm

3. If three capacitors of capacitance C_1, C_2, and C_3, are connected in series, the total capacitance of the circuit, C, is given by

$$\frac{1}{C_1} + \frac{1}{C_2} + \frac{1}{C_3} = \frac{1}{C}.$$

[Capacitance is always positive and is measured in microfarads (μF).] In a certain circuit, C_2 is 5 times as great as C_1, and C_3 is 5 μF less than C_2. If the total capacitance of the circuit is 2 μF, find C_1, C_2, and C_3. $C_1 = 3\ \mu F$, $C_2 = 15\ \mu F$, $C_3 = 10\ \mu F$

4. A copy machine can copy a report in 30 min. When a second copy machine is used at the same time, the copying can be done in 18 min. How long would it take the second copy machine to do the copying alone? 45 min

5. Two harvesting machines can each harvest a field in 15 h. If they are joined by a newer machine and the three machines work together, the job takes 3 h. How long would it take for the newer machine to harvest the field alone? 5 h

6. A new model computer is advertised as being 6 h faster at a certain processing job than an older model. If 3 of the older models and 2 of the new models work simultaneously, they can finish the job in 1 h. How long would it take one of the new machines to finish the job alone? 3 h

7. In a 16 km cross-country ski race Brett skied the first 7 km at a constant speed and then increased his speed by 3 km/h for the last 9 km. If he finished the entire race in 1 h 20 min, what was his speed for the first 7 km? 10.5 km/h

8. Lucy Steele drove 72 mi to a meeting and returned by the same route. She drove 5 mi/h slower and 12 min longer on the return trip. What was her speed on the way to the meeting? 45 mi/h

9. John Dashwood received a 75¢/h raise for his part-time job. He can now earn his former weekly salary of $63 working 2 h less. How many hours a week did he work before he received his raise? 14 h

B 10. A college crew team can maintain a speed equivalent to 15 km/h in still water. If it takes the team 50 min to row 6 km upstream and back, what is the rate of the current in the river? 3 km/h

11. A canoeist can paddle 12 km upstream and 12 km back downstream in the same amount of time as she can paddle 25 km in still water. If the rate of the current is 2 km/h, what is her rate in still water? 10 km/h

12. A day laborer takes three more days to paint a house than an apprentice painter. A master painter takes three days fewer than the apprentice. The master painter can do as much work in seven days as the day laborer and the apprentice working together can accomplish in six days. How long would it take the apprentice alone to paint the house? 9 days

C 13. In a certain city a birth occurs on the average every 24 min and a death every half hour. A resident moves out of the city every 1.5 h, and a new person moves into the city every 4.5 h. How long does it take on the average for the population to increase by 1 person? 18 h

Polynomials and Rational Expressions **237**

Solve.

1. $\dfrac{1}{2} + \dfrac{7}{4a} = \dfrac{15}{4a^2}$ $\left\{-5, \dfrac{3}{2}\right\}$

2. $\dfrac{3z}{z+1} - \dfrac{2}{z-1} = 0$ $\left\{-\dfrac{1}{3}, 2\right\}$

3. $\dfrac{b+4}{b} + \dfrac{3}{b-4} = \dfrac{-16}{b^2 - 4b}$
$\{-3\}$

4. $p - \dfrac{1}{2} = \dfrac{5 + p^2}{3p - 6}$ $\left\{-\dfrac{1}{4}, 4\right\}$

5. If a flight that is being chartered for $19,200 could acquire 4 more passengers, each would pay $20 less for the flight. How many passengers are there?   60 passengers

6. Dave increased his running speed by 40 m/min and completed a 3600 m course in 1 min less than his previous time. What was his original speed? 360 m/min

Mixed Review

1. Solve $|x| < 5$ over $\mathscr{R}$.
$\{x: -5 < x < 5\}$

2. Solve $2p - (5p - 7) = 19$.
$\{-4\}$

3. Simplify $[2 \cdot 7 - 3(-2)] \div (-18 + 13)$. -4

4. The difference of the squares of two consecutive even integers is 22 greater than the smaller of the two integers. What are the integers?   6 and 8

1. Simplify $\dfrac{4y^3 - 36y}{2y^2 + 6y}$.

   $2y - 6$

2. Express

   $\dfrac{6x^3 - 13x^2 + 3x - 20}{2x - 5}$ as a

   sum by using division.
   $3x^2 + x + 4$

3. Simplify $\dfrac{a^2 - a - 2}{(a - 2)^2} \div$

   $\dfrac{a^3 + a^2}{2a^2 - 8} \cdot \dfrac{2(a + 2)}{a^2}$

4. Simplify $\dfrac{x}{x - 4} - \dfrac{16}{x^2 - 16} -$

   $\dfrac{2}{x + 4} \cdot \dfrac{x - 2}{x - 4}$

5. Solve $\dfrac{8}{9 - x^2} - \dfrac{x}{3 - x} -$

   $2 = 0.$ $\{-2, 5\}$

6. Each of three steamrollers can smooth a certain section of road in 21 h working alone. When these three are joined by a faster steamroller, the section of the road can be smoothed in 5 h. How long would it take the faster steamroller to smooth the section of road working alone?
   17.5 h

Self-Test 3

VOCABULARY   rational number (p. 221)
              rational algebraic expression (p. 221)
              rational function (p. 221)
              reduce a fraction to lowest terms (p. 221)
              simplified rational expression (p. 222)
              division (p.224)
              least common denominator (p. 229)
              base (p. 231)
              percent (p. 231)
              percentage (p. 231)
              fractional equation (p. 234)

1. Simplify $\dfrac{3y^2 + y^{-2}}{y^3 - y} \cdot \dfrac{3y^4 + 1}{y^5 - y^3}$ *Obj. 1, p. 221*

   2. $2x^2 + 2x - 5 + \dfrac{13}{3x + 5}$

2. Express $\dfrac{6x^3 + 16x^2 - 5x - 12}{3x + 5}$ as a sum by using division.   *Obj. 2, p. 221*

3. Simplify $\dfrac{2y^2 - 3y - 9}{y^2 - 2y - 3} \div \dfrac{y^3 - 9y}{2y^2 + 9y + 9} \cdot \dfrac{(2y + 3)^2}{y(y + 1)(y - 3)}$ *Obj. 3, p. 221*

4. Simplify $\dfrac{4x + 9}{4x - 5} + \dfrac{x - 3}{x + 1} - \dfrac{34 - 2x}{4x^2 - x - 5} \cdot 2$ *Obj. 4, p. 221*

5. Solve $\dfrac{2x - 32}{16 - x^2} + \dfrac{x + 14}{x + 4} - 3 = 0.$ $\{-2, 6\}$ *Obj. 5, p. 221*

6. A group of college students is planning a ski trip that costs a total of $600, to be shared equally by each of the students. At the last minute, two people decide not to go on the trip and the cost for each of the other students increases by $15. How many students had originally planned to go on the trip? 10 students

Check your answers with those at the back of the book.

Chapter Summary

1. Laws for working with positive integral exponents are extended to any integral exponent by defining $b^0 = 1$ and $b^{-n} = \dfrac{1}{b^n}$ ($b \neq 0$, $n > 0$).

2. To multiply two polynomials, multiply each term of one polynomial by each term of the other polynomial and then add all the monomial products.

238   *Chapter 6*

3. To *factor a polynomial*, express the polynomial as the product of two or more polynomials of lower positive degree. A *reducible* polynomial is one that can be factored; otherwise, the polynomial is *irreducible*. The *greatest monomial factor* of a polynomial is the monomial with greatest numerical coefficient and greatest degree that is a factor of each term of the polynomial. A polynomial is *factored completely* when it is expressed as a product of a constant and one or more irreducible polynomials or powers of irreducible polynomials.

4. To solve quadratic equations by factoring, you may use the following fact. For all a and $b \in R$, $ab = 0$ if and only if $a = 0$ or $b = 0$.

5. To solve quadratic inequalities by factoring, you may use the following fact. For a and $b \in R$, $ab > 0$ if and only if a and b have the same sign, and $ab < 0$ if and only if a and b have opposite signs.

6. To reduce a rational expression to lowest terms, factor the numerator and denominator completely and then divide both by their greatest common factor.

7. By using the division algorithm, you can transform a nonzero rational expression into the sum of a polynomial and a rational expression in which the degree of the numerator is less than the degree of the denominator.

8. Operations can be performed with rational expressions by using the corresponding rules for operations with rational numbers.

Additional Exercises
Written Exercises
(continued from p. 223)

30. 1. $\dfrac{a + c}{b + c} = \dfrac{a}{b}$,

   $c \neq 0, c \in R$
   Hypothesis
2. $b(a + c) = a(b + c)$
   Prod. of means =
   prod. of ex.
3. $ba + bc = ab + ac$
   Dist. ax.
4. $ab + bc = ab + ac$
   Comm. ax. for $\times$
5. $bc = ac$
   Canc. prop. of +
6. $b = a$
   Canc. prop. of $\times$
7. $a = b$
   Symm. prop. of =

Chapter Review

Write the letter of the correct answer.

1. Simplify $(x^{-9}y^5)^{-1}(2^{-1}x^{-4}y)^2$. 6-1

   a. $\dfrac{-2x}{y^3}$ **ⓑ** $\dfrac{x}{4y^3}$ **c.** $\dfrac{1}{2x^{18}y^6}$ **d.** $\dfrac{x}{2y^3}$

2. Express $(2y - 3)(y^2 - 4y + 2)$ as a polynomial in simple form. 6-2
   a. $2y^3 - 8y^2 + 4y - 3$ **b.** $-y^2 + 12y - 6$
   ⓒ $2y^3 - 11y^2 + 16y - 6$ **d.** $2y^3 - 3y^2 + 4y - 6$

3. Factor $24a^3b - 32a^2b^2 - 6ab^3$ completely. 6-3
   a. $2ab(12a^2 - 16ab - 3b^2)$ **b.** $2ab(4a + 3b)(3a - b)$
   c. $2b(12a^3 - 16a^2b - 6ab^2)$ **ⓓ** $2ab(6a + b)(2a - 3b)$

4. Solve $x^3 = 3x^2 + 4x$. 6-4

   a. $\dfrac{3x^2 + 4}{x^2}$ **b.** $\{4, -1\}$

   ⓒ $\{0, 4, -1\}$ **d.** $\varnothing$

Polynomials and Rational Expressions **239**

5. Solve $x^2 - 5x \le 6$. 6-5
   - **(a.)** $\{x: -1 \le x \le 6\}$
   - **b.** $\{x: x \le -3 \text{ or } x \ge -2\}$
   - **c.** $\{x: -3 \le x \le -2\}$
   - **d.** $\{x: x \le -1 \text{ or } x \ge 6\}$

6. Simplify $\dfrac{x^4 - 20x^2 + 64}{(x + 2)(x^2 + 2x - 8)}$. 6-6
   - **a.** $\dfrac{x^4 - 20x^2 + 64}{(x + 2)(x - 2)(x + 4)}$
   - **(b.)** $x - 4$
   - **c.** $x + 4$
   - **d.** $\dfrac{(x^2 - 4)(x^2 - 16)}{(x + 2)(x - 2)(x + 4)}$

7. Express $\dfrac{4y^4 - 68}{y - 2}$ as a sum by using division. 6-7
   - **a.** $4y^3 + 34$
   - **b.** $4y^3 + 8y^2 + 16y + 32$
   - **c.** $4y^3 - 8y^2 + 16y - 32 + \dfrac{-4}{y - 2}$
   - **(d.)** $4y^3 + 8y^2 + 16y + 32 + \dfrac{-4}{y - 2}$

8. Simplify $\dfrac{y^2 - 25}{y^2 + 11y + 30} \cdot \dfrac{y^2 + 3y - 18}{y^2 - 8y + 15}$. 6-8
   - **a.** $\dfrac{(y - 5)(y - 5)}{(y + 6)(y + 6)}$
   - **b.** $\dfrac{y^4 - 8y^3 - 10y^2 + 200y - 375}{y^4 + 14y^3 + 45y^2 + 108y - 540}$
   - **(c.)** 1
   - **d.** 0

9. Simplify $\dfrac{2x - 3}{x^2 - 5x + 4} - \dfrac{x + 1}{x^2 - 3x - 4}$. 6-9
   - **(a.)** $\dfrac{x - 2}{(x - 4)(x - 1)}$
   - **b.** $\dfrac{x^2 - x - 4}{(x - 4)(x - 1)(x + 1)}$
   - **c.** $\dfrac{x - 4}{8 - 2x}$
   - **(d.)** $\dfrac{x - 2}{(x - 4)(x - 1)}$

10. How much salt must be added to 220 g of a 22% salt solution to produce a 34% salt solution? 6-10
    - **a.** 0.36 g
    - **b.** 29 g
    - **c.** 100 g
    - **(d.)** 40 g

11. Solve $\dfrac{x + 1}{x^2 - 4} + \dfrac{x - 1}{x^2 + x - 2} = \dfrac{1}{x + 2}$. 6-11
    - **a.** $\{1, -1\}$
    - **(b.)** $\{-1\}$
    - **c.** $\varnothing$
    - **d.** $\{-2\}$

Chapter Test

Write an equivalent expression using only positive exponents. Assume that no variable that appears in a denominator or with a negative exponent is zero.

1. $(2x^{-3}y^{-1})(4x^{-2}y^4)^{-2}$ $\frac{x}{8y^9}$ 6-1

2. $\dfrac{m^{-2} - n^{-2}}{m^{-1} + n^{-1}}$ $\frac{n - m}{nm}$

Write each product as a polynomial in simple form.

3. $(2x - 1)(3x^2 - 5x + 8)$
$6x^3 - 13x^2 + 21x - 8$

4. $[(3x - 2)(3x + 2)]^2$ 6-2
$81x^4 - 72x^2 + 16$

Factor each polynomial completely.

5. $2g^3h - 6g^2h^2 + 4gh^3$
$2gh(g - h)(g - 2h)$

6. $x^6 - y^6$ 6-3
$(x - y)(x^2 + xy + y^2)(x + y)(x^2 - xy + y^2)$

Solve.

7. $2w^2 + 7w = 15$ $\{\frac{3}{2}, -5\}$ 6-4

8. $(3x + 2)(x - 3) = 10x$ $\{-\frac{1}{3}, 6\}$

9. If the length of one side of a square is increased by 6 cm and the length of an adjacent side of the same square is doubled, a rectangle is formed having an area that is 13 cm^2 greater than the area of the original square. Find the length of a side of the square. 1 cm

10. Solve $4x^2 - 7x - 2 \geq 0$ over $\mathcal{R}$ and graph its solution set.   6-5

11. Simplify $\dfrac{2m^2 - 6m - 20}{4m^2 - 100} \cdot \dfrac{m + 2}{2(m + 5)}$ 6-6

12. Express $\dfrac{8y^4 - 2y^3 + y^2 - 4}{2y + 1}$ as a sum by using division.   6-7
$4y^3 - 3y^2 + 2y - 1 - \dfrac{3}{2y + 1}$

13. Simplify $\left(x - \dfrac{x}{y^2}\right) \div \left(x - \dfrac{x}{y}\right).$ $\frac{y + 1}{y}$ 6-8

14. Simplify $\dfrac{x^2 - 5x}{x - 1} - \dfrac{4x}{1 - x} + 3.$ $x + 3$ 6-9

15. A certain printer can print 1000 pages in 4 h. Another printer takes 6 h to print the same number of pages. If the two machines work together, how long will it take to print 1000 pages? 2 h 24 min   6-10

16. Solve $\dfrac{x^2 - 6x - 11}{x^2 - 2x - 3} + \dfrac{x + 2}{x - 3} = \dfrac{3x - 1}{x + 1}.$ $\{4\}$ 6-11

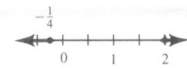
Mixed Review

Simplify.

1. $14 - 8 \div 4 \cdot 3 + (1 + 2)^2$ 17

2. $-[11 + (-z)] - z + (-2)$ -13

3. $-4[-(2x + y) + (-3)(x - y)]$ $20x - 8y$

4. $5s^2 - (3st - 2t)^2 - (-3s^2 + 5st)$
$8s^2 - 9s^2t^2 + 12st^2 - 4t^2 - 5st$

Solve.

5. $-(-x + 4) = 2x - 6$ $\{2\}$

6. $x - 3(x + 1) = x + 6$ $\{-3\}$

7. $2(m - 3) - 4m + 1 < 1$ $\{m : m > -3\}$

8. $-12 < 3(d - 2) < 6$
$\{d : -2 < d < 4\}$

9. $\frac{1}{5}|2x - 3| = 1$ $\{4, -1\}$

10. $|y - 3| \geq 6$ $\{y : y \leq -3 \text{ or } y \geq 9\}$

Polynomials and Rational Expressions   **241**

Solve for b.

11. $4ab - 5b = 3a^2$ $\frac{3a^2}{4a - 5}$

12. $\frac{b}{n} - n = 2c$ $n^2 + 2cn$

13. Solve the system by either the linear combination method or the substitution method.

$$\frac{x + 1}{2} = \frac{y - 3}{4}$$
$$x + 2y = 5 \quad \{(-1, 3)\}$$

14. Determine the equation of a line that passes through the point $(-2, -5)$ and is parallel to the line $x = 4$. $x = -2$

15. If $f: x \rightarrow |x - 3|$ and $g: x \rightarrow x^2 + 4$, find:

  a. $[f \circ g](-2)$ 5
  b. $[g \circ f](-2)$ 29

16. Use Cramer's Rule to solve each system.

  a. $4x - y = 3$
       $2x + 3y = 5$ $\{(1, 1)\}$

  b. $y = \frac{-2}{3}x + 1$
      $12x + 18y = 18$ $\{(x, y): y = \frac{-2}{3}x + 1\}$

17. Determine the equation in the form $y = mx + b$ of a line that passes through the points $(10, -2)$ and $(-4, 5)$. $y = -\frac{1}{2}x + 3$

18. The distance traveled at a constant speed varies directly with the time traveled. If a car can travel 270 km in 3 h, how long would it take to travel 585 km at the same speed?   6.5 h

19. Mary has a total of $4.05 in her bank. There are 39 coins in all; some are dimes and some are quarters. How many coins of each type are in her bank?   1 quarter, 38 dimes

20. A babysitter charges $2 for each hour (or fraction of an hour) that he works before ten o'clock. He increases his fee $.50 for each additional hour he babysits after ten o'clock so that his rate of pay is a function of the hours he works. Graph this function starting at 8 P.M. up to but not including 2 A.M.

Contest Problems

1. Three people, Alvarez, Baird, and Curtis find a box of dollar bills and decide that each person will receive a share of the money based on the amount of work that each contributed to the search. Thus they decide that Alvarez is to receive $\frac{1}{2}$ of the money, Baird $\frac{1}{3}$ of the money, and Curtis $\frac{1}{6}$ of the money. They divide the entire amount of money only to find that each person does not have the correct share. Alvarez returns $\frac{1}{2}$ of the money he took, Baird returns $\frac{1}{3}$ of the money he took, and Curtis returns $\frac{1}{6}$ of the money he took. After this returned money is divided equally among the three people, each person has his rightful share. Determine a *possible* amount of money that was found in the box and the amount that each person took originally.

242   *Chapter 6*

PROGRAMMING IN PASCAL

Problems that involve mechanical activities such as factoring integers or polynomial expressions are ideally solved by computer. Methods of factoring and determining whether or not an integer is prime are *not* remote from real-life concerns. The solutions to such problems are essential to methods of ensuring the security of secret and confidential data. For example, public key encryption and other methods of coding safeguard the access to highly sensitive corporate and government information. The security of public key encryption codes rests upon the enormous difficulty of factoring certain crucial numbers, each of which is the product of two huge (at least 100 digits in length) prime numbers.

The program below determines whether or not a given integer is prime.

```
PROGRAM prime_test(INPUT, OUTPUT);

VAR
   num, factor : integer;
   prime : boolean;

( ****************************************************************)
PROCEDURE get_number;

BEGIN
   writeln;
   writeln('Enter 2 to stop.');
   writeln;
   REPEAT
     write('Enter a positive integer > 1: ');
     readln(num);
   UNTIL (trunc(num) = num) AND (num > 1);
END; (* get_number *)

( ****************************************************************)
PROCEDURE check_it_out;

BEGIN
   prime :=  TRUE;
   factor := 1;
   REPEAT
     factor := factor + 1;
     IF (num MOD factor = 0) AND (num <> 2)   (* 2 is an exception *)
        THEN prime :=  FALSE;
   UNTIL (NOT prime)  OR  (factor > = sqrt(num));
END; (* check_it_out *)
```

(*Program continues on next page.*)

Polynomials and Rational Expressions   **243**

1. $\{x: x \le -3 \text{ or } x \ge 4\}$

2. $\{y: 2 \le y \le 5\}$

3. $\{a: -1 < a < 4\}$

4. $\{x: 2 < x < 7\}$

5. $\{x: x \le 0 \text{ or } x \ge 7\}$

6. $\{b: b < -6 \text{ or } b > 6\}$

7. $\{c: -5 \le c \le 5\}$

8. $\left\{x: -2 < x < \dfrac{1}{3}\right\}$

9. $\{x: -1 < x < 6\}$

10. $\{x: x \le 5 \text{ or } x \ge 7\}$

11. $\{k: k > 3 \text{ or } k < 3\}$

12. $\{-7\}$

13. $\{y: y \ge 3\}$

14. $\{x: x < 3, x \ne 0\}$

15. $\left\{x: x < \dfrac{1}{4}, x \ne 0\right\}$

16. $\{t: t \ge 0\}$

```
( ************************************************************)
PROCEDURE tell_all;
BEGIN
   writeln;
   IF NOT prime
     THEN
       BEGIN
         writeln(num:1, ' is not a prime.');
         writeln;
         writeln(factor:1, ' is a divisor of ',num,'.');
       END
     ELSE writeln(num:1, ' is a prime.');
   writeln;
END;   (* tell_all *)

( ************************************************************)
BEGIN   (* Main *)
   REPEAT
     get_number;
     check_it_out;
     tell_all;
   UNTIL num = 2;
END.
```

Exercises

1. Enter *prime_test* and use it to test whether or not several integers are prime.

2. a. Delete the procedure *tell_all* then modify the procedure *check_it_out* to find and display all the pairs of positive, integral factors of a positive integer. For example, if *num* is 78, the output of the program should be the following.

The pairs of positive integral factors of 78 are:

1	78
2	39
3	26
6	13
13	6
26	3
39	2
78	1

   b. Modify the procedure again, so it displays each pair of factors only once.

   c. Modify the procedure so it counts each pair of different factors and displays the final count.

244   *Chapter 6*

3. a. Write a program similar to *prime_test* that finds and displays all the positive integral primes less 1000.

   b. Modify the program so it counts each prime number, then displays the final count.

   c. Insert code so the user can enter 2 values and the computer will find and display all the prime numbers between the entered values as well as report the number of prime numbers that were displayed.

4. a. Suppose you are testing a positive integer *num* to see if it is prime. Explain why it is sufficient to test only the positive primes less than *num* as possible divisors of *num*. (If none of those primes divide *num*, *num* is itself prime. Why?)

   b. Modify the program in Exercise 3 to store the prime numbers found in an array called *primes*, and use only the elements of *primes* as test divisors in displaying all primes less than 1000. Declare the array *primes* to be a size which is appropriate in view of your solution to Exercise 3. Does the resulting program execute noticeably faster than your solution to Exercise 3?

5. Research Problem

The Sieve of Eratosthenes is an ancient method for determining prime numbers. Find out what it is and convert it to a Pascal program for finding and displaying all primes less than 1000. Such a program is commonly used as a test to gauge the speed of a given programming language on a given computer system. (*Hint*: First create an array that contains all of the integers from 2 to 1000. Next, step through the array in order from 2 to 1000 and replace all of the successive multiples of the current number in the array with 0. When all of the numbers in the array have been processed, the non-zero entries in the array are prime numbers.)

17. $\{z: z > 3\}$

18. $\left\{y: y \le -\dfrac{1}{3} \text{ or } 0 \le y \le \dfrac{1}{3}\right\}$

19. $\{x: x \le 0\}$

20. $\left\{x: x > 0, x \ne \dfrac{5}{2}\right\}$

21. $\{x: x < -2 \text{ or } x > 2\}$

22. $\left\{x: -\dfrac{5}{2} < x < 0 \text{ or } x > \dfrac{5}{2}\right\}$

23. $\{a: -1 \le a \le 1, a \le -3 \text{ or } a \ge 3\}$

24. $\{x: -1 < x < 1\}$

Polynomials and Rational Expressions **245**

The plankton shown have been photographed through a microscope and are between 1×10^{-5} m and 2×10^{-4} m in length. The animal in the center is a copepod whose length can range from 1×10^{-4} m to 1.5×10^{-4} m. Scientific notation is used to express these small numbers and is discussed on pages 257-259.

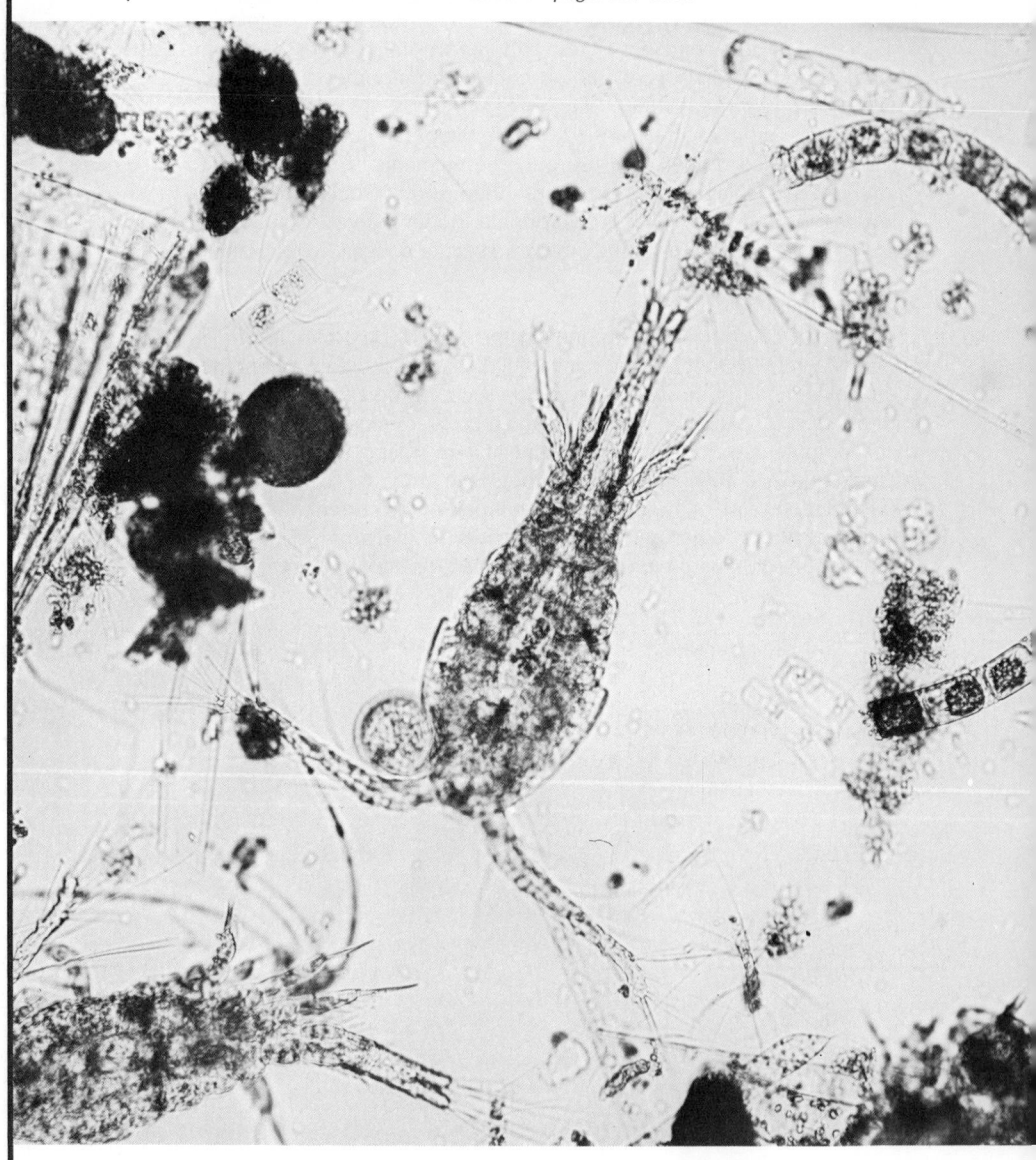

Chapter 7

Radicals and Irrational Numbers

Power Functions, Roots, and Radicals

OBJECTIVES for Sections 7-1 through 7-3:
1. *To use power functions in solving variation problems.*
2. *To determine all real nth roots of a given real number.*
3. *To determine all rational roots of an equation with integral coefficients.*

7–1 Power Functions and Variation

A function f defined by an equation of the form $f(x) = x^n$ is called a
power function. Figure 1 shows the graphs of power functions for the
values $n = 1, 2, 3,$ and 4.

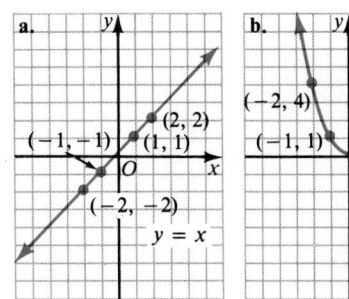
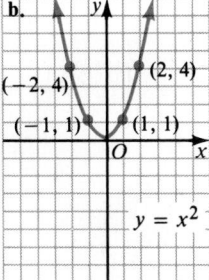
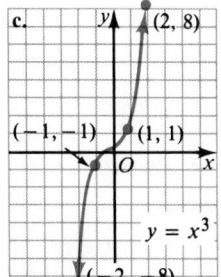
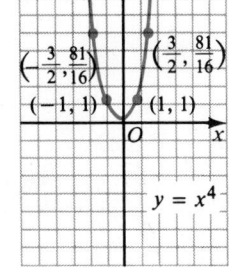

Figure 1

Radicals and Irrational Numbers **247**

Sidebar:

Problem Solving Strategies

Generalizing from Specific
In Sections 7–6 and 7–7
students learn how in
general to rationalize
denominators by examining
several examples.
Looking for a Pattern
Rational numbers are shown
to have repeating patterns in
Section 7–5.

Teaching Suggestions
p. T93

Key Ideas
Solve variation problems by
using power functions.

Chalkboard Examples
Graph each set of equations
on one set of axes.
1. $y = x^2$; $y = \frac{1}{4}x^2$

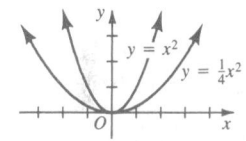

2. $y = \frac{x^4}{2}$; $y = \frac{x^3}{2}$

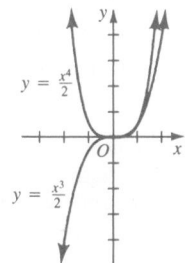

(continued)

3. The distance that a body near Earth's surface will fall from rest varies directly as the square of the number of seconds it has been falling. If a boulder falls from a cliff a distance of 176.4 meters in 6 seconds, how far will it fall in 4 seconds?

Method I
$$d = at^2$$
$$176.4 = a(6)^2$$
$$a = \frac{176.4}{36} = 4.9$$
$$d = 4.9t^2$$
$$= 4.9(4)^2 = 78.4 \text{ m}$$

Method II
$$\frac{d_1}{(t_1)^2} = \frac{d_2}{(t_2)^2}$$
$$\frac{176.4}{6^2} = \frac{d_2}{4^2}$$
$$d_2 = \frac{176.4 \times 16}{36} = 78.4 \text{ m}$$

Common Errors

Students may make a variety of errors when working with graphs of this type. Emphasize that they should graph carefully and neatly, with large enough units to show the graph clearly. You may want to have them practice estimating the most convenient unit size for their graphs.
   Students may tend to connect the few points they have plotted and end up with a graph with "sharp corners" instead of a smooth curvature. Remind the students to make smooth graphs and note carefully where graphs cross each other when they are graphing two functions on one set of axes.

248

A comparison of the graphs in (a) and (c) of Figure 1 indicates the fact that when n is *odd*, the graph of $y = x$ is symmetric with respect to the *origin*. Thus when the function contains the ordered pair (a, b) it also contains the pair $(-a, -b)$; a function with this property is said to be an **odd function**.

When n is *even*, as in (b) and (d) of Figure 1, the graph is symmetric with respect to the *vertical axis*. Thus when the function contains (a, b), it also contains $(-a, b)$; such a function is called an **even function**.

Closely related to the power function is the function defined by

$$y = ax^n, \; n > 0, \; a \neq 0.$$

This function is termed a **variation**. We say that y *varies directly as*, or *is directly proportional to*, the *n*th power of x, and that a is the **constant of variation**, or **proportionality**. When $n = 1$, the function becomes the direct variation $y = ax$ (see page 108).

Figure 2 shows a comparison of the graphs of $y = 3x^2$ and $y = -2x^2$ with that of the power function $y = x^2$.

The concept of variation arises frequently in problems related to the physical world.

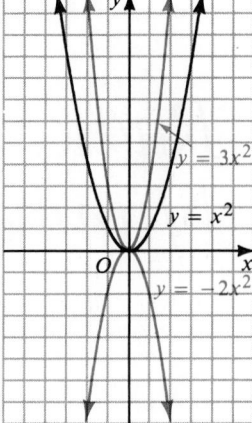

Figure 2

EXAMPLE   The distance that a body near Earth's surface will fall from rest varies directly as the square of the number of seconds it has been falling. If a boulder falls from a cliff a distance of 122.5 m in 5 s, approximately how far will it fall in 8 s?

SOLUTION   Let d = distance (in meters), t = time (in seconds), and a = constant of variation. Then the problem can be solved by either of the following methods.

Method I

$$d = at^2$$
$$122.5 = a(5)^2;$$
$$a = \frac{122.5}{25} = 4.9$$
$$d = 4.9t^2$$

When $t = 8$,

$$d = 4.9(8)^2 = 313.6$$

Method II

$$d_1 = a(t_1)^2; \; d_2 = a(t_2)^2$$
$$\frac{d_1}{(t_1)^2} = \frac{d_2}{(t_2)^2}$$
$$\frac{122.5}{5^2} = \frac{d_2}{8^2}$$
$$d_2 = \frac{122.5 \times 64}{25} = 313.6$$

∴ the boulder falls approximately 313.6 m in 8 s.

Oral Exercises

State an equation that expresses each relationship.

1. y varies directly as x^5. $\quad y = ax^5$

2. The weight w of a solid rubber ball varies directly as the cube of its radius r. $\quad w = ar^3$

3. The period (length of time required for one swing) T of a pendulum varies directly as the square root of the length L of the pendulum. $\quad T = a\sqrt{L}$

4. The kinetic energy E of an object being acted upon by a constantly increasing force varies directly as the fourth power of the time t during which the force has acted. $\quad E = at^4$

Tell whether each function is odd, even, or neither, and whether its graph is symmetric to the y-axis, the origin, or neither.

5. $f(x) = x^6 + 1$	**6.** $f(x) = 2x^3 - 1$	**7.** $f(x) = 2\lvert x \rvert$	**8.** $f(x) = x\lvert x \rvert$
even; symmetric to y-axis	neither; symmetric to neither	even; symmetric to y-axis	odd; symmetric to origin

Written Exercises

Graph each pair of equations on one set of axes, labeling each graph with its equation.

A
1. $y = \frac{1}{2}x^2$, $y = -x^2$

2. $y = \frac{1}{3}x^3$, $y = \frac{3}{2}x^2$

3. $y = \frac{x^3}{2}$, $y = -\frac{x^3}{2}$

4. $y = x^4$, $y = -\frac{x^4}{4}$

5. $y = \frac{x^3}{3}$, $y = -\frac{x^2}{2}$

6. $y = -\frac{x^3}{4}$, $y = \frac{x^2}{2}$

Find the value of k for which the point with the given coordinates lies on the graph of the equation.

7. $(-3, 6)$; $y = kx^2$ $\quad \frac{2}{3}$

8. $(2, 48)$; $y = kx^3$ $\quad 6$

9. $(-3, 54)$; $y = kx^3$ $\quad -2$

10. $\left(\frac{1}{3}, -\frac{5}{6}\right)$; $y = kx^2$ $\quad -\frac{15}{2}$

11. $\left(\frac{1}{4}, \frac{3}{8}\right)$; $y = kx^3$ $\quad 24$

12. $\left(-\frac{1}{2}, \frac{7}{24}\right)$; $y = kx^4$ $\quad \frac{14}{3}$

In Exercises 13–16, assume that y varies directly as x^2.

13. If y is 12 when x is 8, find y when x is 28. 147

14. If y is 36 when x is 9, find x when y is 4. 3, -3

15. If y is $\frac{2}{3}$ when x is $\frac{1}{6}$, find x when y is 24. 1, -1

16. If y is -8 when x is 6, find y when x is 2. $-\frac{8}{9}$

Radicals and Irrational Numbers **249**

Suggested Assignments

Minimum
 249/1–20
 250/*P*: 1–5
Extended Alg. with Trig.
 249/5–21 odd
 250/*P*: 1–9
Enriched Alg.
 249/5–21 odd
 250/*P*: 1–9
Enriched Alg. with Trig.
 249/5–21 odd
 250/*P*: 1–9

Additional Answers
Written Exercises

2.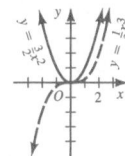

4.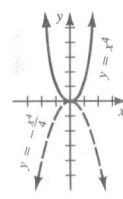

6.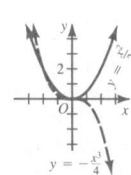

17. The value of the function is multiplied by 8.

18. The value of the function is multiplied by 9.

19. The value of the function is divided by 4.

20. The value of the function is divided by 8.

(continued)

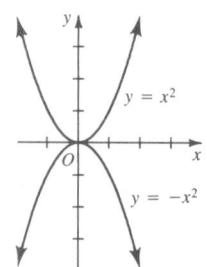
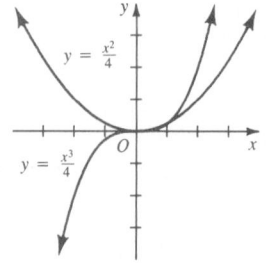
For each function, determine how the value of the function changes under the stated conditions.

EXAMPLE $f(x) = 3x^2$; x is doubled.

SOLUTION $f(2x) = 3(2x)^2 = 3(4x^2) = 4(3x^2) = 4(f(x))$
            The value of the function is multiplied by 4 or quadrupled.

B **17.** $f(x) = 5x^3$; x is doubled.   **18.** $f(x) = 4x^2$; x is tripled.

19. $f(x) = 6x^2$; x is halved.   **20.** $f(x) = 10x^3$; x is halved.

21. $f(x) = 3x^2$; x is quadrupled.   **22.** $f(x) = x^3$; x is multiplied by k.

Problems

A   **1.** In a simple pendulum the length of the pendulum varies directly as the square of the time necessary for one swing (the period). If a pendulum having length 1.2 m has a period of 2.2 s, what is the length of a pendulum that has a period of 1.1 s? 0.3 m

2. In structural engineering the maximum deflection of a flat rectangular plate varies directly as the fourth power of the width of the plate. If the maximum deflection is 0.162 cm for a plate 3 m wide, what is the maximum deflection for a plate 2 m wide? 0.032 cm

3. The power lost by an electrical transmission line varies directly as the square of the current in the line. If the current is 500 A (amperes), the power lost is 100 kW (kilowatts). How much power is lost if the current is lowered to 4 A? 0.0064 kW, or 6.4 W

4. The mass of a spherical drop of mercury that has a diameter 0.2 cm is 0.057 g. If the mass of a drop of mercury varies directly with the cube of its diameter, what is the mass of a drop of diameter 0.4 cm? 0.456 g

5. The power required to overcome air resistance and keep a bicycle moving at a constant speed (neglecting friction) on level ground varies directly as the cube of the speed. If 160 W (watts) is required to maintain a speed of 40 km/h, what power is required to maintain a speed of 30 km/h? 67.5 W

B   **6.** The radiancy of an object that absorbs all the radiation that falls on it (measured in joules per second per square meter), varies directly as the fourth power of the Kelvin temperature of the object. If such an object has a radiancy of 324 (J/s)/m² at 300° K, what is the radiancy of the object at 200° K? 64 (J/s)/m²

7. Energy (measured in joules) that is stored in an electric capacitor varies directly as the square of the electric potential (measured in volts) across the capacitor. **a.** If 6 V produces 0.00009 J of energy, how much energy is produced by 8 V? **b.** What is the constant of variation? a. 0.00016 J   b. 0.0000025

250   *Chapter 7*

8. The force of air resistance on a falling object varies directly as the square of its speed. If the force on an object traveling at 3.6 m/s is 2.7 N (newtons), at what speed will the force on the object be 7.5 N? 6 m/s
9. The square of the period of revolution of an artificial satellite varies directly as the cube of its distance from the center of Earth. A satellite 20,000 km from the center of Earth has a period of revolution of 8 h. What would be the period of a satellite 30,000 km from the center of Earth? 14.70 h

7–2 The Real nth Roots of a Number

By observing the graphs in Figures 3, 4, and 5 of the power functions specified by $f(x) = x^2$, $g(x) = x^3$, and $h(x) = x^4$, you can answer questions such as the following about the functions f, g, and h.

1. How many values in the domain of f are paired with the value $f(x) = 9$ in the range? That is, how many values of x satisfy the equation $x^2 = 9$?

2. How many values in the domain of g are paired with the value $g(x) = -8$ in the range? That is, how many solutions are there of the equation $x^3 = -8$?

3. Are there any real values of x such that the pair $(x, -2) \in h$? That is, are there any real values of x that satisfy $x^4 = -2$?

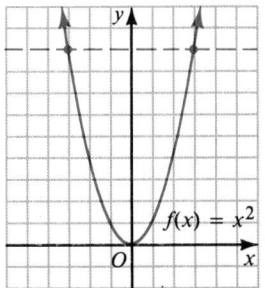

Figure 3

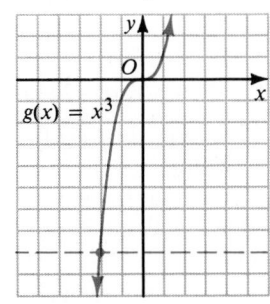

Figure 4

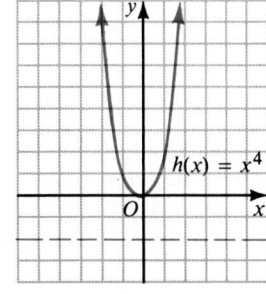

Figure 5

The abscissas of the points shown by red dots on the graphs in Figures 3 and 4 indicate the values of x that satisfy the equation given in each case. In Figure 5 you can see that since there is no pair $(x, -2)$ in h, -2 is not an element of the range, and hence there is no real solution to $x^4 = -2$.

Each solution of the equation $x^n = b$, for n a positive integer, is called an nth root of b. Thus, 3 and -3 are the 2nd roots (or *square roots*) of 9, and -2 is the 3rd (or *cube*) root of -8.

Radicals and Irrational Numbers   **251**

Mixed Review

1. Find the solution over $\mathcal{R}$.
   $$\frac{x + 3}{x - 3} - \frac{x}{2x - 6} = \frac{2}{5} \quad \{-42\}$$

2. Factor $4x^3 + 2x^2 - 30x$ completely.
   $2x(x + 3)(2x - 5)$

3. Solve for a.
   $a : 12 = 3 : 4$.   9

4. Find the slope of the line joining $(-3, 7)$ and $(-2, 4)$. -3

Teaching Suggestions
p. T93

Key Ideas

Find all the real nth roots of a given real number.

Chalkboard Examples

1. Solve $1.44x^2 - 1 = 0$ over $\mathcal{R}$.
   $$1.44x^2 - 1 = 0$$
   $$1.44x^2 = 1$$
   $$x^2 = \frac{1}{1.44}$$
   $$x^2 = \frac{100}{144}$$
   $$x = \pm\frac{10}{12} = \pm\frac{5}{6}$$
   $$\left\{\frac{5}{6}, -\frac{5}{6}\right\}$$

2. Find all real values of the variable for which $\sqrt{(x - 4)^2} = x - 4$ is true.
   $\sqrt{(x - 4)^2}$ is nonnegative
   $\therefore x - 4$ is nonnegative
   $x - 4 \geq 0$, or $x \geq 4$.

Additional A Exercises

Solve over the set of rational numbers.

1. $4x^2 = 36$ $\{-3, 3\}$

2. $1.96x^2 - 1 = 0$ $\left\{\dfrac{5}{7}, -\dfrac{5}{7}\right\}$

3. $3x^3 = 0.024$ $\{0.2\}$

Simplify.

4. $\sqrt[3]{-\dfrac{1}{27}} + \sqrt{\dfrac{4}{9}}$ $\dfrac{1}{3}$

Find all real values of x for which the statement is true.

5. $\sqrt{x - 7} = 0$ $\{7\}$

6. $\sqrt{(x + 2)^2} = x + 2$ $\{x: x \geq -2\}$

Common Errors

Students often dismiss an expression such as $\sqrt[3]{-64}$ as undefined or write "no solution" for an equation such as $x^3 = -64$, not realizing that an *odd* root of a negative number is defined. Remind them that a negative number taken as a factor an odd number of times results in a negative number. Familiarize your students with the notion that an odd root of a negative number is defined. (and is negative), while an even root of a negative number is undefined.

The facts concerning the *real* nth roots of b, as suggested by Figures 3, 4, and 5, are summarized in the following table.

Number and Nature of Real nth Roots of b

	$b > 0$	$b < 0$	$b = 0$
n even	one positive root one negative root	no real roots	one root, namely, 0
n odd	one positive root	one negative root	one root, namely, 0

EXAMPLE 1 Solve $1.21x^2 = 1$ over $\mathcal{R}$.

SOLUTION $1.21x^2 = 1$

$$x^2 = \frac{1}{1.21} = \frac{100}{121}$$

$$x = \frac{10}{11}, \ x = -\frac{10}{11}$$

$\therefore$ the solution set is $\left\{\dfrac{10}{11}, -\dfrac{10}{11}\right\}$

The symbol $\sqrt[n]{b}$ (read "the nth root of b") denotes the principal nth root of b, that is:

1. The *nonnegative* nth root of b if n is even and $b \geq 0$. For example, we write
$$\sqrt[2]{36} = 6, \ -\sqrt[2]{36} = -6, \ \sqrt[4]{0} = 0.$$

2. The single real nth root of b if n is odd. For example,
$$\sqrt[3]{8} = 2, \ -\sqrt[3]{8} = -2.$$

Notice that we are talking about only the *real* nth roots of b; that is, in the set of real numbers, $\sqrt[4]{-16}$ is not defined.

The symbol $\sqrt[n]{b}$ is called a **radical**; b is the **radicand** and n the **index**. The index 2 for the square root is usually omitted. Always be careful to include all of the desired radicand under the radical sign. Thus,
$$\sqrt{36 - 9} = \sqrt{27} = 3\sqrt{3},$$
whereas
$$\sqrt{36} - 9 = 6 - 9 = -3.$$

Notice that $\sqrt[3]{(-2)^3} = -2$, but $\sqrt[4]{(-2)^4} = 2$, because a radical of even index denotes a *nonnegative* root. In general,
$$\sqrt[n]{b^n} = b \text{ if } n \text{ is odd}, \quad \text{and} \quad \sqrt[n]{b^n} = |b| \text{ if } n \text{ is even}.$$

EXAMPLE 2 Find all real values of the variable for which $\sqrt{(x - 3)^2} = x - 3$ is true.

SOLUTION Since $\sqrt{(x - 3)^2}$ denotes a nonnegative number, the equation is true if and only if $x - 3$ is nonnegative; that is $x - 3 \geq 0$, or $x \geq 3$.

252 *Chapter 7*

Oral Exercises

State the value of the indicated principal root, or state that the expression is not defined as a real number.

1. $\sqrt[3]{125}$ 5
2. $\sqrt[3]{-64}$ −4
3. $\sqrt{-49}$ Not defined
4. $-\sqrt[4]{81}$ −3

5. $\sqrt[4]{-16}$ Not defined
6. $\sqrt[5]{1}$ 1
7. $\sqrt[4]{0}$ 0
8. $-\sqrt[3]{-\frac{1}{27}}$ $\frac{1}{3}$

9. $-\sqrt{-\frac{9}{4}}$ Not defined
10. $\sqrt{(-8)^2}$ 8
11. $\sqrt{-(-\frac{1}{2})^2}$ Not defined
12. $-\sqrt[4]{(-2)^4}$ −2

Is the statement true for all real numbers, or false? If the statement is false, give an example that shows this.

13. $\sqrt{a^2} = a$ False, $a = -1$
14. $\sqrt{a^2} = \sqrt{(-a)^2}$ True
15. $\sqrt[3]{a^3} = a$ True
16. $\sqrt[5]{-a} = -\sqrt[5]{a}$ True

Written Exercises

A Solve over the set of rational numbers.

1. $x^2 = 144$ {12, −12}
2. $x^2 - 25 = 0$ {5, −5}
3. $x^4 - 1 = 0$
4. $x^4 = 81$ {3, −3}
5. $x^3 + 8 = 0$ {−2}
6. $27 - x^3 = 0$
7. $x^2 + 9 = 0$ ∅
8. $64x^3 = 1$ {$\frac{1}{4}$}
9. $49 - 16x^2 = 0$
10. $27x^3 + 125 = 0$ {$-\frac{5}{3}$}
11. $32x^5 - 1 = 0$ {$\frac{1}{2}$}
12. $121x^2 = -36$

3. {1, −1}
6. {3}
9. {$\frac{7}{4}$, $-\frac{7}{4}$}
12. ∅

Simplify.

13. $\sqrt[3]{-\frac{1}{8}} + \sqrt{\frac{49}{4}}$ 3
14. $\sqrt[5]{-32} - \sqrt{(-5)^2}$ −7

15. $(-\sqrt[4]{16})(\sqrt[3]{-1})$ 2
16. $\sqrt[3]{\frac{-27}{64}} \div \sqrt{\frac{25}{4}}$ $-\frac{3}{10}$

Find all real values for which the statement is true.

17. $\sqrt{x - 4} = 0$ {4}
18. $\sqrt{x^2} = 5$ {5, −5}
19. $\sqrt[3]{-x^3} = 6$ {−6}
20. $\sqrt{x^2} = -x$ {x: x ≤ 0}
21. $\sqrt{x^2} = \sqrt[3]{x^3}$ {x: x ≥ 0}
22. $\sqrt[3]{-x^3} = -\sqrt{x^2}$ {x: x ≥ 0}

B 23. $\sqrt[3]{(x - 1)^3} = 7$ {8}
24. $\sqrt{(x - 3)^2} = 4$ {−1, 7}
25. $\sqrt{(y + 1)^2} = y + 1$ {y: y ≥ −1}
26. $\sqrt{(y + 1)^4} = (y + 1)^2$ ℛ
27. $\sqrt{(x - 2)^6} = (x - 2)^3$ {x: x ≥ 2}
28. $\sqrt{x - 8} = \sqrt{8 - x}$ {8}

C 29. Prove: If $x \geq 0$, $y \geq 0$, then $\sqrt{x^2 + y^2} \leq x + y$. (*Hint:* Assume $\sqrt{x^2 + y^2} > x + y$ and show that this leads to a contradiction; see Exercise 34 on page 60.)

30. Show that if $\sqrt{x^2 + y^2} = x + y$, then $x = 0$ or $y = 0$.

Radicals and Irrational Numbers **253**

Suggested Assignments

Minimum
253/1–28
Extended Alg. with Trig.
253/3–27 odd
Enriched Alg.
253/7–25 odd, 26–30
Enriched Alg. with Trig.
253/7–25 odd, 26–30

Additional Answers
Written Exercises

30. If $\sqrt{x^2 + y^2} = x + y$, then $(\sqrt{x^2 + y^2})^2 = (x + y)^2$, or, $x^2 + y^2 = x^2 + 2xy + y^2$. Then by the add. prop. of order, $0 = 2xy$. By the mult. prop. of zero, $x = 0$ or $y = 0$.

Mixed Review

1. True or false?
$$\frac{1}{a} + \frac{1}{b} = \frac{1}{a + b}$$ F

2. True or false?
$$\frac{1}{a}(a + b) = \frac{a}{b}$$ F

3. $$\frac{1}{\frac{1}{a} + \frac{1}{b}} = \frac{?}{b + a}$$ ab

4. Solve $\frac{2}{3}(p + 12) = -4$.
{−18}

5. Solve $4[c - 3(2 - c)] = 24$.
{3}

Key Ideas

Find rational roots of equations with integral coefficients.

Chalkboard Examples

1. If $f(x) = 4x^3 + 4x^2 - x - 1$, list the possible roots of $f(x) = 0$; then find which, if any, actually satisfy that equation.

$p \in \{1, -1\}$

$q \in \{1, 2, 4, -1, -2, -4\}$

$\therefore \dfrac{p}{q} : \pm\dfrac{1}{4}, \pm\dfrac{1}{2}, \pm 1$

$f\left(\dfrac{1}{4}\right) = -\dfrac{15}{16} \qquad f\left(\dfrac{1}{2}\right) = 0$

$f\left(-\dfrac{1}{4}\right) = -\dfrac{9}{16} \quad f\left(-\dfrac{1}{2}\right) = 0$

$f(1) = 6 \qquad f(-1) = 0$

$\therefore$ the rational roots are $\dfrac{1}{2}$,

$-\dfrac{1}{2}$, and -1.

Common Errors

Students sometimes miss some of the possible roots of a polynomial equation. Caution the students against overlooking the positive and the negative values of $\dfrac{p}{q}$ that should be tested. You may want to suggest that they write down the factors of an integer in some kind of order to help them remember the possibilities. For example, write the factors of 8 as 1, -1, 2, -2, 4, -4, 8, and -8.

7–3 Roots of a Polynomial Equation

A *rational number* has been defined (page 221) as any number that can be represented in the form $\dfrac{a}{b}$, where a and b are integers, $b \neq 0$.

You know that $\sqrt{4}$ is a rational number because $\sqrt{4} = \dfrac{2}{1}$, but what about $\sqrt{3}$? Since $\sqrt{3}$ is a root of the equation $x^2 - 3 = 0$, we can answer the question by using the following theorem concerning the rational roots, if any, of such a polynomial equation. (The **leading coefficient** of a polynomial is the coefficient of the term of highest degree.)

Rational Root Theorem

Let $f(x)$ be a simplified polynomial with integral coefficients. If the equation $f(x) = 0$ has a rational root $\dfrac{p}{q}$ that is in lowest terms, then p must be an integral factor of the constant term of $f(x)$ and q must be an integral factor of the leading coefficient of $f(x)$.

EXAMPLE 1 If $f(x) = 2x^3 + 7x^2 - 14x + 5$, list the possible roots of $f(x) = 0$; then find which, if any, actually satisfy that equation.

SOLUTION The numerator of any rational root $\dfrac{p}{q}$ must be an integral factor of 5 and the denominator an integral factor of 2. That is,

$$p \in \{1, -1, 5, -5\} \text{ and } q \in \{1, -1, 2, -2\}.$$

Hence the possible rational roots $\dfrac{p}{q}$ are ± 1, ± 5, $\pm\dfrac{1}{2}$, $\pm\dfrac{5}{2}$. Next, check each possible value for $\dfrac{p}{q}$ in $f(x)$ to see which, if any, satisfy $f(x) = 0$.

$$f(1) = 0 \qquad f(5) = 360 \qquad f(\tfrac{1}{2}) = 0 \qquad f(\tfrac{5}{2}) = 45$$
$$f(-1) = 24 \qquad f(-5) = 0 \qquad f(-\tfrac{1}{2}) = \tfrac{27}{2} \qquad f(-\tfrac{5}{2}) = \tfrac{105}{2}$$

$\therefore$ the rational roots of $2x^3 + 7x^2 - 14x + 5 = 0$ are 1, -5, and $\dfrac{1}{2}$.

EXAMPLE 2 If $f(x) = x^2 - 3$, determine the rational roots, if any, of $f(x) = 0$.

SOLUTION A rational root $\dfrac{p}{q}$ must be such that $p \in \{1, -1, 3, -3\}$ and $q \in \{1, -1\}$.

Hence $\dfrac{p}{q} \in \{1, -1, 3, -3\}$. Check in $f(x)$:

$$f(1) = f(-1) = -2 \neq 0 \qquad f(3) = f(-3) = 6 \neq 0$$

$\therefore x^2 - 3 = 0$ has no rational roots.

Real numbers that are not rational are called *irrational numbers.* Example 2 establishes the fact that $\sqrt{3}$ and $-\sqrt{3}$, which are the roots of $x^2 - 3 = 0$, are irrational numbers. In general, for any positive integer b and any integer $n > 1$, you can show, by considering the equation $x^n = b$, that $\sqrt[n]{b}$ is an irrational number unless b is the nth power of an integer.

Example 3 illustrates the following fact: *with the exceptions of $0 \cdot x$ and $\frac{0}{x}$, where x is irrational, the sum, difference, product, or quotient of a rational number and an irrational number is an irrational number.*

EXAMPLE 3 Determine whether $5 + 2\sqrt{3}$ is a rational or an irrational number.

SOLUTION If $5 + 2\sqrt{3}$ is a rational number, there must be integers a and b, $b \neq 0$, such that

$$5 + 2\sqrt{3} = \frac{a}{b}$$

or

$$\sqrt{3} = \frac{a - 5b}{2b}.$$

Since $a - 5b$ and $2b$ are both integers and $b \neq 0$, $\frac{a - 5b}{2b}$ is a rational number. Example 2 showed that $\sqrt{3}$ is an irrational number. Thus there is a contradiction, and accordingly, the hypothesis that $5 + 2\sqrt{3}$ is a rational number must be false. Hence, $5 + 2\sqrt{3}$ is irrational.

Oral Exercises

State all possible rational roots of each equation.

2. $\pm 1, \pm 2, \pm 4, \pm 8$
4. $\pm 1, \pm\frac{1}{2}, \pm\frac{1}{4}$

1. $x^3 - 2x^2 - x + 2 = 0$ $\pm 1, \pm 2$
2. $x^3 - 8 = 0$
3. $x^3 - x^2 - 4x + 4 = 0$ $\pm 1, \pm 2, \pm 4$
4. $4x^3 + 4x^2 - x - 1 = 0$
5. $6x^3 + 7x^2 - 1 = 0$ $\pm 1, \pm\frac{1}{2}, \pm\frac{1}{3}, \pm\frac{1}{6}$
6. $x^3 - 5x^2 + 3x + 9 = 0$
7. $3x^3 + 11x^2 + 8x - 4 = 0$
8. $2x^3 - 9x^2 + 10x - 3 = 0$
9. $x^4 - 26x^2 + 25 = 0$ $\pm 1, \pm 5, \pm 25$
10. $4x^4 - 5x^2 + 1 = 0$
7. $\pm 1, \pm 2, \pm 4, \pm\frac{1}{3}, \pm\frac{2}{3}, \pm\frac{4}{3}$
6. $\pm 1, \pm 3, \pm 9$
8. $\pm 1, \pm 3, \pm\frac{1}{2}, \pm\frac{3}{2}$
10. $\pm 1, \pm\frac{1}{2}, \pm\frac{1}{4}$

Written Exercises

A **1–10.** Find the rational roots of the equations in Oral Exercises 1–10.

Find the rational roots of each equation.

11. $2x^3 - 9x^2 + 3x + 4 = 0$ $1, 4, -\frac{1}{2}$
12. $6x^3 + 7x^2 - 9x + 2 = 0$
13. $4x^3 - 13x + 6 = 0$ $-2, \frac{1}{2}, \frac{3}{2}$
14. $4x^3 - 8x^2 - 7x + 5 = 0$
15. $2x^4 + 3x^3 + 6x^2 + 12x - 8 = 0$ $-2, \frac{1}{2}$
16. $4x^4 - 4x^3 + x^2 - 4x - 3 = 0$

Radicals and Irrational Numbers **255**

Additional A Exercises

Find the rational roots of each equation.

1. $x^3 + x + 2 = 0$ $\{-1\}$
2. $x^3 - 3x^2 + x - 3 = 0$ $\{3\}$
3. $x^3 - 3x + 2 = 0$ $\{1, -2\}$
4. $x^3 - 3x^2 + 4 = 0$ $\{-1, 2\}$
5. $x^4 - 8x^2 - 9 = 0$ $\{3, -3\}$

Suggested Assignments

Minimum
First day
255/1–17 odd, 18–25
Second day
R 256/*Self-Test 1*
S 222/1–11 odd
Extended Alg. with Trig.
255/1–27 odd
R 256/*Self-Test 1*
Enriched Alg.
255/9–27 odd, 28–30
R 256/*Self-Test 1*
Enriched Alg. with Trig.
255/9–27 odd, 28–30
R 256/*Self-Test 1*

Additional Answers
Written Exercises

1. $1, -1, 2$ 2. 2
3. $1, 2, -2$ 4. $-1, \frac{1}{2}, -\frac{1}{2}$
5. $-1, -\frac{1}{2}, \frac{1}{3}$ 6. $-1, 3$
7. $-2, \frac{1}{3}$ 8. $1, 3, \frac{1}{2}$
9. $1, -1, 5, -5$
10. $1, -1, \frac{1}{2}, -\frac{1}{2}$
12. $-2, \frac{1}{2}, \frac{1}{3}$
14. $-1, \frac{1}{2}, \frac{5}{2}$
16. $\frac{3}{2}, -\frac{1}{2}$

(continued on p. 287)

Mixed Review

1. Simplify $\dfrac{\sqrt{36}}{\sqrt{4}}$.   3

2. Simplify $\dfrac{7}{24} + \dfrac{4}{9}$. $\dfrac{53}{72}$

3. Simplify $\dfrac{2d^2 - 6d}{4d - 12} \cdot \dfrac{d}{2}$.

4. Solve $x^2 - 3x - 10 = 0$. $\{5, -2\}$

5. What is the slope of the line $3x + \dfrac{5}{2}y - 2 = 0$?

   $-\dfrac{6}{5}$

6. Find the slope of the line joining $(3, 7)$ and $(3, -7)$. undefined

Quick Quiz

1. If y varies directly as x^3 and y is 3 when x is 2, find y when x is 4.   24

2. Determine the value of **(a)** $\sqrt{(-14)^2}$ and **(b)** $\sqrt[3]{-8}$.
   a) 14 b) -2

3. Solve $5x^3 = 40$ over the set of rational numbers. $\{2\}$

4. Determine all rational roots of the equation $x^3 - 7x - 6 = 0$. $\{-2, -1, 3\}$

256

Write an equation, with integral coefficients, that has the given number as a root. Then use the method of Example 2 to show that the equation has no rational roots and therefore that the given number must be irrational.

B 17. $\sqrt{5}$ 18. $\sqrt[3]{2}$ 19. $\sqrt{7}$

20. $-\sqrt[3]{6}$ 21. $\sqrt[3]{\dfrac{1}{2}}$ 22. $\sqrt{\dfrac{5}{3}}$

Use the method of Example 3 to show that each number is irrational. Use the results of Exercises 17–19.

23. $\sqrt{5} + 9$ 24. $\dfrac{\sqrt{7}}{2}$ 25. $8\sqrt[3]{2}$

26. $6 - \sqrt[3]{2}$ 27. $\dfrac{4 + \sqrt{7}}{3}$ 28. $\dfrac{3\sqrt{5} - 1}{4}$

C 29. This exercise is a partial proof of the Rational Root Theorem on page 254 so the theorem itself cannot be used in the proof.

    a. Prove that if n is an integral root of $ax^3 + bx^2 + cx + d = 0$ ($a, b, c,$ and d are integers) then n must be a factor of d.

    b. Prove that if n is an integer and $\dfrac{1}{n}$ is a root of

$$ax^3 + bx^2 + cx + d = 0$$

    ($a, b, c,$ and d are integers), then n must be a factor of a.

30. Prove that if t is an irrational number, then for any rational numbers a and b, $a \neq 0$, the number $at + b$ is irrational.

Self-Test 1

VOCABULARY   power function (p. 247)   radical (p. 252)
               odd function (p. 248)   radicand (p. 252)
               even function (p. 248)   index of a radical (p. 252)
               variation (p. 248)   leading coefficient (p. 254)
               constant of variation, or   Rational Root Theorem
                  proportionality (p. 248)   (p. 254)
               nth root of b (p. 251)   irrational number (p. 255)

1. If y varies directly as x^3 and y is 5 when x is 2, find y when $x = 4$. 40 *Obj. 1, p. 247*

2. Determine the value of **(a)** $\sqrt{(-12)^2}$ and **(b)** $\sqrt[3]{-27}$. a. 12 b. -3 *Obj. 2, p. 247*

3. Solve $8t^3 + 343 = 0$ over the set of rational numbers. $\{-\frac{7}{2}\}$ *Obj. 3, p. 247*

4. Determine all the rational roots of $x^3 + x + 2 = 0$. -1

Check your answers with those at the back of the book.

---

256   *Chapter 7*

Rational and Irrational Numbers

OBJECTIVES for Sections 7-4 and 7-5:
1. *To represent a number in standard notation.*
2. *To estimate products and quotients.*
3. *To express a rational number as a terminating or repeating decimal, and express a repeating decimal as a fraction.*
4. *To find rational and irrational numbers between any two given real numbers.*

7–4 Standard Notation

It is often convenient to break off, or **round** a lengthy or infinite decimal, leaving an approximation of the number represented by the original decimal. Using $\approx$ to mean "equals approximately" you can write

$$453.182 \approx 453.2 \qquad 453.182 \approx 453.18 \qquad 453.182 \approx 453$$

as approximations of 453.182 to the nearest tenth, the nearest hundredth, or the nearest unit, respectively. In rounding use the following rule.

> To round a decimal, add 1 to the last digit retained if the value of the first digit dropped is 5 or more; otherwise leave the retained digits unchanged.

Using this rule the difference between a number and its approximation, which is the **rounding error,** is *at most* half the unit of the last digit retained. For example, the statement $2.4777 \ldots \approx 2.48$ is equivalent to

$$2.48 - 0.005 < 2.4777 \ldots < 2.48 + 0.005$$

or

$$2.475 < 2.4777 \ldots < 2.485.$$

Measurements almost always produce approximations. The **precision** of a measurement is given by the unit used in making the measurement. For example a physicist reporting the diameter of a piece of wire as 1.79 mm precise to the nearest 0.01 mm means that the true diameter, d, satisfies the inequality

$$1.785 < d < 1.795.$$

The unit or precision of this measurement is 0.01 mm. The **maximum possible error** is half the unit of precision. In this example the maximum

Radicals and Irrational Numbers **257**

Key Ideas

Use standard notation to express very large or very small numbers.
Define significant digits.

Chalkboard Examples

1. Determine the number of significant digits in each of the following:
   a. 0.135   **b.** 3420
   c. 3420.0   **d.** 786
   a. 3, the leading 0 is used to place the decimal.
   b. 3; since no decimal follows, the zero is assumed to be a place holder.
   c. Five; the decimal following the zero makes it significant.
   d. Three

2. Find a one-significant-digit estimate of A if
   $$A = \frac{2120 \times 36.94 \times 194}{365.3}$$
   Round each number to one-significant-digit:
   $$A \approx \frac{2000 \times 40 \times 200}{400}$$
   Express each in standard notation and mult.
   $$A \approx 4 \times 10^4 = 40,000$$

possible error is 0.005 mm. On the other hand, an interior designer may indicate the length of a room as 4.3 m to the nearest 0.1 m. In this measurement the precision is 0.1 m and the maximum possible error is 0.05 m.

To determine which of the measurements in the previous paragraph is more accurate, we need a definition of *accuracy*. The **accuracy** of a measurement is the **relative error,** usually expressed as the ratio of the maximum possible error in the measurement to the measurement itself. Thus the accuracy of the physicist's measurement is

$$\frac{0.005}{1.79} \approx 0.003 \text{ or } 0.3\%.$$

The accuracy of the interior designer's measurement is

$$\frac{0.05}{4.3} \approx 0.012 \text{ or } 1.2\%.$$

Thus, because the relative error is smaller, the physicist made the more accurate measurement.

In a numeral, each digit reporting the number of units of measure contained in a measurement is called a **significant digit.**

EXAMPLE 1   Determine the number of significant digits in each of the following:

    a. 92.3   **b.** 0.0305   **c.** 200   **d.** 2.0

SOLUTION   **a.** three significant digits

    b. The zero between 3 and 5 is significant, but the leading zeros are only used to place the decimal point. Thus 0.0305 has three significant digits.

    c. Without additional information, it cannot be determined which, if any, of the zeros are significant. Thus 200 could have 1, 2, or 3 significant digits.

    d. The zero after the decimal point is significant. Thus 2.0 has two significant digits.

To indicate the accuracy of a measurement we use **standard** or **scientific notation,** that is, a number written in the form $a \times 10^n$ where $1 \le |a| < 10$ and n is an integer.

Some examples of numbers written in ordinary decimal notation and in scientific notation are shown in the table at the right. Notice that a red caret is placed after the first significant digit in each numeral in the numbers written in decimal notation. By counting the number of places *from the caret to the decimal point*, you can determine n. Note that n is positive or negative according as you count to the right or left from the caret.

Decimal notation	Scientific notation
-41.5	-4.15×10^1
0.0058	5.8×10^{-3}
$41,620,000$	4.162×10^7
0.0001	1×10^{-4}

It is easy to tell the number of significant digits if a measurement is given in standard (scientific) notation. In Example 1(c), instead of 200 meters, you would write

$$2 \times 10^2, \quad 2.0 \times 10^2, \quad \text{or} \quad 2.00 \times 10^2,$$

according as the measurement is precise to the nearest hundred meters, the nearest ten meters, or the nearest meter.

For computations in general, we use the following working rules:

To Round the Numerals for Results:

1. Give products, quotients, and powers to the same number of *significant digits* as appear in the *least accurate* approximation involved.

2. Give sums and differences to the same number of decimal places as appear in the approximation with the *least number* of decimal places (the least *precise* measurement).

You can use scientific notation to estimate products and quotients rapidly.

EXAMPLE 2   Find a one-significant-digit estimate of A if

$$A = \frac{3250 \times 61.27 \times 0.076}{4289}.$$

SOLUTION

1. Round each number to its one-significant-digit approximation.

$$A \approx \frac{3000 \times 60 \times 0.08}{4000}$$

2. Express the approximation in scientific notation.

$$A \approx \frac{3 \times 10^3 \times 6 \times 10^1 \times 8 \times 10^{-2}}{4 \times 10^3}$$

3. Compute and round to one significant digit.

$$A \approx \frac{3 \times 6 \times 8 \times 10^2}{4 \times 10^3} = 36 \times 10^{-1} = 3.6$$

$\therefore A \approx 4$ (to four decimal places, A is actually equal to 3.5285).

Oral Exercises

State the number of significant digits in each of the following.

1. 89.4 3
2. 172 3
3. 0.096 2
4. 1.238 4
5. 1.064 4
6. 45.04 4
7. 7.9×10^3 2
8. 1.20×10^4 3

9. 4.3861 4.39   **10.** 317.298 317   **11.** 0.03867 0.0387   **12.** 0.6753 0.675

13. $61,752$ 61,800   **14.** $212,499$ 212,000   **15.** 5.008 5.01   **16.** 42.96 43.0

Written Exercises

Express each of the following in standard notation.

3.82×10^2

A **1.** $34,000$ 3.4×10^4 **2.** 0.0467 4.67×10^{-2} **3.** 0.00051 5.1×10^{-4} **4.** 382

5. 2785 2.785×10^3 **6.** 0.0501 5.01×10^{-2} **7.** 20.23 2.023×10^1 **8.** 0.004006

4.006×10^{-3}

Express each of the following in decimal notation.

287,000

9. 3.5×10^4 35,000   **10.** 4.1×10^{-3} 0.0041   **11.** 9.65×10^{-2} 0.0965   **12.** 2.87×10^5

13. 3.03×10^{-1} **14.** 4.007×10^{-4} **15.** 6.781×10^6 **16.** 7.146×10^{-5}

0.303   0.0004007   6,781,000   0.00007146

Find a one-significant-digit estimate of each of the following.

17. $\dfrac{(40,000)(0.06)}{0.003}$ 8×10^5, or 800,000

18. $\dfrac{(0.07)(9000)}{3,000,000}$ 2×10^{-4}, or 0.0002

B **19.** $\dfrac{(350,000)(6800)}{(0.007)(0.00023)}$ 2×10^{15}, or 2,000,000,000,000,000

20. $\dfrac{(42,000)(0.837)}{(0.512)(2.13)}$ 3×10^4, or 30,000

21. $\dfrac{(72,100)(0.633)}{(0.0929)(20.523)}$ 2×10^4, or 20,000

22. $\dfrac{(43,105)(0.244)}{(0.083)(52.17)}$ 2×10^3, or 2000

23. $\dfrac{(21,051)(0.943)}{(1.82)(0.2816)}$ 3×10^4, or 30,000

24. $\dfrac{(692.3)(1.87)(2397)}{(0.032)(0.83)}$ 1×10^8, or 100,000,000

In Exercises 25 and 26 let $a = 4.135$ and $b = 2.692$.

C **25.** Write a and b as a one-significant-digit approximation to estimate $a + b$, $a - b$, ab, and $a \div b$. Determine the accuracy of each calculation. Which calculation is most accurate?

26. Calculate the value of $a + b$, $a - b$, ab, and $a \div b$ to four significant digits. Determine the accuracy of each calculation. Which calculation is most accurate?

ON THE CALCULATOR

For very large or very small numbers, calculators usually display the answer in standard notation. The readout usually leaves a space before the appropriate exponent of 10 and the 10 is not shown. The number of significant digits in the readout depends on the number of places the calculator has for display.

260 *Chapter 7*

Use a calculator to evaluate each expression. Give your answer in standard notation.

1. $2{,}372{,}000 \times 59{,}000{,}000$ 1.3995×10^{14}
2. $67{,}820 \div 0.00015$ 4.5213×10^{8}
3. $43{,}000{,}000 \div 2{,}745{,}000{,}000$ 1.5665×10^{-2}
4. $0.0000823 \div 210{,}000$ 3.919×10^{-10}
5. 0.000573×0.000056 3.2088×10^{-8}
6. $13{,}478{,}298 \times 9.00036$ 1.2131×10^{8}

7–5 Decimal Numerals for Real Numbers

To find a decimal numeral for a rational number, first express the number as the quotient of two integers and then perform the indicated division.

EXAMPLE 1 Express each of the following as a decimal numeral. Check your answer for part (a).

  a. $2\frac{5}{8}$ **b.** $\frac{3}{22}$

SOLUTION   **a.** $2\frac{5}{8} = \frac{21}{8} = 21 \div 8$ **b.** $\frac{3}{22} = 3 \div 22$

$$
\begin{array}{r}
2.625 \\
8\overline{)21.000} \\
\underline{16} \\
5\,0 \\
\underline{4\,8} \\
20 \\
\underline{16} \\
40 \\
\underline{40} \\
0
\end{array}
\qquad
\begin{array}{r}
.13636 \\
22\overline{)3.00000} \\
\underline{2\,2} \\
80 \\
\underline{66} \\
140 \\
\underline{132} \\
80 \\
\underline{66} \\
140 \\
\underline{132} \\
80
\end{array}
$$

Check:
$$2.625 = 2 + \frac{6}{10} + \frac{2}{100} + \frac{5}{1000}$$
$$= \frac{2000 + 600 + 20 + 5}{1000}$$
$$= \frac{2625}{1000} = \frac{21}{8}$$

$\therefore 2\frac{5}{8} = 2.625$ and $\frac{3}{22} = 0.13636 \ldots$.

You can see that in Example 1(a), the division process effectively terminates when the remainder 0 occurs, since thereafter only 0's appear in the quotient. Accordingly, the decimal numeral for $\frac{21}{8}$ is called a **terminating decimal**.

Radicals and Irrational Numbers   **261**

Teaching Suggestions p. T94

Related Activities p. T94

Supplementary Material
Test 20

Key Ideas
Express rational numbers as decimals.
Express decimal numbers as fractions.
Find a number between two given real numbers.

Chalkboard Examples
1. Express each as a decimal numeral.

  a. $1\frac{3}{8}$ **b.** $\frac{1}{22}$

  a. $1\frac{3}{8} = \frac{11}{8}$
$$
\begin{array}{r}
1.375 \\
8\overline{)11.000} \\
\underline{8} \\
30 \\
\underline{24} \\
60 \\
\underline{56} \\
40
\end{array}
$$

  b. $\frac{1}{22}$
$$
\begin{array}{r}
.0454545 \\
22\overline{)1.0000000} \\
\underline{88} \\
120 \\
\underline{110} \\
100 \\
\underline{88} \\
120 \\
\underline{110} \\
100 \\
\underline{88} \\
120 \\
\underline{110} \\
100 \\
\underline{88} \\
120
\end{array}
$$

$0.0454545 \ldots$ or $0.04\overline{5}$

(continued)

2. Express $0.\overline{378}$ as ratio of two integers.

Let $N = 0.\overline{378}$

$1000N = 378.\overline{378}$

$\underline{N = 0.\overline{378}}$

$999N = 378$

$N = \dfrac{378}{999} = \dfrac{42}{111} = \dfrac{14}{37}$

3. a. Find a rational number between $4.\overline{5211}$ and $4.\overline{521}$.

  b. Find an irrational number between $\sqrt{2}$ and $1.41515515551\ldots$.

    a. $4.\overline{5211} < 4.5212 < 4.\overline{521}$

    b. $\sqrt{2} < 1.41505505550\ldots < 1.41515515551\ldots$

Additional A Exercises

Express each fraction as a decimal.

1. $\dfrac{5}{16}$ 0.3125

2. $\dfrac{5}{6}$ $0.8\overline{3}$

Find a rational number and an irrational number between each pair of numbers.

3. 1.4 and 1.5   1.45, 1.414114111 . . .

4. $-\dfrac{7}{10}$ and $-\dfrac{3}{4}$ -0.725, $-0.717117111\ldots$

Express each decimal as a fraction in lowest terms.

5. 0.325 $\dfrac{13}{40}$

6. $0.\overline{57}$ $\dfrac{19}{33}$

In Example 1(b), however, the remainder 0 never occurs. As a result, the quotient consists of an endlessly *repeating block of digits*, or repetend: 36. The decimal numeral for $\frac{3}{22}$ is an example of a **repeating (or periodic) decimal,** which we usually denote as follows:

$$\tfrac{3}{22} = 0.1\overline{36}$$

The bar indicates the block of digits that repeats without end.

Example 1 illustrates the following facts concerning the result when an integer p is divided by a positive integer q:

1. The remainder at each step in the process may be any element of $\{0, 1, 2, \ldots, q - 1\}$.
2. After only 0's are left in the dividend, within at most $q - 1$ steps either 0 occurs as a remainder and the division process stops, or one of the possible nonzero remainders recurs, initiating a repeating sequence of dividends in the algorithm process, and hence a repeating block of digits in the quotient.
3. If the repetend is $\overline{0}$, the quotient is a terminating decimal; otherwise, the quotient is a repeating decimal.

The decimal representation of any rational number $\dfrac{p}{q}$ either terminates or has a repetend of fewer than q digits. Conversely, every terminating or repeating decimal represents a rational number.

You can use the following method to convert a repeating decimal to a common fraction. This method involves multiplying the repeating decimal by 10^n where n is the number of digits that repeat.

EXAMPLE 2   Express $0.\overline{126}$ as a ratio of two integers.

SOLUTION   Let $N = 0.\overline{126}$.

$$1000N = 126.\overline{126}$$
$$\underline{N = 0.\overline{126}}$$
$$1000N - N = 126$$
$$999N = 126$$
$$N = \dfrac{126}{999} = \dfrac{14}{111}$$

Since the rational numbers are the real numbers named by terminating or repeating decimals, the irrational numbers must be the real numbers represented by the nonterminating, nonrepeating decimals.

It is possible to find successive digits in the infinite decimal representing an irrational number such as $\sqrt{2}$ by various methods. In the geometric method in Figure 6 on the next page, where the interval from

262   *Chapter 7*

1 to 2 is subdivided into tenths, you can see that

$$1.4 < \sqrt{2} < 1.5.$$

You can verify this algebraically as follows:

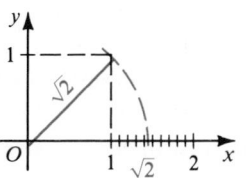
Figure 6

$$(1.4)^2 \overset{?}{<} (\sqrt{2})^2 \overset{?}{<} (1.5)^2$$
$$1.96 < 2 < 2.25$$

Next, if the same interval were divided into hundredths and you could observe the location of $\sqrt{2}$, you would find, and again verify by comparing squared members of the following inequality, that

$$1.41 < \sqrt{2} < 1.42.$$

By further subdividing the unit interval into thousandths, ten-thousandths, and so on, you could obtain as many additional digits in the decimal for $\sqrt{2}$ as you like. *In accordance with the usual rules for rounding, when you write* $\sqrt{2} \approx 1.414$, you mean

$$1.414 - 0.0005 \le \sqrt{2} < 1.414 + 0.0005.$$

Notice that, for example, the decimal numeral $0.2_121_121_2_1112\ldots$, which consists, from the decimal point on, of a succession of 2's, separated first by one 1, then by two 1's, then three 1's, and so on, is neither terminating nor repeating. The numeral, therefore, represents an irrational number. No such pattern is known for the decimal expansion of the irrational number $\sqrt{2}$.

Using decimal representations, you can illustrate the following property of **density** *of the set of real numbers.*

Property of Density of Real Numbers

A set S of real numbers is dense if between every two real numbers there is a member of S.

Both the set of rational numbers and the set of irrational numbers are dense. The set of integers, however, is not dense.

The following example uses decimal representations of numbers to illustrate the property of density of rational numbers and irrational numbers.

EXAMPLE 3   **a.** Find a rational number between $3.\overline{6751}$ and $3.\overline{675}$.
           b. Find an irrational number between $\sqrt{5}$ and $2.2373773777\ldots$.

SOLUTION   **a.** $3.\overline{6751} < 3.6752 < 3.\overline{675}$; thus a rational number between $3.\overline{6751}$ and $3.\overline{675}$ is 3.6752.
           b. Using a calculator or Table 3 shows that $\sqrt{5} < 2.237$.
             Thus $\sqrt{5} < 2.237363663666\ldots < 2.2373773777\ldots$, and
             $2.237363663666\ldots$ is one such irrational number.

Radicals and Irrational Numbers   **263**

Suggested Assignments

Minimum
    264/1–18, 20–28 even
R 265/Self-Test 2
Extended Alg. with Trig.
    264/5–27 odd
R 265/Self–Test 2
Enriched Alg.
    264/7, 11, 13–25 odd, 27–29
R 265/Self-Test 2
Enriched Alg. with Trig.
    264/7, 11, 13–25 odd, 27–29
R 265/Self-Test 2

Additional Answers
Written Exercises

17. $\frac{5}{11}$ 18. $\frac{29}{33}$ 19. $\frac{5}{37}$

20. $\frac{31}{11}$ 25. $\frac{55}{27}$ 26. $\frac{93}{74}$

27. $-\frac{2}{5}$ 28. $\frac{1}{8}$

30. By Ex. 29a, if $\frac{a}{b} < \frac{c}{d}$, then

   $ad < bc$. By the add.
   prop. of order $ad + cd <$
   $bc + cd$, or $d(a + c) <$
   $c(b + d)$. Again by Ex.

   29a, $\frac{a + c}{b + d} < \frac{c}{d}$.

31. Let $0.\overline{abc} = N$. Then
   $1000N = abc.\overline{abc}$, and
   $999N = abc$. Therefore,

   $N = \frac{abc}{999}$, which is in the

   form $\frac{p}{q}$, where q can

   have at most 3 digits.

Oral Exercises

In Exercises 1–8 state the power of ten with which you would multiply each repeating decimal to express it as a ratio of two integers.

1. $0.\overline{79}$ 10^2

2. $0.\overline{2}$ 10^1

3. $0.0\overline{7}$ 10^1

4. $0.0\overline{23}$ 10^2

5. $4.5\overline{45}$ 10^2

6. $0.\overline{496}$ 10^3

7. $0.\overline{1328}$ 10^4

8. $0.0\overline{6391}$ 10^4

In Exercises 9–22 state whether each number is rational or irrational.

9. $2\sqrt{5}$ irrational

10. $3.6\overline{2}$ rational

11. $\sqrt{7} - 0.\overline{27}$ irrational

12. $\frac{\sqrt{5}}{4.19}$ irrational

13. $\frac{7}{41}$ rational

14. $\sqrt[3]{125}$ rational

15. $0.0238238238 \ldots$ rational   16. $6.5151151115 \ldots$ irrational   17. $8.3232232223 \ldots$ irrational

18. the sum of the numbers in Exercises 16 and 17 irrational

19. the product of the numbers in Exercises 9 and 12 rational

20. the decimal consisting of the positive integers written in order:
    $0.12345678910111213 \ldots$ irrational

21. the sum of $7.8181181118 \ldots$ and $4.1818818881 \ldots$ rational

22. the sum of $6.5151151115 \ldots$ and $8.3737737773 \ldots$ rational

Written Exercises

Express each fraction as a decimal.

A   1. $\frac{7}{16}$ 0.4375

2. $\frac{13}{40}$ 0.325

3. $\frac{1}{6}$ $0.1\overline{6}$

4. $\frac{-6}{11}$ $-0.\overline{54}$

5. $\frac{56}{75}$ $0.74\overline{6}$

6. $\frac{-5}{7}$ $-0.\overline{714285}$

7. $\frac{8}{37}$ $0.\overline{216}$

8. $\frac{23}{41}$ $0.\overline{56097}$

Find a rational number and an irrational number between each pair of numbers. Answers may vary. Examples are given.

9. 1.3 and 1.4 1.35;
   1.313113111 . . .

10. $\frac{7}{10}$ and $\frac{5}{8}$ $\frac{2}{3}$;
    0.6252252225 . . .

11. $\frac{7}{9}$ and 0.8 $\frac{71}{90}$;
    0.7891011 . . .

12. $0.\overline{13}$ and $0.1\overline{3}$ 0.132;
    0.131331333 . . .

Express each decimal as a fraction in lowest terms.

13. 0.475 $\frac{19}{40}$

14. 0.0062 $\frac{31}{5000}$

15. 5.072 $\frac{634}{125}$

16. -3.0084 $\frac{7521}{2500}$

17. $0.\overline{45}$

18. $0.\overline{87}$

19. $0.1\overline{35}$

20. $2.8\overline{1}$

B   21. $2.3\overline{8}$ $\frac{43}{18}$

22. $-3.\overline{360}$ $-\frac{373}{111}$

23. $0.41\overline{6}$ $\frac{5}{12}$

24. $-4.5\overline{90}$ $-\frac{101}{22}$

25. $2.\overline{037}$

26. $1.2\overline{567}$

27. $-0.3\overline{9}$

28. $0.124\overline{9}$

C   29. **a.** Show that for any positive integers a, b, c, and d, $\frac{a}{b} < \frac{c}{d}$ if and only if $ad < bc$.

   b. Use the result of part (a) to show that if $\frac{a}{b} < \frac{c}{d}$, then $\frac{a}{b} < \frac{a + c}{b + d}$.

264   *Chapter 7*

30. Use the result of Exercise 29(a) to show that

$$\frac{a + c}{b + d} < \frac{c}{d}.$$

31. Show that any repeating decimal of the form $0.\overline{abc}$ (3-digit repetend) can be represented as a fraction $\frac{p}{q}$ in lowest terms where q is an integer of at most 3 digits.

32. Show that if a fraction in lowest terms can be represented by a terminating decimal, its denominator can have no prime factors other than 2 and 5.

33. If r and s are positive rational numbers such that $r < s$, show the following.

a. $r < r + \frac{\sqrt{2}}{2}(s - r) < s$

b. $r + \frac{\sqrt{2}}{2}(s - r)$ is an irrational number.

34. Prove that the sum of a rational number and an irrational number is always irrational.

32. If $\frac{a}{b}$ is a fraction in lowest terms and can be represented by a terminating decimal, then $\frac{a}{b} = \frac{c}{10^n} = \frac{c}{(2 \cdot 5)^n} = \frac{c}{2^n \cdot 5^n}$. Since a and b are relatively prime, b must be a factor of $2^n \cdot 5^n$. Thus the only possible factors of b are powers of 2 and 5.

(continued)

Self-Test 2

VOCABULARY round (p. 257)
rounding error (p. 257)
precision (p. 257)
maximum possible error (p. 257)
accuracy (p. 258)
relative error (p. 258)

significant digit (p. 258)
standard or scientific notation (p. 258)
terminating decimal (p. 261)
repetend (p. 262)
repeating decimal (p. 262)
property of density (p. 263)

Express in standard notation.

1. 5614 5.614×10^3 **2.** 0.00837 8.37×10^{-3} **3.** 34.92 *Obj. 1, p. 257*

3.492×10^1 *Obj. 2, p. 257*

4. Find a one-significant-digit estimate of

$$\frac{(4126)(49.72)}{(4.11)(0.0071)}. \quad 7 \times 10^6, \text{ or } 7,000,000$$

5. Express $\frac{4}{9}$ as a terminating or repeating decimal. $0.\overline{4}$ *Obj. 3, p. 257*

6. Express $0.\overline{31}$ as a fraction. $\frac{31}{99}$

7. Find a rational and an irrational number between $\frac{5}{11}$ and 0.45. *Obj. 4, p. 257*

Answers may vary. An example is 0.451; 0.454454445 . . .

Check your answers with those at the back of the book.

Quick Quiz

Express in standard notation.
1. 56234 5.6234×10^4
2. 0.04002 4.002×10^{-2}
3. 762.5 7.625×10^2
4. Find a one-significant-digit estimate of
$$\frac{(397)(0.00584)}{(8250)(0.973)} \quad 3 \times 10^{-4}$$
5. Express $\frac{2}{11}$ as a terminating or repeating decimal. $0.\overline{18}$
6. Express $0.\overline{72}$ as a fraction. $\frac{8}{11}$
7. Find a rational and an irrational number between $\frac{4}{9}$ and 0.45. 0.445; 0.44505505550 . . .

33. a. If $r < s$, then $s - r >$

0, so $r + \frac{\sqrt{2}}{2}(s - r) >$

r. Since $r < s$ and $1 -$

$\frac{\sqrt{2}}{2} > 0$, $r\left(1 - \frac{\sqrt{2}}{2}\right) <$

$s\left(1 - \frac{\sqrt{2}}{2}\right)$,

$r - \frac{\sqrt{2}}{2}r < s - \frac{\sqrt{2}}{2}s$,

$r - \frac{\sqrt{2}}{2}r + \frac{\sqrt{2}}{2}s < s$,

$r + \frac{\sqrt{2}}{2}(s - r) < s$.

$\therefore r < r + \frac{\sqrt{2}}{2}(s - r) < s$.

b. $\sqrt{2}$ is irrat. Thus by Ex. 30 p. 256, $r + \frac{\sqrt{2}}{2}(s - r)$ is irrat.

34. Assume the sum of a rat. number $\frac{p}{q}$ and an irrat. number t is a rat. number $\frac{r}{s}$, where r and s are integers, $s \neq 0$. $\frac{p}{q} + t = \frac{r}{s}$, $t = \frac{r}{s} - \frac{p}{q}$, $t = \frac{rq - ps}{qs}$, which is rational. Contradiction of hypoth. $\therefore$ the sum of a rat. number and an irrat. number is irrat.

Careers

Meteorology

Meteorology is the study of atmospheric phenomena. In addition to weather forecasting, there are many other specializations in the field, such as climatology (the study of average weather patterns), the design of meteorological instruments, and the study of the chemical composition and physical properties of the atmosphere.

Meteorologists work in a variety of situations. In many countries government and military weather stations employ a large number of meteorologists. The operations of commercial airlines, aerospace industries, and insurance companies rely upon weather information and other atmospheric data. A number of meteorologists teach and do research at universities. The knowledge and skills of meteorologists are also needed in the field of air pollution control.

EXAMPLE Thermal wind in the atmosphere results from temperature differences within a layer of air. To find the direction and speed of a wind at a particular altitude above the ground, meteorologists find the sum of the *vectors* (see Section 16-6) representing the thermal wind and the wind at the surface.

At the wind at the surface is from the south at 10 km/h and the thermal wind between the surface and an altitude of 5450 m is from the west at 15 km/h, what is the speed of the wind at an altitude of 5450 m?

SOLUTION The diagram at the right is a representation of the speed and direction of the winds. The vector **w** is the resultant vector of the vectors that represent the winds from the south and west. Use the Pythagorean Theorem to find the value of **w**.

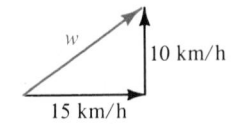

$$\mathbf{w}^2 = 10^2 + 15^2$$
$$\mathbf{w}^2 = 100 + 225$$
$$\mathbf{w}^2 = 325$$
$$\mathbf{w} = \sqrt{325}$$
$$= 5\sqrt{13}$$
$$\mathbf{w} \approx 18.0 \text{ km/h}$$

$\therefore$ the speed of the wind is approximately 18 km/h.

266 *Chapter 7*

Operating with Radicals

OBJECTIVES for Sections 7-6 through 7-8:
1. To use properties of radicals to simplify algebraic expressions.
2. To solve equations involving radicals.

Teaching Suggestions
p. T95

Key Ideas
Simplify radical expressions.

7–6 Properties of Radicals

Notice that $\sqrt[3]{8 \cdot 27} = \sqrt[3]{216} = 6$, and $\sqrt[3]{8} \cdot \sqrt[3]{27} = 2 \cdot 3 = 6$, and therefore $\sqrt[3]{8 \cdot 27} = \sqrt[3]{8} \cdot \sqrt[3]{27}$. Likewise,

$\sqrt{\dfrac{64}{16}} = \sqrt{4} = 2$ and $\dfrac{\sqrt{64}}{\sqrt{16}} = \dfrac{8}{4} = 2$, and therefore $\sqrt{\dfrac{64}{16}} = \dfrac{\sqrt{64}}{\sqrt{16}}$.

These examples illustrate the following theorem.

Theorem. For all a, b, $\sqrt[n]{a}$, and $\sqrt[n]{b} \in \mathcal{R}$:

1. $\sqrt[n]{ab} = \sqrt[n]{a} \cdot \sqrt[n]{b}$ 2. $\sqrt[n]{\dfrac{a}{b}} = \dfrac{\sqrt[n]{a}}{\sqrt[n]{b}}$ $(b \neq 0)$

PROOF OF PART 1

Note that $\sqrt[n]{ab}$ denotes the *principal* nth root of ab. By the laws of exponents $(\sqrt[n]{a} \cdot \sqrt[n]{b})^n = (\sqrt[n]{a})^n \cdot (\sqrt[n]{b})^n = a \cdot b$ and $\sqrt[n]{a} \cdot \sqrt[n]{b}$ is one of the nth roots of ab. To determine if $\sqrt[n]{ab}$ is the principal nth root consider three cases.

1. If n is even, then a and b must both be nonnegative and accordingly the principal nth root of each of these numbers is nonnegative. Hence their product, $\sqrt[n]{a} \cdot \sqrt[n]{b}$, is nonnegative and must therefore represent the *principal* nth root of $a \cdot b$, $\sqrt[n]{ab}$.
2. If n is odd and $a \cdot b \geq 0$, then $\sqrt[n]{a} \cdot \sqrt[n]{b} \geq 0$, and also $\sqrt[n]{ab} \geq 0$.
3. Exercise 35 on page 269 shows the case if n is odd and exactly one of a and b is negative.

Part 2 of the theorem is left as Exercises 36 and 37 on page 269.

If we let $b = a$ in Part 1 we have $\sqrt[n]{a^2} = \sqrt[n]{a} \cdot \sqrt[n]{a} = (\sqrt[n]{a})^2$. This illustrates a special case ($m = 2$) of the following useful theorem.

Theorem. For all b and $\sqrt[n]{b} \in \mathcal{R}$, and m and n positive integers,
$$\sqrt[n]{b^m} = (\sqrt[n]{b})^m.$$

Radicals and Irrational Numbers **267**

Chalkboard Examples

1. Simplify each expression.
   a. $\sqrt[3]{(-64)^2}$ b. $\sqrt[4]{(81)^3}$

   a. $\sqrt[3]{(-64)^2}$ b. $\sqrt[4]{(81)^3}$
   $= (\sqrt[3]{-64})^2$ $= (\sqrt[4]{81})^3$
   $= (-4)^2$ $= (3)^3$
   $= 16$ $= 27$

2. Simplify.

   a. $\sqrt[6]{64x^7}$ b. $\dfrac{2}{\sqrt{2a}}$

   c. $\sqrt[5]{\dfrac{128}{b}}$ d. $\sqrt[3]{x^{-3} + (2y)^{-3}}$

   a. $\sqrt[6]{64x^7}$
   $= \sqrt[6]{2^6 x^6 \cdot x}$
   $= 2x\sqrt[6]{x}$

   b. $\dfrac{2}{\sqrt{2a}}$
   $= \dfrac{2\sqrt{2a}}{\sqrt{2a}\,\sqrt{2a}}$
   $= \dfrac{2\sqrt{2a}}{2a}$
   $= \dfrac{\sqrt{2a}}{a}$

   c. $\sqrt[5]{\dfrac{128}{b}}$
   $= \sqrt[5]{\dfrac{128}{b} \cdot \dfrac{b^4}{b^4}}$
   $= \sqrt[5]{\dfrac{2^5 \cdot 2^2 \cdot b^4}{b^5}}$
   $= \dfrac{2\sqrt[5]{4b^4}}{b}$

   d. $\sqrt[3]{x^{-3} + (2y)^{-3}}$
   $= \sqrt[3]{\dfrac{1}{x^3} + \dfrac{1}{8y^3}}$
   $= \sqrt[3]{\dfrac{8y^3 + x^3}{8x^3y^3}}$
   $= \dfrac{\sqrt[3]{8y^3 + x^3}}{2xy}$

Additional A Exercises

Simplify.

1. $\sqrt{108}$ $6\sqrt{3}$

2. $\sqrt[8]{16} \cdot \sqrt[6]{125}$ $\sqrt[]{10}$

Express each radical in simplest form; then give an approximation correct to the nearest hundredth. Use Tables 3 and 4 as needed.

3. $5\sqrt{111}$ 52.68

4. $\sqrt{1.62}$ 1.27

Express each radical in simplest form. Assume that no denominator is 0.

5. $\dfrac{\sqrt[5]{64c^{11}}}{\sqrt[5]{2c^4}}$ $2c\sqrt[5]{c^2}$

6. $\sqrt{\dfrac{a^4}{b^3}}$ $\dfrac{a^2\sqrt{b}}{b^2}$

Common Errors

Most mistakes here stem from simplifying $\sqrt[mn]{ab}$ to $\sqrt[m]{a} \cdot \sqrt[n]{b}$, or $\sqrt[]{a^n + b^n}$ to $a + b$.

Additional Answers
Written Exercises

21. $6x^8$ **22.** $5x^6y^2\sqrt{3y}$

23. $2|y|^5\sqrt[4]{3x^3}$ **24.** $-2a^2b^3$

25. $\dfrac{|a|^3\sqrt{b}}{b^3}$ **26.** $\dfrac{2y^2\sqrt[3]{3xz}}{z}$

27. $2c\sqrt[5]{c^2}$

28. $\dfrac{(x+3)\sqrt{x^2+9}}{x^2+9}$

29. $\dfrac{\sqrt{25d^2-c^2}}{5|cd|}$

30. $\dfrac{3x^2\sqrt{x^2+y^2}}{|y|}$

31. $\dfrac{\sqrt[3]{x^2y}}{x-y}$ **32.** $\dfrac{4\sqrt{xy}}{xy}$

33. $\dfrac{a}{a+b}$ **34.** $\dfrac{\sqrt[5]{1+729c}}{3c}$

(continued on p. 289)

EXAMPLE 1 Simplify each expression.

  a. $\sqrt[3]{(-27)^2}$ **b.** $\sqrt[5]{(32)^3}$

SOLUTION   **a.** $\sqrt[3]{(-27)^2} = (\sqrt[3]{-27})^2 = (-3)^2 = 9$

   b. $\sqrt[5]{(32)^3} = (\sqrt[5]{32})^3 = 2^3 = 8$

The following theorem enables you to replace the index of a radical with a lesser index.

> *Theorem.* For k and m integers and all b and $\sqrt[km]{b} \in \mathcal{R}$,
> $$\sqrt[km]{b} = \sqrt[k]{\sqrt[m]{b}} = \sqrt[m]{\sqrt[k]{b}}.$$

For example:

$$\sqrt[8]{36} = \sqrt[4\cdot2]{36} = \sqrt[4]{\sqrt[2]{36}} = \sqrt[4]{6}$$

$$\sqrt[4]{x^2} = \sqrt[2\cdot2]{x^2} = \sqrt{\sqrt{x^2}} = \sqrt{|x|}$$

The theorems of this section, together with other number properties, enable you to *simplify* a radical of index n. A radical is in *simplest form* when:

1. the index is as small as possible,
2. there are no radicands containing a fraction or a negative exponent, or radicals appearing in a denominator, and
3. no radicand contains the nth power of an integer or polynomial other than 1.

The phrase rationalizing the denominator is used to describe the process of transforming a fraction that contains a radical in the denominator into an equivalent fraction whose denominator is free of radicals.

EXAMPLE 2 Simplify.

  a. $\sqrt[7]{128x^8}$ **b.** $\dfrac{3}{\sqrt{12b}}$ **c.** $\sqrt[4]{\dfrac{162}{b}}$ $b \neq 0$ **d.** $\sqrt{x^{-2} + (3y)^{-2}}$

SOLUTION   **a.** $\sqrt[7]{128x^8} = \sqrt[7]{128x^7 \cdot x} = 2x\sqrt[7]{x}$

   b. $\dfrac{3}{\sqrt{12b}} = \dfrac{3}{2\sqrt{3b}} \cdot \dfrac{\sqrt{3b}}{\sqrt{3b}} = \dfrac{3\sqrt{3b}}{2 \cdot 3|b|} = \dfrac{\sqrt{3b}}{2|b|}$

   c. $\sqrt[4]{\dfrac{162}{b}} = \dfrac{\sqrt[4]{81} \cdot \sqrt[4]{2}}{\sqrt[4]{b}} \cdot \dfrac{\sqrt[4]{b^3}}{\sqrt[4]{b^3}} = \dfrac{3\sqrt[4]{2b^3}}{b}$

   d. $\sqrt{x^{-2} + (3y)^{-2}} = \sqrt{\dfrac{1}{x^2} + \dfrac{1}{(3y)^2}} = \sqrt{\dfrac{9y^2 + x^2}{9x^2y^2}} = \dfrac{\sqrt{9y^2 + x^2}}{3|xy|}$

Oral Exercises

Express each radical in simplest form. Assume each radical denotes a real number.

1. $\sqrt{50}$ $5\sqrt{2}$ 2. $\sqrt{63}$ $3\sqrt{7}$ 3. $\sqrt{125}$ $5\sqrt{5}$ 4. $\sqrt[3]{40}$ $2\sqrt[3]{5}$ 5. $\sqrt[3]{-250}$ $-5\sqrt[3]{2}$

6. $\sqrt{\dfrac{5y^2}{36}}$ $\dfrac{y\sqrt{5}}{6}$ 7. $\sqrt{\dfrac{18}{x^7}}$ $\dfrac{3\sqrt{2x}}{x^4}$ 8. $\sqrt[6]{9}$ $\sqrt[3]{3}$ 9. $\sqrt[3]{64^2}$ 16   10. $\sqrt{16^{-3}}$ $\dfrac{1}{64}$

Written Exercises

Simplify.

A

1. $\sqrt{192}$ $8\sqrt{3}$ 2. $\sqrt{320}$ $8\sqrt{5}$ 3. $\sqrt{\dfrac{72}{49}}$ $\dfrac{6\sqrt{2}}{7}$ 4. $\sqrt{\dfrac{121}{3}}$ $\dfrac{11\sqrt{3}}{3}$

5. $\sqrt[3]{125^2}$ 25   6. $\sqrt[6]{(-64)^2}$ 4   7. $-\sqrt[3]{27^{-2}}$ $-\dfrac{1}{9}$ 8. $\sqrt[4]{(-81)^{-2}}$ $\dfrac{1}{9}$

9. $\sqrt[6]{\dfrac{27}{64}}$ $\dfrac{\sqrt{3}}{2}$ 10. $\sqrt[4]{\dfrac{36}{25}}$ $\dfrac{\sqrt{30}}{5}$ 11. $\sqrt[3]{\dfrac{27}{125}}$ $\dfrac{3}{5}$ 12. $\sqrt[6]{8}\cdot\sqrt[8]{81}$ $\sqrt{6}$

Express each radical in simplest form; then give an approximation correct to the nearest hundredth. Use Tables 3 and 4 as needed.

13. $7\sqrt{48}$ $28\sqrt{3}$; 48.50   14. $\dfrac{2\sqrt{150}}{5}$ $2\sqrt{6}$; 4.90   15. $\dfrac{2}{3}\sqrt{24}$ $\dfrac{4}{3}\sqrt{6}$; 3.27   16. $\sqrt{0.98}$ $\dfrac{7\sqrt{2}}{10}$; 0.99

17. $\sqrt{\dfrac{27}{125}}\cdot\sqrt{80}$ $\dfrac{12\sqrt{3}}{5}$; 4.16   18. $\sqrt[3]{\dfrac{640}{25}}$ $\dfrac{4\sqrt[3]{50}}{5}$; 2.95   19. $\sqrt{\dfrac{3}{28}}\cdot\sqrt{\dfrac{175}{3}}$ $\dfrac{5}{2}$; 2.50   20. $\sqrt[3]{-72}\cdot\sqrt[3]{375}$ -30

Express each radical in simplest form. Assume that no denominator is 0 and each radical denotes a real number.

21. $\sqrt{36x^{16}}$ 22. $\sqrt{75x^{12}y^5}$ 23. $\sqrt[4]{48x^3y^{20}}$ 24. $\sqrt[5]{-32a^{10}b^{15}}$

25. $\sqrt{\dfrac{a^6}{b^5}}$ 26. $\sqrt[3]{\dfrac{24xy^6}{z^2}}$ 27. $\dfrac{\sqrt[5]{96c^9}}{\sqrt[5]{3c^2}}$ 28. $\dfrac{x+3}{\sqrt{x^2+9}}$

B

29. $\sqrt{c^{-2}-(5d)^{-2}}$ 30. $\sqrt{9x^4+9x^6y^{-2}}$

31. $\sqrt[3]{x^3(x-y)^{-3}-x^2(x-y)^{-2}}$ 32. $\sqrt{16x^{-1}+16y^{-1}}\cdot\sqrt{(x+y)^{-1}}$

33. $\sqrt[8]{a^4(a+b)^{-4}}\cdot\sqrt[6]{a^3(a+b)^{-3}}$ 34. $\sqrt[5]{(3c)^{-5}+3c^{-4}}$

Prove each of the following statements.

C

35. $\sqrt[n]{ab}=\sqrt[n]{a}\sqrt[n]{b}$; n odd, and exactly one of the numbers a, b negative. (*Hint*: First show that $\sqrt[n]{ab}\le 0$ and $\sqrt[n]{a}\sqrt[n]{b}\le 0$.)

36. $\sqrt[n]{\dfrac{a}{b}}=\dfrac{\sqrt[n]{a}}{\sqrt[n]{b}}$; $a\ge 0$, $b>0$.   37. $\sqrt[n]{\dfrac{a}{b}}=\dfrac{\sqrt[n]{a}}{\sqrt[n]{b}}$; n odd; $\dfrac{a}{b}<0$.

38. $\sqrt[n]{b^m}=(\sqrt[n]{b})^m$; $\sqrt[n]{b}$ is a real number, m a positive integer.

Radicals and Irrational Numbers **269**

Mixed Review

1. Simplify $\dfrac{1}{6}\left(-24+\dfrac{27}{2}\right)\cdot\left(-\dfrac{1}{2}\right)$. $\dfrac{7}{8}$

2. True or false?
   $\dfrac{a+b}{a^2+b^2}=\dfrac{1}{a+b}$ False

3. Solve for x.
   $3x+2(5-x)+x=18$
   $\{4\}$

4. Express $0.\overline{7}$ as a ratio of two integers. $\dfrac{7}{9}$

5. If $f(x)=x^2+1$, find $f(\sqrt{2})$. 3

6. Solve the following for a.
   $a:2=7:5$ $\left\{\dfrac{14}{5}\right\}$

Chalkboard Examples

1. Simplify $(\sqrt{3} - \sqrt{5})^2$.
$(\sqrt{3} - \sqrt{5})(\sqrt{3} - \sqrt{5})$
$= (\sqrt{3})^2 - 2\sqrt{3}\sqrt{5} + (\sqrt{5})^2$
$= 3 - 2\sqrt{15} + 5$
$= 8 - 2\sqrt{15}$
$= 2(4 - \sqrt{15})$

2. Rationalize the denominator of $\dfrac{x}{\sqrt{x} - 2}$.

$\dfrac{x}{(\sqrt{x} - 2)} \cdot \dfrac{(\sqrt{x} + 2)}{(\sqrt{x} + 2)}$
$= \dfrac{x(\sqrt{x} + 2)}{x - 4}$

3. Factor $x^2 - 18$ completely over the set of polynomials with real coefficients.
$x^2 - 18 = x^2 - (\sqrt{18})^2$
$= (x + \sqrt{18})(x - \sqrt{18})$
$= (x + \sqrt{9 \cdot 2})(x - \sqrt{9 \cdot 2})$
$= (x + 3\sqrt{2})(x - 3\sqrt{2})$

Common Errors

Students often try to simplify a denominator with two terms (one of which contains a radical) by multiplying by the denominator itself rather than by its conjugate (and then erroneously omitting the middle term that would thereby be produced). Point out this error to the students and it will help them avoid it.

7-7 Operations with Radicals

A sum of radicals can be simplified in accordance with the following rules.

1. Simplify each radical in the sum.
2. Then combine radical terms containing the *same index and radicand*, using the distributive law.

EXAMPLE 1 Simplify $a\sqrt{18a} + \sqrt{50a^3} - a\sqrt{192}$. Assume each radical denotes a real number.

SOLUTION $a\sqrt{18a} + \sqrt{50a^3} - a\sqrt{192} = a\sqrt{3^2 \cdot 2a} + \sqrt{25a^2 \cdot 2a} - a\sqrt{64 \cdot 3}$
$= 3a\sqrt{2a} + 5a\sqrt{2a} - 8a\sqrt{3}$
$= 8a\sqrt{2a} - 8a\sqrt{3} = 8a(\sqrt{2a} - \sqrt{3})$

EXAMPLE 2 Simplify $(\sqrt{3a} - 4)(4\sqrt{3a} + 3a)$. Assume each radical denotes a real number.

SOLUTION $(\sqrt{3a} - 4)(4\sqrt{3a} + 3a) = 4 \cdot 3a + 3a\sqrt{3a} - 16\sqrt{3a} - 12a$
$= 3a\sqrt{3a} - 16\sqrt{3a} = \sqrt{3a}(3a - 16)$

EXAMPLE 3 Rationalize the denominator of $\dfrac{3\sqrt{x}}{\sqrt{x} - 3}$

SOLUTION Use the fact that $(a - b)(a + b) = a^2 - b^2$ so $(\sqrt{x} - 3)(\sqrt{x} + 3) = x - 9$.

$$\frac{3\sqrt{x}}{\sqrt{x} - 3} = \frac{3\sqrt{x}}{(\sqrt{x} - 3)} \cdot \frac{(\sqrt{x} + 3)}{(\sqrt{x} + 3)} = \frac{3x + 9\sqrt{x}}{x - 9}$$

Using radicals, you can factor certain quadratic polynomials that are irreducible over the set of polynomials with *integral* coefficients.

EXAMPLE 4 Factor $x^2 - 50$ completely over the set of polynomials with *real* coefficients.

SOLUTION $x^2 - 50 = x^2 - (\sqrt{50})^2 = (x - \sqrt{50})(x + \sqrt{50})$
$\therefore x^2 - 50 = (x - 5\sqrt{2})(x + 5\sqrt{2})$

Oral Exercises

State the factor by which you would multiply the numerator and denominator of each rational expression in order to rationalize the denominator

1. $\dfrac{1}{9 - \sqrt{2}}$ $\quad 9 + \sqrt{2}$

2. $\dfrac{a^3}{a + \sqrt{14}}$ $\quad a - \sqrt{14}$

3. $\dfrac{12}{5x + 2\sqrt{3}}$ $\quad 5x - 2\sqrt{3}$

4. $\dfrac{\sqrt{q}}{7p - 6\sqrt{q}}$ $\dfrac{7p + 6\sqrt{q}}$

270 *Chapter 7*

Written Exercises

Simplify.

A 1. $3\sqrt{5} + \sqrt{125}$ $8\sqrt{5}$

3. $\frac{1}{3}\sqrt{27} - 2\sqrt{75} + 5\sqrt{12}$ $\sqrt{3}$

5. $5\sqrt[3]{24} - 2\sqrt[3]{54} + \sqrt[3]{3000}$ $20\sqrt[3]{3} - 6\sqrt[3]{2}$

7. $4\sqrt{9b^3} + b\sqrt{49b} - \sqrt{b^3}$ $18b\sqrt{b}$

9. $3x\sqrt{80x^3} - \frac{1}{3}\sqrt{180x^5} + \sqrt{320x}$ $\sqrt{5x}(10x^2 + 8)$

11. $2\sqrt{15}(4\sqrt{3} - 3\sqrt{12})$ $-12\sqrt{5}$

13. $\sqrt{\frac{5}{3}}\left(4\sqrt{3} - \frac{11}{\sqrt{3}}\right)$ $\frac{\sqrt{5}}{3}$

15. $(\sqrt{2} - \sqrt{7})^2$ $9 - 2\sqrt{14}$

17. $(3\sqrt{5} - 4)(3\sqrt{5} + 4)$ 29

19. $(\sqrt[3]{9} + 2\sqrt[3]{3})(\sqrt[3]{3} - 3)$ $3 - \sqrt[3]{9} - 6\sqrt[3]{3}$

21. $\dfrac{4}{\sqrt{10} - 3}$ $4\sqrt{10} + 12$

23. $\dfrac{\sqrt{3} - 5}{\sqrt{3} + 2}$ $7\sqrt{3} - 13$

8. $-42|y|^3\sqrt{2} + 8y^3\sqrt{2}$

2. $4\sqrt{98} - 3\sqrt{72}$ $10\sqrt{2}$

4. $3\sqrt{\frac{9}{5}} - \frac{2}{5}\sqrt{80} + \sqrt{\frac{81}{5}}$ $2\sqrt{5}$

6. $2\sqrt[3]{0.005} + \sqrt[3]{0.135}$ $0.5\sqrt[3]{5}$

8. $\sqrt{18y^6} + y\sqrt{128y^4} - 5y^2\sqrt{162y^2}$

10. $3\sqrt[3]{16r^2} + \sqrt[3]{250r^8} - 5\sqrt[3]{2r^5}$ $\sqrt[3]{2r^2}(6 + 5r^2 - 5r)$

12. $\sqrt{10}\left(\dfrac{5}{\sqrt{2}} - \dfrac{\sqrt{2}}{4}\right)$ $\frac{9\sqrt{5}}{2}$

14. $(\sqrt{5} - \sqrt{13})(\sqrt{5} + \sqrt{13})$ -8

16. $(2\sqrt{3} - \sqrt{6})^2$ $18 - 12\sqrt{2}$

18. $(5\sqrt{2} - 3\sqrt{7})(5\sqrt{2} + 3\sqrt{7})$ -13

20. $(\sqrt[3]{3} - 3\sqrt[3]{2})(\sqrt[3]{4} + 2\sqrt[3]{9})$ $\sqrt[3]{12} - 6\sqrt[3]{18}$

22. $\dfrac{6}{\sqrt{3} + 2}$

24. $\dfrac{4 + \sqrt{5}}{6 - 2\sqrt{5}}$ $\frac{17 + 7\sqrt{5}}{8}$ **22.** $-6\sqrt{3} + 12$

B 25. $(\sqrt[3]{3} - 1)(\sqrt[3]{9} + \sqrt[3]{3} + 1)$ 2

27. $(\sqrt[3]{a} + \sqrt[3]{b})(\sqrt[3]{a^2} - \sqrt[3]{ab} + \sqrt[3]{b^2})$ $a + b$

26. $(\sqrt[3]{2} + \sqrt[3]{5})(\sqrt[3]{4} - \sqrt[3]{10} + \sqrt[3]{25})$ 7

28. $(\sqrt[3]{a} - \sqrt[3]{b})(\sqrt[3]{a^2} + \sqrt[3]{ab} + \sqrt[3]{b^2})$ $a - b$

Use the results of Exercises 27 and 28 to rationalize the denominator of each expression.

29. $\dfrac{\sqrt[3]{4}}{\sqrt[3]{2} - 1}$ $2 + 2\sqrt[3]{2} + \sqrt[3]{4}$

30. $\dfrac{\sqrt[3]{5}}{\sqrt[3]{5} - \sqrt[3]{4}}$ $5 + \sqrt[3]{100} + 2\sqrt[3]{10}$

31. $\dfrac{\sqrt[3]{4}}{\sqrt[3]{3} - \sqrt[3]{2}}$ $\sqrt[3]{36} + 2\sqrt[3]{3} + 2\sqrt[3]{2}$

32. $\dfrac{\sqrt[3]{3}}{\sqrt[3]{6} + \sqrt[3]{3}}$ $\frac{\sqrt[3]{4} - \sqrt[3]{2} + 1}{3}$

Factor completely over the set of polynomials with real coefficients.

33. $3x^2 - y^2$ $(x\sqrt{3} - y)(x\sqrt{3} + y)$

35. $x^2 + 2x\sqrt{5} + 5$ $(x + \sqrt{5})^2$

37. $a^3 - 10$ $(a - \sqrt[3]{10})(a^2 + a\sqrt[3]{10} + \sqrt[3]{100})$

39. $x^2 + 10x\sqrt{2} + 50$ $(x + 5\sqrt{2})^2$

34. $25a^2 - 7$ $(5a - \sqrt{7})(5a + \sqrt{7})$

36. $3x^2 - 2xy\sqrt{6} + 2y^2$ $(x\sqrt{3} - y\sqrt{2})^2$

38. $c^3 + 6$ $(c + \sqrt[3]{6})(c^2 - c\sqrt[3]{6} + \sqrt[3]{36})$

40. $a^2 + 2a\sqrt{6} + 6$ $(a + \sqrt{6})^2$

C 41. Prove that the set of all numbers of the form $a + b\sqrt{2}$, where a and b are rational, is closed under multiplication.

42. Prove that the set of all numbers of the form $a + b\sqrt{2}$, where a and b are rational, is closed under division (except by 0).

43. Factor $x^4 + 1$ over the set of polynomials with real coefficients. [*Hint:* $x^4 + 1 = (x^4 + 2x^2 + 1) - 2x^2$.] $(x^2 - x\sqrt{2} + 1)(x^2 + x\sqrt{2} + 1)$

Radicals and Irrational Numbers   **271**

Suggested Assignments

Minimum
 First day
  271/1–28
 Second day
  272/P: 1–9
S 206/2–18 even
Extended Alg. with Trig.
 271/5–37
 272/P: 1–9
S 206/2–18 even
Enriched Alg.
 271/5–41 odd
 272/P: 1–9
S 206/6–18 even
Enriched Alg. with Trig.
 271/17–41 odd
 272/P: 1–10
S 206/6–18 even

Additional Answers
Written Exercises

41. 1. $a, b, c, d \in \mathcal{R}$
      Hypothesis
   2. $(a + b\sqrt{2})(c + d\sqrt{2}) = ac + (bc + ad)\sqrt{2} + 2bd$
      Distributive axiom
   3. $(a + b\sqrt{2})(c + d\sqrt{2}) = ac + 2bd + (bc + ad)\sqrt{2}$
      Commutative axiom
   4. $ac + 2bd \in \mathcal{R}$, $bc + ad \in \mathcal{R}$
      Closure
   5. The set of numbers of the form $a + b\sqrt{2}$, a, $b \in \mathcal{R}$ is closed under multiplication.
      Definition of closure

42. 1. $a, b, c, d \in \mathcal{R}$, $c \pm d\sqrt{2} \neq 0$
      Hypothesis
   2. $\dfrac{a + b\sqrt{2}}{c + d\sqrt{2}}$
      $= \dfrac{(a + b\sqrt{2})(c - d\sqrt{2})}{(c + d\sqrt{2})(c - d\sqrt{2})}$
      Iden. for mult.

(continued)

Additional Answers
Written Exercises
(continued)

3. $\dfrac{a + b\sqrt{2}}{c + d\sqrt{2}}$

$= \dfrac{ac + (bc - ad)\sqrt{2} - 2bd}{c^2 - 2d^2}$

Distributive axiom

4. $\dfrac{a + b\sqrt{2}}{c + d\sqrt{2}}$

$= \dfrac{ac - 2bd}{c^2 - 2d^2} +$
$\left(\dfrac{bc - ad}{c^2 - 2d^2}\right)\sqrt{2}$

Theorem, p. 228

5. $\dfrac{ac - 2bd}{c^2 - 2d^2} \in \mathscr{R},$

$\dfrac{bc - ad}{c^2 - 2d^2} \in \mathscr{R}$

Closure

6. The set of numbers of the form $a + b\sqrt{2}$, $a, b \in \mathscr{R}$ is closed under division except by 0.
Definition of closure

Additional A Exercises

Simplify.

1. $5\sqrt{3} - \sqrt{27}$ $2\sqrt{3}$

2. $\dfrac{1}{2}\sqrt{20} - 2\sqrt{45} + 3\sqrt{80}$

$7\sqrt{5}$

3. $2\sqrt{\dfrac{25}{3}} - \dfrac{2}{3}\sqrt{48} - \sqrt{\dfrac{49}{3}}$

$-\dfrac{5}{3}\sqrt{3}$

4. $3\sqrt{10}(2\sqrt{5} - 3\sqrt{20})$
$-60\sqrt{2}$

5. $(\sqrt{7} - \sqrt{11})(\sqrt{7} + \sqrt{11})$
-4

6. $\dfrac{\sqrt{6}}{5 + \sqrt{3}}$ $\dfrac{5\sqrt{6} - 3\sqrt{2}}{22}$

Problems

Give each answer to the same number of significant digits as the least accurate measurement.

A 1. The power P in watts of a circuit with a total resistance R is related to the current I by the equation $P = I^2R$. The power is expressed in watts (W), the resistance in ohms (Ω), and the current in amperes (A). If the total resistance is 10Ω, what current will be required to produce 1960 W of power? 14 A

2. An object propelled *horizontally* at a velocity v_0 from a height h will hit the ground $v_0\sqrt{\dfrac{h}{4.9}}$ meters from its starting point if air resistance is neglected. How far will a baseball that is thrown horizontally at a speed of 35 m/s from a height of 10 m travel before hitting the ground? 50 m

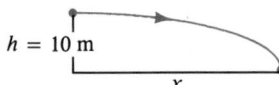

$h = 10\ \text{m}$

x

3. The frequency of f of a string on a musical instrument is given by the equation $f = \dfrac{1}{2L}\sqrt{\dfrac{10^5 F}{m}}$, where L is the length of the string, F is the tension on the string, and m is the mass of the string. F is expressed in newtons (N), and f is expressed in hertz (Hz), or vibrations per second. What is the frequency of a violin string if it has a length of 50.0 cm, a mass of 1.25 g, and is under a tension of 484 N? 62.2 Hz

4. The resonant frequency f of a circuit with inductance L and capacitance C is given by the equation $f = \dfrac{1}{2\pi\sqrt{LC}}$. The resonant frequency is expressed in hertz (Hz), the capacitance in farads (F), and the inductance in henrys (H). Find the resonant frequency of a circuit containing an inductance of 1.2×10^{-2} H and a capacitance of 50×10^{-6} F. 210 Hz

5. What is the radius of a sphere whose volume is 92 cm³? (Use $\pi \approx \frac{22}{7}$.) 2.8 cm

B 6. The speed of sound in air varies directly as the square root of the Kelvin temperature. If the speed of sound is about 320 m/s when the temperature is 256° K, what is its speed at 300° K? 346 m/s

7. When helium gas is compressed using a certain physical process, the product of the Kelvin temperature and the cube root of the volume squared remains the same. Using this process the Kelvin temperature of helium is 300° when the volume is compressed to $\frac{8}{125}$ of its original volume. What was the Kelvin temperature of the helium when it was not compressed? 48°

272   *Chapter 7*

8. Find the side of a cube whose volume is the same as the volume of a sphere of radius 25. (Use $\pi \approx \frac{22}{7}$.) 40

9. According to Kepler's third law of planetary motion, the cubes of the average distances of the planets from the sun are proportional to the squares of their times of one revolution around the sun. If Mars is one-sixth as far from the sun as Saturn, find the ratio of their times of revolution. 0.07

C 10. According to the theory of relativity, at high speeds the mass m of an object is given by the equation

$$m = \frac{m_0}{\sqrt{1 - \dfrac{v^2}{c^2}}},$$

where m_0 is the rest mass of the object, v is the velocity of the object, and c is the speed of light. By what factor is the mass of an object increased at $\frac{2}{3}$ the speed of light? $\dfrac{3\sqrt{5}}{5}$

7–8 Equations Involving Radicals

To solve an equation in which one term contains a variable in a radicand, first isolate that term on one side of the equality sign. Then raise both sides to the power of the radical index, and solve the resulting equation.

EXAMPLE 1   Solve $y + \sqrt{y - 3} - 5 = 0$.

SOLUTION
$$y + \sqrt{y - 3} - 5 = 0$$
$$y - 5 = -\sqrt{y - 3} \qquad (1)$$
$$(y - 5)^2 = (-\sqrt{y - 3})^2 \qquad (2)$$
$$y^2 - 10y + 25 = y - 3$$
$$y^2 - 11y + 28 = 0$$
$$(y - 7)(y - 4) = 0$$
$$y = 7,\ y = 4$$

Check: For $y = 7: 7 + \sqrt{7 - 3} - 5 \overset{?}{=} 0$
$$7 + 2 - 5 \neq 0$$

For $y = 4: 4 + \sqrt{4 - 3} - 5 \overset{?}{=} 0$
$$4 + 1 - 5 = 0 \ \checkmark$$

∴ the solution set is {4}.

Can you explain why an extraneous value appeared, that is, why the squared Equation (2) is not equivalent to the given equation? Think of the equation $x = 4$ and the equation obtained by squaring both sides, $x^2 = 16$; the latter has *two* real solutions, $x = 4$ and $x = -4$, while the former has only *one*, $x = 4$.

Radicals and Irrational Numbers   **273**

Mixed Review

1. Factor $(9a^2 - 16b^2)$.
   $(3a + 4b)(3a - 4b)$
2. Solve $(x - 3)^2 = 4$. {5, 1}
3. Factor $(1 - 10a^3 + 25a^6)$.
   $(1 - 5a^3)^2$
4. If f is a linear function and $f(8) = -2$ and $f(-2) = -7$, find $f(x)$. $f(x) = \frac{1}{2}x - 6$
5. If y varies directly as x, and y is 10 when x is 6, find y when x is 8. $\frac{40}{3}$

Teaching Suggestions
p. T95

Key Ideas

Solve equations involving radical expressions.

Chalkboard Examples

1. Solve $y + \sqrt{y - 2} - 4 = 0$ over $\mathcal{R}$.
   $$y + \sqrt{y - 2} - 4 = 0$$
   $$y - 4 = -\sqrt{y - 2}$$
   $$(y - 4)^2 = (-\sqrt{y - 2})^2$$
   $$y^2 - 8y + 16 = y - 2$$
   $$y^2 - 9y + 18 = 0$$
   $$(y - 6)(y - 3) = 0$$
   $$y = 6,\ y = 3$$
   By checking in the original equation, 6 is not a solution, 3 is a solution. {3}
2. Solve $\sqrt{x - 3} = \sqrt{2} - \sqrt{x}$ over $\mathcal{R}$.
   $$(\sqrt{x - 3})^2 = (\sqrt{2} - \sqrt{x})^2$$
   $$x - 3 = 2 - 2\sqrt{2x} + x$$
   $$-5 = -2\sqrt{2x}$$
   $$(-5)^2 = (-2\sqrt{2x})^2$$
   $$25 = 8x$$
   $$\frac{25}{8} = x$$
   By checking in the original equation, $\frac{25}{8}$ is not a solution. The solution set is $\emptyset$.

(continued on p. 290)

Theorem. For n a positive integer and a and $b \in \mathcal{R}$:

1. If $a = b$, then $a^n = b^n$.
2. If $a^n = b^n$ and n is odd, then $a = b$.
3. If $a^n = b^n$ and n is even, then $a = \pm b$.

If more than one term in an equation contains a variable in a radicand, you may have to repeat the process of isolating a radical term.

EXAMPLE 2 Solve $\sqrt{x-6} = \sqrt{3} - \sqrt{x}$ over $\mathcal{R}$.

SOLUTION
$$\sqrt{x-6} = \sqrt{3} - \sqrt{x}$$
$$(\sqrt{x-6})^2 = (\sqrt{3} - \sqrt{x})^2$$
$$x - 6 = 3 - 2\sqrt{3x} + x$$
$$-9 = -2\sqrt{3x}$$
$$(-9)^2 = (-2\sqrt{3x})^2$$
$$81 = 12x$$
$$x = \frac{81}{12} = \frac{27}{4}$$

Check: $\sqrt{\frac{27}{4} - 6} \stackrel{?}{=} \sqrt{3} - \sqrt{\frac{27}{4}}$

$$\sqrt{\frac{3}{4}} \stackrel{?}{=} \sqrt{3} - \frac{3\sqrt{3}}{2}$$
$$\frac{\sqrt{3}}{2} \stackrel{?}{=} \frac{2\sqrt{3} - 3\sqrt{3}}{2}$$
$$\frac{\sqrt{3}}{2} \neq -\frac{\sqrt{3}}{2}$$

$\therefore$ the solution set is $\emptyset$.

Example 2 illustrates the fact that the solution set of $P(x) = Q(x)$ may be empty even though $[P(x)]^n = [Q(x)]^n$ has one or more real roots.

Oral Exercises

a. State any restrictions on the variable.
b. Explain how you would solve the given equation. Answers may vary.

1. $\sqrt{x+3} = 7$

2. $8 - \sqrt{4-y} = 0$

3. $\sqrt[3]{5n} = 3$

4. $\sqrt[3]{2r-1} = 5$

5. $\sqrt{4c-1} + 2 = 9$

6. $-2 + \sqrt[4]{\frac{a}{3}} = 1$

7. $3 + \sqrt[3]{\frac{x}{2}} = 7$

8. $4 + \sqrt{3n+10} = 9$

9. $\sqrt{k^2+9} - k = 1$

10. $p - \sqrt{p^2-16} = 2$

11. $\frac{5}{2}\sqrt[3]{3m} = 15$

12. $\frac{2}{3}\sqrt[4]{x} = 20$

Written Exercises

Solve over $\mathcal{R}$. **1.** {46} **2.** {−60} **3.** {$\frac{27}{5}$} **4.** {63} **5.** {$\frac{25}{2}$} **6.** {243} **7.** {128}
8. {5} **9.** {4} **10.** {5} **11.** {72} **12.** {810,000}

A **1–12.** Solve the equations in Oral Exercises 1–12.

13. $2\sqrt{u + 5} = u + 2$ {4}   **14.** $6 - \sqrt{5t} = -4$ {20}

15. $7 + \sqrt{8 - r} = 9 - r$ {−1}   **16.** $\sqrt[3]{n^2 - 9} - 6 = 0$ {15, −15}

17. $3 - 2\sqrt[3]{3n + 5} = 0$ {$-\frac{13}{24}$}   **18.** $-2x = \sqrt{6x + 4}$ {$-\frac{1}{2}$}

19. $\sqrt{k + 9} = \sqrt{k} + 1$ {16}   **20.** $\sqrt{2z} - \sqrt{2z - 7} = 1$ {8}

21. $\sqrt{r - 10} = 5 + \sqrt{r}$ ∅   **22.** $\sqrt{c - 6} + 3 = \sqrt{c}$ ∅

23. $\sqrt{x - 1} = \sqrt{x + 4} - 1$ {5}   **24.** $\sqrt{y + 6} - \sqrt{4 - y} = 2$ {3}

25. $\sqrt{2x + 1} = \sqrt{x - 3} + 2$ {4, 12}   **26.** $\sqrt{2x + 3} - \sqrt{x + 1} = 1$ {3, −1}

27. $\sqrt{2x + 8} - \sqrt{x + 2} = 2$ {14, −2}   **28.** $\sqrt{x + 3} + \sqrt{3x + 10} = 3$ {−2}

B **29.** $x\sqrt{x} = \frac{1}{8}$ {$\frac{1}{4}$}   **30.** $x\sqrt{x - 2} = x$ {3}

31. $\dfrac{2x - 3}{\sqrt{x + 2}} = \sqrt{x}$ {9}   **32.** $\dfrac{4x + 2}{\sqrt{x + 1}} = 3\sqrt{x}$ {4, 1}

33. $\dfrac{5x + 5}{\sqrt{2x + 5}} = \sqrt{2x} + 1$ {0, 8}   **34.** $\dfrac{3 - 2x}{\sqrt{2x - 3}} = \sqrt{2x} - 2$ {$\frac{1}{2}, \frac{9}{8}$}

C **35.** Prove: If r and s denote real numbers and $r^3 = s^3$, then:
    a. r and s are both 0, both positive, or both negative.   **b.** $r = s$
    [*Hint*: Use the result of part (a) and $r^3 - s^3 = (r - s)(r^2 + rs + s^2)$.]

36. Prove: If r and s denote real numbers and $r^4 = s^4$, then:
    a. r and s are both 0 or both nonzero.   **b.** $r = s$ or $r = -s$
    [*Hint*: Use the fact that $r^4 - s^4 = (r^2 - s^2)(r^2 + s^2)$.]

Self-Test 3

VOCABULARY rationalizing the denominator (p. 268)

Express in simplest radical form.
                                                              $\sqrt[3]{4xy}(2y - 3x)$

1. $\sqrt{28} - 2\sqrt{63} - 3\sqrt{175}$ $-19\sqrt{7}$ **2.** $\sqrt[3]{32xy^4} - \sqrt[3]{108x^4y}$ *Obj. 1, p. 267*

3. $(2\sqrt{5} - 3)(\sqrt{5} + 4)$ $5\sqrt{5} - 2$ **4.** $\dfrac{3}{6 - \sqrt{7}}$ $\frac{18 + 3\sqrt{7}}{29}$

Solve over $\mathcal{R}$.

5. $\sqrt{\dfrac{2x}{5}} = 4$ {40}   **6.** $\sqrt{6 - x} + 1 = \sqrt{3 - x}$ ∅   *Obj. 2, p. 267*

Check your answers with those at the back of the book.

Radicals and Irrational Numbers **275**

The Set of Complex Numbers

OBJECTIVES for Sections 7-9 through 7-11:
1. To simplify a square-root radical whose radicand is a negative number.
2. To find the sum and difference of two complex numbers.
3. To find the product and quotient (divisor not zero) of two complex numbers.

Teaching Suggestions
p. T95

Key Ideas

Introduce the imaginary
number "*i*". Define *i* as $\sqrt{-1}$.

7–9 The Number *i*

Over the set of positive real numbers, the linear equation $x + 1 = 0$ has no solution. When you extend the replacement set of x to contain negative as well as positive numbers, however, the given equation has a solution, namely, -1.

Over the set $\mathscr{R}$ of all real numbers, the quadratic equation

$$x^2 + 1 = 0$$

has no solution. We can extend the replacement set of x to contain new numbers which will satisfy this equation. About four hundred years ago, mathematicians proposed the introduction of a number, i, with the property that $i^2 + 1 = 0$, or

$$i^2 = -1.$$

Thus, i is a solution of the equation $x^2 + 1 = 0$.

The fact that $i^2 = -1$ suggests that you write

$$\sqrt{-1} = i$$

and call i "a square root of -1."

By requiring that multiplication continue to have the commutative and associative properties, you can discover how to multiply real numbers and i. Study the following examples:

EXAMPLE 1   **a.** $5 \cdot i = 5i$ **b.** $3(4i) = (3 \cdot 4)i = 12i$
c. $4i(-7) = 4(-7)i = -28i$
d. $(6i)(3i) = (6 \cdot 3)(i \cdot i) = 18i^2 = 18(-1) = -18$
e. $(-i)i = -i^2 = -(-1) = 1$
f. $(8i)^2 = (8i)(8i) = 8^2 \cdot i^2 = 64(-1) = -64$

Since for any $r > 0$, $(\sqrt{r}\, i)^2 = (\sqrt{r})^2 i^2 = r(-1) = -r$, it is natural to make this definition:

> For every positive real number r, $\sqrt{-r} = i\sqrt{r}$.

276   *Chapter 7*

EXAMPLE 2 $\sqrt{-20} = \sqrt{20}\,i$
$= 2\sqrt{5}\,i$, or $2i\sqrt{5}$.

The last form of the answer in Example 2 is often used to avoid the error of writing $\sqrt{5i}$ for $\sqrt{5}\,i$.

The preceding suggests the following fact: *For every nonzero real number b, bi is a number whose square is $-b^2$; that is, $(bi)^2 = -b^2$. For $b \neq 0$, we call bi a **pure imaginary number.** The number i is called the **imaginary unit.** We define $0 \cdot i$ to be 0.

Notice that the product of a nonzero real number and a pure imaginary number is a pure imaginary number, but the product of two pure imaginary numbers is a real number. When you simplify successive powers of i, you find the values repeating in cycles of four, according to the pattern $i, -1, -i, 1$.

$i^1 = i$ $\qquad\qquad$ $i^5 = i^4 \cdot i = 1 \cdot i = i$
$i^2 = -1$ $\qquad\qquad$ $i^6 = i^4 \cdot i^2 = 1 \cdot -1 = -1$
$i^3 = i^2 \cdot i = -1 \cdot i = -i$ $\qquad$ $i^7 = i^4 \cdot i^3 = 1 \cdot -i = -i$
$i^4 = i^2 \cdot i^2 = -1 \cdot -1 = 1$ $\qquad$ $i^8 = i^4 \cdot i^4 = 1 \cdot 1 = 1$

Notice also that $i(-i) = (-i)i = 1$. Therefore, i and $-i$ are reciprocals; that is,

$$\frac{1}{i} = -i \quad \text{and} \quad \frac{1}{-i} = i.$$

You can use this fact in computing a quotient in which the divisor is a pure imaginary number.

EXAMPLE 3   **a.** $\dfrac{15}{6i} = \dfrac{15}{6} \cdot \dfrac{1}{i} = \dfrac{5}{2} \cdot -i = -\dfrac{5}{2}i$

b. $\dfrac{10}{i^3} = \dfrac{10}{-i} = 10\,\dfrac{1}{-i} = 10 \cdot i = 10i$

To simplify a square-root radical whose radicand is a negative number, take these steps:

1. Express the radical as the product of a real number and i.
2. Then use the properties of the *real roots of real numbers* to simplify this product.

EXAMPLE 4 $\sqrt{-49} + \sqrt{-12} = i\sqrt{49} + i\sqrt{12}$
$= 7i + 2\sqrt{3}\,i$
$= (7 + 2\sqrt{3})i$

Notice the use of the distributive property in Example 4.

Suggested Assignments

Minimum
  278/1–33
Extended Alg. with Trig.
  278/1–33 odd
Enriched Alg.
  278/1–33 odd
Enriched Alg. with Trig.
  278/1–35 odd

EXAMPLE 5 $\sqrt{-5} \cdot \sqrt{-80} = i\sqrt{5} \cdot i\sqrt{80}$
$$= \sqrt{5 \cdot 80} \cdot i^2 = \sqrt{400} \cdot (-1) = -20.$$

Notice that if you wrote $\sqrt{-5} \cdot \sqrt{-80} = \sqrt{-5 \cdot -80} = \sqrt{400} = 20$, you would be applying properties that have been proved only for radicals denoting *real numbers*, and you would in fact obtain an incorrect result. This is why it is important to follow the order of operations.

EXAMPLE 6 $\dfrac{3}{\sqrt{-75}} = \dfrac{3}{i\sqrt{75}} = -\dfrac{3i}{\sqrt{75}} = -\dfrac{3i}{5\sqrt{3}}$
$$= -\dfrac{3i \cdot \sqrt{3}}{5\sqrt{3} \cdot \sqrt{3}} = -\dfrac{3\sqrt{3}\,i}{5 \cdot 3} = -\dfrac{i\sqrt{3}}{5}$$

Oral Exercises

8. 28*i*   **16.** 15

Express each of the following as a real number or a pure imaginary number.

1. $5i^2$ -5

2. $(-5i)^2$ -25

3. $\sqrt{-9}$ $3i$

4. $\sqrt{-13}$ $i\sqrt{13}$

5. $\sqrt{-12}$ $2i\sqrt{3}$

6. $\sqrt{-48}$ $4i\sqrt{3}$

7. $2i\sqrt{-3}$ $-2\sqrt{3}$

8. $7\sqrt{-16}$

9. $-i(i)$ 1

10. $\dfrac{14i}{-2i}$ -7

11. $\dfrac{-4i}{-5i}$ $\frac{4}{5}$

12. $\dfrac{i}{3} \cdot \dfrac{i}{6}$ $-\frac{1}{18}$

13. $(-2i)(-8i)$ -16

14. $\sqrt{-9}\sqrt{-25}$ -15

15. $\sqrt{(-2)(-18)}$ 6

16. $(3i)(-5i)$

17. $\dfrac{3}{i}$ $-3i$

18. $\dfrac{6}{-i}$ $6i$

19. $\dfrac{15}{-3i}$ $5i$

20. $\dfrac{-20}{4i}$ $5i$

21. i^5 i

22. i^9 i

23. i^{13} i

24. i^{17} i

25. For what values of a and b will the expression i^{an+b} equal i for *all* positive integers n. (See Exercises 21–24.) $a = 4, b = 1$

Written Exercises

Express each of the following in simplest form as a real number or a pure imaginary number.

A **1.** i^{10} -1

2. i^7 $-i$

3. i^{12} 1

4. i^{21} i

5. $\sqrt{-50}$ $5i\sqrt{2}$

6. $\sqrt{-45}$ $3i\sqrt{5}$

7. $3\sqrt{-98}$ $21i\sqrt{2}$

8. $-2\sqrt{-75}$ $-10i\sqrt{3}$

9. $\sqrt{-\dfrac{3}{16}}$ $\frac{\sqrt3}{4}i$

10. $\sqrt{-\dfrac{4}{7}}$ $\frac{2\sqrt7}{7}i$

11. $-\sqrt{-\dfrac{5}{8}}$ $-\frac{\sqrt{10}}{4}i$

12. $\sqrt{-2} \cdot \sqrt{-18}$ -6

13. $3\sqrt{-10} \cdot \sqrt{-15}$ $-15\sqrt{6}$

14. $5\sqrt{-12} \cdot 2\sqrt{-6}$ $-60\sqrt{2}$

15. $\dfrac{\sqrt{-20}}{\sqrt{-5}}$ 2

31. $i\left(\dfrac{2}{5}\sqrt{5} - \dfrac{1}{10}\sqrt{30}\right)$

34. Let $r = -a$, $s = -b$; thus $r > 0$ and $s > 0$. $\sqrt{a} \cdot \sqrt{b} = \sqrt{-r} \cdot \sqrt{-s} = i\sqrt{r} \cdot i\sqrt{s} = i^2\sqrt{rs} = -\sqrt{rs}$. But $rs = (-a)(-b) = ab$. Substituting, $\sqrt{a} \cdot \sqrt{b} = -\sqrt{ab}$.

35. Let $r = -a$, $s = -b$; thus $r > 0$ and $s > 0$. $\dfrac{\sqrt{a}}{\sqrt{b}} = \dfrac{\sqrt{-r}}{\sqrt{-s}} = \dfrac{i\sqrt{r}}{i\sqrt{s}} = \dfrac{\sqrt{r}}{\sqrt{s}}$. But $\dfrac{r}{s} = \dfrac{-a}{-b} = \dfrac{a}{b}$. Substituting, $\dfrac{\sqrt{a}}{\sqrt{b}} = \sqrt{\dfrac{a}{b}}$.

36.

×	1	−1	i	$-i$
1	1	−1	i	$-i$
−1	−1	1	$-i$	i
i	i	$-i$	−1	1
$-i$	$-i$	i	1	−1

yes; yes

16. $\dfrac{8\sqrt{-63}}{\sqrt{-7}}$ 24

17. $\dfrac{6}{\sqrt{-10}}$ $\dfrac{-3\sqrt{10}}{5}i$

18. $\dfrac{5}{\sqrt{-6}}$ $-\dfrac{5\sqrt{6}}{6}i$

19. $\dfrac{3\sqrt{7}}{\sqrt{-28}}$ $-\dfrac{3}{2}i$

20. $\dfrac{\sqrt{5}}{-4i}$ $\dfrac{\sqrt{5}}{4}i$

21. $\dfrac{39i}{-3i}$ -13

22. $\dfrac{4}{i^3}$ $4i$

23. $\dfrac{-8}{i^6}$ 8

24. $\dfrac{9i}{i^{14}}$ $-9i$

B **25.** $3\sqrt{-48} + 5\sqrt{-3}$ $17i\sqrt{3}$ **26.** $2\sqrt{-7} - 3\sqrt{-175}$ $-13i\sqrt{7}$ **27.** $\sqrt{-192} + \sqrt{-300}$ $18i\sqrt{3}$

28. $i^7 + i^8 + i^9$ 1

29. $i^{11} + i^{12} - i^{13} + i^{14}$ $-2i$

30. $\dfrac{1}{i^5} + \dfrac{1}{i^6} + \dfrac{1}{i^7}$ -1

31. $\sqrt{-\frac{4}{5}} - \sqrt{-\frac{3}{10}}$

32. $3\sqrt{-\frac{1}{6}} + \sqrt{-\frac{2}{3}}$ $\dfrac{5\sqrt{6}}{6}i$

33. $3\sqrt{-\frac{1}{48}} - 2\sqrt{-\frac{5}{12}}$

C **34.** Prove that if a and $b < 0$, then $\sqrt{a} \cdot \sqrt{b} = -\sqrt{ab}$.

$i(\frac{1}{4}\sqrt{3} - \frac{1}{3}\sqrt{15})$

35. Prove that if a and $b < 0$, then $\dfrac{\sqrt{a}}{\sqrt{b}} = \sqrt{\dfrac{a}{b}}$.

36. Write a multiplication table for $\{1, -1, i, -i\}$. Is the set closed under multiplication? Does every element have a multiplicative inverse?

37. If a and b are positive, find two pure imaginary numbers that are roots of the equation $ax^2 + b = 0$. $\dfrac{i\sqrt{ab}}{a}$ and $-\dfrac{i\sqrt{ab}}{a}$

7–10 Complex Numbers; Addition and Subtraction

The real numbers together with the pure imaginary numbers form a set in which you can compute products (Section 7-8). But to be able to add two numbers in this set, such as 2 and $7i$, you have to invent another new number, namely, $2 + 7i$, which is neither a real number nor a pure imaginary number. In fact, to assign a sum to any real number a and any pure imaginary number bi, you have to invent a number

$$a + bi,$$

which is called a **complex number**. If $b \neq 0$, $a + bi$ is also called an **imaginary number**. You call a, the *real part*, and b, the *imaginary part*, of $a + bi$.

We shall use the letter $\mathcal{C}$ to refer to the set of all complex numbers. In $\mathcal{C}$, equality of numbers is defined as follows:

If a, b, c, and d are real numbers, then $a + bi = c + di$ if and only if $a = c$ and $b = d$.

Notice that $9 + 6i \neq 6 + 9i$, because $9 \neq 6$.

Radicals and Irrational Numbers **279**

Mixed Review

Solve over $\mathcal{R}$.

1. $\sqrt{2x + 1} + 7 = x$ $\{12\}$

2. $3x\sqrt{3x} = 8$ $\left\{\dfrac{4}{3}\right\}$

3. $\sqrt{x + 2} + \sqrt{x - 6} = \sqrt{3x - 5}$ $\{7\}$

Simplify each expression.

4. $\sqrt{54} - 2\sqrt{24} + 3\sqrt{6}$ $2\sqrt{6}$

5. $2\left(\dfrac{2}{\sqrt{2}} - \dfrac{\sqrt{2}}{3}\right)$ $\dfrac{4\sqrt{2}}{3}$

6. $(\sqrt{2} + 2\sqrt{3})^2$ $14 + 4\sqrt{6}$ or $2(7 + 2\sqrt{6})$

Teaching Suggestions p. T95

Related Activities p. T95

Key Ideas

Add and subtract complex numbers.

EXAMPLE 1   Solve $2y + yi = 3i + xi - x$ for real values of x and y.

SOLUTION   Express each side in $a + bi$ form and use the definition of equality of numbers in $\mathcal{C}$.

$$2y + yi = 3i + xi - x$$
$$2y + yi = -x + (3 + x)i$$
$$2y = -x \quad \text{and} \quad y = 3 + x$$
$$2(3 + x) = -x$$
$$6 = -3x$$
$$x = -2$$
$$y = 1$$

By identifying the real number a with the complex number $a + 0i$, you can say that every real number belongs to $\mathcal{C}$. Similarly, from the agreement $0 + bi = bi$, it follows that every pure imaginary number is also a complex number.

To define the sum $(a + bi) + (c + di)$, we are guided by the requirement that the familiar properties of sums and products in $\mathcal{R}$ continue to be true in $\mathcal{C}$. For example, if the commutative and associative properties of addition are valid in $\mathcal{C}$ and if multiplication in $\mathcal{C}$ is distributive with respect to addition, then

$$(5 + 3i) + (4 + 2i) = (5 + 4) + (3i + 2i)$$
$$= (5 + 4) + (3 + 2)i$$
$$\therefore (5 + 3i) + (4 + 2i) = 9 + 5i.$$

This example suggests the following definition.

Definition of Addition in $\mathcal{C}$

If a, b, c, and d are real numbers, then

$$(a + bi) + (c + di) = (a + c) + (b + d)i.$$

Whenever you operate with complex numbers, express the result in $a + bi$ form unless told to do otherwise. Real numbers should be written as usual and pure imaginary numbers should be written as in the form bi. To simplify notation, the symbol $a + (-b)i$ is often written $a - bi$.

The following facts are true in $\mathcal{C}$. (See Exercises 25 and 26, page 282.)

1. The *additive identity element* is $0 + 0i$, or 0.
   For example, $(6 + 7i) + (0 + 0i) = 6 + 7i$.

2. For all real numbers c and d, the *negative* or *additive inverse*, of $c + di$ is $-c - di$; that is, $-(c + di) = -c - di$.
   For example, $-(2 - 5i) = -2 + 5i$.

280   *Chapter 7*

Because the definition of subtraction (page 22) holds in $\mathcal{C}$, you can use fact 2 to obtain the rule for subtracting one complex number from another. You have:

$$(a + bi) - (c + di) = (a + bi) + (-c - di)$$
$$\therefore (a + bi) - (c + di) = (a - c) + (b - d)i.$$

Complex numbers such as $2 + 3i$ and $2 - 3i$ are called *complex conjugates*. Thus, for any real numbers a and b, the complex conjugate of $a + bi$ is $a - bi$; conversely, the complex conjugate of $a - bi$ is $a + bi$. We will use the notation $\bar{z}$ to denote the conjugate of complex number z.

EXAMPLE 2 Find **(a)** the sum and **(b)** the difference of $4 + 7i$ and its conjugate.

SOLUTION Let $z = 4 + 7i$. Then its conjugate, $\bar{z}$, is $4 - 7i$.
  a. $z + \bar{z} = (4 + 7i) + (4 - 7i) = (4 + 4) + (7 - 7)i = 8$
  b. $z - \bar{z} = (4 + 7i) - (4 - 7i) = (4 - 4) + (7 + 7)i = 14i$

This example suggests the following theorem. Its proof is left as Exercise 27 on page 282.

Theorem. For all real numbers a and b:
$$(a + bi) + (a - bi) = 2a$$
$$(a + bi) - (a - bi) = 2bi$$

Thus, the sum of a complex number $a + bi$ and its complex conjugate is a real number; if $b \neq 0$, their difference is a pure imaginary number.

Oral Exercises

In Exercises 1–8, state:
a. the additive inverse of the number;
b. the complex conjugate of the number;
c. the sum of each number and its complex conjugate; and
d. the difference of each number and its complex conjugate.

1. $4 + 5i$ 2. $7 - 2i$ 3. $-6 + i$ 4. $\sqrt{3} + 12i$
5. $-3 - 9i$ 6. $\sqrt{5} - 8i$ 7. 10 8. $4i$

State each sum or difference as a complex number.

9. $(3 + 8i) + (7 - i)$ $10 + 7i$ 10. $(-6 + 5i) + (2 - 11i)$ $-4 - 6i$
11. $(6 + 13i) - (4 + 9i)$ $2 + 4i$ 12. $(7 - 3i) - (4 + i)$ $3 - 4i$

Radicals and Irrational Numbers **281**

Suggested Assignments
Minimum
  282/1–24
Extended Alg. with Trig.
  282/5–27 odd
Enriched Alg.
  282/9–25 odd, 26–28
Enriched Alg. with Trig.
  282/9–25 odd, 26–28

Additional Answers
Oral Exercises

1. a. $-4 - 5i$ b. $4 - 5i$
   c. 8   d. $10i$
2. a. $-7 + 2i$ b. $7 + 2i$
   c. 14   d. $-4i$
3. a. $6 - i$ b. $-6 - i$
   c. -12 d. $2i$
4. a. $-\sqrt{3} - 12i$
   b. $\sqrt{3} - 12i$
   c. $2\sqrt{3}$ d. $24i$
5. a. $3 + 9i$ b. $-3 + 9i$
   c. -6 d. $-18i$
6. a. $-\sqrt{5} + 8i$ b. $\sqrt{5} + 8i$
   c. $2\sqrt{5}$ d. $-16i$
7. a. -10 b. 10
   c. 20 d. 0
8. a. $-4i$ b. $-4i$
   c. 0 d. $8i$

Mixed Review
Simplify.
1. $\sqrt[3]{24}$ $2\sqrt[3]{3}$
2. $\sqrt[3]{\sqrt{27}}$ $\sqrt{3}$
3. $\sqrt{75x^5}$ $5x^2\sqrt{3x}$
4. $\dfrac{c^2 d}{\sqrt{2c^3}}$ $\dfrac{d\sqrt{2c}}{2}$
5. $\dfrac{\sqrt{5} - 3}{2\sqrt{5} - 3}$ $\dfrac{1 - 3\sqrt{5}}{11}$
6. $\dfrac{3x^2}{\sqrt{8x^3}} - \dfrac{2x}{\sqrt{18x}}$ $\dfrac{5\sqrt{2x}}{12}$

1. $-2 + 2i$ **2.** $7 + i$

3. $-1 - 13i$ **4.** $4 + 13i$

5. $9 - 15i$ **6.** $-3 - 12i$

7. $-\frac{1}{2} + 2i$ **8.** $-\frac{3}{2} - \frac{1}{4}i$

9. $\frac{7}{24} - \frac{3}{4}i$ **10.** $-\frac{1}{3} - \frac{3}{4}i$

11. $1 + \frac{9}{4}i$ **12.** $\frac{7}{6} + i$

13. 6 **14.** $-\frac{5}{3} - i$

15. $3i$ **16.** $-\frac{3}{2}i$

17. $-\frac{11}{24} + \frac{1}{2}i$ **18.** $\frac{2}{3} - \frac{21}{4}i$

23. $z = a + bi; \bar{z} = a - bi.$
But $z = \bar{z}$ so $a + bi = a - bi.$ Thus $bi = -bi, 2bi = 0, b = 0. \therefore z = a + 0i = a,$ a real number.

24. $z = a + bi; \bar{z} = a - bi.$
But $z = -\bar{z}$ so $a + bi = -(a - bi), a + bi = -a + bi, a = -a, 2a = 0, a = 0.$ $\therefore z = 0 + bi = bi,$ a pure imaginary number.

25. $z = a + bi$ so $z + (0 + 0i) = (a + bi) + (0 + 0i) = (a + 0) + (b + 0)i = a + bi = z.$ Also, $(0 + 0i) + z = (0 + 0i) + (a + bi) = (0 + a) + (0 + b)i = a + bi = z.$

26. $z = a + bi; -z = -a - bi.$ $z + (-z) = (a + bi) + (-a - bi) = (a - a) + (b - b)i = 0 + 0i.$

28. Prove the conjugate of $(a + bi) + (c + di)$ is the same as $(a - bi) + (c - di). (a + bi) + (c + di) = (a + c) + (b + d)i;$ thus the conjugate of the sum is $(a + c) - (b + d)i.$ On the other hand, the sum of the conjugates is $(a - bi) + (c - di) = (a + c) + (-b - d)i = (a + c) - (b + d)i.$

Written Exercises

Let $a = 3 - 6i,$ $b = -5 + 8i,$ $c = 4 + 7i,$ $d = -6i,$
 $e = \frac{1}{3} + \frac{3}{2}i,$ $f = -\frac{5}{6} + \frac{1}{2}i,$ $g = \frac{2}{3} - \frac{3}{4}i,$ $h = -\frac{3}{8}.$

Express each of the following as a complex number.

A **1.** $a + b$ **2.** $a + c$ **3.** $a - c$ **4.** $c - d$ **5.** $\bar{c} - b$ **6.** $d - \bar{a}$

7. $e + f$ **8.** $f - \bar{g}$ **9.** $h + g$ **10.** $\bar{e} - g$ **11.** $e + \bar{g}$ **12.** $e - f$

13. $a + \bar{a}$ **14.** $\bar{f} + f$ **15.** $e - \bar{e}$ **16.** $g - \bar{g}$ **17.** $f - \bar{h}$ **18.** $\bar{g} - d$

Solve for x and y, where x and y are *real* numbers.

B **19.** $(x + y) + yi = 2 - 3i$ $x = 5, y = -3$ **20.** $(x - y) + (x + y)i = 5 + 9i$ $x = 7, y = 2$

21. $(2x - y) - (x + 3y)i = 7$ $x = 3, y = -1$ **22.** $(3x + 5y) - (x - 7y)i = -2 + 10i$ $x = -\frac{32}{13}, y = \frac{14}{13}$

Let $z = a + bi, a, b \neq 0$. Prove each of the following statements.

23. If $z = \bar{z}$, then z is a real number.

24. If $z = -\bar{z}$, then z is a pure imaginary number.

25. $z + (0 + 0i) = (0 + 0i) + z = z$

26. $z + (-z) = 0 + 0i$

27. Prove the theorem stated on page 281.

28. Prove that the complex conjugate of the sum of two complex numbers is the sum of their complex conjugates.

7-11 Complex Numbers; Multiplication and Division

Assuming that the commutative, associative, and distributive properties of multiplication are to hold in $\mathcal{C}$, you can compute products of complex numbers in the same way that products of binomials over $\mathcal{R}$ were done on page 205, namely using the distributive axiom and then simplifying. The following example suggests the definition below.

$$(5 + 3i)(4 + 2i) = 5(4 + 2i) + 3i(4 + 2i)$$
$$= (20 + 10i) + (12i + 6i^2)$$
$$= (20 - 6) + (10i + 12i)$$
$$\therefore (5 + 3i)(4 + 2i) = 14 + 22i.$$

Definition of Multiplication in $\mathcal{C}$

If $a, b, c,$ and d are real numbers, then

$$(a + bi)(c + di) = (ac - bd) + (ad + bc)i.$$

In particular, for $b = 0$, this definition gives you the following rule for multiplying a real and a complex number:

$$a(c + di) = ac + adi$$

Thus, $1(c + di) = 1 \cdot c + 1 \cdot di = c + di$, so that 1, or $1 + 0i$, is the *multiplicative identity element* in $\mathcal{C}$.

Notice that, by the definition of multiplication in $\mathcal{C}$,

$$(a + bi)(a - bi) = (a^2 + b^2) + (-ab + ab)i = a^2 + b^2.$$

Hence we have the following result.

Theorem. For all real numbers a and b,

$$(a + bi)(a - bi) = a^2 + b^2.$$

Thus the product of a complex number and its complex conjugate is a real number and this theorem can be used to factor over $\mathcal{C}$.

EXAMPLE 1 Factor $r^2 + 16$ over $\mathcal{C}$.

SOLUTION $r^2 + 16 = (r + 4i)(r - 4i)$

Equality, addition, and multiplication of complex numbers have been defined so that the properties of equality, addition, and multiplication for the set of real numbers (Chapter 1) are also valid when restated for the set of complex numbers. Thus the concepts and methods based on these properties of real numbers apply to complex numbers.

The following examples show how the preceding theorem can be used to express the reciprocal of a complex number and the quotient of two complex numbers (divisor not zero) in the standard form, $a + bi$. (See Exercises 22 and 23 on page 284.)

EXAMPLE 2 Express the reciprocal of $-3 + 2i$ in the form $a + bi$.

SOLUTION Multiply the numerator and denominator of $\dfrac{1}{-3 + 2i}$ by the complex conjugate of the denominator.

$$\frac{1}{-3 + 2i} = \frac{1}{-3 + 2i} \cdot \frac{-3 - 2i}{-3 - 2i} = \frac{-3 - 2i}{9 + 4} = \frac{-3 - 2i}{13} = -\frac{3}{13} - \frac{2}{13}i$$

EXAMPLE 3 Express $\dfrac{6 + i}{5 + 3i}$ in the form $a + bi$.

SOLUTION $\dfrac{6 + i}{5 + 3i} = \dfrac{6 + i}{5 + 3i} \cdot \dfrac{5 - 3i}{5 - 3i} = \dfrac{(30 + 3) + (-18 + 5)i}{25 + 9}$

$$= \frac{33 - 13i}{34} = \frac{33}{34} - \frac{13}{34}i$$

Radicals and Irrational Numbers **283**

Key Ideas

Multiply and divide complex numbers.

Chalkboard Examples

1. Factor $a^2 + 4$ over $\mathcal{C}$.
   $(a + 2i)(a - 2i)$

2. Express the reciprocal of $2 - 5i$ in the form $a + bi$.
   $$\frac{1}{2 - 5i} = \frac{1}{2 - 5i} \cdot \frac{2 + 5i}{2 + 5i}$$
   $$= \frac{2 + 5i}{4 + 25} = \frac{2}{29} + \frac{5}{29}i$$

3. Express $\dfrac{2 + 3i}{7 + 4i}$ in the form $a + bi$.
   $$\frac{2 + 3i}{7 + 4i} \cdot \frac{7 - 4i}{7 - 4i}$$
   $$= \frac{14 + 13i + 12}{49 + 16}$$
   $$= \frac{26}{65} + \frac{13i}{65} = \frac{2}{5} + \frac{1}{5}i$$

Common Errors

When working with quotients and products of complex numbers, students often make the same errors that they make when working with quotients and products of real numbers: erroneous canceling, omission of the middle term when squaring a binomial, and using the wrong binomial when trying to rationalize a denominator. Some errors may also occur when simplifying an expression that involves powers of i.

Suggested Assignments

Minimum
 284/1–21
 R 285/Self-Test 4
Extended Alg. with Trig.
 284/1–21 odd
 R 285/Self-Test 4
Enriched Alg.
 284/5–25 odd
 R 285/Self-Test 4
Enriched Alg. with Trig.
 284/5–25 odd
 R 285/Self-Test 4

Additional A Exercises

Express each of the following in the form $a + bi$. When
possible, give your answer in
the form a, bi, or 0.

1. $(4 - 7i)(3 + i)$ $19 - 17i$

2. $(3 - 4i)^2$ $-7 - 24i$

3. $(-1 + i\sqrt{5})^2$ $-4 - 2\sqrt{5}i$

4. $\dfrac{3 + i}{4 - i}$ $\dfrac{11}{17} + \dfrac{7}{17}i$

5. $\dfrac{8i}{1 + 3i}$ $\dfrac{12}{5} + \dfrac{4}{5}i$

6. $\dfrac{3 - 6i}{-2 - 5i}$ $\dfrac{24}{29} + \dfrac{27}{29}i$

**Additional Answers
Written Exercises**

16. $(x + 3i)(x - 3i)$

17. $(2y + 7i)(2y - 7i)$

18. $(v + i\sqrt{5})(v - i\sqrt{5})$

19. $(4a\sqrt{5} + bi\sqrt{15}) \cdot$
 $(4a\sqrt{5} - bi\sqrt{15})$, or
 $5(4a + bi\sqrt{3})(4a - bi\sqrt{3})$

20. $(5x\sqrt{2} + 2i\sqrt{3}) \cdot$
 $(5x\sqrt{2} - 2i\sqrt{3})$,
 or $2(5x + i\sqrt{6})(5x - i\sqrt{6})$

21. $(z\sqrt{2} + 3i\sqrt{3}) \cdot$
 $(z\sqrt{2} - 3i\sqrt{3})$

Oral Exercises

State each product in the form $a + bi$.

1. $(3 + 7i)(3 - 7i)$ **2.** $(-4 - 4i)(-4 + 4i)$ **3.** $(2 - i\sqrt{3})(2 + i\sqrt{3})$

4. $(-\sqrt{5} + 2i)(-\sqrt{5} - 2i)$ **5.** $-(3 + 5i)(3 - 5i)$ **6.** $(\sqrt{6} - i\sqrt{7})(\sqrt{6} + i\sqrt{7})$

7. $(-8 + i\sqrt{2})(-8 - i\sqrt{2})$ **8.** $(\sqrt{5} - 3i)(\sqrt{5} + 3i)$ **9.** $-(4 + i\sqrt{10})(4 - i\sqrt{10})$

1. $58 + 0i$ **2.** $32 + 0i$ **3.** $7 + 0i$
4. $9 + 0i$ **5.** $-34 + 0i$ **6.** $13 + 0i$
7. $66 + 0i$ **8.** $14 + 0i$ **9.** $-26 + 0i$

Written Exercises

Express each of the following in the form $a + bi$. When possible, you may give your answer in the form a, bi, or 0.

A **1.** $(5 - 9i)(2 + i)$ $19 - 13i$ **2.** $(-7 + 3i)(8 - 2i)$ $-50 + 38i$

3. $(5 - 2i)^2$ $21 - 20i$ **4.** $\left(\dfrac{\sqrt{2}}{2} + \dfrac{\sqrt{2}}{2}i\right)^2$ i

5. $\left(-\dfrac{1}{2} + \dfrac{\sqrt{3}}{2}i\right)\left(-\dfrac{1}{2} - \dfrac{\sqrt{3}}{2}i\right)$ 1 **6.** $(-1 + i\sqrt{7})^2$ $-6 - 2i\sqrt{7}$

7. $\left(-\dfrac{1}{2} + \dfrac{\sqrt{3}}{2}i\right)^2$ $-\dfrac{1}{2} - \dfrac{\sqrt{3}}{2}i$ **8.** $\dfrac{6 + 2i}{5 - i}$ $\dfrac{14}{13} + \dfrac{8}{13}i$

9. $\dfrac{10 - 5i}{1 + 2i}$ $-5i$ **10.** $\dfrac{7 - 5i}{4 + 3i}$ $\dfrac{13}{25} - \dfrac{41}{25}i$

11. $\dfrac{8 - 3i}{-1 - 6i}$ $\dfrac{10}{37} + \dfrac{51}{37}i$ **12.** $\dfrac{7}{5 - i\sqrt{3}}$ $\dfrac{5}{4} + \dfrac{\sqrt{3}}{4}i$

B **13.** $\dfrac{5 + 2i}{4 - i} - \dfrac{3 - i}{2 + 3i}$ $\dfrac{183}{221} + \dfrac{356}{221}i$ **14.** $\left(-\dfrac{1}{2} + \dfrac{\sqrt{3}}{2}i\right)^3$ 1 **15.** $\left(\dfrac{\sqrt{2}}{2} + i\dfrac{\sqrt{2}}{2}\right)^4$ -1

Use the theorem on page 283 to factor each of the following over $\mathcal{C}$.

16. $x^2 + 9$ **17.** $4y^2 + 49$ **18.** $v^2 + 5$

19. $80a^2 + 15b^2$ **20.** $50x^2 + 12$ **21.** $2z^2 + 27$

C **22.** Show that for all real numbers a and b, not both zero,

$$\frac{1}{a + bi} = \frac{a}{a^2 + b^2} - \frac{b}{a^2 + b^2}i.$$

23. Show that for all real numbers a, b, c, d, where c and d are not both zero, $\dfrac{a + bi}{c + di} = \dfrac{ac + bd}{c^2 + d^2} + \dfrac{bc - ad}{c^2 + d^2}i$.

24. Show that the complex conjugate of the product of two complex numbers equals the product of their complex conjugates.

25. Factor $x^4 + 1$ completely over $\mathcal{C}$. (*Hint:* Use the theorem on page 283 first, then use the result of Exercise 4.)

26. Let $z_1 = a + bi$ and $z_2 = c + di$, with b, $d \neq 0$. Prove that if $z_1 + z_2$ and $z_1 z_2$ are both real, then z_1 and z_2 are complex conjugates.

284 *Chapter 7*

Computer Exercises   For students with computer experience

1. Write a program to find a given positive integer power of i. Recall the repeating cycle of powers of i discussed in Section 7-9.

Run the program in Exercise 1 for each of the following positive integer powers of i.

2. i^{29} i **3.** i^{68} 1   **4.** i^{54} -1 **5.** i^{75} $-i$

6. Write a program that will raise a complex number in the form $a + bi$ to a positive integer power n. Use the definition of multiplication of complex numbers on page 282 and repeated multiplication. The program should input the values of a, b, and n.

Run the program in Exercise 6 for each of the following.

7. $(1 + i)^{10}$ 32i **8.** $(-1 + i)^9$ **9.** $(2 - i)^7$ **10.** $(1 + 3i)^8$
$\qquad\qquad\qquad\qquad$ $-16 + 16i$ $\quad$ $-278 + 29i$ $\quad$ $-8432 - 5376i$

11. Write a program that will express a quotient of the form

$$\frac{a + bi}{c + di}$$

in $a + bi$ form. The program should input the values of a, b, c, and d.

Run the program in Exercise 11 for each indicated quotient.

12. $\dfrac{5 - 10i}{3 + 4i}$ $-1 - 2i$ **13.** $\dfrac{7 - 11i}{1 - 3i}$ $4 + i$ **14.** $\dfrac{8 + 15i}{7 - 24i}$ **15.** $\dfrac{-9 - 5i}{6 + 4i}$
$\qquad\qquad\qquad\qquad\qquad\qquad\qquad$ $-0.4864 + 0.4752i$ $\quad$ $-1.423077 + 0.1153846i$

▌ Self-Test 4

VOCABULARY   pure imaginary number (p. 277)   complex number (p. 279)
$\qquad\qquad\qquad$ imaginary unit (p. 277)   imaginary number (p. 279)
$\qquad\qquad\qquad\qquad\qquad\qquad\qquad\qquad\qquad\qquad\qquad\qquad$ complex conjugate (p. 281)

Express as a real or pure imaginary number.

1. $2\sqrt{-25}$ 10i **2.** $3\sqrt{-\frac{64}{3}}$ 8$i\sqrt{3}$ **3.** $\sqrt{-54}$ 3$i\sqrt{6}$ **4.** $\sqrt{-17}$ *Obj. 1, p. 276*
$\qquad\qquad\qquad\qquad\qquad\qquad\qquad\qquad\qquad\qquad\qquad\qquad\qquad\qquad$ $i\sqrt{17}$

Express in the form $a + bi$.

5. $(2 + 3i) + (-7 - 2i)$ $-5 + i$ **6.** $5i - (6 + 4i)$ $-6 + i$ *Obj. 2, p. 276*

7. $(4 - 3i)(5 + 2i)$ 26 $- 7i$ **8.** $\dfrac{3 + 2i}{6 - 5i}$ $\frac{8}{61} + \frac{27}{61}i$ *Obj. 3, p. 276*

9. $(2 - 6i)^2$ $-32 - 24i$ **10.** $(4 - 2i)^{-1}$ $\frac{1}{5} + \frac{1}{10}i$

Check your answers with those at the back of the book.

Radicals and Irrational Numbers **285**

---

22. $\dfrac{1}{a + bi} = \dfrac{1}{a + bi} \cdot \dfrac{a - bi}{a - bi}$

$\qquad = \dfrac{a - bi}{a^2 + b^2}$

$\qquad = \dfrac{a}{a^2 + b^2} - \dfrac{b}{a^2 + b^2}i$

24. Prove the conjugate of $(a + bi)(c + di) = (a - bi)(c - di)$. $(a + bi) \cdot (c + di) = (ac - bd) + (ad + bc)i$; thus the conjugate of the product is $(ac - bd) - (ad + bc)i$. The product of the conjugates is $(a - bi)(c - di) = (ac - bd) + (-ad - bc)i = (ac - bd) - (ad + bc)i$.

26. If $z_1 + z_2 = (a + c) + (b + d)i$ is real, $b + d = 0$, or $b = -d$. If $z_1z_2 = (ac - bd) + (ad + bc)i$ is real, $ad + bc = 0$. Since $b = -d$, $ad - dc = 0$, or $d(a - c) = 0$. Since $d \neq 0$, $a - c = 0$, or $a = c$. ∴ $z_2 = c + di = a - bi$, the conjugate of z_1.

Quick Quiz

Express as a real or pure imaginary number.

1. $3\sqrt{-16}$ 12i

2. $4\sqrt{\dfrac{16}{-2}}$ 8$i\sqrt{2}$

3. $\sqrt{-72}$ 6$i\sqrt{2}$

4. $\sqrt{-13}$ $i\sqrt{13}$

Express in the form $a + bi$.

5. $(3 + 2i) + (-2 - 7i)$
$\qquad$ $1 - 5i$

6. $3i - (2 + 13i)$ $-2 - 10i$

7. $(2 - 3i)(5 + 8i)$ $34 + i$

8. $\dfrac{2 - i}{3 + i}$ $\frac{1}{2} - \frac{1}{2}i$

9. $(2 - 3i)^2$ $-5 - 12i$

10. $\dfrac{4}{1 - i}$ $2 + 2i$

285

Simplify.

1. $(\sqrt{5} - 2\sqrt{3})(\sqrt{5} + \sqrt{3})$
$-1 - \sqrt{15}$

2. $\dfrac{\sqrt{5} - 3}{2\sqrt{5} - 3} \quad \dfrac{1 - 3\sqrt{5}}{11}$

3. $(\sqrt{3} - 2\sqrt{5})^2 \quad 23 - 4\sqrt{15}$

4. $\dfrac{x^2 - 9}{(x - 3)^2} \quad \dfrac{x + 3}{x - 3}$

5. $\dfrac{a^2 - 2ab - 3b^2}{a^2 - 4ab + 3b^2} \quad \dfrac{a + b}{a - b}$

6. $\dfrac{3cd^4}{(3c^2d)^{-2}} \quad 27c^5d^6$

7-10. Prove that if z_1 and z_2 are nonreal numbers and $z_1 + z_2$ and z_1z_2 are both real, then z_1 and z_2 are complex conjugates. (*Hint*: Start by letting $z_1 = a + bi$, $z_2 = c + di$, with $b \neq 0$, $d \neq 0$.)
Since $z_1 + z_2 \in \mathcal{R}$, $b + d = 0$, so $d = -b$.
Since $z_1z_2 \in \mathcal{R}$, $ad + bc = 0$, and hence, $a(-b) + bc = 0$, and $bc = ab$.
$\therefore c = a$ (since $b \neq 0$), and we get $z_2 = a - bi$.

Chapter Summary

1. The graph of the *power function* defined by $p(x) = x^n$ with n an *even* positive integer, is *symmetric with respect to the vertical axis*. If n is an *odd* positive integer, the graph is *symmetric with respect to the origin*.

2. Whenever a function is specified by an equation of the form $y = ax^n$, $a \neq 0$, we say that y *varies directly as* x^n or that y *is directly proportional to the nth power of x*.

3. For every positive integer n, any solution of $x^n = b$ is an *nth root of b*. The radical $\sqrt[n]{b}$ denotes the *principal nth root of b*. If n is odd, $\sqrt[n]{b^n} = b$; if n is even, $\sqrt[n]{b^n} = |b|$.

4. If a rational root of a *polynomial equation in simple form* with integral coefficients is expressed in lowest terms $\dfrac{p}{q}$, with $q \neq 0$, then p must be an integral factor of the constant term and q must be an integral factor of the leading coefficient. Any other real root of the equation is an *irrational number*.

5. A numeral is given in *standard notation* when it is expressed in the form $a \times 10^n$ where $1 \leq |a| < 10$ and n is an integer.

6. A number can be represented by a terminating or repeating decimal if and only if it is a rational number.

7. If n is a positive integer and a, b, and $\sqrt[n]{b}$ denote real numbers, then:

$$\sqrt[n]{ab} = \sqrt[n]{a}\sqrt[n]{b} \quad \text{and} \quad \sqrt[n]{\dfrac{a}{b}} = \dfrac{\sqrt[n]{a}}{\sqrt[n]{b}}, b \neq 0$$

If m is also a positive integer, $\sqrt[n]{b^m} = (\sqrt[n]{b})^m$.

8. You can write the *sum or difference of radicals having the same index and the same radicand* as a single term by using the distributive property. You can write the *product or quotient of radicals having the same index* as a single term by applying the product or quotient property of radicals.

9. To solve equations involving radicals, isolate a radical as one side and raise each side to the power corresponding to the root index.

10. A *complex number*, $a + bi$, where a and b are real numbers and i, or $\sqrt{-1}$, is the *imaginary unit*, is a real number if $b = 0$, is an *imaginary number* if $b \neq 0$, and is a *pure imaginary number* if $a = 0$, $b \neq 0$. If a, b, c, and d are real numbers, then $a + bi = c + di$ if and only if $a = c$ and $b = d$.

11. Complex numbers may be added and also multiplied: for all real numbers a, b, c, and d, $(a + bi) + (c + di) = (a + c) + (b + d)i$; and $(a + bi)(c + di) = (ac - bd) + (ad + bc)i$. The closure, associative, commutative, and distributive properties hold for the set $\mathcal{C}$ of complex numbers. The *additive identity element* in $\mathcal{C}$ is $0 + 0i$, or 0; for all real numbers c and d, the *additive inverse* of $c + di$ is $-c - di$. The *multiplicative identity element* in $\mathcal{C}$ is $1 + 0i$, or 1.

Chapter Review

Write the letter of the correct answer.

1. For what value of k will the point $(-2, 28)$ lie on the graph of $y = kx^2$?   *7-1*

   a. 14   **(b.)** 7   **c.** -14 **d.** $\frac{1}{7}$

2. Simplify $\dfrac{\sqrt{27}}{-3}$.   *7-2*

   (a.) $-\sqrt{3}$ **b.** $-3\sqrt{3}$ **c.** -1 **d.** $\dfrac{3}{\sqrt{3}}$

3. Determine the rational roots, if any, of $x^3 - 2x - 4 = 0$.   *7-3*

   a. -2 **b.** 2, 4   **(c.)** 2   **d.** 1, -4

4. Express 4.126×10^4 in decimal notation.   *7-4*

   a. 4126   **b.** 0.0004126   **(c.)** 41,260   **d.** 4.1260

5. Find a one-significant-digit estimate of $\dfrac{390 \times 1.86}{1513}$.

   a. 2×10^{-1} **b.** 8×10^1 **c.** 4×10^1 **(d.)** 4×10^{-1}

6. Which number represents an irrational number?   *7-5*

   a. $3.\overline{16}$ **(b.)** $3.16116\ldots$ **c.** 3.16 **d.** $\frac{1}{3}$

7. Express $0.\overline{923}$ as a ratio of two integers.

   (a.) $\frac{923}{999}$ **b.** $\frac{922}{1000}$ **c.** $\frac{923}{1000}$ **d.** 0.923×10^3

8. Simplify $\sqrt[6]{\dfrac{27}{b^3}}$. Assume each radical represents a real number.   *7-6*

   a. $b\sqrt{3}$ **b.** $\dfrac{\sqrt{3}}{\sqrt[3]{b}}$ **c.** $\dfrac{\sqrt{3}}{b}$ **(d.)** $\dfrac{\sqrt{3b}}{b}$

9. Simplify $\dfrac{\sqrt{a} + 2}{\sqrt{a} - 5}$. Assume each radical represents a real number.   *7-7*

   a. $\dfrac{a + 10}{a - 25}$ **(b.)** $\dfrac{a + 7\sqrt{a} + 10}{a - 25}$ **c.** $\dfrac{a + 7\sqrt{a} + 10}{a + 25}$ **d.** $\dfrac{7\sqrt{a} + 2}{5}$

10. Solve $6 - \sqrt{5y} = 1$ over $\mathcal{R}$.   *7-8*

    (a.) $\{5\}$ **b.** $\{-5\}$ **c.** $\{1, 5\}$ **d.** $\{1\}$

11. Express $\dfrac{\sqrt{-20}}{i^3\sqrt{5}}$ in simplest form as a real or imaginary number.   *7-9*

    a. $10i$ **(b.)** -2 **c.** $-2i$ **d.** -10

12. Solve $(x + y)i + 3x = 3i + 6$ for x and y, where x and y are real numbers.   *7-10*

    a. $x = 0, y = 3$ **b.** $x = 3, y = 0$
    c. $x = 1, y = 2$ **(d.)** $x = 2, y = 1$

Radicals and Irrational Numbers **287**

26. If $6 - \sqrt[3]{2}$ is rat., $6 - \sqrt[3]{2} = \frac{a}{b}$ where a, b are integers, $b \neq 0$. $\sqrt[3]{2} = \frac{6b - a}{b}$, which is rat. Ex. 18 shows that $\sqrt[3]{2}$ is irrat. Contradiction; $\therefore 6 - \sqrt[3]{2}$ is irrat.

27. If $\frac{4 + \sqrt{7}}{3}$ is rat., $\frac{4 + \sqrt{7}}{3} = \frac{a}{b}$ where a, b are integers, $b \neq 0$. $\sqrt{7} = \frac{3a - 4b}{b}$, which is rat. Ex. 19 shows that $\sqrt{7}$ is irrat. Contradiction; $\therefore \frac{4 + \sqrt{7}}{3}$ is irrat.

28. If $\frac{3\sqrt{5} - 1}{4}$ is rat., $\frac{3\sqrt{5} - 1}{4} = \frac{a}{b}$ where a, b are integers, $b \neq 0$. $\sqrt{5} = \frac{4a + b}{3b}$, which is rat. Ex. 17 shows that $\sqrt{5}$ is irrat. Contradiction; $\therefore \frac{3\sqrt{5} - 1}{4}$ is irrat.

30. If $at + b$ is rat., then $at + b = \frac{c}{d}$ where c, d are integers, $d \neq 0$. $t = \frac{c - bd}{ad}$, which is rat. Contradiction of hypoth. that t is irrat. $\therefore at + b$ is irrat.

13. Express $\frac{5 + 2i}{6 + 3i} - \frac{1 + i}{15}$ in the form $a + bi$. *7-11*
    a. $\frac{11}{15} - \frac{2}{15}i$ **b.** $\frac{4}{5} - \frac{1}{15}i$ **c.** $\frac{7}{15} - \frac{-2}{15}i$ **d.** $\frac{11}{15}$

14. Express $(2 + i\sqrt{5})^2$ in the form $a + bi$.
    a. $9 + 10i$ **b.** $-1 + 10i$ **c.** $9 + 5i$ **(d.)** $-1 + 4i\sqrt{5}$

Chapter Test

1. If y is directly proportional to x^2 and y is 48 when x is 4, determine y when x is 3. 27 *7-1*

2. The distance that a ball will roll down an inclined plane is directly proportional to the square of the time it rolls. If a ball rolls 54 ft in 3 s, how far will it roll in 4 s? 96 ft

3. Solve $x^2 - 1.21 = 0$ over $\mathcal{R}$. {1.1, -1.1} *7-2*

4. Find the rational roots, if any, of $y^3 + 3y^2 + y - 2 = 0$. -2 *7-3*

5. Express 390,700 in standard notation. 3.907×10^5 *7-4*

6. Find a one-significant-digit estimate of $\frac{190 \times 0.083}{3.9}$. 4

7. Express $\frac{3}{7}$ as a decimal. $0.\overline{428571}$ *7-5*

8. Express $1.6\overline{2}62$ as a fraction in lowest terms. $\frac{161}{99}$

9. Find a rational number and an irrational number between $\frac{3}{5}$ and $0.\overline{727}$. Answers may vary. An example is $\frac{2}{3}$; 0.61661666 . . .

Simplify. Assume all radicals denote real numbers.

10. $\frac{2\sqrt{147}}{\sqrt[4]{9}}$ 14

11. $\frac{\sqrt[3]{81a^4}}{\sqrt[3]{30a^3}}$ $\frac{3\sqrt[3]{100a}}{10}$, $a \neq 0$ *7-6*

12. $\sqrt[4]{\frac{(64a)^2}{c^{10}}}$ $\frac{8\sqrt{|ac|}}{|c|^3}$

13. $\sqrt{25a^2b + 25b^2} \cdot \sqrt[4]{16a^6}$ $10|a|\sqrt{|a^3b| + |ab^2|}$

14. $b\sqrt[3]{27} + \sqrt{b^2a} + 2b\sqrt{144a}$ $3b + |b|\sqrt{a} + 24b\sqrt{a}$

15. $\frac{\sqrt{5} + 3}{\sqrt{5} - 2}$ $11 + 5\sqrt{5}$ *7-7*

16. Solve $x - 4 - \sqrt{x - 4} = 0$ over $\mathcal{R}$. {4, 5} *7-8*

17. Express $\frac{2i}{\sqrt{-16}} + i\sqrt{-27}$ in simplest form as a real number or pure imaginary number. $\frac{1}{2} - 3\sqrt{3}$ *7-9*

18. Express $\bar{a} + b$ as a complex number, given $a = 2 - 3i$ and $b = \frac{2}{3}i$. $2 + \frac{11}{3}i$ *7-10*

19. Express $\frac{6 - 4i}{1 + 2i}$ in the form $a + bi$. $-\frac{2}{5} - \frac{16}{5}i$ *7-11*

288 *Chapter 7*

PREPARING FOR
COLLEGE ENTRANCE EXAMS

Strategy for Success: Be prepared for the test by becoming familiar with the format and organization of the test. Study and complete some sample tests ahead of the test date. This will enable you to be familiar with the directions, explanations, and types of questions on the test.

Decide which is the best of the choices given and write the corresponding letter on your answer sheet.

1. $(a^{5p} + 4a^{-3p})(a^{5p} - 4a^{-3p})$ is equivalent to C
   (A) $a^{25p} - 4^{9p}$ **(B)** $a^{25p} - 4^{-9p}$ **(C)** $a^{10p} - 16a^{-6p}$
   (D) $a^{10p} - 8a^{15p} + 16a^{6p}$ **(E)** $a^{10p} + 16a^{-6p}$

2. Which of the following must be true for the polynomial $x^3 - 3x^2 + 4x - 12$? E
   I. It is irreducible. II. Its factors are $(x - 4)(x - 3)$.
   III. Its factors are $(x - 2)(x + 2)(x - 3)$.
   (A) I only **(B)** II only **(C)** III only **(D)** II and III only **(E)** None

3. If the square of a number exceeds 10 times the number by 416, what is the number? D
   (A) 16 **(B)** -21 **(C)** 15 **(D)** -16 **(E)** 38

4. What is the solution set of the following system? C
$$\frac{2x + y}{x - y} = \frac{3}{4}$$
$$\frac{x}{y} = -2$$
   (A) $\{0\}$ **(B)** $\{\frac{4}{9}\}$ **(C)** $\emptyset$ **(D)** $\{-\frac{9}{4}\}$ **(E)** $\{-\frac{4}{9}\}$

5. Which of the following must be true for the function $f(x) = x|x|$? D
   I. The function is odd.
   II. The graph of the function is symmetric with respect to the origin.
   III. The function is neither even nor odd.
   (A) I only **(B)** II only **(C)** III only **(D)** I and II only **(E)** II and III only

6. If y varies directly as x^2 and z^3, what happens to the value of y when the values of both x and z are doubled? A
   (A) It is multiplied by 32. **(B)** It is divided by 8. **(C)** It is divided by 2.
   (D) It is multiplied by 4. **(E)** It is multiplied by 2.

7. Which of the following must be rational roots of $2x^3 + x^2 - 2x - 1 = 0$? E
   I. 1 II. -1 III. $-\frac{1}{2}$
   (A) I only **(B)** II only **(C)** III only **(D)** I and II only **(E)** I, II, and III

8. What is the solution set of $\sqrt{2x + 7} - \sqrt{x + 3} = 1$ over $\mathcal{R}$? A
   (A) $\{-3, 1\}$ **(B)** $\{-3\}$ **(C)** $\{1\}$ **(D)** $\emptyset$ **(E)** $\{3, -1\}$

Radicals and Irrational Numbers **289**

Additional Answers
Oral Exercises

(continued from p. 274)

3. **a.** no restrictions
   b. Cube both sides.

4. **a.** no restrictions
   b. Cube both sides.

5. **a.** $c \geq \dfrac{1}{4}$
   b. Add -2 to both sides; square both sides.

6. **a.** $a \geq 0$
   b. Add 2 to both sides; raise both sides to the fourth power.

7. **a.** no restrictions
   b. Add -3 to both sides; cube both sides.

8. **a.** $n \geq -\dfrac{10}{3}$
   b. Add -4 to both sides; square both sides.

9. **a.** no restrictions
   b. Add k to both sides; square both sides.

10. **a.** $p \leq -4$ or $p \geq 4$
    b. Add $\sqrt{p^2 - 16}$ to both sides; add -2 to both sides; square both sides.

11. **a.** no restrictions
    b. Multiply both sides by $\dfrac{2}{5}$; cube both sides.

12. **a.** $x \geq 0$
    b. Multiply both sides by $\dfrac{3}{2}$; raise both sides to the fourth power.

Contest Problems

1. In the long division shown below, each X represents one digit: 0, 1, 2, 3, 4, 5, 6, 7, 8, or 9. Find the digit represented by each X so that there is a remainder of 0.

```
            XX8XX                    80809
     XXX) XXXXXXXX          124)10020316
          XXX                      992
          XXXX                    1003
           XXX                     992
          XXXX                    1116
          XXXX                    1116
             0                       0
```

PROGRAMMING IN PASCAL

There are many ways in which output can be formatted in Pascal. Integer and real values are right aligned with a field width of 10 spaces when they are printed as output in a Pascal program. The *field width* is the number of spaces allotted for the print-out of a value. This format in Pascal is most obvious when printing integers. For example each integer below occupies a different number of spaces, but each integer field occupies a full 10 spaces.

10 spaces	10 spaces	10 spaces
876321	9136	125

Sometimes this automatic allotment of spaces, called the *default* field width, causes output to have an awkward appearance. For example, the statement

writeln('The number is', num);

where *num* is of type integer and of value 3, would output the following:

The number is   3

Pascal prints real values in *floating point notation* (what you know as standard or scientific notation). This, too, can create an awkward appearance. If we enter a real value 652.9 for the identifier *rval*, the output for *rval* would be 6.52900000000000E + 02, a rather lengthy representation of a 4 digit number. The number of digits in the output also implies an accuracy which is not actually present. Note that every system has its own default field widths. The most common are 10 spaces for integers and 22 spaces for real, but you should be familiar with your own system's values.

In order to overcome these format problems, Pascal allows the user to define field widths in output statements, thereby giving the user the option of deciding what format the output will take. The user can delete extra space

290 *Chapter 7*

by making the field width smaller, or add space by making the field width larger. To change the field width, follow the value to be output by a colon, and then the positive integer number of spaces desired in the new field width. Neither digits of an integer variable nor digits to the left of a decimal point of a real variable will be lost. Extra spaces that occur if the field width is greater than the number of spaces needed to represent the value will appear on the left. If the field width is too small for an integer or char value, it will automatically expand to make space for the full value. The final digit of real values is rounded to fit the desired field width.

With real values, the programmer is allowed to determine not only the field width, but also the decimal accuracy that should appear. For example, the field width allotted for each expression below is the same, but the number of decimal places of accuracy allowed increases.

$$\text{writeln}(126/12{:}10{:}1,\ 126/12{:}10{:}2,\ 126/12{:}10{:}5);$$

The accompanying output would be:

$$10.5 \qquad\qquad 10.50 \qquad\qquad 10.50000$$

Exercises

1. a. Write a function with the heading

$$\text{FUNCTION power}(x : \text{real}; \ n : \text{integer}) : \text{real};$$

that computes the nth power of x for a specified real number x, and integral exponent n. (If $n \geq 0$, find the power using one method; if $n < 0$, find the power using another method.

b. Write a Pascal program *integral_powers* that uses the function *power* to compute the integral power of the real number specified by the user. Remember that 0^n is undefined for $n < 0$.

c. Use your program to compute some values of exponentiated real numbers.

d. Compare the values you obtain in (c) by the computer to the values you obtain by using a scientific calculator. Are there any differences? If so, suggest a possible explanation.

2. a. Write a Pascal program that displays all possible rational roots of a polynomial $P(x)$ in fractional form and with integral coefficients. (*Hint*: Use either a two-dimensional array or two parallel one-dimensional arrays to store the coefficient and exponent of each term.)

b. Enhance the program in (a) to test each possible root to see if it is a root of $P(x)$. You may not get exactly zero when you evaluate the polynomial at a possible root. If, however, the result you obtain is very close to zero, you may want to consider this possible root an *approximation* of an actual root. Install a check in your program that finds the difference $P(x) - 0$ where x is a possible rational root you are testing. If the absolute value of this difference is less than 0.000005 (or a tolerance you choose) the program should declare x as an approximation of a root.

Radicals and Irrational Numbers　**291**

This High-Speed Systems of Transport (HSST) car, introduced at the Expo '85 in Tsukuba, Japan, can reach speeds of 250 km/h. The magnetic field in the track and the magnet on the train are oppositely charged to produce changes in magnetism that propel the train.

Chapter 8

Sequences and Series

Arithmetic Sequences and Series

OBJECTIVES for Sections 8-1 through 8-3:
1. *To specify a sequence recursively or explicitly.*
2. *To find a specified term of an arithmetic sequence when two terms, or one term and the common difference, are given.*
3. *To insert any number of arithmetic means between two given numbers.*
4. *To find the sum of a given arithmetic series.*
5. *To solve problems involving arithmetic sequences and series.*

8–1 Sequences

Margaret has $23 in her checking account and plans to deposit $10 every week for the next five weeks. The number of dollars in her account each week will form the following *sequence*:

$$23, 33, 43, 53, 63, 73$$

The numbers in a sequence are called the terms of the sequence. In the preceding sequence the first term is 23, the second term is 33, and so on. A sequence that has a last term is called a finite sequence. A sequence that has no last term, such as the odd positive integers

$$1, 3, 5, 7, 9, \ldots ,$$

is called an infinite sequence.

An arithmetic sequence, or an arithmetic progression, is any sequence in which each term after the first is obtained by adding a fixed number, called the common difference, to the preceding term. In the sequence 23, 33, 43, 53, 63, 73 the common difference is 10. The terms in an arithmetic sequence are said to be in arithmetic progression.

Sequences and Series **293**

Problem Solving Strategies

Looking for a Pattern
Throughout the chapter students are challenged to look for numerical patterns. See especially Sections 8-1 and 8-4.
Deduction
This method is used to establish formulas for sums of arithmetic series on page 301 and geometric series on page 316.
Generalizing from Specific
In Section 8-7 students use specific examples to help generalize the concept of the limit of a sequence.

Teaching Suggestions
p. T96

Related Activities p. T96

Key Ideas

Define and specify an arithmetic sequence.

1. $-8, -4, 0, 4$
   $(-4) - (-8) = 4$
   $0 - (-4) = 4$
   $4 - 0 = 4$ Yes, the sequence is arithmetic with the common difference 4.

2. 5, 10, 20, 40 $10 - 5 = 5$ $20 - 10 = 10$ No, the sequence is not arithmetic.

3. Specify **(a)** recursively and **(b)** explicitly the sequence 10, 5, 0, $-5, \ldots$.
   a. $a_1 = 10$; $a_{n+1} = a_n + (-5)$, or $a_{n+1} = a_n - 5$
   b. $a_n = 10 - (n-1)5$, or $a_n = 10 - 5n + 5 = 15 - 5n$

Give the first four terms of each sequence. If the sequence is arithmetic give the common difference.

1. $a_1 = 4$, $a_{n+1} = 4a_n$ 4, 16, 64, 256; not arithmetic

2. $a_1 = -4$, $a_{n+1} = a_n + 4$
   $-4, 0, 4, 8$; common difference: 4

3. $a_1 = -7$, $a_{n+1} = a_n - 7$
   $-7, -14, -21, -28$; common difference: -7

4. Specify the sequence recursively.
   $-3, -6, -9, -12, \ldots$
   $a_{n+1} = a_n + (-3)$, or $a_{n+1} = a_n - 3$

5. Specify the sequence explicitly.
   1, 5, 9, 13, $\ldots$ $a_n = 1 + (n-1)(4)$ or $a_n = 4n - 3$

EXAMPLE 1   Determine whether each sequence is arithmetic. If it is, give the common difference.

a. $-7, 1, 9, 17$ **b.** 6, 12, 24, 48, . . .

SOLUTION   To determine whether the sequence is arithmetic, subtract each term from its successor.

a. $1 - (-7) = 8$ $9 - 1 = 8$ $17 - 9 = 8$

Since these differences are all equal, the sequence is arithmetic with common difference 8

b. $12 - 6 = 6$ $24 - 12 = 12$ $48 - 24 = 24$

Since these differences are not equal, the sequence is not arithmetic.

It is convenient to use subscript notation to refer to the terms of a sequence. For example, the first term of a sequence is denoted a_1, the second term a_2, and so on, with a_n denoting the nth term. This notation is used to specify a sequence in either of two distinct ways:

1. A sequence can be specified *recursively* by giving the value of a_1 and stating the relationship between any subsequent term and the term (or terms) before it in the sequence. For example, the *recursive* rule for the sequence in Example 1(a) is shown below.

$$a_1 = -7, \qquad a_{n+1} = a_n + 8, \qquad n \in \{1, 2, 3\}$$

That is, the first term is -7, and any subsequent term can be found by adding 8 to the preceding term.

2. A sequence can be specified by giving the value of any term as a function of its *position* in the sequence. For example, the *explicit* rule for the sequence in Example 1(b) is shown below.

$$a_n = 3 \cdot 2^n, \qquad n \in \{1, 2, 3, \ldots\}$$

Notice that in (1) above, a finite sequence with four terms is specified. In (2) an infinite sequence is specified. In general, this book will omit "$n \in \{1, 2, 3, \ldots\}$" when giving a rule for an infinite sequence. Unless otherwise specified, a sequence in this book may be assumed to be infinite.

Any *arithmetic* sequence can be defined recursively as follows.

> If the first term of an arithmetic sequence is a_1, and d is the common difference, then
> $$a_{n+1} = a_n + d.$$

EXAMPLE 2   Specify **(a)** recursively and **(b)** explicitly the following sequence:

$$7, 3, -1, -5, -9, \ldots$$

SOLUTION   **a.** $a_1 = 7$; $a_{n+1} = a_n + (-4)$, or $a_{n+1} = a_n - 4$

b. Since $a_1 = 7 - 0 \cdot 4$, $a_2 = 7 - 1 \cdot 4$, $a_3 = 7 - 2 \cdot 4$, and so on,

$$a_n = 7 - (n - 1)4, \text{ or } a_n = 11 - 4n.$$

Oral Exercises

State the second and third terms of the arithmetic sequence whose first term is given.

1. $a_1 = 5$, $d = 2$ 7, 9

2. $a_1 = 7$, $d = -6$ 1, −5

3. $a_1 = -8$, $d = 3$ −5, −2

4. $a_1 = -\frac{1}{2}$, $d = 1$ $\frac{1}{2}, \frac{3}{2}$

5. $a_1 = 5$, $d = -\frac{5}{2}$ $\frac{5}{2}$, 0

6. $a_1 = -9$, $d = -\frac{3}{2}$
$-\frac{21}{2}, -12$

State whether each of the following is an arithmetic sequence. If it is, give the next term. Y is an arithmetic sequence, N is not an arithmetic sequence.

7. 5, 12, 19, 26, . . . Y, 33

8. 3, 1, −1, −3, . . . Y, −5

9. 1, 4, 9, 16, . . . N

10. 2, 4, 8, 16, . . . N

11. $5\frac{1}{2}$, 4, $2\frac{1}{2}$, 1, . . . Y, $-\frac{1}{2}$

12. −13, −9, −5, −3, . . .
N

State the first four terms of each sequence.

13. $a_1 = 3$, $a_{n+1} = a_n + 8$ 3, 11, 19, 27

14. $a_1 = 5$, $a_{n+1} = 2a_n$ 5, 10, 20, 40

15. $a_1 = 80$, $a_{n+1} = \frac{1}{2}a_n$ 80, 40, 20, 10

16. $a_1 = 7$, $a_{n+1} = a_n - 10$

17. $a_n = 3 + n$ 4, 5, 6, 7

18. $a_n = 7 - 2n$ 5, 3, 1, −1

19. $a_n = n^3$ 1, 8, 27, 64

20. $a_n = 5^n$ 5, 25, 125, 625

21. $a_n = 5n + 1$ 6, 11, 16, 21

22. $a_n = 2^{n-1}$ 1, 2, 4, 8
16. 7, −3, −13, −23

Written Exercises

Give the first four terms of each sequence. If the sequence is arithmetic, give the common difference.

−3, 4, 11, 18; $d = 7$

A **1.** $a_1 = 2$, $a_{n+1} = -3a_n$ 2, −6, 18, −54

2. $a_1 = -3$, $a_{n+1} = a_n + 7$

3. $a_1 = \frac{5}{2}$, $a_{n+1} = a_n - 1$ $\frac{5}{2}$, 2, $\frac{3}{2}$, 1; $d = -\frac{1}{2}$

4. $a_1 = 2$, $a_{n+1} = (a_n)^2$ 2, 4, 16, 256

5. $a_1 = -4$, $a_{n+1} = a_n + k$

6. $a_1 = 8$, $a_{n+1} = 3 - a_n$

7. $a_n = n^2 + 1$ 2, 5, 10, 17

8. $a_n = 4n - 5$

9. $a_n = 8 - 3n$

10. $a_n = \dfrac{3 - 2n}{3}$ $\frac{1}{3}, -\frac{1}{3}, -1, -\frac{5}{3}$; $d = -\frac{2}{3}$

11. $a_n = 5 \cdot 2^{n-1}$ 5, 10, 20, 40

12. $a_n = (-1)^n n$ −1, 2, −3, 4

Specify each sequence recursively.

13. 2, −4, −10, −16, . . . $a_1 = 2$, $a_{n+1} = a_n - 6$

14. $a_1 = 7$, $a_{n+1} = 2a_n$

16. $a_1 = 6$, $a_{n+1} = \frac{1}{2}a_n$

14. 7, 14, 28, 56, . . .

15. k, k^3, k^5, k^7, . . . $a_1 = k$, $a_{n+1} = a_n k^2$

16. 6, 3, $\frac{3}{2}$, $\frac{3}{4}$, . . .

17. $-\frac{1}{2}$, $\frac{5}{2}$, $\frac{11}{2}$, $\frac{17}{2}$, . . . $a_1 = -\frac{1}{2}$, $a_{n+1} = a_n + 3$

18. −9, 9, −9, 9, . . .
$a_1 = -9$, $a_{n+1} = -a_n$

Sequences and Series **295**

Suggested Assignments

Minimum
  295/1–27
Extended Alg. with Trig.
  295/3–29 odd
Enriched Alg.
  295/9–31 odd
Enriched Alg. with Trig.
  295/9–31 odd

Common Errors

Students sometimes mistakenly think of a subscript as having the value of the term. Point out that the sequence is a list of terms, and the subscript *gives the position of each term* on the list. Remind them that because sequences are generated according to a pattern, one can use the position subscript of a term to describe *(explicitly)* the value of the term.

Mixed Review

Simplify.

1. $3p + 2b - 3(-8p + 4b)$
   $27p - 10b$

2. $\dfrac{4s - 8}{s} \cdot \dfrac{s^2 + 2s}{s^2 - 4}$ 4

3. $(k^2)(k^2/^3)$ $k^4/^3$

4. $\dfrac{12m}{6n} \cdot \dfrac{n^2 p}{p}$ $2mn$

5. $\dfrac{3}{c^2 d} - \dfrac{4}{cd} + \dfrac{5}{c^2 d}$ $\dfrac{8 - 4c}{c^2 d}$

6. If $2p^2 - 5 = 45$, then $p = $
   _?_ 5 or −5

Additional Answers
Written Exercises

5. −4, −4 + k, −4 + 2k,
   −4 + 3k; $d = k$

6. 8, −5, 8, −5

8. −1, 3, 7, 11; $d = 4$

9. 5, 2, −1, −4; $d = -3$

Specify each sequence explicitly.

19. 1, 4, 9, 16, . . . $a_n = n^2$
21. 8, $4\frac{1}{2}$, 1, $-2\frac{1}{2}$, . . . $a_n = 11\frac{1}{2} - \frac{7}{2}n$
23. 2, 10, 50, 250, . . . $a_n = 2 \cdot 5^{n-1}$

22. $a_n = c + nb^2$
20. 4, 7, 10, 13, . . . $a_n = 3n + 1$
22. $c + b^2, c + 2b^2, c + 3b^2$, . . .
24. a, ar, ar^2, ar^3, . . . $a_n = ar^{n-1}$

Specify each sequence both explicitly and recursively. (*Hint*: For the recursive rule, consider $a_{n+1} - a_n$.)

B 25. 1, 2, 4, 7, 11, . . .   26. 1, 4, 9, 16, 25, . . .   27. $\frac{1}{2}, \frac{2}{3}, \frac{3}{4}, \frac{4}{5}, \frac{5}{6}$, . . .

28. Specify the first two terms and give a recursive rule for subsequent terms of the sequence 1, 1, 2, 3, 5, 8, 13, ...

29. Show that if a_1, a_2 and a_3 are the first three terms of an arithmetic sequence, then:

$$\frac{a_1 + a_3}{2} = a_2$$

C 30. Show that if a_1, a_2, a_3, and a_4 are the first four terms of an arithmetic sequence, then:

$$\frac{2a_1 + a_4}{3} = a_2$$

31. Find the value of x if it is known that the sequence $2x - 1$, $5x - 3$, $4x + 3$ is an arithmetic progression. $x = 2$

8–2 Arithmetic Sequences and Arithmetic Means

To determine a particular term of an arithmetic sequence, $a_1, a_2, a_3, \ldots$, such as a_{20}, it is not necessary to compute each preceding term. Notice that you have

$$a_1$$
$$a_2 = a_1 + d = a_1 + 1d$$
$$a_3 = a_2 + d = (a_1 + d) + d = a_1 + (d + d) = a_1 + 2d$$
$$a_4 = a_3 + d = (a_1 + 2d) + d = a_1 + (2d + d) = a_1 + 3d$$

and so on. This suggests the following fact.

> The nth term of an arithmetic sequence whose first term is a_1 and whose common difference is d is
> $$a_n = a_1 + (n - 1)d.$$

EXAMPLE 1   Find the thirtieth term in the arithmetic sequence $-4, -1, 2, \ldots$.

SOLUTION $a_1 = -4, d = 2 - (-1) = 3$. Since $a_n = a_1 + (n-1)d$,
$a_{30} = -4 + (30 - 1) \cdot 3 = -4 + 29 \cdot 3 = 83$.

The terms between two given terms of an arithmetic sequence are called **arithmetic means** between the given terms. For example, the three arithmetic means between 18 and 98 are 38, 58, 78 because

$$18, 38, 58, 78, 98$$

is an arithmetic sequence.

A single arithmetic mean inserted between two numbers is *the* **arithmetic mean,** or the **average,** of the two numbers.

EXAMPLE 2   Insert four arithmetic means between 7 and 37.

SOLUTION   Use the formula $a_n = a_1 + (n-1)d$ to determine d for an arithmetic sequence whose first term is 7 and whose sixth term is 37.

$$37 = 7 + (6-1)d$$
$$37 = 7 + 5d$$
$$d = 6$$

The required means are found by successive additions of 6.

$$7, 13, 19, 25, 31, 37$$

EXAMPLE 3   What is the first term of an arithmetic sequence whose fourth term is 5 and whose eighth term is 11?

SOLUTION   To find the value of d, use the formula $a_n = a_1 + (n-1)d$ twice, first with $n = 4$, then with $n = 8$.

Since the fourth term is 5,

$$5 = a_1 + (4-1)d.$$

Since the eighth term is 11,

$$11 = a_1 + (8-1)d.$$

Solve the system: $5 = a_1 + 3d$
$11 = a_1 + 7d$

Subtracting, you obtain the following: $-6 = -4d$
$$d = \tfrac{3}{2}$$

To find the value of a_1, substitute d in the first equation and solve for a_1.

$$5 = a_1 + 3(\tfrac{3}{2})$$
$$a_1 = \tfrac{1}{2}$$

$\therefore$ the first term of the arithmetic sequence is $\dfrac{1}{2}$.

Chalkboard Examples

1. Find the ninth term in the arithmetic sequence $-2, 0, 2, \ldots$ $a_1 = -2, d = 0 - (-2) = 2$. Since $a_n = a_1 + (n-1)d$, $a_9 = -2 + (9-1)2 = -2 + 8 \cdot 2 = 14$

2. Insert five arithmetic means between 3 and 39.
$a_n = a_1 + (n-1)d$
$39 = 3 + (7-1)d$
$39 = 3 + 6d$
$36 = 6d$
$d = 6$
3, 9, 15, 21, 27, 33, 39

3. What is the first term of an arithmetic sequence whose fifth term is 5 and whose eighth term is 9?
$5 = a_1 + (5-1)d$
$9 = a_1 + (8-1)d$
$-4 = -3d$
$$d = \frac{4}{3}$$
$$5 = a_1 + 4\left(\frac{4}{3}\right)$$
$$a_1 = 5 - \frac{16}{3} = -\frac{1}{3}$$

Suggested Assignments

Minimum
298/2–36 even, 37–42
299/*P*: 1–8
Extended Alg. with Trig.
298/13–41 odd
299/*P*: 1–8
Enriched Alg.
298/17–43 odd
299/*P*: 3–8
Enriched Alg. with Trig.
298/17–43 odd
299/*P*: 3–8

Additional A Exercises

Find the specified term of the arithmetic sequence.

1. 18, 14, 10, 6, . . . ; a_{10}
 -18

2. $-7, -3, 1, 5, . . . ; a_{15}$ 49

3. 1, 26, 51, . . . ; a_7 151

4. 0.1, 0.6, 1.1, . . . ; a_9 4.1

Find the missing value, using the given values for an arithmetic sequence.

5. $a_1 = 12, a_7 = 90, d =$
 ? $d = 13$

6. $a_{10} = 53, d = 6, a_1 =$
 ? $a_1 = -1$

Additional Answers
Written Exercises

1. -28 **2.** 703 **3.** 47

4. -79 **5.** 77 **6.** 0.9

7. $-\dfrac{65}{6}$ **8.** 1718

43. 1. $c - a = b - c$
  Def. of arith. seq.
  2. $2c = a + b$
   Add. prop. of =

  3. $c = \dfrac{1}{2}(a + b)$

   Mult. Prop. of =

Oral Exercises

State the values you would use for a_1, n, and d in order to use the formula on page 296 to find the specified term of each given arithmetic sequence.

1. 17, 14, 11, . . . ; a_{16} 17, 16, -3
2. 3, 53, 103, . . . ; a_{15} 3, 15, 50
3. $-8, -3, 2, . . . ; a_{12}$ -8, 12, 5
4. 1, $-3, -7, . . . ; a_{21}$ 1, 21, -4
5. $-4, \frac{1}{2}, 5, . . . ; a_{19}$ -4, 19, $\frac{9}{2}$
6. 0.1, 0.15, 0.2, . . . ; a_{17} 0.1, 17, 0.05
7. $\frac{1}{6}, -\frac{1}{3}, -\frac{5}{6}, . . . ; a_{23}$ $\frac{1}{6}$, 23, $-\frac{1}{2}$
8. 1, 102, 203, . . . ; a_{18} 1, 18, 101

State the arithmetic mean of the given numbers.

9. 7 and 11 9 **10.** 1 and 23 12 **11.** -5 and 9 2 **12.** -2 and -12 -7

13. 3.2 and -1.8 0.7 **14.** -7.5 and 3.5 -2 **15.** $a + b$ and $a - b$ a **16.** $3c$ and $c + 2d$ $2c + d$

Written Exercises

A **1–8.** Find the specified term of the arithmetic sequences in Oral Exercises 1–8.

Find the missing value, using the given values for an arithmetic sequence.

9. $a_1 = 4, d = 8, a_{51} =$? 404
10. $a_1 = 11, a_{15} = 95, d =$? 6
11. $a_{26} = 472, d = 19, a_1 =$? -3
12. $a_1 = 172, a_{51} = -28, d =$? -4
13. $a_1 = -2, a_{17} = -82, d =$? -5
14. $a_1 = \frac{3}{2}, d = -\frac{5}{2}, a_{32} =$? -76
15. $a_{29} = 63, d = 4, a_1 =$? -49
16. $a_{45} = -27, d = -0.75, a_1 =$? 6
17. $a_1 = 12, d = -7, a_n = -114, n =$? 19
18. $a_1 = 19, d = -\frac{2}{3}, a_n = 5, n =$? 22
19. $a_1 = -44, d = \frac{3}{2}, a_n = -5, n =$? 27
20. $a_1 = 1.8, d = 3.2, a_n = 117, n =$? 37

24. 0.75, 2.5, 4.25, 6, 7.75, 9.5, 11.25
Insert the stated number of arithmetic means between the given numbers.

21. three between 7 and 91 28, 49, 70
22. five between -5 and 37 2, 9, 16, 23, 30
23. three between 3 and 21 7.5, 12, 16.5
24. seven between -1 and 13
25. nine between -18 and -3
 $-16.5, -15, -13.5, -12, -10.5, -9, -7.5, -6, -4.5$
26. five between $\frac{1}{5}$ and 2 $\frac{1}{2}, \frac{4}{5}, \frac{11}{10}, \frac{7}{5}, \frac{17}{10}$

Given the specified values for the terms of an arithmetic sequence, find a_1 and d.

27. $a_{12} = 8, a_{17} = 23$ $-25, 3$
28. $a_7 = 3, a_{16} = -15$ 15, -2
29. $a_{10} = 25, a_{16} = 67$ $-38, 7$
30. $a_5 = 9, a_{13} = 21$ 3, 1.5
31. $a_7 = -1, a_{11} = 9$ $-16, 2.5$
32. $a_{10} = -2, a_{16} = -5$ 2.5, -0.5

Find the specified term of each arithmetic sequence.

B **33.** $a_2 = -3, a_7 = -18, a_3 =$? -6
34. $a_6 = 18, a_{10} = 6, a_9 =$? 9
35. $a_{13} = 4, a_{21} = 8, a_5 =$? 0
36. $a_4 = 2, a_{25} = 65, a_{30} =$? 80

298 Chapter 8

Find the value of x.

37. The arithmetic mean of $x + 4$ and $4x + 5$ is 12. 3

38. The arithmetic mean of $x - 5$ and $2x + 1$ is $x + 2$. 8

39. The first term of an arithmetic sequence is $x - 2$, the fifth term is $2x$, and the common difference is $x - 7$. 10

40. The first two terms of an arithmetic sequence are $x + 2$ and $3x - 2$, and the sixth term is $5x$. 3

41. Which term of the arithmetic sequence $-1, 5, 11, 17, \ldots$ is 245? 42 nd

42. Which term of the arithmetic sequence $-3, -10, -17, -24, \ldots$ is -143? 21 st

C **43.** Prove that if a, c, b is an arithmetic sequence, then $c = \frac{1}{2}(a + b)$. Thus, the arithmetic mean of a and b is $\frac{1}{2}(a + b)$.

44. Prove that in an arithmetic sequence, a_n is the arithmetic mean of a_1 and a_{2n-1}, for any integer n greater than 1.

Problems

A **1.** For each of the first twelve years of its fruitful life a certain apple tree is expected to produce eight more apples than it had the preceding year. If the tree produces seventeen apples the first year, how many apples can the tree be expected to produce during the twelfth year? 105 apples

2. The first tread of a flight of stairs is 24 cm above the ground. If each tread after the first is 16.5 cm above the level of the previous step, how high above the ground is the tread of the fifteenth step? 255 cm

3. A child's toy consists of eight rings. The first ring has a diameter of 12.4 cm and the eighth ring has a diameter of 4 cm. What is the diameter of the second ring if the difference between the diameters of any two consecutive rings is constant? 11.2 cm

4. In a certain school system, the starting salary for a teacher is $12,000 with an annual raise of $1150 each year until a maximum salary of $28,100 is reached. How many years must a person teach in order to reach the maximum salary? 15 yr

5. Between March 1 and March 31, the sunrise at 40° north latitude occurs about 1.6 min earlier each day than the preceding day. If the sun rose at 6:33 A.M. on March 1, at what time did it rise on March 26? On what day did the sun rise at 6:09 A.M.? 5:53 A.M.; March 16

B **6.** Because of gravity, an object given an initial upward velocity *loses* 9.8 m/s of that velocity every second, or *gains* 9.8 m/s every second it falls in a downward direction. If an object has an upward velocity of 49 m/s at the end of 1 s, what will be its velocity at the end of 8 s? 19.6 m/s downward; After how many seconds will it have a *downward* velocity of 49 m/s? 11 s

44.
1. $a_n - a_1 = (n - 1)d$
   Def. of arith. seq.
2. $a_{2n-1} - a_n = (2n - 1 - n)d$
   Def. of arith. seq.
3. $a_{2n-1} - a_n = (n - 1)d$
   Simplifying
4. $a_n - a_1 = a_{2n-1} - a_n$
   Trans. prop. of =
5. $2a_n = a_1 + a_{2n-1}$
   Add. prop. of =
6. $a_n = \frac{1}{2}(a_1 + a_{2n-1})$

   Mult. prop. of =

Mixed Review

Simplify.

1. $3xy + 2x(9xy^2 - y) - (x + y)$ $18x^2y^2 + xy - x - y$

2. $\dfrac{3m^2 + 9m}{(m - 3)} \cdot \dfrac{m}{(m^2 - 9)}$ $\dfrac{3m^2}{m^2 - 6m + 9}$

3. $(3a^2b^3c^4)^2$ $9a^4b^6c^8$

4. $\dfrac{21q^3r^2s^2}{35q^4r^5s}$ $\dfrac{3s}{5qr^3}$

5. $\dfrac{6t}{t + 3} + \dfrac{2}{2t + 6}$ $\dfrac{6t + 1}{t + 3}$

6. Write an equation of a line having a slope of 3 and a y-intercept of -7. $y = 3x - 7$

7. Luke Rowan started working in 1980 at an annual salary of $14,000 with a raise of $550 each subsequent year. Rachel Ray started at the same office in 1984 at an annual salary of $15,000 with a raise of $700 each subsequent year. In what year will they earn the same salary? 1992

8. In its original form, the width of the Great Pyramid at Giza decreased by 1.57 m for each successive 1 m of height. If the width at a height of 1 m was 229.22 m, at what height was the width 103.62 m? How high was the Great Pyramid? 81 m; 146 m

Teaching Suggestions
p. T97

Supplementary Material
Test 23

Related Activities p. T97

Key Ideas

Use the summation symbol. Find the sum of an arithmetic series.

Reading Algebra

To make the use and meaning of the summation symbol more clear to the students, spend extra time on examples that use the summation notation. You may want to point out that the symbol comes from the Greek letter sigma, and carefully read aloud an example like $\sum_{i=1}^{4} (2i + 3)$ as "the summation of $2i + 3$ from $i = 1$ to $i = 4$." Be sure it is clear to the student that the i below the sigma is used to count the terms of the sum, four terms in all.

8–3 Arithmetic Series

From the terms of any given sequence, such as

$$4, 7, 10, 13, 16,$$

you can construct an associated sequence S_1, S_2, S_3, S_4, S_5 of *sums*:

$$S_1 = 4$$
$$S_2 = 4 + 7 = 11$$
$$S_3 = 4 + 7 + 10 = 21$$
$$S_4 = 4 + 7 + 10 + 13 = 34$$
$$S_5 = 4 + 7 + 10 + 13 + 16 = 50$$

Each of the indicated sums $S_1, S_2, S_3, S_4,$ and S_5 is called a *series* associated with the sequence 4, 7, 10, 13, 16.

In general, given any sequence

$$a_1, a_2, a_3, \ldots$$

with n or more terms, the associated series of n terms, S_n, is

$$S_n = a_1 + a_2 + a_3 + \cdots + a_n.$$

The Greek letter Σ (*sigma*), called the **summation sign,** is used to abbreviate the writing of a series. For example, to abbreviate the series

$$S_5 = 4 + 7 + 10 + 13 + 16,$$

first observe that 4, 7, 10, 13, 16 is an arithmetic sequence with nth term, or general term, $4 + (n - 1)3$. Therefore the series can be denoted by the symbol

$$\sum_{n=1}^{5} [4 + (n - 1)3]$$

which is read "the summation of $4 + (n - 1)3$ from $n = 1$ to $n = 5$." This means that you successively replace n with 1, 2, 3, 4, and 5, and then write an expression denoting the sum of the resulting values. The letter n is called the **index** (plural, *indexes* or *indices*) and the replacement

300   *Chapter 8*

set of n is the **range of summation**. (Note that the index need not be n; any convenient letter may be chosen as the index.)

EXAMPLE 1 Write $\sum_{k=1}^{5} 3k$ in expanded form.

SOLUTION Replace k with the numerals 1, 2, 3, 4, and 5 in turn and write the sum.

$$\sum_{k=1}^{5} 3k = 3(1) + 3(2) + 3(3) + 3(4) + 3(5)$$

EXAMPLE 2 Use summation notation to write the series

$$3 + 7 + 11 + 15 + 19 + 23 + 27.$$

SOLUTION Each term is of the form $3 + 4(n - 1)$, or $4n - 1$. Since there are seven terms,

$$3 + 7 + 11 + 15 + 19 + 23 + 27 = \sum_{n=1}^{7} (4n - 1).$$

A series, such as $3 + 7 + 11 + 15 + 19 + 23 + 27$, whose terms are in arithmetic progression is called an **arithmetic series**. Because

$$3 + 7 + 11 + 15 + 19 + 23 + 27 = 105$$

we say that the **sum** of this series is 105.

You can find a formula for the sum S_n of any arithmetic series by noticing that there are two ways to obtain the terms of the series:

(1) Start with the first term a_1 and successively add the common difference d.

(2) Start with the nth term a_n and successively subtract d.

Thus, (1) $S_n = a_1 + (a_1 + d) + (a_1 + 2d) + \cdots + [a_1 + (n - 1)d]$;
   (2) $S_n = a_n + (a_n - d) + (a_n - 2d) + \cdots + [a_n - (n - 1)d]$.

Adding the corresponding sides of these equations, you obtain

$$2S_n = (a_1 + a_n) + (a_1 + a_n) + (a_1 + a_n) + \cdots + (a_1 + a_n),$$

where $a_1 + a_n$ occurs n times. Hence,

$$2S_n = n(a_1 + a_n), \quad \text{or} \quad S_n = \frac{n}{2}(a_1 + a_n).$$

EXAMPLE 3 Find the sum of the first two hundred positive odd integers.

SOLUTION $a_1 = 1$, $d = 2$, $n = 200$.

$$a_n = a_1 + (n - 1)d$$
$$a_{200} = 1 + (200 - 1)2 = 399$$

and

$$S_{200} = \frac{200}{2}(1 + 399) = 40,000$$

Sequences and Series **301**

Suggested Assignments

Minimum
  First day
    303/1–34; *P:* 1–8
  Second day
    R 305/*Self-Test 1*
    S 298/1–35 odd
Extended Alg. with Trig.
  First day
    303/5–33 odd; *P:* 1–8
  Second day
    R 305/*Self-Test 1*
    S 298/14–32 even
Enriched Alg.
  First day
    303/7–37 odd; *P:* 5–12
  Second day
    R 305/*Self-Test 1*
    S 298/18–32 even
Enriched Alg. with Trig.
  First day
    303/7–37 odd; *P:* 5–12
  Second day
    R 305/*Self-Test 1*
    S 298/18–32 even

Additional Answers
Oral Exercises

9. $a_1 = 9$, $n = 11$, $a_n =$
   -37; $S_n = \frac{n}{2}(a_1 + a_n)$

10. $a_1 = 3$, $d = 2$, $n = 21$;
    $S_n = \frac{n}{2}[2a_1 + (n - 1)d]$

11. $a_1 = 4$, $d = 4$, $a_n = 72$,
    $n = 18$; either formula

12. $a_1 = 31$, $n = 25$, $a_n =$
    -19; $S_n = \frac{n}{2}(a_1 + a_n)$

13. $a_1 = \frac{1}{2}$, $d = \frac{1}{3}$, $n = 10$;
    $S_n = \frac{n}{2}[2a_1 + (n - 1)d]$

14. $a_1 = 15$, $d = 3$, $a_n = 48$;
    find n; either formula

15. $a_1 = 41$, $d = 2$, $a_n = 63$;
    find n; either formula

16. $a_1 = -13$, $d = 6$, $n = 15$;
    $S_n = \frac{n}{2}[2a_1 + (n - 1)d]$

Example 3 illustrates the following interesting fact: *The sum of the first n positive odd integers is n^2* (see Exercise 37, page 303).

In applying the formula for S_n in Example 3, we used the fact that

$$a_n = a_1 + (n - 1)d.$$

Thus

$$S_n = \frac{n}{2}(a_1 + a_n) = \frac{n}{2}[a_1 + a_1 + (n - 1)d], \text{ or } S_n = \frac{n}{2}[2a_1 + (n - 1)d].$$

In finding the two formulas for S_n for an arithmetic series, we have proved the following theorem.

Theorem. If S_n is the sum of the first n terms of an arithmetic sequence whose first term is a_1, whose common difference is d, and whose nth term is a_n, then

$$S_n = \frac{n}{2}(a_1 + a_n) \qquad \text{and} \qquad S_n = \frac{n}{2}[2a_1 + (n - 1)d].$$

Oral Exercises

State each series in expanded form.

2. $2 + 3 + 4 + 5 + 6 + 7$
3. $1 + 3 + 5 + 7 + 9 + 11 + 13$
4. $-2 - 4 - 6 - 8$

1. $\sum_{i=1}^{5} 3i$
   $3 + 6 + 9 + 12 + 15$

2. $\sum_{k=1}^{6} (k + 1)$

3. $\sum_{j=1}^{6} (2j - 1)$

4. $\sum_{n=1}^{4} -2n$

5. $\sum_{m=2}^{6} (3m + 1)$
   $7 + 10 + 13 + 16 + 19$

6. $\sum_{i=4}^{7} (5 - i)$
   $1 + 0 - 1 - 2$

7. $\sum_{i=8}^{11} (4i - 5)$
   $27 + 31 + 35 + 39$

8. $\sum_{i=2}^{5} (1 - 5i)$
   $-9 - 14 - 19 - 24$

For each arithmetic series state the values of whichever of the constants a_1, d, n, and a_n can be determined by inspection. State which of the formulas above you would use to find the sum of the series.

9. a series with first term 9 and -37 as the eleventh and last term

10. a series of twenty-one terms beginning $3 + 5 + 7 + \cdots$

11. the sum of all the positive multiples of 4 up to, and including, 72

12. a series of twenty-five terms beginning with 31 and ending with -19

13. a series of ten terms beginning $\frac{1}{2} + \frac{5}{6} + \frac{7}{6} + \cdots$

14. the sum of all the multiples of 3 between 15 and 48, inclusive

15. the sum of all the odd integers between 41 and 63, inclusive

16. a series of fifteen terms beginning $-13 - 7 - 1 + \cdots$

17. a series of fourteen terms beginning $9 + 4 - 1 - \cdots$

18. $\sum_{n=1}^{17} (2n + 3)$

19. $\sum_{k=1}^{12} (7k - 2)$

20. $\sum_{i=2}^{11} (9 - 4i)$

302 *Chapter 8*

Written Exercises

Find the sum of each series.

A
1. $a_1 = 3, a_7 = 21, n = 7$ 84
3. $a_1 = -86, a_{32} = -14, n = 32$ −1600
5. $a_1 = -12, d = 3, n = 52$ 3354
7. $a_1 = 14, a_2 = 23, n = 11$ 649

2. $a_1 = -7, a_{15} = 29, n = 15$ 165
4. $a_1 = 33, d = -4, n = 14$ 98
6. $a_1 = 9, a_{27} = 45, n = 27$ 729
8. $a_1 = 68, a_5 = 59, n = 17$ 850

9–20. Find the sum of each series in Oral Exercises 9–20. **9.** −154 **10.** 483 **11.** 684
12. 150 **13.** 20 **14.** 378 **15.** 624 **16.** 435 **17.** −329 **18.** 357 **19.** 522

Write each series using summation notation. **20.** −170

21. $9 + 11 + 13 + 15$
23. $-5 - 1 + 3 + 7 + 11$
25. $4 + 6 + 8 + \cdots + 40$

22. $1 + 8 + 15 + 22$
24. $11 + 2 - 7 - 16 - 25$
26. $-9 - 2 + 5 + \cdots + 54$

Use the given data about an arithmetic series to find the required value.

B
27. $a_1 = 2, S_{17} = -170, d = \underline{\ ?\ }$ $-\frac{3}{2}$
28. $a_1 = -21, S_{16} = -288, a_{16} = \underline{\ ?\ }$ −15
29. $a_1 = 42, a_n = -26, S_n = 360, n = \underline{\ ?\ }$ 45
30. $a_{12} = 59, S_{12} = 324, a_1 = \underline{\ ?\ }$ −5
31. $d = 6, S_{11} = 198, a_1 = \underline{\ ?\ }$ −12
32. $a_1 = -15, d = 4, S_n = -8, n = \underline{\ ?\ }$ 8

Find the sum of each series.

33. $3 + 6 + 9 + \cdots + 210$ 7455
34. $100 + 95 + 90 + \cdots + 10$ 1045

C
35. The sum of the first n positive integers is 4950. What is the value of n? 99
36. The sum of the first n positive multiples of 3 is 315. What is the value of n? 14
37. Show that the sum of the first n positive odd integers is n^2.
38. Show that if m is a positive integer, then the sum of all the integers between m and m^2, inclusive, is $\dfrac{m(m^3 + 1)}{2}$.

Problems

A
1. In January of a certain year Will Belton deposited $120 in his savings account. For each subsequent month of the same year and each month of the following year his monthly deposit increased by $5. How much did he deposit in the savings account altogether during the two years? $4260

Sequences and Series **303**

---

17. $a_1 = 9, d = -5, n = 14;$
$$S_n = \frac{n}{2}[2a_1 + (n - 1)d]$$
18. $a_1 = 5, d = 2, n = 17,$
$a_n = 37;$ either formula
19. $a_1 = 5, d = 7, n = 12,$
$a_n = 82;$ either formula
20. $a_1 = 1, d = -4, n = 10,$
$a_n = -35;$ either formula

---

21. $\sum_{i=1}^{4} 2i + 7$ **22.** $\sum_{i=1}^{4} 7i - 6$

23. $\sum_{i=1}^{5} 4i - 9$ **24.** $\sum_{i=1}^{5} 20 - 9i$

25. $\sum_{i=1}^{19} 2i + 2$ **26.** $\sum_{i=1}^{10} 7i - 16$

37. $a_1 = 1, d = 2, S_n =$
$$\frac{n}{2}[2 \cdot 1 + (n - 1)2] =$$
$$\frac{n}{2}(2 + 2n - 2) = n^2$$

38. $a_1 = m, a_n = m^2, d = 1$
so $m^2 = m + (n - 1)1$ or
$n = m^2 - m + 1.$ Thus
$$S_n = \frac{m^2 - m + 1}{2}.$$
$$(m + m^2) = \frac{m(m^3 + 1)}{2}.$$

Mixed Review

Simplify.
1. $(2p - 3)(3p - 2)$ $6p^2 - 13p + 6$
2. $\dfrac{(5z + 10)}{2z} \div \dfrac{(z + 2)}{z}$ $\dfrac{5}{2}$
3. $(pk^{-2})(p^2k^3)$ p^3k
4. $\dfrac{16x^2y^2z}{4x^7z^4}$ $\dfrac{4y^2}{x^5z^3}$
5. $\dfrac{c}{(c + d)} - \dfrac{2d}{(c - d)} + \dfrac{4c}{(c - d)}$ $\dfrac{5c^2 + cd - 2d^2}{c^2 - d^2}$
6. If $4r^2 - 4 = 60$, then $r = \underline{\ ?\ }$ 4 or −4

303

11. T_n is an arithmetic series with $a_1 = 1$, $d = 1$, and n $= n$. So $T_n = \frac{n}{2}[2 \cdot 1 +$ $(n - 1)1] = \frac{n^2 + n}{2}$. For the arithmetic series T_{n-1}, n is replaced by $n - 1$. So $T_{n-1} =$ $\frac{n-1}{2}[2 \cdot 1 + (n - 2)1] =$ $\frac{n^2 - n}{2}$. $\therefore T_n + T_{n-1} =$ $\frac{n^2 + n}{2} + \frac{n^2 - n}{2} = n^2$

12. $T_{2n+1} - 2T_n = \frac{2n + 1}{2}$. $[2 \cdot 1 + (2n)1] -$ $2\left(\frac{n^2 + n}{2}\right) = (n + 1)^2$

In the diagram, the number of dots in each of the small triangles represents T_5. Furthermore, the number of dots in the large triangle represents $T_{2(5)+1}$. The difference between the number of dots in the large triangle and the number of dots in the two small triangles is the number of dots in the square. So $T_{11} - 2T_5 = (5 + 1)^2$.

2. A potato farmer gathers 35 bu of potatoes on the first day of the harvest. The farmer estimates that on each successive day of the harvest, the amount gathered can be 4 bu more than the preceding day. If the harvest lasts fourteen days, what is the total number of bushels that the farmer can expect to collect? 854 bu

3. Jack Neville has a job that pays $15,000 for the first year with a raise of $1000 at the end of each year thereafter. Kate O'Hara has a job that pays $7500 for the first six months with a raise of $250 at the end of every six months thereafter. Who has earned the greater total income after ten years? after twenty years? Kate with $197,500; Kate with $495,000

4. The unit used to measure the fuel requirements of buildings is the *degree-day*. Each degree that the average daily temperature is below 65°F is one degree-day. If the average temperature was 42°F on December 1 and fell by 1°F for each subsequent day up to and including December 16, how many degree-days were there from December 1 to December 16? 488 degree-days

B   5. A free-falling object falls 4.9 m during the first second of its descent. During each subsequent second it falls 9.8 m farther than it fell during the preceding second. Neglecting air resistance, how long would it take an object to reach the ground if it were dropped from the top of a building 313.6 m high? 8 s

6. Amelia Roper wants to invest a total of $40,000 in her credit union over the next sixteen years. She wants to invest $1600 the first year and increase her deposit by a fixed amount each year. What should this fixed amount be? $120

7. Imagine 26 points, A, B, C, . . . , Z, equally spaced along a line, as shown. How many line segments are there whose endpoints are any two of the 26 points? 325 segments

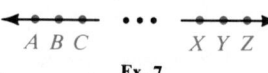
$A\ B\ C$ $X\ Y\ Z$
Ex. 7

8. A spiral maze has paths that are 1 cm wide and an initial divider segment that is 2 cm long, as shown. The outer dimensions of the maze are 25 cm by 23 cm. What is the total length of the dividers? 623 cm

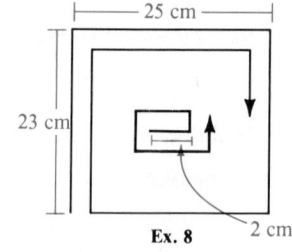
25 cm
23 cm
Ex. 8   2 cm

C   9. A chemical storage tank holds 180,000 L and is supplied by nine inlet pipes, each having a flow of 2500 L/h. If the pipes are opened at 30 min intervals, how long after the first pipe is opened will the tank be filled? 10 h

10. One day during the month of May, Grace Crawley made a charitable contribution in dollars that was equal to the number of the day of the month. She continued to contribute in this manner on successive days up to, and including, the day of the month whose number was twice that of the day on which she made her first contribution. If her total contribution was $165, what day of the month did she begin her contributions? May 10

304   *Chapter 8*

11. A number of the form $T_n = 1 + 2 + 3 + 4 + \ldots + n$ is called a *triangular number* because it can be represented by dots arranged to form a triangle. Show that for any positive integer n, $T_n + T_{n-1} = n^2$.

12. Using the definition of T_n given in Problem 11, show that for any positive integer n,

$$T_{2n+1} - 2T_n = (n + 1)^2.$$

Use the diagram to interpret this result for $n = 5$.

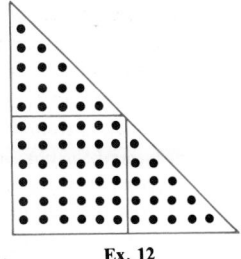

Ex. 12

Self-Test 1

VOCABULARY

terms (p. 293)
finite sequence (p. 293)
infinite sequence (p. 293)
arithmetic sequence (p. 293)
arithmetic progression (p. 293)
common difference (p. 293)
arithmetic means (p. 297)
average (p. 297)

summation sign (p. 300)
index (p. 300)
range of summation (p. 301)
arithmetic series (p. 301)
sum (p. 301)

Specify each sequence (a) recursively and (b) explicitly.

1. 4, 8, 12, 16, . . . **a.** $a_1 = 4$,
   $a_{n+1} = a_n + 4$ **b.** $a_n = 4n$

2. 3, 9, 27, 81, . . . **a.** $a_1 = 3$,   *Obj. 1, p. 293*
   $a_{n+1} = 3a_n$ **b.** $a_n = 3^n$

Find the specified term of each arithmetic sequence.

3. $a_{41} = -22$, $d = -3$, $a_1 = $ __?__ 98

4. $a_{15} = 9$, $a_{57} = 30$, $a_1 = $ __?__   *Obj. 2, p. 293*
   $\frac{2}{}$

Insert the stated number of arithmetic means between the given numbers.

5. one between -2 and 10 4

6. five between -2 and 10   *Obj. 3, p. 293*
   0, 2, 4, 6, 8

Find the sum of each arithmetic series.

7. $a_1 = 2$, $a_2 = -3$, $n = 12$ -306

8. $a_1 = 4$, $a_{10} = 58$, $n = 10$ *Obj. 4, p. 293*
   310

9. On March 2 Mary jogged for fifteen minutes. If each day thereafter   *Obj. 5, p. 293*
   she jogged two minutes longer than she had the previous day, on
   what day did she jog forty-five minutes? March 17

10. If a person deposits $200 in a savings account one month and then
    increases the deposit by $10.00 each month thereafter, how much
    money will the person have deposited altogether after 3 years? $13,500

Check your answers with those at the back of the book.

Quick Quiz

Specify each sequence **(a)** recursively and **(b)** explicitly.

1. 1, 9, 17, . . .
   a. $a_1 = 1$, $a_{n+1} = a_n + 8$
   b. $a_n = 8n - 7$

2. 2, 4, 8, 16, . . .
   a. $a_1 = 2$, $a_{n+1} = 2a_n$
   b. $a_n = 2^n$

Find the specified term of each arithmetic sequence.

3. $a_{10} = 20$, $d = 4$, $a_1 = $ __?__ -16

4. $a_3 = -5$, $a_{14} = 28$, $a_1 = $ __?__ -11

Insert the stated number of arithmetic means between the given numbers.

5. two between 2 and 20
   8, 14

6. four between -12 and 13 $-7, -2, 3, 8$

Find the sum of each arithmetic series.

7. $a_1 = a_2 = -8$, $n = 6$
   -48

8. $a_1 = 3$, $a_9 = 75$, $n = 9$
   351

9. On January 1 Joan spent 20 minutes studying Latin. Each day thereafter she studied 5 minutes longer than she had on the previous day. On what day did she study 65 minutes? January 10

10. A young man decides to make weekly deposits to a money market account, increasing the amount of deposit by $3 each week. If he deposits $50 the first week, how much will he deposit in the 12th week? How much will he have deposited at the end of 12 weeks?
    $83, $798

Geometric Sequences and Series

OBJECTIVES for Sections 8-4 through 8-6:
1. To find a specified term of a geometric sequence when two terms, or one term and the common ratio, are given.
2. To insert any number of geometric means between two given numbers.
3. To find the sum of a given geometric series.
4. To solve problems involving geometric sequences and series.

Teaching Suggestions
p. T98

Key Ideas

Define a geometric sequence.
Find a specified term within a geometric sequence.

8–4 Geometric Sequences

In Section 8-1 you learned that in an arithmetic sequence each term after the first is obtained by adding a fixed number, the common difference, to the preceding term. Notice that in the sequence

$$2, 10, 50, 250$$

each term after the first is obtained by *multiplying* the preceding term by a fixed number, namely 5.

$$a_2 = 10 = 2 \cdot 5 = a_1 \cdot 5$$
$$a_3 = 50 = 10 \cdot 5 = a_2 \cdot 5$$
$$a_4 = 250 = 50 \cdot 5 = a_3 \cdot 5$$

Thus:

$$a_{n+1} = a_n \cdot 5 \text{ for } n = 1, 2, 3$$

Any sequence in which each term after the first is the product of the preceding term and a fixed number is called a **geometric sequence** or a **geometric progression**. The fixed number is called the **common ratio**. In general:

If the first term of a geometric sequence is a_1 and r is the common ratio, then the successive terms are obtained from the rule

$$a_{n+1} = a_n \cdot r.$$

The terms of a geometric sequence are said to be in geometric progression.

EXAMPLE 1   The first three terms of a geometric sequence are 96, 24, 6.
a. Find the common ratio.
b. Find the fourth term of this sequence.

306   *Chapter 8*

SOLUTION   **a.** To find the common ratio, divide any one term into its successor.

$$24 \div 96 = \frac{1}{4}$$

$$\text{or } 6 \div 24 = \frac{1}{4}$$

$\therefore$ the common ratio is $\frac{1}{4}$.

b. To find the fourth term, multiply the third term by the common ratio.

$$6 \times \frac{1}{4} = \frac{3}{2}$$

$\therefore$ the fourth term is $\frac{3}{2}$.

To discover a formula for the general term in any geometric sequence, let us continue the equations for the terms of the geometric sequence

$$2, 10, 50, 250, \ldots$$

in the following way.

$$a_2 = a_1 \cdot 5 = a_1 \cdot 5^1$$
$$a_3 = a_2 \cdot 5 = a_1 \cdot 5^2$$
$$a_4 = a_3 \cdot 5 = a_1 \cdot 5^3$$

You can see that the nth term ($n > 1$) of the sequence is given by the formula

$$a_n = a_1 \cdot 5^{n-1}.$$

Now consider a geometric sequence whose first term is a_1 and whose common ratio is r. The first few terms of the sequence are

$$a_1, a_1 r^1, a_1 r^2, a_1 r^3, a_1 r^4,$$

so that the nth term is given by

$$a_n = a_1 r^{n-1}, \qquad n > 1.$$

Notice that this formula applies, whatever value r may have. Of course, if the value of r is 0, all terms after the first are also 0.

The preceding formula also applies when $n = 1$ and $r \neq 0$, since we have defined (page 202) r^0 to be 1 whenever $r \neq 0$. Thus we have the following fact.

The nth term of a geometric sequence whose first term is a_1 and whose common ratio is a nonzero number r is

$$a_n = a_1 r^{n-1}.$$

Chalkboard Examples

1. The first three terms of a geometric sequence are 245, 35, 5. Find the common ratio and the fourth term of the sequence.

   $5 \div 35 = \frac{1}{7}$: the common ratio is $\frac{1}{7}$. $5 \cdot \frac{1}{7} = \frac{5}{7}$: the fourth term is $\frac{5}{7}$.

2. Find the seventh term of the geometric progression $-6, -24, -96, \ldots$.
   $a_1 = -6; r = -24 \div (-6) = 4$
   $a_7 = -6(4)^{7-1}$
   $= -6(4)^6$
   $= -6(4096)$
   $= -24{,}576$

3. Find the value after three years of an investment of \$1500, compounded semi-annually (twice a year) at a rate of 7%.

   $$A = P\left(1 + \frac{r}{n}\right)^{nt}$$
   $$= 1500\left(1 + \frac{0.07}{2}\right)^{2 \cdot 3}$$
   $$= 1500(1.035)^6$$
   $$= 1500(1.229)$$
   $$= \$1843.50 \text{ (to the nearest cent)}$$

Find the first four terms of the geometric sequence with the given first term and common ratio.

1. $a_1 = 4, r = 3$ 4, 12, 36, 108

2. $a_1 = -1, r = -2$ −1, 2, −4, 8

3. $a_1 = 10, r = \dfrac{1}{4}$ $10, 2\dfrac{1}{2}, \dfrac{5}{8},$

$\dfrac{5}{32}$, or 10, 2.5, 0.625, 0.15625

Find the specified term of the geometric sequence with the given first term and common ratio.

4. $a_1 = 25, r = \dfrac{1}{5}, a_6 = \underline{\ ?\ }$

$\dfrac{1}{125}$

5. $a_1 = -\dfrac{1}{2}, r = -4, a_6 = \underline{\ ?\ }$

512

Find the specified term of the geometric sequence.

6. $\dfrac{1}{2}, \dfrac{1}{4}, \dfrac{1}{8}, \ldots ; a_{10}$ $\dfrac{1}{1024}$

In finding the terms of a geometric progression, you may use the following table or a calculator.

Short Table of Powers

N	N^2	N^3	N^4	N^5
1	1	1	1	1
2	4	8	16	32
3	9	27	81	243
4	16	64	256	1,024
5	25	125	625	3,125
6	36	216	1,296	7,776
7	49	343	2,401	16,807
8	64	512	4,096	32,768
9	81	729	6,561	59,049
10	100	1,000	10,000	100,000
11	121	1,331	14,641	161,051
12	144	1,728	20,736	248,832
13	169	2,197	28,561	371,293
14	196	2,744	38,416	537,824
15	225	3,375	50,625	759,375
16	256	4,096	65,536	1,048,576
17	289	4,913	83,521	1,419,857
18	324	5,832	104,976	1,889,568
19	361	6,859	130,321	2,476,099
20	400	8,000	160,000	3,200,000

EXAMPLE 2   Find the sixth term of the geometric progression $-3, -15, -75, \ldots$.

SOLUTION
$$a_1 = -3; \ r = -15 \div (-3) = 5$$
$$a_n = a_1 r^{n-1}$$
$$a_6 = -3(5)^{6-1}$$
$$= -3(5)^5$$
$$= -3(3125)$$
$$= -9375$$

An important application of geometric sequences is the computation of *compound interest*. If P dollars are invested at an annual rate of 10%, at the end of the year the value of the investment is 10%, more than the value of the investment at the beginning of the year. After one year the value of the investment is $P + 0.10(P) = P(1.10)$. After two years the value of the investment is $P(1.10)(1.10) = P(1.10)^2$. Continuing this pattern, you can see that the values of the investment for successive years form a geometric sequence with $a_1 = P(1.10)$ and $r = 1.10$. After t years the value of the investment will be

$$a_1 r^{t-1} = [P(1.10)](1.10)^{t-1} = P(1.10)^t.$$

308   *Chapter 8*

In general, if interest on an investment is compounded n times per year at an annual rate r (expressed as a decimal), then the value A of the investment after t years is given by

$$A = P\left(1 + \frac{r}{n}\right)^{nt}.$$

EXAMPLE 3 Find the value after two years of an investment of $1250, compounded semiannually (twice a year) at an annual rate of 8%.

SOLUTION Use the formula for value of an investment. Substitute for P, r, n, and t.

$$A = P\left(1 + \frac{r}{n}\right)^{nt}$$
$$= 1250\left(1 + \frac{0.08}{2}\right)^{2 \cdot 2}$$
$$= 1250(1.04)^4$$
$$= \$1462.32 \qquad \text{(to the nearest cent)}$$

Oral Exercises

State whether each sequence is arithmetic, geometric, or neither. If it is geometric, state the common ratio.

1. 25, 5, 1, $\frac{1}{5}$ geometric; $\frac{1}{5}$ **2.** 1, 4, 9, 16 neither   **3.** -4, 0, 4, 8 arithmetic

4. 9, 3, -3, -9 arithmetic   **5.** 54, 18, 6, 2 geometric; $\frac{1}{3}$ **6.** -1, 3, -1, 3 neither

7. 7, -14, 28, -56 geometric; -2 **8.** 5, -5, 5, -5 geometric; -1 **9.** 0.3, 0.33, 0.333, 0.3333 neither

10. 26, 2.6, 0.26, 0.026 geometric; 0.1   **11.** $\frac{a}{b}$, $\frac{a}{b^2}$, $\frac{a}{b^3}$, $\frac{a}{b^4}$ geometric; $\frac{1}{b}$ **12.** $x + y$, x, $x - y$, $x - 2y$ arithmetic

Written Exercises

Find the first four terms of the geometric sequence with the given first term and common ratio.

A **1.** $a_1 = 5$, $r = 5$ 5, 25, 125, 625   **2.** $a_1 = 3$, $r = -3$ 3, -9, 27, -81 **3.** $a_1 = 40$, $r = \frac{1}{2}$ 40, 20, 10, 5

4. $a_1 = \frac{9}{25}$, $r = -\frac{5}{3}$ $\frac{9}{25}$, $-\frac{3}{5}$, 1, $-\frac{5}{3}$ **5.** $a_1 = \frac{c^2}{d}$, $r = \frac{2d}{c}$ $\frac{c^2}{d}$, $2c$, $4d$, $\frac{8d^2}{c}$ **6.** $a_1 = p$, $r = \frac{q}{p}$ p, q, $\frac{q^2}{p}$, $\frac{q^3}{p^2}$

Find the specified term of the geometric sequence with the given first term and common ratio.

7. $a_1 = 20$, $r = \frac{1}{4}$, $a_4 = \underline{\quad?\quad}$ $\frac{5}{16}$ **8.** $a_1 = -\frac{1}{9}$, $r = -3$, $a_5 = \underline{\quad?\quad}$ -9

9. $a_1 = 18$, $r = -\frac{1}{3}$, $a_5 = \underline{\quad?\quad}$ $\frac{2}{9}$ **10.** $a_1 = \frac{1}{3}$, $r = \frac{3}{2}$, $a_6 = \underline{\quad?\quad}$ $\frac{81}{32}$

11. $a_1 = \frac{b}{c^2}$, $r = \frac{c}{b}$, $a_6 = \underline{\quad?\quad}$ $\frac{c^3}{b^4}$ **12.** $a_1 = \frac{1}{4n^3}$, $r = 2n$, $a_7 = \underline{\quad?\quad}$ $16n^3$

Suggested Assignments

Minimum
  First day
    309/1–26 odd
  Second day
    310/P: 1–11
Extended Alg. with Trig.
  309/1–25 odd
  310/P: 1–11
Enriched Alg.
  309/5, 9, 15–31 odd
  310/P: 1–11
Enriched Alg. with Trig.
  309/5, 9, 15–31 odd
  311/P: 5–11

Find the specified term of each geometric sequence.

13. $\frac{8}{3}, \frac{4}{3}, \frac{2}{3}, \ldots$; $a_8 = $ __?__ $\frac{1}{48}$

14. $\frac{1}{20}, \frac{1}{10}, \frac{1}{5}, \ldots$; $a_7 = $ __?__ $\frac{16}{5}$

15. $250, 50, 10, \ldots$; $a_6 = $ __?__ $\frac{2}{25}$

16. $\frac{27}{2}, -\frac{9}{2}, \frac{3}{2}, \ldots$; $a_7 = $ __?__ $\frac{1}{54}$

17. $333\frac{1}{3}, 33\frac{1}{3}, 3\frac{1}{3}, \ldots$; $a_8 = $ __?__ $\frac{1}{30{,}000}$

18. $0.0008, 0.008, 0.08, \ldots$; $a_6 = $ __?__ $\frac{?}{80}$

B **19.** Find the first term of a geometric sequence whose fifth and sixth terms are $\frac{3}{4}$ and $\frac{3}{2}$, respectively. $\frac{3}{64}$

20. Find the second term of a geometric sequence whose fourth and fifth terms are $-\frac{5}{2}$ and $\frac{1}{2}$, respectively. $-\frac{125}{2}$

Determine whether each sequence is arithmetic or geometric and then find the specified term.

21. $-16, -4, 8, \ldots$; $a_7 = $ __?__ Arithmetic; 56

22. $-4\frac{1}{2}, -1\frac{1}{2}, -\frac{1}{2}, \ldots$; $a_6 = $ __?__ Geometric; $-\frac{1}{54}$

23. $222\frac{2}{9}, 22\frac{2}{9}, 2\frac{2}{9}, \ldots a_6 = $ __?__ Geometric; $\frac{1}{450}$

24. $\frac{3}{2}, \frac{9}{4}, 3, \ldots$; $a_7 = $ __?__ Arithmetic; 6

25. $-0.64, -0.16, 0.32, \ldots$; $a_8 = $ __?__ Arithmetic; 2.72

26. $a^7 b^6, a^6 b^4, a^5 b^2, \ldots$; $a_7 = $ __?__ Geometric; $\frac{a}{b^6}$

C **27.** Show that in any geometric sequence,

$$a_{n-1} a_{n+1} = (a_n)^2 \qquad \text{for any integer } n > 1.$$

28. Show that if a sequence of at least three terms is both arithmetic and geometric, then its terms must all be equal. (*Hint*: Assume that the sequence is arithmetic and then set two ratios of successive terms equal.)

If the sequence $a_1, a_2, a_3, \ldots, a_n$ is geometric with common ratio r, then each of the following sequences is also geometric. Give the common ratio of each in terms of r.

29. $a_n, a_{n-1}, a_{n-2}, \ldots, a_1 \quad \frac{1}{r}$

30. $2a_1, 2a_2, 2a_3, \ldots, 2a_n \quad r$

31. $a_1, a_3, a_5, \ldots, a_{2n+1} \quad r^2$

32. $a_1, 3a_2, 9a_3, \ldots, 3^{n-1}a_n \quad 3r$

Problems

A **1.** When a certain atom is struck by a neutron, it undergoes fission, producing three more neutrons in the first stage of a chain reaction. Suppose each of these neutrons in turn strikes another atom and produces three more neutrons in the second stage. How many neutrons would be produced in the fifth stage, if the reaction continues in this way? 243 neutrons

310 *Chapter 8*

2. Legend says that the inventor of the game of chess wanted to be rewarded with 1 grain of wheat for the first square of the chessboard, 2 grains for the second square, 4 grains for the third, and so on. Approximately how many grains of wheat would the inventor receive for the 64th square? (Use the approximation $2^{10} \approx 1000$.)
approximately 8×10^{18}

3. A population of insects being observed in an experiment grows by 20% every two weeks. In other words, at the end of two weeks there will be 1.2 times the original number of insects. If there are 5000 insects at the beginning of the experiment, how many will there be at the end of eight weeks? 10,368 insects

4. Each year the value of a certain car is 80% of what it was the previous year. If its value was $10,000 at the end of the first year, what was the value at the end of the fifth year? $4096

5. How many great-great-grandparents do six people have, assuming that they have no common ancestors in the previous four generations? 96 great-great-grandparents

6. During a special sale a certain camera was priced at $128 for the first week and each succeeding week the price was reduced by 25%. What was the price of the camera during the fourth week? What was the price of the camera during the fifth week? $54; $40.50

B 7. The *half-life* of nitrogen-13, an isotope of the gas nitrogen, is about 10 min; that is, at the end of any 10 min period, half of the amount present at the beginning of the period will remain. During a laboratory experiment, 16 mg of nitrogen-13 is observed to be present at 3:10 P.M. How much will be present at 4:20 P.M.? 0.125 mg

8. A side of an equilateral triangle is 96 cm long. A second equilateral triangle is inscribed in it by joining the midpoints of the sides of the first triangle. The process is continued, as suggested by the diagram. Find the perimeter of the seventh *inscribed* equilateral triangle. 2.25 cm

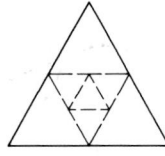

9. Find the value of $8000 invested for one year at an annual rate of 20%, if interest is compounded:
   a. annually $9600   **b.** semiannually $9680   **c.** quarterly $9724.05

10. The interest rate in a certain bank is 8%, compounded semiannually. If $5000 is deposited in an account and no other deposits or withdrawals are made, how much money will be in the account at the end of $1\frac{1}{2}$ years? $5624.32

11. Find the amount of compound interest at the end of five years if $2000 is invested at 10%, compounded annually. $1221.02

Simplify.

1. $3x(x + y)^2$ $3x^3 + 6x^2y + 3xy^2$

2. $\dfrac{k - 4}{k + 4} \div \dfrac{k}{k^2 - 16}$
$\dfrac{k^2 - 8k + 16}{k}$

3. $(2w^2x^3y^2) \cdot (x^2y)$ $2w^2x^5y^3$

4. $\dfrac{16}{p^2 + 4p} + \dfrac{2p + 8}{p}$
$\dfrac{2p^2 + 16p + 48}{p^2 + 4p}$

5. Solve the open sentence.
$|2t - 5| < 1$
$\{t: 2 < t < 3\}$

6. Write the equation $3x + 4y = 8$ in slope-intercept form. $y = -\dfrac{3}{4}x + 2$

Chalkboard Examples

1. Insert two geometric means between 9 and 576.
   $576 = 9 \cdot r^{4-1} = 9 \cdot r^3$
   $64 = r^3$
   $r = 4$
   $9 \times 4 = 36$; $36 \times 4 = 144$
   ∴ the required means are 36 and 144.

2. Find the second term of a geometric sequence whose third term is 243 and whose seventh term is 3.
   Since the third term is 243,
   $243 = a_1 r^{n-1} = a_1 r^2$.
   Since the seventh term is 3,
   $3 = a_1 r^{7-1} = a_1 r^6$.
   From the first equation,
   $$a_1 = \frac{243}{r^2}.$$
   Substitute in the second equation.
   $$r = \pm\frac{1}{3}$$
   When $r = \frac{1}{3}$, the sequence is 2187, 729, 243, 81, 27, 9, 3.
   When $r = -\frac{1}{3}$, the sequence is 2187, −729, 243, −81, 27, −9, 3.
   ∴ the second term is either 729 or −729.

3. Find the geometric mean of 8 and 392.
   $m^2 = 8 \cdot 392$
   $m^2 = 3136$
   $m = 56$ or $m = -56$
   ∴ the geometric mean of 8 and 392 is 56.

8–5 Geometric Means

The terms between two given terms of a geometric sequence are called *geometric means* between the given terms. In the sequence

$$-2, \; 6, \; -18, \; 54, \; -162,$$

6, −18, and 54 are the three geometric means between −2 and −162.

EXAMPLE 1 Insert two geometric means between 7 and 189.

SOLUTION There are four terms in the sequence represented schematically by 7, $\underline{\;?\;}$, $\underline{\;?\;}$, 189. To determine r, use the formula $a_n = a_1 r^{n-1}$ and substitute 7 for a_1, 189 for a_n, and 4 for n.

$$189 = 7 \cdot r^{4-1}$$
$$27 = r^3$$
$$r = 3$$

To complete the sequence, multiply 7 by 3 and then multiply the result by 3.

$$7, \; 21, \; 63, \; 189$$

∴ the required means are 21 and 63.

EXAMPLE 2 Find the second term of a geometric sequence whose third term is 48 and whose seventh term is 3. Give all possible correct answers.

SOLUTION Use the formula $a_n = a_1 r^{n-1}$.
Since the third term is 48,

$$48 = a_1 r^{3-1} = a_1 r^2.$$

Since the seventh term is 3,

$$3 = a_1 r^{7-1} = a_1 r^6.$$

Solve the first equation for a_1.

$$a_1 = \frac{48}{r^2}$$

Substitute in the second equation.

$$3 = \left(\frac{48}{r^2}\right) r^6$$
$$3 = 48 r^4$$
$$r^4 = \frac{3}{48} = \frac{1}{16}$$
$$r = \pm\frac{1}{2}$$

When $r = \frac{1}{2}$, the sequence is $192, 96, 48, 24, 12, 6, 3$.

When $r = -\frac{1}{2}$, the sequence is $192, -96, 48, -24, 12, -6, 3$.

$\therefore$ the second term is either 96 or -96.

A single geometric mean inserted between two numbers is called *the geometric mean* or the **mean proportional** of the numbers. If the real number m is the geometric mean of two nonzero real numbers a and b, then a, m, b is a geometric sequence and

$$\frac{a}{m} = \frac{m}{b}.$$
$$\therefore m^2 = ab.$$

Since m^2 is positive, a and b must be both positive or both negative. If both a and b are positive, then $\sqrt{ab}$ is generally considered to be the geometric mean of the numbers. If a and b are negative, then $-\sqrt{ab}$ is generally considered to be the geometric mean of the numbers.

EXAMPLE 3 Find the geometric mean of 5 and 80.

SOLUTION The required mean satisfies the equation $m^2 = 5 \cdot 80$.

$$m^2 = 400$$
$$m = 20 \quad \text{or} \quad m = -20$$

$\therefore$ the geometric mean of 5 and 80 is 20.

Oral Exercises

State the geometric mean of each pair of numbers.

1. $9, 4$ 6
2. $1, 25$ 5
3. $-2, -50$ -10
4. $3, 75$ 15
5. $-\frac{1}{2}, -128$ -8
6. $\frac{2}{3}, 48$ $4\sqrt{2}$
7. $c, c^3; c > 0$ c^2
8. $a, ab^2; a, b > 0$ ab

State all real roots of each equation.

9. $x^3 = 64$ 4
10. $x^4 = 625$ 5, -5
11. $x^2 = 49$ 7, -7
12. $x^5 = -243$ -3
13. $x^4 = -10,000$ No real roots
14. $x^3 = -216$ -6
15. $x^2 = 121$ 11, -11
16. $x^6 = 64$ 2, -2

In Exercises 17–19 classify each statement as true or false.

17. If all the terms of a geometric sequence are negative, then the common ratio is negative. False
18. If $a_1 > 0$ and $a_4 < 0$ in a geometric sequence, then $a_2 < 0$. True
19. If $ab < 0$, then a and b have no geometric mean. True
20. Name two unequal integers whose geometric mean is 12. Ans. may vary.
    Examples are 2, 72; 4, 36

Sequences and Series **313**

Additional A Exercises

Find the common ratio of a geometric sequence having the given terms.

1. $a_1 = 3$, $a_4 = 1029$ $r = 7$
2. $a_1 = 2$, $a_4 = 1458$ $r = 9$

Give the first two terms of a geometric sequence having the given terms.

3. $a_3 = 24$, $a_6 = 192$ 6, 12
4. $a_3 = 2.5$, $a_5 = 0.625$ 10, 5 or 10, -5

Mixed Review

Simplify.

1. $-[2a + 3b - (-8a + 3b)]$ $-10a$
2. $\dfrac{j + 7}{(j^2 - 49)} \cdot \dfrac{j}{(j + 7)} \div \dfrac{7}{(j - 7)}$
   $\dfrac{j}{7(j + 7)}$
3. $(d^3)(d^2e^3)(d^2e^3f^4)$ $d^7e^6f^4$
4. $\dfrac{4x^2y^2z^3}{12x^5y^2z^7} \cdot 3z$ $\dfrac{1}{x^3z^3}$
5. $\dfrac{w}{v^2w} + \dfrac{v}{vw} + \dfrac{5v}{2w}$
   $\dfrac{2w + 2v^2 + 5v^3}{2v^2w}$
6. If $4h^2 - 100 = 0$, then $h = \underline{\ ?\ }$ 5 or -5

313

Suggested Assignments

Minimum
  314/1–30
Extended Alg. with Trig.
  314/1–29 odd
Enriched Alg.
  314/5, 11, 17–31 odd, 32–36
Enriched Alg. with Trig.
  314/5, 11, 17–31 odd, 32–36

Additional Answers
Written Exercises

16. $\frac{3}{2}$, -6, 24, -96

17. $\frac{7}{4}, \frac{7}{2}$, 7, 14, 28; $\frac{7}{4}$, $-\frac{7}{2}$, 7, -14, 28

18. $-\frac{25}{3}$, -5, -3, $-\frac{9}{5}$, $-\frac{27}{25}$; $-\frac{25}{3}$, 5, -3, $\frac{9}{5}$, $-\frac{27}{25}$

20. -125, -50, -20, -8, $-\frac{16}{5}$; -125, 50, -20, 8, $-\frac{16}{5}$

31. The common ratio of the sequence $\{(a_1 r^n)^2\}$ or a_1^2, $a_1^2 r^2$, $a_1^2 r^4$, $a_1^2 r^6$, ... is r^2 so the sequence is geometric.

32. The common ratio of the sequence $\left\{\frac{1}{a_1 r^n}\right\}$ or $\frac{1}{a_1}$, $\frac{1}{a_1 r}$, $\frac{1}{a_1 r^2}$, $\frac{1}{a_1 r^3}$, ... is $\frac{1}{r}$ so the sequence is geometric.

33. Since $\frac{t^b}{t^a} = \frac{t^c}{t^b} = \frac{t^d}{t^c} = \cdots$, $b - a = c - b = d - c = \cdots$. Then $2b = a + c$, $2c = b + d$, ... and $b = \frac{a + c}{2}$, $c = \frac{b + d}{2}$, ...

Written Exercises

Find the common ratio of a geometric sequence having the given terms.

A   **1.** $a_1 = 4$, $a_4 = 108$ 3   **2.** $a_1 = -2$, $a_4 = 250$ -5

3. $a_1 = 72$, $a_4 = -9$ $-\frac{1}{2}$ **4.** $a_1 = 27$, $a_4 = 8$ $\frac{2}{3}$

5. $a_3 = 7$, $a_5 = 112$ 4 or -4 **6.** $a_2 = -162$, $a_6 = -2\frac{1}{3}$ or $-\frac{1}{3}$

Give the first two terms of a geometric sequence having the given terms.

7. $a_3 = 15$, $a_6 = 120$ $\frac{15}{4}, \frac{15}{2}$ **8.** $a_3 = -2$, $a_6 = 54$ $-\frac{2}{9}, \frac{2}{3}$ **9.** $a_4 = -10$, $a_7 = -80$ $-\frac{5}{4}, -\frac{5}{2}$

10. $a_3 = 6$, $a_6 = \frac{3}{4}$ 24, 12 **11.** $a_5 = 20$, $a_8 = \frac{4}{25}$ $12{,}500$; 2500 **12.** $a_3 = -18$, $a_6 = \frac{16}{3}$ $-\frac{81}{2}, 27$

Insert the stated number of geometric means between the given numbers and write the resulting geometric sequence. Give all possible correct answers.

13. two between 4 and 500   4, 20, 100, 500   **14.** two between -22 and -176 $-22, -44, -88, -176$

15. two between -5 and 135 -5, 15, -45, 135   **16.** two between $\frac{3}{2}$ and -96

17. three between $\frac{7}{4}$ and 28   **18.** three between $-\frac{25}{3}$ and $-\frac{27}{25}$

19. three between 4 and $\frac{81}{4}$ 4, 6, 9, $\frac{27}{2}, \frac{81}{4}$; 4, -6, 9, $-\frac{27}{2}, \frac{81}{4}$ **20.** three between -125 and $-\frac{16}{5}$

Find the specified value using the data given about a geometric sequence.

B   **21.** $a_1 = 2$, $a_2 = 5$, $a_n = \frac{625}{8}$, $n = \underline{\ ?\ }$ 5   **22.** $a_1 = \frac{81}{4}$, $a_2 = \frac{27}{2}$, $a_n = 4$, $n = \underline{\ ?\ }$ 5   2 or -2

23. $a_5 = \frac{1}{2}$, $a_8 = -\frac{1}{128}$, $a_2 = \underline{\ ?\ }$ -32 **24.** $a_6 = 250$, $a_8 = 6250$, $a_3 = \underline{\ ?\ }$

25. $a_4 = \frac{x^3}{y^2}$, $a_7 = \frac{y}{x^3}$, $a_1 = \underline{\ ?\ }$ $\frac{x^9}{y^5}$ **26.** $a_5 = \frac{k^2}{n^4}$, $a_8 = \frac{k^5}{n^{10}}$, $a_1 = \underline{\ ?\ }$ $\frac{n^4}{k^2}$

27. In a laboratory experiment there were 4800 microorganisms present in a culture at the end of one hour. At the end of three hours there were 7500 microorganisms. If the population increased geometrically, by what factor did it increase each hour? $\frac{5}{4}$

28. According to Newton's Law of Cooling, when the *difference* in the temperature of a warm object and its cooler surroundings is measured at equal time intervals, the differences form a geometric sequence. If the temperature of a piece of toast was 26°C at 8:00 A.M. and 22°C at 8:10 A.M., what was its temperature when the toast was taken out of the toaster at 7:40 A.M.? Assume that the air temperature remained constant at 20°C.   74°C

29. Find all possible values of x that make the sequence $x - 1$, $2x$, $5x + 3$ a geometric progression.   3 and -1

30. Find all possible values of x that make $x + 2$ the geometric mean of $x - 2$ and $3x - 2$.   6 and 0

314   *Chapter 8*

For Exercises 31 and 32 let $a_1, a_1r, a_1r^2, a_1r^3, \ldots$ represent a geometric sequence.

C   31. Show that the sequence formed by squaring each term of the sequence above is also a geometric sequence by finding the common ratio.

32. Assuming that $a_1r \neq 0$, show that the sequence formed by taking the reciprocal of each term is also a geometric sequence by finding the common ratio.

33. Prove that for $t \neq 0, t \neq 1, t \neq -1$, if the sequence $t^a, t^b, t^c, t^d, \ldots$ is geometric, then the sequence $a, b, c, d, \ldots$ is an arithmetic sequence.

34. Show that if $a_1, a_2, a_3, \ldots$ and $b_1, b_2, b_3, \ldots$ are geometric sequences, then the sequence $a_1b_1, a_2b_2, a_3b_3, \ldots$ is also a geometric sequence.

35. Let $a, b \geq 0$. Show that the arithmetic mean of a^2 and b^2 is greater than or equal to the geometric mean of a^2 and b^2. (*Hint:* Start with the fact that $(a - b)^2 \geq 0$ for any real numbers a and b.)

36. Show that for any geometric sequence if $r > 0$, then a_n is the geometric mean of a_1 and a_{2n-1}, for any positive integer n.

Leonhard Euler
1707–1783

The son of a Swiss clergyman, Leonhard Euler tried to follow in his father's footsteps, but the lure of mathematics and science was too strong to resist.

Throughout his long life Euler was immensely productive. The total number of books and articles attributed to him is staggering. He contributed to the founding of several areas of mathematics, among them the calculus of variations, the theory of differential equations, and the theory of special functions. Many mathematical symbols originated with Euler, including i to represent $\sqrt{-1}$, Σ to indicate summation, and $f(x)$ to denote the function of x.

Euler was always concerned with the practical application of mathematical principles. He used them to solve problems in such diverse fields as optics, astronomy, hydraulics, mapmaking, navigation, shipbuilding, insurance, and population studies.

Sequences and Series   **315**

Thus, b is the arithmetic mean of a and c, c is the arithmetic mean of b and d, and so on. $\therefore a, b, c, d, \ldots$ is an arithmetic sequence.

34. Let r_1 = the common ratio of $a_1, a_2, a_3, \ldots$ and r_2 = the common ratio of $b_1, b_2, b_3, \ldots$. So $r_1 = \dfrac{a_2}{a_1} = \dfrac{a_3}{a_2} = \cdots$ and $r_2 = \dfrac{b_2}{b_1} = \dfrac{b_3}{b_2} = \cdots$. Then

$$r_1r_2 = \frac{a_2b_2}{a_1b_1} = \frac{a_3b_3}{a_2b_2} = \cdots.$$

Thus r_1r_2 is the common ratio of $a_1b_1, a_2b_2, a_3b_3, \ldots$, and $a_1b_1, a_2b_2, a_3b_3, \ldots$ is a geometric sequence.

35. Since the square of any real number is nonnegative, $(a - b)^2 \geq 0$. $a^2 - 2ab + b^2 \geq 0$ means that $a^2 + b^2 \geq 2ab$ or $\dfrac{a^2 + b^2}{2} \geq ab$. Now, $\dfrac{a^2 + b^2}{2}$ is the arithmetic mean of a^2 and b^2 and ab is the geometric mean of a^2 and $b^2(a, b \geq 0)$. $\therefore$ the arithmetic mean of a^2 and b^2 is greater than or equal to the geometric mean of a^2 and b^2.

36. For $n > 0$, $a_n = a_1r^{n-1}$ and $a_{2n-1} = a_1r^{2n-2}$. Hence, $\dfrac{a_{2n-1}}{a_n} = \dfrac{a_1r^{2n-2}}{a_1r^{n-1}} = r^{n-1}$. Also $\dfrac{a_n}{a_1} = r^{n-1}$. Thus $\dfrac{a_{2n-1}}{a_n} = \dfrac{a_n}{a_1}$. $\therefore a_n$ is the geometric mean of a_1 and a_{2n-1}.

8–6 Geometric Series

A series whose terms are in geometric progression, such as

$$3 + 6 + 12 + 24,$$

is called a **geometric series**. To find a formula for the sum S_n of a geometric series of n terms, first write S_n in expanded form and below it write the product of $-r$ and S_n. Then add the corresponding sides.

$$S_n = a_1 + a_1 r + a_1 r^2 + \cdots + a_1 r^{n-2} + a_1 r^{n-1}$$
$$-rS_n = \qquad\; - a_1 r - a_1 r^2 - a_1 r^3 - \cdots - a_1 r^{n-1} - a_1 r^n$$
$$\overline{S_n - rS_n = a_1 + \; 0 \; + \; 0 \; + \; 0 \; + \cdots + \quad 0 \quad - a_1 r^n}$$
$$S_n - rS_n = a_1 - a_1 r^n$$
$$(1 - r)S_n = a_1 - a_1 r^n$$

$$\therefore S_n = \frac{a_1 - a_1 r^n}{1 - r}, \qquad r \neq 1.$$

EXAMPLE 1   Find the sum of the first six terms of the geometric sequence

$$7, -14, 28, \ldots .$$

SOLUTION   Replace a_1 with 7, r with -2, and n with 6 in $S_n = \frac{a_1 - a_1 r^n}{1 - r}$:

$$S_6 = \frac{7 - 7(-2)^6}{1 - (-2)} = \frac{7 - 7(64)}{3} = \frac{-441}{3}$$
$$= -147$$

$\therefore$ the sum is -147.

Note that $a_1 r^n = r(a_1 r^{n-1}) = r a_n$. Hence, from $S_n = \frac{a_1 - a_1 r^n}{1 - r}$,

$$S_n = \frac{a_1 - r a_n}{1 - r}, \qquad r \neq 1.$$

EXAMPLE 2   Find the sum of a geometric series whose first term is 3, whose last (nth) term is 9375, and whose common ratio is 5.

SOLUTION   Replace a_1 with 3, a_n with 9375 and r with 5 in $S_n = \frac{a_1 - r a_n}{1 - r}$.

$$S_n = \frac{3 - 5(9375)}{1 - (5)} = \frac{3 - 46875}{-4} = 11{,}718$$

$\therefore$ the sum is $11{,}718$.

The theorem on the next page summarizes the results obtained in this section.

8. $a_1 = 3^1, r = 3, n = 5;$

$$S_n = \frac{a_1 - a_1 r^n}{1 - r}$$

9. $a_1 = 5, r = 2, a_n = 640;$

$$S_n = \frac{a_1 - ra_n}{1 - r}$$

10. $a_1 = 405, r = -\frac{1}{3}, n = 5;$

$$S_n = \frac{a_1 - a_1 r^n}{1 - r}$$

For Exercises 11–18, use

$$S_n = \frac{a_1 - a_1 r^n}{1 - r}.$$

11. $a_1 = 2, r = 3, n = 6$

12. $a_1 = \frac{1}{2}, r = 6, n = 5$

13. $a_1 = -4, r = \frac{1}{2}, n = 7$

14. $a_1 = \frac{1}{6}, r = -2, n = 8$

15. $a_1 = \frac{2}{9}, r = 3, n = 5$

16. $a_1 = 100, r = \frac{1}{5}, n = 5$

17. $a_1 = 27, r = \frac{2}{3}, n = 6$

18. $a_1 = 8, r = \frac{3}{4}, n = 5$

19. S_n becomes closer to

$$\frac{1}{1 - r}.$$

Theorem. If S_n is the sum of the first n terms of a geometric sequence whose first term is a_1, whose common ratio is r, and whose nth term is a_n, then

$$S_n = \frac{a_1 - a_1 r^n}{1 - r} \quad \text{and} \quad S_n = \frac{a_1 - ra_n}{1 - r}, \quad r \neq 1.$$

Oral Exercises

State whether each series is arithmetic or geometric.

1. $\displaystyle\sum_{n=1}^{7} 4n$ Arithmetic **2.** $\displaystyle\sum_{n=1}^{6} (3^n \cdot 5)$ Geometric **3.** $\displaystyle\sum_{n=1}^{7} \frac{8}{2^{n+1}}$ Geometric **4.** $\displaystyle\sum_{n=1}^{14} 5(3n + 2)$ Arithmetic

5. the sum of the powers of 2 between 8 and 128, inclusive Geometric
6. the sum of the multiples of 3 between 6 and 27, inclusive Arithmetic

For each geometric series state the values of whichever of the constants a_1, r, n, and a_n can be determined by inspection. State which of the formulas above you would use to find the sum of the series.

7. the sum of all the integral powers of 4 between 1 and 256, inclusive
8. the sum of the first 5 integral powers of 3, beginning with 3^1
9. $5 + 10 + \cdots + 640$ **10.** the first 5 terms of $405 - 135 + 45 - \cdots$

11. $\displaystyle\sum_{n=1}^{6} 2(3)^{n-1}$ **12.** $\displaystyle\sum_{k=1}^{5} \frac{1}{2}(6)^{k-1}$ **13.** $\displaystyle\sum_{i=1}^{7} -4(\tfrac{1}{2})^{i-1}$ **14.** $\displaystyle\sum_{j=1}^{8} \frac{1}{6}(-2)^{j-1}$

15. $\displaystyle\sum_{m=1}^{5} \frac{2}{9}(3)^{m-1}$ **16.** $\displaystyle\sum_{n=1}^{5} 100(\tfrac{1}{5})^{n-1}$ **17.** $\displaystyle\sum_{i=1}^{6} 27(\tfrac{2}{3})^{i-1}$ **18.** $\displaystyle\sum_{n=1}^{5} 8\left(\frac{3^{n-1}}{2^{2n-2}}\right)$

19. In a geometric series, if $a_1 = 1$ and $0 < r < 1$, describe (by examining the first formula on page 316) what happens to S_n as more and more terms are added, that is, as n becomes greater and greater.

Written Exercises

Find the sum S_n of each geometric series described.

A **1.** $a_1 = \frac{3}{8}, r = 2, n = 7$ $47\frac{5}{8}$ **2.** $a_1 = 1, r = -3, n = 6$ -182
3. $a_1 = 324, r = \frac{1}{3}, n = 6$ $485\frac{1}{3}$ **4.** $a_1 = 128, r = \frac{3}{4}, n = 5$ $390\frac{1}{2}$
5. $-64 + 32 - 16 + \cdots - 1$ -43 **6.** $100 + 20 + 4 + \cdots + \frac{4}{125}$ $124\frac{124}{125}$

7–18. Find the sums of the geometric series in Oral Exercises 7–18.

Sequences and Series **317**

Suggested Assignments

Minimum
First day   317/1–28
Second day 318/P: 1–7
              R 320/*Self-Test 2*
Extended Alg. with Trig.
First day   317/5–27 odd
Second day 318/P: 1–7
              R 320/*Self-Test 2*
Enriched Alg.
  317/17–27 odd
  318/P: 1–7
R 320/*Self-Test 2*
Enriched Alg. with Trig.
  317/15–33 odd
  319/P: 4–8
R 320/*Self-Test 2*

In Exercises 19–28 use the given data about a geometric series to find the specified value.

B   19. $r = 2$, $S_6 = 393$, $a_1 = $ ___?___ $\frac{131}{21}$

20. $r = -\frac{1}{4}$, $S_5 = 410$, $a_1 = $ ___?___ 512

21. $r = -\frac{1}{3}$, $a_n = \frac{5}{3}$, $S_n = \frac{305}{3}$, $a_1 = $ ___?___ 135

22. $a_1 = 4$, $r = 3$, $S_n = 484$, $n = $ ___?___ 5

23. $a_1 = 32$, $r = \frac{5}{2}$, $S_n = 812$, $n = $ ___?___ 4

24. $r = \frac{1}{2}$, $a_n = \frac{3}{2}$, $S_n = 94\frac{1}{2}$, $n = $ ___?___ 6

25. $r = 0.25$, $S_4 = 148.75$, $a_4 = $ ___?___ 1.75

26. $r = -0.5$, $S_6 = 126$, $a_6 = $ ___?___ −6

27. $a_1 = 10$, $S_3 = \frac{95}{2}$, $r = $ ___?___ $\frac{3}{2}$ or $-\frac{5}{2}$

28. $a_1 = 6$, $S_3 = \frac{38}{3}$, $r = $ ___?___ $\frac{2}{3}$ or $-\frac{5}{3}$

(*Hint*: For Exercises 27 and 28, use the fact that $\frac{1 - r^3}{1 - r} = 1 + r + r^2$.)

C   29. For a certain geometric series in which $a_1 = 2$ and r is the common ratio, the sum of the first three terms is $2r^2 - r + 3$. Find the value of r. $\frac{1}{3}$

30. The sum of the first two terms of a geometric series is 90. The sum of the sixth and seventh terms is $-\frac{10}{27}$. Find the sum of the first seven terms. $101\frac{8}{27}$

31. Prove: $\sum\limits_{k=1}^{n} (2^{-k}) = 1 - \frac{1}{2^n}$

32. Prove the first formula of the theorem on page 317 by writing

$$S_n = \frac{a_1 - a_1 r^n}{1 - r} = \frac{a_1(1 - r^n)}{1 - r} = a_1\left(\frac{r^n - 1}{r - 1}\right)$$

and using long division.

33. Show that if $a_1 + a_1 r + a_1 r^2 + \cdots + a_1 r^{n-1} = S_n$,

then $a_1 + \frac{a_1}{r} + \frac{a_1}{r^2} + \cdots + \frac{a_1}{r^{n-1}} = \frac{1}{r^{n-1}} \cdot S_n$ $(r \neq 1, 0)$.

Problems

A   1. How many ancestors from parents to great-great-great-great grandparents do you have? (Assume that there is no duplication.) 126

2. A basketball player's contract guarantees a $100,000 salary for each of four successive years. If inflation reduces the buying power of each year's salary to $\frac{9}{10}$ of what it was the preceding year, what will be the total buying power of the four years' income in terms of the first year's prices? $343,900

3. Each coil of a tapering spring is to be 0.8 times the length of the preceding one. If the wire out of which the spring is to be formed is 21.01 cm long and there are to be five coils, how long should the first coil be? 6.25 cm

4. At the beginning of the first year of observation the fish population of a lake is found to be 200. At the beginning of each of the next three years the lake is stocked with 200 additional fish. At the end of each of the four years the fish population had grown to 1.1 times the population at the beginning of the year. What will the fish population be at the end of four years? (*Hint*: The original fish population will grow to $200(1.1)^4$ by the end of the four years. Form a geometric series of such terms.) about 1021

B 5. Every minute, a certain vacuum pump removes $\frac{1}{4}$ of the air remaining in a closed container.
   a. Show that the amounts of air remaining in the container after each minute form a geometric *sequence*, and find the fraction of the original amount remaining in the container after 5 min.
   b. The total amount of air removed from the container after any given number of minutes is the sum of a geometric *series*. Find a_1 and r for this series and use these values to find the fraction of the original amount of air that is removed in 5 min. $a_1 = \frac{1}{4}x, r = \frac{3}{4}, \frac{781}{1024}$

6. The winner of a contest will receive $1280 the first year, with a 25% increase over the preceding year's payment for each subsequent year until a total of $7380 is paid. How many years will it take the winner to collect the total prize? 4 yr

7. When a certain circuit is closed, the charge (measured in coulombs, C) on a capacitor changes from 0 C to $\frac{1}{3}$ C in the first second. The increase in the charge for each succeeding second is $\frac{2}{3}$ of the increase for the preceding second. In how many seconds will the capacitor have a charge of at least $\frac{4}{5}$ C? 4 s

C 8. An *amortized loan*, such as a mortage, is a loan that is paid off in equal monthly installments. Each monthly installment consists of a partial repayment of the principal (money lent) plus the interest on the unpaid balance. Each month the amount repaid on the principal increases and the amount of interest decreases. It can be shown that the amount repaid on the principal for successive months is given by the sequence

$$\frac{m}{(1 + r)^k}, \frac{m}{(1 + r)^{k-1}}, \frac{m}{(1 + r)^{k-2}}, \ldots, \frac{m}{1 + r},$$

where m is the entire monthly installment, r is the monthly interest rate, and k is the number of monthly installments. Show that if P is the total principal, then the monthly installment is:

$$m = \frac{rP}{1 - (1 + r)^{-k}}$$

Sequences and Series **319**

Mixed Review
Simplify.
1. $x(x - y) - (x + y)$
   $x^2 - xy - x - y$
2. $\frac{9}{k + 3} \div \frac{k}{k^2 + 6k + 9}$
   $\frac{9(k + 3)}{k}$
3. $(h^2 j^3 k^4)^2(k)$ $h^4 j^6 k^9$
4. $\frac{300q^3 r^4}{500q^2 r^8} \cdot \frac{s}{t}$ $\frac{3qs}{5r^4 t}$
5. $\frac{6}{r - 2} - \frac{10r + 12}{r^2 - 4}$ $\frac{-4r}{r^2 - 4}$

319

Find the specified term of each geometric sequence.

1. $a_1 = 3$, $r = 4$, $a_5 =$ _?_ 768

2. $a_3 = 252$, $a_5 = 9072$, $a_1 =$ _?_ 7

Insert the stated number of geometric means between the given numbers. Give all possible correct answers.

3. one between 12 and 588. 84

4. three between 5 and 3125. 25, 125, 625 or -25, 125, -625

Find the sum of each geometric series.

5. $a_1 = 6$, $r = 2$, $n = 7$ 762

6. $1 - 3 + 9 - \ldots + 729$ 547

7. The number of rabbits in a rabbit family increases each year by 50%. If there are 1000 rabbits in the family this year, how many rabbits will there be three years from now? 3375 rabbits

8. As a result of improving her sales technique, a salesperson who worked on commission was able to double her earnings each week. If she earned $75 the first week, what were her total earnings after 8 weeks? $19,125

Computer Exercises   For students with computer experience

1. Write a program to find the sum of a geometric series, given the first term a_1, the common ratio r, and the number of terms n, using the formula $a_{n+1} = a_n \cdot r$.

Run the program in Exercise 1 for a series with the given constants.

2. $a_1 = 0.3$, $r = 0.1$, $n = 10$ .333333333

3. $a_1 = 0.75$, $r = 2$, $n = 12$ 3071.25

4. $a_1 = 50$, $r = -0.5$, $n = 10$ 33.3007813

5. $a_1 = -625$, $r = -0.2$, $n = 9$ -520.8336

Self-Test 2

VOCABULARY   geometric sequence (p. 306)
geometric progression (p. 306)
common ratio (p. 306)
geometric means (p. 312)
mean proportional (p. 313)
geometric series (p. 316)

Find the specified term of each geometric sequence.

1. $a_1 = 6$, $r = 2$, $a_{10} =$ _?_ 3072   **2.** $a_4 = -9$, $a_7 = 243$, $a_1 = \dfrac{1}{3}$ _?_   *Obj. 1, p. 306*

Insert the stated number of geometric means between the given numbers. Give all possible correct answers.

3. one between 100 and 10,000 1000 or -1000

4. four between 15 and 480 30, 60, 120, 240

Obj. 2, p. 306

Find the sum of each geometric series.

5. $a_1 = 20$, $r = -4$, $n = 5$ 4100   **6.** $1 - \dfrac{1}{3} + \dfrac{1}{9} - \ldots - \dfrac{1}{243}$ $\dfrac{182}{243}$ *Obj. 3, p. 306*

7. The population of a certain town increases by 10% each year. What will be the population of the town four years from now if it is now 20,000? 29,282   *Obj. 4, p. 306*

8. If you worked at a certain job that paid wages of $1 the first day, $2 the second day, $4 the third day, with your earnings doubling each successive day, what would your total earnings be after working fourteen days? $16,383

Check your answers with those at the back of the book.

Infinite Sequences and Series

OBJECTIVES for Section 8-7 and Section 8-8:
1. *To find the absolute value of the difference between the limit of a convergent sequence and a term in the sequence.*
2. *To find the sum of a convergent geometric series.*

8–7 Limit of a Sequence

Figure 1 pictures the first few terms of the infinite sequence

$$1, 1\tfrac{1}{2}, 1\tfrac{3}{4}, 1\tfrac{7}{8}, \ldots , 2 - (\tfrac{1}{2})^{n-1}, \ldots$$

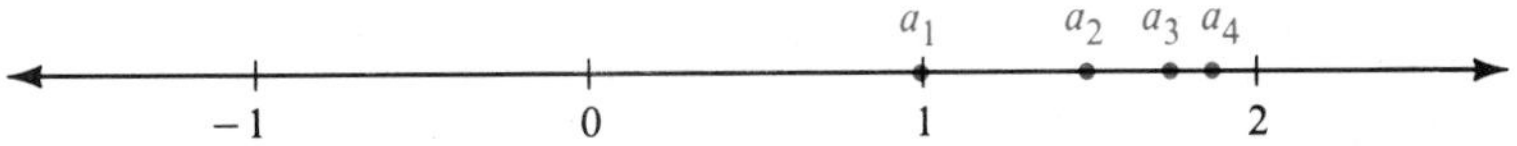

Figure 1

This figure suggests that the graphs of the terms of the given sequence eventually crowd in on the graph of 2. Notice that if you think of each term of the sequence as an approximation of 2, then the error of approximation, that is, the absolute value of the error that you make in considering a_n to be 2, is $(\tfrac{1}{2})^{n-1}$. As shown in the table at the right, this error is halved each time n is increased by 1. Therefore, by choosing n great enough, you can make the error less than any given positive number, however small. For this reason, we say that the *limit* of the sequence is 2, and we write

n	a_n	$\lvert 2 - a_n \rvert$
1	1	1
2	$1\tfrac{1}{2}$	$\tfrac{1}{2}$
3	$1\tfrac{3}{4}$	$\tfrac{1}{4}$
4	$1\tfrac{7}{8}$	$\tfrac{1}{8}$

$$\lim_{n\to\infty} [2 - (\tfrac{1}{2})^{n-1}] = 2,$$

read, "the limit of $2 - (\tfrac{1}{2})^{n-1}$ as n increases without bound is 2." An infinite sequence having a limit is said to **converge** or be **convergent**.

EXAMPLE Find the limit of the sequence $4, 3\tfrac{1}{2}, 3\tfrac{1}{3}, 3\tfrac{1}{4}, \ldots , 3 + \tfrac{1}{n}, \ldots$

SOLUTION Show the first few terms of the sequence on a number line.

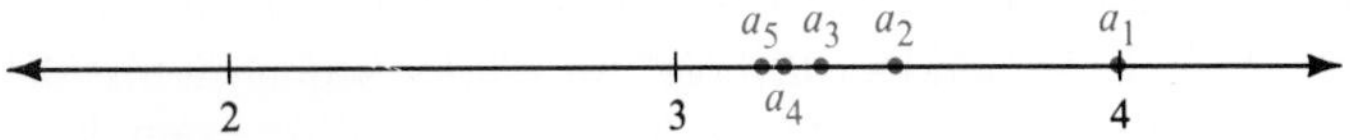

The diagram suggests that the limit is 3; in fact, the error made in considering the nth term a_n to be 3 is $\lvert a_n - 3 \rvert$, or $\tfrac{1}{n}$, and $\tfrac{1}{n}$ is as small a positive number as you like if n is great enough.

$$\therefore \lim_{n\to\infty} \left(3 + \tfrac{1}{n}\right) = 3$$

Sequences and Series **321**

Related Activities p. T99

Key Ideas

Describe the limit of nondecreasing or nonincreasing sequences.

Chalkboard Examples

Find the limit of each sequence.

1. $4, 4\tfrac{1}{2}, 4\tfrac{2}{3}, 4\tfrac{3}{4}, \ldots , 5 - \dfrac{1}{n}$

$$\lim_{n\to\infty} \left(5 - \tfrac{1}{n}\right) = 5$$

2. $\dfrac{4}{2}, \dfrac{5}{3}, \dfrac{6}{4}, \dfrac{7}{5}, \ldots , \dfrac{3+n}{n+1}$

$$\lim_{n\to\infty} \left(\dfrac{3+n}{n+1}\right) = 1$$

Common Errors

Just as some students find it difficult to conceptualize the idea of infinity, some may also have difficulty with the idea of a limit. When they are told that a sequence such as the one defined by $a_n = 1 - \dfrac{1}{n}$ has a limit if we make n great enough (in this case it is 1), they may try to argue that none of the terms ever really do equal 1. Explain that "the limit of the sequence equals 1" means that we can get *as close to the limit as we want* by making n great enough. No matter how close we want the term to get to the limit we can get closer by increasing n.

In general, an infinite sequence has a *limit* L if you can make the error of approximation, $|L - a_n|$, less than any positive number, however small, by choosing n great enough. Notice how the successive terms of the sequence

$$1, 1\tfrac{1}{2}, 1\tfrac{3}{4}, 1\tfrac{7}{8}, \ldots, 2 - (\tfrac{1}{2})^{n-1}, \ldots$$

compare with each other. Because

$$1 \leq 1\tfrac{1}{2} \leq 1\tfrac{3}{4} \leq 1\tfrac{7}{8} \ldots,$$

the sequence is *nondecreasing*. Any sequence in which each term is less than or equal to the following term is called a **nondecreasing sequence**.

The sequence $4, 3\tfrac{1}{2}, 3\tfrac{1}{3}, 3\tfrac{1}{4}, \ldots, 3 + \tfrac{1}{n}, \ldots$, on the other hand, is *nonincreasing*, because

$$4 \geq 3\tfrac{1}{2} \geq 3\tfrac{1}{3} \geq 3\tfrac{1}{4} \geq \ldots.$$

Any sequence in which each term is greater than or equal to the following term is called a **nonincreasing sequence**.

Notice also that each term of the sequence

$$1, 1\tfrac{1}{2}, 1\tfrac{3}{4}, 1\tfrac{7}{8}, \ldots, 2 - (\tfrac{1}{2})^{n-1}, \ldots$$

is less than 2 in absolute value, and that each term of the sequence

$$4, 3\tfrac{1}{2}, 3\tfrac{1}{3}, 3\tfrac{1}{4}, \ldots, 3 + \tfrac{1}{n}, \ldots$$

is less than or equal to 4 in absolute value. Whenever there exists a number that equals or exceeds the *absolute value* of *every* term of a sequence, the sequence is **bounded** by the number. Thus, the sequence $1, 1\tfrac{1}{2}, 1\tfrac{3}{4}, 1\tfrac{7}{8}, \ldots, 2 - (\tfrac{1}{2})^{n-1}, \ldots$ is bounded by 2, and the sequence in the example on page 321 is bounded by 4. Of course, each of the sequences is also bounded by 10, for example.

The fact that both of the sequences are convergent illustrates the final axiom needed to characterize the set $\mathcal{R}$ of real numbers.

Axiom of Completeness

Every bounded, nondecreasing (or nonincreasing) sequence of real numbers converges, and its limit is a real number.

Not all infinite sequences converge. An infinite sequence that does not have a limit is said to **diverge** or to be **divergent**. For example, the sequence $3, 9, 27, \ldots, 3^n, \ldots$ diverges because it contains terms that are arbitrarily great in absolute value. Although this sequence is nondecreasing, it is not bounded. The sequence $1, -1, 1, -1, \ldots,$ $(-1)^{n-1}, \ldots$ is also divergent because its terms are alternately 1 and -1. Although this sequence is bounded, it is neither nondecreasing nor nonincreasing.

Oral Exercises

For each of the following sequences: (a) state the first four terms; (b) tell whether the sequence is nondecreasing, nonincreasing, or neither; (c) tell whether the sequence is bounded; and (d) tell whether the sequence seems to be convergent, and if so make a reasonable guess of the limit.

1. $a_n = (\frac{1}{3})^n$

2. $a_n = 3 + (\frac{1}{2})^n$

3. $a_n = (-\frac{1}{4})^n$

4. $a_n = (-1)^n \cdot 5$

5. $a_n = (-1)^{n-1} \frac{1}{n}$

6. $a_n = \frac{n-1}{n}$

7. $a_n = \frac{n^2}{n^2 + 3}$

8. $a_n = (-1)^n 2n$

9. $a_n = 2 + (\frac{1}{5})^n$

10. $a_n = (-1)^n \frac{n}{n^2 + 1}$

11. $a_n = 4 - \frac{1}{n^2}$

12. $a = 3 + (-1)^n \frac{1}{2^n}$

13. Give an example of a sequence that is neither nondecreasing nor nonincreasing but is convergent. Examples are Exercises 3, 5, 10, 12

14. If the sequence $a_1, a_2, a_3, \ldots$ is convergent, must the sequence

$$a_2, a_4, a_6, \ldots$$

(consisting of just the even terms of the first sequence) also be convergent? Explain. Yes, if the first sequence converges to L, then $\lim_{n \to \infty} |L - a_n| < x$ where x is any positive number. It will also be true that $\lim_{2n \to \infty} |L - a_{2n}| < x$. ∴ L will be the limit of the second sequence.

Written Exercises

Write the first four terms of the sequence whose nth term is given and make a reasonable guess of the limit or state that the sequence is not convergent.

A **1.** $a_n = \frac{3n + 1}{n}$ 4, $3\frac{1}{2}$, $3\frac{1}{3}$, $3\frac{1}{4}$; 3

2. $a_n = \frac{n^2 + 1}{2n^2}$ 1, $\frac{5}{8}$, $\frac{5}{9}$, $\frac{17}{32}$; $\frac{1}{2}$

3. $a_n = \frac{n^2}{20}$ $\frac{1}{20}$, $\frac{1}{5}$, $\frac{9}{20}$, $\frac{4}{5}$; not convergent

4. $a_n = 2 - \frac{1}{3^n}$ $1\frac{2}{3}$, $1\frac{8}{9}$, $1\frac{26}{27}$, $1\frac{80}{81}$; 2

5. $a_n = (-1)^n \frac{2^n}{2^n + 1}$ $-\frac{2}{3}$, $\frac{4}{5}$, $-\frac{8}{9}$, $\frac{16}{17}$; not convergent

6. $a_n = 5(1 - \frac{1}{n})$ 0, $2\frac{1}{2}$, $3\frac{1}{3}$, $3\frac{3}{4}$; 5

In Exercises 7–12 a formula for the nth term of an infinite sequence and its limit L are given. For $n = 1, 2, 3,$ and 4, find $|L - a_n|$, and then give the general formula for $|L - a_n|$.

7. $a_n = \frac{4n - 1}{n}$; $L = 4$ 1, $\frac{1}{2}$, $\frac{1}{3}$, $\frac{1}{4}$; $\frac{1}{n}$

8. $a_n = -1 + (\frac{1}{2})^n$; $L = -1$ $\frac{1}{2}$, $\frac{1}{4}$, $\frac{1}{8}$, $\frac{1}{16}$; $(\frac{1}{2})^n$

9. $a_n = \frac{1 + 3n}{2n}$; $L = \frac{3}{2}$ $\frac{1}{2}$, $\frac{1}{4}$, $\frac{1}{6}$, $\frac{1}{8}$; $\frac{1}{2n}$

10. $a_n = 5 + (-1)^n \frac{1}{n^2}$; $L = 5$ 1, $\frac{1}{4}$, $\frac{1}{9}$, $\frac{1}{16}$; $\frac{1}{n^2}$

B **11.** $a_n = \frac{2n}{4n + 1}$; $L = \frac{1}{2}$

$\frac{1}{10}$, $\frac{1}{18}$, $\frac{1}{26}$, $\frac{1}{34}$; $\frac{1}{8n + 2}$

12. $a_n = \frac{6n^2 + 1}{2n^2 - 1}$; $L = 3$ 4, $\frac{4}{7}$, $\frac{4}{17}$, $\frac{4}{31}$; $\frac{4}{2n^2 - 1}$

Sequences and Series **323**

---

6. a. $0, \frac{1}{2}, \frac{2}{3}, \frac{3}{4}$
 b. nondecreasing
 c. bounded
 d. converges to 1

7. a. $\frac{1}{4}, \frac{4}{7}, \frac{3}{4}, \frac{16}{19}$
 b. nondecreasing
 c. bounded
 d. converges to 1

8. a. $-2, 4, -6, 8$
 b. neither
 c. not bounded
 d. not convergent

9. a. $2\frac{1}{5}, 2\frac{1}{25}, 2\frac{1}{125}, 2\frac{1}{625}$
 b. nonincreasing
 c. bounded
 d. converges to 2

10. a. $-\frac{1}{2}, \frac{2}{5}, -\frac{3}{10}, \frac{4}{17}$
 b. neither
 c. bounded
 d. converges to 0

11. a. $3, 3\frac{3}{4}, 3\frac{8}{9}, 3\frac{15}{16}$
 b. nondecreasing
 c. bounded
 d. converges to 4

12. a. $2\frac{1}{2}, 3\frac{1}{4}, 2\frac{7}{8}, 3\frac{1}{16}$
 b. neither
 c. bounded
 d. converges to 3

Suggested Assignments

Minimum
 323/1–14
 S 264/21–27 odd
Extended Alg. with Trig.
 323/1–12
 S 264/18–28 even
Enriched Alg.
 323/1–19
 S 264/20–28 even
Enriched Alg. with Trig.
 323/1–20
 S 264/20–28 even

13. $n = 11; \dfrac{1}{11}$ **14.** $n = 4; \dfrac{1}{16}$

15. $n = 6; \dfrac{1}{12}$ **16.** $n = 4; \dfrac{1}{16}$

17. $n = 2; \dfrac{1}{18}$ **18.** $n = 5; \dfrac{4}{49}$

19. $2 - \dfrac{1}{n} < a_n$ and $a_n < 2.$

Thus, $2 - a_n < \dfrac{1}{n}$ and

$2 - a_n > 0.$ That is,

$0 < 2 - a_n < \dfrac{1}{n}$, or

$|2 - a_n| < \dfrac{1}{n}.$ Since

$\dfrac{1}{n}$ is a positive number

that is as small as you
wish by choosing

n great enough, $\lim\limits_{n \to \infty} a_n = 2.$

(continued on p. 332)

Mixed Review

Simplify.

1. $(7x - 2)(4x + 3)$ $28x^2 + 13x - 6$

2. $\dfrac{3w + 4}{w} + \dfrac{z + 2}{z}$

$\dfrac{4wz + 4z + 2w}{wz}$

3. $((st^2)(s^3t^2))^2 \cdot 4st$ $4s^9t^9$

4. $\dfrac{p^2 + p - 6}{p - 2} \cdot \dfrac{2p - 6}{p + 3}$

$2(p - 3)$

5. $\dfrac{f}{f^2 + 2fg + g^2} \div \dfrac{f}{g + f}$

$\dfrac{1}{f + g}$

6. If $x^3 - x^2 = 20x$, then $x =$
$\underline{\quad ?\quad}$ $5, -4,$ or 0

13–18. Find a value of n that will make $|L - a_n| < \frac{1}{10}$ for each of Written Exercises 7–12. Give the numerical value of $|L - a_n|$ for this n.

19. Suppose $a_1, a_2, a_3, \ldots$ is an infinite sequence such that

$$2 - \frac{1}{n} < a_n < 2.$$

Explain why $\lim\limits_{n \to \infty} a_n = 2.$

C **20.** Suppose the sequence $a_1, a_2, a_3, \ldots$ has limit L and the sequence

$$b_1, b_2, b_3, \ldots$$

is defined by the relation $b_n = ka_n$. Show that the second sequence has limit kL by showing that the expression $|kL - b_n|$ has limit 0. (*Hint:* Recall that $|ab| = |a| \cdot |b|$.)

8–8 Infinite Geometric Series

Figure 2 pictures the numbers

$$1, \ 1 + \tfrac{1}{2}, \ 1 + \tfrac{1}{2} + \tfrac{1}{4}, \ 1 + \tfrac{1}{2} + \tfrac{1}{4} + \tfrac{1}{8},$$

and suggests three facts:

1. The more terms you add in the infinite series

$$1 + \tfrac{1}{2} + \tfrac{1}{4} + \tfrac{1}{8} + \tfrac{1}{16} + \cdots + (\tfrac{1}{2})^{n-1} + \cdots,$$

   the greater is the sum obtained.
2. The sum never exceeds 2, no matter how many terms you add.
3. If enough terms are added, the sum will approximate 2 as closely as you may demand.

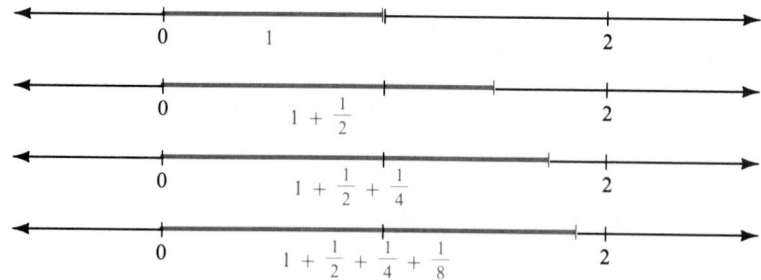

Figure 2

You can never add all the terms, so that you cannot refer to the "sum

of the infinite series" without first defining what you mean by such a
sum. To define the sum of the *infinite geometric series*

$$1 + \tfrac{1}{2} + \tfrac{1}{4} + \tfrac{1}{8} + \tfrac{1}{16} + \cdots + (\tfrac{1}{2})^{n-1} + \cdots,$$

consider the sequence of *partial sums*

$$S_1 = 1, \quad S_2 = 1 + \tfrac{1}{2}, \quad S_3 = 1 + \tfrac{1}{2} + \tfrac{1}{4},$$

and, in general,

$$S_n = 1 + \tfrac{1}{2} + \tfrac{1}{4} + \cdots + (\tfrac{1}{2})^{n-1} = \sum_{i=1}^{n} (\tfrac{1}{2})^{i-1}, \qquad n \geq 1.$$

Since S_n is the sum of the first n terms of a geometric sequence whose
first term is 1 and whose common ratio is $\tfrac{1}{2}$, you have:

$$S_n = \frac{[1 - 1(\tfrac{1}{2})^n]}{1 - \tfrac{1}{2}} = 2 - \frac{1}{2^{n-1}} = 2 - \left(\frac{1}{2}\right)^{n-1}$$

Hence (see page 321), $\lim\limits_{n \to \infty} S_n = 2$. Accordingly, we *define* the sum of this
infinite series to be 2, and we write

$$1 + \tfrac{1}{2} + \tfrac{1}{4} + \cdots + (\tfrac{1}{2})^{n-1} + \cdots = 2, \qquad \text{or} \qquad \sum_{i=1}^{\infty} (\tfrac{1}{2})^{i-1} = 2.$$

In general, for any infinite series $a_1 + a_2 + \cdots + a_n + \cdots$,

$$S_n = \sum_{i=1}^{n} a_i$$

is called a **partial sum**. If the sequence $S_1, S_2, \ldots, S_n, \ldots$ of partial
sums converges and if $\lim\limits_{n \to \infty} S_n = S$, then the **sum of the infinite series**

$$a_1 + a_2 + \cdots + a_n + \cdots$$

is defined to be S. You write

$$\sum_{k=1}^{\infty} a_k = S,$$

and you say that the series **converges** or **is convergent**.

On the other hand, if the sequence of partial sums diverges, then the
series **diverges**, or **is divergent**, and its sum is *not* defined. For example,
the series

$$1 + 2 + 3 + 4 + \cdots + n + \cdots$$

diverges, because the sequence of partial sums

$$1, 3, 6, 10, \ldots, \frac{n(n+1)}{2}, \ldots$$

is divergent. The series $1 - 1 + 1 - 1 + \cdots$ also is divergent because
the sequence of partial sums $1, 0, 1, 0, \ldots$ is a divergent sequence.

Sequences and Series **325**

Teaching Suggestions
p. T99

Supplementary Material
Test 24

Key Ideas

Find the sum of an infinite
geometric series.

Chalkboard Examples

1. Find the sum of the infinite geometric series.

$$\frac{2}{5} - \frac{1}{5} + \frac{1}{10} - \cdots \qquad a_1 = \frac{2}{5},$$

$$r = -\frac{1}{2}, \ |r| < 1$$

$$S = \frac{a_1}{1 - r} = \frac{\frac{2}{5}}{1 - \left(-\frac{1}{2}\right)} = \frac{4}{15}$$

2. Express $0.\overline{620}$ as a fraction
in lowest terms.
Since $0.\overline{620} = 0.620 +$
$0.620(0.001) + 0.620(0.001)^2$
$+ \cdots,$
$a_1 = 0.620$ and $r = 0.001$.

$$S = \frac{a_1}{1 - r} = \frac{0.620}{1 - 0.001} =$$

$$\frac{0.620}{0.999} = \frac{620}{999}$$

Additional A Exercises

Find the sum of each infinite
geometric series, if it converges. If it diverges, so state.

1. $72 + 24 + 8 + \cdots$ 108
2. $-2 + 2 - 2 + \cdots$ div.
3. $6 + 4 + 2\frac{2}{3} + \cdots$ 18
4. $0.3 + 0.03 + 0.003 + \cdots$

$$\frac{1}{3}$$

5. $\frac{5}{3} + 1 + \frac{3}{5} + \cdots$ $\frac{25}{6}$

1. a. $\frac{1}{2}$ b. converges
   c. 28, 42, 49
2. a. -1
   b. does not converge
   c. $-3, 0, -3$
3. a. $\frac{1}{5}$ b. converges
   c. 5, 6, $6\frac{1}{5}$
4. a. $\frac{1}{3}$ b. converges
   c. $1\frac{4}{5}, 2\frac{2}{5}, 2\frac{3}{5}$
5. a. -3
   b. does not converge
   c. $\frac{2}{27}, -\frac{4}{27}, \frac{14}{27}$
6. a. 1
   b. does not converge
   c. 8, 16, 24
7. a. $-\frac{1}{2}$ b. converges
   c. 6, 3, $4\frac{1}{2}$
8. a. $\frac{1}{10}$ b. converges
   c. $\frac{6}{10}, \frac{66}{100}, \frac{666}{1000}$
9. a. 0.1   b. converges
   c. 0.4, 0.44, 0.444
10. a. 0.01   b. converges
    c. 0.27, 0.2727, 0.272727
11. 0.7, 0.1
12. 0.07, 0.01
13. 0.07, 0.1
14. 0.63, 0.01
15. 0.063, 0.01
16. 0.630, 0.001

Consider each of the following cases for any infinite geometric series,
$$a_1 + a_1 r + a_1 r^2 + \cdots .$$

Case 1. $a_1 = 0$. In this case, every term of the series is 0, so that every partial sum is 0. The series is $0 + 0 + 0 + \cdots$, which has the sum 0.

Case 2. $a_1 \neq 0$ and $r = 1$. In this case, every term of the series is a_1, so that the series is $a_1 + a_1 + a_1 + \cdots$. This series diverges because the sequence of partial sums $a_1, 2a_1, \ldots$ diverges.

Case 3. $a_1 \neq 0$ and $r = -1$. In this case, the terms of the series are alternately a_1 and $-a_1$, so that the series is $a_1 - a_1 + a_1 - a_1 + a_1 \cdots$. This series diverges because the sequence of partial sums $a_1, 0, a_1, 0, \ldots$ diverges.

Case 4. $a_1 \neq 0$ and $|r| \neq 1$. In this case, the nth partial sum is

$$S_n = \frac{a_1 - a_1 r^n}{1 - r}, \quad \text{or} \quad S_n = \frac{a_1}{1 - r} - \frac{a_1}{1 - r} r^n .$$

If $|r| < 1$, then $r^n \left(\text{and therefore, } \left| \frac{a_1}{1 - r} r^n \right| \right)$ can be made to approximate 0 as closely as you wish by taking n great enough. Because

$$\left| S_n - \frac{a_1}{1 - r} \right| = \left| \frac{a_1}{1 - r} r^n \right| ,$$

it follows that $S = \lim_{n \to \infty} S_n = \frac{a_1}{1 - r}$, $|r| < 1$.

If $|r| > 1$, $|r^n|$ increases without bound as n increases, so r^n does not have a limit as n increases without bound and neither does S_n. The following theorem summarizes these cases.

Theorem. The infinite geometric series
$$a_1 + a_1 r + a_1 r^2 + \cdots$$
converges and has the sum $\frac{a_1}{1 - r}$ if $|r| < 1$. If $a_1 = 0$, then the series converges and has the sum 0. If $|r| \geq 1$ and $a_1 \neq 0$, then the series diverges.

EXAMPLE 1   Find the sum of the infinite geometric series $\frac{3}{8} - \frac{3}{16} + \frac{3}{32} - \cdots$.

SOLUTION $a_1 = \frac{3}{8}, r = -\frac{1}{2}, |r| < 1$

$$S = \frac{a_1}{1 - r} = \frac{\frac{3}{8}}{1 - (-\frac{1}{2})} = \frac{\frac{3}{8}}{\frac{3}{2}} = \frac{1}{4}$$

You can use the formula for the sum of an infinite series to convert a repeating decimal to a fraction in lowest terms.

EXAMPLE 2   Express $0.\overline{270}$ as a fraction in lowest terms.

SOLUTION   Since $0.\overline{270} = 0.270 + 0.270(0.001) + 0.270(0.001)^2 + \cdots$,
$a_1 = 0.270$ and $r = 0.001$.

$$S = \frac{a_1}{1 - r} = \frac{0.270}{1 - 0.001} = \frac{0.270}{0.999} = \frac{30}{111}$$

Oral Exercises

For each of the following infinite geometric series, (a) state the common ratio, (b) tell whether the series converges, and (c) state the value of S_1, S_2, and S_3.

1. $28 + 14 + 7 + \cdots$

2. $-3 + 3 - 3 + \cdots$

3. $5 + 1 + \frac{1}{5} + \cdots$

4. $\frac{9}{5} + \frac{3}{5} + \frac{1}{5} + \cdots$

5. $\frac{2}{27} - \frac{2}{9} + \frac{2}{3} - \cdots$

6. $8 + 8 + 8 + \cdots$

7. $6 - 3 + \frac{3}{2} - \cdots$

8. $\frac{6}{10} + \frac{6}{100} + \frac{6}{1000} + \cdots$

9. $0.4 + 0.04 + 0.004 + \cdots$

10. $0.27 + 0.0027 + 0.000027 + \cdots$

Each of the following repeating decimals can be represented as the sum of an infinite geometric series. State the first term of the series and the common ratio.

11. $0.\overline{7}$ 12. $0.0\overline{7}$ 13. $0.\overline{07}$ 14. $0.\overline{63}$ 15. $0.0\overline{63}$ 16. $0.\overline{630}$

Give an example to show that each of the following statements is *false*.

17. Every infinite geometric series converges. $1 + 2 + 4 + \ldots + 2^{n+1} + \cdots$

18. A convergent geometric series with a negative common ratio has a negative sum. Oral Exercise 7

Written Exercises

Find the sum of each infinite geometric series, if it converges. If it is divergent, so state.

A   1. $36 + 12 + 4 + \cdots$ 54

2. $250 - 50 + 10 - \cdots$ $208\frac{1}{3}$

3. $6 + 5 + \frac{25}{6} + \cdots$ 36

4. $7 - 7 + 7 - \cdots$ Divergent

5. $\frac{4}{3} - 1 + \frac{3}{4} - \cdots$ $\frac{16}{21}$

6. $\frac{8}{3} + \frac{4}{9} + \frac{2}{27} + \cdots$ $3\frac{1}{5}$

7. $\frac{1}{48} - \frac{1}{24} + \frac{1}{12} - \cdots$ Divergent

8. $14 + 6 + \frac{18}{7} + \cdots$ $24\frac{1}{2}$

9. $0.8 + 0.08 + 0.008 + \cdots$ $0.\overline{8}$ or $\frac{8}{9}$

10. $0.6 + 0.006 + 0.00006 + \cdots$ $0.\overline{60}$ or $\frac{20}{33}$

Sequences and Series   **327**

Suggested Assignments
Minimum
    First day
    327/1–32
    Second day
    328/*P*: 1–6
R 330/*Self-Test 3*
Extended Alg. with Trig.
    First day
    327/1–31 odd
    Second day
    328/*P*: 1–6
R 330/*Self-Test 3*
Enriched Alg.
    First day
    327/7–33 odd
    Second day
    328/*P*: 1–6
R 330/*Self-Test 3*
Enriched Alg. with Trig.
    First day
    327/7–21 odd
    Second day
    328/*P*: 5–10
R 330/*Self-Test 3*

Find the sum of each infinite geometric series, if it converges. If it is divergent, so state.

11. $\displaystyle\sum_{n=1}^{\infty} 4(\tfrac{1}{3})^{n-1}$ 6

12. $\displaystyle\sum_{n=1}^{\infty} 3(-\tfrac{2}{5})^{n-1}$ $2\tfrac{1}{7}$

13. $\displaystyle\sum_{n=1}^{\infty} \tfrac{5}{6}(-\tfrac{3}{4})^{n-1}$ $\tfrac{10}{21}$

14. $\displaystyle\sum_{n=1}^{\infty} \tfrac{1}{10}(\tfrac{6}{5})^{n-1}$ Divergent

Find the specified value for the infinite geometric series that is described.

15. $r = \tfrac{1}{3}$, $S = 15$, $a_1 = \underline{\ ?\ }$ 10

16. $r = -\tfrac{5}{6}$, $S = 9$, $a_1 = \underline{\ ?\ }$ $16\tfrac{1}{2}$

17. $r = -\tfrac{1}{4}$, $S = 24$, $a_1 = \underline{\ ?\ }$ 30

18. $a_1 = 12$, $S = 16$, $r = \underline{\ ?\ }$ $\tfrac{1}{4}$

19. $a_1 = 35$, $S = 30$, $r = \underline{\ ?\ }$ $-\tfrac{1}{6}$

20. $S = \tfrac{4}{3}a_1$, $r = \underline{\ ?\ }$ $-\tfrac{1}{4}$

Convert each nonterminating decimal to a fraction in lowest terms by rewriting the decimal as an infinite geometric series and finding the sum of the series.

21. $0.\overline{5}$

22. $0.\overline{36}$

23. $0.\overline{21}$

24. $0.\overline{162}$

25. $0.\overline{117}$

26. $0.\overline{108}$

27. $0.0\overline{3}$

28. $0.00\overline{72}$

B

29. The sum of an infinite geometric series whose first term is 3 and whose common ratio is x is $\dfrac{2}{x}$. Find the value of x. $\tfrac{2}{5}$

30. The sum of an infinite geometric series whose first term is -10 and whose common ratio is x is $9x$. Find the value of x. $-\tfrac{2}{3}$

31. For what value of x will the sum of the infinite geometric series $x + x^2 + x^3 + \cdots$ be 5? $\tfrac{5}{6}$

32. Find the value of x so that the sum of the infinite geometric series $x + x^3 + x^5 + \cdots$ will be $\dfrac{4}{15}$. $\tfrac{1}{4}$

C

33. Find all the possible values of x for which the sum of the series $5 + 15x^2 + 45x^4 + \cdots$ is $\dfrac{5}{2x}$. $\tfrac{1}{3}$

34. Suppose that the two infinite geometric series $a_1 + a_1 r + a_1 r^2 + \cdots$ and $2a_1 + 2a_1 s + 2a_1 s^2 + \cdots$ have the same sum for some $r \neq \tfrac{1}{2}$, $0 < r < 1$. Find s in terms of r and give an example of two series with this relationship. $s = 2r - 1$, Example: $a_1 = 1$, $r = \tfrac{1}{4}$ and $s = -\tfrac{1}{2}$. The series are $1 + \tfrac{1}{4} + \tfrac{1}{16} + \cdots$ and $2 - 1 + \tfrac{1}{2} - \cdots$; $S = \tfrac{4}{3}$.

Problems

A

1. Friction and air resistance cause each swing (after the first) of a pendulum bob to be 0.9 times as long as the preceding swing. If the length of the first swing is 16 cm, find the total distance traveled by the bob before coming to rest. 160 cm

328   *Chapter 8*

2. A square piece of paper whose sides are each 60 cm long is cut into four smaller squares, each 30 cm on a side. One of these squares is cut into four smaller squares, each 15 cm on a side. One of these squares is then cut into four smaller squares, each 7.5 cm on a side. If this process is repeated infinitely many times, what is the sum of the perimeters of all the squares produced? 720 cm

3. A point on a violin string moves 2.4 mm in one direction as the string begins to vibrate. Each subsequent one-way movement of the point is 2% shorter than the previous one. What is the total distance traveled by the point before the string comes to rest? 120 mm

4. A ball is dropped straight down from a height of 120 cm and rebounds $\frac{3}{8}$ of its preceding maximum height on each successive bounce. How far will the ball travel before coming to rest? (*Hint*: Consider the upward distances and the downward distances separately.) 264 cm

B 5. Seven congruent smaller circles are inscribed in one large circle, as shown. Six of these circles are shaded, and seven smaller circles are inscribed in the seventh unshaded circle. Again six of these are shaded. If this procedure is repeated infinitely many times, what fraction of the area of the original large circle will be shaded? $\frac{3}{4}$

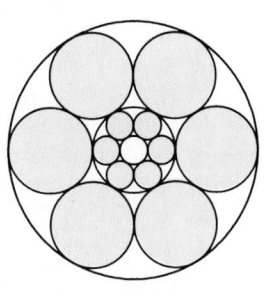

Ex. 5

6. In a square whose sides each have length 7 is inscribed a smaller square whose vertices divide the sides of the larger square in a 4:3 ratio, as shown. A third square is inscribed in the second square, again dividing the sides in a 4:3 ratio. If this process is repeated infinitely many times, what will be the sum of the perimeters of all the squares? 98

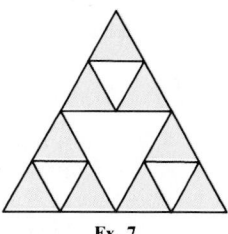

Ex. 6

C 7. The segments joining the midpoints of the sides of an equilateral triangle are drawn, and the interior of the triangle they form is removed from the interior of the original triangle. The segments connecting the midpoints of the remaining triangles are joined, and the interiors of the triangles they form are removed from the interior of the original triangle. (The remaining interior is the shaded region in the diagram.) If this process is repeated infinitely many times, how much of the original interior will be left? none

Ex. 7

Simplify.

1. $3c(b + c) - [2(-4b + 7c + bc)]$ $3c^2 + bc + 8b - 14c$

2. $\dfrac{q + 3}{q^2 - 2q - 15} - \dfrac{5}{2q + 6}$

$\dfrac{-3q + 31}{2q^2 - 4q - 30}$

3. $(5t^3u^2v^2) \cdot (u^2v) \cdot \dfrac{1}{u^2v}$

$5t^3u^2v^2$

4. $-|-3 - 5 + 28|$ -20

5. Solve the open sentence.

$|4u - 3| = 9$ $\left\{3, -\dfrac{3}{2}\right\}$

6. Write the equation $2x - y = -8$ in slope-intercept form. $y = 2x + 8$

8. A car that is 60 km from point P approaches P traveling at 40 km/h. At the same moment the car starts traveling, a bee leaves the front of the car flying at 80 km/h towards P. When the bee reaches P, it instantaneously reverses direction and continues flying at the same speed. When it reaches the car, it reverses direction again, continuing to fly back and forth until the car reaches P. Use an infinite geometric series to find the total distance traveled by the bee. Is there a simpler way to find this total distance? 120 km

9. A "snowflake" curve is constructed as follows: The sides of an equilateral triangle are trisected. A new equilateral triangle is placed on the middle third of each trisection. The sides common to the previous figure and the new triangles are then removed. This process continues indefinitely using the sides of the last figure obtained. If the side of the original equilateral triangle is of length 1, what is the area enclosed by the snowflake curve if the process is continued without end? $\frac{2\sqrt{3}}{5}$ sq. units

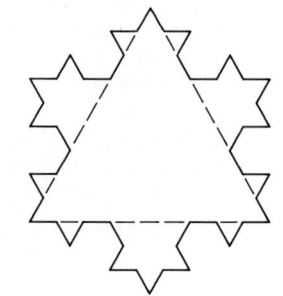

10. Show that the figure described in Problem 9 has no perimeter, that is, that the perimeters of successive polygons formed increase without bound.

Computer Exercises For students with computer experience

1. Write a program to find the sum of an infinite geometric series. Make provisions so that the program will not run if $|r| \geq 1$, where r is the common ratio.

Run the program in Exercise 1 for a series with the given constants.

2. $a_1 = 0.3, r = 0.1$
   0.3

3. $a_1 = 0.75, r = 2$
   $|r| \geq 1$

4. $a_1 = 50, r = -0.5$
   $33.\overline{3}$

▉ Self-Test 3

VOCABULARY convergent infinite sequence
(p. 321)
limit of an infinite sequence
(p. 322)
nondecreasing sequence
(p. 322)
nonincreasing sequence
(p. 322)
bounded sequence (p. 322)

divergent infinite sequence
(p. 322)
partial sum (p. 325)
sum of an infinite series
(p. 325)
convergent infinite series
(p. 325)
divergent infinite series
(p. 325)

In Exercises 1 and 2 a formula for the nth term of an infinite sequence is given. Write the first four terms of each sequence and make a reasonable guess of the limit L. Give the general formula for $|L - a_n|$.

1. $a_n = \dfrac{n + 2}{n}$ 2. $a_n = 3 - \dfrac{1}{3^n}$ *Obj. 1, p. 321*

Find the sum of each infinite geometric series.

3. $9 - 3 + 1 - \cdots$ $\dfrac{27}{4}$ 4. $\displaystyle\sum_{n=1}^{\infty} \frac{2}{5}\left(\frac{1}{2}\right)^{n-1}$ $\dfrac{4}{5}$ *Obj. 2, p. 321*

5. Convert $0.\overline{213}$ to a fraction in lowest terms by rewriting the decimal as an infinite geometric series and finding the sum of the series.
$0.213 + 0.213\,(0.001) + 0.213\,(0.001)^2 + \cdots ;\ \dfrac{71}{333}$
Check your answers with those at the back of the book.
1. $3, 2, \frac{5}{3}, \frac{3}{2}; 1; \frac{2}{n}$

2. $2\frac{2}{3}, 2\frac{8}{9}, 2\frac{26}{27}, 2\frac{80}{81}; 3; \frac{1}{3^n}$

Chapter Summary

1. An *arithmetic sequence*, or *arithmetic progression*, is any sequence in which each term after the first is obtained by adding a fixed number, d, called the *common difference*, to the preceding term. For an arithmetic sequence, you have
$$a_{n+1} = a_n + d \quad \text{and} \quad a_n = a_1 + (n - 1)d.$$

2. The terms between two given terms of an arithmetic sequence are called *arithmetic means* between the given terms; a single arithmetic mean between two numbers is *the arithmetic mean*, or the *average*, of the two numbers.

3. The sum of the first n terms of a given sequence is the associated series, S_n. For an arithmetic series, you have
$$S_n = \frac{n}{2}(a_1 + a_n) = \frac{n}{2}[2a_1 + (n - 1)d].$$

4. You can use the *summation sign* to abbreviate the writing of a series, using an *index* to indicate the *range of summation*. Thus,
$$a_1 + a_2 + \cdots + a_n = \sum_{i=1}^{n} a_i.$$

5. A *geometric sequence*, or *geometric progression*, is any sequence in which each term after the first is the product of the preceding term and a fixed number, r, called the *common ratio*. For a geometric sequence, you have
$$a_{n+1} = a_n \cdot r \text{ and } a_n = a_1 r^{n-1}.$$

Sequences and Series **331**

331

20. Since $|0 - (L - a_n)|$ is less than some positive number for n as great as you wish, then the limit of $|a_n - L|$ is 0. Also $|k| \cdot 0 = 0$, so $\lim_{n \to \infty} |k| \cdot |a_n - L| = 0$. Furthermore, $|k| \, |a_n - L| = |ka_n - kL| = |b_n - kL|$. $\therefore |b_n - kL|$ has limit 0 and the sequence $b_1, b_2, b_3, \ldots$ has limit kL.

6. The terms between two given terms of a geometric sequence are called *geometric means* between the given terms. A single geometric mean between two numbers is called *the geometric mean* or *mean proportional* of the two numbers; the geometric mean of two positive numbers a and b is $\sqrt{ab}$.

7. A series whose terms are in geometric progression is called a geometric series. For a geometric series, you have the sum

$$S_n = \frac{a_1 - a_1 r^n}{1 - r} = \frac{a_1 - ra_n}{1 - r}, \qquad r \neq 1.$$

8. An infinite sequence has a *limit L* if you can make the error of approximation, $|L - a_n|$, less than any positive number, however small, by choosing n great enough. Any infinite sequence that has a limit is said to *converge*, or to be *convergent*: if the sequence does not converge, it is said to *diverge*, or to be *divergent*.

9. *Axiom of Completeness:* Every bounded, nondecreasing (or nonincreasing) sequence of real numbers converges, and its limit is a real number.

10. For any infinite series $a_1 + a_2 + \cdots + a_n + \ldots$, $S_n = \sum_{i=1}^{n} a_i$ is called a *partial sum*. If the sequence $S_1, S_2, \ldots, S_n, \ldots$ of partial sums converges to S, then the infinite series is said to *converge*, or to be *convergent*, and its *sum* is defined to be S.

11. For an infinite geometric series with $|r| < 1$, you have the sum

$$S = \frac{a_1}{1 - r}.$$

Chapter Review

Write the letter of the correct answer.

1. Specify the sequence 8, 4, 0, . . . recursively.   *8-1*
    a. $a_1 = 8$; $a_n = 12 - 4n$ **b.** $a_1 = 8$; $a_{n+1} = a_n - 4$
    c. $a_1 = 8$; $a_{n+1} = \frac{1}{2}a_n$ **d.** $a_1 = 8$; $a_n = 2^{4-n}$

2. Find a_1 for an arithmetic sequence in which $a_{24} = -18$ and $d = -1$.   *8-2*
    a. -41 **b.** -5 **c.** 18   **d.** 5

3. In an arithmetic sequence, $a_4 = -2$ and $a_9 = 13$. Find the value of a_7.
    a. 7   **b.** 8   **c.** 4   **d.** 5

4. Find the sum of an arithmetic series in which $a_1 = 12$, $a_4 = -6$, and $n = 10$.   *8-3*
    a. 30   **b.** -42 **c.** -150 **d.** 28

5. Find the value of $\sum\limits_{i=1}^{6} (13 + 2i)$.

    a. 25   **b.** 156   **c.** 36   (**d.**) 120

6. Find the eighth term of the geometric sequence 100, 50, ...   *8-4*

    a. -250 (**b.**) $\frac{25}{32}$ **c.** $199\frac{7}{32}$ **d.** -1000

7. Insert two geometric means between 81 and -3.   *8-5*

    a. 53, 25   **b.** 27, -9 (**c.**) -27, 9   **d.** 27, 9

8. In a geometric sequence, $a_2 = \frac{3}{4}$ and $a_7 = -\frac{3}{128}$. Find the value of a_4.

    (**a.**) $\frac{3}{16}$ **b.** $-\frac{3}{2}$ **c.** $\frac{3}{32}$ **d.** $\frac{3}{2}$

9. Find the sum S_5 for a geometric series in which $a_1 = -2$ and $r = -\frac{1}{3}$.   *8-6*

    a. $-\frac{2}{81}$ **b.** $-\frac{40}{3}$ (**c.**) $-\frac{122}{81}$ **d.** $-\frac{10}{3}$

10. Find the value of $\sum\limits_{n=1}^{4} 100(\frac{1}{5})^{n-1}$.

    a. $\frac{4}{5}$ (**b.**) $124\frac{4}{5}$ **c.** $100\frac{1}{125}$ **d.** $401\frac{1}{5}$

11. If $a_n = \frac{2n + 1}{n}$ for a certain infinite sequence, what is the limit of the   *8-7*
sequence?

    a. 0   **b.** 1   (**c.**) 2   **d.** no limit

12. Find the sum of the infinite geometric series $180 - 60 + 20 - \cdots$.   *8-8*

    (**a.**) 135   **b.** 0   **c.** 270   **d.** series is divergent

13. Find the value of $\sum\limits_{n=1}^{\infty} 2(\frac{1}{4})^{n-1}$.

    a. 2   **b.** $\frac{3}{2}$ **c.** 0   (**d.**) $\frac{8}{3}$

Chapter Test

1. Give the first four terms of an arithmetic sequence in which $a_1 = -1$ *8-1*
and $a_{n+1} = a_n + \frac{1}{4}$. $-1, -\frac{3}{4}, -\frac{1}{2}, -\frac{1}{4}$

2. Specify the sequence 5, 3, 1, -1, . . . both explicitly and recursively. $a_n = 7 - 2n$;
$a_1 = 5, a_{n+1} = a_n - 2$

3. Find the first term of an arithmetic sequence in which the fifth term   *8-2*
is 5 and the eighth term is -13. 29

4. Find the sum of the series $10 + 25 + 40 + \cdots + 100$. 385   *8-3*

5. Find the common difference in an arithmetic series in which $S_6 = 27$
and $a_1 = 2$. 1

6. Find the tenth term of the geometric sequence $-120, 60, -30, \ldots$. $\frac{15}{64}$ 8-4

7. The owners of a certain store reduce the price of their items at the end of each week. If the original price of a blouse is \$25.00 and its price each week is $\frac{4}{5}$ of what it was the previous week, what will the blouse cost at the beginning of the fifth week? \$10.24

8. Find the geometric mean of 3 and 48. 12   8-5

9. Find the value of $\sum_{n=1}^{6} 81(\frac{1}{3})^{n-1}$. $121\frac{1}{3}$ 8-6

10. Give the general formula for $|L - a_n|$ for an infinite sequence with   8-7
$a_n = \dfrac{n^2 - n}{2n^2}$ and $L = \dfrac{1}{2}$. $\frac{1}{2n}$

11. Convert $0.\overline{24}$ to a fraction in lowest terms. $\frac{8}{33}$ 8-8

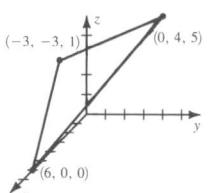
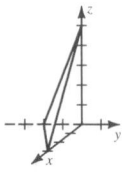
Cumulative Review

Chapter 5

1. Sketch the triangle in space whose vertices have coordinates $(-3, -3, 1)$, $(6, 0, 0)$, and $(0, 4, 5)$.

2. Graph part of the plane having xy-trace: $2x - 3y = 6, z = 0$
yz-trace: $5y - 2z = -10, x = 0$
xz-trace: $10x + 6z = 30, y = 0$

Solve each system.

3. $\begin{aligned} x - 2y + 3z &= 6 \\ 2x + 5y - z &= 10 \\ 3x + 3y + 2z &= 12 \end{aligned}$ $\varnothing$

4. $\begin{aligned} -x + 2y + 3z &= -12 \\ 2x + 7y + 5z &= -2 \\ 3x + 4y - 4z &= 4 \end{aligned}$ $\{(\frac{492}{55}, \frac{-222}{55}, \frac{92}{55})\}$

5. Evaluate. $\begin{vmatrix} -3 & 2 & 1 & 1 \\ 2 & 1 & 2 & 5 \\ 4 & -3 & -1 & 2 \\ 0 & 3 & 0 & 1 \end{vmatrix}$ -68

6. In a child's piggy bank, there is a total of 46 coins, some are nickels, some are dimes, and some are quarters. The total value of the coins in the bank is \$5.50. If there 5 more dimes than nickels, find the number of each type of coin in the bank. 15 nickels, 20 dimes, 11 quarters

Chapter 6

Simplify.

7. $(-r)(rs^2)(r^2s)^3$ $-r^8s^5$

8. $(5m - 3)(3m - 5)$ $15m^2 - 34m + 15$

9. $\dfrac{(-2x^2)^3(7x)^2}{(4x)^2(7x)(x^2)}$ $-\frac{7}{2}x^3$

10. $\dfrac{c^2 + 5c - 6}{c^2 + 7c + 12} \cdot \dfrac{c^2 - c - 20}{c^2 - 6c + 5} \div \dfrac{c^2 - 36}{c^2 - 9}$

11. $\dfrac{12w^2 + 4w - 18}{2w + 3}$ $\frac{12w^2 + 4w - 18}{2w + 3}$

12. $\dfrac{11r^3s^3 - 33rs + 66r^2s^2}{11r^2s^2}$ $\frac{r^2s^2 + 6\,rs - 3}{rs}$

10. $\dfrac{c - 3}{c - 6}$

13. $\dfrac{9}{2 + a} - \dfrac{7}{a - 2} - \dfrac{3a - 1}{4 - a^2}$ $\frac{33 - 5a}{(2 + a)(2 - a)}$

14. $(x^{-1} + y^{-1}) \div \left(\dfrac{x}{y} - \dfrac{y}{x}\right)$ $\frac{1}{x - y}$

Solve.

15. $2x^2 + 3 = -5x$ $\{-\frac{3}{2}, -1\}$

16. $-12 = y^2 - 3y - 10$ $\{1, 2\}$

17. $\dfrac{1}{d + 5} = \dfrac{17}{25 - d^2} + \dfrac{d}{d - 5}$ $\{-6, 2\}$

18. $4y^2 - 3y > 1$ $\{y : y < -\frac{1}{4} \text{ or } y > 1\}$

19. The length of a rectangle is 5 cm less than twice its width, and the area of the rectangle is 187 cm². What are the dimensions of the rectangle?   11 cm by 17 cm

Chapter 7

20. Determine the value of k so that $(2, 28)$ lies on the graph of $y = kx^3$. $\frac{7}{2}$

21. Determine all the rational roots of $x^3 - x^2 - 4x - 6 = 0$.   3

22. Express $0.\overline{37}$ as a ratio of two integers. $\frac{37}{99}$

Simplify.

23. $\sqrt[3]{-\dfrac{1}{27}} + \sqrt{\dfrac{81}{4}}$ $\frac{25}{6}$

24. $\sqrt{cd^2} - d\sqrt{c} + \sqrt{16cd^2}$ $5|d|\sqrt{c} - d\sqrt{c}$

25. $\dfrac{2}{1 - \sqrt{3}}$ $-1 - \sqrt{3}$

26. $(\sqrt{10} - 4)(2\sqrt{10} + 7)$ $-8 - \sqrt{10}$

Solve over $\mathcal{R}$.

27. $81x^4 - 16 = 0$ $\{\frac{2}{3}, -\frac{2}{3}\}$

28. $2n^3 + 16 = 0$ $\{-2\}$

29. $b - 3 - \sqrt{6b + 9} = 0$ $\{12\}$

30. $\sqrt{v + 5} + \sqrt{8 - v} = 5$ $\{4, -1\}$

Express in the form $a + bi$.

31. $(4 - 9i) + (4 - i)$ $8 - 10i$

32. $(5 - 8i) - (2 + 6i)$ $3 - 14i$

33. $(7 - i)(-6 + 2i)$ $-40 + 20i$

34. $\dfrac{2 - 3i}{7 + 4i}$ $\frac{2}{65} - \frac{29}{65}i$

Chapter 8

Give the first four terms of each sequence.

35. $a_1 = -1,\ a_{n+1} = -2a_n$ $-1, 2, -4, 8$

36. $a_n = 7 - 4n$ $3, -1, -5, -9$

37. Find the value of n in the arithmetic sequence where $a_1 = 12$, $d = -\frac{1}{3}$, and $a_n = 5$.   22

38. Find the value of a_1 in the arithmetic sequence where $a_{10} = 7$ and $a_{14} = 15$. -11

39. Find the value of n in the arithmetic series where $a_1 = 2$, $a_n = 44$, and $S_n = 345$. 15

40. Find the value of x if the arithmetic mean of $x + 3$ and $2x + 5$ is 10. 4

41. Find the value of a_4 in the geometric sequence where $a_1 = 32$ and $r = \frac{1}{2}$. 4

42. Find the value of r in the geometric sequence where $a_1 = 54$ and $a_6 = -\frac{2}{9}$. $-\frac{1}{3}$

43. Find the value of S_n in the geometric series where $a_1 = 36$, $r = -\frac{1}{3}$, and $n = 5$. $27\frac{1}{9}$

44. Find the value of a_1 in the geometric series where $n = 6$, $r = \frac{3}{2}$, and $S_n = 498\frac{3}{4}$. 24

45. A ball dropped straight down from a height of 128 ft rebounds $\frac{1}{2}$ of its preceding maximum height on each successive bounce. What is the distance the ball bounces on the ninth rebound? $\frac{1}{4}$ft

APPLICATION

Camera Settings

If you have ever used an adjustable camera, you may have noticed that the available shutter speed settings form a nearly perfect geometric sequence. "Shutter speed" actually refers to the *time* during which the shutter is open (which depends on the speed at which it opens and closes.) Shutter speeds are usually given in seconds. The sequence of shutter speeds given on a typical camera might be: 1/500, 1/250, 1/125, 1/60, 1/30, ... As you can see, the common ratio is approximately 2.

The longer the shutter is open, the greater the exposure of the film in the camera. Thus the exposure is directly proportional to the shutter speed. The amount of exposure is also controlled by the size of the aperture, or opening, through which light passes. The aperture size is controlled by a diaphragm. The

amount of exposure is directly proportional to the *square* of the effective diameter of the lens. Thus if the effective diameter is made $\sqrt{2}$ times as long (by opening up the aperture), the amount of exposure is doubled.

The f-stops of a camera relate the effective diameter of the lens to its focal length, the distance from the center of the lens to the point at which it brings light into focus. At an f-stop of f/2, a lens has an effective diameter that is half as long as its focal length. Notice that the f-stop number is inversely proportional to the lens diameter. Thus, making an f-stop number $\dfrac{1}{\sqrt{2}}$ times as large doubles the exposure. A sequence of f-stop numbers found on a typical camera might be: 22, 16, 11, 8, 5.6, 4, 2.8, 2, · · · · . Note that the common ratio is approximately $\dfrac{1}{\sqrt{2}}$ and that, in both the sequences of f-stops and shutter speeds, the exposure changes by a factor of about two from term to term. Thus, a change in shutter speed can compensate for a change in f-stop, and conversely.

Exercises

1. How do you affect the exposure if you (a) change the shutter speed from 1/250 to 1/125, (b) change the f-stop from f/16 to f/22, (c) do both? **a**. Exposure is doubled.   **b**. Exposure is halved.   **c**. no change

2. Find the geometric mean between 1/500 and 1/30 in the sequence of shutter speeds. What number in the sequence does it approximate?
$\dfrac{1}{122.5}$; $\dfrac{1}{125}$

PROGRAMMING IN PASCAL

The following Pascal program inserts arithmetic means between two numbers specified by the user.

```
PROGRAM arith_means (INPUT, OUTPUT);
VAR
   first, last, common_difference : real;
   num_of_means : integer;


( **************************************************************)
PROCEDURE obtain_first_and_last_terms;
BEGIN
   writeln('Enter the first and last terms with a space between them. ');
   write('If you want to stop, enter the same value for both: ');
   readln(first, last);
END;   (* obtain_first_and_last_terms *)
```

(Program continues on next page.)

```
( ***********************************************************)
PROCEDURE obtain_num_of_means;
BEGIN
  writeln;
  writeln('Enter the number of arithmetic means to be inserted');
  write('between ' ,first:1,' and ' ,last:1,' : ');
  read(num_of_means);
  writeln;
END;   (* obtain_num_of_means *)


( ***********************************************************)
PROCEDURE compute_the_common_difference;
BEGIN
  common_difference := (last − first) / (num_of_means + 1);
END;   (* compute_the_common_difference *)


( ***********************************************************)
PROCEDURE display_the_result;
VAR
  index : integer;
  term : real;
BEGIN
  term := first;
  FOR index := 1 TO num_of_means + 2 DO
    BEGIN
      writeln(term);
      term := term + common_difference;
    END;
END;   (* display_the_result *)


( ***********************************************************)
BEGIN (* Main Program *)
  obtain_first_and_last_terms;
  WHILE first <> last DO
    BEGIN
      obtain_num_of_means;
      compute_the_common_difference;
      display_the_result;
      obtain_first_and_last_terms;
    END;   (* while *)
END.
```

Exercises

1. Enter and run *arith_means*. Use it to check your solutions to Exercises 21-26 in Section 8-2 on page 298.

2. Write a Pascal program to find a user-specified term (the nth term) of
   a. an arithmetic sequence, given the first term and the common difference.
   b. a geometric sequence, given the first term and the common ratio.
3. Write a Pascal program to display a user-specified number of terms (*num*) of an arithmetic sequence or geometric sequence once the user has specified the first term and the respective common difference or common ratio.
4. In a sense, finding the sum of the first n terms of an arithmetic or geometric sequence is trivial because, as you have seen, there exist simple formulas (called *closed forms*) for such sums. For arithmetic sequences the closed form is

$$S_n = \frac{n}{2}(a_1 + a_n)$$

and for geometric sequences the closed form is

$$S_n = \frac{(a_1 - a_1 r^n)}{1 - r}, \quad r \neq 1.$$

Consider, however, the problem of finding the sum of the first n terms of a sequence such as the one below.

$$\frac{1}{1}, \frac{1}{8}, \frac{1}{27}, \frac{1}{64}, \ldots, \frac{1}{n^3}, \ldots$$

There is no known closed form for such a sum. Write a program to find the sum of the first n terms of this sequence after the user specifies n. Impose some reasonable limit on n.
5. Write a program to print the first 50 terms of the sequence in Exercises 2, 3, and 6 on page 323. Use a function with heading

FUNCTION term (n : integer) : real;

to compute each term.

The photo shows a collection of ancient musical instruments. The Application on pages 386 and 387 discusses the "tempered scale," which is fundamental in the design of modern instruments such as the piano.

Chapter 9

Polynomial Functions

Problem Solving Strategies

Generalizing from Specific
Students learn the general procedures for synthetic substitution from specific examples in Section 9-7.
Making an Organized List
In Section 9-10 an organized list proves useful in locating possible real roots of a polynomial equation.
Trial and Error
The use of diagrams and iteration is helpful in estimating roots in Section 9-11.

Solving Quadratic Equations

OBJECTIVES for Sections 9-1 through 9-3:
1. *To solve quadratic equations by completing the square and by using the quadratic formula.*
2. *To use the discriminant of a quadratic equation to determine the nature of the roots of the equation.*
3. *To use the relationship between the roots and coefficients of a quadratic equation to determine the sum and product of the roots when the equation is given, and vice versa.*

9–1 Completing the Square and the Quadratic Formula

Teaching Suggestions
p. T100

Key Ideas

Solve equations by completing the square.
Solve equations by using the quadratic formula.

You learned in Section 6-4 how to solve a quadratic equation of the form $ax^2 + bx + c = 0$, $a \neq 0$, in which the left side can be factored over the integers. For example, you can find the roots of $2x^2 - 5x + 2 = 0$ by factoring:

$$2x^2 - 5x + 2 = 0$$
$$(2x - 1)(x - 2) = 0$$
$$x = \frac{1}{2} \quad \text{or} \quad x = 2$$

Many quadratic equations, such as $x^2 - 8x - 1 = 0$, cannot be factored over the integers. However, an equation of the form

$$(x + d)^2 = e$$

can always be solved over $\mathcal{C}$, as shown on the following page.

Polynomial Functions **341**

1. Solve $2x^2 - 5x + 2 = 0$ by
   a) completing the square,
   b) using the quadratic formula.

   a. $\dfrac{2x^2}{2} - \dfrac{5x}{2} + \dfrac{2}{2} = 0$

   $x^2 - \dfrac{5}{2}x = -1$

   $x^2 - \dfrac{5}{2}x + \dfrac{25}{16} = -1 + \dfrac{25}{16}$

   $\left(x - \dfrac{5}{4}\right)^2 = \dfrac{9}{16}$

   $x - \dfrac{5}{4} = \pm\dfrac{3}{4}$

   $x = \dfrac{5}{4} \pm \dfrac{3}{4}$

   $\therefore$ the solution set is

   $\left\{2, \dfrac{1}{2}\right\}$

   b. $2x^2 - 5x + 2 = 0$
   $a = 2, b = -5, c = 2$

   $x = \dfrac{-b \pm \sqrt{b^2 - 4ac}}{2a}$

   $= \dfrac{-(-5) \pm \sqrt{(-5)^2 - 4(2)(2)}}{2(2)}$

   $= \dfrac{5 \pm \sqrt{25 - 16}}{4}$

   $= \dfrac{5 \pm \sqrt{9}}{4}$

   $= \dfrac{5 \pm 3}{4}$

   $\therefore$ the solution set is $\left\{2, \dfrac{1}{2}\right\}$

2. Solve $x^2 - 6x + 13 = 0$ over $\mathcal{C}$.
   $a = 1, b = -6, c = 13$

   $x = \dfrac{-b \pm \sqrt{b^2 - 4ac}}{2a}$

   $= \dfrac{+6 \pm \sqrt{36 - 52}}{2}$

   $= \dfrac{6 \pm \sqrt{-16}}{2}$

   $= \dfrac{6 \pm 4i}{2}$

   $= 3 \pm 2i$

   $\therefore$ the solution set is
   $\{3 + 2i, 3 - 2i\}$

$$(x + d)^2 = e \qquad (x + 2)^2 = 5 \qquad (x - 3)^2 = -4$$
$$x + d = \pm\sqrt{e} \qquad x + 2 = \pm\sqrt{5} \qquad x - 3 = \pm\sqrt{-4}$$
$$x = -d \pm \sqrt{e} \qquad x = -2 \pm \sqrt{5} \qquad x = 3 \pm 2i$$

We could therefore solve the equation

$$x^2 - 8x - 1 = 0$$

if we could put it in the form $(x + d)^2 = e$. This can be done by the following procedure, called **completing the square**:

$$x^2 - 8x - 1 = 0$$
$$x^2 - 8x = 1$$
$$x^2 - 8x + 16 = 1 + 16$$
$$(x - 4)^2 = 17$$
$$x - 4 = \pm\sqrt{17}$$
$$x = 4 \pm \sqrt{17}$$

In general, given an equation of the form

$$ax^2 + bx + c = 0, \quad a \neq 0,$$

we can apply the same procedure after first dividing both sides by a:

$$x^2 + \dfrac{b}{a}x + \dfrac{c}{a} = 0$$
$$x^2 + \dfrac{b}{a}x = -\dfrac{c}{a}$$

In order to determine what constant term must be added to the left side to make that side the square of a binomial, note that

$$(x + d)^2 = x^2 + 2dx + d^2.$$

The constant term, d^2, is thus the square of one-half the coefficient of the linear term:

$$d^2 = [\tfrac{1}{2}(2d)]^2$$

That is, in order to complete the square of the left side, we should add the square of one-half the coefficient of x to both sides:

$$x^2 + \dfrac{b}{a}x + \left(\dfrac{1}{2} \cdot \dfrac{b}{a}\right)^2 = \left(\dfrac{1}{2} \cdot \dfrac{b}{a}\right)^2 - \dfrac{c}{a}$$
$$x^2 + \dfrac{b}{a}x + \dfrac{b^2}{4a^2} = \dfrac{b^2}{4a^2} - \dfrac{c}{a}$$
$$\left(x + \dfrac{b}{2a}\right)^2 = \dfrac{b^2 - 4ac}{4a^2}$$

Thus, taking the square root of both sides, we obtain

$$x + \dfrac{b}{2a} = \pm\sqrt{\dfrac{b^2 - 4ac}{4a^2}} = \pm\dfrac{\sqrt{b^2 - 4ac}}{2a}.$$

Finally, we have

$$x = \frac{-b \pm \sqrt{b^2 - 4ac}}{2a}.$$

This last equation is called the **quadratic formula**.

EXAMPLE 1 Solve $2y^2 + 6y + 3 = 0$ by **(a)** completing the square, and **(b)** using the quadratic formula.

SOLUTION **a.** $y^2 + \dfrac{6y}{2} + \dfrac{3}{2} = 0$

$$y^2 + 3y = -\frac{3}{2}$$

$$y^2 + 3y + \left(\frac{3}{2}\right)^2 = -\frac{3}{2} + \left(\frac{3}{2}\right)^2$$

$$\left(y + \frac{3}{2}\right)^2 = \frac{-6 + 9}{4} = \frac{3}{4}$$

$$y + \frac{3}{2} = \pm\sqrt{\frac{3}{4}} = \pm\frac{1}{2}\sqrt{3}$$

$$y = -\frac{3}{2} \pm \frac{1}{2}\sqrt{3}$$

$\therefore$ the solution set is $\left\{-\dfrac{3}{2} + \dfrac{1}{2}\sqrt{3}, -\dfrac{3}{2} - \dfrac{1}{2}\sqrt{3}\right\}$.

b. $a = 2, b = 6, c = 3$

$$y = \frac{-b \pm \sqrt{b^2 - 4ac}}{2a}$$

$$= \frac{-6 \pm \sqrt{6^2 - 4(2)(3)}}{2(2)}$$

$$= \frac{-6 \pm \sqrt{12}}{4}$$

$$= -\frac{3}{2} \pm \frac{1}{2}\sqrt{3}$$

$\therefore$ the solution set is $\left\{-\dfrac{3}{2} + \dfrac{1}{2}\sqrt{3}, -\dfrac{3}{2} - \dfrac{1}{2}\sqrt{3}\right\}$.

EXAMPLE 2 Solve $-x^2 + 8x - 25 = 0$ over $\mathcal{C}$.

SOLUTION $a = -1, b = 8, c = -25$
Therefore by the quadratic formula, we have:

$$x = \frac{-b \pm \sqrt{b^2 - 4ac}}{2a} = \frac{-8 \pm \sqrt{8^2 - 4(-1)(-25)}}{2(-1)}$$

$$= \frac{-8 \pm \sqrt{-36}}{-2} = 4 \pm 3i$$

$\therefore$ the solution set is $\{4 + 3i, 4 - 3i\}$.

Polynomial Functions **343**

3. Solve $x^{-2} - 8x^{-1} + 3 = 0$.

$$\frac{1}{x^2} - \frac{8}{x} + 3 = 0$$

$$\left(\frac{1}{x}\right)^2 - 8\left(\frac{1}{x}\right) + 3 = 0$$

$$\left(\frac{1}{x}\right)^2 - 8\left(\frac{1}{x}\right) = -3$$

$$\left(\frac{1}{x}\right)^2 - 8\left(\frac{1}{x}\right) + 16 = -3 + 16$$

$$\left(\frac{1}{x} - 4\right)^2 = 13$$

$$\frac{1}{x} - 4 = \pm\sqrt{13}$$

$$\frac{1}{x} = 4 \pm \sqrt{13}$$

$$x = \frac{1}{4 \pm \sqrt{13}}$$

Rationalizing the denominator, we have:

$$x = \frac{4 \pm \sqrt{13}}{3}$$

$\therefore$ the solution set is

$$\left\{\frac{4 + \sqrt{13}}{3}, \frac{4 - \sqrt{13}}{3}\right\}$$

Additional A Exercises

Solve over $\mathcal{C}$ by completing the square.

1. $y^2 + 10y = 3$
$\{-5 + 2\sqrt{7}, -5 - 2\sqrt{7}\}$

2. $n^2 - 3n - 10 = 0$
$\{5, -2\}$

Solve over $\mathcal{C}$ by using the quadratic formula.

3. $3x^2 + 5x - 4 = 0$
$\left\{\dfrac{-5 + \sqrt{73}}{6}, \dfrac{-5 - \sqrt{73}}{6}\right\}$

4. $4x^2 = 11 + 4x$
$\left\{\dfrac{1 + 2\sqrt{3}}{2}, \dfrac{1 - 2\sqrt{3}}{2}\right\}$

Solve over $\mathcal{C}$ by any method.

5. $3(x + 1)^2 - 5 = 0$
$\left\{-1 + \dfrac{\sqrt{15}}{3}, -1 - \dfrac{\sqrt{15}}{3}\right\}$

6. $x^2 + 10x = -3$
$\{-5 + \sqrt{22}, -5 - \sqrt{22}\}$

Reading Algebra

The quadratic formula may be used to solve an equation if and only if the equation can be written in the form $ax^2 + bx + c = 0$. To help students remember this, ask them to write a paragraph, describing in detail how to solve an equation in the form $ax^2 + bx + c = 0$ using the quadratic formula. Students may need a start such as: "The solution to a quadratic (2nd degree) equation in the form $ax^2 + bx + c = 0$ can be found by taking the opposite of the coefficient of the linear term plus or minus the square root of . . ."

Common Errors

Before using the quadratic formula, emphasize that the equation must be written in the form $ax^2 + bx + c = 0$. The coefficients a, b, and c are often incorrectly identified if the equation is written in another form.

Equations such as $x^4 + 5x^2 + 4 = 0$ and $\left(\frac{1}{x}\right)^2 + 6\left(\frac{1}{x}\right) - 4 = 0$ are said to be *in quadratic form*. Since $x^4 + 5x^2 + 4 = 0$ can be written as

$$(x^2)^2 + 5x^2 + 4 = 0,$$

you can solve first for x^2 and then for x:

$$(x^2 + 1)(x^2 + 4) = 0$$
$$x^2 = -1 \quad \text{or} \quad x^2 = -4$$
$$x = \pm i \quad \text{or} \quad x = \pm 2i$$

Thus the roots over $\mathcal{C}$ of $x^4 + 5x^2 + 4 = 0$ are $i, -i, 2i, -2i$.

EXAMPLE 3   Solve $\left(\frac{1}{x}\right)^2 + 18\left(\frac{1}{x}\right) + 68 = 0$.

SOLUTION   This equation can be solved by first multiplying by x^2 in order to rewrite the equation in the form $ax^2 + bx + c = 0$. We can also apply the methods of this section immediately by treating $\frac{1}{x}$ as the variable.

$$\left(\frac{1}{x}\right)^2 + 18\left(\frac{1}{x}\right) = -68$$

Completing the square, we obtain:

$$\left(\frac{1}{x}\right)^2 + 18\left(\frac{1}{x}\right) + 81 = -68 + 81$$

$$\left(\frac{1}{x} + 9\right)^2 = 13$$

$$\frac{1}{x} + 9 = \pm\sqrt{13}$$

$$\frac{1}{x} = -9 \pm \sqrt{13}$$

$$x = \frac{1}{-9 \pm \sqrt{13}}$$

Rationalizing the denominator, we have:

$$\frac{1}{-9 + \sqrt{13}} = \frac{1(-9 - \sqrt{13})}{(-9 + \sqrt{13})(-9 - \sqrt{13})}$$

$$= \frac{-9 - \sqrt{13}}{68}$$

and

$$\frac{1}{-9 - \sqrt{13}} = \frac{1(-9 + \sqrt{13})}{(-9 - \sqrt{13})(-9 + \sqrt{13})}$$

$$= \frac{-9 + \sqrt{13}}{68}$$

$\therefore$ the solution set is $\left\{\dfrac{-9 - \sqrt{13}}{68}, \dfrac{-9 + \sqrt{13}}{68}\right\}$

Oral Exercises

State the number that must be added to both sides of the equation in order to complete the square.

1. $x^2 - 10x = 2$ 25

2. $x^2 + 12x = -4$ 36

3. $y^2 - 5y = -7$ $\frac{25}{4}$

4. $t^2 - \frac{4}{3}t = -1$ $\frac{4}{9}$

5. $r^2 - \frac{2}{5}r = \frac{1}{5}$ $\frac{1}{25}$

6. $n^2 - \frac{5}{2}n = -\frac{5}{4}$ $\frac{25}{16}$

Tell which method—factoring, completing the square, or using the quadratic formula—you would use to solve each equation. Answers may vary.

7. $x^2 - 6x = 1$ Comp. the square

8. $x^2 - 8x + 15 = 0$ factoring

9. $5x^2 + 3x - 7 = 0$ formula

10. $15x^2 + 8x = 0$ factoring

State values of *a, b,* and *c* that would be used in the quadratic formula.

$a = 1, b = -6, c = 4$ $a = 3, b = -4, c = -1$ $a = 2, b = -8, c = 11$

11. $k^2 - 6k = -4$

12. $3y^2 = 4y + 1$

13. $11 = 8x - 2x^2$

14. $8v^2 = 3 - 12v$

15. $7z^2 = 6z$

16. $9x^2 + 4 = 8x$

$a = 8, b = 12, c = -3$ $a = 7, b = -6, c = 0$ $a = 9, b = -8, c = 4$

Written Exercises

Solve over $\mathcal{C}$ by the indicated method; give irrational answers in simple radical form. In Exercises 1–10 solve by completing the square.

A   **1–6.** Solve the equations in Oral Exercises 1–6.

7. $2u^2 - 8u + 9 = 0$ $\{\frac{4 + i\sqrt{2}}{2}, \frac{4 - i\sqrt{2}}{2}\}$

8. $8v^2 + 4v - 3 = 0$ $\{\frac{-1 \pm \sqrt{7}}{4}\}$

9. $3y^2 + 16y + 8 = 0$ $\{\frac{-8 \pm 2\sqrt{10}}{3}\}$

10. $2x^2 - 6x + 5 = 0$ $\{\frac{3 \pm i}{2}\}$

Solve by using the quadratic formula.

11–16. Solve the equations in Oral Exercises 11–16.

17. $5 + 2k^2 = 10k$ $\{\frac{5 \pm \sqrt{15}}{2}\}$

18. $6v^2 + 16v = 3$ $\{\frac{-8 \pm \sqrt{82}}{6}\}$

19. $\frac{1}{9}n^2 = \frac{2}{9}n - 5$ $\{1 \pm 2i\sqrt{11}\}$

20. $\frac{5}{3}(t^2 - 4t) = -7$ $\{2 \pm \frac{i\sqrt{5}}{5}\}$

Solve by any method.

21. $2(x + 3)^2 + 7 = 0$ $\{-3 \pm \frac{i\sqrt{14}}{2}\}$

22. $\frac{2}{5}x^2 - 3x = 0$ $\{0, \frac{15}{2}\}$

23. $7x^2 - 12x + 4 = 0$

24. $\frac{(3x - 1)^2}{2} = 10$

25. $3x^2 - 2x + 1 = 0$ $\{\frac{1 \pm i\sqrt{2}}{3}\}$

26. $-2x^2 + 2x - 5 = 0$ $\{\frac{1 \pm 3i}{2}\}$

27. $4x^2 - 12x + 7 = 0$

28. $2x^2 + 15x + 29 = 0$

B   **29.** $x^2\sqrt{3} + 6x + 7\sqrt{3} = 0$ $\{-\sqrt{3} \pm 2i\}$

30. $8x^2 - 12\sqrt{5}x + 9 = 0$

31. $\frac{29}{x + 3} + 2x = 10$ $\{1 \pm \frac{\sqrt{6}}{2}\}$

32. $\frac{x + 4}{x - 2} = \frac{3x - 4}{x + 8}$ $\{12, -1\}$

Polynomial Functions   **345**

23. $\left\{\dfrac{6 \pm 2\sqrt{2}}{7}\right\}$

24. $\left\{\dfrac{1 \pm 2\sqrt{5}}{3}\right\}$

27. $\left\{\dfrac{3 \pm \sqrt{2}}{2}\right\}$

28. $\left\{\dfrac{-15 \pm i\sqrt{7}}{4}\right\}$

30. $\left\{\dfrac{3\sqrt{5} \pm 3\sqrt{3}}{4}\right\}$

37. $\left\{2, -1, -1 \pm i\sqrt{3}, \right.$

$\left. \dfrac{1 \pm i\sqrt{3}}{2}\right\}$

38. $\left\{-4, -1, -2 \pm 2i\sqrt{3}, \right.$

$\left. \dfrac{1 \pm i\sqrt{3}}{2}\right\}$

40. $\left\{\dfrac{-11 \pm \sqrt{21}}{10}\right\}$

Mixed Review

Simplify each expression.

1. $2\sqrt{18} + 3\sqrt{50}$ $21\sqrt{2}$

2. $\dfrac{3}{\sqrt{5} - 2}$ $3\sqrt{5} + 6$

3. $(2p^2q^{-3})(2pq^3)^{-2}$ $\dfrac{1}{2q^9}$

4. $(n^3 - y)(n^3 + y)$ $n^6 - y^2$

5. $\left[8\left(3 - \dfrac{1}{2}\right) - 12\left(\dfrac{2}{3} + \dfrac{3}{4}\right)\right]7$

   21

Solve by any method.

33. $\dfrac{5}{x - 3} + \dfrac{4}{x} + 3 = 0$ $\{2, -2\}$

34. $\dfrac{7}{x - 1} - 3 = \dfrac{2}{x + 1}$ $\{3, -\frac{4}{3}\}$

35. $x^4 - 5x^2 - 36 = 0$ $\{\pm 2i, \pm 3\}$

36. $x^4 - 6x^2 + 5 = 0$ $\{\pm 1, \pm\sqrt{5}\}$

37. $x^6 - 7x^3 - 8 = 0$

38. $x^6 + 65x^3 + 64 = 0$

39. $\left(\dfrac{1}{x}\right)^2 - 2\left(\dfrac{1}{x}\right) - 1 = 0$ $\{-1 \pm \sqrt{2}\}$

40. $\left(\dfrac{1}{x + 1}\right)^2 - \left(\dfrac{1}{x + 1}\right) = 5$

41. $x^2 + ix + 6 = 0$ $\{2i, -3i\}$

42. $x^2 - 2ix + 3 = 0$ $\{3i, -i\}$

43. $x^2 - (2 + 3i)x + (3 + 3i) = 0$ $\{1 + \frac{3 \pm \sqrt{17}}{2}i\}$

44. $ix^2 + (1 - 2i)x - (1 - 3i) = 0$
$\{1 + 2i, 1 - i\}$

C 45. **a.** Factor $x^4 + 4$ over the real numbers. $(x^2 + 2x + 2)(x^2 - 2x + 2)$
(*Hint*: You can rewrite $x^4 + 4$ as $x^4 + 4x^2 + 4 - 4x^2$)
   b. Use the same method to factor $x^4 + 1$. $(x^2 + x\sqrt{2} + 1)(x^2 - x\sqrt{2} + 1)$

46. Solve the equation $\sqrt{2x - x^2} = (x - 1)^2$.
(*Hint*: $2x - x^2 = 1 - (1 - 2x + x^2)$) $\{1 \pm \frac{\sqrt{-2 + 2\sqrt{5}}}{2}\}$

Problems

Express each irrational answer **(a)** in simple radical form and **(b)** using a calculator or Table 3, as a decimal approximation to the nearest tenth. Give *only* answers that are physically possible (e.g., distances cannot be negative).

A 1. At time t (in seconds) the height y above the ground (in meters) of an object thrown straight up with an initial velocity v_0 is given by

$$y = -4.9t^2 + v_0t.$$

How many seconds will it take an object thrown upward at a velocity of 14 m/s to reach a height of 7 m? **a.** $\frac{10 + \sqrt{30}}{7}$ s, $\frac{10 - \sqrt{30}}{7}$ s   **b.** 2.2 s, 0.6 s

2. When two resistors of resistances R_1 and R_2 are connected in series, their combined resistance is $R_1 + R_2$. When they are connected in parallel, their combined resistance is $\dfrac{R_1R_2}{R_1 + R_2}$. Two resistors are found to have a combined resistance of 10 Ω (ohms) when connected in series and 2 Ω when connected in parallel. What is the resistance of each? **a.** $(5 + \sqrt{5})$ Ω, $(5 - \sqrt{5})$ Ω **b.** 7.2 Ω, 2.8 Ω

3. A U.S. Postal Service regulation requires that a rectangular package have a "length-plus-girth" measurement of at most 108 in., where the girth is the perimeter of a cross-section perpendicular to the length. If a package with the maximum length-plus-girth allowable has a length of 40 in. and a volume of 11,000 in.^{3}, what are its height and width? **a.** $(17 + \sqrt{14})$ in., $(17 - \sqrt{14})$ in.   **b.** 20.7 in., 13.3 in.

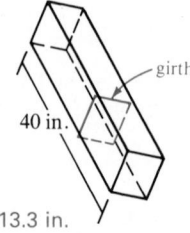

girth

40 in.

346   *Chapter 9*

4. Two cones have the same volume. One of them has a radius 3 cm less than the other and a height twice that of the first. What are the radii of the two cones? **a.** $(6 + 3\sqrt{2})$ cm and $(3 + 3\sqrt{2})$ cm   **b.** 10.2 cm and 7.2 cm

5. Two cubical boxes have the same surface area although one box is open at the top and the other is closed. Each edge of the closed box is 1 cm shorter than each edge of the open box. Find the length of an edge of each box. **a.** $(6 + \sqrt{30})$ cm and $(5 + \sqrt{30})$ cm   **b.** 11.5 cm and 10.5 cm

6. A frame of uniform width surrounds a 16 cm by 20 cm picture so that the total area of the framed picture is 464 cm². What is the width of the frame? **a.** $(-9 + 3\sqrt{13})$ cm   **b.** 1.8 cm

B 7. A machine part, in the shape of a cone on top of a cylinder, has a rim whose width is 2 cm, as shown. The cone and the cylinder each have height 6 cm, and the volume of the entire part is 102π cm³. What is the radius of the cylinder? **a.** $\dfrac{1 + 4\sqrt{3}}{2}$ cm   **b.** 4.0 cm

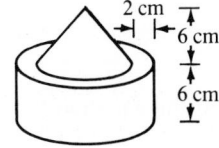

8. A potter uses 55π cm³ of clay to produce a cylindrical container that has a bottom, but no top. The thickness of the walls and base is 1 cm. What is the outside radius of the container if the height and radius are equal? **a.** $\dfrac{1 + \sqrt{73}}{2}$ cm   **b.** 4.8 cm

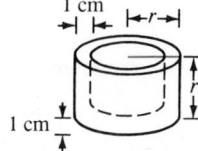

9. A rectangular pen is subdivided by two dividers that meet at the midpoint P of the base of the rectangle, as shown. Each divider is 10 m long, and the perimeter of the pen is 30 m. What are the dimensions of the rectangle? **a.** $(3 + 2\sqrt{11})$ m by $(12 - 2\sqrt{11})$ m   **b.** 9.6 m by 5.4 m

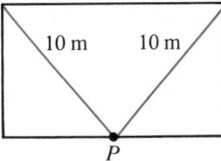

10. A right triangle with a hypotenuse 10 cm long is to have an area of 15 cm². What should be the lengths of the legs? **a.** $\sqrt{10}$ cm, $3\sqrt{10}$ cm   **b.** 3.2 cm, 9.5 cm

---

9–2 The Discriminant

The number $b^2 - 4ac$ (*not* including the radical sign) in the quadratic formula is called the **discriminant** of the corresponding quadratic equation, $ax^2 + bx + c = 0$, because it enables us to discriminate among the possible natures of the roots of such an equation. If we denote the discriminant by D, then the quadratic formula can be written as

$$x = \frac{-b \pm \sqrt{D}}{2a}.$$

We can then distinguish the three cases on the following page.

Polynomial Functions **347**

Teaching Suggestions
p. T100

Related Activities p. T100

Key Ideas

Use the discriminant to determine the nature of the roots of a quadratic equation.

Chalkboard Examples

1. Determine the nature of the roots of $3x^2 + 2x - 6 = 0$.
   $a = 3, b = 2, c = -6$
   $D = b^2 - 4ac =$
   $4 - 4(3)(-6) = 76$
   Since $D > 0$, there are two unequal real roots, which are irrational since D is not the square of a rational number.

2. What is the nature of the zeros of $\{(x, y): y = x^2 - 2x\sqrt{5} + 5\}$?
   $x^2 - 2x\sqrt{5} + 5 = 0$
   $a = 1, b = -2\sqrt{5}, c = 5$
   $D = b^2 - 4ac =$
   $(-2\sqrt{5})^2 - 4 \cdot 1 \cdot 5 =$
   $20 - 20 = 0$
   Since $D = 0$, there is one real zero (a double zero). Although D is the square of a rational number $(0 = 0^2)$, the real zero is an irrational number because b is irrational.

Additional A Exercises

For each equation, find the value of the discriminant, D, and tell how many real and how many imaginary roots the equation has. If it has any real roots, tell whether they are rational.

1. $x^2 - 2x + 2 = 0$
   $D = -4$, 2 imag. roots

2. $2x^2 + 7x + 4 = 0$
   $D = 17$, 2 real irrat. roots

3. $7x^2 - 9x + 2 = 0$
   $D = 25$, 2 real rat. roots

4. $-3x^2 - 12 = 0$
   $D = -144$, 2 imag. roots

5. $-2x^2 + 8x - 9 = 0$
   $D = -8$, 2 imag. roots

6. $9x^2 + 42x + 49 = 0$
   $D = 0$, 1 double real rational root

Case 1: $D > 0$

In this case, since $\sqrt{D}$ is a positive real number, the quadratic formula gives two distinct real solutions,

$$x = \frac{-b + \sqrt{D}}{2a} \quad \text{and} \quad x = \frac{-b - \sqrt{D}}{2a}.$$

Case 2: $D = 0$

In this case, since $\sqrt{D} = \sqrt{0} = 0$, the two solutions, $x = \frac{-b + \sqrt{D}}{2a}$ and $x = \frac{-b - \sqrt{D}}{2a}$, both reduce to

$$x = -\frac{b}{2a}$$

and the equation has a *double root* (see page 214).

Case 3: $D < 0$

In this case, since D is negative, $\sqrt{D}$ is imaginary and

$$D = i\sqrt{|D|}.$$

The solutions are given by

$$x = \frac{-b + i\sqrt{|D|}}{2a} \quad \text{and} \quad x = \frac{-b - i\sqrt{|D|}}{2a}.$$

The results of Cases 1–3 are summarized in the following box.

Theorem. For all real numbers b and c, and all nonzero real numbers a, the quadratic equation $ax^2 + bx + c = 0$ has

1. two different real roots if $b^2 - 4ac > 0$;
2. one double real root if $b^2 - 4ac = 0$;
3. two imaginary complex conjugate roots if $b^2 - 4ac < 0$.

The discriminant also enables you to tell whether the roots of a quadratic equation with *rational coefficients* are rational numbers. If a, b, and c are rational numbers, $a \ne 0$, then $\frac{-b \pm \sqrt{b^2 - 4ac}}{2a}$ denotes a rational number if and only if $\sqrt{b^2 - 4ac}$ is rational (Section 7-3). But $\sqrt{b^2 - 4ac}$ is rational if and only if $b^2 - 4ac$ is the square of a rational number. Thus the following is true:

A quadratic equation with rational coefficients has rational roots if and only if its discriminant is the square of a rational number.

---

348 *Chapter 9*

EXAMPLE 1 Determine the nature of the roots of $4x^2 + 5x - 3 = 0$.

SOLUTION $a = 4$, $b = 5$, $c = -3$
$D = b^2 - 4ac = 5^2 - 4(4)(-3) = 73$
Since $D > 0$, there are two unequal real roots.
Since D is not the square of a rational number, the roots are irrational numbers.

In the domain of a function f, any value of x that satisfies the equation $f(x) = 0$ is said to be a **zero** of f. Notice that "zeros of the function f" and "roots of the equation $f(x) = 0$" are just different ways of referring to the same numbers.

EXAMPLE 2 What is the nature of the zeros of $4x^2 - 4x\sqrt{3} + 3$?

SOLUTION $a = 4$, $b = -4\sqrt{3}$, $c = 3$
$D = b^2 - 4ac = (-4\sqrt{3})^2 - 4 \cdot 4 \cdot 3 = 48 - 48 = 0$
Since $D = 0$, there is one real zero (a double zero).
Although D is the square of a rational number ($0 = 0^2$), the real zero is an irrational number because b is irrational.

Oral Exercises

Find the value of the discriminant D of each equation.

1. $x^2 - 3x + 7 = 0$ -19

2. $10x^2 - 8x - 1 = 0$ 104

3. $5x^2 - 6x = 0$ 36

4. $8x^2 + 1 = 0$ -32

5. $2x^2 + 12x + 18 = 0$ 0

6. $3x^2 - 7x + 4 = 0$ 1

Describe the nature of the roots of a quadratic equation having the given value of the discriminant.

7. $D = 12$
two unequal real roots

8. $D = -4$
two complex conjugate roots

9. $D = 0$
one real root

10. $D = 81$
two unequal rational roots

Written Exercises

For each equation, find the value of the discriminant D, and tell how many real and how many imaginary roots the equation has. If it has any real roots, tell whether they are rational.

A

1. $x^2 + 3x + 3 = 0$ -3; 2 imaginary roots

2. $3x^2 + 10x - 2 = 0$

3. $5x^2 - 11x + 6 = 0$ 1; 2 real rational roots

4. $7x^2 - 15x = 0$

5. $2x^2 + 5x - 8 = 0$ 89; 2 real irrational roots

6. $-4x^2 + 4x - 5 = 0$

7. $8x^2 - 12x + \frac{9}{2} = 0$ 0; one double root

8. $5x^2 - 11x + 2 = 0$

9. $-6x^2 + 72 = 0$ 1728; 2 real irrational roots

10. $\frac{3}{2}x^2 + x - 20 = 0$

Polynomial Functions **349**

Mixed Review

1. Solve by Cramer's Rule.
$2x + y - z = 3$
$x - 2y - z = -3$
$-x + y - 3z = 5$
$\{(0, 2, -1)\}$

2. Solve $x^2 - 4x + 1 = 0$. $\{2 \pm \sqrt{3}\}$

3. Simplify $\sqrt{120}$. $2\sqrt{30}$

4. Factor $3x^2 + 18x - 120$.
$3(x - 4)(x + 10)$

5. Simplify $2\sqrt{24} - \sqrt{54} + 5\sqrt{150}$. $26\sqrt{6}$

Suggested Assignments

Minimum
  349/2–16 even, 17, 18
Extended Alg. with Trig.
  349/1–11 odd, 12–18
Enriched Alg.
  349/7–13 odd, 14–20
Enriched Alg. with Trig.
  349/7–13 odd, 14–20

Additional Answers
Written Exercises

2. 124; 2 real irrational roots

4. 225; 2 real rational roots

6. -64; 2 imaginary roots

8. 81; 2 real rational roots

10. 121; 2 real rational roots

Teaching Suggestions
p. T100

Key Ideas

Find the sum and the product of the roots of a quadratic equation.
Given the roots, find the quadratic equation to which they belong.

Determine the value(s) of k for which the given equation will have exactly one real root.

B 11. $2x^2 - 10x + 5k = 0$ $2\frac{1}{2}$

     13. $x^2 - 2kx + k + 6 = 0$ $3, -2$

     12. $5x^2 - kx + 8 = 0$ $\pm 4\sqrt{10}$

     14. $x^2 + kx + 3 - 2k = 0$ $-4 \pm 2\sqrt{7}$

Determine the value(s) of k for which the given equation will have two distinct real roots.

15. $3x^2 - 6x + k = 0$ $\{k: k < 3\}$

16. $5kx^2 + 4x + 2 = 0$ $\{k: k < 0.4\}$

Determine the value(s) of k for which the given equation will have imaginary (conjugate) roots.

17. $(2k - 1)x^2 - 8x + 1 = 0$ $\{k: k > \frac{17}{2}\}$

18. $x^2 + 4(k + 1)x + 4k^2 = 0$ $\{k: k < -\frac{1}{2}\}$

Exercises 19–20 refer to a quadratic equation of the form $ax^2 + bx + c = 0$. Prove each statement.

C 19. If a and b are nonzero rational numbers and $c = 0$, then the equation has two rational roots.

     20. If a and c have opposite signs, then the equation has two real roots.

9–3 Roots and Coefficients of a Quadratic Equation

Recall (page 213) that

$$x - 3 = 0 \qquad \text{or} \qquad x + 2 = 0$$

is equivalent to the equation $(x - 3)(x + 2) = 0$. In general, if any real or complex numbers r_1 and r_2 are the roots of a quadratic equation in x, then the quadratic equation must be equivalent to $(x - r_1)(x - r_2) = 0$:

$$x^2 - (r_1 + r_2)x + r_1 r_2 = 0$$

By transforming the equation $ax^2 + bx + c = 0$, $a \neq 0$, to

$$x^2 + \frac{b}{a}x + \frac{c}{a} = 0,$$

you can deduce (Exercise 19, page 352) the following theorem.

> *Theorem.* The roots of the equation $ax^2 + bx + c = 0$, $a \neq 0$, are r_1 and r_2, if and only if $r_1 + r_2 = -\frac{b}{a}$ and $r_1 r_2 = \frac{c}{a}$.

350   *Chapter 9*

EXAMPLE 1   Find the values of b and c if the equation $3x^2 + bx + c = 0$ has as its solution set $\left\{\dfrac{1 + 2\sqrt{7}}{3}, \dfrac{1 - 2\sqrt{7}}{3}\right\}$.

SOLUTION

$$r_1 + r_2 = -\frac{b}{a} \qquad\qquad r_1 r_2 = \frac{c}{a}$$

$$\frac{1 + 2\sqrt{7}}{3} + \frac{1 - 2\sqrt{7}}{3} = -\frac{b}{3} \qquad\qquad \left(\frac{1 + 2\sqrt{7}}{3}\right)\left(\frac{1 - 2\sqrt{7}}{3}\right) = \frac{c}{3}$$

$$\frac{2}{3} = -\frac{b}{3} \qquad\qquad \frac{-27}{9} = \frac{c}{3}$$

$$b = -2 \qquad\qquad c = -9$$

$$\therefore b = -2 \text{ and } c = -9$$

EXAMPLE 2   Find a quadratic equation whose roots are $\dfrac{1}{2} + i\sqrt{5}$ and $\dfrac{1}{2} - i\sqrt{5}$.

SOLUTION $\left(\dfrac{1}{2} + i\sqrt{5}\right) + \left(\dfrac{1}{2} - i\sqrt{5}\right) = 1 = -\dfrac{b}{a}$

$\left(\dfrac{1}{2} + i\sqrt{5}\right)\left(\dfrac{1}{2} - i\sqrt{5}\right) = \dfrac{1}{4} + 5 = \dfrac{21}{4} = \dfrac{c}{a}$

By letting $a = 4$, we can avoid fractions as coefficients:

$$1 = -\frac{b}{a} \qquad\qquad \frac{21}{4} = \frac{c}{a}$$

$$1 = -\frac{b}{4} \qquad\qquad \frac{21}{4} = \frac{c}{4}$$

$$b = -4 \qquad\qquad c = 21$$

$\therefore$ an equation with the given roots is $4x^2 - 4x + 21 = 0$.

Check: Using the quadratic formula to show that the roots of $4x^2 - 4x + 21 = 0$ are $\dfrac{1}{2} \pm i\sqrt{5}$ is left to you.

Oral Exercises

For each equation state (a) the sum of the roots and (b) the product of the roots.

1. $x^2 - 6x + 2 = 0$ a. 6   b. 2   2. $2x^2 + 5x = 0$ a. $-\frac{5}{2}$ b. 0   3. $-2x^2 + 7x - 3 = 0$ a. $\frac{7}{2}$ b. $\frac{3}{2}$

4. $6x^2 - 8x - 3 = 0$ 5. $3x^2 - 15 = 0$ 6. $-x^2 - 4x + 7 = 0$

    a. $\frac{4}{3}$ b. $-\frac{1}{2}$ a. 0   b. -5 a. -4 b. -7

Find a, b, or c if the given equation has the given roots.

7. $2x^2 + bx + 3 = 0; \ 3, \dfrac{1}{2}$ $b = -7$ 8. $2x^2 + x + c = 0; \ \dfrac{5}{2}, -3$ $c = -15$

    $a = 2$

9. $ax^2 - 8x + 26 = 0; \ 2 + 3i, \ 2 - 3i$ 10. $3x^2 - bx - 48 = 0; \ 4, -4$ $b = 0$

11. $x^2 - 6x + c = 0; \ 3 + \sqrt{2}, \ 3 - \sqrt{2}$ 12. $ax^2 + 6x - 12 = 0; \ 1 + i\sqrt{3}, \ 1 - i\sqrt{3}$

    $c = 7$ $a = -3$

Polynomial Functions   **351**

Suggested Assignments

Minimum
  First day
    352/1–18
  Second day
    R 353/Self-Test 1
    S 349/1–15 odd
Extended Alg. with Trig.
    352/1–13 odd, 15–18
    R 353/Self-Test 1
    S 349/2–14 even
Enriched Alg.
    352/7–15 odd, 16–20
    R 353/Self-Test 1
    S 349/1–12 even
Enriched Alg. with Trig
    352/7–15 odd, 16–20
    R 353/Self-Test 1
    S 349/1–12 even

Additional A Exercises

Write a quadratic equation having the given solution set.

1. $\{3, -5\}$ $x^2 + 2x - 15 = 0$

2. $\left\{\dfrac{5}{2}\right\}$ $x^2 - 5x + \dfrac{25}{4} = 0$

3. $\{1 + \sqrt{3}, 1 - \sqrt{3}\}$
   $x^2 - 2x - 2 = 0$

4. $\{2 + 6i, 2 - 6i\}$
   $x^2 - 4x + 40 = 0$

5. $\left\{\dfrac{1}{2} + \dfrac{1}{4}i\sqrt{7}, \dfrac{1}{2} - \dfrac{1}{4}i\sqrt{7}\right\}$
   $x^2 - x + \dfrac{11}{16} = 0$

6. $\left\{\dfrac{4 \pm 2i\sqrt{5}}{3}\right\}$
   $3x^2 - 8x + 12 = 0$

Additional Answers
Written Exercises
(See p. 385.)

Written Exercises

Write a quadratic equation having the given solution set.

A 1. $\{-2, 7\}$ $x^2 - 5x - 14 = 0$

2. $\{-3, -5\}$ $x^2 + 8x + 15 = 0$

3. $\left\{-\dfrac{3}{2}\right\}$ $4x^2 + 12x + 9 = 0$

4. $\{1 + \sqrt{7}, 1 - \sqrt{7}\}$ $x^2 - 2x - 6 = 0$

5. $\{6\sqrt{2}, -6\sqrt{2}\}$ $x^2 - 72 = 0$

6. $\{1 + 3i, 1 - 3i\}$ $x^2 - 2x + 10 = 0$

7. $\{3 + i\sqrt{5}, 3 - i\sqrt{5}\}$ $x^2 - 6x + 14 = 0$

8. $\{5 + \sqrt{3}, 5 - \sqrt{3}\}$ $x^2 - 10x + 22 = 0$

9. $\left\{\dfrac{3}{2} - \dfrac{\sqrt{5}}{2}i, \dfrac{3}{2} + \dfrac{\sqrt{5}}{2}i\right\}$ $2x^2 - 6x + 7 = 0$

10. $\left\{\dfrac{4}{3} - \dfrac{2}{3}i, \dfrac{4}{3} + \dfrac{2}{3}i\right\}$ $9x^2 - 24x + 20 = 0$

11. $\{7 - 2\sqrt{3}, 7 + 2\sqrt{3}\}$ $x^2 - 14x + 37 = 0$

12. $\{\sqrt{2} + 3i, \sqrt{2} - 3i\}$
       $x^2 - 2x\sqrt{2} + 11 = 0$

In Exercises 13–18 determine a value for k so that the given conditions are satisfied.

B 13. One root of $6x^2 + x + k = 0$ is $\dfrac{1}{2}$. -2

14. One root of $x^2 - 4x + k = 0$ is $2 + \sqrt{11}$. -7

15. One root of $2x^2 + kx - 21 = 0$ is $-\dfrac{3}{2}$. -11

16. One root of $4x^2 + kx + 25 = 0$ is $\dfrac{3}{2} + 2i$. -12

17. One root of $kx^2 - 3x + 10 = 0$ is twice the other. $\frac{1}{5}$

18. One root of $16x^2 - 9x + k = 0$ is three times the other. $\frac{243}{256}$

Exercises 19–23 refer to an equation of the form $ax^2 + bx + c = 0$, in which a, b, and c are real numbers and $a \neq 0$.

C 19. Use the quadratic formula to prove that the roots of such an equation must have $-\dfrac{b}{a}$ as their sum and $\dfrac{c}{a}$ as their product.

20. Prove that the sum of the *reciprocals* of the roots of such an equation is $-\dfrac{b}{c}$.

21. Prove that the sum of the *squares* of the roots of such an equation is $\dfrac{b^2 - 2ac}{a^2}$. (*Hint:* Consider $(r_1 + r_2)^2$. Derive an expression for $r_1^2 + r_2^2$ from this.)

22. *Without* using the quadratic formula, prove that if such an equation has two imaginary roots, then these roots must be complex conjugates. (*Hint:* See Exercise 26 on page 284.)

23. Prove that if such an equation has *rational* coefficients (that is, if a, b, and c are rational) and has two roots of the form $p + q\sqrt{2}$ and $r + s\sqrt{2}$, with p, q, r, and s nonzero rational numbers, then $p = r$ and $q = -s$.

352   *Chapter 9*

Computer Exercises For students with computer experience

1. Write a program to find the roots of a quadratic equation of the form

$$ax^2 + bx + c = 0,$$

given the values of a, b, and c. If there is one double root, the output should print its value only once. If the roots are nonreal complex conjugates, they should be printed in the form $a + bi$.

Run the program in Exercise 1 for each equation.

2. $x^2 - 7x + 12 = 0$ {3, 4}

3. $4x^2 + 20x + 25 = 0$ {−2.5}

4. $x^2 - 6x + 13 = 0$ {3 + 2i, 3 − 2i}

5. $-3x^2 - 7x - 2 = 0$ {−0.3333, −2}

6. Given two real numbers r_1 and r_2 or the coefficients m and n of a pair of complex conjugates $m + ni$ and $m - ni$, write a program that will print a quadratic equation of the form

$$x^2 + bx + c = 0$$

that has the given numbers as roots. Have the program give the choice of entering two real numbers or the real and imaginary parts of a pair of complex conjugates.   7. $x^2 + 2x - 15 = 0$ 8. $x^2 - 2x + 5 = 0$

Run the program in Exercise 6 for each pair of numbers.   9. $x^2 + 5x + 6.25 = 0$

7. $-5, 3$ 8. $1 \pm 2i$ 9. $-2.5, -2.5$ 10. $0, -6$ 11. $-3 \pm 4i$ 12. $\pm 8i$

13. Write a program that will allow you to input four integers p, q, r, and s and give you the option to treat them either as two rational numbers $\frac{p}{q}$ and $\frac{r}{s}$ or as the real and imaginary parts of a pair of complex conjugates $\frac{p}{q} + \frac{r}{s}i$. The program should then print an equation of the form $ax^2 + bx + c = 0$, with a, b, and c *integers*, that has the given numbers as roots.

Run the program in Exercise 13 for each pair.   17. $100x^2 + 100x + 61 = 0$

14. $\frac{2}{3}, \frac{1}{4}$
$12x^2 - 11x + 2 = 0$

15. $\frac{3}{2} \pm i$
$4x^2 - 12x + 13 = 0$

16. $-5, \frac{3}{4}$
$4x^2 + 17x - 15 = 0$

17. $-\frac{1}{2} \pm \frac{3}{5}i$

10. $x^2 + 6x = 0$ 11. $x^2 + 6x + 25 = 0$ 12. $x^2 + 64 = 0$

▌ Self-Test 1

VOCABULARY   completing the square (p. 342)   discriminant (p. 347)
quadratic formula (p. 343)   zero of a function (p. 349)

1. Solve $2x^2 - 6x + 9 = 0$ by completing the square. {$\frac{3 \pm 3i}{2}$}   *Obj. 1, p. 341*

Polynomial Functions   **353**

353

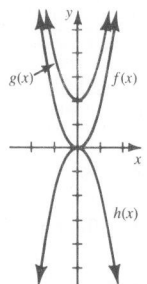
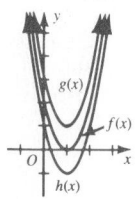
Solve the equation over $\mathcal{C}$ by using the quadratic formula.

2. $5x^2 + 10x - 2 = 0$
3. $x^2 - 6x + 29 = 0$ $\{3 \pm 2i\sqrt{5}\}$

Without solving the equation, determine the nature of its roots.

4. $4x^2 + 8x + 5 = 0$
5. $9x^2 - 30x + 25 = 0$ *Obj. 2, p. 341*

6. Give **(a)** the sum and **(b)** the product of the roots of the equation   *Obj. 3, p. 341*
   $2x^2 - 5x + 3 = 0$. a. $\frac{5}{2}$ b. $\frac{3}{2}$

7. Write a quadratic equation having solution set $\{2 + 5i, 2 - 5i\}$. $x^2 - 4x + 29 = 0$

Check your answers with those at the back of the book.

Quadratic Functions and Their Graphs

OBJECTIVES for Sections 9-4 through 9-6:
1. *To find an equation of the axis of symmetry and the coordinates of the vertex of the graph of an equation of the form $y = a(x - h)^2 + k$, and to determine whether the vertex is a maximum or a minimum point.*
2. *To sketch the graph of a given quadratic function.*
3. *To solve extreme-value problems involving quadratic functions.*
4. *To solve quadratic inequalities.*

9–4 The Graph of $y = a(x - h)^2 + k$

The function $f(x) = x^2$ is called a *quadratic function* or a *polynomial function of degree two*. A **quadratic function** f over $\mathcal{R}$ is a function with domain $\mathcal{R}$ and values given by a quadratic polynomial; that is,

$$f(x) = ax^2 + bx + c,$$

where a, b, and c are real numbers and $a \neq 0$. The graph of a quadratic function over $\mathcal{R}$ is called a **parabola**.

The parabolas shown in Figure 1 are the graphs of the two quadratic functions

$$f(x) = x^2 \qquad \text{and} \qquad g(x) = x^2 + 3.$$

Both parabolas are symmetric with respect to the y-axis. Notice that the **minimum** (lowest) **point** of the graph of g, which is $(0, 3)$, is 3 units above that of f, which is $(0, 0)$.

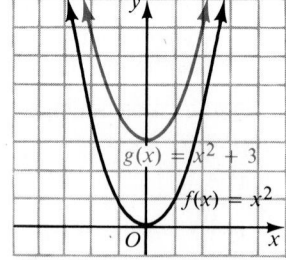

Figure 1

x	$f(x)$		x	$g(x)$
-1	1		-1	4
0	0		0	3
1	1		1	4

354   *Chapter 9*

Now compare the graphs of f and g with those of the quadratic functions F and G in Figure 2, where

$$F(x) = -\tfrac{1}{2}x^2 \quad \text{and} \quad G(x) = -\tfrac{1}{2}x^2 - 4.$$

Again, both parabolas are symmetric with respect to the line $x = 0$ (the y-axis). But in this case, the graph of G has a **maximum** (highest) **point,** that is, a point with greatest ordinate, which is 4 units below the maximum of F.

In general, the graphs of

$$f(x) = ax^2 \quad \text{and} \quad g(x) = ax^2 + k$$

are both symmetric with respect to the line $x = 0$. Moreover, if $k > 0$, then each point of the graph of g is k units above the corresponding point of the graph of f; and if $k < 0$, it is $|k|$ units below.

Next let us compare the graph of $f(x) = ax^2$ with those of two other general quadratic functions, of the form

$$p(x) = a(x - h)^2, a \neq 0,$$

and

$$q(x) = a(x - h)^2 + k, a \neq 0.$$

If we let $a = \tfrac{1}{2}$, $h = 2$, and $k = 3$, we obtain the graphs in Figure 3.

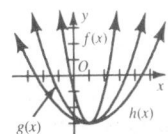

Figure 2

x	F(x)		x	G(x)
−2	−2		−2	−6
0	0		0	−4
2	−2		2	−6

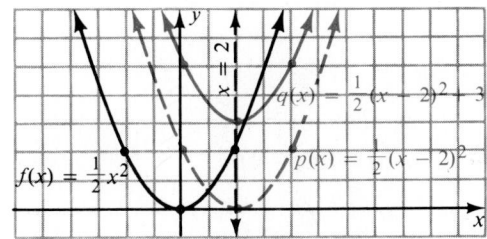

Figure 3

x	f(x)		x	p(x)		x	q(x)
−2	2		0	2		0	5
0	0		2	0		2	3
2	2		4	2		4	5

By observing these three graphs, we can summarize the facts concerning the **axis of symmetry** and the **vertex** (the point of the graph that lies on the axis of symmetry) of the graph of $y = a(x - h)^2 + k$.

The graph of the function

$$y = a(x - h)^2 + k, a \neq 0$$

over $\mathcal{R}$ is a parabola with the line $x = h$ as an axis of symmetry and the point (h, k) as vertex. If $a > 0$, then (h, k) is a minimum point, and the parabola opens upward; if $a < 0$, then (h, k) is a maximum point, and the parabola opens downward.

Polynomial Functions **355**

Suggested Assignments

Minimum
First day
356/1–12
Second day
356/13–22
Extended Alg. with Trig.
356/5–21 odd
Enriched Alg.
356/7–27 odd
Enriched Alg. with Trig.
356/7–27 even

**Additional Answers
Oral Exercises**

1. a. $x = 2$ b. $(2, 9)$
   c. down

2. a. $x = -3$ b. $(-3, -5)$
   c. up

3. a. $x = 0$ b. $(0, 8)$
   c. down

4. a. $x = -2$ b. $(-2, 0)$
   c. up

**Additional Answers
Written Exercises**

2.

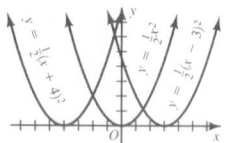

3.

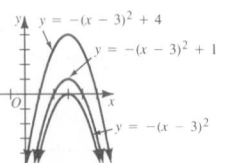

4.

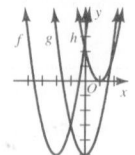

6.

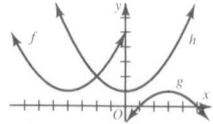

(continued on page 387)

356

Oral Exercises

For the graph of each of the following functions, state (a) an equation of the axis of symmetry, (b) the coordinates of the vertex, and (c) whether the graph opens upward or downward.

1. $y = -3(x - 2)^2 + 9$

2. $y = 4(x + 3)^2 - 5$

3. $f(x) = 8 - 5x^2$

4. $g(x) = \frac{1}{3}(x + 2)^2$

In Exercises 5–10 the first pair of coordinates is the vertex of the parabola and the second pair of coordinates is another point on the parabola. Give the equation of the axis of symmetry and name a third point on the parabola.

5. $(0, -3)$; $(4, 0)$ $x = 0, (-4, 0)$

6. $(0, 2)$; $(6, 1)$ $x = 0, (-6, 1)$

7. $(3, -1)$; $(4, 1)$ $x = 3, (2, 1)$

8. $\left(-\frac{3}{2}, 0\right)$; $(1, 4)$
   $x = -\frac{3}{2}, (-4, 4)$

9. (p, q); $(p - r, s)$
   $x = p, (p + r, s)$

10. $(a + b, c)$; (a, d)
   $x = a + b, (a + 2b, d)$

Written Exercises

In Exercises 1–8 sketch the graphs of all three of the given functions on one set of axes. Be sure to label each graph with its equation.

A

1. $y = x^2$; $y = x^2 + 3$; $y = x^2 - 2$

2. $y = \frac{1}{2}x^2$; $y = \frac{1}{2}(x + 4)^2$; $y = \frac{1}{2}(x - 3)^2$

3. $y = -(x - 3)^2$; $y = -(x - 3)^2 + 1$; $y = -(x - 3)^2 + 4$

4. $f(x) = 2(x + 2)^2 - 5$; $g(x) = 2x^2 - 5$; $h(x) = 2(x - 1)^2$

5. $f(x) = -(x - 2)^2 + 3$; $g(x) = \frac{1}{2}(x - 2)^2 + 3$; $h(x) = -2(x - 2)^2 + 3$

6. $f(x) = \frac{1}{4}(x + 4)^2 + 1$; $g(x) = -\frac{1}{4}(x - 3)^2 + 1$; $h(x) = \frac{1}{4}x^2 + 1$

7. $f(x) = 3(x - 4)^2 - 8$; $g(x) = \frac{1}{2}(x - 4)^2 - 1$; $h(x) = -\frac{1}{5}(x - 4)^2 + 2$

8. $f(x) = -\left(x + \frac{5}{2}\right)^2 + 6$; $g(x) = -\left(x + \frac{5}{2}\right)^2 + 2$; $h(x) = -\left(x - \frac{5}{2}\right)^2 + 3$

In Exercises 9–22 find the function of the form $y = a(x - h)^2 + k$ whose graph satisfies the given conditions.

9. Has vertex $(1, 3)$ and passes through the point $(4, 21)$ $y = 2(x - 1)^2 + 3$

10. Has vertex $(-2, -5)$ and passes through the point $(-3, 0)$ $y = 5(x + 2)^2 - 5$

11. Has vertex $(-1, 4)$ and passes through the point $(5, -14)$ $y = -\frac{1}{2}(x + 1)^2 + 4$

12. Has vertex $(\frac{1}{2}, -\frac{3}{2})$ and passes through the point $(-1, 12)$ $y = 6(x - \frac{1}{2})^2 - \frac{3}{2}$

13. Has $a = 3$, has $x = -3$ as its axis of symmetry, and passes through the point $(-3, -4)$ $y = 3(x + 3)^2 - 4$

14. Has $a = -1$, has $x = \frac{1}{2}$ as its axis of symmetry, and passes through the point $(3, -8)$ $y = -(x - \frac{1}{2})^2 - \frac{7}{4}$
15. Passes through the points $(-2, 1)$, $(0, -5)$, and $(2, 1)$ $y = \frac{3}{2}x^2 - 5$
16. Passes through the points $(-3, 2)$, $(-2, 8)$, and $(1, 2)$
    $y = -2(x + 1)^2 + 10$

B 17. Has an equation of the form $y = (x - h)^2 + 4$ and passes through the point $(2, 13)$ (There is more than one answer.)
18. Has an equation of the form $y = \frac{1}{2}(x - h)^2 - 6$ and passes through the point $(1, 26)$ $y = \frac{1}{2}(x - 9)^2 - 6$ or $y = \frac{1}{2}(x + 7)^2 - 6$
19. Has $x = 3$ as its axis of symmetry and passes through the points $(2, 5)$ and $(-1, -25)$ $y = -2(x - 3)^2 + 7$
    (*Hint*: Solve a system of equations in a and k.)
20. Has $x = -2$ as its axis of symmetry and passes through the points $(-3, 1)$ and $(-5, 13)$ $y = \frac{3}{2}(x + 2)^2 - \frac{1}{2}$
21. Has the y-axis as its axis of symmetry and passes through the points $(1, -13)$ and $(-3, 11)$ $y = 3x^2 - 16$
22. Has $x = -\frac{1}{2}$ as its axis of symmetry and passes through the points $(\frac{3}{2}, 6)$ and $(-1, -9)$ $y = 4(x + \frac{1}{2})^2 - 10$

C 23. Show that the point $(h + r, s)$ is on the graph of $y = a(x - h)^2 + k$, $a \neq 0$, if and only if the point $(h - r, s)$ is also on the graph.
24. Show that the point (r, s) is on the graph of $y = a(x - h)^2 + k$, $a \neq 0$, if and only if the point $(2h - r, s)$ is also on the graph.
25. Show that if the two distinct points $(r_1, 0)$ and $(r_2, 0)$ lie on the graph of $y = a(x - h)^2 + k$, $a \neq 0$, then $\dfrac{r_1 + r_2}{2} = h$.
26. If the parabola with equation $y = a(x - h)^2 + k$ intersects the x-axis in two distinct points, what must be true of a and k?
27. Graph the function $f(x) = |x^2 - 4|$.

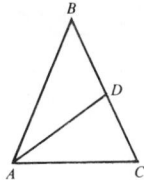

9–5 The Graph of a Quadratic Function

To graph a general quadratic function over $\mathscr{R}$ with equation

$$y = ax^2 + bx + c, \ a \neq 0, \tag{1}$$

it would be helpful if we could express the equation equivalently in the form described in the preceding section:

$$y = a(x - h)^2 + k, \ a \neq 0 \tag{2}$$

We could then specify the vertex, (h, k), and the axis of symmetry, $x = h$, in terms of the coefficients, a, b, c, in (1). Then by locating two or three more points we could draw the graph with reasonable accuracy. First let us consider the example at the top of the following page.

Polynomial Functions **357**

1. Specify the vertex and the axis of symmetry of the graph of $y = 3x^2 + 6x + 1$.

$$\frac{y - 1}{3} = x^2 + 2x$$

$$\frac{y - 1}{3} + 1 = x^2 + 2x + 1$$

$$\frac{y - 1}{3} = (x + 1)^2 - 1$$

Rewrite in the form
$y = a(x - h)^2 + k$.
$y = 3[(x + 1)^2 - 1] + 1$
$y = 3(x + 1)^2 - 2$
axis of symmetry: $x = -1$
vertex: $(-1, -2)$

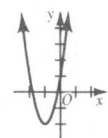

2. Specify the axis of symmetry and the vertex of $f(x) = 2x^2 - 4x + 1$.
$a = 2, b = -4, c = 1$.
Equation of axis of symmetry:

$$x = -\frac{b}{2a} = \frac{4}{4} = 1$$

vertex: $\left(-\dfrac{b}{2a}, \ -\dfrac{b^2 - 4ac}{4a}\right)$,

or $\left(\dfrac{4}{4}, \ -\dfrac{16 - 8}{8}\right)$, or $(1, -1)$

3. Find two real numbers whose sum is 18 and whose product is as great as possible.
Let $x = $ one number.
Then $18 - x = $ other num.
$f(x) = x(18 - x)$
$\qquad = -x^2 + 18x$

$x = -\dfrac{b}{2a}$ at its maximum

$x = -\dfrac{18}{-2} = 9$

$18 - x = 9$
∴ the numbers are 9 and 9.

358 **358**

EXAMPLE 1 Specify the vertex and the axis of symmetry of the graph of $y = 3x^2 - 12x + 11$ and sketch the graph.

SOLUTION 1. Rewrite the given equation in the form $\dfrac{y - c}{a} = x^2 + \dfrac{b}{a}x$.

$$y = 3x^2 - 12x + 11$$

$$\frac{y - 11}{3} = x^2 - \frac{12}{3}x$$

$$\frac{y - 11}{3} = x^2 - 4x$$

2. Add to both sides of the equation the number $\dfrac{b^2}{4a^2}$ in order to complete the square on the right.

$$\frac{y - 11}{3} + 4 = x^2 - 4x + 4$$

3. Then transform the equation into the desired form (2).

$$\frac{y - 11}{3} = (x - 2)^2 - 4$$

$$y - 11 = 3(x - 2)^2 - 12$$

$$y = 3(x - 2)^2 - 1$$

Thus the axis of symmetry is $x = 2$, and the vertex is $(2, -1)$. Substituting 1 and 3 for x in the original equation, you find that $(1, 2)$ and $(3, 2)$ also are on the graph, which is sketched at the right.

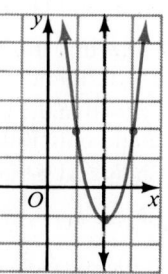

Using the method of Example 1 you can find the coordinates of the vertex and an equation of the axis of symmetry for the graph of

$$y = ax^2 + bx + c, \ a \neq 0.$$

First rewrite the equation as

$$\frac{y - c}{a} = x^2 + \frac{b}{a}x.$$

Completing the square, you obtain:

$$\frac{y - c}{a} + \frac{b^2}{4a^2} = x^2 + \frac{b}{a}x + \frac{b^2}{4a^2} = \left(x + \frac{b}{2a}\right)^2$$

$$y - c = a\left[\left(x + \frac{b}{2a}\right)^2 - \frac{b^2}{4a^2}\right]$$

$$y = a\left(x + \frac{b}{2a}\right)^2 - \frac{b^2}{4a} + c$$

$$y = a\left(x - \left(\frac{-b}{2a}\right)\right)^2 + \left(-\frac{b^2 - 4ac}{4a}\right) \qquad (3)$$

358 Chapter 9

Comparing Equation (3) with $y = a(x - h)^2 + k$, you have for the equation $x = h$ of the axis of symmetry: $x = -\dfrac{b}{2a}$; and for the coordinates (h, k) of the vertex: $\left(-\dfrac{b}{2a}, -\dfrac{b^2 - 4ac}{4a}\right)$. The vertex is a minimum point when $a > 0$ and a maximum point when $a < 0$.

EXAMPLE 2 Specify the axis of symmetry and the vertex of $f(x) = 2x^2 - 20x + 35$.

SOLUTION $a = 2, b = -20, c = 35$.

Equation of axis of symmetry: $x = -\dfrac{b}{2a} = \dfrac{20}{4}$, or $x = 5$

Vertex: $\left(-\dfrac{b}{2a}, -\dfrac{b^2 - 4ac}{4a}\right)$, or $\left(5, -\dfrac{400 - 280}{8}\right)$, or $(5, -15)$

The fact that the quadratic function $f(x) = ax^2 + bx + c$ has exactly one *extreme value* (a minimum when $a > 0$, a maximum when $a < 0$) is useful in practical applications. The ordered pair $\left(-\dfrac{b}{2a}, -\dfrac{b^2 - 4ac}{4a}\right)$ specifies the value of $(x, f(x))$ at the extreme point of the graph of f. Thus the ordinate, $-\dfrac{b^2}{4a} + c$, at the extreme point is the extreme value, $f\left(-\dfrac{b}{2a}\right)$, of the function.

EXAMPLE 3 Find two real numbers with sum 14 and product as great as possible.

SOLUTION Let x be one number, and $14 - x$ be the other. We want to find the value of x for which the function

$$f(x) = x(14 - x), \quad \text{or} \quad f(x) = -x^2 + 14x,$$

assumes its maximum value. Here $a = -1$ and $b = 14$. The maximum occurs when $x = -\dfrac{b}{2a} = -\dfrac{14}{-2}$, or 7. Then the other number is $14 - x = 14 - 7$, or 7.

$\therefore$ the two numbers are 7 and 7.

As you saw in Section 9-3, the sum of the roots of the quadratic equation $ax^2 + bx + c = 0$ is $-\dfrac{b}{a}$. It follows that the *average* of the roots, given by

$$\frac{\text{sum of roots}}{2} = \frac{-\dfrac{b}{a}}{2} = -\frac{b}{2a},$$

is the x-coordinate of the vertex of the graph of

$$y = ax^2 + bx + c.$$

This fact is used in the next example.

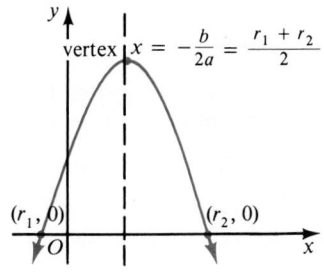

Figure 4

Polynomial Functions **359**

Additional A Exercises

Give the equation of the axis of symmetry and the coordinates of the vertex of the graph of each function and sketch the graph.

1. $y = x^2 + 5x$ $x = -\dfrac{5}{2}$; $\left(-\dfrac{5}{2}, -\dfrac{25}{4}\right)$
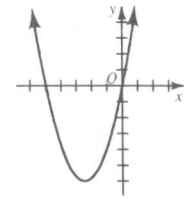

2. $y = 2x^2 - 4$ $x = 0$; $(0, -4)$
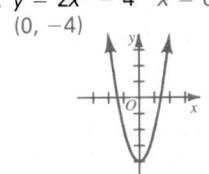

3. $y = \dfrac{1}{2}x^2 - 3$ $x = 0$; $(0, -3)$
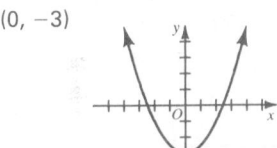

4. $f(x) = x^2 - 8x + 15$ $x = 4$; $(4, -1)$
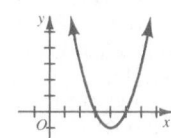

5. $f(x) = -2x^2 + 8x - 5$ $x = 2$; $(2, 3)$
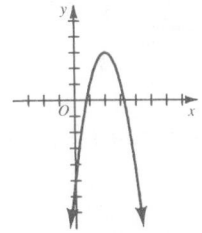

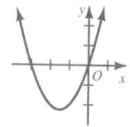
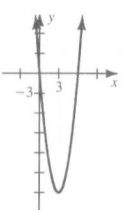
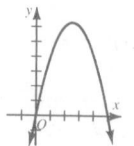
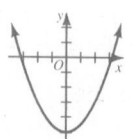
EXAMPLE 4 A potato farmer has 400 bushels of potatoes that she can now sell for $3.20 per bushel. For every week she waits, the price per bushel will drop by 10¢ but she will harvest 20 more bushels. How many weeks should she wait in order to maximize her income?

SOLUTION Let w = the number of weeks after the present.
Then $320 - 10w$ = price per bushel (in cents) after w weeks;
$\quad 400 + 20w$ = number of bushels to be sold after w weeks.

Thus we have:

$$\text{income} = (\text{price/bushel}) \times (\text{number of bushels})$$
$$= (320 - 10w)(400 + 20w)$$

It is easy to see that the roots of the corresponding equation

$$0 = (320 - 10w)(400 + 20w)$$

are -20 and 32.

Therefore the x-coordinate of the vertex of the graph of

$$r = (320 - 10w)(400 + 20w)$$

is $\frac{-20 + 32}{2}$, or 6.

Check: The check is left to you.

∴ She should wait 6 weeks.

Oral Exercises

Tell whether the graph of the equation will have a minimum or maximum point.

1. $y = x^2 - 10x$ min

2. $y = x^2 + 3x$ min

3. $y = 2x^2 - 12x$ min

4. $y = -x^2 + 5x$ max

5. $y = -2x^2 - 6$ max

6. $y = \frac{1}{2}x^2 - 5$ min

Written Exercises

Give the equation of the axis of symmetry and the coordinates of the vertex of the graph of each function and sketch the graph.

A 1–6. Use the equations in Oral Exercises 1–6.

7. $f(x) = x^2 - 6x + 7$ $x = 3; (3, -2)$

8. $f(x) = x^2 + 4x + 1$ $x = -2; (-2, -3)$

9. $f(x) = 2x^2 - 8x - 3$ $x = 2; (2, -11)$

10. $g(x) = 3x^2 - 6x + 1$ $x = 1; (1, -2)$

11. $f(x) = -2x^2 + 12x - 9$ $x = 3; (3, 9)$

12. $g(x) = \frac{1}{2}x^2 - 3x + 4$ $x = 3; (3, -\frac{1}{2})$

Find the quadratic function whose graph contains the given ordered pairs. (*Hint*: Determine a, b, and c so that the given ordered pairs satisfy the equation $y = ax^2 + bx + c$.)

$y = -x^2 - 2x + 5$

B 13. $(0, 2)$, $(-1, 9)$, $(2, 6)$ $\quad y = 3x^2 - 4x + 2$ 14. $(-3, 2)$, $(-1, 6)$, $(1, 2)$
15. $(-2, 9)$, $(1, -3)$, $(2, 3)$ $\qquad\qquad$ 16. $(1, 2)$, $(-2, 1)$, $(3, 6)$
$\qquad\qquad y = \frac{5}{2}x^2 - \frac{3}{2}x - 4 \qquad\qquad\qquad\qquad y = \frac{1}{3}x^2 + \frac{2}{3}x + 1$

Put each equation in the form $y = a(x - h)^2 + k$ and give the equation of the axis of symmetry and the coordinates of the vertex of the graph in terms of r and s.

17. $y = r(x^2 - s)$ $\qquad\qquad\qquad$ 18. $y = rx^2 + 2rx$

19. $y = rx^2 - 2rsx - 2s^2$ $\qquad\qquad$ 20. $y = \frac{s^2}{2} + rsx - rx^2$

C 21. Prove that if the graphs of two quadratic functions have the same vertex and one other common point, then the functions are the same.

22. Show that if $\left(\dfrac{-b}{2a} + r, s\right)$ is a point on the graph of $ax^2 + bx + c = y$,

then so is $\left(\dfrac{-b}{2a} - r, s\right)$

Problems

A 1. Find two numbers with sum 24 and product as great as possible. 12, 12

2. Find the dimensions of the rectangle of greatest area whose perimeter is 36 m. 9 m by 9 m

3. A pen is to be built along a wall from 72 m of fencing. Determine the value of x that maximizes the area that will be enclosed. 18

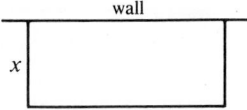

4. In a 120-volt circuit having a resistance of 12 Ω (ohms), the power W in watts when a current I (in amperes) is flowing is given by $W = 120I - 12I^2$. Determine the maximum power that can be delivered in this circuit. 300 W

B 5. If an object is projected vertically upward with a velocity of 29.4 m/s from a height of 30 m, then its height h above the ground in meters after t seconds is given by $h = 30 + 29.4t - 4.9t^2$. What is the maximum height the object will reach? 74.1 m

6. Sixty meters of fencing material is to be used to build a play yard around the corner of a house, as shown. What should dimensions x and y be in order that the area of the yard be a maximum? 18.75 m, 18.75 m

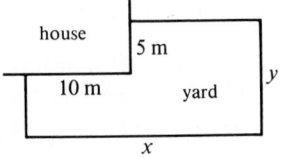

Polynomial Functions **361**

7.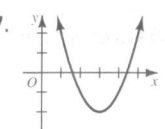

8.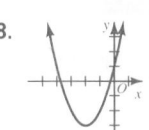

9.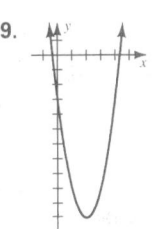

10.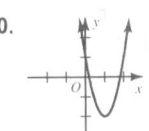

11.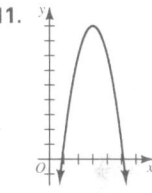

12.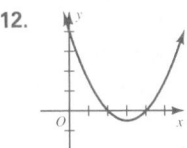

17. $y = rx^2 - rs$; $x = 0$; $(0, -rs)$

18. $y = r(x + 1)^2 - r$; $x = -1$; $(-1, -r)$

19. $y = r(x - s)^2 - (rs^2 + 2s^2)$; $x = s$; $(s, -rs^2 - 2s^2)$

20. $y = -r\left(x - \dfrac{s}{2}\right)^2 + \dfrac{s^2}{2} + \dfrac{rs^2}{4}$; $x = \dfrac{s}{2}$; $\left(\dfrac{s}{2}, \dfrac{s^2}{2} + \dfrac{rs^2}{4}\right)$

(continued on p. 389)

1. Express in lowest terms:

$$\frac{x^2 - 3x}{2} \cdot \frac{2x - 4}{x^3 - 9x} \cdot \frac{x - 2}{x + 3}$$

2. Solve $\frac{x + 3}{x - 3} - \frac{x}{2x - 6} = \frac{2}{5}$

over $\mathcal{R}$. $\{-42\}$

3. Find the solution set of
$x^2 - 2x - 15 \le 0$.
$\{x: -3 \le x \le 5\}$

4. Rewrite $(3a^{-3}a^{-5}b^3) \cdot$
$(3a^4a^{-2}b^{-1})$ using only
positive exponents. $\frac{9b^2}{a^6}$

Teaching Suggestions
p. T102

Supplementary Material

Test 28

Key Ideas

Solve quadratic inequalities.

Chalkboard Examples

1. Find the solution set over
$\mathcal{R}$ of $-x^2 + 3x + 5 > 0$.
Determine the roots of
$-x^2 + 3x + 5 = 0$. Use
the quadratic formula.
$a = -1$, $b = 3$, $c = 5$

$$x = \frac{-3 \pm \sqrt{3^2 - 4(-1)(5)}}{2(-1)}$$

$$= \frac{-3 \pm \sqrt{29}}{-2}$$

$$= \frac{3}{2} \pm \frac{\sqrt{29}}{2}$$

$$\left\{ x: \frac{3}{2} - \frac{\sqrt{29}}{2} < x < \frac{3}{2} + \frac{\sqrt{29}}{2} \right\}$$

---

7. The perimeter of the two adjacent squares shown is to
be 30 m. What should dimension x be in order for the
area to be a minimum? 6 m

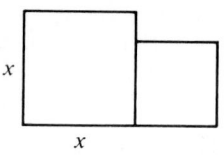

8. The velocity v of the object in Problem 5 after t seconds is given by $v = 29.4 - 9.8t$. Solve this equation for t, and show that h is a maximum when $v = 0$.

C 9. A wire 60 cm long is cut into two pieces, and each piece is bent into a
square. Where should the wire be cut if the total area of the two
squares is to be a minimum? At its midpoint.

10. A limousine shuttle service operating between an airport and the
center of a city charges a fare of $12 and carries 400 persons per day.
The firm estimates that business will decrease by 10 passengers per
day for each increase of $1 in the fare. Find the most profitable fare
to charge for the service. $26

11. In the diagram at the right, $AFED$ is a rectangle, $AC = 8$, and $AB = 12$.
   a. Use similar triangles to find an expression for y in
      terms of x.
   b. Give an expression for the area of rectangle $AFED$
      in terms of x alone.
   c. Find the value of x that makes this area a maximum.
   a. $y = \frac{24 - 2x}{3}$ b. $\frac{24x - 2x^2}{3}$ c. 6

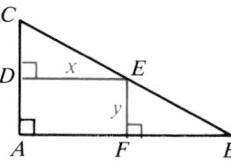

9–6 Quadratic Inequalities

Figure 5 shows the graph of the function with equation

$$y = x^2 - 4x + 1.$$

From the graph, you can see that the points with coordinates $2 - \sqrt{3}$ and $2 + \sqrt{3}$ (the zeros of the function)
separate the other points on the x-axis into the following
three sets:

$$A = \{x: x < 2 - \sqrt{3}\}$$
$$B = \{x: 2 - \sqrt{3} < x < 2 + \sqrt{3}\}$$
$$C = \{x: x > 2 + \sqrt{3}\}$$

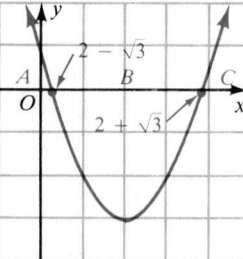

Figure 5

Do you see that over any one of these subsets the value of $x^2 - 4x + 1$ is
always positive (the graph is above the x-axis) or *always negative* (the
graph is below the x-axis)?

$x \in A$	$x \in B$	$x \in C$
$x^2 - 4x + 1 > 0$	$x^2 - 4x + 1 < 0$	$x^2 - 4x + 1 > 0$

Thus, to test whether $x^2 - 4x + 1$ denotes a positive number or whether it denotes a negative number for *every* value of x in one of these subsets, you need only determine the sign of $x^2 - 4x + 1$ for *any one* value of x in that subset.

EXAMPLE   Find the solution set over $\mathscr{R}$ of $-x^2 + 2x + 9 > 0$.

SOLUTION   1. Determine the roots of $-x^2 + 2x + 9 = 0$; $a = -1$, $b = 2$, $c = 9$.

$$x = \frac{-2 \pm \sqrt{2^2 - 4(-1)(9)}}{2(-1)} = \frac{-2 \pm \sqrt{40}}{-2} = 1 \pm \frac{2\sqrt{10}}{-2} = 1 \pm \sqrt{10}$$

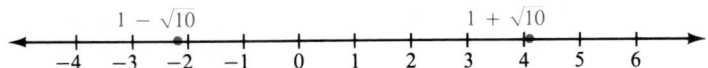

2. From each subset into which these numbers separate the set of real numbers, choose a number for which to evaluate $-x^2 + 2x + 9$.

Subset	Particular Number	$-x^2 + 2x + 9$
$x < 1 - \sqrt{10}$	-4	-15, negative
$1 - \sqrt{10} < x < 1 + \sqrt{10}$	0	9, positive
$x > 1 + \sqrt{10}$	5	-6, negative

$$\therefore \{x: -x^2 + 2x + 9 > 0\} = \{x: 1 - \sqrt{10} < x < 1 + \sqrt{10}\}$$

Note that you could avoid testing particular values of x in the given quadratic expression by simply finding the roots of the corresponding equation and then determining (from the sign of a) whether the graph of the corresponding quadratic function opens up or down.

Oral Exercises

State whether the given value of x satisfies the given inequality.

1. $x^2 - 3x + 4 \le 0$; 2 yes ✓
2. $3x^2 + 2x - 5 > 0$; 4 yes
3. $x^2 - \frac{5}{4} > 0$; $\frac{\sqrt{5}}{2}$ no
4. $x^2 - 9x \ge -3$; -1 yes
5. $0 < x^2 + 6x + 8$; -5 yes
6. $6x - 10 > x^2$; $\sqrt{3}$ no

Polynomial Functions   **363**

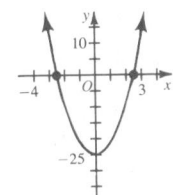
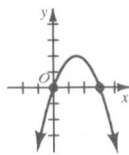

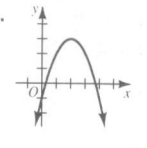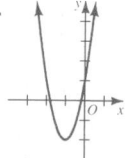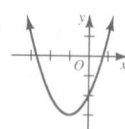

Quick Quiz

1. For the graph of the equation $y = -(x - 3)^2 + 4$ find an equation for the axis of symmetry and the coordinates of the vertex. State whether the vertex is a maximum or a minimum.
   $x = 3$
   $(3, 4)$
   maximum

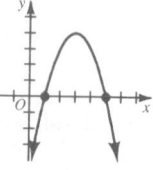

2. Sketch the graph of $y = 2x^2 - 8x + 5$.
   $y = 2(x - 2)^2 - 3$

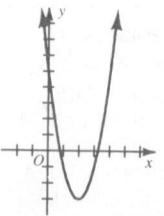

3. Find the area of the right triangle of largest area such that the sum of the lengths of its legs is 16 cm. 32 cm^2

Written Exercises

In Exercises 1–9
a. rewrite the inequality as one in which 0 is the right side;
b. find the roots of the related equation;
c. sketch the graph of the related quadratic function; and
d. solve the inequality by noting which part(s) of the graph lie above or below the x-axis.

A 1. $x^2 - 16 > 0$ 2. $9x^2 - 4 \leq 0$ 3. $5x - x^2 < 0$
    4. $3x^2 - 8x < 0$ 5. $x^2 - x \geq 12$ 6. $2x^2 + 7 \leq 9x$

B 7. $x^2 - 6x + 4 \leq 0$ 8. $2x^2 \geq 4x + 1$ 9. $x^2 + 4x > 1$

Write the inequality as a conjunction of simpler inequalities. Then sketch the graphs of the related quadratic functions to determine the solution set of the original inequality.

C 10. $-5 < x^2 - 9 < 0$ 11. $0 \geq x^2 - 5x \geq -6$

■ Self-Test 2

VOCABULARY   quadratic function (p. 354)
            parabola (p. 354)
            minimum point (p. 354)
            maximum point (p. 355)
            axis of symmetry (p. 355)
            vertex (p. 355)

Exercises 1–3 refer to the graph of $y = -(x - 2)^2 + 3$.

1. Find an equation for the axis of symmetry and the coordinates   *Obj. 1, p. 354*
   of the vertex. $x = 2; (2, 3)$

2. State whether the vertex is a maximum or a minimum. maximum

3. Sketch the graph.   *Obj. 2, p. 354*

4. Rewrite $y = 3x^2 + 6x + 1$ in the form $y = a(x - h)^2 + k$, and
   sketch its graph. $y = 3(x + 1)^2 - 2$

5. Find the area of the right triangle of greatest area such that the   *Obj. 3, p. 354*
   sum of the lengths of the legs is 12 cm. 18 cm^2

6. Solve the inequality $x^2 + 2x - 2 > 0$ by drawing a sketch of the   *Obj. 4, p. 354*
   graph of the related quadratic function. $\{x: x > -1 + \sqrt{3} \text{ or } x < -1 - \sqrt{3}\}$

Check your answers with those at the back of the book.

364   *Chapter 9*

Polynomial Functions and Equations

OBJECTIVES for Sections 9-7 through 9-9:
1. *To use synthetic substitution to find the value of a given polynomial function at a given domain value.*
2. *To use synthetic division to find the partial quotient and the remainder when a polynomial is divided by $x - c$, where $c \in \mathcal{C}$.*
3. *To use synthetic division and depressed equations to factor polynomials and to find rational zeros of polynomial functions.*
4. *To use the fact that imaginary roots of polynomial equations with real coefficients occur in complex conjugate pairs to solve such equations.*

9-7 Values of Polynomial Functions

Functions whose values are given by polynomials, such as

$$y = 3x - \sqrt{2}, \qquad y = x^2 - 2ix + 3, \qquad y = 4x^3 + 2x^2 - 1,$$

are called **polynomial functions.** You know how to evaluate a polynomial function by direct substitution. For example, if a_0, a_1, a_2, a_3 denote complex numbers and P is the function

$$P(x) = a_0x^3 + a_1x^2 + a_2x + a_3,$$

then to evaluate P at 7, you write

$$P(7) = a_0(7^3) + a_1(7^2) + a_2(7) + a_3.$$

To find the value of $P(7)$, you might compute as follows:

(1) 7^2
(2) 7^3
(3) $a_0(7^3)$
(4) $a_1(7^2)$
(5) $a_2(7)$
(6) $a_0(7^3) + a_1(7^2) + a_2(7) + a_3$

However, if you wished to program a computer to find $P(7)$, this method would not be a good one, since exponentiation (raising a number to a power) often produces inexact results on a computer (and some computer languages, such as Pascal, do not have a built-in exponentiation function). Even using a hand calculator, you would find the method indicated above inefficient. To discover a more efficient way to compute $P(7)$, study the sequence of operations on the following page.

Polynomial Functions **365**

Mixed Review
1. Simplify $\sqrt{-45}$. $3i\sqrt{5}$
2. Solve.
   $7(x - 5) - 4x = 28$ $\{21\}$
3. Solve the system:
   $3x - 6y = 18$
   $5x + 3y = 17$ $\{(4, -1)\}$
4. Rewrite $(8a^3b^{-2}c^{-3}) \cdot$
   $(2a^2b^{-1}c)^{-2}$ using only
   positive exponents. $\dfrac{2}{ac^5}$
5. Find the product.
   $(x - 2)(x^2 + 2x - 3)$
   $x^3 - 7x + 6$

Teaching Suggestions
p. T102

Key Ideas
Find the value of a polynomial function at a given domain value by using synthetic substitution.

Chalkboard Examples
1. If $Q(x) = 2x^4 - x^3 + 2x - 1$, find $Q(3)$ and $Q(3i)$.
   Write the coefficients of $Q(x)$ in order, using 0 where necessary. Then use synthetic substitution.

$$\begin{array}{r|rrrrr} 3 & 2 & -1 & 0 & 2 & -1 \\ & & 6 & 15 & 45 & 141 \\ \hline & 2 & 5 & 15 & 47 & 140 \end{array}$$

   $\therefore Q(3) = 140$;
   $Q(3i) = 161 + 33i$.

365

Additional A Exercises

Use synthetic substitution to find the given values over $\mathcal{C}$ of the polynomial function given directly above each exercise. If the substituted value is a zero, so state.

$P(x) = x^3 - 3x^2 - 4x + 12$

1. $P(-1)$ 12

2. $P(-2)$ 0; a zero

$Q(x) = x^4 - x^3 + 2x^2 - 3x - 10$

3. $Q(-1)$ -3 **4.** $Q(3)$ 53

$S(x) = x^4 - 3x^3 - 12x + 16$

5. $S(2)$ -16

6. $S(3i)$ $97 + 45i$

Mixed Review

Find the sum of the arithmetic series.

1. $a_1 = 10$, $a_7 = 16$, $n = 7$ 91

2. $\sum_{i=6}^{15} (2i + 1)$ 220

3. Simplify $\sqrt[3]{\sqrt{27}}$. $\sqrt{3}$

4. Solve $\sqrt{x - 1} + 3 = x$ over $\mathcal{R}$. $\{5\}$

5. Solve $x^2 - 6x + 1 = 0$ over $\mathcal{R}$. $\{3 \pm 2\sqrt{2}\}$

Suggested Assignments

Minimum
  First day
  367/2–24 even
  Second day
  367/23, 25–28
Extended Alg. with Trig.
  367/7–27 odd
Enriched Alg.
  367/11–33 odd
Enriched Alg. with Trig.
  367/11–33 odd

Additional Answers
Written Exercises (p. 367)

31.

0	a	b	c	d
	0	0	0	
a	b	c	d	

366

---

1. Multiply a_0 by 7: $a_0 \cdot 7$
2. Add a_1: $a_0 \cdot 7 + a_1$
3. Multiply the result of Step 2 by 7: $(a_0 \cdot 7 + a_1) \cdot 7$
4. Add a_2: $(a_0 \cdot 7 + a_1) \cdot 7 + a_2$
5. Multiply the result of Step 4 by 7: $[(a_0 \cdot 7 + a_1) \cdot 7 + a_2] \cdot 7$
6. Add a_3: $[(a_0 \cdot 7 + a_1) \cdot 7 + a_2] \cdot 7 + a_3$

By simplifying the expression in Step 6, you can verify that

$$[(a_0 \cdot 7 + a_1) \cdot 7 + a_2] \cdot 7 + a_3 = a_0 \cdot 7^3 + a_1 \cdot 7^2 + a_2 \cdot 7 + a_3 = P(7).$$

Notice that in this sequence of steps, each result is computed directly from the preceding result. If you followed the sequence using a calculator, you would not need to use the calculator's memory.

You can use this second method even if you do not have a computer or a calculator. Steps 1–6 can be arranged conveniently as shown below. The circled numerals designate each of the steps.

$$
\begin{array}{c|cccc}
7 & a_0 & a_1 & a_2 & a_3 \\
  & & ① \ a_0 \cdot 7 & ③ \ (a_0 \cdot 7 + a_1) \cdot 7 & ⑤ \ [(a_0 \cdot 7 + a_1) \cdot 7 + a_2] \cdot 7 \\
\hline
  & a_0 & a_0 \cdot 7 + a_1 & (a_0 \cdot 7 + a_1) \cdot 7 + a_2 & [(a_0 \cdot 7 + a_1) \cdot 7 + a_2] \cdot 7 + a_3 \\
  & & ② & ④ & ⑥
\end{array}
$$

If $P(x) = 5x^3 - 12x^2 - 20x + 1$, you can find $P(7)$ by following Steps 1–6, using 5, -12, -20, and 1 in place of a_0, a_1, a_2, and a_3, respectively:

$$
\begin{array}{c|cccc}
7 & 5 & -12 & -20 & 1 \\
  &   & 35 & 161 & 987 \\
\hline
  & 5 & 23 & 141 & 988 \\
  &   &   &   & \overset{\shortmid}{P(7)}
\end{array}
$$

Thus $P(7) = 988$. This process, called **synthetic substitution**, applies to polynomials of any degree. Notice that $P(x)$ must be written in descending powers of x. Also, if a power is missing, 0 must be written in the corresponding place.

EXAMPLE   If $Q(x) = 3x^4 - x^3 + 5x - 7$, find $Q(-2)$ and $Q(2i)$.

SOLUTION   Write the coefficients of $Q(x)$ in order, using 0 where necessary. Then use synthetic substitution.

$$
\begin{array}{c|ccccc}
-2 & 3 & -1 & 0 & 5 & -7 \\
   &   & -6 & 14 & -28 & 46 \\
\hline
   & 3 & -7 & 14 & -23 & 39
\end{array}
$$

$$
\begin{array}{c|ccccc}
2i & 3 & -1 & 0 & 5 & -7 \\
   &   & 0 + 6i & -12 - 2i & 4 - 24i & 48 + 18i \\
\hline
   & 3 & -1 + 6i & -12 - 2i & 9 - 24i & 41 + 18i
\end{array}
$$

$\therefore\ Q(-2) = 39$; $Q(2i) = 41 + 18i$

Oral Exercises

Find the requested values of $P(x) = 2x^3 - 5x^2 - 6x - 9$ for the given values by considering the equation in the rewritten form shown:

$$P(x) = [(2x - 5)x - 6]x - 9$$

1. $P(0)$ _−9_ **2.** $P(1)$ _−18_ **3.** $P(-1)$ _−10_ **4.** $P(-3)$ _−90_ **5.** $P(4)$ _15_

Written Exercises

In Exercises 1–20 use synthetic substitution to find the given values over $\mathcal{C}$ of the polynomial function given directly above each exercise. If the substituted value is a zero, so state.

$P(x) = x^3 - 2x^2 - 9x + 18$

A **1.** $P(2)$ _0; zero_ **2.** $P(-1)$ _24_ **3.** $P(-3)$ _0; zero_ **4.** $P(3)$ _0; zero_ **5.** $P(-2)$ _20_

$Q(x) = x^4 + x^3 - 3x^2 - 5x - 6$

6. $Q(-1)$ _−4_ **7.** $Q(-2)$ _0; zero_ **8.** $Q(2)$ _−4_ **9.** $Q(3)$ _60_ **10.** $Q(-3)$ _36_

$R(x) = 3x^5 - 4x^3 - 10x^2 - 7x - 10$

11. $R(-1)$ _−12_ **12.** $R(i)$ _0; zero_ **13.** $R(-i)$ _0; zero_ **14.** $R(2)$ _0; zero_ **15.** $R(2i)$ _30 + 114i_

$S(x) = 2x^4 - 3x^3 - 12x - 32$

16. $S(3)$ _13_ **17.** $S(-2)$ _48_ **18.** $S(2i)$ _0; zero_ **19.** $S(-2i)$ _0; zero_ **20.** $S(i)$ _−30 − 9i_

B **21.** Determine m so that $G(1) = 2$ for $G(x) = 8x^4 + mx - 5$. _−1_
  22. Determine m so that $F(3) = -1$ for $F(x) = 2x^3 - 7x^2 + 5x + m$. _−7_
  23. Determine m so that -2 is a zero of $H(x) = 3x^3 - x^2 + mx + 10$. _−9_
  24. Determine m so that 3 is a zero of $K(x) = 2x^3 + mx^2 - 8x + 6$. _−4_

Use the Rational Root Theorem (page 254) to locate possible rational roots of each polynomial equation. Then test each possible root by synthetic substitution to find the actual roots.

25. $x^3 + 2x^2 - 13x + 10 = 0$ _{−5, 1, 2}_
26. $x^3 - 8x^2 + 5x + 14 = 0$ _{−1, 2, 7}_
27. $2x^3 - x^2 - 7x + 6 = 0$ _{−2, 1, $\frac{3}{2}$}_
28. $10x^3 + 3x^2 - 6x + 1 = 0$ _{−1, $\frac{1}{5}$, $\frac{1}{2}$}_

C **29.** Determine a and b so that i is a root of
$$5x^4 - 3x^3 + ax^2 + bx - 7 = 0.$$ _$a = -2$, $b = -3$_

  30. Determine a and b so that $-2i$ is a root of
$$2x^5 + 6x^4 + ax^3 + bx^2 + 4x - 36 = 0.$$ _$a = 9$, $b = 15$_

  31. Use synthetic substitution to show that if $P(x) = ax^3 + bx^2 + cx + d$, then $P(0) = d$.

Polynomial Functions **367**

Additional Answers
Written Exercises (p. 364)

2. a. $9x^2 - 4 \leq 0$

b. $\left\{\frac{2}{3}, -\frac{2}{3}\right\}$

c.
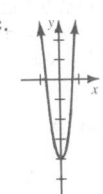

d. $\left\{x: -\frac{2}{3} \leq x \leq \frac{2}{3}\right\}$

4. a. $3x^2 - 8x < 0$

b. $\left\{0, \frac{8}{3}\right\}$

c.
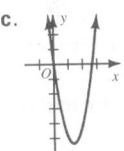

d. $\left\{x: 0 < x < \frac{8}{3}\right\}$

5. a. $x^2 - x - 12 \geq 0$
b. $\{-3, 4\}$
c.
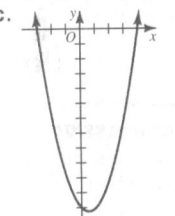

d. $\{x: x \leq -3\} \cup \{x: x \geq 4\}$

6. a. $2x^2 - 9x + 7 \leq 0$

b. $\left\{1, \frac{7}{2}\right\}$

c.
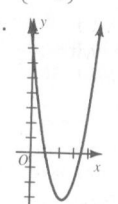

d. $\left\{x: 1 \leq x \leq \frac{7}{2}\right\}$

(continued on p. 373)

368

---

32. Use synthetic substitution to show that if $P(x) = ax^3 + bx^2 + cx + d$ has 2 as one of its zeros, then $8a + 4b + 2c + d = 0$.

33. Let $P(x) = ax^3 + bx^2 + cx + d$ with a, b, c, and d all real numbers.
    a. Show that if i is a zero of $P(x)$, then $a = c$ and $b = d$.
    b. Show that if $a = c$ and $b = d$, then $-i$ is a zero of $P(x)$.
    c. What conclusion can be drawn from (a) and (b)?

Computer Exercises   For students with computer experience

1. Write a program that will evaluate a given polynomial $P(x)$ with real coefficients for a given real value of x.

Run the program in Exercise 1 for each polynomial $P(x)$ and each value of x.

2. $x^3 - 5x^2 - 10x + 7$;   3 _−41
3. $2x^4 + 7x^3 - 9x^2 - 17x + 10$;   −4 _−2
4. $6x^4 + 13x^3 - 9x^2 + 2x + 25$;   2.5 _411.25

5. Modify the program in Exercise 1 so it evaluates a given polynomial $P(x)$ with real coefficients when x is a complex number, $a + bi$, and a and b are input. (*Hint*: Keep separate real and imaginary parts.)

Run the program in Exercise 5 for each polynomial for each given value of x.

6. $x^3 + 5x^2 + 4x + 7$ **a.** $x = 2i$ _−13   **b.** $x = -i$ _2 − 3i
7. $x^4 - 7x^3 + 7x^2 - 25x + 20$ **a.** $x = 3 + 2i$ **b.** $x = 3 - 2i$
_−76 − 168i   _−76 + 168i

9–8 Remainder and Factor Theorems

By dividing $2x^2 - 4x - 5$, $x^2 - x - 6$, $x^3 - 4x^2 + 5x - 6$, and $x^3 + 2x^2 - 7x - 4$, respectively, by $x - 3$, you obtain the *partial quotients* (Q) and *remainders* (R) shown in the following chart.

$P(x)$	$Q(x)$	R	$P(3)$
$2x^2 - 4x - 5$	$2x + 2$	1	1
$x^2 - x - 6$	$x + 2$	0	0
$x^3 - 4x^2 + 5x - 6$	$x^2 - x + 2$	0	0
$x^3 + 2x^2 - 7x - 4$	$x^2 + 5x + 8$	20	20

Notice that in every case the remainder R equals $P(3)$.

368   *Chapter 9*

Let us now take a more formal approach and divide the general third degree polynomial by $x - 7$ using the following steps:

$$P(x) = a_0x^3 + a_1x^2 + a_2x + a_3$$
$$P(7) = a_0(7^3) + a_1(7^2) + a_2(7) + a_3$$
$$\begin{aligned}
P(x) - P(7) &= a_0(x^3 - 7^3) + a_1(x^2 - 7^2) + a_2(x - 7) + (a_3 - a_3) \\
&= a_0(x - 7)(x^2 + 7x + 49) + a_1(x - 7)(x + 7) + a_2(x - 7) \\
&= (x - 7)[a_0(x^2 + 7x + 49) + a_1(x + 7) + a_2] \\
&= (x - 7)[a_0x^2 + [a_0(7) + a_1]x + [a_0(7)^2 + a_1(7) + a_2]]
\end{aligned}$$

Thus, if you let $Q(x) = a_0x^2 + [a_0(7) + a_1]x + [a_0(7^2) + a_1(7) + a_2]$, you have

$$P(x) = (x - 7)Q(x) + P(7), \quad \text{or} \quad \frac{P(x)}{x - 7} = Q(x) + \frac{P(7)}{x - 7}.$$

You can use this argument to prove the following theorem for any polynomial of any positive degree in x and any divisor $x - r$.

Remainder Theorem

For every polynomial $P(x)$ of positive degree n over the set of complex numbers, and for every complex number r, there exists a polynomial $Q(x)$ of degree $n - 1$, such that

$$P(x) = (x - r)Q(x) + P(r).$$

Did you recognize the coefficients of $Q(x)$ in the discussion preceding the Remainder Theorem? They are the first three expressions in the last line of the substitution process shown on page 366. Because for any polynomial $P(x)$ you can use the synthetic-substitution process to find the partial quotient $Q(x)$ and the remainder $P(r)$ that is obtained on dividing $P(x)$ by $(x - r)$, synthetic substitution is often called synthetic division.

EXAMPLE 1 Use synthetic division to divide $x^3 - 3x^2 + 4x - 12$ by $x - 3$.

SOLUTION

$$\begin{array}{r|rrrr}
3 & 1 & -3 & 4 & -12 \\
  &   & 3 & 0 & 12 \\
\hline
  & 1 & 0 & 4 & 0
\end{array}$$

$Q(x) = x^2 + 4; R = 0$

$$\therefore \frac{x^3 - 3x^2 + 4x - 12}{x - 3} = x^2 + 4 + \frac{0}{x - 3} = x^2 + 4$$

A corollary of the Remainder Theorem is stated on the following page.

Use synthetic division to write the given polynomial in the form $P(x) = (x - r)Q(x) + P(r)$ for the given values of r.

1. $P(x) = 2x^3 + 5x^2 - 3x + 2$
   a. $r = 2$ b. $r = 3$
   a. $(x - 2)(2x^2 + 9x + 15) + 32$
   b. $(x - 3)(2x^2 + 11x + 30) + 92$

2. $P(x) = 4x^3 - 4x^2 + 5x - 8$
   a. $r = -2$ b. $r = \dfrac{1}{2}$
   a. $(x + 2)(4x^2 - 12x + 29) - 66$
   b. $\left(x - \dfrac{1}{2}\right)(4x^2 - 2x + 4) - 6$

Use the Factor Theorem to decide whether the binomial given in the form $x - r$ is a factor of the polynomial given as $P(x)$. If it is not, give the remainder when $P(x)$ is divided by $x - r$.

3. $x + 2$; $P(x) = 3x^3 - 7x^2 - 20x + 3$ $R = -\dfrac{9}{x + 2}$

4. $x - 2i$; $P(x) = x^4 - x^3 - 4x - 16$ $x - 2i$ is a factor of $P(x)$

5. $x - 3$; $P(x) = 2x^3 + x^2 - 3x - 9$ $R = \dfrac{45}{x - 3}$

6. If one root of $x^3 - 5x^2 - 2x + 24 = 0$ is 3, find the other roots. $4, -2$

Factor Theorem

Over the set of complex numbers, $x - r$ is a factor of a polynomial $P(x)$ if and only if r is a root of $P(x) = 0$.

PROOF

If r is a root of $P(x) = 0$, then by the definition of root, $P(r) = 0$. Therefore, by the Remainder Theorem, we have:

$$P(x) = (x - r)Q(x) + P(r)$$
$$= (x - r)Q(x) + 0$$
$$= (x - r)Q(x)$$

Thus $(x - r)$ is a factor of $P(x)$. Conversely, if $(x - r)$ is a factor of $P(x)$, then

$$P(x) = (x - r)Q(x),$$

so that we have:

$$P(r) = (r - r)Q(r)$$
$$= 0 \cdot Q(r)$$
$$= 0$$

This theorem can help you identify factors of polynomials and zeros of polynomial functions.

EXAMPLE 2 Is $x - 6$ a factor of $P(x) = x^4 - 4x^3 - 13x^2 + 3x + 18$?

SOLUTION If $P(6) = 0$, then $x - 6$ is a factor. Use synthetic substitution.

$$
\begin{array}{r|rrrrr}
6 & 1 & -4 & -13 & 3 & 18 \\
  &   & 6 & 12 & -6 & -18 \\
\hline
  & 1 & 2 & -1 & -3 & 0
\end{array}
$$

$\therefore x - 6$ is a factor.

Whenever r is a root of the polynomial equation $P(x) = 0$, you find the remaining roots by solving the **depressed equation**

$$P(x) \div (x - r) = 0.$$

EXAMPLE 3 Find the zeros of the function P, where $P(x) = 2x^3 - 11x^2 + 3$.

SOLUTION *Plan:* Solve the equation $P(x) = 0$.

1. Use the Rational Root Theorem (page 254) to identify each possible rational root, $\dfrac{p}{q}$.

$$2x^3 - 11x^2 + 3 = 0$$

$$\dfrac{p}{q} \in \left\{ \dfrac{1}{2}, -\dfrac{1}{2}, \dfrac{3}{2}, -\dfrac{3}{2}, 1, -1, 3, -3 \right\}$$

370 *Chapter 9*

2. Use the Factor Theorem and synthetic substitution to test each possibility. By mentally doing the addition steps in the process, you can arrange the work conveniently, as shown.

x				$P(x)$
	2	-11	0	3
$\frac{1}{2}$	2	-10	-5	$\frac{1}{2}$
$-\frac{1}{2}$	2	-12	6	0

$\therefore P(x) = \left(x + \frac{1}{2}\right)(2x^2 - 12x + 6) = 0.$

3. Solve the depressed equation $2x^2 - 12x + 6 = 0$, or $x^2 - 6x + 3 = 0$: Using the quadratic formula, you find that $x = 3 \pm \sqrt{6}$.

$\therefore$ the set of zeros of P is $\left\{-\frac{1}{2}, 3 + \sqrt{6}, 3 - \sqrt{6}\right\}.$

Oral Exercises

Exercises 1–7 refer to the following synthetic division problem in which $P(x)$ is divided by $x - r$.

$$\begin{array}{r|rrrr} -3 & 2 & 1 & 0 & 32 \\ & & -6 & 15 & -45 \\ \hline & 2 & -5 & 15 & -13 \end{array}$$

1. State $P(x)$ as a polynomial. $P(x) = 2x^3 + x^2 + 32$
2. State the value of r and $x - r$ as a polynomial. $r = -3, x - r = x + 3$
3. State the partial quotient. $2x^2 - 5x + 15$
4. State the remainder. -13
5. State the value of $P(r)$. -13 $2x^3 + x^2 + 32 = (x + 3)(2x^2 - 5x + 15) - 13$
6. State the division as an equation in the form

$$P(x) = (x - r)Q(x) + P(r).$$

7. State the division as an equation with $\dfrac{P(x)}{x - r}$ as its left side.

$$\frac{2x^3 + x^2 + 32}{x + 3} = 2x^2 - 5x + 15 - \frac{13}{x + 3}$$

Written Exercises

Use synthetic division to write the given polynomial in the form $P(x) = (x - r)Q(x) + P(r)$ for the given values of r.

A
1. $P(x) = 3x^3 - 8x^2 + 5x + 6$ a. $r = 2$ b. $r = -1$
2. $P(x) = 2x^3 - 7x^2 - x + 10$ a. $r = 3$ b. $r = 4$
3. $P(x) = 4x^3 + 8x^2 - 7x + 5$ a. $r = -3$ b. $r = \frac{1}{2}$
4. $P(x) = 8x^3 - 12x - 9$ a. $r = \frac{3}{2}$ b. $r = i$

Polynomial Functions **371**

Suggested Assignments

Minimum
First day
371/1–8
Second day
372/9–28
Extended Alg. with Trig.
371/1–31 odd
Enriched Alg.
371/3, 7, 11–33 odd
Enriched Alg. with Trig.
371/3, 7, 11–33 odd

Additional Answers
Written Exercises

1. a. $3x^3 - 8x^2 + 5x + 6 = (x - 2)(3x^2 - 2x + 1) + 8$
   b. $3x^3 - 8x^2 + 5x + 6 = (x + 1)(3x^2 - 11x + 16) - 10$

2. a. $2x^3 - 7x^2 - x + 10 = (x - 3)(2x^2 - x - 4) - 2$
   b. $2x^3 - 7x^2 - x + 10 = (x - 4)(2x^2 + x + 3) + 22$

3. a. $4x^3 + 8x^2 - 7x + 5 = (x + 3)(4x^2 - 4x + 5) - 10$
   b. $4x^3 + 8x^2 - 7x + 5 = \left(x - \frac{1}{2}\right)(4x^2 + 10x - 2) + 4$

4. a. $8x^3 - 12x - 9 = \left(x - \frac{3}{2}\right)(8x^2 + 12x + 6)$
   b. $8x^3 - 12x - 9 = (x - i) \cdot (8x^2 + 8ix - 20) - 9 - 20i$

6. a. $\dfrac{-2x^3 + 3x^2 - 2x + 3}{x - 3}$

$= -2x^2 - 3x - 11 + \dfrac{-30}{x - 3}$

b. $\dfrac{-2x^3 + 3x^2 - 2x + 3}{x + i} =$

$-2x^2 + (3 + 2i)x - 3i$

8. a. $\dfrac{8x^3 + 10x^2 - 7x - 5}{x + \dfrac{3}{2}}$

$= 8x^2 - 2x - 4 + \dfrac{1}{x + \dfrac{3}{2}}$

18. $-3, 4$ **20.** $2 \pm i$

26. $\left\{2, \dfrac{1}{3}, -4\right\}$

33. Let $P(x) = 2x^{18} + mx^2 - 1$. Since $(x - i)$ is a factor of $P(x)$, $P(i) = 0 = 2i^{18} + mi^2 - 1$, $0 = -2 - m - 1$, $m = -3$. Substituting for m and evaluating $P(-i)$ gives $2(-i)^{18} - 3(-i)^2 - 1 = 0$. Therefore, $x + i$ is a factor of $P(x)$.

Mixed Review

State the possible rational roots of the given equation.

1. $2x^3 - x + 1 = 0$ $\pm 1, \pm \dfrac{1}{2}$

2. $4x^3 - 3x^2 - 1 = 0$ $\pm 1,$ $\pm \dfrac{1}{2}, \pm \dfrac{1}{4}$

3. Find the rational roots of $x^3 - 3x + 2 = 0$. $\{1, -2\}$

4. If y varies directly as x^3 and y is 3 when x is 2, find y when $x = 4$. 24

Evaluate. Express your answer in standard notation.

5. $1,234,000 \times 62,000,000$
    7.6508×10^{13}

Use synthetic division to divide the given polynomial $P(x)$ by the given polynomials $x - r$. Express the result as an equation whose left side is $\dfrac{P(x)}{x - r}$.

8. b. $\dfrac{8x^3 + 10x^2 - 7x - 5}{x - \frac{1}{2}} = 8x^2 + 14x + \dfrac{-5}{x - \frac{1}{2}}$

5. $P(x) = x^3 - 13x + 18$ **a.** $x + 4$ **b.** $x - 2$

6. $P(x) = -2x^3 + 3x^2 - 2x + 3$ **a.** $x - 3$ **b.** $x + i$

7. $P(x) = x^4 - 4x^3 - 3x - 9$ **a.** $x - 5$ **b.** $x - 2i$

8. $P(x) = 8x^3 + 10x^2 - 7x - 5$ **a.** $x + \frac{3}{2}$ **b.** $x - \frac{1}{2}$

Use the Factor Theorem to decide whether the binomial given in the form $x - r$ is a factor of the polynomial given as $P(x)$. If it is not, give the remainder when $P(x)$ is divided by $x - r$.

9. $x + 2$; $P(x) = x^4 + x^3 + 3x^2 + 7x - 6$ yes

10. $x - 3$; $P(x) = x^4 - 5x^3 + 10x^2 - 6x - 18$ yes

11. $x + i$; $P(x) = x^4 + 7x^2 - 4$ no; -10

12. $x - 2i$; $P(x) = 2x^3 - 5x^2 + 8x + 20$ no; 40

13. $x - \frac{1}{2}$; $P(x) = 2x^3 - 7x^2 + 5x - 3$ no; -2

14. $x - \frac{1}{3}$; $P(x) = 3x^4 - 7x^3 + x - 2$ no; $-\frac{17}{9}$

For each of the polynomials below, one root is given. Find the other roots.

15. $x^3 - 3x^2 - 10x + 24$; $r_1 = 2$ $-3, 4$

16. $x^3 + 3x^2 - 2x - 6$; $r_1 = -3 \pm\sqrt{2}$

17. $2x^3 + 3x^2 - 2x - 3$; $r_1 = 1$ $-1, -\frac{3}{2}$

18. $2x^3 - x^2 - 25x - 12$; $r_1 = -\frac{1}{2}$

19. $x^3 + 5x^2 + 9x + 45$; $r_1 = -5$ $\pm 3i$

20. $x^3 - 7x^2 + 17x - 15$; $r_1 = 3$

B   **21.** $x^4 + x^3 - 11x^2 - 9x + 18$; $r_1 = -2$ $1, \pm 3$

22. $x^4 + 2x^3 - 2x^2 + 2x - 3$; $r_1 = 1$ $-3, \pm i$

Find the zeros of each function.

23. $P(x) = x^3 + x^2 - 4x - 4$ $\{-1, \pm 2\}$

24. $P(x) = x^3 + 4x^2 + x - 6$ $\{1, -2, -3\}$

25. $P(x) = x^3 - 5x^2 + 10x - 8$ $\left\{2, \frac{3 \pm i\sqrt{7}}{2}\right\}$

26. $P(x) = 3x^3 + 5x^2 - 26x + 8$

27. $P(x) = x^4 - x^3 - 5x^2 - x - 6$ $\{-2, 3, \pm i\}$

28. $P(x) = 2x^4 + 3x^3 + 6x^2 + 12x - 8$ $\left\{-2, \frac{1}{2}, \pm 2i\right\}$

29. Find m such that $x - 3$ will be a factor of $x^3 - 4x^2 + mx - 9$. 6

30. Find m such that -2 will be a zero of $x^3 + 3x^2 - 7x + m$. -18

31. Find m such that $x^3 + 2x^2 + mx - 3$ will leave a remainder of -3 when divided by $x + 3$. -3

C   **32.** Find m such that $x + 1$ will be a factor of $x^{85} + mx - 4$. (Hint: Use the Factor Theorem.) -5

33. Show that if $x - i$ is a factor of $2x^{18} + mx^2 - 1$, then so is $x + i$.

34. Show that if $P(x) = ax^3 + bx^2 + cx + d$, with a, b, c, and d real numbers, then $P(ki)$ is the complex conjugate of $P(-ki)$, for any real number k.

READING ALGEBRA   *Notation*

The abbreviations that mathematicians use are the set of symbols which we call mathematical notation. They provide us with a brief and concise way of expressing ideas which would often be cumbersome if we tried to say them in words.

Look, for example, at the Remainder Theorem on page 369 and the Factor Theorem on page 370. Try and write an understandable version of these theorems in words. The task is difficult even if you are allowed to use variables.

When you read $P(-1)$ you know that what is referred to is "the value of the polynomial function, $P(x)$, when x is equal to negative one". The abbreviated form, $P(-1)$, can easily be used in equations such as $P(-1) = 5$ or $P(-1) = P(3)$.

If you are careful to learn the precise meaning of each of the symbols and the abbreviations used, you will appreciate their usefulness. To help yourself to learn the meaning of new notation, look for a logical link between the symbol and its meaning. The inequality symbols, $<$ and $>$, for example, are not confusing if you remember that the smaller side points toward the smaller quantity. For example, $3 < 4$.

A symbol may have an obvious meaning and even look like the concept that it denotes. An example of this is the symbol for triangle ($\triangle$). Other symbols may be conventions that have developed over time, such as $+$, the symbol for addition, and $=$, the symbol for is equal to. If an association is not obvious for a particular symbol, try to invent something which will help you to remember its meaning. Such a memory aid is called a mnemonic device.

As you complete each chapter in your book and are preparing to study for a chapter test, make a list of the new symbols notation, and terms found in the chapter. If you need help in locating the page where a symbol is first used, consult the list of symbols in the front of your book. There you will find each symbol, how it is read, and the number of the page on which it was introduced.

Exercises

Translate each symbol into words.

1. $\{ \ \}$ set
2. Σ summation
3. $\approx$ approximately equal
4. $\mathcal{R}$ set of real numbers

Translate each phrase into mathematical notation.

5. the absolute value of c $|c|$
6. the set of complex numbers $\mathcal{C}$
7. the limit as n increases without bound $\lim\limits_{n \to \infty}$

***Additional Answers
Written Exercises***
(continued from p. 367)

7. a. $x^2 - 6x + 4 \le 0$
   b. $\{3 + \sqrt{5}, 3 - \sqrt{5}\}$
   c.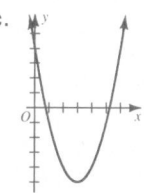
   d. $\{x: 3 - \sqrt{5} \le x \le 3 + \sqrt{5}\}$

8. a. $2x^2 - 4x - 1 \ge 0$
   b. $\left\{\dfrac{2 + \sqrt{6}}{2}, \dfrac{2 - \sqrt{6}}{2}\right\}$
   c.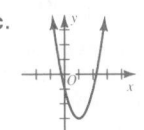
   d. $\left\{x: x \le \dfrac{2 - \sqrt{6}}{2}\right\} \cup$ $\left\{x: x \ge \dfrac{2 + \sqrt{6}}{2}\right\}$

9. a. $x^2 + 4x - 1 > 0$
   b. $\{-2 + \sqrt{5}, -2 - \sqrt{5}\}$
   c.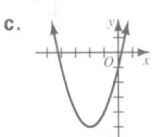
   d. $\{x: x < -2 - \sqrt{5}\} \cup$ $\{x: x > -2 + \sqrt{5}\}$

10. $x^2 - 4 > 0$ and $x^2 - 9 < 0$
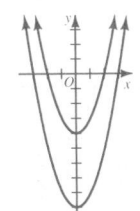
$\{x: -3 < x < -2\} \cup$ $\{x: 2 < x < 3\}$

373

Key Ideas

Apply theorems about the number of complex roots of a polynomial equation and conjugate pairs of roots.

Chalkboard Examples

1. Find a polynomial equation having the roots $\sqrt{2}$, $-\sqrt{2}$, -4. If the roots are $\sqrt{2}, -\sqrt{2}, -4$, the factors of the polynomial are $(x - \sqrt{2})(x + \sqrt{2})(x + 4)$. Multiply to find $P(x)$.
$x^3 + 4x^2 - 2x - 8 = 0$

2. If $1 - 2i$ is a root of $x^3 + x^2 - x + 15 = 0$, find the remaining roots. Since the equation has real coefficients, the conjugate of $1 - 2i$ is also a root.
∴ $[x - (1 - 2i)][x - (1 + 2i)] = x^2 - 2x + 5$ is a factor of $x^3 + x^2 - x + 15 = 0$.
$x^3 + x^2 - x + 15 \div (x^2 - 2x + 5) = x + 3$. ∴ $x + 3$ is a factor.
∴ the roots are $1 - 2i$, $1 + 2i$ and -3.

Additional A Exercises

Find a polynomial equation having the given roots.

1. 2, −2, 3
$x^3 - 3x^2 - 4x + 12 = 0$

2. $\sqrt{3}, -\sqrt{3}, -1$
$x^3 + x^2 - 3x - 3 = 0$

3. $-2, 3 + i, 3 - i$
$x^3 - 4x^2 - 2x + 20 = 0$

Find the other root(s) of the given equation and factor the polynomial completely.

4. $x^3 + 2x^2 + 9x + 18 = 0$;
$3i$ $-3i, -2; (x + 3i) \cdot (x - 3i)(x + 2)$

9–9 The Fundamental Theorem of Algebra

Over the set of real numbers a polynomial equation may have no solution. For example, $x^2 + 1 = 0$ has no *real* root. But over the set of complex numbers it has two roots, namely i and $-i$. The German mathematician C. F. Gauss in 1799 first proved that *every polynomial equation with complex coefficients has at least one root.* This result, called the Fundamental Theorem of Algebra, leads to the following assertion, which we will accept without proof.

> *Theorem.* Every polynomial equation with complex coefficients and positive degree n has exactly n complex roots.

In applying this theorem, you may have to count the same number as a root more than once. For example, 5 is a *double* root of the equation $x^2 - 10x + 25 = 0$.

Recall (page 348) that the imaginary roots of a quadratic equation with real coefficients occur in conjugate pairs. Thus the fact that $2 - 3i$ is a root of $x^2 - 4x + 13 = 0$ implies that $2 + 3i$ is also a root. This property is typical of all polynomial equations with *real* coefficients.

> *Theorem.* If a polynomial equation with real coefficients has $a + bi$ as a root (a and b real, $b \neq 0$), then $a - bi$ is also a root.

The proof of this theorem is left as Exercises 17–19, page 375.

If you know that

$$P(x) = x^4 + 4x^3 + 12x^2 + 28x + 35 = 0$$

has the root $-2 + i$, then the preceding theorem and the Factor Theorem enable you to solve the equation. Since the equation has real coefficients, *both* $-2 + i$ and $-2 - i$ are roots. Thus both $x - (-2 + i)$ and $x - (-2 - i)$ are factors of $P(x)$. Since neither of these polynomials is a factor of the other, their product

$$[(x + 2) - i][(x + 2) + i],$$

or $x^2 + 4x + 5$, must be a factor of $P(x)$. But

$$P(x) \div (x^2 + 4x + 5) = x^2 + 7.$$

The roots of the depressed equation are $i\sqrt{7}$ and $-i\sqrt{7}$; hence the solution set of $P(x) = 0$ is

$$\{-2 + i, -2 - i, i\sqrt{7}, -i\sqrt{7}\}.$$

374 *Chapter 9*

Oral Exercises

1. Two roots of $x^3 - 3x^2 + 2x - 6 = 0$ are 3 and $i\sqrt{2}$. What is the other root? $-i\sqrt{2}$

2. Two roots of $x^3 + 6x + 20 = 0$ are -2 and $1 + 3i$. What is the other root? $1 - 3i$

3. A cubic equation with real coefficients has roots 9 and $5 - 2i$. What is the third root? $5 + 2i$

4. One root of a quadratic equation is $-4i$. What is the other root? Find such an equation. $4i; x^2 + 16 = 0$

5. One root of a quadratic equation is $5 + i$. What is the other root? Find such an equation. $5 - i; x^2 - 10x + 26 = 0$

6. A cubic equation with real coefficients has roots -1 and $3 - 2i$. What is the third root? Find such an equation. $3 + 2i; x^3 - 5x^2 + 7x + 13 = 0$

Written Exercises

Find a polynomial equation having the given roots.

A

1. $3, -1, 2$

2. $\sqrt{6}, -\sqrt{6}, -2$

3. $3i, -3i, 4$
   $x^3 - 4x^2 + 9x - 36 = 0$

4. $-5, 2 + i, 2 - i$
   $x^3 + x^2 - 15x + 25 = 0$

5. $2, -2, i\sqrt{3}, -i\sqrt{3}$
   $x^4 - x^2 - 12 = 0$

6. $6, 1 + i\sqrt{5}, 1 - i\sqrt{5}$
   $x^3 - 8x^2 + 18x - 36 = 0$

In Exercises 7–14 you are given a polynomial equation and one or more of its roots. Find the other root(s) and factor the polynomial completely.

7. $x^3 - 3x^2 + 9x + 13 = 0; 2 + 3i, -1$

8. $x^3 - 2x^2 - 4x - 16 = 0; 4, -1 + i\sqrt{3}$

9. $x^3 - 3x^2 + 5x - 15 = 0; i\sqrt{5}$

10. $x^4 + 11x^2 + 18 = 0; -3i$

11. $x^3 - 2x^2 - 14x + 40 = 0; 3 + i$

12. $x^4 + 4x^3 + 2x^2 - 12x - 15 = 0; -2 - i$

13. $x^4 + 4x^3 + 11x^2 + 24x + 30 = 0; -2 + i$

14. $x^4 - 6x^3 + 27x^2 - 18x + 72 = 0; -i\sqrt{3}$

B

15. Prove that every polynomial of *odd* degree with real coefficients has at least one real root. (*Hint*: Use the theorems on page 374.)

16. Prove that every polynomial of *even* degree with real coefficients has an even number of real roots or no real roots.

17. For any positive integer k and real numbers a, b, and c, show that $c(a + bi)^k$ is the conjugate of $c(a - bi)^k$. (*Hint*: Use the fact that the conjugate of a real number is itself, in conjunction with the results of Exercise 24 on page 284.)

C

18. Show that if $P(x)$ is a polynomial with real coefficients, then $P(a + bi)$ is the conjugate of $P(a - bi)$, for any real numbers a and b. (*Hint*: Use Exercise 17 above, and Exercise 28 on page 282.)

19. Show that if $P(x)$ is a polynomial with real coefficients and $P(a + bi) = 0$, then $P(a - bi) = 0$, for any real numbers a and b.

Polynomial Functions **375**

---

5. $x^3 - x^2 - 7x + 15 = 0$;
   $2 + i$, $2 - i$, -3; $(x + 3) \cdot (x - (2 - i))(x - (2 + i))$

6. $x^4 + 9x^2 + 20 = 0$; $-i\sqrt{5}$
   $i\sqrt{5}, \pm 2i; (x + i\sqrt{5})(x - i\sqrt{5})(x - 2i)(x + 2i)$

Suggested Assignments

Minimum
First day
375/1–15
Second day
R 376/*Self-Test 3*
S 367/1–23 odd
Extended Alg. with Trig.
375/1–19
R 376/*Self-Test 3*
S 367/2–22 even
Enriched Alg.
375/5–13 odd, 14–19
R 376/*Self-Test 3*
S 367/2–22 even
Enriched Alg. with Trig.
375/5–13 odd, 14–19
R 376/*Self-Test 3*
S 367/2–22 even

Additional Answers
Written Exercises
(See p. 388.)

Mixed Review

Evaluate.

1. $\sum_{n=1}^{9} (3n - 5)$ 90

2. $\sum_{j=1}^{14} \left(\frac{1}{3}j + 4\right)$ 91

Express as a fraction in lowest terms.

3. $0.4\overline{09}$ $\frac{9}{22}$ 4. $0.19\overline{3}$ $\frac{29}{150}$

Solve over $\mathcal{R}$.

5. $\sqrt{2x + 1} - \sqrt{x - 3} = 2$
   $\{4, 12\}$

6. $\sqrt{x - 2} - \sqrt{x + 3} = 1$ $\varnothing$

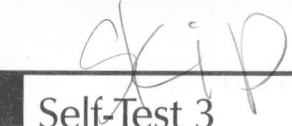

Quick Quiz

1. If $P(x) = 2x^4 + 6x^3 - x^2 + 7x + 4$, use synthetic substitution to find: **a.** $P(2)$ and **b.** $P(i)$
 a. 94   **b.** $7 + i$

2. For $P(x) = x^4 - x^3 - 4x - 20$, use synthetic division to divide $P(x)$ by $x + 3$. Express the result as an equation whose left side is $\dfrac{P(x)}{x + 3}$.

$$\frac{x^4 - x^3 - 4x - 20}{x + 3} = $$
$$x^3 - 4x^2 + 12x - 40 + \frac{100}{x + 3}$$

3. Find all the roots of $x^3 - 5x^2 - 2x + 24 = 0$.
 3, 4, −2

4. Find all the roots of $x^3 - 3x + 52 = 0$ given that one root is $2 - 3i$.
 $\{2 - 3i, 2 + 3i, -4\}$

Teaching Suggestions
p. T103

Key Ideas

Determine the possible number of positive and negative roots by using Descartes' Rule of Signs.
Determine the upper and lower bounds of the roots.

Self-Test 3

VOCABULARY   polynomial function (p. 365)   Factor Theorem (p. 370)
synthetic substitution (p. 366)   depressed equation (p. 370)
Remainder Theorem (p. 369)   Fundamental Theorem of
synthetic division (p. 369)   Algebra (p. 374)

1. If $P(x) = x^4 - 5x^3 + 3x^2 - 4x - 1$, use synthetic substitution to find **(a)** $P(-2)$ and **(b)** $P(i)$. **a.** 75   **b.** $-3 + i$ *Obj. 1, p. 365*

2. For $P(x) = 2x^3 - 5x^2 + 3x + 4$, use synthetic division to divide $P(x)$ by $x - 3$. Express the result as an equation whose left side is $\dfrac{P(x)}{x - 3}$. $\dfrac{2x^3 - 5x^2 + 3x + 4}{x - 3} = 2x^2 + x + 6 + \dfrac{22}{x - 3}$ *Obj. 2, p. 365*

3. Find all the roots of $2x^3 - 7x^2 + x + 10 = 0$. $\{-1, 2, \frac{5}{2}\}$ *Obj. 3, p. 365*

4. Find all the roots of $x^3 - 8x^2 + 25x - 26 = 0$ given that one root is $3 - 2i$. $\{3 \pm 2i, 2\}$ *Obj. 4, p. 365*

Check your answers with those at the back of the book.

Estimating Real Roots

OBJECTIVES for Sections 9-10 and 9-11:
1. *To determine the possible number of positive and negative roots of a polynomial equation.*
2. *To determine upper and lower bounds for the roots of a polynomial equation.*
3. *To estimate the real roots of a polynomial equation.*

9–10 Locating Possible Real Roots

Using synthetic division and the Factor Theorem, you can check the possible rational roots of a polynomial equation with integral coefficients to see which are in fact roots of the equation. Even if the roots of a polynomial equation are irrational, however, the theorem below, named after the French mathematician and philosopher René Descartes (1596–1650), provides some information as to the *number* of real roots and how many of these are positive and how many are negative. In order to state the theorem, we need the following definition: If the terms of a polynomial $P(x)$ are written in decreasing order according to the powers of x (ignoring missing terms), each pair of

376   *Chapter 9*

successive coefficients with *opposite signs* is called a **variation of sign**. For example, the polynomial

$$5x^5 - 3x^3 - 7x^2 + 6x - 2$$

$$\underbrace{\qquad}_{1} \quad \underbrace{\qquad}_{2} \quad \underbrace{\qquad}_{3}$$

contains the three indicated variations of sign.

We can now state the theorem.

Descartes' Rule of Signs

If $P(x)$ is a polynomial with real coefficients, then

1. the number of positive roots of $P(x) = 0$ is either equal to the number of variations of sign of $P(x)$ or is less than this number by a positive even integer;

2. the number of negative roots of $P(x) = 0$ is either equal to the number of variations of sign of $P(-x)$ or is less than this number by a positive even integer.

EXAMPLE 1   Determine the possible number of **(a)** positive roots and **(b)** negative roots of $x^4 + 3x^3 - 2x^2 - x - 5 = 0$.

SOLUTION   If

$$P(x) = x^4 + 3x^3 - 2x^2 - x - 5,$$

then $P(x)$ has only the one indicated variation of sign. Therefore, according to Descartes' Rule of Signs, $P(x) = 0$ must have exactly 1 positive root.

$$P(-x) = (-x)^4 + 3(-x)^3 - 2(-x)^2 - (-x) - 5$$
$$= x^4 - 3x^3 - 2x^2 + x - 5$$

Thus $P(-x)$ has three variations of sign, and hence, by Descartes' Rule, $P(x)$ has either 3 negative roots or 1 negative root. The following table summarizes the possibilities (in view of the second theorem on page 374).

positive roots	negative roots	imaginary roots
1	1	2
1	3	0

Descartes' Rule of Signs provides a way of determining the possible number of positive and negative roots of a polynomial equation. The next theorem gives us a way to narrow the region of possible real roots to an interval on the number line.

1. Determine the number of **(a)** positive roots and **(b)** negative roots of $x^4 + 8x^3 - 2x^2 - 3x + 8 = 0$. If $P(x) = x^4 + 8x^3 - 2x^2 - 3x + 8$, then $P(x)$ has two sign changes, therefore by Descartes' Rule of Signs, $P(x) = 0$ must have either two or zero positive roots. If $P(-x) = x^4 - 8x^3 - 2x^2 + 3x + 8$, therefore $P(x) = 0$ has two or zero negative roots.

pos. roots	neg. roots	imag. roots
2	2	0
0	0	4
2	0	2
0	2	2

2. Find an upper and a lower bound for the real roots of $x^3 - x - 1 = 0$. Use synthetic division for $M = 1, 2, \ldots$ and $L = -1, -2, \ldots$ until numbers are found that satisfy the conditions of the theorem. The results of the synthetic divisions are shown in the table below.

x	coef. of quotient			$P(x)$
1	1	1	0	-1
2	1	2	3	5
-1	1	-1	0	-1

∴ the upper bound is 2; the lower bound is -1.

Have your students explain in their own words what it means to have an upper bound or a lower bound on the roots of a polynomial equation.

Theorem.   Let $P(x)$ be a polynomial with real coefficients and with positive leading coefficient.

1. If $M \geq 0$ and the coefficients of the quotient and remainder obtained on dividing $P(x)$ by $x - M$ are all greater than or equal to 0, then $P(x)$ has no roots greater than M.

2. If $L \leq 0$ and the coefficients obtained on dividing $P(x)$ by $x - L$ are alternately greater than or equal to 0 and less than or equal to 0, then $P(x)$ has no roots less than L.

A proof of this theorem is suggested in Exercises 15 and 16 on page 379. The numbers M and L are called **upper** and **lower bounds**, respectively, for the roots of the given polynomial.

EXAMPLE 2   Find an upper and a lower bound for the real roots of

$$x^3 - 3x^2 + 2x + 9 = 0.$$

SOLUTION   Use synthetic division for

$$M = 1, 2, 3, \ldots \quad \text{and} \quad L = -1, -2, -3, \ldots$$

until numbers are found that satisfy the conditions of the theorem. The results of the synthetic divisions are shown in the table below.

x	coefficients of quotient			P(x)	
1	1	−2	0	9	
2	1	−1	0	9	
3	1	0	2	15	← all ≥ 0
−1	1	−4	6	3	⎧ alternately ≥ 0
−2	1	−5	12	−15	⎨ and ≤ 0

Thus, by the theorem above, there is no root greater than 3 or less than −2. The real roots must lie in the interval graphed below.

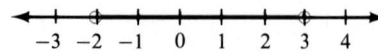

Thus, by the theorem above, there is no root greater than 3 or less than −2. The real roots must lie in the interval graphed below.

Oral Exercises

State the possible number of positive roots of each equation.

1. $x^3 - 2x^2 + 5x - 11 = 0$ 3 or 1

2. $-3x^4 - 7x^3 + 8x^2 - x + 2 = 0$ 3 or 1

3. $x^5 - x^3 + x^2 - x + 7 = 0$ 4, 2, or 0

4. $-x^5 + 2x^4 - 4x^3 + x^2 - 9 = 0$ 4, 2, o

378   *Chapter 9*

By testing to see if the conditions of the theorem on page 378 are satisfied, determine whether the number M is an upper bound for the real roots or whether the number L is a lower bound for the real roots of the equation $x^3 - 2x^2 - x - 3 = 0$.

5. $M = 1$ no

6. $M = 2$ no

7. $M = 3$ yes

8. $M = 4$ yes

9. $L = -1$ yes

10. $L = -2$ yes

Written Exercises

Determine the possible number of positive roots and the possible number of negative roots of each equation. Summarize the possibilities (including the number of imaginary roots) in a table like the one shown in Example 1 on page 377.

A
1. $x^3 - 5x^2 + x + 2 = 0$
2. $-x^3 - 4x^2 + 6x - 3 = 0$
3. $-2x^4 + x^3 - 7x^2 + 4x + 1 = 0$
4. $x^4 + 5x^3 - x^2 - 7x + 8 = 0$
5. $y^5 + y^4 - 2y^3 - y^2 + y - 10 = 0$
6. $v^6 - 2v^5 - 4v^4 - v^3 + v + 1 = 0$

Determine the least positive integral upper bound and the greatest negative integral lower bound for the roots of each equation that you can find using the theorem on page 378.

7. $x^3 - x^2 + 2x + 9 = 0$ $M = 1; L = -2$
8. $x^3 - 2x^2 - x - 3 = 0$
9. $y^3 - y^2 - 5y + 2 = 0$ $M = 3; L = -2$
10. $2z^3 - z^2 - 4z + 3 = 0$

                                              $M = 3; L = -2$

B
11. $u^4 - 2u^3 - 3u^2 + 4u + 8 = 0$
12. $2w^4 + w^3 - 3w^2 - w + 5 = 0$
13. $x^4 - 3x^2 - 7x + 6 = 0$ $M = 3; L = -2$
14. $v^4 + 2v^3 - v^2 + 7v - 5 = 0$

Exercises 15 and 16 suggest a proof of the theorem on page 378. Let $P(x)$ be a polynomial of degree at least 1, with real coefficients.

C
15. Suppose that $M \geq 0$ and $P(x) = (x - M)Q(x) + R$, where the coefficients of $Q(x)$ and the number R are all greater than or equal to 0. Prove that if $x_0 > M$, then x_0 cannot be a root of $P(x) = 0$. (*Hint:* Suppose that $x_0 > M$ and $P(x_0) = 0$. Note that since $P(x)$ has degree greater than or equal to 1, $Q(x)$ cannot be the zero polynomial.)

16. Suppose that $L \leq 0$ and $P(x) = (x - L)Q(x) + R$, where the coefficients of $Q(x)$ and the number R are alternately greater than or equal to zero and less than or equal to zero. Prove that if $x_0 < L$, then x_0 cannot be a root of $P(x) = 0$. (*Hint:* Suppose that $x_0 < L$ and $P(x_0) = 0$. Consider two cases: (1) The constant term of $Q(x) \geq 0$; and (2) The constant term of $Q(x) \leq 0$.)

Polynomial Functions   **379**

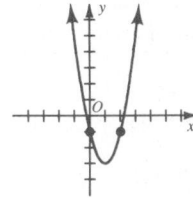

Key Ideas

Approximate real roots of polynomial functions.

Chalkboard Examples

1. Graph P if $P(x) = x^3 - 2x^2 + x + 1$. Use synthetic substitution.

x	$P(x)$
1	1
2	3
-1	-3
0	1

2. Find all intervals of length one-half unit that contain a zero of $P(x) = x^3 - 2x^2 - 3x + 1$. By Descartes' Rule of Signs there are two or zero positive zeros and one negative zero. Find values of $P(x)$ for integral values of x. When you find that $P(x)$ changes sign, evaluate $P(x)$ at the halfway point.

x	$P(x)$
-2	-9
-1.5	-2.375
-1	1
0	1
0.5	-0.875
1	-3
2	-5
2.5	-3.375
3	1

∴ the zeros r_1, r_2, and r_3 are such that
$-1.5 < r_1 < -1.0$
$0.0 < r_2 < 0.5$
$2.5 < r_3 < 3.0$.

9–11 Estimates of Real Roots

Estimates of the real roots of a polynomial equation (or the real zeros of a polynomial function) can be found by several methods. First, you can consider the graph of the polynomial function. You have sketched the graphs of many first- and second-degree polynomial functions over $\mathcal{R}$, as well as graphs of some power functions (Section 7-1). The following example suggests how to get some idea of the shape of the graph of a polynomial function of higher degree over $\mathcal{R}$.

EXAMPLE 1   Graph P if $P(x) = x^3 - 3x^2 + 3$.

SOLUTION   Use synthetic substitution to find some values for $P(x)$. When these points are connected by a smooth curve, you obtain the graph shown at the right.

x	$P(x)$
-1	-1
0	3
1	1
2	-1
3	3

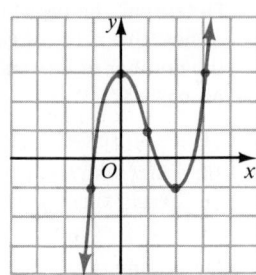

In drawing the graphs of polynomial functions as smooth unbroken curves, you assume the following:

> ### Intermediate Value Property
> If P is a polynomial function with real coefficients, and if m is any number between $P(a)$ and $P(b)$, then there is at least one number c between a and b for which $P(c) = m$.

In other words, P takes on every value between any two of its values.

In Example 1, since $P(2) = -1$ and $P(3) = 3$, so that $P(2) < 0 < P(3)$, there must be a value of x between 2 and 3 for which $P(x) = 0$. By inspecting the diagram, you can see that the graph crosses the x-axis at a point whose x-coordinate is approximately 2.5. Thus, an estimate of one root of $P(x) = 0$ is 2.5.

You can also see that since $P(1) = 1$ and $P(2) = -1$, there must be a root between 1 and 2 (approximately 1.4), and since $P(-1) = -1$ and $P(0) = 3$, there must be a root between -1 and 0 (approximately -0.9). Thus, the equation $x^3 - 3x^2 + 3 = 0$ has three real roots.

EXAMPLE 2   Find all intervals of length one-half unit that contain a zero of $P(x) = x^3 + x^2 - 3x - 1$.

SOLUTION By Descartes' Rule of Signs there is 1 positive zero and either 0 or 2 negative zeros. Find values of $P(x)$ for integral values of x. When you find that $P(x)$ changes sign between $x = -3$ and $x = -2$, for example, evaluate $P(x)$ for $x = -2.5$.

Since the changes of sign of $P(x)$ occur between

x	$P(x)$
-3	-10
-2.5	-2.875
-2	1
-1	2
-0.5	0.625
0	-1
1	-2
1.5	0.125
2	5

$$x = -2.5 \quad \text{and} \quad x = -2.0,$$
$$x = -0.5 \quad \text{and} \quad x = 0.0,$$
$$x = 1.0 \quad \text{and} \quad x = 1.5,$$

the zeros r_1, r_2, and r_3 are such that

$$-2.5 < r_1 < -2.0, \quad -0.5 < r_2 < 0.0, \quad \text{and} \quad 1.0 < r_3 < 1.5.$$

After you have located a root by estimating its position on a graph or by using the method of Example 2, you can obtain a closer approximation by computation. For example, consider the root of $x^3 - 3x^2 + 3 = 0$ that we estimated as 2.5 from the graph of $P(x) = x^3 - 3x^2 + 3$. By direct or synthetic substitution you can show that

$$P(2.5) = -0.125 < 0 \quad \text{and} \quad P(2.6) = 0.296 > 0,$$

so that $2.5 < r_1 < 2.6$. Now look at Figure 6, which shows the part of the graph of P over the interval $2.5 \le x \le 2.6$. Notice that the line segment joining $A(2.5, -0.125)$ and $B(2.6, 0.296)$ crosses the x-axis at C, which is near the point where the graph itself crosses. This suggests that the x-coordinate of C is a fairly good approximation of r_1. Denoting the coordinates of C by $(2.5 + h, 0)$, you have the following:

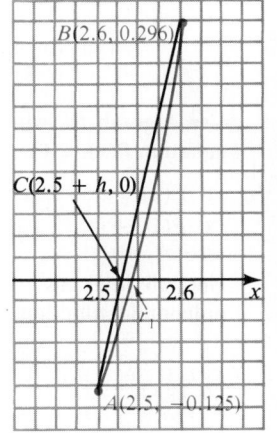

Figure 6

$$\text{slope of } \overline{AC} = \text{slope of } \overline{AB}$$
$$\frac{0.125}{h} \approx \frac{0.421}{0.1} = 4.21$$
$$\therefore h \approx \frac{0.125}{4.21} \approx 0.03$$

Thus, $r_1 \approx 2.5 + 0.03$, or 2.53. As a matter of fact, $P(2.53) \approx -0.008423$, which is fairly close to 0. This process of approximating a value of $P(x)$ by using a line segment is called **linear interpolation**.

You can check that $P(2.54) \approx 0.032264$, so that $2.53 < r_1 < 2.54$. To obtain an even better approximation of r_1, you can repeat the interpolation over this shorter interval. The process can be repeated as many times as desired, until you have an approximation of a root to any number of decimal places. Such repetitive procedures can be programmed for a computer.

Polynomial Functions **381**

---

3. Find all intervals of length one-half unit that contain a zero of $P(x) = x^3 + 5x - 1$. By Descartes' Rule of Signs, there is 1 positive zero and 0 negative zeros.

x	$P(x)$
0	-1
0.5	1.625
1	5

$\therefore$ the zero r is such that $0.0 < r < 0.5$.

Additional A Exercises

1. Find all intervals of length one-half unit that contain a zero of $P(x) = 2x^3 + 2x^2 - 2x - 1$. By Descartes' Rule of Signs there is 1 positive zero and either 0 or 2 negative zeros.

x	$P(x)$
-2	-5
-1.5	-0.25
-1	1
-0.5	0.25
0	-1
0.5	-1.25
1	1

$\therefore$ the zeros r_1, r_2, and r_3 are such that
$-1.5 < r_1 < -1.0$,
$-0.5 < r_2 < 0.0$, and
$0.5 < r_3 < 1.0$.

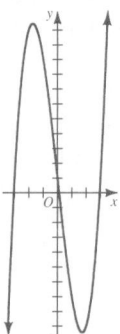
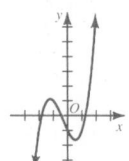
Oral Exercises

1. A table of values is given for $P(x) = x^3 + 2x^2 - 3x - 3$. List all intervals of length one-half unit that contain a zero of the polynomial function. $-3 < r_1 < -2.5;\ -1 < r_2 < -0.5;\ 1.5 < r_3 < 2$

x	-3	-2.5	-2	-1.5	-1	-0.5	0	0.5	1	1.5	2
$P(x)$	-3	1.375	3	2.625	1	-1.125	-3	-5.625	-3	-0.375	7

Exercises 2–4 refer to $P(x) = x^3 + 2x - 7$.

2. State the possible number of negative and positive roots of $P(x) = 0$. 1 pos., 0 neg.
3. Locate between two consecutive integers each real root of $P(x) = 0$. $1 < r < 2$
4. Find all intervals of length one-half unit that contain a root of $P(x) = 0$. $1.5 < r < 2$

Written Exercises

Use a graph to estimate to the nearest half unit the real roots of the equation $P(x) = 0$.

A
1. $P(x) = x^3 - 9x + 1$ $r_1 \approx -3;\ r_2 \approx 0;\ r_3 \approx 3$
2. $P(x) = x^3 + x^2 - 2x - 1$ $r_1 \approx 1;\ r_2 \approx -0.5;\ r_3 \approx -2$
3. $P(x) = -x^3 + x^2 + 6x - 2$ $r_1 \approx -2;\ r_2 \approx \frac{1}{2};\ r_3 \approx 3$
4. $P(x) = -x^3 + 5x^2 - 7$ $r_1 \approx 4.5;\ r_2 \approx 1.5;\ r_3 \approx -1$

Find all intervals of length one-half unit that contain a zero of the polynomial function.

5. $P(x) = x^3 - 3x + 1$
6. $P(x) = x^3 - 4x - 2$
7. $P(x) = x^3 - x^2 - 2x + 1$
8. $P(x) = 2x^3 - x^2 - 7x + 1$
9. $P(x) = x^3 - x^2 - 4x + 3$
10. $P(x) = x^3 - 2x^2 - 4x + 7$

B
11. $P(x) = x^3 - x^2 - 6$ $2 < r < \frac{5}{2}$
12. $P(x) = 2x^3 + 5x + 1$ $-\frac{1}{2} < r < 0$

Use linear interpolation to estimate to the nearest hundredth each zero found in the given exercise.

13. Exercise 11 $r \approx 2.22$
14. Exercise 9 $r_1 \approx -1.91;\ r_2 \approx 0.71;\ r_3 \approx 2.20$
15. Exercise 1 $r_1 \approx -3.05;\ r_2 \approx 0.11;\ r_3 \approx 2.94$
16. Exercise 5 $r_1 \approx -1.88;\ r_2 \approx 0.35;\ r_3 \approx 1.53$

C
17. Find all intervals of length one-half unit that contain a zero of $P(x) = 16x^4 + 16x^3 - 31x^2 - 32x - 2$. (*Hint:* You will need to evaluate P for certain half-unit values even when the sign does not change at the adjacent integral values.) $(-\frac{3}{2}, -1), (-1, -\frac{1}{2}), (-\frac{1}{2}, 0), (1, \frac{3}{2})$

18. Write a polynomial of the third degree with integral coefficients that has three real zeros all of which are greater than 0.3 and less than 1.
Answers will vary. $50x^3 - 75x^2 + 37x - 6 = 0$

382 Chapter 9

Self-Test 4

VOCABULARY   variation of sign (p. 377)
Descartes' Rule of Signs
(p. 377)
upper and lower bounds
(p. 378)

Intermediate Value Property
(p. 380)
linear interpolation (p. 381)

1. Determine the possible number of positive roots and negative roots of $x^5 - 2x^4 - 3x^3 - x^2 + x - 2 = 0$. Summarize the possibilities in a table. *Obj. 1, p. 376*

2. Determine the least positive integral upper bound and the greatest negative integral lower bound for the roots of $y^3 - 3y^2 + 2y - 2 = 0$ that you can find using the theorem on page 378. $M = 3; L = -1$ *Obj. 2, p. 376*

3. Find all intervals of length one-half unit that contain a zero of $P(x) = x^3 - 2x^2 - 3x + 2$. $-\frac{3}{2} < r_1 < -1; \frac{1}{2} < r_2 < 1; \frac{5}{2} < r_3 < 3$ *Obj. 3, p. 376*

Check your answers with those at the back of the book.

Chapter Summary

1. The *quadratic formula*,
$$x = \frac{-b \pm \sqrt{b^2 - 4ac}}{2a},$$
enables you to solve any quadratic equation of the form $ax^2 + bx + c = 0$, $a \neq 0$.

2. From the value of the *discriminant*, $b^2 - 4ac$, of a quadratic equation with real coefficients, you can tell whether its roots are real roots or imaginary, complex conjugate roots and whether its real roots are rational or irrational, and equal or unequal.

3. The sum of the roots of a quadratic equation is $-\frac{b}{a}$; the product of the roots is $\frac{c}{a}$.

4. The graph of the equation $y = a(x - h)^2 + k$ is a *parabola* having a vertical *axis* of symmetry, the line $x = h$. The point (h, k), where it crosses its axis, is the *vertex* of the parabola.

5. The graph of the *quadratic function* $f(x) = ax^2 + bx + c$ is a parabola. The function f takes on its *extreme value* when $x = -\frac{b}{2a}$. This value is a *maximum* if $a < 0$, a *minimum* if $a > 0$.

Polynomial Functions   **383**

3.

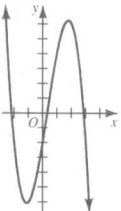

4.

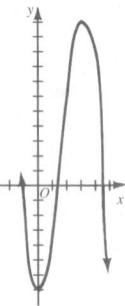

5. $-2 < r_1 < -\dfrac{3}{2}$,

$0 < r_2 < \dfrac{1}{2}, \dfrac{3}{2} < r_3 < 2$

6. $-2 < r_1 < -\dfrac{3}{2}$,

$-1 < r_2 < -\dfrac{1}{2}$,

$2 < r_3 < 2.5$

7. $-\dfrac{3}{2} < r_1 < -1$,

$0 < r_2 < \dfrac{1}{2}, \dfrac{3}{2} < r_3 < 2$

8. $-2 < r_1 < -\dfrac{3}{2}$,

$0 < r_2 < \dfrac{1}{2}, 2 < r_3 < \dfrac{5}{2}$

9. $-2 < r_1 < -\dfrac{3}{2}$,

$\dfrac{1}{2} < r_2 < 1, 2 < r_3 < \dfrac{5}{2}$

10. $-2 < r_1 < -\dfrac{3}{2}$,

$1 < r_2 < \dfrac{3}{2}, 2 < r_3 < \dfrac{5}{2}$

6. To solve a *quadratic inequality* such as $ax^2 + bx + c > 0$, $a \neq 0$, you first find roots, r_1 and r_2, $r_1 \leq r_2$, of $ax^2 + bx + c = 0$, and then consider the intervals $x < r_1$, $r_1 < x < r_2$, and $x > r_2$.

7. For any polynomial function P, *synthetic substitution* (or *synthetic division*) may be used in finding values of $P(x)$ for given values of x.

8. The *Remainder Theorem* states that for every polynomial $P(x)$, of degree n $(n \geq 1)$, and every complex number r, there is a polynomial $Q(x)$, of degree $n - 1$, such that $P(x) = (x - r)Q(x) + P(r)$. This leads to the *Factor Theorem*, which states that $x - r$ is a factor of $P(x)$ if and only if r is a root of $P(x) = 0$. The coefficients of $Q(x)$, as well as $P(r)$, can be determined by synthetic division.

9. From the *Fundamental Theorem of Algebra* it can be proved that every polynomial equation of degree n $(n \geq 1)$ with complex coefficients has exactly n complex roots.

10. If a polynomial with real coefficients has $a + bi$ as a root (a and b real, $b \neq 0$), then $a - bi$ is also a root.

11. *Descartes' Rule of Signs* can be used to determine the possible number of positive and negative roots of a polynomial equation.

12. The *Intermediate Value Property* states that if P is a polynomial function, and if m is any number between $P(a)$ and $P(b)$, then there is a number c between a and b for which $P(c) = m$. In particular, if a and b are real numbers and you find either $P(a) > 0$ and $P(b) < 0$ or $P(a) < 0$ and $P(b) > 0$, then there is at least one root between a and b.

Chapter Review

Write the letter of the correct answer.

1. Solve $4x^2 - 4x + 3 = 0$ over $\mathcal{C}$.
   a. $\{1 + i\sqrt{2}, 1 - i\sqrt{2}\}$
   b. $\{-1 + i\sqrt{2}, -1 - i\sqrt{2}\}$
   c. $\left\{\dfrac{1}{2} + \dfrac{\sqrt{2}}{2}i, \dfrac{1}{2} - \dfrac{\sqrt{2}}{2}i\right\}$
   d. $\left\{-\dfrac{1}{2} + \dfrac{\sqrt{2}}{2}i, -\dfrac{1}{2} - \dfrac{\sqrt{2}}{2}i\right\}$

   9-1

2. What is the nature of the roots of the equation $3z^2 - 11z - 4 = 0$?
   a. real, irrational
   b. imaginary
   c. real, rational, unequal
   d. a double root

   9-2

3. Give the sum of the roots of the equation $3x^2 - 5x + 1 = 0$.
   a. $-\dfrac{5}{3}$
   b. $-\dfrac{1}{3}$
   c. $\dfrac{1}{3}$
   d. $\dfrac{5}{3}$

   9-3

4. Give the product of the roots of the equation $3x^2 - 5x + 1 = 0$.
   a. $-\dfrac{5}{3}$
   b. $-\dfrac{1}{3}$
   c. $\dfrac{1}{3}$
   d. $\dfrac{5}{3}$

5. Give an equation for the axis of symmetry of the graph of $y = \frac{1}{2}(x + 6)^2 + 3$.

    a. $x = 3$ **b.** $x = -3$ **c.** $x = 6$ **(d.)** $x = -6$

6. Find the vertex of the function $y = x^2 - 6x + 9$.

    a. $(0, 9)$ **(b.)** $(3, 0)$ **c.** $(6, 9)$ **d.** $(1, 4)$

7. Find the solution set of $6 \geq x^2 - 5x$.

    a. $\{-1, 6\}$ **b.** $\{2, 3\}$

    (c.) $\{x: -1 \leq x \leq 6\}$ **d.** $\{x: x \leq 2\} \cup \{x: 3 \leq x\}$

8. If $P(x) = 2x^3 - 3x^2 - 5x - 1$, use synthetic substitution to find $P(3)$.

    a. 119 **(b.)** 11 **c.** -7 **d.** -67

9. If $P(x) = 3x^3 - 5x + 2$, use synthetic division to write $P(x)$ in the form $(x + 2)Q(x) + P(-2)$.

    a. $(x + 2)(3x^2 + 6x + 7) + 16$ **b.** $(x + 2)(3x^2 + x) + 4$

    (c.) $(x + 2)(3x^2 - 6x + 7) - 12$ **d.** $(x + 2)(3x^2 - 11x) + 24$

10. Given that $-2 + i$ is a solution of $x^3 + 3x^2 + x - 5 = 0$, find its solution set.

    (a.) $\{-2 + i, -2 - i, 1\}$ **b.** $\{-2 + i, 2 - i, 1\}$

    c. $\{-2 + i, 2 + i, -1\}$ **d.** $\{-2 + i, 2 - i, -1\}$

11. The equation $3x^5 + 2x^4 + x^3 - 4x^2 + x - 1$ cannot have which of the following combinations of roots?

    a. 3 positive, 2 negative   **b.** 1 positive, 0 negative

    (c.) 1 positive, 3 negative   **d.** 3 positive, 0 negative

12. Which interval must contain a root of $x^3 - 3x^2 - x + 2 = 0$?

    a. $-3 < x < -2$ **b.** $-2 < x < -1$

    c. $2 < x < 3$ **(d.)** $3 < x < 4$

Chapter Test

3. 2 unequal, real, rational roots

1. Solve $x^2 - 10x = -8$ over $\mathcal{C}$ by completing the square. $\{5 \pm \sqrt{17}\}$

2. Solve $4y^2 - 4y = -5$ over $\mathcal{C}$ by using the quadratic formula. $\{\frac{1}{2} \pm i\}$

3. How many real or complex roots does the equation $4x^2 + 4x - 15 = 0$ have? Explain whether any real roots are rational or irrational.

4. Give the sum and product of the roots of $3x^2 + 15x + 1 = 0$. $-5; \frac{1}{3}$

5. Write a quadratic equation with integral coefficients whose solution set is $\{1 + \sqrt{3}, 1 - \sqrt{3}\}$. $x^2 - 2x - 2 = 0$

6. Find a function in the form $y = a(x - h)^2 + k$ whose graph has vertex $(2, 5)$ and passes through the point $(-1, 2)$. $y = -\frac{1}{3}(x - 2)^2 + 5$

Polynomial Functions **385**

Additional Answers
Written Exercises (p. 352)

19. According to the quadratic formula, the roots of such an equation are

$$\frac{-b + \sqrt{b^2 - 4ac}}{2a} \text{ and }$$

$$\frac{-b - \sqrt{b^2 - 4ac}}{2a}. \text{ Let } D =$$

$\sqrt{b^2 - 4ac}$. Adding the two roots yields

$$\frac{-b + D - b - D}{2a} = \frac{-2b}{2a} =$$

$-\frac{b}{a}$. Multiplying the two roots yields

$$\frac{(-b + D)(-b - D)}{4a^2} =$$

$$\frac{b^2 - D^2}{4a^2}; \text{ substituting}$$

then gives

$$\frac{b^2 - (b^2 - 4ac)}{4a^2} =$$

$$\frac{b^2 - b^2 + 4ac}{4a^2} = \frac{4ac}{4a^2} = \frac{c}{a}.$$

20. 1. $r_1 = \dfrac{-b + \sqrt{b^2 - 4ac}}{2a}$,

$r_2 = \dfrac{-b - \sqrt{b^2 - 4ac}}{2a}$

Def. r_1, r_2

2. $\dfrac{1}{r_1} + \dfrac{1}{r_2} = \dfrac{r_1 + r_2}{r_1 r_2}$

Add. of fractions

3. $r_1 + r_2 = \dfrac{-b}{a}, r_1 r_2 = \dfrac{c}{a}$

Results of Ex. 19

4. $\dfrac{r_1 + r_2}{r_1 r_2} = \dfrac{-\frac{b}{a}}{\frac{c}{a}} = -\dfrac{b}{c}$

Substitution and simplifying

385

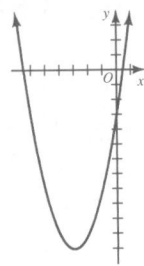
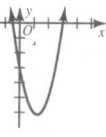
7. Give the equation of the axis of symmetry and the coordinates of the vertex of the graph of $x^2 + 6x - 3 = y$. Sketch the graph. *9-5*

8. a. Sketch the graph of the function related to $2x^2 - 5x < 3$. *9-6*

    b. Give the solution set of the inequality. $\{x: -\frac{1}{2} < x < 3\}$

9. For $P(x) = x^3 + 2x^2 - 5x + 3$, use synthetic substitution to find (a) $P(-1)$ and (b) $P(2i)$. **a.** 9   **b.** $-5 - 18i$ *9-7*

10. For $P(x) = 3x^3 - 7x^2 - 20x + 3$, use synthetic division to divide $P(x)$ by $x - 3$. Express the result as an equation whose left member is $\dfrac{P(x)}{x - 3}$. $3x^2 + 2x - 14 + \frac{-39}{x - 3}$ *9-8*

11. Solve $x^4 - 5x^3 + 8x^2 - 10x + 12 = 0$ over $\mathcal{C}$ given that $i\sqrt{2}$ is a root. $\{3, 2, i\sqrt{2}, -i\sqrt{2}\}$ *9-9*

12. Determine the possible number of positive roots and negative roots of $2x^4 + 3x^3 - x^2 - 2x + 1 = 0$. Summarize the possibilities in a table. *9-10*

13. Find all intervals of length one-half unit that contain a zero of $P(x) = x^3 - 5x^2 + 3x + 5$. $-1 < r_1 < -\frac{1}{2}, \frac{3}{2} < r_2 < 2, \frac{7}{2} < r_3 < 4$ *9-11*

    7. axis: $x = 3$ vertex: $(-3, -12)$

APPLICATION

Musical Scales

Over the years, musicians have discovered many ways to produce sounds. All of these ways involve the vibration of something, such as a

string or a column of air. The number of vibrations per second that occur is called the frequency of the sound and is measured in hertz (Hz). For example, the sound made by middle C on the piano has a frequency of 261.6 vibrations per second, or 261.6 Hz.

Musicians do not utilize all possible frequencies in making music. They normally use only those frequencies that comprise a special series known as a scale. The equal-tempered scale emerged after many years of musical experimentation.

The frequency produced by a string of length s is proportional to $\frac{1}{s}$.

About 2500 years ago, the Greeks were comparing tones produced when strings of different lengths were vibrated simultaneously. They found that harmonious tones were produced when the lengths of the strings were in whole number ratios, such as $\frac{1}{2}$, $\frac{2}{3}$, and $\frac{3}{4}$.

The interval between a given frequency, f, and twice that frequency, $2f$, eventually became divided into twelve unequal intervals. This interval is called an octave because eight of the twelve notes are sounded in playing a standard scale.

Nearly all of the tones in the scale had frequencies that could be expressed in terms of whole number ratios. However in order to allow tuned instruments such as pipe organs and pianos to be played with equally good results in any key, it was necessary to make the intervals equal. That is, the interval between the two frequencies would have to consist of twelve equally spaced tones, or notes. Consequently, the frequency following the frequency f would have to be greater than f by a factor a, such that

$$a^{12}f = 2f$$
$$a^{12} = 2$$
$$a = \sqrt[12]{2}$$
$$= 2^{\frac{1}{12}}$$
$$\approx 1.059$$

On the equal-tempered scale, then, the frequency of each tone is 1.059 times as great as the preceding tone. The frequencies on this scale very closely approximate those corresponding to the traditional whole number values.

Exercises

1. String s is twice as long as string t. In all other respects the strings are identical. Express the frequency of string t in terms of the frequency of string s. The frequency of string t is twice the frequency of string s.

2. The keys on a piano produce frequencies ranging from 27.5 Hz to 4186 Hz. To the nearest integer, how many octaves does a piano keyboard contain? 7

7.

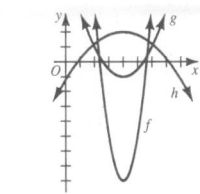

8.

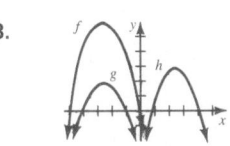

17. $y = (x - 5)^2 + 4$ or $y = (x + 1)^2 + 4$

24. Assume (r, s) is on the graph of $y = a(x - h)^2 + k$, $a \neq 0$. Therefore:
$$\begin{aligned} s &= a(r - h)^2 + k \\ &= a(-1)^2(r - h)^2 + k \\ &= a(h - r)^2 + k \\ &= a(2h - r - h)^2 + k \end{aligned}$$
This shows that $(2h - r, s)$ is also on the graph.

25. Since $(r_1, 0)$ and $(r_2, 0)$ lie on the graph, $0 = a(r_1 - h)^2 + k = a(r_2 - h)^2 + k$. Simplifying, $a(r_1 - h)^2 = a(r_2 - h)^2$. Since $a \neq 0$, $(r_1 - h)^2 = (r_2 - h)^2$ and $r_1 - h = \pm(r_2 - h)$. If $r_1 - h = r_2 - h$, that would imply $r_1 = r_2$. But $r_1 \neq r_2$. Therefore $r_1 - h = -(r_2 - h)$, $r_1 - h = -r_2 + h$,
$$r_1 + r_2 = 2h,$$
or $\dfrac{r_1 + r_2}{2} = h$.

26. $k < 0$ and $a > 0$, or $k > 0$ and $a < 0$.

27.

Additional Answers

1. $x^3 - 4x^2 + x + 6 = 0$

2. $x^3 + 2x^2 - 6x - 12 = 0$

7. $2 - 3i$; $(x + 1)(x - 2 - 3i)(x - 2 + 3i)$

8. $-1 - i\sqrt{3}$; $(x - 4)(x + 1 - i\sqrt{3})(x + 1 + i\sqrt{3})$

9. $-i\sqrt{5}, 3$; $(x + i\sqrt{5})(x - i\sqrt{5})(x - 3)$

10. $3i, \pm i\sqrt{2}$; $(x + 3i)(x - 3i)(x + i\sqrt{2})(x - i\sqrt{2})$

11. $3 - i, -4$; $(x - 3 - i)(x - 3 + i)(x + 4)$

12. $-2 + i, \pm\sqrt{3}$; $(x + 2 + i)(x + 2 - i)(x + \sqrt{3}) \cdot (x - \sqrt{3})$

13. $-2 - i, \pm i\sqrt{6}$; $(x + 2 + i)(x + 2 - i)(x + i\sqrt{6}) \cdot (x - i\sqrt{6})$

14. $i\sqrt{3}, 3 + i\sqrt{15}, 3 - i\sqrt{15}$; $(x - i\sqrt{3})(x + i\sqrt{3})(x - 3 - i\sqrt{15})(x - 3 + i\sqrt{15})$

15. 1. $P(x)$ is a polynomial of degree $2n - 1$, $n > 0$, with real coefficients. Hypothesis
   2. Suppose $P(x)$ has no real roots. Hypothesis
   3. $P(x)$ has $2n - 1$ complex roots. First theorem, p. 374.
   4. All roots are of the form $a + bi$, $b \neq 0$. Definition of imaginary number
   5. If $a + bi$ is a root, $a - bi$ is a root. Second theorem, p. 374.
   6. There must be an even number of roots. All the roots occur in pairs.
   7. $P(x)$ has at least one real root. Since (6) contradicted (3), the assumption (2) must be false.

(continued on p. 390)

PREPARING FOR COLLEGE ENTRANCE EXAMS

Strategy for Success: You are not expected to solve the test problems in your head. You can do calculations on your test booklet and cross off answer choices that you can eliminate. Be sure, however, not to make extra marks on your answer sheet, since that could interfere with the scoring on your test.

Decide which is the best of the choices given and write the correct letter on your answer sheet.

1. Which of the following must be true for the arithmetic sequence $\frac{5}{6}, \frac{1}{3}, -\frac{1}{6}, \ldots$? E

   I. The common difference is $\frac{1}{2}$ II. $a_{17} = 16$ III. $a_{16} = 8$

   (A) I only   **(B)** II only   **(C)** III only   **(D)** I, II, and III   **(E)** None

2. In an arithmetic series, $a_{10} = 79$ and the sum of the first ten terms is 430. What is the value of a_1? A

   (A) 7   **(B)** 8   **(C)** −7   **(D)** −8   **(E)** None of these

3. Which of the following values of x make the sequence $x - 1, x + 2, 3x$ a geometric sequence? D

   I. $-\frac{1}{2}$ II. 4   III. 0

   (A) I only   **(B)** II only   **(C)** III only   **(D)** I and II   **(E)** None

4. What is the value of $\sum\limits_{n=1}^{\infty} 4\left(\frac{4}{3}\right)^{n-1}$? E

   (A) 24   **(B)** 8   **(C)** 12   **(D)** 64   **(E)** None of these

5. Express $2\sqrt{-18} + 3\sqrt{-2}$ in simplest form. B

   (A) $-9\sqrt{2}$ **(B)** $9i\sqrt{2}$ **(C)** $3i\sqrt{2}$ **(D)** $-3i\sqrt{2}$ **(E)** $3\sqrt{2}i$

6. Solve $9x^2 - 6\sqrt{2}x + 2 = 0$ over $\mathcal{C}$. A

   (A) $\frac{\sqrt{2}}{3}$ **(B)** $3i\sqrt{2}$ **(C)** $\frac{3\sqrt{2}}{2}$ **(D)** $-\frac{\sqrt{2}}{3}$ **(E)** $\frac{-3\sqrt{2}}{2}$

7. For which value(s) of k will the equation $a^2 - ka + k = -8$ have exactly one real root? E

   I. 4   II. −4   III. 8

   (A) I only   **(B)** II only   **(C)** III only   **(D)** I and II only   **(E)** II and III only

8. What is the minimum value of $y - 2x$ under the following constraints? B

   $$x \geq 0$$
   $$y \geq 0$$
   $$2x + y \leq 4$$

   (A) −2   **(B)** −4   **(C)** 4   **(D)** 0   **(E)** 2

388   *Chapter 9*

PROGRAMMING IN PASCAL

The program *complex_arith*, below, multiplies and divides complex numbers using the Pascal data structure called the *record*. The use of records is similar to that of arrays, but the data to be stored in a record are not necessarily all of the same type. Records must be defined by the user in the TYPE section of the program.

The type identifier *complex*, in the program *complex_arith*, is a record and has two fields called *a* and *b* (to hold the real and imaginary parts of a complex number). Each *field* of a record must have its type declared; in this case both *a* and *b* are of type real.

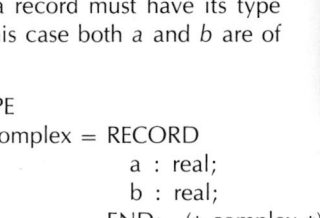

```
TYPE
    complex = RECORD
                a : real;
                b : real;
              END;   (* complex *)
```

Given the above TYPE section in a program, the variable identifiers shown below are of type *complex*.

```
VAR
    z1, z2 : complex;
```

Both *z1* and *z2* have the two fields *a* and *b*. To access a field of one of these identifiers, you must name the identifier, follow it by a period, then name the field. For example, the statement

```
z2.a := 2.6;
```

assigns the *a* field of the record *z2* the value 2.6.

Consider the following program.

```
PROGRAM complex_arith(INPUT, OUTPUT);
TYPE
    complex = RECORD
                a : real;
                b : real;
              END;   (* complex RECORD *)
VAR
    z1, z2 : complex;
    operation : char;
```

(Program continues on next page.)

Polynomial Functions   **389**

389

16. 1. $P(x)$ is a polynomial of degree $2n$, $n \geq 0$, with real coefficients. Hypothesis

2. Suppose $P(x)$ has at least one real root, r_1. Hypothesis

3. $P_1(x) = \dfrac{P(x)}{x - r_1}$ is a polynomial with real coefficients of degree $2n - 1$. Factor theorem, cancellation property of multiplication

4. $P_1(x)$ has at least one real root, r_2. Ex. 15, p. 375

5. $P_2(x) = \dfrac{P_1(x)}{x - r_2}$ is a polynomial with real coefficients and degree $2n - 2$. Factor theorem, cancellation property of multiplication

6. Suppose $P_2(x)$ has at least one real root, r_3. Hypothesis

7. $P_3(x) = \dfrac{P_2(x)}{x - r_3}$ is a polynomial with degree $2(n - 1) - 1$. Factor theorem, cancellation

8. $P_3(x)$ has at least one real root, r_4. Ex. 15, p. 375

9. $P(x)$ has no roots or it has an even number of roots. Continuing as above, for each root found for an even-degree polynomial, by Ex. 15, the resulting depressed equation of odd degree has at least one real root. This continues until an

```
(*************************************************************)
PROCEDURE obtain_nums;
BEGIN
   writeln('Enter the real and imaginary parts of the first');
   write('complex number, with a space between them: ');
   readln (z1.a, z1.b);
   writeln('Enter the real and imaginary parts of the second');
   write('complex number, with a space between them: ');
   readln(z2.a, z2.b);
END;   (* obtain_nums *)
(*************************************************************)
PROCEDURE which_operation;
BEGIN
   writeln('Do you wish to multiply <*>');
   write('or divide </> the first by the second: ');
   readln(operation);
END;   (* which_operation *);

(*************************************************************)
PROCEDURE do_operation;
VAR
   denominator : real;
BEGIN
   IF operation = '*'
      THEN BEGIN
              z3.a := (z1.a * z2.a) − (z1.b * z2.b);
              z3.b := (z1.a * z2.b) + (z1.b * z2.a);
           END   (* THEN *)
      ELSE   BEGIN
              denominator : = z2.a * z2.a + z2.b * z2.b;
              z3.a := (z1.a * z2.a + z1.b * z2.b)/denominator;
              z3.b := (z1.b * z2.a − z1.a * z2.b)/denominator;
           END;   (* else *)
END;   (* do_operation *)
(*************************************************************)
PROCEDURE display_result;
BEGIN
   writeln(z3.a :1 :4, ' + ', z3.b :1 :4, 'i')
END;   (* display_result *)
(*************************************************************)
BEGIN   (* Main Program *)
   obtain_nums;
   which_operation;
   do_operation;
   display_result
END.
```

Exercises

1. **a.** Carefully study the program *complex_arith* then run the program to produce some listings.

   b. The fields of the record *complex* are both of the same type (*real*). You can therefore replace the record with a one-dimensional array of reals of length two. Rewrite *complex_arith* using arrays instead of records to store the complex number.

   c. Modify *complex_arith* so that *which_operation* offers the user the option of choosing one of $+$, $-$, $*$, $/$ as the operation. Alter *do_operation* accordingly. Run, your program, testing each operation with various numbers.

2. Given the polynomial

$$P(x) = 5x^3 - 12x^2 - 20x + 1,$$

   write a program *synthetic_substitution* which displays the value of $P(a)$ where a is a real number specified by the user. The program should, of course, use synthetic substitution to compute the value of $P(a)$.

3. Write a Pascal program to approximate the real roots of a polynomial by using the idea of linear interpolation described in Section 9-11. (*Hint:* The loop approximating the root should terminate when $|P(x) - 0| < 0.000005$.)

1.b. In the TYPE declaration, make the following change:
   complex = ARRAY [1 . . 2] OF real;

   In the procedures and the main program, make the following replacements

   z1.a replace with z1[1]
   z1.b replace with z1[2]
   z2.a replace with z2[1]
   z2.b replace with z2[2]
   z3.a replace with z3[1]
   z3.b replace with z3[2]

---

even-degree polynomial with no roots occurs, leaving a set of an even number of real roots.

17. From Ex. 24, p. 284,

$$\overline{c(a + bi)^k} \cdot \overline{c(a - bi)^k} =$$
$$\overline{[c(a + bi)^k c(a - bi)^k]} =$$
$$\overline{c^2[(a + bi)(a - bi)]^k} =$$
$$\overline{c^2(a^2 + b^2)^k} =$$
$$c^2(a^2 + b^2)^k. \text{ Then}$$
$$\overline{c(a - bi)^k} =$$
$$\frac{c^2(a^2 + b^2)^k}{\overline{c(a + bi)^k}} =$$
$$\frac{c^2}{c}\left[\frac{a^2 + b^2}{a + bi}\right]^k =$$
$$c\left[\frac{(a^2 + b^2)(a + bi)}{(a - bi)(a + bi)}\right]^k =$$
$$c\left[\frac{(a^2 + b^2)(a + bi)}{a^2 + b^2}\right]^k =$$
$$c(a + bi)^k$$

18. $P(x)$ is a polynomial with real coefficients, say, $P(x) = a_0x^k + a_1x^{k-1} + a_2x^{k-2} + \cdots + a_{k-1}x + a_k$. Then the conjugate of $P(a - bi)$ equals the conjugate of $a_0(a - bi)^k + a_1(a - bi)^{k-1} + \cdots + a_{k-1}(a - bi) + a_k$, and by Ex. 17 above and by Ex. 28 on page 282 this is equal to $a_0(a + bi)^k + a_1(a + bi)^{k-1} + \cdots + a_{k-1}(a + bi) + a_k = P(a + bi)$.

19. $P(x)$ is a polynomial with real coefficients and $P(a + bi) = 0$. By Ex. 18 above, $P(a + bi)$ is the complex conjugate of $P(a - bi)$. But $P(a + bi)$ is a real number, since 0 is real, therefore the complex conjugate of $P(a + bi)$, $P(a - bi)$ is equal to $P(a + bi) = 0$.

The experimental solar generator shown in the photo is designed to supply power for irrigation pumps. The mirrors, which are shaped like part of a parabola, concentrate the sun's energy on the central tube.

Chapter 10

Quadratic Relations and Systems

Coordinates and Distances in a Plane

OBJECTIVES for Sections 10-1 and 10-2:
1. *To determine the distance between two points in a plane.*
2. *To determine the midpoint of a line segment.*
3. *To determine an equation of the line perpendicular to a given line and passing through a given point.*

10–1 Distance between Points

On page 69 you saw that the distance between two points on a number line having coordinates a and b is $|b - a|$. You can use this same method to find the distance between two points in a coordinate plane if they lie on the same vertical or horizontal line. To determine the distance between *any* two points in a coordinate plane with equal units on both axes, you use the following familiar theorem.

Pythagorean Theorem

In a right triangle, the square of the length c of the hypotenuse is equal to the sum of the squares of the lengths a and b of the other two sides: $c^2 = a^2 + b^2$.

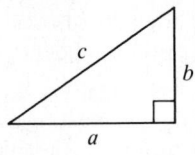

Quadratic Relations and Systems   **393**

Problem Solving Strategies

Deduction
Deduction is used in Section 10-1 to establish the distance formula on page 394, and in Sections 10-4 through 10-6 to obtain general forms for equations of parabolas, ellipses, and hyperbolas.
Drawing a Diagram
Various diagrams in Section 10-8 (page 423) show the possible number of solutions of quadratic systems.
Simulation
Actually using string, tacks, and a board might be useful in teaching ellipses in Section 10-5 on page 408.

Teaching Suggestions
p. T104

Key Ideas

Derive the Distance Formula. Introduce the Midpoint Formula.

Suggested Assignments

Minimum
    396/1–22
Extended Alg. with Trig.
    396/7–27 odd
Enriched Alg.
    396/15, 17, 19–29
Enriched Alg. with Trig.
    396/15, 17, 19–29

1. Determine the distance between $P_1(-2, 6)$ and $P_2(3, 7)$.

P_1P_2
$= \sqrt{(x_2 - x_1)^2 + (y_2 - y_1)^2}$
$= \sqrt{[3 - (-2)]^2 + (7 - 6)^2}$
$= \sqrt{5^2 + 1^2} = \sqrt{26}$

2. Determine the perimeter of the triangle whose vertices are $(6, 4)$, $(-3, 1)$, and $(9, -5)$. Determine whether the triangle is isosceles.

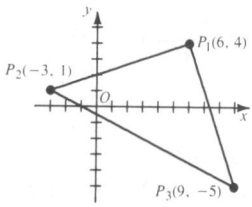

Use the distance formula to find the length of each side.

P_2P_1
$= \sqrt{[6 - (-3)]^2 + (4 - 1)^2}$
$= \sqrt{9^2 + 3^2} = \sqrt{90}$
$= 3\sqrt{10} \approx 9.49$

P_1P_3
$= \sqrt{(9 - 6)^2 + (-5 - 4)^2}$
$= \sqrt{3^2 + (-9)^2}$
$= \sqrt{90}$
$= 3\sqrt{10} \approx 9.49$

P_2P_3
$= \sqrt{[9 - (-3)]^2 + (-5 - 1)^2}$
$= \sqrt{12^2 + (-6)^2}$
$= \sqrt{180}$
$= 6\sqrt{5} \approx 13.42$

∴ the perimeter of the triangle is $3\sqrt{10} + 3\sqrt{10} + 6\sqrt{5} =$
$6\sqrt{10} + 6\sqrt{5} \approx 32.40$
Since $P_2P_1 = P_1P_3$, the triangle is isos.

The converse of this theorem is also true.

Converse of the Pythagorean Theorem

If a, b, and c are the lengths of the sides of a triangle, and if

$$c^2 = a^2 + b^2,$$

then the triangle is a right triangle with hypotenuse of length c.

Figure 1 shows two points $P_1(x_1, y_1)$ and $P_2(x_2, y_2)$ in the plane, where segment $\overline{P_1P_2}$ is not parallel to a coordinate axis. You can construct a right triangle such that $\overline{P_1P_2}$ is the hypotenuse of right $\triangle P_1P_2T$ and T has coordinates (x_2, y_1) as shown in Figure 1. Since $\overline{P_1T}$ and $\overline{P_2T}$ are parallel to the coordinate axes, their lengths are $|x_2 - x_1|$ and $|y_2 - y_1|$, respectively. Then the distance from P_1 to P_2, denoted P_1P_2, can be found using the Pythagorean Theorem.

$$(P_1P_2)^2 = (P_1T)^2 + (P_2T)^2$$
$$= |x_2 - x_1|^2 + |y_2 - y_1|^2$$
$$= (x_2 - x_1)^2 + (y_2 - y_1)^2$$

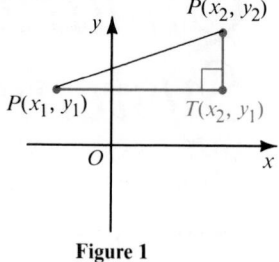

Figure 1

Since distance is a nonnegative number, take the principal square root of each side to obtain the following formula.

Distance Formula

$$P_1P_2 = \sqrt{(x_2 - x_1)^2 + (y_2 - y_1)^2}$$

where $P_1(x_1, y_1)$ and $P_2(x_2, y_2)$ are points in the plane.

EXAMPLE 1 Determine the distance between $P_1(-3, 5)$ and $P_2(2, 8)$.

SOLUTION $P_1P_2 = \sqrt{[2 - (-3)]^2 + (8 - 5)^2}$
$= \sqrt{5^2 + 3^2} = \sqrt{34}$.

EXAMPLE 2 Determine the perimeter of the triangle whose vertices are $(6, 5)$, $(-3, 2)$, and $(9, -4)$. Determine whether the triangle is isosceles and whether it is a right triangle.

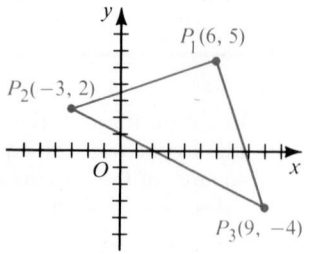

SOLUTION First, make a sketch showing the triangle. Next, use the distance formula to find the lengths of the sides.

$$P_2P_1 = \sqrt{[6 - (-3)]^2 + (5 - 2)^2} = \sqrt{9^2 + 3^2} = \sqrt{90} = 3\sqrt{10}$$
$$P_1P_3 = \sqrt{(9 - 6)^2 + (-4 - 5)^2} = \sqrt{3^2 + (-9)^2} = \sqrt{90} = 3\sqrt{10}$$
$$P_2P_3 = \sqrt{[9 - (-3)]^2 + (-4 - 2)^2} = \sqrt{12^2 + (-6)^2} = \sqrt{180} = 6\sqrt{5}$$

Then the perimeter is

$$3\sqrt{10} + 3\sqrt{10} + 6\sqrt{5} = 6\sqrt{10} + 6\sqrt{5}.$$

Since $P_2P_1 = P_1P_3$, the triangle is isosceles.

Also, since $(3\sqrt{10})^2 + (3\sqrt{10})^2 = (6\sqrt{5})^2$, the converse of the Pythagorean Theorem assures us that the triangle is a right triangle.

$\therefore$ the perimeter is $6\sqrt{10} + 6\sqrt{5}$, and the triangle is an isosceles right triangle.

An immediate consequence of the distance formula is the following formula for the midpoint of any line segment.

Midpoint Formula

The midpoint of the segment joining points $P_1(x_1, y_1)$ and $P_2(x_2, y_2)$ is

$$M\left(\frac{x_1 + x_2}{2}, \frac{y_1 + y_2}{2}\right).$$

You can prove this formula as follows:

If $P_1M = MP_2 = \frac{1}{2}(P_1P_2)$, then M is the midpoint of P_1P_2.
Using the distance formula you find that:

$$P_1M = \sqrt{\left(\frac{x_1 + x_2}{2} - x_1\right)^2 + \left(\frac{y_1 + y_2}{2} - y_1\right)^2}$$
$$= \sqrt{\left(\frac{x_2 - x_1}{2}\right)^2 + \left(\frac{y_2 - y_1}{2}\right)^2}$$
$$= \frac{1}{2}\sqrt{(x_2 - x_1)^2 + (y_2 - y_1)^2}$$

$$MP_2 = \sqrt{\left(x_2 - \frac{x_1 + x_2}{2}\right)^2 + \left(y_2 - \frac{y_1 + y_2}{2}\right)^2}$$
$$= \sqrt{\left(\frac{x_2 - x_1}{2}\right)^2 + \left(\frac{y_2 - y_1}{2}\right)^2}$$
$$= \frac{1}{2}\sqrt{(x_2 - x_1)^2 + (y_2 - y_1)^2}$$

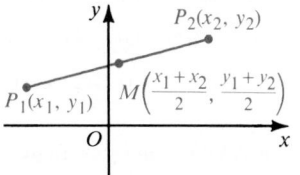

Since $P_1P_2 = \sqrt{(x_2 - x_1)^2 + (y_2 - y_1)^2}$,

$$P_1M = \frac{1}{2}(P_1P_2) \quad \text{and} \quad MP_2 = \frac{1}{2}(P_1P_2).$$

$\therefore$ M is the midpoint of $\overline{P_1P_2}$.

Quadratic Relations and Systems **395**

Additional Answers
Written Exercises

10.
$$\sqrt{(4-3)^2+(5-1)^2}=\sqrt{17};$$
$$\sqrt{(4-5)^2+(5-9)^2}=\sqrt{17}$$

11.
$$\sqrt{(0+2)^2+(4-7)^2}=\sqrt{13};$$
$$\sqrt{(0-2)^2+(4-1)^2}=\sqrt{13}$$

12.
$$\sqrt{\left(\frac{7}{2}-3\right)^2+\left(-\frac{1}{2}+3\right)^2}=\sqrt{\frac{13}{2}}$$
$$\sqrt{\left(\frac{7}{2}-4\right)^2+\left(-\frac{1}{2}-2\right)^2}=\sqrt{\frac{13}{2}}$$

14.

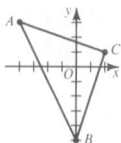

15.

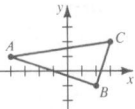

16.

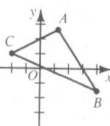

23. P_1P_2
$$=\sqrt{(b-a)^2+(c-c)^2}$$
$$=\sqrt{(b-a)^2}=|a-b|$$

24. $P_1P_2=$
$$\sqrt{(b-a)^2+(2\sqrt{ab}-0)^2}$$
$$=\sqrt{b^2-2ab+a^2+4ab}$$
$$=\sqrt{a^2+2ab+b^2}$$
$$=\sqrt{(a+b)^2}=|a+b|$$

25. Let $B=(a,b)$.
$$(p,q)=\left(\frac{r+a}{2},\frac{s+b}{2}\right);$$
thus $p=\dfrac{r+a}{2}$, $q=\dfrac{s+b}{2}$.
$a=2p-r$, $b=2q-s$.
$\therefore B=(2p-r,2q-s)$.

Oral Exercises

Give the coordinates of the midpoint of the line segment joining the given points.

1. (3, 1), (5, 9) (4, 5)   **2.** (−2, 7), (2, 1) (0, 4)   **3.** (3, −3), (4, 2) $\left(\frac{7}{2},-\frac{1}{2}\right)$

4. If the midpoint of a line segment is (0, 0) and one endpoint has coordinates (a, b), determine the coordinates of the other endpoint. $(-a,-b)$

Written Exercises

Determine the length of the line segment joining the given points. Express all radicals in simple form.

A

1. (4, 3), (−1, 3) 5   **2.** (5, −9), (5, −2) 7   **3.** (4, 2), (7, 6) 5

4. (−2, 1), (10, −4) 13   **5.** (−2, 6), (−3, 8) $\sqrt{5}$ **6.** (5, −2), (−4, 1) $3\sqrt{10}$

7. $\left(5,\frac{1}{2}\right)$, (−1, −4) $\frac{15}{2}$ **8.** $(-\sqrt{3},-2)$, $(5\sqrt{3},-8)$ 12   **9.** (a, b), (b, a) $\sqrt{2}|a-b|$

10–12. Verify that the midpoints calculated in Oral Exercises 1–3 are equidistant from the endpoints of the segments.

In Exercises 13–16:
a. sketch the triangle whose coordinates are given;
b. state whether the triangle is isosceles and, if so, give the lengths of the equal sides; I = isosceles; R = right
c. state whether the triangle is a right triangle and, if so, write an equation relating the lengths of the legs to that of the hypotenuse.

13. A(0, 0), B(6, 8), C(2, −2) not I; not R   **14.** A(−4, 3), B(0, −5), C(2, 1) I, $BC=AC=2\sqrt{10}$; R, $(AC)^2+(BC)^2=(AB)^2$

15. A(−4, 1), B(2, −1), C(3, 2) not I; R, $(AB)^2+(BC)^2=(AC)^2$ **16.** $A\left(1,\frac{5}{2}\right)$, $B\left(4,-\frac{3}{2}\right)$, C(−2, 1) not I; not R

Find the coordinates of F if M is the midpoint of $\overline{FG}$.

17. M(2, 3); G(7, 4) (−3, 2)   **18.** M(−2, 5); G(3, −1) (−7, 11)

19. $M\left(\frac{3}{2},-\frac{1}{2}\right)$; G(2, −6) (1, 5)   **20.** M(h, k); G(h + m, k − n) (h − m, k + n)

In △XYZ, whose vertices are given, find the length of the median to side $\overline{XY}$. Recall that the median to a side of a triangle is the segment joining the midpoint of the side to the opposite vertex.

B **21.** X(−1, 2); Y(5, 8); Z(−3, −7) 13   **22.** X(9, −3); Y(5, −15); Z(−1, 6) 17

23. Show that the distance between $P_1(a, c)$ and $P_2(b, c)$ is $|a - b|$.

24. Show that the distance between $P_1(a, 0)$ and $P_2(b, 2\sqrt{ab})$ is $|a + b|$.

25. Show that if M(p, q) is the midpoint of $\overline{AB}$, with A(r, s), then the coordinates of B are (2p − r, 2q − s).

396 Chapter 10

26. Determine all values of x such that the distance between $A(3, -4)$ and $B(x, 8)$ is 13 units. $\{-2, 8\}$

27. Determine an equation whose solution set consists of those points (x, y) equidistant from $(-3, 4)$ and $(5, -2)$. $4x - 3y = 1$

C 28. The coordinates of the vertices of a parallelogram are $A(0, 0)$, $B(a, 0)$, $C(a + b, c)$ and $D(b, c)$. Show that the diagonals of the parallelogram bisect each other.

29. Show that if $A(r, mr + b)$, $B(s, ms + b)$, and $C(t, mt + b)$ are three points on the graph of $y = mx + b$, with $r < s < t$, then

$$AB + BC = AC.$$

10–2 Perpendicular Lines

You should recall that two lines intersecting at right angles are called **perpendicular lines**. Every horizontal line in a plane is perpendicular to every vertical line in the same plane. For example, in Figure 2, the graph of $y = 2$ is perpendicular to the graph of $x = -3$.

If neither of two perpendicular lines is vertical, you can use the Pythagorean Theorem and its converse to establish an interesting relationship between their slopes. Figure 3 shows two lines L_1 and L_2, with equations

$$y = m_1 x + b_1,$$
$$y = m_2 x + b_2,$$

intersecting at $P(x_1, y_1)$. Since $P(x_1, y_1)$ lies on both lines, the points

$$T_1(x_1 + 1, y_1 + m_1) \quad \text{and} \quad T_2(x_1 + 1, y_1 + m_2)$$

must lie on L_1 and L_2, respectively. The points P, T_1, and T_2 are then the vertices of a triangle and the distances T_1T_2, PT_1, and PT_2 can be determined as follows.

$$T_1T_2 = |m_1 - m_2| \quad PT_1 = \sqrt{1 + m_1^2} \quad PT_2 = \sqrt{1 + m_2^2}$$

If the lines L_1 and L_2 are perpendicular then $\triangle PT_1T_2$ is a right triangle, with right angle at P. Thus, by the Pythagorean Theorem,

$$(T_1T_2)^2 = (PT_1)^2 + (PT_2)^2$$
$$(m_1 - m_2)^2 = (1 + m_1^2) + (1 + m_2^2)$$
$$m_1^2 - 2m_1m_2 + m_2^2 = 2 + m_1^2 + m_2^2$$
$$-2m_1m_2 = 2$$
$$m_1m_2 = -1$$

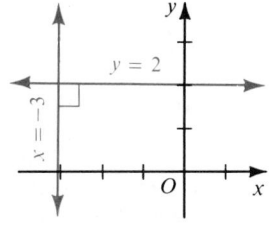

Figure 2

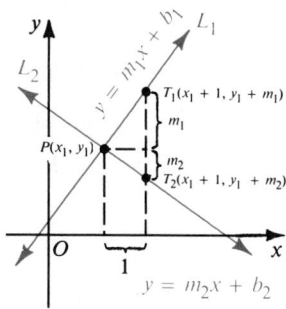

Figure 3

Quadratic Relations and Systems **397**

28. midpt. of $\overline{AC} =$

$\left(\dfrac{a + b}{2}, \dfrac{c}{2}\right)$, midpt. of

$\overline{BD} = \left(\dfrac{a + b}{2}, \dfrac{c}{2}\right)$; since

diag. have same midpt., they bisect each other.

29. $AB = (s - r)\sqrt{1 + m^2}$;
$BC = (t - s)\sqrt{1 + m^2}$;
$AC = (t - r)\sqrt{1 + m^2}$;
$\therefore AB + BC = AC$

Teaching Suggestions
p. T104

Key Ideas

Find the equation of a line perpendicular to a given line.

Reading Algebra

Have the students carefully read the theorem on p. 398 and rewrite it using their own words as two if-then statements. Ask them why this theorem excludes vertical lines. (The slope of a vertical line is undefined and does not have a negative reciprocal.)

Common Errors

Students may forget to take the negative as well as the reciprocal of the slope of a given line when they are looking for the slope of a line perpendicular to it. Remind them to check the slopes of the two lines to be sure their product is -1, that is, to be sure the lines are perpendicular.

Determine an equation of the line passing through the given point and perpendicular to the given line.

1. $(-2, 5); x - 3y = 7$

$y = \frac{1}{3}x - \frac{7}{3}$, so $m = \frac{1}{3}$.

Slope of any line perpendicular to the given line is -3.

$y = mx + b$
$5 = -3(-2) + b, -1 = b$
∴ the equation is
$y = -3x - 1$.

2. $(-1, -3); 2x - y = 4$
$y = 2x - 4$, so $m = 2$.
Slope of any line perpendicular to the given line is $-\frac{1}{2}$.

$y = mx + b$
$-3 = \left(-\frac{1}{2}\right)(-1) + b$
$-3 = \frac{1}{2} + b, b = -\frac{7}{2}$
∴ the equation is
$y = -\frac{1}{2}x - \frac{7}{2}$.

Suggested Assignments

Minimum
  First day
  399/1–14
  Second day
R 400/Self-Test 1
S 345/2–10 even
Extended Alg. with Trig.
  399/1–15 odd
R 400/Self-Test 1
S 345/2–10 even
Enriched Alg.
  399/9, 11–18
R 400/Self-Test 1
S 345/2–10 even
Enriched Alg. with Trig.
  399/9, 11–18
R 400/Self-Test 1
S 345/2–10 even

Conversely, suppose that, for nonvertical lines L_1 and L_2, the slopes m_1 and m_2 are such that $m_1 m_2 = -1$. In this case, L_1 and L_2 cannot be parallel, because for parallel lines, $m_1 = m_2$, and $m_1 m_2 = -1$ would imply $m_1{}^2 = -1$, a statement that is false for every $m_1 \in \mathcal{R}$. Hence, L_1 and L_2 must intersect at some point $P(x_1, y_1)$. If points T_1 and T_2 are now determined as on the preceding page,

$$(T_1 T_2)^2 = m_1{}^2 - 2m_1 m_2 + m_2{}^2 = m_1{}^2 - 2(-1) + m_2{}^2$$
$$= m_1{}^2 + 2 + m_2{}^2 = (1 + m_1{}^2) + (1 + m_2{}^2).$$
$$\therefore (T_1 T_2)^2 = (PT_1)^2 + (PT_2)^2$$

By the converse of the Pythagorean Theorem, then, L_1 and L_2 must be perpendicular.

The following theorem summarizes the previous arguments.

Theorem. Two nonvertical lines are perpendicular if and only if the product of their slopes is -1.

Notice that $m_1 m_2 = -1$ implies that

$$m_1 = -\frac{1}{m_2} \quad \text{and} \quad m_2 = -\frac{1}{m_1}.$$

The slopes of perpendicular lines are *negative reciprocals* of each other.

EXAMPLE Determine an equation of the line passing through $(-3, 7)$ that is perpendicular to the graph of $x - 2y = 5$.

SOLUTION Transform the equation to slope-intercept form (page 104):

$$y = \tfrac{1}{2}x - \tfrac{5}{2}$$

Thus the slope of its graph is $\frac{1}{2}$. Any line perpendicular to the graph of the given equation will have slope $-\dfrac{1}{\frac{1}{2}}$, or -2. Using the point-slope form of an equation of a line:

$$y - 7 = -2(x + 3)$$
$$y = -2x + 1$$

∴ an equation of the line is $y = -2x + 1$.

Oral Exercises

State the slope of a line perpendicular to the line whose equation is given.

1. $y = 4x - 5 \text{ -}\frac{1}{4}$ **2.** $y = \frac{3}{2}x + 7 \text{ -}\frac{2}{3}$ **3.** $y = -\frac{3}{4}x + 6\frac{4}{3}$ **4.** $x + 3y = 8 \text{ } 3$ **5.** $5x - y = 2 \text{ -}\frac{1}{5}$

Written Exercises

Determine an equation of the line passing through the given point and perpendicular to the line with the given equation.

$5x - 3y = 18$

A 1. $3y = x - 5$; $(-2, 7)$ $y = -3x + 1$ 2. $5y + 3x = 6$; $(3, -1)$

   3. $4x + 5 = 3y$; $(8, -11)$ $3x + 4y = -20$ 4. $4y = 7x + 1$; $(-\frac{1}{3}, -\frac{2}{3})$

$4x + 7y = -6$

Determine an equation of the line containing C and perpendicular to the line passing through points A and B.

$3y = 4x + 7$

5. $A(7, 6)$, $B(-1, 4)$, $C(-3, 3)$ $y = -4x - 9$ 6. $A(6, -1)$, $B(-2, 5)$, $C(-4, -3)$

7. $A(-8, -7)$, $B(2, 5)$, $C(\frac{2}{3}, -3)$ $15x + 18y = -44$ 8. $A(\frac{2}{5}, \frac{1}{4})$, $B(-\frac{1}{5}, \frac{7}{4})$, $C(\frac{3}{2}, 4)$

$5y = 2x + 17$

Determine an equation for the perpendicular bisector of the line segment with the given endpoints. Recall that the perpendicular bisector of a segment is a line passing through the midpoint of the segment and perpendicular to the line containing the segment.

9. $(1, -5)$, $(3, 7)$ 10. $(6, -7)$, $(10, -13)$ 11. $(2, 1)$, $(-3, 4)$ 12. $(-1, \frac{1}{2})$, $(\frac{1}{3}, \frac{5}{6})$

   $x + 6y = 8$ $2x - 3y = 46$ $5x - 3y = -10$ $12x + 3y = -2$

Prove that the quadrilateral with vertices A, B, C, and D is a rectangle by showing that its opposite sides are of equal length and that one pair of adjacent sides is perpendicular.

B 13. $A(-5, 1)$, $B(1, 4)$, $C(-1, 8)$, $D(-7, 5)$

   14. $A(-4, -2)$, $B(2, -6)$, $C(4, -3)$, $D(-2, 1)$

   15. Consider the parallelogram with vertices $(0, 0)$, $(\sqrt{a^2 + b^2}, 0)$, $(\sqrt{a^2 + b^2} + b, a)$, and (b, a).

   a. Show that it is a rhombus by showing that its sides are of equal length.

   b. Show that its diagonals are perpendicular.

   16. Determine an equation of the perpendicular bisector of the line segment joining $(2a, 0)$ and $(0, 2b)$. $ax - by = a^2 - b^2$

C 17. Show that the median to side $\overline{BC}$ in the triangle with vertices $A(a, a)$, $B(a + 2b, a + 2c)$, and $C(a + 2c, a + 2b)$ is perpendicular to $\overline{BC}$.

   18. Show that if the triangle with vertices $A(a, 0)$, $B(-a, 0)$, and $C(r, s)$, $a > 0$, is a right triangle with right angle at C, then the distance from $(0, 0)$ to C is a. (*Hint*: Use the slope relationship for perpendicular lines to show that $r^2 + s^2 = a^2$.)

Computer Exercises   For students with computer experience

1. Write a program that will compare the slopes of the sides of a quadrilateral when its vertices are input. The output should tell if the quadrilateral is a rectangle, parallelogram, trapezoid, or quadrilateral.

Quadratic Relations and Systems   **399**

17. midpt., M, of $\overline{BC} =$
$(a + b + c, a + b + c)$.

Slope of $\overline{BC} = \dfrac{2b - 2c}{2c - 2b} =$

-1; slope of $\overline{AM} =$

$\dfrac{b + c}{b + c} = 1; \therefore \overline{BC} \perp \overline{AM}$.

18. Since $\overline{AC} \perp \overline{BC}$,

$\dfrac{s - 0}{r - a} \cdot \dfrac{s - 0}{r + a} = -1$.

Thus $s^2 = -1(r^2 - a^2)$,
or $r^2 + s^2 = a^2$. $OC =$
$\sqrt{(r - 0)^2 + (s - 0)^2} =$
$\sqrt{r^2 + s^2} = \sqrt{a^2} = a$.

Quick Quiz

1. Determine the distance between $P_1(4, -2)$ and $P_2(7, 3)$. $\sqrt{34}$

2. Determine the midpoint of the segment joining $(-2, 4)$ and $(-4, -8)$. $(-3, -2)$

3. Determine an equation of the line containing $(-1, 3)$ and perpendicular to the line through the points $(2, 0)$ and $(3, -2)$.

$y = \dfrac{1}{2}x + \dfrac{7}{2}$

Mixed Review

1. Simplify $\sqrt{-32}$. $4i\sqrt{2}$

2. Give the sum and product of the roots of $3x^2 + 18x +$

$2 = 0$. $-6, \dfrac{2}{3}$

3. Find the vertex of the graph of the function $y = x^2 - 6x + 9$. $(3, 0)$

4. Find $P(2)$ if $P(x) = 3x^3 - 2x^2 - 4x - 3$.
5

Run the program in Exercise 1 for each set of vertices.

parallelogram
2. $(-1, -3)$, $(4, 0)$, $(5, 4)$, $(0, 1)$

rectangle
3. $(0, 6)$, $(-2, 3)$, $(1, 1)$, $(3, 4)$

4. $(-5, 7)$, $(4, 1)$, $(-1, 2)$, $(5, -2)$
trapezoid

5. $(3, 8)$, $(-2, 5)$, $(-3, -1)$, $(6, -1)$
quadrilateral

 ## Self-Test 1

VOCABULARY   Pythagorean Theorem (p. 393)
distance formula (p. 394)
midpoint formula (p. 395)
perpendicular lines (p. 397)

2. $\left(\dfrac{9}{2}, 3\right)$

1. Determine the distance between $P_1(-3, 4)$ and $P_2(7, 3)$. $\sqrt{101}$ *Obj. 1, p. 393*

2. Determine the midpoint of the segment joining $(3, 5)$ and $(6, 1)$.   *Obj. 2, p. 393*

3. Determine an equation of the line containing $(4, 2)$ and perpendicular to the line through the points $(0, 2)$ and $(-3, 8)$. $y = \dfrac{1}{2}x$ *Obj. 3, p. 393*

Check your answers with those at the back of the book.

Emmy Noether
1882–1935

When Emmy Noether entered the University of Erlanger in 1900, she was one of only two women among a thousand enrolled students. In 1907 she completed her doctoral dissertation and then occasionally substituted as a university lecturer for her mathematician father, Max Noether. From 1922 to 1933 she taught mathematics at the University of Göttingen. In 1933 Emmy Noether came to the United States, where she taught at Bryn Mawr College and the Institute of Advanced Study in Princeton.

Noether's strength as a mathematician lay in her ability to think in abstractions rather than to resort to concrete examples, and to visualize remote connections.

Her contributions to mathematics are centered on noncommutative algebras and their transformations. Her impact, however, extended far beyond her own work. Her insight, advice, and encouragement affected the research of her associates and students.

400   *Chapter 10*

Graphing Quadratic Relations

OBJECTIVES for Sections 10-3 through 10-7:
1. To sketch graphs for second-degree equations in two variables.
2. To write equations for circles, parabolas, ellipses, and hyperbolas, given appropriate properties of these curves.
3. To determine a set of values in an inverse variation, given appropriate information.
4. To apply inverse variations to solve word problems.

10–3 Circles

Recall that in a plane, a *circle* is the set of points at a given distance, called the *radius*, from a given fixed point, called the *center*. To find an equation of the circle with the center $(3, 2)$ and radius 5, use the distance formula. For each point (x, y) on the circle (Figure 4), you have

$$\sqrt{(x - 3)^2 + (y - 2)^2} = 5, \qquad (1)$$

or

$$(x - 3)^2 + (y - 2)^2 = 25. \qquad (2)$$

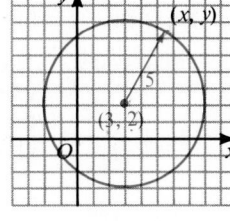

Figure 4

Conversely, if (2) is satisfied then so is (1), and the distance from $(3, 2)$ to (x, y) is 5. Hence (2) is an equation of the circle.

In general, you have:

> An equation of the circle with center (h, k) and radius r $(r > 0)$ is
> $$(x - h)^2 + (y - k)^2 = r^2.$$
> If the center is the origin, then an equation of the circle is
> $$x^2 + y^2 = r^2.$$

Notice that $(x - 3)^2 + (y - 2)^2 = 25$ is equivalent to

$$x^2 + y^2 - 6x - 4y - 12 = 0.$$

This is an example of the fact that $(x - h)^2 + (y - k)^2 = r^2$ is equivalent to an equation of the form $x^2 + y^2 + ax + by + c = 0$, where a, b, and c are real-number constants. If you are given an equation of a circle in this form, you can transform it to an equivalent equation in the form $(x - h)^2 + (y - k)^2 = r^2$ by completing the square (page 342) twice, once for x and once for y.

Quadratic Relations and Systems   **401**

Teaching Suggestions
p. T104

Key Ideas

Identify the center and radius of a circle from the equation. Write an equation of the circle with given center and radius.

Chalkboard Examples

1. Sketch the graph of
   $x^2 + y^2 + 6x - 2y - 6 = 0$.
   $(x^2 + 6x) + (y^2 - 2y) = 6$
   $(x^2 + 6x + 9) + (y^2 - 2y + 1)$
   $= 6 + 9 + 1$
   $(x + 3)^2 + (y - 1)^2 = 16$
   center: $(-3, 1)$; radius: 4

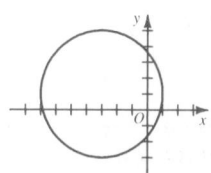

2. Sketch the graph of
   $x^2 + y^2 \leq 4$.
   center: $(0, 0)$; radius: 2;
   solid line $\leq$; shade interior.

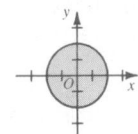

Determine an equation of the form $x^2 + y^2 + ax + by + c = 0$ for the circle with the given center and radius.

1. $(0, 0)$; 7
$x^2 + y^2 - 49 = 0$

2. $\left(-\dfrac{1}{2}, 1\right)$; $\dfrac{3}{2}$
$x^2 + y^2 + x - 2y - 1 = 0$

3. (a, b); a $\quad x^2 + y^2 - 2ax - 2by + b^2 = 0$

Write the given equation in the form $(x - h)^2 + (y - k)^2 = r^2$, give the center and radius of the circle defined by the equation, and sketch its graph.

4. $2x^2 + 2y^2 = 18$
$(x - 0)^2 + (y - 0)^2 = 3^2$
$(0, 0)$; $r = 3$

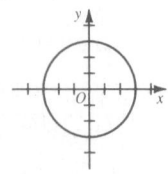

5. $x^2 + y^2 + 6y = 0$
$(x - 0)^2 + (y + 3)^2 = 3^2$
$(0, -3)$; 3

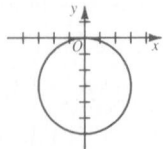

6. $x^2 + y^2 + 6x + 2y + 6 = 0$
$(x + 3)^2 + (y + 1)^2 = 2^2$
$(-3, -1)$; 2

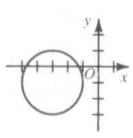

EXAMPLE 1 Sketch the graph of $x^2 + y^2 - 8x + 2y + 8 = 0$.

SOLUTION Add -8 to each side and group the terms involving x and y.

$$(x^2 - 8x \quad) + (y^2 + 2y \quad) = -8$$

Next, complete the square in x by adding $\left(\dfrac{-8}{2}\right)^2 = 16$ to each side and the square in y by adding $\left(\dfrac{2}{2}\right)^2 = 1$ to each side.

$$\underbrace{(x^2 - 8x + 16)}_{(x - 4)^2} + \underbrace{(y^2 + 2y + 1)}_{(y + 1)^2} = \underbrace{-8 + 16 + 1}_{9}$$

From the resulting equation, it is evident by inspection that the graph is a circle with center $(4, -1)$ and radius 3, as shown at the right.

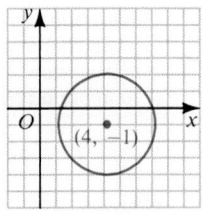

EXAMPLE 2 Sketch the graph of $x^2 + y^2 > 16$.

SOLUTION The graph of $x^2 + y^2 = 16$ separates the plane into two sets of points, those inside the circle and those outside the circle. The graph of $x^2 + y^2 > 16$ is the shaded part of the plane outside the circle. The dashed line of the circle shows the circle itself is not included.

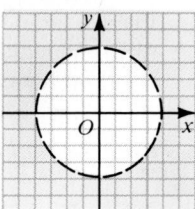

Oral Exercises

State the radius and the center of a circle with the given equation.

1. $(x - 3)^2 + (y - 1)^2 = 81$ $\quad (3, 1)$; 9

2. $(x - 7)^2 + (y - 2)^2 = 25$ $\quad (7, 2)$; 5

3. $x^2 + (y - 6)^2 = 49$ $\quad (0, 6)$; 7

4. $x^2 + (y + 3)^2 = 12$ $\quad (0, -3)$; $2\sqrt{3}$

5. $x^2 + y^2 = 121$ $\quad (0, 0)$; 11

6. $(x + 5)^2 + (y + 9)^2 = 3$ $\quad (-5, -9)$; $\sqrt{3}$

Written Exercises

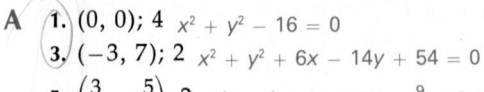

Determine an equation of the form $x^2 + y^2 + ax + by + c = 0$ for the circle with the given center and radius.

A

1. $(0, 0)$; 4 $\quad x^2 + y^2 - 16 = 0$

2. $(1, 5)$; 6 $\quad x^2 + y^2 - 2x - 10y - 10 = 0$

3. $(-3, 7)$; 2 $\quad x^2 + y^2 + 6x - 14y + 54 = 0$

4. $(-2, -4)$; 7 $\quad x^2 + y^2 + 4x + 8y - 29 = 0$

5. $\left(\dfrac{3}{2}, -\dfrac{5}{2}\right)$; 2 $\quad x^2 + y^2 - 3x + 5y + \dfrac{9}{2} = 0$

6. $\left(-3, \dfrac{1}{2}\right)$; $\dfrac{3}{2}$ $\quad x^2 + y^2 + 6x - y + 7 = 0$

7. (a, b); b $\quad x^2 + y^2 - 2ax - 2by + a^2 = 0$

8. $(-c, -d)$; $c + d$ $\quad x^2 + y^2 + 2cx + 2dy - 2cd = 0$

Write the given equation in the form $(x - h)^2 + (y - k)^2 = r^2$, give the center and radius of the circle defined by the equation, and sketch its graph. 9. $(x - 0)^2 + (y - 0)^2 = 6^2$; $C(0, 0)$; $r = 6$ 10. $(x - 4)^2 + (y - 0)^2 = 4^2$; $C(4, 0)$; $r = 4$

9. $2x^2 + 2y^2 = 72$ 10. $x^2 + y^2 - 8x = 0$

11. $x^2 + y^2 + 4y = 0$ 12. $x^2 + y^2 + 2x + 6y = 15$

13. $x^2 + y^2 + 4x - 12y - 9 = 0$ 14. $x^2 + y^2 - 4x - 3y = 0$

15. $2x^2 + 2y^2 - 10x + 2y - 5 = 0$ 16. $4x^2 + 4y^2 - 24x - 12y + 29 = 0$
                                  $(x - 3)^2 + (y - \frac{3}{2})^2 = 2^2$; $C(3, \frac{3}{2})$; $r = 2$

Sketch the graph of each inequality.

17. $x^2 + y^2 \geq 9$ 18. $x^2 + y^2 < 4$

B   19. $x^2 + y^2 + 4y \leq 0$ 20. $x^2 + y^2 > 6x$

Determine an equation for a circle with a diameter having the given endpoints. (*Hint*: The midpoint of a diameter is the center of any circle.)
                                        $(x + 3)^2 + (y - 2)^2 = 20$

21. $(5, 9)$, $(5, 3)$ 22. $(-2, 4)$, $(8, 4)$ 23. $(-4, 1)$, $(2, 9)$ 24. $(-5, 6)$, $(-1, -2)$

$(x - 5)^2 + (y - 6)^2 = 9$ $(x - 3)^2 + (y - 4)^2 = 25$ $(x + 1)^2 + (y - 5)^2 = 25$

Determine an equation of the form $x^2 + y^2 + ax + by + c = 0$ for the circle with center C and passing through the point P.
                                   $x^2 + y^2 + 2x - 10y + 17 = 0$

25. $C(3, -2)$; $P(7, 1)$ 26. $C(-1, 5)$; $P(-1, 8)$

27. $C(k, 0)$; $P(2k, 0)$ $x^2 + y^2 - 2kx = 0$ 28. $C(r, s)$; $P(0, 0)$ $x^2 + y^2 - 2rx - 2sy = 0$

25. $x^2 + y^2 - 6x + 4y - 12 = 0$

Sketch the graph of each equation. Recall that $\sqrt{a}$ is nonnegative.

29. $x = \sqrt{9 - y^2}$ 30. $y = -\sqrt{16 - x^2}$

C   31. $y = 1 + \sqrt{6x - x^2}$ 32. $x = \sqrt{4y - y^2}$

33. Determine an equation for the circle with center at $(4, -1)$ and tangent to the y-axis. (*Hint*: The tangent line to a circle is perpendicular to the radius drawn to the point of tangency.) $(x - 4)^2 + (y + 1)^2 = 16$

34. A circle with center at the origin passes through $(-4, 3)$. Find an equation of the tangent line at this point. $4x - 3y = -25$

35. The circle defined by $x^2 + y^2 - 8y - 9 = 0$ passes through $(3, 8)$ and $(5, 4)$.
   a. Prove that the chord with the given endpoints is perpendicular to the radius that passes through the midpoint of the chord.
   b. Prove that the radius perpendicular to the chord passes through its midpoint.   36. $x^2 + y^2 - 8y - 9 = 0$

36. Determine an equation of the circle passing through $(3, 0)$, $(-3, 0)$, and $(0, 9)$, using the general equation $x^2 + y^2 + ax + by + c = 0$.

37. Show that the set of all points (x, y) whose distance from the point $(6, 0)$ is twice as great as their distance from $(3, 0)$ is a circle, and give the center and radius of the circle.

38. Prove that if $x^2 + y^2 + ax + by + c = 0$ has a graph in the coordinate plane, then $a^2 + b^2 - 4c \geq 0$.

Quadratic Relations and Systems   **403**

Suggested Assignments

Minimum
   First day
   402/1–24
   Second day
   403/25–30
S 356/1–22 even
Extended Alg. with Trig.
   402/3–29 odd
S 356/1–22 even
Enriched Alg.
   402/5, 15, 19–37 odd
Enriched Alg. with Trig.
   402/5, 15, 19–37 odd

Additional Answers
Written Exercises

10.

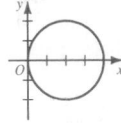

11. $(x - 0)^2 + (y + 2)^2 = 2^2$; $C(0, -2)$; $r = 2$

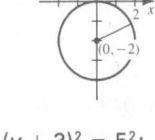

12. $(x + 1)^2 + (y + 3)^2 = 5^2$; $C(-1, -3)$; $r = 5$

13. $(x + 2)^2 + (y - 6)^2 = 7^2$; $C(-2, 6)$; $r = 7$

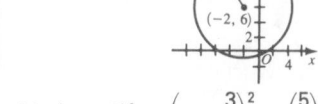

14. $(x - 2)^2 + \left(y - \frac{3}{2}\right)^2 = \left(\frac{5}{2}\right)^2$; $C\left(2, \frac{3}{2}\right)$; $r = \frac{5}{2}$

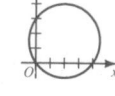

(continued on p. 428)

1. Solve.

$$4a - 3(7 - 6a) = 12 \quad \left\{\tfrac{3}{2}\right\}$$

2. Find the sum:

$$\sum_{n=1}^{8} \tfrac{1}{4}(2)^{n-1} \quad 63\tfrac{3}{4}$$

3. Simplify $\sqrt[3]{(-8)^4}$.   16

4. Solve $\sqrt{x - 4} + x = 6$
over $\mathcal{R}$.   {5}

Teaching Suggestions
p. T104

Related Activities p. T105

Key Ideas

Identify the parts of a parabola using its equation.

Common Errors

Students may sometimes state the vertex of the parabola whose equation is $x = 2(y - 4)^2 + 3$ to be (4, 3) when in fact it is (3, 4). Make sure students understand that although both equations
$y = a(x - h)^2 + k$ and
$x = a(y - k)^2 + h$
describe parabolas, the vertex in each case is (h, k). (You may have to specifically point out to the students that the positions of the h and k are switched in those forms.)

Computer Exercises   <small>For students with computer experience</small>

1. Write a program that will tell whether the graph of an equation of the form $x^2 + y^2 + ax + by + c = 0$ is a circle, a single point, or the empty set. If the graph is a circle, the output should give the circle's center and radius. (*Hint*: Complete the square in x and y by adding the same number to both sides of the equation. The graph is then a circle, a single point, or the empty set as this number is positive, zero, or negative, respectively.)

Run the program in Exercise 1 for each equation.

2. $x^2 + y^2 - 4x + 6y - 12 = 0$ Circle; (2, −3); $r = 5$ **3.** $x^2 + y^2 - 10x = 0$ Circle; (5, 0); $r = 5$

4. $x^2 + y^2 + 2x - 12y - 37 = 0$ Circle; (−1, 6); $r \approx 8.6$ **5.** $x^2 + y^2 + 8x + 2y + 20 = 0$ Empty set

---

10–4 Parabolas

In the preceding section, you saw that an equation can be found for the set of points called a *circle*. Equations can be found for sets of points that satisfy other geometric conditions. Consider, for example, the set consisting of every point P whose perpendicular distance from a fixed line is equal to the distance from a fixed point not on the line. A curve consisting of a set of points satisfying these conditions is called a **parabola**. The fixed line is called the **directrix** and the fixed point is called the **focus**.

To find an equation for the parabola with focus $F(0, 4)$ and directrix the line L with equation $y = -2$ (Figure 5), you use the distance formula. If D is any point on line L, then

$$FP = \sqrt{(x - 0)^2 + (y - 4)^2},$$
$$PD = |y - (-2)|,$$

from which:

$$|y + 2| = \sqrt{x^2 + (y - 4)^2}$$
$$|y + 2|^2 = (\sqrt{x^2 + (y - 4)^2})^2$$
$$(y + 2)^2 = x^2 + (y - 4)^2$$
$$y^2 + 4y + 4 = x^2 + y^2 - 8y + 16$$
$$12y = x^2 + 12$$
$$y = \tfrac{1}{12}x^2 + 1$$

Thus, the set of points described above is the graph of the quadratic equation

$$y = \tfrac{1}{12}x^2 + 1.$$

Using the methods of Section 9-4 you can plot its graph as pictured in Figure 5.

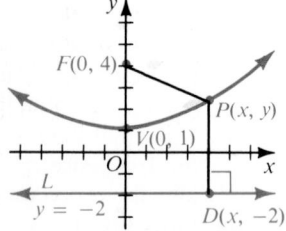

Figure 5

404   *Chapter 10*

The table at the right suggests that the graph is symmetric with respect to the line with equation $x = 0$, that is, the y-axis. In fact, since $(-r, t)$ satisfies the equation of the function whenever (r, t) does, the y-axis is the **axis of symmetry**, or simply the **axis**, of this parabola. The point $V(0, 1)$ where the parabola intersects its axis is called the **vertex** of the parabola.

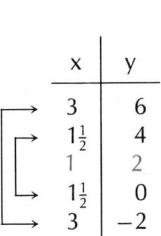

Similarly, you can show that an equation for the parabola with focus $F(3, 2)$ and directrix the line L with equation $x = -1$ is $x = \frac{1}{8}(y - 2)^2 + 1$, whose graph is shown in Figure 6. The vertex of this parabola is the point $V(1, 2)$, and the axis is the line with equation $y = 2$, as suggested by the table.

From the definition of a parabola and Figures 5 and 6, you can see that *the vertex is the midpoint of the segment of the axis between the focus and the directrix.*

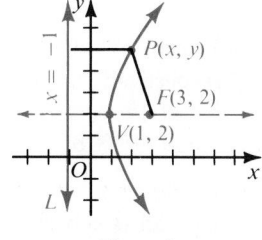

Figure 6

The parabola with vertex $V(h, k)$ and directrix the line L with equation $y = k - m$, $m \neq 0$ (where m is the distance between the focus and the vertex) is the graph of the relation

$$y - k = \frac{1}{4m}(x - h)^2.$$

Similarly, the parabola with vertex $V(h, k)$ and directrix the line L with equation $x = h - m$, $m \neq 0$ (where m is the distance between the focus and the vertex) is the graph of the relation

$$x - h = \frac{1}{4m}(y - k)^2.$$

(See Exercises 28–30 on page 407.)

A parabola has a simple equation if its focus is on a coordinate axis and its vertex is at the origin.

EXAMPLE 1 Determine an equation of the parabola whose focus F is at $(0, 3)$ and whose vertex is at the origin.

SOLUTION The equation of the directrix is $y = -3$. If P is any point on the parabola and D is a point on the directrix as shown, then by the definition of a parabola:

$$PF = PD$$
$$\sqrt{(x - 0)^2 + (y - 3)^2} = \sqrt{(x - x)^2 + (y + 3)^2}$$
$$x^2 + (y - 3)^2 = (y + 3)^2$$
$$x^2 + y^2 - 6y + 9 = y^2 + 6y + 9$$
$$x^2 = 12y$$
$\therefore x^2 = 12y$ is an equation of the parabola.

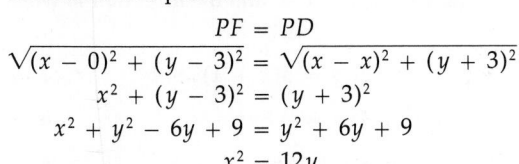

Quadratic Relations and Systems **405**

Chalkboard Examples

1. Determine an equation of the parabola whose focus F is at $(0, 2)$ and whose vertex is at the origin.
Equation of directrix: $y = -2$.
If P is any point on the parabola and D is a point on the directrix, then by definition of the parabola:
$$PF = PD$$
Using the distance formula:
$$\sqrt{(x - 0)^2 + (y - 2)^2} = \sqrt{(x - x)^2 + (y + 2)^2}$$
$$x^2 + (y - 2)^2 = (y + 2)^2$$
$$x^2 + y^2 - 4y + 4 = y^2 + 4y + 4$$
$$x^2 = 8y$$
$\therefore x^2 = 8y$ is an equation of the parabola.

2. Sketch the graph of $x - 3 = 2y - y^2$.
Complete the square in y:
$$-x + 3 = y^2 - 2y$$
$$-x + 3 + 1 = y^2 - 2y + 1$$
$$-x + 4 = (y - 1)^2$$
$$x - 4 = -(y - 1)^2$$
Vertex: $(4, 1)$; axis: $y = 1$; $a < 0$
$\therefore$ graph of parabola opens to left.

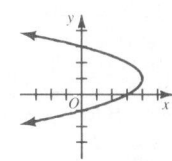

Suggested Assignments

Minimum
407/1–24
Extended Alg. with Trig.
407/1–27 odd
Enriched Alg.
407/11–31 odd
Enriched Alg. with Trig.
407/11–31 odd

Additional A Exercises

Determine the equation of a parabola with vertex at the origin and satisfying the given conditions.

1. focus: (2, 0) $y^2 = 8x$

2. focus: (0, −1) $x^2 = -4y$

3. focus: (0, 3) $x^2 = 12y$

4. directrix: $x = 4$
$y^2 = -16x$

Rewrite each equation in the form $y = a(x - h)^2 + k$ or $x = a(y - k)^2 + h$ and sketch its graph.

5. $y = 2x^2 - 12x + 14$
$y = 2(x - 3)^2 - 4$

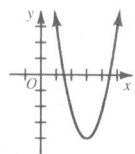

6. $x = 3y^2 + 6y - 2$
$x = 3(y + 1)^2 - 5$

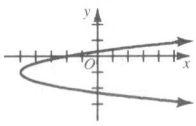

Mixed Review

1. Factor $36c^2 - d^6$ completely.
$(6c - d^3)(6c + d^3)$

2. Determine the solution set of $2x^2 - 3 = x.$ $\left\{-1, \frac{3}{2}\right\}$

3. Simplify $\dfrac{x^2 - 9}{(x - 3)^2} \cdot \dfrac{x + 3}{x - 3}.$

4. Simplify $\dfrac{x}{x - 2} - \dfrac{2x - 1}{x^2 - 4}.$
$\dfrac{x^2 + 1}{x^2 - 4}$

5. In an arithmetic series, find a_{13} if $a_1 = 2$, and $d = -\dfrac{1}{3}.$ -2

In general, a parabola with vertex (0, 0) and focus (0, p) has the equation $x^2 = 4py$. Similarly, a parabola with vertex (0, 0) and focus (p, 0) has the equation $y^2 = 4px$. If $p > 0$, then the parabola opens upward or to the right. If $p < 0$, then the parabola opens downward or to the left.

A parabola whose equation is of the form

$$y = a(x - h)^2 + k \quad \text{or} \quad x = a(y - k)^2 + h$$

has vertex $V(h, k)$ and axis of symmetry

$$x = h \quad \text{or} \quad y = k,$$

respectively. If $a > 0$, the graph opens upward or to the right; if $a < 0$, the graph opens downward or to the left.

If you are given an equation of the form

$$y = ax^2 + bx + c \quad \text{or} \quad x = ay^2 + by + c,$$

you can sketch the curve more readily by completing the square in the variable that is squared and comparing the resulting equation with the information above.

EXAMPLE 2 Sketch the graph of $x + 1 = 4y - y^2.$

SOLUTION Complete the square in y:

$$-x - 1 = y^2 - 4y$$
$$-x - 1 + 4 = y^2 - 4y + 4$$
$$-x + 3 = (y - 2)^2$$
$$x = -(y - 2)^2 + 3$$

x	y
−1	0
2	1
3	2
2	3
−1	4

Comparing this with the relation

$$x = a(y - k)^2 + h,$$

you see that the vertex is $V(3, 2)$, and the axis is the graph of $y = 2$. Since $a = -1$, and $-1 < 0$, the equation has a graph that is a parabola opening to the left.

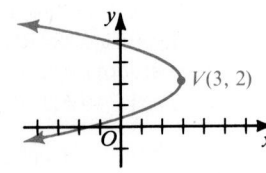

Oral Exercises

For each parabola, state the vertex, the axis, and the direction in which it opens. D = down; U = up; L = left; R = right

1. $y = -3(x - 7)^2 + 2$
(7, 2); $x = 7$; D

2. $y = 5(x + 1)^2 + 6$
(−1, 6); $x = -1$; U

3. $y = 2x^2 - 9$
(0, −9); $x = 0$; U

4. $x = 3(y - 1)^2 + 4$
(4, 1); $y = 1$; R

5. $x = -8y^2 - 7$
(−7, 0); $y = 0$; L

6. $x = \frac{1}{4}(y + 5)^2 + 10$
(10, −5); $y = -5$; R

7. $x^2 = 4y$
(0, 0); $x = 0$; U

8. $y^2 = 8x$
(0, 0); $y = 0$; R

9. $x^2 = -12y$
(0, 0); $x = 0$; D

406 Chapter 10

Written Exercises

Determine the equation of a parabola with vertex at the origin and satisfying the given conditions.

$x^2 = -2y$

A

1. focus: (10, 0) $y^2 = 40x$ **2.** focus: (-2, 0) $y^2 = -8x$ **3.** focus: $\left(0, -\frac{1}{2}\right)$

4. focus: $\left(0, \frac{2}{3}\right)$ $x^2 = \frac{8}{3}y$ **5.** directrix: $y = -2$ $x^2 = 8y$ **6.** directrix: $x = 5$
$y^2 = -20x$

Rewrite each equation in either $y = a(x - h)^2 + k$ or $x = a(y - k)^2 + h$ form, and sketch its graph.

$y = \frac{1}{2}(x - 0)^2 - 3$ $x = 2(y - 0)^2 - 8$ $x = (y - 1)^2 - 4$

B **7.** $y = \frac{1}{2}(x^2 - 6)$ **8.** $x = -2(4 - y^2)$ **9.** $x + 3 = y^2 - 2y$

$y = 2(x - 3)^2 - 18$ $y = \frac{1}{4}(x + 6)^2 - 9$

10. $y = 2x^2 - 12x$ **11.** $y = \frac{1}{4}x^2 + 3x$ **12.** $x = \frac{1}{2}y^2 + 4y$
$y = 2(x - 2)^2 - 3$ $x = 3(y - 1)^2 - 8$

13. $y = 2x^2 - 8x + 5$ **14.** $x = 3y^2 - 6y - 5$ **15.** $3x = y^2 - 2y + 10$

16. $x = \sqrt{y} + 3$ **17.** $x = 3 + \sqrt{y - 2}$ **18.** $y = -4 + \sqrt{x - 1}$
18. $x = (y + 4)^2 + 1, y \geq -4$

(*Hint:* In Exercises 16–18, restrict one of the variables.)

16. $y = x^2 - 3, x \geq 0$ **17.** $y = (x - 3)^2 + 2, x \geq 3$

Determine an equation of the form $y = ax^2 + bx + c$ or $x = ay^2 + by + c$ for the parabola with focus F and directrix having the given equation.
$x = \frac{1}{8}y^2 - y + 3$

C **19.** $F(0, 1); y = -5$ $y = \frac{1}{12}x^2 - 2$ **20.** $F(-2, 6); y = 4$ **21.** $F(3, 4); x = -1$

22. $F(-3, 3); x = -1$ **23.** $F\left(4, \frac{5}{2}\right); y = \frac{3}{2}$ **24.** $F\left(\frac{7}{8}, 3\right); x = \frac{9}{8}$
$x = -\frac{1}{4}y^2 + \frac{3}{2}y - \frac{17}{4}$ $y = \frac{1}{2}x^2 - 4x + 10$ $x = -2y^2 + 12y - 17$

25. Determine an equation of the parabola with vertex $\left(-\frac{1}{2}, \frac{1}{4}\right)$ and directrix having the equation $y = \frac{1}{2}$. $y = -(x + \frac{1}{2})^2 + \frac{1}{4}$

26. Determine an equation of the parabola with vertex $\left(-\frac{9}{4}, \frac{3}{2}\right)$ and focus $(-2, \frac{3}{2})$. $x = (y - \frac{3}{2})^2 - \frac{9}{4}$

27. Express the constant p in the form $x^2 = 4py$ in terms of the constant a in the form $y = a(x + h)^2 + k$. $p = \frac{1}{4a}$

28. Verify that the parabola with focus $F(0, m)$ and directrix with equation $y = -m$ is the graph of $y = \frac{1}{4m}x^2$.

29. Verify that the parabola with focus $F(m, k)$ and directrix with equation $x = -m$ is the graph of $x = \frac{1}{4m}(y - k)^2$.

30. Verify that the parabola with vertex (h, k) and directrix with equation $y = k - m$ is the graph of $y = \frac{1}{4m}(x - h)^2 + k$. $\left(\text{This shows that if } m \text{ is the distance between the focus and the vertex or between the vertex and the directrix of a parabola, then } a = \frac{1}{4m}.\right)$

407

Additional Answers
Written Exercises

7.
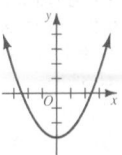

8.
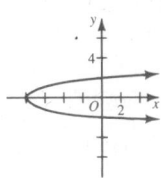

9.
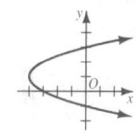

10.
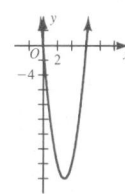

11.
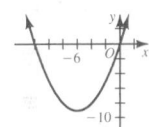

12. $x = \frac{1}{2}(y + 4)^2 - 8$

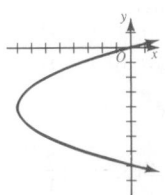

13.
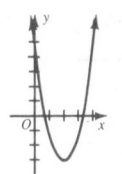

(continued on p. 434)

31. Show that the parabola with focus $(h, k + m)$ and directrix with equation $y = k - m$ passes through the points $(h + 2m, k + m)$ and $(h - 2m, k + m)$, and that the line segment joining these points has length $4|m|$.

Key Ideas

Sketch the graph of an ellipse.
Write an equation of an ellipse given conditions on the curve.

Chalkboard Examples

1. Sketch the graph of $4x^2 + 36y^2 = 144$.
   Rewrite in the form
   $$\frac{x^2}{a^2} + \frac{y^2}{b^2} = 1: \frac{x^2}{36} + \frac{y^2}{4} = 1$$
   By inspection, the graph is an ellipse, and:
   1. Is symmetric with respect to both axes.
   2. Has x-intercepts 6 and -6.
   3. Has y-intercepts 2 and -2.
   Find first-quadrant points.
   $$y = \frac{1}{3}\sqrt{36 - x^2}$$

x	y
0	2
2	$\frac{\sqrt{32}}{3} \approx 1.885$

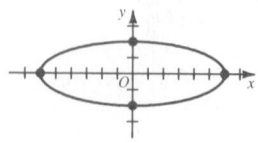

10–5 Ellipses

The path traversed by a planet as it revolves about the sun is a plane curve called an *ellipse*. In a plane an **ellipse** is defined as the set of all points P for each of which the sum of the distances from P to two fixed points is a constant. Each of the fixed points is a **focus** (plural: **foci**) of the ellipse, and the distances from the foci to a point P on the curve are called **focal radii** of P (Figure 7). The midpoint C of $\overline{F_1F_2}$ is called the **center** of the ellipse. This definition suggests that a drawing of an ellipse may be made by fastening a loop of string at the points F_1 and F_2, stretching it taut with a pencil at point P, and drawing the curve as shown in Figure 8.

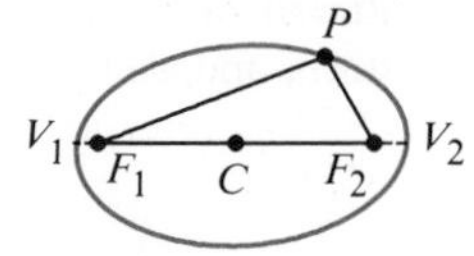

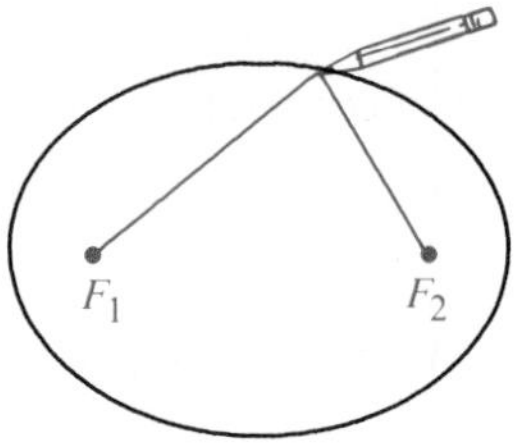

To find an equation for an ellipse with foci $F_1(-4, 0)$ and $F_2(4, 0)$ and $PF_1 + PF_2 = 10$, you have:

$$PF_1 = \sqrt{[x - (-4)]^2 + (y - 0)^2} = \sqrt{(x + 4)^2 + y^2}$$
$$PF_2 = \sqrt{(x - 4)^2 + (y - 0)^2} = \sqrt{(x - 4)^2 + y^2}$$

Since $PF_1 + PF_2 = 10$, it follows that

$$\sqrt{(x + 4)^2 + y^2} + \sqrt{(x - 4)^2 + y^2} = 10,$$

or

$$\sqrt{(x + 4)^2 + y^2} = 10 - \sqrt{(x - 4)^2 + y^2}.$$

Squaring each side and simplifying, you obtain

$$(x + 4)^2 + y^2 = 100 - 20\sqrt{(x - 4)^2 + y^2} + (x - 4)^2 + y^2,$$
$$4x - 25 = -5\sqrt{(x - 4)^2 + y^2}.$$

Again squaring and simplifying, you find that:

$$16x^2 - 200x + 625 = 25[(x - 4)^2 + y^2]$$
$$9x^2 + 25y^2 = 225$$
$$\frac{x^2}{25} + \frac{y^2}{9} = 1$$

Thus, the ellipse described is the graph of the equation

$$\frac{x^2}{25} + \frac{y^2}{9} = 1.$$

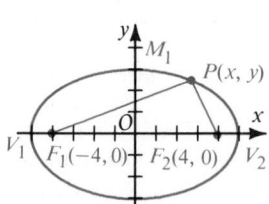

Figure 7

Figure 8

Figure 9

408 *Chapter 10*

You can verify that the equation is satisfied by the coordinates of these points shown in Figure 9: $V_1(-5, 0)$, $V_2(5, 0)$, $M_1(0, 3)$, $M_2(0, -3)$. Thus, *the x-intercepts are -5 and 5; the y-intercepts are 3 and -3.* Also, since $(r, -t)$ and $(-r, t)$ satisfy the equation whenever (r, t) does,

the curve is symmetric with respect to both coordinate axes.

The graph of $\dfrac{x^2}{25} + \dfrac{y^2}{9} = 1$ can be sketched by plotting the intercepts and a few additional points obtained from $y = \pm\frac{3}{5}\sqrt{25 - x^2}$, as shown in the adjoining table. The graph is pictured in Figure 9. Note that the curve contains only points for which $|x| \le 5$ and $|y| \le 3$.

In general if the foci are on the x-axis, you can show (Exercise 35, page 412) that:

x	y
3	$\pm\frac{12}{5}$
4	$\pm\frac{9}{5}$
-3	$\pm\frac{12}{5}$
-4	$\pm\frac{9}{5}$

An equation of the ellipse with center at the origin, foci $(c, 0)$ and $(-c, 0)$, and the sum of the focal radii for each of its points the constant $2a$ $(a > c)$ is

$$\frac{x^2}{a^2} + \frac{y^2}{b^2} = 1, \qquad (1)$$

where $b^2 = a^2 - c^2$. The graph has x-intercepts a and $-a$ and y-intercepts b and $-b$.

If the foci are on the y-axis, you can verify (Exercise 36, page 412) the following.

An equation of the ellipse with center at the origin, foci $(0, c)$ and $(0, -c)$, and the sum of the focal radii for each of its points the constant $2a$ $(a > c)$ is

$$\frac{x^2}{b^2} + \frac{y^2}{a^2} = 1, \qquad (2)$$

where again $b^2 = a^2 - c^2$. The graph has x-intercepts b and $-b$, and y-intercepts a and $-a$.

In each case, the ellipse is symmetric with respect to both the x-axis and the y-axis. Notice also that $a > b$.

The **major axis** is the segment of length $2a$ cut off on the axis by the ellipse. The **minor axis** is the segment of length $2b$ that is perpendicular to the major axis at the center. The points where the ellipse cuts its major axis are called vertices. In Figure 9, the major axis is $\overline{V_1V_2}$, the minor axis is $\overline{M_1M_2}$, and the vertices are V_1 and V_2.

Quadratic Relations and Systems **409**

2. Determine an equation of an ellipse with axes on the coordinate axes and with major axis of length 8, foci at (3, 0) and (−3, 0).
major axis $2a = 8$
$\qquad\qquad a = 4$
a focus at (3, 0) yields:
$\qquad c = 3$.
Since $b^2 = a^2 - c^2$:
$\qquad\qquad b^2 = 16 - 9$
$\qquad\qquad b^2 = 7$
$\therefore \dfrac{x^2}{16} + \dfrac{y^2}{7} = 1$

is an equation of the ellipse.

Mixed Review

1. Express in lowest terms:
$$\frac{a^2 - 2a + 1}{a^2 - a} \div \frac{a^2 - 1}{a^2 - a - 2}$$
$$\frac{a - 2}{a}$$

2. Find the equation of a circle with center (1, −4) and radius 7.
$(x - 1)^2 + (y + 4)^2 = 49$

3. Simplify $\dfrac{\sqrt{-16}}{i^3}$. -4

4. Solve
$$\frac{a - 1}{a - 2} - \frac{3a - 2}{a^2 - 4} = \frac{3}{a + 2}.$$
$\{3\}$

5. Factor $54c^3 + 2d^3$ completely. $2(3c + d)(9c^2 - 3cd + d^2)$

Suggested Assignments

Minimum
 411/1–24
Extended Alg. with Trig.
 411/7–31 odd
Enriched Alg.
 411/11–35 odd
Enriched Alg. with Trig.
 411/11–35 odd

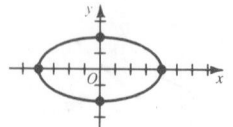
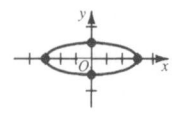
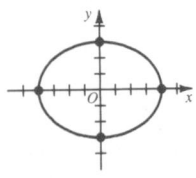
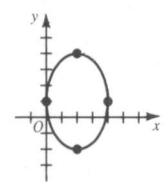
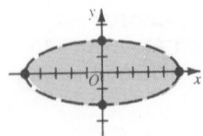
EXAMPLE 1 Sketch the graph of $4x^2 + 25y^2 = 100$.

SOLUTION Divide both sides of $4x^2 + 25y^2 = 100$ by 100 to obtain $\dfrac{x^2}{25} + \dfrac{y^2}{4} = 1$. Since $25 > 4$, compare this with Equation (1). By inspection, the graph is an ellipse, and:

1. It is symmetric with respect to both axes.
2. The x-intercepts are 5 and -5.
3. The y-intercepts are 2 and -2.

x	y
0	2
3	1.6
4	1.2
5	0

From $y = \frac{1}{5}\sqrt{100 - 4x^2}$ construct a table of points in the first quadrant and use symmetry to determine corresponding points in other quadrants. Sketch the graph as shown.

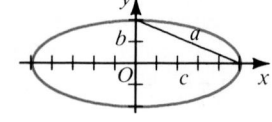

In the preceding example, the foci are on the x-axis. Since you know that $c^2 = a^2 - b^2 = 25 - 4 = 21$, it follows that $c = \sqrt{21}$, and the foci are $(\sqrt{21}, 0)$ and $(-\sqrt{21}, 0)$. The foci need not be used in sketching the graph. The right triangle with sides a, b, and c, shown in the figure for the example above, illustrates the relationship between a, b, and c in an ellipse.

EXAMPLE 2 Determine an equation for an ellipse with axes on the coordinate axes, with major axis of length 10, foci at $(3, 0)$ and $(-3, 0)$.

SOLUTION Since the major axis has length $2a$, $2a = 10$, or $a = 5$. Since one focus is at $(3, 0)$, $c = 3$. Then $b^2 = a^2 - c^2$; so $b^2 = 25 - 9$, and $b = 4$.
$\therefore \dfrac{x^2}{25} + \dfrac{y^2}{16} = 1$ is an equation of the ellipse.

An ellipse can have its center at a point other than the origin, and its axes need not lie on the coordinate axes. In fact, the graph of an equation of the form

$$\frac{(x-h)^2}{a^2} + \frac{(y-k)^2}{b^2} = 1 \quad \text{or} \quad \frac{(x-h)^2}{b^2} + \frac{(y-k)^2}{a^2} = 1,$$

where $a > b$, is an ellipse with center (h, k), whose axes are *parallel* to the coordinate axes. As with ellipses centered at the origin, the sum of the focal radii for each point is $2a$, and the lengths of the major and minor axes are $2a$ and $2b$, respectively. The foci of such an ellipse are located c units to either side of the center along the major axis; where as before, $b^2 = a^2 - c^2$. For example, the graph of

$$\frac{(x-2)^2}{9} + \frac{(y+1)^2}{4} = 1$$

is shown in Figure 10.

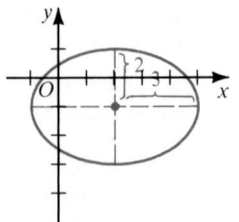

Figure 10

Oral Exercises

Find the coordinates of the foci of the ellipse whose equation is given.

1. $\dfrac{x^2}{16} + \dfrac{y^2}{25} = 1$ $(0, \pm3)$

2. $\dfrac{x^2}{25} + \dfrac{y^2}{9} = 1$ $(\pm4, 0)$

3. $x^2 + \dfrac{y^2}{9} = 1$ $(0, \pm2\sqrt{2})$

4. $\dfrac{x^2}{8} + y^2 = 1$ $(\pm\sqrt{7}, 0)$

5. $\dfrac{x^2}{169} + \dfrac{y^2}{144} = 1$ $(\pm5, 0)$

6. $\dfrac{x^2}{81} + \dfrac{y^2}{225} = 1$ $(0, \pm12)$

Written Exercises

Graph each equation and label the foci.

A **1–6.** Use the equations in Oral Exercises 1–6.

7. $4x^2 + 9y^2 = 144$

8. $x^2 + 9y^2 = 225$

9. $9x^2 + 3y^2 = 900$

10. $16x^2 + 25y^2 = 400$

11. $9x^2 + 25y^2 = 100$

12. $16x^2 + 25y^2 = 144$

13. $\dfrac{(x - 3)^2}{16} + \dfrac{(y - 2)^2}{9} = 1$

14. $\dfrac{(x + 1)^2}{4} + \dfrac{(y - 6)^2}{36} = 1$

15. $\dfrac{(x + 3)^2}{9} + (y - 1)^2 = 1$

16. $\dfrac{(x - 5)^2}{25} + \dfrac{(y + 3)^2}{4} = 1$

Sketch the graph of each inequality.

B **17.** $\dfrac{x^2}{4} + \dfrac{y^2}{36} < 1$

18. $\dfrac{x^2}{25} + \dfrac{y^2}{9} > 1$

19. $\dfrac{(x + 1)^2}{16} + \dfrac{(y - 3)^2}{4} \geq 1$

20. $\dfrac{(x - 2)^2}{4} + \dfrac{(y + 1)^2}{25} \leq 1$

Sketch the graph of each equation. Recall that $\sqrt{a}$ is nonnegative.

21. $y = 3\sqrt{1 - x^2}$

22. $x = 6\sqrt{1 - y^2}$

23. $x^2 + 6x + 25y^2 - 100y + 9 = 0$

24. $9x^2 + 36x + 4y^2 - 8y + 4 = 0$

In Exercises 25–30 determine an equation for the ellipse with axes on the coordinate axes and satisfying the given conditions.

25. foci at $(5, 0)$ and $(-5, 0)$; y-intercepts 12 and -12

26. major axis of length 26; x-intercepts 9 and -9

27. minor axis of length 6; foci at $(0, 4)$ and $(0, -4)$

28. sum of focal radii 20; y-intercepts 7 and -7

29. sum of focal radii 14; foci at $(0, \sqrt{13})$ and $(0, -\sqrt{13})$

30. major axis of length 22, foci at $(4\sqrt{6}, 0)$ and $(-4\sqrt{6}, 0)$

25. $\dfrac{x^2}{169} + \dfrac{y^2}{144} = 1$

26. $\dfrac{x^2}{81} + \dfrac{y^2}{169} = 1$

27. $\dfrac{x^2}{9} + \dfrac{y^2}{25} = 1$

28. $\dfrac{x^2}{100} + \dfrac{y^2}{49} = 1$

29. $\dfrac{x^2}{36} + \dfrac{y^2}{49} = 1$

30. $\dfrac{x^2}{121} + \dfrac{y^2}{25} = 1$

Quadratic Relations and Systems **411**

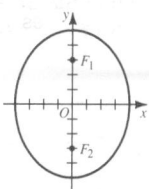
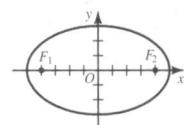
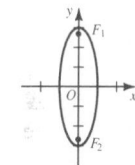
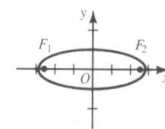
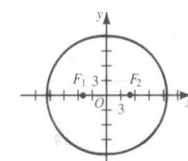
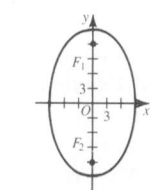
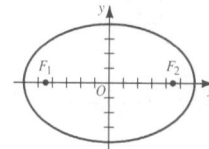
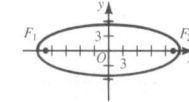

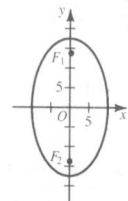
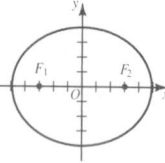
31. Determine an equation of the ellipse with major axis of length 10, and with foci $(0, -7)$ and $(8, -7)$. $\dfrac{(x-4)^2}{25} + \dfrac{(y+7)^2}{9} = 1$

32. Determine an equation of the ellipse with minor axis of length 8, and with foci $(-3, 5)$ and $(-3, -1)$. $\dfrac{(x+3)^2}{25} + \dfrac{(y-2)^2}{16} = 1$

C 33. Determine the equation of the ellipse with foci $(-3, 0)$ and $(3, 0)$ that passes through the point $(2, 2\sqrt{6})$. (*Hint*: Find the sum of the focal radii.) $\dfrac{x^2}{36} + \dfrac{y^2}{27} = 1$

34. Consider the set of all points (x, y) whose distance from the line $x = 8$ is twice their distance from the point $(2, 0)$. Show that this set is an ellipse by determining an equation for this set of points. $\dfrac{x^2}{16} + \dfrac{y^2}{12} = 1$

35. Use the definition of an ellipse to derive an equation of the ellipse with foci $(c, 0)$ and $(-c, 0)$ and sum of focal radii $2a$.

36. Repeat Exercise 35 for the ellipse with foci $(0, c)$ and $(0, -c)$ with sum of focal radii $2a$.   35. $\dfrac{x^2}{a^2} + \dfrac{y^2}{a^2 - c^2} = 1$ 36. $\dfrac{x^2}{a^2 - c^2} + \dfrac{y^2}{a^2} = 1$

10–6 Hyperbolas

Just as in Section 10-5 you used the sum of distances between points to define an ellipse, so do you use the *difference* of such distances to define another curve. Consider the set of all points in a plane such that, for each point P, the absolute value of the difference of the distance from P to two fixed points is a constant. Such a set of points is a two-branched curve called a **hyperbola**. As with ellipses, each fixed point is called a **focus**, and the distances from the foci to a point P on the curve are called **focal radii**.

To obtain an equation for the hyperbola with foci at $F_1(-5, 0)$ and $F_2(5, 0)$ and with focal radii differing by 8 (Figure 11), you can begin by expressing the fact that for any point $P(x, y)$ on the hyperbola, either

$$\sqrt{[x - (-5)]^2 + y^2} - \sqrt{(x - 5)^2 + y^2} = 8$$

or

$$\sqrt{[x - (-5)]^2 + y^2} - \sqrt{(x - 5)^2 + y^2} = -8.$$

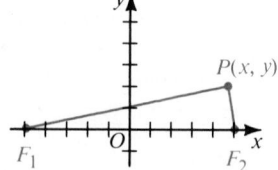

Figure 11

In either case, the squaring and simplifying process you used to obtain the equation for an ellipse (page 408) can be applied to obtain

$$\dfrac{x^2}{16} - \dfrac{y^2}{9} = 1.$$

By inspection, you can see that the graph of this equation:

1. Is symmetric with respect to both axes.
2. Has x-intercepts 4 and -4.

412   *Chapter 10*

It has no y-intercepts, since if you solve the equation for y in terms of x, you obtain

$$y = \pm\tfrac{3}{4}\sqrt{x^2 - 16},$$

from which it is evident that the curve has no points for which $|x| < 4$.

Using these facts, and constructing a table of first-quadrant values as shown, you can sketch the hyperbola in Figure 12.

x	$\tfrac{3}{4}\sqrt{x^2 - 16}$	y
4	$\tfrac{3}{4}\sqrt{16 - 16}$	0
5	$\tfrac{3}{4}\sqrt{25 - 16}$	2.3
6	$\tfrac{3}{4}\sqrt{36 - 16}$	3.3

As you can see in Figure 12, the graph lies entirely within two of the regions determined by the diagonals of the rectangle that is bounded by segments of the lines with equations

$$x = -4, \quad x = 4, \quad y = 3, \quad \text{and} \quad y = -3.$$

These diagonals are called *asymptotes* of the hyperbola and have equations

$$y = \tfrac{3}{4}x \quad \text{and} \quad y = -\tfrac{3}{4}x.$$

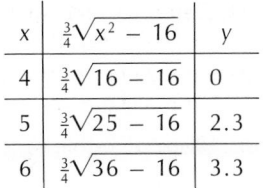

Figure 12

An **asymptote** for a graph is a line such that the distance between this line and a point P on the graph goes to 0 as the distance between P and the origin becomes greater and greater. For example, Figure 13 shows the first-quadrant graph of the hyperbola and asymptote drawn in Figure 12. For a given value of x, the difference between the y-coordinate of a point on the asymptote in the first quadrant and the y-coordinate of the corresponding point on the curve is given by

$$\tfrac{3}{4}x - \tfrac{3}{4}\sqrt{x^2 - 16}.$$

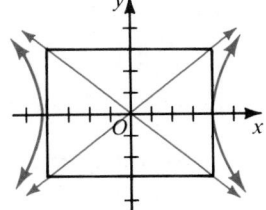

Figure 13

As you can see, as x increases, the difference between the y-coordinates decreases and for increasing values of x, the curve approaches closer and closer to the asymptote. By symmetry you can see that the corresponding situation holds in each of the remaining quadrants.

In general, you can show (Exercise 33, page 417) that:

> If $(-c, 0)$ and $(c, 0)$ are foci of the hyperbola for which the absolute value of the difference of the focal radii is the constant $2a > 0$, then the hyperbola is the graph of the equation
>
> $$\frac{x^2}{a^2} - \frac{y^2}{b^2} = 1, \qquad (1)$$
>
> where $b^2 = c^2 - a^2$. The equations for the asymptotes are
>
> $$y = \frac{b}{a}x \quad \text{and} \quad y = -\frac{b}{a}x.$$

Quadratic Relations and Systems **413**

2. Sketch the graph of $x^2 - 4y^2 = 36$.

Divide both sides by 36.

$$\frac{x^2}{36} - \frac{y^2}{9} = 1; \ a = 6, \ b = 3$$

By inspection the graph is a hyperbola and:
1. Symmetric to both axes.
2. Has x-intercepts 6 and -6; no y-intercepts
3. Has asymptotes:

$$y = \frac{1}{2}x; \ y = -\frac{1}{2}x$$

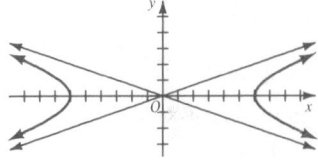

Additional A Exercises

In Exercises 1–6:

a. Draw the rectangle whose diagonals are the asymptotes of the hyperbola and draw the asymptotes themselves.

b. Sketch the hyperbola.

c. Give the coordinates of the foci.

1. $\dfrac{x^2}{4} - y^2 = 1$

foci: $(\sqrt{5}, 0), (-\sqrt{5}, 0)$

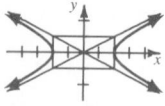

2. $\dfrac{y^2}{4} - \dfrac{x^2}{4} = 1$

foci: $(0, 2\sqrt{2}), (0, -2\sqrt{2})$

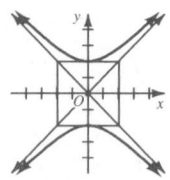

3. $\dfrac{x^2}{4} - \dfrac{y^2}{4} = 1$

foci: $(2\sqrt{2}, 0), (-2\sqrt{2}, 0)$

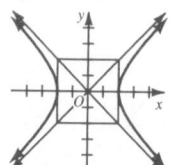

4. $\dfrac{(x - 1)^2}{4} - \dfrac{(y - 1)^2}{4} = 1$

foci: $(1 + 2\sqrt{2}, 1),$
$(1 - 2\sqrt{2}, 1)$

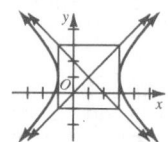

Similarly, you can show (Exercise 34, page 417) that:

> If $(0, -c)$ and $(0, c)$ are foci of the hyperbola for which the absolute value of the difference of the focal radii is the constant $2a > 0$, then the hyperbola is the graph of the equation
>
> $$\frac{y^2}{a^2} - \frac{x^2}{b^2} = 1, \qquad (2)$$
>
> where $b^2 = c^2 - a^2$. The equations for the asymptotes are
>
> $$y = \frac{a}{b}x \qquad \text{and} \qquad y = -\frac{a}{b}x.$$

The **transverse axis** of a hyperbola is the line segment of length $2a$ that intersects the hyperbola in two points called the **vertices** of the hyperbola. The midpoint of the line segment joining the foci is called the **center** of the hyperbola.

EXAMPLE Sketch the graph of $9x^2 - 25y^2 = 225$.

SOLUTION Divide each side by 225 to obtain

$$\frac{x^2}{25} - \frac{y^2}{9} = 1.$$

Comparing this with Equation (1) on page 413, $a = 5$ and $b = 3$. Then, by inspection, the graph is a hyperbola and:

1. The graph is symmetric with respect to both axes.
2. The x-intercepts are 5 and -5. There are no y-intercepts.
3. The asymptotes are the graphs of $y = \frac{3}{5}x$ and $y = -\frac{3}{5}x$.

Use the asymptotes and the intercepts to sketch the graph shown at the right. A right triangle with sides measuring a, b, and c is also shown in the figure, and illustrates the relationship between a, b, and c in a hyperbola.

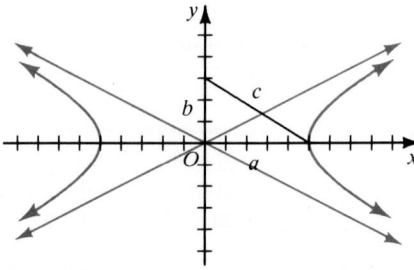

Like ellipses, hyperbolas can have a center other than the origin. A hyperbola with center (h, k) and transverse axis parallel to the x- or y-axis can be defined by an equation of the form

$$\frac{(x - h)^2}{a^2} - \frac{(y - k)^2}{b^2} = 1 \quad \text{or} \quad \frac{(y - k)^2}{a^2} - \frac{(x - h)^2}{b^2} = 1.$$

414 *Chapter 10*

Figure 14 shows the hyperbola defined by

$$\frac{(x-2)^2}{9} - \frac{(y-1)^2}{4} = 1.$$

The circle, ellipse, parabola, and hyperbola are called **conic sections** because each can be formed as the intersection of a plane with a *conical surface of two nappes* (see Figure 15). (It is assumed that such a conical surface extends indefinitely.) A point, a line, and a pair of intersecting lines are sometimes called **degenerate conic sections**.

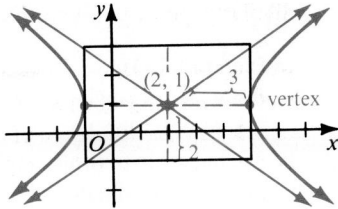

Figure 14

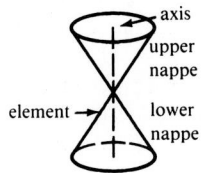

Right circular conical surface (of two nappes)

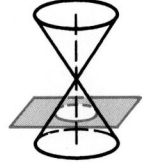

Circle
(Plane perpendicular to the axis, cutting one nappe)

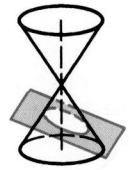

Ellipse
(Plane oblique to the axis, cutting one nappe)

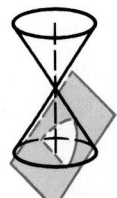

Parabola
(Plane parallel to an element)

Hyperbola
(Plane cutting both nappes)

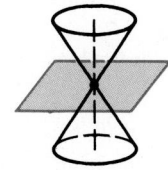

Point
(Plane cutting only at the vertex)

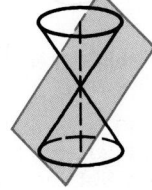

Single line
(Plane tangent along an element)

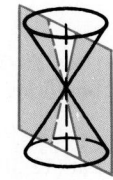

Pair of intersecting lines
(Plane through the axis)

Figure 15

Oral Exercises

State the vertices and the equations of the asymptotes for the hyperbola whose equation is given.

1. $\frac{x^2}{9} - \frac{y^2}{16} = 1$ $(\pm3, 0)$; $y = \pm\frac{4}{3}x$

2. $\frac{x^2}{64} - \frac{y^2}{36} = 1$ $(\pm8, 0)$; $y = \pm\frac{3}{4}x$

3. $\frac{y^2}{25} - \frac{x^2}{100} = 1$ $(0, \pm5)$; $y = \pm\frac{1}{2}x$

4. $\frac{x^2}{9} - y^2 = 1$ $(\pm3, 0)$; $y = \pm\frac{1}{3}x$

5. $\frac{y^2}{49} - \frac{x^2}{49} = 1$ $(0, \pm7)$; $y = \pm x$

6. $\frac{x^2}{4} - \frac{y^2}{36} = 1$ $(\pm2, 0)$; $y = \pm3x$

Quadratic Relations and Systems **415**

5. $\frac{(y-1)^2}{4} - \frac{(x-1)^2}{4} = 1$

foci: $(1, 1 + 2\sqrt{2})$, $(1, 1 - 2\sqrt{2})$

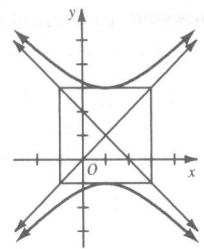

6. $\frac{x^2}{4} - y^2 > 1$

foci: $(\sqrt{5}, 0)$, $(-\sqrt{5}, 0)$

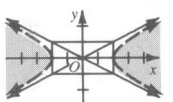

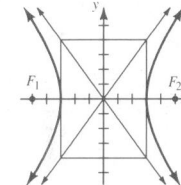
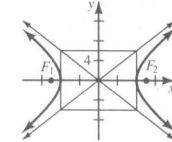
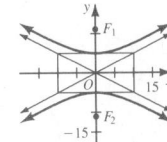
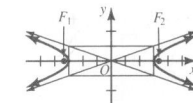

5.

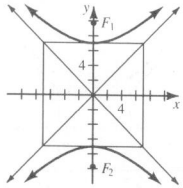

6.

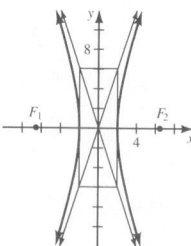

7.

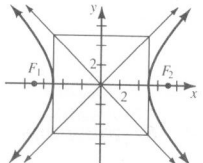

8.

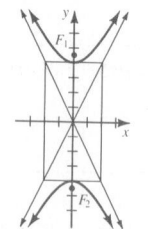

9.

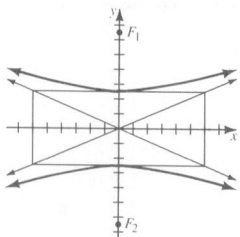

10.

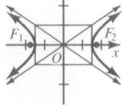

Written Exercises

In Exercises 1–16:

a. draw the rectangle whose diagonals are the asymptotes of the hyperbola and draw the asymptotes themselves;

b. sketch the hyperbola; and

c. give the coordinates of the foci.

1. $(\pm5, 0)$ **2.** $(\pm10, 0)$ **3.** $(0, \pm5\sqrt{5})$

4. $(\pm\sqrt{10}, 0)$ **5.** $(0, \pm7\sqrt{2})$ **6.** $(\pm2\sqrt{10}, 0)$

A **1–6.** Use the equations in Oral Exercises 1–6.

7. $2x^2 - 2y^2 = 50$ $(\pm5\sqrt{2}, 0)$ **8.** $4y^2 - 16x^2 = 64$ $(0, \pm2\sqrt{5})$

9. $144y^2 - 25x^2 = 900$ $(0, \pm\frac{13}{2})$ **10.** $9x^2 - 16y^2 = 16$ $(\pm\frac{5}{3}, 0)$

11. $6x^2 - 30y^2 = 180$ $(\pm6, 0)$ **12.** $16y^2 - 20x^2 = 400$ $(0, \pm3\sqrt{5})$

13. $\dfrac{(x-2)^2}{16} - \dfrac{(y+1)^2}{25} = 1$ $(2 \pm \sqrt{41}, -1)$ **14.** $\dfrac{(y-3)^2}{9} - (x-1)^2 = 1$ $(1, 3 \pm \sqrt{10})$

15. $\dfrac{(y-3)^2}{4} - \dfrac{(x+4)^2}{9} = 1$ $(-4, 3 \pm \sqrt{13})$ **16.** $x^2 - y^2 + 6y = 10$ $(\pm\sqrt{2}, 3)$

Sketch the graph of each inequality.

17. $y^2 - x^2 \geq 9$ **18.** $y^2 - x^2 \leq 4$ **19.** $x^2 < y^2 + 4$ **20.** $y^2 < x^2 - 1$

Sketch the graph of each equation. Recall that $\sqrt{a}$ is nonnegative.

B **21.** $x = \sqrt{y^2 + 1}$ **22.** $y = \sqrt{x^2 - 9}$ **23.** $y = \sqrt{x^2 - 16}$ **24.** $x = \sqrt{y^2 + 25}$

In Exercises 25–28 determine an equation for the hyperbola satisfying the given conditions.

25. foci at $(0, 10)$ and $(0, -10)$; one vertex at $(0, 8)$ **25.** $\dfrac{y^2}{64} - \dfrac{x^2}{36} = 1$ **26.** $\dfrac{x^2}{81} - \dfrac{y^2}{63} = 1$

26. foci at $(12, 0)$ and $(-12, 0)$; absolute value of the difference of focal radii 18

C **27.** foci at $(-2, 1)$ and $(8, 1)$; transverse axis of length 6   **27.** $\dfrac{(x-3)^2}{9} - \dfrac{(y-1)^2}{16} = 1$

28. foci at $(3, 11)$ and $(3, -5)$; vertices at $(3, 7)$ and $(3, -1)$ **28.** $\dfrac{(y-3)^2}{16} - \dfrac{(x-3)^2}{48} = 1$

29. Determine an equation of the hyperbola whose asymptotes are given by $y = \frac{4}{3}x$ and $y = -\frac{4}{3}x$, and with foci at $(15, 0)$ and $(-15, 0)$; use the following steps. **a.** Use the relationships given on page 413 to determine two equations involving a and b. **b.** Solve these equations simultaneously by substitution for a^2 and b^2. $\dfrac{x^2}{81} - \dfrac{y^2}{144} = 1$

30. Determine an equation of the hyperbola with asymptotes $y = 2x$ and $y = -2x$, and foci at $(0, 5)$ and $(0, -5)$. $\dfrac{y^2}{20} - \dfrac{x^2}{5} = 1$

31. Determine an equation of the hyperbola with foci at $(0, 7)$ and $(0, -7)$ and passing through the point $(12, 2)$. $y^2 - \dfrac{x^2}{48} = 1$

32. Consider the set of points (x, y) whose distance from the point $(8, 0)$ is twice their distance from the line $x = 2$. Show that this set is a hyperbola by determining an equation for this set of points. $\dfrac{x^2}{16} - \dfrac{y^2}{48} = 1$

416   *Chapter 10*

33. Use the definition of a hyperbola to derive an equation for a hyperbola with foci $(c, 0)$ and $(-c, 0)$ and absolute value of difference of focal radii $2a > 0$ where $a < c$.

34. Repeat Exercise 33 for the hyperbola with foci $(0, c)$ and $(0, -c)$ and absolute value of the difference of focal radii $2a > 0$, where $a < c$.

33. $\dfrac{x^2}{a^2} - \dfrac{y^2}{c^2 - a^2} = 1$ 34. $\dfrac{y^2}{a^2} - \dfrac{x^2}{c^2 - a^2} = 1$

Computer Exercises For students with computer experience

1. The graph of an equation of the form $ax^2 + by^2 = c$ $(a, b \neq 0)$ is an ellipse, a circle, a hyperbola, a pair of lines, one point, or the empty set depending on the signs of a, b, and c. Write a program that will identify the graph when a, b, and c are input. Have the output give the coordinates of the foci if the graph is an ellipse or a hyperbola, or the length of the radius if the graph is a circle.

Run the program in Exercise 1 for each equation.

2. $9x^2 - 16y^2 = 144$ Hyperbola; $(\pm5, 0)$ 3. $25x^2 + 16y^2 = 400$ Ellipse; $(0, \pm3)$

4. $-x^2 - 4y^2 = 25$ Empty set   5. $x^2 + 25y^2 = 0$ Point

6. $4x^2 - 9y^2 = 0$ A pair of lines   7. $25x^2 - 144y^2 = -3600$

8. $5x^2 + 5y^2 = 125$ Circle; $r = 5$ 9. $-7x^2 - 7y^2 = 13$ Empty set

7. Hyperbola; $(0, \pm13)$

11.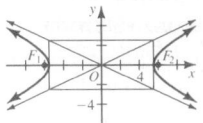

12.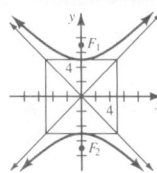

13.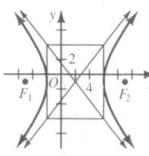

(continued on p. 438)

10–7 Inverse and Other Kinds of Variation

When two pulleys are connected (Figure 16), the one with smaller diameter revolves more rapidly. If d_1 represents the diameter of one pulley and n_1 the number of revolutions per minute (r/min) of the pulley, and if d_2 and n_2 represent the corresponding numbers for the other pulley, then $d_1 n_1 = d_2 n_2$, or $\dfrac{d_1}{d_2} = \dfrac{n_2}{n_1}$. In the pulleys shown in Figure 16, $d_1 = 6$ cm and $d_2 = 10$ cm. If the smaller pulley revolves at 120 r/min, you can compute the revolutions per minute of the other pulley:

$$\frac{6}{10} = \frac{n_2}{120}$$
$$n_2 = 72$$

For this set of pulleys, $dn = 720$. Because this implies that $n = \dfrac{720}{d}$ or $d = \dfrac{720}{n}$, you say that n and d vary inversely as each other, or are *inversely proportional* to each other.

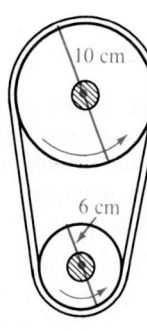

Figure 16

Teaching Suggestions p. T106

Supplementary Material

Test 31

Key Ideas

Use inverse variation to solve problems.

Quadratic Relations and Systems   **417**

1. Determine an equation in x and y in which y varies inversely as x and passes through the point (3, 9).
   Since y varies inversely as x, the equation is of the form
   $$xy = k, k \neq 0$$
   Substitute x and y:
   $$(3)(9) = k$$
   $$27 = k$$
   ∴ the equation is $xy = 27$.

2. If 20 m of wire of diameter 1.5 mm has a resistance of 12 Ω, what is the resistance of 20 m of the same type of wire if the diameter is increased to 2 mm?
   $$R = \frac{kl}{d^2}$$
   $R = 12, l = 20, d = 1.5$
   $$12 = \frac{k(20)}{(1.5)^2}$$
   $1.35 = k$
   $$\therefore R = \frac{1.35l}{d^2}$$
   Now: $l = 20, d = 2$
   $$R = \frac{1.35(20)}{(2)^2} = 6.75$$
   ∴ the resistance is 6.75 Ω.

In general, any function defined by an equation of the form

$$xy = k,$$

where k is a nonzero constant, is called an **inverse variation,** and k is the **constant of variation.** The graph of such a function is of the form shown in Figure 17 if $k > 0$, and of the form shown in Figure 18 if $k < 0$. It can be shown that these graphs are hyperbolas, with foci on the lines with equations $y = x$ and $y = -x$, respectively, and with the coordinate axes as asymptotes. Ordinarily, in practical situations, you have $x > 0$, $y > 0$, and $k > 0$, and the graph is limited to one such as the first-quadrant branch of Figure 17.

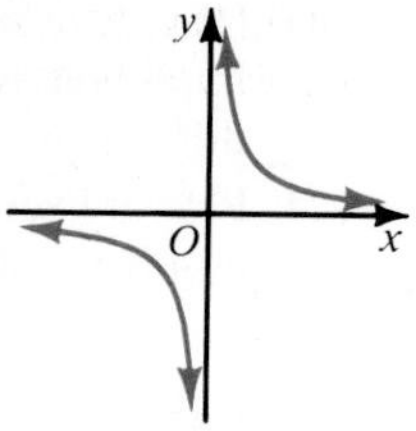

As with direct variation (recall page 108), there is an important relationship among the coordinates of two ordered pairs (x_1, y_1) and (x_2, y_2) of the inverse variation specified by the equation $xy = k$, $k \neq 0$. You have

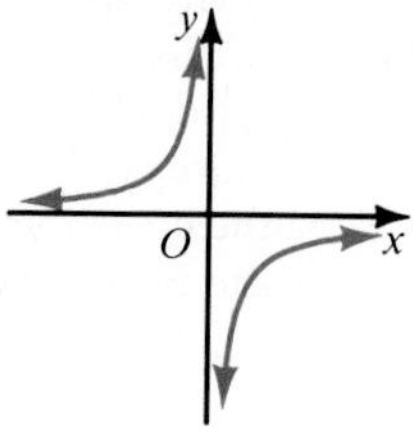

$$x_1 y_1 = k \quad \text{and} \quad x_2 y_2 = k, \quad \text{and so} \quad x_1 y_1 = x_2 y_2.$$

Hence, $x_1 \neq 0$, $x_2 \neq 0$, $y_1 \neq 0$, and $y_2 \neq 0$, and you have

$$\frac{x_1}{x_2} = \frac{y_2}{y_1}.$$

We state this result as a theorem.

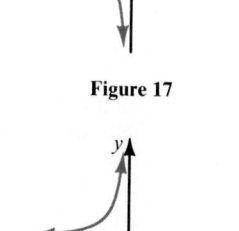

Figure 17

Figure 18

Theorem. For all real numbers x_1, y_1, x_2, y_2, $(x_1 \neq 0, x_2 \neq 0, y_1 \neq 0,$ and $y_2 \neq 0)$, if (x_1, y_1) and (x_2, y_2) are ordered pairs of an inverse variation, then

$$\frac{x_1}{x_2} = \frac{y_2}{y_1}.$$

The following list gives some everyday examples of inverse variation.

1. For a fixed distance, the greater the speed, the proportionately less time needed to cover it (and vice versa).

   $rt = d$

2. For a stated income, the greater the rate of interest, the proportionately less the amount of principal needed.

   $pr = I$

3. For a given area, the greater the length of a rectangle, the proportionately less the width.

   $lw = A$

4. Boyle's Law in physics: If the temperature is kept constant, the greater the pressure on a gas, the proportionately less the volume.

   $pv = K$

In Section 7-1, the idea of direct variation was extended to include variation defined by $y = ax^2$, that is, "y varies directly as x^2." (The graph

418 *Chapter 10*

of that variation is a parabola.) Similarly, the idea of inverse variation is extended to include variation defined, for example, by

$$y = \frac{k}{x^2};$$

that is, y *varies inversely as* x^2. (The graph of such a variation is not a curve with which you are familiar.) Corresponding extensions involving other powers may also be made.

Still another form of variation is typified by the relationship between the electrical resistance of a wire and the length and diameter of the wire. The electrical resistance R of a wire varies *directly* as the length l of the wire and *inversely* as the square of its diameter d. An equation expressing this fact is $R = \frac{kl}{d^2}$. You call such a variation a **combined variation**. If z varies directly as x and also directly as y, then the equation relating these variables is of the form $z = kxy$, and you say z **varies jointly** as x and y.

EXAMPLE   If 18 m of high-resistance wire of diameter 1.2 mm has a resistance of 10 Ω, what is the resistance of 27 m of the same type of wire if the diameter is increased to 1.5 mm?

SOLUTION   Use the combined variation $R = \frac{kl}{d^2}$.

Since $R = 10$ when $l = 18$ and $d = 1.2$,

$$10 = \frac{k(18)}{(1.2)^2}, \quad \text{or} \quad k = 0.8.$$

$$\therefore R = \frac{0.8l}{d^2}$$

Replacing l with 27 and d with 1.5, you have

$$R = \frac{0.8(27)}{(1.5)^2} = 9.6$$

$\therefore$ the resistance is 9.6 Ω.

Oral Exercises

To describe the relation of the variable in the left side of the equation to the other variables, state whether the equation is a direct, inverse, combined, or joint variation.

4. inverse   8. joint

1. $y = \frac{x}{9}$ direct
2. $z = 5xy$ joint
3. $y = \frac{8x}{z}$ combined
4. $y = \frac{6}{x}$

5. $z = \frac{x}{12y}$ combined
6. $y = \frac{1}{7x}$ inverse
7. $y = 6.3x$ direct
8. $z = \frac{xy}{15}$

Quadratic Relations and Systems   **419**

Reading Algebra

Students have already learned that direct variation is represented by an equation in the form $y = mx$. In this chapter students will become familiar with the general equation $xy = k$ which represents inverse variation. Briefly discuss the differences between these two types of variation. Be sure your students understand that the phrases "inverse variation", "vary inversely", and "inversely proportional" are all different ways of stating the same relationship.

Mixed Review

1. Simplify $\sqrt[3]{(-27)^2}$.   9
2. Simplify $-2\sqrt{48x^5} + 3\sqrt{27x^5}$. $x^2\sqrt{3x}$
3. Solve
   $\sqrt{x-2} - \sqrt{x+3} = 1$
   over $\mathcal{R}$.   Ø
4. Solve $5x^2 - 4x = 3$.
   $\left\{ \dfrac{2 \pm \sqrt{19}}{5} \right\}$
5. Find the sum of the series
   $500 - 100 + 20 - \ldots$
   $416\frac{2}{3}$
6. True or false: The graph of the equation $x^2 - 4y^2 = 16$ is an ellipse.   False.

Suggested Assignments

Minimum
  First day
    420/1–20
  Second day
    421/*P*: 1–7
  R 422/*Self-Test 2*
Extended Alg. with Trig.
  First day
    420/1–19 odd
  Second day
    421/*P*: 1–7
  R 422/*Self-Test 2*
Enriched Alg.
  420/5–21 odd
  421/*P*: 1–7
  R 422/*Self-Test 2*
Enriched Alg. with Trig.
  420/5–21 odd
  421/*P*: 1–7
  R 422/*Self-Test 2*

Additional Answers
Written Exercises

20.

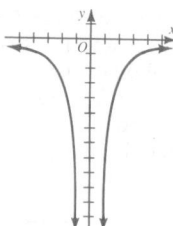

Written Exercises

Determine an equation in x and y that defines the given variation and contains the given ordered pair.

A   **1.** y varies inversely as x; $(7, 4)$ $xy = 28$

     2. y varies inversely as x; $(6, -\frac{1}{3})$ $xy = -2$

     3. y varies inversely as x^2; $(1, 5)$ $x^2 y = 5$

     4. y varies inversely as x^3; $(-\frac{3}{4}, 6)$ $x^3 y = -\frac{81}{32}$

     5. y varies inversely as $\sqrt{x}$; $(25, \frac{-3}{2})$ $y\sqrt{x} = -\frac{15}{2}$

     6. y varies inversely as $\sqrt[3]{x}$; $(0.027, 250)$ $y\sqrt[3]{x} = 75$

In Exercises 7–9, y varies inversely as x. Determine the value of x_2 or y_2.

   7. $(\frac{1}{5}, 5)$, $(x_2, 4)$ $\frac{1}{4}$ **8.** $(-5, 2)$, $(3, y_2)$ $-\frac{10}{3}$ **9.** $(\frac{4}{5}, 30)$, $(16, y_2)$ $\frac{3}{2}$

In Exercises 10–12, y varies inversely as x^2. Determine the value of x_2 or the value of y_2.

   10. $(2, 9)$, $(x_2, 4)$ $3, -3$ **11.** $(-\frac{2}{3}, 27)$, $(4, y_2)$ $\frac{3}{4}$ **12.** $(3\sqrt{3}, \frac{1}{4})$, $(x_2, 12)$ $\frac{3}{4}, -\frac{3}{4}$

   13. If z varies jointly as x and y, and $z = 12$ when $x = 9$ and $y = 4$, find z when $x = 7$ and $y = 2$. $\frac{14}{3}$

   14. If z varies jointly as x and y, and $z = \frac{2}{3}$ when $x = 27$ and $y = \frac{1}{2}$, find z when $x = 9$ and $y = 18$.   8

B   **15.** If z varies directly as x and inversely as y, and $z = 6$ when $x = 4$ and $y = 5$, find z when $x = 8$ and $y = 3$.   20

     16. If z varies directly as x and inversely as the cube of y, and $z = 9$ when $x = 2$ and $y = \frac{1}{3}$, find z when $x = 3$ and $y = \frac{1}{2}$.   4

     17. In a certain variation z varies jointly as x and y and inversely as w, and $z = 5$ when $x = \frac{1}{2}$, $y = 3$, and $w = 6$. Find z when $x = 2$, $y = \frac{1}{3}$, and $w = 8$. $\frac{5}{3}$

     18. In a certain variation, z varies jointly as x and y and inversely as w^2, and $z = 72$ when $x = 80$, $y = 30$, and $w = 5$. Find z when $x = 20$, $y = 60$, and $w = 9$. $\frac{100}{9}$

     19. Sketch the graph of $y = \dfrac{6}{x^2}$.

     20. Sketch the graph of $y = \dfrac{-12}{x^2}$.

C   **21.** Use the definition on page 412 to find an equation for the hyperbola with foci at $(2, 2)$ and $(-2, -2)$ and with difference of focal radii 4; draw its graph. $xy = 2$

     22. Find an equation of the hyperbola that is identical to the graph in Exercise 21, but rotated $45°$ clockwise. That is, its asymptotes are the lines $y = x$ and $y = -x$. $\frac{x^2}{4} - \frac{y^2}{4} = 1$

420   *Chapter 10*

Problems

Give each answer to the same number of significant digits as the least accurate measurement.

A 1. The number of oscillations per minute made by an object suspended from a given spring varies inversely as the square root of the object's mass. If an object with a mass of 400 g makes 60 oscillations per minute, how many oscillations per minute will be made by an object with a mass of 625 g? 48 oscillations per minute

2. The slope at which a curve in a road should be banked varies inversely with the radius of the curve and directly with the square of the maximum speed of the cars that will use it. If a curve of radius 1250 m carrying cars traveling at a maximum speed of 24.5 m/s has a slope of 0.049, what slope should the road have on a curve of radius 765 m if the maximum speed is 15 m/s? approx. 0.030

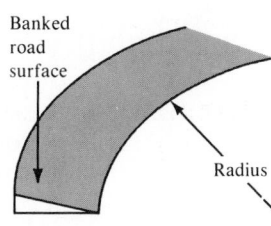
Banked road surface

Radius

3. The volume of a gas varies directly with the Kelvin temperature and inversely with the pressure. If a certain gas has a volume of 342 m³ at a temperature of 300° K under a pressure of 200 kPa (kilopascals), what will be the volume of the same gas at a temperature of 320° K under a pressure of 400 kPa? 182 m³

4. The speed of a satellite orbiting Earth is inversely proportional to the square root of its distance from Earth's center. If an artificial satellite 16,000 km from Earth's center travels at 1.58×10^5 m/s, what is the speed of the moon, which is 400,000 km from the center of Earth?
3.16×10^4 m/s

5. The power produced by an electric circuit varies directly as the square of the voltage and inversely as the resistance. If a voltage of 120 V in a circuit with a resistance of 25 Ω produces 576 W, what voltage applied to a circuit with a resistance of 20 Ω will produce 605 W? 110 V

B 6. The speed of sound in a gas at 0° C varies inversely as the square root of the density of the gas. If sound travels at 2904 m/s in carbon dioxide, how fast will it travel in helium, whose density is 0.09 times that of carbon dioxide? 9680 m/s

Quadratic Relations and Systems **421**

Additional A Exercises

Determine an equation in x and y that defines the given variation and contains the given ordered pair.

1. y varies inversely as x; (3, 6). $xy = 18$

2. y varies inversely as x^2; $\left(4, -\dfrac{1}{4}\right)$. $x^2y = -4$

3. y varies inversely as $\sqrt{x}$; $\left(9, -\dfrac{5}{4}\right)$. $y\sqrt{x} = -\dfrac{15}{4}$

In Exercises 4 and 5, y varies inversely as x. Find x_2.

4. $\left(\dfrac{1}{2}, 6\right)$, $(x_2, 2)$ $x_2 = \dfrac{3}{2}$

5. $(3, 4)$, $(x_2, 9)$ $x_2 = \dfrac{4}{3}$

6. The resistance required in an electrical circuit to produce a given amount of power varies inversely with the square of the current. If a current of 0.8 amperes requires a resistance of 50 ohms, what resistance will be required by a current of 0.5 amperes? 128 ohms.

Additional Answers Self-Test 2

1.
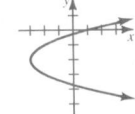

2.
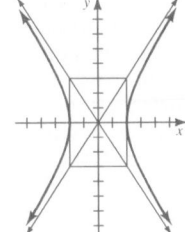

Sketch the graph of each equation.

1. $2x = y^2 - 2y - 3$

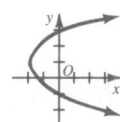

2. $4x^2 - 9y^2 = 36$

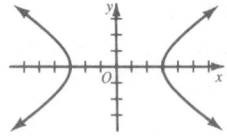

3. Write an equation of the form $x^2 + y^2 + ax + by + c = 0$ for the circle with center $(2, -1)$ and radius 5.
$x^2 + y^2 - 4x + 2y - 20 = 0$

4. Write an equation of the set of points (x, y) the sum of whose distances from $(0, 5)$ and $(0, -5)$ is 26.
$\dfrac{x^2}{144} + \dfrac{y^2}{169} = 1$

5. If z varies directly as x^2 and inversely as y, and $z = \dfrac{3}{4}$ when $x = 9$ and $y = 2$, find the value of z when $x = 12$ and $y = \dfrac{1}{2}$. $\dfrac{16}{3}$

6. The illuminance expressed in lux (lx), of an unshaded electric bulb directly above a flat surface is inversely proportional to the square of the distance to the surface. If the illuminance is 4.5 lx when the bulb is 4 m from the surface, what is the illuminance when the bulb is 3 m from the surface? 8 lx

7. Newton's law of gravitation states that two objects attract each other with a force that varies jointly with the masses of the two objects and inversely with the square of the distance between them. If two 30 kg masses 1 m apart are found to attract each other with a force of 6×10^{-8} N, and a 1 kg mass on the surface of Earth (6×10^6 m from its center of gravity) is attracted by Earth with a force of 9.8 N, compute the mass of Earth. 5×10^{24} kg

Self-Test 2

VOCABULARY
circle (p. 401)
radius (p. 401)
center of a circle (p. 401)
parabola (p. 404)
directrix (p. 404)
focus of a parabola (p. 404)
axis of symmetry (p. 405)
vertex of a parabola (p. 405)
ellipse (p. 408)
focus of an ellipse (p. 408)
focal radii of an ellipse
  (p. 408)
center of an ellipse (p. 408)
major axis of an ellipse
  (p. 409)
minor axis of an ellipse
  (p. 409)

hyperbola (p. 412)
focus of a hyperbola (p. 412)
focal radii of a hyperbola
  (p. 412)
asymptotes (p. 413)
transverse axis (p. 414)
vertices of a hyperbola (p. 414)
center of a hyperbola (p. 414)
conic sections (p. 415)
degenerate conic sections
  (p. 415)
inverse variation (p. 418)
constant of variation (p. 418)
combined variation (p. 419)
joint variation (p. 419)

Sketch the graph of each equation. (See p. 421 for the graphs.)

1. $x = y^2 + 4y + 1$ **2.** $9x^2 - 4y^2 = 36$ *Obj. 1, p. 401*

3. Write an equation of the form $x^2 + y^2 + ax + by + c = 0$ for the circle   *Obj. 2, p. 401* with center $(1, -2)$ and radius 3. $x^2 + y^2 - 2x + 4y - 4 = 0$

4. Write an equation for the set of points (x, y), the sum of whose distances from $(0, 5)$ and $(0, -5)$ is 26. $\frac{x^2}{144} + \frac{y^2}{169} = 1$

5. If a varies directly as b^2 and inversely as c, and $a = \frac{9}{2}$ when $b = 3$ and   *Obj. 3, p. 401* $c = 4$, find the value of a when $b = 6$ and $c = 12$. $a = 6$

6. The speed of a gear (in revolutions per minute) varies inversely as the   *Obj. 4, p. 401* number of teeth on the gear. If a gear that has 36 teeth makes 30 rev/min, what is the speed of a gear that has 90 teeth? 12 rev/min

Check your answers with those at the back of the book.

Solving Quadratic Systems

OBJECTIVES for Sections 10-8 through 10-10:
1. *To solve simple quadratic systems graphically.*
2. *To solve simple quadratic systems by substitution.*

10–8 Graphic Solutions of Systems

Graphical methods can be used to determine the number of real-number solutions of systems of equations in two variables in which one or both of the equations are quadratic. These methods can also be used to estimate the solutions.

EXAMPLE   Use a graph to estimate the solution of the system over $\mathcal{R}$.

$$9x^2 + 4y^2 = 100$$
$$25x^2 - 4y^2 = 36$$

SOLUTION   Graph both equations on the same coordinate plane. There are 4 points of intersection that appear to be $(2, 4)$, $(-2, 4)$, $(-2, -4)$, and $(2, -4)$. Check these points by substituting in both equations.

Check: $9(2)^2 + 4(4)^2 \overset{?}{=} 100$
$36 + 64 = 100 \checkmark$
$25(2)^2 - 4(4)^2 \overset{?}{=} 36$
$100 - 64 = 36 \checkmark$

Actually $(2, 4)$ is an *exact* solution as the check shows. Since

$$(-2)^2 = 2^2$$

and

$$(-4)^2 = 4^2,$$

all values will check in a similar way.

$\therefore$ the solution set is $\{(2, 4), (-2, 4), (-2, -4), (2, -4)\}$.

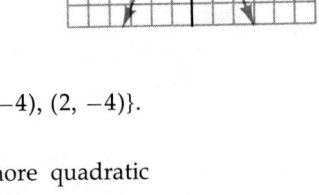

In solving graphically systems containing one or more quadratic equations, you will discover that:
1. A system consisting of a linear equation and a quadratic equation may have no real solutions or as many as two.
2. A system consisting of two quadratic equations may have no real solutions or as many as four.

Teaching Suggestions
p. T106

Key Ideas

Use graphs to solve simple quadratic equations.

Chalkboard Examples

1. Use a graph to estimate the solution of the system over $\mathcal{R}$.
$$x^2 + y^2 = 25$$
$$y = x + 1$$
Sketch the graph of each equation on the same set of coordinate axes. Find the points of intersection. Check them in both equations.

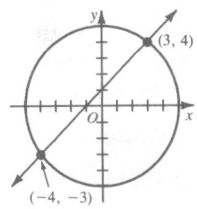

Additional A Exercises

Solve each system over $\mathcal{R}$ by graphing. Then check the coordinates of each graphical solution by substituting in both equations.
1. $y = 4x - x^2$
$y = 2x - 3$
$\{(3, 3), (-1, -5)\}$
2. $x = 1 - y^2$
$x = 2y + 2$ $\{(0, -1)\}$

423

Suggested Assignments

Minimum
  First day
    424/1–12
  Second day
  S 345/30–44 even
    131/1–11 odd
Extended Alg. with Trig.
  424/1–12
  S 345/30–44 even
    131/2–14 even
Enriched Alg.
  424/1–15 odd
  S 345/36–44 even
    131/2–14 even
Enriched Alg. with Trig.
  424/1–15 odd
  S 345/36–44 even
    131/2–14 even

Mixed Review

In $\triangle ABC$ with vertices $A(11, -4)$, $B(-1, 1)$, and $C(3, 9)$, find each of the following:

1. the length of side $\overline{AB}$.   13
2. the length of the median to side $\overline{BC}$. $\sqrt{181} \approx 13.45$
3. the equation of the altitude to side $\overline{BC}$.
$$y = -\frac{1}{2}x + \frac{3}{2}$$

Teaching Suggestions
p. T107

Key Ideas

Solve linear-quadratic systems by using substitution.

Oral Exercises

Identify the graphs of the equations.

EXAMPLE $y = 3x^2 + 6$
               $y = -6x$

SOLUTION   The graph of $y = 3x^2 + 6$ is a parabola. The graph of $y = -6x$ is a line.

6. ellipse; ellipse
parabola

1. $y = x^2 - 3$ parabola   2. $y = 6x - x^2$ parabola 3. $x = y^2 - 5y + 4$
   $y = 2x$ line $y = 3x - 4$ line $x + y = 1$ line
4. $x = 9 - y^2$ parabola   5. $x^2 + 4y^2 = 25$ ellipse 6. $x^2 + 9y^2 = 25$
   $x = -4y + 13$ line $x^2 + y^2 = 13$ circle $x^2 + 36y^2 = 100$

1. $\{(-1, -2), (3, 6)\}$ 2. $\{(-1, -7), (4, 8)\}$ 3. $\{(-2, 3), (0, 1)\}$
Written Exercises 4. $\{(5, 2)\}$ 5. $\{(-3, -2), (-3, 2), (3, -2), (3, 2)\}$ 6. $\{(0, \frac{5}{3}), (0, -\frac{5}{3})\}$

Solve each system over $\mathcal{R}$ by graphing. Then check the coordinates of each graphical solution by substituting in both equations.

A **1–6.** Use Oral Exercises 1–6.

                                      $\{(-4, 3), (-4, -3), (5, 0)\}$ $\{(8, 6), (-8, 6)\}$
   7. $x^2 + 9y^2 = 9$ **8.** $x = 5 - y^2$ **9.** $x^2 + y^2 = 100$
       $4x^2 - y^2 = 36$ $\{(3, 0), (-3, 0)\}$ $x^2 + y^2 = 25$ $x^2 - 10 = 9y$

B **10.** $x^2 - 2x + y^2 - 4y - 5 = 0$ **11.** $x^2 - y^2 = 9$ **12.** $y^2 - 4x^2 = 9$
        $x^2 + y^2 = 25$ $4y = x^2 - 9$ $y = 3 - 2x^2$
           $\{(0, 5), (4, 3)\}$ **11.** $\{(-5, 4), (5, 4), (3, 0), (-3, 0)\}$ $\{(-2, -5), (2, -5), (0, 3)\}$

Estimate the solution(s) of each system over $\mathcal{R}$ to the nearest $\frac{1}{2}$ unit by graphing. Check that the solution satisfies the two given equations to the nearest whole number.

                                              $\{(4.5, 3), (-4.5, 3), (0, -2)\}$
C **13.** $x^2 + y^2 = 25$ $\{(4, 3), (4, -3),$ **14.** $4x^2 + y^2 = 16$ **15.** $4y = x^2 - 8$
        $4x^2 + 9y^2 = 144$ $(-4, -3), (-4, 3)\}$ $y^2 - 8x^2 = 4$ $4y^2 - x^2 = 16$
                                 $\{(1, 3.5), (1, -3.5), (-1, -3.5), (-1, 3.5)\}$

10–9 Linear-Quadratic Systems: Substitution

If a system of equations involves a linear equation and a quadratic equation, you can solve the system by substitution. In this process the quadratic equation is replaced by one involving a single variable.

EXAMPLE 1   Find the solution set of the system:

$$x^2 + 4y^2 = 29$$
$$x - 3y = 2$$

SOLUTION

1. Transform the *linear equation* to express x in terms of y.

$$x - 3y = 2$$
$$x = 3y + 2$$

2. Replace the given quadratic equation with the equation obtained from it by replacing x with "$3y + 2$."

$$x^2 + 4y^2 = 29$$
$$(3y + 2)^2 + 4y^2 = 29$$
$$9y^2 + 12y + 4 + 4y^2 = 29$$

3. Solve the new quadratic equation.

$$13y^2 + 12y - 25 = 0$$
$$(13y + 25)(y - 1) = 0$$
$$y = -\frac{25}{13}, \ y = 1$$

4. Solve two linear systems:

$x = 3y + 2$	$x = 3y + 2$
$y = -\frac{25}{13}$	$y = 1$
$x = 3(-\frac{25}{13}) + 2$	$x = 3(1) + 2$
$x = -\frac{49}{13}$	$x = 5$

∴ the solutions are $(-\frac{49}{13}, -\frac{25}{13})$ and $(5, 1)$.

5. Check each ordered pair in both equations. Checking the solutions in the first equation is done below. Checking the solutions in the second equation is left to you.

$$(-\tfrac{49}{13})^2 + 4(-\tfrac{25}{13})^2 \overset{?}{=} 29 \qquad\qquad (5)^2 + 4(1)^2 \overset{?}{=} 29$$
$$\tfrac{2401}{169} + \tfrac{2500}{169} \overset{?}{=} 29 \qquad\qquad\qquad 25 + 4 \overset{?}{=} 29$$
$$\tfrac{4901}{169} \overset{?}{=} 29 \qquad\qquad\qquad\qquad\qquad 29 = 29 \ \checkmark$$
$$29 = 29 \ \checkmark$$

∴ the solution set is $\{(-\frac{49}{13}, -\frac{25}{13}), (5, 1)\}$.

Figure 19 depicts the graphical situation in the preceding example. Figure 19a shows the graph of the original system of equations, while Figure 19b shows the result after Step 3, when the ellipse has been replaced with the two horizontal lines with equations $y = 1$ and $y = -\frac{25}{13}$.

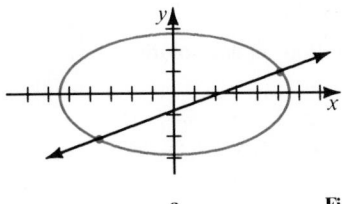

a   **Figure 19**   b

Notice also, in Example 1, that the linear equation was transformed to express x in terms of y rather than y in terms of x in order to make the resulting computation simpler.

Quadratic Relations and Systems   **425**

Suggested Assignments

Minimum
  First day
  427/1–14
  Second day
  427/P: 1–9
S 131/13–17 odd
Extended Alg. with Trig.
  First day
  424/1–14
  Second day
  427/P: 1–9
Enriched Alg.
  427/1–13 odd
  427/P: 1–11
Enriched Alg. with Trig.
  427/1–13 odd
  427/P: 1–11

Additional A Exercises

Find the solution set of each system over $\mathcal{C}$.

1. $y = x^2 + 4$
   $y = 5x$ $\{(1, 5), (4, 20)\}$

2. $x^2 - y^2 = 8$
   $x - 3y = 0$
   $\{(3, 1), (-3, -1)\}$

3. $x^2 + y^2 = 13$
   $2x + y = 4$
   $\left\{\left(\dfrac{1}{5}, \dfrac{18}{5}\right), (3, -2)\right\}$

4. $x^2 + y^2 = 6$
   $x + y = -2\sqrt{3}$
   $\{(-\sqrt{3}, -\sqrt{3})\}$

5. $x^2 - y^2 = 16$
   $3x + y = 0$
   $\{(i\sqrt{2}, -3i\sqrt{2}), (-i\sqrt{2}, 3i\sqrt{2})\}$

6. Find the dimensions of a rectangle whose area is 48 cm² and whose perimeter is 38 cm.
   16 cm × 3 cm

It is important to be aware of the fact that systems of equations involving one or more quadratic equations may have complex as well as real solutions. While graphing may be used to identify such real solutions as exist, substitution will yield complex as well as real solutions.

EXAMPLE 2   Find the solution set of the system over $\mathcal{C}$:

$$x^2 - 3y^2 = 8$$
$$x - y = 2$$

SOLUTION   Solve the linear equation for y in terms of x:

$$y = x - 2$$

Substitute $x - 2$ for y in the quadratic equation and simplify:

$$x^2 - 3(x - 2)^2 = 8$$
$$x^2 - 3x^2 + 12x - 12 = 8$$
$$2x^2 - 12x + 20 = 0$$
$$x^2 - 6x + 10 = 0$$

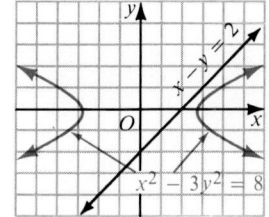

Solve for x using the quadratic formula:

$$x = \frac{-(-6) \pm \sqrt{(-6)^2 - 4(1)(10)}}{2(1)} = \frac{6 \pm 2i}{2} = 3 \pm i$$

Then since $y = x - 2$:

$$
\begin{array}{ll}
x = 3 + i & \text{or} \quad x = 3 - i \\
y = 3 + i - 2 & \quad\quad\; y = 3 - i - 2 \\
y = 1 + i & \quad\quad\; y = 1 - i
\end{array}
$$

$\therefore$ the solution set over $\mathcal{C}$ is $\{(3 + i, 1 + i), (3 - i, 1 - i)\}$.

Note that, as in Example 2, when a system has no real solutions, the graphs of the equations do not intersect and the solution set over $\mathcal{R}$ is $\emptyset$.

Oral Exercises

State whether each ordered pair in the proposed solution set is a solution of the system when it is solved over $\mathcal{C}$.

1. $x^2 + y^2 = 5;$ $\{(2i, 3), (-2i, 3), (2i, -3), (-2i, -3)\}$
   $\quad\quad y = -3$ No   No   Yes   Yes

2. $x^2 - y^2 = -24;$ $\{(5i, i), (-5i, -i), (-5i, i), (5i, -i)\}$
   $\quad\quad x = 5y$ Yes   Yes   No   No

3. $x^2 - 9y^2 = -7;$ $\{(4i, i), (-4i, i), (4i, -i), (-4i, -i)\}$
   $\quad\quad x + 4y = 0$ No   Yes   Yes   No

Written Exercises

Find the solution set of each system over $\mathcal{C}$.

A 1. $y = x^2 + 3$
$\quad y = 4x$ {(3, 12), (1, 4)}

2. $x^2 - y^2 = 15$
$\quad x - 4y = 0$ {(4, 1), (−4, −1)}

3. $x^2 + y^2 = 40$
$\quad 3x - y = -20$ {(−6, 2)}

4. $x^2 + y^2 = 41$
$\quad x + 2y = 6$

5. $x^2 - y^2 = 72$ {($i\sqrt{3}$, −5$i\sqrt{3}$),
$\quad 5x + y = 0$ (−$i\sqrt{3}$, 5$i\sqrt{3}$)}

6. $39 - x^2 = 6y$
$\quad 2x + y = 12$

B 7. $\sqrt{x^2 + y^2} = y + 2$ {(0, −1),
$\quad 2y = x - 2$ (2, 0)}

8. $\sqrt{x^2 + y^2} = 1 - y$ {(2, −$\frac{3}{2}$),
$\quad 2y + x = -1$ (−1, 0)}

9. $\dfrac{4x}{y} + \dfrac{y}{x} = 7$
$\quad x - y = 3$

10. $\dfrac{x}{y} + \dfrac{6y}{x} = \dfrac{-5}{xy}$
$\quad x - 3y = 1$

11. $x^2 + y^2 - 6xy = 28$
$\quad y + 10 = 3x$
$\quad$ {(3, −1), (−3, −19)}

12. $y^2 + x^2 - 2xy = 1$
$\quad 5x - y = 5$

13. $5\sqrt{x^2 + y^2} - \dfrac{17}{5\sqrt{x^2 + y^2}} = 0$
$\quad x + y = 1$ {($\frac{1}{5}$, $\frac{4}{5}$), ($\frac{4}{5}$, $\frac{1}{5}$)}

14. $\dfrac{13}{\sqrt{x^2 - 3y^2}} - \sqrt{x^2 - 3y^2} = 0$
$\quad 3y - x = 7$
$\quad$ {(11, 6), (−4, 1)}

C 15. Prove that the solution set of any system

$$y = ax^2 + bx + c$$
$$y = mx + k,$$

with a, b, c, m, and k all real numbers, must consist of one or two ordered pairs all of whose components are real or all of whose components are imaginary.

16. Prove that the solution set of any system

$$ax^2 + by^2 = c$$
$$y = mx + k,$$

with a, b, c, m, and k all real numbers, must consist of one or two ordered pairs all of whose components are real or all of whose components are imaginary.

Problems

3. Door: 36 cm × 36 cm; panel: 12 cm × 12 cm

A 1. Find the dimensions of a rectangle whose area is 72 cm² and whose perimeter is 44 cm. 18 cm × 4 cm

2. Find the lengths of the sides of an isosceles triangle if 10, 10, 16 the perimeter is 36 and the altitude to the base is 6.

3. A square cupboard door contains four square panels, as shown. The border around each square panel has a uniform width of 4 cm. If the total area of the door *excluding* the panels is 720 cm², what are the dimensions of the door and each panel?

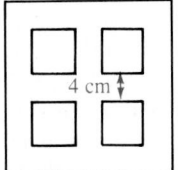

4 cm

Mixed Review

1. Find $[f \circ g](2)$ if $f(x) = x^2 + 1$ and $g(x) = 3x$. 37

2. Express $2\sqrt[3]{x} + \sqrt[6]{64x^2}$ in simplest radical form.
$2\sqrt[3]{x} + 2\sqrt[6]{x^2}$

3. Find an equation of the line perpendicular to the graph of $3x - 4y = 12$ and passing through the point (−2, 3).
$4x + 3y = 1$

Determine the solution set.

4. $x^2 - 3x - 10 = 0$ {5, −2}

5. $2x^2 - 3 = x$ $\left\{-1, \dfrac{3}{2}\right\}$

6. $(x - 3)^2 = 4$ {5, 1}

Additional Answers
Written Exercises

4. $\left\{\left(\dfrac{32}{5}, -\dfrac{1}{5}\right), (-4, 5)\right\}$

6. {(6 + $\sqrt{3}$, −2$\sqrt{3}$),
(6 − $\sqrt{3}$, 2$\sqrt{3}$)}

9. $\left\{\left(\dfrac{15 + 3\sqrt{33}}{4}, \dfrac{3 + 3\sqrt{33}}{4}\right),\right.$
$\left.\left(\dfrac{15 - 3\sqrt{33}}{4}, \dfrac{3 - 3\sqrt{33}}{4}\right)\right\}$

10. $\left\{\left(\dfrac{2}{5} + \dfrac{9}{5}i, -\dfrac{1}{5} + \dfrac{3}{5}i\right),\right.$
$\left.\left(\dfrac{2}{5} - \dfrac{9}{5}i, -\dfrac{1}{5} - \dfrac{3}{5}i\right)\right\}$

12. $\left\{\left(\dfrac{3}{2}, \dfrac{5}{2}\right), (1, 0)\right\}$

15. By subst., $mx + k = ax^2 + bx + c$, or $ax^2 + (b - m)x + (c - k) = 0$, a quad. eq. with at most two roots; 2 real, 2 imag., or 1 real. The solution of $y = mx + k$ is real if x is real, and is imag. if x is imag. ∴ x, y are both real or both imag. and system has 1 or 2 roots.

(continued)

427

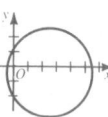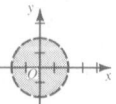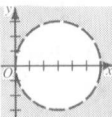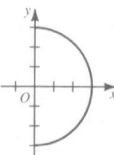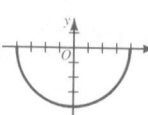
4. A square piece of metal is cut to form an isosceles trapezoid with legs of length 25 cm, as shown. If the perimeter of the trapezoid is 84 cm, find the length of a side of the original square. 24 cm

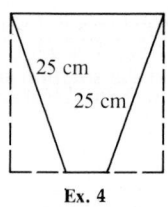

Ex. 4

5. If an object moves with a constant acceleration a, then the distance d that it travels in any interval of time is given by

$$d = \frac{v_2^2 - v_1^2}{2a},$$

where v_1 is its initial (starting) velocity and v_2 is its terminal (final) velocity. The object's average velocity $\bar{v}$ is given by

$$\bar{v} = \frac{v_1 + v_2}{2}.$$

Determine the initial and terminal velocities of a ball that travels 20 m with an average velocity of 14 m/s, accelerating at 9.8 m/s² (the acceleration of gravity on Earth). $v_1 = 7$ m/s, $v_2 = 21$ m/s

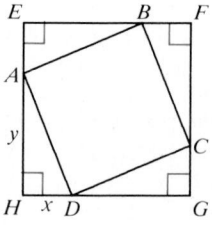

Ex. 6

6. Square $ABCD$ is inscribed in square $EFGH$ forming four congruent right triangles, as shown. If the perimeter of the square $EFGH$ is 64 and the area of square $ABCD$ is 130, determine the values of x and y. 7 and 9

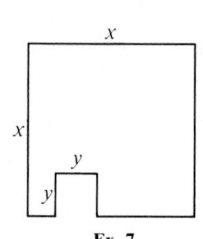

Ex. 7

B

7. An architectural plan calls for a square room with a square cutout in one wall to create space for a closet in an adjoining room. The room shown has 21 m² of floor space and a perimeter of 24 m. Determine the values of x and y. $x = 5$ m, $y = 2$ m

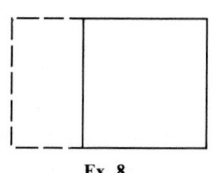

Ex. 8

8. A rectangle is cut to form a square, as shown. The perimeter of the rectangle is 54 cm, and the area of the cut-off piece is 36 cm². What are the dimensions of the original rectangle? 12 cm × 15 cm or 1.5 cm × 25.5 cm

9. A fishing pier is to be built in the shape of a square surmounted by a right triangle as shown in the diagram. If the area of the pier is to be 1050 m² and the part of its perimeter along the water is to be 100 m long, what are the dimensions of the pier?
$x = 30$ m, $y = 10$ m

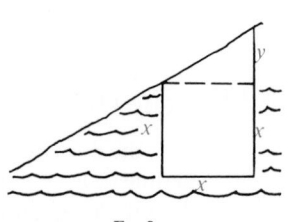

Ex. 9

C

10. A parabola that has its vertex on the line $y = 3x + 1$ and whose equation is of the form $y = 3(x - h)^2 + k$ passes through the point $(1, 10)$. What are the equations of all such parabolas?

11. An isosceles triangle has a perimeter of 50 cm. The sum of the length of the base and the height of the triangle is 31 cm. Find the area of the triangle. 105 cm² or 120 cm²

10. $y = 3(x - 2)^2 + 7$ or $y = 3(x + 1)^2 - 2$

Careers

Civil Engineering

Civil engineers use mathematics and engineering concepts to design roads, bridges, airports, water systems and other similar structures.

The three major areas of specialization in civil engineering are transportation, hydraulic, and structural engineering. Transportation engineers concentrate on the various structures used in transporting goods and people, such as roads, railways, and bridges. Hydraulic engineers work on flood control structures, irrigation systems, and canal and water systems. Structural engineers specialize in building tunnels and structures where a study of stress factors, and cost and safety is necessary.

EXAMPLE   A civil engineer wants to design an underpass in the form of a parabolic arch that is 18 ft high and 40 ft wide, as shown. What is the maximum height of a truck that can drive under the arch? Allow a tolerance of 1 ft for the height, and assume that a truck is 8 ft wide and stays 1 ft to the right of the center line.

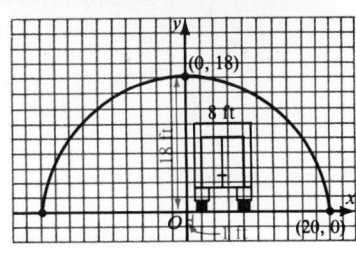

SOLUTION   Set a coordinate system on the arch such that the point on the road directly below the vertex of the parabola is the origin. Substitute the points $(20, 0)$ and $(0, 18)$ into the equation $y = a(x - h)^2 + k$ to obtain

$$a = -\frac{9}{200}.$$

Thus the equation of the parabola is

$$y = -\frac{9}{200}x^2 + 18$$

and the height of the underpass 9 ft from the center line is

$$y = -\frac{9}{200}(9)^2 + 18$$
$$\approx 14.36$$

∴ allowing a tolerance of 1 ft, the maximum height of a truck that can drive under the arch is 13.36 ft.

Quadratic Relations and Systems   **429**

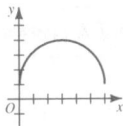

31.

32.

35. The circle has center $(0, 4)$. Midpt. of chord is $(4, 6)$.
   a. Slope of radius is
   $$\frac{6 - 4}{4 - 0} = \frac{1}{2}; \text{ slope of}$$
   chord is $\frac{4 - 8}{5 - 3} = -2;$
   $$\frac{1}{2}(-2) = -1; \therefore \perp.$$
   b. Slope of chord $= -2$.
   The radius passes through $(0, 4)$ and has slope $\frac{1}{2}$; its equation
   is $x - 2y = -8$.
   $(4, 6)$ satisfies the equation; ∴ radius passes through midpt.

37. $\sqrt{(x - 6)^2 + y^2} = 2\sqrt{(x - 3)^2 + y^2};$
   squaring and simplifying gives $x^2 + y^2 - 4x = 0$, or $(x - 2)^2 + y^2 = 4$, which is a circle. Center is $(2, 0)$; $r = 2$.

38. $x^2 + y^2 + ax + by + c = 0$ is equiv. to $\left(x + \frac{a}{2}\right)^2 +$
   $\left(y + \frac{b}{2}\right)^2 = -c + \frac{a^2}{4} + \frac{b^2}{4}.$
   Thus $r^2 = -c + \frac{a^2}{4} + \frac{b^2}{4} =$
   $\frac{a^2 + b^2 - 4c}{4}$, and
   $r = \sqrt{\frac{a^2 + b^2 - 4c}{4}}$, a
   nonneg. number.
   ∴ $a^2 + b^2 - 4c \geq 0.$

Supplementary Material

Test 32

Key Ideas

Use substitution to solve
quadratic-quadratic systems
of equations.

Chalkboard Examples

1. Find the solution set of
   the system:
   $$x^2 + 4y^2 = 17$$
   $$3x^2 - y^2 = -1$$
   Solve for y^2 in terms of x.
   $$y^2 = 3x^2 + 1$$
   Substitute.
   $$x^2 + 4(3x^2 + 1) = 17$$
   $$x^2 + 12x^2 + 4 = 17$$
   $$13x^2 = 13$$
   $$x^2 = 1$$
   $$x = \pm1$$
   $$y^2 = 3x^2 + 1$$
   Using $x = 1$: $y^2 = 3(1)^2 + 1$
   $$y^2 = 4$$
   $$y = \pm2$$
   Using $x = -1$: $y^2 = 3(1)^2 + 1$
   $$y^2 = 4$$
   $$y = \pm2$$
   $$\{(1, 2), (1, -2), (-1, 2), (-1, -2)\}$$

Mixed Review

1. Factor $9x^2 - 30xy + 25y^2$
   completely. $(3x - 5y)^2$
2. Solve $x^2 + 6x + 12 = 0$
   over $\mathcal{C}$. $\{-3 \pm i\sqrt{3}\}$
3. Simplify $\sqrt{-12}\sqrt{-3}$. -6
4. Solve $\sqrt{3 - 2x} = \sqrt{3} + x +$
   3 over $\mathcal{R}$. $\{-3\}$
5. Evaluate:
$$\begin{vmatrix} 3 & 0 & 0 \\ 0 & 0 & -2 \\ 0 & -4 & 1 \end{vmatrix} \quad -24$$

10–10 Quadratic-Quadratic Systems

Substitution often provides a means of solving systems of two quadratic
equations in two variables.

EXAMPLE 1 Find the solution set of the system:

$$4x^2 + y^2 = 25$$
$$2x^2 - y^2 = -1$$

SOLUTION

1. Solve the first equation for y^2
   in terms of x.

   $$4x^2 + y^2 = 25$$
   $$y^2 = 25 - 4x^2$$

2. Replace y^2 in the second equation with "$25 - 4x^2$" and
   simplify.

   $$2x^2 - (25 - 4x^2) = -1$$
   $$2x^2 + 4x^2 - 25 = -1$$
   $$6x^2 = 24$$
   $$x^2 = 4$$

3. Solve the resulting equation. $x = 2$ or $x = -2$

4. Solve two simple systems.

$y^2 = 25 - 4x^2$	$y^2 = 25 - 4x^2$
$x = 2$	$x = -2$
$y^2 = 25 - 4(2)^2$	$y^2 = 25 - 4(-2)^2$
$y^2 = 9$	$y^2 = 9$
$y = 3$ or $y = -3$	$y = 3$ or $y = -3$
$\{(2, 3), (2, -3)\}$	$\{(-2, 3), (-2, -3)\}$

5. Checking the solutions in the original system is left to you.

   $\therefore$ the solution set is $\{(2, 3), (2, -3), (-2, 3), (-2, -3)\}$.

Figure 20 shows that the result of Step 3 is the replacement of the
hyperbola having equation $2x^2 - y^2 = -1$ with the two straight lines
having equations $x = 2$ and $x = -2$.

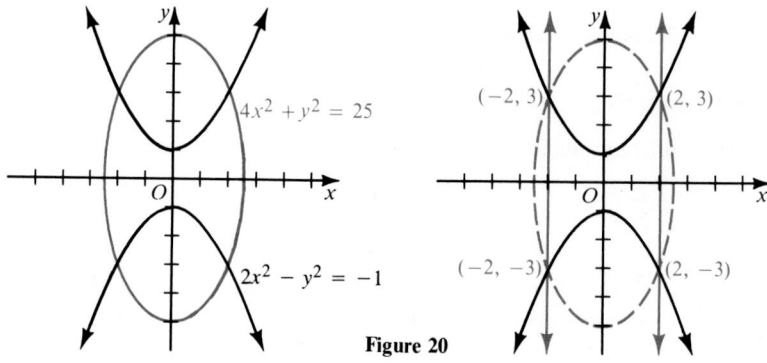

Figure 20

Linear combinations (Section 4-2) of quadratic equations can also be used to find solution sets of systems of such equations.

EXAMPLE 2   Find the solution set of the system:

$$x^2 + 9y^2 = 25 \qquad (1)$$
$$x^2 + y^2 = 17 \qquad (2)$$

SOLUTION   Subtracting the equations produces the following:

$$8y^2 = 8$$
$$y^2 = 1.$$
$$y = \pm 1$$

This equation can be used together with Equation (2) to form the equivalent system:

$$x^2 + y^2 = 17 \qquad \text{or} \qquad x^2 + y^2 = 17$$
$$y = 1 \qquad\qquad\qquad y = -1$$

Substituting 1 and −1 for y in turn in Equation (2) then produces values for x.

$x^2 + 1^2 = 17$	$x^2 + (-1)^2 = 17$
$x^2 = 16$	$x^2 = 16$
$x = 4$ or $x = -4$	$x = 4$ or $x = -4$
$\{(4, 1), (-4, 1)\}$	$\{(4, -1), (-4, -1)\}$

Checking the solutions in both equations is left to you.

∴ the solution set is $\{(4, 1), (-4, 1), (4, -1), (-4, -1)\}$.

Oral Exercises

Express y^2 in terms of x in each of the following equations.

1. $x^2 + y^2 = 22$ $y^2 = 22 - x^2$

2. $5x - 3y^2 = 2$ $y^2 = \frac{5x - 2}{3}$

3. $9y^2 - x^2 = 15$ $y^2 = \frac{x^2 + 15}{9}$

4. $x^2 + 6y^2 - 5 = 0$ $y^2 = \frac{5 - x^2}{6}$

5. $4x^2 + 8y^2 = 2$ $y^2 = \frac{1 - 2x^2}{4}$

6. $-5x^2 - 7y^2 = 36$ $y^2 = \frac{-5x^2 - 36}{7}$

Written Exercises

Find the solution set over $\mathcal{C}$.

A   1. $x^2 + 16y^2 = 25$
       $x^2 - 2y^2 = 7$

2. $2x^2 + y^2 = 33$
    $x^2 + 3y^2 = 79$

3. $x^2 - 9y^2 = 7$
    $3x^2 + 4y^2 = 52$

4. $x^2 + 4y^2 = 16$
    $4x^2 - 5y^2 = 1$

5. $x^2 + y^2 = 4$
    $5x^2 + 6y^2 = 8$

6. $x^2 - 4y^2 = 40$
    $x^2 + 9y^2 = 66$

7. $y^2 - 9x^2 = 36$
    $2x^2 + y^2 = 3$

8. $y^2 - 3x^2 = 23$
    $x^2 - 16y^2 = 8$

9. $7x^2 + 9y^2 = 22$
    $3x^2 - 7y^2 = -15$

Quadratic Relations and Systems   **431**

Suggested Assignments
Minimum
  First day
    431/1–15
  Second day
    432/P: 1–7
R 433/*Self-Test 3*
Extended Alg. with Trig.
  First day
    431/1–16
  Second day
    432/P: 1–7
R 433/*Self-Test 3*
Enriched Alg.
    431/1–13 odd, 14–16
    432/P: 3–7
R 433/*Self-Test 3*
Enriched Alg. with Trig.
    431/1–13 odd, 14–16
    432/P: 3–7
R 433/*Self-Test 3*

Additional Answers
Written Exercises

1. $\{(3, 1), (3, -1), (-3, 1), (-3, -1)\}$

2. $\{(2, 5), (2, -5), (-2, 5), (-2, -5)\}$

3. $\{(4, 1), (4, -1), (-4, 1), (-4, -1)\}$

4. $\{(2, \sqrt{3}), (2, -\sqrt{3}), (-2, \sqrt{3}), (-2, -\sqrt{3})\}$

5. $\{(4, 2i\sqrt{3}), (4, -2i\sqrt{3}), (-4, 2i\sqrt{3}), (-4, -2i\sqrt{3})\}$

6. $\{(4\sqrt{3}, \sqrt{2}), (4\sqrt{3}, -\sqrt{2}), (-4\sqrt{3}, \sqrt{2}), (-4\sqrt{3}, -\sqrt{2})\}$

7. $\{(i\sqrt{3}, 3), (i\sqrt{3}, -3), (-i\sqrt{3}, 3), (-i\sqrt{3}, -3)\}$

8. $\{(2i\sqrt{2}, i), (2i\sqrt{2}, -i), (-2i\sqrt{2}, i), (-2i\sqrt{2}, -i)\}$

9. $\left\{\left(\frac{1}{2}, \frac{3}{2}\right), \left(\frac{1}{2}, -\frac{3}{2}\right), \left(-\frac{1}{2}, \frac{3}{2}\right), \left(-\frac{1}{2}, -\frac{3}{2}\right)\right\}$

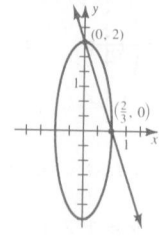
10. $25x^2 + 4y^2 = 99$
$3x^2 - 5y^2 = -21$

11. $x^2 + y^2 = 10$
$y = x^2 - 4$

12. $9x^2 + y^2 = 27$
$y = 3x^2 - 3$

B **13.** $x^2 + y^2 = 12$
$x = 3y^2 + 16$

14. $y^2 - 8x^2 = 9$
$y = 4x^2 - 3$

15. $x^2 + y^2 = 15$
$xy = 6$

C **16.** Use determinants (or linear combinations) to show that if the system

$$ax^2 + by^2 = 1 \qquad (ad - bc \neq 0)$$
$$cx^2 + dy^2 = 1$$

has four real solutions, then they lie on the graph of the circle

$$x^2 + y^2 = \frac{a + d - (b + c)}{ad - bc}.$$

Problems

A **1.** A rectangular flower bed that measures 13 m along a diagonal has an area of 60 m². What are its dimensions? 5 m × 12 m

2. What are the dimensions of a rectangular television screen with a diagonal of length 50 cm and an area of 1200 cm²? 40 cm × 30 cm

3. Find the coordinates of all points that are 10 units from the origin and 17 units from the point $(0, -9)$.
(8, 6) and (−8, 6)

4. In the diagram, point A on the circle with center at O and point B on tangent $\overline{BC}$ are equidistant from point C. If $AD = 6\sqrt{2}$ and $BD = 3\sqrt{10}$, find AC and DC.
$AC = 3$, $DC = 9$

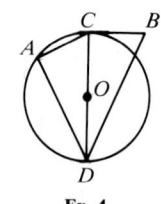

Ex. 4

B **5.** Find the dimensions of a right triangle whose area is 60 cm² and whose perimeter is 40 cm. 8 cm, 15 cm, 17 cm

6. A building complex consisting of three square buildings is to be built around a courtyard in the shape of an isosceles triangle of altitude 60 m. If the combined area of the ground floors of the three buildings is to be 45,600 m², what should the dimensions x and y be?
$x = 80$ m, $y = 100$ m

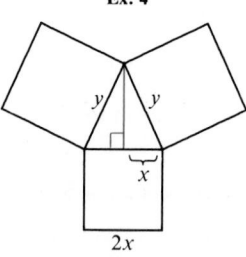

Ex. 6

7. Each lateral edge of a square pyramid is twice as long as the altitude of the pyramid. The slant height of the pyramid is 5 cm. Find the length of the altitude and a base edge. Altitude: $\sqrt{10}$ cm; base edge: $2\sqrt{15}$ cm

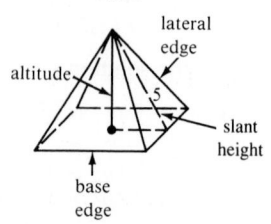

Ex. 7

432 Chapter 10

Self-Test 3

1. Solve the system over $\mathcal{R}$ by graphing: $\quad 9x^2 + y^2 = 4$ $\{(\frac{2}{3}, 0), (0, 2)\}$ $\qquad$ *Obj. 1, p. 423*
$$3x + y = 2$$

Solve each system over $\mathcal{C}$.

2. $4x^2 + 9y^2 = 100$ $\qquad\qquad$ 3. $x^2 + 3y^2 = 4$ $\qquad\qquad$ *Obj. 2, p. 423*
$\quad\; x^2 - y^2 = 12$ $\qquad\qquad\qquad\;\; x - y = 10$
$\{(4, 2), (4, -2), (-4, 2), (-4, -2)\}$

Check your answers with those at the back of the book.

3. $\{(\frac{15 + i\sqrt{71}}{2}, \frac{-5 + i\sqrt{71}}{2}), (\frac{15 - i\sqrt{71}}{2}, \frac{-5 - i\sqrt{71}}{2})\}$

Chapter Summary

1. The distance from point $P_1(x_1, y_1)$ to point $P_2(x_2, y_2)$ is given by the formula
$$P_1P_2 = \sqrt{(x_2 - x_1)^2 + (y_2 - y_1)^2}.$$

2. The midpoint (M) of the segment with endpoints $P_1(x_1, y_1)$ and $P_2(x_2, y_2)$ is
$$M\left(\frac{x_1 + x_2}{2}, \frac{y_1 + y_2}{2}\right).$$

3. If m_1 and m_2 are the slopes of nonvertical lines L_1 and L_2, respectively, then L_1 and L_2 are perpendicular if and only if
$$m_1m_2 = -1.$$

4. An equation of the circle with center (h, k) and radius r is
$$(x - h)^2 + (y - k)^2 = r^2.$$

5. A *parabola* whose equation is of the form
$$y = a(x - h)^2 + k \qquad \text{or} \qquad x = a(y - k)^2 + h$$
has vertex $V(h, k)$ and axis of symmetry
$$x = h \qquad \text{or} \qquad y = k.$$

6. The graph of an equation of the form
$$\frac{(x - h)^2}{a^2} + \frac{(y - k)^2}{b^2} = 1 \qquad \text{or} \qquad \frac{(x - h)^2}{b^2} + \frac{(y - k)^2}{a^2} = 1$$
(where $a > b$) will be an *ellipse* with center at (h, k) and whose axes are parallel to the coordinate axes.

Quadratic Relations and Systems **433**

Additional A Exercises
Find the solution set over $\mathcal{C}$.

1. $x^2 + 4y^2 = 13$
$\quad x^2 - y^2 = 8$
$\{(3, 1), (-3, 1), (3, -1),$
$(-3, -1)\}$

2. $y^2 - 4x^2 = 16$
$\quad 2x^2 + y^2 = 16$
$\{(0, 4), (0, -4)\}$

3. $\quad x^2 - 4y^2 = 9$
$\quad 4x^2 + 9y^2 = 36$
$\{(3, 0), (-3, 0)\}$

4. $\quad y^2 - x^2 = 3$
$\quad x^2 - 9y^2 = 5$
$\{(2i, i), (2i, -i),$
$(-2i, i), (-2i, -i)\}$

5. $3x^2 + 2y^2 = 21$
$\quad 2x^2 - 5y^2 = -24$
$\{(\sqrt{3}, \sqrt{6}), (-\sqrt{3}, \sqrt{6}),$
$(\sqrt{3}, -\sqrt{6}), (-\sqrt{3}, -\sqrt{6})\}$

6. A rectangular flower bed that measures 5 m along a diagonal has an area of 12 m². What are its dimensions? $\quad$ 3 m × 4 m

Quick Quiz

1. Solve the system over $\mathcal{R}$ by graphing:
$\quad 4x^2 + y^2 = 16$
$\quad\; 2x + y = 4$

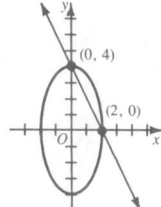

Solve each system over $\mathcal{C}$.

2. $x^2 + 5y^2 = 45$
$\qquad\quad 2y = 1 - x$
$\left\{(5, -2), \left(-\frac{35}{9}, \frac{22}{9}\right)\right\}$

3. $9x^2 + 4y^2 = 100$
$\quad\; y^2 - x^2 = 12$
$\{(2, 4), (2, -4),$
$(-2, 4), (-2, -4)\}$

14.

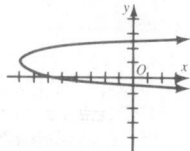

15. $x = \frac{1}{3}(y - 1)^2 + 3$

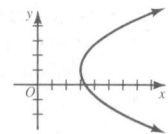

16.

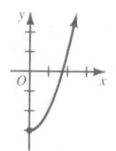

18.

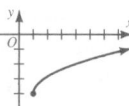

20. $y = \frac{1}{4}x^2 + x + 6$

28. $P(x, y)$ on parabola;
$D(x, -m)$ on directrix.
$PF = PD$; thus
$\sqrt{(x - 0)^2 + (y - m)^2} = |y - (-m)|$, $x^2 + (y - m)^2 = (y + m)^2$; simplifying gives $x^2 = 4my$,
or $y = \frac{1}{4m}x^2$.

29. $P(x, y)$ on parabola;
$D(-m, y)$ on directrix.
$PF = PD$; thus
$\sqrt{(x - m)^2 + (y - k)^2} = |x - (-m)|$, $(x - m)^2 + (y - k)^2 = (x + m)^2$;
simplifying gives $(y - k)^2 = 4mx$, or $x = \frac{1}{4m}(y - k)^2$.

30. $V(h, k)$ is vertex;
$F(h, k + m)$ is focus; $P(x, y)$ on parabola; $D(x, k - m)$ on directrix. $PF = PD$; thus
$\sqrt{(x - h)^2 + (y - (k + m))^2} = |y - (k - m)|$. Squar-

434

---

7. The graph of an equation of the form

$$\frac{(x - h)^2}{a^2} - \frac{(y - k)^2}{b^2} = 1 \quad \text{or} \quad \frac{(y - k)^2}{a^2} - \frac{(x - h)^2}{b^2}$$

will be a *hyperbola* with center at (h, k).

8. A function specified by an equation of the form $xy = k$, $k \neq 0$, is called an *inverse variation*. You say that x and y *vary inversely* as each other or are *inversely proportional* to each other.

  In a function specified by an equation of the form $z = kxy$, you say that z *varies jointly* as x and y.

9. The points of intersection of the graphs of the equations of a system represent the real solutions of the system. A linear-quadratic system may have as many as two real solutions; a quadratic-quadratic system may have as many as four real solutions. These systems may also have complex solutions.

Chapter Review

Write the letter of the correct answer.

1. Determine the distance between $(4, -1)$ and $(3, 7)$.   *10-1*
   a. $\sqrt{5}$ (b.) $\sqrt{65}$ c. $\sqrt{113}$ d. $\sqrt{37}$

2. Find the coordinates of the midpoint of the line segment with endpoints $(-a, -b)$ and $(a, -b)$.
   a. $\left(-\frac{a}{2}, \frac{b}{2}\right)$ b. $(0, b)$ (c.) $(0, -b)$ d. $\left(\frac{a}{2}, \frac{b}{2}\right)$

3. Determine an equation of the line containing $(3, -2)$ and perpendicular to the graph of $5x + 2y = 7$.   *10-2*
   a. $2x - 5y = -19$ b. $5x + 2y = -4$
   c. $2x + 5y = -4$ (d.) $2x - 5y = 16$

4. Determine an equation of the circle with center at $(1, -2)$ and containing the point $(5, 1)$.   *10-3*
   a. $x^2 + y^2 - 8x + 2y = -8$ b. $x^2 + y^2 + 2x - 4y = 25$
   c. $x^2 + y^2 + 2x - 8y = -4$ (d.) $x^2 + y^2 - 2x + 4y = 20$

5. Determine an equation of the parabola with focus $F(0, 2)$ and directrix $y = -2$.   *10-4*
   (a.) $x^2 = 8y$ b. $x^2 = -8y$ c. $x^2 - 9x = 3y$ d. $y^2 = -8x$

6. Determine an equation of an ellipse with major axis of length 8 and y-intercepts 2 and -2.   *10-5*
   (a.) $\frac{x^2}{16} + \frac{y^2}{4} = 1$ b. $\frac{x^2}{4} + \frac{y^2}{16} = 1$ c. $\frac{x^2}{16} - \frac{y^2}{4} = 1$ d. $\frac{y^2}{16} - \frac{x^2}{16} = 1$

7. Determine the equations of the asymptotes of the hyperbola described by $36x^2 - 4y^2 = 144$.   10-6

(a.) $y = 3x$ $y = -3x$ b. $y = \frac{1}{9}x$ $y = -\frac{1}{9}x$

c. $y = \frac{1}{4}x$ $y = -\frac{1}{4}x$ d. $y = 4x$ $y = -4x$

8. If x varies inversely as y^2, and $x = 3$ when $y = 2$ find x when $y = \frac{1}{4}$.   10-7

a. 96   b. 24   c. 48   (d.) 192

9. Determine the number of points of intersection.   10-8
$$y = x^2 + 4$$
$$8y = x^2 + y^2$$

a. 0   b. 1   (c.) 2   d. 4

10. Solve over $\mathcal{C}$: $y = x^2 + 5$ 10-9
$$y = 4x$$

(a.) $\{(2 + i, 8 + 4i), (2 - i, 8 - 4i)\}$

b. $\{(2, 8), (-2, -8)\}$

c. $\{(3, 12), (-3, -12)\}$

d. $\{(3 + i, 12 + 4i), (3 - i, 12 - 4i)\}$

11. Solve over $\mathcal{C}$: $x^2 + y^2 = 25$ 10-10
$$x^2 + 4y^2 = 46$$

a. $\{(3\sqrt{2}, 3), (-3\sqrt{2}, 3), (3\sqrt{2}, -3), (-3\sqrt{2}, -3)\}$

b. $\{(3, \sqrt{7}), (-3, \sqrt{7}), (3, -\sqrt{7}), (-3, -\sqrt{7})\}$

(c.) $\{(3\sqrt{2}, \sqrt{7}), (-3\sqrt{2}, \sqrt{7}), (3\sqrt{2}, -\sqrt{7}), (-3\sqrt{2}, -\sqrt{7})\}$

d. $\{(3, 3), (3, -3), (-3, 3), (-3, -3)\}$

Chapter Test

1. Determine the coordinates of A if the midpoint of $\overline{AB}$ is the point $(4, 5)$ and B is the point $(6, 1)$. (2, 9)   10-1

2. Determine an equation of the line containing $(2, 3)$ and perpendicular to the line passing through $(3, 4)$ and $(5, 6)$. $x + y = 5$ 10-2

3. Determine the center and radius of the circle with equation $x^2 + y^2 - 6x + 2y = 6$. $C(3, -1); r = 4$ 10-3

4. Sketch the graph of $x = y^2 + 6y + 11$. Identify the vertex and the axis of symmetry. $V(2, -3)$; axis: $y = -3$ 10-4

5. Sketch the graph of $\dfrac{(x + 2)^2}{25} + \dfrac{y^2}{9} = 1$.   10-5

6. Sketch the graph of $9x^2 - 4y^2 = 36$. Label the asymptotes on the graph and give the equations of the asymptotes. $y = \frac{3}{2}x$, $y = -\frac{3}{2}x$ 10-6

7. If x varies directly as y^2 and inversely as z, and $x = 2$ when $y = 3$ and $z = 9$, find the value of x when $y = 2$ and $z = 8$. 1   10-7

Quadratic Relations and Systems   **435**

ing both sides and simplifying gives $(x - h)^2 = 4my - 4mk$, or $y = \dfrac{1}{4m}(x - h)^2 + k$.

31. By Ex. 30, the equation of the parabola is $y = \dfrac{1}{4m}(x - h)^2 + k$. Subst. in $(h + 2m, k + m)$ gives $k + m = m + k$. Subst. in $(h - 2m, k + m)$ gives $k + m = m + k$. ∴ points are on parabola $P_1P_2 = |(h + 2m) - (h - 2m)| = |4m| = 4|m|$.

Additional Answers
Chapter Test

4.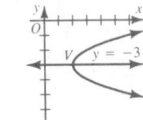

5.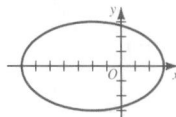

6.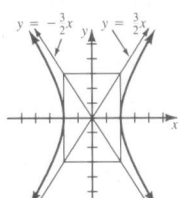

435

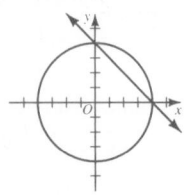

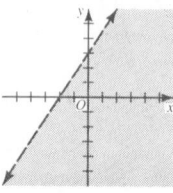
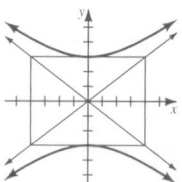
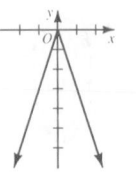
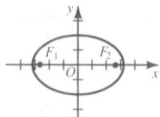
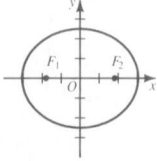
8. Solve over $\mathcal{R}$ by graphing: $x^2 + y^2 = 16$ $\qquad$ *10-8*
$x + y - 4 = 0$ {(4, 0), (0, 4)}

9. Solve over $\mathcal{C}$ by substitution: $x^2 - 2y^2 = 7$ $\qquad$ *10-9*
$x - y = 1$ {(2 + $i\sqrt{5}$, 1 + $i\sqrt{5}$), (2 − $i\sqrt{5}$, 1 − $i\sqrt{5}$)}

10. Solve over $\mathcal{C}$: $4x^2 + y^2 = 52$ $\qquad$ *10-10*
$x^2 - y^2 = -7$ {(3, 4), (3, −4), (−3, 4), (−3, −4)}

Mixed Review

Simplify.

2. $-\dfrac{m^8 n^7}{2a^6 b^3}$ **4.** $\dfrac{7x + 2y}{3(x - y)^2}$ **10.** $\varnothing$

1. $[(5 \cdot 3)^2 + (4 \cdot 3)^2 - (2 \cdot 3)^2] \div [3^2 \cdot 37]$ 1

2. $\dfrac{(-3mn^2)^2(-2m^2n)^3}{9a^2b(-4a^2b)^2}$

3. $8(u - \frac{1}{4}v) - 4(\frac{1}{2}v + 2u) - (-v + 2)$ $-3v - 2$

4. $\dfrac{x + 2y}{x^2 - 2xy + y^2} + \dfrac{4}{3x - 3y}$

5. $(2 + 7i) - (3 + 4i) + (7 - 6i)$ $6 - 3i$

6. $\dfrac{\sqrt{x} + 3\sqrt{y}}{\sqrt{x} - 3\sqrt{y}}$ $\dfrac{x + 6\sqrt{xy} + 9y}{x - 9y}$

Solve over $\mathcal{R}$.

7. $-(1 - 4c) = 2(2c - 1) + 1$ $\mathcal{R}$

8. $|2n - 3| > 7$ {$n: n < -2$ or $n > 5$}

9. $6x = x^2 - 6$ {3 + $\sqrt{15}$, 3 − $\sqrt{15}$}

10. $\dfrac{4}{a - 3} - \dfrac{2a}{9 - a^2} = \dfrac{1}{a + 3}$

11. Find all the rational roots of $x^3 - 3x^2 + x - 4 = 0$. none

12. Solve $2x^2 = -9$ over $\mathcal{C}$. {$\dfrac{3i\sqrt{2}}{2}, -\dfrac{3i\sqrt{2}}{2}$}

Graph.

13. $8 + 5t \geq 2$ or $-(8 + 5t) \geq 2$

14. $7x > 5y - 15$

15. $16y^2 = 9x^2 + 144$

16. $y = -3|x|$

17. Determine the maximum and minimum values of $2x - y$ in the region R. max.: 8; min.: -3

18. Determine the coordinates of F if $M(2, -1)$ is the midpoint of $\overline{FG}$ and G has the coordinates (4, 2). (0, −4)

19. Find the sum of the arithmetic series such that $a_1 = 3$, $a_n = -217$, and $n = 45$. -4815

20. Find two geometric means between 9 and $\frac{1}{3}$. 3, 1

21. Find the sum of the infinite geometric series $8 + 4 + 2 + 1 + \ldots$. 16

22. What is the radius of the circle with equation $4x^2 + 4y^2 = 17$? $\dfrac{\sqrt{17}}{2}$

23. Find the negative integer whose square is 45 more than 4 times the integer. -5

24. Express $\dfrac{6x^4 - 2x^3 - 6x^2 + 8x - 6}{2x - 2}$ as a sum by using division. $3x^3 + 2x^2 - x + 3$

25. Each year the value of the Thomas' truck is 80% of the value of the previous year. If the truck's value was \$9,000 at the end of the first year, what will be its value at the end of the fourth year? \$4608

436   Chapter 10

Contest Problems

1. What is the length to the nearest cm of the longest stick that can be placed inside a box that is 60 cm wide, 75 cm long and 45 cm high? 106 cm

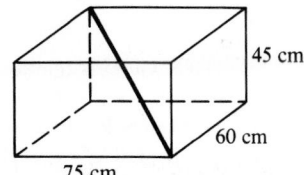

PROGRAMMING IN PASCAL

Exercises

1. Carefully study the (incomplete) program *conics*. Use *ck_ellipse* as a model and insert code for **(a)** the procedure *ck_circle* **(b)** the procedure *ck_parabola* and **(c)** the procedure *ck_hyperbola*.

(∗ This program analyzes the graph of an equation of the form ∗)
(∗ $ax^2 + cy^2 + dx + ey + f = 0$ where not both a and c are 0. ∗)

```
PROGRAM conics (INPUT, OUTPUT);

TYPE
  str9 = PACKED ARRAY[1 . . 9] OF char;

VAR
  a, c, d, e, f : real;
  kind : str9;

( ****************************************************************)
PROCEDURE get_coefficients;

BEGIN
  REPEAT
    write('Enter the values of a: '); readln(a);
    write('                    c: '); readln(c);
  UNTIL abs(a) + abs(c) > 0;
    write('                    d: '); readln(d);
    write('                    e: '); readln(e);
    write('                    f: '); readln(f);
  IF a < 0
    THEN BEGIN
      a := -a; c := -c; d := -d; e := -e; f := -f;
      END;
END;
```

(Program continues on next page.)

Quadratic Relations and Systems **437**

13.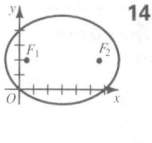
14.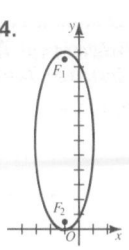

15.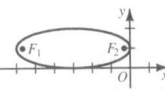

16.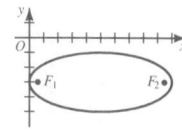

18.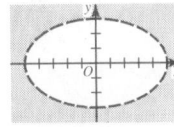

19.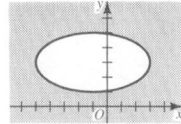

20.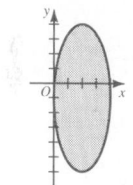

22.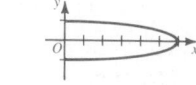

23.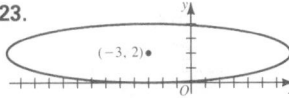

24.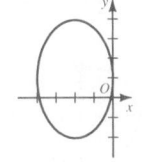

14.

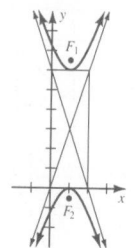

15.

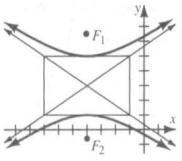

16.

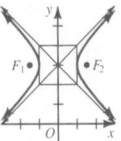

18.

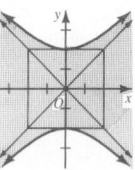

19.

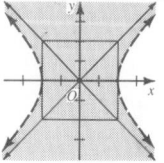

20.

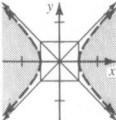

21.

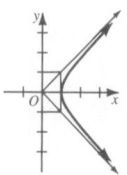

```
( ************************************************************************)
PROCEDURE determine_conic;

BEGIN
  If a = c
    THEN kind := 'circle   '
    ELSE IF a * c = 0                          (* either a or c is 0 *)
        THEN kind := 'parabola '
        ELSE IF a * c > 0                      (* a and c have the same sign *)
            THEN kind := 'ellipse '
            ELSE kind := 'hyperbola';
END;

( ************************************************************************)
PROCEDURE ck_ellipse;

VAR
  k, dist_to_focus, maj_axis_len, min_axis_len, x_ctr, y_ctr,
  top_x_vertex, top_y_vertex, btm_x_vertex, btm_y_vertex,
  left_x_vertex, left_y_vertex, rt_x_vertex, rt_y_vertex,
  x_f1, y_f1, x_f2, y_f2 : real;


  BEGIN
    x_ctr := -d / (2 * a); (* found by completing the square for x *)
    y_ctr := -e / (2 * c); (* found by completing the square for y *)

(* general equation ax² + cy² + dx + ey + f = 0 is transformed to *)
(*     a(x + d / 2a)² + c(y + e / 2c)² = -f + d² / 4a + e² / 4 c      *)
(*       let k represents the quantity to the right of the equal sign    *)

    k := sqr(x_ctr) * a + sqr(y_ctr) * c - f;
IF k < 0
  THEN writeln('No ellipse.')
  ELSE BEGIN
        IF a < c
          THEN BEGIN
                    maj_axis_len := 2 * sqrt(k / a);
                    min_axis_len := 2 * sqrt(k / c);
                    dist_to_focus := sqrt(k * (1 / a - 1 / c));

                    (* the coordinates of the vertices *)
                    top_x_vertex := x_ctr;
                    top_y_vertex := y_ctr + 0.5 * min_axis_len;
                    btm_x_vertex := x_ctr;
                    btm_y_vertex := y_ctr - 0.5 * min_axis_len;
                    left_x_vertex := x_ctr - 0.5 * maj_axis_len;
                    left_y_vertex := y_ctr;
```

438   *Chapter 10*

```
                        rt_x_vertex := x_ctr + 0.5 * maj_axis_len;
                        rt_y_vertex := y_ctr;

                        (* the coordinates of the foci *)
                        x_f1 := x_ctr − dist_to_focus;
                        y_f1 := y_ctr;
                        x_f2 := x_ctr + dist_to_focus;
                        y_f2 := y_ctr;
                   END
              ELSE BEGIN
                        maj_axis_len := 2 * sqrt(k / c);
                        min_axis_len := 2 * sqrt(k / a);
                        dist_to_focus := sqrt(k * (1 / c − 1 / a));

                        (* the coordinates of the vertices *)
                        top_x_vertex := x_ctr;
                        top_y_vertex := y_ctr + 0.5 * maj_axis_len;
                        btm_x_vertex := x_ctr;
                        btm_y_vertex := y_ctr − 0.5 * maj_axis_len;
                        left_x_vertex := x_ctr − 0.5 * min_axis_len;
                        left_y_vertex := y_ctr;
                        rt_x_vertex := x_ctr + 0.5 * min_axis_len;
                        rt_y_vertex := y_ctr;

                        (* the coordinates of the foci *)
                        x_f1 := x_ctr;
                        y_f1 := y_ctr + dist_to_focus;
                        x_f2 := x_ctr;
                        y_f2 := y_ctr − dist_to_focus;
                    END;
            writeln('Center of ellipse: (',x_ctr:3:2,',',y_ctr:3:2,')');
            writeln('Vertices:  (',top_x_vertex:3:2,',',top_y_vertex:3:2,')');
            writeln('            (',btm_x_vertex:3:2,',',btm_y_vertex:3:2,')');
            writeln('            (',left_x_vertex:3:2,',',left_y_vertex:3:2,')');
            writeln('            (',rt_x_vertex:3:2,',',rt_y_vertex:3:2,')');
            writeln('Foci:  (',x_f1:3:2,',',y_f1:3:2,')');
            writeln('        (',x_f2:3:2,',',y_f2:3:2,')');
         END;
END;


( ******************************************************************)
BEGIN (* main *)
  get_coefficients;
  determine_conic;
  IF kind = 'ellipse  '
     THEN ck_ellipse;
END.
```

22.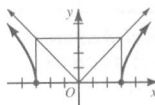

23.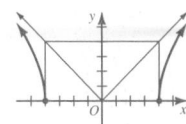

24.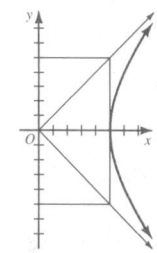

Quadratic Relations and Systems   **439**

This Egyptian mural from the twelfth dynasty (1991-1786 B.C.) has an unpainted wooden background. The approximate ages of artifacts made from organic materials can be determined using carbon dating as discussed on pages 476-478.

440   *Chapter 11*

Chapter 11

Exponents and Logarithms

Extending the Laws of Exponents

OBJECTIVES for Sections 11-1 and 11-2:
1. *To simplify an expression involving rational and real-number exponents.*
2. *To solve an equation containing a variable in an exponent.*

11–1 Rational Exponents

In Section 6-1 negative and zero exponents were defined in such a way that the laws of exponents hold for any integral exponent. Thus,

$$4^{-1} \cdot 4^2 = 4^{-1+2} = 4^1 = 4 \quad \text{and} \quad 3^{-2} \cdot 3^0 = 3^{-2+0} = 3^{-2} = \tfrac{1}{9}.$$

Similarly, we can extend the definition of an exponent to include any rational number in such a way that the laws of exponents hold. Consider, for example, $3^{\frac{1}{2}}$. For $3^{\frac{1}{2}}$ to have a meaning consistent with the familiar laws of exponents (page 200), it should be true that

$$\left(3^{\frac{1}{2}}\right)^2 = 3^{\left(\frac{1}{2} \cdot 2\right)} = 3^1 = 3.$$

Since $3^{\frac{1}{2}}$ is to denote a number whose square is 3, we define it to be $\sqrt{3}$. We choose the positive square root, $\sqrt{3}$, rather than $-\sqrt{3}$ so that the inequality

$$3^0 < 3^{\frac{1}{2}} < 3^1$$

will be true. Similar reasoning requires that

$$3^{\frac{5}{2}} = \left(3^{\frac{1}{2}}\right)^5 = (\sqrt{3})^5 \quad \text{and} \quad 3^{-\frac{5}{2}} = \left(3^{\frac{1}{2}}\right)^{-5} = (\sqrt{3})^{-5}.$$

These observations suggest the definition on the following page.

Exponents and Logarithms **441**

Problem Solving Strategies

Looking for a Pattern
Tables and graphs suggest the continuous pattern of the graph of $y = 2^x$ in Section 11-2 (page 445). In Section 11-9 students see the pattern in $\left(1 + \dfrac{1}{k}\right)^k$ as k increases and learn the meaning of *e* on page 468.
Recognizing Problem Types
Even though population and bacteria problems are different from compound interest problems, students learn in Section 11-8 that they are of the same type.
Generalizing from Specific
Students learn to abstract the notion of rational exponents in Section 11-1 (page 442) from specific examples.

Teaching Suggestions
p. T107

Key Ideas

Define and apply the relationship between rational exponents and expressions in radical form.

Reading Algebra

Students should be encouraged to verbalize exponential and radical expressions. For example, the expression "$x^{\frac{2}{3}}$" can be verbalized as "x to the two-thirds power" and "$\sqrt[5]{x^3}$" can be verbalized as "the fifth root of x to the third power."

1. Evaluate $\sqrt[3]{625} \cdot \sqrt[6]{625}$.
$$\sqrt[3]{625} \cdot \sqrt[6]{625} = (5^4)^{\frac{1}{3}} \cdot (5^4)^{\frac{1}{6}}$$
$$= 5^{\frac{4}{3}} \cdot 5^{\frac{4}{6}}$$
$$= 5^{\frac{4}{3} + \frac{2}{3}}$$
$$= 25$$

2. Evaluate $\left(-\dfrac{8}{125}\right)^{-\frac{4}{3}}$.
$$\left(-\frac{8}{125}\right)^{-\frac{4}{3}} = \left(-\frac{125}{8}\right)^{\frac{4}{3}}$$
$$= \left[\left(-\frac{125}{8}\right)^{\frac{1}{3}}\right]^4$$
$$= \left(-\frac{5}{2}\right)^4$$
$$= \frac{625}{16}$$

3. Convert $\sqrt[5]{243a^6b^{-15}}$ to exponential form and then write in simplified radical form.
$$\sqrt[5]{243a^6b^{-15}}$$
$$= 243^{\frac{1}{5}}(a^6)^{\frac{1}{5}}(b^{-15})^{\frac{1}{5}}$$
$$= 3a^{\frac{6}{5}}b^{-3}$$
$$= 3a^{\frac{5}{5}}a^{\frac{1}{5}}b^{-3}$$
$$= \frac{3a}{b^3}\sqrt[5]{a}$$

Common Errors

When evaluating expressions with rational exponents, students sometimes multiply the base by the exponent. Encourage students to work neatly so they can distinguish (on paper as well as in their minds) between multiplying *a* numbers together and raising a base to a power.

Reinforce the meaning of the denominator of the exponent (root) and the numerator of the exponent (power). Stress conversion to radical form to help students remember the correct definitions.

If p denotes an *integer*, r a *positive integer*, and b a *positive real number*, then

$$b^{\frac{p}{r}} = (\sqrt[r]{b})^p.$$

When p *as well as* r is a positive integer, we define $0^{\frac{p}{r}} = 0$.

In particular, if $p = 1$,

$$b^{\frac{1}{r}} = \sqrt[r]{b}.$$

The fact (page 267) that $\sqrt[r]{b^p} = (\sqrt[r]{b})^p$ implies that

$$(b^p)^{\frac{1}{r}} = (b^{\frac{1}{r}})^p,$$

and either side of this latter equation is thus equal to $b^{\frac{p}{r}}$.

Using powers with rational exponents, you can write radical expressions in **exponential form,** that is, as powers or products of powers. Then, because the laws of exponents apply to these powers (Exercises 45–48, page 444), you can use the laws to simplify the exponential expressions.

EXAMPLE 1 Evaluate.

a. $\sqrt[12]{8} \cdot \sqrt[4]{8}$

b. $\left(\dfrac{1}{81}\right)^{-\frac{3}{4}}$

SOLUTION a. $\sqrt[12]{8} \cdot \sqrt[4]{8} = (2^3)^{\frac{1}{12}} \cdot (2^3)^{\frac{1}{4}}$
$$= 2^{\frac{3}{12}} \cdot 2^{\frac{3}{4}}$$
$$= 2^{\frac{3}{12} + \frac{3}{4}}$$
$$= 2$$

b. $\left(\dfrac{1}{81}\right)^{-\frac{3}{4}} = (81)^{\frac{3}{4}}$
$$= (81^{\frac{1}{4}})^3$$
$$= 3^3$$
$$= 27$$

Note that in extending powers to include all rational exponents, we have restricted the base b to be a *positive* real number. Without that restriction, we could not always define b^r to be a real number. For example, $(-2)^{\frac{1}{2}}$ could not represent a real number, since there is no real number whose square is negative. In the examples and exercises that follow, assume that all variables denote positive real numbers, unless otherwise specified.

EXAMPLE 2 Convert to exponential form and then write in simplified radical form.

a. $\sqrt[4]{16x^{11}y^{-4}}$

b. $\sqrt{x} \cdot \sqrt[3]{x^2}$

SOLUTION a. $\sqrt[4]{16x^{11}y^{-4}} = (16)^{\frac{1}{4}}(x^{11})^{\frac{1}{4}}(y^{-4})^{\frac{1}{4}} = 2x^{\frac{11}{4}}y^{-1} = 2(x^{\frac{8}{4}} \cdot x^{\frac{3}{4}})y^{-1}$
$$= 2x^2y^{-1}x^{\frac{3}{4}} = \frac{2x^2}{y}\sqrt[4]{x^3}$$

b. $\sqrt{x} \cdot \sqrt[3]{x^2} = x^{\frac{1}{2}} \cdot x^{\frac{2}{3}} = x^{\frac{3}{6} + \frac{4}{6}} = x^{\frac{7}{6}} = x^{\frac{6}{6}} \cdot x^{\frac{1}{6}} = x\sqrt[6]{x}$

442 *Chapter 11*

The theorems on pages 267 and 268 can be used to develop another useful fact about radical expressions. Let b denote a positive real number, r and s be positive integers, and p and q be integers. Then the following is true.

$$\sqrt[rs]{b^{ps}} = (\sqrt[rs]{b})^{ps} = \left[\left(\sqrt[s]{\sqrt[r]{b}}\right)^s\right]^p = (\sqrt[r]{b})^p = \sqrt[r]{b^p}$$

That is,

$$\sqrt[rs]{b^{ps}} = \sqrt[r]{b^p}.$$

For example,

$$\sqrt[12]{81x^8} = \sqrt[3\cdot4]{3^{1\cdot4}x^{2\cdot4}} = \sqrt[3\cdot4]{(3x^2)^4} = \sqrt[3]{3x^2}.$$

Oral Exercises

State each expression in exponential form.

1. $\sqrt{7}$ $7^{\frac{1}{2}}$ **2.** $\sqrt[3]{x^2}$ $x^{\frac{2}{3}}$ **3.** $\sqrt[4]{y^5}$ $y^{\frac{5}{4}}$ **4.** $\dfrac{1}{\sqrt[3]{5}}$ $5^{-\frac{1}{3}}$

5. $\sqrt[5]{3x^4}$ $3^{\frac{1}{5}}x^{\frac{4}{5}}$ **6.** $\sqrt{(3x)^7}$ $3^{\frac{7}{2}}x^{\frac{7}{2}}$ **7.** $\sqrt{c+d}$ $(c+d)^{\frac{1}{2}}$ **8.** $\sqrt[4]{3(x+y)}$ $3^{\frac{1}{4}}(x+y)^{\frac{1}{4}}$

State each expression in radical form.

9. $7^{\frac{1}{3}}$ $\sqrt[3]{7}$ **10.** $2^{\frac{3}{2}}$ $\sqrt{2^3}$ **11.** $3^{\frac{1}{4}}$ $\sqrt[4]{3}$ **12.** $5^{-\frac{1}{2}}$ $\dfrac{1}{\sqrt{5}}$

13. $(6x)^{\frac{1}{2}}$ $\sqrt{6x}$ **14.** $3x^{-\frac{1}{5}}$ $\dfrac{3}{\sqrt[5]{x}}$ **15.** $(a+5)^{\frac{4}{7}}$ $\sqrt[7]{(a+5)^4}$ **16.** $(u^2+v^2)^{\frac{1}{2}}$ $\sqrt{u^2+v^2}$

17. Is it true that $(a+b)^{\frac{1}{2}} \neq a^{\frac{1}{2}} + b^{\frac{1}{2}}$ for all positive real numbers a and b? Explain. Yes

Written Exercises

Evaluate.

A **1.** $27^{\frac{1}{3}}$ 3 **2.** $8^{\frac{2}{3}}$ 4 **3.** $36^{-\frac{1}{2}}$ $\frac{1}{6}$ **4.** $64^{\frac{4}{3}}$ 256

5. $16^{\frac{3}{4}}$ 8 **6.** $32^{-\frac{3}{5}}$ $\frac{1}{8}$ **7.** $\left(\frac{1}{27}\right)^{\frac{2}{3}}$ $\frac{1}{9}$ **8.** $\left(\frac{125}{64}\right)^{-\frac{1}{3}}$ $\frac{4}{5}$

9. $0.25^{\frac{3}{2}}$ $\frac{1}{8}$ **10.** $0.008^{\frac{5}{3}}$ 0.00032 **11.** $(25+144)^{\frac{1}{2}}$ 13 **12.** $(10^2-6^2)^{-\frac{1}{2}}$ $\frac{1}{8}$

Convert to exponential form and then write in simplified radical form.

13. $\sqrt[6]{x^9}$ $x^{\frac{9}{6}}$; $x\sqrt{x}$ **14.** $\sqrt[4]{3^2y^6}$ $3^{\frac{2}{4}}y^{\frac{6}{4}}$; $y\sqrt{3y}$ **15.** $\sqrt[4]{49n^2}$ $49^{\frac{1}{4}}n^{\frac{2}{4}}$; $\sqrt{7n}$ **16.** $\sqrt[6]{\dfrac{1}{27r^{24}}}$

17. $\sqrt[8]{81}$ $81^{\frac{1}{8}}$; $\sqrt{3}$ **18.** $\sqrt[12]{27}$ $27^{\frac{1}{12}}$; $\sqrt[4]{3}$ **19.** $\sqrt{\sqrt[3]{125}}$ $125^{\frac{1}{6}}$; $\sqrt{5}$ **20.** $\sqrt[4]{\sqrt{625}}$

21. $\sqrt[6]{\dfrac{1}{1000}}$ $10^{-\frac{1}{2}}$; $\dfrac{\sqrt{10}}{10}$ **22.** $\sqrt[9]{\dfrac{125}{8}}$ $\left(\frac{125}{8}\right)^{\frac{1}{9}}$; $\dfrac{\sqrt[3]{20}}{2}$ **23.** $\sqrt[3]{\left(\dfrac{64}{27}\right)^2}$ $\left(\frac{64}{27}\right)^{\frac{2}{3}}$; $\frac{16}{9}$ **24.** $\sqrt[4]{\left(\dfrac{49}{144}\right)^{-2}}$

Exponents and Logarithms **443**

Additional A Exercises
Evaluate.

1. $49^{-\frac{1}{2}}$ $\frac{1}{7}$

2. $64^{\frac{2}{3}}$ 16

3. $0.0001^{-\frac{1}{4}}$ 10

Convert to exponential form and then write in simplified radical form.

4. $\sqrt[8]{64}$ $\sqrt[4]{8}$

5. $\sqrt[5]{-32x^{10}y^{12}}$ $-2x^2y^2\sqrt[5]{y^2}$

6. $\sqrt[6]{8}$ $\sqrt[3]{2}$

Mixed Review
Simplify.

1. $(-6)^{-2}$ $\frac{1}{36}$

2. $\left(\dfrac{3}{2}\right)^{-4}$ $\frac{16}{81}$

3. $(0.7)^2$ 0.49

4. $(-7)\frac{1}{3} + \frac{3}{4}$ $-\frac{19}{12}$

5. $\dfrac{\sqrt{162}}{6}$ $\dfrac{3\sqrt{2}}{2}$

6. $\sqrt[3]{54} + \sqrt[3]{686}$ $10\sqrt[3]{2}$

Additional Answers
Written Exercises

16. $\left(\dfrac{1}{27r^{24}}\right)^{\frac{1}{6}}$; $\dfrac{\sqrt{3}}{3r^4}$

20. $625^{\frac{1}{8}}$; $\sqrt{5}$

24. $\left(\dfrac{49}{144}\right)^{-\frac{1}{2}}$; $\dfrac{12}{7}$

Express in simplest radical form.

25. $\sqrt[6]{36} \cdot \sqrt[3]{36}$ 6   **26.** $\sqrt[8]{25} \cdot \sqrt[4]{125}$ 5   **27.** $\sqrt[3]{4} \cdot \sqrt[6]{32}$ $2\sqrt{2}$ **28.** $\sqrt[12]{3} \cdot \sqrt[12]{27}$ $\sqrt[3]{3}$

29. $\dfrac{\sqrt[8]{81}}{\sqrt[6]{3}}$ $\sqrt[3]{3}$ **30.** $\dfrac{\sqrt[4]{27}}{\sqrt[4]{3}}$ $\sqrt{3}$ **31.** $\dfrac{\sqrt[6]{25}}{\sqrt[3]{625}}$ $\frac{1}{5}$ **32.** $\dfrac{\sqrt[3]{81}}{\sqrt[12]{81}}$ 3

B   **33.** $\sqrt{\sqrt[3]{9}}$ $\sqrt[6]{3}$ **34.** $\sqrt{\sqrt[4]{27}} \cdot \sqrt[4]{3}$ 3   **35.** $\sqrt{10 \cdot \sqrt[3]{10}}$ $\sqrt[3]{100}$

36. $\sqrt[3]{\sqrt[4]{5}} \cdot \sqrt[4]{25}$ $\sqrt[4]{5}$ **37.** $\sqrt{\sqrt[6]{7} \cdot \sqrt{7}}$ $\sqrt[3]{7}$ **38.** $\sqrt{\sqrt[3]{27}} \cdot \sqrt{\sqrt{27}}$ 3

Solve over $\mathcal{R}$.

EXAMPLE $x^{\frac{2}{3}} + x^{\frac{1}{3}} - 20 = 0$

SOLUTION   Use the fact that $x^{\frac{2}{3}} = (x^{\frac{1}{3}})^2$ to factor the left side of the equation.

$$x^{\frac{2}{3}} + x^{\frac{1}{3}} - 20 = 0$$
$$(x^{\frac{1}{3}} - 4)(x^{\frac{1}{3}} + 5) = 0$$
$$x^{\frac{1}{3}} = 4 \quad \text{or} \quad x^{\frac{1}{3}} = -5$$
$$x = 64 \quad \text{or} \quad x = -125$$

We reject -125 because $(-125)^{\frac{2}{3}}$ is not defined.

$\therefore$ the solution set is $\{64\}$.

39. $\frac{1}{2}z^{-\frac{4}{5}} = 8$ $\{\frac{1}{32}\}$ **40.** $8k^{-\frac{3}{4}} = 125$ $\{\frac{16}{625}\}$ **41.** $t - 4t^{\frac{1}{2}} + 3 = 0$ $\{1, 9\}$

42. $27x^3 - 28x^{\frac{3}{2}} + 1 = 0$ $\{\frac{1}{9}, 1\}$ **43.** $u^{\frac{4}{3}} - 5u^{\frac{2}{3}} + 4 = 0$ $\{8, 1\}$ **44.** $9x^{\frac{4}{5}} - 10x^{\frac{2}{5}} + 1 = 0$ $\{\frac{1}{243}, 1\}$

Let a and b denote positive real numbers, r and s be positive integers, and p and q be integers. Use the definition of a rational exponent, the laws of integral exponents, and the theorems on pages 267 and 268 to prove each statement.

C   **45.** $b^{\frac{p}{r}} \cdot b^{\frac{q}{s}} = b^{\frac{ps+rq}{rs}}$ **46.** $(b^{\frac{p}{r}})^{\frac{q}{s}} = b^{\frac{pq}{rs}}$ **47.** $a^{\frac{p}{r}} \cdot b^{\frac{p}{r}} = (ab)^{\frac{p}{r}}$ **48.** $\dfrac{a^{\frac{p}{r}}}{b^{\frac{p}{r}}} = \left(\dfrac{a}{b}\right)^{\frac{p}{r}}$

11–2 Real-Number Exponents

In Chapter 1 we defined powers with natural-number exponents, in Chapter 6 we extended this definition to integral powers, and then in Section 11-1 we extended it to powers with rational-number exponents. Thus, we have defined such powers as

$$2^3 = 2 \cdot 2 \cdot 2 = 8, \quad 2^0 = 1, \quad 2^{-3} = \frac{1}{2^3} = \frac{1}{8}, \quad \text{and} \quad 2^{\frac{1}{2}} = \sqrt{2}.$$

Also,

$$2^{\frac{3}{2}} = \sqrt{2^3} \quad \text{and} \quad 2^{1.7} = 2^{\frac{17}{10}} = \sqrt[10]{2^{17}}.$$

444   *Chapter 11*

Now we will consider the problem of defining powers, such as $2^{\sqrt{3}}$, in which the exponent is irrational. The graph of $y = 2^x$ for selected rational values of x is shown in Figure 1.

In order for the graph of $y = 2^x$ to be represented by a smooth unbroken curve (Figure 2), it must be true that powers such as $2^{\sqrt{3}}$, in which the exponents are irrational, exist. You can see that since

$$1.5 < \sqrt{3} < 2,$$

you have

$$2^{\frac{3}{2}} < 2^{\sqrt{3}} < 2^2.$$

The power $2^{\sqrt{3}}$ can be approximated by the successive powers

$$2^1,\ 2^{1.7},\ 2^{1.73},\ 2^{1.732},\ \ldots$$

in which the exponents are rational numbers represented by taking more and more places in the decimal representing $\sqrt{3}$. Since these powers steadily increase but remain less than 2^2, it follows from the Axiom of Completeness (page 322) that they converge to a certain positive real number, called $2^{\sqrt{3}}$. To four decimal places $2^{\sqrt{3}} \approx 3.3220$.

Similar reasoning leads to the definition of b^x where b is any *positive* real number and x any irrational number. Furthermore, it can be proved that the laws of exponents continue to hold for these powers. For example,

$$(3^{\sqrt{2}})^{\sqrt{2}} = 3^{\sqrt{2}\cdot\sqrt{2}} = 3^2 = 9;$$
$$2^{1-\pi} \cdot 2^{\pi} = 2^{(1-\pi)+\pi} = 2^1 = 2.$$

The curve shown in Figure 2 continuously rises with increasing abscissa and is typical of the graph of every function of the form

$$y = b^x, \text{ where } b > 1.$$

On the other hand, the graph of

$$y = b^x, \text{ where } 0 < b < 1,$$

falls with increasing abscissa, as illustrated by the graph of the function $y = (\frac{1}{2})^x$, shown in Figure 3. Notice that this can also be described as $y = 2^{-x}$.

In either Figure 2 or Figure 3, any vertical line and any horizontal line above the x-axis intersects the graph in exactly one point. In general, you have the result shown at the top of the following page.

x	2^x
-3	$\frac{1}{8}$
-2	$\frac{1}{4}$
-1	$\frac{1}{2}$
0	1
$\frac{1}{2}$	$1.4\ldots$
1	2
$\frac{3}{2}$	$2.8\ldots$
2	4
$\frac{5}{2}$	$5.6\ldots$

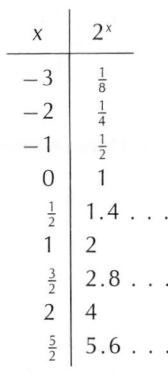

Figure 1

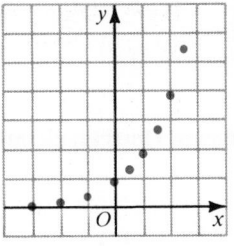

Figure 2

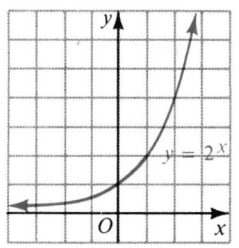

Figure 3

Exponents and Logarithms **445**

Suggested Assignments

Minimum
  First day
    446/1–20
  Second day
    R 447/*Self-Test 1*
Extended Alg. with Trig.
  First day
    446/1–19 odd
  Second day
    R 447/*Self-Test 1*
    S 111/8–18 even
Enriched Alg.
  First day
    446/7–17 odd, 18–22
  Second day
    R 447/*Self-Test 1*
    S 111/8–18 even
Enriched Alg. with Trig.
  First day
    446/7–17 odd, 18–22
  Second day
    R 447/*Self-Test 1*
    S 111/8–18 even

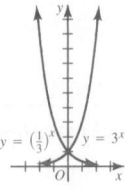
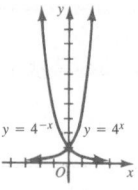
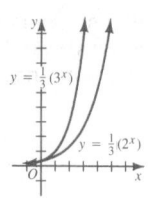
For $b > 0$, $b \neq 1$, $b^{x_1} = b^{x_2}$ if and only if $x_1 = x_2$.

EXAMPLE   Solve $4^{3r+1} = \left(\frac{1}{2}\right)^{r-5}$ over $\mathcal{R}$.

SOLUTION   1. First express each side as a power of the same base, 2.

2. Simplify each side.

3. *Equate exponents* and solve for r.

4. The check is left for you.

∴ the solution set is $\{\frac{3}{7}\}$.

$$4^{3r+1} = \left(\frac{1}{2}\right)^{r-5}$$
$$(2^2)^{3r+1} = (2^{-1})^{r-5}$$
$$2^{6r+2} = 2^{-r+5}$$
$$6r + 2 = -r + 5$$
$$7r = 3$$
$$r = \frac{3}{7}$$

Oral Exercises

State whether the inequality is true or false.

1. $3^{1.4} < 3^{\sqrt{2}}$ T

2. $7^{\pi} > 7^{4.1}$ F

3. $11^{\sqrt{5}} < 11^{\frac{5}{2}}$ T

4. $5^{\sqrt{6}} > 5^{2.9}$ F

5. $4^{3.4} > 16^{\sqrt{3}}$ F

6. $6^{2\sqrt{2}} > 36^{1.4}$ T

7. $9^{2\sqrt{2}} < 81^{1.5}$ T

8. $3^{3\sqrt{5}} < 27^{2.1}$ F

Written Exercises

Use the facts that $3^{\sqrt{2}} \approx 4.7288$ and $3^{\sqrt{3}} \approx 6.7050$ to express each of the following in the form 3^x.

A 1. $(4.7288)(6.7050)$ $3^{\sqrt{2}+\sqrt{3}}$

2. $\dfrac{6.7050}{4.7288}$ $3^{\sqrt{3}-\sqrt{2}}$

3. $(4.7288)^3$ $3^{3\sqrt{2}}$

4. $\dfrac{1}{6.7050}$ $3^{-\sqrt{3}}$

5. $(4.7288)(6.7050)^2$ $3^{\sqrt{2}+2\sqrt{3}}$

6. $\sqrt{4.7288}$ $3^{\frac{\sqrt{2}}{2}}$

7. $\dfrac{(4.7288)^3}{27}$ $3^{3\sqrt{2}-3}$

8. $\dfrac{9}{\sqrt[3]{6.7050}}$ $3^{2-\frac{\sqrt{3}}{3}}$

Solve over $\mathcal{R}$.

9. $4^x = 2^{x+3}$ $\{3\}$

10. $5^{3x-1} = 25^{x+4}$ $\{9\}$

11. $3^{x-1} = 27^{2x+3}$ $\{-2\}$

12. $125^{2x-2} = 25^{3-x}$

13. $(\frac{1}{6})^{x-3} = 6^{x-1}$ $\{2\}$

14. $10^{2x+4} = (\frac{1}{100})^{x-3}$ $\{\frac{1}{2}\}$

15. $8^{x+3} = (\frac{1}{4})^{3x-6}$ $\{\frac{1}{3}\}$

16. $(\frac{1}{32})^{x+2} = (\frac{1}{8})^{x-4}$

12. $\{\frac{3}{2}\}$

16. $\{-11\}$

Graph each pair of functions on one set of axes. Label each graph.

B 17. $y = 3^x$
       $y = (\frac{1}{3})^x$

18. $y = 4^x$
    $y = 4^{-x}$

19. $y = \frac{1}{3}(3^x)$
    $y = \frac{1}{3}(2^x)$

20. $y = 9(\frac{1}{3})^x$
    $y = 6(\frac{1}{2})^x$

C 21. Find positive irrational numbers a and b such that $2^a \cdot 2^b$ is a rational number. Answers may vary. $a = \sqrt{2}, b = 2 - \sqrt{2}$

22. Solve the system at the right graphically, approximating the solution to the nearest half unit.

$$y = 2^x$$
$$x + y = 8$$

$\{(2.5, 5.5)\}$

Self-Test 1

VOCABULARY exponential form (p. 442)

Evaluate.

1. $10,000^{\frac{3}{4}}$ 1000

2. $(\frac{9}{25})^{\frac{1}{2}}$ $\frac{3}{5}$

3. $64^{-\frac{2}{3}}$ $\frac{1}{16}$

Obj. 1, p. 441

Express in simplest radical form.

4. $\sqrt[4]{\sqrt[3]{625}}$ $\sqrt[4]{5}$

5. $\sqrt[8]{49} \cdot \sqrt[4]{7}$ $\sqrt{7}$

6. $\dfrac{\sqrt[6]{36}}{\sqrt[6]{9}}$ $\sqrt[3]{2}$

Solve over $\mathcal{R}$.

7. $81^{2x-3} = 27^{7-x}$ {3}

8. $(\frac{1}{25})^{2x+1} = (125)^{x+4}$ {−2}

Obj. 2, p. 441

Check your answers with those at the back of the book.

From Exponents to Logarithms

***OBJECTIVES** for Sections 11-3 through 11-5:*
1. To identify inverses of functions.
2. To convert equations from exponential to logarithmic form and vice versa.
3. To evaluate certain logarithms and to solve equations involving logarithms.
4. To apply the laws of logarithms.

11–3 The Inverse of a Function

In Section 3-1 you saw that two functions f and g can often be combined to form *composite* functions:

$$[f \circ g](x) = f(g(x)) \qquad [g \circ f](x) = g(f(x))$$

When f and g are related in a special way so that

$$f(g(x)) = x \text{ for all } x \text{ in the domain of } g$$

and

$$g(f(x)) = x \text{ for all } x \text{ in the domain of } f,$$

f and g are called **inverse functions**.

Exponents and Logarithms **447**

20.
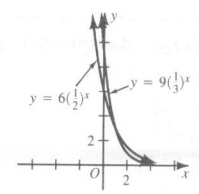
$y = 6(\frac{1}{2})^x$ $y = 9(\frac{1}{3})^x$

Quick Quiz
Evaluate.

1. $\left(-\dfrac{8}{125}\right)^{\frac{2}{3}}$ $\dfrac{4}{25}$

2. $(81)^{-\frac{3}{4}}$ $\dfrac{1}{27}$

3. $0.0064^{-\frac{1}{2}}$ $\dfrac{25}{2}$

Express in simplest radical form.

4. $\dfrac{\sqrt[4]{625}}{\sqrt[4]{25}}$ $\sqrt{5}$

5. $(\sqrt[3]{11})(\sqrt[3]{121})$ 11

6. $\sqrt[3]{\sqrt{32}}$ $\sqrt[6]{32}$

Solve over $\mathcal{R}$.

7. $32^{2x} = \left(\dfrac{1}{16}\right)^{6-3x}$ {12}

8. $7^{1-2x} = \left(\dfrac{1}{49}\right)^{5x+3}$ $\left\{-\dfrac{7}{8}\right\}$

Teaching Suggestions
p. T108

Key Ideas
Find the inverse of a function.

Reading Algebra
Be sure that students verbalize the expression $f^{-1}(x)$ as "f inverse of x" or "the inverse of f of x."

1. Show that $f(x) = -27x^3$ and $g(x) = -\frac{1}{3}\sqrt[3]{x}$ are inverse functions.

$f(g(x)) = -27\left(-\frac{1}{3}\sqrt[3]{x}\right)^3$

$= -27\left(-\frac{1}{27}x\right)$

$= x$

$g(f(x)) = -\frac{1}{3}\sqrt[3]{-27x^3}$

$= -\frac{1}{3}(-3x)$

$= x$

2. Find the inverse of the function defined by $y = \frac{3}{4}x - 6$.

Interchange x and y.

$x = \frac{3}{4}y - 6$

Solve for y.

$4x = 3y - 24$

$3y = 4x + 24$

$y = \frac{4}{3}x + 8$

Common Errors

Because the topic of inverse functions follows the study of rational exponents, it is important to clarify the meaning of "-1" in the expression "$f^{-1}(x)$". Remind students that although $x^{-1} = \frac{1}{x}$, $f^{-1}(x)$ denotes the inverse of a function, not the reciprocal of a number.

EXAMPLE 1 Show that $f(x) = 5x + 1$ and $g(x) = \frac{x-1}{5}$ are inverse functions.

SOLUTION

$$f(g(x)) = f\left(\frac{x-1}{5}\right) \qquad g(f(x)) = g(5x + 1)$$

$$= 5\left(\frac{x-1}{5}\right) + 1 \qquad = \frac{(5x+1) - 1}{5}$$

$$= (x - 1) + 1 \qquad = \frac{5x}{5}$$

$$= x \qquad = x$$

The inverse of a function f is denoted by f^{-1}, read "f inverse" or "the inverse of f." The domain and range of f^{-1} are the range and domain of f, respectively.

Not all functions have inverse functions. For example, consider the function $f(x) = x^2 + 1$. Since $f(2) = 5$ and $f(-2) = 5$, an inverse function of f would have to map 5 onto both 2 and -2. But that is impossible since a function maps each element of its domain onto exactly one element of its range.

If a function f is to have an inverse function, then not only must each element in the domain of f be paired with exactly one element in the range, but also each element in the range must be paired with exactly one element in the domain. Such a function is called a **one-to-one function.** There is a simple geometric test to determine whether a function is one-to-one.

> A function is one-to-one if and only if no horizontal line intersects the graph more than once.

Since the dashed vertical line in Figure 4 intersects the graph in more than one point, the graph does not represent a function (page 88). Although the graph in Figure 5 represents a function, the dashed horizontal line shows that the function is not one-to-one. The graph in Figure 6 represents a function that is one-to-one, since no horizontal or vertical line intersects it more than once.

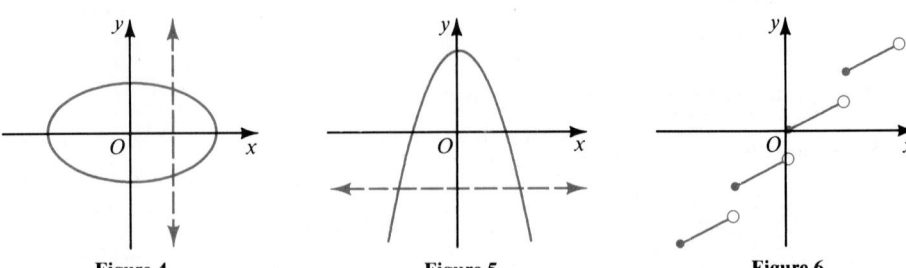

Figure 4　　　　　Figure 5　　　　　Figure 6

A function f and its inverse f^{-1} are related in the following way:

$$f^{-1}(b) = a \text{ if and only if } f(a) = b$$

Thus, when a function is defined by an equation involving x and y, you can find the rule for its inverse by interchanging x and y and then solving for y.

EXAMPLE 2   Find the inverse of the function f defined by $y = 3x - 2$.

SOLUTION   Interchange x and y. $\qquad x = 3y - 2$
Solve for y. $\qquad\qquad\qquad 3y = x + 2$
$$y = \tfrac{1}{3}(x + 2)$$

$\therefore f^{-1}(x) = \tfrac{1}{3}(x + 2)$

Figure 7 shows the graphs of

$$f(x) = 3x - 2$$

and

$$f^{-1}(x) = \tfrac{1}{3}(x + 2),$$

together with the line having equation $y = x$. Notice that f and f^{-1} appear to be symmetric with respect to the graph of $y = x$. This is because (a, b) lies on the graph of f if and only if (b, a) lies on the graph of f^{-1}.

The line with equation $y = x$ is the *line of symmetry* for the points (a, b) and (b, a) because it is the perpendicular bisector of the segment with endpoints (a, b) and (b, a). (See Written Exercise 16.) The point (b, a) is the *reflection* of (a, b) in the line with equation $y = x$. The graph of f^{-1} is the reflection of f in the line with equation $y = x$.

Although the function with rule $y = x^2 + 1$ is not one-to-one and hence does not have an inverse function, you can work with a related one-to-one function by *restricting the domain* so that either $x \geq 0$ or $x \leq 0$. Figure 8 shows the graph of the function g with rule $y = x^2 + 1$ and domain $\{x: x \geq 0\}$ and the graph of its inverse g^{-1} with rule $y = \sqrt{x - 1}$ and domain $\{x: x \geq 1\}$.

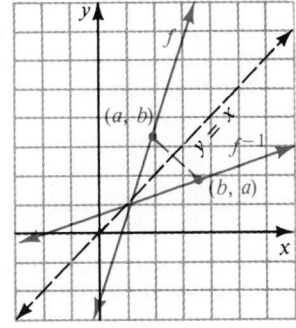

Figure 7

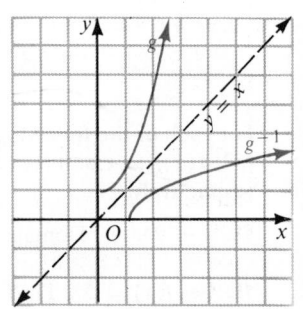

Figure 8

Oral Exercises

State the inverse of the function.

1. $f(x) = 3x$ $f^{-1}(x) = \frac{x}{3}$

4. $h(x) = \sqrt[3]{x}$ $h^{-1}(x) = x^3$

2. $g(x) = \frac{x}{7}$ $g^{-1}(x) = 7x$

5. $f(x) = x^3$ $f^{-1}(x) = \sqrt[3]{x}$

3. $g^{-1}(x) = x + 8$

6. $h^{-1}(x) = -\frac{1}{x}$

3. $g(x) = x - 8$

6. $h(x) = -\frac{1}{x}$

Exponents and Logarithms   **449**

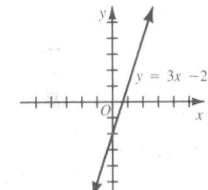
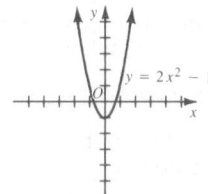
449

Suggested Assignments

Minimum
450/1–15
Extended Alg. with Trig.
450/1–9 odd, 10–16
Enriched Alg.
450/5–11 odd, 13–18
Enriched Alg. with Trig.
450/5–11 odd, 13–18

Additional Answers
Written Exercises

1. $f^{-1}(f(x)) = \dfrac{(4x - 4) + 4}{4} = \dfrac{4x}{4} = x$; $f(f^{-1}(x)) = 4\left(\dfrac{x + 4}{4}\right) - 4 = x + 4 - 4 = x$

2. $f^{-1}(f(x)) = \dfrac{(3x + 12) - 12}{3} = \dfrac{3x}{3} = x$; $f(f^{-1}(x)) = 3\left(\dfrac{x - 12}{3}\right) + 12 = x - 12 + 12 = x$

3. $f^{-1}(f(x)) = -3\left(-\dfrac{1}{3}x + 7\right) + 21 = x - 21 + 21 = x$; $f(f^{-1}(x)) = -\dfrac{1}{3}(-3x + 21) + 7 = x - 7 + 7 = x$

4. $f^{-1}(f(x)) = -4\left(-\dfrac{1}{4}x - 5\right) - 20 = x + 20 - 20 = x$; $f(f^{-1}(x)) = -\dfrac{1}{4}(-4x - 20) - 5 = x + 5 - 5 = x$

5. $f^{-1}(f(x)) = 2\sqrt[3]{\dfrac{x^3}{8}} = 2 \cdot \dfrac{x}{2} = x$; $f(f^{-1}(x)) = \dfrac{(2\sqrt[3]{x})^3}{8} = \dfrac{8x}{8} = x$

6. $f^{-1}(f(x)) = (\sqrt[3]{x + 1})^3 - 1 = x + 1 - 1 = x$; $f(f^{-1}(x)) = \sqrt[3]{(x^3 - 1) + 1} = \sqrt[3]{x^3} = x$

(continued on p. 476)

State whether the graph is the graph of a one-to-one function, a function that is not one-to-one, or a relation that is not a function.

7.
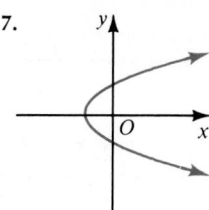
a relation that is not a function

8.
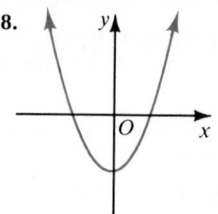
a function that is not one-to-one

9.
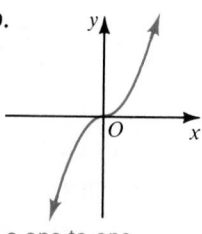
a one-to-one function

Written Exercises

For each of the following, find $f^{-1}(x)$ and show that $f^{-1}(f(x)) = x$ and $f(f^{-1}(x)) = x$.

$f^{-1}(x) = -3x + 21$

A

1. $y = 4x - 4$ $f^{-1}(x) = \dfrac{x + 4}{4}$

2. $y = 3x + 12$ $f^{-1}(x) = \dfrac{x - 12}{3}$

3. $y = -\dfrac{1}{3}x + 7$

4. $y = -\dfrac{1}{4}x - 5$ $f^{-1}(x) = -4x - 20$

5. $y = \dfrac{x^3}{8}$ $f^{-1}(x) = 2\sqrt[3]{x}$

6. $y = \sqrt[3]{x + 1}$ $f^{-1}(x) = x^3 - 1$

In Exercises 7–12 find the inverse of the given function and graph both the function and its inverse on one set of axes. **10.** $f^{-1}(x) = \sqrt{x - 3}, x \geq 3$

7. $y = 3x + 2$ $f^{-1}(x) = \dfrac{x - 2}{3}$

8. $y = \dfrac{6}{x}, x \neq 0$ $f^{-1}(x) = \dfrac{6}{x}, x \neq 0$

9. $y = x^3 - 1$ $f^{-1}(x) = \sqrt[3]{x + 1}$

10. $y = x^2 + 3$; Domain: $\{x: x \geq 0\}$

B

11. $y = \sqrt{x^2 + 9}$; Domain: $\{x: x \leq 0\}$
$f^{-1}(x) = -\sqrt{x^2 - 9}, x \geq 3$

12. $y = \sqrt{x - 2}$; Domain: $\{x: x \geq 2\}$
$f^{-1}(x) = x^2 + 2, x \geq 0$

For each of the following, find $f^{-1}(x)$, specify its domain, and show that $f^{-1}(f(x)) = x$ and $f(f^{-1}(x)) = x$ for each x in the domains of f and f^{-1}, respectively.

13. $y = 2\sqrt{x}$; Domain: $\{x: x \geq 0\}$
$f^{-1}(x) = \dfrac{x^2}{4}, x \geq 0$

14. $y = \sqrt{x - 1}$; Domain: $\{x: x \geq 1\}$
$f^{-1}(x) = x^2 + 1, x \geq 0$

15. Let $f(x) = 2^x$. Find $f^{-1}(1)$, $f^{-1}(2)$, $f^{-1}(4)$, $f^{-1}(8)$, $f^{-1}(\frac{1}{2})$, and $f^{-1}(\frac{1}{4})$ and sketch the graphs of f and f^{-1} on one set of axes. 0; 1; 2; 3; −1; −2

16. Show that the line $y = x$ is the perpendicular bisector of the segment joining (a, b) and (b, a).

In Exercises 17 and 18 draw the graph of the given relation. Does the rule specify a one-to-one function, a function that is not one-to-one, or a relation that is not a function? If it is not a one-to-one function, restrict the domain so that it will be and give a rule for the inverse.

C

17. $y = x^2 - 2x$

18. $25x^2 + 9y^2 = 225$

450 Chapter 11

11–4 The Logarithmic Function

A function defined by an equation of the form

$$y = b^x \quad (b > 0, b \neq 1)$$

is called an **exponential function with base** b. Its domain is $\mathcal{R}$ and its range is $\{y: y > 0\}$.

For example, the graph of the function f with equation

$$y = 2^x$$

is pictured in Figure 9. As x increases, y increases, and thus f is one-to-one. Hence its inverse, f^{-1}, with equation

$$x = 2^y$$

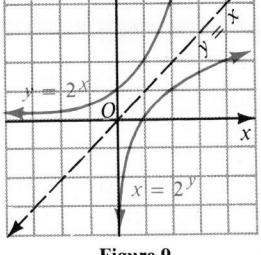

Figure 9

is also a function. Because the domain and range of f^{-1} are the range and domain, respectively, of f, f^{-1} has $\{x: x > 0\}$ as domain and $\mathcal{R}$ as range. Its graph is also shown in Figure 9.

Each exponential function with equation

$$y = b^x \quad (b > 0, b \neq 1)$$

has an inverse function with equation

$$x = b^y \quad (b > 0, b \neq 1).$$

In order to express y in terms of x, we adopt the notation $\log_b x$, read "the **logarithm** of x to the base b," for the exponent y in this equation. Thus the inverse of the exponential function with base b has equation

$$y = \log_b x \quad (b > 0, b \neq 1, x > 0)$$

and is called the **logarithmic function with base** b.

The fact that, for inverses, $f^{-1}(f(x)) = x$ and $f(f^{-1}(x)) = x$ implies the following:

$$b^{\log_b x} = x \quad (b > 0, b \neq 1, x > 0)$$
$$\log_b b^x = x \quad (b > 0, b \neq 1)$$

Another important fact is the following:

$$y = \log_b x \text{ if and only if } x = b^y.$$

That is, these two equations are equivalent. By definition, since $b^0 = 1$, $\log_b 1 = 0$ whenever $b > 0$ and $b \neq 1$.

Teaching Suggestions
p. T108

Key Ideas

Define the logarithmic function.
Convert equations from exponential form to logarithmic form and from logarithmic form to exponential form.

Reading Algebra

Encourage students to read logarithmic expressions such as "$\log_b x = y$" as "the logarithm of x to the base b is y". You should also have them explain what this expression means. One might say "the expression $\log_b x = y$ means y is the power to which b is raised to get x."

Common Errors

When converting the expression $\log_b x = y$ into exponential form, students may confuse x and y. Emphasize to the students that this logarithmic notation is used to express an exponential relationship where b is the base, and y is the power to which b is raised to get x.

Chalkboard Examples

1. Express $3^{-3} = \frac{1}{27}$ in logarithmic form.

$\log_3 \frac{1}{27} = -3$

2. Express $\log_{625} 5 = \frac{1}{4}$ in exponential form. $625^{\frac{1}{4}} = 5$

3. Evaluate $\log_2 \frac{1}{32}$.

Let $\log_2 \frac{1}{32} = x$.

Then $2^x = \frac{1}{32}$.

Since $\frac{1}{32} = 2^{-5}$, $2^x = 2^{-5}$.

$\therefore x = -5$.

4. Solve for x: $\log_x \sqrt[3]{9} = 2$
If $\log_x \sqrt[3]{9} = 2$,
then $x^2 = \sqrt[3]{9} = 9^{\frac{1}{3}}$.
Raise both sides to the $\frac{1}{2}$ power:
$(x^2)^{\frac{1}{2}} = (9^{\frac{1}{3}})^{\frac{1}{2}} = (9^{\frac{1}{2}})^{\frac{1}{3}}$
$x = 3^{\frac{1}{3}} = \sqrt[3]{3}$
$\therefore$ the solution set is $\{\sqrt[3]{3}\}$.

Mixed Review

Evaluate.

1. $\sqrt[3]{\frac{1}{27}}$ $\frac{1}{3}$

2. 5^4 625

3. 10^{-3} 0.001

Solve and graph.

4. $-3x - 2 \le 7$ $\{x: x \ge -3\}$

5. $7x - 4 > 10$ $\{x: x > 2\}$

6. $|x - 2| \ge 3$
$\{x: x \le -1\} \cup \{x: x \ge 5\}$

EXAMPLE 1 Express each equation in logarithmic form.
 a. $2^4 = 16$ b. $4^{-2} = \frac{1}{16}$

SOLUTION a. $\log_2 16 = 4$ b. $\log_4 \frac{1}{16} = -2$

EXAMPLE 2 Express each equation in exponential form.
 a. $\log_{10} 1000 = 3$ b. $\log_8 4 = \frac{2}{3}$

SOLUTION a. $10^3 = 1000$ b. $8^{\frac{2}{3}} = 4$

When you evaluate an expression or solve an equation involving a logarithm, it is often helpful to use the equivalent exponential form.

EXAMPLE 3 Evaluate $\log_4 64$.

SOLUTION Let $\log_4 64 = x$; then $4^x = 64$.
Since $64 = 4^3$, $4^x = 4^3$ and $x = 3$.
$\therefore \log_4 64 = 3$.

EXAMPLE 4 Solve for x: $\log_x \sqrt{7} = \frac{3}{2}$

SOLUTION If $\log_x \sqrt{7} = \frac{3}{2}$, then $x^{\frac{3}{2}} = \sqrt{7} = 7^{\frac{1}{2}}$.

Raise both sides to the $\frac{2}{3}$ power:
$$(x^{\frac{3}{2}})^{\frac{2}{3}} = (7^{\frac{1}{2}})^{\frac{2}{3}}$$
$$x = 7^{\frac{1}{3}} = \sqrt[3]{7}$$

$\therefore$ the solution set is $\{\sqrt[3]{7}\}$.

The relationship between exponential and logarithmic functions produces two additional useful facts about the latter functions. From the boxed statement on page 446 you know that for $b > 0$, $b \ne 1$, $b^{y_1} = b^{y_2}$ if and only if $y_1 = y_2$. It follows that:

For $b > 0$, $b \ne 1$,
$$\log_b x_1 = \log_b x_2 \text{ if and only if } x_1 = x_2 \ (x_1, x_2 > 0).$$

EXAMPLE 5 Evaluate $\log_5 (\sqrt{5})^6$.

SOLUTION $\log_5 (\sqrt{5})^6 = \log_5 (5^{\frac{1}{2}})^6$
$= \log_5 5^3$
$= 3$

452 *Chapter 11*

Oral Exercises

Express each equation in logarithmic form.

1. $10^5 = 100{,}000$ $\log_{10} 100{,}000 = 5$
2. $3^2 = 9$ $\log_3 9 = 2$
3. $5^{-3} = \frac{1}{125}$ $\log_5 \frac{1}{125} = -3$
4. $b^0 = 1$ $\log_b 1 = 0$
5. $32^{\frac{3}{5}} = 8$ $\log_{32} 8 = \frac{3}{5}$
6. $(\frac{1}{3})^{-4} = 81$ $\log_{\frac{1}{3}} 81 = -4$
7. $10^{-3} = 0.001$ $\log_{10} 0.001 = -3$
8. $(\sqrt{2})^6 = 8$ $\log_{\sqrt{2}} 8 = 6$

Express each equation in exponential form.

9. $\log_2 16 = 4$ $2^4 = 16$
10. $\log_6 \frac{1}{36} = -2$ $6^{-2} = \frac{1}{36}$
11. $\log_{10} 0.01 = -2$ $10^{-2} = 0.01$
12. $\log_{\frac{1}{3}} \frac{1}{27} = 3$ $(\frac{1}{3})^3 = \frac{1}{27}$
13. $\log_b 1 = 0$ $b^0 = 1$
14. $\log_5 \sqrt[4]{5} = \frac{1}{4}$ $5^{\frac{1}{4}} = \sqrt[4]{5}$

Written Exercises

Evaluate.

A

1. $\log_3 81$ 4
2. $\log_{10} 0.0001$ −4
3. $\log_2 \frac{1}{8}$ −3
4. $\log_7 1$ 0
5. $\log_{49} 7$ $\frac{1}{2}$
6. $\log_{\frac{1}{2}} 128$ −7
7. $\log_{\frac{1}{20}} 8000$ −3
8. $\log_{15} 15$ 1
9. $\log_{\sqrt{3}} 9$ 4
10. $\log_{25} 125$ 1.5
11. $\log_{27} 9$ $\frac{2}{3}$
12. $\log_{\sqrt{6}} 36\sqrt{6}$ 5

Solve for x by first converting each equation to exponential form.

13. $\log_x 64 = 6$ {2}
14. $\log_x 16 = \frac{1}{2}$ {256}
15. $\log_8 x = 3$ {512}
16. $\log_{1000} x = -\frac{2}{3}$ {0.01}
17. $\log_x 81 = \frac{4}{3}$ {27}
18. $\log_x 7 = -2$ $\{\frac{\sqrt{7}}{7}\}$
19. $\log_{\sqrt{3}} \frac{1}{9} = x$ {−4}
20. $\log_{100} 1000 = x$ {1.5}
21. $\log_x \sqrt{7} = \frac{1}{4}$ {49}
22. $\log_x 125 = 6$ $\{\sqrt{5}\}$
23. $\log_x \frac{1}{36} = -\frac{2}{3}$ {216}
24. $\log_{32} 8 = x$ $\{\frac{3}{5}\}$

Evaluate.

B

25. $\log_7 7^{10}$ 10
26. $\log_3 9^7$ 14
27. $\log_{1000} 10^{12}$ 4
28. $\log_5 25^9$ 18

Solve for x.

EXAMPLE $\log_{2x} 27 = 3$

SOLUTION Since $\log_{2x} 27 = 3$, $(2x)^3 = 27$.

$$8x^3 = 27$$
$$x^3 = \frac{27}{8}$$
$$x = \frac{3}{2}$$

29. $\log_{4x} 36 = 2$ $\{\frac{3}{2}\}$
30. $\log_5 (2x + 5) = 3$ {60}
31. $\log_2 32 = 3x - 1$ {2}
32. $\log_3 (x^2 + 17) = 4$ {±8}
33. $\log_2 32 = 3x + 1$ $\{\frac{4}{3}\}$
34. $\log_{x^3} 64 = \frac{2}{3}$ {8}

C

35. $\log_2 (\log_2 16) = \log_7 x$ {49}
36. $\log_2 (\log_9 3) = \log_x 6$ $\{\frac{1}{6}\}$
37. $\log_{10} [\log_2 (\log_3 9)] = x$ {0}
38. $\log_4 (\log_8 64) = \log_5 x$ $\{\sqrt{5}\}$

Suggested Assignments

Minimum
First day
453/1–19 odd
Second day
453/21–34
Extended Alg. with Trig.
453/1–33 odd
Enriched Alg.
453/9–37 odd
Enriched Alg. with Trig.
453/9–37 odd

Additional A Exercises

Evaluate.

1. $\log_{64} 4$ $\frac{1}{3}$
2. $\log_{\frac{1}{4}} 16$ −2
3. $\log_{36} \frac{1}{6}$ $-\frac{1}{2}$

Solve for x by first converting each equation to exponential form.

4. $\log_x 27 = -3$ $\left\{\frac{1}{3}\right\}$
5. $\log_2 x = -4$ $\left\{\frac{1}{16}\right\}$
6. $\log_{\frac{1}{3}} x = -2$ {9}

Supplementary Material

Test 34

Key Ideas

Apply the laws of logarithms to evaluate logarithmic expressions and solve logarithmic equations.

Common Errors

Students may reverse the laws of logarithms; that is they may write $\dfrac{\log a}{\log b} =$ $\log (a - b)$ or $(\log a)(\log b)$ $= \log (a + b)$. Clarify the laws by showing the students these incorrect examples, and also by reminding students of the laws of exponents on which these laws of logarithms are based.

Chalkboard Examples

1. Express $\log_a \dfrac{y^2\sqrt{y}}{x^5z}$ in terms of $\log_a x$, $\log_a y$, and $\log_a z$.

$\log_a \dfrac{y^2\sqrt{y}}{x^5z} = \log_a \dfrac{y^2 y^{\frac{1}{2}}}{x^5z}$

$= \log_a \dfrac{y^{\frac{5}{2}}}{x^5z}$

$= \log_a (y^{\frac{5}{2}}) - \log_a (x^5z)$

$= \log_a (y^{\frac{5}{2}}) -$
$(\log_a x^5 + \log_a z)$

$= \dfrac{5}{2} \log_a y - 5 \log_a x -$

$\log_a z$

11–5 Laws of Logarithms

Logarithms were first developed at the end of the sixteenth century as an aid to carrying out lengthy arithmetic processes such as long division and extraction of roots. Because logarithms are exponents, you can multiply (or divide) positive numbers by adding (or subtracting) the logarithms of the numbers and then finding the number that has the result as its logarithm. Although the introduction of calculators and computers has reduced the use of logarithms as a computational tool, the logarithmic function is indispensable in many scientific applications. There are also certain kinds of equations whose solution still requires the use of logarithms, as we will see in Section 11-7. It is therefore important to understand the laws that govern their use.

Laws of Logarithms

Let b be a positive number not equal to 1. Let x_1 and x_2 be any positive numbers and n be any real number. Then:

1. $\log_b x_1 x_2 = \log_b x_1 + \log_b x_2$

2. $\log_b \dfrac{x_1}{x_2} = \log_b x_1 - \log_b x_2$

3. $\log_b (x_1)^n = n \log_b x_1$

To prove Law 1, start with the fact (page 451) that:

$$x_1 = b^{\log_b x_1} \quad \text{and} \quad x_2 = b^{\log_b x_2}$$

Then:

$$x_1 x_2 = b^{\log_b x_1} \cdot b^{\log_b x_2}$$
$$= b^{\log_b x_1 + \log_b x_2} \text{ by Law 1 for exponents (page 200)}$$

Thus:

$$\log_b x_1 x_2 = \log_b x_1 + \log_b x_2$$

Proofs of the second and third laws, which are similar, are left as Exercises 35 and 36.

EXAMPLE 1   Express $\log_b \dfrac{x\sqrt{y}}{z^3}$ in terms of $\log_b x$, $\log_b y$, and $\log_b z$.

SOLUTION $\log_b \dfrac{x\sqrt{y}}{z^3} = \log_b \dfrac{xy^{\frac{1}{2}}}{z^3}$

$\qquad\qquad = \log_b (xy^{\frac{1}{2}}) - \log_b z^3$ by Law 2

$\qquad\qquad = (\log_b x + \log_b y^{\frac{1}{2}}) - \log_b z^3$ by Law 1

$\qquad\qquad = \log_b x + \frac{1}{2} \log_b y - 3 \log_b z$ by Law 3

454   *Chapter 11*

EXAMPLE 2 Evaluate.

    a. $\log_2 40 - \log_2 5$ **b.** $(5^{\log_{25} 7})^2$ **c.** $8^{\log_2 5}$

SOLUTION   **a.** $\log_2 40 - \log_2 5 = \log_2 \dfrac{40}{5} = \log_2 8 = 3$

    b. $(5^{\log_{25} 7})^2 = 5^{2 \log_{25} 7} = (5^2)^{\log_{25} 7} = 25^{\log_{25} 7} = 7$

    c. $8^{\log_2 5} = (2^3)^{\log_2 5} = 2^{3 \log_2 5} = 2^{\log_2 5^3} = 2^{\log_2 125} = 125$

EXAMPLE 3   Solve the equation $\log_4 x + \log_4 (x - 6) = 2$.

SOLUTION $\log_4 x + \log_4 (x - 6) = 2$

                $\log_4 x(x - 6) = 2$

                $\log_4 (x^2 - 6x) = 2$

    Convert to exponential form:

                $x^2 - 6x = 4^2$

          $x^2 - 6x - 16 = 0$

        $(x - 8)(x + 2) = 0$

           $x = 8$ or $x = -2$

    Check: $x = 8$: $\log_4 8 + \log_4 (8 - 6) \overset{?}{=} 2$

                                 $\log_4 8 + \log_4 2 \overset{?}{=} 2$

                                        $\frac{3}{2} + \frac{1}{2} = 2$ ✓

        $x = -2$:   Since $\log_4 (-2)$ is not defined, -2 is not a solution.

    ∴ the solution set is $\{8\}$.

Oral Exercises

Express in terms of $\log_b x$ and $\log_b y$.

1. $\log_b x^2 y$ $2 \log_b x + \log_b y$ **2.** $\log_b \dfrac{x}{y^4}$ $\log_b x - 4 \log_b y$ **3.** $\log_b (\sqrt[3]{x})^4$ $\frac{4}{3} \log_b x$

4. $\log_b \sqrt{xy^5}$ $\frac{1}{2} \log_b x + \frac{5}{2} \log_b y$ **5.** $\log_b \left(\dfrac{x}{y}\right)^7$ $7(\log_b x - \log_b y)$ **6.** $\log_b \dfrac{1}{(xy)^3}$ $-3(\log_b x + \log_b y)$

Express as the logarithm of a single number.

7. $\log_8 5 + 2 \log_8 3$ $\log_8 45$ **8.** $\frac{1}{2} \log_6 100 - \log_6 2$ $\log_6 5$

9. $\frac{1}{3}(\log_5 56 - \log_5 7)$ $\log_5 2$ **10.** $\log_7 24 - (\log_7 2 + \log_7 3)$ $\log_7 4$

Give values of x and y for which each statement is *false*.

11. $(\log_2 x)(\log_2 y) = \log_2 x + \log_2 y$ $x = 1, y = 2$

12. $\dfrac{\log_3 x}{\log_3 y} = \log_3 x - \log_3 y$ $x = 9, y = 3$

13. $(\log_2 x)^3 = 3 \log_2 x$ $x = 2$

                                      Exponents and Logarithms **455**

---

2. Evaluate.

    a. $\log_{10} 200 + \log_{10} 5$

    b. $(9^{\log_3 11})^{\frac{1}{2}}$

    c. $25^{\log_5 3}$

    a. $\log_{10} 200 + \log_{10} 5 =$
        $\log_{10} (200 \cdot 5) =$
        $\log_{10} (1000) = 3$

    b. $(9^{\log_3 11})^{\frac{1}{2}} = 9^{\frac{1}{2} \log_3 11} =$
        $(9^{\frac{1}{2}})^{\log_3 11} = 3^{\log_3 11} = 11$

    c. $25^{\log_5 3} = (5^2)^{\log_5 3} = 5^{2 \log_5 3}$
        $= 5^{\log_5 3^2} = 5^{\log_5 9} = 9$

Mixed Review

Solve.

1. $x^2 - 4x = 12$ $\{6, -2\}$

2. $x^2 + 2x - 8 = 0$ $\{2, -4\}$

Simplify.

3. $(x^{5y})(x^{2y})(x^{-7y})$ 1

4. $\left(\dfrac{2z^4}{xy^7}\right)^{-3}$ $\dfrac{x^3 y^{21}}{8z^{12}}$

Solve each system of equations.

5. $3y = -x - 2$;
    $5y + 2x = -5$ $\{(-5, 1)\}$

6. $y = 9 - 3x$;
    $2x - 2y = 6$ $\{(3, 0)\}$

Additional A Exercises

Evaluate.

1. $\log_9 108 - \log_9 4 + \log_9 3$
    2

2. $\frac{1}{3}(\log_4 10 + 5 \log_4 2 -$
    $\log_4 5)$ 1

3. $4(\log_5 2 - \dfrac{1}{2} \log_5 100)$
    -4

Solve for x.

4. $3 \log_5 2 + \log_5 4 = \log_5 x$
    $\{32\}$

5. $\log_3 36 + 2 \log_3 x =$
    $\log_3 18 + \log_3 8$ $\{2\}$

Evaluate.

6. $7^{\log_7 10 + \log_7 5}$ 50

Find the inverse of each function and graph both the function and its inverse.

1. $y = 5x - 2$ $\quad y = \dfrac{x + 2}{5}$

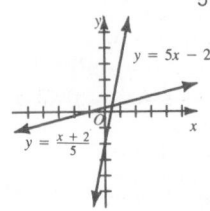

2. $y = 2x^2 + 1$;
Domain: $\{x: x \leq 0\}$

$y = -\sqrt{\dfrac{x - 1}{2}}$;

Domain: $\{x: x \geq 1\}$

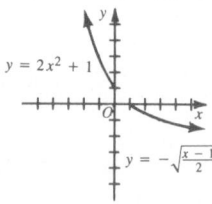

Express each equation in logarithmic form.

3. $\left(\dfrac{1}{2}\right)^{-5} = 32$ $\quad \log_{\frac{1}{2}} 32 = -5$

4. $625^{\frac{1}{2}} = 25$ $\quad \log_{625} 25 = \dfrac{1}{2}$

Express each equation in exponential form.

5. $\log_3 243 = 5$ $\quad 3^5 = 243$

6. $\log_{49} 7 = \dfrac{1}{2}$ $\quad 49^{\frac{1}{2}} = 7$

Evaluate.

7. $\log_{11}\left(\dfrac{1}{121}\right)$ $\quad -2$

8. $\log_4 \dfrac{1}{2}$ $\quad -\dfrac{1}{2}$

Solve for x.

9. $\log_x\left(\dfrac{1}{10{,}000}\right) = 2$ $\quad \left\{\dfrac{1}{100}\right\}$

10. $\log_7 x = \dfrac{1}{2}\log_7 144 - \log_7 12$ $\quad \{1\}$

Written Exercises

Evaluate.

A   **1.** $\log_3 45 + \log_3 6 - \log_3 10$ 3

2. $\log_6 4 + 2\log_6 3$ 2

3. $\dfrac{1}{2}\log_3 144 - 2\log_3 6$ -1

4. $2(\log_2 20 - \log_2 5)$ 4

5. $\dfrac{1}{3}(\log_2 12 + \log_2 16 - \log_2 3)$ 2

6. $\dfrac{1}{2}\log_5 16 - 2\log_5 10$ -2

Solve for x.

10. $\{\pm 4\}$

7. $\log_7 x = 2\log_7 6 - \log_7 4$ $\{9\}$

8. $\log_{10} x = \dfrac{1}{3}\log_{10} 8 + \dfrac{1}{2}\log_{10} 81$ $\{18\}$

9. $\log_5 x = \dfrac{2}{3}\log_5 64$ $\{16\}$

10. $\log_3 x^2 = \log_3 8 + \log_3 10 - \log_3 5$

11. $2\log_{10} x = \log_{10} 12 + \log_{10} 3$ $\{6\}$

12. $\log_3 x - \log_3 4 = 2\log_3 5$ $\{100\}$

13. $\log_2 6 + \log_2 x = \log_2 3 + \log_2 10$ $\{5\}$

14. $\dfrac{1}{2}\log_7 x = \log_7 20 - 2(\log_7 2 + \log_7 5)$

15. $\log_b 5 + 2\log_b x = \log_b 45$ $\{3\}$

16. $\log_5 2 - \dfrac{1}{2}\log_5 x = \log_5 3$ $\left\{\dfrac{4}{9}\right\}$ $\left\{\dfrac{1}{25}\right\}$

Evaluate.

17. $3^{2\log_9 5}$ 5

18. $5^{\log_5 2 + \log_5 3}$ 6

19. $25^{\log_5 7}$ 49

20. $2^{\frac{1}{3}\log_2 27}$ 3

Let b be a number such that $\log_b 3 = 1.5$ and $\log_b 5 = 2.2$. Evaluate each of the following.

EXAMPLE $\quad \log_b 15$

SOLUTION $\quad \log_b 15 = \log_b (3 \cdot 5) = \log_b 3 + \log_b 5 = 1.5 + 2.2 = 3.7$

B   **21.** $\log_b \dfrac{5}{3}$ 0.7

22. $\log_b 0.6$ -0.7

23. $\log_b 45$ 5.2

24. $\log_b 125$ 6.6

25. $\log_b \sqrt{5}$ 1.1

26. $\log_b (\sqrt[3]{3})^7$ 3.5

27. $\log_b 3b^2$ 3.5

28. $\log_b \dfrac{b}{15}$ -2.7

Solve for x.

29. $\log_{10} x + \log_{10}(x + 3) = 1$ $\{2\}$

30. $\log_6 x + \log_6 (x + 5) = 2$ $\{4\}$

31. $\log_2 (x + 6) - \log_2 (x - 1) = 3$ $\{2\}$

32. $\log_3 2x^2 - \log_3 (5x - 9) = 1$ $\{3, 4.5\}$

33. $\log_8 (x^2 - 1) - \log_8 (3x + 9) = 0$ $\{5, -2\}$

34. $\log_7 (x^3 + 27) - \log_7 (x + 3) = 2$ $\{8\}$

C   **35.** Prove Law 2 on page 454.

36. Prove Law 3 on page 454.

▍Self-Test 2

VOCABULARY   inverse functions (p. 447)   logarithm (p. 451)
one-to-one function (p. 448)   logarithmic function (p. 451)
exponential function (p. 451)

Find the inverse of the given function and graph both the function and its inverse.

1. $y = \frac{1}{3}x + 1$
   $f^{-1}(x) = 3x - 3$

2. $y = x^2 - 2;$ Domain: $\{x: x \geq 0\}$
   $f^{-1}(x) = \sqrt{x + 2}; x \geq -2$

Obj. 1, p. 447

Express each equation in logarithmic form.

3. $(\frac{1}{2})^{-6} = 64$
   $\log_{\frac{1}{2}} 64 = -6$

4. $36^{\frac{1}{2}} = 6$ $\log_{36} 6 = \frac{1}{2}$

Obj. 2, p. 447

Express each equation in exponential form.

5. $\log_5 \frac{1}{125} = -3$ $5^{-3} = \frac{1}{125}$

6. $\log_{10} 1000 = 3$ $10^3 = 1000$

Evaluate.

7. $\log_4 256$ 4

8. $\log_{27} 81$ $\frac{4}{3}$

Obj. 3, p. 447

Solve for x.

9. $\log_x \sqrt{2} = \frac{1}{4}$ $\{4\}$

10. $\log_{\frac{1}{125}} x = -\frac{4}{3}$ $\{625\}$

11. $\log_5 x = 3 \log_5 4 - \frac{1}{3} \log_5 64$ $\{16\}$

12. $2 \log_{12} x = \frac{1}{2} \log_{12} 9 + 3 \log_{12} 3$ $\{9\}$

Obj. 4, p. 447

Check your answers with those at the back of the book.

Caroline Herschel

1750–1848

Trained to be a concert singer, Caroline Herschel gave up this career for a consuming interest in astronomy. Lacking formal education in mathematics, Herschel taught herself the geometric and logarithmic concepts that were necessary for her work. In addition to doing the detailed calculations and reductions, she catalogued 2500 nebulae and reorganized a listing of approximately 3000 stars.

In 1783, Herschel discovered three new nebulae. Between 1786 and 1797 she is credited with discovering no less than eight new comets. An appointment by the king as assistant court astronomer, the first position of this kind for a woman, granted her a salary for her work.

In 1828, after she had completed a catalog of 1500 nebulae, the Royal Astronomical Society awarded her a gold medal as an "extraordinary monument to the unextinguished ardor of a lady of seventy-five to the cause of science."

Exponents and Logarithms   **457**

Suggested Assignments

Minimum
  First day
  456/1–20
  Second day
  456/21–34
  Third day
R 456/*Self-Test 2*
S 453/8–20 even
Extended Alg. with Trig.
  First day
  456/1–15 odd
  Second day
  456/17–33
  Third day
R 456/*Self-Test 2*
S 453/14–24 even
Enriched Alg.
  First day
  456/1–35 odd
  Second day
R 456/*Self-Test 2*
S 453/14–24 even
Enriched Alg. with Trig.
  First day
  456/1–35 odd
  Second day
R 456/*Self-Test 2*
S 453/14–24 even

Additional Answers
Written Exercises

(See p. 479.)

Additional Answers
Self-Test 2

1.
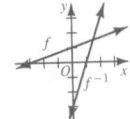

2.
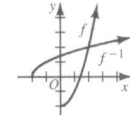

Using Logarithms

OBJECTIVES for Sections 11-6 through 11-9:
1. *To use a calculator or a table to find the common logarithm of a number or to find the number that has a given common logarithm.*
2. *To use logarithms to solve exponential equations.*
3. *To solve problems involving interest and exponential growth and decay.*
4. *To use the number e and natural logarithms.*

Teaching Suggestions
p. T109

Key Ideas

Find the common logarithm of a number or a number from a given common logarithm.
Use common logarithms in arithmetic calculations.

Common Errors

Students may forget to re-write a negative logarithm in the correct form before they use a table to find its antilogarithm. Remind them that the values in the table are positive, and that a negative logarithm must be rewritten as a combination of a positive mantissa and a negative characteristic before finding the antilogarithm.

11–6 Common Logarithms

Until relatively recently, logarithms were used primarily as an aid to arithmetic computation. Since such computation is carried out in the decimal system of notation, it was found convenient to adopt the number 10 as the base for most logarithmic work.

Logarithms to the base 10 are called **common logarithms.** In addition to being used for computation, common logarithms are the basis for *logarithmic scales,* such as the Richter scale used for measuring the magnitude of an earthquake and the decibel scale that measures the intensity of a sound. Logarithmic scales are particularly useful for measuring quantities that vary over very large ranges.

It is customary to omit the subscript 10 in writing a common logarithm. Thus we write

$$\log_{10} x \qquad \text{as simply} \qquad \log x.$$

Many calculators have keys that give the values of the common logarithms of positive numbers. Tables of common logarithms, such as Table 5 of this book, give approximate values of the common logarithms of numbers greater than or equal to 1 and less than 10. To find the logarithms of positive numbers outside this range, such as 1240 and 0.0124, you can write the numbers in scientific notation and apply the laws of logarithms. For example, you can determine from the table that the logarithm of 1.24 is approximately 0.0934. Although the value found in the table is an approximation, it is customary to write = rather than $\approx$. Thus we write:

$$\log 1.24 = 0.0934$$

Since $1240 = 1.24 \times 10^3$, we have:
$$\begin{aligned}\log 1240 &= \log 1.24 + \log 10^3\\ &= 0.0934 + 3\\ &= 3.0934\end{aligned}$$

Since $0.0124 = 1.24 \times 10^{-2}$, we have:
$$\begin{aligned}\log 0.0124 &= \log 1.24 + \log 10^{-2}\\ &= 0.0934 + (-2)\\ &= -1.9066\end{aligned}$$

458   *Chapter 11*

Notice that the common logarithm of a number can be written as the sum of an integer, called the **characteristic,** and a nonnegative number less than 1, called the **mantissa.** The common logarithm of a positive number that is less than 1 has a negative characteristic. A calculator combines the characteristic and mantissa to give one number, such as 3.0934 or -1.9066, and, in general, we will do the same.

If $\log x = a$, then x is sometimes called the **antilogarithm** of a, and we write:

$$x = \text{antilog } a$$

It follows that

$$\text{antilog } a = 10^a,$$

so that antilogarithms can be found on a calculator by merely raising 10 to the given power. For convenience, some calculators have a key marked 10^x. Appendix A discusses how to use a table of logarithms to find antilogarithms.

The following example uses the fact that $\log_b x_1 = \log_b x_2$ if and only if $x_1 = x_2$ (page 452).

EXAMPLE   Solve: $143x^{\frac{3}{7}} = 65.2$

SOLUTION $143x^{\frac{3}{7}} = 65.2$

$$x^{\frac{3}{7}} = \frac{65.2}{143}$$

Take logarithms of both sides:

$$\tfrac{3}{7} \log x = \log\left(\frac{65.2}{143}\right)$$

$$\log x = \tfrac{7}{3} \log\left(\frac{65.2}{143}\right)$$

$$= \tfrac{7}{3}(\log 65.2 - \log 143)$$

$$= \tfrac{7}{3}(1.8142 - 2.1553)$$

$$= -0.7959 \text{ (or } 0.2041 - 1)$$

$$x = 0.160 \text{ (to three significant digits)}$$

Notice that you can use a calculator to find the antilogarithm of -0.7959 in the example. However, if you use a table of logarithms you will have to write -0.7959 in the form $0.2041 - 1$ so you can look up a *positive* mantissa in the table. In general, we will round answers to three significant digits.

Oral Exercises

Given that $\log 4.28 = 0.6314$, state the value of each expression.

1. $\log 428$ 2.6314   **2.** $\log 0.428$ -0.3686 **3.** $\log 428{,}000$ 5.6314   **4.** $\log 0.000428$ -3.3686

Exponents and Logarithms   **459**

459

Chalkboard Examples

1. Solve: $20.5x^{-\frac{1}{2}} = 3.58$

$$20.5x^{-\frac{1}{2}} = 3.58$$

$$x^{-\frac{1}{2}} = \frac{3.58}{20.5}$$

Take logarithms of both sides.

$$-\frac{1}{2} \log x = \log\left(\frac{3.58}{20.5}\right)$$

$$\log x = -2 \log\left(\frac{3.58}{20.5}\right)$$

$$= -2(\log 3.58 - \log 20.5)$$

$$= -2(0.5539 - 1.3118)$$

$$= -2(-0.7579)$$

$$= 1.5158$$

Take the antilogarithm.
$x \approx 32.8$ (to three significant digits)

2. Solve: $\sqrt{62.4x} = 5.19$

$$\sqrt{62.4x} = 5.19$$

$$62.4x = (5.19)^2$$

$$x = \frac{(5.19)^2}{62.4}$$

Take logarithms of both sides.

$$\log x = \log \frac{(5.19)^2}{62.4}$$

$\log x =$
$2 \log 5.19 - \log 62.4$
$= 2(0.7152) - 1.7952$
$= -0.3648 \text{ (or } 0.6352 - 1)$
Take the antilogarithm.
$x \approx 0.432$

Suggested Assignments

Minimum
  460/1–24
Extended Alg. with Trig.
  460/1, 2, 3–27 odd
Enriched Alg.
  460/2–4, 7–31 odd
Enriched Alg. with Trig.
  460/2–4, 7–31 odd

Additional A Exercises

Evaluate.

1. a. log 5.07   0.7050
   b. log 50,700   4.7050
   c. log 0.00507   −2.2950
   d. log $\dfrac{1}{507}$ −2.7050

2. a. log 368   2.5658
   b. log 0.368   −0.4342
   c. log 0.0368   −1.4342
   d. log $\dfrac{1}{368,000}$ −5.5658

Solve for x without using a table or a calculator.

3. log x = log 4.07 − 3
   {0.00407}

4. −7 + log x = log 5.62
   {56,200,000}

Evaluate.

5. log ($\sqrt[5]{32}$ · 6)   1.0792

6. log $\dfrac{\sqrt{64}}{2^2}$ 0.3010

Use Table 5 or a calculator to find each value.

5. antilog 3.8525 7120   **6.** antilog (0.6902 − 2) 0.049   **7.** antilog (−1.5391)   0.0289
8. $10^{1.7193}$ 52.4   **9.** $10^{-0.1296}$ 0.742   **10.** $10^{-2.8}$ 0.00158

Given that $\log_3 2 = 0.6309$, state the value of each expression. (*Hint*: Apply the laws of logarithms.)

11. $\log_3 6$ 1.6309   **12.** $\log_3 18$ 2.6309   **13.** $\log_3 54$ 3.6309   **14.** $\log_3 \frac{2}{3}$ −0.3691

Complete each statement. Assume that $x > 0$.

15. log x + 2 = log (___?___) 100x

16. log x + ___?___ = log $\dfrac{x}{10}$ −1

17. $10^{3 + \log x}$ = ___?___ 1000x

1.d. −0.3729
2.d. 0.1186
3.c. 0.256

Written Exercises

Evaluate.

A **1. a.** log 23.6 1.3729   **b.** log 23,600 4.3729   **c.** log 0.00236 −2.6271   **d.** log $\dfrac{1}{2.36}$

2. a. log 761 2.8814   **b.** log 0.0761 −1.1186   **c.** log 761,000 5.8814   **d.** log $\dfrac{100}{76.1}$

3. a. antilog 3.4082 2560   **b.** antilog 5.4082 256,000   **c.** antilog (−0.5918)

4. a. $10^{2.9248}$ 841   **b.** $10^{0.9248 - 2}$ 0.0841   **c.** $10^{0.0752 - 3}$ 0.00119

Solve for x without using a table or a calculator.

5. log x = log 3.62 + 4 {36,200}   **6.** 1 + log x = log 2.69 {0.269}
7. log 7520 = log 7.52 + x {3}   **8.** log $\dfrac{831}{10^5}$ = x + log 8.31 {−3}
9. log 9.23 − log 0.00923 = x {3}   **10.** log 852 = log 0.0852 + x {4}

Evaluate.

0.7782
11. log (5^2 · 3) 1.8751   **12.** log $\left(\dfrac{3}{2}\right)^2$ 0.3522   **13.** log $\sqrt{\dfrac{144}{4^2}}$ 0.4771   **14.** log $\dfrac{12}{\sqrt[3]{8}}$

15. log $\left(\dfrac{\sqrt[4]{81}}{15}\right)$ −0.6990   **16.** log $\left(\dfrac{56}{\sqrt{64}}\right)$ 0.8451   **17.** log $\sqrt[3]{\dfrac{36^2}{3^3 \cdot 6}}$ 0.3010   **18.** log $\left(\dfrac{7^2}{98 \cdot 5}\right)$
−1

Solve for x.

{0.327}
B **19.** $8.3x^5 = 47.9$ {1.42}   **20.** $634x^4 = 58$ {0.550}   **21.** $\dfrac{4.72}{x^3} = 135$

22. $x^{\frac{2}{3}} = \dfrac{726}{68.1}$ {34.8}   **23.** $\sqrt[3]{87x^5} = 391$ {14.7}   **24.** $11.6 = \dfrac{2700}{x^{\frac{4}{3}}}$
{59.6}

460   *Chapter 11*

In Exercises 25–28 use the facts that log 2 = 0.3010 and log 3 = 0.4771 to find an approximation of each of the following without using a table or a calculator.

1.1761

25. **a.** log 0.2 −0.6990 **b.** log 5 0.6990 **c.** log 0.25 −0.6020 **d.** log 15

26. Given that $7^2 \approx 48$ and $11^2 \approx 120$, find approximations of log 7 and log 11. log 7 ≈ 0.8406; log 11 ≈ 1.0396

27. Given that $13^3 \approx 3^7$, find an approximation of log 13. 1.1132

28. Given that $3 \cdot 5 \cdot 17 \approx 2^8$, find an approximation of log 17. 1.2319

Exercises 29–32 deal with the relative intensity of sound and the decibel scale, which are described in Exercise 29.

C 29. The decibel scale measures the relative intensity of a sound. Let I_0 represent the intensity of a sound that is barely audible. Then the decibel level of a sound with intensity I is defined to be:

$$10 \log\left(\frac{I}{I_0}\right)$$

a. What is the decibel level of a sound that is barely audible? 0 decibels

b. The sound intensity of a jet take-off at close range is $10^{14}I_0$, or 100,000,000,000,000 times the intensity of a barely audible sound. What is the decibel level of the jet take-off? 140 decibels

c. The decibel level of loud rock music with amplifiers is 120. How many times more intense is this sound than a barely audible sound? 10^{12}

30. A whisper has a decibel level of 20 and ordinary conversation has a decibel level of 60. How many times more intense is the sound of an ordinary conversation than a whisper? 10,000

31. A telephone bell has a decibel level of 70. What is the decibel level of two telephone bells ringing simultaneously? approx. 73 decibels

32. A softly played flute has a decibel level of 41. What is the decibel level of three such flutes playing together? approx. 46 decibels

11–7 Solving Exponential Equations

An **exponential equation** is an equation in which a variable appears in an exponent. You learned in Section 11-2 that certain exponential equations, such as $2^{x+1} = 8^{2x-1}$, can be solved by expressing both sides as powers of a common base. If a common base cannot be found easily, however, logarithms provide the only practical method of solving the equation.

Exponents and Logarithms **461**

Mixed Review
Express in scientific notation.
1. 0.000193 1.93×10^{-4}
2. 5,648,000 5.648×10^6
Evaluate to four significant digits.
3. $\sqrt{0.44}$ 0.6633
4. $\sqrt{97}$ 9.849
5. $(8.5)^2$ 72.25
6. 54^2 2916

Teaching Suggestions
p. T109

Key Ideas
Use logarithms to solve exponential equations in which a common base cannot be easily found.

1. Solve: $11^{3x-6} = 94$

$$11^{3x-6} = 94$$
$$\log 11^{3x-6} = \log 94$$
$$(3x - 6)\log 11 = \log 94$$
$$3x - 6 = \frac{\log 94}{\log 11}$$
$$3x = \frac{\log 94}{\log 11} + 6$$
$$x = \frac{\log 94}{3 \log 11} + 2$$

Substituting by means of a table or a calculator, you have:

$$x = \frac{1.9731}{3(1.0414)} + 2$$

$x = 2.63$ (to three significant digits)

$\therefore$ the solution set is $\{2.63\}$.

Additional A Exercises

Solve each equation.
a. Give the answer in calculation-ready form.
b. Give the solution to three significant digits.

1. $38^{2x} = 109$

   a. $x = \dfrac{\log 109}{2 \log 38}$

   b. $x \approx 0.645$

2. $428^{-x} = 37.5$

   a. $x = -\dfrac{\log 37.5}{\log 428}$

   b. $x \approx -0.598$

3. $56.1^{\frac{x}{2}} = 1260$

   a. $x = \dfrac{2 \log 1260}{\log 56.1}$

   b. $x \approx 3.55$

4. $278^{4x} = 37,800$

   a. $x = \dfrac{\log 37,800}{4 \log 278}$

   b. $x \approx 0.468$

Approximate to three significant digits.

5. $\log_7 648$ 3.33

6. $\log_{6.2} 5.07$ 0.889

EXAMPLE 1   Solve: $5^{3x} = 87$

SOLUTION
$$5^{3x} = 87$$
$$\log 5^{3x} = \log 87$$
$$3x \log 5 = \log 87$$
$$x = \frac{\log 87}{3 \log 5}$$

Substituting by means of a table or a calculator, you have:

$$x = \frac{1.9395}{3(0.6990)} = 0.925 \quad \text{(to three significant digits)}$$

$\therefore$ the solution set is $\{0.925\}$.

In Example 1 the equation $x = \dfrac{\log 87}{3 \log 5}$ gives the solution in **calculation-ready form,** where the next step is to obtain a decimal approximation using a table or a calculator.

EXAMPLE 2   Find $\log_7 12$.

a. Give the answer in calculation-ready form.
b. Give the answer as a decimal correct to three significant digits.

SOLUTION   a. Let $x = \log_7 12$. Convert this equation to exponential form:

$$7^x = 12$$
$$x \log 7 = \log 12$$
$$x = \frac{\log 12}{\log 7}$$

b.
$$x = \frac{1.0792}{0.8451} \approx 1.28$$

$\therefore \log_7 12 = 1.28$ (to three significant digits).

The preceding example suggests the following formula for finding $\log_x y$ when the logarithms of both x and y to a common base (for example, base 10) are available.

Change of Base Formula

$$\log_x y = \frac{\log_b y}{\log_b x} \qquad (y > 0, b > 0, b \neq 1, x > 0, x \neq 1)$$

This formula is especially useful in constructing tables of common logarithms, since the logarithm of a given number can be found (by advanced methods) most easily to the base $e = 2.71828\ldots$. We will study e in Section 11-9. Once $\log_e n$ and $\log_e 10$ are known, the formula above can be used to find $\log_{10} n$.

Oral Exercises

Express x in terms of common logarithms.

1. $7^x = 13$ $\dfrac{\log 13}{\log 7}$ **2.** $23^{-x} = 8$ $-\dfrac{\log 8}{\log 23}$ **3.** $5^{2x} = 79$ $\dfrac{\log 79}{2\log 5}$ **4.** $8^{\frac{x}{3}} = 156$ $\dfrac{3\log 156}{\log 8}$

5. Explain how the Change of Base Formula can be used to show that

$$\log_x y = \frac{1}{\log_y x}. \qquad \log_x y = \frac{\log_y y}{\log_y x} = \frac{1}{\log_y x}$$

6. Replace the "?" below with one of the symbols $>$, $=$, or $<$:

$$\frac{\log_2 5}{\log_2 11} \; ? \; \frac{\log_3 5}{\log_3 11} \qquad \frac{\log_2 5}{\log_2 11} = \log_{11} 5 = \frac{\log_3 5}{\log_3 11}$$

Explain your answer.

Written Exercises

Solve each equation.
a. Give the solution in calculation-ready form.
b. Give the solution to three significant digits.

9. $\{-5.78\}$

12. $\{5.40\}$

A
1. $1.59^x = 4.02$ $\{3.00\}$ **2.** $7.4^{-x} = 18.6$ $\{-1.46\}$ **3.** $19^{\frac{x}{2}} = 145$ $\{3.38\}$

4. $54^{3x} = 19$ $\{0.246\}$ **5.** $9.12^{4x} = 46$ $\{0.433\}$ **6.** $128^{\frac{x}{8}} = 33.7$ $\{5.80\}$

7. $6(2^x) = 50.1$ $\{3.06\}$ **8.** $0.76(5^{2x}) = 29.3$ $\{1.13\}$ **9.** $8.35(3^{-x}) = 4780$

10. $4.1(8.2)^{\frac{x}{2}} = 132$ $\{3.30\}$ **11.** $\sqrt[3]{25.3}\,(6^x) = 122$ $\{2.08\}$ **12.** $\sqrt[5]{203}\,(7^{\frac{x}{3}}) = 96$

13. $15^{\frac{3}{4}} \cdot 9^{-x} = 18$ $\{-0.391\}$ **14.** $12^{x+1} = \sqrt[4]{56.6}$ $\{-0.594\}$ **15.** $(0.027)^{x-2} = \sqrt[3]{94}$

     $\{1.58\}$

Approximate to three significant digits.

16. $\log_5 11$ 1.49   **17.** $\log_{23} 6.2$ 0.582   **18.** $\log_{7.3} 85.9$ 2.24

Solve each equation.
a. Give the solution in calculation-ready form.
b. Give the solution to three significant digits.

B
19. $3^x = 5^{x-1}$ $\{3.15\}$ **20.** $3^{x-1} = 2^{x+1}$ $\{4.42\}$ **21.** $6.7^{2x} = 15^{x+1}$ $\{2.47\}$

22. $6^{3x-1} = 28^x$ $\{0.877\}$ **23.** $8^{2x-1} = 39^{x+1}$ $\{11.6\}$ **24.** $42^{x-1} = 17^{3x-1}$

     $\{-0.190\}$

Use the Change of Base Formula to prove each of the following. Do not use a table of logarithms.

C
25. $\log_{25} 2 = \dfrac{\log_5 2}{2}$ **26.** $\log_a 9 = \dfrac{2\log 3}{\log a}$ **27.** $\log_b x = 2\log_{b^2} x$

28. $\log_b x = n\log_{b^n} x$ **29.** $\log_{ab} x = \dfrac{\log_a x}{1 + \log_a b}$ **30.** $\log_{\frac{a}{b}} x = \dfrac{\log_a x}{1 - \log_a b}$

31. $(\log_a x)(\log_b y) = (\log_b x)(\log_a y)$ (*Hint:* Consider $\log_x y$.)

Suggested Assignments

Minimum
  First day
  463/1–10
  Second day
  463/11–24
Extended Alg. with Trig.
  First day
  463/1–15 odd
  Second day
  463/17–25 odd
Enriched Alg.
  463/7–31 odd
Enriched Alg. with Trig.
  463/7–31 odd

Mixed Review

Give all answers in slope-intercept form.

1. Find an equation of the line with slope $-\dfrac{3}{5}$, passing through $(7, -10)$.

$y = -\dfrac{3}{5}x - \dfrac{29}{5}$

2. Find an equation of the line passing through $(-1, 3)$ and $(5, 7)$. $y = \dfrac{2}{3}x + \dfrac{11}{3}$

3. Find an equation of the line with x-intercept 8 and y-intercept -3. $y = \dfrac{3}{8}x - 3$

4. Find an equation of the line with y-intercept 7 and having the same slope as the line passing through $(-4, -3)$ and $(6, 9)$.

$y = \dfrac{6}{5}x + 7$

Additional Answers
Written Exercises

1. a. $\dfrac{\log 4.02}{\log 1.59}$

2. a. $\dfrac{-\log 18.6}{\log 7.4}$

(continued on p. 480)

Key Ideas

Use logarithms to solve exponential growth and decay problems.

Chalkboard Examples

1. A colony of bacteria consists of 3.9×10^6 bacteria at 3:30 P.M. If the population quadruples every 30 min, how many bacteria will there be in the colony at 5:15 P.M.?
Choose 3:30 P.M. as $t = 0$. Using the formula with $N_0 = 3.9 \times 10^6$, $b = 4$, $t = 105$, $k = 30$, you have
$N = (3.9 \times 10^6)4^{\frac{105}{30}}$
$= (3.9 \times 10^6)4^{\frac{7}{2}}$
Solving by logarithms,
$\log N = \log (3.9 \times 10^6) +$
$\frac{7}{2} \log 4$
$= 6.5911 + \frac{7}{2}(0.6021)$
$= 8.69845$
Take the antilogarithm of both sides:
$N = 4.99 \times 10^8$
There will be approximately 4.99×10^8 bacteria at 5:15 P.M.

2. An isotope of a certain unstable element has a half-life of about 28 min. How long would it take for a given sample to decay to 37% of its original mass?

11–8 Exponential Growth and Decay

In many practical situations a certain quantity changes so that in a fixed length of time it grows (or diminishes) by a fixed percent. As an example of such a process, which is called exponential growth (or decay), suppose the population of a certain town doubles every 30 years. That is, the population is multiplied by 2 in any 30-year period. The table on the left below gives some values for the population P of such a town t years after a certain year that is arbitrarily designated as $t = 0$. For comparison, a table illustrating a *constant* increase in population (by 5000 every 30 years) is shown at the right. (Such growth is said to be *linear*.)

Exponential Growth

t	P
0	1,000
30	2,000
60	4,000
90	8,000
120	16,000
150	32,000

Linear Growth

t	P
0	1,000
30	6,000
60	11,000
90	16,000
120	21,000
150	26,000

By observing that the values of P in the table on the left form a geometric sequence, you may be able to guess that a formula for P as a function of t is

$$P = 1000 \cdot 2^{\frac{t}{30}}.$$

Written in the form

$$P = 1000 \cdot (2^{\frac{1}{30}})^t,$$

the formula resembles the one for the general term of a geometric sequence (page 307). In general, a quantity N that grows (or diminishes) by a factor b in any period of fixed length k can be expressed as a function of time t by

$$N = N_0 \cdot b^{\frac{t}{k}},$$

where N_0 is the value of N when $t = 0$.

EXAMPLE 1 A colony of bacteria consists of 1.6×10^5 bacteria at noon. If the bacteria population triples every 40 min, how many bacteria will there be in the colony at 1:40 P.M.?

SOLUTION Choose noon as $t = 0$. Using the formula above with $N_0 = 1.6 \times 10^5$, $b = 3$, $t = 100$, and $k = 40$, you have:

$$N = (1.6 \times 10^5) \cdot 3^{\frac{100}{40}} = (1.6 \times 10^5) \cdot 3^{\frac{5}{2}}$$

464   *Chapter 11*

You can find N directly using a scientific calculator. Otherwise, using logarithms, you have:

$$\log N = \log (1.6 \times 10^5) + \tfrac{5}{2} \log 3$$
$$= 5.2041 + \tfrac{5}{2}(0.4771)$$
$$= 6.3969$$
$$N = 2.49 \times 10^6 \quad \text{(or 2,490,000)}$$

There will be approximately 2.49×10^6 bacteria at 1:40 P.M.

Often in describing exponential decay we state the **half-life** of a particular substance. This is the length of time that must elapse before only half of the original amount of the substance is left unchanged (by some physical process). If the half-life of a given substance is k, the amount N remaining after time t is given by the formula on page 464, with $b = \tfrac{1}{2}$:

$$N = N_0 \cdot \left(\frac{1}{2}\right)^{\frac{t}{k}}$$

EXAMPLE 2   An isotope of the synthetic element Californium has a half-life of about 45 min. How long would it take for a given sample to decay to 15% of its original mass?

SOLUTION   Let N_0 = the original mass of the sample and let t = the desired time (in minutes). Then the amount left at time t will be 15% of N_0, or $0.15N_0$. Thus, by the formula above:

$$0.15N_0 = N_0 \cdot \left(\frac{1}{2}\right)^{\frac{t}{45}}$$
$$0.15 = \left(\frac{1}{2}\right)^{\frac{t}{45}}$$
$$\log 0.15 = \frac{t}{45} \log \frac{1}{2}$$
$$t = \frac{45 \log 0.15}{\log \frac{1}{2}}$$
$$= \frac{45(-0.8239)}{(-0.3010)}$$
$$= 123 \quad \text{(to three significant digits)}$$

∴ it would take about 123 min.

Another important instance of exponential growth is the accumulation of money in a bank account that earns compound interest. The balance in a bank account that earns interest compounded, say, semiannually (twice a year) at an annual rate of 6% is increased by 3% every half-year and is thus multiplied by 1.03 in any half-year period. The amount A in the account after t years is thus given by the formula

Let N_0 = the original mass of the sample and let t = the desired time (in minutes). Then the amount left at time t will be 37% of N_0, or $0.37N_0$. Thus, by the formula on p. 465

$$0.37N_0 = N_0\left(\frac{1}{2}\right)^{\frac{t}{28}}$$
$$0.37 = \left(\frac{1}{2}\right)^{\frac{t}{28}}$$
$$\log 0.37 = \frac{t}{28} \log \frac{1}{2}$$
$$t = \frac{28 \log 0.37}{\log \frac{1}{2}}$$
$$= \frac{28(-0.4318)}{-0.3010}$$
$$= 40.2 \text{ (to three significant digits)}$$

∴ it would take 40.2 min.

3. How many years would it take for $2000 to grow to $18,900 in a savings certificate that pays interest at an annual rate of 10%, compounded semiannually?

Use the compound interest formula for $P = 2000$, $A = 18,900$, $r = 0.10$, and $n = 2$.

$$A = P\left(1 + \frac{r}{n}\right)^{nt}$$
$$18,900 = 2000\left(1 + \frac{0.10}{2}\right)^{2t}$$
$$18.9 = 2(1.05)^{2t}$$
$$\log 18.9 = \log 2 + 2t(\log 1.05)$$
$$t = \frac{\log 18.9 - \log 2}{2 \log 1.05}$$
$$= \frac{1.2765 - 0.3010}{2(0.0212)}$$
$$= 23.0$$

∴ it would take 23 years.

Additional A Exercises

1. How long will it take for an amount of money invested at an annual rate of 12%, compounded semiannually, to quadruple? 11.9 years

2. An isotope of the synthetic element Californium has a half-life of about 45 min. How long will it take for a given sample to decay to 40% of its original mass? 59.5 minutes

3. A bacteria culture is found to double in size every 19 min. How many minutes will it take for a culture of 1.7×10^{10} bacteria to grow to 2.8×10^{11}? 76.9 minutes

4. The population of a certain city grows 11% every year. The population was 400,000 six years ago. What is it now? 748,000

$N = N_0 \cdot b^{\frac{t}{k}}$, with $N_0 = P$, the original amount invested, or *principal*, $b = 1.03$, and $k = \frac{1}{2}$. Then

$$\frac{t}{k} = \frac{t}{\frac{1}{2}} = 2t,$$

and

$$A = P(1.03)^{2t}$$

This suggests that if a principal P is deposited in an account that earns interest compounded n times a year at an annual rate r (expressed as a decimal), the amount in the account A after t years is given by

$$A = P\left(1 + \frac{r}{n}\right)^{nt}.$$

This formula is often referred to as the **compound interest formula.**

EXAMPLE 3 How many years would it take for $1000 to grow to $7450 in a savings certificate that pays interest at an annual rate of 12%, compounded quarterly?

SOLUTION Use the compound interest formula for $P = 1000$, $A = 7450$, $r = 0.12$, and $n = 4$.

$$A = P\left(1 + \frac{r}{n}\right)^{nt}$$

$$7450 = 1000\left(1 + \frac{0.12}{4}\right)^{4t}$$

$$7.45 = (1.03)^{4t}$$

$$\log 7.45 = 4t \log 1.03$$

$$t = \frac{\log 7.45}{4 \log 1.03} = \frac{0.8722}{4(0.0128)}$$

$$= 17.0 \quad \text{(to three significant digits)}$$

∴ it would take 17.0 years.

Problems

A **1.** Find the amount in an account after 15 years, if $5000 was originally invested and the account earns 8% annual interest, compounded quarterly. $16,400

2. How long would it take for an amount of money invested at an annual rate of 8%, compounded semiannually, to triple? 14.0 yr

3. Eighty-eight grams of the synthetic element Berkelium is found to decay radioactively to 56 g in 3 h. If this decay is exponential, what is the half-life of Berkelium? approx. 4.6 h

466 *Chapter 11*

4. A bacteria culture is found to double in size every 24 min. How many minutes would it take for a culture of 3.2×10^5 bacteria to grow to 7.61×10^5 bacteria? approx. 30 min

5. The population of Hope Springs grows by 5% each year. The population is now 3470. How many years ago was it 1000? approx. 25.5 yr

6. Forty years ago the population of a certain town was 2000. It is now 12,000. Assuming exponential growth, in how many years will the population be 45,000?
approx. 29.5 yr

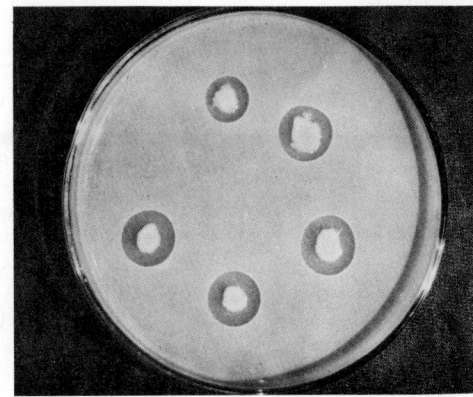

B

7. The charge on a capacitor decays exponentially with time so that after 1 second the charge is 62% of its original value. After how many seconds will it be 5.68% of its original value? approx. 6 s

8. Carbon-14, which has a half-life of 5730 years, is used to calculate the age of fossilized animal remains. If 4.5% of the animal's original amount of carbon-14 is found in its remains, how long ago did it die?
approx. 25,600 yr

C

9. The monthly payment M on a mortgage is given by

$$M = \frac{rP}{1 - (1 + r)^{-n}},$$

where r is the *monthly* rate of interest, P is the principal, and n is the number of months in the life of the mortgage. If a mortgage at 12% *annual* interest on a principal of $62,000 has a monthly payment of $653, what is the life of the mortgage? 25 yr

10. Atmospheric pressure is approximately halved for each 4.8 km increase in elevation above sea level. What is the difference in elevation of two points at which the atmospheric pressures are 50 kPa (kilopascals) and 2.21 kPa, respectively? 21.6 km

Computer Exercises   For students with computer experience

1. Given an annual rate of interest r and the number of times per year n that this interest is to be compounded, write a program that will compute the value of an investment at the end of each interest period and report the time in years that must elapse before this value is at least double the original value.

Mixed Review

Use Table 1 on page 746, if necessary, to solve the problems below.

1. Find the area of a rectangle with a base of length 10 ft and a diagonal of length 26 ft. 240 ft^2

2. A sphere has a diameter of 7 in. Find the volume of the sphere.
$\frac{343\pi}{6}$ in.^{3}

3. A trapezoid of area 16 m^2 has an altitude of measure 2 m and a base of length 10 m. Find the length of the other base.   6 m

Exponents and Logarithms   **467**

Run the program in Exercise 1 for each rate _r_ and number of interest periods per year _n_.

2. $r = 6\%$, $n = 2$ 12 yr

3. $r = 8\%$, $n = 4$ 9 yr

4. $r = 10\%$, $n = 4$ 7.25 yr

5. $r = 15\%$, $n = 12$ 4.7 yr

6. Modify the program in Exercise 1 so that when the initial value and the final value of an investment are input, the output will report how long it will take for the given initial investment to grow to at least the given final value.

Run the program in Exercise 6 for each initial value _P_, final value _A_, rate _r_, and number of interest periods per year _n_.

7. $P = \$3000$, $A = \$5500$, $r = 8\%$, $n = 4$ 7.75 yr

8. $P = \$5000$, $A = \$12,000$, $r = 10\%$, $n = 2$ 9 yr

9. $P = \$8000$, $A = \$20,000$, $r = 12\%$, $n = 12$ 7.75 yr

Teaching Suggestions
p. T110

Related Activities p. T110

Supplementary Material

Test 35

Key Ideas

Introduce the number _e_ and natural logarithms.
Apply natural logarithms to compound interest problems.

11–9 Natural Logarithms

Although common (base 10) logarithms are the most convenient for numerical computation, most theoretical mathematics involving logarithms uses the irrational number _e_ for the base. This number is named _e_ in honor of the Swiss mathematician Leonhard Euler (1707–1783), who investigated the number and discovered many of its properties. The number _e_ is defined (using the notation of Section 8-7) by the following equation.

$$e = \lim_{k \to \infty} \left(1 + \frac{1}{k}\right)^k$$

Euler was the first to prove that this limit exists. The table below shows some approximations for $\left(1 + \frac{1}{k}\right)^k$ as _k_ gets larger.

k	10	100	1000	10,000	100,000
$\left(1 + \frac{1}{k}\right)^k$	2.5937	2.7048	2.7169	2.7181	2.7183

In fact,

$$e = 2.718281828459045 \ldots.$$

While this may seem a cumbersome number for the base of a system of logarithms, theoretical work with logarithms is much simpler with base e than with any other base. You will also learn in calculus that the shaded region in Figure 10 has area exactly equal to

$$\log_e r.$$

Logarithms to the base e are called **natural logarithms**. Just as we use the notation $\log x$ to mean $\log_{10} x$, we usually write

$$\log_e x \quad \text{as} \quad \ln x.$$

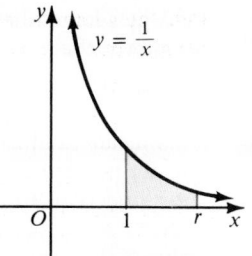

Figure 10

EXAMPLE 1 Express each statement in logarithmic form.
    a. $e^2 = 7.389$ **b.** $e^x = 12$

SOLUTION   **a.** $\ln 7.389 = 2$ **b.** $\ln 12 = x$

EXAMPLE 2 Express each statement in exponential form.
    a. $\ln 148 = 5$ **b.** $\ln x = 3$

SOLUTION   **a.** $e^5 = 148$ **b.** $e^3 = x$

EXAMPLE 3 Write as a single logarithm.
    a. $3 \ln x - (\ln y + \frac{1}{2} \ln z)$ **b.** $\ln 20 + 4 - 2 \ln 2$

SOLUTION   **a.** $3 \ln x - (\ln y + \frac{1}{2} \ln z) = \ln x^3 - (\ln y + \ln \sqrt{z})$
$$= \ln x^3 - \ln(y\sqrt{z})$$
$$= \ln \frac{x^3}{y\sqrt{z}}$$

    b. $\ln 20 + 4 - 2 \ln 2 = (\ln 20 + \ln e^4) - \ln 2^2$
$$= \ln 20e^4 - \ln 4$$
$$= \ln \frac{20e^4}{4}$$
$$= \ln 5e^4$$

The number e also plays a role in the calculation of compound interest. Recall (page 466) that after one year the amount A in a bank account begun with a principal P and earning interest at an annual rate r, compounded n times per year, is given by:

$$A = P\left(1 + \frac{r}{n}\right)^n$$
$$= P\left[\left(1 + \frac{r}{n}\right)^{\frac{n}{r}}\right]^r$$

If we let $k = \frac{n}{r}$, this formula can be written:

$$A = P\left[\left(1 + \frac{1}{k}\right)^k\right]^r.$$

Chalkboard Examples

1. Express each statement in logarithmic form.
  a. $e^6 = 403.43$
    $\ln 403.43 = 6$
  b. $e^{3x} = 908$ $\ln 908 = 3x$

2. Express each statement in exponential form.
  a. $\ln 17 = 2.833$
    $e^{2.833} = 17$
  b. $\ln 5 = 7x$ $e^{7x} = 5$

3. Write as a single natural logarithm.
  a. $\ln a - (5 \ln b + 7 \ln c)$
    $= \ln a - (\ln b^5 + \ln c^7)$
    $= \ln a - \ln(b^5 c^7)$
    $= \ln\left(\frac{a}{b^5 c^7}\right)$

  b. $7 - \left(\ln 6 + \frac{1}{2} \ln 54\right)$
    $= \ln e^7 - (\ln 6 + \ln \sqrt{54})$
    $= \ln e^7 - \ln(6\sqrt{54})$
    $= \ln\left(\frac{e^7}{6\sqrt{54}}\right)$

Additional A Exercises

Express each statement in logarithmic form.

1. $e^{-3} = 0.0498$
In 0.0498 = −3

2. $e^{\frac{3}{8}} = 1.455$ In 1.455 = $\frac{3}{8}$

Express each statement in exponential form.

3. ln 123 = 4.812
$e^{4.812} = 123$

4. ln 0.09 = −2.408
$e^{-2.408} = 0.09$

5. Simplify the expression
$\frac{1}{2}$ ln 2 + ln 8 − (5 ln y + ln 6).   In $\frac{8\sqrt{2}}{6y^5}$

6. Use the table on page 470 to find the value of the investment after one year if the interest is compounded continuously at the rate of 10% annually, on a principal of $3000.
A = $3315

As k gets large, the number inside the brackets approaches e, and it can be shown that:

$$\lim_{k \to \infty} P\left[\left(1 + \tfrac{1}{k}\right)^k\right]^r = Pe^r$$

This means that as the compounding of interest occurs a greater number of times per year, the value of the account after one year approaches Pe^r. We say that if interest is compounded continuously (or instantaneously), then the value of the bank account equals this number. That is, after one year:

$$A = Pe^r$$

The table at the right, which gives values of e^r for selected values of r, can be used to compute continuously compounded interest.

r	e^r
0.01	1.094
0.02	1.020
0.03	1.030
0.04	1.041
0.05	1.051
0.06	1.062
0.07	1.073
0.08	1.083
0.09	1.094
0.10	1.105
0.11	1.116
0.12	1.127
0.13	1.139
0.14	1.150
0.15	1.162

EXAMPLE 4   Find the value of an investment of $8000 after one year if interest is compounded continuously at 10%.

SOLUTION $A = Pe^r = 8000e^{0.10}$
= 8000(1.105)   (from the table)
= 8840   (to three significant digits)

Note that the continuously compounded interest accumulated in one year in the example above is $840, about 5% more than the $800 that would have been earned by the same amount of money if interest had been compounded *annually*.

If a principal P is invested at an annual rate r, compounded continuously for t years, then the value A of the investment is given by

$$A = Pe^{rt}.$$

Oral Exercises

Evaluate each expression.

1. ln e 1   **2.** ln 1 0   **3.** ln e^4 4   **4.** ln $\sqrt{e}$ $\frac{1}{2}$ **5.** ln $\frac{1}{e}$ -1

6. $e^{\ln 2}$ 2   **7.** e^0 1   **8.** $e^{\ln 2 + \ln 5}$ 10   **9.** $e^{2 \ln 3}$ 9   **10.** $(e^{\ln 5})^3$ $\frac{1}{125}$

11. State an expression for the amount of money in a bank account at the end of one year if interest is compounded daily (365 times per year) at an annual rate of 100% on a principal of $1. Do you think that the value of this expression is close to e? $(1 + \frac{1}{365})^{365} \approx 2.71$; $e \approx 2.72$

12. For any positive number x, give a formula for ln x in terms of common logarithms. $\frac{\log x}{\log e}$

470   *Chapter 11*

Written Exercises

Express each statement in logarithmic form.

A **1.** $e^{2.5} = 12.18$ $\ln 12.18 = 2.5$

3. $e^{\frac{1}{5}} = 1.221$ $\ln 1.221 = \frac{1}{5}$

2. $e^{-2} = 0.135$ $\ln 0.135 = -2$

4. $\sqrt{e} = 1.649$ $\ln 1.649 = \frac{1}{2}$

Express each statement in exponential form.

5. $\ln 8 = 2.079$ $e^{2.079} = 8$

7. $\ln 0.1 = -2.303$ $e^{-2.303} = 0.1$

6. $\ln 0.5 = -0.693$ $e^{-0.693} = 0.5$

8. $\ln 0.01 = -4.605$ $e^{-4.605} = 0.01$

Simplify each expression.

9. $\ln 48 - 4 \ln 2$ $\ln 3$

11. $\frac{1}{2}(\ln 45 + \ln 5) - 2 \ln 3$ $\ln \frac{5}{3}$

13. $e^{\ln 6 + \ln 7}$ 42

15. $e^{\ln 8 - \ln 6}$ $\frac{4}{3}$

10. $\frac{1}{2} \ln 9 + \ln 12 - 2 \ln 3$ $\ln 4$

12. $\ln 6 + \ln 30 - (\ln 5 + 3 \ln 2)$ $\ln \frac{9}{2}$

14. $e^{2 \ln 7}$ 49

16. $e^{\frac{1}{2} \ln 3}$ $\sqrt{3}$

Use the table on page 470 to find the value of each investment after one year if interest is compounded continuously at the given annual rate r on the principal P.

17. $r = 8\%$; $P = \$7000$ $\$7581$

19. $r = 13\%$; $P = \$5500$ $\$6264.50$

18. $r = 11\%$; $P = \$4500$ $\$5022$

20. $r = 9\%$; $P = \$2200$ $\$2406.80$

Find the value of each investment after t years if interest is compounded continuously at the given annual rate r on the principal P.

21. $t = 2$ yr; $r = 7\%$; $P = \$6000$ $\$6900$

22. $t = 2\frac{1}{2}$ yr; $r = 6\%$; $P = \$7500$ $\$8715$

Use the table on page 470 and the given data to solve each equation for an approximate value of x. (Hint: Take natural logarithms of both sides.)

B **23.** $x^{10} = 3$; $\ln 3 = 1.099$ $\{1.12\}$

25. $x^{44} = 9$; $\ln 9 = 2.197$ $\{1.05\}$

24. $x^{20} = 6$; $\ln 6 = 1.792$ $\{1.09\}$

26. $x^{33} = 10$; $\ln 10 = 2.30$ $\{1.07\}$

27. Graph $y = e^x$ and $y = \ln x$ on the same set of axes.

In Exercises 28 and 29 you can use the facts that

$$\lim_{k \to \infty} f(k)g(k) = \lim_{k \to \infty} f(k) \cdot \lim_{k \to \infty} g(k) \text{ and } \lim_{k \to \infty} \frac{1}{f(k)} = \frac{1}{\lim_{k \to \infty} f(k)}.$$

C **28.** Show that $\lim_{k \to \infty} \left(1 + \frac{1}{k}\right)^{k+1} = e$.

29. Show that $\lim_{n \to \infty} \left(1 - \frac{1}{n}\right)^n = \frac{1}{e}$. (Hint: Simplify the expression inside parentheses, let $k = n - 1$, and use the result of Exercise 28.)

Exponents and Logarithms **471**

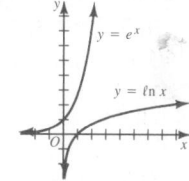

Computer Exercises For students with computer experience

1. Write a program that will use the formula

$$e = 1 + \frac{1}{1} + \frac{1}{2 \cdot 1} + \frac{1}{3 \cdot 2 \cdot 1} + \frac{1}{4 \cdot 3 \cdot 2 \cdot 1} + \cdots$$

to approximate the value of e by finding the partial sum of a given number of terms.

5. approx. 2.7182818

Run the program in Exercise 1 for the given number of terms.

2. 5   **3.** 8   **4.** 10   **5.** 15
approx. 2.7083   approx. 2.718254   approx. 2.7182815

ON THE CALCULATOR

In order to facilitate comparison of rates that are compounded in different ways, an interest rate is often accompanied by the corresponding *effective annual rate*, or *annual yield*. The *effective annual rate* is the rate that would have to be paid at the end of one year to yield the same amount of annual interest.

Exercises

In the following exercises use a scientific calculator and give each answer to four decimal places.

1. Find the effective annual rate corresponding to each continuously compounded rate.
   a. 7% 7.2508%   **b.** 8% 8.3287%   **c.** 11% 11.6278%   **d.** 14% 15.0274%

2. Find the continuously compounded rate corresponding to each effective annual rate.
   a. 8% 7.6961%   **b.** 9% 8.6178%   **c.** 14% 13.1028%   **d.** 10.5% 9.9845%

3. Complete the table with each effective annual rate.

compounded	9%	9.25%	9.5%	9.75%	10%
quarterly	?	?	?	?	?
monthly	?	?	?	?	?
continuously	?	?	?	?	?

4. What continuously compounded interest rate would be needed in order to double your money in 5 years? 13.8629%

5. What continuously compounded interest rate would be needed to triple your money in 9 years? 12.2068%

Self-Test 3

VOCABULARY   common logarithm (p. 458)   half-life (p. 465)
characteristic (p. 459)   compound interest formula
mantissa (p. 459)   (p. 466)
antilogarithm (p. 459)   natural logarithm (p. 469)
exponential equation (p. 461)   compounded continuously
calculation-ready form (p. 462)   (p. 470)
exponential growth (or decay)
(p. 464)

Evaluate.

1. $\log 0.00152$ -2.8182 **2.** $\log \sqrt{\dfrac{48 \cdot 12}{30^2}}$ -0.0969 *Obj. 1, p. 458*

Solve for x.

3. $5.2x^3 = 13.4$ $\{1.37\}$ **4.** $x^{\frac{2}{5}} = \dfrac{421}{26.3}$ $\{1030\}$

Solve each equation.
a. Give the solution in calculation-ready form.
b. Give the solution to three significant digits.

5. $4(2.9)^{2x} = \sqrt[3]{34.6}$ $\{-0.0963\}$ **6.** $7^{x-1} = 3^{4x}$ $\{-0.795\}$ *Obj. 2, p. 458*

7. If the population of a certain town grows by 5% each year, how long   *Obj. 3, p. 458*
will it take the population to triple? approx. $22\frac{1}{2}$ yr

8. How many years ago was $5000 invested in an account paying 8%
annual interest compounded quarterly, if the amount presently in
the account is $11,500? approx. $10\frac{1}{2}$ yr

Simplify each expression.

9. $e^{\ln 6 - \ln 18}$ $\frac{1}{3}$ **10.** $\ln 6 + \ln \frac{2}{3} - (\ln \frac{1}{8} + \ln 32)$ 0   *Obj. 4, p. 458*

11. Find the value of an investment of $6000 after $1\frac{1}{2}$ years if the interest
is compounded continuously at 8%. $6765

Check your answers with those at the back of the book.

Chapter Summary

1. Radical expressions may be written equivalently in exponential
form: $(\sqrt[r]{b})^p = b^{\frac{p}{r}}$, provided $\sqrt[r]{b}$ and p are real numbers.
2. The laws of exponents apply to real-number exponents for positive
bases.

Exponents and Logarithms **473**

Quick Quiz
Evaluate.
1. $\log 381$ $\quad 2.5089$
2. $\log 14^5 \sqrt[3]{\dfrac{29}{315}}$ $\quad 5.3853$
Solve for x.
3. $68x^{-2} = 29.9$ $\quad \{1.51\}$
4. $x^{\frac{3}{7}} = \dfrac{902}{17.8}$ $\quad \{9502\}$
Solve each equation.
a. Give the answer in
calculation-ready form.
b. Give the answer to three
significant digits.
5. $17(8)^{3x+1} = \sqrt[5]{6.5}$
a. $x =$
$$\dfrac{\frac{1}{5}\log 6.5 - \log 17}{3 \log 8} - \dfrac{1}{3}$$
b. $x = -0.727$
6. $5^{2x+9} = 2^{4x-2}$
a. $x = -\dfrac{9 \log 5 + 2 \log 2}{2 \log 5 - 4 \log 2}$
b. $x = -35.6$
7. A bacteria culture triples
every 20 min. At 10 A.M.
the culture contains
5.8×10^7 bacteria. How
many bacteria are in the
culture at 11:10 A.M.?
2.71×10^9
8. In seven years, what will
be the value of a $45,000
savings certificate with an
annual interest rate of
13%, compounded
quarterly? $110,188
Simplify each expression.
9. $e^{\ln 18 + \ln 9}$ $\quad 162$
10. $\ln 14 + \ln \dfrac{1}{7} - \left(\dfrac{1}{3} \ln 24 + \right.$
$\left. \ln 48 \right)$ $\ln \dfrac{1}{48\sqrt[3]{3}}$
11. Find the value of an in-
vestment of $5,400 after
2 years if the interest is
compounded continuous-
ly, at a rate of 7%.
$6211

3. Every one-to-one function has an *inverse* function. If f^{-1} is the inverse of a one-to-one function f, then $f^{-1}(f(x)) = x$ for each x in the domain of f, and $f(f^{-1}(x)) = x$ for each x in the domain of f^{-1}.

4. The inverse of the *exponential function* with equation $y = b^x$, $b > 0$, $b \neq 1$ is the *logarithmic function* with equation $y = \log_b x$, $b > 0$, $b \neq 1$, $x > 0$. If $x \in R$, $x > 0$, $b > 0$, $b \neq 1$, then $b^{\log_b x} = x$; if $x \in R$, $b > 0$, $b \neq 1$, then $\log_b b^x = x$.

5. The laws of exponents are the basis for the *laws of logarithms*:

$$\log_b x_1 x_2 = \log_b x_1 + \log_b x_2 \qquad \log_b \frac{x_1}{x_2} = \log_b x_1 - \log_b x_2$$

$$\log_b x_1{}^n = n \log_b x_1$$

6. The *characteristic* of the common logarithm of a number may be found by inspection of the number in scientific notation; the *mantissa* is determined from a table.

7. To find the logarithm of a number to another base, you use the Change of Base Formula:

$$\log_x y = \frac{\log_b y}{\log_b x}$$

8. There are certain types of equations for which logarithms provide the best, or the only, method of solution. Logarithms are used in studying exponential equations and their applications to such phenomena as radioactive decay, population growth, and compound interest.

9. Logarithms to the base $e = 2.71828 \ldots$ are *natural logarithms*.

Chapter Review

Write the letter of the correct answer.

1. Express $\sqrt[5]{4\sqrt[10]{32}}$ in simplest radical form.   11-1
   a. $\sqrt[50]{128}$ (b.) $\sqrt{2}$ c. $\sqrt{128}$ d. $\sqrt[10]{2^{21}}$

2. Solve $3^{2x+3} = (\frac{1}{27})^{4+x}$ over R.   11-2
   a. $\{-9\}$ b. $\{1\}$ c. $\{\frac{1}{27}\}$ (d.) $\{-3\}$

3. Which of the following functions is *not* one-to-one?   11-3
   a. $y = -2x^3$ (b.) $y = \frac{x^2}{4}$ c. $y = \frac{2}{x}$, $x \neq 0$ d. $y = -\frac{3}{4}x - 2$

4. Evaluate $\log_{\sqrt{5}} 25$.   11-4
   a. $\frac{1}{2}$ b. 2   (c.) 4   d. 8

5. Solve $\log_x \frac{2}{5} = \frac{1}{3}$.
   a. $\{\frac{4}{25}\}$ b. $\{\frac{2}{15}\}$ (c.) $\{\frac{8}{125}\}$ d. $\{\frac{6}{5}\}$

474   *Chapter 11*

6. Evaluate $\frac{1}{3} \log_8 8 - 4 \log_8 2$. — *11-5*

    a. $-\frac{2}{3}$ **b.** $\frac{3}{2}$ **c.** 1   **(d.)** -1

7. Solve $\log_2 (x^2) - \log_2 (x + 3) = 2$.

    (a.) $\{6, -2\}$ **b.** $\{6\}$ **c.** $\varnothing$ **d.** $\{-6, 2\}$

8. Evaluate $\log \left(\dfrac{20\sqrt{3.24}}{\sqrt[3]{1000}} \right)$. — *11-6*

    (a.) 0.5563   **b.** -0.4437 **c.** 0.8116   **d.** -1.4437

9. Solve $8^{1+x} = 5^{3-2x}$ for x. — *11-7*

    a. $\{\frac{2}{3}\}$ **b.** $\{0.143\}$ **(c.)** $\{0.519\}$ **d.** $\{0.627\}$

10. Find the amount in an account after 9 years if \$2600 was invested — *11-8*
originally and the account earns $7\frac{1}{2}\%$ annual interest, compounded
quarterly.

    (a.) \$5070   **b.** \$2800   **c.** \$3070   **d.** \$4980

11. Simplify $\frac{1}{3} (\ln 48 - \frac{1}{2} \ln 36) + \ln 3$. — *11-9*

    a. $\ln 13$ **b.** e^6 **c.** $\ln 5$ **(d.)** $\ln 6$

Chapter Test

Express in simplest radical form.

1. $\dfrac{\sqrt[15]{125}}{\sqrt[10]{25}}$ 1   **2.** $\sqrt[8]{64} \cdot \sqrt[12]{64}$ $2\sqrt[4]{2}$ — *11-1*

3. Solve $(\frac{1}{7})^{1-3x} = (49)^{x+3}$. $\{7\}$ — *11-2*

4. Find the inverse of the function $y = \sqrt{3 - x}$ with domain $f^{-1}(x) = 3 - x^2$, — *11-3*
$\{x: x \le 3\}$. Graph both the function and its inverse. $x \ge 0$

Solve.

5. $\log_9 x = \frac{3}{2}$ $\{27\}$ **6.** $\log_x 32 = -5$ $\{\frac{1}{2}\}$ — *11-4*

7. $\log_3 2x = 2 \log_3 4 - 3 \log_3 2$ $\{1\}$ **8.** $\log_4 x + \log_4 (x - 3) = 1$ $\{4\}$ — *11-5*

9. $24x^9 = 144$ $\{1.22\}$ **10.** $\sqrt[4]{31x^2} = 103$ $\{1910\}$ — *11-6*

11. $2^{3x} = 34$ $\{1.70\}$ **12.** $72^{x-1} = 19^x$ $\{3.21\}$ — *11-7*

13. If the population of a certain town grows exponentially and was — *11-8*
12,000 in 1967 and 21,000 in 1982, in what year will the population
reach 36,750? 1997

14. Find the rate of interest on an account in which a \$1200 investment — *11-9*
has grown to \$1326 in 1 year, if the interest on the account is
compounded continuously. 10%

Exponents and Logarithms **475**

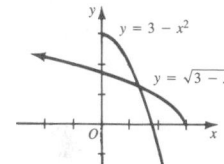

7.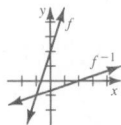

8.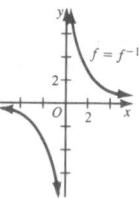

9.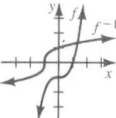

10.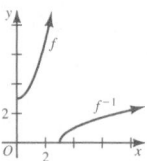

11.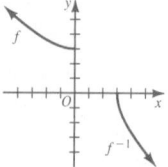

12.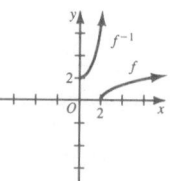

13. $f^{-1}(f(x)) = \dfrac{(2\sqrt{x})^2}{4} = \dfrac{4x}{4} =$

$x; \; f(f^{-1}(x)) = 2\sqrt{\dfrac{x^2}{4}} =$

$2 \cdot \dfrac{x}{2} = x$

APPLICATION

Carbon Dating

The radioactive decay of the unstable nucleus of an uncommon form of carbon permits scientists to determine the approximate age of archaeological objects. Atomic nuclei are composed largely of *protons* (positively charged particles) and *neutrons* (electrically neutral particles). The most common form of carbon has 6 protons and 6 neutrons in each nucleus; it is called carbon-12, or C-12, where the 12 refers to the total number of protons and neutrons. In the atmosphere there is a small amount of another form of carbon, C-14, with 6 protons and 8 neutrons, that is formed from the bombardment of nitrogen by neutrons that originate in outer space.

C-14 has an unstable nucleus that achieves stability over long periods of time by emitting a negatively charged particle, or *electron*. This radioactive decay occurs when one of the neutrons is transformed into a proton, which remains in the nucleus, and an electron, which is emitted. The end product is a stable nitrogen atom whose nucleus has 7 protons and 7 neutrons.

Living plants absorb C-14 from the atmosphere in the form of carbon dioxide. While the plant is alive, equilibrium is reached between the rate at which C-14 is absorbed and the rate at which it decays. When the plant dies, the intake of C-14 ceases, but the C-14 that is present continues to decay. Scientists can measure the amount of C-14 remaining in the plant and use that information to estimate the time that has elapsed since the plant's death. Since plants form the basis of the food chain for animals, leather and bone can be dated in the same way.

Carbon dating is based on an assumption that the level of atmospheric C-14 is the same today as it was when the organism was alive. However, it is now known that variations in atmospheric C-14 exist, with

the most major ones occurring before 1000 B.C. and after 1900. For example, rings of bristlecone pine trees have been used to show that a sample with a carbon dating age of 7500 years has a true age of 8200 years. With appropriate corrections for carbon dating ages obtained, carbon dating remains an important tool.

The number n of C-14 nuclei per gram of carbon extracted from a piece of wood or bone at time t can be written as

$$N = N_0 e^{-\lambda t},$$

where N_0 is the number of such nuclei that were present when the object was part of a living organism and λ (the Greek letter *lambda*) is the *decay constant* for C-14. Assuming a half-life of 5730 years for C-14, you can find the value of λ.

$$\tfrac{1}{2} N_0 = N_0 e^{-\lambda(5730)}$$
$$\tfrac{1}{2} = e^{-5730\lambda}$$
$$e^{5730\lambda} = 2$$
$$5730\,\lambda = \ln 2$$
$$\lambda = \frac{\ln 2}{5730} \approx 1.21 \times 10^{-4}$$

The value of λ that corresponds to a time in years is 1.21×10^{-4}, to three significant digits.

EXAMPLE 1   Three quarters of the C-14 from an old bison bone has decayed since the bison died. About how old is the bone?

SOLUTION $\tfrac{1}{4} N_0 = N_0 e^{-\lambda t}$

$$\tfrac{1}{4} = e^{-\lambda t}$$
$$e^{\lambda t} = 4$$
$$\lambda t = \ln 4$$
$$t = \frac{\ln 4}{1.21 \times 10^{-4}} \approx 11{,}500 \text{ (to three significant digits)}$$

∴ the bone has a carbon dating age of 11,500 years.

Carbon dating usually involves measuring the *rate of decay* of C-14. In calculus the rate of decay is written as $\dfrac{\Delta N}{\Delta t}$, where ΔN represents the small number of atoms that decay in an extremely short time interval, Δt. The rate of decay can be written as

$$\frac{\Delta N}{\Delta t} = -\lambda N,$$

and thus

$$\frac{\Delta N}{\Delta t} = -\lambda N_0 e^{-\lambda t}.$$

14. $f^{-1}(f(x)) = (\sqrt{x-1})^2 + 1 = x - 1 + 1 = x$; $f(f^{-1}(x)) = \sqrt{(x^2 + 1) - 1} = \sqrt{x^2} = x$

15.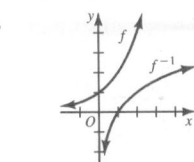

16. The segment joining (a, b) and (b, a) has slope $\dfrac{b-a}{a-b} = -1$ and midpoint $\left(\dfrac{a+b}{2}, \dfrac{a+b}{2}\right)$. Since $y = x$ passes through this point and has slope 1, it is the perpendicular bisector of the segment.

17. A function that is not one-to-one; $\{x: x \geq 1\}$; $f^{-1}(x) = \sqrt{x+1} + 1$; $\{x: x \geq -1\}$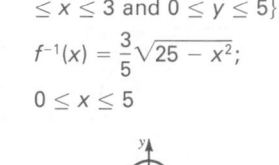

18. Not a function; $\{(x, y): 0 \leq x \leq 3 \text{ and } 0 \leq y \leq 5\}$; $f^{-1}(x) = \dfrac{3}{5}\sqrt{25 - x^2}$; $0 \leq x \leq 5$

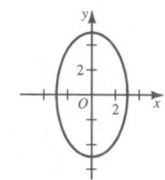

EXAMPLE 2   **a.** The rate of decay is measured to be 15.3 nuclei per minute for 1 g of carbon from a living tree. How many C-14 atoms are contained in this gram of carbon?

b. The rate of decay per gram of carbon from a chariot spoke is 12.0 nuclei per minute. Estimate the time that has elapsed since the wood from which the spoke was made was cut.

SOLUTION   **a.** Since we have calculated λ in terms of years, we multiply 15.3 by 60 × 24 × 365 to determine the decay rate in nuclei per year. Since C-14 is decaying, the rate is negative. Finally, since the carbon is from a living tree, $t = 0$.

$$\frac{\Delta N}{\Delta t} = -\lambda N_0 e^{-\lambda t}$$

$$-(15.3 \times 60 \times 24 \times 365) = -1.21 \times 10^{-4} N_0 e^{-0}$$

$$N_0 = 6.646 \times 10^{10}$$

∴ the gram of carbon contains about 6.65×10^{10} atoms of C-14.

b. Since $-12.0 \times (60 \times 24 \times 365) = -\lambda N_0 e^{-\lambda t}$
and $-15.3 \times (60 \times 24 \times 365) = -\lambda N_0 e^0$,
we obtain the following by substituting 1 for e^0, substituting for $-\lambda N_0$ in the first equation, and then solving for t:

$$\frac{12.0}{15.3} = e^{-\lambda t}.$$

$$\ln\left(\frac{12.0}{15.3}\right) = -\lambda t$$

$$t = -\frac{1}{1.21 \times 10^{-4}} \ln\left(\frac{12.0}{15.3}\right)$$

$$\approx 2010$$

∴ the wood for the chariot spoke was cut about 2010 years ago.

Exercises

1. Solve for t in terms of the other variables: $N = N_0 e^{-\lambda t}$ $t = \frac{\ln N_0 - \ln N}{\lambda}$

2. How are the equations $N = N_0 e^{-\lambda t}$ and $N = N_0 \cdot b^{\frac{t}{k}}$ (page 464) related? Explain.

3. Forty percent of the C-14 from an Egyptian flute has decayed since the flute was made. About how old is the flute? 4220 yr

4. The rate of decay per gram of carbon from a leather sandal is 13.9 nuclei per minute. Could it have been made in ancient Rome? Explain. No

PREPARING FOR ▆▆▆▆▆▆▆▆
COLLEGE ENTRANCE EXAMS

Strategy for Success: Be sure to read the directions, questions, and answer choices carefully. You may wish to underline important words such as *not*, *exactly*, *false*, *never*, and *except* and to cross out answer choices that are clearly incorrect.

Decide which is the best of the choices given and write the correct letter on your answer sheet.

1. What is the distance between point $P(x, z)$ and point $Q(y, z)$? C

   (A) $2z$ **(B)** $x - y$ **(C)** $|x - y|$ **(D)** $y - x$ **(E)** $\dfrac{x - y}{2}$

2. Which of the following is an equation of the perpendicular bisector of the line segment with endpoints $(4, -3)$ and $(8, -11)$? A
   (A) $x - 2y = 20$ **(B)** $2y = x + 20$ **(C)** $4x - 2y = 20$
   (D) $2y = -4x + 20$ **(E)** $4x + 2y = -20$

3. Which of the following is the focus of the parabola with directrix $y = -1$ and whose equation is $y = \frac{1}{8}x^2 + 1$? E

   (A) $(3, 0)$ **(B)** $(0, 0)$ **(C)** $(0, 2)$
   (D) $(0, -3)$ **(E)** None of these

4. Which of the following must be true for the function $f(x) = x^2 + 6x + 7$? B
   I. The equation of the axis of symmetry is $x = 3$.
   II. The vertex of the graph is $(-3, -2)$.
   III. The graph has a maximum point.
   (A) I only   **(B)** II only   **(C)** III only
   (D) I and II only   **(E)** II and III only

5. If x and y are both nonnegative real numbers, which of the following must be true? C
   I. $\sqrt{x^2 + y^2} > x + y$ II. $\sqrt{x^2 + y^2} \neq x + y$ III. $\sqrt{x^2 + y^2} \leq x + y$
   (A) I only   **(B)** II only   **(C)** III only   **(D)** I and II only   **(E)** None

6. Which of the following must be true for the sequence $a_n = (-1)^n 3n$? B
   I. $a_1 = 3$ II. The sequence diverges.   III. The sequence is nondecreasing.
   (A) I only   **(B)** II only   **(C)** III only
   (D) I and II only   **(E)** I, II, and III

7. What is the solution of $\log_{10} x + \log_{10} (x - 3) = 1$? A
   (A) $\{5\}$ **(B)** $\{-5\}$ **(C)** $\{2\}$ **(D)** $\{6\}$ **(E)** None of these

8. Which of the following must be true for any positive number x? A
   I. $\log x^2 = 2 \log x$ II. $\sqrt{x} = \frac{1}{2} \log x$ III. $\log x^2 = 2 \log \sqrt{x}$
   (A) I only   **(B)** II only   **(C)** III only
   (D) I and II only   **(E)** I and III only

Exponents and Logarithms   **479**

<table>
<tr><td colspan="2">

</td></tr>
</table>

Additional Answers
Written Exercises

(continued from p. 463)

3. a. $\dfrac{2 \log 145}{\log 19}$

4. a. $\dfrac{\log 19}{3 \log 54}$

5. a. $\dfrac{\log 46}{4 \log 9.12}$

6. a. $\dfrac{8 \log 33.7}{\log 128}$

7. a. $\dfrac{\log 50.1 - \log 6}{\log 2}$

8. a. $\dfrac{\log 29.3 - \log 0.76}{2 \log 5}$

9. a. $\dfrac{\log 8.35 - \log 4780}{\log 3}$

10. a. $\dfrac{2(\log 132 - \log 4.1)}{\log 8.2}$

11. a. $\dfrac{\log 122 - \frac{1}{3} \log 25.3}{\log 6}$

12. a. $\dfrac{3\left(\log 96 - \frac{1}{5} \log 203\right)}{\log 7}$

13. a. $\dfrac{\frac{3}{4} \log 15 - \log 18}{\log 9}$

14. a. $\dfrac{\frac{1}{4} \log 56.6}{\log 12} - 1$

15. a. $\dfrac{\frac{1}{3} \log 94}{\log 0.027} + 2$

19. a. $\dfrac{\log 5}{\log 5 - \log 3}$

20. a. $\dfrac{\log 2 + \log 3}{\log 3 - \log 2}$

21. a. $\dfrac{\log 15}{2 \log 6.7 - \log 15}$

22. a. $\dfrac{\log 6}{3 \log 6 - \log 28}$

23. a. $\dfrac{\log 39 + \log 8}{2 \log 8 - \log 39}$

24. a. $\dfrac{\log 17 - \log 42}{3 \log 17 - \log 42}$

25. $\log_{25} 2 = \dfrac{\log_5 2}{2 \log_5 5} = \dfrac{\log_5 2}{2}$

PROGRAMMING IN PASCAL

As you learned in Chapter 7, standard Pascal lacks an exponential operator. You wrote a function in Chapter 7 that evaluates x^n for x a real number and n an integer. Now you can use your understanding of exponents and logarithms to extend that function to

FUNCTION power (x, y : real) : real;

a function that computes x^y for x and y both of type real, when x^y is defined.

Standard Pascal contains two functions exp and ln which are inverses of each other. It can be shown that

$$\lim_{n \to \infty}\left(1 + \frac{1}{n}\right)^n \approx 2.7182818284 \ldots$$

The name for the value of this limit is e. This number e is an irrational number which is used as the base of the natural logarithm and exponential functions. So $\exp(x) = e^x$ for all x, and $\ln(x) = \log_e x$ for $x > 0$. These functions are extremely important because they occur frequently in formulas such as those used for continuous compounding of interest, population growth, radioactive decay, and others. Calculus makes use of the natural exponential and logarithmic functions because they have particularly nice properties.

Exercises

1. Write a program which displays a table of values for $n = 1$ to 20 as follows:

n	$\exp(n)$	$\ln(n)$
1	2.718281	0
2	7.389056	0.693147
3	20.085540	1.098612
.		
.		
.		

2. a. Use the laws of exponents and logarithms to show that $a^x = e^{x \ln a}$. What restrictions, if any, must be imposed on a and x so that this equation can be defined?
   b. Using the idea of part (a), write a function with the heading

FUNCTION power (a, x : real) : real;

   that computes a^x whenever it is defined.
   c. Write a program that allows the user to enter the values of a and x, then calls the function *power* to compute a^x. When the function is not defined, the program should inform the user, then ask the user to re-enter the values of a and x.

3. Use the program to solve the equation $x^{\frac{4}{3}} = 28.9$. $\{12.464458\}$

4. Research Problem

a. The relationship between x^y and y^x is difficult to predict. For example,

$$2^3 < 3^2, \quad 2^4 = 4^2, \quad 5^2 < 2^5$$

Use the ideas developed in this chapter to examine enough cases to attempt a generalization about the relationship e^x and x^e.

b. Write a Pascal program that calls the function *power* when the user enters a value of x to attempt to find a value for y with $x \neq y$, such that $x^y = y^x$. The program should print x and y as an ordered pair (x,y).

c. Using the ideas of this chapter, write a program that will find the value to which the sequence

$$Z, \; Z^Z, \; Z^{(Z^Z)}$$

converges for $Z = 0.06, 0.1, 0.25, 0.4, 0.77$.

For values of Z that are greater than 1, the sequence above will diverge. However, for some values very close to 1, for example 1.1, the sequence diverges so slowly that the program might interpret that the sequence converges.

4. a. When $x > 0$: $e^x \geq x^e$. In fact, the two expressions are equal only when $x = e$.

26. $\log_a 9 = \dfrac{\log 9}{\log a} = \dfrac{2 \log 3}{\log a}$

27. $\log_b x = \dfrac{\log x}{\log b} = \dfrac{2 \log x}{2 \log b} =$

$\dfrac{2 \log x}{\log b^2} = 2 \log_{b^2} x$

28. $\log_b x = \dfrac{\log x}{\log b} = \dfrac{n \log x}{n \log b} =$

$\dfrac{n \log x}{\log b^n} = n \log_{b^n} x$

29. $\log_{ab} x = \dfrac{\log_a x}{\log_a a + \log_a b} =$

$\dfrac{\log_a x}{1 + \log_a b}$

30. $\log_{\frac{a}{b}} x = \dfrac{\log_a x}{\log_a a - \log_a b} =$

$\dfrac{\log_a x}{1 - \log_a b}$

31. $\log_x y = \dfrac{\log_a y}{\log_a x} = \dfrac{\log_b y}{\log_b x}$

$\therefore (\log_a x)(\log_b y) = (\log_b x)(\log_a y)$

Many of the plants in this photo have been grown from seeds. When seeds are planted, some of the seeds sprout, or germinate, while others do not. The science of probability may be used in studying plant germination.

482 *Chapter 12*

Chapter 12

Permutations, Combinations, and Probability

Problem Solving Strategies
Making an Organized List
This technique is useful in Sections 12-1 (page 483) and 12-2 (page 486) in solving counting problems.
Generalizing from Specific
In Section 12-6 the Binomial Theorem is developed from specific examples.
Looking for a Pattern.
Students are encouraged to look for a pattern in developing Pascal's triangle in Section 12-7.

Permutations

OBJECTIVES for Sections 12-1 through 12-3:
1. *To apply fundamental counting principles.*
2. *To find the number of permutations of the elements of a set.*
3. *To find the number of permutations of the elements of an r-element subset of an n-element set.*
4. *To find the number of permutations of elements that are not all different.*

12–1 Two Fundamental Counting Principles

Teaching Suggestions
p. T111

Key Ideas
Use two fundamental counting principles.

The number of elements in the union of two finite sets is related to the number of elements in each set and the number of elements in their intersection. For example, if $A = \{1, 2, 3, 4\}$ and $B = \{2, 4, 6\}$, then

$$A \cup B = \{1, 2, 3, 4, 6\} \quad \text{and} \quad A \cap B = \{2, 4\}.$$

Notice that the union of A and B contains five elements rather than seven because 2 and 4 are elements of both A and B. In general, we have the following fundamental counting principle.

> If set A contains r elements, set B contains s elements, and $A \cap B$ contains t elements, then $A \cup B$ contains $r + s - t$ elements.

If $A \cap B = \emptyset$, then $t = 0$ and the number of elements in $A \cup B$ is $r + s$.

Permutations, Combinations, and Probability **483**

1. In a certain high school, 310 students are registered for an algebra class, 200 are registered for a biology class, and 120 are registered for both subjects. If all students attend, how many are taking algebra or biology?

To determine the number N of students in the union of two sets whose intersection contains 120 members, let $N = 310 + 200 - 120 = 390$
$\therefore$ 390 students are taking algebra or biology.

2. How many odd, three-digit natural numbers can be formed using only the digits 1, 2, 3, 4, 5?

The hundreds digit and the tens digit may each be any of the 5 digits 1, 2, 3, 4, 5. The units digit may be 1, 3, or 5, but not 2 or 4. Write the appropriate numbers of possibilities in each space.

<u>5 5 3</u>

The second counting principle tells you that there are $5 \times 5 \times 3$, or 75 ways of forming the required odd integers.

EXAMPLE 1 The Mathematics Club at East High School has 36 members, the Spanish Club has 25 members, and 7 students are members of both organizations. If all members attended, what would be the attendance at a joint meeting of the two clubs?

SOLUTION You want to determine the number N of members in the union of two sets whose intersection contains 7 members. You have:

$$N = 36 + 25 - 7 = 54$$

$\therefore$ the attendance at the joint meeting would be 54.

To discover a second counting principle, consider this problem: If A is the set of *two* integers $\{1, 2\}$ and B is the set of *three* integers $\{4, 5, 6\}$, how many different ordered pairs (a, b) are there with $a \in A$ and $b \in B$?

For each of the *two* ways that you can choose the first entry, a, there are *three* ways you can choose the second entry, b. Thus the set of all such ordered pairs (a, b) is

$$\{(1, 4), (1, 5), (1, 6), (2, 4), (2, 5), (2, 6)\},$$

which contains $2 \cdot 3$, or 6, elements. This set of ordered pairs is called the **Cartesian product** of A and B and is denoted by $A \times B$.

This result can be generalized as follows.

If set A contains r elements and set B contains s elements, then there are rs different ordered pairs (a, b) with $a \in A$ and $b \in B$ (that is, $A \times B$ contains rs elements).

This principle can be extended to any number of sets and applied in many counting situations.

EXAMPLE 2 How many four-digit even integers can be formed using the digits 0, 1, 2, 3, 4, 5?

SOLUTION To help you think through such problems, it is useful to employ a diagram such as this: □□□□ or this: __ __ __ __ .

For the thousands digit, you can use any of the five digits 1, 2, 3, 4, or 5, but not 0. Therefore, you write 5 in the first space: | 5 | | | |

For the hundreds and tens digits you can use any one of the given six digits, so write 6 in each of these places: | 5 | 6 | 6 | |

In the units place, you can use any one of the three digits 0, 2, or 4 but not 1, 3, or 5. Therefore, write 3 in the units place: | 5 | 6 | 6 | 3 |

The second counting principle tells you that there are $5 \times 6 \times 6 \times 3$, or 540, ways of forming the required even integers.

484 *Chapter 12*

Oral Exercises

If *A* and *B* are as given, state the number of elements in $A \cap B$, $A \cup B$, and $A \times B$.

1; 5; 9

1. $A = \{3, 5, 7\}$, $B = \{6, 8, 10\}$ 0; 6; 9
2. $A = \{2, 4, 7\}$, $B = \{5, 7, 9\}$
3. $A = \{4\}$, $B = \{2, 3, 4, 5, 6\}$ 1; 5; 5
4. $A = \{8\}$, $B = \{3\}$ 0; 2; 1
5. $A = B = \{4, 5, 6, 7, 8\}$ 5; 5; 25
6. $A = \{4, 5, 6\}$, $B = \emptyset$

0; 3; 0

In Exercises 7–8 the numbers of elements in three of the sets *A*, *B*, $A \cup B$, $A \cap B$, and $A \times B$ are given. State the numbers of elements in the other two.

1

7. *A*: 4, *B*: 6, $A \cup B$: 9, $A \cap B$: ?, $A \times B$: ? 24
8. *A*: 5, *B*: ?, $A \cup B$: 8, $A \cap B$: 4, $A \times B$: ? 35

7

Written Exercises

A
1. How many ordered pairs of letters are there that use only the letters L, M, N, O, and P? 25
2. How many three-digit integers can be formed that do not contain a 4? 648
3. How many sequences of 4 letters can be formed from the letters A, B, C, D, E, and F if no letter may be used more than once? 360
4. How many different sequences of heads or tails are possible if a coin is flipped 8 times? (If you let H stand for heads and T for tails, one such sequence could be HTHHTHTH.) 256
5. Out of a faculty of 76 at Central High School, 52 teach at least one eleventh-grade class, 45 teach at least one twelfth-grade class, and 3 teach no eleventh- or twelfth-grade classes. How many on the faculty teach both eleventh- and twelfth-grade classes? 24
6. In a survey of 375 dog and cat owners there were 215 dog owners and 193 cat owners. How many in the survey own a dog and no cat? 182
7. There are 9 different routes between cities A and B, and 8 different routes between cities B and C. How many different routes are there from city A to city C by way of city B? 72
8. How many different employee identification symbols are possible if each symbol consists of two letters of the alphabet followed by four digits? 6,760,000

B
9. In Exercise 8 how many symbols are possible if the two letters must be different and all four of the digits cannot be zero? 6,499,350
10. How many seven-digit phone numbers are possible if 0 and 1 cannot be used as the first digit and the first three digits cannot be 555, 411, or 936? 7,970,000

Permutations, Combinations, and Probability **485**

Suggested Assignments

Enriched Alg.
 485/1–13 odd, 14
Enriched Alg. with Trig.
 485/1–13 odd, 14

Additional A Exercises

1. How many ordered pairs of letters are there that use only the letters A, B, C, D, and E? 25
2. How many three-digit integers can be formed that do not contain a 6? 648
3. How many sequences of 4 letters can be formed from the letters J, K, L, M, N, and O if no letter may be used more than once? 360
4. How many different sequences of heads or tails are possible if a coin is flipped 7 times? 128

Mixed Review

1. Find log 0.0167.
 -1.7773
2. Find antilog 3.6395. 4360
3. Find an equation of the circle that has a diameter with endpoints $(3, -1)$ and $(7, -3)$.
 $(x - 5)^2 + (y + 2)^2 = 5$
4. Simplify $\sqrt{-45}$. $3i\sqrt{5}$
5. Find *x* if $3x$, $2x - 1$, $5x + 4$ is an arithmetic sequence. $-\dfrac{3}{2}$

11. How many positive odd integers less than 1000 can be formed from the digits 0, 1, 2, 3, 4, 5, and 6? 147

12. How many numbers between 450 and 700 can be formed using only the digits 3, 4, 5, 6, 7, and 8? 96

13. How many of the numbers in Exercise 12 will be odd? 48

C 14. How many multiples of 3 between 100 and 1000 can be formed from the digits 1, 4, 5, 6, and 8? (*Hint:* The sum of the digits of any multiple of 3 is also a multiple of 3.) 41

12–2 Linear and Circular Permutations

You can list the members of {*a*, *b*, *c*} in six different orders:

$$abc \qquad acb \qquad bac \qquad bca \qquad cab \qquad cba$$

Each ordering, or arrangement, of the letters is called a (*linear*) *permutation* of {*a*, *b*, *c*}. A **permutation** is any arrangement of the elements of a set in a definite order.

Notice that the first letter listed can be any member of {*a*, *b*, *c*}. This means there are *three* choices for first place, so we write 3 in the first space of a diagram: 3☐☐. *After a letter has been selected for first place,* the choice for second place is made from the set of *two* letters remaining. Therefore, we write 2 in the second space: 3|2☐. *After letters have been assigned to both the first and the second places,* there is only *one* choice for third place; so we write 1 in the third space: 3|2|1. Thus, the number of permutations of the elements of {*a*, *b*, *c*} is

$$3 \times 2 \times 1.$$

The product 3 × 2 × 1 can be written in brief **factorial notation** as 3! (read "three factorial" or "factorial three"). Thus seven factorial is

$$7! = 7 \cdot 6 \cdot 5 \cdot 4 \cdot 3 \cdot 2 \cdot 1 = 5040,$$

and in general,

$$n! = n \cdot (n-1) \cdot \cdots \cdot 3 \cdot 2 \cdot 1,$$

where *n* is any natural number.

The preceding discussion illustrates the following fact.

> The number of permutations of the members of a set containing *n* different elements is *n*!.

486 *Chapter 12*

EXAMPLE 1   How many different signals can be made using the five flags pictured at the right if all the flags must be used in each signal?

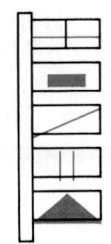

SOLUTION   You want to determine the number of permutations of 5 things:

$$5! = 5 \cdot 4 \cdot 3 \cdot 2 \cdot 1 = 120$$

∴ 120 signals can be made using the five flags.

Now suppose you are asked to find the number of permutations of five different letters taken three at a time. In the diagram ☐ ☐ ☐, the first space could be filled in *five* ways, the second in *four* ways, and the last in *three*. Thus $\boxed{5\ 4\ 3}$ would represent the situation. From the fundamental counting principle on page 484 there are $5 \cdot 4 \cdot 3$, or 60, ways in which the letters could be arranged.

The number of permutations of n different elements taken r at a time is denoted by $_nP_r$. Other representations are $P(n, r)$ and P_r^n. To obtain a formula for $_nP_r$, notice that the diagram representing the situation contains r spaces to be filled as shown below.

$$\boxed{n\ \ n-1\ \ n-2\ \cdots\ n-(r-1)}$$

Thus, you have the following result:

The number of permutations of r members of a set containing n different elements is

$$_nP_r = n(n-1)(n-2) \cdots (n-r+1).$$

Note that if $r = n$, then $_nP_n = n!$.

EXAMPLE 2   How many different three-letter sequences can you form from the first 15 letters of the alphabet if no two letters in a sequence are the same?

SOLUTION   You want to determine the number of permutations of 3 elements from a set of 15 elements:

$$_{15}P_3 = 15 \cdot 14 \cdot 13 = 2730$$

There is a special type of permutation, called a **circular permutation**, that is an arrangement of objects in a circular pattern. A common example is the seating of people around a circular table. In such an arrangement there is no first place, so that if each person shifts position by one place counterclockwise (or clockwise) the relative positions are not changed. In fact, if there are n people at the table, each person can shift position n times and return to his or her original position without disturbing the arrangement. Therefore, if you use the formula for a

Permutations, Combinations, and Probability   **487**

Students are introduced here to the meaning of permutations. Tell the students that the word permute means "to change the order of." Careful reading is necessary to determine what is actually asked for in each problem. Tell students that if they find key words such as "arrangement" and "order" in a problem, it is probably a situation where order matters, and permutations are therefore used to solve the problem.

Be sure students understand that the different notations below represent the same thing,

$$_nP_r \qquad P(n, r) \qquad P_r^n$$

and should be read "the number of permutations of n objects taken r at a time." Also be sure students read factorial notation correctly.

Additional A Exercises

1. In how many ways can 5 people be seated on a bench?   120

2. How many arrangements are there of the letters in the word MEAL?   24

3. In how many different ways can 5 children be seated on a merry-go-round?   24

4. How many 3-digit numbers can be formed if the first digit must be a 5?   100

5. How many 2-letter permutations are there of the letters in the word DISH?   12

**Additional Answers
Written Exercises**

16. $n\left(_{n-1}P_{r-1}\right) \overset{?}{=} {_nP_r}$
$7\left(_6P_3\right) \overset{?}{=} {_7P_4}$
$7 \cdot 6 \cdot 5 \cdot 4 \overset{?}{=} 7 \cdot 6 \cdot 5 \cdot 4$
$840 = 840$

18. $_nP_{n-r} \overset{?}{=} \dfrac{n!}{r!}$

$_7P_3 \overset{?}{=} \dfrac{7!}{4!}$

$7 \cdot 6 \cdot 5 \overset{?}{=}$

$\dfrac{7 \cdot 6 \cdot 5 \cdot 4 \cdot 3 \cdot 2 \cdot 1}{4 \cdot 3 \cdot 2 \cdot 1}$

$7 \cdot 6 \cdot 5 \overset{?}{=} 7 \cdot 6 \cdot 5$
$210 = 210$

19. $\left(_nP_r\right)\left(_{n-r}P_{n-r}\right) \overset{?}{=} {_nP_n}$
$\left(_7P_4\right)\left(_3P_3\right) \overset{?}{=} {_7P_7}$
$7 \cdot 6 \cdot 5 \cdot 4 \cdot 3 \cdot 2 \cdot 1 \overset{?}{=}$
$7 \cdot 6 \cdot 5 \cdot 4 \cdot 3 \cdot 2 \cdot 1$
$5040 = 5040$

20. $n\left(_{n-1}P_{r-1}\right)$
$= n((n-1)(n-2)\cdots$
$((n-1)-(r-1)+1))$
$= n(n-1)(n-2)\cdots$
$(n-1-r+1+1)$
$= n(n-1)(n-2)\cdots$
$(n-r+1)$
$= {_nP_r}$

22. $_nP_{n-r} = n(n-1)(n-2)\cdots$
$(n-(n-r)+1)$
$= n(n-1)(n-2)\cdots$
$(n-n+r+1)$
$= n(n-1)(n-2)\cdots$
$(r+1)$
$= \dfrac{n(n-1)(n-2)\cdots(r+1)}{1} \cdot$
$\dfrac{r(r-1)\cdots 1}{r(r-1)\cdots 1} = \dfrac{n!}{r!}$

---

linear permutation to find the number of possible arrangements, you will have counted each different arrangement n times. Thus, there are

$$\frac{n!}{n} = \frac{n \cdot (n-1) \cdot \ldots \cdot 3 \cdot 2 \cdot 1}{n} = (n-1)!$$

distinguishable permutations.

The diagram below shows the $(3-1)!$, or 2, circular permutations of the 3-element set $\{a, b, c\}$ and the corresponding 3!, or 6, linear permutations.

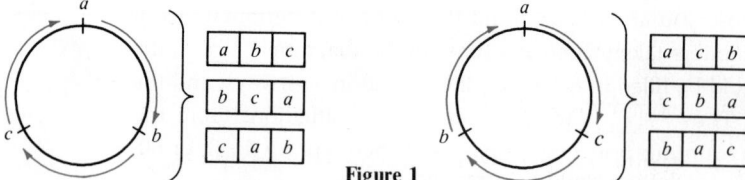

Figure 1

EXAMPLE 3 In how many ways can seven persons be seated around a circular table?

SOLUTION Since this is a circular permutation of 7 things, there are $(7-1)!$, or 720, possible different seating arrangements.

You might instead think of this in a slightly different manner. Since a rotation of any permutation does not produce a new permutation, one of the positions can be considered fixed, and $\boxed{1\,|\,6\,|\,5\,|\,4\,|\,3\,|\,2\,|\,1}$ describes the situation. We see again that there are 720 different arrangements.

The analysis of problems involving circular permutations of objects that do not have a definite top or bottom, such as bracelets or key rings, is somewhat different. In these cases it seems reasonable to consider that flipping an arrangement over does not change the arrangement. Thus, since flipping over the first arrangement in Figure 1 yields the second arrangement, and vice versa, there is only $\dfrac{(3-1)!}{2}$, or 1, permutation of 3 objects about a key ring or bracelet. In general, for n objects, there are $\dfrac{(n-1)!}{2}$ such permutations, provided $n > 2$.

EXAMPLE 4 In how many ways can 6 keys be arranged on a key ring?

SOLUTION $\dfrac{(n-1)!}{2} = \dfrac{5!}{2} = \dfrac{5 \cdot 4 \cdot 3 \cdot 2 \cdot 1}{2} = 60$

Oral Exercises

Evaluate.

1. $_5P_2$ 20

2. $_4P_3$ 24

3. $_6P_3$ 120

4. $_9P_1$ 9

5. $_8P_8$ 40,320

6. $_{11}P_2$ 110

Written Exercises

A

1. **a.** In how many ways can 9 books be arranged on a shelf? 362,880
   b. In how many ways can 9 books be arranged on a shelf if a particular book must occupy the central position? 40,320

2. Given 9 starting batters on a baseball team, how many batting orders are possible if the shortstop must bat first and the right fielder must bat fourth? 5040

3. How many arrangements are there of the letters in the word CURTAIN? 5040

4. How many 3-letter permutations are there of the letters in the word COMPUTER? 336

5. In how many ways can the 6 positions on a hockey team be assigned among 10 players? 151,200

6. In how many ways can a president, a vice-president, a secretary, and a treasurer be chosen from the 13 members of a club? 17,160

7. In how many ways can 8 persons be seated in a row of 9 chairs? 362,880

8. In how many ways can 5 persons be seated in a row of 9 chairs? 15,120

9. In how many ways can 8 distinct dishes be arranged around a revolving platter at a buffet dinner? 5040

10. In how many different ways can 9 people be seated at a round table? 40,320

11. In how many ways can 7 keys be arranged on a key ring? 360

12. In how many different ways can 8 different colored beads be arranged on a bracelet? 2520

B

13. How many arrangements of the letters in the word TRIANGLE begin with three vowels? 720

14. How many arrangements of the letters in the word GRACIOUS both begin and end with a vowel? 8640

15. **a.** In how many ways can 7 students be seated in a row if 2 particular students must be seated next to each other? 1440
   b. In how many ways can 7 students be seated in a row if 2 particular students must not be seated next to each other? 3600

Show that each of the following is true when $n = 7$ and $r = 4$.

16. $n\left({}_{n-1}P_{r-1}\right) = {}_nP_r$

17. $(n - r)\,{}_nP_r = {}_nP_{r+1}$

18. ${}_nP_{n-r} = \dfrac{n!}{r!}$

19. $\left({}_nP_r\right)\left({}_{n-r}P_{n-r}\right) = {}_nP_n$

C

20–23. Show that the statements in Exercises 16–19 are true for all positive integers n and r, where $n > r$.

Show that each of the following is true for all positive integers n, r, and s, where $n > r + s$.

24. ${}_nP_r - {}_nP_{r-1} = (n - r)\,{}_nP_{r-1}$

25. $\left({}_nP_r\right)\left({}_{n-r}P_s\right) = {}_nP_{r+s}$

23. $\left({}_nP_r\right)\left({}_{n-r}P_{n-r}\right)$
$= n(n - 1)(n - 2) \cdots$
$(n - r + 1)(n - r)(n - r - 1) \cdots (n - r - n + r + 1)$
$= n(n - 1)(n - 2) \cdots$
$(n - r + 1)(n - r)(n - r - 1) \cdots 1$
$= {}_nP_n$

24. ${}_nP_r - {}_nP_{r-1}$
$= n(n - 1)(n - 2) \cdots$
$(n - r + 1) - n(n - 1)(n - 2) \cdots (n - r + 2)$
$= n(n - 1)(n - 2) \cdots$
$(n - r + 2)(n - r + 1 - 1)$
$= n(n - 1)(n - 2) \cdots$
$(n - r + 2)(n - r)$
$= (n - r)(n(n - 1)(n - 2) \cdots (n - (r - 1) + 1))$
$= (n - r)\,{}_nP_{r-1}$

25. ${}_nP_r\left({}_{n-r}P_s\right)$
$= n(n - 1)(n - 2) \cdots$
$(n - r + 1)(n - r)(n - r - 1) \cdots (n - r - s + 1)$
$= n(n - 1)(n - 2) \cdots$
$(n - (r + s) + 1) = {}_nP_{r+s}$

1. Find all the rational roots of $x^3 + x^2 - 7x - 3 = 0$.
   -3

2. Simplify $(3\sqrt{2} - 5)^2$.
   $43 - 30\sqrt{2}$

3. Give the equation of the axis of symmetry, and the coordinates of the vertex of the graph of $y = -x^2 + 6x - 7$.
   equation of axis of symmetry: $x = 3$
   vertex: $(3, 2)$

4. If x varies inversely as y^3 and $x = 6$ when $y = 2$, find x when $y = \dfrac{1}{2}$. 384

12–3 Permutations with Repeated Elements

Finding the number of distinguishable permutations of a set of elements that are not all different involves an extension of the method used in Section 12-2. For example, to find the number of distinguishable permutations of the letters in the word

REVERSE,

we must consider the fact that the letter E occurs three times and the letter R twice in the word, and that simply interchanging any of the three E's or the two R's with each other does not produce a distinguishable permutation.

To analyze the situation in detail, let us label the R's and E's with subscripts:

$$R_1 \; E_1 \; V \; E_2 \; R_2 \; S \; E_3$$

There are, of course, $_7P_7 = 7!$ permutations of these 7 letters. If we let P denote the number of *distinguishable* permutations, then for each of these P permutations, there are $3!$ permutations of the E's (E_1, E_2, and E_3) and $2!$ permutations of the R's (R_1 and R_2). It follows that

$$2! \cdot 3! \cdot P = {}_7P_7 = 7!, \text{ so that}$$

$$P = \frac{7!}{2! \cdot 3!} = 420.$$

Using similar reasoning, we can assert that, in general, the following is true.

The number of distinguishable permutations of n elements taken n at a time, with n_1 elements alike, n_2 of another kind alike, and so on, is

$$\frac{n!}{n_1! n_2! \cdot \cdot \cdot}.$$

Oral Exercises

Find the number of distinguishable permutations of the letters in the given word.

1. EGG 3   **2.** AREA 12   **3.** GEESE 20   **4.** DITTO 60   **5.** ROTOR 30

Written Exercises

Find the number of distinguishable permutations of the letters in the given word.

A **1.** TURRET 180   **2.** REFEREES 840   **3.** MINIMUM 420   **4.** ALFALFA 210

5. CALCULATOR 453,600   **6.** MILLIMETERS 2,494,800   **7.** CURRICULUM 151,200   **8.** INTERFERENCE 4,989,600

How many different numerals can be formed using all the digits of the given numeral?

9. 526218 360

10. 7373473 140

B **11.** 8106929595 302,400

12. 5630103050 37,800

13. How many distinguishable circular permutations of the letters in the word CONIC can be formed? 12

14. How many distinguishable circular permutations of the letters in the word DEGREE can be formed? 20

C **15.** Find the number of distinguishable five-letter permutations that can be formed from the letters in the word WINDOW. Use the following steps:
  a. Find the number of distinguishable five-letter sequences containing exactly one w. 120
  b. Find the number of distinguishable five-letter sequences containing two w's. 240
  c. Add the results of steps (a) and (b). 360

Self-Test 1

VOCABULARY   Cartesian product (p. 484)   factorial notation (p. 486)
              permutation (p. 486)   circular permutation (p. 487)

1. How many three-digit odd integers can be formed using only   *Obj. 1, p. 483*
the digits 3, 4, 5, 6, and 7? 75

2. In how many ways can 6 names be arranged on a ballot if each   *Obj. 2, p. 483*
name must be used exactly once? 720

3. In how many ways can 4 countries on a map be colored if 6   *Obj. 3, p. 483*
colors are available and each country must be a different color? 360

4. How many distinguishable permutations are there of the letters   *Obj. 4, p. 483*
in the word DIFFERENCE? 302,400

Check your answers with those at the back of the book.

Permutations, Combinations, and Probability   **491**

Suggested Assignments

Enriched Alg.
  First day
    491/5–15 odd
  Second day
R 491/*Self-Test 1*
S 489/2–12 even
Enriched Alg. with Trig.
  491/5–15 odd
R 491/*Self-Test 1*

Mixed Review

1. Find x so that the sum of the infinite geometric series $x + 3x^2 + 9x^3 + \dots$ is $\frac{1}{2}$. $\frac{1}{5}$

2. Write $(3x - y)^2(3x + y)^2$ as a polynomial in simple form. $81x^4 - 18x^2y^2 + y^4$

3. Solve.
$3x - 6y = 18$
$5x + 3y = 17$ $\{(4, -1)\}$

4. Solve.
$3x + 4y = 17$
$2x - y = 4$ $\{(3, 2)\}$

Quick Quiz

1. How many three-digit even integers can be formed using only the digits 2, 3, 4, 5, and 6? 75

2. In how many ways can 5 colors be arranged in a design if each color must be used exactly once? 120

3. How many 3-letter sequences of letters can be made up from the letters of the word TRIANGLE, if no letter may be used more than once? 336

4. How many distinguishable permutations are there of the letters in the word INTUITION? 15,120

Key Ideas

Determine the number of combinations of n elements taken r at a time.

Chalkboard Examples

1. How many different debating teams of 4 members can be chosen from a 9-member debating club?
You are asked to find the number of 4-member team subsets of a 9-member set. Letting $n = 9$ and $r = 4$,

$$_9C_4 = \frac{9 \cdot 8 \cdot 7 \cdot 6}{1 \cdot 2 \cdot 3 \cdot 4} = 126.$$

∴ there are 126 possible 4-member debating teams.

2. How many combinations of 6 cassettes can be chosen from a group of 10 cassettes?
You are asked to find the number of 6-cassette subsets of a 10-cassette set. Letting $n = 10$ and $r = 6$,

$$_{10}C_6 = \frac{10 \cdot 9 \cdot 8 \cdot 7 \cdot 6 \cdot 5}{1 \cdot 2 \cdot 3 \cdot 4 \cdot 5 \cdot 6}$$
$$= 210.$$

∴ there are 210 possible 6-cassette subsets.

Reading Algebra

Students should read the notation $_nC_r$, $C(n, r)$, or C_r^n as "the number of combinations of n elements taken r at a time," or as "the number of r-element subsets of a set with n elements." Students should remember that the word "combine" implies

Combinations

OBJECTIVES for Sections 12-4 and 12-5:
1. *To find the number of combinations of* n *elements taken* r *at a time.*
2. *To find the number of ways in which specified subsets can be selected from two or more given sets.*

12–4 Counting Subsets

Can you list the three-element subsets of the set $T = \{a, b, c, d\}$? To obtain any of these subsets, all you have to do is remove one of the members of the original set. Thus the three-element subsets are:

$$\{a, b, c\} \qquad \{a, b, d\} \qquad \{a, c, d\} \qquad \{b, c, d\}$$

Hence, denoting the number of three-element subsets of a four-element set by $_4C_3$, you have $_4C_3 = 4$.

You can classify the *permutations* of T's elements *taken three at a time* according to the three-element subset involved in each permutation. For example, the $3!$, or 6, arrangements

$$abd \qquad adb \qquad bad \qquad bda \qquad dab \qquad dba$$

are the permutations of the subset $\{a, b, d\}$. Similarly, each of the other three-element subsets yields $3!$ other permutations of the letters $a, b, c,$ and d taken three at a time. Thus, it is true that

$$_4C_3 \times 3! = {}_4P_3, \qquad \text{or} \qquad _4C_3 = \frac{_4P_3}{3!}.$$

This formula is consistent with our observation that $_4C_3 = 4$, since

$$\frac{_4P_3}{3!} = \frac{4 \cdot 3 \cdot 2}{1 \cdot 2 \cdot 3} = 4.$$

Moreover, the formula suggests the following general relationship between $_nC_r$, the number of r-element subsets of a set with n elements, and $_nP_r$, the number of permutations of the n elements taken r at a time for $0 < r < n$:

$$_nC_r = \frac{_nP_r}{r!}$$

Since $_nP_r = n(n-1)(n-2) \cdots (n-r+1)$, the following is true.

The number of r-element subsets of a set containing n elements is

$$_nC_r = \frac{n(n-1)(n-2) \cdots (n-r+1)}{r!}.$$

Note that the numerator and the denominator of the expression on the right in the box are both products of r factors.

An r-element subset of a set with n elements is often called a **combination** of n elements taken r at a time. Thus, $_nC_r$, also denoted by $\binom{n}{r}$, $C(n, r)$, or C_r^n, is the number of combinations of n elements taken r at a time.

EXAMPLE How many 5-card hands that contain only hearts can be dealt from a standard bridge deck?

SOLUTION You are asked for the number of 5-card subsets of a 13-card set: the set of hearts. Letting $r = 5$ and $n = 13$, you can begin by noting that the denominator of

$$_{13}C_5 = \frac{?}{5!}$$

contains 5 factors. Therefore the required numerator contains 5 descending factors starting with 13. Thus

$$_{13}C_5 = \frac{13 \cdot 12 \cdot 11 \cdot 10 \cdot 9}{1 \cdot 2 \cdot 3 \cdot 4 \cdot 5} = 1287.$$

$\therefore$ there are 1287 possible 5-card hands containing only hearts.

If you multiply the numerator and denominator of the expression for $_nC_r$ by $(n - r)!$, you obtain the equivalent expression:

$$_nC_r = \frac{n(n - 1)(n - 2) \cdots (n - r + 1)(n - r)(n - r - 1) \cdots 3 \cdot 2 \cdot 1}{r!(n - r)!}$$

or

$$_nC_r = \frac{n!}{r!(n - r)!}$$

The symbol $_nC_0$ denotes the number of subsets with no elements in a set having n elements. There is just one such subset, namely the empty set, $\emptyset$. Thus, $_nC_0 = 1$ and you have:

$$_nC_0 = \frac{n!}{0! \, n!} = \frac{1}{0!}, \qquad \text{or} \qquad 1 = \frac{1}{0!}$$

Therefore, for the formula to hold when r is a whole number, we must *define* $0!$ to be 1. You should verify that this definition also makes the formula for $_nC_r$ valid in the case $r = n$.

You can discover a useful fact about $_nC_r$ by noticing that whenever r elements are selected from a set of n elements, $n - r$ elements are left behind. Therefore, the combinations of r elements selected, and the combinations of $n - r$ elements left, are paired one-to-one and are consequently the same in number; that is,

$$_nC_r = {_nC_{n-r}}.$$

grouping together and *not* ordering or arranging. Tell students to look for key words such as "group," "team," and "subset." When these words appear, the problem very likely involves combinations.

___

Common Errors

Students tend to confuse permutations with combinations. Remind them that since order does matter in finding permutations, but does not matter in combinations, there are always a number of permutations for each combination. For example, the combination of letters ABC has 6 different permutations.

___

Additional A Exercises

1. How many combinations of 3 books can be chosen from a shelf containing 10 books?   120

2. How many ways can a debating team consisting of 4 members be chosen from a group of 7 people?   35

3. In how many ways can a committee consisting of 11 members choose a subcommittee of 4?   330

4. In a group of 9 points, no 3 are collinear. How many triangles are there that have 3 of these points as vertices?   84

5. How many ways are there of scoring exactly 70% on a 10-question true-false test?   120

$$_{50}C_{48} = {_{50}C_2} = \frac{50 \cdot 49}{1 \cdot 2} = 1225.$$

Mixed Review

1. Use Cramer's Rule to solve:
   $3x - 4y = 5$
   $-4x + 6y = -7$
   $\left\{\left(1, -\frac{1}{2}\right)\right\}$

2. Find the slope of the line through the points (1, 4) and (5, 4).   0

3. Find the slope of the line with the equation $5x - 3y = 7$. $\frac{5}{3}$

4. Write an equation of a line that passes through the point (0, 3), and has a slope of -2. $y = -2x + 3$

5. Write an equation of a line that passes through the points (8, −1) and (6, 5). $y = -3x + 23$

Oral Exercises

Evaluate.

1. $_5C_1$ 5
2. $_6C_2$ 15
3. $_{10}C_3$ 120
4. $_7C_5$ 21
5. $_{15}C_{15}$ 1
6. $_{17}C_0$ 1

Written Exercises

A

1. How many combinations of 4 records can be chosen from 14 records offered by a record club? 1001

2. How many steering subcommittees of 6 people can be formed from a committee consisting of 10 people? 210

3. How many roads are needed to connect 9 cities if there is to be exactly one road between any two of the cities? 36

4. At a chess tournament with 13 players, each player played exactly one game with each of the other players. How many games of chess were played at the tournament? 78

5. How many lines are determined by 7 points, no 3 of which are collinear? 21

6. Of 10 points, no 3 are collinear. How many triangles are there that have 3 of these points as vertices? 120

7. How many ways are there of scoring exactly 80% on a 10-question true-false test? 45

8. In how many ways can 12 people be divided into hockey teams of 6 players each? 924

B

9. In a club consisting of 16 members, how many ways are there of choosing a committee of 6 if the club president must be on the committee and another member of the club is not able to serve on the committee? 2002

10. How many diagonals are there in a regular ten-sided polygon? (Note that a diagonal cannot connect two *adjacent* vertices.) 35

11. How many committees of 5 can be chosen from a management group of 11 people if the president and first vice-president are not to serve on the same committee? 378

12. Of 10 points, 6 are collinear, but no other 3 are. How many triangles are there that have 3 of these points as vertices? 100

Additional Answers
Written Exercises

17. $_nC_r = \dfrac{n!}{r!(n-r)!}; \; _nC_{n-r} =$

   $\dfrac{n!}{(n-r)![n-(n-r)]!} =$

   $\dfrac{n!}{(n-r)!r!}; \; \dfrac{n!}{r!(n-r)!} =$

   $\dfrac{n!}{(n-r)!r!},$ so $_nC_r = {_nC_{n-r}}$

494   *Chapter 12*

13. How many different sums of money can be made using at most one bill of each of the following denominations? $1, $5, $10, $20, $50 31

14. Find n if: **a.** $_nC_2 = {}_{100}C_{98}$ **b.** $_nC_4 = {}_nC_3$ 100; 7

15. At a party each guest shook hands with every other guest exactly once. There were a total of 105 handshakes. How many guests were there? 15

16. At Herb's Deli there are 21 different meat sandwiches, each made up of one kind or two different kinds of meat. How many different kinds of meat does Herb's Deli have? 6

Use the formula $_nC_r = \dfrac{n!}{(n-r)!r!}$ to show that each of the following is true.

17. $_nC_r = {}_nC_{n-r}$

18. $_nC_r = \dfrac{n}{r}({}_{n-1}C_{r-1})$

19. $(n-r)({}_nC_r) = n({}_{n-1}C_r)$

C 20. Show that the total number of subsets of a set with n elements is 2^n. (*Hint*: Each member of the set either is or is not selected in forming a subset.)

Computer Exercises   For students with computer experience

1. Write a program that will compute $n!$ for a given integer value of n, $n \geq 0$. (If you did Computer Exercise 1, page 472, you may wish to refer to part of that program.)

Run the program in Exercise 1 for each value of n.

2. 7 5040

3. 10 3,628,800

4. 14
   87,178,291,200

5. 18

6. Use the program of Exercise 1 to help you write a program to compute $_nC_r$ for given values of n and r. Make sure your program accepts only positive integer values of n and r with $n \geq r$. Use the formula for $_nC_r$ that is given in the box on page 492.

Run the program of Exercise 6 to find each value.

7. $_{10}C_7$ 120

8. $_{13}C_5$ 1287

9. $_{52}C_5$
   2,598,960

10. $_{52}C_{10}$

11. Write a program that uses the formula on page 490 to compute the number of distinguishable permutations of n elements takes n at a time, with n_1 alike, n_2 of another kind alike, and so on. Have the computer first ask the user of the program how many *sets* of like elements there are that consist of more that one element. Then have the computer ask how many elements are in each set.

Run the program in Exercise 11 for each set of data.

12. $n = 10$, $n_1 = 2$, $n_2 = 3$

13. $n = 11$, $n_1 = 2$, $n_2 = 2$, $n_3 = 3$

Permutations, Combinations, and Probability **495**

18. $_nC_r = \dfrac{n!}{r!(n-r)!}$;

$\dfrac{n}{r}({}_{n-1}C_{r-1}) =$

$\dfrac{n}{r}\left[\dfrac{(n-1)!}{(r-1)!(n-1-(r-1))!}\right]$

$= \dfrac{n}{r}\left[\dfrac{(n-1)!}{(r-1)!(n-r)!}\right] =$

$\dfrac{n(n-1)!}{r(r-1)!(n-r)!} =$

$\dfrac{n!}{r!(n-r)!}$; $\dfrac{n!}{r!(n-r)!} =$

$\dfrac{n!}{r!(n-r)!}$, so $_nC_r =$

$\dfrac{n}{r}({}_{n-1}C_{r-1})$

19. $(n-r)({}_nC_r) = \dfrac{(n-r)n!}{r!(n-r)!}$;

$n({}_{n-1}C_r) = \dfrac{n(n-1)!}{r!(n-1-r)!}$

$= \dfrac{n!}{r!(n-r-1)!} =$

$\dfrac{n!(n-r)}{r!(n-r)(n-r-1)!} =$

$\dfrac{n!(n-r)}{r!(n-r)!} = (n-r)({}_nC_r)$

20. Represent the n elements of a set by n boxes:

☐ ☐ ☐ ☐ . . .

To indicate a particular subset, assign T or F to the boxes to show whether or not that element is in the subset. Since there are 2^n different sequences of T or F, then there are 2^n subsets.

Additional Answers
Computer Exercises

5. 6,402,373,705,728,000

10. 15,820,024,220

12. 302,400

13. 1,663,200

495

Chalkboard Examples

1. In how many ways can a 4-card combination of 2 aces and 2 face cards be drawn from a deck of playing cards which contains 4 aces and 12 face cards?

For the aces, you compute

$$_4C_2 = \frac{4 \cdot 3}{1 \cdot 2} = 6.$$

For the face cards, you compute

$$_{12}C_2 = \frac{12 \cdot 11}{1 \cdot 2} = 66.$$

Then the number of ways both selections can be made is $6 \times 66 = 396$.

2. In how many ways can a 5-letter combination consisting of 2 vowels and 3 consonants be drawn from a word containing 5 vowels and 7 consonants?

For the vowels, you compute

$$_5C_2 = \frac{5 \cdot 4}{1 \cdot 2} = 10.$$

For the consonants, you compute

$$_7C_3 = \frac{7 \cdot 6 \cdot 5}{1 \cdot 2 \cdot 3} = 35.$$

Then the number of ways that both selections can be made is $10 \cdot 35 = 350$.

12–5 Combinations and Products

You can use the counting principle on page 484 to help determine the number of ways specified subsets can be selected from two or more given sets.

EXAMPLE   How many 7-marble combinations drawn from an urn containing 10 white and 7 red marbles consist of 4 white and 3 red marbles?

SOLUTION   For the white marbles, you compute

$$_{10}C_4 = \frac{10 \cdot 9 \cdot 8 \cdot 7}{1 \cdot 2 \cdot 3 \cdot 4} = 210.$$

For the red marbles, you compute

$$_7C_3 = \frac{7 \cdot 6 \cdot 5}{1 \cdot 2 \cdot 3} = 35.$$

Then the number of ways both selections can be made is

$$35 \times 210 = 7350.$$

Oral Exercises

An urn contains 10 red marbles and 8 blue marbles. State an expression for the number of ways of drawing each of the following combinations.

EXAMPLE   2 red, 3 blue   SOLUTION $_{10}C_2 \times {}_8C_3$

1. 4 red, 6 blue $_{10}C_4 \times {}_8C_6$
2. 5 red, 7 blue $_{10}C_5 \times {}_8C_7$
3. 7 red, 4 blue $_{10}C_7 \times {}_8C_4$
4. 9 red, no blue $_{10}C_9 \times {}_8C_0$
5. 2 red, 8 blue $_{10}C_2 \times {}_8C_8$
6. 10 red, 2 blue $_{10}C_{10} \times {}_8C_2$

Written Exercises
1. 5880   2. 2016   3. 8400   4. 10
5. 45   6. 28

A   1–6. Find the number of combinations described in Oral Exercises 1–6.

7. A family dinner special at a certain restaurant allows a choice of any 2 out of 6 dishes from Group A and 3 out of 7 dishes from Group B. How many different combinations are possible? 525

8. Parts A, B, and C of a test contain 4, 5, and 6 questions respectively, and a student must choose 2, 3, and 4 questions respectively from the three parts. In how many ways can this be done? 900

Exercises 9–18 refer to a standard bridge deck containing 52 cards with 13 cards in each of four suits (clubs, diamonds, hearts, and spades) and 4 cards in each denomination (two, three, . . . , ace). In each case, tell how many combinations of the specified cards there are.

9. 2 tens and 3 jacks   24   10. 2 clubs and 3 diamonds   22,308

496   Chapter 12

11. 4 face cards (jacks, queens, or kings) and 2 aces 2970

12. 2 black cards and 2 hearts 25,350

13. 1 nine, 1 ten, 1 jack, 1 queen, and 1 king 1024

14. 1 heart, 1 club, 1 diamond, 1 spade 28,561

15. 3 kings and 2 cards neither of which is a king 4512

16. 2 aces, 2 threes, and a card that is neither an ace nor a three 1584

B **17.** 2 cards of one denomination and 1 card of another denomination (*Hint*: First find the number of possible combinations of 2 denominations.) 3744

18. 3 cards of one denomination, 2 cards of another denomination, and a card that is of a third denomination 164,736

19. How many 5-letter sequences containing exactly 2 vowels and 3 consonants can be made from the letters in the word COMPUTERS? (*Hint*: Find the number of possible *combinations* of letters first, then the number of permutations of each combination.) 7200

20. How many 5-letter sequences can be formed using the letters in the word MATRICES if each arrangement has 2 vowels and 3 consonants? 3600

C **21.** There are 10 empty seats in a row of a theater. In how many ways can 3 people be seated in the row? In how many ways can they be seated next to each other in the row? 720; 48

22. Find the number of distinguishable 5-letter sequences that can be made from the letters in the word MANTISSA. (*Hint*: Consider separately the cases where there are no duplications, both A and S are duplicated, or exactly one letter is duplicated.) 2040

Self-Test 2

VOCABULARY   combination (p. 493)

1. How many different combinations of 4 books can be chosen from a summer reading list of 10 books? 210   *Obj. 1, p. 492*

2. How many different committees of 5 people can be formed from a group of 11 people? 462

3. In how many ways can you select 2 blue marbles and 3 red marbles from an urn containing 6 blue marbles and 6 red marbles? 300   *Obj. 2, p. 492*

4. Parts A and B of a test contain 6 and 8 questions respectively. In how many ways can a student select 4 questions from part A and 6 questions from part B? 420

Check your answers with those at the back of the book.

Permutations, Combinations, and Probability **497**

Suggested Assignments
Enriched Alg.
 First day
 496/3–15 odd, 16–22
 Second day
R 497/*Self-Test 2*
S 206/24, 28, 30
Enriched Alg. with Trig.
 496/3–15 odd, 16–22
R 497/*Self-Test 2*
S 206/24, 28, 30

Additional A Exercises

A bag contains 8 red beads and 6 blue beads. Tell how many ways there are of drawing each of the following combinations.

1. 3 red, 2 blue 840

2. 4 red, 5 blue 420

Exercises 3–4 refer to a standard bridge deck. Tell how many combinations of the specified cards there are.

3. 2 hearts and 3 clubs 845,000

4. 3 spades and 4 hearts 204,490

Quick Quiz

1. How many different combinations of 5 names can be chosen from a list of 13 names? 1287

2. How many different 5-member basketball teams can be chosen from a group of 9 players? 126

3. In how many ways can you select 3 green beads and 4 white beads from a box containing 12 green beads and 12 white beads? 108,900

4. Parts *A* and *B* of a test contain 10 and 12 questions respectively. In how many ways can a student select 7 questions from part *A* and 8 questions from part *B*? 59,400

Chalkboard Examples

1. Expand $(3n - 2)^5$.
$(3n)^5 + 5(3n)^4(-2) +$
$10(3n)^3(-2)^2 + 10(3n)^2 \cdot$
$(-2)^3 + 5(3n)(-2)^4 +$
$(-2)^5 = 243n^5 - 810n^4 +$
$1080n^3 - 720n^2 + 240n -$
32

2. Find the fifth term in the
expansion of $(2x + y)^{10}$.
You have $n = 10$ and $r =$
5.
1. The exponent of y is
$5 - 1$ or 4. y^4
2. The exponent of $(2x)$ is
$10 - 4$, or 6. $(2x)^6 y^4$
3. The denominator of the
coefficient is $(5 - 1)!$,
or $4 \cdot 3 \cdot 2 \cdot 1$.

$$\frac{}{4 \cdot 3 \cdot 2 \cdot 1}(2x)^6 y^4$$

4. The numerator of the
coefficient contains 4
factors starting with 10.

$$\frac{10 \cdot 9 \cdot 8 \cdot 7}{4 \cdot 3 \cdot 2 \cdot 1}(2x)^6 y^4$$

Simplifying, you have
$210(2)^6 x^6 y^4$, or
$13{,}440 x^6 y^4$.

Binomial Expansions

OBJECTIVES for Sections 12-6 and 12-7:
1. *To expand a binomial.*
2. *To find a given term in a binomial expansion.*
3. *To use Pascal's triangle to expand a binomial.*

12–6 The Binomial Theorem

When you expand powers of binomials, you discover an interesting pattern:

$$(a + b)^1 = a + b$$
$$(a + b)^2 = a^2 + 2ab + b^2$$
$$(a + b)^3 = a^3 + 3a^2b + 3ab^2 + b^3$$
$$(a + b)^4 = a^4 + 4a^3b + 6a^2b^2 + 4ab^3 + b^4$$

This suggests the following:

1. The number of terms in the expansion of $(a + b)^n$ is $n + 1$.

2. The coefficient of the first term is 1.

3. The coefficient of any other term is the product of the coefficient of the preceding term and the exponent of a in the preceding term divided by the number of the preceding term.

4. The exponent of a in any term after the first is one less than the exponent of a in the preceding term.

5. The exponent of b in any term after the first is one greater than the exponent of b in the preceding term.

6. The sum of the exponents of a and b in each term is n.

EXAMPLE 1   Expand $(2n + 3)^5$.

SOLUTION

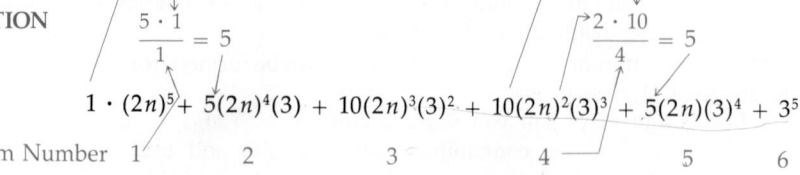

Term Number   1   2   3   4   5   6

The arrows show how the numerical coefficients of the second and fifth terms are computed. (You should be able to explain the others.) In simplified form,

$$(2n + 3)^5 = 32n^5 + 240n^4 + 720n^3 + 1080n^2 + 810n + 243.$$

498   *Chapter 12*

The pattern displayed by the expansion of binomials suggests the **Binomial Theorem**, which states that for any positive integer n the expansion of $(a + b)^n$ is

$$a^n + \frac{n}{1}a^{n-1}b^1 + \frac{n}{1} \cdot \frac{n-1}{2}a^{n-2}b^2 + \frac{n}{1} \cdot \frac{n-1}{2} \cdot \frac{n-2}{3}a^{n-3}b^3 + \cdots$$

$$+ \frac{n}{1} \cdot \frac{n-1}{2} \cdot \frac{n-2}{3} \cdots \frac{n-(r-2)}{r-1}a^{n-(r-1)}b^{r-1} + \cdots + b^n.$$

The rth term is

$$\frac{n(n-1)(n-2)\cdots(n-r+2)}{(r-1)!}a^{n-r+1}b^{r-1}, \qquad r > 1.$$

Observe that in the rth term, the exponent of b is $r - 1$, the exponent of a is $n - r + 1$, the denominator of the coefficient is $(r - 1)!$, and the numerator of the coefficient is $n(n - 1) \ldots (n - r + 2)$, which consists of $r - 1$ consecutive integral factors decreasing from n. Thus, the numerator and denominator of the coefficient contain the same number of factors.

EXAMPLE 2 Find the fourth term in the expansion of $(x - 2y)^{10}$.

SOLUTION You have $n = 10$ and $r = 4$.

1. Find the exponent of b. Since $x - 2y$ is a way of writing $x + (-2y)$, in this case b is $-2y$. The exponent is $4 - 1$, or 3. $(-2y)^3$

2. The exponent of a (in this case, x) is then $10 - 3$, or 7. $x^7(-2y)^3$

3. The denominator of the coefficient is then $(4 - 1)!$, or $3 \cdot 2 \cdot 1$. $\dfrac{1}{3 \cdot 2 \cdot 1}x^7(-2y)^3$

4. The numerator of the coefficient contains 3 factors, starting with 10. $\dfrac{10 \cdot 9 \cdot 8}{3 \cdot 2 \cdot 1}x^7(-2y)^3$

Simplifying, you have $120(-2)^3x^7y^3$, or $-960x^7y^3$.

Oral Exercises

Find the requested parts of the binomial expansion of $(5x - 2)^4$.

$$1 \cdot (5x)^4 + ?(5x)^3(-2) + ?(5x)^2(-2)^2 + 4(5x)(-2)^3 + ?$$

1. The numerical coefficient of the second term. 4
2. The exponent of $5x$ in the second term. 3
3. The numerical coefficient of the third term. 6
4. The exponent of -2 in the third term. 2
5. The fifth term. 16

Additional Answers
Written Exercises

1. $b^5 + 10b^4 + 40b^3 + 80b^2$
   $+ 80b + 32$

2. $y^4 - 12y^3 + 54y^2 - 108y$
   $+ 81$

3. $625 - 500r + 150r^2 -$
   $20r^3 + r^4$

4. $x^6 + 12x^5y + 60x^4y^2 +$
   $160x^3y^3 + 240x^2y^4 +$
   $192xy^5 + 64y^6$

5. $243c^5 + 135c^4 + 30c^3 +$
   $\dfrac{10}{3}c^2 + \dfrac{5}{27}c + \dfrac{1}{243}$

6. $\dfrac{1}{16}x^4 - \dfrac{1}{2}x^3 + \dfrac{3}{2}x^2 -$
   $2x + 1$

7. $\dfrac{1}{64} - \dfrac{3}{8}b + \dfrac{15}{4}b^2 - 20b^3$
   $+ 60b^4 - 96b^5 + 64b^6$

8. $p^{21} + 7p^{18}q + 21p^{15}q^2 +$
   $35p^{12}q^3 + 35p^9q^4 +$
   $21p^6q^5 + 7p^3q^6 + q^7$

9. $y^{32} - 8y^{28} + 28y^{24} -$
   $56y^{20} + 70y^{16} - 56y^{12} +$
   $28y^8 - 8y^4 + 1$

10. $x^7 + 7x^6y^3 + 21x^5y^6 +$
    $35x^4y^9 + 35x^3y^{12} +$
    $21x^2y^{15} + 7xy^{18} + y^{21}$

11. $2187 + 10,206b +$
    $20,412b^2 + 22,680b^3 +$
    $15,120b^4 + 6048b^5 +$
    $1344b^6 + 128b^7$

12. $c^{32} + 8c^{28}d^3 + 28c^{24}d^6 +$
    $56c^{20}d^9 + 70c^{16}d^{12} +$
    $56c^{12}d^{15} + 28c^8d^{18} +$
    $8c^4d^{21} + d^{24}$

13. $210a^6b^4$

14. $66x^{20}$

15. $672p^6q^3$

16. $-112,266n^{18}$

17. $160\dfrac{7}{8}c^5d^8$

500

Written Exercises

Expand each binomial and express the result in simplified form.

A
1. $(b + 2)^5$
2. $(y - 3)^4$
3. $(5 - r)^4$
4. $(x + 2y)^6$

5. $\left(3c + \dfrac{1}{3}\right)^5$
6. $\left(\dfrac{1}{2}x - 1\right)^4$
7. $\left(\dfrac{1}{2} - 2b\right)^6$
8. $(p^3 + q)^7$

9. $(y^4 - 1)^8$
10. $(x + y^3)^7$
11. $(3 + 2b)^7$
12. $(c^4 + d^3)^8$

Find and simplify the specified term.

13. $(a + b)^{10}$; fifth
14. $(x^2 - 1)^{12}$; third
15. $(p + 2q)^9$; fourth

16. $(n^3 - 3)^{11}$; sixth
17. $\left(2c - \dfrac{d}{2}\right)^{13}$; ninth
18. $(1 - u^2)^{14}$; sixth

Find the first four terms in each expansion.

B
19. $(x^2 + 2)^{15}$
20. $(2z^2 - 1)^{10}$

21. $\left(x^2 - \dfrac{1}{2}y^3\right)^{24}$
22. $(x + x^{\frac{3}{2}})^{40}$

Assume that the Binomial Theorem as stated on page 499 holds when n is a positive _rational number_ (and the expansion is an infinite series). Give the first three terms of each expansion in simplified form.

C
23. $(64 + 1)^{\frac{1}{2}}$
24. $(81 - 1)^{\frac{1}{2}}$

25. $(16 - r)^{\frac{1}{2}}$
26. $(27a^3 + 1)^{\frac{2}{3}}$

Computer Exercises   For students with computer experience

1. Write a program that will print the binomial expansion of $(a + b)^n$, given a positive integer n. Use the fact that the binomial coefficients, C_k, can be defined recursively by

$$C_0 = 1;\ C_k = C_{k-1} \cdot \dfrac{n - k + 1}{k}.$$

Run the program in Exercise 1 for each value of n.

2. $n = 7$
3. $n = 10$
4. $n = 12$
5. $n = 20$

6. Modify the program in Exercise 1 so that when numerical _values_ of a and b, in addition to the value of n, are input, the output will give the numerical value of the expansion.

Run the program in Exercise 6 for each expression.

7. $(1 + 0.06)^{12}$
8. $(2 - 0.5)^6$
9. $(1 - 0.08)^9$

Careers

Computer Programming

Computers are useful in many different fields. With the ability to store large amounts of information and make quick calculations, computers aid people in tasks such as keeping track of business finances and guiding rockets. In order to be useful, computers must be programmed by people to perform these tasks.

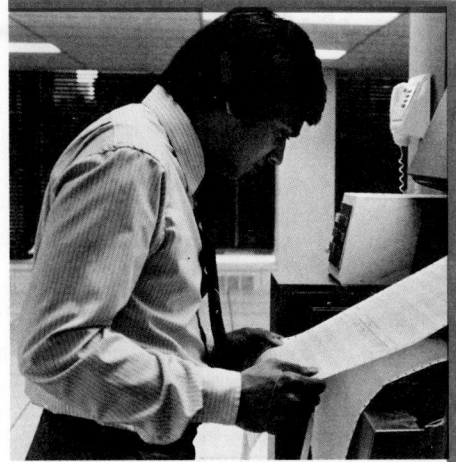

A systems analyst begins programming a computer by analyzing the problem to be solved. The analyst must choose a language and verify that the computer to be used is equipped to solve the problem. The analyst then lists the general steps to be taken by the program and often uses diagrams and flowcharts to help show the relationships between these steps. Sections of the solution are assigned to programmers who each write a detailed version for their section and translate it into code. The program is then run on a computer, with revisions if necessary.

EXAMPLE A computer programmer wants to write an outline for a program that computes and prints the new daily balance of a savings account when deposits and withdrawals are made and interest is earned.

SOLUTION The outline below describes the steps taken by the program.

RETRIEVE program.

  Are there deposits or withdrawals?
    No   Yes ⟶ Function: compute balance using deposits and withdrawals.

  Is there interest earned?
    No   Yes ⟶ Function: compute interest and new balance.

  Is a printed copy wanted?
    No   Yes ⟶ Procedure: print out information.

END

Permutations, Combinations, and Probability **501**

Key Ideas

Use Pascal's triangle to expand a binomial.

Chalkboard Examples

1. Use Pascal's triangle to expand $(2m - 1)^6$.
   The seventh row is:
   1 6 15 20 15 6 1
   $\therefore (2m - 1)^6 = 1(2m)^6 +$
   $6(2m)^5(-1) + 15(2m)^4 \cdot$
   $(-1)^2 + 20(2m)^3(-1)^3 +$
   $15(2m)^2(-1)^4 + 6(2m) \cdot$
   $(-1)^5 + (-1)^6 = 64m^6 -$
   $192m^5 + 240m^4 -$
   $160m^3 + 60m^2 - 12m + 1$.

2. Find the seventh term in the expansion of $(t + 2)^{10}$.
   The seventh term is:
   $_{10}C_6 t^{10-7+1}(2)^{7-1} = 210t^4(2)^6$
   $= 210(64)t^4$
   $= 13,440t^4$

Mixed Review

1. Solve $16x^2 = 81$. $\left\{\dfrac{9}{4}, -\dfrac{9}{4}\right\}$

2. Express 0.000503 in standard notation.
   5.03×10^{-4}

3. Solve $\sqrt{3x + 4} + 2 = x$.
   $\{7\}$

4. Solve by substitution.
   $4x^2 + y^2 = 16$
   $y = 2x - 4$
   $\{(0, -4), (2, 0)\}$

5. Simplify $\sqrt[6]{4} \cdot \sqrt[3]{4}$. 2

12–7 Pascal's Triangle

You can look at the expansion of a positive integral power of a binomial from the point of view of combinations of terms selected from each of the binomial factors. Consider the following expansion:

$$(a + b)^3 = (a + b)(a + b)(a + b)$$
$$= aaa + baa + aab + aba + bba + bab + abb + bbb$$

You obtain each product shown in the expansion by multiplying three variables, one from each of the binomial factors of $(a + b)^3$. The term baa, for example, is the result of choosing b from the first binomial factor, a from the second, and a from the third. If you combine similar terms in the expansion to obtain

$$(a + b)^3 = a^3 + 3a^2b + 3ab^2 + b^3,$$

then 3, the coefficient of a^2b, is the number of ways of selecting one b from the three factors, that is, $_3C_1$. Similarly, because you obtain a^3 by choosing no b from the three factors, the coefficient of a^3 is 1, or $_3C_0$. In fact, you can rewrite the expansion as follows:

$$(a + b)^3 = {}_3C_0a^3 + {}_3C_1a^2b + {}_3C_2ab^2 + {}_3C_3b^3$$

The reasoning used in determining the coefficients in the expansion of $(a + b)^3$ can be extended to determining the coefficients in the expansion of $(a + b)^n$. Thus,

$$(a + b)^n = {}_nC_0a^n + {}_nC_1a^{n-1}b + {}_nC_2a^{n-2}b^2 + \cdots + {}_nC_nb^n,$$

where the rth term is

$$_nC_{r-1}a^{n-(r-1)}b^{r-1}.$$

If you write the expansions of $(a + b)^n$ for successive values of n in the form of a triangle, you have the following:

$$
\begin{array}{rl}
(a + b)^0 = & 1 \\
(a + b)^1 = & a + b \\
(a + b)^2 = & a^2 + 2ab + b^2 \\
(a + b)^3 = & a^3 + 3a^2b + 3ab^2 + b^3 \\
(a + b)^4 = & a^4 + 4a^3b + 6a^2b^2 + 4ab^3 + b^4
\end{array}
$$

Now, looking only at the coefficients, you see the following triangle:

$$
\begin{array}{ccccccccc}
 & & & & 1 & & & & \\
 & & & 1 & & 1 & & & \\
 & & 1 & & 2 & & 1 & & \\
 & 1 & & 3 & & 3 & & 1 & \\
1 & & 4 & & 6 & & 4 & & 1
\end{array}
$$

502   *Chapter 12*

Notice that each term other than 1 is the sum of the term to the right and the term to the left of it in the row directly above. Thus, the sixth row is obtained from the fifth row as follows:

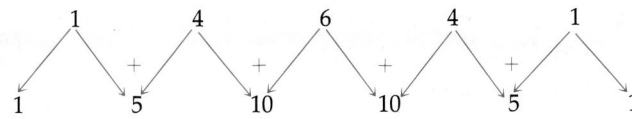

This array, known as **Pascal's triangle**, is named after the French mathematician and philosopher Blaise Pascal.

EXAMPLE 1 Use Pascal's triangle to expand $(2a + b)^6$.

SOLUTION You use the seventh row of Pascal's triangle, which you can obtain from the sixth row.

Sixth Row: 1   5   10   10   5   1

Seventh Row: 1   6   15   20   15   6   1

$$\therefore (2a + b)^6 = 1(2a)^6 + 6(2a)^5b + 15(2a)^4b^2 + 20(2a)^3b^3$$
$$+ 15(2a)^2b^4 + 6(2a)b^5 + 1 \cdot b^6$$
$$= 64a^6 + 192a^5b + 240a^4b^2 + 160a^3b^3 + 60a^2b^4 + 12ab^5 + b^6$$

EXAMPLE 2 Find the seventh term in the expansion of $(t - 2)^9$.

SOLUTION The rth term is given by $_nC_{r-1}(t)^{n-r+1}(2)^{r-1}$.
The seventh term is $_9C_6 t^{9-7+1}(2)^{7-1}$.

Since $_9C_6 = _9C_3 = \dfrac{9 \cdot 8 \cdot 7}{1 \cdot 2 \cdot 3} = 84$, the seventh term (which is positive) is

$$84t^3(2)^6 = 84(64)t^3 = 5376t^3.$$

Oral Exercises

1. Use the seventh row of Pascal's triangle shown below to find the eighth and ninth rows of the triangle.

         1   6   15   20   15   6   1

2. State the first four terms of $(a + b)^7$. $a^7 + 7a^6b + 21a^5b^2 + 35a^4b^3$
3. State the last four terms of $(c - d)^8$. $-56c^3d^5 + 28c^2d^6 - 8cd^7 + d^8$
4. State the constant term of $\left(x + \dfrac{1}{x}\right)^8$. fifth term: 70

Permutations, Combinations, and Probability **503**

Suggested Assignments

Enriched Alg.
First day
504/1–15 odd, 16–19
Second day
R 504/*Self-Test 3*
S 485/2–12 even
Enriched Alg. with Trig.
First day
504/1–15 odd, 16–19
Second day
R 504/*Self-Test 3*
S 485/2–12 even

Additional Answers
Written Exercises

2. $r^7 - 14r^6 + 84r^5 - 280r^4$
$+ 560r^3 - 672r^2 + 448r$
$- 128$

3. $243x^5 + 405x^4y +$
$270x^3y^2 + 90x^2y^3 +$
$15xy^4 + y^5$

4. $\frac{1}{64}a^6 + \frac{3}{16}a^5b + \frac{15}{16}a^4b^2$
$+ \frac{5}{2}a^3b^3 + \frac{15}{4}a^2b^4 + 3ab^5$
$+ b^6$

5. $1 - 7y^3 + 21y^6 - 35y^9 +$
$35y^{12} - 21y^{15} + 7y^{18} - y^{21}$

6. $\frac{1}{243}a^5 + \frac{5}{27}a^4 + \frac{10}{3}a^3 +$
$30a^2 + 135a + 243$

7. $r^{18} - 6r^{15}t^2 + 15r^{12}t^4 -$
$20r^9t^6 + 15r^6t^8 - 6r^3t^{10} +$
t^{12}

Quick Quiz

1. Expand $(x^2 - 2)^4$.
$x^8 - 8x^6 + 24x^4 - 32x^2 +$
16

2. Find the fourth term in the expansion of $(2k + n^2)^6$.
$160k^3n^6$

3. Use Pascal's triangle to expand $\left(\frac{a}{2} + 2\right)^5$.

$\frac{a^5}{32} + \frac{5a^4}{8} + 5a^3 + 20a^2 +$
$40a + 32$

Written Exercises

Use Pascal's triangle to expand each binomial. Express answers in simplified form. 1. $c^6 + 6c^5 + 15c^4 + 20c^3 + 15c^2 + 6c + 1$

A 1. $(c + 1)^6$ 2. $(r - 2)^7$ 3. $(3x + y)^5$ 4. $\left(\frac{a}{2} + b\right)^6$

5. $(1 - y^3)^7$ 6. $\left(\frac{1}{3}a + 3\right)^5$ 7. $(r^3 - t^2)^6$ 8. $(k^2 + 2n)^8$

8. $k^{16} + 16k^{14}n + 112k^{12}n^2 + 448k^{10}n^3 + 1120k^8n^4 + 1792k^6n^5 + 1792k^4n^6 + 1024k^2n^7 + 256n^8$

For each expansion, express in simplified form the term that contains the given expression.

B 9. $(1 + m)^8$; m^4 $70m^4$ 10. $(x + \frac{1}{2})^9$; x^6 $\frac{21}{2}x^6$

11. $\left(\frac{p}{2} + q\right)^{10}$; q^7 $15p^3q^7$ 12. $(a + \frac{1}{3}b)^{11}$; a^7 $\frac{110}{27}a^7b^4$

13. $(1 - 2x)^{12}$; $(2x)^4$ $7920x^4$ 14. $(r^2 - t^3)^{14}$; $(t^3)^5$ $-2002r^{18}t^{15}$

15. The eighth term of $(x - y)^{15}$ is $-6435x^8y^7$.
   a. What is the ninth term? $6435x^7y^8$
   b. What is the ninth term of $(x - y)^{16}$? $12,870x^8y^8$

16. The fifteenth term of $(z + 1)^{30}$ is $145,422,675z^{16}$.
   a. What is the seventeenth term? $145,422,675z^{14}$
   b. What is the sixteenth term? $155,117,520z^{15}$

C 17. Prove: $_nC_3 = {}_{n-1}C_2 + {}_{n-1}C_3$, for $n \geq 4$

18. Prove: $_nC_r = {}_{n-1}C_{r-1} + {}_{n-1}C_r$, for $n \geq r + 1$

19. Add the entries in each of the first few rows of Pascal's triangle and make a conjecture about the sum of the entries in the nth row as a function of n. Test the conjecture for $n = 8$.
$f(n) = 2^{n-1}$; the sum of the entries in the eighth row is $2^7 = 128$.

Self-Test 3

VOCABULARY   Binomial Theorem (p. 499)
               Pascal's triangle (p. 503)

1. Expand $(a^2 - 2)^5$. $a^{10} - 10a^8 + 40a^6 - 80a^4 + 80a^2 - 32$ *Obj. 1, p. 498*

2. Find the seventh term in the expansion of $(2p + q)^{10}$. $3360p^4q^6$ *Obj. 2, p. 498*

3. Use Pascal's triangle to expand $\left(x - \frac{y}{2}\right)^6$.   *Obj. 3, p. 498*

$x^6 - 3x^5y + \frac{15}{4}x^4y^2 - \frac{5}{2}x^3y^3 + \frac{15}{16}x^2y^4 - \frac{3}{16}xy^5 + \frac{1}{64}y^6$

Check your answers with those at the back of the book.

---

504   Chapter 12

Probability

OBJECTIVES for Sections 12-8 through 12-11:
1. To list a sample space for an experiment, and list an event.
2. To find the probability of an event and of its complement.
3. To find the probability of mutually exclusive events.
4. To find the probability of the occurrence of a second event given the occurrence of a first event.

12–8 Sample Spaces and Events

An experiment in which you do not necessarily get the same *outcome* when you repeat the experiment under essentially the same conditions is called a **random experiment**. The set of all possible outcomes of a random experiment is known as the **sample space** for the experiment. For example, suppose you conduct a random experiment by tossing three coins—a nickel, a dime, and a quarter. If H represents heads and T tails, then one possible outcome of the experiment is (H, H, T), where the components of the ordered triple represent in order the result of tossing the nickel, the dime, and the quarter. The sample set of the experiment is

{(H, H, H), (H, H, T), (H, T, H), (H, T, T), (T, H, H), (T, H, T), (T, T, H), (T, T, T)}.

Any subset of possible outcomes for an experiment is known as an event. When an event involves a single element of the sample space, it is often called a *simple* event.

EXAMPLE 1   A die is rolled and the number of spots on its top face when it comes to rest is observed. Specify the following.
a. the sample space
b. the event that a number greater than 5 results
c. the event that a number less than 4 results
d. the event that an even number results

SOLUTION   **a.** {1, 2, 3, 4, 5, 6}   **b.** {6}   **c.** {1, 2, 3}   **d.** {2, 4, 6}

It is convenient to discuss the sample space for an experiment involving two distinct occurrences, such as throwing two dice, as the Cartesian product of the sample spaces of each of the occurrences. Since the sample space for each die is

$$A = \{1, 2, 3, 4, 5, 6\},$$

the sample space for the experiment is $A \times A$. A *lattice* that portrays this sample space is shown in Figure 2.

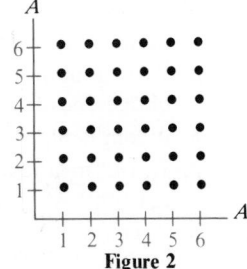
Figure 2

Permutations, Combinations, and Probability   **505**

Related Activities p. T113

Key Ideas
List sample spaces and events.

Chalkboard Examples
1. A die is rolled and the number of spots on its top face when it comes to rest is observed. Specify the following:
   a. the sample space {1, 2, 3, 4, 5, 6}
   b. the event that a number greater than 3 results   {4, 5, 6}
   c. the event that a number less than 5 results {1, 2, 3, 4}
   d. the event that an odd number results {1, 3, 5}
2. Two dice are thrown. Specify the event that the sum of the spots showing is:
   a. 5   {(1, 4), (2, 3), (3, 2), (4, 1)}
   b. 3   {(1, 2), (2, 1)}

Additional A Exercises
In Exercises 1–6 specify the **(a)** sample space for each experiment, and **(b)** the event.
1. A number is chosen from the set of even integers 0 through 10. Event: The number is a multiple of 2.
   a. {0, 2, 4, 6, 8, 10}
   b. {0, 2, 4, 6, 8, 10}

(continued)

EXAMPLE 2 Two dice are thrown. Specify the event that the sum of the spots showing is:

a. 4 **b.** 6

SOLUTION **a.** {(1, 3), (2, 2), (3, 1)} **b.** {(1, 5), (2, 4), (3, 3), (4, 2), (5, 1)}

Additional A Exercises
(continued)

2. A marble is drawn from a bag containing 4 red marbles, numbered 1 through 4, and 3 blue marbles numbered 1 through 3. Event: The marble is red.
   a. $\{R_1, R_2, R_3, R_4, B_1, B_2, B_3\}$ **b.** $\{R_1, R_2, R_3, R_4\}$
3. A coin is flipped twice. Event: Tails appear once.
   a. {(H, H), (H, T), (T, H), (T, T)}
   b. {(H, T), (T, H)}
4. Two cards are drawn randomly from the 4 kings of a standard card deck. Event: The cards are both black.
   a. {(C, H), (C, D), (C, S), (H, C), (H, D), (H, S), (D, C), (D, S), (D, H), (S, C), (S, H), (S, D)}
   b. {(C, S), (S, C)}
5. A block is drawn from a box containing 3 blocks labeled A, B, and C. The block is returned to the box and another block is drawn. Event: Both blocks have the same letter.
   a. {(A, A), (A, B), (A, C), (B, A), (B, B), (B, C), (C, A), (C, B), (C, C)}
   b. {(A, A), (B, B), (C, C)}
6. Two different digits are chosen randomly from the number 3247. Event: The two digits are both even.
   a. {(3, 2), (3, 4), (3, 7), (2, 3), (2, 4), (2, 7), (4, 3), (4, 2), (4, 7), (7, 3), (7, 2), (7, 4)}
   b. {(2, 4), (4, 2)}

Oral Exercises

For the experiment in which a letter is selected at random from those in the word PLANET, **specify the following.**

1. the sample space {P, L, A, N, E, T}
2. the event that the letter selected is a vowel {A, E}
3. the event that the letter selected is a consonant {P, L, N, T}
4. the event that the letter selected is the fifth letter of the alphabet {E}

For the experiment in which two different letters are selected at random from those in the word ARC, **specify the following.**

5. the sample space
   {(A, R), (A, C), (R, C), (R, A), (C, A), (C, R)}
6. the event that both letters selected are consonants
   {(R, C), (C, R)}

Written Exercises

In Exercises 1–8, (a) specify the sample space for the given experiment and (b) specify the given event.

A 1. A number is chosen from the integers 1 through 14. Event: The number is a multiple of 4.
   2. A sock is drawn from a bag containing 2 white socks, numbered 1 and 2, and 5 black socks, numbered 1 through 5. Event: The sock is black.
   3. A tag is drawn from a box containing three tags numbered 1, 2, and 3. The tag is returned to the box, and another tag is drawn. Event: Both tags have the same number.
   4. Same experiment as in Exercise 3 except that the first tag is not returned before the second is drawn. Event: The sum of the two numbers is even.
   5. Two cards are drawn randomly from the four aces of a standard bridge deck. Event: The cards are both red.
   6. A coin is flipped twice. Event: Heads appears at least once.
   7. Two *different* letters are chosen randomly from the letters A, B, C, and E. Event: At least one of the two letters is a vowel.
   8. Same experiment as in Exercise 7. Event: The two letters are not both vowels.

Suggested Assignments

Enriched Alg.
  506/1–15 odd, 16
S 494/2–10 even
Enriched Alg. with Trig.
  506/1–15 odd, 16
S 494/6, 8, 10

In Exercises 9–16 the experiment is drawing the given number of cards from a standard 52-card bridge deck. Tell *how many* elements are in the sample space and *how many* elements are in the given event. Recall that the sample space consists of all the possible outcomes of the experiment.

B   **9.** A single card is drawn; it is a heart.   **10.** Of two cards drawn, both are kings.

11. Of two cards drawn, both are red. 1326; 325

12. Of two cards drawn, both are face cards. 1326; 66

13. Of two cards drawn, both are clubs. 1326; 78

14. Of three cards drawn, all are spades. 22,100; 286

C   **15.** Of five cards drawn, three are diamonds and two are spades.

16. Five cards are drawn, and all four suits are represented.

15. 2,598,960; 22,308   **16.** 2,598,960; 685,464

ON THE CALCULATOR

Certain sample spaces and events contain a great number of elements. Counting these elements can be a long and difficult process without the aid of a calculator.

Some calculators have a factorial key ($x!$) that can calculate factorials of large numbers automatically. If your calculator has a factorial key, use the formula

$$_nC_r = \frac{n!}{r!(n-r)!}$$

to calculate $_nC_r$. If your calculator does not have a factorial key, use the formula

$$_nC_r = \frac{n(n-1)(n-2) \cdots (n-r+1)}{r(r-1)(r-2) \cdots 1}.$$

If the numbers are too great to fit on your calculator screen, the screen will print out the results in scientific notation (see On the Calculator, page 260). If so, your result will be only an approximation.

Exercises

In an experiment of drawing the given number of cards from a standard 52-card bridge deck, tell how many elements are in the sample space and how many elements are in the given event. (Aces are considered high.)

270,725; 14,950   2,598,960; 1287

1. Of four cards drawn, all are black.   **2.** Of five cards drawn, all are diamonds.

3. Of five cards drawn, all have denomination less than eight. 2,598,960; 42,504

4. Of six cards drawn, four are hearts and two are clubs. 20,358,520; 55,770

5. Of fifteen cards drawn, all are red. 4.4814×10^{12}; 7,726,160

6. Of twelve cards drawn, eight are face cards and the remaining four have denomination less than 5. 2.1×10^{11}; 245,025

Permutations, Combinations, and Probability **507**

Mixed Review

1. Solve $\frac{1}{2}x^{-\frac{2}{3}} = 8$. $\left\{\frac{1}{64}\right\}$

2. Solve $\log_x \frac{1}{9} = -4$. $\{\sqrt{3}\}$

3. Solve the equation $x^2 + 2x + 5 = 0$.
$\{-1 + 2i, -1 - 2i\}$

4. Express $\sqrt{\dfrac{50x^5}{x-1}}$ in simple radical form. $\dfrac{5x^2\sqrt{2x^2 - 2x}}{x-1}$

5. Simplify $\dfrac{x^2 - 4y^2}{(x-2y)^2} \cdot \dfrac{x+2y}{x-2y}$

Determine the mathematical
probability of an event and
its complement.

1. If two cards are drawn at
random from a standard
52-card bridge deck, what
is the probability that both
are face cards?
Since there are 12 face
cards in the deck, there
are $_{12}C_2$ ways in which
two face cards can be
drawn. There are 52 cards
all together, so there are
$_{52}C_2$ possible ways in
which two cards can be
drawn. If E represents
drawing two face cards,
then

$$P(E) = \frac{_{12}C_2}{_{52}C_2} = \frac{\dfrac{12 \cdot 11}{1 \cdot 2}}{\dfrac{52 \cdot 51}{1 \cdot 2}}$$

$$= \frac{66}{1326} = \frac{11}{221}$$

2. Two marbles are drawn at
random from a bag con-
taining 12 blue and 14 red
marbles.
   a. What is the probability
   that at least one marble
   is blue?
   b. What are the odds that
   at least one marble is
   blue?
   a. The probability that at
   least one marble is
   blue is just the probability that not both
   marbles are red. Let E
   represent the event
   that both are red.

12–9 Mathematical Probability

Consider the following experiment: From a bag containing 5 blue and 12
white marbles, one marble is drawn. If the experiment is designed so
that each marble is just as likely to be drawn as any other, we say that
the experiment has 17 *equally likely* outcomes. The event that the marble
drawn is white consists of 12 outcomes. Therefore, if you replace the
marble drawn and repeat the experiment many times, it seems reasonable to expect that about $\frac{12}{17}$ of the time you will find that you have drawn
a white marble. This ratio, $\frac{12}{17}$, is called the *probability* that the outcome of
any single trial of the experiment will be the drawing of a white marble.
This example suggests the following.

Let S be the sample space of an experiment in which there are n
possible outcomes, each equally likely. If an event E with h
elements is a subset of S, then the probability of E, denoted by
$P(E)$, is given by

$$P(E) = \frac{h}{n}.$$

EXAMPLE 1 If two cards are drawn at random from a standard 52-card bridge deck,
what is the probability that both cards are hearts?

SOLUTION Since there are 13 hearts in the deck, there are $_{13}C_2$ ways in which two of
them can be drawn. There are 52 cards altogether, so there are $_{52}C_2$
possible ways in which two of them can be drawn. If E represents
drawing 2 hearts, then

$$P(E) = \frac{_{13}C_2}{_{52}C_2} = \frac{\dfrac{13 \cdot 12}{1 \cdot 2}}{\dfrac{52 \cdot 51}{1 \cdot 2}} = \frac{78}{1326} = \frac{1}{17}.$$

In the foregoing example, the answer $\frac{1}{17}$ does not tell you anything
certain about what is going to happen. It does not, for example, tell you
that you will get exactly one pair of hearts out of 17 draws. You might get
one such draw, or you might get none, or you might even get 17.
However, if you perform the experiment a very large number of times,
the ratio of the number of times you draw 2 hearts to the total number of
draws will probably come close to $\frac{1}{17}$.

An event E in the sample space S is called *certain* if $E = S$; it is called
impossible if $E = \emptyset$. Since $P(S) = \frac{n}{n} = 1$, while $P(\emptyset) = \frac{0}{n} = 0$, the
probability is 1 for a certain event and 0 for an impossible one. Notice
that the probability of an event that is neither certain nor impossible is a
number between 0 and 1.

508 *Chapter 12*

By the symbol $\overline{E}$, read "the complement of E," we mean the set of the elements of S that are not members of E. If S has n members and E has h members, then $\overline{E}$ contains $n - h$ elements. Therefore, $P(\overline{E})$ is the probability that E does not occur, and

$$P(\overline{E}) = \frac{n - h}{n} = 1 - \frac{h}{n} = 1 - P(E).$$

The odds that the event E will occur are given by

$$\frac{P(E)}{P(\overline{E})}, \quad \text{or} \quad \frac{h}{n - h}, \quad \text{or} \quad h \text{ to } n - h.$$

Thus, in the original experiment the odds are *12 to 5 in favor of* drawing a white marble or *5 to 12 against* drawing a white marble.

EXAMPLE 2   Two marbles are drawn at random from an urn containing 12 red and 10 blue marbles.
    a. What is the probability that at least one marble is blue?
    b. What are the odds that at least one marble is blue?

SOLUTION   **a.** The probability that at least one marble is blue is just the probability that *not both* marbles are red. Let E represent the event that both *are* red. Then

$$P(E) = \frac{{}_{12}C_2}{{}_{22}C_2} = \frac{\frac{12 \cdot 11}{1 \cdot 2}}{\frac{22 \cdot 21}{1 \cdot 2}} = \frac{66}{231} = \frac{2}{7},$$

and the probability that at least one marble is blue is just

$$P(\overline{E}) = 1 - \frac{2}{7} = \frac{5}{7}.$$

b. The odds that at least one marble is blue are $\dfrac{5}{7 - 5}$, or 5 to 2.

Oral Exercises

A single die is tossed. What is the probability that the number of spots showing is

1. five? $\frac{1}{6}$ **2.** odd? $\frac{1}{2}$ **3.** one or four? $\frac{1}{3}$ **4.** less than 5? $\frac{2}{3}$

5–8. Give the odds in favor of each of the events in Exercises 1–4 above. $\frac{1}{5}; \frac{1}{1}; \frac{1}{2}; \frac{2}{1}$

9. A nickel and a dime are tossed.
    a. What is the sample space of the experiment? {(H, H), (H, T), (T, H), (T, T)}
    b. What is the probability of getting exactly 1 head? $\frac{1}{2}$
    c. What is the probability of getting 2 heads? $\frac{1}{4}$

10. A coin is tossed 4 times.
    a. How many elements are there in the sample space? 16
    b. What is the probability of getting exactly 1 head? $\frac{1}{4}$

Permutations, Combinations, and Probability **509**

Then

$$P(E) = \frac{{}_{14}C_2}{{}_{26}C_2} = \frac{14 \cdot 13}{26 \cdot 25} = \frac{7}{25}$$

and the probability that at least one marble is blue is

$$P(\overline{E}) = 1 - \frac{7}{25} = \frac{18}{25}.$$

b. The odds that at least one marble is blue are $\dfrac{18}{25 - 18}$, or 18 to 7.

Reading Algebra

Be sure students read the symbol $P(E)$ correctly as "the probability of event E" where E is an event or a subset of the sample space. The symbol $\overline{E}$ is read "the complement of E" and $P(\overline{E})$ is read "the probability of the complement of E" or "the probability that E does not occur."

Common Errors

Students sometimes forget that the probability of an event will always be a fraction between 0 and 1, inclusive. To help them remember this, remind them that if they know for certain that an event A will occur, then $P(A) = 1$. If they know for certain that an event B will *not* occur, then $P(B) = 0$. All other probabilities must fall between 0 and 1. It might be useful at this point to contrast the "number of ways" an event can occur, and the probability of an event occurring.

Additional A Exercises

1. One letter is selected at random from the first 10 letters of the alphabet. What is the probability that the letter is

   a. a vowel? $\frac{3}{10}$

   b. after the letter F?
      $\frac{4}{10}$, or $\frac{2}{5}$

2. What are the odds for each event in Exercise 1?
   a. 3 to 7   b. 2 to 3

3. Two dice are thrown. What is the probability of each of the following events?
   a. The sum of the numbers showing is greater than 10? $\frac{3}{36}$, or $\frac{1}{12}$

   b. The sum of the numbers showing is 13? 0

4. What are the odds for each event in Exercise 3?
   a. 1 to 11   b. 0 to 1

5. Two marbles are drawn from a bag containing 3 red, 5 green, and 4 blue marbles. What is the probability of each of the following events?

   a. Both are red. $\frac{1}{22}$

   b. Neither is green. $\frac{7}{22}$

6. A coin is tossed 4 times. What is the probability of getting exactly 2 heads?
   $\frac{6}{16}$, or $\frac{3}{8}$

Written Exercises

A 1. One letter is selected at random from the letters in the word LOGARITHM. What is the probability that the letter is
   a. a vowel? $\frac{1}{3}$ b. a consonant? $\frac{2}{3}$
   c. before J in the alphabet? $\frac{4}{9}$ d. in the word PARTICLE? $\frac{5}{9}$

2. Two dice are thrown. What is the probability of each of the following events? (If necessary, refer to Figure 2, page 505.)
   a. The sum of the numbers showing is 9. $\frac{1}{9}$
   b. Both dice show the same number. $\frac{1}{6}$
   c. The dice show different numbers. $\frac{5}{6}$
   d. The sum of the numbers showing is 3 or 7. $\frac{2}{9}$

3. In Exercise 1 what are the odds for each event? $\frac{1}{2}; \frac{2}{1}; \frac{4}{5}; \frac{5}{4}$

4. In Exercise 2 what are the odds for each event? $\frac{1}{8}; \frac{1}{5}; \frac{5}{1}; \frac{2}{7}$

5. Two socks are drawn at random from a drawer containing 4 red, 5 green, and 7 blue socks. What is the probability of each of the following events?
   a. Both are red. $\frac{1}{20}$ b. Both are green. $\frac{1}{12}$
   c. Both are blue. $\frac{7}{40}$ d. Neither is red. $\frac{11}{20}$
   e. Neither is green. $\frac{11}{24}$ f. Neither is blue. $\frac{3}{10}$

6. Are the events of parts (a) and (d) in Exercise 5 complements of each other? Give a reason for your answer. no; $\frac{1}{20} + \frac{11}{20} \neq 1$

7. A coin is tossed 5 times. What is the probability of getting
   a. all heads? $\frac{1}{32}$ b. exactly 1 tail? $\frac{5}{32}$
   c. exactly 3 heads? $\frac{5}{16}$ d. exactly 2 tails? $\frac{5}{16}$
   e. at least 1 head? (*Hint:* Consider the complement.) $\frac{31}{32}$

8. On a 10-question true-false test:
   a. How many possible ways are there of answering all the questions? 1024
   b. If the questions are answered at random, what is the probability of answering exactly 9 questions correctly? $\frac{5}{512}$
   c. If the questions are answered at random, what is the probability of answering exactly 7 questions correctly? $\frac{15}{128}$

B 9. A three-digit numeral with no repeated digits is made from the digits 1 through 6. What is the probability that the number indicated is
   a. odd? $\frac{1}{2}$ b. a multiple of 5? $\frac{1}{6}$
   c. between 100 and 400? $\frac{1}{2}$ d. between 200 and 550? $\frac{19}{30}$

10. Four cards are drawn at random from the 13 hearts in a standard deck. What is the probability that the selection contains
    a. no face cards (jack, queen, or king)? $\frac{42}{143}$ b. both the queen and the king? $\frac{1}{13}$
    c. the queen and not the king? $\frac{3}{13}$ d. at least two face cards? $\frac{29}{143}$

11. A committee of 5 is to be chosen by lot from a group of 11 people. If Herb and Nora are among those being considered, what is the

probability of each of these events?
a. Both Herb and Nora will be on the committee. $\frac{2}{11}$
b. Neither Herb nor Nora will be on the committee. $\frac{3}{11}$
c. Nora but not Herb will be on the committee. $\frac{3}{11}$
d. At least one of the two will be on the committee. (*Hint:* See part (b).) $\frac{8}{11}$

12. The letters of the word QUIET are rearranged at random. What is the probability of each of the following events?
a. Q will be followed directly by U. $\frac{1}{5}$
b. Q and U will be together in either order. $\frac{2}{5}$

13. Two different letters are chosen at random from the word NUMERI-CAL. What is the probability of each of the following events?
a. Both are vowels. $\frac{1}{6}$ b. Both are consonants. $\frac{5}{18}$

14. A committee of 5 is to be chosen at random from 4 juniors and 7 seniors. What is the probability that the committee will contain exactly 3 seniors and 2 juniors? $\frac{5}{11}$

15. If 4 boys and 5 girls are seated at random at 9 desks in a row, what is the probability that the boys and girls are in alternate seats? $\frac{1}{126}$

C 16. In Exercise 15 what is the probability that the 4 boys are in 4 adjacent seats? $\frac{1}{21}$

12–10 Mutually Exclusive Events

The diagram in Figure 3 shows the sample space S for the experiment of drawing a number at random from {1, 2, 3, 4, 5, 6, 7, 8}. (Such diagrams are known as *Venn diagrams*.) The diagram also shows events A and B in S. Event A corresponds to the drawing of a number less than 4, so that $A = \{1, 2, 3\}$. Event B corresponds to the drawing of an even number; that is, $B = \{2, 4, 6, 8\}$. Therefore,

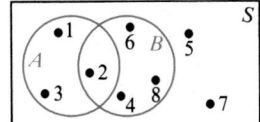

Figure 3

$$P(A) = \frac{3}{8} \quad \text{and} \quad P(B) = \frac{4}{8} = \frac{1}{2}.$$

What is the probability that either A or B (or both) will occur? This is equivalent to $P(A \cup B)$. Since $A \cup B = \{1, 2, 3, 4, 6, 8\}$, $P(A \cup B) = \frac{6}{8} = \frac{3}{4}$. To see the relationship between $P(A \cup B)$, $P(A)$, and $P(B)$, notice that $A \cap B = \{2\}$ and thus $P(A \cap B) = \frac{1}{8}$.

$$P(A \cup B) = \frac{6}{8} = \frac{3 + 4 - 1}{8} = \frac{3}{8} + \frac{4}{8} - \frac{1}{8}$$
$$P(A \cup B) = P(A) + P(B) - P(A \cap B)$$

Permutations, Combinations, and Probability   **511**

Mixed Review

1. Solve $\sqrt{2x + 6} - \sqrt{x - 1} = 2$. {5}

2. Find a if $2 - i$ is a root of the equation $ax^2 - 4ix + 10 = 0$. -2

3. The difference of the squares of two consecutive even integers is 22 greater than the smaller of the two integers. What are the integers? 6 and 8

4. Simplify $\dfrac{21}{x^2 - 49} + \dfrac{3}{x + 7} \cdot \dfrac{3x}{x^2 - 49}$

5. Solve the equation $x^3 - x^2 - 2x = 0$ by factoring. $\{-1, 0, 2\}$

Teaching Suggestions p. T114

Related Activities p. T114

Key Ideas

Determine the probability of mutually exclusive events.

Reading Algebra

To reinforce the meaning of *union* and *intersection* be sure students read the notation correctly. The probability of a union, denoted $P(A \cup B)$ is read as "the probability of A or B (or both)." The probability of an intersection, denoted $P(A \cap B)$, is read as "the probability of A and B."

1. Five beads are drawn at random from a bag containing 4 yellow, 8 white, and 6 red beads. What is the probability that at least 3 of them are white?
   The desired probability is the sum of the probabilities that exactly three beads are white (A), that exactly four beads are white (B), or that exactly five beads are white (C).

   $P(A) = \dfrac{_8C_3 \times {}_{10}C_2}{_{18}C_5} = \dfrac{2520}{8568}$

   $P(B) = \dfrac{_8C_4 \times {}_{10}C_1}{_{18}C_5} = \dfrac{700}{8568}$

   $P(C) = \dfrac{_8C_5}{_{18}C_5} = \dfrac{56}{8568}$

   $\therefore P(A) + P(B) + P(C) =$
   $\dfrac{3276}{8568} = \dfrac{91}{238}$

2. A change purse contains 3 nickels and 5 dimes. Two coins are drawn at random. Give the probability that at least one of the two coins is a nickel.
   The probability is the sum of the probabilities that exactly one coin is a nickel (A) or that exactly 2 coins are nickels (B).

   $P(A) = \dfrac{_3C_1 \times {}_5C_1}{_8C_2} = \dfrac{15}{28}$

   $P(B) = \dfrac{_3C_2}{_8C_2} = \dfrac{3}{28}$

   $P(A \cup B) = \dfrac{15}{28} + \dfrac{3}{28} = \dfrac{9}{14}$

Common Errors

Stress that an "or" means the "union," and an "and" means the "intersection." Students will sometimes interpret the "and" as addition and mistakenly use the union formula.

It can be proved that this relationship is true in general.

> For any two events A and B in a sample space,
> $$P(A \cup B) = P(A) + P(B) - P(A \cap B).$$

If the events A and B have no outcome in common, that is, if $A \cap B = \emptyset$, we say that the events are **mutually exclusive**. For mutually exclusive events, $P(A \cap B) = P(\emptyset) = 0$. Thus we have the following:

> If A and B are mutually exclusive events,
> $$P(A \cup B) = P(A) + P(B).$$

EXAMPLE Four marbles are drawn at random from an urn containing 2 white, 4 red, and 6 blue marbles. What is the probability that at least 2 of them are red?

SOLUTION The desired probability is the sum of the probabilities that *exactly* two marbles are red (A), that *exactly* three marbles are red (B), or that *exactly* four marbles are red (C).

$$P(A) = \dfrac{_4C_2 \times {}_8C_2}{_{12}C_4} = \dfrac{6 \times 28}{495} = \dfrac{168}{495}$$

$$P(B) = \dfrac{_4C_3 \times {}_8C_1}{_{12}C_4} = \dfrac{4 \times 8}{495} = \dfrac{32}{495}$$

$$P(C) = \dfrac{_4C_4}{_{12}C_4} = \dfrac{1}{495}$$

$$\therefore P(A \cup B \cup C) = \dfrac{168}{495} + \dfrac{32}{495} + \dfrac{1}{495} = \dfrac{201}{495} = \dfrac{67}{165}$$

Oral Exercises

A single die is tossed. State whether the events are mutually exclusive.

1. The number of spots showing is less than 4; the number of spots showing is even. No
2. The number of spots showing is less than 3; the number of spots showing is greater than 2. Yes
3. Five spots are showing; the number of spots showing is odd. No
4. The number of spots showing is a prime number; the number of spots showing is greater than or equal to 3. No

Written Exercises

A
1. Six coins are flipped simultaneously. What is the probability there are
   a. at least 5 heads? $\frac{7}{64}$
   b. at least 4 heads? $\frac{11}{32}$
   c. at least 3 tails? $\frac{21}{32}$
   d. no heads? $\frac{1}{64}$
   e. at least 4 heads or at least 3 tails? 1

2. Two marbles are drawn at random from a bag containing 3 red, 5 blue, and 6 green marbles. What is the probability of drawing
   a. at least one red marble? $\frac{36}{91}$
   b. at least one green marble? $\frac{9}{13}$
   c. two marbles of the same color? $\frac{4}{13}$
   d. two marbles of different colors? $\frac{9}{13}$

3. What is the probability of getting 8 or more questions correct on a 10-question true-false test if the questions are answered at random? $\frac{7}{128}$

4. Two cards are drawn from a standard 52-card deck. What is the probability that the cards are
   a. both spades? $\frac{1}{17}$ b. of the same suit? $\frac{4}{17}$
   c. both jacks? $\frac{1}{221}$ d. a pair? $\frac{1}{17}$
   e. both spades or both jacks? $\frac{14}{221}$ f. both of the same suit or a pair? $\frac{5}{17}$

5. Three letters are chosen at random from the word POSTER. What is the probability that the selection will contain
   a. E or O but not both? $\frac{3}{5}$ b. E or O or both? $\frac{4}{5}$

B
6. Three dice are thrown. What is the probability of the following events?
   a. All 3 dice show the same number. $\frac{1}{36}$
   b. Exactly 2 of the dice show the same number. $\frac{5}{12}$
   c. Two or more of the dice show the same number. $\frac{4}{9}$

7. Two letters are chosen at random from the word DELETE and two are chosen at random from the word ENTER. What is the probability that the selection contains 4 E's or no E's? $\frac{2}{25}$

8. Five letters are chosen at random from the first 10 letters of the alphabet. Find the probability that the selection contains A or B but not both. $\frac{5}{9}$

C
9. Four beads are drawn at random from a box containing 4 black, 4 white, and 2 red beads. What is the probability that the selection will contain exactly 2 black beads or exactly 2 white beads or both? (Note that the two events of drawing exactly 2 white beads and of drawing exactly 2 black beads are not mutually exclusive.) $\frac{24}{35}$

10. In Exercise 8 what is the probability that the selection contains A or E or both, but not I? $\frac{5}{12}$

Permutations, Combinations, and Probability **513**

Additional A Exercises

1. Five coins are flipped simultaneously. What is the probability there are
   a. at least 4 heads? $\frac{3}{16}$
   b. no tails? $\frac{1}{32}$

2. Three marbles are drawn at random from an urn containing 4 white, 3 red, and 5 blue marbles. What is the probability that at least one of them is red? $\frac{34}{55}$

3. What is the probability of getting 7 or more questions correct on a 10-question true-false test if the questions are answered at random? $\frac{11}{64}$

Mixed Review

1. Solve $1 + \frac{2}{x} - \frac{x+5}{x^2+x} = 0$.
   $\{1, -3\}$

2. Simplify $\frac{1 - \sqrt{7}}{3 - \sqrt{7}}$.
   $-2 - \sqrt{7}$

3. Solve $4x^2 - 4x = 5$.
   $\left\{\frac{1}{2} + \frac{\sqrt{6}}{2}, \frac{1}{2} - \frac{\sqrt{6}}{2}\right\}$

4. Write a quadratic equation with integral coefficients whose solutions set is $\{-1 + i\sqrt{3}, -1 - i\sqrt{3}\}$.
   $x^2 + 2x + 4 = 0$

Key Ideas

Determine conditional probabilities.

Chalkboard Examples

1. If two cards are drawn at random from a 52-card deck, what is the probability that both cards are aces?

Let A be the event that the first card is an ace, and B be the event that the second card is an ace. Then

$P(A) = \frac{4}{52}, P(B|A) = \frac{3}{51},$

$P(A \cap B) = P(A) \cdot P(B|A)$

$\quad = \frac{4}{52} \cdot \frac{3}{51} = \frac{1}{221}.$

2. A red die and a green die are thrown. What is the probability that the sum of the numbers shown is 8 and that the green die shows a number greater than 3? Are these events independent?
Graph the sample space.

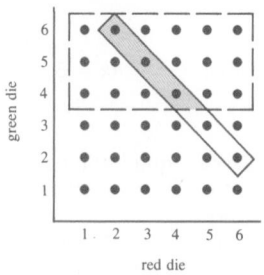

red die

Let A be the event that the sum of the numbers is 8.

12–11 Conditional Probability

Suppose 2 balls are drawn at random from a bag containing 4 red and 3 black balls. (1) If the first ball is replaced before the second is drawn, what is the probability that both balls are red? (2) If the first ball is not replaced before the second is drawn, what is the probability that both balls are red?

Let A be the event of drawing a red ball the first time, and B be the event of drawing a red ball the second time. The sample space S of the experiment depends on whether or not the first ball drawn is returned to the bag before the second ball is chosen. Then in each case we want to find $P(A \cap B)$.

Case 1: The first ball is replaced before the second is drawn.

The sample space S consists of all ordered pairs (x, y) where both x and y denote elements of a set of 7 outcomes (4 red, 3 black).

Figure 4 shows the sample space containing 7×7, or 49, ordered pairs. We use r_i and b_i to designate the drawing of red and black balls, respectively. The dashed rectangle outlines all possible experiment outcomes in which the first ball is red, and the solid rectangle outlines all possible outcomes in which the second ball is red. Since $A \cap B$ consists of the ordered pairs of the form (red, red), there are $4 \times 4 = 16$ elements in $A \cap B$. Therefore,

$$P(A \cap B) = \tfrac{16}{49} = \tfrac{4}{7} \cdot \tfrac{4}{7}.$$

Notice that

$$P(A \cap B) = P(A) \cdot P(B).$$

The outcomes in $A \cap B$ are in the region that is shaded in Figure 4.

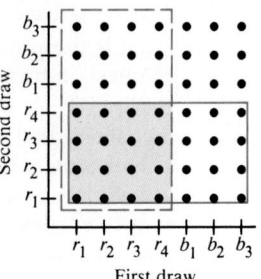

First draw

Figure 4

Case 2: The first ball is not replaced before the second is drawn.

Any one of the 7 balls may be selected on the first draw, but since this ball will not be replaced, there remain only 6 balls for the second draw. All ordered pairs with equal components must be deleted from the sample space. Note in Figure 5 that one diagonal will not be in the new sample space. Therefore, there are 7×6, or 42, ordered pairs possible. The number of these that are of the form (red, red) is 4×3, or 12, because any of the 4 red balls can be the first, but there are only 3 choices for the second red ball.

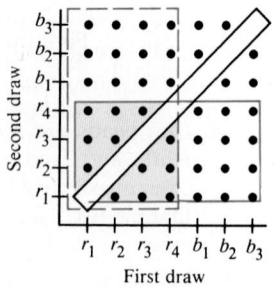

First draw

Figure 5

514   *Chapter 12*

Thus,

$$P(A \cap B) = \tfrac{12}{42} = \tfrac{2}{7}.$$

Analyzing this result as

$$P(A \cap B) = \tfrac{12}{42} = \tfrac{4}{7} \cdot \tfrac{3}{6},$$

you can see that

$$\tfrac{4}{7} = P(A).$$

We can interpret the second factor, $\tfrac{3}{6}$, as the *probability that the second ball drawn is red under the condition that the first ball drawn was red*; we will denote this probability by $P(B|A)$, read "the probability of B given A."

This discussion suggests the following general rule for conditional probability, the probability of one event under the condition that another event has occurred.

Let $P(A)$ denote the probability of an event A, and $P(B|A)$ denote the conditional probability of an event B given that event A has occurred. If $P(A \cap B)$ is the probability that A and B occur, then

$$P(A \cap B) = P(A) \cdot P(B|A).$$

We can use this rule to solve Example 1, page 508, in a different way.

EXAMPLE 1 If two cards are drawn at random from a 52-card deck, what is the probability that both cards are hearts?

SOLUTION Let A be the event that the first card is a heart and B be the event that the second card is a heart. Then $P(A) = \tfrac{13}{52}$ and $P(B|A) = \tfrac{12}{51}$.

$$P(A \cap B) = P(A) \cdot P(B|A)$$
$$= \tfrac{13}{52} \cdot \tfrac{12}{51} = \tfrac{1}{17}.$$

We say that events A and B are *independent* when the probability of one does not depend on the occurrence of the other. For example, in Case 1 where the balls are drawn *with replacement*, the events A and B are independent because the outcome on the first draw does not affect the outcome on the second draw. Thus, we define two events A and B as independent events if and only if

$$P(A \cap B) = P(A) \cdot P(B).$$

For independent events,

$$P(A|B) = P(A) \quad \text{and} \quad P(B|A) = P(B).$$

Two events that are not independent are said to be dependent.

This is shown by the solid rectangle.

Let B be the event that the green die shows a number greater than 3. This is shown by the dashed rectangle.

Then $A \cap B$ is shown by the shaded region. You have

$$P(A) = \frac{5}{36}, \; P(B) = \frac{18}{36} = \frac{1}{2},$$

and

$$P(A \cap B) = \frac{3}{36} = \frac{1}{12}.$$

Since $\dfrac{1}{12} \neq \dfrac{5}{36} \cdot \dfrac{1}{2}$,

$$P(A \cap B) \neq P(A) \cdot P(B).$$

Hence A and B are dependent events.

3. A die is thrown and a coin is tossed. What is the probability that the die shows an odd number, and the coin shows a tail? Let A be the event that the die shows an odd number, and B be the event that the coin shows a tail.

Then $P(A) = \dfrac{1}{2}$ and

$$P(B) = \frac{1}{2}.$$

These events are clearly independent. Hence,
$$P(A \cap B) = P(A) \cdot P(B)$$
$$= \frac{1}{2} \cdot \frac{1}{2} = \frac{1}{4}.$$

1. Given that $P(C) = 0.3$ and $P(D|C) = 0.15$, find $P(C \cap D)$. 0.045

2. Given that $P(K) = \frac{1}{3}$ and

   $P(J \cap K) = \frac{1}{6}$, find

   $P(J|K)$. $\frac{1}{2}$

A coin is tossed twice and a die is rolled once. Find the probability of each event in Exercises 3 and 4.

3. Two heads are tossed and a five shows on the die.

   $\frac{1}{24}$

4. At least one head is tossed, and a two does

   not show on the die. $\frac{5}{8}$

5. Two marbles are drawn at random from a bag containing 3 green, 4 orange, and 5 red marbles. What is the probability of drawing:

   a. 1 green and 1 red? $\frac{5}{22}$

   b. 1 orange but no

   red? $\frac{2}{11}$

EXAMPLE 2　A red die and a green die are thrown. What is the probability that the sum of the numbers shown is 6 and that the green die shows a number less than 3? Are these events independent?

SOLUTION　Graph the sample space. Let A be the event that the sum of the numbers is 6. This is shown by the solid rectangle.

Let B be the event that the green die shows a number less than 3. This is shown by the dashed rectangle.

Then $A \cap B$ is shown by the shaded region. You have

$$P(A) = \tfrac{5}{36}, \; P(B) = \tfrac{12}{36} = \tfrac{1}{3}, \text{ and}$$

$$P(A \cap B) = \tfrac{2}{36} = \tfrac{1}{18}.$$

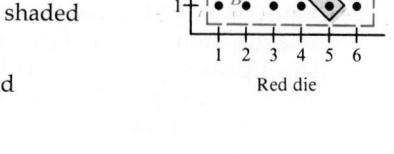

Since $\tfrac{1}{18} \neq \tfrac{5}{36} \cdot \tfrac{1}{3}$, $P(A \cap B) \neq P(A) \cdot P(B)$. Hence A and B are dependent events.

In many cases, rather than use the relationship

$$P(A \cap B) = P(A) \cdot P(B)$$

to determine the independence of events A and B, you use the fact that A and B are obviously independent events and employ the relationship to find $P(A \cap B)$.

EXAMPLE 3　A die is thrown and a coin is tossed. What is the probability that the die shows a 2 or a 3 and the coin shows a head?

SOLUTION　Let A be the event that the die shows a 2 or a 3, and B be the event that the coin shows a head. These events are clearly independent. Hence, knowing that $P(A) = \tfrac{2}{6} = \tfrac{1}{3}$, and $P(B) = \tfrac{1}{2}$, you have

$$P(A \cap B) = \tfrac{1}{3} \cdot \tfrac{1}{2} = \tfrac{1}{6}.$$

Oral Exercises

In each exercise use the probabilities given to tell whether A and B are dependent or independent events.

1. $P(A) = \tfrac{1}{3}$, $P(B) = \tfrac{1}{4}$, $P(A \cap B) = \tfrac{1}{12}$ independent
2. $P(A) = \tfrac{1}{4}$, $P(B) = \tfrac{1}{8}$, $P(A \cap B) = \tfrac{1}{8}$ dependent
3. $P(A) = 0.2$, $P(B) = 0.4$, $P(A \cap B) = 0.15$ dependent
4. $P(A) = 0.35$, $P(B) = 0.4$, $P(A \cap B) = 0.14$ independent
5. $P(A) = 0.1$, $P(B) = 0.2$, $P(A \cap B) = 0.1$ dependent
6. $P(A) = 0.8$, $P(B) = 0.8$, $P(A \cap B) = 0.8$ dependent

516　*Chapter 12*

Written Exercises

A 1. Given that $P(C) = 0.4$ and $P(D|C) = 0.55$, find $P(C \cap D)$. 0.22

2. Given that $P(K) = \frac{1}{4}$ and $P(J \cap K) = \frac{1}{12}$, find $P(J|K)$. $\frac{1}{3}$

3. A coin is tossed twice and a die is rolled once. Find the probability.
   a. Two tails are tossed and a 6 shows on the die. $\frac{1}{24}$
   b. Exactly 1 head is tossed and a 3 or a 4 shows on the die. $\frac{1}{6}$
   c. At least 1 head is tossed, and a 5 does not show on the die. $\frac{5}{8}$
   d. At most 1 head is tossed and a number less than 4 shows. $\frac{3}{8}$

4. Two marbles are drawn at random from a bag containing 2 green, 3 orange, and 6 red marbles. What is the probability of drawing
   a. 1 green and 1 red? $\frac{12}{55}$ b. 1 orange and 1 green? $\frac{6}{55}$
   c. 1 red and 1 non-orange? $\frac{27}{55}$ d. 1 orange but no red? $\frac{9}{55}$

B 5. One card is drawn at random from the 13 hearts in a standard deck and then replaced; two more are then drawn. What is the probability of each event?
   a. The first is the ace and the second two are face cards (jacks, queens, and kings). $\frac{1}{338}$
   b. All three are face cards. $\frac{3}{338}$

6. Two *different* letters are chosen at random from the word MODULAR and three numbers are chosen at random from the numbers 1, 2, 3, 4, 5, and 6 (*with* repetition allowed). What is the probability of getting
   a. A and M and three of the same digit? b. two vowels and three odd digits?
   c. two consonants and no 3's among the digits? $\frac{1}{756}; \frac{1}{56}; \frac{125}{756}$

7. A single die is rolled twice. If A is the event that the sum of the numbers showing is 7, and B is the event of at least one 3 showing, calculate $P(A)$, $P(B)$, $P(A|B)$, and $P(B|A)$, and show that
$$P(A) \cdot P(B|A) = P(B) \cdot P(A|B).$$

8. Two cards are drawn at random from a 52-card deck. Calculate the probabilities of the following events to determine which is more likely. A: Both cards are of the same color. B: One card is red and the other is black. $P(A) = \frac{25}{51}$ $P(B) = \frac{26}{51}$

In Exercises 9–13 a cube with two adjacent faces painted green, two adjacent faces painted yellow, and two adjacent faces painted red is tossed and lands on one of its faces. The exercises refer to the following events:
A: **The cube lands on a green face.** B: **The top face is red.**
C: **Two of the vertical faces are green.** D: **Exactly one of the vertical faces is red.**

9. Find the probabilities of events A, B, C, and D. $\frac{1}{3}; \frac{1}{3}; \frac{1}{3}; \frac{2}{3}$
10. Are events A and B independent? Explain your answer.
11. Are events B and C independent? Explain your answer.
12. Are events B and D independent? Explain your answer.
13. Which pair(s) of events are mutually exclusive?

Permutations, Combinations, and Probability **517**

Suggested Assignments
Enriched Alg.
   517/1–16
R 518/*Self-Test 4*
Enriched Alg. with Trig.
   517/1–15 odd
R 518/*Self-Test 4*

Additional Answers
Written Exercises

7. $P(A) = \frac{1}{6}$

$P(B) = \frac{11}{36}$

$P(A|B) = \frac{2}{11}$

$P(B|A) = \frac{1}{3}$

$\frac{1}{6} \cdot \frac{1}{3} \stackrel{?}{=} \frac{11}{36} \cdot \frac{2}{11}$

$\frac{1}{18} = \frac{2}{36}$

10. No; since $P(A|B) \neq P(A)$.
11. No; since $P(B|C) \neq P(B)$.
12. No; since $P(B|D) \neq P(B)$.
13. A and C

Mixed Review

1. Name two consecutive integers between which a root of $x^3 + x - 11 = 0$ exists. 2 and 3
2. Find the fourth term of $(x - 2y)^7$. $-280x^4y^3$
3. How many distinguishable permutations exist of the letters in the word DEPENDENT? 15,120
4. A student must answer any 6 out of 8 questions on an exam. In how many ways might she complete the exam? 28
5. How many different 3-digit even numbers can be written? 450

1. A letter is selected at random from the word CAT, and a number is selected from the numeral 325. List a sample space for this experiment and list the event that the letter is a vowel and the number is even. $S = \{(c, 3), (c, 2), (c, 5), (a, 3), (a, 2), (a, 5), (t, 3), (t, 2), (t, 5)\}$
$E = \{(a, 2)\}$

An integer from 1 to 6 inclusive is selected at random.

2. a. What is the probability that the integer is a multiple of 2? $\frac{1}{2}$

   b. What is the probability that the integer is greater than 2? $\frac{2}{3}$

3. What is the probability that the integer is a multiple of 2 or else it is 1 or 5? $\frac{5}{6}$

4. Two dice are tossed. Let A be the event one die shows a 2. Let B be the event that the sum of the numbers shown is 4. Are the events independent? Why or why not? No. Because $P(A \cap B) \neq P(A) \cdot P(B)$.

Additional Answers
Self-Test 4

1. $\{(A, 1), (A, 2), (A, 3), (B, 1), (B, 2), (B, 3), (C, 1), (C, 2), (C, 3)\}; \{(A, 1), (A, 3), (B, 1), (B, 3)\}$

14. The student council consists of 6 girls and 4 boys, of whom 3 girls and 3 boys are seniors. If a 4-person dance committee is to be chosen by lot, what is the probability that the committee will consist of all seniors, of whom at least 2 are girls? $\frac{2}{35}$

C 15. Box X contains p red marbles and q green marbles. Box Y contains r red marbles. One box is chosen at random and one marble is chosen at random from that box. What is the probability that the chosen marble is
   a. red and from Box X? $\frac{p}{2(p + q)}$ b. red? $\frac{2p + q}{2(p + q)}$

16. Consider the experiment described in Exercise 15 with $p = 4$, $q = 4$, and $r = 1$. Suppose you know only that the chosen marble is red. What is the probability that it was picked from Box Y? $\frac{2}{3}$

Self-Test 4

VOCABULARY   random experiment (p. 505)
             sample space (p. 505)
             event (p. 505)
             complement (p. 509)
             odds (p. 509)
             mutually exclusive (p. 512)
             conditional probability (p. 515)
             independent events (p. 515)
             dependent events (p. 515)

1. A letter is selected at random from the letters A, B, and C, and a number is selected at random from the numbers 1, 2, and 3. List a sample space for this experiment and list the event that the letter is not C and the number is odd.   *Obj. 1, p. 505*

In Exercises 2 and 3 an integer from 1 to 8 inclusive is selected at random.

2. a. What is the probability that the integer is greater than 6? $\frac{1}{4}$ *Obj. 2, p. 505*
   b. What is the probability that the integer is less than 7? $\frac{3}{4}$

3. What is the probability that the integer is greater than 6 or else is 3 or 5? $\frac{1}{2}$ *Obj. 3, p. 505*

4. Two dice are tossed. Let A be the event one die shows a 1. Let B be the event that the sum of the numbers shown is 8. Are the events independent? Why or why not? No; if the sum of the numbers equals 8, it is impossible for one die to show a 1.   *Obj. 4, p. 505*

Check your answers with those at the back of the book.

Chapter Summary

1. If sets A, B, and $A \cap B$ contain r, s, and t elements, respectively, then $A \cup B$ contains $r + s - t$ elements and $A \times B$ contains rs elements. From these fundamental counting principles you can derive a formula for the number of *permutations* of n elements, r at a time.

$$_nP_r = n(n - 1)(n - 2) \cdots (n - (r - 1)).$$

2. The number of *combinations* of n things, r at a time, is given by

$$_nC_r = \frac{_nP_r}{r!}.$$

Also, $_nC_r = {_nC_{n-r}}$.

3. The Binomial Theorem gives the expansion of $(a + b)^n$. (See page 499.) Pascal's triangle can also be used to find binomial coefficients.

4. If the sample space for an experiment consists of n equally likely outcomes and event E consists of h outcomes, then the *probability* of E is $P(E) = \frac{h}{n}$. The *odds* in favor of E are $\frac{h}{n - h}$.

5. The probability that at least *one* of the events A and B will occur is given by $P(A \cup B) = P(A) + P(B) - P(A \cap B)$. When A and B are *mutually exclusive*, $P(A \cup B) = P(A) + P(B)$.

6. The probability that *two* events A and B will occur is given by $P(A \cap B)$ where $P(A \cap B) = P(A) \cdot P(B|A)$. $P(B|A)$ is the *conditional probability* that B will occur given that A has occurred.

7. Two events are *independent* if and only if

$$P(A \cap B) = P(A) \cdot P(B).$$

Chapter Review

Write the letter of the correct answer.

1. $A = \{1, 2, 3\}$, $B = \{2, 4\}$. State the number of elements in $A \times B$.   *12-1*
   a. 1   b. 4   c. 5   (d.) 6

2. A contains 4 elements, B contains 3 elements, and $A \cap B$ contains 2 elements. State the number of elements in $A \cup B$.
   a. 7   (b.) 5   c. 9   d. 12

3. In how many ways can 4 books be arranged on a shelf?   *12-2*
   (a.) 24   b. 6   c. 4   d. 10

Permutations, Combinations, and Probability   **519**

4. In how many ways can 5 people be seated at a circular table?
   a. 120   b. 15   (c.) 24   d. 5

5. State the number of distinguishable permutations of the letters in
   the word MISSISSIPPI.   *12-3*
   a. 69,300   b. 3150   c. 138,600   (d.) 34,650

6. Evaluate $_{36}C_{33}$.   *12-4*
   a. 3570   (b.) 7140   c. 14,280   d. 21,420

7. How many combinations of 3 people can be selected for the math
   team from a group of 7 people?
   (a.) 35   b. 840   c. 210   d. 21

8. How many 7-marble combinations drawn from an urn containing 6   *12-5*
   blue and 5 red marbles include 3 blue and 4 red marbles?
   a. 20   b. 25   c. 330   (d.) 100

9. Use the Binomial Theorem to find the fourth term in the expansion   *12-6*
   $(3x + 2)^7$.
   a. $840x^4$ b. $2923x^4$ (c.) $22,680x^4$ d. $210x^3$

10. Use Pascal's triangle to find the third term in the expansion $\left(\frac{x}{2} + 3\right)^5$.   *12-7*

    (a.) $\frac{45}{4}x^3$ b. $45x^3$ c. $\frac{19}{2}x^3$ d. $\frac{27}{4}x^3$

11. Four coins are tossed. How many elements are in the sample space?   *12-8*
    a. 4   (b.) 16   c. 15   d. 24

12. An integer between 20 and 35 inclusive is selected at random. What   *12-9*
    is the probability that it is a multiple of 3?
    a. $\frac{3}{8}$ b. $\frac{1}{3}$ (c.) $\frac{5}{16}$ d. $\frac{5}{14}$

13. A marble is selected at random from an urn containing 6 black and 4
    white marbles. What are the odds in favor of the marble being
    white?
    a. 2 to 5   (b.) 2 to 3   c. 3 to 2   d. 1 to 3

14. Five marbles are drawn at random from an urn containing 6 black, 8   *12-10*
    white, and 7 red marbles. What expression represents the probabili-
    ty that at least 3 of the marbles are red?

    a. $\dfrac{_7C_3 \times\, _{14}C_2}{_{21}C_5}$

    b. $1 - \dfrac{_7C_3 \times\, _{14}C_2}{_{21}C_5}$

    c. $\dfrac{_7C_3 \times\, _{14}C_2}{_{21}C_5} + \dfrac{_7C_4 \times\, _{14}C_1}{_{21}C_5}$

    (d.) $\dfrac{_7C_3 \times\, _{14}C_2}{_{21}C_5} + \dfrac{_7C_4 \times\, _{14}C_1}{_{21}C_5} + \dfrac{_7C_5}{_{21}C_5}$

15. A die is rolled and a coin is tossed. What is the probability that the *12-11*
    die shows a 3 and the coin lands heads?
    a. $\frac{1}{6}$ b. $\frac{2}{3}$ c. $\frac{1}{4}$ (d.) $\frac{1}{12}$

520   *Chapter 12*

Chapter Test

1. How many 3-letter code words can be formed from the letters in the word COMPLEX? 343

2. Evaluate $_{10}P_4$. 5040

3. How many different permutations exist of all the letters in the word COEFFICIENT? 2,494,800

4. Given a set with 10 elements. How many 4-element subsets are there? 210

5. How many different hands consisting of 2 aces and 3 queens can be chosen from a standard 52-card bridge deck? 24

6. Find the eighth term of $(\frac{3}{4}x - 1)^{13}$. $-\frac{312741}{1024}x^6$

7. Use Pascal's triangle to expand $(y - 3)^5$.

8. A coin is tossed and a die is rolled. List the sample space.

Three marbles are drawn at random from an urn containing 4 black, 7 yellow, and 5 green marbles.

9. What is the probability that none of them are yellow? $\frac{3}{20}$

10. What is the probability that at least 2 of them are black? $\frac{19}{140}$

11. A drawer contains 7 blue socks, 3 green socks, and 6 brown socks. If you choose 2 socks at random, what is the probability that they are both blue? $\frac{7}{40}$

12-1

12-2

12-3

12-4

12-5

12-6

12-7

12-8

12-9

12-10

12-11

Cumulative Review

Chapter 9

Solve over $\mathcal{C}$.

1. $x^2 + 4x + 20 = 0$ $\{-2 \pm 4i\}$

2. $x^2 - 2x + 4 = 0$ $\{1 \pm i\sqrt{3}\}$

3. Determine the value of k for which $4x^2 - 12x + 3k = 0$ will have exactly one real root. 3

4. Write the quadratic equation whose solution set is $\{-2 + i\sqrt{2}, -2 - i\sqrt{2}\}$. $x^2 + 4x + 6 = 0$

5. Write the equation of the axis of symmetry for the graph of $y = x^2 - 3x + 1$. $x = \frac{3}{2}$

6. Find the function of the form $y = a(x - h)^2 + k$ whose graph has vertex (4, 1) and passes through the point (3, 0). $y = -(x - 4)^2 + 1$

Permutations, Combinations, and Probability   **521**

7. Find the quadratic function whose graph contains the ordered pairs $(1, -1)$, $(0, 5)$, and $(-1, 15)$. $y = 2x^2 - 8x + 5$

8. If $Q(x) = 12x^3 + 7x^2 - 14x + 3$, find $Q(-2)$. -37

9. Find the zeroes of $P(x) = x^3 + 3x^2 - x - 3$. $\pm 1; -3$

10. Determine the number of possible positive roots and the number of possible negative roots of $x^3 - 2x^2 - x + 7 = 0$. 2 or 0; 1

Chapter 10

11. Determine an equation of the perpendicular bisector of the line segment with endpoints $(1, 0)$ and $(5, -2)$. $y = 2x - 7$

12. Determine an equation of the circle with center $(0, 0)$ and passing through $(-2, -3)$. $x^2 + y^2 = 13$

13. Determine an equation of the parabola with focus $(0, -3)$ and directrix $y = 3$. $y = -\frac{1}{12}x^2$

14. Determine an equation of the ellipse with major axis of length 12 and y-intercepts 3 and -3. $\frac{x^2}{36} + \frac{y^2}{9} = 1$

15. Determine an equation of the hyperbola with foci at $(0, 5)$, and $(0, -5)$ and one vertex at $(0, -3)$. $\frac{y^2}{9} - \frac{x^2}{16} = 1$

16. If y varies inversely as x^2 and $x = 3\sqrt{2}$ when $y = \frac{1}{2}$ find the value of y when $x = \frac{2}{3}$. $\frac{81}{4}$

Solve over $\mathcal{C}$.

17. $2x^2 - y^2 = 28$
    $4x + y = 0$

18. $x^2 + y = 0$
    $2x - y - 3 = 0$

19. $x^2 + y^2 = 6$
    $x + y = -2\sqrt{3}$

20. $3x^2 + 2y^2 = 21$
    $2x^2 - 5y^2 = -24$

21. Find the dimensions of a right triangle whose area is 30 m^2 and whose perimeter is 30 m. 5 m, 12 m, 13 m

Chapter 11

22. Express $\dfrac{\sqrt[4]{8}}{\sqrt[4]{2}}$ in simplest radical form. $\sqrt{2}$

Solve over $\mathcal{R}$.

23. $8^{2x-2} = 4^{2-x}$ $\{\frac{5}{4}\}$

24. $\log_{25} x = \frac{3}{2}$ $\{125\}$

25. $\log_3 63 - \log_3 7x = \log_3 2$ $\{\frac{9}{2}\}$

26. $125^{x-6} = 25^{2x-3}$ $\{-12\}$

27. Determine the solution of $2^{3x-2} = 64$ to three significant digits. $\{2.67\}$

28. If \$3000 is invested at 12% annual interest compounded quarterly, what is the amount in the account after 11 yr? \$11,014.36

29. Express $e^7 = 1097$ in logarithmic form. $\ln 1097 = 7$

522 *Chapter 12*

30. How many two-digit integers can be formed using the digits 1, 2, 3, and 4? 16

31. In how many ways can 6 books be arranged on a shelf? 720

32. How many different integers can be formed using all the digits in the numeral 55232332? 560

Evaluate.

33. $_7P_2$ 42   **34.** $_9P_9$ 362,880   **35.** $_7C_2$ 21   **36.** $_{12}C_0$ 1

37. In how many different ways can a committee of 4 be selected from a group of 9 people? 126

38. In how many ways can 3 jacks and 3 fours be chosen from a standard bridge deck? 16

39. Expand $(2x + 3b)^4$. $16x^4 + 96x^3b + 216x^2b^2 + 216xb^3 + 81b^4$

40. Find the fifth term in the expansion of $(3a - b)^{11}$. $721{,}710a^7b^4$

41. If a coin is tossed 5 times, what is the probability of getting exactly 3 heads? $\frac{5}{16}$

42. Five digits are chosen at random from the digits 0, 1, 2, 3, . . . , 9. What is the probability that a 7 or an 8, but not both, will be chosen? $\frac{5}{9}$

Contest Problems

1. A certain coin dealer received nine coins that were supposedly solid gold. The coin dealer, however, believed that one of the coins was counterfeit and was either *heavier or lighter* than the others. The coin dealer had only a very old balance scale that would fall apart after three weighings. Tell how the balance scale can be used to determine (in three weighings) which is the counterfeit coin and whether it is heavier or lighter than the others.

2. Two boats are at points *A* and *B* across a river from each other as shown. At the same instant, each boat starts traveling toward the opposite shore (at right angles to the shore). Each boat travels at a different, but constant speed. At a point 720 ft from *A* the boats pass each other and continue to travel to the opposite shore. Each boat spends exactly 20 min unloading supplies and then starts back toward its original starting point. The boats pass each other again 400 ft from point *B*. How wide is the river? 1760 ft

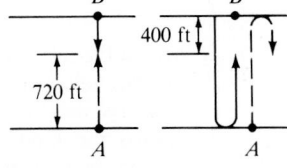

3. Determine the pattern for the following set of two-digit numbers:

$$11, 31, 71, 91, 32, 92, 13, 73, 14, 34$$

What are the next three numbers in this pattern? 74, 35, 95

Permutations, Combinations, and Probability   **523**

PROGRAMMING IN PASCAL

In this chapter you learned the definition of "factorial". You know that

$$n! = n \times (n - 1) \times (n - 2) \times \ldots \times 2 \times 1, \quad \text{for } n \geq 1$$
$$0! = 1$$

You can also think of n! as being n × (n − 1)! Thus, n! could be defined as

$$n! = \begin{cases} n \times (n - 1)!, & n \geq 1 \\ 1, & n = 0 \end{cases}$$

This type of definition, where a function is defined in terms of itself, is called a *recursive* function. Notice that if some restriction were not placed on the definition, an infinite regression would occur. For example, 3 factorial would be defined in the following manner:

$$3! = 3*2! = 3*2*1! = 3*2*1*0! = 3*2*1*0*(-1!), \quad \text{etc.}$$

There would be nothing to stop the function from continuing on and on.

The recursive definition of factorial is useful in illustrating the use of recursion in Pascal. The Pascal function below is recursive; that is, it calls upon a simpler case of itself during its execution.

```
FUNCTION factorial(n : integer) : real;
BEGIN
  IF n = 0
    THEN factorial := 1
    ELSE factorial := n * factorial (n−1)
END;   (* factorial *)
```

Although recursion in Pascal is both time- and money-consuming, it is often the simplest way to solve a problem and the most easily understood process. If you wish to investigate recursion further, almost any Pascal text will provide you with more information.

Exercises

1. **a.** Write a Pascal function, *factorial*, that computes $n!$ for non-negative, integral values of n. Use a loop containing a running product (*prod* := *prod* * *j*, for example) for the actual computation.
   b. Discuss the limits on n and *factorial(n)* that are imposed on the program by your system or implementation of Pascal.
   c. Compare the speed of the iterative function *factorial* to that of the recursive function *factorial* given earlier in this section. Use each of them to display a table of factorials of the form

Number	Factorial
n	$n!$

   for $1 \leq n \leq 30$. You should find the recursive method to be noticably slower than iteration (using a loop). In addition, recursion uses up a lot of memory. When you are considering using recursion to solve a problem by computer, you must decide whether it is more important to solve the problem with simple code or more important to save time and memory space.

2. a. Write a Pascal function, with the header

   FUNCTION permutations(n, r : integer) : integer;

   This function will calculate the number of permutations that n objects taken r at a time can have. Use a loop similar to that suggested in Exercise 1. Discuss any restrictions on n, r, and $_nP_r$.

   b. Write a Pascal program, *evaluate*, that uses the function you wrote in part (a) and may be used to solve problems such as those in Section 12-2. The user of your program must be able to specify n and r during the execution of *evaluate*.

3. a. Write a Pascal function with the header

   FUNCTIONS combinations (n, r : integer) : integer ;

   This function will calculate the number of combinations that n objects taken r at a time can have. Use the fact that

   $$_nC_r = \frac{_nP_r}{r!}.$$

   Discuss any restrictions on n, r, and $_nC_r$.

   b. Modify the program *evaluate* to include the function *combinations*. This function must be positioned after the function *permutations*. The main program should ask the user for the values of n and r then check them for validity. Next, the program should ask the user whether they want to find the number of permutations or the number of combinations of those objects and then make the call to the appropriate function. An example of what the computer should print is:

   > The number of combinations of
   > 5 items taken 3 at a time
   > is 10.

   c. How can you further modify the program *evaluate* so the function *combinations* uses the fact that $_nC_r = {}_nC_{n-r}$ to increase the efficiency of the computer when $r > \frac{n}{2}$?

4. a. Write a Pascal program, *pascal_triangle*, that displays the first n rows of Pascal's triangle (say for $1 \le n \le 10$) after n is specified by the user.

   b. Enhance the program in part (a) so that it offers the user the option of displaying the first n rows or the nth row only of Pascal's triangle where n is specified by the user.

The Harvard Women's Crew is shown practicing on the Charles River in Cambridge, Massachusetts. The Application on pages 567–569 shows how matrices can be used to determine the rankings of teams in a competitive sport.

Chapter 13

Matrices

Problem Solving Strategies
Drawing a Diagram
Communication networks are presented as directed diagrams and matrices are formulated from examination of the networks in the Application on page 567.
Deduction
Proof techniques are used in Section 13–6 to express the formula for the inverse of a 2×2 matrix (page 548).

Basic Properties of Matrices

OBJECTIVES for Sections 13-1 through 13-5:
1. To find sums and differences of matrices.
2. To find the product of a scalar and a matrix.
3. To solve certain matrix equations.
4. To find the product of two matrices.

13–1 Matrices and Their Sums

Teaching Suggestions
p. T114

In Chapter 4 you saw that it is sometimes convenient to name a number by means of a square array of numerals. Thus the *determinant* $\begin{vmatrix} a & b \\ c & d \end{vmatrix}$ names the number $ad - bc$:

$$\begin{vmatrix} a & b \\ c & d \end{vmatrix} = ad - bc$$

Key Ideas
Find sums and differences of matrices.

The array $\begin{bmatrix} a & b \\ c & d \end{bmatrix}$ itself is called a *matrix*. Notice that brackets [ ] are used to distinguish the matrix from the determinant. In general, a rectangular array such as that shown in red at the right below is a **matrix** (plural, **matrices**).

Each numeral in the array is an **entry** of the matrix. The number of (horizontal) rows and the number of (vertical) columns of entries in the matrix are its **dimensions**. For example, the matrix displayed has three rows and two columns and is called a 3×2 (read "three by two") matrix.

$$\begin{array}{c} \text{column} \\ \begin{array}{cc} 1 & 2 \end{array} \\ \begin{array}{c} \text{row } 1 \\ 2 \\ 3 \end{array} \begin{bmatrix} 2 & 1 \\ -3 & 0 \\ 5 & 2 \end{bmatrix} \end{array}$$

Common Errors
Students sometimes confuse the dimensions of a matrix to read the number of columns first, and then the number of rows. Emphasize the correct usage.

Matrices **527**

1. Find the values of x and y
so that:
$[x - 2 \quad 3 \quad x - z] =$
$[3 \quad x + y \quad 4]$
This matrix equation is
true if and only if
$x - 2 = 3$, $3 = x + y$, and
$x - z = 4$.
This system can be solved
to obtain the equivalent
system
$x = 5$, $y = -2$, and $z = 1$.
$\therefore$ the value for x is 5, for y
is -2, and for z is 1.

2. Find $\begin{bmatrix} 2 & 3 \\ -4 & 5 \end{bmatrix} + \begin{bmatrix} -3 & 7 \\ -6 & 1 \end{bmatrix}$.
$\begin{bmatrix} 2 & 3 \\ -4 & 5 \end{bmatrix} + \begin{bmatrix} -3 & 7 \\ -6 & 1 \end{bmatrix} =$
$\begin{bmatrix} 2 + (-3) & 3 + 7 \\ -4 + (-6) & 5 + 1 \end{bmatrix} =$
$\begin{bmatrix} -1 & 10 \\ -10 & 6 \end{bmatrix}$

3. Find $\begin{bmatrix} 0 & 5 & -6 \\ 3 & 1 & -2 \end{bmatrix} -$
$\begin{bmatrix} -3 & 2 & 0 \\ 5 & -1 & 3 \end{bmatrix}$.
$\begin{bmatrix} 3 & 3 & -6 \\ -2 & 2 & -5 \end{bmatrix}$

Notice that the number of rows is given first and then the number of columns. In this book, we shall use only real numbers for entries of a matrix. An entry of a matrix is referred to by giving its row and column numbers. In matrix A, below, the entry in the third row and second column is 3.

As was just indicated, capital letters, such as A, B, and C may be used to denote matrices. Sometimes subscripts, as in $A_{3 \times 2}$, are also used to represent the dimensions of a matrix. Thus, for

$$A = \begin{bmatrix} 4 & 1 & 7 \\ 2 & 0 & 6 \\ 5 & 3 & 2 \end{bmatrix}, \quad B = [1 \quad 5], \quad \text{and} \quad C = \begin{bmatrix} -1 \\ 7 \\ 8 \end{bmatrix},$$

you might write $A_{3 \times 3}$, $B_{1 \times 2}$, and $C_{3 \times 1}$. Matrices, such as B and C, that have only one row or one column are called *row matrices* and *column matrices* (or *row vectors* and *column vectors*), respectively. Since the matrix $A_{3 \times 3}$ has 3 rows and 3 columns, it is said to be a *square matrix of order* 3. Similarly, we have square matrices of order 2, square matrices of order 4, and so on.

For two matrices of the same dimensions, the entries in the same row and same column are said to be *corresponding entries*.

Two matrices are *equal* if and only if they have the same dimensions and all their corresponding entries are equal. Notice that matrices of different dimensions are never said to be equal, even if corresponding entries, as far as they extend, are equal. For instance,

$$[4 \quad 5] \neq \begin{bmatrix} 4 \\ 5 \end{bmatrix} \quad \text{and} \quad \begin{bmatrix} 0 & 0 \\ 0 & 0 \end{bmatrix} \neq \begin{bmatrix} 0 & 0 & 0 \\ 0 & 0 & 0 \end{bmatrix}.$$

EXAMPLE 1 Find the values of w, x, y, and z so that:

a. $\begin{bmatrix} x - 3 & y + 4 \\ z + 5 & w \end{bmatrix} = \begin{bmatrix} 2 & 4 \\ 0 & 3 \end{bmatrix}$ b. $[x + y \quad z - x \quad y] = [-\frac{3}{2} \quad 9 \quad \frac{1}{2}]$

SOLUTION a. By the definition of equal matrices, this matrix equation is true if and only if

$$x - 3 = 2, \quad y + 4 = 4, \quad z + 5 = 0, \quad \text{and} \quad w = 3 \text{ or}$$
$$x = 5, \quad y = 0, \quad z = -5, \quad \text{and} \quad w = 3.$$

$\therefore$ the value for w is 3, for x is 5, for y is 0, and for z is -5.

b. The given matrix equation is true if and only if

$$x + y = -\tfrac{3}{2}$$
$$z - x = 9$$
$$y = \tfrac{1}{2}$$

The system can be solved to obtain the equivalent system: $x = -2$
$$y = \tfrac{1}{2}$$
$$z = 7$$

$\therefore$ the value for x is -2, for y is $\frac{1}{2}$, and for z is 7.

528 *Chapter 13*

If two matrices A and B have the same dimensions, then their sum, denoted by

$$A + B,$$

is a matrix of the same dimensions, whose entries are the sums of the corresponding entries of A and B.

EXAMPLE 2 Find each of the following.

    a. $[2 \quad -3] + [-5 \quad 4]$

    b. $\begin{bmatrix} 1 & 4 \\ -3 & 2 \end{bmatrix} + \begin{bmatrix} -6 & 2 \\ -5 & 7 \end{bmatrix}$

SOLUTION   **a.** $[2 \quad -3] + [-5 \quad 4] = [2 + (-5) \quad -3 + 4]$

$$= [-3 \quad 1]$$

    b. $\begin{bmatrix} 1 & 4 \\ -3 & 2 \end{bmatrix} + \begin{bmatrix} -6 & 2 \\ -5 & 7 \end{bmatrix} = \begin{bmatrix} 1 + (-6) & 4 + 2 \\ -3 + (-5) & 2 + 7 \end{bmatrix}$

$$= \begin{bmatrix} -5 & 6 \\ -8 & 9 \end{bmatrix}$$

For any given natural numbers m and n, suppose that $S_{m \times n}$ represents the set of all $m \times n$ matrices $A_{m \times n}$ with real number entries. Then the **zero matrix,** denoted by

$$O_{m \times n} \text{ (or simply } O\text{)},$$

is the $m \times n$ matrix each of whose entries is 0. For example, some zero matrices are:

$$O_{2 \times 2} = \begin{bmatrix} 0 & 0 \\ 0 & 0 \end{bmatrix}, \quad O_{2 \times 3} = \begin{bmatrix} 0 & 0 & 0 \\ 0 & 0 & 0 \end{bmatrix}, \quad \text{and} \quad O_{3 \times 3} = \begin{bmatrix} 0 & 0 & 0 \\ 0 & 0 & 0 \\ 0 & 0 & 0 \end{bmatrix}$$

Because the sum of the zero matrix $O_{m \times n}$ and any other matrix $A_{m \times n}$ is $A_{m \times n}$, the zero matrix $O_{m \times n}$ is also called the **identity matrix for addition** in the set $S_{m \times n}$. For example,

$$[a \quad b] + [0 \quad 0] = [a + 0 \quad b + 0] = [a \quad b].$$

The **opposite** of the $m \times n$ matrix A, denoted by

$$-A \text{ (read "the opposite of } A\text{"}),$$

is the $m \times n$ matrix whose entries are the opposites of the corresponding entries in A. For example, if

$$A = \begin{bmatrix} 1 \\ -3 \end{bmatrix}, \quad \text{then} \quad -A = \begin{bmatrix} -1 \\ 3 \end{bmatrix},$$

and if

$$B = \begin{bmatrix} 3 & -7 \\ -4 & 2 \end{bmatrix}, \quad \text{then} \quad -B = \begin{bmatrix} -3 & 7 \\ 4 & -2 \end{bmatrix}.$$

Matrices   **529**

Additional A Exercises

Find the indicated sum or difference.

1. $\begin{bmatrix} 3 \\ -2 \end{bmatrix} + \begin{bmatrix} -5 \\ 3 \end{bmatrix} \quad \begin{bmatrix} -2 \\ 1 \end{bmatrix}$

2. $[3 \quad -1 \quad 7] - [-2 \quad -3 \quad 4]$
   $[5 \quad 2 \quad 3]$

3. $\begin{bmatrix} 8 & 0 \\ 0 & -3 \end{bmatrix} - \begin{bmatrix} 6 & -5 \\ 2 & 4 \end{bmatrix}$
   $\begin{bmatrix} 2 & 5 \\ -2 & -7 \end{bmatrix}$

4. $\begin{bmatrix} 11 & 5 & -6 \\ -4 & 0 & 3 \\ 2 & 12 & 10 \end{bmatrix} +$
   $\begin{bmatrix} 6 & 20 & -3 \\ 4 & -1 & -5 \\ 9 & -12 & 4 \end{bmatrix}$
   $\begin{bmatrix} 17 & 25 & -9 \\ 0 & -1 & -2 \\ 11 & 0 & 14 \end{bmatrix}$

Find the values of the variables for which each statement is true.

5. $\begin{bmatrix} w+2 & 1-x \\ 5 & 2y \end{bmatrix} =$
   $\begin{bmatrix} -1 & 4 \\ z+3 & 12 \end{bmatrix}$
   $w = -3, x = -3, y = 6,$
   $z = 2$

6. $[w \quad y \quad x] - [x \quad -w \quad y] =$
   $[1 \quad -1 \quad -4]$
   $w = -2, x = -3, y = 1$

Mixed Review

1. Find the slope of the line through the points $(-5, 3)$ and $(-5, 2)$. undefined

2. If $f(x) = -3x^2 - x + 2$ find $f(1)$. -2

3. Find the value of $\begin{vmatrix} 2 & -1 \\ 6k & -3k \end{vmatrix}$. 0

4. Solve $x^2 - 3x - 40 = 0$ over $\mathcal{R}$ by factoring. $\{8, -5\}$

5. Simplify $\dfrac{3}{x^2-9} + \dfrac{2}{x+3} - \dfrac{2x-3}{x^2-9}$

Because the sum of $A_{m \times n}$ and $-A_{m \times n}$ is $O_{m \times n}$, we call $-A_{m \times n}$ the *additive inverse* of $A_{m \times n}$. For example, if

$$A = \begin{bmatrix} a & b \\ c & d \end{bmatrix}, \quad \text{then} \quad -A = \begin{bmatrix} -a & -b \\ -c & -d \end{bmatrix},$$

and

$$A + (-A) = \begin{bmatrix} a-a & b-b \\ c-c & d-d \end{bmatrix} = \begin{bmatrix} 0 & 0 \\ 0 & 0 \end{bmatrix}.$$

If two matrices have the same dimensions, then they have a *difference*, denoted by

$$A - B,$$

which is defined to be the sum $A + (-B)$ and is therefore a matrix. Thus, to obtain the difference of two matrices of the same dimensions, you simply subtract their corresponding entries.

EXAMPLE 3 Find $\begin{bmatrix} 2 & 8 \\ -1 & 4 \end{bmatrix} - \begin{bmatrix} -3 & 2 \\ -5 & -1 \end{bmatrix}$.

SOLUTION $\begin{bmatrix} 2 & 8 \\ -1 & 4 \end{bmatrix} - \begin{bmatrix} -3 & 2 \\ -5 & -1 \end{bmatrix} = \begin{bmatrix} 2-(-3) & 8-2 \\ -1-(-5) & 4-(-1) \end{bmatrix} = \begin{bmatrix} 5 & 6 \\ 4 & 5 \end{bmatrix}$

Oral Exercises

State the dimensions of each matrix.

1. $\begin{bmatrix} 3 & -2 & 0 \\ 1 & 0 & 4 \end{bmatrix}$
   2×3

2. $\begin{bmatrix} 6 & 2 \\ -1 & 4 \\ 0 & -5 \end{bmatrix}$
   3×2

3. $\begin{bmatrix} 0 \\ 8 \end{bmatrix}$
   2×1

4. $\begin{bmatrix} 7 & 9 & -1 & 3 \\ -2 & 1 & 0 & 5 \\ -6 & 4 & 0 & -3 \end{bmatrix}$
   3×4

State the specified entry in the sum.

$$\begin{bmatrix} 2 & -3 \\ 0 & -4 \end{bmatrix} + \begin{bmatrix} 8 & -4 \\ 7 & -1 \end{bmatrix} = \begin{bmatrix} a & b \\ c & d \end{bmatrix}.$$

5. a 10
6. b -7
7. c 7
8. d -5

If the second matrix is **subtracted** from the first in the addition example above, state the value of the specified entry in the *difference*.

9. a -6
10. b 1
11. c -7
12. d -3

State the value of the specified variable so that

$$\begin{bmatrix} 0 & x+3 \\ 5 & z-2 \end{bmatrix} = \begin{bmatrix} w & -1 \\ y+4 & -3 \end{bmatrix}.$$

13. w 0
14. x -4
15. y 1
16. z -1

17. Explain why there is not just one identity matrix for addition. In other words, explain why there is not just one zero matrix. Matrices can have different dimensions.

Chapter 13

530

Written Exercises

Find the indicated sum or difference.

A

1. $\begin{bmatrix} 2 \\ -3 \end{bmatrix} + \begin{bmatrix} -6 \\ 2 \end{bmatrix} \begin{bmatrix} -4 \\ -1 \end{bmatrix}$

2. $[-4 \quad 0] + [2 \quad -3] \, [-2 \ -3]$

3. $\begin{bmatrix} 9 & -3 \\ 8 & 0 \end{bmatrix} - \begin{bmatrix} 5 & -2 \\ -4 & -1 \end{bmatrix} \begin{bmatrix} 4 & -1 \\ 12 & 1 \end{bmatrix}$

4. $\begin{bmatrix} 6 & 0 \\ 0 & 5 \end{bmatrix} - \begin{bmatrix} -2 & 7 \\ 4 & -1 \end{bmatrix} \begin{bmatrix} 8 & -7 \\ -4 & 6 \end{bmatrix}$

5. $\begin{bmatrix} 4 & 0 & -3 \\ -7 & 2 & 5 \end{bmatrix} + \begin{bmatrix} -2 & 8 & -5 \\ 3 & 0 & 1 \end{bmatrix} \begin{bmatrix} 2 & 8 & -8 \\ -4 & 2 & 6 \end{bmatrix}$

6. $[6 \quad -3 \quad 8] - [-1 \quad -2 \quad 5]$ $[7 \ -1 \ 3]$

7. $\begin{bmatrix} -1 & 6 \\ 2 & -4 \\ 5 & 0 \end{bmatrix} - \begin{bmatrix} 8 & 1 \\ 1 & 3 \\ -2 & 6 \end{bmatrix} \begin{bmatrix} -9 & 5 \\ 1 & -7 \\ 7 & -6 \end{bmatrix}$

8. $\begin{bmatrix} 12 & 7 & -5 \\ -1 & 0 & 4 \\ 3 & 10 & 11 \end{bmatrix} + \begin{bmatrix} 8 & 15 & -6 \\ 5 & -2 & -7 \\ 7 & -10 & 5 \end{bmatrix}$ $\begin{bmatrix} 20 & 22 & -11 \\ 4 & -2 & -3 \\ 10 & 0 & 16 \end{bmatrix}$

Find the values of the variables for which the given statement is true.

9. $\begin{bmatrix} w + 3 & 2 - x \\ 6 & 2y \end{bmatrix} = \begin{bmatrix} -2 & 5 \\ z + 1 & 10 \end{bmatrix}$ $w = -5; x = -3; y = 5; z = 5$

10. $\begin{bmatrix} w - 6 & 11 \\ -2 & 2z + 3 \end{bmatrix} = \begin{bmatrix} 2 & 3x - 1 \\ y + 7 & -5 \end{bmatrix}$ $w = 8; x = 4; y = -9; z = -4$

11. $[x + 3 \quad x - y] = [-2 \quad 3]$ $x = -5; y = -8$

12. $\begin{bmatrix} 17 \\ 7 \end{bmatrix} = \begin{bmatrix} x - y \\ x + y \end{bmatrix}$ $x = 12; y = -5$

13. $[w - x \quad -1 \quad 2] = [5 \quad x - y \quad w + y]$ $w = 3; x = -2; y = -1$

14. $\begin{bmatrix} w + x & w - y \\ y + 2 & z + y \end{bmatrix} = \begin{bmatrix} 7 & -1 \\ 0 & 2 \end{bmatrix}$ $w = -3; x = 10; y = -2; z = 4$

15. $[w \quad y \quad x] - [x \quad -w \quad y] = [11 \quad 6 \quad -5]$ $w = 6; x = -5; y = 0$

16. $\begin{bmatrix} w + 3 & x \\ y - 4 & z \end{bmatrix} + \begin{bmatrix} x - 2 & -z \\ 3 & 3z \end{bmatrix} = \begin{bmatrix} 5 & 2 \\ 1 & -2 \end{bmatrix}$ $w = \frac{5}{2}; x = \frac{3}{2}; y = 2; z = -\frac{1}{2}$

B

17. Let $A = \begin{bmatrix} 2 & 3 \\ -1 & 6 \end{bmatrix}$ and $B = \begin{bmatrix} -4 & 5 \\ 2 & -7 \end{bmatrix}$. Then $|A| = \begin{vmatrix} 2 & 3 \\ -1 & 6 \end{vmatrix}$ denotes the determinant of A, and similarly for B and $A + B$. Find $|A|, |B|$, and $|A + B|$ and show that $|A| + |B| \neq |A + B|$.

18. Show that if $A = \begin{bmatrix} a & b \\ c & d \end{bmatrix}$ and $B = \begin{bmatrix} rc & rd \\ ta & tb \end{bmatrix}$, then it will be true that $|A| + |B| = |A + B|$. For the meaning of $|A|, |B|$, and $|A + B|$, see Exercise 17.

C

19. Show that if $A = \begin{bmatrix} a & b \\ c & d \end{bmatrix}$ and $B = \begin{bmatrix} w & x \\ y & z \end{bmatrix}$, then $|A + B| = |A| + |B| + \begin{vmatrix} a & b \\ y & z \end{vmatrix} + \begin{vmatrix} w & x \\ c & d \end{vmatrix}$. For the meaning of $|A|, |B|$, and $|A + B|$, see Exercise 17.

Suggested Assignments

Enriched Alg.
  531/3–17 odd, 18, 19

Additional Answers
Written Exercises

17. $|A| = 15, |B| = 18,$
$A + B = \begin{bmatrix} -2 & 8 \\ 1 & -1 \end{bmatrix}$
$|A + B| = -6$
$\therefore |A| + |B| = 33 \neq |A + B|.$

18. $|A| = ad - bc.$
$|B| = rctb - rdta.$
$A + B = \begin{bmatrix} a & b \\ c & d \end{bmatrix} + \begin{bmatrix} rc & rd \\ ta & tb \end{bmatrix}$
$= \begin{bmatrix} a + rc & b + rd \\ c + ta & d + tb \end{bmatrix}.$
$|A + B| = (a + rc)(d + tb) - (b + rd)(c + ta) =$
$ad + rcd + atb + rctb - (bc + rdc + bta + rdta) =$
$(ad - bc) + (rctb - rdta) =$
$|A| + |B|$

19. $|A| = ad - bc.$
$|B| = wz - xy.$
$A + B = \begin{bmatrix} a & b \\ c & d \end{bmatrix} + \begin{bmatrix} w & x \\ y & z \end{bmatrix} =$
$\begin{bmatrix} a + w & b + x \\ c + y & d + z \end{bmatrix}$
$|A + B| = (a + w)(d + z) - (b + x)(c + y)$
$= ad + dw + az + wz - bc - cx - by - xy.$
$|A| + |B| + \begin{vmatrix} a & b \\ y & z \end{vmatrix} + \begin{vmatrix} w & x \\ c & d \end{vmatrix}$
$= ad - bc + wz - xy + az - by + wd - cx = |A + B|$

Chalkboard Examples

1. Solve over $S_{2\times3}$.
$$X + \begin{bmatrix} 4 & 2 & 0 \\ 3 & -1 & 7 \end{bmatrix} = \begin{bmatrix} -3 & 6 & 2 \\ -1 & 5 & 4 \end{bmatrix}$$

Assume that there is a matrix $X \in S_{2\times3}$ such that
$$X + \begin{bmatrix} 4 & 2 & 0 \\ 3 & -1 & 7 \end{bmatrix} = \begin{bmatrix} -3 & 6 & 2 \\ -1 & 5 & 4 \end{bmatrix}.$$

Then
$$X + \begin{bmatrix} 4 & 2 & 0 \\ 3 & -1 & 7 \end{bmatrix} + \begin{bmatrix} -4 & -2 & 0 \\ -3 & 1 & -7 \end{bmatrix} = \begin{bmatrix} -3 & 6 & 2 \\ -1 & 5 & 4 \end{bmatrix} + \begin{bmatrix} -4 & -2 & 0 \\ -3 & 1 & -7 \end{bmatrix}$$

and
$$X + \begin{bmatrix} 0 & 0 & 0 \\ 0 & 0 & 0 \end{bmatrix} = \begin{bmatrix} -7 & 4 & 2 \\ -4 & 6 & -3 \end{bmatrix}$$
$$X = \begin{bmatrix} -7 & 4 & 2 \\ -4 & 6 & -3 \end{bmatrix}$$

∴ the solution set is
$$\left\{ \begin{bmatrix} -7 & 4 & 2 \\ -4 & 6 & -3 \end{bmatrix} \right\}.$$

Mixed Review

1. Simplify
$$\frac{4x^2 - 9y^2}{(2x + 3y)^2} \cdot \frac{2x - 3y}{2x + 3y}$$

13–2 Properties of Matrix Addition

With the definitions of the equality and the sum of two matrices (pages 528 and 529), for any given natural numbers m and n we can state some addition properties of the set $S_{m \times n}$ of all $m \times n$ matrices with real-number entries as follows:

Theorem. For any given natural numbers m and n, let A, B, and C be members of the set $S_{m \times n}$ of all $m \times n$ matrices with real-number entries. Then:

I. $A + B \in S_{m \times n}$.   **Closure Property**

II. $A + B = B + A$.   **Commutative Property**

III. $(A + B) + C = A + (B + C)$.   **Associative Property**

IV. There exists in $S_{m \times n}$ a unique element O such that for each $A \in S_{m \times n}$, $O + A = A$ and $A + O = A$.   **Identity Property for Addition**

V. For each $A \in S_{m \times n}$ there exists a unique element $-A$ in $S_{m \times n}$ such that $A + (-A) = O$ and $(-A) + A = O$.   **Property of Additive Inverses**

Compare the properties listed in the preceding theorem with the axioms of addition in $\mathcal{R}$ given in Chapter 1. We shall consider the proofs of these various properties only for the representative case $m = 2$, $n = 2$. (The proofs for all other cases are similar.) All the properties are direct applications of the definitions that we made for the equality and the sum of two matrices together with the corresponding properties of real numbers.

PROOF OF CLOSURE FOR $S_{2\times2}$

By definition of the sum of two matrices, the entries of $A + B$ are the sums of the corresponding entries in A and B. Since the sum of two real numbers is a real number, the entries of $A + B$ are real numbers. Therefore $A + B$ is a 2 × 2 matrix with real-number entries and thus a member of $S_{2\times2}$.

PROOF OF THE COMMUTATIVE PROPERTY FOR $S_{2\times2}$

Let $A = \begin{bmatrix} a_1 & b_1 \\ a_2 & b_2 \end{bmatrix}$ and $B = \begin{bmatrix} c_1 & d_1 \\ c_2 & d_2 \end{bmatrix}$. Then

$$A + B = \begin{bmatrix} a_1 + c_1 & b_1 + d_1 \\ a_2 + c_2 & b_2 + d_2 \end{bmatrix} \quad \text{and} \quad B + A = \begin{bmatrix} c_1 + a_1 & d_1 + b_1 \\ c_2 + a_2 & d_2 + b_2 \end{bmatrix}.$$

532   Chapter 13

But by the commutative property of addition for real numbers,

$$a_1 + c_1 = c_1 + a_1, \qquad b_1 + d_1 = d_1 + b_1,$$
$$a_2 + c_2 = c_2 + a_2, \qquad b_2 + d_2 = d_2 + b_2.$$

Each entry of the matrix $A + B$ is equal to the corresponding entry of the matrix $B + A$. Thus, by the definition of equality for matrices you have $A + B = B + A$.

Proofs of Parts III, IV, and V, for the representative case $m = 2, n = 2$, are similar and are left as Exercises 11–13 on page 534. By using Part II, you need to establish only one of the equations in IV and one in V.

By using the substitution principle and the properties of equality (Chapter 1), you can also solve certain matrix equations.

EXAMPLE Solve $X + \begin{bmatrix} 5 & 3 & 0 \\ 4 & -2 & 8 \end{bmatrix} = \begin{bmatrix} -2 & 7 & 1 \\ -1 & 4 & 5 \end{bmatrix}$ over $\mathcal{S}_{2 \times 3}$.

SOLUTION Assume that there is a matrix $X \in \mathcal{S}_{2 \times 3}$ such that

$$X + \begin{bmatrix} 5 & 3 & 0 \\ 4 & -2 & 8 \end{bmatrix} = \begin{bmatrix} -2 & 7 & 1 \\ -1 & 4 & 5 \end{bmatrix}.$$

Then by the addition property of equality:

$$\left(X + \begin{bmatrix} 5 & 3 & 0 \\ 4 & -2 & 8 \end{bmatrix} \right) + \begin{bmatrix} -5 & -3 & 0 \\ -4 & 2 & -8 \end{bmatrix}$$
$$= \begin{bmatrix} -2 & 7 & 1 \\ -1 & 4 & 5 \end{bmatrix} + \begin{bmatrix} -5 & -3 & 0 \\ -4 & 2 & -8 \end{bmatrix}$$

By the associative property, additive inverse property, and additive identity property of matrices:

$$X + \left(\begin{bmatrix} 5 & 3 & 0 \\ 4 & -2 & 8 \end{bmatrix} + \begin{bmatrix} -5 & -3 & 0 \\ -4 & 2 & -8 \end{bmatrix} \right)$$
$$= \begin{bmatrix} -2 & 7 & 1 \\ -1 & 4 & 5 \end{bmatrix} + \begin{bmatrix} -5 & -3 & 0 \\ -4 & 2 & -8 \end{bmatrix}$$
$$X + \begin{bmatrix} 0 & 0 & 0 \\ 0 & 0 & 0 \end{bmatrix} = \begin{bmatrix} -7 & 4 & 1 \\ -5 & 6 & -3 \end{bmatrix}$$
$$X = \begin{bmatrix} -7 & 4 & 1 \\ -5 & 6 & -3 \end{bmatrix}$$

Check: Replace X with $\begin{bmatrix} -7 & 4 & 1 \\ -5 & 6 & -3 \end{bmatrix}$ in the original equation.

$$\begin{bmatrix} -7 & 4 & 1 \\ -5 & 6 & -3 \end{bmatrix} + \begin{bmatrix} 5 & 3 & 0 \\ 4 & -2 & 8 \end{bmatrix} = \begin{bmatrix} -7 + 5 & 4 + 3 & 1 + 0 \\ -5 + 4 & 6 - 2 & -3 + 8 \end{bmatrix}$$
$$= \begin{bmatrix} -2 & 7 & 1 \\ -1 & 4 & 5 \end{bmatrix} \checkmark$$

$\therefore$ the solution set is $\left\{ \begin{bmatrix} -7 & 4 & 1 \\ -5 & 6 & -3 \end{bmatrix} \right\}$.

Matrices **533**

2. Solve $\sqrt{x - 2} + 4 = x$.
   {6}

3. Give the equation for the axis of symmetry and the coordinates of the vertex of the graph of $y = x^2 + 8x - 3$. $x = -4$; $(-4, -19)$

4. Graph the solution set of $y = |x| - 2$ and state whether it is a function. yes

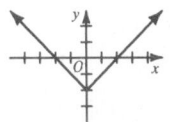

Additional A Exercises

Solve for the matrix X.

1. $X + \begin{bmatrix} 3 & 0 \\ -2 & 4 \end{bmatrix} = \begin{bmatrix} 5 & -2 \\ 3 & -1 \end{bmatrix}$
   $X = \begin{bmatrix} 2 & -2 \\ 5 & -5 \end{bmatrix}$

2. $X - \begin{bmatrix} 1 & 3 \\ -2 & -5 \end{bmatrix} = \begin{bmatrix} 4 & -3 \\ 1 & -2 \end{bmatrix}$
   $X = \begin{bmatrix} 5 & 0 \\ -1 & -7 \end{bmatrix}$

3. $X + \begin{bmatrix} -6 & 5 \\ 2 & 8 \end{bmatrix} = \begin{bmatrix} 7 & -2 \\ -3 & 0 \end{bmatrix}$
   $X = \begin{bmatrix} 13 & -7 \\ -5 & -8 \end{bmatrix}$

4. $\begin{bmatrix} 4 & 7 \\ -3 & 8 \end{bmatrix} - X = \begin{bmatrix} 2 & 9 \\ 5 & -1 \end{bmatrix}$
   $X = \begin{bmatrix} 2 & -2 \\ -8 & 9 \end{bmatrix}$

5. $\begin{bmatrix} 4 & -3 & 7 \\ 8 & 4 & -5 \end{bmatrix} + X =$
   $\begin{bmatrix} -1 & 0 & 5 \\ 2 & -3 & 0 \end{bmatrix}$
   $X = \begin{bmatrix} -5 & 3 & -2 \\ -6 & -7 & 5 \end{bmatrix}$

6. $\begin{bmatrix} 8 & -11 & 0 \\ -7 & 6 & 2 \end{bmatrix} - X =$
   $\begin{bmatrix} 6 & -4 & 2 \\ 3 & 0 & 5 \end{bmatrix}$
   $X = \begin{bmatrix} 2 & -7 & -2 \\ -10 & 6 & -3 \end{bmatrix}$

(continued on page 564)

534

Oral Exercises

Exercises 1–5 list the steps you might use to solve the matrix equation
$X - \begin{bmatrix} 3 & -4 & 6 \end{bmatrix} = \begin{bmatrix} 5 & -2 & -1 \end{bmatrix}$. **Give a reason for each step.**

1. $X + (-\begin{bmatrix} 3 & -4 & 6 \end{bmatrix}) = \begin{bmatrix} 5 & -2 & -1 \end{bmatrix}$ Def. of subtraction

Addition property of equality

2. $(X + (-\begin{bmatrix} 3 & -4 & 6 \end{bmatrix})) + \begin{bmatrix} 3 & -4 & 6 \end{bmatrix} = \begin{bmatrix} 5 & -2 & -1 \end{bmatrix} + \begin{bmatrix} 3 & -4 & 6 \end{bmatrix}$

3. $X + (-\begin{bmatrix} 3 & -4 & 6 \end{bmatrix} + \begin{bmatrix} 3 & -4 & 6 \end{bmatrix}) = \begin{bmatrix} 8 & -6 & 5 \end{bmatrix}$ Associative property

4. $X + \begin{bmatrix} 0 & 0 & 0 \end{bmatrix} = \begin{bmatrix} 8 & -6 & 5 \end{bmatrix}$ Additive inverse property

5. $X = \begin{bmatrix} 8 & -6 & 5 \end{bmatrix}$ Additive identity property

Suppose the addition of 1×2 matrices were defined as follows:

$$[a \quad b] + [c \quad d] = [a \quad b + d]$$

Which of the following properties in the theorem on page 532 would hold?

6. I Yes   **7.** II No   **8.** III Yes   **9.** IV No   **10.** V
No

Written Exercises

7. $\begin{bmatrix} -6 & 3 & -3 \\ 0 & -8 & 6 \end{bmatrix}$ **8.** $\begin{bmatrix} 2 & -4 & -2 \\ -10 & 7 & 7 \end{bmatrix}$

Solve for the matrix X.

1. $X + \begin{bmatrix} 2 & 0 \\ -3 & 1 \end{bmatrix} = \begin{bmatrix} 7 & -4 \\ 2 & -5 \end{bmatrix}\begin{bmatrix} 5 & -4 \\ 5 & -6 \end{bmatrix}$ **2.** $X + \begin{bmatrix} -1 & 7 \\ 5 & 8 \end{bmatrix} = \begin{bmatrix} 8 & -6 \\ -2 & 0 \end{bmatrix}\begin{bmatrix} 9 & -13 \\ -7 & -8 \end{bmatrix}$

3. $X - \begin{bmatrix} 2 & 1 \\ -3 & -9 \end{bmatrix} = \begin{bmatrix} 6 & -2 \\ 3 & -4 \end{bmatrix}\begin{bmatrix} 8 & -1 \\ 0 & -13 \end{bmatrix}$ **4.** $X - \begin{bmatrix} 8 & -3 \\ 5 & -4 \end{bmatrix} = \begin{bmatrix} 1 & 4 \\ -11 & 0 \end{bmatrix}\begin{bmatrix} 9 & 1 \\ -6 & -4 \end{bmatrix}$

5. $\begin{bmatrix} 3 & 8 \\ -6 & 7 \end{bmatrix} - X = \begin{bmatrix} 4 & 5 \\ 2 & -1 \end{bmatrix}\begin{bmatrix} -1 & 3 \\ -8 & 8 \end{bmatrix}$ **6.** $\begin{bmatrix} -4 & 0 \\ 2 & 5 \end{bmatrix} - X = \begin{bmatrix} -7 & 3 \\ -2 & 1 \end{bmatrix}\begin{bmatrix} 3 & -3 \\ 4 & 4 \end{bmatrix}$

7. $\begin{bmatrix} 4 & -3 & 9 \\ 2 & 5 & -6 \end{bmatrix} + X = \begin{bmatrix} -2 & 0 & 6 \\ 2 & -3 & 0 \end{bmatrix}$ **8.** $\begin{bmatrix} 6 & -10 & 0 \\ -9 & 7 & 4 \end{bmatrix} - X = \begin{bmatrix} 4 & -6 & 2 \\ 1 & 0 & -3 \end{bmatrix}$

Solve for X in terms of a, b, c, and d and give a reason for each step.

9. $[a \quad b] - X = [c \quad d]$ $X = [a - c \quad b - d]$ **10.** $\begin{bmatrix} a \\ b \end{bmatrix} + X = \begin{bmatrix} c \\ d \end{bmatrix}$ $X = \begin{bmatrix} c - a \\ d - b \end{bmatrix}$

In Exercises 11–16, prove the given statement when A, B, and C are 2×2 matrices.

11. $(A + B) + C = A + (B + C)$ **12.** $A + O = A$

13. $A + (-A) = O$ **14.** If $A = -A$, then $A = O$.

15. $(A + B) - A = B$ **16.** If $A + B = A$, then $B = O$.

17. Find all ordered triples (x, y, z) for which the following is true.

$(-3, 5, -4), (-3, 5, -2), (-3, -2, -4),$
$(-3, -2, -2)$

$$\begin{bmatrix} 4 & 3x \\ y^2 - 3y & 6z \end{bmatrix} - \begin{bmatrix} 4 & -9 \\ 10 & -(8 + z^2) \end{bmatrix} = O_{2 \times 2}$$

13–3 Product of a Scalar and a Matrix

In dealing with matrices, we often refer to real numbers as **scalars**.
We define the **product** of a scalar c and a matrix A, denoted by

$$cA,$$

as the matrix of the same dimensions as A whose entries are the products of c and the corresponding entries of A. For example, if

$$A = \begin{bmatrix} 2 & 3 \\ 1 & 0 \\ -4 & 7 \end{bmatrix},$$

then

$$5A = 5 \begin{bmatrix} 2 & 3 \\ 1 & 0 \\ -4 & 7 \end{bmatrix} = \begin{bmatrix} 5(2) & 5(3) \\ 5(1) & 5(0) \\ 5(-4) & 5(7) \end{bmatrix} = \begin{bmatrix} 10 & 15 \\ 5 & 0 \\ -20 & 35 \end{bmatrix}$$

and

$$-2A = -2 \begin{bmatrix} 2 & 3 \\ 1 & 0 \\ -4 & 7 \end{bmatrix} = \begin{bmatrix} -2(2) & -2(3) \\ -2(1) & -2(0) \\ -2(-4) & -2(7) \end{bmatrix} = \begin{bmatrix} -4 & -6 \\ -2 & 0 \\ 8 & -14 \end{bmatrix}.$$

Notice that the product of a *scalar* and a *matrix* is a *matrix*.

Products of scalars and matrices have a number of basic properties that follow from the definition above and the properties of real numbers. These basic properties are given in the following theorem.

Theorem. If $A \in \mathcal{S}_{m \times n}$ and $B \in \mathcal{S}_{m \times n}$, where m and n are given natural numbers, and if $c \in \mathcal{R}$ and $d \in \mathcal{R}$, then:

I. $cA \in \mathcal{S}_{m \times n}$	**Closure Property**
II. $c(dA) = (cd)A$	**Associative Property**
III. $(c + d)A = cA + dA$	**Distributive Properties**
IV. $c(A + B) = cA + cB$	
V. $1 \cdot A = A$	**Identity Property for Multiplication**
VI. $(-1)A = -A$	**Multiplicative Property of -1**
VII. $0 \cdot A = O$	**Zero Properties**
VIII. $cO = O$	

As in Section 13-2, we shall consider proofs only for the representative case $m = 2$, $n = 2$. For the sample proof on the next page we let

$$A = \begin{bmatrix} a_1 & b_1 \\ a_2 & b_2 \end{bmatrix}.$$

Matrices **535**

Key Ideas

Find the product of a scalar and a matrix.

Chalkboard Examples

1. Solve over $\mathcal{S}_{3 \times 2}$:

$$5X - 3 \begin{bmatrix} 0 & -2 \\ 1 & 2 \\ -4 & 7 \end{bmatrix} = 2 \begin{bmatrix} 5 & -7 \\ 1 & -3 \\ -9 & 2 \end{bmatrix}$$

First simplify the products; thus,

$$5X - \begin{bmatrix} 0 & -6 \\ 3 & 6 \\ -12 & 21 \end{bmatrix} = \begin{bmatrix} 10 & -14 \\ 2 & -6 \\ -18 & 4 \end{bmatrix}.$$

Then add $\begin{bmatrix} 0 & -6 \\ 3 & 6 \\ -12 & 21 \end{bmatrix}$ to each side.

$$5X - \begin{bmatrix} 0 & -6 \\ 3 & 6 \\ -12 & 21 \end{bmatrix} + \begin{bmatrix} 0 & -6 \\ 3 & 6 \\ -12 & 21 \end{bmatrix} = \begin{bmatrix} 10 & -14 \\ 2 & -6 \\ -18 & 4 \end{bmatrix} + \begin{bmatrix} 0 & -6 \\ 3 & 6 \\ -12 & 21 \end{bmatrix}$$

$$5X + O = \begin{bmatrix} 10 & -20 \\ 5 & 0 \\ -30 & 25 \end{bmatrix}$$

$$X = \frac{1}{5} \begin{bmatrix} 10 & -20 \\ 5 & 0 \\ -30 & 25 \end{bmatrix}$$

$$= \begin{bmatrix} 2 & -4 \\ 1 & 0 \\ -6 & 5 \end{bmatrix}$$

∴ the solution set is

$$\left\{ \begin{bmatrix} 2 & -4 \\ 1 & 0 \\ -6 & 5 \end{bmatrix} \right\}.$$

Additional A Exercises

In Exercises 1–3 find the 2 × 2 matrix X that satisfies the equation, if

$A = \begin{bmatrix} 3 & -9 \\ 21 & 0 \end{bmatrix}$ and $B = \begin{bmatrix} 7 & -11 \\ -1 & 10 \end{bmatrix}$.

1. $X = 2A + B$ $\quad \begin{bmatrix} 13 & -29 \\ 41 & 10 \end{bmatrix}$

2. $3X = A$ $\quad \begin{bmatrix} 1 & -3 \\ 7 & 0 \end{bmatrix}$

3. $X - 3A = B$ $\quad \begin{bmatrix} 16 & -38 \\ 62 & 10 \end{bmatrix}$

In Exercises 4–6 find the 2 × 2 matrix X that satisfies the given equation, if

$A = \begin{bmatrix} 2 & -1 \\ 5 & -4 \end{bmatrix}$ and $B = \begin{bmatrix} 4 & 13 \\ -5 & -8 \end{bmatrix}$.

4. $5A - X = 2B$ $\quad \begin{bmatrix} 2 & -31 \\ 35 & -4 \end{bmatrix}$

5. $\frac{1}{2}X + 4A = B$ $\quad \begin{bmatrix} -8 & 34 \\ -50 & 16 \end{bmatrix}$

6. $2X + 5A = 3B$ $\quad \begin{bmatrix} 1 & 22 \\ -20 & -2 \end{bmatrix}$

Reading Algebra

Have your students verbalize the different parts of the theorem on page 535. This will help make the theorem more meaningful. For example, read part V as "the product of the scalar 1 and the matrix A is equal to the matrix A."

Suggested Assignments

Enriched Alg.
537/7–25 odd, 26–28

$$(-1)A = (-1)\begin{bmatrix} a_1 & b_1 \\ a_2 & b_2 \end{bmatrix}$$
$$= \begin{bmatrix} (-1)a_1 & (-1)b_1 \\ (-1)a_2 & (-1)b_2 \end{bmatrix}$$
$$= \begin{bmatrix} -a_1 & -b_1 \\ -a_2 & -b_2 \end{bmatrix} = -A$$

You should be able to supply a reason for each step given above.

Proofs of the remaining parts of this theorem for the representative case $m = 2$, $n = 2$ are similar to the one we have given. Writing them is left to you (Exercises 19–25 on pages 537 and 538).

You can use parts of the preceding theorem to help you solve some equations involving matrices.

EXAMPLE Solve over $\mathcal{S}_{3\times2}$: $5X - 2\begin{bmatrix} 0 & -3 \\ 1 & 4 \\ -6 & 7 \end{bmatrix} = 3\begin{bmatrix} 5 & -8 \\ 1 & -6 \\ 4 & 2 \end{bmatrix}$

SOLUTION Assuming that there is a 3 × 2 matrix X satisfying the equation, first simplify the products; thus,

$$5X - \begin{bmatrix} 0 & -6 \\ 2 & 8 \\ -12 & 14 \end{bmatrix} = \begin{bmatrix} 15 & -24 \\ 3 & -18 \\ 12 & 6 \end{bmatrix}.$$

Then add $\begin{bmatrix} 0 & -6 \\ 2 & 8 \\ -12 & 14 \end{bmatrix}$ to each side, as shown below.

$$5X - \begin{bmatrix} 0 & -6 \\ 2 & 8 \\ -12 & 14 \end{bmatrix} + \begin{bmatrix} 0 & -6 \\ 2 & 8 \\ -12 & 14 \end{bmatrix} = \begin{bmatrix} 15 & -24 \\ 3 & -18 \\ 12 & 6 \end{bmatrix} + \begin{bmatrix} 0 & -6 \\ 2 & 8 \\ -12 & 14 \end{bmatrix}$$

$$5X + O = \begin{bmatrix} 15 & -30 \\ 5 & -10 \\ 0 & 20 \end{bmatrix}$$

$$X = \frac{1}{5}\begin{bmatrix} 15 & -30 \\ 5 & -10 \\ 0 & 20 \end{bmatrix}$$

$$= \begin{bmatrix} 3 & -6 \\ 1 & -2 \\ 0 & 4 \end{bmatrix}$$

The check is left to you.

$\therefore$ the solution set is $\left\{ \begin{bmatrix} 3 & -6 \\ 1 & -2 \\ 0 & 4 \end{bmatrix} \right\}$.

536   Chapter 13

Oral Exercises

In Exercises 1–14, find the value of the variable so that the equation in which it appears is true.

$$2\begin{bmatrix} 3 & -1 \\ 5 & 0 \end{bmatrix} = \begin{bmatrix} a & b \\ c & d \end{bmatrix} \qquad w\begin{bmatrix} -4 & x \\ y & 7 \end{bmatrix} = \begin{bmatrix} -12 & 15 \\ -6 & z \end{bmatrix}$$

1. a 6   **2.** b −2   **3.** c 10   **4.** d 0   **5.** w 3   **6.** x 5   **7.** y −2   **8.** z 21

$$3\begin{bmatrix} -1 & 2 & 0 \\ 3 & 4 & -5 \end{bmatrix} + \begin{bmatrix} 7 & -6 & -1 \\ k & m & n \end{bmatrix} = \begin{bmatrix} r & p & q \\ -2 & 5 & 9 \end{bmatrix}$$

9. k −11   **10.** m −7   **11.** n 24   **12.** r 4   **13.** p 0   **14.** q −1

In Exercises 15–18, $A = \begin{bmatrix} -1 & 2 \\ 5 & 7 \end{bmatrix}$ and $B = \begin{bmatrix} 3 & 0 \\ 2 & -5 \end{bmatrix}$.
Tell whether each statement is true or false.

15. $-2A = \begin{bmatrix} 2 & -4 \\ -10 & -14 \end{bmatrix}$ true

16. $3B = \begin{bmatrix} 9 & 3 \\ 6 & -15 \end{bmatrix}$ false

17. $A - B = \begin{bmatrix} 2 & 2 \\ 3 & -12 \end{bmatrix}$ false

18. $2B - 3A = \begin{bmatrix} 9 & -6 \\ -11 & -31 \end{bmatrix}$ true

Written Exercises

In Exercises 1–12, find the 2 × 2 matrix X that satisfies the equation, if

$$A = \begin{bmatrix} 9 & -3 \\ -21 & 0 \end{bmatrix} \quad \text{and} \quad B = \begin{bmatrix} 11 & -7 \\ 1 & 10 \end{bmatrix}.$$

A   **1.** $X = 3A$ **2.** $X = -4B$ **3.** $X = 2A + B$

4. $X = A - 2B$ **5.** $3X = A$ **6.** $\frac{1}{2}X = B$

7. $X + A = 2B$ **8.** $X - B = 3A$ **9.** $\frac{1}{3}X = B - A$

10. $3X = A - 3B$ **11.** $5X = 3A - 2B$ **12.** $2X - 3A = B$

In Exercises 13–18, find the 2 × 2 matrix X that satisfies the given equation
if $A = \begin{bmatrix} 1 & -2 \\ 4 & -5 \end{bmatrix}$ and $B = \begin{bmatrix} -13 & -4 \\ 8 & 5 \end{bmatrix}$.

13. $X + 2A = B$ **14.** $4A - X = 2B$ **15.** $\frac{1}{2}X + 5A = B$

16. $3X - 5A = 2B$ **17.** $4A - 5X = -3B$ **18.** $2X + 3B = 5A$

In Exercises 19–27, prove the statement for all 2 × 2 matrices A, B, and X
and all real numbers c and d.

B   **19.** $cA \in \mathcal{S}_{2 \times 2}$ **20.** $c(dA) = (cd)A$ **21.** $(c + d)A = cA + dA$

22. $c(A + B) = cA + cB$ **23.** $1 \cdot A = A$ **24.** $0 \cdot A = O_{2 \times 2}$

Matrices   **537**

Mixed Review

1. Solve $\log_{27} x = -\dfrac{2}{3}$ over
$\mathcal{R}$. $\left\{\dfrac{1}{9}\right\}$

2. Write an equation of a line
through $(-7, 2)$ and
$(-3, 1)$. $y = -\dfrac{x}{4} + \dfrac{1}{4}$

3. Evaluate $\begin{vmatrix} 5 & 3 \\ 6 & -1 \end{vmatrix}$.   −23

4. Simplify $(\sqrt{2} + 4)(3\sqrt{2} - 5)$. $7\sqrt{2} - 14$

5. Solve $x^2 - 2x + 5 = 0$
over $\mathcal{C}$. $\{1 + 2i, 1 - 2i\}$

Additional Answers
Written Exercises

1. $\begin{bmatrix} 27 & -9 \\ -63 & 0 \end{bmatrix}$

2. $\begin{bmatrix} -44 & 28 \\ -4 & -40 \end{bmatrix}$

3. $\begin{bmatrix} 29 & -13 \\ -41 & 10 \end{bmatrix}$

4. $\begin{bmatrix} -13 & 11 \\ -23 & -20 \end{bmatrix}$

5. $\begin{bmatrix} 3 & -1 \\ -7 & 0 \end{bmatrix}$

6. $\begin{bmatrix} 22 & -14 \\ 2 & 20 \end{bmatrix}$

7. $\begin{bmatrix} 13 & -11 \\ 23 & 20 \end{bmatrix}$

8. $\begin{bmatrix} 38 & -16 \\ -62 & 10 \end{bmatrix}$

9. $\begin{bmatrix} 6 & -12 \\ 66 & 30 \end{bmatrix}$

10. $\begin{bmatrix} -8 & 6 \\ -8 & -10 \end{bmatrix}$

11. $\begin{bmatrix} 1 & 1 \\ -13 & -4 \end{bmatrix}$

12. $\begin{bmatrix} 19 & -8 \\ -31 & 5 \end{bmatrix}$

13. $\begin{bmatrix} -15 & 0 \\ 0 & 15 \end{bmatrix}$

14. $\begin{bmatrix} 30 & 0 \\ 0 & -30 \end{bmatrix}$

(continued)

25. $cO_{2\times2} = O_{2\times2}$

26. If $cX = B$, $c \neq 0$, then $X = \dfrac{1}{c}B$.

27. *If $cX + B = A$, then $X = \dfrac{1}{c}(A - B)$.*

C **28.** Find all ordered pairs (x, y) for which the following matrix equation is true:

$(2 + 2\sqrt{3}, 8)$, $(2 + 2\sqrt{3}, -3)$,
$(2 - 2\sqrt{3}, 8)$, $(2 - 2\sqrt{3}, -3)$

$$\begin{bmatrix} x^2 \\ y^2 \end{bmatrix} - \begin{bmatrix} 4x \\ 5y \end{bmatrix} - 8\begin{bmatrix} 1 \\ 3 \end{bmatrix} = O_{2\times1}$$

13–4 Product of Two Matrices

Before introducing the definition of the product of two matrices in general, let us consider the two matrices

$$A_{1\times3} = \begin{bmatrix} a & b & c \end{bmatrix} \quad \text{and} \quad B_{3\times2} = \begin{bmatrix} x_1 & x_2 \\ y_1 & y_2 \\ z_1 & z_2 \end{bmatrix}.$$

Suppose that the numbers a, b, and c represent the price per package of frozen strawberries, peaches, and apricots, respectively, charged by a supermarket, while x_1, y_1, and z_1 represent the number of packages of each sold the first week, and x_2, y_2, z_2 the numbers sold the second week. How much money would the supermarket collect for these items? For the first week, the total amount collected for the strawberries is $a \cdot x_1$, for the peaches $b \cdot y_1$, and for the apricots $c \cdot z_1$. By adding these products, you obtain the total amount collected, $ax_1 + by_1 + cz_1$. For the second week, you have a total collection of $ax_2 + by_2 + cz_2$.

The process of adding the products obtained by multiplying the elements of a row in one matrix by the corresponding elements of a column in another matrix suggests a fruitful way of defining the product of two matrices. We may say that the product of $A_{1\times3}$ and $B_{3\times2}$ shown above is

$$C_{1\times2} = \begin{bmatrix} ax_1 + by_1 + cz_1 & ax_2 + by_2 + cz_2 \end{bmatrix}.$$

The **product** of two 2×2 matrices A and B, denoted by

$$A \times B \quad \text{or} \quad A \cdot B \quad \text{or} \quad AB,$$

is defined as follows.

If $A = \begin{bmatrix} a_1 & b_1 \\ a_2 & b_2 \end{bmatrix}$ and $B = \begin{bmatrix} c_1 & d_1 \\ c_2 & d_2 \end{bmatrix}$, then:

$$A \times B = AB = \begin{bmatrix} a_1 & b_1 \\ a_2 & b_2 \end{bmatrix} \times \begin{bmatrix} c_1 & d_1 \\ c_2 & d_2 \end{bmatrix} = \begin{bmatrix} a_1c_1 + b_1c_2 & a_1d_1 + b_1d_2 \\ a_2c_1 + b_2c_2 & a_2d_1 + b_2d_2 \end{bmatrix}$$

538   *Chapter 13*

Notice that the entries of a given row of A, say the ith row, are multiplied by the entries of a given column of B, say the jth column, in order, and these products are then added to obtain the entry in the ith row and jth column of $A \times B$. Thus, the multiplication of two matrices can be described as "row by column" multiplication.

As an example, let us find the product

$$\begin{bmatrix} 3 & 2 \\ 5 & 4 \end{bmatrix} \times \begin{bmatrix} -2 & 3 \\ 1 & 6 \end{bmatrix}$$

by displaying the computation of each entry, one at a time, in red. Notice that we may omit the times sign between the matrices.

1. First row, first column:

$$\begin{bmatrix} 3 & 2 \end{bmatrix}\begin{bmatrix} -2 \\ 1 \end{bmatrix} = \begin{bmatrix} 3 \times (-2) + 2 \times 1 \end{bmatrix} = \begin{bmatrix} -6 + 2 \end{bmatrix} = \begin{bmatrix} -4 \end{bmatrix}$$

2. First row, second column:

$$\begin{bmatrix} 3 & 2 \end{bmatrix}\begin{bmatrix} 3 \\ 6 \end{bmatrix} = \begin{bmatrix} -4 & 3 \times 3 + 2 \times 6 \end{bmatrix} = \begin{bmatrix} -4 & 9 + 12 \end{bmatrix} = \begin{bmatrix} -4 & 21 \end{bmatrix}$$

3. Second row, first column:

$$\begin{bmatrix} 5 & 4 \end{bmatrix}\begin{bmatrix} -2 \\ 1 \end{bmatrix} = \begin{bmatrix} -4 & & 21 \\ 5 \times (-2) + 4 \times 1 & \end{bmatrix} = \begin{bmatrix} -4 & 21 \\ -10 + 4 \end{bmatrix} = \begin{bmatrix} -4 & 21 \\ -6 \end{bmatrix}$$

4. Second row, second column.

$$\begin{bmatrix} 5 & 4 \end{bmatrix}\begin{bmatrix} 3 \\ 6 \end{bmatrix} = \begin{bmatrix} -4 & & 21 \\ -6 & 5 \times 3 + 4 \times 6 \end{bmatrix} = \begin{bmatrix} -4 & 21 \\ -6 & 15 + 24 \end{bmatrix} = \begin{bmatrix} -4 & 21 \\ -6 & 39 \end{bmatrix}$$

Putting Steps 1–4 together, we have

$$\begin{bmatrix} 3 & 2 \\ 5 & 4 \end{bmatrix}\begin{bmatrix} -2 & 3 \\ 1 & 6 \end{bmatrix} = \begin{bmatrix} -4 & 21 \\ -6 & 39 \end{bmatrix}.$$

Ordinarily, of course, the steps are not all shown in detail.

In general, the product AB of any two matrices A and B, where A has the same number of *columns* as B has *rows*, can be defined through "row by column" multiplication. The number of rows in the product matrix AB will be the same as the number of rows in A, and the number of columns in AB will be the same as the number of columns in B. Thus,

$$A_{m \times p} \times B_{p \times n} = C_{m \times n}.$$

Find the given product, if it is defined.

1. $\begin{bmatrix} 4 & -2 \\ 5 & 0 \\ -1 & 3 \end{bmatrix}\begin{bmatrix} 2 & 6 \\ 0 & -3 \end{bmatrix}$

$\begin{bmatrix} 4 & -2 \\ 5 & 0 \\ -1 & 3 \end{bmatrix}\begin{bmatrix} 2 & 6 \\ 0 & -3 \end{bmatrix} =$

$\begin{bmatrix} 8 & 24 + 6 \\ 10 + 0 & 30 - 0 \\ -2 + 0 & -6 - 9 \end{bmatrix} =$

$\begin{bmatrix} 8 & 30 \\ 10 & 30 \\ -2 & -15 \end{bmatrix}$

2. $\begin{bmatrix} 3 \\ 0 \\ -4 \end{bmatrix}\begin{bmatrix} 3 & 6 & 2 \\ -1 & 0 & 4 \end{bmatrix}$

The number of columns in the first matrix is 1 and the number of rows in the second matrix is 2. Since $1 \neq 2$, the given product is not defined.

Mixed Review

1. Factor $2x^3 - x^2 - 7x + 6$ completely.
    $(x + 2)(x - 1)(2x - 3)$
2. If y varies directly as x, and y is 8 when x is 6, find x when y is 15. $\dfrac{45}{4}$
3. State whether this system has none, one, or infinitely many solutions.
    $\dfrac{2}{3}x - y = 7$ and $3y - 2x = 6$ none
4. Solve by the linear combination method.
    $8x + 3y = 5$
    $5x + 4y = 18$ $\{(-2, 7)\}$

Additional A Exercises

Find the given product, if it is defined. Otherwise write "not defined."

1. $\begin{bmatrix} -2 & 5 \\ 3 & 1 \end{bmatrix}\begin{bmatrix} 4 \\ -2 \end{bmatrix}$ $\begin{bmatrix} -18 \\ 10 \end{bmatrix}$

2. $\begin{bmatrix} 4 \\ 3 \end{bmatrix}\begin{bmatrix} -5 & 0 \\ 5 & 2 \end{bmatrix}$ not defined

3. $\begin{bmatrix} 5 & -1 \\ 3 & 0 \\ -2 & 4 \end{bmatrix}\begin{bmatrix} -3 \\ -4 \end{bmatrix}$ $\begin{bmatrix} -11 \\ -9 \\ -10 \end{bmatrix}$

For $A = \begin{bmatrix} 3 & -4 \\ 2 & 1 \end{bmatrix}$ and $B = \begin{bmatrix} 1 & -3 \\ -3 & 1 \end{bmatrix}$, find a 2 × 2 matrix equal to the given product.

4. AB $\begin{bmatrix} 15 & -13 \\ -1 & -5 \end{bmatrix}$

5. BA $\begin{bmatrix} -3 & -7 \\ -7 & 13 \end{bmatrix}$

6. A^2 $\begin{bmatrix} 1 & -16 \\ 8 & -7 \end{bmatrix}$

Suggested Assignments

Enriched Alg.
 541/7–21 odd, 22–30

Additional Answers
Written Exercises

13. $\begin{bmatrix} -7 & 9 \\ -9 & 14 \end{bmatrix}$ 14. $\begin{bmatrix} 2 & 9 \\ 3 & 5 \end{bmatrix}$

15. $\begin{bmatrix} 1 & 0 \\ 0 & 1 \end{bmatrix}$ 16. $\begin{bmatrix} 13 & -18 \\ 24 & -11 \end{bmatrix}$

17. $\begin{bmatrix} 9 & 0 \\ 18 & 9 \end{bmatrix}$ 18. $\begin{bmatrix} 21 & -18 \\ 36 & -21 \end{bmatrix}$

19. $AB = \begin{bmatrix} ax + 0z & ay + 0w \\ 0x + az & 0y + aw \end{bmatrix}$

$= \begin{bmatrix} ax & ay \\ az & aw \end{bmatrix}$

$BA = \begin{bmatrix} ax + 0y & 0x + ay \\ az + 0w & 0z + aw \end{bmatrix}$

$= \begin{bmatrix} ax & ay \\ az & aw \end{bmatrix} = AB$

EXAMPLE 1 Find the indicated product.

a. $\begin{bmatrix} -2 & 8 & 1 \\ 0 & -7 & 3 \end{bmatrix}\begin{bmatrix} 0 \\ 2 \\ -6 \end{bmatrix}$ b. $\begin{bmatrix} 4 & 8 \\ 0 & -3 \\ 1 & -4 \end{bmatrix}\begin{bmatrix} 1 & 0 \\ -2 & 5 \end{bmatrix}$

SOLUTION a. $\begin{bmatrix} -2 & 8 & 1 \\ 0 & -7 & 3 \end{bmatrix}\begin{bmatrix} 0 \\ 2 \\ -6 \end{bmatrix} = \begin{bmatrix} -2 \times 0 + & 8 \times 2 + 1 \times (-6) \\ 0 \times 0 + (-7) \times 2 + 3 \times (-6) \end{bmatrix}$

$= \begin{bmatrix} 0 + 16 - 6 \\ 0 - 14 - 18 \end{bmatrix} = \begin{bmatrix} 10 \\ -32 \end{bmatrix}$

b. $\begin{bmatrix} 4 & 8 \\ 0 & -3 \\ 1 & -4 \end{bmatrix}\begin{bmatrix} 1 & 0 \\ -2 & 5 \end{bmatrix} = \begin{bmatrix} 4 \times 1 + & 8 \times (-2) & 4 \times 0 + & 8 \times 5 \\ 0 \times 1 + (-3) \times (-2) & 0 \times 0 + (-3) \times 5 \\ 1 \times 1 + (-4) \times (-2) & 1 \times 0 + (-4) \times 5 \end{bmatrix}$

$= \begin{bmatrix} 4 - 16 & 0 + 40 \\ 0 + 6 & 0 - 15 \\ 1 + 8 & 0 - 20 \end{bmatrix}$

$= \begin{bmatrix} -12 & 40 \\ 6 & -15 \\ 9 & -20 \end{bmatrix}$

If the number of *columns* in A is not equal to the number of rows in B, then the product of A and B is not defined. Notice that $B_{p \times n} \times A_{m \times p}$ is not defined unless $n = m$.

In particular, the product of a 2 × 2 and a 2 × 1 matrix is defined as follows.

If $A = \begin{bmatrix} a_1 & b_1 \\ a_2 & b_2 \end{bmatrix}$ and $B = \begin{bmatrix} x \\ y \end{bmatrix}$, then

$$AB = \begin{bmatrix} a_1 & b_1 \\ a_2 & b_2 \end{bmatrix}\begin{bmatrix} x \\ y \end{bmatrix} = \begin{bmatrix} a_1x + b_1y \\ a_2x + b_2y \end{bmatrix}.$$

EXAMPLE 2 Find the indicated product, if it is defined.

a. $\begin{bmatrix} 1 & 2 \\ 3 & -4 \end{bmatrix}\begin{bmatrix} 1 \\ 0 \end{bmatrix}$ b. $\begin{bmatrix} 1 \\ 0 \end{bmatrix}\begin{bmatrix} 1 & 2 \\ 3 & -4 \end{bmatrix}$

SOLUTION a. $\begin{bmatrix} 1 & 2 \\ 3 & -4 \end{bmatrix}\begin{bmatrix} 1 \\ 0 \end{bmatrix} = \begin{bmatrix} 1 \times 1 + & 2 \times 0 \\ 3 \times 1 + (-4) \times 0 \end{bmatrix} = \begin{bmatrix} 1 \\ 3 \end{bmatrix}$

b. The number of columns in $\begin{bmatrix} 1 \\ 0 \end{bmatrix}$ is 1, while the number of rows in $\begin{bmatrix} 1 & 2 \\ 3 & -4 \end{bmatrix}$ is 2. Since $1 \neq 2$, the given product is not defined.

As you can see, the product of a 2 × 2 matrix and a 2 × 1 matrix is a 2 × 1 matrix. Later in this chapter, we shall use products of 2 × 2 and 2 × 1 matrices in connection with systems of equations.

As with numbers, for square matrices A we use the symbol A^2 to mean $A \times A$.

540 *Chapter 13*

Oral Exercises

State the value of the specified letter in the computation of the product.

$$\begin{bmatrix} -4 & 6 & 3 \\ 5 & 0 & -2 \end{bmatrix} \begin{bmatrix} 3 \\ 1 \\ -4 \end{bmatrix} = \begin{bmatrix} (-4) \cdot 3 + 6 \cdot 1 + a(-4) \\ 5 \cdot b + c \cdot 1 + (-2)d \end{bmatrix} = \begin{bmatrix} e \\ f \end{bmatrix}$$

1. a 3 **2.** b 3 **3.** c 0 **4.** d −4 **5.** e −18 **6.** f 23

Tell whether the given product is defined.

7. $\begin{bmatrix} 1 & 2 \\ 6 & 1 \end{bmatrix} \begin{bmatrix} 1 & 0 \\ 0 & -1 \end{bmatrix}$ yes

8. $\begin{bmatrix} -1 & 3 \\ 2 & 0 \end{bmatrix} \begin{bmatrix} 7 \\ 4 \end{bmatrix}$ yes

9. $\begin{bmatrix} 6 \\ 6 \end{bmatrix} \begin{bmatrix} -1 & 3 \\ 3 & 4 \end{bmatrix}$ no

10. $\begin{bmatrix} 3 & 4 \end{bmatrix} \begin{bmatrix} -4 \\ 3 \end{bmatrix}$ yes

11. $\begin{bmatrix} -3 \\ 2 \end{bmatrix} \begin{bmatrix} 4 & 0 \end{bmatrix}$ yes

12. $\begin{bmatrix} 1 \\ 0 \end{bmatrix} \begin{bmatrix} 0 \\ 1 \end{bmatrix}$ no

Written Exercises

Find the indicated product, if it is defined. Otherwise write "not defined."

A **1.** $\begin{bmatrix} 3 & 4 \\ -2 & 1 \end{bmatrix} \begin{bmatrix} 5 \\ -1 \end{bmatrix} \begin{bmatrix} 11 \\ -11 \end{bmatrix}$

2. $\begin{bmatrix} -1 & 11 \\ 2 & -6 \end{bmatrix} \begin{bmatrix} -6 \\ 1 \end{bmatrix} \begin{bmatrix} 17 \\ -18 \end{bmatrix}$

3. $\begin{bmatrix} 6 & 0 & -4 \\ -7 & -1 & 2 \end{bmatrix} \begin{bmatrix} 2 \\ -5 \\ 3 \end{bmatrix} \begin{bmatrix} 0 \\ -3 \end{bmatrix}$

4. $\begin{bmatrix} 8 & -2 \\ 5 & 0 \\ -6 & 1 \end{bmatrix} \begin{bmatrix} -2 \\ -9 \end{bmatrix} \begin{bmatrix} 2 \\ -10 \\ 3 \end{bmatrix}$

5. $\begin{bmatrix} -1 & 5 \\ 3 & 2 \end{bmatrix} \begin{bmatrix} 2 & -2 \\ 0 & 1 \end{bmatrix} \begin{bmatrix} -2 & 7 \\ 6 & -4 \end{bmatrix}$

6. $\begin{bmatrix} -5 & 3 \\ 4 & 2 \end{bmatrix} \begin{bmatrix} 0 & 3 \\ -2 & 7 \end{bmatrix} \begin{bmatrix} -6 & 6 \\ -4 & 26 \end{bmatrix}$

7. $\begin{bmatrix} 4 & 3 \\ -7 & -5 \end{bmatrix} \begin{bmatrix} -5 & -3 \\ 7 & 0 \end{bmatrix} \begin{bmatrix} 1 & -12 \\ 0 & 21 \end{bmatrix}$

8. $\begin{bmatrix} 6 \\ 1 \end{bmatrix} \begin{bmatrix} -1 & 0 \\ 2 & 3 \end{bmatrix}$ not defined

9. $\begin{bmatrix} 0 \\ -2 \end{bmatrix} \begin{bmatrix} 5 \\ 1 \end{bmatrix}$ not defined

10. $\begin{bmatrix} 1 & 0 \\ 0 & 1 \end{bmatrix} \begin{bmatrix} -5 & 2 \\ 3 & 6 \end{bmatrix} \begin{bmatrix} -5 & 2 \\ 3 & 6 \end{bmatrix}$

11. $\begin{bmatrix} 1 & 6 \\ 2 & -4 \\ 0 & 3 \end{bmatrix} \begin{bmatrix} 7 & 0 & 4 \\ -2 & 1 & 1 \end{bmatrix} \begin{bmatrix} -5 & 6 & 10 \\ 22 & -4 & 4 \\ -6 & 3 & 3 \end{bmatrix}$

12. $\begin{bmatrix} 7 & -2 & 5 \\ 4 & -1 & -6 \end{bmatrix} \begin{bmatrix} 2 & -1 & 0 \\ 0 & 6 & -3 \\ 3 & 5 & 2 \end{bmatrix} \begin{bmatrix} 29 & 6 & 16 \\ -10 & -40 & -9 \end{bmatrix}$

For $A = \begin{bmatrix} 5 & -3 \\ 4 & 1 \end{bmatrix}$ and $B = \begin{bmatrix} -2 & 3 \\ -1 & 2 \end{bmatrix}$, find a 2 × 2 matrix equal to the given product.

13. AB **14.** BA **15.** B^2 **16.** A^2 **17.** $(A + B)^2$ **18.** $(A + B)(A - B)$

19. Show that if $A = \begin{bmatrix} a & 0 \\ 0 & a \end{bmatrix}$ and $B = \begin{bmatrix} x & y \\ z & w \end{bmatrix}$, then $AB = BA$.

20. Show that if $A = \begin{bmatrix} a & 0 \\ 0 & b \end{bmatrix}$ and $B = \begin{bmatrix} x & 0 \\ 0 & y \end{bmatrix}$, then $AB = BA$.

Matrices **541**

---

20. $AB = \begin{bmatrix} ax + 0 & 0a + 0y \\ 0x + 0b & 0 + by \end{bmatrix}$

$= \begin{bmatrix} ax & 0 \\ 0 & by \end{bmatrix}$

$BA = \begin{bmatrix} ax + 0 & 0x + 0b \\ 0a + 0y & 0 + by \end{bmatrix}$

$= \begin{bmatrix} ax & 0 \\ 0 & by \end{bmatrix} = AB$

27. $AB = \begin{bmatrix} a & b \\ c & d \end{bmatrix} \begin{bmatrix} d & -b \\ -c & a \end{bmatrix} =$

$\begin{bmatrix} ad + (-c)b & a(-b) + ab \\ cd + (-c)d & (-b)c + ad \end{bmatrix}$

$= \begin{bmatrix} ad - bc & 0 \\ 0 & ad - bc \end{bmatrix}$

which is in the form $\begin{bmatrix} x & 0 \\ 0 & x \end{bmatrix}$.

28. $BA = \begin{bmatrix} d & -b \\ -c & a \end{bmatrix} \begin{bmatrix} a & b \\ c & d \end{bmatrix} =$

$\begin{bmatrix} ad - bc & bd - bd \\ -ac + ac & -bc + ad \end{bmatrix}$

$= \begin{bmatrix} ad - bc & 0 \\ 0 & ad - bc \end{bmatrix}$

$\therefore AB = BA$

29. $AB =$

$\begin{bmatrix} 17 & 21 \\ 2 & 4 \end{bmatrix} \begin{bmatrix} 20 & 15 & 10 \\ 30 & 40 & 50 \end{bmatrix} =$

$\begin{bmatrix} 970 & 1095 & 1220 \\ 160 & 190 & 220 \end{bmatrix}$

The entries in the top row represent the number of jewels used in producing both models on Monday, Tuesday, and Wednesday, respectively. The entries in the second row represent the number of straps used in producing both models on Monday, Tuesday, and Wednesday, respectively.

30. $ABC =$

$\begin{bmatrix} 125 & 15 \end{bmatrix} \begin{bmatrix} 4 & 0.6 \\ 8 & 1.0 \end{bmatrix} \begin{bmatrix} 12.50 \\ 14.00 \end{bmatrix}$

$= \begin{bmatrix} 620 & 90 \end{bmatrix} \begin{bmatrix} 12.50 \\ 14.00 \end{bmatrix} =$

$\begin{bmatrix} 9010 \end{bmatrix}$

The entry in ABC represents the total labor cost of producing the tennis rackets.

Solve for x and y, or for x, y, and z.

B 21. $\begin{bmatrix} 3 & 0 \\ -3 & 1 \end{bmatrix}\begin{bmatrix} x \\ y \end{bmatrix} = \begin{bmatrix} 6 \\ -7 \end{bmatrix}$ $x = 2, y = -1$ 22. $\begin{bmatrix} 4 & -7 \\ 2 & -3 \end{bmatrix}\begin{bmatrix} x \\ y \end{bmatrix} = \begin{bmatrix} -3 \\ 1 \end{bmatrix}$ $\begin{array}{l} x = 8 \\ y = 5 \end{array}$

23. $\begin{bmatrix} -5 & -2 \\ 4 & 3 \end{bmatrix}\begin{bmatrix} x \\ y \end{bmatrix} = \begin{bmatrix} -8 \\ 12 \end{bmatrix}$ $x = 0, y = 4$ 24. $[x \quad y]\begin{bmatrix} 4 & -5 \\ -2 & 6 \end{bmatrix} = [2 \quad 15]$ $x = 3, y = 5$

25. $\begin{bmatrix} -1 & 0 & 1 \\ 1 & -1 & 0 \\ 0 & 1 & 1 \end{bmatrix}\begin{bmatrix} x \\ y \\ z \end{bmatrix} = \begin{bmatrix} -4 \\ 2 \\ 4 \end{bmatrix}$ 26. $\begin{bmatrix} 2 & 0 & -1 \\ 0 & 3 & 0 \\ 1 & 1 & 0 \end{bmatrix}\begin{bmatrix} x \\ y \\ z \end{bmatrix} = \begin{bmatrix} 6 \\ 9 \\ 5 \end{bmatrix}$

$x = 5, y = 3, z = 1$ $\qquad$ $x = 2, y = 3, z = -2$

In Exercises 27 and 28, let $A = \begin{bmatrix} a & b \\ c & d \end{bmatrix}$ **and** $B = \begin{bmatrix} d & -b \\ -c & a \end{bmatrix}$.

27. Show that AB is in the form $\begin{bmatrix} x & 0 \\ 0 & x \end{bmatrix}$.

28. What is the relationship between AB and BA? $AB = BA$

C 29. Let the matrix $A = \begin{bmatrix} 17 & 21 \\ 2 & 4 \end{bmatrix}$ represent the fact that wristwatch Model I has 17 jewels and 2 straps, while Model II has 21 jewels and 4 straps. Let the matrix $B = \begin{bmatrix} 20 & 15 & 10 \\ 30 & 40 & 50 \end{bmatrix}$ represent the fact that a factory produced 20 Model I wristwatches on Monday, 15 on Tuesday, and 10 on Wednesday, and on the same days produced 30, 40, and 50 sets of the Model II wristwatch, respectively. Simplify the product AB and tell what its entries represent.

30. Let the matrix $A = [125 \quad 15]$ represent the fact that a sporting-goods company produces 125 standard and 15 deluxe tennis rackets each week. Let the matrix $B = \begin{bmatrix} 4 & 0.6 \\ 8 & 1.0 \end{bmatrix}$ represent the fact that the standard racket requires 4 h of labor for assembly and 0.6 h of labor for lamination whereas the deluxe model requires 8 h of labor for assembly and 1 h of labor for lamination. Let the matrix $C = \begin{bmatrix} 12.50 \\ 14.00 \end{bmatrix}$ represent the fact that the costs per hour of labor are \$12.50 for assembly and \$14 for lamination. Simplify the product ABC and specify what each entry represents.

Computer Exercises For students with computer experience

1. Write a program that will compute A^2, A^3, and A^4 when the entries of a square matrix A are input.

Run the program in Exercise 1 for each matrix.

2. $\begin{bmatrix} 3 & -2 \\ 1 & 0 \end{bmatrix}$ 3. $\begin{bmatrix} -1 & -4 \\ 5 & 8 \end{bmatrix}$ 4. $\begin{bmatrix} 1 & -1 & -3 \\ -5 & 2 & -2 \\ 3 & 4 & 0 \end{bmatrix}$ 5. $\begin{bmatrix} 1 & 0 & 0 \\ 0 & 1 & 0 \\ 0 & 0 & 1 \end{bmatrix}$

542 *Chapter 13*

13-5 Properties of Matrix Multiplication

Although multiplication of matrices with real-number entries has some of the properties of multiplication of real numbers, there are some important differences. We shall illustrate these by considering multiplication in the set $S_{2\times 2}$.

EXAMPLE 1 For $A = \begin{bmatrix} 6 & 4 \\ -9 & -6 \end{bmatrix}$ and $B = \begin{bmatrix} -2 & -4 \\ 3 & 6 \end{bmatrix}$, find:

    a. AB **b.** BA

SOLUTION   **a.** $AB = \begin{bmatrix} 6 & 4 \\ -9 & -6 \end{bmatrix}\begin{bmatrix} -2 & -4 \\ 3 & 6 \end{bmatrix} = \begin{bmatrix} -12+12 & -24+24 \\ 18-18 & 36-36 \end{bmatrix} = \begin{bmatrix} 0 & 0 \\ 0 & 0 \end{bmatrix}$

             b. $BA = \begin{bmatrix} -2 & -4 \\ 3 & 6 \end{bmatrix}\begin{bmatrix} 6 & 4 \\ -9 & -6 \end{bmatrix} = \begin{bmatrix} -12+36 & -8+24 \\ 18-54 & 12-36 \end{bmatrix}$

                                 $= \begin{bmatrix} 24 & 16 \\ -36 & -24 \end{bmatrix}$

Notice in Example 1 (a) that $A \neq O$ and $B \neq O$ but $AB = O$. Thus, in $S_{2\times 2}$ the product of two matrices can be the zero matrix without either factor being the zero matrix!

Notice also in Example 1 that $BA \neq AB$. Therefore, multiplication in $S_{2\times 2}$ *is not commutative*. For this reason, you must be careful to specify the order of the factors in matrix multiplication. To specify the product AB, for example, you say that you **right-multiply** A by B or that you **left-multiply** B by A. Some special matrix products, however, do not depend on the order of the factors.

EXAMPLE 2 If $I_{2\times 2} = \begin{bmatrix} 1 & 0 \\ 0 & 1 \end{bmatrix}$ and $A = \begin{bmatrix} a_1 & b_1 \\ a_2 & b_2 \end{bmatrix}$, simplify:

    a. $I_{2\times 2}A$ **b.** $AI_{2\times 2}$

SOLUTION   **a.** $I_{2\times 2}A = \begin{bmatrix} 1 & 0 \\ 0 & 1 \end{bmatrix}\begin{bmatrix} a_1 & b_1 \\ a_2 & b_2 \end{bmatrix} = \begin{bmatrix} a_1+0 & b_1+0 \\ 0+a_2 & 0+b_2 \end{bmatrix} = \begin{bmatrix} a_1 & b_1 \\ a_2 & b_2 \end{bmatrix}$

             b. $AI_{2\times 2} = \begin{bmatrix} a_1 & b_1 \\ a_2 & b_2 \end{bmatrix}\begin{bmatrix} 1 & 0 \\ 0 & 1 \end{bmatrix} = \begin{bmatrix} a_1+0 & 0+b_1 \\ a_2+0 & 0+b_2 \end{bmatrix} = \begin{bmatrix} a_1 & b_1 \\ a_2 & b_2 \end{bmatrix}$

As Example 2 suggests, not only do the products of *some* matrices AB in $S_{2\times 2}$ commute, but also the product of any 2×2 matrix A and

$$I_{2\times 2} = \begin{bmatrix} 1 & 0 \\ 0 & 1 \end{bmatrix}$$

always is the matrix A. Thus, $I_{2\times 2}$ is an **identity matrix for multiplication** in $S_{2\times 2}$ and

$$I_{2\times 2}A = AI_{2\times 2} = A.$$

Teaching Suggestions
p. T115

Supplementary Material

Test 40

Key Ideas

Use properties of matrix multiplication.

Chalkboard Examples

1. For $A = \begin{bmatrix} 3 & -1 \\ 2 & 4 \end{bmatrix}$ and $B = \begin{bmatrix} -1 & 5 \\ 2 & -3 \end{bmatrix}$ find:

  a. $(A + B)^2$
  b. $A^2 + 2AB + B^2$

  a. $A + B =$

$$\begin{bmatrix} 3 & -1 \\ 2 & 4 \end{bmatrix} + \begin{bmatrix} -1 & 5 \\ 2 & -3 \end{bmatrix}$$
$$= \begin{bmatrix} 2 & 4 \\ 4 & 1 \end{bmatrix}$$

  Then

$$(A + B)^2 = \begin{bmatrix} 2 & 4 \\ 4 & 1 \end{bmatrix}\begin{bmatrix} 2 & 4 \\ 4 & 1 \end{bmatrix}$$
$$= \begin{bmatrix} 20 & 12 \\ 12 & 17 \end{bmatrix}$$

  b. $A^2 = \begin{bmatrix} 7 & -7 \\ 14 & 14 \end{bmatrix}$

$$B^2 = \begin{bmatrix} 11 & -20 \\ -8 & 19 \end{bmatrix}$$

$$2AB = 2\begin{bmatrix} 3 & -1 \\ 2 & 4 \end{bmatrix}\begin{bmatrix} -1 & 5 \\ 2 & -3 \end{bmatrix}$$
$$= 2\begin{bmatrix} -3-2 & 15+3 \\ -2+8 & 10-12 \end{bmatrix}$$
$$= 2\begin{bmatrix} -5 & 18 \\ 6 & -2 \end{bmatrix} = \begin{bmatrix} -10 & 36 \\ 12 & -4 \end{bmatrix}$$

  so that $A^2 + 2AB + B^2 =$

$$\begin{bmatrix} 7 & -7 \\ 14 & 14 \end{bmatrix} + \begin{bmatrix} -10 & 36 \\ 12 & -4 \end{bmatrix} +$$
$$\begin{bmatrix} 11 & -20 \\ -8 & 19 \end{bmatrix} = \begin{bmatrix} 8 & 9 \\ 18 & 29 \end{bmatrix}$$

Reading Algebra

Read the matrix product *AB* as "matrix *A* times matrix *B*" to emphasize the order in which the matrices are placed. Make sure the students read it correctly so that it will help them remember that this is matrix multiplication, which is *not* commutative. Read A^{-1} as the "inverse of matrix *A*".

Mixed Review

1. Use Cramer's Rule to solve the given system.

$4x - 5y = 2$

$8x + 10y = 4$ $\quad \left\{ \left(\dfrac{1}{2}, 0 \right) \right\}$

2. Graph the solution set of the system:

$2 \le y < 5$

$y \ge -x - 3$

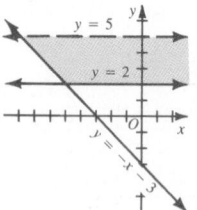

3. Evaluate $\begin{vmatrix} 1 & 3 & 6 \\ -2 & 2 & -4 \\ -1 & 5 & 2 \end{vmatrix}$. $\quad 0$

4. Simplify

$(2x + 2)(4x^2 - 5x + 6)$.

$8x^3 - 2x^2 + 2x + 12$

5. Expand $(x + y)^2$.

$x^2 + 2xy + y^2$

Suggested Assignments

Enriched Alg.
First day
    545/5–17 odd, 19–27
Second day
R 546/Self–Test 1
S 531/2–16 even
    541/8–20 even

544

This property and some other properties of multiplication of 2×2 matrices are listed in the following theorem.

Theorem. If $A \in \mathcal{S}_{2 \times 2}$, $B \in \mathcal{S}_{2 \times 2}$, $C \in \mathcal{S}_{2 \times 2}$, and $a \in \mathcal{R}$, then:

I. $AB \in \mathcal{S}_{2 \times 2}$ $\qquad\qquad\qquad$ **Closure Property**

II. $(AB)C = A(BC)$ $\qquad\qquad$ **Associative Property**

III. $A(B + C) = AB + AC$ $\qquad$ **Distributive Properties**

IV. $(B + C)A = BA + CA$

V. $I_{2 \times 2}A = AI_{2 \times 2} = A$ $\qquad$ **Identity Property for Multiplication**

VI. $a(AB) = (aA)B = A(aB)$ $\quad$ **Associative Property for Scalar Multiplication**

VII. $O_{2 \times 2}A = AO_{2 \times 2} = O_{2 \times 2}$ $\quad$ **Multiplicative Property of the Zero Matrix**

EXAMPLE 3 Show that if $A = \begin{bmatrix} 6 & 4 \\ -9 & -6 \end{bmatrix}$ and $B = \begin{bmatrix} -2 & -4 \\ 3 & 6 \end{bmatrix}$, then:

a. $(A + B)(A - B) \ne A^2 - B^2$

b. $(A + B)(A - B) = A^2 - AB + BA - B^2$.

SOLUTION **a.** First:

$$A + B = \begin{bmatrix} 6 & 4 \\ -9 & -6 \end{bmatrix} + \begin{bmatrix} -2 & -4 \\ 3 & 6 \end{bmatrix} = \begin{bmatrix} 4 & 0 \\ -6 & 0 \end{bmatrix}$$

$$A - B = \begin{bmatrix} 6 & 4 \\ -9 & -6 \end{bmatrix} - \begin{bmatrix} -2 & -4 \\ 3 & 6 \end{bmatrix} = \begin{bmatrix} 8 & 8 \\ -12 & -12 \end{bmatrix}$$

$$(A + B)(A - B) = \begin{bmatrix} 4 & 0 \\ -6 & 0 \end{bmatrix}\begin{bmatrix} 8 & 8 \\ -12 & -12 \end{bmatrix} = \begin{bmatrix} 32 & 32 \\ -48 & -48 \end{bmatrix}$$

Next:

$$A^2 = \begin{bmatrix} 6 & 4 \\ -9 & -6 \end{bmatrix}\begin{bmatrix} 6 & 4 \\ -9 & -6 \end{bmatrix}$$

$$= \begin{bmatrix} 0 & 0 \\ 0 & 0 \end{bmatrix}$$

$$B^2 = \begin{bmatrix} -2 & -4 \\ 3 & 6 \end{bmatrix}\begin{bmatrix} -2 & -4 \\ 3 & 6 \end{bmatrix}$$

$$= \begin{bmatrix} -8 & -16 \\ 12 & 24 \end{bmatrix}$$

$$A^2 - B^2 = \begin{bmatrix} 0 & 0 \\ 0 & 0 \end{bmatrix} - \begin{bmatrix} -8 & -16 \\ 12 & 24 \end{bmatrix}$$

$$= \begin{bmatrix} 8 & 16 \\ -12 & -24 \end{bmatrix}$$

Since $\begin{bmatrix} 32 & 32 \\ -48 & -48 \end{bmatrix} \ne \begin{bmatrix} 8 & 16 \\ -12 & -24 \end{bmatrix}$, $(A + B)(A - B) \ne A^2 - B^2$.

b. From Example 1 on page 543,

$$-AB + BA = -\begin{bmatrix} 0 & 0 \\ 0 & 0 \end{bmatrix} + \begin{bmatrix} 24 & 16 \\ -36 & -24 \end{bmatrix} = \begin{bmatrix} 24 & 16 \\ -36 & -24 \end{bmatrix}$$

so that $A^2 - AB + BA - B^2 = A^2 - B^2 - AB + BA$

$$= \begin{bmatrix} 8 & 16 \\ -12 & -24 \end{bmatrix} + \begin{bmatrix} 24 & 16 \\ -36 & -24 \end{bmatrix} = \begin{bmatrix} 32 & 32 \\ -48 & -48 \end{bmatrix}$$

Also, since $(A + B)(A - B) = \begin{bmatrix} 32 & 32 \\ -48 & -48 \end{bmatrix}$,

$$(A + B)(A - B) = A^2 - AB + BA - B^2.$$

To understand the results in Example 3, notice that by the distributive property you have

$$(A + B)(A - B) = A(A - B) + B(A - B) = A^2 - AB + BA - B^2.$$

Oral Exercises

In Exercises 1–5, let A be any 2×2 matrix. State a simpler name for each of these.

1. $AI_{2\times 2}$ A **2.** $A + O_{2\times 2}$ A **3.** $A - A$ $O_{2\times 2}$ **4.** $A^2 O_{2\times 2}$ $O_{2\times 2}$ **5.** $I^2_{2\times 2}A$ A

In Exercises 6–10, let $A = \begin{bmatrix} 0 & -1 \\ 1 & 0 \end{bmatrix}$, $B = \begin{bmatrix} 0 & 1 \\ 1 & 0 \end{bmatrix}$, and $C = \begin{bmatrix} 0 & 1 \\ -1 & 0 \end{bmatrix}$

Find each product.

6. AB $\begin{bmatrix} -1 & 0 \\ 0 & 1 \end{bmatrix}$ **7.** BA $\begin{bmatrix} 1 & 0 \\ 0 & -1 \end{bmatrix}$ **8.** AC $\begin{bmatrix} 1 & 0 \\ 0 & 1 \end{bmatrix}$ **9.** CA $\begin{bmatrix} 1 & 0 \\ 0 & 1 \end{bmatrix}$ **10.** CB $\begin{bmatrix} 1 & 0 \\ 0 & -1 \end{bmatrix}$

11. In Exercises 6–10, which, if any, of the products AB, BA, AC, CA, and CB are equal? $BA = CB$, $AC = CA$

Written Exercises

In Exercises 1–10 let $A = \begin{bmatrix} 4 & 2 \\ -8 & -4 \end{bmatrix}$, $B = \begin{bmatrix} -1 & 3 \\ 2 & -6 \end{bmatrix}$, $C = \begin{bmatrix} 8 & 2 \\ 4 & 6 \end{bmatrix}$, and $D = \begin{bmatrix} 3 & -1 \\ -2 & 4 \end{bmatrix}$. Compute the value of the expressions and state whether or not they are equal.

A

1. AB and $BA \neq$ **2.** AC and $CA \neq$ **3.** AD and $DA \neq$

4. BC and $CB \neq$ **5.** CD and $DC =$ **6.** BD and $DB \neq$

7. $A(B + C)$ and $AB + AC =$ **8.** $A(BC)$ and $(AB)C =$

9. $B(AC)$ and $(BA)C =$ **10.** $B(A + C)$ and $BA + BC =$

---

1. $AB = \begin{bmatrix} 0 & 0 \\ 0 & 0 \end{bmatrix}$; $BA = \begin{bmatrix} -28 & -14 \\ 56 & 28 \end{bmatrix}$; $AB \neq BA$

2. $AC = \begin{bmatrix} 40 & 20 \\ -80 & -40 \end{bmatrix}$; $CA = \begin{bmatrix} 16 & 8 \\ 32 & -16 \end{bmatrix}$; $AC \neq CA$

3. $AD = \begin{bmatrix} 8 & 4 \\ -16 & -8 \end{bmatrix}$; $DA = \begin{bmatrix} 20 & 10 \\ -40 & -20 \end{bmatrix}$; $AD \neq DA$

4. $BC = \begin{bmatrix} 4 & 16 \\ -8 & -32 \end{bmatrix}$; $CB = \begin{bmatrix} -4 & 12 \\ 8 & -24 \end{bmatrix}$; $BC \neq CB$

5. $CD = DC = \begin{bmatrix} 20 & 0 \\ 0 & 20 \end{bmatrix}$

6. $BD = \begin{bmatrix} -9 & 13 \\ 18 & -26 \end{bmatrix}$; $DB = \begin{bmatrix} -5 & 15 \\ 10 & -30 \end{bmatrix}$; $BD \neq DC$

7. $A(B + C) = AB + AC = \begin{bmatrix} 40 & 20 \\ -80 & -40 \end{bmatrix}$

8. $A(BC) = (AB)C = \begin{bmatrix} 0 & 0 \\ 0 & 0 \end{bmatrix}$

9. $B(AC) = (BA)C = \begin{bmatrix} -280 & -140 \\ 560 & 280 \end{bmatrix}$

10. $B(A + C) = BA + BC = \begin{bmatrix} -24 & 2 \\ 48 & -4 \end{bmatrix}$

11. $A(BC) = \begin{bmatrix} 0 & 0 \\ 0 & 0 \end{bmatrix}$; $(BC)A = \begin{bmatrix} -112 & -56 \\ 224 & 112 \end{bmatrix}$; $A(BC) \neq (BC)A$

12. $A(CD) = (CD)A = \begin{bmatrix} 80 & 40 \\ -160 & -80 \end{bmatrix}$

13. $C(A + B) = \begin{bmatrix} 12 & 20 \\ -24 & -40 \end{bmatrix}$; $(A + B)C = \begin{bmatrix} 44 & 36 \\ -88 & -72 \end{bmatrix}$; $C(A + B) \neq (A + B)C$

14. $(AB)(CD) = (CD)(AB) = \begin{bmatrix} 0 & 0 \\ 0 & 0 \end{bmatrix}$

Additional A Exercises

In Exercises 1–3 let

$A = \begin{bmatrix} 2 & 1 \\ -6 & -3 \end{bmatrix}$, $B = \begin{bmatrix} -1 & 4 \\ 2 & -8 \end{bmatrix}$,

$C = \begin{bmatrix} 6 & 2 \\ 10 & 4 \end{bmatrix}$, $D = \begin{bmatrix} 2 & -1 \\ -5 & 3 \end{bmatrix}$.

Compute the value of the expressions and state whether or not they are equal.

1. AB and BA no

$AB = \begin{bmatrix} 0 & 0 \\ 0 & 0 \end{bmatrix}$; $BA =$

$\begin{bmatrix} -26 & -13 \\ 52 & 26 \end{bmatrix}$

2. AD and DA no

$AD = \begin{bmatrix} -1 & 1 \\ 3 & -3 \end{bmatrix}$; $DA =$

$\begin{bmatrix} 10 & 5 \\ -28 & -14 \end{bmatrix}$

3. $A(B + C)$ and $AB + AC$ yes

$A(B + C) = \begin{bmatrix} 22 & 8 \\ -66 & -24 \end{bmatrix}$;

$AB + AC = \begin{bmatrix} 22 & 8 \\ -66 & -24 \end{bmatrix}$

4. $A(CD)$ and $(CD)A$ yes

$A(CD) = \begin{bmatrix} 4 & 2 \\ -12 & -6 \end{bmatrix}$;

$(CD)A = \begin{bmatrix} 4 & 2 \\ -12 & -6 \end{bmatrix}$

5. $C(A + B)$ and $(A + B)C$

no

$C(A + B) = \begin{bmatrix} -2 & 8 \\ -6 & 6 \end{bmatrix}$;

$(A + B)C = \begin{bmatrix} 56 & 22 \\ -134 & -52 \end{bmatrix}$

6. $(AB)(CD)$ and $(CD)(AB)$

yes

$(AB)(CD) = \begin{bmatrix} 0 & 0 \\ 0 & 0 \end{bmatrix}$;

$(CD)(AB) = \begin{bmatrix} 0 & 0 \\ 0 & 0 \end{bmatrix}$

Quick Quiz

Find the indicated sum or difference.

1. $\begin{bmatrix} 2 & -4 \\ -5 & 7 \end{bmatrix} + \begin{bmatrix} -3 & 1 \\ -8 & 2 \end{bmatrix}$

$\begin{bmatrix} -1 & -3 \\ -13 & 9 \end{bmatrix}$

In Exercises 11–18 let $A = \begin{bmatrix} 4 & 2 \\ -8 & -4 \end{bmatrix}$, $B = \begin{bmatrix} -1 & 3 \\ 2 & -6 \end{bmatrix}$, $C = \begin{bmatrix} 8 & 2 \\ 4 & 6 \end{bmatrix}$, and

$D = \begin{bmatrix} 3 & -1 \\ -2 & 4 \end{bmatrix}$. Compute the value of the expressions and state whether or not they are equal.

11. $A(BC)$ and $(BC)A$ $\neq$

12. $A(CD)$ and $(CD)A$ $=$

13. $C(A + B)$ and $(A + B)C$ $\neq$

14. $(AB)(CD)$ and $(CD)(AB)$ $=$

Compute each of the three expressions and tell which of the three are the same. **15.** a and c   **16.** a and b   **17.** a and b   **18.** a, b, and c

B **15. a.** $(A + B)^2$ **b.** $A^2 + 2AB + B^2$ **c.** $A^2 + AB + BA + B^2$

16. a. $(A - B)^2$ **b.** $A^2 - AB - BA + B^2$ **c.** $A^2 - 2AB + B^2$

17. a. $(A + B)(C + D)$ **b.** $AC + BC + AD + BD$ **c.** $C(A + B) + D(A + B)$

18. a. $(A - B)(C + D)$ **b.** $AC - BC - BD + AD$ **c.** $A(C + D) - B(C + D)$

Prove each of the following for all 2×2 matrices A, B, and C, and all real numbers a.

19. $AB \in \mathcal{S}_{2 \times 2}$ **20.** $(AB)C = A(BC)$

21. $A(B + C) = AB + AC$ **22.** $(B + C)A = BA + CA$

23. $a(AB) = (aA)B$ **24.** $a(AB) = A(aB)$

25. $O_{2 \times 2}A = O_{2 \times 2}$ **26.** $AO_{2 \times 2} = O_{2 \times 2}$

C **27.** Let $|A|$ denote the determinant of A. Show that for all 2×2 matrices A and B and all real numbers c:

     a. $|cA| = c^2|A|$ **b.** $|AB| = |A| \cdot |B|$

▮ Self-Test 1

VOCABULARY   matrix (p. 527)

                  entry of a matrix (p. 527)

                  dimensions of a matrix (p. 527)

                  zero matrix (p. 529)

                  identity matrix for addition (p. 529)

                  opposite of a matrix (p. 529)

                  scalar (p. 535)

                  product of a scalar and a matrix (p. 535)

                  product of two matrices (p. 538)

                  right- or left-multiplication of a matrix (p. 543)

                  identity matrix for multiplication (p. 543)

Find the indicated sum or difference.

1. $\begin{bmatrix} 3 & 6 \\ 4 & -2 \end{bmatrix} + \begin{bmatrix} 6 & 7 \\ -1 & 9 \end{bmatrix} \begin{bmatrix} 9 & 13 \\ 3 & 7 \end{bmatrix}$ 2. $\begin{bmatrix} 0 & 3 & -4 \\ -1 & 7 & 8 \end{bmatrix} - \begin{bmatrix} 7 & -4 & 0 \\ 5 & 1 & -3 \end{bmatrix}$

3. Simplify $5A - 4B$ if $A = \begin{bmatrix} -4 & -1 \\ 8 & 5 \end{bmatrix}$ and $B = \begin{bmatrix} -4 & 2 \\ 4 & 7 \end{bmatrix} . \begin{bmatrix} -4 & -13 \\ 24 & -3 \end{bmatrix}$

$\begin{bmatrix} -7 & 7 & -4 \\ -6 & 6 & 11 \end{bmatrix}$ *Obj. 1, p. 527*

Obj. 2, p. 527

Using the matrices A and B in Exercise 3, find the 2 × 2 matrix X that satisfies the given equation.

4. $X - B = A$ $\begin{bmatrix} -8 & 1 \\ 12 & 12 \end{bmatrix}$ 5. $2X + 5B = 3A$ $\begin{bmatrix} 4 & -\frac{13}{2} \\ 2 & -10 \end{bmatrix}$ *Obj. 3, p. 527*

Find the indicated product.

6. $\begin{bmatrix} 1 & 5 \\ -6 & 0 \end{bmatrix} \begin{bmatrix} 9 & 2 \\ -1 & -3 \end{bmatrix} \begin{bmatrix} 4 & -13 \\ -54 & -12 \end{bmatrix}$ 7. $\begin{bmatrix} 1 & 2 & -4 \\ 6 & 0 & 3 \end{bmatrix} \begin{bmatrix} 5 \\ 0 \\ 1 \end{bmatrix} \begin{bmatrix} 1 \\ 33 \end{bmatrix}$ *Obj. 4, p. 527*

8. Use the matrices A and B in Exercise 3 to compute AB and BA. Then state whether or not AB and BA are equal. $AB = \begin{bmatrix} 12 & -15 \\ -12 & 51 \end{bmatrix}$, $BA = \begin{bmatrix} 32 & 14 \\ 40 & 31 \end{bmatrix}$; $AB \neq BA$

Check your answers with those at the back of the book.

Matrices and Linear Systems

OBJECTIVES for Section 13-6:
1. To determine the inverse of a nonsingular square matrix of order 2.
2. To use the inverse of a matrix to solve a matrix equation.
3. To write a linear system of equations in matrix form.
4. To use a matrix equation to solve a linear system.

13–6 Matrix Solution of a Linear System

Every nonzero real number r has a multiplicative inverse. That is, for each $r \in \mathcal{R}$, $r \neq 0$, there is a number $r^{-1} \in \mathcal{R}$ for which

$$rr^{-1} = 1 \quad \text{and} \quad r^{-1}r = 1.$$

Does every nonzero 2 × 2 matrix A with real-number entries also have a multiplicative inverse? That is, for each $A \in \mathcal{S}_{2 \times 2}$, $A \neq 0$, is there a matrix $A^{-1} \in \mathcal{S}_{2 \times 2}$ for which

$$AA^{-1} = I \quad \text{and} \quad A^{-1}A = I?$$

2. $\begin{bmatrix} 0 & 5 \\ -1 & 6 \end{bmatrix} - \begin{bmatrix} 3 & -7 \\ -2 & 1 \end{bmatrix}$

$\begin{bmatrix} -3 & 12 \\ 1 & 5 \end{bmatrix}$

3. Simplify $3A - 2B$ if

$A = \begin{bmatrix} 5 & -2 \\ -1 & 3 \end{bmatrix}$ and $B =$

$\begin{bmatrix} 4 & 2 \\ -5 & -3 \end{bmatrix} . \begin{bmatrix} 7 & -10 \\ 7 & 15 \end{bmatrix}$

Using the matrices A and B in Quiz Item 3, find the 2 × 2 matrix X that satisfies the given equation.

4. $X - A = B$ $\begin{bmatrix} 9 & 0 \\ -6 & 0 \end{bmatrix}$

5. $3X + 4B = 5A$ $\begin{bmatrix} 3 & -6 \\ 5 & 9 \end{bmatrix}$

6. Find the given product.
$\begin{bmatrix} -4 & 3 \\ 0 & -2 \end{bmatrix} \begin{bmatrix} 1 & -3 \\ 2 & -4 \end{bmatrix} \begin{bmatrix} 2 & 0 \\ -4 & 8 \end{bmatrix}$

7. $\begin{bmatrix} 2 & 1 & -3 \\ 5 & 0 & 4 \end{bmatrix} \begin{bmatrix} 6 \\ 0 \\ 1 \end{bmatrix} \begin{bmatrix} 9 \\ 34 \end{bmatrix}$

8. Use the matrices A and B in Quiz Item 3 to compute AB and BA. Then state whether or not AB and BA are equal.

$AB = \begin{bmatrix} 30 & 16 \\ -19 & -11 \end{bmatrix}$

$BA = \begin{bmatrix} 18 & -2 \\ -22 & 1 \end{bmatrix}$

$AB \neq BA$

Teaching Suggestions p. T115

Related Activities p. T116

Matrices **547**

Key Ideas

Use the inverse of a matrix to solve a matrix equation. Use a matrix equation to solve a linear system in two variables.

Chalkboard Examples

1. If $A = \begin{bmatrix} 3 & 6 \\ 2 & 5 \end{bmatrix}$, find A^{-1}.

Note first that
$\det A = 3 \cdot 5 - 2 \cdot 6 = 3$,
so that $\det A \neq 0$. Then

$A^{-1} = \dfrac{1}{3} \begin{bmatrix} 5 & -6 \\ -2 & 3 \end{bmatrix}$

$= \begin{bmatrix} \frac{5}{3} & -2 \\ -\frac{2}{3} & 1 \end{bmatrix}$

2. Solve the following equation for the matrix X.

$\begin{bmatrix} 2 & -3 \\ 4 & -5 \end{bmatrix} X = \begin{bmatrix} 8 & -6 \\ 14 & -8 \end{bmatrix}$

First, find the inverse of $\begin{bmatrix} 2 & -3 \\ 4 & -5 \end{bmatrix}$ and note that its determinant is $-10 + 12$, or 2.

Then $\begin{bmatrix} 2 & -3 \\ 4 & -5 \end{bmatrix}^{-1} =$

$\dfrac{1}{2} \begin{bmatrix} -5 & 3 \\ -4 & 2 \end{bmatrix} = \begin{bmatrix} -\frac{5}{2} & \frac{3}{2} \\ -2 & 1 \end{bmatrix}$.

Next, left-multiply each member of the given equation by this inverse.

$X = \begin{bmatrix} 1 & 3 \\ -2 & 4 \end{bmatrix}$

You can readily see, for example, that the matrix $A = \begin{bmatrix} 3 & 0 \\ 0 & 3 \end{bmatrix}$ has a multiplicative inverse $A^{-1} = \begin{bmatrix} \frac{1}{3} & 0 \\ 0 & \frac{1}{3} \end{bmatrix}$, since

$$\begin{bmatrix} 3 & 0 \\ 0 & 3 \end{bmatrix}\begin{bmatrix} \frac{1}{3} & 0 \\ 0 & \frac{1}{3} \end{bmatrix} = \begin{bmatrix} 1 & 0 \\ 0 & 1 \end{bmatrix} \quad \text{and} \quad \begin{bmatrix} \frac{1}{3} & 0 \\ 0 & \frac{1}{3} \end{bmatrix}\begin{bmatrix} 3 & 0 \\ 0 & 3 \end{bmatrix} = \begin{bmatrix} 1 & 0 \\ 0 & 1 \end{bmatrix}.$$

In general, for 2×2 matrices, given $A = \begin{bmatrix} a_1 & b_1 \\ a_2 & b_2 \end{bmatrix}$, let us try to find $A^{-1} = \begin{bmatrix} x_1 & y_1 \\ x_2 & y_2 \end{bmatrix}$ such that

$$AA^{-1} = \begin{bmatrix} a_1 & b_1 \\ a_2 & b_2 \end{bmatrix}\begin{bmatrix} x_1 & y_1 \\ x_2 & y_2 \end{bmatrix} = \begin{bmatrix} a_1x_1 + b_1x_2 & a_1y_1 + b_1y_2 \\ a_2x_1 + b_2x_2 & a_2y_1 + b_2y_2 \end{bmatrix} = I_{2\times2} = \begin{bmatrix} 1 & 0 \\ 0 & 1 \end{bmatrix}.$$

This is true if and only if

$$a_1x_1 + b_1x_2 = 1, \quad a_1y_1 + b_1y_2 = 0,$$
$$a_2x_1 + b_2x_2 = 0, \quad a_2y_1 + b_2y_2 = 1.$$

Recall that in Chapter 4 we used determinants to solve systems of equations. If the determinant of coefficients in this system, $\begin{vmatrix} a_1 & b_1 \\ a_2 & b_2 \end{vmatrix}$, is not equal to zero, that is, if $a_1b_2 - a_2b_1 \neq 0$, then these equations can be solved for x_1, x_2, y_1, and y_2 to produce

$$x_1 = \frac{b_2}{a_1b_2 - a_2b_1}, \ y_1 = \frac{-b_1}{a_1b_2 - a_2b_1}, \text{ and } x_2 = \frac{-a_2}{a_1b_2 - a_2b_1}, \ y_2 = \frac{a_1}{a_1b_2 - a_2b_1}.$$

You can check that with these values for x_1, x_2, y_1, and y_2, not only do you have $AA^{-1} = I_{2\times2}$, but also (somewhat surprisingly, since products AB of matrices do not ordinarily commute) you have $A^{-1}A = I_{2\times2}$. (See Exercises 25 and 26, page 552.)

With each square matrix $\begin{bmatrix} a_1 & b_1 \\ a_2 & b_2 \end{bmatrix}$ we associate a particular real number, namely, the determinant $\begin{vmatrix} a_1 & b_1 \\ a_2 & b_2 \end{vmatrix}$, and we write

$$\det \begin{bmatrix} a_1 & b_1 \\ a_2 & b_2 \end{bmatrix} = \begin{vmatrix} a_1 & b_1 \\ a_2 & b_2 \end{vmatrix} = a_1b_2 - a_2b_1.$$

The pairing of each square matrix with its determinant constitutes a function, since associated with each such matrix is one and only one real number. The symbol "det A" (read "determinant of A") represents the element in the range of the function det associated with the matrix A in its domain.

Because each denominator in the preceding equations for x_1, x_2, y_1, and y_2 is det A, you can see that

$$A^{-1} = \frac{1}{\det A} \begin{bmatrix} b_2 & -b_1 \\ -a_2 & a_1 \end{bmatrix}.$$

548 *Chapter 13*

If det $A = 0$, then the equations for x_1, x_2, y_1, and y_2 have no solution and so such a matrix A (called a **singular matrix**) has no inverse (see Exercise 27, page 552). If det $A \neq 0$, then matrix A is said to be **nonsingular**, or **invertible**.

Notice that, to find A^{-1} from A, you interchange a_1 and b_2, replace a_2 and b_1 with their opposites, and multiply by the reciprocal of det A.

EXAMPLE 1 If $A = \begin{bmatrix} 2 & 3 \\ 4 & 7 \end{bmatrix}$, find A^{-1}.

SOLUTION Note first that det $A = 2 \times 7 - 4 \times 3 = 14 - 12 = 2$, so that det $A \neq 0$. Then

$$A^{-1} = \frac{1}{2}\begin{bmatrix} 7 & -3 \\ -4 & 2 \end{bmatrix} = \begin{bmatrix} \frac{7}{2} & -\frac{3}{2} \\ -2 & 1 \end{bmatrix}.$$

The check is left to you.

$$\therefore A^{-1} = \begin{bmatrix} \frac{7}{2} & -\frac{3}{2} \\ -2 & 1 \end{bmatrix}$$

Matrix equations in the form $AX = B$ may be solved using inverse matrices in the following way:

$$AX = B$$
$$A^{-1}AX = A^{-1}B$$
$$IX = A^{-1}B$$
$$X = A^{-1}B$$

EXAMPLE 2 Solve $\begin{bmatrix} 6 & -3 \\ 5 & -2 \end{bmatrix}X = \begin{bmatrix} 9 & -6 \\ 12 & -3 \end{bmatrix}$ for the matrix X.

SOLUTION 1. First, to find the inverse of $\begin{bmatrix} 6 & -3 \\ 5 & -2 \end{bmatrix}$, note that its determinant is $-12 + 15$, or 3. Then the inverse is

$$\begin{bmatrix} 6 & -3 \\ 5 & -2 \end{bmatrix}^{-1} = \frac{1}{3}\begin{bmatrix} -2 & 3 \\ -5 & 6 \end{bmatrix} = \begin{bmatrix} -\frac{2}{3} & 1 \\ -\frac{5}{3} & 2 \end{bmatrix}.$$

2. Next, left-multiply each side of the given equation by this inverse.

$$\begin{bmatrix} -\frac{2}{3} & 1 \\ -\frac{5}{3} & 2 \end{bmatrix}\begin{bmatrix} 6 & -3 \\ 5 & -2 \end{bmatrix}X = \begin{bmatrix} -\frac{2}{3} & 1 \\ -\frac{5}{3} & 2 \end{bmatrix}\begin{bmatrix} 9 & -6 \\ 12 & -3 \end{bmatrix}$$
$$\begin{bmatrix} 1 & 0 \\ 0 & 1 \end{bmatrix}X = \begin{bmatrix} 6 & 1 \\ 9 & 4 \end{bmatrix}$$
$$X = \begin{bmatrix} 6 & 1 \\ 9 & 4 \end{bmatrix}$$

The check is left to you.

$$\therefore X = \begin{bmatrix} 6 & 1 \\ 9 & 4 \end{bmatrix}$$

3. Use matrices to find the solution set of the system:
$$-x + 2y = -6$$
$$3x + 4y = 8$$

The matrix equation is:
$$\begin{bmatrix} -1 & 2 \\ 3 & 4 \end{bmatrix}\begin{bmatrix} x \\ y \end{bmatrix} = \begin{bmatrix} -6 \\ 8 \end{bmatrix}$$
Then find the inverse of the coefficient matrix:
$$\begin{bmatrix} -1 & 2 \\ 3 & 4 \end{bmatrix}^{-1} = -\frac{1}{10}\begin{bmatrix} 4 & -2 \\ -3 & -1 \end{bmatrix}$$
$$= \begin{bmatrix} -\frac{4}{10} & \frac{2}{10} \\ \frac{3}{10} & \frac{1}{10} \end{bmatrix}$$
Then: $\begin{bmatrix} x \\ y \end{bmatrix} = \begin{bmatrix} 4 \\ -1 \end{bmatrix}$
$x = 4$ and $y = -1$
$\therefore$ the solution set is
$\{(4, -1)\}$.

Additional A Exercises

Find the inverse of the given matrix if it is nonsingular. If the matrix is singular, so state.

1. $\begin{bmatrix} 5 & 3 \\ 3 & 2 \end{bmatrix}$ $\begin{bmatrix} 2 & -3 \\ -3 & 5 \end{bmatrix}$

2. $\begin{bmatrix} 3 & 2 \\ 9 & 6 \end{bmatrix}$ singular

3. $\begin{bmatrix} 0 & 1 \\ 1 & 0 \end{bmatrix}$ $\begin{bmatrix} 0 & 1 \\ 1 & 0 \end{bmatrix}$

Let $A = \begin{bmatrix} 2 & 3 \\ 1 & 2 \end{bmatrix}$, $B = \begin{bmatrix} 1 & 2 \\ 3 & 7 \end{bmatrix}$, $C = \begin{bmatrix} -2 & 0 \\ 0 & -2 \end{bmatrix}$, and $D = \begin{bmatrix} 4 & 3 \\ -2 & -2 \end{bmatrix}$.

Solve each of the following matrix equations for X.

4. $AX = B$ $\begin{bmatrix} -7 & -17 \\ 5 & 12 \end{bmatrix}$

5. $BX = D$ $\begin{bmatrix} 32 & 25 \\ -14 & -11 \end{bmatrix}$

6. $DX = C$ $\begin{bmatrix} -2 & -3 \\ 2 & 4 \end{bmatrix}$

Mixed Review

1. Determine the solution set of the equation.
$x^3 + 8x^2 = 20x$
$\{0, -10, 2\}$

2. Simplify $\dfrac{x^2 + 2}{x^2 - 3x} + \dfrac{2x}{2x - 6}$.

$\dfrac{2x^2 + 2}{x^2 - 3x}$

3. Solve $\dfrac{8}{9 - x^2} - 2 = \dfrac{x}{3 - x}$.

$\{-2, 5\}$

4. Find the solution set for this system of equations.
$3x + y = 8$
$2x + y = 5$ $\{(3, -1)\}$

5. Evaluate $\begin{vmatrix} 3 & 6 \\ 2 & 4 \end{vmatrix}$. 0

We shall now consider a method of solving the system

$$a_1x + b_1y = c_1$$
$$a_2x + b_2y = c_2$$

by using matrices. By the definition of matrix equality, this system may be written in matrix notation as

$$\begin{bmatrix} a_1x + b_1y \\ a_2x + b_2y \end{bmatrix} = \begin{bmatrix} c_1 \\ c_2 \end{bmatrix}.$$

You saw on page 540 that

$$\begin{bmatrix} a_1x + b_1y \\ a_2x + b_2y \end{bmatrix} = \begin{bmatrix} a_1 & b_1 \\ a_2 & b_2 \end{bmatrix}\begin{bmatrix} x \\ y \end{bmatrix}.$$

Therefore, you can rewrite the matrix equation as

$$\begin{bmatrix} a_1 & b_1 \\ a_2 & b_2 \end{bmatrix}\begin{bmatrix} x \\ y \end{bmatrix} = \begin{bmatrix} c_1 \\ c_2 \end{bmatrix},$$

which represents the linear system in the simple matrix form

$$AX = B,$$

where $A = \begin{bmatrix} a_1 & b_1 \\ a_2 & b_2 \end{bmatrix}$ is called the **coefficient matrix**, and both X and B are 2×1 matrices.

If the coefficient matrix is invertible, the components of the single member of the solution set are the entries in $A^{-1}B$; if it is not invertible, the equations in the system are either dependent or inconsistent.

Cramer's Rule for solving equations using determinants, which was presented in Chapter 5, may be derived from this method.

EXAMPLE 3 Use matrices to find the solution set of the system:

$$-4x + y = -11$$
$$-5x + 3y = -19$$

SOLUTION 1. First, write the matrix equation:

$$\begin{bmatrix} -4 & 1 \\ -5 & 3 \end{bmatrix}\begin{bmatrix} x \\ y \end{bmatrix} = \begin{bmatrix} -11 \\ -19 \end{bmatrix}$$

2. Next, find the inverse of the coefficient matrix:

$$\begin{bmatrix} -4 & 1 \\ -5 & 3 \end{bmatrix}^{-1} = -\frac{1}{7}\begin{bmatrix} 3 & -1 \\ 5 & -4 \end{bmatrix} = \begin{bmatrix} -\frac{3}{7} & \frac{1}{7} \\ -\frac{5}{7} & \frac{4}{7} \end{bmatrix}$$

3. Then: $\begin{bmatrix} x \\ y \end{bmatrix} = \begin{bmatrix} -\frac{3}{7} & \frac{1}{7} \\ -\frac{5}{7} & \frac{4}{7} \end{bmatrix}\begin{bmatrix} -11 \\ -19 \end{bmatrix} = \begin{bmatrix} 2 \\ -3 \end{bmatrix}$

A direct check shows that the values $x = 2$, $y = -3$ satisfy the given equations.

$\therefore$ the solution set is $\{(2, -3)\}$.

550 *Chapter 13*

Systems of three linear equations in three variables, and also larger systems, can similarly be represented in the simple matrix form $AX = B$, with the unique solution $X = A^{-1}B$ in the case where A is nonsingular.

Oral Exercises

Let $A = \begin{bmatrix} 8 & 5 \\ 4 & 3 \end{bmatrix}$. If $A^{-1} = \frac{1}{a}\begin{bmatrix} b & c \\ d & e \end{bmatrix}$, give the value of the specified variable.

1. a 4   **2.** b 3   **3.** c −5   **4.** d −4   **5.** e 8

6. State the matrix equation $\begin{bmatrix} -5 & -2 \\ 3 & 4 \end{bmatrix}\begin{bmatrix} x \\ y \end{bmatrix} = \begin{bmatrix} 1 \\ -9 \end{bmatrix}$ as a system of linear equations. $-5x - 2y = 1$; $3x + 4y = -9$

If the linear system $\begin{array}{r} 2x - y = 7 \\ -3x + 4y = 8 \end{array}$ is written as a matrix equation $\begin{bmatrix} a & b \\ c & d \end{bmatrix}\begin{bmatrix} x \\ y \end{bmatrix} = \begin{bmatrix} e \\ f \end{bmatrix}$, state the value of the specified letter.

7. a 2   **8.** b −1   **9.** c −3   **10.** d 4   **11.** e 7   **12.** f 8

Written Exercises

Find the inverse of the given matrix if it is nonsingular. If the matrix is singular, so state.

A   **1.** $\begin{bmatrix} 5 & 2 \\ 7 & 3 \end{bmatrix}$ $\begin{bmatrix} 3 & -2 \\ -7 & 5 \end{bmatrix}$ **2.** $\begin{bmatrix} 9 & 2 \\ 5 & 1 \end{bmatrix}$ **3.** $\begin{bmatrix} 8 & 4 \\ 6 & 3 \end{bmatrix}$ singular **4.** $\begin{bmatrix} 0 & 2 \\ 2 & 0 \end{bmatrix}$ $\begin{bmatrix} 0 & \frac{1}{2} \\ \frac{1}{2} & 0 \end{bmatrix}$ **5.** $\begin{bmatrix} 3 & 0 \\ 0 & 3 \end{bmatrix}$

6. $\begin{bmatrix} 4 & -7 \\ 3 & -5 \end{bmatrix}$ **7.** $\begin{bmatrix} 4 & 3 \\ 6 & 5 \end{bmatrix}$ **8.** $\begin{bmatrix} 9 & -6 \\ -3 & 2 \end{bmatrix}$ **9.** $\begin{bmatrix} 9 & 4 \\ -6 & -3 \end{bmatrix}$ **10.** $\begin{bmatrix} -11 & 5 \\ -4 & 2 \end{bmatrix}$

Let $A = \begin{bmatrix} 3 & 4 \\ 2 & 3 \end{bmatrix}$, $B = \begin{bmatrix} 2 & 1 \\ 5 & 3 \end{bmatrix}$, $C = \begin{bmatrix} 2 & 0 \\ 0 & 2 \end{bmatrix}$, and $D = \begin{bmatrix} 5 & 4 \\ -2 & -2 \end{bmatrix}$.

Solve each of the following matrix equations for X.

11. $AX = B$ **12.** $AX = C$ **13.** $BX = A$ **14.** $BX = D$

15. $CX = D$ **16.** $CX = A$ **17.** $DX = C$ **18.** $DX = A$

15. $\begin{bmatrix} \frac{5}{2} & 2 \\ -1 & -1 \end{bmatrix}$ **16.** $\begin{bmatrix} \frac{3}{2} & 2 \\ 1 & \frac{3}{2} \end{bmatrix}$ **17.** $\begin{bmatrix} 2 & 4 \\ -2 & -5 \end{bmatrix}$

Find the solution set of the given system by using a matrix equation.

B   **19.** $x + 3y = 0$ {(3, −1)}   **20.** $3x + 2y = -3$ {(7, −12)}   **21.** $4x + 3y = -5$ {(−2, 1)}
        $x + 2y = 1$ $7x + 4y = 1$ $-5x - 4y = 6$

22. $4x - 2y = 8$ **23.** $6x - 2y = 3$ **24.** $3x + 2y = -6$
        $5x - 2y = 11$ {(3, 2)} $10x - 2y = 5$ {($\frac{1}{2}$, 0)} $6x + 5y = -9$

Matrices   **551**

Suggested Assignments

Enriched Alg.
First day
    551/1–27 odd
Second day
R 552/Self–Test 2
S 534/2–12 even

Additional Answers
Written Exercises

2. $\begin{bmatrix} -1 & 2 \\ 5 & -9 \end{bmatrix}$

5. $\begin{bmatrix} \frac{1}{3} & 0 \\ 0 & \frac{1}{3} \end{bmatrix}$

6. $\begin{bmatrix} -5 & 7 \\ -3 & 4 \end{bmatrix}$

7. $\begin{bmatrix} \frac{5}{2} & \frac{3}{2} \\ -3 & 2 \end{bmatrix}$

8. singular

9. $\begin{bmatrix} 1 & \frac{4}{3} \\ -2 & -3 \end{bmatrix}$

10. $\begin{bmatrix} -1 & \frac{5}{2} \\ -2 & \frac{11}{2} \end{bmatrix}$

11. $\begin{bmatrix} -14 & -9 \\ 11 & 7 \end{bmatrix}$

12. $\begin{bmatrix} 6 & -8 \\ -4 & 6 \end{bmatrix}$

13. $\begin{bmatrix} 7 & 9 \\ -11 & -14 \end{bmatrix}$

14. $\begin{bmatrix} 17 & 14 \\ -29 & -24 \end{bmatrix}$

18. $\begin{bmatrix} 7 & 10 \\ -8 & -\frac{23}{2} \end{bmatrix}$

24. {(−4, 3)}

551

27. Assume A^{-1} exists. Then $AA^{-1} = I$ and det $AA^{-1} =$ det $I = 1$. By Exercise 27(b) on page 546, det $AA^{-1} =$ det A det A^{-1}. But if det $A = 0$, det $A \cdot$ det $A^{-1} = 0$, and $0 \neq 1$ so A^{-1} cannot exist.

For $A = \begin{bmatrix} a & b \\ c & d \end{bmatrix}$, prove the given statement.

C **25.** $A \cdot \dfrac{1}{\det A}\begin{bmatrix} d & -b \\ -c & a \end{bmatrix} = \begin{bmatrix} 1 & 0 \\ 0 & 1 \end{bmatrix}$

26. $\dfrac{1}{\det A}\begin{bmatrix} d & -b \\ -c & a \end{bmatrix} \cdot A = \begin{bmatrix} 1 & 0 \\ 0 & 1 \end{bmatrix}$

27. Use the result of Written Exercise 27(b) on page 546 and the fact that $AA^{-1} = I$ to show that if det $A = 0$, then A has no inverse. (*Hint*: Assume A has an inverse and show that this leads to a contradiction.)

Computer Exercises For students with computer experience

1. Write a program that will compute the inverse of a given 2×2 matrix when the entries of the matrix are input. If the matrix does not have an inverse, the output should state this.

Run the program in Exercise 1 for each matrix.

2. $\begin{bmatrix} 4 & 5 \\ 0 & 1 \end{bmatrix}$
$\begin{bmatrix} 0.25 & -1.25 \\ 0 & 1 \end{bmatrix}$

3. $\begin{bmatrix} 3 & -1 \\ 9 & -3 \end{bmatrix}$
no inverse

4. $\begin{bmatrix} 2 & 0 \\ 0 & 2 \end{bmatrix}$
$\begin{bmatrix} 0.5 & 0 \\ 0 & 0.5 \end{bmatrix}$

5. $\begin{bmatrix} 6 & 4 \\ -5 & -3 \end{bmatrix}$
$\begin{bmatrix} -1.5 & -2 \\ 2.5 & 3 \end{bmatrix}$

.6. $\begin{bmatrix} 4 & 5 \\ 20 & 25 \end{bmatrix}$
no inverse

Quick Quiz

1. Find the inverse of
$\begin{bmatrix} 3 & 5 \\ -3 & -4 \end{bmatrix}$.
$\begin{bmatrix} -\frac{4}{3} & -\frac{5}{3} \\ 1 & 1 \end{bmatrix}$

2. Solve for X.
$\begin{bmatrix} 6 & 5 \\ 4 & 3 \end{bmatrix}X = \begin{bmatrix} 2 & -4 \\ -6 & 4 \end{bmatrix}$
$X = \begin{bmatrix} -18 & 16 \\ 22 & -20 \end{bmatrix}$

3. Write in the matrix form $AX = B$: $3x - 4y = 2$
$5x + y = 11$
$\begin{bmatrix} 3 & -4 \\ 5 & 1 \end{bmatrix}\begin{bmatrix} x \\ y \end{bmatrix} = \begin{bmatrix} 2 \\ 11 \end{bmatrix}$

4. Use a matrix equation to find the solution set of the system:
$2x - 5y = 3$
$x - 3y = 1$ $\{(4, 1)\}$

Self-Test 2

VOCABULARY   singular matrix (p. 549)
nonsingular, or invertible
matrix (p. 549)
coefficient matrix of a linear
system (p. 550)

1. Find the inverse of $\begin{bmatrix} -2 & 1 \\ 5 & 7 \end{bmatrix}$. $\begin{bmatrix} \frac{-7}{19} & \frac{1}{19} \\ \frac{5}{19} & \frac{2}{19} \end{bmatrix}$ *Obj. 1, p. 547*

2. Solve for X: $\begin{bmatrix} 4 & 9 \\ 2 & 3 \end{bmatrix}X = \begin{bmatrix} -4 & 5 \\ -2 & 7 \end{bmatrix}$ $\begin{bmatrix} -1 & 8 \\ 0 & -3 \end{bmatrix}$ *Obj. 2, p. 547*

3. Write in the matrix form $AX = B$: $2x - y = -3$ $\begin{bmatrix} 2 & -1 \\ 7 & 8 \end{bmatrix}\begin{bmatrix} x \\ y \end{bmatrix} = \begin{bmatrix} -3 \\ 1 \end{bmatrix}$ *Obj. 3, p. 547*
$7x + 8y = 1$

4. Use a matrix equation to find the solution set of the system: *Obj. 4, p. 547*

$$4x + 3y = -10$$
$$2x - 4y = 17 \quad \{(\tfrac{1}{2}, -4)\}$$

Check your answers with those at the back of the book.

---

552 *Chapter 13*

Transformations of the Plane

OBJECTIVES for Sections 13-7 and 13-8:
1. *To determine an equation of a translation of the plane when the image of one point under the translation is given.*
2. *To find the coordinates of the image of a point, and also the coordinates of the preimage of a point, under a given translation of the plane.*
3. *To find the coordinates of the image of a point under a given linear transformation of the plane.*
4. *To find the coordinates of the preimage of a point under a given nonsingular transformation of the plane.*

13–7 Transformations by Matrix Addition

Teaching Suggestions
p. T116

Key Ideas

Use matrix addition to find a matrix equation of a transformation.
Use matrix addition to find the coordinates of the image of a point and the coordinates of the preimage of a point under a transformation.

Imagine a triangular piece of cardboard resting on a coordinate plane and having vertices at $A(-4, -3)$, $B(3, -1)$, and $C(2, 3)$, as shown in Figure 1. You can think of sliding the cardboard 2 units in the x-direction and 1 unit in the y-direction to the position shown in Figure 2. In their new positions, the vertices will be at $A'(-4 + 2, -3 + 1)$, $B'(3 + 2, -1 + 1)$, and $C'(2 + 2, 3 + 1)$, or $A'(-2, -2)$, $B'(5, 0)$, and $C'(4, 4)$. By such a sliding, you can think of each point $P(x, y)$ of the plane as being *mapped* onto a corresponding point $P'(x + 2, y + 1)$.

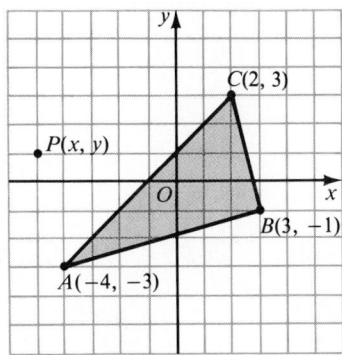

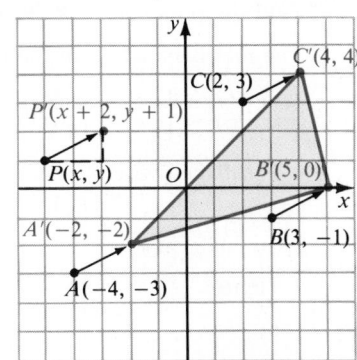

Figure 1   **Figure 2**

The sliding described above is an example of a **transformation,** or **mapping,** of the plane called a **translation.** In the translation of amount h in the x-direction and k in the y-direction, the point $P(x, y)$ is mapped onto the point $P'(x', y')$, where

$$x' = x + h \quad \text{and} \quad y' = y + k.$$

We say that P' is the **image** of P, and that P is the **preimage** of P', under the mapping.

It is convenient to use matrices in working with transformations of the plane and to designate the coordinates (x, y) by the matrix $\begin{bmatrix} x \\ y \end{bmatrix}$. Thus the translation given on page 553 is written simply as a *matrix equation*

$$\begin{bmatrix} x' \\ y' \end{bmatrix} = \begin{bmatrix} x \\ y \end{bmatrix} + \begin{bmatrix} h \\ k \end{bmatrix}. \tag{1}$$

If h and k are given, then Equation (1) can be considered as defining a function,

$$\begin{bmatrix} x \\ y \end{bmatrix} \longrightarrow \begin{bmatrix} x' \\ y' \end{bmatrix}$$

with domain the set of 2×1 matrices $\begin{bmatrix} x \\ y \end{bmatrix}$, where $x, y \in \mathcal{R}$, and range the set of corresponding matrices $\begin{bmatrix} x' \\ y' \end{bmatrix}$.

Of course, the function $(x, y) \longrightarrow (x', y')$, whose domain and range are sets of ordered pairs, describes the same transformation. Matrices are used because they are very convenient in studying linear transformations (Section 13-8).

EXAMPLE In a translation of the plane, the image of $P(3, 1)$ is $P'(5, -4)$.

    a. Find a matrix equation of the transformation.

    b. Find the image of $Q(-3, 0)$ under the transformation.

    c. Find the preimage of $R'(9, 8)$ under the transformation.

SOLUTION **a.** The matrix equation of the translation,

$$\begin{bmatrix} x' \\ y' \end{bmatrix} = \begin{bmatrix} x \\ y \end{bmatrix} + \begin{bmatrix} h \\ k \end{bmatrix}$$

is satisfied when

$$\begin{bmatrix} x \\ y \end{bmatrix} = \begin{bmatrix} 3 \\ 1 \end{bmatrix} \quad \text{and} \quad \begin{bmatrix} x' \\ y' \end{bmatrix} = \begin{bmatrix} 5 \\ -4 \end{bmatrix}.$$

Hence:

$$\begin{bmatrix} 5 \\ -4 \end{bmatrix} = \begin{bmatrix} 3 \\ 1 \end{bmatrix} + \begin{bmatrix} h \\ k \end{bmatrix}$$

$$\begin{bmatrix} 5 \\ -4 \end{bmatrix} - \begin{bmatrix} 3 \\ 1 \end{bmatrix} = \begin{bmatrix} 3 \\ 1 \end{bmatrix} + \begin{bmatrix} h \\ k \end{bmatrix} - \begin{bmatrix} 3 \\ 1 \end{bmatrix} = \begin{bmatrix} h \\ k \end{bmatrix}$$

$$\begin{bmatrix} h \\ k \end{bmatrix} = \begin{bmatrix} 5 - 3 \\ -4 - 1 \end{bmatrix} = \begin{bmatrix} 2 \\ -5 \end{bmatrix}$$

$\therefore$ the transformation is given by $\begin{bmatrix} x' \\ y' \end{bmatrix} = \begin{bmatrix} x \\ y \end{bmatrix} + \begin{bmatrix} 2 \\ -5 \end{bmatrix}$

554 *Chapter 13*

b. To find the image of $Q(-3, 0)$, substitute $\begin{bmatrix} -3 \\ 0 \end{bmatrix}$ for $\begin{bmatrix} x \\ y \end{bmatrix}$ in the equation of the transformation:

$$\begin{bmatrix} x' \\ y' \end{bmatrix} = \begin{bmatrix} -3 \\ 0 \end{bmatrix} + \begin{bmatrix} 2 \\ -5 \end{bmatrix}$$

$$= \begin{bmatrix} -1 \\ -5 \end{bmatrix}$$

$\therefore$ the image of $Q(-3, 0)$ under the transformation is $Q'(-1, -5)$.

c. To find the preimage of $R'(9, 8)$, substitute $\begin{bmatrix} 9 \\ 8 \end{bmatrix}$ for $\begin{bmatrix} x' \\ y' \end{bmatrix}$ in the equation of the transformation:

$$\begin{bmatrix} 9 \\ 8 \end{bmatrix} = \begin{bmatrix} x \\ y \end{bmatrix} + \begin{bmatrix} 2 \\ -5 \end{bmatrix}$$

$$\begin{bmatrix} 9 \\ 8 \end{bmatrix} - \begin{bmatrix} 2 \\ -5 \end{bmatrix} = \begin{bmatrix} x \\ y \end{bmatrix} + \begin{bmatrix} 2 \\ -5 \end{bmatrix} - \begin{bmatrix} 2 \\ -5 \end{bmatrix} = \begin{bmatrix} x \\ y \end{bmatrix}$$

$$\begin{bmatrix} x \\ y \end{bmatrix} = \begin{bmatrix} 9 - 2 \\ 8 - (-5) \end{bmatrix}$$

$$= \begin{bmatrix} 7 \\ 13 \end{bmatrix}$$

$\therefore$ the preimage of $R'(9, 8)$ under the transformation is $R(7, 13)$.

Parts (b) and (c) of the Example illustrate the fact that under a translation every point of the plane has a unique image, and also every point has a unique preimage. Therefore such a transformation is a one-to-one mapping of the entire plane onto itself.

Oral Exercises

Find the images of the given points under the following translation:

$$\begin{bmatrix} x' \\ y' \end{bmatrix} = \begin{bmatrix} x \\ y \end{bmatrix} + \begin{bmatrix} 4 \\ -1 \end{bmatrix}$$

1. $P(5, 3)$
(9, 2)

2. $Q(4, -2)$
(8, -3)

3. $R(-7, 1)$
(-3, 0)

4. $S(-6, -5)$
(-2, -6)

5. $T(0, 0)$
(4, -1)

Find the preimages of the given points under the following translation:

$$\begin{bmatrix} x' \\ y' \end{bmatrix} = \begin{bmatrix} x \\ y \end{bmatrix} + \begin{bmatrix} -2 \\ 3 \end{bmatrix}$$

6. $P'(4, -1)$
(6, -4)

7. $Q'(5, 6)$
(7, 3)

8. $R'(-1, -1)$
(1, -4)

9. $S'(-3, 0)$
(-1, -3)

10. $T'(0, 0)$
(2, -3)

Additional A Exercises

Find the coordinates of the image P' of the given point P under the translation $\begin{bmatrix} x' \\ y' \end{bmatrix} = \begin{bmatrix} x \\ y \end{bmatrix} + \begin{bmatrix} h \\ k \end{bmatrix}$ for the given matrix $\begin{bmatrix} h \\ k \end{bmatrix}$.

1. $P(-5, 2)$; $\begin{bmatrix} 4 \\ 1 \end{bmatrix}$ (-1, 3)

2. $P(a, b)$; $\begin{bmatrix} -a \\ b \end{bmatrix}$ (0, 2b)

Find the coordinates of the preimage P of the given point P' under the translation $\begin{bmatrix} x' \\ y' \end{bmatrix} = \begin{bmatrix} x \\ y \end{bmatrix} + \begin{bmatrix} h \\ k \end{bmatrix}$ for the given matrix $\begin{bmatrix} h \\ k \end{bmatrix}$.

3. $P'(3, -5)$; $\begin{bmatrix} 1 \\ -3 \end{bmatrix}$ (2, -2)

4. $P'(0, 0)$; $\begin{bmatrix} a \\ b \end{bmatrix}$ (-a, -b)

Find a matrix equation of the translation of the plane that transforms the given point P into the given point P'.

5. $P(3, 7)$; $P'(5, 1)$ $\begin{bmatrix} x' \\ y' \end{bmatrix} = \begin{bmatrix} x \\ y \end{bmatrix} + \begin{bmatrix} 2 \\ -6 \end{bmatrix}$

6. $P(-4, 6)$; $P'(-6, 2)$ $\begin{bmatrix} x' \\ y' \end{bmatrix} = \begin{bmatrix} x \\ y \end{bmatrix} + \begin{bmatrix} -2 \\ -4 \end{bmatrix}$

Mixed Review

1. Find a_7 for the geometric sequence whose first term is $\frac{3}{2}$ and whose second term is 3. **96**

2. How many three-digit numerals can be formed that do not contain a 2? **648**

3. In how many ways can a committee of 3 be chosen from a group 8 people? **56**

4. If two cards are drawn from a standard 52 card deck, what is the probability that the cards are both spades? **$\frac{1}{17}$**

5. Find x if $3x$, $2x - 1$, $5x + 4$ is an arithmetic sequence. **$-\frac{3}{2}$**

Written Exercises

Find the coordinates of the image P' of the given point P under the translation $\begin{bmatrix} x' \\ y' \end{bmatrix} = \begin{bmatrix} x \\ y \end{bmatrix} + \begin{bmatrix} h \\ k \end{bmatrix}$ for the given matrix $\begin{bmatrix} h \\ k \end{bmatrix}$.

A 1. $P(-6, 1)$; $\begin{bmatrix} 5 \\ 2 \end{bmatrix}$ (−1, 3) 2. $P(-4, 9)$; $\begin{bmatrix} 7 \\ -6 \end{bmatrix}$ (3, 3) 3. $P(-3, -8)$; $\begin{bmatrix} -1 \\ 11 \end{bmatrix}$ (−4, 3)

4. $P(0, 10)$; $\begin{bmatrix} -8 \\ -1 \end{bmatrix}$ (−8, 9) 5. $P(a, b)$; $\begin{bmatrix} a \\ -b \end{bmatrix}$ (2a, 0) 6. $P(a, a - b)$; $\begin{bmatrix} a \\ a + b \end{bmatrix}$ (2a, 2a)

Find the coordinates of the preimage P of the given point P' under the translation $\begin{bmatrix} x' \\ y' \end{bmatrix} = \begin{bmatrix} x \\ y \end{bmatrix} + \begin{bmatrix} h \\ k \end{bmatrix}$ for the given matrix $\begin{bmatrix} h \\ k \end{bmatrix}$.

7. $P'(2, -7)$; $\begin{bmatrix} 4 \\ -5 \end{bmatrix}$ (−2, −2) 8. $P'(0, 4)$; $\begin{bmatrix} -3 \\ 9 \end{bmatrix}$ (3, −5) 9. $P'(-1, -9)$; $\begin{bmatrix} 3 \\ -4 \end{bmatrix}$ (−4, −5)

10. $P'(6, -2)$; $\begin{bmatrix} -5 \\ 8 \end{bmatrix}$ (11, −10) 11. $P'(0, 0)$; $\begin{bmatrix} a \\ -b \end{bmatrix}$ (−a, b) 12. $P'(a, -b)$; $\begin{bmatrix} a - b \\ a + b \end{bmatrix}$ (b, −a − 2b)

Find a matrix equation of the translation of the plane that transforms the given point P into the given point P'.

13. $P(2, 8)$; $P'(3, 4)$

14. $P(-3, 7)$; $P'(4, -1)$

15. $P(2, -4)$; $P'(7, -1)$

16. $P(-1, -8)$; $P'(-7, 2)$

17. $P(a, b)$; $P'(c, d)$

18. $P(a - b, a + b)$; $P'(a, b)$

Find the image of the point X under the translation that transforms the given point P into the given point P'.

B 19. $X(6, 1)$; $P(0, -3)$; $P'(7, -8)$ (13, −4) 20. $X(-5, 3)$; $P(-2, 1)$; $P'(9, -4)$ (6, −2)

21. $X(8, 3)$; $P(-6, -7)$; $P'(-6, 4)$ (8, 14) 22. $X(a, b)$; $P(c, d)$; $P'(-a, b)$ (−c, 2b − d)

Find the preimage of X' under the translation that transforms the given point P into the given point P'.

23. $X'(6, -2)$; $P(4, 0)$; $P'(9, -8)$ (1, 6) 24. $X'(8, 2)$; $P(-1, -6)$; $P'(-3, 7)$ (10, −11)

25. $X'(-1, -6)$; $P(5, 9)$; $P'(1, -2)$ (3, 5) 26. $X'(a, b)$; $P(c, d)$; $P'(2a, b)$ (c − a, d)

C 27. Let $\begin{bmatrix} x' \\ y' \end{bmatrix} = \begin{bmatrix} x \\ y \end{bmatrix} + \begin{bmatrix} h_1 \\ k_1 \end{bmatrix}$ and $\begin{bmatrix} x' \\ y' \end{bmatrix} = \begin{bmatrix} x \\ y \end{bmatrix} + \begin{bmatrix} h_2 \\ k_2 \end{bmatrix}$ define two translations of the plane. What is the image of $P(a, b)$ under the translations applied in succession? $(a + h_1 + h_2, b + k_1 + k_2)$

28. If the order of the translations in Exercise 27 is reversed, what is the image of $P(a, b)$? $(a + h_2 + h_1, b + k_2 + k_1)$

29. On the basis of Exercises 27 and 28, is the application of successive translations of the plane commutative? yes

Careers

Geology

Geologists can work either in a laboratory or outdoors in remote areas. The laboratory scientists analyze rock and mineral samples to learn the age and composition of each specimen. In the field, geologists test rocks in their natural form, identify minerals, and draw maps.

There is more than one branch of geology and many types of geologists work within each branch. The geologists who study the composition of the earth include petroleum geologists, who seek oil and natural gas, and economic geologists, who search for minerals and solid fuel. There are also mineralogists who study the structure of rocks and minerals and geochemists who determine the chemical makeup of rocks and minerals.

EXAMPLE   Geologists can measure the age of rocks by using isotopes. A radioactive form of potassium, called potassium-40, slowly undergoes radioactive decay, and turns into the nonradioactive element argon-40. By measuring the amounts of potassium-40 and argon-40 in a rock, geologists can determine the age of the rock. The formula that is used is

$$t = T \cdot \frac{\log\left(1 + 8.33\frac{A}{K}\right)}{\log 2}$$

where A and K are the number of atoms of argon-40 and potassium-40 in the specimen, respectively, T is the half-life of potassium-40, and t is the age of the specimen in years. If the half-life of potassium-40 is approximately 1.26×10^9 years, how old, to three significant digits, would a sample of rock be if tests showed that the ratio $\frac{A}{K}$ is 0.212?

SOLUTION   Substitute the given numbers in the formula to obtain the following:

$$t \approx 1.26 \times 10^9 \cdot \frac{\log(1 + 8.33(0.212))}{\log 2}$$

$$\approx 1.26 \times 10^9 \cdot \frac{\log(2.76596)}{\log 2}$$

$$\approx 1.26 \times 10^9 \cdot \frac{0.4425}{0.3010}$$

$$\approx 1.85 \times 10^9$$

∴ the sample is approximately 1.85×10^9 years old.

Matrices   **557**

Chalkboard Examples

1. Under the transformation $\begin{bmatrix} x' \\ y' \end{bmatrix} = \begin{bmatrix} 4 & 1 \\ -2 & -3 \end{bmatrix}\begin{bmatrix} x \\ y \end{bmatrix}$ determine:
   a. the image of $P(3, -5)$
   b. the preimage of $Q'(7, -1)$

a. When $\begin{bmatrix} 3 \\ -5 \end{bmatrix}$ is substituted for $\begin{bmatrix} x \\ y \end{bmatrix}$ in the equation for the transformation, the result is:
$$\begin{bmatrix} x' \\ y' \end{bmatrix} = \begin{bmatrix} 4 & 1 \\ -2 & -3 \end{bmatrix}\begin{bmatrix} 3 \\ -5 \end{bmatrix}$$
$$\begin{bmatrix} x' \\ y' \end{bmatrix} = \begin{bmatrix} 7 \\ 9 \end{bmatrix}$$
∴ the image of $P(3, -5)$ is $P'(7, 9)$.

b. When $\begin{bmatrix} 7 \\ -1 \end{bmatrix}$ is substituted for $\begin{bmatrix} x' \\ y' \end{bmatrix}$ in the equation for the transformation, the result is:
$$\begin{bmatrix} 7 \\ -1 \end{bmatrix} = \begin{bmatrix} 4 & 1 \\ -2 & -3 \end{bmatrix}\begin{bmatrix} x \\ y \end{bmatrix}$$

13–8 Transformations by Matrix Multiplication

The system of linear equations

$$x' = 3x - 2y$$
$$y' = 7x - 5y$$

can be written in the matrix form $X' = AX$ as:

$$\begin{bmatrix} x' \\ y' \end{bmatrix} = \begin{bmatrix} 3 & -2 \\ 7 & -5 \end{bmatrix}\begin{bmatrix} x \\ y \end{bmatrix} \qquad (1)$$

For each $x, y \in \mathcal{R}$, this matrix equation yields just one $\begin{bmatrix} x' \\ y' \end{bmatrix}$. For example, if $\begin{bmatrix} x \\ y \end{bmatrix} = \begin{bmatrix} 3 \\ 5 \end{bmatrix}$, then

$$\begin{bmatrix} x' \\ y' \end{bmatrix} = \begin{bmatrix} 3 & -2 \\ 7 & -5 \end{bmatrix}\begin{bmatrix} 3 \\ 5 \end{bmatrix}$$
$$= \begin{bmatrix} -1 \\ -4 \end{bmatrix}.$$

Accordingly, Equation (1) can be considered as defining a mapping function, or transformation of the plane,

$$\begin{bmatrix} x \\ y \end{bmatrix} \rightarrow \begin{bmatrix} x' \\ y' \end{bmatrix},$$

with domain the set of 2×1 matrices $\begin{bmatrix} x \\ y \end{bmatrix}$, where $x, y \in \mathcal{R}$, and range the set of corresponding matrices $\begin{bmatrix} x' \\ y' \end{bmatrix}$. (See Figure 3.)

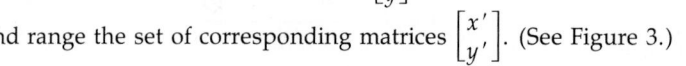

Figure 3

EXAMPLE 1   Under the transformation $\begin{bmatrix} x' \\ y' \end{bmatrix} = \begin{bmatrix} 4 & 2 \\ -7 & -5 \end{bmatrix}\begin{bmatrix} x \\ y \end{bmatrix}$ determine:
   a. the image of $P(-1, 3)$
   b. the preimage of $Q'(-4, 1)$

SOLUTION   **a.** When $\begin{bmatrix} -1 \\ 3 \end{bmatrix}$ is substituted for $\begin{bmatrix} x \\ y \end{bmatrix}$ in the equation for the transformation, the result is:

$$\begin{bmatrix} x' \\ y' \end{bmatrix} = \begin{bmatrix} 4 & 2 \\ -7 & -5 \end{bmatrix}\begin{bmatrix} -1 \\ 3 \end{bmatrix}$$
$$\begin{bmatrix} x' \\ y' \end{bmatrix} = \begin{bmatrix} 2 \\ -8 \end{bmatrix}$$

∴ the image of $P(-1, -3)$ is $P'(2, -8)$.

558   *Chapter 13*

b. When $\begin{bmatrix} -4 \\ 1 \end{bmatrix}$ is substituted for $\begin{bmatrix} x' \\ y' \end{bmatrix}$ in the equation for the transforma-

tion, the result is: $\begin{bmatrix} -4 \\ 1 \end{bmatrix} = \begin{bmatrix} 4 & 2 \\ -7 & -5 \end{bmatrix}\begin{bmatrix} x \\ y \end{bmatrix}$

Find the multiplicative inverse of $\begin{bmatrix} 4 & 2 \\ -7 & -5 \end{bmatrix}$.

$$\frac{1}{-20 + 14}\begin{bmatrix} -5 & -2 \\ 7 & 4 \end{bmatrix} = -\frac{1}{6}\begin{bmatrix} -5 & -2 \\ 7 & 4 \end{bmatrix}$$

$$= \begin{bmatrix} \frac{5}{6} & \frac{2}{6} \\ -\frac{7}{6} & -\frac{4}{6} \end{bmatrix}$$

When both sides of the equation are multiplied by the multiplicative inverse, the result is:

$$\begin{bmatrix} \frac{5}{6} & \frac{2}{6} \\ -\frac{7}{6} & -\frac{4}{6} \end{bmatrix}\begin{bmatrix} -4 \\ 1 \end{bmatrix} = \begin{bmatrix} \frac{5}{6} & \frac{2}{6} \\ -\frac{7}{6} & -\frac{4}{6} \end{bmatrix}\begin{bmatrix} 4 & 2 \\ -7 & -5 \end{bmatrix}\begin{bmatrix} x \\ y \end{bmatrix}$$

$$\begin{bmatrix} -3 \\ 4 \end{bmatrix} = \begin{bmatrix} 1 & 0 \\ 0 & 1 \end{bmatrix}\begin{bmatrix} x \\ y \end{bmatrix}$$

$$\begin{bmatrix} x \\ y \end{bmatrix} = \begin{bmatrix} -3 \\ 4 \end{bmatrix}$$

A direct check of these values shows that they satisfy the original equation.

∴ the preimage of $Q'(-4, 1)$ is $Q(-3, 4)$.

In general, any transformation of the form

$$X' = AX$$

where

$$X' = \begin{bmatrix} x' \\ y' \end{bmatrix}, \qquad A = \begin{bmatrix} a_1 & b_1 \\ a_2 & b_2 \end{bmatrix}, \qquad X = \begin{bmatrix} x \\ y \end{bmatrix}$$

and $a_1, b_1, a_2, b_2 \in \mathcal{R}$, is called a **linear transformation of the plane.**

Each point in the plane has an image under such a transformation, as illustrated in Example 1(a). In particular, since

$$AO = O,$$

the image of the origin is the origin under every linear transformation.

Further, as illustrated in Example 1(b), if $\det A \neq 0$ then each point also has a unique preimage because the matrix A has a unique inverse. In this case, the transformation is a one-to-one mapping of the entire plane onto itself, and is said to be **nonsingular.**

Matrices **559**

The multiplicative inverse of

$$\begin{bmatrix} 4 & 1 \\ -2 & -3 \end{bmatrix} \text{ is}$$

$$\begin{bmatrix} \frac{3}{10} & \frac{1}{10} \\ -\frac{2}{10} & \frac{4}{10} \end{bmatrix}.$$

When both sides of the equation are multiplied by the multiplicative inverse, the result is:

$$\begin{bmatrix} x \\ y \end{bmatrix} = \begin{bmatrix} 2 \\ -1 \end{bmatrix}$$

∴ the preimage of $Q'(7, -1)$ is $Q(2, -1)$.

2. Describe the mapping of the plane onto itself under the linear transformation of the plane $X' = AX$, for which $A = \begin{bmatrix} 1 & 0 \\ 0 & -1 \end{bmatrix}$.

When $\begin{bmatrix} 1 & 0 \\ 0 & -1 \end{bmatrix}$ is substituted for $\begin{bmatrix} a_1 & b_1 \\ a_2 & b_2 \end{bmatrix}$ in

$$\begin{bmatrix} x' \\ y' \end{bmatrix} = \begin{bmatrix} a_1 & b_1 \\ a_2 & b_2 \end{bmatrix}\begin{bmatrix} x \\ y \end{bmatrix},$$

the result is:

$$\begin{bmatrix} x' \\ y' \end{bmatrix} = \begin{bmatrix} 1 & 0 \\ 0 & -1 \end{bmatrix}\begin{bmatrix} x \\ y \end{bmatrix},$$

$$\begin{bmatrix} x' \\ y' \end{bmatrix} = \begin{bmatrix} x \\ -y \end{bmatrix}, \text{ or } (x', y') = (x, -y).$$

Thus for each $P(a, b)$ the image is the reflection $P'(a, -b)$ of P in the x-axis as shown in the figure below.

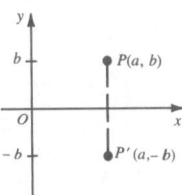

∴ the transformation is a reflection in the x-axis.

(continued)

3. Describe the mapping of the plane under the singular linear transformation for which $A = \begin{bmatrix} 2 & 1 \\ 4 & 2 \end{bmatrix}$.

For any x and $y \in R$,
$$\begin{bmatrix} x' \\ y' \end{bmatrix} = \begin{bmatrix} 2x + y \\ 4x + 2y \end{bmatrix},$$
so that $x' = 2x + y$, $y' = 4x + 2y$. Since $4x + 2y = 2(2x + y)$, $y' = 2x'$. Therefore the map of the entire plane lies on the line $y' = 2x'$. Further, each point, say $(c, 2c)$, on this line is the image of many different points; for example, $(c, 2c)$ is the image of both $\left(\dfrac{c}{2}, 0\right)$ and $\left(\dfrac{c}{3}, \dfrac{c}{3}\right)$.

∴ under this transformation the plane is mapped onto the line $y' = 2x'$.

Mixed Review

1. Find the solution set of the linear system by substitution.
   $2x + y = 2$
   $5x + 3y = -1$ $\{(7, -12)\}$

2. Given that $P(C) = \dfrac{1}{4}$ and
   $P(D \cap C) = \dfrac{1}{8}$, find
   $P(D|C)$. $\dfrac{1}{2}$

3. Find log 14.70.   1.1673

4. Find $f^{-1}(x)$ if $f(x) = x^2 + 2$, $x \geq 0$.
   $f^{-1}(x) = \sqrt{x - 2}$, $x \geq 2$

5. Graph $3x - 2y = 8$.

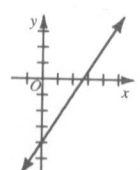

EXAMPLE 2   Describe the mapping of the plane onto itself under the linear transformation of the plane, $X' = AX$, for which:

a. $A = \begin{bmatrix} -1 & 0 \\ 0 & 1 \end{bmatrix}$

b. $A = \begin{bmatrix} 0 & 1 \\ 1 & 0 \end{bmatrix}$

SOLUTION   a. When $\begin{bmatrix} -1 & 0 \\ 0 & 1 \end{bmatrix}$ is substituted for $\begin{bmatrix} a_1 & b_1 \\ a_2 & b_2 \end{bmatrix}$ in
$$\begin{bmatrix} x' \\ y' \end{bmatrix} = \begin{bmatrix} a_1 & b_1 \\ a_2 & b_2 \end{bmatrix}\begin{bmatrix} x \\ y \end{bmatrix},$$

the result is:
$$\begin{bmatrix} x' \\ y' \end{bmatrix} = \begin{bmatrix} -1 & 0 \\ 0 & 1 \end{bmatrix}\begin{bmatrix} x \\ y \end{bmatrix}$$
$$\begin{bmatrix} x' \\ y' \end{bmatrix} = \begin{bmatrix} -x \\ y \end{bmatrix}, \quad \text{or} \quad (x', y') = (-x, y)$$

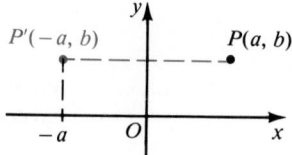

Thus, for each $P(a, b)$ the image is the reflection $P'(-a, b)$ of P in the y-axis, as shown in the figure.

∴ the transformation is a reflection in the y-axis.

b. When $\begin{bmatrix} 0 & 1 \\ 1 & 0 \end{bmatrix}$ is substituted for $\begin{bmatrix} a_1 & b_1 \\ a_2 & b_2 \end{bmatrix}$, the result is:
$$\begin{bmatrix} x' \\ y' \end{bmatrix} = \begin{bmatrix} 0 & 1 \\ 1 & 0 \end{bmatrix}\begin{bmatrix} x \\ y \end{bmatrix}$$
$$\begin{bmatrix} x' \\ y' \end{bmatrix} = \begin{bmatrix} y \\ x \end{bmatrix} \quad \text{or} \quad (x', y') = (y, x)$$

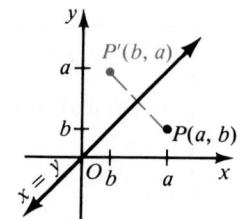

Thus for each $P(a, b)$ the image is the reflection $P'(b, a)$ of P in the line $y = x$, as discussed on page 449 and as shown in the figure.

∴ the transformation is a reflection in the line $y = x$.

If $\det A = 0$ then the linear transformation $X' = AX$ is said to be **singular.**

EXAMPLE 3   Describe the mapping of the plane under the singular linear transformation for which:

a. $A = \begin{bmatrix} 0 & 0 \\ 0 & 0 \end{bmatrix}$

b. $A = \begin{bmatrix} 2 & 1 \\ 6 & 3 \end{bmatrix}$

560   *Chapter 13*

SOLUTION   **a.** For all $x, y \in \mathcal{R}$,

$$\begin{bmatrix} x' \\ y' \end{bmatrix} = \begin{bmatrix} 0 & 0 \\ 0 & 0 \end{bmatrix} \begin{bmatrix} x \\ y \end{bmatrix}$$

$$\begin{bmatrix} x' \\ y' \end{bmatrix} = \begin{bmatrix} 0 \\ 0 \end{bmatrix}.$$

Thus the image of each $P(a, b)$ is $P'(0, 0)$.

$\therefore$ each point of the plane is mapped onto the origin under this transformation.

b. For any x and $y \in \mathcal{R}$,

$$\begin{bmatrix} x' \\ y' \end{bmatrix} = \begin{bmatrix} 2 & 1 \\ 6 & 3 \end{bmatrix} \begin{bmatrix} x \\ y \end{bmatrix} = \begin{bmatrix} 2x + y \\ 6x + 3y \end{bmatrix},$$

so that $x' = 2x + y$, and $y' = 6x + 3y$. Since $6x + 3y = 3(2x + y)$, $y' = 3x'$. Thus the map of the entire plane lies on the line $y' = 3x'$. Further, each point, say $(c, 3c)$ on this line is the image of many different points; for example, $(c, 3c)$ is the image of $\left(\frac{c}{2}, 0\right)$ and also of $\left(\frac{c}{3}, \frac{c}{3}\right)$.

$\therefore$ under this transformation the plane is mapped onto the line $y' = 3x'$.

Examples 1, 2, and 3 illustrate the theorem below.

Theorem.   Suppose that $X' = AX$ is a linear transformation, where $A = \begin{bmatrix} a_1 & b_1 \\ a_2 & b_2 \end{bmatrix}$, $X = \begin{bmatrix} x \\ y \end{bmatrix}$, $X' = \begin{bmatrix} x' \\ y' \end{bmatrix}$, and $a_1, b_1, a_2, b_2 \in \mathcal{R}$.

Then:

If $\det A \neq 0$, then the transformation is a one-to-one mapping of the plane onto itself.

If $A = O$, then the transformation maps the entire plane onto the origin.

If $A \neq O$ but $\det A = 0$, then the transformation maps the entire plane onto a line through the origin.

Oral Exercises **1.** $\begin{bmatrix} x' \\ y' \end{bmatrix} = \begin{bmatrix} 1 & 1 \\ 1 & -1 \end{bmatrix} \begin{bmatrix} x \\ y \end{bmatrix}$ **2.** $\begin{bmatrix} x' \\ y' \end{bmatrix} = \begin{bmatrix} 2 & -1 \\ 1 & 0 \end{bmatrix} \begin{bmatrix} x \\ y \end{bmatrix}$ **3.** $\begin{bmatrix} x' \\ y' \end{bmatrix} = \begin{bmatrix} 3 & 2 \\ 0 & 1 \end{bmatrix} \begin{bmatrix} x \\ y \end{bmatrix}$

State each system of linear equations in matrix form.

1. $x' = x + y$ **2.** $x' = 2x - y$ **3.** $x' = 3x + 2y$
    $y' = x - y$ $y' = x$ $y' = y$

Additional A Exercises

Find the coordinates of the image P' of the given point P under the linear transformation $\begin{bmatrix} x' \\ y' \end{bmatrix} = \begin{bmatrix} a_1 & b_1 \\ a_2 & b_2 \end{bmatrix} \begin{bmatrix} x \\ y \end{bmatrix}$ for the given matrix $\begin{bmatrix} a_1 & b_1 \\ a_2 & b_2 \end{bmatrix}$.

1. $P(1, 2)$; $\begin{bmatrix} 3 & 0 \\ 0 & 3 \end{bmatrix}$ $(3, 6)$

2. $P(-2, 3)$; $\begin{bmatrix} 0 & 2 \\ 2 & 0 \end{bmatrix}$ $(6, -4)$

3. $P(3, 5)$; $\begin{bmatrix} -4 & 3 \\ 5 & -2 \end{bmatrix}$ $(3, 5)$

Find the coordinates of the preimage P of the given point P' under the linear transformation $\begin{bmatrix} x' \\ y' \end{bmatrix} = \begin{bmatrix} a_1 & b_1 \\ a_2 & b_2 \end{bmatrix} \begin{bmatrix} x \\ y \end{bmatrix}$ for the given matrix $\begin{bmatrix} a_1 & b_1 \\ a_2 & b_2 \end{bmatrix}$.

4. $P'(3, 12)$; $\begin{bmatrix} -3 & 0 \\ 0 & -3 \end{bmatrix}$
$(-1, -4)$

5. $P'(4, -1)$; $\begin{bmatrix} 5 & 3 \\ 2 & 1 \end{bmatrix}$
$(-7, 13)$

6. $P'(-2, 5)$; $\begin{bmatrix} 2 & 3 \\ 1 & 2 \end{bmatrix}$
$(-19, 12)$

Matrices **561**

Suggested Assignments

Enriched Alg.
First day
    562/1–29 odd, 30
Second day
R 563/Self–Test 3

State each matrix equation as a system of linear equations.

4. $\begin{bmatrix} x' \\ y' \end{bmatrix} = \begin{bmatrix} 2 & -1 \\ 6 & 4 \end{bmatrix}\begin{bmatrix} x \\ y \end{bmatrix}$

$x' = 2x - y$
$y' = 6x + 4y$

5. $\begin{bmatrix} x' \\ y' \end{bmatrix} = \begin{bmatrix} 1 & 0 \\ 0 & 1 \end{bmatrix}\begin{bmatrix} x \\ y \end{bmatrix}$

$x' = x$
$y' = y$

6. $\begin{bmatrix} x' \\ y' \end{bmatrix} = \begin{bmatrix} 3 & 0 \\ -2 & 0 \end{bmatrix}\begin{bmatrix} x \\ y \end{bmatrix}$

$x' = 3x$
$y' = -2x$

Written Exercises

Find the coordinates of the image P' of the given point P under the linear transformation $\begin{bmatrix} x' \\ y' \end{bmatrix} = \begin{bmatrix} a_1 & b_1 \\ a_2 & b_2 \end{bmatrix}\begin{bmatrix} x \\ y \end{bmatrix}$ for the given matrix $\begin{bmatrix} a_1 & b_1 \\ a_2 & b_2 \end{bmatrix}$.

A
1. $P(2, 1)$; $\begin{bmatrix} 5 & 0 \\ 0 & 5 \end{bmatrix}$ (10, 5)

2. $P(-3, 2)$; $\begin{bmatrix} 0 & 4 \\ 4 & 0 \end{bmatrix}$ (8, −12)

3. $P(7, -2)$; $\begin{bmatrix} 3 & 5 \\ 0 & 2 \end{bmatrix}$ (11, −4)

4. $P(-1, 2)$; $\begin{bmatrix} 0 & 3 \\ -3 & 0 \end{bmatrix}$ (6, 3)

5. $P(-3, -1)$; $\begin{bmatrix} 4 & -5 \\ -1 & 0 \end{bmatrix}$ (−7, 3)

6. $P(4, 3)$; $\begin{bmatrix} -1 & 2 \\ 5 & -3 \end{bmatrix}$ (2, 11)

Find the coordinates of the preimage P of the given point P' under the linear transformation $\begin{bmatrix} x' \\ y' \end{bmatrix} = \begin{bmatrix} a_1 & b_1 \\ a_2 & b_2 \end{bmatrix}\begin{bmatrix} x \\ y \end{bmatrix}$ for the given matrix $\begin{bmatrix} a_1 & b_1 \\ a_2 & b_2 \end{bmatrix}$.

7. $P'(6, 10)$; $\begin{bmatrix} -2 & 0 \\ 0 & -2 \end{bmatrix}$ (−3, −5)

8. $P'(9, -3)$; $\begin{bmatrix} 0 & 3 \\ -3 & 0 \end{bmatrix}$ (1, 3)

9. $P'(-1, 3)$; $\begin{bmatrix} 3 & 4 \\ 2 & 3 \end{bmatrix}$ (−15, 11)

10. $P'(-3, -1)$; $\begin{bmatrix} 4 & 3 \\ 3 & 2 \end{bmatrix}$ (3, −5)

11. $P'(-5, 9)$; $\begin{bmatrix} 7 & 4 \\ 5 & 6 \end{bmatrix}$ (−3, 4)

12. $P'(4, -3)$; $\begin{bmatrix} 2 & 5 \\ 3 & 9 \end{bmatrix}$ (17, −6)

For each of the following linear transformations the image P' of each point $P(x, y)$ lies on a line with equation $y' = mx'$. Find x' and y' in terms of x and y, and find m.

B
13. $\begin{bmatrix} x' \\ y' \end{bmatrix} = \begin{bmatrix} 2 & 3 \\ 4 & 6 \end{bmatrix}\begin{bmatrix} x \\ y \end{bmatrix}$ $x' = 2x + 3y$; $y' = 4x + 6y$; $m = 2$

14. $\begin{bmatrix} x' \\ y' \end{bmatrix} = \begin{bmatrix} 3 & 1 \\ -6 & -2 \end{bmatrix}\begin{bmatrix} x \\ y \end{bmatrix}$

15. $\begin{bmatrix} x' \\ y' \end{bmatrix} = \begin{bmatrix} -10 & -12 \\ -5 & -6 \end{bmatrix}\begin{bmatrix} x \\ y \end{bmatrix}$ $x' = -10x - 12y$; $y' = -5x - 6y$; $m = \frac{1}{2}$

16. $\begin{bmatrix} x' \\ y' \end{bmatrix} = \begin{bmatrix} 6 & 15 \\ 4 & 10 \end{bmatrix}\begin{bmatrix} x \\ y \end{bmatrix}$

Find the image of the square with vertices (0, 0), (1, 0), (1, 1), and (0, 1) under the linear transformation $\begin{bmatrix} x' \\ y' \end{bmatrix} = \begin{bmatrix} a_1 & b_1 \\ a_2 & b_2 \end{bmatrix}\begin{bmatrix} x \\ y \end{bmatrix}$ for the given matrix.

17. $\begin{bmatrix} 3 & 0 \\ 0 & 3 \end{bmatrix}$

18. $\begin{bmatrix} 4 & 0 \\ 0 & -4 \end{bmatrix}$

19. $\begin{bmatrix} 0 & -1 \\ -1 & 0 \end{bmatrix}$

20. $\begin{bmatrix} 1 & 0 \\ 2 & 1 \end{bmatrix}$

Describe the mapping of the plane onto itself under the linear transformation $\begin{bmatrix} x' \\ y' \end{bmatrix} = \begin{bmatrix} a_1 & b_1 \\ a_2 & b_2 \end{bmatrix}\begin{bmatrix} x \\ y \end{bmatrix}$ **for the given matrix** $\begin{bmatrix} a_1 & b_1 \\ a_2 & b_2 \end{bmatrix}$.

21–24. Use the matrices in Exercises 17–20 on page 562.

25. $\begin{bmatrix} 0 & 0 \\ 0 & 1 \end{bmatrix}$ **26.** $\begin{bmatrix} 1 & 0 \\ 1 & 0 \end{bmatrix}$ **27.** $\begin{bmatrix} 0 & 0 \\ 3 & 0 \end{bmatrix}$ **28.** $\begin{bmatrix} 0 & 0 \\ 0 & -1 \end{bmatrix}$

C **29.** Show that the images of the points $(0, 0)$, $(1, 0)$, $(1, 1)$, and $(0, 1)$ are the vertices of a parallelogram under any nonsingular linear transformation $\begin{bmatrix} x' \\ y' \end{bmatrix} = \begin{bmatrix} a & b \\ c & d \end{bmatrix}\begin{bmatrix} x \\ y \end{bmatrix}$.

30. Show that if $X' = AX$ and $X' = BX$ are two linear transformations, then the image of point $P(x, y)$ under the transformations applied in succession (B, then A) is the same as the image of $P(x, y)$ under the product transformation AB. That is, show that $A(BX) = (AB)X$.
$\left(\text{Use } A = \begin{bmatrix} a & b \\ c & d \end{bmatrix} \text{ and } B = \begin{bmatrix} e & f \\ g & h \end{bmatrix}.\right)$

■ Self-Test 3

VOCABULARY   transformation or mapping (p. 553)
nonsingular linear transformation (p. 559)
translation (p. 553)
singular linear transformation (p. 560)
image (p. 554)
preimage (p. 554)
linear transformation of the plane (p. 559)   1. $\begin{bmatrix} x' \\ y' \end{bmatrix} = \begin{bmatrix} x \\ y \end{bmatrix} + \begin{bmatrix} 1 \\ 7 \end{bmatrix}$

1. Find a matrix equation of the translation of the plane that transforms the point $P(3, -1)$ into the point $P'(4, 6)$.   *Obj. 1, p. 553*

2. Find the coordinates of the image P' of the point $P(-3, 5)$ under the translation of the plane given by $\begin{bmatrix} x' \\ y' \end{bmatrix} = \begin{bmatrix} x \\ y \end{bmatrix} + \begin{bmatrix} -2 \\ 2 \end{bmatrix}$. $(-5, 7)$ *Obj. 2, p. 553*

3. Find the coordinates of the preimage P of the point $P'(-5, 8)$ under the translation given in Exercise 2. $(-3, 6)$

4. Find the coordinates of the image P' of the point $P(7, -2)$ under the linear transformation $\begin{bmatrix} x' \\ y' \end{bmatrix} = \begin{bmatrix} 6 & 1 \\ 4 & 5 \end{bmatrix}\begin{bmatrix} x \\ y \end{bmatrix}$. $(40, 18)$ *Obj. 3, p. 553*

5. Find the coordinates of the preimage P of the point $P'(8, 3)$ under the linear transformation given in Exercise 4. $(\frac{37}{26}, \frac{-7}{13})$ *Obj. 4, p. 553*

Check your answers with those at the back of the book.

Quick Quiz

1. Find a matrix equation of the translation of the plane that transforms the point $P(-1, 4)$ into the point $P'(5, 3)$.
$\begin{bmatrix} x' \\ y' \end{bmatrix} = \begin{bmatrix} x \\ y \end{bmatrix} + \begin{bmatrix} 6 \\ -1 \end{bmatrix}$

2. Find the coordinates of the image P' of the point $P(-2, 7)$ under the translation of the plane given by
$\begin{bmatrix} x' \\ y' \end{bmatrix} = \begin{bmatrix} x \\ y \end{bmatrix} + \begin{bmatrix} -2 \\ 5 \end{bmatrix}$.
$(-4, 12)$

3. Find the preimage P of the point $P'(4, -9)$ under the translation given in Quiz Item 2 above. $(6, -14)$

4. Find the coordinates of the image P' of the point $P(-1, -6)$ under the linear transformation
$\begin{bmatrix} x' \\ y' \end{bmatrix} = \begin{bmatrix} 5 & 2 \\ 7 & 3 \end{bmatrix}\begin{bmatrix} x \\ y \end{bmatrix}$.
$(-17, -25)$

5. Find the coordinates of the preimage P of the point $P'(2, 4)$ under the linear transformation given in Quiz Item 4 above. $(-2, 6)$

563

16. If $A + B = A$, then
$$\begin{bmatrix} a_1 + c_1 & b_1 + d_1 \\ a_2 + c_2 & b_2 + d_2 \end{bmatrix} =$$
$\begin{bmatrix} a_1 & b_1 \\ a_2 & b_2 \end{bmatrix}$, and $a_1 + c_1 =$
$a_1, b_1 + d_1 = b_1, a_2 + c_2$
$= a_2, b_2 + d_2 = b_2. \therefore c_1$
$= d_1 = c_2 = d_2 = 0$, and
$B = O$

Chapter Summary

1. Two matrices are *equal* if and only if they have the same dimensions and all their corresponding entries are equal.

2. If two matrices have the same dimensions, then their *sum* is a matrix of the same dimensions. The entries of the new matrix are the sums of the corresponding entries of the given matrices.

3. In the set $\mathcal{S}_{m \times n}$ of all $m \times n$ matrices with real-number entries, the *identity matrix for addition* is the zero matrix $O_{m \times n}$ all of whose entries are zero. The *additive inverse*, or *opposite*, of the matrix $A_{m \times n}$ is the $m \times n$ matrix $-A_{m \times n}$ each of whose entries is the opposite of the corresponding entry of $A_{m \times n}$. The *difference* $A_{m \times n} - B_{m \times n}$ is defined to be the sum $A_{m \times n} + (-B_{m \times n})$.

4. The set $\mathcal{S}_{m \times n}$ of all $m \times n$ matrices with real-number entries has the same addition properties as the set of all real numbers: closure, commutative, associative, additive-identity, and additive-inverse.

5. In work with matrices, real numbers are often referred to as *scalars*. The *product* of a scalar c and a matrix A is denoted by cA; it is the matrix of the same dimensions as A whose entries are the products of c and the corresponding entries of A. Basic properties of these products follow from the definition and the properties of real numbers.

6. The *product* AB of the matrices A and B can be described as "row by column" multiplication. The product is defined only if the number of columns in A is equal to the number of rows in B. The product matrix then has the same number of rows as A and the same number of columns as B. Thus $A_{m \times p} \times B_{p \times n} = C_{m \times n}$.

7. The set $\mathcal{S}_{2 \times 2}$ is closed under matrix multiplication, and the associative and distributive properties hold for matrix multiplication. The *identity matrix for multiplication* is

$$I_{2 \times 2} = \begin{bmatrix} 1 & 0 \\ 0 & 1 \end{bmatrix}.$$

Multiplication in $\mathcal{S}_{2 \times 2}$ is *not commutative*, though certain products do commute. The product of two matrices can be the zero matrix without either factor being the zero matrix.

8. The determinant having the same entries as a given square matrix A is the *determinant* of the matrix. The matrix A is *singular* if det $A = 0$; otherwise A is *nonsingular*, or *invertible*. A has a *multiplicative inverse*, that is, there is a square matrix A^{-1} such that $AA^{-1} = I$ and $A^{-1}A = I$, if and only if A is nonsingular.

564   *Chapter 13*

9. A *transformation*, or *mapping* of the plane defined by an equation of the form $\begin{bmatrix} x' \\ y' \end{bmatrix} = \begin{bmatrix} x \\ y \end{bmatrix} + \begin{bmatrix} h \\ k \end{bmatrix}$, where $h, k \in \mathcal{R}$, is called a *translation*. Under the transformation, the point $P'(x', y')$ is called the *image* of the point $P(x, y)$, and P is called the *preimage* of P'.

10. An equation of the form $\begin{bmatrix} x' \\ y' \end{bmatrix} = \begin{bmatrix} a_1 & b_1 \\ a_2 & b_2 \end{bmatrix}\begin{bmatrix} x \\ y \end{bmatrix}$, where $a_1, b_1, a_2, b_2 \in \mathcal{R}$

defines a *linear transformation of the plane*. If $\det \begin{bmatrix} a_1 & b_1 \\ a_2 & b_2 \end{bmatrix} \neq 0$, then the transformation is a one-to-one mapping of the plane onto itself. Such a transformation is said to be *nonsingular*.

Chapter Review

1. Find the dimensions of the matrix $\begin{bmatrix} 2 & 0 \\ 1 & 1 \\ -6 & 5 \end{bmatrix}$.   *13-1*

(a.) 3×2 **b.** 2×3 **c.** 1×3 **d.** 3×1

2. Find the sum: $\begin{bmatrix} 3 & -1 \\ 4 & 0 \end{bmatrix} + \begin{bmatrix} -3 & 6 \\ -2 & 8 \end{bmatrix}$

(a.) $\begin{bmatrix} 0 & 5 \\ 2 & 8 \end{bmatrix}$ **b.** $\begin{bmatrix} 0 & 5 \\ -2 & 8 \end{bmatrix}$ **c.** $\begin{bmatrix} 0 & -5 \\ 2 & 8 \end{bmatrix}$ **d.** $\begin{bmatrix} 0 & 5 \\ 2 & -8 \end{bmatrix}$

In Exercises 3 and 4, solve for the matrix X.

3. $X + \begin{bmatrix} 3 & 0 \\ 4 & 7 \end{bmatrix} = \begin{bmatrix} 4 & 1 \\ 4 & 9 \end{bmatrix}$ *13-2*

a. $\begin{bmatrix} 7 & 3 \\ 8 & 16 \end{bmatrix}$ **b.** $\begin{bmatrix} 1 & -1 \\ 0 & 2 \end{bmatrix}$ (c.) $\begin{bmatrix} 1 & 1 \\ 0 & 2 \end{bmatrix}$ **d.** $\begin{bmatrix} 1 & 3 \\ 0 & 2 \end{bmatrix}$

4. $X - \begin{bmatrix} 2 & 6 \\ 0 & 3 \end{bmatrix} = \begin{bmatrix} 1 & -8 \\ 2 & 7 \end{bmatrix}$

a. $\begin{bmatrix} -1 & -14 \\ 2 & 4 \end{bmatrix}$ **b.** $\begin{bmatrix} 1 & 14 \\ -2 & -4 \end{bmatrix}$ **c.** $\begin{bmatrix} 3 & -14 \\ 2 & 10 \end{bmatrix}$ (d.) $\begin{bmatrix} 3 & -2 \\ 2 & 10 \end{bmatrix}$

In Exercises 5 and 6, find the 2 × 2 matrix X that satisfies the equation for $A = \begin{bmatrix} 8 & -6 \\ 7 & -2 \end{bmatrix}$ **and** $B = \begin{bmatrix} -3 & -4 \\ 1 & 2 \end{bmatrix}$.

5. $X = A - 2B$ *13-3*

a. $\begin{bmatrix} 19 & -8 \\ 13 & -6 \end{bmatrix}$ **b.** $\begin{bmatrix} -2 & 14 \\ -9 & -2 \end{bmatrix}$ **c.** $\begin{bmatrix} 22 & -4 \\ 12 & -8 \end{bmatrix}$ (d.) $\begin{bmatrix} 14 & 2 \\ 5 & -6 \end{bmatrix}$

19. Let $A = \begin{bmatrix} a_1 & b_1 \\ a_2 & b_2 \end{bmatrix}$. Then

$cA = c\begin{bmatrix} a_1 & b_1 \\ a_2 & b_2 \end{bmatrix} =$

$\begin{bmatrix} ca_1 & cb_1 \\ ca_2 & cb_2 \end{bmatrix}$, by definition

of the product of a scalar
and a matrix; so $cA \in$
$\mathcal{S}_{2\times2}$ because the product
of two real numbers is a
real number.

20. Let $A = \begin{bmatrix} a_1 & b_1 \\ a_2 & b_2 \end{bmatrix}$. $c(dA)$

$= c\begin{bmatrix} da_1 & db_1 \\ da_2 & db_2 \end{bmatrix} =$

$\begin{bmatrix} cda_1 & cdb_1 \\ cda_2 & cdb_2 \end{bmatrix}$. $(cd)A =$

$(cd)\begin{bmatrix} a_1 & b_1 \\ a_2 & b_2 \end{bmatrix} =$

$\begin{bmatrix} cda_1 & cdb_1 \\ cda_2 & cdb_2 \end{bmatrix}$, $\therefore c(dA) =$
$(cd)A$

21. Let $A = \begin{bmatrix} a_1 & b_1 \\ a_2 & b_2 \end{bmatrix}$.

$(c + d)A =$

$\begin{bmatrix} (c+d)a_1 & (c+d)b_1 \\ (c+d)a_2 & (c+d)b_2 \end{bmatrix} =$

$\begin{bmatrix} ca_1 + da_1 & cb_1 + db_1 \\ ca_2 + da_2 & cb_2 + db_2 \end{bmatrix}$;

$cA + dA = c\begin{bmatrix} a_1 & b_1 \\ a_2 & b_2 \end{bmatrix} +$

$d\begin{bmatrix} a_1 & b_1 \\ a_2 & b_2 \end{bmatrix} = \begin{bmatrix} ca_1 & cb_1 \\ ca_2 & cb_2 \end{bmatrix} +$

$\begin{bmatrix} da_1 & db_1 \\ da_2 & db_2 \end{bmatrix} =$

$\begin{bmatrix} ca_1 + da_1 & cb_1 + db_1 \\ ca_2 + da_2 & cb_2 + db_2 \end{bmatrix}$;
$\therefore (c + d)A = cA + dA$

22. Let $A = \begin{bmatrix} a_1 & b_1 \\ a_2 & b_2 \end{bmatrix}$ and B

$= \begin{bmatrix} c_1 & d_1 \\ c_2 & d_2 \end{bmatrix}$. $c(A + B) =$

$c\begin{bmatrix} a_1 + c_1 & b_1 + d_1 \\ a_2 + c_2 & b_2 + d_2 \end{bmatrix} =$

$\begin{bmatrix} c(a_1 + c_1) & c(b_1 + d_1) \\ c(a_2 + c_2) & c(b_2 + d_2) \end{bmatrix} =$

$\begin{bmatrix} ca_1 + cc_1 & cb_1 + cd_1 \\ ca_2 + cc_2 & cb_2 + cd_2 \end{bmatrix}$;

(continued)

---

6. $X - B = A$

   (a.) $\begin{bmatrix} 5 & -10 \\ 8 & 0 \end{bmatrix}$ **b.** $\begin{bmatrix} -5 & 10 \\ -8 & 0 \end{bmatrix}$ **c.** $\begin{bmatrix} 5 & -2 \\ 8 & 0 \end{bmatrix}$ **d.** $\begin{bmatrix} 5 & -10 \\ 6 & 0 \end{bmatrix}$

7. Find the product: $\begin{bmatrix} 2 & -6 \\ 1 & 0 \end{bmatrix}\begin{bmatrix} -2 \\ 5 \end{bmatrix}$ *13-4*

    a. $\begin{bmatrix} -4 & -30 \\ -2 & 0 \end{bmatrix}$ **b.** $\begin{bmatrix} 26 \\ -2 \end{bmatrix}$ (c.) $\begin{bmatrix} -34 \\ -2 \end{bmatrix}$ **d.** $\begin{bmatrix} -34 \\ 3 \end{bmatrix}$

8. Find the 2×2 matrix equal to $\begin{bmatrix} 5 & -2 \\ 0 & 3 \end{bmatrix} \times \begin{bmatrix} -1 & 0 \\ 2 & -4 \end{bmatrix}$.

    a. $\begin{bmatrix} -9 & -8 \\ 5 & -12 \end{bmatrix}$ **b.** $\begin{bmatrix} -9 & 8 \\ 6 & 0 \end{bmatrix}$ (c.) $\begin{bmatrix} -9 & 8 \\ 6 & -12 \end{bmatrix}$ **d.** $\begin{bmatrix} -9 & -8 \\ 6 & 0 \end{bmatrix}$

9. If $A \in \mathcal{S}_{2\times2}$, $B \in \mathcal{S}_{2\times2}$, and $C \in \mathcal{S}_{2\times2}$, then $(AB)C = \underline{}$ *13-5*
    a. $C(AB)$ (b.) $A(BC)$ **c.** $(BA)C$ **d.** $C(BA)$

10. If $A \in \mathcal{S}_{2\times2}$ and $B \in \mathcal{S}_{2\times2}$, then $(B + A)(B - A) = \underline{}$
    a. $B^2 - A^2$ (b.) $B^2 - BA + AB - A^2$ **c.** $B^2 - 2BA - A^2$ **d.** $A^2 + B^2$

11. Find the inverse of $\begin{bmatrix} 5 & 9 \\ 4 & 8 \end{bmatrix}$.   *13-6*

    a. $\begin{bmatrix} -2 & \frac{9}{4} \\ 1 & -\frac{5}{4} \end{bmatrix}$ **b.** $\begin{bmatrix} \frac{5}{4} & -\frac{9}{4} \\ -1 & 2 \end{bmatrix}$ **c.** $\begin{bmatrix} \frac{5}{4} & -\frac{9}{4} \\ -1 & 2 \end{bmatrix}$ (d.) $\begin{bmatrix} 2 & -\frac{9}{4} \\ -1 & \frac{5}{4} \end{bmatrix}$

12. If $AX = B$, then $X = \underline{}$
    a. AB (b.) $A^{-1}B$ **c.** BA^{-1} **d.** $B^{-1}A$

In Exercises 13–14, use the translation

$$\begin{bmatrix} x' \\ y' \end{bmatrix} = \begin{bmatrix} x \\ y \end{bmatrix} + \begin{bmatrix} 3 \\ -5 \end{bmatrix}.$$

13. Find the coordinates of the image P' of the point $P(-6, 1)$.   *13-7*
    (a.) $P'(-3, -4)$ **b.** $P'(9, -4)$ **c.** $P'(-3, 6)$ **d.** $P'(9, 6)$

14. Find the coordinates of the preimage Q of the point $Q'(-2, 4)$.
    (a.) $Q(-5, 9)$ **b.** $Q(1, -1)$ **c.** $Q(1, 9)$ **d.** $Q(-5, -1)$

In Exercises 15–16, use the linear transformation

$$\begin{bmatrix} x' \\ y' \end{bmatrix} = \begin{bmatrix} 2 & 2 \\ 1 & -3 \end{bmatrix}\begin{bmatrix} x \\ y \end{bmatrix}.$$

15. Find the coordinates of the image R' of the point $R(3, -4)$.   *13-8*
    a. $R'(-2, -15)$ **b.** $R'(-4, 3)$ **c.** $R'(4, -3)$ (d.) $R'(-2, 15)$

16. Find the coordinates of the preimage S of the point $S'(-8, -16)$.
    a. $S(3, -7)$ (b.) $S(-7, 3)$ **c.** $S(-16, -8)$ **d.** $S(8, 16)$

7. $\begin{bmatrix} x' \\ y' \end{bmatrix} = \begin{bmatrix} x \\ y \end{bmatrix} + \begin{bmatrix} 5 \\ -9 \end{bmatrix}$

1. Find the values of x and y for which the following statement is true: 13-1
   $[2x \quad 3y] - [3y \quad -x] = [5 \quad -11]$ $x = -2; y = -3$

2. Solve for the matrix X: $X + \begin{bmatrix} 2 & 0 & 1 \\ -1 & 6 & 4 \end{bmatrix} = \begin{bmatrix} -3 & 9 & 8 \\ 2 & 4 & 0 \end{bmatrix} \begin{bmatrix} -5 & 9 & 7 \\ 3 & -2 & -4 \end{bmatrix}$ 13-2

3. Find the 2×2 matrix X that satisfies the following equation: 13-3
   $2X - 2\begin{bmatrix} 0 & -6 \\ 7 & 1 \end{bmatrix} = 3\begin{bmatrix} 4 & 0 \\ -2 & 2 \end{bmatrix} \begin{bmatrix} 6 & -6 \\ 4 & 4 \end{bmatrix}$

4. Find the product: $\begin{bmatrix} 1 & 0 & 2 \\ 0 & -3 & 1 \end{bmatrix} \begin{bmatrix} -1 & 8 \\ 3 & 1 \\ 7 & 0 \end{bmatrix}$. $\begin{bmatrix} 13 & 8 \\ -2 & -3 \end{bmatrix}$ 13-4

5. For $A = \begin{bmatrix} -3 & 0 \\ 2 & 4 \end{bmatrix}$ and $B = \begin{bmatrix} 5 & 0 \\ 1 & 1 \end{bmatrix}$, compute $(2A)B$ and $A(2B)$ 13-5
   and state whether or not they are equal. $(2A)B = A(2B) = \begin{bmatrix} -30 & 0 \\ 28 & 8 \end{bmatrix}$

6. Solve for X: $\begin{bmatrix} 5 & -4 \\ 1 & -1 \end{bmatrix} X = \begin{bmatrix} 3 & 0 \\ 1 & -12 \end{bmatrix} \begin{bmatrix} -1 & 48 \\ -2 & 60 \end{bmatrix}$ 13-6

7. Find a matrix equation of the translation of the plane that transforms 13-7
   the point $P(-1, 7)$ into the point $P'(4, -2)$.

8. Use the matrix equation of Exercise 7 to find the image of $Q(9, -6)$. $(14, -15)$

9. Find the coordinates of the image P' of the point $P(3, -4)$ under the 13-8
   linear transformation $\begin{bmatrix} x' \\ y' \end{bmatrix} = \begin{bmatrix} 6 & -2 \\ 0 & 1 \end{bmatrix} \begin{bmatrix} x \\ y \end{bmatrix}$. $(26, -4)$

10. Find the coordinates of the preimage P of the point $P'(-5, 2)$ under
    the linear transformation $\begin{bmatrix} x' \\ y' \end{bmatrix} = \begin{bmatrix} 1 & 3 \\ 2 & 7 \end{bmatrix} \begin{bmatrix} x \\ y \end{bmatrix}$. $(-41, 12)$

APPLICATION

Communication and Dominance Matrices

Matrices can be used to represent communications between groups of human beings. Consider the relationships shown below.

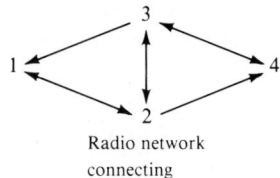

Radio network
connecting
four ships

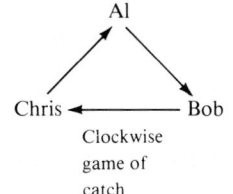

Clockwise
game of
catch

$cA + cB = c\begin{bmatrix} a_1 & b_1 \\ a_2 & b_2 \end{bmatrix} +$
$c\begin{bmatrix} c_1 & d_1 \\ c_2 & d_2 \end{bmatrix} = \begin{bmatrix} ca_1 & cb_1 \\ ca_2 & cb_2 \end{bmatrix} +$
$\begin{bmatrix} cc_1 & cd_1 \\ cc_2 & cd_2 \end{bmatrix} =$
$\begin{bmatrix} ca_1 + cc_1 & cb_1 + cd_1 \\ ca_2 + cc_2 & cb_2 + cd_2 \end{bmatrix}$;
$\therefore c(A + B) = cA + cB$

23. Let $A = \begin{bmatrix} a_1 & b_1 \\ a_2 & b_2 \end{bmatrix}$. $1 \cdot A =$
    $1\begin{bmatrix} a_1 & b_1 \\ a_2 & b_2 \end{bmatrix} =$
    $\begin{bmatrix} 1 \cdot a_1 & 1 \cdot b_1 \\ 1 \cdot a_2 & 1 \cdot b_2 \end{bmatrix} =$
    $\begin{bmatrix} a_1 & b_1 \\ a_2 & b_2 \end{bmatrix} = A$

24. Let $A = \begin{bmatrix} a_1 & b_1 \\ a_2 & b_2 \end{bmatrix}$.
    $0 \cdot A = 0\begin{bmatrix} a_1 & b_1 \\ a_2 & b_2 \end{bmatrix} =$
    $\begin{bmatrix} 0 \cdot a_1 & 0 \cdot b_1 \\ 0 \cdot a_2 & 0 \cdot b_2 \end{bmatrix} =$
    $\begin{bmatrix} 0 & 0 \\ 0 & 0 \end{bmatrix} = O.$

25. $O_{2\times2} = \begin{bmatrix} 0 & 0 \\ 0 & 0 \end{bmatrix}$. $cO_{2\times2} =$
    $\begin{bmatrix} c \cdot 0 & c \cdot 0 \\ c \cdot 0 & c \cdot 0 \end{bmatrix} = \begin{bmatrix} 0 & 0 \\ 0 & 0 \end{bmatrix} =$
    $O_{2\times2}$

26. Let $B = \begin{bmatrix} c_1 & d_1 \\ c_2 & d_2 \end{bmatrix}$ and $X = \begin{bmatrix} x_1 & y_1 \\ x_2 & y_2 \end{bmatrix}$. Since $cX = B$,
    $\begin{bmatrix} cx_1 & cy_1 \\ cx_2 & cy_2 \end{bmatrix} = \begin{bmatrix} c_1 & d_1 \\ c_2 & d_2 \end{bmatrix}$,
    and $cx_1 = c_1$, $cy_1 = d_1$,
    $cx_2 = c_2$, $cy_2 = d_2$.
    $\therefore x_1 = \frac{1}{c}c_1, y_1 = \frac{1}{c}d_1, x_2 = \frac{1}{c}c_2, y_2 = \frac{1}{c}d_2$, and
    $X = \begin{bmatrix} \frac{1}{c}c_1 & \frac{1}{c}d_1 \\ \frac{1}{c}c_2 & \frac{1}{c}d_2 \end{bmatrix} = \frac{1}{c}B.$

27. If $cX + B = A$, then $cX + B + (-B) = A + (-B)$
    and $cX = A - B$. From the previous exercise $X = \frac{1}{c}(A - B)$.

In each of these relationships, there is a group of people and a network of one-way or two-way communications lines that connect at least some of the people. Such a collection of people and communication lines is called a **communication network.** To each such network there corresponds a **communication matrix.** The communication matrices for the above examples are shown below.

$$
\text{From Ship }
\begin{array}{c}
\\
1 \\ 2 \\ 3 \\ 4
\end{array}
\overset{\displaystyle \text{To Ship}}{\overset{\displaystyle 1 \;\; 2 \;\; 3 \;\; 4}{
\begin{bmatrix}
0 & 1 & 0 & 0 \\
1 & 0 & 1 & 1 \\
1 & 1 & 0 & 1 \\
0 & 0 & 1 & 0
\end{bmatrix}}} = A
\qquad
\begin{array}{c}
\\
\text{Al} \\ \text{Bob} \\ \text{Chris}
\end{array}
\overset{\displaystyle \text{Al} \;\; \text{Bob} \;\; \text{Chris}}{
\begin{bmatrix}
0 & 1 & 0 \\
0 & 0 & 1 \\
1 & 0 & 0
\end{bmatrix}} = B
$$

Radio network
connecting
four ships

A clockwise
game of
catch

In a communication matrix, the element "1" indicates that direct communication *to* another point is possible. The element "0" indicates that direct communication *to* another point is not possible. For example, the "1" in row one, column two of matrix A indicates that Ship 1 can send a message *to* Ship 2. The "0" in row four, column one of the same matrix indicates that Ship 4 may not send a message *to* Ship 1. Rows, therefore, represent the possible directions in which a member of the communication network may transmit a message. Note that the principal diagonal of the matrix (running from the upper left to the lower right) consists entirely of zeros. This represents the assumption that a member of a communication network does not send a message *to* himself or herself.

A special kind of communication matrix, called the domination matrix, is represented by matrix B in the example. A **dominance relation** is said to exist between members of a group when, between any two members, exactly one may send a message to the other. In the game of clockwise catch, each pair of players consists of one who only throws the ball to the other while the other player only catches. Thus, the thrower in the pair "dominates" the other player.

A member x of a group is said to have a **two-stage dominance** over a member z when x dominates y and y dominates z. (The simpler dominance relation between two members of a group that was described earlier can now be referred to as a **one-stage dominance.**) It can be shown that if M is the matrix for a dominance relation, then M^2 represents the matrix for a two-stage dominance relation.

EXAMPLE Show that in the example of clockwise catch, each player has a two-stage dominance over exactly one other player.

568 *Chapter 13*

SOLUTION   Calculate the value of B^2.

$$B^2 = \begin{bmatrix} 0 & 1 & 0 \\ 0 & 0 & 1 \\ 1 & 0 & 0 \end{bmatrix} \begin{bmatrix} 0 & 1 & 0 \\ 0 & 0 & 1 \\ 1 & 0 & 0 \end{bmatrix}$$

$$= \begin{bmatrix} 0 & 0 & 1 \\ 1 & 0 & 0 \\ 0 & 1 & 0 \end{bmatrix}$$

Each row of B^2 contains exactly one "1." Recall that rows represent the possible directions in which a member of a communication network may transmit a message. Since a dominance relation is a special kind of communication, it follows that each player has a two-stage dominance over exactly one other player.

1. $\begin{bmatrix} 0 & 1 & 0 \\ 1 & 0 & 1 \\ 1 & 0 & 0 \end{bmatrix}$ **2.** $\begin{bmatrix} 0 & 1 & 0 & 0 \\ 1 & 0 & 1 & 1 \\ 0 & 0 & 0 & 1 \\ 1 & 0 & 0 & 0 \end{bmatrix}$ **3.** $\begin{bmatrix} 0 & 1 & 0 & 0 \\ 1 & 0 & 1 & 0 \\ 0 & 0 & 0 & 1 \\ 0 & 1 & 0 & 0 \end{bmatrix}$

Exercises

In Exercises 1–3, write the communication matrix that corresponds to the given communication network.

1. 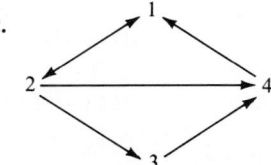   **2.**   **3.**

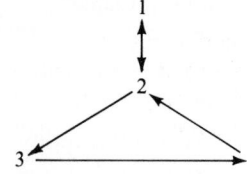

In Exercises 4–6, draw a communication network that corresponds to the given communication matrix.

4. $\begin{bmatrix} 0 & 0 & 0 \\ 1 & 0 & 0 \\ 0 & 1 & 0 \end{bmatrix}$ **5.** $\begin{bmatrix} 0 & 0 & 1 \\ 1 & 0 & 0 \\ 0 & 1 & 0 \end{bmatrix}$ **6.** $\begin{bmatrix} 0 & 1 & 0 & 0 \\ 1 & 0 & 0 & 0 \\ 0 & 1 & 0 & 0 \\ 0 & 1 & 0 & 0 \end{bmatrix}$

The dominance matrix, M, at the right, represents the results of a chess tournament.

	Ed	Tod	Pat	Dot
Ed	0	1	1	0
Tod	0	0	1	1
Pat	0	0	0	0
Dot	1	0	1	0

7. Which player(s) exert(s) a one-stage dominance over Ed? Dot

8. Over which player(s) does Tod exert a one-stage dominance? Pat, Dot

9. Which player(s) exert(s) a two-stage dominance over Tod? Dot

10. Which player(s) exert(s) both a one-stage dominance and a two-stage dominance over another player? Ed, Tod, Dot

569

 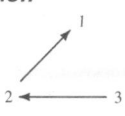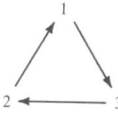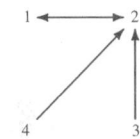

PREPARING FOR ▮▮▮▮▮▮
COLLEGE ENTRANCE EXAMS

Strategy for Success: On multiple-choice exams, you are often asked to determine the *best* answer to a question. For any given question, more than one answer may seem right to some extent, and you should be careful not to choose the first answer that seems reasonable. Be sure to evaluate all the possible choices before you determine which answer is best.

Decide which is the best of the choices given and write the correct letter on your answer sheet.

1. Which of the following must be true if set A contains 6 elements, $A \cup B$ contains 8 elements, and $A \cap B$ contains 5 elements? E
   I. Set B contains 2 elements.   II. $A \times B$ contains 12 elements.   III. $B \subset A$
   (A) I only   **(B)** II only   **(C)** III only   **(D)** I and II only   **(E)** None

2. In how many different ways can 2 adults and 6 teenagers be seated at a round table?   C
   (A) 1260   **(B)** 2520   **(C)** 5040   **(D)** 630   **(E)** 1360

3. What is the sixth term of $(x^2 - 1)^{12}$?   C
   (A) $210\ x^{12}$ **(B)** $792\ x^7$ **(C)** $-792x^{14}$ **(D)** $-210x^{14}$ **(E)** $-792x^7$

4. What is the vertex of the parabola with equation $x = 2y^2 + 6y + 1$?   A
   (A) $\left(-\frac{7}{2}, -\frac{3}{2}\right)$ **(B)** $\left(-\frac{7}{2}, \frac{5}{2}\right)$ **(C)** $\left(\frac{29}{2}, \frac{3}{2}\right)$ **(D)** $\left(-\frac{3}{2}, 1\right)$ **(E)** $\left(\frac{5}{2}, \frac{7}{2}\right)$

5. Which of the following must be true for the circle with equation $x^2 + y^2 - 6x + 4y + 13 = 9$?   C
   I. The circle has center $(3, 2)$.   II. The diameter of the circle is 9.
   III. The circle is tangent to the y-axis at $(0, -2)$.
   (A) I only   **(B)** II only   **(C)** III only   **(D)** I, II, and III   **(E)** None

6. Which of the following is a zero of the function $P(x) = x^4 - x^3 + 2x^2 - 4x - 8$?   A
   (A) $-2i$ **(B)** i **(C)** $\frac{1}{2}$
   (D) $i\sqrt{2}$ **(E)** None of these

7. For which values of a and b is the statement $\begin{bmatrix} 16 \\ 6 \end{bmatrix} = \begin{bmatrix} a - b \\ a + b \end{bmatrix}$ true?   A
   (A) $a = 11, b = -5$ **(B)** $a = -11, b = 5$ **(C)** $a = 9, b = -6$
   (D) $a = -5, b = -1$ **(E)** $a = 5, b = 11$

8. For the matrices $A = \begin{bmatrix} 2 & 1 \\ -4 & -2 \end{bmatrix}$ and $B = \begin{bmatrix} -1 & -3 \\ 2 & 6 \end{bmatrix}$ which of the following expressions are equivalent?   B
   I. $(A + B)^2$ II. $A^2 + 2AB + B^2$ III. $A^2 + AB + BA + B^2$
   (A) I and II   **(B)** I and III   **(C)** II and III   **(D)** I, II, and III   **(E)** None

570   *Chapter 13*

PROGRAMMING IN PASCAL

Standard Pascal does not provide a built-in procedure for manipulating matrices. If your system is equipped with a matrix algebra package, please do not use the package to solve the exercises on pages 572 and 573. These exercises will allow you to use some of the mathematical concepts you learned in this chapter and develop the coding necessary to perform some of the matrix operations described in this chapter.

Vectors and matrices are best represented in Pascal with arrays. Recall that standard Pascal does not allow you to choose the dimensions of arrays during the execution of the program, but rather, you must declare the dimensions of the arrays in the declaration part of the program. In these exercises, the size of the arrays is limited to 2.

There is an advantage to using the constant *dim* throughout the program instead of the value 2. That is, if need be, the value of *dim* can be changed, and the program therefore used for larger sized matrices, without going through and changing the program wherever a 2 appears.

Consider the declaration section below.

```
CONST dim = 2;   (* or 3 or whatever *)

TYPE
   scalar = real;
   vector = ARRAY[1. .dim] OF real;
   matrix = ARRAY[1..dim, 1..dim] OF real;

VAR
   a,b : scalar;
   u,v : vector;
   m,n, mat : matrix;
```

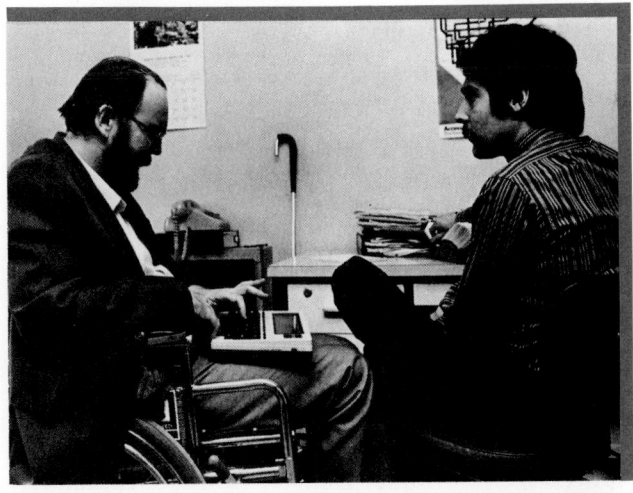

The variables of type *vector* are arrays with two elements; the variables of type *matrix* are two by two arrays with a total of four elements. Refer to the elements of a two-dimensional array just as you would a matrix, but remain consistent with Pascal notation. The elements of *m* are referred to as $m[1,1]$, $m[1,2]$, $m[2,1]$, $m[2,2]$ in Pascal. The procedure below uses a nested FOR loop structure to add matrices. Assume the declaration section on page 571 appears in the same program.

```
PROCEDURE add_matrices (m,n : matrix; VAR sum : matrix);
VAR
   row, col : 1..dim;
BEGIN
  FOR row := 1 TO dim DO
    FOR col := 1 TO dim DO
      sum[row, col] := m[row, col] + n[row, col];
END;   (* add_matrices *)
```

To follow the execution of this procedure, enter the first FOR loop (which sets *row* equal to 1). Within the first iteration, the second FOR loop is entered where *col* is set equal to 1. Then $sum[1,1]$ (since *row* and *col* are both equal to 1 at that point) is assigned the value $m[1,1] + n[1,1]$. Next, the second iteration of the FOR loop sets *col* equal to 2, and $sum[1,2]$ is assigned the value $m[1,2] + n[1,2]$. Now the inside loop is finished, and control returns to the outside FOR loop where *row* is incremented by 1 (set equal to 2). The inside FOR loop is now entered again, and the second row of *sum* is assigned its respective values. Study this process carefully so that you can use this type of technique for manipulating matrices throughout the exercises.

Exercises

1. a. Write a procedure with heading

    PROCEDURE get_matrix (VAR m : matrix);

    in which the user enters the elements of a *dim* × *dim* matrix.
   b. Write a procedure with heading

    PROCEDURE print_matrix (m : matrix);

    that displays the elements of a *dim* × *dim* matrix in matrix form.
   c. Use *get_matrix*, *add_matrices*, and *print_matrix* to write a Pascal program that prompts the user for the elements of two *dim* × *dim* matrices, adds the matrices, and displays the result.

2. a. Supply the missing lines of code for the procedure *mult_matrices* on the following page that multiplies two 2 × 2 matrices and stores the result in the 2 × 2 matrix, *prod*.

```
PROCEDURE mult_matrices (m,n : matrix; VAR prod : matrix);

VAR
  row, col, j : 1..dim;

BEGIN
  FOR row := 1 TO dim DO
    FOR col := 1 TO dim DO
      FOR j := 1 TO dim DO
        (* student supplies code *);
END;
```

b. Use *get_matrix, mult_matrices,* and *print_matrix* to write a Pascal program that prompts the user for the elements of two *dim* × *dim* matrices, multiplies the two matrices together, and displays the result.

3. a. Use the formula and ideas on A^{-1} in Section 13-6 to write a procedure with the heading

```
PROCEDURE inv (a : matrix; VAR b : matrix);
```

that computes the inverse of a 2 × 2 matrix A and stores the result in a matrix B. Assume the inverse of A exists.

b. Write a program that solves any system of two linear equations in two variables by using matrix operations. If there is no solution or there are infinitely many solutions, the program should say which is the case.

4. The transpose of a matrix, A, is the matrix A^t, having the rows of A as its corresponding columns and the columns of A as its corresponding rows. For example, if

$$A = \begin{vmatrix} 5 & 2 & 1 \\ 3 & 4 & 6 \\ 8 & 0 & 7 \end{vmatrix}$$

then

$$A^t = \begin{vmatrix} 5 & 3 & 8 \\ 2 & 4 & 0 \\ 1 & 6 & 7 \end{vmatrix}$$

Write a procedure *transpose* that computes the transpose of A and stores the result in a matrix T. Use the heading:

```
PROCEDURE transpose (a : matrix; VAR t : matrix);
```

Pictured below is Hydro Quebec in Canada. Civil engineering, fluid mechanics, and electrical engineering are some of the special disciplines required in the development of such a project. All of these depend upon mathematics.

574 *Chapter 14*

Chapter 14

Trigonometric and Circular Functions

Problem Solving Strategies

Drawing a Diagram
Graphs are used extensively in Section 14–6 to illustrate periodic functions. In Sections 14–3 and 14–8 diagrams and tables are used to help students remember graphs of trigonometric functions and variations of them.
Word Problem Plan
The five-step plan is used implicitly in Section 14–9 to solve problems involving right triangles.

Angles and Their Measurement

OBJECTIVES for Sections 14-1 and 14-2:
1. *To find the distance traveled during a given number of revolutions by a point on the rim of a wheel.*
2. *To convert between radian measure and degree measure.*

14–1 Rotations and Angles

How can you measure the distance traveled by a wheeled vehicle in going from one point to another? One way may be described as follows: First, count the number of times a wheel on which the vehicle travels has rotated through a complete revolution in making the trip. Count partial revolutions as fractions of complete revolutions. Then multiply this number of revolutions by the circumference of the wheel. The result will be (in the absence of slippage, of course) the distance the wheel, and hence the vehicle, has moved.

  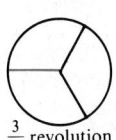 

start $\frac{1}{3}$ revolution $\frac{3}{4}$ revolution   1 revolution

Figure 1

Teaching Suggestions
p. T118

Key Ideas
Redefine angles on the Cartesian plane.

Trigonometric and Circular Functions **575**

1. A coin has a diameter of 1.5 cm. How far will it roll on its edge for 8 revolutions?
   $C \approx 3.14(1.5) = 4.71$ cm
   $s \approx 4.71(8) \approx 37.7$ cm

2. A car wheel has a diameter of 0.7 m. How far will the car travel if the wheel makes 25.5 revolutions?
   $C \approx 3.14(0.7) \approx 2.198$ m
   $s \approx 2.198(25.5) \approx 56.0$ m

Reading Algebra

When reading quantities, the distinction between approximate values and exact values should be made clear. The symbol $\approx$, read "is approximately equal to," is used whenever a value is calculated where an approximation is used for one of the quantities involved. Frequently, in this chapter, 3.14 is used as an approximation to π. (In actuality π is a non-repeating, non-terminating decimal.) Any quantity calculated using this approximation is an approximation itself. For example, when using $\pi \approx 3.14$, the circumference of a circle with diameter 1 cm is "approximately equal to 3.14 cm" not "equal to 3.14 cm."

Suggested Assignments

Extended Alg. with Trig.
  578/3–23 odd, 4, 6, 22
Enriched Alg.
  578/3, 7, 11, 13–25 odd
Enriched Alg. with Trig.
  578/3, 7, 11, 13–25 odd

EXAMPLE A tractor wheel has a diameter of 2 m. How far will the tractor travel as the wheel makes 5.4 revolutions?

SOLUTION To find the circumference C of the wheel, we use $C = \pi d$ where $\pi \approx 3.14$.
Thus $C \approx 3.14(2) = 6.28$.
Then the distance, s, that the tractor travels in 5.4 revolutions is

$$s \approx 6.28(5.4) \approx 33.9.$$

$\therefore$ the distance traveled is approximately 33.9 m.

To study rotating objects, we need to reconsider the concept of *angle*. An *angle* may have been defined in your geometry course as the union of two noncollinear rays that have the same endpoint, as suggested by $\angle AOB$ pictured in Figure 2. We shall now define a **directed angle** as an *ordered pair* of rays with a common endpoint, one ray called the *initial* side of the angle and the other called the *terminal* side of the angle, together with a *rotation* from the initial side to the terminal side (see Figure 3). For this definition we drop the restriction that the rays be noncollinear.

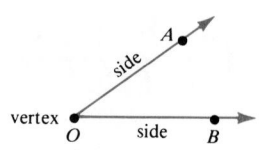

Figure 2

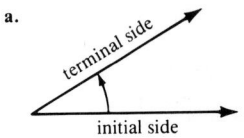
a.

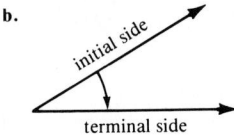
b.

Figure 3

The rotation of the directed angle pictured in Figure 3b is *clockwise;* the rotation pictured in Figure 3a is *counterclockwise.* The angle in Figure 4 has a rotation of $\frac{1}{4}$ of a revolution *counterclockwise.* An angle having a rotation of $\frac{1}{4}$ of a revolution is a **right angle**.

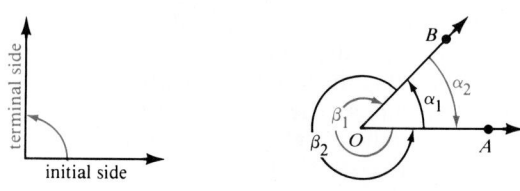

Figure 4

Figure 5

The sides of any geometric angle form the sides of many different directed angles. The angles are often named by Greek letters such as α (alpha) or β (beta). In Figure 5, the sides of $\angle AOB$ are those of angle α_1, which has a counterclockwise rotation of $\frac{1}{8}$ of a revolution; angle α_2, which has a clockwise rotation of $\frac{1}{8}$ of a revolution; angle β_1, which has a

576   *Chapter 14*

clockwise rotation of $\frac{7}{8}$ of a revolution; and angle β_2, which has a counterclockwise rotation of $\frac{7}{8}$ of a revolution.

Angles that have the same initial side and the same terminal side are called **coterminal angles.** In Figure 5, α_1 and β_1 are coterminal angles, as are α_2 and β_2.

To study an angle, it is convenient to consider it as placed on a rectangular coordinate system with the vertex, O, of the angle at the origin and the initial side of the angle as the positive part of the horizontal axis, as shown in Figure 6. The angle is then said to be in **standard position.**

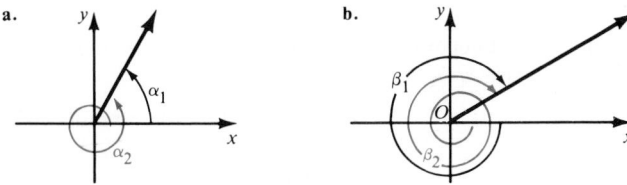

Figure 6

In Figure 6a, angle α_1 and angle α_2 are coterminal. Angle α_1 has a counterclockwise rotation of $\frac{1}{6}$ of a revolution, and angle α_2 has a counterclockwise rotation of $1\frac{1}{6}$ revolutions.

In Figure 6b, angles β_1 and β_2 are coterminal, with angle β_1 having a clockwise rotation of $\frac{11}{12}$ of a revolution and angle β_2 having a clockwise rotation of $1\frac{11}{12}$ of a revolution.

If the terminal side of an angle in standard position lies in a given quadrant, then the angle is said to lie *in* that quadrant. If the terminal side of an angle coincides with a coordinate axis, then the angle is called a **quadrantal angle.** The angles in Figure 6 all lie in the first quadrant. Angle α_1 in Figure 7a lies in the second quadrant; angle α_2 lies in the third quadrant. Angle β_1 in Figure 7b is a quadrantal angle having $\frac{1}{2}$ of a revolution; its terminal side lies in the negative x-axis. It is called a **straight angle.** Angle β_2 is a quadrantal angle having $1\frac{3}{4}$ revolutions; its terminal side lies in the negative y-axis.

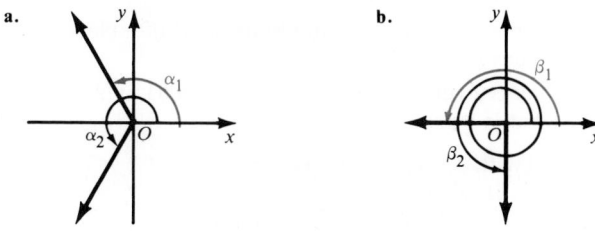

Figure 7

9.

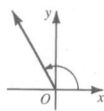

10.

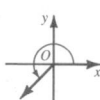

11.

12.

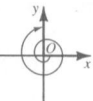

13.

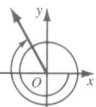

14.

15–20. Answers may vary.

15. $\frac{2}{3}$ revolution clockwise

16. $\frac{3}{8}$ revolution clockwise

17. $\frac{1}{8}$ revolution clockwise

18. $\frac{1}{4}$ revolution counter-clockwise

19. $\frac{1}{3}$ revolution counter-clockwise

20. $\frac{7}{8}$ revolution clockwise

Oral Exercises

A circular flywheel has a radius of 1. How many revolutions does the wheel make when a fixed point on its circumference travels the given distance?

1. 2π 1

2. π $\frac{1}{2}$

3. $\frac{\pi}{4}$ $\frac{1}{8}$

4. $\frac{\pi}{3}$ $\frac{1}{6}$

5. $\frac{3\pi}{2}$ $\frac{3}{4}$

6. 5π $\frac{5}{2}$

Written Exercises

Find the distance traveled by a wheel of the given diameter as it makes the given number of revolutions. Use $\pi \approx 3.14$.

A

1. 25 cm; 1.6 revolutions 126 cm

2. 20 m; 3.5 revolutions 220 m

3. 12 m; 6.9 revolutions 260 m

4. 10 m; 0.6 revolutions 18.8 m

Find the radius of a wheel that will travel the given distance in the given number of revolutions. Use $\pi \approx 3.14$.

5. 942 cm; 4 revolutions 37.5 cm

6. 15.7 m; 8 revolutions 0.313 m

7. 62.8 m; 6 revolutions 1.67 m

8. 25.12 m; 6 revolutions 0.667 m

Sketch the following angles in standard position, indicating the rotation with a curved arrow.

9. $\frac{1}{3}$ revolution counterclockwise

10. $\frac{5}{8}$ revolution counterclockwise

11. $\frac{7}{8}$ revolution counterclockwise

12. $1\frac{3}{4}$ revolutions clockwise

13. $\frac{5}{3}$ revolution clockwise

14. $2\frac{1}{8}$ revolutions counterclockwise

15–20. Express each angle in Exercises 9–14 as a number of revolutions in the opposite direction, clockwise or counterclockwise.

In Exercises 21–23, use $\pi \approx 3.14$.

B

21. How far does a point on the outer tip of the minute hand of a clock travel in 5 min if the hand is 42 cm long? 22.0 cm

22. How far does a point on the outer tip of the hour hand of a clock travel in 5 min if the hand is 7.5 cm long? 0.327 cm

23. A satellite circles Earth once every 6 h. If the satellite is 6800 km from the center of Earth, what is its speed? 7120 km/h

C

24. The radius of the wheel with center O is 4 times that of the wheel with center P. If wheel O is stationary and wheel P rolls around it once, how many revolutions does radius $\overline{PA}$ make? 5

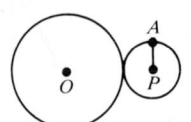

25. In Exercise 24, if wheel P were rolling *inside* wheel O, how many revolutions would radius $\overline{PA}$ make? 3

578   *Chapter 14*

14–2 Measurement of Angles

In Section 14-1, angles were described in terms of complete revolutions and fractions of revolutions. However, in order to make effective use of angles, we need some system of measurement that is based on a smaller unit than a complete revolution.

In one system of measurement a complete revolution is divided into 360 equal parts, each of which is called a degree of rotation or simply a degree. If the rotation is counterclockwise, the measure is ordinarily taken as positive. If the rotation is clockwise, the measure is negative. In Figure 8, the measure of angle α is negative 40 degrees, written as $-40°$, and the measure of angle β is 320°.

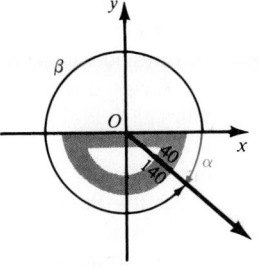

Figure 8

For more precise measurements, degrees can be subdivided into decimal degrees as is done by most calculators, or into *minutes* and *seconds*. Each degree can be divided into 60 equal parts called minutes (denoted by ') and each minute can be divided into 60 equal parts called seconds (denoted by ''). Thus,

$$1° = 60' \quad \text{and} \quad 1' = 60''.$$

The equals sign is used to indicate that we have written two names for the same amount of rotation.

EXAMPLE 1 Find the degree measure of an angle formed by a rotation of:

a. $1\frac{3}{4}$ revolutions clockwise   b. $\frac{17}{32}$ revolution counterclockwise

SOLUTION a. $1\frac{3}{4}(-360°) = -630°$

b. $\frac{17}{32}(360°) = 191.25°$, or $191°15'$

Several notations exist for recording angle measurements. If angle α defines a rotation of 35° we write $m(\alpha) = 35°$. This is read "the measure of α is 35°."

If α and β are coterminal angles, you can write

$$m(\beta) = m(\alpha) + k \cdot 360°, \text{ where } k \text{ is an integer.}$$

For another system of measuring angles, consider $\angle AOB$ placed in standard position on the *uv*-coordinate axes (Figure 9). On the same set of axes, picture the unit circle, that is, the circle with center at (0, 0) and radius 1. Notice that any angle in standard position is determined by the point of intersection of its terminal side with the unit circle. We will measure angles by measuring the length of the arc of the unit circle intercepted by the angle.

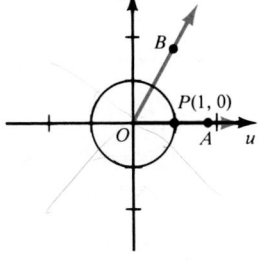

Figure 9

Trigonometric and Circular Functions **579**

Key Ideas

Find degree and radian measures of angles.

Chalkboard Examples

1. Find the degree measure of an angle formed by a rotation of $\frac{4}{5}$ revolutions counterclockwise.

   $\frac{4}{5}(360°) = 288°$

2. Express 450° in radians.

   $\frac{\pi}{180}(450) = \frac{5\pi}{2}$

   $\therefore 450° = \frac{5\pi^{R}}{2}$

3. Find the length of the arc intercepted by a central angle with measure $\frac{8\pi^{R}}{9}$ in a circle whose radius is 36 cm. Use $\pi \approx \frac{22}{7}$.

   $C = 2\pi r \approx 2\left(\frac{22}{7}\right)(36)$

   ≈ 226

   The angle makes $\frac{\frac{8\pi}{9}}{2\pi}$, or $\frac{4}{9}$, of a revolution, so the length of the arc is $\frac{4}{9}$ of the circumference, or

   $\frac{4}{9}(226) \approx 100.$

   $\therefore$ the length of the arc is approximately 100 cm.

Imagine a flexible x number line tangent to the unit circle at $P(1, 0)$, with origin at P and the same scale, or unit length, as on the u and v axes. Think of the number line as being wound around the unit circle as a thread would be wound on a spool. The positive ray would be wound in the counterclockwise direction, and the negative ray in the clockwise direction (Figure 10). This winding procedure pairs each real number (with graph on the flexible number line) with one and only one point on the unit circle. For example, 2 is paired with point S.

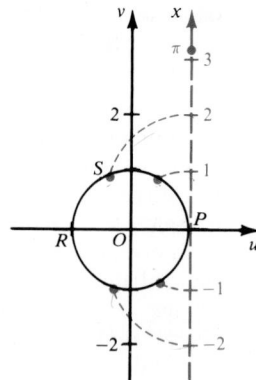

Figure 10

As the flexible number line is wound around the unit circle, more than one number will be paired with the same point. Thus, both π and $-\pi$ are paired with point R. Moreover, the unit circle has a circumference of 2π so that S is paired with $2, 2 + 2\pi, 2 - 2\pi$, and so on, and P with $0, 2\pi, -2\pi, 4\pi$, and so on.

Now we can assign a measure to any angle in standard position as follows: If the *length of an arc* measured from point P is a units, then the *measure of the angle* intercepting that arc is said to be a radians. For example, the measure of the angle α in Figure 11 is 2 radians, written 2^R. The angle for a complete revolution is measured by $2\pi^R$.

A notation similar to that for degree measure is used. Thus in Figure 11, $m(\alpha) = 2^R$. In general, if α and β are coterminal angles, you can write

$$m(\beta) = m(\alpha) + 2k\pi^R, \text{ where } k \text{ is an integer.}$$

Thus, in Figure 11

$$m(\beta) = 2^R + 2(-1)\pi^R = (2 - 2\pi)^R.$$

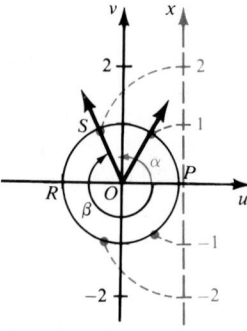

Figure 11

Since an angle for a complete revolution is also measured by 360°, you have

$$360° = 2\pi^R, \quad \text{or} \quad 180° = \pi^R.$$

Thus,

$$1° = \frac{\pi}{180}^R \quad \text{and} \quad 1^R = \frac{180°}{\pi}.$$

To change from degrees to radians, or from radians to degrees, we use the equations above to get *conversion rules*.

---

1. Multiply $\frac{180}{\pi}$ times the measure of the angle in radians to find the angle's measure in degrees.

2. Multiply $\frac{\pi}{180}$ times the measure of the angle in degrees to find the angle's measure in radians.

Since $\pi \approx 3.14159$,

$$\frac{\pi}{180}(1) = 0.01745, \quad \text{or} \quad 1° = 0.01745^R$$

and

$$\frac{180}{\pi}(1) = 57.3, \quad \text{or} \quad 1^R = 57.3°.$$

EXAMPLE 2   **a.** Convert $\frac{7\pi}{6}^R$ to degrees.   **b.** Convert $120°$ to radians.

c. Convert 4^R to the nearest tenth of a degree.

SOLUTION   **a.** $\frac{180}{\pi}\left(\frac{7\pi}{6}\right) = 210$

$$\therefore \frac{7\pi}{6}^R = 210°.$$

b. $\frac{\pi}{180}(120) = \frac{2\pi}{3}$

$$\therefore 120° = \frac{2\pi}{3}^R.$$

c. $\frac{180}{\pi}(4) = \frac{720}{\pi} \approx \frac{720}{3.14} \approx 229.3$

$$\therefore 4^R \approx 229.3°.$$

Whichever system of angle measure you choose, decimal degree, degree-minute-second, or radian, the properties of the angle remain the same. Each system is convenient for different situations. Surveyors use the degree-minute-second system, calculators usually use the decimal degree system, and advanced mathematics usually use radians.

The next two examples use radian measure.

EXAMPLE 3   Find the length of the arc intercepted by a central angle with measure $\frac{7\pi}{4}^R$ in a circle whose radius is 15 cm. Use $\pi \approx \frac{22}{7}$.

SOLUTION $C = 2\pi r = 2\pi(15) \approx 2(\frac{22}{7})(15) = \frac{660}{7} \approx 94.3$

The angle makes $\dfrac{\frac{7\pi}{4}}{2\pi}$, or $\frac{7}{8}$ of a revolution, so the length of the arc is $\frac{7}{8}$ of the circumference, or $\frac{7}{8}(94.3) \approx 82.5$.

$\therefore$ the length of the arc is approximately 82.5 cm.

In general, you can show (Exercise 33, page 583) that in a circle of radius r, the length, s, of the arc intercepted by a central angle α, measured in radians, is given by the formula

$$s = r \cdot m(\alpha).$$

Trigonometric and Circular Functions   **581**

Suggested Assignments

Extended Alg. with Trig.
  First day
  582/3–33 odd
  Second day
R 584/*Self-Test 1*
S 578/10–20 even
Enriched Alg.
  First day
  582/7–31 odd, 32–34
  Second day
R 584/*Self-Test 1*
S 578/6–12 even
Enriched Alg. with Trig.
  First day
  582/7–31 odd, 32–34
  Second day
R 584/*Self-Test 1*
S 578/6–12 even

___

Additional A Exercises

Express each degree measure as a radian measure using π.

1. $125°$ $\dfrac{25\pi}{36}^R$

2. $-50°$ $-\dfrac{5\pi}{18}^R$

3. $400°$ $\dfrac{20\pi}{9}^R$

Express each radian measure as a degree measure.

4. $\dfrac{4\pi^R}{15}$ $48°$

5. $\dfrac{7\pi^R}{5}$ $252°$

6. $\dfrac{\pi^R}{8}$ $22.5°$

EXAMPLE 4 Find the radius of a circle in which arc AB, denoted $\overarc{AB}$, measures 16π and intercepts a central angle of $240°$.

SOLUTION Convert $240°$ to radian measure:

$$\frac{\pi}{180} \cdot 240 = \frac{4\pi}{3}$$

By substitution in the formula on page 581:

$$s = r \cdot m(\alpha)$$
$$16\pi = r \cdot \frac{4\pi}{3}$$
$$12 = r$$

$\therefore$ the radius of the circle is 12.

Oral Exercises

Find the degree measure of an angle of rotation through the given number of revolutions in the given direction.

4. $840°$

1. $\frac{5}{6}$ revolution counterclockwise $300°$

2. $2\frac{1}{4}$ revolutions clockwise $-810°$

3. $2\frac{1}{2}$ revolutions counterclockwise $900°$

4. $\frac{7}{3}$ revolutions counterclockwise

5. $3\frac{2}{5}$ revolutions counterclockwise $1224°$

6. $1\frac{3}{8}$ revolutions clockwise $-495°$

Written Exercises

15. $150°$

Express each degree measure as a radian measure using π.

A **1.** $240°$ $\frac{4\pi}{3}^R$ **2.** $-225°$ $-\frac{5\pi}{4}^R$ **3.** $150°$ $\frac{5\pi}{6}^R$ **4.** $-60°$ $-\frac{\pi}{3}^R$ **5.** $330°$ $\frac{11\pi}{6}^R$

6. $-300°$ $-\frac{5\pi}{3}^R$ **7.** $315°$ $\frac{7\pi}{4}^R$ **8.** $270°$ $\frac{3\pi}{2}^R$ **9.** $-108°$ $-\frac{3\pi}{5}^R$ **10.** $144°$ $\frac{4\pi}{5}^R$

Express each radian measure as a degree measure.

11. $\dfrac{3\pi^R}{4}$ $135°$ **12.** $\dfrac{5\pi^R}{3}$ $300°$ **13.** $-\dfrac{7\pi^R}{4}$ $-315°$ **14.** $\dfrac{3\pi^R}{2}$ $270°$ **15.** $\dfrac{5\pi^R}{6}$

16. $-\dfrac{3\pi^R}{2}$ $-270°$ **17.** $-\dfrac{11\pi^R}{6}$ $-330°$ **18.** $\dfrac{7\pi^R}{3}$ $420°$ **19.** $-\dfrac{8\pi^R}{9}$ $-160°$ **20.** $\dfrac{7\pi^R}{12}$

$105°$

Find the length of the arc on a circle with the given radius that is intercepted by a central angle of the given measure. Use $\pi \approx \frac{22}{7}$.

275 cm

21. 35 cm; $72°$ 44 cm   **22.** 2.8 cm; $330°$ 16.1 cm   **23.** 105 cm; $150°$

24. 630 mm; $\dfrac{5\pi^R}{6}$ 1650 mm   **25.** 56 cm; $\dfrac{\pi^R}{8}$ 22 cm   **26.** 0.42 cm; $\dfrac{9\pi^R}{2}$

5.94 cm

582 *Chapter 14*

Find the radius of a circle in which the arc of given length is intercepted by the given central angle.

27. $\overset{\frown}{AB}$: 30π; $m(\alpha) = 150°$ 36

28. $\overset{\frown}{AB}$: 35π; $m(\alpha) = 315°$ 20

29. $\overset{\frown}{AB}$: $\dfrac{7\pi}{4}$; $m(\alpha) = 210°$ $\frac{3}{2}$

30. $\overset{\frown}{AB}$: $\dfrac{25\pi}{9}$; $m(\alpha) = 300°$ $\frac{5}{3}$

31. $\overset{\frown}{AB}$: $\dfrac{15\pi}{8}$; $m(\alpha) = 225°$ $\frac{3}{2}$

32. $\overset{\frown}{AB}$: $\dfrac{j\pi}{k}$; $m(\alpha) = c°$ $\frac{180j}{ck}$

B **33.** Derive the formula $s = r \cdot m(\alpha)$, given on page 581.

C **34.** Write an arithmetic sequence such that each of its terms expresses the measure in radians of an angle coterminal with $\dfrac{\pi}{2}$.

$$\ldots, -\frac{7\pi}{2}, -\frac{3\pi}{2}, \frac{\pi}{2}, \frac{5\pi}{2}, \frac{9\pi}{2}, \ldots$$

Computer Exercises For students with computer experience

1. Write a program that will convert a positive angle measure given in radians to degrees, minutes, and seconds, giving the result to the nearest whole number of seconds.

Run the program in Exercise 1 for each radian measure.

2. 0.68^R
38°57′40″

3. 1.14^R
65°19′2″

4. 2.56^R
146°40′38″

5. 3.63^R
207°59′2″

ON THE CALCULATOR

Although most calculators cannot automatically convert the measure of an angle from degrees to radians or from radians to degrees, you can use a calculator to simplify the computations of the conversion rules on page 580. Many calculators have a π key that further simplifies the calculations.

Exercises

Use a calculator to convert each degree measure to radian measure. Round your answer to the nearest ten thousandth.

1. 56° 0.9774^R

2. 32.8° 0.5725^R

3. $-24.1°$ -0.4206^R

4. $-480°$ -8.3776^R

5. 13°25′ 0.2342^R

6. 45.6° 0.7959^R

7. 28°15′3″ 0.4931^R

8. $-2°43′19″$ -0.0475^R

Use a calculator to convert each radian measure to decimal degrees. Round your answer to the nearest tenth.

9. 0.4521 25.9°

10. 0.3670 21.0°

11. 0.7291 41.8°

12. -0.8619 $-49.4°$

13. 2.421 138.7°

14. 6.2832 360.0°

15. -4.5236 $-259.2°$

16. 1.3264 76.0°

Trigonometric and Circular Functions **583**

Mixed Review

1. How far will a car travel if its wheel of diameter 37.2 cm makes 5.6 revolutions? Use $\pi \approx 3.14$
654 cm

2. Find the circumference of a circle with radius 12 cm. Use $\pi \approx 3.14$. 75.36 cm

3. Find the fifth term of the geometric sequence 5, 15, 45, ... 405

4. Solve $64 = 2^{2x}$. {3}

Additional Answers Written Exercises

33. Let s = length of arc intercepted by α.

$$\frac{m(\alpha)}{2\pi} = \frac{s}{2\pi r}$$
$$2\pi s = m(\alpha) \cdot 2\pi r$$
$$s = r \cdot m(\alpha)$$

1. How far has a wagon traveled if its wheel with a radius of 56 cm has made 6.5 revolutions? Use $\pi \approx \frac{22}{7}$. 2288 cm

2. Convert $-288°$ to radian measure. $-\frac{8\pi^R}{5}$

3. Convert $\frac{13\pi^R}{9}$ to degree measure. 260°

Teaching Suggestions
p. T118

Key Ideas

Define the trigonometric functions sine and cosine.

Reading Algebra

Be sure students understand that the notation sin x (or cos x) involves no multiplication, but rather is function notation, similar to the notation log x, where the parentheses have been omitted. Although this notation is generally read "sine x," the reading changes when specific angles are used. For example, sin 45° is read "sine of 45 degrees."

Self-Test 1

VOCABULARY   directed angle (p. 576)   degree (p. 579)
right angle (p. 576)   decimal degrees (p. 579)
coterminal angles (p. 577)   minutes (p. 579)
standard position (p. 577)   seconds (p. 579)
quadrantal angle (p. 577)   unit circle (p. 579)
straight angle (p. 577)   radians (p. 580)

1. How far will a point on the rim of a wheel of radius 28 cm travel in 3.2 revolutions? Use $\pi \approx \frac{22}{7}$. 563.2 cm   *Obj. 1, p. 575*

2. Convert 480° to radian measure. $\frac{8\pi^R}{3}$ *Obj. 2, p. 575*

3. Convert $\frac{11\pi^R}{6}$ to degree measure. 330°

Check your answers with those at the back of the book.

The Sine and Cosine Functions

OBJECTIVES for Sections 14-3 through 14-7:
1. To determine the sine and cosine of an angle in standard position given the coordinates of a point other than the origin on the terminal side of the angle.
2. To determine one of the values sin α or cos α given the other value and the quadrant in which α lies
3. To find values for cos α and sin α for specified angles α.
4. To sketch the graphs of $y = A \sin Bx$ and $y = A \cos Bx$ for given constants A and B.

14–3 The Sine and Cosine Functions

Figure 12 shows an angle α in standard position on a uv-coordinate system. $P(u_1, v_1)$ and $R(u_2, v_2)$ are two distinct points different from the origin on the terminal side of α. $\overline{PA}$ and $\overline{RB}$ are perpendicular to the u-axis. Since right triangles OAP and OBR share a common acute angle, α, they are similar. Therefore their sides are proportional, that is:

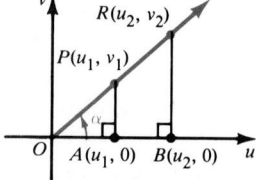

Figure 12

$$\frac{v_1}{\sqrt{u_1{}^2 + v_1{}^2}} = \frac{v_2}{\sqrt{u_2{}^2 + v_2{}^2}} \quad \text{and} \quad \frac{u_1}{\sqrt{u_1{}^2 + v_1{}^2}} = \frac{u_2}{\sqrt{u_2{}^2 + v_2{}^2}}.$$

Although the terminal side of angle α is pictured in Quadrant I in Figure 12, the preceding proportions hold for an angle in any quadrant (see Figure 13, for example) and for quadrantal angles as well. The ratios on page 584 are independent of the points chosen on the terminal side of α, and so for each angle α there are unique values

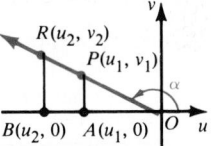

Figure 13

$$\frac{v_1}{\sqrt{u_1^2 + v_1^2}} \quad \text{and} \quad \frac{u_1}{\sqrt{u_1^2 + v_1^2}}.$$

Thus, we can define two functions as follows:

If α is an angle in standard position, and $P(u, v)$ is any point other than the origin on the terminal side of α, and if $\sqrt{u^2 + v^2} = r$, then:

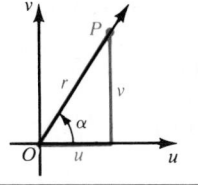

$$\text{sine: } \alpha \longrightarrow \frac{v}{r} \qquad \text{cosine: } \alpha \longrightarrow \frac{u}{r}$$

Values of these functions are denoted by:

sine of α, written $\sin \alpha$ cosine of α, written $\cos \alpha$

The preceding definitions tell you what is meant by the sine and the cosine of any angle *in standard position*. But any angle can be put into standard position, and so *the domain of each of these functions is the set of all angles*. Because of their relation to triangles, these functions are called **trigonometric functions**, where the word "trigonometric" comes from the Greek (*trigonon*, meaning *triangle*, and *metron*, meaning *measure*). There are several other trigonometric functions, which will be defined later in this chapter.

EXAMPLE 1   If α is an angle in standard position and its terminal side contains the point $(-3, 2)$, find $\sin \alpha$ and $\cos \alpha$.

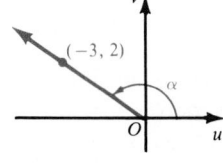

SOLUTION   Substituting -3 for u and 2 for v in the Pythagorean Theorem and the definitions of the functions you have

$$\sqrt{u^2 + v^2} = \sqrt{(-3)^2 + 2^2} = \sqrt{13}$$

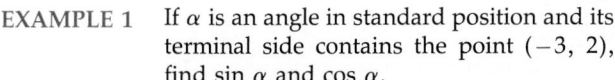

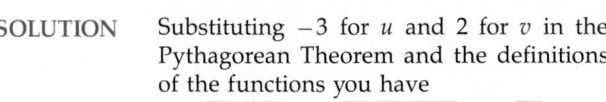

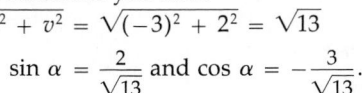

$$\sin \alpha = \frac{2}{\sqrt{13}} \text{ and } \cos \alpha = -\frac{3}{\sqrt{13}}.$$

By computation, you find that for the angle α in the example above,

$$\sin \alpha \approx 0.5547 \quad \text{and} \quad \cos \alpha \approx -0.8321.$$

Trigonometric and Circular Functions   **585**

Suggested Assignments

Extended Alg. with Trig.
  588/1–23 odd
Enriched Alg.
  588/5–25 odd, 24
Enriched Alg. with Trig.
  588/5–25 odd, 24

Additional A Exercises

Find $\sin \alpha$ and $\cos \alpha$ if α is in standard position and passes through the given point. Leave your answers in fractional form.

1. (5, 7)

$$\sin \alpha = \frac{7}{\sqrt{74}}; \ \cos \alpha = \frac{5}{\sqrt{74}}$$

2. (−6, 4)

$$\sin \alpha = \frac{2}{\sqrt{13}};$$

$$\cos \alpha = -\frac{3}{\sqrt{13}}$$

3. (−9, −3)

$$\sin \alpha = -\frac{1}{\sqrt{10}};$$

$$\cos \alpha = -\frac{3}{\sqrt{10}}$$

Mixed Review

1. A group of five judges can be arranged in how many ways along a bench?
 5! = 120

2. State the discriminant and the nature of the roots of the equation $x^2 + 4 = 0$.
 −16; two complex roots

3. Is the relation {(1, 3), (2, 4), (3, 5), (2, 6)} a function? No

4. Solve $\log_2 8 + \log_2 4 = \log_2 x$. {32}

The definitions of $\sin \alpha$ and $\cos \alpha$ make it clear that these values are positive or negative depending on the quadrant in which the terminal side of α lies. Thus, since $u > 0$ and $v < 0$ in the fourth quadrant (and you always have $r > 0$ except at the origin), it follows that in the fourth quadrant,

$$\sin \alpha = \frac{v}{r} < 0$$

and

$$\cos \alpha = \frac{u}{r} > 0.$$

These inequalities and corresponding ones for the other quadrants are shown in Figure 14.

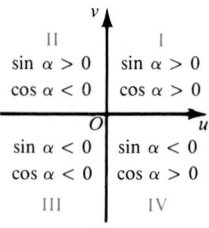

Figure 14

Let us now find the ranges of the sine and cosine functions. Suppose that $R(u, v)$ is the point of intersection of the terminal side of an angle α in standard position with the unit circle (Figure 15). Then you have

$$u^2 + v^2 = 1.$$

$$\sin \alpha = \frac{v}{\sqrt{u^2 + v^2}} = \frac{v}{1} = v,$$

$$\cos \alpha = \frac{u}{\sqrt{u^2 + v^2}} = \frac{u}{1} = u.$$

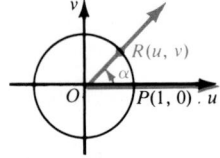

Figure 15

This means that the coordinates of R are $(\cos \alpha, \sin \alpha)$. Thus, the ranges of the sine and cosine functions are given, respectively, by

$$-1 \le \sin \alpha \le 1 \quad \text{and} \quad -1 \le \cos \alpha \le 1.$$

Since the coordinates of R must satisfy $u^2 + v^2 = 1$, you have

$$(\sin \alpha)^2 + (\cos \alpha)^2 = 1,$$

which is usually written

$$\sin^2 \alpha + \cos^2 \alpha = 1.$$

This statement is one of the *fundamental trigonometric identities*. Recall that an identity is an equation that is true for all real values of the variable for which both sides of the equation are defined.

> **Theorem.**   For every angle α,
>
> $$\sin^2 \alpha + \cos^2 \alpha = 1.$$

EXAMPLE 2   Find $\cos \alpha$ when α is in the second quadrant and $\sin \alpha = \frac{3}{5}$.

586   *Chapter 14*

SOLUTION   Since $\sin^2\alpha + \cos^2\alpha = 1$ for every angle α, $(\frac{3}{5})^2 + \cos^2\alpha = 1$. Therefore $\cos^2\alpha = 1 - \frac{9}{25} = \frac{16}{25}$ and $\cos\alpha = \frac{4}{5}$ or $\cos\alpha = -\frac{4}{5}$. Since α is in the second quadrant, choose $\cos\alpha < 0$.

$\therefore \cos\alpha = -\frac{4}{5}$.

EXAMPLE 3   Find $\sin\alpha$ and $\cos\alpha$ when $R(u, v)$ is the point where the terminal side of α (in standard position) intersects the unit circle and $u = v$, $u > 0$.

SOLUTION   Since $R(u, v)$ is on the unit circle, $u = \cos\alpha$ and $v = \sin\alpha$. Substituting u for v in $u^2 + v^2 = 1$ yields the following:

$$u^2 + u^2 = 1$$
$$2u^2 = 1$$
$$u^2 = \frac{1}{2}$$
$$u = \frac{1}{\sqrt{2}} \quad \text{or} \quad u = -\frac{1}{\sqrt{2}}$$

Since $u > 0$, choose the positive value.

$\therefore \sin\alpha = \frac{1}{\sqrt{2}}$, or $\frac{\sqrt{2}}{2}$, and $\cos\alpha = \frac{1}{\sqrt{2}}$, or $\frac{\sqrt{2}}{2}$.

There is an alternative way to define sine and cosine functions, a way in which the elements of the domains of the functions are real numbers instead of angles. Recall that when we set up the system of radian measure, each real number was paired with one and only one point on the unit circle. Thus, with each real number x, you can pair a value $\sin x$ and a value $\cos x$ to define functions

sine: $x \longrightarrow \sin x$ cosine: $x \longrightarrow \cos x$

which each have $\mathcal{R}$ as domain. Each range is, as before,

$$-1 \leq \sin x \leq 1 \quad \text{and} \quad -1 \leq \cos x \leq 1.$$

Because, in this context, values in the domains of these functions can be pictured as lengths of arcs on the unit circle (Figure 16), these functions are sometimes called **circular functions** to distinguish them from the trigonometric functions with the set of angles as domains.

The fact that each angle α has one and only one measure in radians ensures that the circular functions and angle functions have identical properties. Since it is sometimes convenient (and easier) to study properties of one of these kinds of functions in preference to the other, we shall, hereafter, use whichever seems best suited for our purposes, with the understanding that fundamental properties developed using one kind are equally applicable to the other.

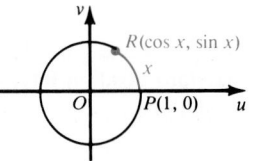

Figure 16

Trigonometric and Circular Functions   **587**

 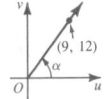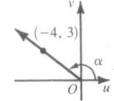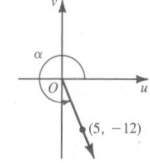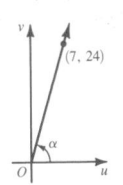

5.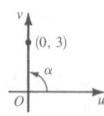

6.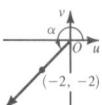

7.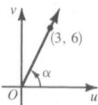

8.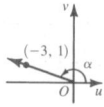

9.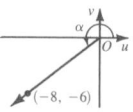

10.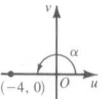

Oral Exercises

State the sine and cosine of the angle.

1. $\angle AOB$ $\frac{1}{2}; \frac{\sqrt{3}}{2}$ 2. $\angle AOC$ $\frac{4}{5}; \frac{3}{5}$

3. $\angle AOD$ $1; 0$ 4. $\angle AOE$ $\frac{\sqrt{2}}{2}; -\frac{\sqrt{2}}{2}$

5. $\angle AOF$ $\frac{3}{5}; -\frac{4}{5}$ 6. $\angle AOG$ $0; -1$

7. $\angle AOH$ 8. $\angle AOT$

$-\frac{\sqrt{2}}{2}; -\frac{\sqrt{2}}{2}$ $-\frac{\sqrt{3}}{2}; \frac{1}{2}$

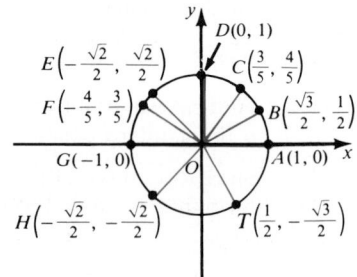

Written Exercises

Sketch the angle α whose terminal side in standard position passes through the given point, and find $\sin \alpha$ and $\cos \alpha$. Leave your answers in fractional form.

$\frac{4}{5}; \frac{3}{5}$ $\frac{3}{5}; -\frac{4}{5}$ $-\frac{12}{13}; \frac{5}{13}$ $\frac{24}{25}; \frac{7}{25}$ $1; 0$

A 1. $(9, 12)$ 2. $(-4, 3)$ 3. $(5, -12)$ 4. $(7, 24)$ 5. $(0, 3)$

6. $(-2, -2)$ 7. $(3, 6)$ 8. $(-3, 1)$ 9. $(-8, -6)$ 10. $(-4, 0)$

$-\frac{1}{\sqrt{2}}; -\frac{1}{\sqrt{2}}$ $\frac{2}{\sqrt{5}}; \frac{1}{\sqrt{5}}$ $\frac{1}{\sqrt{10}}; -\frac{3}{\sqrt{10}}$ $-\frac{3}{5}; -\frac{4}{5}$ $0; -1$

In Exercises 11–18, find $\sin \alpha$ or $\cos \alpha$, whichever is not given, for α in the given quadrant.

11. $\sin \alpha = -\frac{12}{13}$; III $\cos \alpha = -\frac{5}{13}$ 12. $\sin \alpha = -\frac{5}{13}$; III $\cos \alpha = -\frac{12}{13}$

13. $\cos \alpha = \frac{15}{17}$; I $\sin \alpha = \frac{8}{17}$ 14. $\cos \alpha = -\frac{21}{25}$; II $\sin \alpha = \frac{2\sqrt{46}}{25}$

15. $\cos \alpha = \frac{1}{4}$; IV $\sin \alpha = -\frac{\sqrt{15}}{4}$ 16. $\sin \alpha = -\frac{\sqrt{3}}{2}$; IV $\cos \alpha = \frac{1}{2}$

17. $\sin \alpha = \frac{2\sqrt{6}}{5}$; II $\cos \alpha = -\frac{1}{5}$ 18. $\cos \alpha = \frac{5}{7}$; I $\sin \alpha = \frac{2\sqrt{6}}{7}$

Find $\sin \alpha$ and $\cos \alpha$ if $R(u, v)$ is the point where the terminal side of α in standard position intersects the unit circle and u and v satisfy the given conditions.

$-\frac{2}{\sqrt{5}}; -\frac{1}{\sqrt{5}}$ $-\frac{4}{5}; \frac{3}{5}$

B 19. $2u = v, u < 0$ 20. $4u = -3v, u > 0$

21. $u = -3v, u < 0$ $\frac{1}{\sqrt{10}}; -\frac{3}{\sqrt{10}}$ 22. $v = u\sqrt{3}, u < 0$ $-\frac{\sqrt{3}}{2}; -\frac{1}{2}$

C 23. $v = 2u^2 - 2, u < 0, v < 0$ 24. $20v^2 = 9u, v > 0$

25. Show that if $P(a, b)$ is a point on the terminal side of an angle α in standard position, then its image P' under the linear transformation
$$\begin{bmatrix} x' \\ y' \end{bmatrix} = \begin{bmatrix} \cos \alpha & \sin \alpha \\ -\sin \alpha & \cos \alpha \end{bmatrix} \begin{bmatrix} x \\ y \end{bmatrix}$$
is on the x-axis the same distance from the origin as P itself.

23. $-\frac{1}{2}; -\frac{\sqrt{3}}{2}$ 24. $\frac{3}{5}; \frac{4}{5}$

588   *Chapter 14*

588

14–4 Special Values of Sine and Cosine

We used shortened notation to refer to the values of trigonometric functions for specific angles. The notation "sin 30°" means "the value of the sine of the angle whose measure is 30°." The notation "$\cos \frac{\pi^R}{4}$" means "the value of the cosine of the angle whose measure is $\frac{\pi^R}{4}$."

In geometry, you learned that if one acute angle of a right triangle measures 30° $\left(\text{or } \frac{\pi^R}{6}\right)$, then the lengths of the sides of the triangle are in the ratio

$$1 : \sqrt{3} : 2,$$

and that in a right triangle with an acute angle measuring 45° $\left(\text{or } \frac{\pi^R}{4}\right)$, the lengths of the sides are in the ratio

$$1 : 1 : \sqrt{2}.$$

You can use these facts to find sin α and cos α when the measure of α is a multiple of 30° $\left(\text{or } \frac{\pi^R}{6}\right)$ or a multiple of 45° $\left(\text{or } \frac{\pi^R}{4}\right)$.

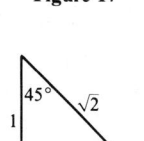

Figure 17

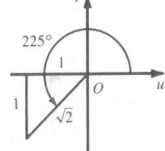

Figure 18

EXAMPLE 1 Find **(a)** sin 60° and **(b)** $\cos \frac{3\pi}{4}^R$.

SOLUTION In each case, sketch an angle α in standard position with the given measure.

a. Since the measure of the angle is 60°, locate point $A(1, 0)$ on the u-axis and complete the right triangle AOP with dimensions as shown at the right.

$$\therefore \sin 60° = \frac{v}{r} = \frac{\sqrt{3}}{2}.$$

b. Since the measure of the given angle is $\frac{3\pi}{4}^R$, the measure of angle β in the diagram is

$$\left(\pi - \frac{3\pi}{4}\right)^R, \text{ or } \frac{\pi^R}{4}.$$

Locate point $A(-1, 0)$ on the u-axis and complete the right triangle AOP with the dimensions as shown at the right. In the second quadrant cos α is negative.

$$\therefore \cos \frac{3\pi}{4}^R = -\frac{1}{\sqrt{2}}.$$

Trigonometric and Circular Functions **589**

Teaching Suggestions
p. T119

Key Ideas

Use right triangles to find trigonometric values for angles of 30°, 45°, and their multiples.

Chalkboard Examples

1. Find cos 225°.

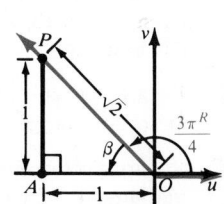

$$\cos 225° = \frac{u}{r} = -\frac{1}{\sqrt{2}}$$

2. Find $\sin \frac{5\pi^R}{6}$.

$$\sin \frac{5\pi^R}{6} = \frac{v}{r} = \frac{1}{2}$$

3. Find all values of α for which the terminal side of α in standard position passes through (0, −5). Give your answer in degrees and radians.

$$r = \sqrt{u^2 + v^2} = \sqrt{0^2 + (-5)^2} = 5$$

$$\sin \alpha = \frac{-5}{5} = -1; \cos \alpha =$$

$$\frac{0}{5} = 0$$

From the table,
$$\alpha = 270° + k \cdot 360°$$

$$\alpha = \frac{3\pi}{2} + 2k\pi, \text{ where } k \text{ is}$$

an integer

Additional A Exercises

Find the given function values.

1. $\sin \dfrac{14\pi^R}{3}$ $\dfrac{\sqrt{3}}{2}$

2. $\cos 3\pi^R$ -1

3. $\sin 570°$ $-\dfrac{1}{2}$

4. $\cos(-135°)$ $-\dfrac{1}{\sqrt{2}}$

Find all values of α for which the terminal side of α in standard position passes through the given point. Give your answer (a) in degrees, and (b) in radians.

5. $(-\sqrt{2}, \sqrt{2})$
  a. $135° + k \cdot 360°$
  b. $\dfrac{3\pi^R}{4} + 2k\pi^R$, where k is
     an integer

6. $(-3, -3\sqrt{3})$
  a. $240° + k \cdot 360°$
  b. $\dfrac{4\pi^R}{3} + 2k\pi^R$, where k is
     an integer

For quadrantal angles, that is, angles with measures which are multiples of $90°$ $\left(\text{or } \dfrac{\pi^R}{2}\right)$, the definitions of $\sin \alpha$ and $\cos \alpha$ are sufficient to provide these values by inspection. Thus, you can see from a particular point P on the unit circle (Figure 19) that:

$\cos 0° = 1$	$\cos 90° = 0$	$\cos 180° = -1$	$\cos 270° = 0$
$\sin 0° = 0$	$\sin 90° = 1$	$\sin 180° = 0$	$\sin 270° = -1$

Using these function values for quadrantal angles and procedures such as those supplied in Example 1, you can construct the table shown below.

You can use this table for circular functions also. For example,

$$\text{if } x = \frac{\pi}{3}, \text{ then } \sin x = \frac{\sqrt{3}}{2} \text{ and } \cos x = \frac{1}{2}.$$

On the other hand, for values of x between 0 and 2π,

$$\text{if } \sin x = \frac{1}{\sqrt{2}}, \text{ then } x = \frac{\pi}{4} \text{ or } x = \frac{3\pi}{4}.$$

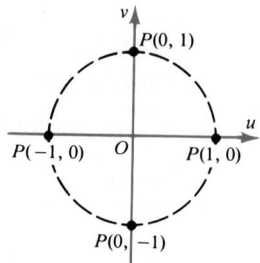

Figure 19

measure of α		$\sin \alpha$	$\cos \alpha$	measure of α		$\sin \alpha$	$\cos \alpha$
$0°$	0^R	0	1	$180°$	π^R	0	-1
$30°$	$\dfrac{\pi^R}{6}$	$\dfrac{1}{2}$	$\dfrac{\sqrt{3}}{2}$	$210°$	$\dfrac{7\pi^R}{6}$	$-\dfrac{1}{2}$	$-\dfrac{\sqrt{3}}{2}$
$45°$	$\dfrac{\pi^R}{4}$	$\dfrac{1}{\sqrt{2}}$	$\dfrac{1}{\sqrt{2}}$	$225°$	$\dfrac{5\pi^R}{4}$	$-\dfrac{1}{\sqrt{2}}$	$-\dfrac{1}{\sqrt{2}}$
$60°$	$\dfrac{\pi^R}{3}$	$\dfrac{\sqrt{3}}{2}$	$\dfrac{1}{2}$	$240°$	$\dfrac{4\pi^R}{3}$	$-\dfrac{\sqrt{3}}{2}$	$-\dfrac{1}{2}$
$90°$	$\dfrac{\pi^R}{2}$	1	0	$270°$	$\dfrac{3\pi^R}{2}$	-1	0
$120°$	$\dfrac{2\pi^R}{3}$	$\dfrac{\sqrt{3}}{2}$	$-\dfrac{1}{2}$	$300°$	$\dfrac{5\pi^R}{3}$	$-\dfrac{\sqrt{3}}{2}$	$\dfrac{1}{2}$
$135°$	$\dfrac{3\pi^R}{4}$	$\dfrac{1}{\sqrt{2}}$	$-\dfrac{1}{\sqrt{2}}$	$315°$	$\dfrac{7\pi^R}{4}$	$-\dfrac{1}{\sqrt{2}}$	$\dfrac{1}{\sqrt{2}}$
$150°$	$\dfrac{5\pi^R}{6}$	$\dfrac{1}{2}$	$-\dfrac{\sqrt{3}}{2}$	$330°$	$\dfrac{11\pi^R}{6}$	$-\dfrac{1}{2}$	$\dfrac{\sqrt{3}}{2}$
$180°$	π^R	0	-1	$360°$	$2\pi^R$	0	1

Since coterminal angles differ in measure by multiples of $360°$, or 2π radians, it follows that for any angle α

$$\sin(\alpha + k \cdot 360°) = \sin \alpha, \qquad \sin(\alpha + 2k\pi^R) = \sin \alpha,$$
$$\cos(\alpha + k \cdot 360°) = \cos \alpha, \qquad \cos(\alpha + 2k\pi^R) = \cos \alpha,$$

where k is an integer.

590   *Chapter 14*

EXAMPLE 2 Find all values of $m(\alpha)$ for which the terminal side of α in standard position passes through $(-4, 4)$.
  a. Give your answer in degrees.   **b.** Give your answer in radians.

SOLUTION Use $u = -4$ and $v = 4$ to obtain r.

$$r = \sqrt{u^2 + v^2} = \sqrt{(-4)^2 + 4^2} = \sqrt{32} = 4\sqrt{2}$$

Then $\sin \alpha = \dfrac{4}{4\sqrt{2}} = \dfrac{1}{\sqrt{2}}$ and $\cos \alpha = \dfrac{-4}{4\sqrt{2}} = -\dfrac{1}{\sqrt{2}}$.

Using the table on page 590 gives the following:

a. $m(\alpha) = 135° + k \cdot 360°$, where k is an integer.

b. $m(\alpha) = \dfrac{3\pi}{4}^{R} + 2k\pi^{R}$, where k is an integer.

Sine and cosine are examples of *periodic* functions. A function f is **periodic** if there is some positive constant p such that

$$f(x + p) = f(x) = f(x - p)$$

for all x in the domain of f. The value p is called a **period** of the function. If there is a smallest positive constant p for which f is periodic, then p is called the **fundamental period** of f. Sine and cosine both have fundamental periods of $360°$ or $2\pi^{R}$. If their domains are each $\mathcal{R}$, the fundamental period is 2π.

Oral Exercises

Find the given function values.

1. $\sin 30°$ $\frac{1}{2}$

2. $\cos 420°$ $\frac{1}{2}$

3. $\cos (-60°)$ $\frac{1}{2}$

4. $\sin \dfrac{7\pi}{2}^{R}$ -1

Written Exercises

Find the given function values.

8. $-\dfrac{\sqrt{3}}{2}$

A
1. $\cos 495°$ $-\frac{1}{\sqrt{2}}$ 2. $\sin (-210°)$ $\frac{1}{2}$ 3. $\cos 765°$ $\frac{1}{\sqrt{2}}$ 4. $\sin 600°$ $-\dfrac{\sqrt{3}}{2}$

5. $\sin \dfrac{9\pi}{4}^{R}$ $\frac{1}{\sqrt{2}}$ 6. $\sin \dfrac{5\pi}{2}^{R}$ 1 7. $\cos \left(-\dfrac{7\pi}{3}^{R}\right)$ $\frac{1}{2}$ 8. $\cos \left(-\dfrac{7\pi}{6}^{R}\right)$

9. $\sin \left(-\dfrac{4\pi}{3}^{R}\right)$ $\dfrac{\sqrt{3}}{2}$ 10. $\cos \dfrac{19\pi}{6}^{R}$ $-\dfrac{\sqrt{3}}{2}$ 11. $\cos (-5\pi^{R})$ -1 12. $\sin \dfrac{15\pi}{4}^{R}$ $-\frac{1}{\sqrt{2}}$

Find all values of $m(\alpha)$ for which the terminal side of α in standard position passes through the given point.
a. Give your answer in degrees.   **b.** Give your answer in radians.

13. $(1, \sqrt{3})$ 14. $(5, -5)$ 15. $(-2\sqrt{3}, -2)$ 16. $(9, 9)$

17. $(-3, -3)$ 18. $(\sqrt{5}, -\sqrt{15})$ 19. $(-\sqrt{3}, -\sqrt{3})$ 20. $(-\frac{1}{3}, \frac{1}{3}\sqrt{3})$

Trigonometric and Circular Functions **591**

1. Insert three arithmetic means between 3 and 47. 14, 25, 36

2. Determine the xy-trace of the equation $2x - y + 3z = 8$. $2x - y = 8$

3. Simplify the expression $2\sqrt[3]{8x^7y^{-3}}z$. $\dfrac{4x^2}{y}\sqrt[3]{xz}$

4. Find the distance between the two points $(-5, 4)$ and $(7, -1)$. 13

Express each radian measure as a degree measure.

5. $-\dfrac{3\pi}{20}^{R}$ $-27°$

6. $\dfrac{7\pi}{9}^{R}$ $140°$

Additional Answers
Written Exercises

13. **a.** $60° + k \cdot 360°$
    b. $\dfrac{\pi}{3}^{R} + 2k\pi^{R}$

14. **a.** $315° + k \cdot 360°$
    b. $\dfrac{7\pi}{4}^{R} + 2k\pi^{R}$

15. **a.** $210° + k \cdot 360°$
    b. $\dfrac{7\pi}{6}^{R} + 2k\pi^{R}$

16. **a.** $45° + k \cdot 360°$
    b. $\dfrac{\pi}{4}^{R} + 2k\pi^{R}$

17. **a.** $225° + k \cdot 360°$
    b. $\dfrac{5\pi}{4}^{R} + 2k\pi^{R}$

18. **a.** $300° + k \cdot 360°$
    b. $\dfrac{5\pi}{3}^{R} + 2k\pi^{R}$

19. **a.** $225° + k \cdot 360°$
    b. $\dfrac{5\pi}{4}^{R} + 2k\pi^{R}$

20. **a.** $120° + k \cdot 360°$
    b. $\dfrac{2\pi}{3}^{R} + 2k\pi^{R}$

(continued)

21. The line defined by $y = \sqrt{3}x$ goes through $(0, 0)$ and $(1, \sqrt{3})$. Since $\sqrt{(1)^2 + (\sqrt{3})^2} = \sqrt{4} = 2$, $\sin \alpha = \frac{\sqrt{3}}{2}$ and $\cos \alpha = \frac{1}{2}$. From the table on page 590, $m(\alpha) = 60°$.

Key Ideas

Use reference angles to find the sine and cosine of an angle in any quadrant.

Chalkboard Examples

Find a four-place decimal approximation for each.

1. $\sin 215°$
$m(\theta) = m(\alpha) - 180°$
$\qquad = 215° - 180° = 35°$
$\sin 215° = -\sin 35°$
$\qquad = -0.5736$
$\therefore \sin 215° = -0.5736$

2. $\cos 332°$
$m(\theta) = 360° - m(\alpha)$
$\qquad = 360° - 332° = 28°$
$\cos 332° = \cos 28°$
$\qquad = 0.8829$
$\therefore \cos 332° = 0.8829$

B **21.** Show that the line defined by $y = \sqrt{3}x$ makes an angle of $60°$ with the positive x-axis.

22. What angle does the line $y = -\dfrac{x}{\sqrt{3}}$ make with the positive x-axis?
$330°$ or $150°$

C **23.** What is the fundamental period of the function $y = \sin nx$ where n is a positive integer? $\frac{2\pi}{n}$

Computer Exercises For students with computer experience

1. Write a program to find $\sin x$ (x in radians) using the series $\sin x = \dfrac{x}{1!} - \dfrac{x^3}{3!} + \dfrac{x^5}{5!} - \dfrac{x^7}{7!} + \ldots$, when the number of terms to be evaluated, n, is given.

Run the program in Exercise 1 for each value of n and x. Answers may vary.

2. $n = 2$, $x = 0.2$
0.1986667

3. $n = 5$, $x = 1.47$
0.9949261

4. $n = 10$, $x = 1.47$
0.9949244

14–5 Reference Angles and Arcs

In the preceding section you used some facts of geometry to find the sine and cosine of certain angles. In order to find the sine or cosine of an angle that is not a multiple of $30°$ or $45°$, you can use a table or a calculator. Tables 6 and 7 in the back of the book give approximate values for the sine and cosine functions of angles whose measures are given from $0°$ to $90°$. The values are given in multiples of tenths and $10'$, respectively. For example, $\sin 28°50' = 0.4823$ (from Table 7) and $\cos 72.8° = 0.2957$ (from Table 6).

To find the sine or cosine of an angle outside the range $0° \leq m(\alpha) \leq 90°$, consider an angle in each of Quadrants II, III, and IV, as shown in Figure 20. Let $P(a, b)$ be any point other than the origin on the terminal side of α. Let T be the point with coordinates $(|a|, |b|)$; thus, T lies in the first quadrant. Then let θ (Greek *theta*) be the angle whose terminal side contains T.

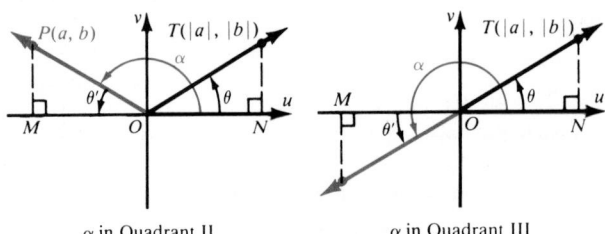

α in Quadrant II $\qquad$ α in Quadrant III $\qquad$ α in Quadrant IV

Figure 20

592 *Chapter 14*

Notice that for α in Quadrant II, right triangles OMP and ONT are congruent and $\theta' \cong \theta$. Thus, the v-coordinates of P and T are equal, but their u-coordinates are opposites of each other; that is, $a = -|a|$. This means

$$\sin \alpha = \frac{b}{\sqrt{a^2 + b^2}} = \frac{|b|}{\sqrt{|a|^2 + |b|^2}} = \sin \theta;$$

$$\cos \alpha = \frac{a}{\sqrt{a^2 + b^2}} = \frac{-|a|}{\sqrt{|a|^2 + |b|^2}} = -\cos \theta.$$

By observing similar relationships between a and $|a|$ and b and $|b|$ in each quadrant, and noting that $\sqrt{a^2 + b^2} = \sqrt{|a|^2 + |b|^2}$, you can see that the sines and cosines of α and θ are related as shown in the table below. Either θ or θ' is called a **reference angle** for α, because you can use it to determine values for trigonometric functions of α. In this book the reference angle will hereafter be denoted by θ.

function value	quadrant in which α lies			
	I	II	III	IV
$\sin \alpha$	$\sin \theta$	$\sin \theta$	$-\sin \theta$	$-\sin \theta$
$\cos \alpha$	$\cos \theta$	$-\cos \theta$	$-\cos \theta$	$\cos \theta$

EXAMPLE   Find a four-decimal place approximation for each of the following.
   a. $\cos 129°$ **b.** $\sin 307.2°$

SOLUTION   Use a sketch to help picture the angle.
   a. From the sketch it can be seen that

$$m(\theta) = 180° - m(\alpha) = 180° - 129° = 51°.$$

From the table above, $\cos 129° = -\cos 51°$.
Then, from Table 7, $\cos 51° = 0.6293$.

$$\therefore \cos 129° = -0.6293.$$

   b. From the sketch it can be seen that

$$m(\theta) = 360° - m(\alpha) = 360° - 307.2°$$
$$= 52.8°.$$

From the table above, $\sin 307.2° = -\sin 52.8°$.
Then, from Table 7 $\sin 52.8° = 0.7965$.

$$\therefore \sin 307.2° = -0.7965.$$

For angles with measures outside the range $0° \le m(\alpha) < 360°$, first use the periodic properties of the trigonometric functions on page 591 to

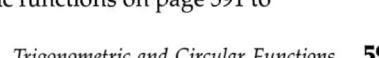

Trigonometric and Circular Functions   **593**

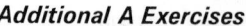

Additional A Exercises

For each function value (a) find the reference angle of the given angle, and (b) evaluate the function using Table 6, 7, or 8.

1. $\sin 261°$
   a. $81°$ b. -0.9877
2. $\cos 364°$
   a. $4°$ b. 0.9976
3. $\sin 702°$
   a. $18°$ b. -0.3090
4. $\cos 148°20'$
   a. $31°40'$ b. -0.8511
5. $\sin 281°$
   a. $79°$ b. -0.9816
6. $\cos 5.07^R$
   a. 1.21^R b. 0.3530

Mixed Review

1. Determine an equation of a line perpendicular to the line $y = 2x - 3$ and passing through the point (3, 3).

$$y = -\frac{1}{2}x + \frac{9}{2}$$

2. If 3 is a root of the polynomial equation $x^3 + 4x^2 - 9x - 36 = 0$, find the other two roots. $-4, -3$

3. Insert two geometric means between 12 and 324.   36, 108

4. Find the product AB of the two matrices below.

$$A = \begin{bmatrix} 3 & 2 \\ 4 & 0 \\ 1 & 3 \end{bmatrix} \quad B = \begin{bmatrix} 2 & -1 \\ 2 & -2 \end{bmatrix}$$

$$AB = \begin{bmatrix} 10 & -7 \\ 8 & -4 \\ 8 & -7 \end{bmatrix}$$

5. Describe fully the graph of the equation $\frac{x^2}{36} + \frac{y^2}{4} = 1$.

An ellipse with x-intercepts 6 and -6, and y-intercepts 2 and -2. The foci are $(4\sqrt{2}, 0)$ and $(-4\sqrt{2}, 0)$.

Suggested Assignments

Extended Alg. with Trig.
  594/7–33 odd
Enriched Alg.
  594/7–43 odd
Enriched Alg. with Trig.
  594/9–47 odd

**Additional Answers
Written Exercises**

1. a. 75°   c. 0.9659
   b.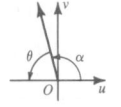

2. a. 55°   c. −0.5736
   b.

3. a. 12°   c. 0.9781
   b.

4. a. 80°   c. −0.9848
   b.

5. a. 15°   c. −0.9659
   b.

6. a. 23°   c. 0.3907
   b.

7. a. 64.2°   c. −0.9003
   b.

find an equivalent value for an angle with measure that is in the range $0° \leq m(\alpha) < 360°$. Then use an appropriate reference angle. For example, to find sin 672°, you would first write

$$\sin 672° = \sin(672° - 360°) = \sin 312°,$$

and then proceed as in part b of the example on the previous page.

When angle measures are given in radians, measures for reference angles can be approximated by using

$$\frac{\pi}{2} \approx 1.57, \qquad \pi \approx 3.14, \qquad \frac{3\pi}{2} \approx 4.71, \qquad \text{and} \qquad 2\pi \approx 6.28.$$

When finding values of circular functions, you may want to think of the real number x as the measure of an arc on the unit circle, and use a *reference arc* with measure x'. For example,

$$\cos 4^R = -\cos(4 - \pi)^R$$
$$\approx -\cos(4.00 - 3.14)^R$$
$$\approx -\cos 0.86^R.$$

Then, from Table 8, $\cos 0.86^R \approx 0.6524$.

$$\therefore \cos 4^R \approx -0.6524.$$

Oral Exercises

8. 36.5°

State the measure of the reference angle θ you would use to evaluate each function.

1. sin 105° 75°   2. cos 235° 55°   3. cos 348° 12°   4. sin 260° 80°
5. cos 195° 15°   6. sin 157° 23°   7. sin 295.8° 64.2°   8. cos 216.5°
9. cos 4.23^R 1.09^R 10. sin 5.78^R 0.50^R 11. cos 1.86^R 1.28^R 12. sin 2.69^R
                                                                    0.45^R

Written Exercises

A 1–12. For each of the function values in Oral Exercises 1–12, **(a)** find the reference angle of the given angle, **(b)** make a sketch showing the given angle (labeled α) and the reference angle (labeled θ), and **(c)** evaluate the function using Table 6, 7, or 8.

17. −0.7880   18. −0.3256   20. −0.7071   24. 0.5299
Evaluate each function using Table 6, 7, or 8.
                                −0.7096   −0.6862   −0.4566
13. sin 136.7° 0.6858   14. sin 314.8°   15. cos 226°40′   16. cos 117°10′
17. sin (−128°)   18. cos (−251°)   19. cos 339° 0.9336   20. sin 675°
21. cos 812° −0.0349   22. sin 542° −0.0349   23. sin 963° −0.8910   24. cos 1022°
25. cos (−3.4^R)   26. sin (−2.7^R)   27. sin 8.7^R 28. cos 11.6^R
−0.9664   −0.4259   0.6594   0.5735
594   *Chapter 14*

In Exercises 29–36 determine the measure of α such that α satisfies the given conditions and $0° \le m(\alpha) < 360°$. Give your answer in degrees and minutes to the nearest 10′ or in decimal degrees to the nearest tenth of a degree.

29. $\sin \alpha = -0.2079$; $\cos \alpha < 0$ 192° **30.** $\cos \alpha = 0.9239$; $\sin \alpha < 0$

31. $\cos \alpha = -0.1132$; $\sin \alpha < 0$ **32.** $\sin \alpha = 0.7716$; $\cos \alpha < 0$

33. $\cos \alpha = -0.9150$; $\sin \alpha > 0$ **34.** $\sin \alpha = -0.7408$; $\cos \alpha < 0$

35. $\sin \alpha = 0.8854$; $\cos \alpha < 0$ **36.** $\cos \alpha = -0.2538$; $\sin \alpha < 0$
117°40′, or 117.7° 255°20′, or 255.3°

In Exercises 37–40 determine the measure of angle α such that the terminal side of α (in standard position) passes through the given point and $0° \le m(\alpha) < 360°$. Give your answer in decimal degrees to the nearest tenth of a degree.

129.5°

B **37.** $(-20, 21)$ 133.6° **38.** $(-15, -8)$ 208.1° **39.** $(2, -\sqrt{5})$ 311.8° **40.** $(-7, 6\sqrt{2})$

In Exercises 41–44 determine the measure of angle α such that the terminal side of α (in standard position) passes through the given point and $0° \le m(\alpha) < 360°$. Give your answer in degrees and minutes to the nearest 10′.

257°20′ 295°20′

41. $(3, -4)$ 306°50′ **42.** $(-5, 12)$ 112°40′ **43.** $(-9, -40)$ **44.** $(3, -2\sqrt{10})$

In Exercises 45–48 determine the measure of angle α such that the terminal side of α (in standard position) passes through the given point and $0^R \le m(\alpha) < 2\pi^R$. Give your answer in radians to the nearest hundredth of a radian.

5.60^R

C **45.** $(15, -8)$ 5.79^R **46.** $(-\sqrt{7}, 3)$ 2.29^R **47.** $(-2, -3\sqrt{5})$ 4.42^R **48.** $(7, -4\sqrt{2})$

Computer Exercises For students with computer experience

1. Write a program that will find the degree measure of the reference angle associated with an angle whose degree measure is given.

Run the program in Exercise 1 for each of the following.

2. 465° 75° **3.** 683° 37° **4.** −520° 20° **5.** −79° 79°

14–6 Graphs of Sine and Cosine I

The sine and cosine circular functions, when specified as sets of ordered pairs of real numbers, are represented respectively by

$$(x, \sin x) \quad \text{and} \quad (x, \cos x).$$

Trigonometric and Circular Functions **595**

---

8. a. 36.5° **c.** −0.8039
b.

9. a. 1.09^R **c.** −0.4625
b.

10. a. 0.50^R **c.** −0.4794
b.

11. a. 1.28^R **c.** −0.2867
b.

12. a. 0.45^R **c.** 0.4350
b.

31. 263°30′, or 263.5°
32. 129°30′, or 129.5°
33. 156°10′, or 156.2°
34. 227°50′, or 227.8°

Teaching Suggestions
p. T119

Graph sine and cosine functions, and use maximum and
minimum values of the sine
and cosine functions to
graph related functions with
different amplitudes.

Chalkboard Examples

1. Sketch the graph of
   $y = 5 \sin x$ on the interval
   $-2\pi \le x \le 2\pi$.

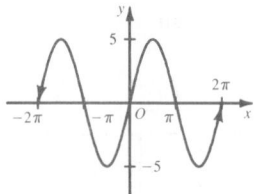

2. Sketch the graph of
   $y = \dfrac{3}{2} \cos x$ on the interval

   $-2\pi \le x \le 2\pi$.

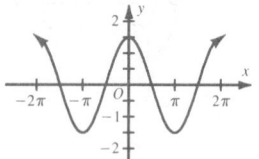

Common Errors

When graphing sine and cosine and related functions,
some students may use
straight lines to connect
those points that are to be
used as guides in graphing.
This results in a jagged
graph instead of a smooth
curve through the points.

Each function has $\mathcal{R}$ as its domain and as its range the set of real numbers $\{y : |y| \le 1\}$. Therefore, each can be graphed in the coordinate plane. Because both functions are periodic with fundamental period 2π, we need to determine their graphs only over the interval $0 \le x \le 2\pi$; the pattern over this interval then repeats endlessly in both directions along the x-axis.

To graph the function $y = \sin x$ you can use the table on page 590 to determine the ordered pairs

$$(x, \sin x)$$

that have multiples of $\dfrac{\pi}{6}$ or $\dfrac{\pi}{4}$ as first components.

When all such ordered pairs with first components that are in the interval $0 \le x \le 2\pi$ are graphed, you obtain Figure 21. Assuming that the graph of sine is a smooth unbroken curve (as it is), you can connect the points shown in Figure 21 to produce the graph shown in Figure 22, which represents one fundamental period.

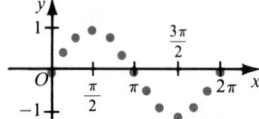

Figure 21

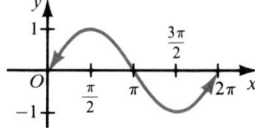

Figure 22

Now you need only to duplicate the pattern shown in Figure 22 over successive intervals of length 2π along the x-axis to obtain as much of the graph of the sine function as you wish. Figure 23 pictures the graph over $-4\pi \le x \le 4\pi$. This graph is called a **sine wave**.

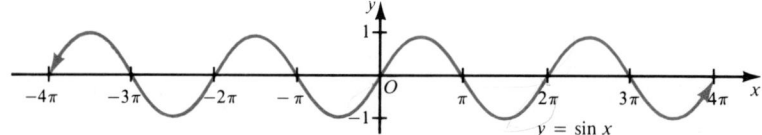

Figure 23

To graph the function $y = \cos x$ you can use the same procedure as for the sine function. Graphing points for ordered pairs from the table on page 590, you obtain the pattern shown in Figure 24, and, upon connecting the points, you have the graph of one fundamental period of the cosine function (Figure 25).

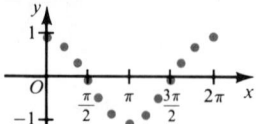

Figure 24

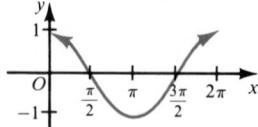

Figure 25

596   *Chapter 14*

Drawing three more periods produces the graph of cosine over $-4\pi \le x \le 4\pi$, as shown in Figure 26. This graph is also an example of a sine wave, but one displaced along the x-axis $\frac{\pi}{2}$ to the left.

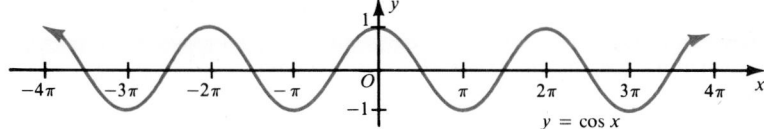

Figure 26

Notice that both curves (Figures 23 and 26) lie between the graphs of the *maximum* y-coordinate, 1, and the *minimum* y-coordinate, -1. When a periodic function attains a maximum value M and a minimum value m, you say that the function has an **amplitude** of

$$\frac{M - m}{2}.$$

Thus, the amplitude of sine and cosine is $\dfrac{1 - (-1)}{2}$, or 1.

Notice that the maximum y-coordinate in the graph of $y = \sin x$ occurs at $\frac{1}{4}$ of the distance across the fundamental periodic interval, which is $0 \le x \le 2\pi$ (Figure 22). That is,

$$\sin \frac{\pi}{2} = 1,$$

and the minimum y-coordinate occurs at $\frac{3}{4}$ of the distance across that interval, that is,

$$\sin \frac{3\pi}{2} = -1.$$

Correspondingly, the maximum y-coordinates in the graph of $y = \cos x$ occur at the beginning and the end of the interval (see Figure 25), that is

$$\cos 0 = 1 \quad \text{and} \quad \cos 2\pi = 1,$$

and the minimum at the midpoint of the interval, that is,

$$\cos \pi = -1.$$

Next consider the graph of $y = 2\cos x$ shown in Figure 27. Comparing this graph to the graph of $y = \cos x$ in Figure 26, you should notice that the y-coordinate of each point on the graph of $y = 2\cos x$ is 2 times the y-coordinate of each point on the graph of $y = \cos x$. For $J = \{\text{integers}\}$ the maximum values on the graph of $y = 2\cos x$ occur at the points $(2k\pi, 2)$, $k \in J$, and the minimum values occur at the points $(\pi + 2k\pi, -2)$, $k \in J$.

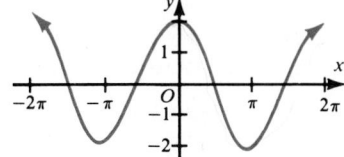

Figure 27

Trigonometric and Circular Functions **597**

Suggested Assignments

Extended Alg. with Trig.
598/1–13 odd
Enriched Alg.
598/1–19 odd
Enriched Alg. with Trig.
598/1–19 odd

Additional A Exercises

Sketch the graph of each function on the interval $-2\pi \le x \le 2\pi$.

1. $y = -4 \sin x$

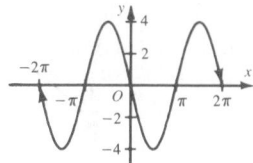

2. $y = 3 \sin x$

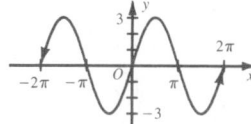

3. $y = 4 \cos x$

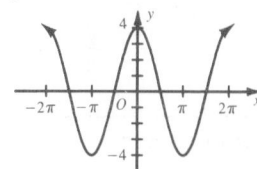

Determine the measure of α such that $0° \leq m(\alpha) \leq 360°$ and α satisfies the conditions in each exercise. Give your answer in degrees and minutes to the nearest 10' or in decimal degrees to the nearest tenth of a degree.

1. $\sin \alpha = 0.8339$; $\cos \alpha < 0$
   123°30' or 123.5°

2. $\cos \alpha = 0.9986$; $\sin \alpha < 0$
   357°

3. $\sin \alpha = 0.9563$; $\cos \alpha > 0$
   73°

4. A card is chosen at random from a standard deck of playing cards. What is the probability that the card is a red 4? $\frac{1}{26}$

5. Simplify the expression $(3 + 2i)(\sqrt{-64} + 2i)(1 - i)$.
   $50i + 10$

6. Express 420° as a radian measure using π. $\frac{7\pi^R}{3}$

Additional Answers
Written Exercises

1.

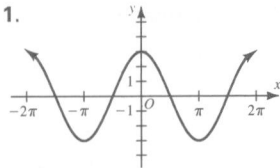

2.

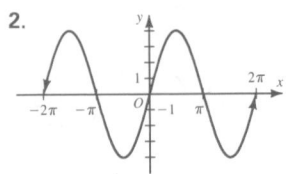

Similarly in Figure 28, the y-coordinate of each point on the graph of $y = -2 \sin x$ is -2 times the y-coordinate of each point on the graph of $y = \sin x$ shown in Figure 23. Note also that the graph of

$$y = -2 \sin x$$

is a reflection over the x-axis of the graph of

$$y = 2 \sin x.$$

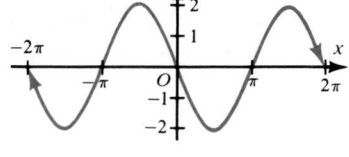

Figure 28

The maximum values occur at the points $\left(\frac{3\pi}{2} + 2k\pi, 2\right)$, $k \in J$ and the minimum values occur at the points $\left(\frac{\pi}{2} + 2k\pi, -2\right)$, $k \in J$.

In general the functions $y = A \sin x$ and $y = A \cos x$ $(A \neq 0)$, have fundamental period 2π and amplitude $|A|$.

Oral Exercises

a. **State the maximum and minimum values of each function.**
b. **State the amplitude of each function.**

1. $y = 3 \cos x$ a. 3; −3   b. 3
2. $y = 4 \sin x$ a. 4; −4   b. 4
3. $y = -\sin x$ a. 1; −1   b. 1
4. $y = \frac{1}{2} \cos x$ a. $\frac{1}{2}$; −$\frac{1}{2}$ b. $\frac{1}{2}$
5. $y = -\frac{3}{2} \sin x$ a. $\frac{3}{2}$; −$\frac{3}{2}$ b. $\frac{3}{2}$
6. $y = -3 \cos x$ a. 3; −3   b. 3

Written Exercises

Sketch the graph of each function on the interval $-2\pi \leq x \leq 2\pi$.

A 1–6. Sketch the graphs of the functions in Oral Exercises 1–6.

7. $y = -\frac{2}{3} \cos x$ 8. $y = -\frac{4}{3} \sin x$ 9. $y = \frac{5}{2} \sin x$

In Exercises 10–13, give the maximum and minimum values of the function and state the amplitude.

10. $y = 2 \sin x + 2$ 4; 0; 2   11. $y = 3 \cos x - 1$ 2; −4; 3
12. $y = \frac{1}{2} \cos x + 1$ $\frac{3}{2}$; $\frac{1}{2}$; $\frac{1}{2}$ 13. $y = -4 \sin x + 3$ 7; −1; 4

B 14–17. Graph the functions in Exercises 10–13 above on the interval $-2\pi \leq x \leq 2\pi$.

C 18. Graph $y = \cos\left(x - \frac{\pi}{2}\right)$ over the interval $-2\pi \leq x \leq 2\pi$. Can you recognize this graph as the graph of an equivalent function? $y = \sin x$

598   *Chapter 14*

19. On one set of axes, sketch the graphs of $y = \sin x$, $y = -\sin x$, and $y = \sin(-x)$ on the interval $-2\pi \leq x \leq 2\pi$. Which graphs are the same? $y = -\sin x$ and $y = \sin(-x)$

20. On one set of axes, sketch the graphs of $y = \cos x$, $y = -\cos x$, and $y = \cos(-x)$ on the interval $-2\pi \leq x \leq 2\pi$. Which graphs are the same? $y = \cos x$ and $y = \cos(-x)$

14–7 Graphs of Sine and Cosine II

In Section 14-6 you saw that the graphs of

$$y = A \sin x \qquad \text{and} \qquad y = A \cos x$$

were sine waves with amplitude $|A|$ and fundamental period 2π.

Next consider the graph of $y = \sin 2x$ shown in Figure 29. Notice that as x varies from 0 to π, $2x$ varies from 0 to 2π. Hence, $\sin 2x$ will assume all the values of sine while in the interval $0 \leq x \leq \pi$. Thus, the fundamental period of this function is π rather than 2π.

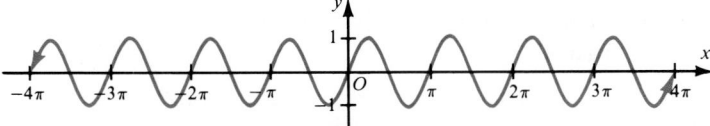

Figure 29

In general, the graphs of the equations

$$y = \sin Bx \qquad \text{and} \qquad y = \cos Bx \ (B \neq 0)$$

have fundamental period $\dfrac{2\pi}{|B|}$ and amplitude 1.

Combining these results with those in Section 14-6, you have the following result:

The graphs of the equations

$$y = A \sin Bx \tag{1}$$
$$y = A \cos Bx \tag{2}$$

each have amplitude $|A|$ and fundamental period $\dfrac{2\pi}{|B|}$. These graphs are sine waves.

These facts can be used to make a rapid sketch of such a graph.

Trigonometric and Circular Functions **599**

3.
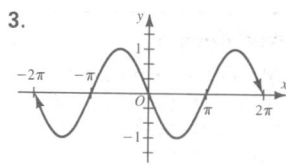

(continued on p. 613)

Teaching Suggestions
p. T119

Supplementary Material
Test 45

Key Ideas

Use the graphs of the sine and cosine functions to graph related functions with different periods.

Chalkboard Examples

Sketch the graph over one fundamental period.

1. $y = -3 \cos \dfrac{1}{2}x$

Amplitude: $|A| = 3$; fundamental period:

$$\dfrac{2\pi}{\frac{1}{2}} = 4\pi;$$

maximum: $(2\pi, 3)$; minimum: $(0, -3)$, $(4\pi, -3)$; intercepts: $(\pi, 0)$, $(3\pi, 0)$.

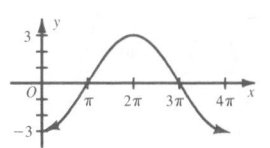

600

Additional A Exercises

Sketch the graph of the given function over the interval $0 \le x \le 2\pi$ if its fundamental period p is less than or equal to 2π. Otherwise, graph the function over the interval $0 \le x \le p$.

$y = 2 \sin \frac{1}{2}x$

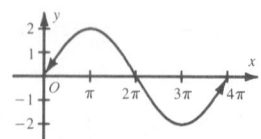

Mixed Review

Find the reference angle of α, and $\cos \alpha$ and $\sin \alpha$, if α is in standard position and the terminal side passes through the given point.

1. $(-2, 0)$ $0°$, $\sin \alpha = 0$, $\cos \alpha = -1$

EXAMPLE Sketch the graph of $y = 2 \cos \frac{1}{2}x$ over one fundamental period.

SOLUTION The amplitude is 2 and the fundamental period is $\frac{2\pi}{\frac{1}{2}} = 4\pi$. Then over the interval $0 \le x \le 4\pi$:

1. The maximum points are $(0, 2)$ and $(4\pi, 2)$.
2. The minimum point occurs when $x = \frac{1}{2} \cdot 4\pi = 2\pi$; it is $(2\pi, -2)$.
3. The intercepts are midway between the maximum and minimum points, that is at $(\pi, 0)$ and $(3\pi, 0)$.

These facts yield the curve shown.

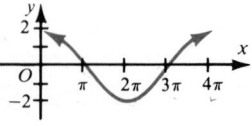

To graph the trigonometric functions sine and cosine, each having as domain the set of all angles, we represent each angle by its measure. If the angles are measured in radians, the graphs are the same as those of the associated circular functions. If the angles are measured in the degree system, the scale on the horizontal axis must be labeled accordingly. At the right is the graph of one fundamental period of

$$y = \cos \alpha,$$

where the measure of α in radians is shown in red along the horizontal axis, and the measure of α in degrees is shown in black along the horizontal axis.

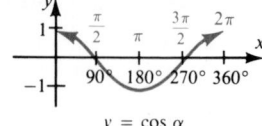

$y = \cos \alpha$

Figure 30

Oral Exercises

State the amplitude, $|A|$, and the fundamental period, p, of each function.

1. $y = \cos 3x$ $1; \frac{2\pi}{3}$

2. $y = \sin 2x$ $1; \pi$

3. $y = 3 \sin \frac{1}{2}x$ $3; 4\pi$

4. $y = \frac{3}{2} \cos 2x$ $\frac{3}{2}; \pi$

5. $y = 4 \cos \frac{1}{3}x$ $4; 6\pi$

6. $y = -3 \sin \frac{1}{4}x$ $3; 8\pi$

Written Exercises

Sketch the graph of the given function over the interval $0 \le x \le 2\pi$ if its fundamental period $p \le 2\pi$. Otherwise, graph the function over the interval $0 \le x \le p$.

A **1–6.** Graph the functions in Oral Exercises 1–6.

7. $y = -\frac{1}{2} \cos 3x$

8. $y = 3 \sin \frac{3}{2}x$

9. $y = 2 \sin \frac{3}{4}x$

10. $y = 2 \cos(-4x)$

11. $y = 4 \sin(-3x)$

12. $y = \cos \pi x$

600 *Chapter 14*

Graph each pair of functions on one set of axes over the given interval.

B 13. $y = 2 \cos 2x$; $y = \frac{1}{3}\cos 2x$; $0 \le x \le 2\pi$

14. $y = 3 \cos \frac{1}{2}x$; $y = 3 \cos \frac{1}{4}x$; $0 \le x \le 8\pi$

15. $y = \sin \pi x$; $y = \sin \frac{\pi}{2}x$; $0 \le x \le 4$

16. $y = 2 \sin 2x$; $y = 2 \sin 3x$; $0 \le x \le 2\pi$

C 17. $y = \sin 4x$; $y = \sin 4x + 1$; $0 \le x \le 2\pi$

18. $y = \cos 2x$; $y = \cos 2x - 2$; $0 \le x \le 2\pi$

19. $y = \sin \frac{1}{2}x$; $y = \sin \left(\frac{1}{2}x - \frac{\pi}{2}\right)$; $0 \le x \le 4\pi$

20. $y = \sin 3x$; $y = |\sin 3x|$; $0 \le x \le 2\pi$

◼ Self-Test 2

VOCABULARY   sine of an angle (p. 585)
cosine of an angle (p. 585)
trigonometric functions
    (p. 585)
circular functions (p. 587)
periodic function (p. 591)
period of a function (p. 591)
fundamental period (p. 591)
reference angle (p. 593)
sine wave (p. 596)
amplitude (p. 597)

1. Determine $\sin \alpha$ and $\cos \alpha$ if the terminal side of angle α *Obj. 1, p. 584*
in standard position passes through $(-4, 4\sqrt{3})$. $\frac{\sqrt{3}}{2}$; $-\frac{1}{2}$

2. Determine $\cos \alpha$ if $\sin \alpha = -\frac{15}{17}$ and the terminal side of α *Obj. 2, p. 584*

lies in Quadrant III. $-\frac{8}{17}$

Evaluate each function using a calculator or Table 6, 7, or 8.

3. $\sin 83.7°$ 0.9940   4. $\cos 242°$ −0.4695   *Obj. 3, p. 584*

5. $\sin 1.3^R$ 0.9636   6. $\cos (-7.8^R)$ 0.0508

7. Sketch the graph of $y = 3 \sin \frac{1}{4}x$ over one fundamental   *Obj. 4, p. 584*
period.

Check your answers with those at the back of the book.

Trigonometric and Circular Functions   **601**

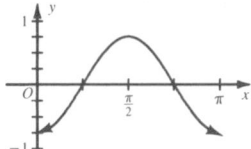

Other Trigonometric and Circular Functions and Applications

OBJECTIVES for Sections 14-8 and 14-9:
1. *To find values for tan x, sec x, cot x, and csc x for values of x.*
2. *To solve and apply solutions of right triangles.*

Teaching Suggestions
p. T120

Related Activities p. T120

Key Ideas

Extend the trigonometric functions to include the tangent function and the reciprocal functions cosecant, secant, and cotangent.

Chalkboard Examples

If the terminal side of α, in standard position, passes through the given point, find tan α, cot α, sec α, and csc α.

1. (5, −12)

Since $u = 5$ and $v = -12$:
$$r = \sqrt{u^2 + v^2} = \sqrt{5^2 + 12^2}$$
$$= 13$$

$$\tan \alpha = \frac{\sin \alpha}{\cos \alpha} = \frac{\frac{-12}{13}}{\frac{5}{13}}$$

$$= -\frac{12}{5}$$

$$\cot \alpha = \frac{\cos \alpha}{\sin \alpha} = \frac{\frac{5}{13}}{\frac{-12}{13}} = -\frac{5}{12}$$

14–8 The Tangent, Cotangent, Secant, and Cosecant Functions

Several other trigonometric functions in common use are defined in terms of sine and cosine as follows:

tangent: $\alpha \longrightarrow \dfrac{\sin \alpha}{\cos \alpha}$, cos $\alpha \neq 0$ secant: $\alpha \longrightarrow \dfrac{1}{\cos \alpha}$, cos $\alpha \neq 0$

cotangent: $\alpha \longrightarrow \dfrac{\cos \alpha}{\sin \alpha}$, sin $\alpha \neq 0$ cosecant: $\alpha \longrightarrow \dfrac{1}{\sin \alpha}$, sin $\alpha \neq 0$

Values of these functions are denoted by:

tangent of α, written tan α secant of α, written sec α
cotangent of α, written cot α cosecant of α, written csc α

Notice that tangent and secant are not defined when $m(\alpha)$ is 90°, 270°, and so forth. $\left(\text{or } \dfrac{\pi}{2}^{\text{R}}, \text{ etc.}\right)$, and cotangent and cosecant are not defined when $m(\alpha)$ is 0°, 180°, etc. (or π^{R}, etc.).

EXAMPLE   If the terminal side of α, in standard position, passes through the point $(-3, 4)$, find each of the following.

     a. tan α **b.** cot α **c.** sec α **d.** csc α

SOLUTION   Since $u = -3$ and $v = 4$,

$$r = \sqrt{u^2 + v^2} = \sqrt{(-3)^2 + 4^2} = 5.$$

$$\text{Thus } \sin \alpha = \frac{v}{r} = \frac{4}{5} \text{ and } \cos \alpha = \frac{u}{r} = -\frac{3}{5}.$$

a. $\tan \alpha = \dfrac{\sin \alpha}{\cos \alpha} = \dfrac{\frac{4}{5}}{-\frac{3}{5}}$ **b.** $\cot \alpha = \dfrac{\cos \alpha}{\sin \alpha} = \dfrac{-\frac{3}{5}}{\frac{4}{5}}$

$\therefore \tan \alpha = -\dfrac{4}{3}$ $\therefore \cot \alpha = -\dfrac{3}{4}$

c. $\sec \alpha = \dfrac{1}{\cos \alpha} = \dfrac{1}{-\frac{3}{5}}$ **d.** $\csc \alpha = \dfrac{1}{\sin \alpha} = \dfrac{1}{\frac{4}{5}}$

$\therefore \sec \alpha = -\dfrac{5}{3}$ $\therefore \csc \alpha = \dfrac{5}{4}$

602   *Chapter 14*

By using the definitions of sine and cosine on page 585,

$$\text{sine: } \alpha \longrightarrow \frac{v}{r}, \qquad \text{cosine: } \alpha \longrightarrow \frac{u}{r},$$

you can verify the following theorem.

> **Theorem.** If α is an angle in standard position, with $P(u, v)$ any point other than the origin on the terminal side of α, and if $\sqrt{u^2 + v^2} = r$, then:
>
> $$\text{tangent: } \alpha \longrightarrow \frac{v}{u}, \, u \neq 0 \qquad \text{secant: } \alpha \longrightarrow \frac{r}{u}, \, u \neq 0$$
>
> $$\text{cotangent: } \alpha \longrightarrow \frac{u}{v}, \, v \neq 0 \qquad \text{cosecant: } \alpha \longrightarrow \frac{r}{v}, \, v \neq 0$$

From the theorem you can see why tangent and cotangent are called **reciprocal functions.** Sine and cosecant are also reciprocal functions, as are cosine and secant.

Although the theorem above is stated for angles in standard position, the domain of each of the functions is the set of all angles, since any angle can be put into standard position.

We can now extend Figure 14 to Figure 31, shown below.

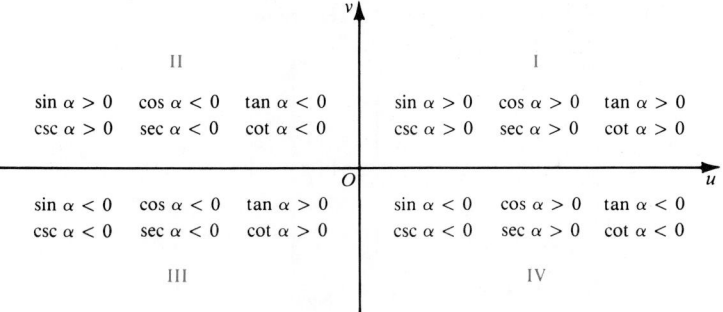

Figure 31

The circular functions tangent, cotangent, secant, and cosecant are defined in terms of arc length on the unit circle (Figure 32); since $\sin x = v$ and $\cos x = u$:

$$\tan x = \frac{v}{u}, \, u \neq 0 \qquad \sec x = \frac{1}{u}, \, u \neq 0$$

$$\cot x = \frac{u}{v}, \, v \neq 0 \qquad \csc x = \frac{1}{v}, \, v \neq 0$$

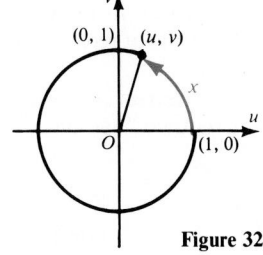

Figure 32

Trigonometric and Circular Functions **603**

$$\sec \alpha = \frac{1}{\cos \alpha} = \frac{1}{\dfrac{5}{13}} = \frac{13}{5}$$

$$\csc \alpha = \frac{1}{\sin \alpha} = \frac{1}{\dfrac{-12}{13}}$$

$$= -\frac{13}{12}$$

2. $(-6, -8)$

Since $u = -6$ and $v = -8$,
$r = \sqrt{(-6)^2 + (-8)^2} = 10$

$$\tan \alpha = \frac{\sin \alpha}{\cos \alpha} = \frac{\dfrac{-8}{10}}{\dfrac{-6}{10}} = \frac{4}{3}$$

$$\cot \alpha = \frac{\cos \alpha}{\sin \alpha} = \frac{\dfrac{-6}{10}}{\dfrac{-8}{10}} = \frac{3}{4}$$

$$\sec \alpha = \frac{1}{\cos \alpha} = \frac{1}{\dfrac{-6}{10}} = -\frac{5}{3}$$

$$\csc \alpha = \frac{1}{\sin \alpha} = \frac{1}{\dfrac{-8}{10}} = -\frac{5}{4}$$

Common Errors

Students may incorrectly determine the sign of the function value. Use a mnemonic device like: **A**ll **S**tudents **T**ake **C**alculus. This phrase can help students remember which functions are positive in a given quadrant:
Quadrant I: **A**ll
Quadrant II: **S**ine
Quadrant III: **T**angent
Quadrant IV: **C**osine

Suggested Assignments

Extended Alg. with Trig.
  605/1–47 odd
Enriched Alg.
  605/1–51 odd
Enriched Alg. with Trig.
  605/1–51 odd

Special values for tan α, cot α, sec α, and csc α can be computed from the values for sin α and cos α given in the table on page 590. You can use these values in sketching graphs of these functions, as shown in Figures 33–36 below.

Notice that the fundamental period of tangent and cotangent is π rather than 2π, and that the tangent, cotangent, secant, and cosecant functions each have certain real numbers excluded from their respective domains.

Furthermore, as x increases from 0 to $\frac{\pi}{2}$, the function values vary as follows:

sin x increases, tan x increases, sec x increases,
cos x decreases, cot x decreases, csc x decreases.

Values of tangent, cotangent, secant, and cosecant can be found from Tables 6, 7, and 8 at the back of the book.

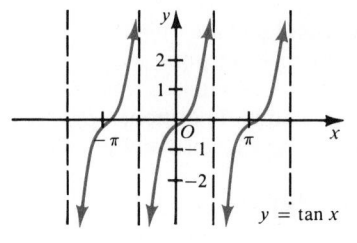

Figure 33

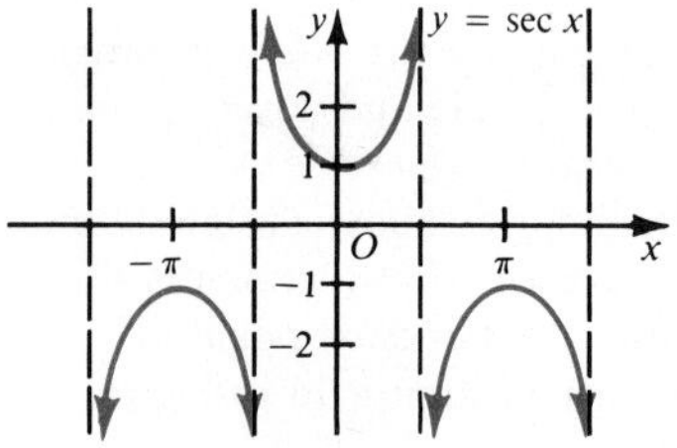

Figure 34

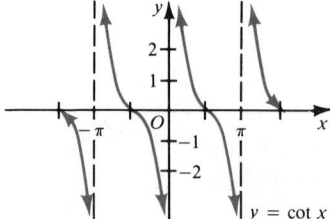

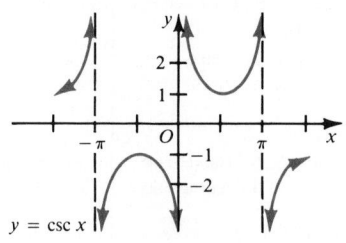

Figure 35

Figure 36

Compare the graph in Figure 33 with the graph of its reciprocal function in Figure 35. The graphs in Figure 34 and Figure 26 are also graphs of reciprocal functions, as are the graphs in Figure 36 and Figure 23.

We can now summarize the properties of the six circular functions that we have discussed (J = {the integers}).

sine: Domain: $\mathcal{R}$
Range: $\{y: y \in \mathcal{R}, |y| \leq 1\}$
Fundamental period: 2π

Mixed Review

1. How many different sequences of heads or tails are possible if a coin is flipped 6 times? 64
2. In how many ways can 7 people be seated at a round table? 720
3. How many 2-letter permutations are there in the word CLOTH? 20
4. Determine the slope of the line which passes through the points (−4, −6) and (6, −1). $\frac{1}{2}$
5. If y varies jointly as x^2 and z, and y is 36 when x is 2 and z is 3, what is the value of y when x is 3 and z is 4? 108

Additional A Exercises

Use Table 6, 7, or 8 to find the given values to four significant digits.

1. tan 217° 0.7536
2. csc (−170°) −5.759
3. sec 297° 2.203
4. cot 1.75^R −0.1828

Additional Answers Written Exercises

25. $\sin\alpha = -\frac{4}{5}$; $\cos\alpha = \frac{3}{5}$;
$\tan\alpha = -\frac{4}{3}$; $\cot\alpha = -\frac{3}{4}$;
$\sec\alpha = \frac{5}{3}$; $\csc\alpha = -\frac{5}{4}$
26. $\sin\alpha = \frac{12}{13}$; $\cos\alpha = -\frac{5}{13}$;
$\tan\alpha = -\frac{12}{5}$; $\csc\alpha = \frac{13}{12}$;
$\sec\alpha = -\frac{13}{5}$;
$\cot\alpha = -\frac{5}{12}$

cosine:	Domain: $\mathcal{R}$
	Range: $\{y: y \in \mathcal{R}, |y| \leq 1\}$
	Fundamental period: 2π

tangent: Domain: $\left\{x: x \in \mathcal{R}, x \neq \dfrac{(2k+1)\pi}{2}, k \in J\right\}$

Range: $\mathcal{R}$
Fundamental period: π

cotangent: Domain: $\{x: x \in \mathcal{R}, x \neq k\pi, k \in J\}$
Range: $\mathcal{R}$
Fundamental period: π

secant: Domain: $\left\{x: x \in \mathcal{R}, x \neq \dfrac{(2k+1)\pi}{2}, k \in J\right\}$

Range: $\{y: y \in \mathcal{R}, |y| \geq 1\}$
Fundamental period: 2π

cosecant: Domain: $\{x: x \in \mathcal{R}, x \neq k\pi, k \in J\}$
Range: $\{y: y \in \mathcal{R}, |y| \geq 1\}$
Fundamental period: 2π

Oral Exercises

Use Figure 31 on page 603 to tell whether the given function will be positive, negative, or zero.

1. tan 132° neg. **2.** sec 305° pos. **3.** csc 213°10′ neg. **4.** tan 325°40′ neg.

5. cot 247.3° pos. **6.** csc 289.7° neg. **7.** sec 2.46^R neg. **8.** cot 3.75^R pos.

1. −1.111 **2.** 1.743 **3.** −1.828 **4.** −0.6830 **5.** 0.4183
6. −1.062 **7.** −1.286 **8.** 1.431

Written Exercises

Use Table 7 or 8 to find the given values to four significant digits. Make a sketch showing each angle in standard position.

A **1–8.** Find the values of the functions in Oral Exercises 1–8.

9. tan 5.87^R **10.** csc 6.02^R **11.** cot 3.5^R **12.** sec 0.74^R
−0.4346   −3.890   2.657   1.354

Use Table 6 to find α in decimal degrees to the nearest tenth of a degree so that $0 \leq m(\alpha) \leq 90°$.

49.6°

13. tan α = 1.904 62.3° **14.** sec α = 1.080 22.2° **15.** csc α = 1.313

16. tan α = 5.000 78.7° **17.** sec α = 1.271 38.1° **18.** cot α = 4.372

12.9°

27. $\sin \alpha = -1$; $\cos \alpha = 0$; $\tan \alpha$ not defined; $\cot \alpha = 0$; $\sec \alpha$ not defined; $\csc \alpha = -1$

28. $\sin \alpha = -\dfrac{24}{25}$; $\cos \alpha = -\dfrac{7}{25}$; $\tan \alpha = \dfrac{24}{7}$; $\csc \alpha = -\dfrac{25}{24}$; $\sec \alpha = -\dfrac{25}{7}$; $\cot \alpha = \dfrac{7}{24}$

29. $\sin \alpha = 0$; $\cos \alpha = -1$; $\tan \alpha = 0$; $\cot \alpha$ not defined; $\sec \alpha = -1$; $\csc \alpha$ not defined

30. $\sin \alpha = \dfrac{2}{\sqrt{5}}$; $\cos \alpha = \dfrac{1}{\sqrt{5}}$; $\tan \alpha = 2$; $\csc \alpha = \dfrac{\sqrt{5}}{2}$; $\sec \alpha = \sqrt{5}$; $\cot \alpha = \dfrac{1}{2}$

31. $\sin \alpha = -\dfrac{5}{7}$; $\cos \alpha = \dfrac{2\sqrt{6}}{7}$; $\tan \alpha = -\dfrac{5}{2\sqrt{6}}$; $\cot \alpha = -\dfrac{2\sqrt{6}}{5}$; $\sec \alpha = \dfrac{7}{2\sqrt{6}}$; $\csc \alpha = -\dfrac{7}{5}$

32. $\sin \alpha = \dfrac{\sqrt{5}}{3}$; $\cos \alpha = -\dfrac{2}{3}$; $\tan \alpha = -\dfrac{\sqrt{5}}{2}$; $\cot \alpha = -\dfrac{2}{\sqrt{5}}$; $\sec \alpha = -\dfrac{3}{2}$; $\csc \alpha = \dfrac{3}{\sqrt{5}}$

(continued)

Additional Answers
Written Exercises
(continued)

33. $\cos \alpha = \dfrac{1}{2}$; $\tan \alpha = -\sqrt{3}$;

   $\cot \alpha = -\dfrac{1}{\sqrt{3}}$;

   $\sec \alpha = 2$; $\csc \alpha = -\dfrac{2}{\sqrt{3}}$

34. $\cos \alpha = -\dfrac{1}{\sqrt{2}}$;

   $\tan \alpha = 1$; $\cot \alpha = 1$;
   $\sec \alpha -\sqrt{2}$;
   $\csc \alpha = -\sqrt{2}$

35. $\cos \alpha = -\dfrac{2}{3}$; $\tan \alpha =$

   $-\dfrac{\sqrt{5}}{2}$; $\cot \alpha = -\dfrac{2}{\sqrt{5}}$;

   $\sec \alpha = -\dfrac{3}{2}$; $\csc \alpha = \dfrac{3}{\sqrt{5}}$

36. $\sin \alpha = -\dfrac{1}{3}$;

   $\cos \alpha = -\dfrac{2\sqrt{2}}{3}$;

   $\tan \alpha = \dfrac{1}{2\sqrt{2}}$;

   $\cot \alpha = 2\sqrt{2}$; $\csc \alpha = -3$

37. $\sin \alpha = \dfrac{\sqrt{15}}{7}$;

   $\cos \alpha = -\dfrac{\sqrt{34}}{7}$;

   $\tan \alpha = -\dfrac{\sqrt{510}}{34}$;

   $\cot \alpha = -\dfrac{\sqrt{510}}{15}$;

   $\sec \alpha = -\dfrac{7}{\sqrt{34}}$

38. $\sin \alpha = \dfrac{3}{\sqrt{13}}$;

   $\tan \alpha = -\dfrac{3}{2}$;

   $\cot \alpha = -\dfrac{2}{3}$; $\sec \alpha = \dfrac{\sqrt{13}}{2}$;

   $\csc \alpha = \dfrac{\sqrt{13}}{3}$

(continued on p. 721)

606

Find α in decimal degrees to the nearest tenth of a degree so that α lies in the given quadrant.

19. IV: $\tan \alpha = -0.7265$ 324°
20. II: $\csc \alpha = 1.244$ 126.5°
21. III: $\sec \alpha = -1.108$ 205.5°
22. II: $\cot \alpha = -0.3899$ 111.3°
23. IV: $\csc \alpha = -2.572$ 337.1°
24. III: $\tan \alpha = 8.000$ 262.9°

Find the values of the six trigonometric functions of the angle α whose terminal side in standard position passes through the given point. If the function is not defined for the angle, so state.

25. $(6, -8)$
26. $(-5, 12)$
27. $(0, -3)$
28. $(-7, -24)$
29. $(-2, 0)$
30. $(3, 6)$
31. $(2\sqrt{6}, -5)$
32. $(-4, 2\sqrt{5})$

Find the values of the other five trigonometric functions of the angle α in the given quadrant and having the given function value.

33. IV: $\sin \alpha = -\dfrac{\sqrt{3}}{2}$
34. III: $\sin \alpha = -\dfrac{1}{\sqrt{2}}$
35. II: $\sin \alpha = \dfrac{\sqrt{5}}{3}$
36. III: $\sec \alpha = -\dfrac{3}{2\sqrt{2}}$
37. II: $\csc \alpha = \dfrac{7}{\sqrt{15}}$
38. IV: $\cos \alpha = \dfrac{2}{\sqrt{13}}$

B 39. III: $\tan \alpha = \dfrac{1}{5}$
40. II: $\cot \alpha = -\dfrac{3\sqrt{5}}{2}$

(*Hint*: In Exercises 39 and 40, let $u = \sin \alpha$ and $v = \cos \alpha$ and solve two simultaneous equations involving u and v.)

41. Use the theorem on page 603 to prove that for any angle α:
   $1 + \tan^2 \alpha = \sec^2 \alpha$.
42. Use the theorem on page 603 to prove that for any angle α:
   $1 + \cot^2 \alpha = \csc^2 \alpha$.

Graph each of the following functions over the given interval.

43. $y = \tan \frac{1}{2}x \; (-\pi < x < \pi)$
44. $y = 4 \csc x \; (-\pi < x < \pi)$
45. $y = -\tan x \; (-\pi < x < \pi)$
46. $y = -3 \sec x \; (0 < x < 2\pi)$
47. $y = \frac{3}{2} \csc 2x \; (0 < x < \pi)$
48. $y = 2 \cot \frac{1}{2}x \; (0 < x < 2\pi)$

Use the diagram at the right, in which $P(u, v)$ is a point on the unit circle O, and $\overline{AB}$ and $\overline{CD}$ are tangents, to prove each of the following.

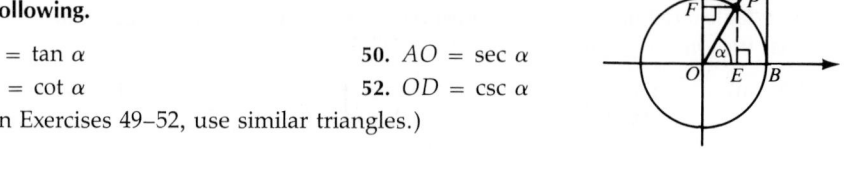

C 49. $AB = \tan \alpha$
50. $AO = \sec \alpha$
51. $CD = \cot \alpha$
52. $OD = \csc \alpha$

(*Hint*: In Exercises 49–52, use similar triangles.)

606   *Chapter 14*

14-9 Solving Right Triangles

To solve a right triangle means to find the measures (or approximations of these) of various parts of the triangle (angles and sides) when measures of other parts of the triangle are given. In working with right triangles, it is customary to use either capital letters or Greek letters to identify the angles, and the corresponding lower-case Roman letters to represent the lengths of the sides opposite these angles (see Figure 37). It is also common practice to label the vertex of the right angle as C. The trigonometric functions discussed earlier in this chapter can be used to solve right triangles.

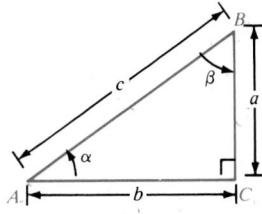

Figure 37   **Figure 38**

If angle α of right triangle ABC is placed in standard position with side $\overline{AC}$ along the positive u-axis (Figure 38), then the definitions of the six trigonometric functions can be interpreted as follows.

If α is an acute angle of a right triangle, then:

$$\sin \alpha = \frac{\text{length of side opposite } \alpha}{\text{length of hypotenuse}} \qquad \cos \alpha = \frac{\text{length of side adjacent to } \alpha}{\text{length of hypotenuse}}$$

$$\tan \alpha = \frac{\text{length of side opposite } \alpha}{\text{length of side adjacent to } \alpha} \qquad \cot \alpha = \frac{\text{length of side adjacent to } \alpha}{\text{length of side opposite } \alpha}$$

$$\sec \alpha = \frac{\text{length of hypotenuse}}{\text{length of side adjacent to } \alpha} \qquad \csc \alpha = \frac{\text{length of hypotenuse}}{\text{length of side opposite } \alpha}$$

From these statements, you can see that for right triangle ABC in Figure 37:

$$\sin \beta = \frac{b}{c} = \cos \alpha \qquad \cos \beta = \frac{a}{c} = \sin \alpha$$

$$\tan \beta = \frac{b}{a} = \cot \alpha \qquad \cot \beta = \frac{a}{b} = \tan \alpha$$

$$\sec \beta = \frac{c}{a} = \csc \alpha \qquad \csc \beta = \frac{c}{b} = \sec \alpha$$

Trigonometric and Circular Functions   **607**

Teaching Suggestions
p. T120

Related Activities p. T121

Supplementary Material
Test 46

Key Ideas

Use trigonometric functions to find the measures of various parts of a right triangle when either the measures of two sides or the measures of one side and one acute angle are known.

Chalkboard Examples

1. Solve the right triangle shown.

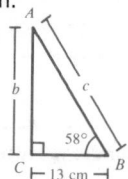

Solution: $m(A) + m(B) = 90°$
$m(A) = 90° - m(B)$
$ = 90° - 58° = 32°$

Method 1

$\sec B = \dfrac{c}{13}$

$c = 13 \sec 58°$
$c \approx 13(1.887) \approx 24.5$

Method 2

$\cos B = \dfrac{13}{c}$

$c = \dfrac{13}{\cos 58°}$

$c \approx \dfrac{13}{0.5299} \approx 24.5$

$\tan B = \dfrac{b}{13}$

$b = 13 \tan 58°$
$b \approx 13(1.6003)$
$b \approx 20.8$
$\therefore m(A) = 32°,\ c \approx 24.5$ cm,
$b \approx 20.8$ cm

2. Solve the right triangle shown.

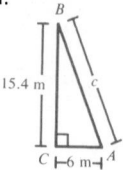

$\cot A = \dfrac{6}{15.4} \approx 0.3896$

$m(A) \approx 68.7°$
$m(B) = 90° - m(A)$
$m(B) \approx 90° - 68.7$
$m(B) \approx 21.3°$

$\sec A = \dfrac{c}{6}$

Angles α and β are complementary angles, and the pairs of functions sine and cosine, tangent and cotangent, secant and cosecant are called cofunctions.

EXAMPLE 1 Solve the right triangle shown, stating lengths of sides correct to three significant digits and angle measures to the nearest degree.

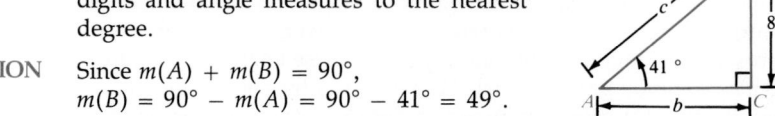

SOLUTION Since $m(A) + m(B) = 90°$,
$m(B) = 90° - m(A) = 90° - 41° = 49°$.

To determine c, use one of the following methods.

Method 1	Method 2
$\csc A = \dfrac{c}{8}$	$\sin A = \dfrac{8}{c}$
$c = 8 \csc 41°$	$c = \dfrac{8}{\sin 41°}$
$c \approx 8(1.524)$	$c \approx \dfrac{8}{0.6561}$
$c \approx 12.2$	$c \approx 12.2$

To determine b, use the tangent function.

$$\tan B = \frac{b}{8}$$

$$b = 8 \tan 49° \approx 8(1.150) \approx 9.2$$

$$\therefore m(B) = 49°,\ c \approx 12.2,\ \text{and}\ b \approx 9.20.$$

In theoretical examples like Example 1, the given measures (in this case, the measures of side $\overline{BC}$, angle α, and the right angle at C) are taken to be exact. Notice also, that in Method 1 the cosecant was used instead of the sine to avoid long division. Method 2 is usually the better choice when a calculator can be used.

When you use Table 6 or 7 in the solution of triangles or in solving practical problems, lengths need be stated to no more than four significant digits, since this is the limit of the accuracy of the entries in these tables. The following relationships between angle measure and length can be used as a guide to the accuracies of length and angle measurements.

angle measured to nearest	corresponds to	lengths measured to
1°		2 significant digits
10′ or 0.1°		3 significant digits
1′ or 0.01°		4 significant digits

EXAMPLE 2 Solve the right triangle shown. Give the lengths of the sides to three significant digits and the measure of the angles to the nearest tenth of a degree.

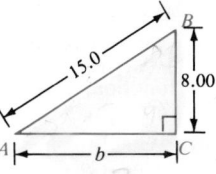

SOLUTION Since a and c are known, use $\sin A$ to determine $m(A)$.

$$\sin A = \frac{8}{15}$$

$$\approx 0.5333$$

Use Table 6 or a calculator to determine that

$$m(A) \approx 32.2°.$$

Since $m(A) + m(B) = 90°$:

$$m(B) = 90° - m(A)$$
$$m(B) \approx 90° - 32.2°$$
$$m(B) \approx 57.8°$$

To determine b, use the cosine function.

$$\cos A = \frac{b}{15}$$

$$\cos 32.2° = \frac{b}{15}$$

$$15(0.8462) \approx b$$

$$12.7 \approx b$$

$$\therefore m(A) \approx 32.2°, \ m(B) \approx 57.8°, \text{ and } b \approx 12.7.$$

In many practical problems applying trigonometric function values, an angle is described as an *angle of elevation* or an *angle of depression* (see Figure 39). Since the point B is elevated with respect to the observer at A, $\angle CAB$, the angle between the horizontal ray AC through A and the line of sight, is called an **angle of elevation**. The point T is depressed with respect to the observer at R; therefore, $\angle TRS$, the angle between the line of sight and the horizontal ray RS, is called an **angle of depression**.

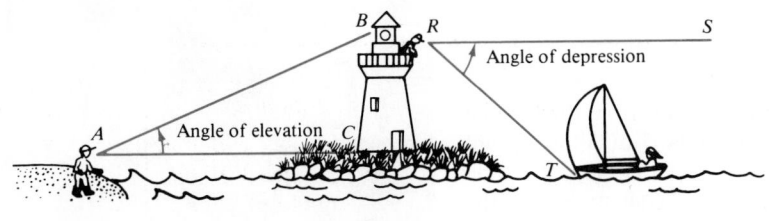

Figure 39

Trigonometric and Circular Functions **609**

$$\sec 68.7° \approx \frac{c}{6}$$
$$c \approx (2.753)(6)$$
$$c \approx 16.5$$
$$\therefore m(A) \approx 68.7°, \ m(B) \approx$$
$$21.3°, \ c \approx 16.5 \text{ m}$$

Suggested Assignments

Extended Alg. with Trig.
  First day
  610/1–19 odd
  Second day
  610/P: 1–6
R 612/*Self-Test 3*
Enriched Alg.
  First day
  610/1–19 odd
  Second day
  610/P: 1–6
R 612/*Self-Test 3*
Enriched Alg. with Trig.
  610/1–19 odd
  610/P: 1–6
R 612/*Self-Test 3*

Mixed Review

Solve.

1. $25x^2 = 36$ $\left\{\dfrac{6}{5}, -\dfrac{6}{5}\right\}$

2. $\sqrt{4x + 25} = x - 5$ $\{14\}$

3. $5x^2 - 6x + 4 = 0$
$\left\{\dfrac{3 + i\sqrt{11}}{5}, \dfrac{3 - i\sqrt{11}}{5}\right\}$

Simplify.

4. $\dfrac{4 + \sqrt{5}}{1 + \sqrt{5}}$ $\dfrac{1 + 3\sqrt{5}}{4}$

5. $\dfrac{1 - \dfrac{3}{x} + \dfrac{2}{x^2}}{1 - \dfrac{4}{x^2}}$ $\dfrac{x - 1}{x + 2}$

Additional A Exercises

Solve the right triangle with the given parts.

1. $m(A) = 47.4°$; $b = 12.0$
   $m(B) = 42.6°$; $a \approx 13.0°$;
   $c \approx 17.7$

2. $m(B) = 36.9°$; $a = 8.0$
   $m(A) = 53.1°$; $b \approx 6.01$;
   $c \approx 10.0$

3. $a = 10.0$; $b = 24.0$
   $m(A) \approx 22.6°$; $m(B) \approx$
   $67.4°$; $c \approx 26.0$

4. $b = 18.7$; $c = 26.8$
   $m(B) \approx 44.3°$; $m(A) \approx$
   $45.7°$; $a \approx 19.2$

5. A tree stands 3.60 m tall and casts a shadow. If the angle of depression from the top of the tree to the top of the shadow is 56°, what is the length of the shadow? 2.43 m

---

Additional Answers
Written Exercises

1. $a \approx 52.9$; $b \approx 28.4$;
   $m(A) = 61.8°$

2. $a \approx 17.8$; $b \approx 46.7$;
   $m(B) = 69.2°$

3. $a \approx 23.4$; $b \approx 70.8$;
   $m(B) = 71.7°$

4. $a \approx 20.7$; $b \approx 27.9$;
   $m(A) = 36.6°$

5. $a \approx 16.0$; $c \approx 29.7$;
   $m(B) = 57.3°$

6. $a \approx 106$; $c \approx 337$;
   $m(A) = 18.3°$

7. $b \approx 32.3$; $c \approx 57.8$;
   $m(B) = 33.9°$

8. $b \approx 13.8$; $c \approx 18.3$;
   $m(A) = 41.1°$

9. $b \approx 84.6$; $c \approx 93.6$;
   $m(A) = 25.3°$

10. $a \approx 184$; $c \approx 290$;
    $m(B) = 50.8°$

11. $c \approx 28.0$; $m(A) \approx 55.2°$;
    $m(B) \approx 34.8°$

12. $b \approx 67.1$; $m(A) \approx 16.6°$;
    $m(B) \approx 73.4°$

Oral Exercises

Use the right triangle shown to give the value of each function. Give your answer as a fraction.

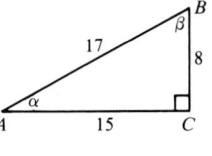

1. $\sin \alpha$ $\frac{8}{17}$
2. $\cos \alpha$ $\frac{15}{17}$
3. $\tan \alpha$ $\frac{8}{15}$
4. $\csc \alpha$ $\frac{17}{8}$
5. $\sin \beta$ $\frac{15}{17}$
6. $\cos \beta$ $\frac{8}{17}$
7. $\sec \beta$ $\frac{17}{8}$
8. $\tan \beta$ $\frac{15}{8}$

Written Exercises

Solve the right triangle with the given parts. In each triangle angle C is a right angle. Make a sketch with sides and angles labeled according to the data. State the lengths of sides to three significant digits and the measures of angles to the nearest tenth of a degree.

A
1. $m(B) = 28.2°$; $c = 60.0$
2. $m(A) = 20.8°$; $c = 50.0$
3. $m(A) = 18.3°$; $c = 74.6$
4. $m(B) = 53.4°$; $c = 34.8$
5. $m(A) = 32.7°$; $b = 25.0$
6. $m(B) = 71.7°$; $b = 320$
7. $m(A) = 56.1°$; $a = 48.0$
8. $m(B) = 48.9°$; $a = 12.0$
9. $m(B) = 64.7°$; $a = 40.0$
10. $m(A) = 39.2°$; $b = 225$
11. $a = 23.0$; $b = 16.0$
12. $a = 20.0$; $c = 70.0$
13. $a = 60.0$; $c = 65.0$
14. $a = 19.5$; $c = 41.0$
15. $a = 28.2$; $b = 44.8$
16. $b = 11.4$; $c = 61.5$
17. $a = 12.4$; $c = 25.1$
18. $a = 103$; $b = 496$
19. $b = 17.5$; $c = 29.3$

Problems

Give the measures of angles to the nearest tenth of a degree and lengths to three significant digits unless specified otherwise.

A
1. A ladder 32 ft long leans against the side of a building and makes an angle of 75° with the ground. How high above the ground is the top of the ladder? 30.9 ft

2. The angle of elevation from an observer on the ground to an airplane flying at an altitude of 8.3 km is 22.5°. How far is the observer in a straight line from the plane? 21.7 km

3. A support wire runs from the top of a utility pole that is 17 m tall to a point on the ground that is 11 m from the base of the pole. What angle does the wire make with the ground? 57.1°

4. The angle of depression from a ship to a wreck on the ocean floor is 10.9°. If the depth of the ocean at the wreck is 156 m, how far should the ship sail so that it is directly over the wreck? 810 m

5. Determine the length of the base of isosceles triangle PQR shown below. 11.5

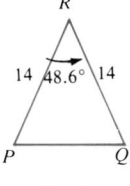 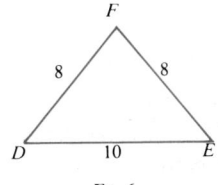 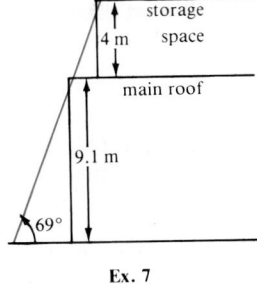

Ex. 5   Ex. 6   Ex. 7

6. Determine the measure of angle D of isosceles triangle DEF shown above. 51.3°

7. The top of a recessed storage space is 4 m above the main roof of a house as shown. If the main roof is 9.1 m above the ground and a ladder that reaches the top of the storage space makes an angle of 69° with the ground, how long is the ladder? 14.0 m

B 8. When a rocket that has taken off vertically from a point on a straight line between two observers reaches a height of 15 km, the angles of elevation of the lines of sight of the observers are 82.7° and 64.1°, respectively. How far apart are the two observers? 9.2 km

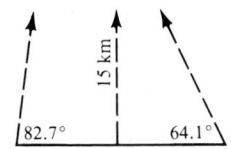

9. One ship starts at point P and sails on a course that makes an angle measuring 54.3° east of north as shown. Another ship also starts at point P and sails on a course making an angle of 60.7° west of north. After the first ship has sailed 25 km, it is exactly east of the second ship. How far apart are the two ships? 46.3 km

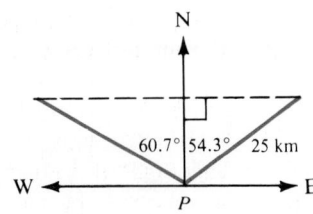

10. At 6:00 A.M. on a certain day a tree casts a shadow that is 12.1 m long, and the angle of elevation from the tip of the shadow to the top of the tree is 63.7°. At 6:00 P.M. the tree casts a shadow that is 8.5 m long in the other direction. What is the angle of elevation from the tip of this shadow to the top of the tree? 70.9°

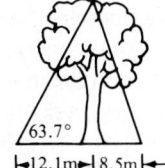

13. $b \approx 25.0$; $m(A) \approx 67.4°$; $m(B) \approx 22.6°$

14. $b \approx 36.1$; $m(A) \approx 28.4°$; $m(B) \approx 61.6°$

15. $c \approx 52.9$; $m(A) \approx 32.2°$; $m(B) \approx 57.8°$

16. $a \approx 60.4$; $m(A) \approx 79.3°$; $m(B) \approx 10.7°$

17. $b \approx 21.8$; $m(A) \approx 29.6°$; $m(B) \approx 60.4°$

18. $c \approx 508$; $m(A) \approx 11.7°$; $m(B) \approx 78.3°$

19. $a \approx 23.5$; $m(A) \approx 53.3°$; $m(B) \approx 36.7°$

Trigonometric and Circular Functions **611**

11. The angle of elevation to the top of the lighthouse from two ships at points A and B, which are on a straight line with the base of a lighthouse 160 m high, are 10.2° and 11.3°, respectively. How far apart are the ships? Give your answer to 2 significant digits. 89 m

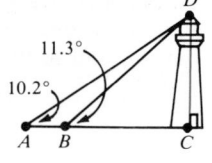

Ex. 11

C 12. In Exercise 11, if the ships are exactly 15 m apart and the angles of elevation are the same as above, how tall is the lighthouse to the nearest tenth of a meter? (*Hint*: Let $BC = x$ and $CD = y$, and solve two equations in two variables.) 27.1 m

Quick Quiz

1. Determine the values of the six trigonometric functions of an angle in standard position whose terminal side passes through the point $(5, -2\sqrt{14})$.

$\sin \alpha = -\dfrac{2\sqrt{14}}{9}$, $\cos \alpha = \dfrac{5}{9}$, $\tan \alpha = -\dfrac{2\sqrt{14}}{5}$,

$\csc \alpha = -\dfrac{9}{2\sqrt{14}}$, $\sec \alpha = \dfrac{9}{5}$,

$\cot \alpha = -\dfrac{5}{2\sqrt{14}}$

2. Solve the right triangle ABC if $m(B) = 26.8°$ and $a = 11$. $m(A) = 63.2°$; $b = 5.55$; $c = 12.3$

Additional Answers
Self-Test 3

1. $\sin \alpha = \dfrac{2\sqrt{10}}{7}$

$\cos \alpha = -\dfrac{3}{7}$

$\tan \alpha = -\dfrac{2\sqrt{10}}{3}$

$\cot \alpha = -\dfrac{3}{2\sqrt{10}}$

$\sec \alpha = -\dfrac{7}{3}$

$\csc \alpha = \dfrac{7}{2\sqrt{10}}$

Self-Test 3

VOCABULARY   tangent of an angle (p. 602)
secant of an angle (p. 602)
cotangent of an angle (p. 602)
cosecant of an angle (p. 602)
reciprocal functions (p. 603)
angle of elevation (p. 609)
angle of depression (p. 609)

1. Determine the values of the six trigonometric functions of an angle in standard position whose terminal side passes through the point $(-3, 2\sqrt{10})$. *Obj. 1, p. 602*

2. Solve the right triangle ABC in which $m(C) = 90°$, $m(A) = 37.2°$, and $a = 7$. Give angle measures to the nearest tenth and lengths to three significant digits. $b = 9.22$; $c = 11.6$; $m(B) = 52.8°$ *Obj. 2, p. 602*

Check your answers with those at the back of the book.

Chapter Summary

1. A *directed angle* is an ordered pair of rays with a common endpoint, one ray called the *initial side* and the other the *terminal side* of the angle, together with a rotation from the initial side to the terminal side. Angle measures can be expressed in either *degrees* or *radians*.

2. If α is an angle in standard position, with $P(u, v)$ any point other than the origin on the terminal side of α, and if $\sqrt{u^2 + v^2} = r$, then the six *trigonometric functions* are listed on the following page.

612   *Chapter 14*

sine: $\alpha \longrightarrow \sin\alpha = \dfrac{v}{r}$ \qquad cosine: $\alpha \longrightarrow \cos\alpha = \dfrac{u}{r}$

tangent: $\alpha \longrightarrow \tan\alpha = \dfrac{v}{u},\ u \neq 0$ \qquad cotangent: $\alpha \longrightarrow \cot\alpha = \dfrac{u}{v},\ v \neq 0$

secant: $\alpha \longrightarrow \sec\alpha = \dfrac{r}{u},\ u \neq 0$ \qquad cosecant: $\alpha \longrightarrow \csc\alpha = \dfrac{r}{v},\ v \neq 0$

3. Trigonometric and circular functions are *periodic*, and values for these functions can be found using *reference angles* or *reference arcs*.

4. Graphs of periodic functions consist of basic patterns repeated over each interval having a length of one fundamental period. The graphs of sine and cosine are *sine waves*. For a sine wave, the absolute value of one-half the difference of the minimum and maximum ordinates on the curve is called the *amplitude* of the wave.

5. Many practical problems can be solved by applying trigonometric function values.

Chapter Review

Write the letter of the correct answer.

1. A wheel has a radius of 3.2 cm. Find the distance traveled by the wheel when it makes 2.3 revolutions. Use $\pi \approx 3.14$. \hfill *14-1*

   (a.) 46.2 cm \qquad **b.** 23.1 cm \qquad **c.** 10.0 cm \qquad **d.** 20.1 cm

2. Express $\dfrac{5\pi}{4}^{\mathrm{R}}$ as a degree measure. \hfill *14-2*

   a. 45° \qquad **b.** 300° \qquad **(c.)** 225° \qquad **d.** 60°

3. Express $-330°$ as a radian measure using π.

   a. $-\dfrac{2\pi}{3}^{\mathrm{R}}$ \qquad **(b.)** $-\dfrac{11\pi}{6}^{\mathrm{R}}$ \qquad **c.** $\dfrac{11\pi}{6}^{\mathrm{R}}$ \qquad **d.** $\dfrac{5\pi}{6}^{\mathrm{R}}$

4. If α is an angle in standard position and its terminal side contains the point $(3, -4)$, find $\cos\alpha$. \hfill *14-3*

   a. $\dfrac{4}{5}$ \qquad **b.** $-\dfrac{4}{5}$ \qquad **c.** $-\dfrac{3}{5}$ \qquad **(d.)** $\dfrac{3}{5}$

5. Find $\sin\alpha$, if $\cos\alpha = \dfrac{3}{4}$ and $\sin\alpha > 0$.

   a. $\dfrac{7}{16}$ \qquad **b.** $\dfrac{5}{16}$ \qquad **c.** $\dfrac{1}{4}$ \qquad **(d.)** $\dfrac{\sqrt{7}}{4}$

6. Find $\sin\dfrac{19\pi}{6}^{\mathrm{R}}$. \hfill *14-4*

   a. $\dfrac{1}{2}$ \qquad **b.** $-\dfrac{1}{\sqrt{2}}$ \qquad **(c.)** $-\dfrac{1}{2}$ \qquad **d.** $-\dfrac{\sqrt{3}}{2}$

Trigonometric and Circular Functions **613**

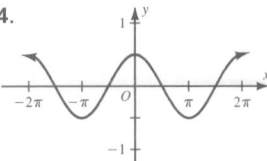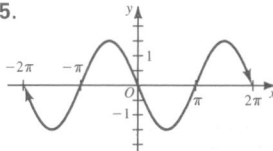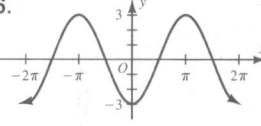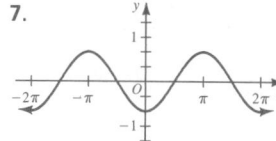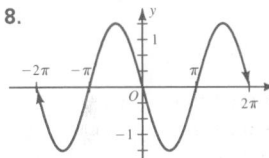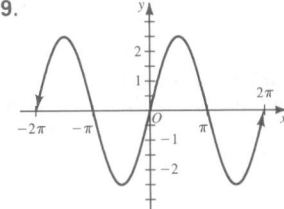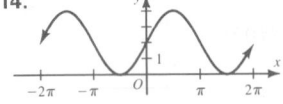

15.

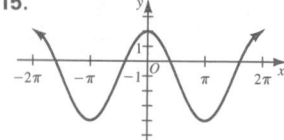

16.

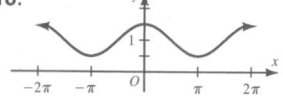

17.

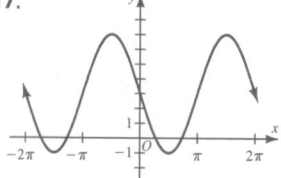

18.

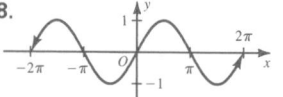

19.

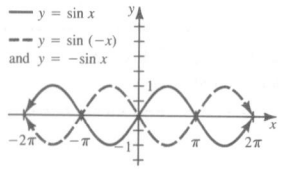

20.

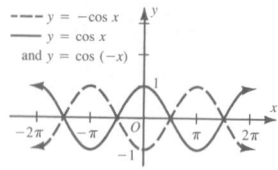

7. Find all values of $m(\alpha)$ for which the terminal side of α in standard position passes through $(-\sqrt{6}, -\sqrt{6})$.

    a. $135° + k \cdot 360°$ **b.** $45° + k \cdot 360°$

    c. $240° + k \cdot 360°$ **(d.)** $225° + k \cdot 360°$

8. Use Table 6 to find cos 143.7°.

    a. 0.8059   **(b.)** -0.8059 **c.** 0.5920   **d.** -0.6336

14-5

9. Use Table 8 to find $m(\alpha)$ if sin $\alpha = 0.9320$.

    (a.) 1.2^R **b.** 2.1^R **c.** 6.8^R **d.** 1.8^R

10. What is the amplitude of $y = -3 \sin x$?

    a. 1   **b.** $\frac{1}{3}$ **(c.)** 3   **d.** -3

14-6

11. What is the fundamental period of $y = 2 \cos 5x$?

    a. 2π **b.** 5   **(c.)** $\frac{2\pi}{5}$ **d.** $\frac{5}{2\pi}$

14-7

12. If $\cos \alpha = -\frac{\sqrt{5}}{3}$ and $\sin \alpha > 0$, find tan α.

14-8

    (a.) $-\frac{2}{\sqrt{5}}$ **b.** $\frac{2}{\sqrt{5}}$ **c.** $-\frac{\sqrt{5}}{2}$ **d.** $\frac{\sqrt{5}}{3}$

13. If $\triangle ABC$ is a right triangle, $m(B) = 42°$, $m(C) = 90°$, and the length of $\overline{AC} = 10$, find the length of $\overline{BC}$.

14-9

    a. 14.9   **(b.)** 11.1   **c.** 9.2   **d.** 8.9

Chapter Test

1. If a wheel travels 4.2 m as it makes 2.4 revolutions, find the diameter of the wheel to the nearest 0.01 m. Use $\pi \approx 3.14$. 0.56 m

14-1

2. Find the length of the arc on a circle with radius 25 cm that is intercepted by a central angle with measure $\frac{2\pi}{5}^R$. Use $\pi \approx 3.14$. 31.4 cm

14-2

3. Express 108° as a radian measure using π. $\frac{3\pi}{5}^R$

4. Express $\frac{4\pi}{5}^R$ as a degree measure. 144°

5. If $\sin \alpha = \frac{-15}{17}$ and $\cos \alpha > 0$, find $\cos \alpha$. $\frac{8}{17}$

14-3

6. If $\frac{\pi}{2} < x < \pi$, and $\sin x = \frac{1}{2}$, what is the value of x? $\frac{5\pi}{6}^R$

14-4

7. Find all the values of $m(\alpha)$ for which the terminal side of α in standard position passes through $(2\sqrt{3}, -2)$. $m(\alpha) = 330° + k \cdot 360°$, or $\frac{11\pi}{6}^R + 2k\pi^R$

Find a four-decimal place approximation for each of the following.

8. cos 672° 0.6691   **9.** sin (-4.2^R) 0.8724

14-5

614   *Chapter 14*

10. Sketch the graph of $y = 3 \sin x + 1$ over the interval $-2\pi \le x \le 2\pi$.   *14-6*

11. Sketch the graph of $y = 2 \cos 4x$ over one fundamental period.   *14-7*

12. Find the values of the other five trigonometric functions of α if   *14-8*
    $\sin \alpha = \frac{1}{2}$ and $\cos \alpha < 0$.

13. If $\triangle ABC$ is a right triangle with $m(C) = 90°$, $m(B) = 37.2°$ and the   *14-9*
    length of $\overline{AB} = 121$, find the length of $\overline{AC}$ to three significant digits.
    73.2

Contest Problems

1. On school mornings, a girl walks from her home at point P in a path perpendicular to a canal, arriving at point Q on the canal. At point Q she boards a boat that travels downstream to her school. One morning she hears the boat's whistle upstream as she leaves her house and realizes that if she wants to ride the boat she must meet it farther down the canal. If her house is 1 mi from point Q, and she walks at a rate that is half the rate of the boat, how far downstream from point Q should she head in order to have the best chance of catching the boat?
   about 0.58 mi

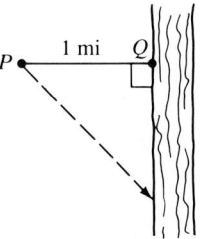

Mixed Review

4. $-3y^3 - y^2 - 3y + 7$

Simplify.

1. $\left(-\frac{1}{4}c + 7c\right) \div \left(\frac{-3}{4}\right)$ $-9c$

2. $\dfrac{-3x^6y^3}{x^2y^4}$ $-3x^4y^{-1}$

3. $(w - 3)(w^2 - 2w + 6)$
   $w^3 - 5w^2 + 12w - 18$

4. $(-3y^3 + y^2 - 8) - (2y^2 + 3y - 15)$

5. $\sqrt[3]{(-8)^4}$
   16

6. $\sqrt[3]{\dfrac{125}{4}}$ $\dfrac{5\sqrt[3]{2}}{2}$

Solve.

7. $\dfrac{3d}{5} - \dfrac{7d}{5} = 16$ $\{-20\}$

8. $x^2 - 13x = 40$ $\{\dfrac{13 \pm \sqrt{329}}{2}\}$

9. $2y^2 + 5y - 3 < 0$ $\{y: -3 < y < \frac{1}{2}\}$

10. $y - \sqrt{y + 5} = 7$ $\{11\}$

11. $y^2 - 6x^2 = 4$
    $y = 3x^2 - 2$
    $\{(\sqrt{2}, 4), (-\sqrt{2}, 4), (0, -2)\}$

12. $3x + 5y + z = 3$
    $-2x - y + 3z = 7$
    $x + 4y - 2z = -6$ $\{(\frac{2}{3}, -\frac{1}{3}, \frac{8}{3})\}$

Graph.

13. $|3x + 2| > 1$

14. $\dfrac{x^2}{16} - \dfrac{y^2}{9} = 1$

Trigonometric and Circular Functions   **615**

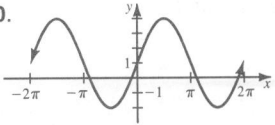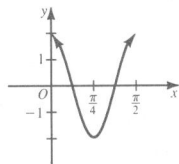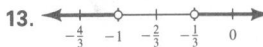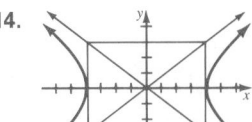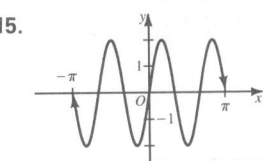

2.
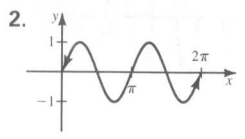

4.
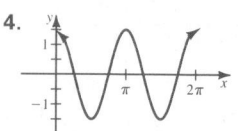

5.
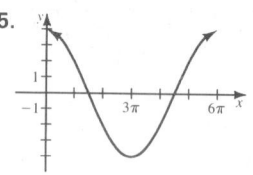

6.
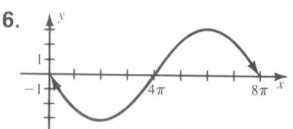

7.
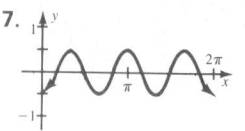

8.
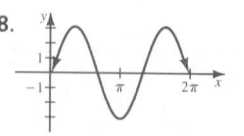

9.
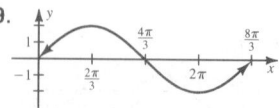

10.
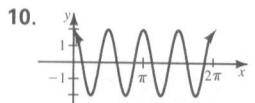

15. Graph $y = 2 \sin 3x$ over the interval $-\pi \le x \le \pi$.

16. Write the equation of the line passing through the points (2, 5) and (4, −1). $y = -3x + 11$

17. Write the equation of the axis of symmetry of the parabola defined by the equation $y = -x^2 - 4x + 4$. $x = -2$

Find the sum for each series.

18. $\displaystyle\sum_{i=1}^{9} (2i - 3)$ 63

19. $\displaystyle\sum_{k=1}^{8} \frac{1}{3}(2)^{k-1}$ 85

20. If y varies jointly as x and z, and $y = 30$ when $x = 3$ and $z = 15$, find the value of y when $x = 5$ and $z = 2$. $\frac{20}{3}$

21. What is the third term in the expansion of $(2x + 3y)^5$? $720x^3y^2$

22. Find $\cos \frac{13\pi}{3}^{R}$. $\frac{1}{2}$

23. In $\triangle ABC$, $m(C) = 90°$, $a = 12$, and $\sin A = \frac{3}{4}$. Find c and b.   16, 10.6

24. Find the coordinates of the preimage P of the point $P'(8, -3)$ under the translation
$$\begin{bmatrix} x' \\ y' \end{bmatrix} = \begin{bmatrix} x \\ y \end{bmatrix} + \begin{bmatrix} 9 \\ -3 \end{bmatrix}.$$
(−1, 0)

25. If $A = \begin{bmatrix} 2 & 1 \\ -4 & -2 \end{bmatrix}$ and $B = \begin{bmatrix} -1 & -3 \\ 2 & 6 \end{bmatrix}$, find $(A + B)(A - B)$. $\begin{bmatrix} 15 & 20 \\ -30 & -40 \end{bmatrix}$

26. The difference of the areas of two squares is 385. The length of a side of one square is 1 less than twice the length of a side of the other square. What is the length of a side of each square?   12, 23

27. How many years will it take $2500 invested at 8% and compounded quarterly, to grow to $7000?   13 years

28. In a survey of 400 car and truck owners, there were 230 car owners and 185 truck owners. How many people in the survey own a car, but no truck?   215

PROGRAMMING IN PASCAL

Standard Pascal implementations provide the functions sin, cos, and arctan (atan in some systems). It is standard to use radian measure when computations involve angles. In this context, it will often be useful to make a constant declaration such as:

```
CONST
    pi = 3.141593;
```

so that *pi* is a reasonable approximation to π.

616   *Chapter 14*

Exercises

1. In a right triangle ABC, $\angle C$ is the right angle and the measures of the other angles are in degrees. Write a Pascal program *solve_rt_triangle* that solves the right triangle when the user enters any one of the following sets of data:

$$m(\angle A), a; \quad m(\angle A), b; \quad m(\angle A), c;$$
$$m(\angle B), a; \quad m(\angle B), b; \quad m(\angle B), c;$$

   where a is the length of the side opposite angle A, etc. Reducing some of the cases to others already coded will minimize the amount of coding needed to solve all the cases. For example, given the $m(\angle B)$ and the length of side a, you could use the code for solving the triangle given $m(\angle A)$ and the length of side a, since $m(\angle A) = 90° - m(\angle B)$. Your program should output a menu of the six options from which the user may choose.

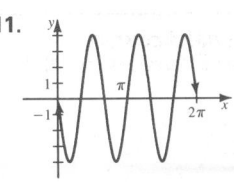

2. The function *arctan*(x) takes in the argument x and returns the measure in radians $\left(\text{usually between } -\frac{\pi}{2} \text{ and } \frac{\pi}{2}\right)$ of the angle having tangent x. Thus, *arctan*(x) can be read "the angle having tangent x." For example,

$$arctan(1) = 0.785398$$
$$arctan(0) = 0$$
$$arctan(-1) = -0.785398$$

   Extend *solve_rt_triangle* from Exercise 1 to cover these additional cases of user input:

$$a, c; \quad a, b; \quad b, c$$

   Use *solve_rt_triangle* to solve Written Exercises 1–19 in Section 14-10.

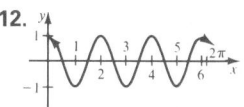

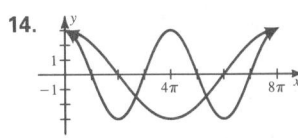

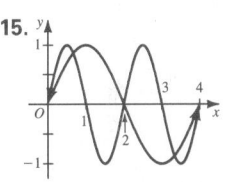

3. Write a program that computes and displays $\frac{\sin x}{x}$. Use your program to evaluate $\frac{\sin x}{x}$ for $x = 1, 0.1, 0.01, 0.001, 0.0001, \ldots$ (you decide where to stop). Also compute and display $\frac{\sin x}{x}$ for $x = -1, -0.1, -0.01, -0.001, -0.0001 \ldots$ etc. Why is it unnecessary to compute $\frac{\sin x}{x}$ for the negative values once it has been done for the positive ones?

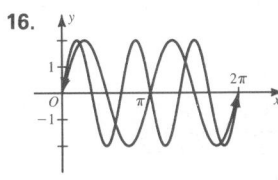

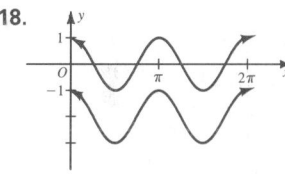

4. Research Problem
   a. Write a program to compare the value of the series
$$x - \frac{x^3}{3!} + \frac{x^5}{5!} - \frac{x^7}{7!} + \frac{x^9}{9!} - \ldots$$
   with the computer's approximation of sin x.

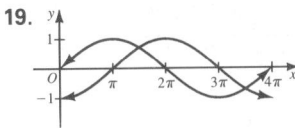

   b. Write a program to compare the value of the series
$$1 - \frac{x^2}{2!} + \frac{x^4}{4!} - \frac{x^6}{6!} + \frac{x^8}{8!} - \ldots$$
   with the computer's approximation of cos x.

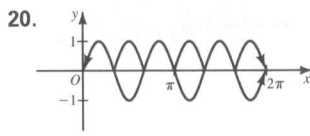

The cloud-filtered daylight dramatically illuminates the palm trees shown in the photo. See pages 656–657 for a discussion of using trigonometric functions to predict the number of hours of daylight at a given latitude on a given day of the year.

Chapter 15

Trigonometric Identities

Problem Solving Strategies

Deduction
Sections 15-2 through 15-5 rely heavily on deductive proof to develop and establish important trigonometric identities. Proof is used to establish a variety of theorems beginning in Section 15-7 with the Pythagorean Theorem and continuing through the law of cosines on page 645 and the law of sines on page 648.
Word Problem Plan
The five-step plan is used implicitly in this chapter along with diagrams and trigonometric equations.

Identities

OBJECTIVE for Sections 15-1 and 15-2:
1. To prove simple trigonometric identities.

15–1 Fundamental Identities

Following are listed the eight *fundamental trigonometric identities,* some of which you have already seen. You saw in Section 14-3 that the values of the sine and cosine functions are related by the identity

$$\sin^2 \alpha + \cos^2 \alpha = 1. \qquad (1)$$

In Section 14-8 the following statements were made as definitions:

$$\tan \alpha = \frac{\sin \alpha}{\cos \alpha}, \quad \cos \alpha \neq 0 \quad (2) \qquad \cot \alpha = \frac{\cos \alpha}{\sin \alpha}, \quad \sin \alpha \neq 0 \quad (3)$$

$$\sec \alpha = \frac{1}{\cos \alpha}, \quad \cos \alpha \neq 0 \quad (4) \qquad \csc \alpha = \frac{1}{\sin \alpha}, \quad \sin \alpha \neq 0 \quad (5)$$

Since values of trigonometric functions are real numbers, the properties of real numbers can be used to find other relationships between the function values. For example, since

$$\frac{\sin \alpha}{\cos \alpha} = \frac{1}{\frac{\cos \alpha}{\sin \alpha}}, \quad \sin \alpha \neq 0, \cos \alpha \neq 0,$$

from (2) and (3) you have

$$\tan \alpha = \frac{1}{\cot \alpha}, \quad \tan \alpha \neq 0, \cot \alpha \neq 0. \qquad (6)$$

Reading Algebra

Be sure students understand the distinction between $\sin^2 \alpha$ and $\sin \alpha^2$. The quantity $\sin^2 \alpha = (\sin \alpha)(\sin \alpha)$, whereas $\sin \alpha^2 = \sin (\alpha^2)$. If these quantities are read correctly in class and in study, students will more easily remember the distinction. Read $\sin^2 \alpha$ as "sine-squared alpha" and read $\sin \alpha^2$ as "sine alpha-squared".

Trigonometric Identities **619**

1. Express $\csc \alpha \tan \alpha$ in terms of $\cos \alpha$, noting any restrictions.

$\csc \alpha \tan \alpha = \dfrac{1}{\sin \alpha} \dfrac{\sin \alpha}{\cos \alpha}$

$= \dfrac{1}{\cos \alpha};\ \sin \alpha \neq 0,$

$\cos \alpha \neq 0$

2. Express $\sin \alpha \cot \alpha$ in terms of $\cos \alpha$.

$\sin \alpha \cot \alpha = \sin \alpha \dfrac{\cos \alpha}{\sin \alpha} =$

$\cos \alpha;\ \sin \alpha \neq 0$

3. Express $\sec^2 \alpha \cos \alpha$ in terms of $\sin \alpha$.

$\sec^2 \alpha \cos \alpha = \dfrac{1}{\cos^2 \alpha} \cos \alpha$

$= \dfrac{1}{\cos \alpha} = \pm \dfrac{1}{\sqrt{1 - \sin^2 \alpha}};$

$\dfrac{1}{\cos \alpha} = \dfrac{1}{\sqrt{1 - \sin^2 \alpha}}$ if α

lies in Quadrant I or IV;

$\dfrac{1}{\cos \alpha} = \dfrac{-1}{\sqrt{1 - \sin^2 \alpha}}$ if α

lies in Quadrant II or III;

$\cos \alpha \neq 0$.

Additional A Exercises

Write an equivalent expression in simplest form, using only the sine function and/or cosine function. Note any restrictions.

1. $\dfrac{\tan^2 \alpha}{\sec^2 \alpha}$ $\sin^2 \alpha,\ \sec \alpha \neq 0$

2. $\csc^2 \alpha\ (\sin \alpha + \tan \alpha)$

$\dfrac{\cos \alpha + 1}{\sin \alpha \cos \alpha},$

$\sin \alpha \neq 0,\ \cos \alpha \neq 0$

3. $\cot \alpha\ (\tan \alpha + \cot \alpha)$

$\dfrac{1}{\sin^2 \alpha},\ \sin \alpha \neq 0$

4. $\dfrac{1 - \cos^2 \alpha}{\tan \alpha}$

$\sin \alpha \cos \alpha,\ \tan \alpha \neq 0$

The last two of the eight fundamental trigonometric identities can be obtained from (1) as follows. For each angle α for which $\cos \alpha \neq 0$, you have

$$\frac{\sin^2 \alpha}{\cos^2 \alpha} + \frac{\cos^2 \alpha}{\cos^2 \alpha} = \frac{1}{\cos^2 \alpha},$$

or

$$\tan^2 \alpha + 1 = \sec^2 \alpha, \quad \cos \alpha \neq 0. \tag{7}$$

Similarly, for each angle α for which $\sin \alpha \neq 0$, you have

$$\frac{\sin^2 \alpha}{\sin^2 \alpha} + \frac{\cos^2 \alpha}{\sin^2 \alpha} = \frac{1}{\sin^2 \alpha},$$

or

$$1 + \cot^2 \alpha = \csc^2 \alpha, \quad \sin \alpha \neq 0. \tag{8}$$

Of course, Identities (1)–(8) are equally valid for the circular functions. You know that for every real number x,

$$\sin^2 x + \cos^2 x = 1,$$

because $\sin x$ and $\cos x$ are coordinates of points on the unit circle. The fact that the remaining circular functions are defined in precisely the same way as the corresponding trigonometric functions of angles guarantees that every trigonometric identity is as valid for values of circular functions as it is for values of those functions with the same names whose domains are the set of angles.

EXAMPLE 1   Express $\cot \alpha \sec \alpha$ in terms of $\sin \alpha$, noting any restrictions.

SOLUTION   The given expression is defined provided $\cos \alpha \neq 0$ and $\sin \alpha \neq 0$. Thus, from Identities (3) and (4), you have:

$$\cot \alpha\ \sec \alpha = \frac{\cos \alpha}{\sin \alpha} \cdot \frac{1}{\cos \alpha} = \frac{1}{\sin \alpha}, \quad \cos \alpha \neq 0,\ \sin \alpha \neq 0$$

EXAMPLE 2   Express $\sin \alpha$ in terms of $\cos \alpha$.

SOLUTION   From Identity (1), you have:

$$\sin^2 \alpha + \cos^2 \alpha = 1$$
$$\sin^2 \alpha = 1 - \cos^2 \alpha$$
$$\therefore \sin \alpha = \pm\sqrt{1 - \cos^2 \alpha}$$

The quadrant in which α lies determines whether $\sin \alpha$ is positive or negative. Thus:

$\sin \alpha = \sqrt{1 - \cos^2 \alpha}$ if α lies in Quadrant I or Quadrant II.

$\sin \alpha = -\sqrt{1 - \cos^2 \alpha}$ if α lies in Quadrant III or Quadrant IV.

620   *Chapter 15*

Oral Exercises

State an equivalent expression in terms of the cosine function only. Note any restrictions.

1. $1 - \sin^2 \alpha$ $\cos^2 \alpha$

2. $\sin \alpha \cot \alpha$ $\cos \alpha; \sin \alpha \neq 0$

3. $\sec^2 \alpha$

4. $\dfrac{\cos \alpha}{1 - \sin^2 \alpha}$ $\dfrac{1}{\cos \alpha}; \cos \alpha \neq 0$

5. $\csc \alpha \tan \alpha$

6. $\sin \alpha \tan \alpha$

State an equivalent expression in terms of $\sin x$ only. Note any restrictions.

7. $\cos^2 x$ $1 - \sin^2 x$

8. $\sec x \tan x$

9. $\dfrac{\cos^2 x}{\cot^2 x}$

10. $\sin x \cot^2 x$ $\dfrac{1 - \sin^2 x}{\sin x}; \sin x \neq 0$

11. $\dfrac{1 - \cos^2 x}{\sin x}$ $\sin x; \sin x \neq 0$

12. $\sec^2 x$ $\dfrac{1}{1 - \sin^2 x}; \cos x \neq 0$

Written Exercises

Write an equivalent expression in simplest form, using only the sine function and/or the cosine function. Note any restrictions.

A

1. $\cos^2 \alpha \tan^2 \alpha$ $\sin^2 \alpha; \cos \alpha \neq 0$

2. $(1 - \sin^2 \alpha)\cot \alpha$

3. $\tan^2 \alpha \csc^2 \alpha \cos \alpha$ $\dfrac{1}{\cos \alpha}; \sin \alpha \neq 0, \cos \alpha \neq 0$

4. $\tan \alpha(\cot \alpha + \tan \alpha)$

5. $\sin \alpha \tan \alpha + \cos \alpha$ $\dfrac{1}{\cos \alpha}; \cos \alpha \neq 0$

6. $\cot \alpha(\sec^2 \alpha - 1)$

7. $\sec \alpha(\cos \alpha + \sin \alpha \tan \alpha)$ $\dfrac{1}{\cos^2 \alpha}; \cos \alpha \neq 0$

8. $(\csc \alpha - \sin \alpha)\sec^2 \alpha$

9. $\dfrac{1 + \tan^2 \alpha}{\tan^2 \alpha}$ $\dfrac{1}{\sin^2 \alpha}; \sin \alpha \neq 0, \cos \alpha \neq 0$

10. $\dfrac{1 - \sin^2 \alpha}{\cot \alpha}$

11. $\dfrac{\sec^2 x - 1}{1 - \cos^2 x}$ $\dfrac{1}{\cos^2 x}; \sin x \neq 0, \cos x \neq 0$

12. $\dfrac{1 + \cot^2 x}{\cot x \csc x}$

13. $\cos^2 x \csc x + \sin x$ $\dfrac{1}{\sin x}; \sin x \neq 0$

14. $\dfrac{\sec^2 x - \tan^2 x}{\tan x \csc x}$

15. $(\sec x - \cos x)\cot^2 x$ $\cos x; \sin x \neq 0, \cos x \neq 0$

16. $\cos x(\sin x + \cos x \cot x)$

17. $\dfrac{\cos x + \sin x \tan x}{1 + \tan^2 x}$ $\cos x; \cos x \neq 0$

18. $\tan x(1 - \sin x)(1 + \sin x)$

Write an equivalent expression in simplest form that involves only a single function (e.g., $\tan x$) or a constant. Note any restrictions.

B

19. $\dfrac{\sin x}{1 + \cos x} - \dfrac{\sin x}{1 - \cos x}$ $-2 \cot x; \sin x \neq 0$

20. $\dfrac{1}{\sin x \cos x} - \tan x$

21. $\dfrac{\sec^2 \alpha - \tan^2 \alpha}{1 - \sin^2 \alpha}$ $\sec^2 \alpha; \cos \alpha \neq 0$

22. $\dfrac{\csc^2 \alpha - \cot^2 \alpha}{1 + \tan^2 \alpha}$

23. $\sec x \csc x - \cot x$ $\tan x; \sin x \neq 0, \cos x \neq 0$

24. $\sin x(\tan x + \cot x)$

25. $\dfrac{\tan x}{1 - \cos x} - \dfrac{\tan x}{1 + \cos x}$ $2\csc x; \sin x \neq 0, \cos x \neq 0$

26. $\dfrac{\cos \alpha}{\sec \alpha + 1} + \dfrac{\cos \alpha}{\sec \alpha - 1}$

27. $\dfrac{(\tan \alpha + 1)^2}{\sec \alpha} - \sec \alpha$ $2 \sin \alpha; \cos \alpha \neq 0$

28. $\dfrac{1 + \tan x}{\sin x + \cos x}$

28. $\sec x; \cos x \neq 0, \sin x \neq -\cos x$

Trigonometric Identities   **621**

Suggested Assignments

Extended Alg. with Trig.
  621/9–31 odd
Enriched Alg. with Trig.
  621/7–35 odd

Additional Answers
Oral Exercises

3. $\dfrac{1}{\cos^2 \alpha}; \cos \alpha \neq 0$

5. $\dfrac{1}{\cos \alpha}; \sin \alpha \neq 0, \cos \alpha \neq 0$

6. $\dfrac{1 - \cos^2 \alpha}{\cos \alpha}; \cos \alpha \neq 0$

8. $\dfrac{\sin x}{1 - \sin^2 x}; \cos x \neq 0$

9. $\sin^2 x; \sin x \neq 0, \cos x \neq 0$

Additional Answers
Written Exercises

2. $\dfrac{\cos^3 \alpha}{\sin \alpha}; \sin \alpha \neq 0$

4. $\dfrac{1}{\cos^2 \alpha}; \sin \alpha \neq 0,$
  $\cos \alpha \neq 0$

6. $\dfrac{\sin \alpha}{\cos \alpha}; \sin \alpha \neq 0,$
  $\cos \alpha \neq 0$

8. $\dfrac{1}{\sin \alpha}; \sin \alpha \neq 0,$
  $\cos \alpha \neq 0$

10. $\sin \alpha \cos \alpha; \cos \alpha \neq 0,$
  $\sin \alpha \neq 0$

12. $\dfrac{1}{\cos x}; \sin x \neq 0, \cos x \neq 0$

14. $\cos x; \sin x \neq 0, \cos x \neq 0$

16. $\dfrac{\cos x}{\sin x}; \sin x \neq 0$

18. $\sin x \cos x; \cos x \neq 0$

20. $\cot x; \sin x \neq 0, \cos x \neq 0$

22. $\cos^2 \alpha; \sin \alpha \neq 0,$
  $\cos \alpha \neq 0$

24. $\sec x; \cos x \neq 0,$
  $\sin x \neq 0$

26. $2 \cot^2 \alpha; \cos \alpha \neq 0,$
  $\sin \alpha \neq 0$

Mixed Review

1. Find the slope of the line passing through the points (2, 3) and (−1, −5).

$\frac{8}{3}$

2. Express 5.127×10^7 in decimal notation.

51,270,000

3. Express $0.\overline{477}$ as a fraction in lowest terms. $\frac{53}{111}$

Teaching Suggestions
p. T122

Key Ideas

Prove trigonometric identities using algebraic transformations.

Chalkboard Examples

Prove the identity. Note any restrictions.

1. $\dfrac{1}{2 \sec x - 2} - \dfrac{1}{2 \sec x + 2} = \cot^2 x$

$= \dfrac{\dfrac{1}{2 \sec x - 2} - \dfrac{1}{2 \sec x + 2}}{1}$

$= \dfrac{(2 \sec x + 2) - (2 \sec x - 2)}{4 \sec^2 x - 4}$

$= \dfrac{4}{4(\sec^2 x - 1)}$

$= \dfrac{1}{\tan^2 x}$

$= \cot^2 x, \sin x \neq 0$

Suggested Assignments

Extended Alg. with Trig.
623/1–9, 11–19 odd, 21–26
R 624/Self-Test 1
Enriched Alg. with Trig.
623/1–9, 11–19 odd, 21–28
R 624/Self-Test 1

Write an equivalent expression in simplest form that involves only a single function (e.g., tan x) or a constant. Note any restrictions.

$\tan x$; $\sin x \neq 0$, $\cos x \neq 0$

29. $\dfrac{\sin \alpha}{1 + \cos \alpha} + \cot \alpha \csc \alpha$; $\sin \alpha \neq 0$

30. $(\sec x + 1)(\csc x - \cot x)$

31. $\dfrac{\cos x}{1 + \sin x} + \dfrac{1 + \sin x}{\cos x}$ $2 \sec x$; $\cos x \neq 0$

32. $\dfrac{\tan \alpha \csc \alpha}{\cos \alpha} - \dfrac{\sec \alpha - \cos \alpha}{\sin \alpha \tan \alpha}$

32. $\tan^2 \alpha$; $\sin \alpha \neq 0$, $\cos \alpha \neq 0$

C 33. If α is an angle in Quadrant II, express $\cos \alpha$ in terms of $\tan \alpha$.

34. If α is an angle in Quadrant III, express $\csc \alpha$ in terms of $\tan \alpha$.

35. If α is a positive acute angle and $\sin \alpha = \dfrac{1}{r}$, express $\tan \alpha$ in terms of r.

33. $\dfrac{1}{-\sqrt{\tan^2 \alpha + 1}}$ **34.** $-\dfrac{\sqrt{\tan^2 \alpha + 1}}{\tan \alpha}$ **35.** $\dfrac{1}{\sqrt{r^2 - 1}}$

15–2 Proving Identities

Is the equation

$$\frac{1}{1 + \sin x} + \frac{1}{1 - \sin x} = 2 \sec^2 x$$

true for every real number x for which both sides are defined? That is, is it an identity? One method of proving that an equation is an identity is to transform the more complicated side to the form of the simpler side.

EXAMPLE 1 Prove that $\dfrac{1}{1 + \sin x} + \dfrac{1}{1 - \sin x} = 2 \sec^2 x$ is an identity, noting any restrictions on values for x.

SOLUTION First notice that the left side is not defined for any number x for which $\sin x = 1$ or $\sin x = -1$, and the right side is not defined for those x for which $\cos x = 0$. For all other values of x we can simplify the left side to obtain the right side.

$$\frac{1(1 - \sin x)}{(1 + \sin x)(1 - \sin x)} + \frac{1(1 + \sin x)}{(1 - \sin x)(1 + \sin x)} = \frac{(1 - \sin x) + (1 + \sin x)}{1 - \sin^2 x}$$

$$= \frac{2}{1 - \sin^2 x}$$

$$= \frac{2}{\cos^2 x}$$

$$= 2 \sec^2 x$$

Since no new restrictions on x were introduced in the process, the original sentence is an identity when $\cos x \neq 0$. This restriction is sufficient to ensure that $\sin x \neq -1$ and $\sin x \neq 1$.

You can also prove that a given equation is an identity by showing that both sides are equal to the same expression.

622 *Chapter 15*

EXAMPLE 2   Prove that sec α(sec α − cos α) = (sec α csc α − cot α)^2, noting any restrictions on values for α.

SOLUTION   Simplifying the left side, you obtain the following:

$$\sec \alpha(\sec \alpha - \cos \alpha) = \sec^2 \alpha - \sec \alpha \cos \alpha$$

$$= \sec^2 \alpha - \left(\frac{1}{\cos \alpha}\right) \cos \alpha$$

$$= \sec^2 \alpha - 1$$

$$= \tan^2 \alpha$$

The left side is not defined for cos α = 0.
Simplifying the right side, you obtain the following:

$$(\sec \alpha \csc \alpha - \cot \alpha)^2 = \left(\frac{1}{\cos \alpha} \cdot \frac{1}{\sin \alpha} - \frac{\cos \alpha}{\sin \alpha}\right)^2$$

$$= \left(\frac{1 - \cos^2 \alpha}{\cos \alpha \sin \alpha}\right)^2$$

$$= \left(\frac{\sin^2 \alpha}{\cos \alpha \sin \alpha}\right)^2$$

$$= \left(\frac{\sin \alpha}{\cos \alpha}\right)^2$$

$$= \tan^2 \alpha$$

The right side is not defined for cos α = 0 or sin α = 0.

Since both sides equal tan^2 α, the identity is proved for all values of α for which cos α ≠ 0 and sin α ≠ 0.

Strategies for Proving Identities

1. Simplify the more complicated side of the equation first.
2. Look for ways to use the eight fundamental trigonometric identities.
3. If no other approach suggests itself, express all function values in terms of values of sine and cosine.
4. After simplifying one side as much as you can, work on the other side until the two sides are equal.

Written Exercises

Prove each identity. Note any restrictions.

A 1. cos α(sec α − cos α) = sin^2 α
   3. tan α + cot α = sec α csc α
   5. cos x(sec x + cos x csc^2 x) = csc^2 x

   2. csc α − sin α = cos α cot α
   4. cos x + sin x tan x = sec x
   6. $\dfrac{1 + \tan^2 x}{1 + \cot^2 x} = \tan^2 x$

Trigonometric Identities   **623**

(continued on p. 654)

Prove each identity. Note any restrictions.

1. $\csc^2 x(1 - \cos^2 x)$
$= \sec^2 x(1 - \sin^2 x)$
Left side:
$= \csc^2 x - \cos^2 x \csc^2 x$
$= \dfrac{1}{\sin^2 x} - \dfrac{\cos^2 x}{\sin^2 x}$
$= \dfrac{1 - \cos^2 x}{\sin^2 x} = \dfrac{\sin^2 x}{\sin^2 x} = 1$
Right side:
$= \sec^2 x - \sin^2 x \sec^2 x$
$= \dfrac{1}{\cos^2 x} - \dfrac{\sin^2 x}{\cos^2 x}$
$= \dfrac{1 - \sin^2 x}{\cos^2 x} = \dfrac{\cos^2 x}{\cos^2 x} = 1,$
$\sin x \neq 0, \cos x \neq 0$

2. $2 \csc x - 2 \cot x \cos x$
$= \dfrac{2}{\csc x}$ Left side:
$= \dfrac{2}{\sin x} - \dfrac{2 \cos^2 x}{\sin x}$
$= \dfrac{2 - 2\cos^2 x}{\sin x}$
$= \dfrac{2\sin^2 x}{\sin x} = 2 \sin x =$
$\dfrac{2}{\csc x}, \sin x \neq 0$

3. $\sec x(\cos x - \sec x)$
$= -\sec^2 x \sin^2 x$
Left side:
$= \sec x \cos x - \sec^2 x$
$= 1 - \sec^2 x = -\tan^2 x$
Right side: $= \dfrac{-1}{\cos^2 x} \cdot \sin^2 x$
$= \dfrac{-\sin^2 x}{\cos^2 x}$
$= -\tan^2 x, \cos x \neq 0$

Mixed Review

Factor completely.

1. $x^2 - 9xy - 22y^2$
$(x + 2y)(x - 11y)$

2. $3x^2 + 19x - 14$
$(3x - 2)(x + 7)$

3. $x^3 + 4x^2 - 32x$
$x(x + 8)(x - 4)$

Prove each identity. Note any restrictions.

7. $\dfrac{(\cos \alpha - \sin \alpha)^2}{\sin \alpha} = \csc \alpha - 2 \cos \alpha$

8. $\dfrac{\cos x}{1 - \cos x} - \dfrac{\cos x}{1 + \cos x} = 2 \cot^2 x$

9. $\dfrac{\sin \alpha}{\sec \alpha - 1} + \dfrac{\sin \alpha}{\sec \alpha + 1} = 2 \cot \alpha$

10. $\sec \alpha(\csc \alpha - \cot \alpha \cos \alpha) = \tan \alpha$

11. $\sec^2 x + \csc^2 x = \sec^2 x \csc^2 x$

12. $\sec x(\csc x - \sin x) = \cot x$

13. $\dfrac{1 + \cot \alpha}{\sin \alpha + \cos \alpha} = \csc \alpha$

14. $\dfrac{\sec \alpha + \csc \alpha}{1 + \cot \alpha} = \sec \alpha$

15. $(\sec x + 1)(\csc x - \cot x) = \tan x$

16. $(\tan x + \cos x)(\cot x + \tan x) = \csc x + \sec^2 x$

B 17. $(\tan \alpha + \cot \alpha)^2 = \csc^2 \alpha \sec^2 \alpha$

18. $(\csc \alpha - \sin \alpha)^2 = \cot^2 \alpha - \cos^2 \alpha$

19. $\dfrac{1}{1 - \cos x} - \dfrac{1}{1 + \sec x} = \csc^2 x + \cot^2 x$

20. $\dfrac{\sec \alpha}{\sec \alpha - 1} - \dfrac{\sec \alpha + 1}{\tan^2 \alpha} = 1$

21. $\sec^4 \alpha - \tan^4 \alpha = \sec^2 \alpha + \tan^2 \alpha$

22. $\dfrac{\sec x - \tan x}{\cos x} - \dfrac{\cos x}{\sec x + \tan x} = \dfrac{\sin^2 x}{1 + \sin x}$

23. $\dfrac{\cos \alpha}{1 - \sin \alpha} + \dfrac{1 - \sin \alpha}{\cos \alpha} = 2 \sec \alpha$

24. $\dfrac{\sec x - 1}{\tan x} + \dfrac{\tan x}{\sec x + 1} = \dfrac{2 \sin x}{1 + \cos x}$

25. $\dfrac{\tan x + 1}{\tan x} - \dfrac{\sec x \csc x + 1}{\tan x + 1} = \dfrac{\cos x}{\sin x + \cos x}$

26. $\dfrac{1 - \sin^4 x}{\sec^2 x + \tan^2 x} = \cos^4 x$

C 27. $\dfrac{\sec \alpha - \tan \alpha}{\sec \alpha + \tan \alpha} = \left(\dfrac{1 - \sin \alpha}{\cos \alpha}\right)^2$

28. $\dfrac{\tan x + \sin x}{\tan x - \sin x} = \left(\dfrac{1 + \cos x}{\sin x}\right)^2$

Self-Test 1

Prove each identity. Note any restrictions.

1. $\sec \alpha(\cos \alpha + \sin \alpha \tan \alpha) = \sec^2 \alpha$ *Obj. 1, p. 619*

2. $\dfrac{\cos x}{\sec x} + \dfrac{\sin x}{\csc x} = 1$

3. $\dfrac{\sin^2 \alpha}{1 - \cos \alpha} = 1 + \cos \alpha$

4. $\sec \alpha \csc \alpha - \tan \alpha = \cot \alpha$

Check your answers with those at the back of the book.

Functions: Sums and Differences

OBJECTIVES for Sections 15-3 through 15-6:
1. *To use formulas for the sine, cosine, and tangent of a sum or difference.*
2. *To use the double-angle and half-angle formulas for the sine, cosine, and tangent.*
3. *To prove trigonometric identities involving double- and half-angle formulas.*

15–3 The Cosine of a Sum or a Difference

Teaching Suggestions
p. T122

Key Ideas

Develop and use formulas for the cosine of a sum or the cosine of a difference of two angles.

If $m(\alpha_1) = x_1{}^R$ and $m(\alpha_2) = x_2{}^R$, then the **sum** of α_1 and α_2, denoted by $\alpha_1 + \alpha_2$, is an angle whose measure in radians is $x_1 + x_2$:

$$m(\alpha_1 + \alpha_2) = (x_1 + x_2)^R$$

Similarly, the **difference** of α_1 and α_2, denoted by $\alpha_1 - \alpha_2$, has measure $x_1 - x_2$:

$$m(\alpha_1 - \alpha_2) = (x_1 - x_2)^R$$

Angles α_1, α_2, $\alpha_1 + \alpha_2$, $\alpha_1 - \alpha_2$, where all are in the first quadrant, are pictured in Figure 1.

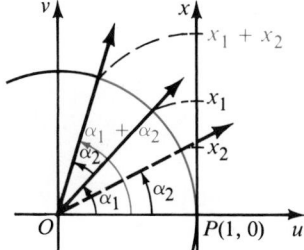

Figure 1

The **opposite** of α_1, denoted by $-\alpha_1$, has measure $-x_1$:

$$m(-\alpha_1) = (-x_1)^R$$

Angle α_1 (in the first quadrant) and angle $-\alpha_1$ (in the fourth quadrant) are pictured in Figure 2.

Figure 2 can be used to deduce two very important properties of sin $(-x)$ and cos $(-x)$ and of sin $(-\alpha)$ and cos $(-\alpha)$. Because of the symmetry of the unit circle with respect to the axes, the points R and S have *equal abscissas*, but *ordinates* that are the *opposites* of each other. Since a similar situation exists if α is in any of the other quadrants, you can assert that for each angle α,

$$\sin{(-\alpha)} = -\sin{\alpha} \quad \text{and} \quad \cos{(-\alpha)} = \cos{\alpha}.$$

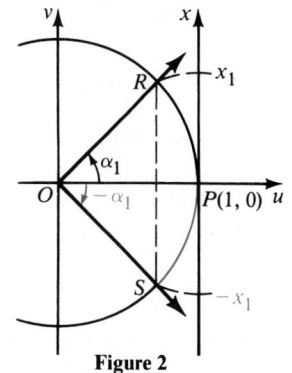

Figure 2

Common Errors

Many students will confuse the sum and difference formulas with the distributive property. To prevent this confusion, use an example to show that the cosine function is not distributive. For example, show that cos (60° − 30°) ≠ cos 60° − cos 30°. Also, remind students that trigonometric notation is *functional* notation, not multiplication.

1. Find $\cos \dfrac{\pi}{12}$.

$$\dfrac{\pi}{12} = \dfrac{4\pi}{12} - \dfrac{3\pi}{12} = \dfrac{\pi}{3} - \dfrac{\pi}{4}$$

$$\therefore \cos \dfrac{\pi}{12} = \cos \left(\dfrac{\pi}{3} - \dfrac{\pi}{4} \right) =$$

$$\cos \dfrac{\pi}{3} \cos \dfrac{\pi}{4} + \sin \dfrac{\pi}{3} \sin \dfrac{\pi}{4}$$

$$\cos \dfrac{\pi}{3} = \dfrac{1}{2}, \ \cos \dfrac{\pi}{4} = \dfrac{1}{\sqrt{2}},$$

$$\sin \dfrac{\pi}{3} = \dfrac{\sqrt{3}}{2}, \ \sin \dfrac{\pi}{4} = \dfrac{1}{\sqrt{2}}$$

$$\cos \dfrac{\pi}{12} = \cos \left(\dfrac{\pi}{3} - \dfrac{\pi}{4} \right)$$

$$= \left(\dfrac{1}{2} \right) \left(\dfrac{1}{\sqrt{2}} \right) + \left(\dfrac{\sqrt{3}}{2} \right) \left(\dfrac{1}{\sqrt{2}} \right)$$

$$= \dfrac{\sqrt{2}}{4} + \dfrac{\sqrt{6}}{4}$$

$$= \dfrac{\sqrt{2}(1 + \sqrt{3})}{4}$$

2. Find $\cos (x - y)$, if

$$\sin x = \dfrac{-2}{3}, \dfrac{3\pi}{2} < x < 2\pi,$$

and $\sin y = \dfrac{5}{9}, 0 < y < \dfrac{\pi}{2}$.

$$\left(\dfrac{-2}{3} \right)^2 + \cos^2 x = 1$$

$$\cos^2 x = \dfrac{5}{9}$$

$$\cos x = \pm \dfrac{\sqrt{5}}{3}$$

Since $\dfrac{3\pi}{2} < x < 2\pi$,

$$\cos x = \dfrac{\sqrt{5}}{3}.$$

$$\left(\dfrac{5}{9} \right)^2 + \cos^2 y = 1$$

$$\cos^2 y = \dfrac{56}{81}$$

$$\cos y = \pm \dfrac{2\sqrt{14}}{9}$$

(continued)

Similarly, for each real number x,

$$\sin (-x) = -\sin x \qquad \text{and} \qquad \cos (-x) = \cos x.$$

A function for which

$$f(-x) = -f(x)$$

for every x in the domain of f is called an *odd function*, while a function for which

$$f(-x) = f(x)$$

for every x in the domain of f is called an *even function* (see Section 7-1). Thus, sine is an *odd* function, and cosine is an *even* function.

Figure 3 shows $x_1 - x_2$ when x_1 corresponds to a second-quadrant angle and x_2 to a first-quadrant angle.

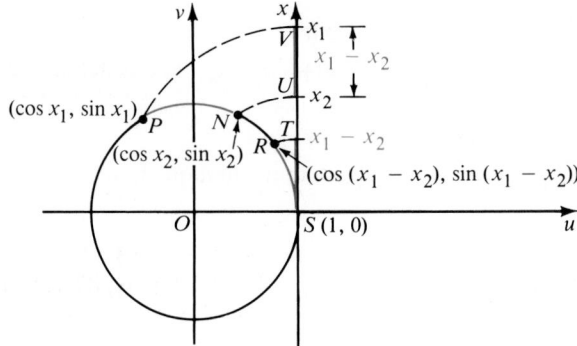

Figure 3

Observe that $\overline{UV}$, whose endpoints are the graphs of x_2 and x_1 on the x number line, is congruent to $\overline{ST}$, with endpoints corresponding to 0 and $x_1 - x_2$. Therefore, the arc of the unit circle from P to R is congruent to the arc from N to S, and consequently chord $\overline{PR}$ is congruent to chord $\overline{NS}$. Using the distance formula (page 394), you have the following:

$$\sqrt{[\cos (x_1 - x_2) - 1]^2 + [\sin (x_1 - x_2) - 0]^2}$$
$$= \sqrt{(\cos x_1 - \cos x_2)^2 + (\sin x_1 - \sin x_2)^2}$$

Squaring both sides of this equation, you obtain:

$$\cos^2 (x_1 - x_2) - 2 \cos (x_1 - x_2) + 1 + \sin^2 (x_1 - x_2)$$
$$= \cos^2 x_1 - 2 \cos x_1 \cos x_2 + \cos^2 x_2$$
$$+ \sin^2 x_1 - 2 \sin x_1 \sin x_2 + \sin^2 x_2$$

On rearranging terms, you have the equation at the top of the following page.

626 *Chapter 15*

$$\cos^2 (x_1 - x_2) + \sin^2 (x_1 - x_2) - 2 \cos (x_1 - x_2) + 1$$
$$= \cos^2 x_1 + \sin^2 x_1 - 2 \cos x_1 \cos x_2 +$$
$$\cos^2 x_2 + \sin^2 x_2 - 2 \sin x_1 \sin x_2$$

Because $\cos^2 x + \sin^2 x = 1$ for every real number x, you can replace each of the three pairs of terms shown in red in the preceding equation with "1" and simplify the result to obtain the following:

$$\cos (x_1 - x_2) = \cos x_1 \cos x_2 + \sin x_1 \sin x_2$$

Since the development given is not limited by the particular values of x_1 and x_2 shown in Figure 3, the preceding equation is an identity. It is sometimes called the formula for the cosine of a difference. In terms of angles, this formula could be written:

$$\cos (\alpha_1 - \alpha_2) = \cos \alpha_1 \cos \alpha_2 + \sin \alpha_1 \sin \alpha_2$$

EXAMPLE 1   Find $\cos \dfrac{5\pi}{12}$.

SOLUTION   Express $\dfrac{5\pi}{12}$ as the difference of two numbers for which you know the sine and cosine. (Refer to the table on page 590.) Thus,

$$\frac{5\pi}{12} = \frac{8\pi}{12} - \frac{3\pi}{12} = \frac{2\pi}{3} - \frac{\pi}{4}.$$

Replacing x_1 and x_2 in the formula for the cosine of a difference with $\dfrac{2\pi}{3}$ and $\dfrac{\pi}{4}$, respectively, you get

$$\cos \left(\frac{2\pi}{3} - \frac{\pi}{4} \right) = \cos \frac{2\pi}{3} \cos \frac{\pi}{4} + \sin \frac{2\pi}{3} \sin \frac{\pi}{4}.$$

From the table, you have

$$\cos \frac{2\pi}{3} = -\frac{1}{2}, \quad \cos \frac{\pi}{4} = \frac{1}{\sqrt{2}}, \quad \sin \frac{2\pi}{3} = \frac{\sqrt{3}}{2}, \quad \sin \frac{\pi}{4} = \frac{1}{\sqrt{2}}.$$

Therefore:

$$\cos \frac{5\pi}{12} = \cos \left(\frac{2\pi}{3} - \frac{\pi}{4} \right)$$
$$= \left(-\frac{1}{2} \right)\left(\frac{1}{\sqrt{2}} \right) + \left(\frac{\sqrt{3}}{2} \right)\left(\frac{1}{\sqrt{2}} \right)$$
$$= -\frac{1}{2\sqrt{2}} + \frac{\sqrt{3}}{2\sqrt{2}}$$
$$= -\frac{\sqrt{2}}{4} + \frac{\sqrt{6}}{4}$$
$$= \frac{-\sqrt{2} + \sqrt{6}}{4}$$

Since $0 < y < \dfrac{\pi}{2}$,

$$\cos y = \frac{2\sqrt{14}}{9}.$$

$$\therefore \cos (x - y)$$
$$= \cos x \cos y + \sin x \sin y$$
$$= \frac{\sqrt{5}}{3} \cdot \frac{2\sqrt{14}}{9} + \frac{-2}{3} \cdot \frac{5}{9}$$
$$= \frac{2\sqrt{70} - 10}{27}$$

Mixed Review

Simplify.

1. $\sqrt{\dfrac{x^9}{z^4}} \quad \dfrac{x^4 \sqrt{x}}{z^2}$

2. $\sqrt{\dfrac{a^5}{b^2 c^3}} \quad \dfrac{a^2 \sqrt{ac}}{|b| c^2}$

3. $\sqrt[4]{9 a^6} \quad a\sqrt{3a}$

4. $\sqrt[8]{2} \sqrt[4]{32} \quad 2\sqrt[8]{8}$

Use special right triangles to find the values of each. Leave in radical form.

5. sec 30° $\dfrac{2}{\sqrt{3}}$

6. csc 45° $\sqrt{2}$

Trigonometric Identities   **627**

627

By replacing x_2 with $-x_2$ in the difference formula, you obtain

$$\cos[x_1 - (-x_2)] = \cos x_1 \cos(-x_2) + \sin x_1 \sin(-x_2).$$

Then, since $\cos(-x) = \cos x$ and $\sin(-x) = -\sin x$ (page 626), you have the **formula for the cosine of a sum:**

$$\cos(x_1 + x_2) = \cos x_1 \cos x_2 - \sin x_1 \sin x_2$$

In terms of angles, this formula could be written:

$$\cos(\alpha_1 + \alpha_2) = \cos \alpha_1 \cos \alpha_2 - \sin \alpha_1 \sin \alpha_2$$

You can develop a number of other valuable identities by using the formulas for the cosine of a difference and the cosine of a sum. For example, if you let $x_1 = \pi$ and $x_2 = x$ in the difference formula, you have the following:

$$\cos(\pi - x) = \cos \pi \cos x + \sin \pi \sin x$$
$$= (-1)\cos x + (0)\sin x = -\cos x$$

Similarly, if $m(\alpha_1) = 180°$ and $\alpha_2 = \alpha$, you find the following from the difference formula:

$$\cos(180° - \alpha) = \cos 180° \cos \alpha + \sin 180° \sin \alpha$$
$$= (-1)\cos \alpha + (0)\sin \alpha = -\cos \alpha$$

You can also show (Exercise 22, page 630) that

$$\cos(\pi + x) = -\cos x \qquad \text{and} \qquad \cos(180° + \alpha) = -\cos \alpha.$$

Now if you let $x_1 = \frac{\pi}{2}$ and $x_2 = x$ in the difference formula, you have the following:

$$\cos\left(\frac{\pi}{2} - x\right) = \cos\frac{\pi}{2}\cos x + \sin\frac{\pi}{2}\sin x$$
$$= (0)\cos x + (1)\sin x = \sin x$$

Notice that since the measures of the acute angles of a right triangle can be expressed as x and $\frac{\pi}{2} - x$ radians, the preceding formula is a generalization of the corresponding one given on page 607.

You can also show (Exercise 25, page 630) that

$$\cos\left(\frac{\pi}{2} + x\right) = -\sin x.$$

By letting $x_1 = \frac{3\pi}{2}$ in the sum and difference formulas, you can find two more identities (Exercises 23 and 24, page 630). Because all these formulas can be used to "reduce" a given circular or trigonometric function value in any quadrant to a function value in Quadrant I, they are called **reduction formulas**. Here is a list of the most useful reduction

Additional A Exercises

Express each of the following in terms of a single cosine and evaluate that cosine.

1. $\cos\dfrac{13\pi}{18}\cos\dfrac{\pi}{18} +$
$\sin\dfrac{13\pi}{18}\sin\dfrac{\pi}{18}$
$\cos\dfrac{2\pi}{3} = -\dfrac{1}{2}$

2. $\cos\dfrac{7\pi}{16}\cos\dfrac{\pi}{16} -$
$\sin\dfrac{7\pi}{16}\sin\dfrac{\pi}{16}$
$\cos\dfrac{\pi}{2} = 0$

3. $\cos 160° \cos 25° +$
$\sin 160° \sin 25°$
$\cos 135° = -\dfrac{1}{\sqrt{2}}$

4. $\cos 100° \cos 80° -$
$\sin 100° \sin 80°$
$\cos 180° = -1$

formulas for cosine. From now on we shall work primarily with formulas stated in terms of x, where x is a real number, with the understanding that corresponding formulas hold in terms of angles.

$$\cos\left(\frac{\pi}{2} + x\right) = -\sin x \qquad \cos\left(\frac{\pi}{2} - x\right) = \sin x$$

$$\cos(\pi + x) = -\cos x \qquad \cos(\pi - x) = -\cos x$$

$$\cos\left(\frac{3\pi}{2} + x\right) = \sin x \qquad \cos\left(\frac{3\pi}{2} - x\right) = -\sin x$$

EXAMPLE 2 Find $\cos(x - y)$, given that $\sin x = \frac{3}{5}$, $0 < x < \frac{\pi}{2}$, and $\sin y = \frac{7}{25}$, $\frac{\pi}{2} < y < \pi$.

SOLUTION Since $\sin^2 x + \cos^2 x = 1$, you have the following:

$$\left(\frac{3}{5}\right)^2 + \cos^2 x = 1$$

$$\cos^2 x = \frac{16}{25}$$

$$\cos x = \pm\sqrt{\frac{16}{25}} = \pm\frac{4}{5}$$

But since $0 < x < \frac{\pi}{2}$, $\cos x = \frac{4}{5}$.

Similar reasoning shows that

$$\cos y = \pm\frac{24}{25}.$$

But since $\frac{\pi}{2} < y < \pi$, $\cos y = -\frac{24}{25}$.

Using the formula for the cosine of a difference, you obtain the following:

$$\cos(x - y) = \cos x \cos y + \sin x \sin y$$

$$= \frac{4}{5}\left(-\frac{24}{25}\right) + \frac{3}{5} \cdot \frac{7}{25} = -\frac{75}{125} = -\frac{3}{5}$$

Oral Exercises

Express each of the following in terms of a single cosine.

1. $\cos 45° \cos 60° - \sin 45° \sin 60°$ $\cos 105°$

2. $\cos \frac{\pi}{4} \cos \frac{\pi}{6} + \sin \frac{\pi}{4} \sin \frac{\pi}{6}$ $\cos \frac{\pi}{12}$

3. $\cos 20° \cos 35° + \sin 20° \sin 35°$ $\cos 15°$

4. $\cos \frac{\pi}{3} \cos \frac{\pi}{4} - \sin \frac{\pi}{3} \sin \frac{\pi}{4}$ $\cos \frac{7\pi}{12}$

5. $\cos \frac{4\pi}{9} \cos \frac{\pi}{9} - \sin \frac{4\pi}{9} \sin \frac{\pi}{9}$ $\cos \frac{5\pi}{9}$

6. $\sin 65° \sin 15° - \cos 65° \cos 15°$ $-\cos 80°$

Trigonometric Identities **629**

Express each of the following as the cosine of a sum or a difference of two of the angles listed in the table on page 590. Then apply a sum or difference formula to evaluate the given cosine. Express your answer in radical form.

5. $\cos 165°$
$120° + 45°$; $\cos 165° =$
$\cos 120° \cos 45° -$
$\sin 120° \sin 45° =$
$\frac{-1}{2}\frac{1}{\sqrt{2}} - \frac{\sqrt{3}}{2}\frac{1}{\sqrt{2}} =$
$\frac{-\sqrt{2}}{4} - \frac{\sqrt{6}}{4} = \frac{-\sqrt{2}(1 + \sqrt{3})}{4}$

6. $\cos(-75°)$
$135° - 210°$;
$\cos(-75°) =$
$\cos 135° \cos 210° +$
$\sin 135° \sin 210° =$
$\left(-\frac{1}{\sqrt{2}}\right)\left(-\frac{\sqrt{3}}{2}\right) +$
$\left(\frac{1}{\sqrt{2}}\right)\left(-\frac{1}{2}\right)$
$= \frac{\sqrt{6}}{4} - \frac{\sqrt{2}}{4} = \frac{\sqrt{6} - \sqrt{2}}{4}$

Additional Answers
Written Exercises

4. $\cos 150° = -\frac{\sqrt{3}}{2}$

5. $\cos 1.5\pi = 0$

21. $\cos\left(x - \frac{\pi}{2}\right) = \cos x \cos \frac{\pi}{2}$
$+ \sin x \sin \frac{\pi}{2} = \cos x(0)$
$+ \sin x(1) = \sin x$

22. $\cos(\pi + x) =$
$\cos \pi \cos x - \sin \pi \sin x =$
$(-1)\cos x - (0)\sin x$
$= -\cos x$

(continued)

23. $\cos\left(\dfrac{3\pi}{2} + x\right) = \cos\dfrac{3\pi}{2} \cdot$

$\cos x - \sin\dfrac{3\pi}{2} \sin x =$

$(0)\cos x - (-1)\sin x =$
$\sin x$

24. $\cos\left(\dfrac{3\pi}{2} - x\right) = \cos\dfrac{3\pi}{2} \cdot$

$\cos x + \sin\dfrac{3\pi}{2} \sin x =$

$(0)\cos x + (-1)\sin x =$
$-\sin x$

25. $\cos\left(\dfrac{\pi}{2} + x\right) = \cos\dfrac{\pi}{2} \cos x$

$- \sin\dfrac{\pi}{2} \sin x = (0)\cos x$

$- (1)\sin x = -\sin x$

26. $\cos\left(x - \dfrac{\pi}{6}\right) = \cos x \cdot$

$\cos\dfrac{\pi}{6} + \sin x \sin\dfrac{\pi}{6} =$

$\cos x\left(\dfrac{\sqrt{3}}{2}\right) + \sin x\left(\dfrac{1}{2}\right) =$

$\dfrac{\sqrt{3}\cos x + \sin x}{2}$

27. $\cos\left(\dfrac{\pi}{3} - x\right) = \cos\dfrac{\pi}{3} \cdot$

$\cos x + \sin\dfrac{\pi}{3} \sin x =$

$\left(\dfrac{1}{2}\right)\cos x + \left(\dfrac{\sqrt{3}}{2}\right)\sin x =$

$\dfrac{\cos x + \sqrt{3}\sin x}{2}$

28. $\cos\left(\dfrac{3\pi}{4} + x\right) = \cos\dfrac{3\pi}{4} \cdot$

$\cos x - \sin\dfrac{3\pi}{4} \sin x =$

$\left(-\dfrac{1}{\sqrt{2}}\right)\cos x - \left(\dfrac{1}{\sqrt{2}}\right)\sin x$

$= -\dfrac{1}{\sqrt{2}}(\cos x + \sin x)$

33. $\sin x = \dfrac{\sqrt{5}}{3}$; $\sin y = -\dfrac{1}{2}$;

$\cos(x + y) = \dfrac{2\sqrt{3} + \sqrt{5}}{6}$

Written Exercises

Express each of the following in terms of a single cosine and evaluate that cosine. $\cos\dfrac{3\pi}{4} = -\dfrac{1}{\sqrt{2}}$

A

1. $\cos\dfrac{7\pi}{12} \cos\dfrac{\pi}{12} - \sin\dfrac{7\pi}{12} \sin\dfrac{\pi}{12}$ $\cos\dfrac{2\pi}{3} = -\dfrac{1}{2}$ 2. $\cos\dfrac{\pi}{8} \cos\dfrac{7\pi}{8} + \sin\dfrac{\pi}{8} \sin\dfrac{7\pi}{8}$

3. $\cos\dfrac{5\pi}{9} \cos\dfrac{2\pi}{9} + \sin\dfrac{5\pi}{9} \sin\dfrac{2\pi}{9}$ $\cos\dfrac{\pi}{3} = \dfrac{1}{2}$ 4. $\cos 175° \cos 25° + \sin 175° \sin 25°$

5. $\cos 1.2\pi \cos 0.3\pi - \sin 1.2\pi \sin 0.3\pi$ 6. $\cos 185° \cos 40° - \sin 185° \sin 40°$

7. $\cos 135° \cos(-15°) + \sin 135° \sin(-15°)$ $\cos 150° = -\dfrac{\sqrt{3}}{2}$ $\cos 225° = -\dfrac{1}{\sqrt{2}}$

8. $\cos(-5°) \cos 245° - \sin(-5°) \sin 245°$ $\cos 240° = -\dfrac{1}{2}$

Express each of the following as the cosine of a sum or difference of two of the angles listed in the table on page 590. Then apply a sum or difference formula to evaluate the given cosine. Express your answer in radical form. $\dfrac{-\sqrt{6} - \sqrt{2}}{4}$

9. $\cos 15°$ $\dfrac{\sqrt{6} + \sqrt{2}}{4}$ 10. $\cos 75°$ $\dfrac{\sqrt{6} - \sqrt{2}}{4}$ 11. $\cos 165°$ $\dfrac{-\sqrt{6} - \sqrt{2}}{4}$ 12. $\cos 195°$

13. $\cos 345°$ $\dfrac{\sqrt{2} + \sqrt{6}}{4}$ 14. $\cos 285°$ $\dfrac{-\sqrt{2} + \sqrt{6}}{4}$ 15. $\cos(-15°)$ $\dfrac{\sqrt{6} + \sqrt{2}}{4}$ 16. $\cos(-105°)$ $\dfrac{\sqrt{2} - \sqrt{6}}{4}$

Express each of the following as the cosine of a sum or difference you can evaluate using the table on page 590. Then evaluate it, expressing your answer in radical form.

17. $\cos\dfrac{\pi}{12}$ $\dfrac{\sqrt{6} + \sqrt{2}}{4}$ 18. $\cos\dfrac{13\pi}{12}$ $\dfrac{-\sqrt{6} - \sqrt{2}}{4}$ 19. $\cos\dfrac{17\pi}{12}$ $\dfrac{-\sqrt{6} + \sqrt{2}}{4}$ 20. $\cos\left(-\dfrac{5\pi}{12}\right)$ $\dfrac{\sqrt{6} - \sqrt{2}}{4}$

Use the formulas for the cosine of a sum or a difference to prove each of the following identities.

21. $\cos\left(x - \dfrac{\pi}{2}\right) = \sin x$ 22. $\cos(\pi + x) = -\cos x$

23. $\cos\left(\dfrac{3\pi}{2} + x\right) = \sin x$ 24. $\cos\left(\dfrac{3\pi}{2} - x\right) = -\sin x$

25. $\cos\left(\dfrac{\pi}{2} + x\right) = -\sin x$ 26. $\cos\left(x - \dfrac{\pi}{6}\right) = \dfrac{\sqrt{3}\cos x + \sin x}{2}$

27. $\cos\left(\dfrac{\pi}{3} - x\right) = \dfrac{\cos x + \sqrt{3}\sin x}{2}$ 28. $\cos\left(\dfrac{3\pi}{4} + x\right) = -\dfrac{1}{\sqrt{2}}(\cos x + \sin x)$

Find sin x and sin y and use these values to find a value for cos $(x - y)$ under the given conditions.

B

29. $\cos x = \dfrac{4}{5}$; $0 < x < \dfrac{\pi}{2}$; $\cos y = \dfrac{12}{13}$; $0 < y < \dfrac{\pi}{2}$ $\dfrac{63}{65}$

30. $\cos x = -\dfrac{3}{5}$; $\dfrac{\pi}{2} < x < \pi$; $\cos y = \dfrac{5}{13}$; $0 < y < \dfrac{\pi}{2}$ $\dfrac{33}{65}$

31. $\cos x = -\dfrac{7}{25}$; $\pi < x < \dfrac{3\pi}{2}$; $\cos y = -\dfrac{4}{5}$; $\dfrac{\pi}{2} < y < \pi$ $-\dfrac{44}{125}$

32. $\cos x = \dfrac{15}{17}$; $\dfrac{3\pi}{2} < x < 2\pi$; $\cos y = -\dfrac{3}{5}$; $\pi < y < \dfrac{3\pi}{2}$ $-\dfrac{13}{85}$

630 Chapter 15

Find sin x and sin y and use these values to find a value for cos $(x + y)$ under the given conditions.

33. $\cos x = \frac{2}{3}$; $\ 0 < x < \frac{\pi}{2}$; $\quad \cos y = \frac{\sqrt{3}}{2}$; $\ \frac{3\pi}{2} < y < 2\pi$

34. $\cos x = -\frac{3}{4}$; $\ \pi < x < \frac{3\pi}{2}$; $\quad \cos y = \frac{2\sqrt{2}}{3}$; $\ 0 < y < \frac{\pi}{2}$

35. $\cos x = -\frac{\sqrt{15}}{4}$; $\ \frac{\pi}{2} < x < \pi$; $\quad \cos y = \frac{\sqrt{5}}{3}$; $\ \frac{3\pi}{2} < y < 2\pi$

C 36. $\cos x = a$; $\ 0 < x < \frac{\pi}{2}$; $\quad \cos y = \sqrt{1 - b^2}$; $\ 0 < y < \frac{\pi}{2}$

15–4 The Sine and Tangent of a Sum or a Difference

To find formulas for the sine of a sum and of a difference, first recall (page 629) that

$$\cos\left(\frac{\pi}{2} - x\right) = \sin x. \qquad (1)$$

If you replace x with $\frac{\pi}{2} - x$, you obtain

$$\cos\left(\frac{\pi}{2} - \left(\frac{\pi}{2} - x\right)\right) = \sin\left(\frac{\pi}{2} - x\right),$$

or

$$\sin\left(\frac{\pi}{2} - x\right) = \cos x.$$

To find a formula for $\sin(x_1 + x_2)$, use (1) and the formula for the cosine of a difference to get:

$$\sin(x_1 + x_2) = \cos\left(\frac{\pi}{2} - (x_1 + x_2)\right)$$

$$= \cos\left(\left(\frac{\pi}{2} - x_1\right) - x_2\right)$$

$$= \cos\left(\frac{\pi}{2} - x_1\right)\cos x_2 + \sin\left(\frac{\pi}{2} - x_1\right)\sin x_2$$

Since $\cos\left(\frac{\pi}{2} - x_1\right) = \sin x_1$ and $\sin\left(\frac{\pi}{2} - x_1\right) = \cos x_1$, you have the formula for the sine of a sum:

$$\sin(x_1 + x_2) = \sin x_1 \cos x_2 + \cos x_1 \sin x_2$$

In terms of angles, you have:

$$\sin(\alpha_1 + \alpha_2) = \sin \alpha_1 \cos \alpha_2 + \cos \alpha_1 \sin \alpha_2$$

Trigonometric Identities **631**

---

34. $\sin x = -\frac{\sqrt{7}}{4}$; $\sin y = \frac{1}{3}$;
$\cos(x + y) = \frac{-6\sqrt{2} + \sqrt{7}}{12}$

35. $\sin x = \frac{1}{4}$; $\sin y = -\frac{2}{3}$;
$\cos(x + y) = \frac{-5\sqrt{3} + 2}{12}$

36. $\sin x = \sqrt{1 - a^2}$;
$\sin y = |b|$; $\cos(x + y) = a\sqrt{1 - b^2} - |b|\sqrt{1 - a^2}$

Teaching Suggestions
p. T122

Key Ideas

Develop and use formulas for the sine and tangent of a sum or a difference of two angles.

1. Find $\sin \dfrac{7\pi}{12}$.

$$\dfrac{7\pi}{12} = \dfrac{3\pi}{12} + \dfrac{4\pi}{12} = \dfrac{\pi}{4} + \dfrac{\pi}{3}$$

$$\therefore \sin \dfrac{7\pi}{12} = \sin \left(\dfrac{\pi}{4} + \dfrac{\pi}{3} \right)$$

$$= \sin \dfrac{\pi}{4} \cos \dfrac{\pi}{3} + \cos \dfrac{\pi}{4} \sin \dfrac{\pi}{3}$$

$$= \left(\dfrac{1}{\sqrt{2}} \right)\left(\dfrac{1}{2} \right) + \left(\dfrac{1}{\sqrt{2}} \right)\left(\dfrac{\sqrt{3}}{2} \right)$$

$$= \dfrac{\sqrt{2}}{4} + \dfrac{\sqrt{6}}{4} = \dfrac{\sqrt{2} + \sqrt{6}}{4}$$

2. Find $\sin \dfrac{\pi}{12}$.

$$\dfrac{\pi}{12} = \dfrac{4\pi}{12} - \dfrac{3\pi}{12} = \dfrac{\pi}{3} - \dfrac{\pi}{4}$$

$$\therefore \sin \dfrac{\pi}{12} = \sin \left(\dfrac{\pi}{3} - \dfrac{\pi}{4} \right)$$

$$= \sin \dfrac{\pi}{3} \cos \dfrac{\pi}{4} - \cos \dfrac{\pi}{3} \sin \dfrac{\pi}{4}$$

$$= \left(\dfrac{\sqrt{3}}{2} \right)\left(\dfrac{1}{\sqrt{2}} \right) - \left(\dfrac{1}{2} \right)\left(\dfrac{1}{\sqrt{2}} \right)$$

$$= \dfrac{\sqrt{6}}{4} - \dfrac{\sqrt{2}}{4} = \dfrac{\sqrt{6} - \sqrt{2}}{4}$$

3. Find $\tan \dfrac{5\pi}{12}$.

$$\dfrac{5\pi}{12} = \dfrac{3\pi}{12} + \dfrac{2\pi}{12} = \dfrac{\pi}{4} + \dfrac{\pi}{6}$$

$$\therefore \tan \dfrac{5\pi}{12} = \dfrac{\tan \dfrac{\pi}{4} + \tan \dfrac{\pi}{6}}{1 - \tan \dfrac{\pi}{4} \tan \dfrac{\pi}{6}}$$

$$= \dfrac{1 + \dfrac{1}{\sqrt{3}}}{1 - \dfrac{1}{\sqrt{3}}}$$

$$= 2 + \sqrt{3}$$

If, in the formula for the sine of a sum, you replace x_2 with $-x_2$, you obtain:

$$\sin (x_1 + (-x_2)) = \sin x_1 \cos (-x_2) + \cos x_1 \sin (-x_2)$$

Then, because $\sin (-x) = -\sin x$ and $\cos (-x) = \cos x$ (page 626), you have the formula for the sine of a difference:

$$\sin (x_1 - x_2) = \sin x_1 \cos x_2 - \cos x_1 \sin x_2$$

In terms of angles, you have:

$$\sin (\alpha_1 - \alpha_2) = \sin \alpha_1 \cos \alpha_2 - \cos \alpha_1 \sin \alpha_2$$

EXAMPLE 1 Find $\sin \dfrac{5\pi}{12}$.

SOLUTION $\dfrac{5\pi}{12}$ can be expressed as the sum of two numbers for which you know the sine and cosine, namely,

$$\dfrac{5\pi}{12} = \dfrac{\pi}{6} + \dfrac{\pi}{4}.$$

Then, you can replace x_1 and x_2 in the sum formula with $\dfrac{\pi}{6}$ and $\dfrac{\pi}{4}$, respectively, to obtain

$$\sin \dfrac{5\pi}{12} = \sin \left(\dfrac{\pi}{6} + \dfrac{\pi}{4} \right) = \sin \dfrac{\pi}{6} \cos \dfrac{\pi}{4} + \cos \dfrac{\pi}{6} \sin \dfrac{\pi}{4}.$$

Since

$$\sin \dfrac{\pi}{6} = \dfrac{1}{2}, \ \cos \dfrac{\pi}{4} = \dfrac{1}{\sqrt{2}}, \ \cos \dfrac{\pi}{6} = \dfrac{\sqrt{3}}{2}, \ \text{and} \ \sin \dfrac{\pi}{4} = \dfrac{1}{\sqrt{2}},$$

you have:

$$\sin \dfrac{5\pi}{12} = \sin \left(\dfrac{\pi}{6} + \dfrac{\pi}{4} \right)$$

$$= \left(\dfrac{1}{2} \right)\left(\dfrac{1}{\sqrt{2}} \right) + \left(\dfrac{\sqrt{3}}{2} \right)\left(\dfrac{1}{\sqrt{2}} \right)$$

$$= \dfrac{1}{2\sqrt{2}} + \dfrac{\sqrt{3}}{2\sqrt{2}}$$

$$= \dfrac{\sqrt{2}}{4} + \dfrac{\sqrt{6}}{4}$$

$$= \dfrac{\sqrt{2} + \sqrt{6}}{4}$$

Following are the reduction formulas for sine corresponding to those for cosine given on page 629. The second of these was derived at the beginning of this section. The others are left as Exercises 27–31, page 635.

632 *Chapter 15*

$$\sin\left(\frac{\pi}{2} + x\right) = \cos x \qquad\qquad \sin\left(\frac{\pi}{2} - x\right) = \cos x$$

$$\sin(\pi + x) = -\sin x \qquad\qquad \sin(\pi - x) = \sin x$$

$$\sin\left(\frac{3\pi}{2} + x\right) = -\cos x \qquad\qquad \sin\left(\frac{3\pi}{2} - x\right) = -\cos x$$

Sum, difference, and reduction formulas for the functions tangent, cotangent, secant, and cosecant can be obtained by using appropriate formulas established for sine and cosine, in conjunction with the definitions of the other functions. Although the formulas for sine and cosine hold for all values of the variable, there are certain restrictions placed on the domains of the variables in formulas for the other functions.

In practice, the only ones of these additional formulas that are of much concern to us are those for tangent, because to find values for cotangent, secant, or cosecant, you need only take the reciprocal of the corresponding values of tangent, cosine, or sine.

To start, notice that tangent, like sine, is an odd function (page 626), since

$$\tan(-x) = \frac{\sin(-x)}{\cos(-x)} = \frac{-\sin x}{\cos x} = -\frac{\sin x}{\cos x}.$$

Thus,

$$\tan(-x) = -\tan x, \quad \cos x \neq 0.$$

To obtain a difference formula for the tangent function, you can replace x with $(x_1 - x_2)$ in the definition of $\tan x$ to obtain:

$$\tan(x_1 - x_2) = \frac{\sin(x_1 - x_2)}{\cos(x_1 - x_2)}, \quad \cos(x_1 - x_2) \neq 0$$

From this, you find:

$$\tan(x_1 - x_2) = \frac{\sin x_1 \cos x_2 - \cos x_1 \sin x_2}{\cos x_1 \cos x_2 + \sin x_1 \sin x_2}$$

Then, if $\cos x_1 \cos x_2 \neq 0$, you can divide the numerator and denominator of the right side of this identity by this product to obtain:

$$\tan(x_1 - x_2) = \frac{\dfrac{\sin x_1 \cos x_2}{\cos x_1 \cos x_2} - \dfrac{\cos x_1 \sin x_2}{\cos x_1 \cos x_2}}{\dfrac{\cos x_1 \cos x_2}{\cos x_1 \cos x_2} + \dfrac{\sin x_1 \sin x_2}{\cos x_1 \cos x_2}}, \quad \cos x_1 \cos x_2 \neq 0$$

From this and the definition of $\tan x$, you have the formula for the tangent of a difference:

$$\tan(x_1 - x_2) = \frac{\tan x_1 - \tan x_2}{1 + \tan x_1 \tan x_2}, \quad \cos(x_1 - x_2) \neq 0, \cos x_1 \neq 0, \cos x_2 \neq 0$$

Trigonometric Identities   **633**

Additional A Exercises

Use an identity to express each of the following as a function of a single angle and then evaluate that function. Express your answer in radical form.

1. $\sin 160° \cos 50° + \cos 160° \sin 50°$

$\sin 210° = -\dfrac{1}{2}$

2. $\sin 350° \cos 35° - \cos 350° \sin 35°$

$\sin 315° = -\dfrac{1}{\sqrt{2}}$

3. $\dfrac{\tan 85° + \tan 65°}{1 - \tan 85° \tan 65°}$

$\tan 150° = -\dfrac{1}{\sqrt{3}}$

4. $\dfrac{\tan 320° - \tan 20°}{1 + \tan 320° \tan 20°}$

$\tan 300° = -\sqrt{3}$

Use the formulas for the sine and tangent of a sum or difference, and the chart on page 590, to evaluate each function. Express your answer in radical form.

5. $\sin 195°$

$\sin 150° \cos 45° + \cos 150° \sin 45°$

$= \left(\dfrac{1}{2}\right)\left(\dfrac{1}{\sqrt{2}}\right) + \left(-\dfrac{\sqrt{3}}{2}\right)\left(\dfrac{1}{\sqrt{2}}\right)$

$= \dfrac{\sqrt{2}}{4} + \left(-\dfrac{\sqrt{6}}{4}\right) =$

$\dfrac{\sqrt{2} - \sqrt{6}}{4}$

6. $\tan 105°$

$\dfrac{\tan 60° + \tan 45°}{1 - \tan 60° \tan 45°}$

$= \dfrac{\sqrt{3} + 1}{1 - \sqrt{3}} = -2 - \sqrt{3}$

Using the fact that tangent is an odd function, you can replace x_2 with $-x_2$ in the formula for the tangent of a difference to deduce that

$$\tan [x_1 - (-x_2)] = \frac{\tan x_1 - \tan (-x_2)}{1 + \tan x_1 \tan (-x_2)},$$

from which you have the **formula for the tangent of a sum:**

$$\tan (x_1 + x_2) = \frac{\tan x_1 + \tan x_2}{1 - \tan x_1 \tan x_2}, \quad \cos (x_1 + x_2) \neq 0, \cos x_1 \neq 0, \cos x_2 \neq 0$$

You saw from the graph of tangent on page 604 that the fundamental period of tangent is π (or 180°) rather than 2π (or 360°). You can readily verify that π is a period by replacing x_1 with π and x_2 with x in the formula for $\tan (x_1 + x_2)$ to obtain:

$$\tan (\pi + x) = \frac{\tan \pi + \tan x}{1 - \tan \pi \tan x} = \frac{0 + \tan x}{1 - (0)(\tan x)} = \tan x, \quad \cos x \neq 0$$

This latter relationship can then be used to obtain a reduction formula for $\tan (\pi - x)$. If you replace x with $-x$, you have

$$\tan (\pi - x) = \tan (-x),$$

or

$$\tan (\pi - x) = -\tan x, \quad \cos x \neq 0.$$

Notice that the formula for $\tan (x_1 + x_2)$ cannot be used to produce reduction formulas for tangent for values $\dfrac{\pi}{2} \pm x$ or $\dfrac{3\pi}{2} \pm x$, because $\tan \dfrac{\pi}{2}$ and $\tan \dfrac{3\pi}{2}$ are not defined. Such formulas, however, for tangent can be found directly from the definition of $\tan x$.

EXAMPLE 2 Prove that $\tan \left(\dfrac{\pi}{2} + x\right) = -\cot x, \sin x \neq 0.$

SOLUTION $\tan \left(\dfrac{\pi}{2} + x\right) = \dfrac{\sin \left(\dfrac{\pi}{2} + x\right)}{\cos \left(\dfrac{\pi}{2} + x\right)} = \dfrac{\cos x}{-\sin x} = -\cot x, \quad \sin x \neq 0$

$\therefore \tan \left(\dfrac{\pi}{2} + x\right) = -\cot x, \quad \sin x \neq 0$

Following are reduction formulas for tangent. Three were derived above. The others are left as Exercises 35–37, page 636.

$\tan \left(\dfrac{\pi}{2} + x\right) = -\cot x, \sin x \neq 0 \qquad \tan \left(\dfrac{\pi}{2} - x\right) = \cot x, \sin x \neq 0$

$\tan (\pi + x) = \tan x, \cos x \neq 0 \qquad \tan (\pi - x) = -\tan x, \cos x \neq 0$

$\tan \left(\dfrac{3\pi}{2} + x\right) = -\cot x, \sin x \neq 0 \qquad \tan \left(\dfrac{3\pi}{2} - x\right) = \cot x, \sin x \neq 0$

634 *Chapter 15*

Oral Exercises

State the requested value.

1. If $\sin x_1 = 0.6923$, then $\sin(-x_1) = \underline{\ ?\ }$. -0.6923
2. If $\cos x_2 = -0.8015$, then $\cos(-x_2) = \underline{\ ?\ }$. -0.8015
3. If $\tan x_3 = 1.665$, then $\tan(-x_3) = \underline{\ ?\ }$. -1.665
4. If $\sin x_4 = -0.7882$, then $\sin(\pi + x_4) = \underline{\ ?\ }$. 0.7882
5. If $\cos x_5 = -0.1834$, then $\sin\left(\dfrac{\pi}{2} + x_5\right) = \underline{\ ?\ }$. -0.1834

1. $\sin 60° = \dfrac{\sqrt{3}}{2}$
2. $\sin 240° = -\dfrac{\sqrt{3}}{2}$
3. $\sin 330° = -\dfrac{1}{2}$
4. $\sin 225° = -\dfrac{1}{\sqrt{2}}$

Written Exercises

Use an identity to express each of the following as a function of a single angle and then evaluate that function. Express your answer in radical form.

A
1. $\sin 80° \cos 20° - \cos 80° \sin 20°$
2. $\sin 70° \cos 170° + \cos 70° \sin 170°$
3. $\sin 275° \cos 55° + \cos 275° \sin 55°$
4. $\sin 335° \cos 110° - \cos 335° \sin 110°$
5. $\dfrac{\tan 95° + \tan 25°}{1 - \tan 95° \tan 25°}$ $\tan 120° = -\sqrt{3}$
6. $\dfrac{\tan 155° - \tan 20°}{1 + \tan 155° \tan 20°}$ $\tan 135° = -1$
7. $2 \sin 112.5° \cos 112.5°$ $\sin 225° = -\dfrac{1}{\sqrt{2}}$
8. $2 \sin 165° \cos 165°$ $\sin 330° = -\dfrac{1}{2}$

(*Hint for Exercises 7 and 8:* $2 \sin x \cos x = \sin x \cos x + \cos x \sin x$.)

9. $\dfrac{2 \tan 105°}{1 - \tan^2 105°}$ $\tan 210° = \dfrac{1}{\sqrt{3}}$
10. $\dfrac{2 \tan 67.5°}{1 - \tan^2 67.5°}$ $\tan 135° = -1$

Use the formulas for the sine and tangent of a sum or difference, and the table on page 590, to evaluate each function. Express your answer in radical form.

14. $\dfrac{-\sqrt{6} - \sqrt{2}}{4}$
18. $\dfrac{\sqrt{2} - \sqrt{6}}{4}$
22. $-2 + \sqrt{3}$

11. $\sin 15°$ $\dfrac{\sqrt{6} - \sqrt{2}}{4}$
12. $\sin 105°$ $\dfrac{\sqrt{6} + \sqrt{2}}{4}$
13. $\sin \dfrac{11\pi}{12}$ $\dfrac{\sqrt{6} - \sqrt{2}}{4}$
14. $\sin \dfrac{17\pi}{12}$

15. $\sin(-75°)$ $\dfrac{-\sqrt{6} - \sqrt{2}}{4}$
16. $\sin 165°$ $\dfrac{\sqrt{6} - \sqrt{2}}{4}$
17. $\sin\left(-\dfrac{7\pi}{12}\right)$ $\dfrac{-\sqrt{6} - \sqrt{2}}{4}$
18. $\sin \dfrac{13\pi}{12}$

19. $\tan 75°$ $2 + \sqrt{3}$
20. $\tan 165°$ $-2 + \sqrt{3}$
21. $\tan 195°$ $2 - \sqrt{3}$
22. $\tan(-15°)$

23. $\tan \dfrac{7\pi}{12}$ $-2 - \sqrt{3}$
24. $\tan\left(-\dfrac{5\pi}{12}\right)$ $-2 - \sqrt{3}$
25. $\tan\left(-\dfrac{11\pi}{12}\right)$ $2 - \sqrt{3}$
26. $\tan\left(\dfrac{17\pi}{12}\right)$ $2 + \sqrt{3}$

Use the formulas for the sine and cosine of the sum or difference to prove each identity.

B
27. $\sin\left(\dfrac{\pi}{2} + x\right) = \cos x$
28. $\sin(\pi + x) = -\sin x$
29. $\sin(\pi - x) = \sin x$
30. $\sin\left(\dfrac{3\pi}{2} + x\right) = -\cos x$
31. $\sin\left(\dfrac{3\pi}{2} - x\right) = -\cos x$
32. $\sin\left(x - \dfrac{\pi}{2}\right) = -\cos x$

Trigonometric Identities **635**

Additional Answers
Written Exercises

27. $\sin\left(\dfrac{\pi}{2} + x\right) =$

$\sin \dfrac{\pi}{2} \cos x + \cos \dfrac{\pi}{2} \sin x =$

$(1) \cos x + (0) \sin x = \cos x$

28. $\sin(\pi + x) =$
$\sin \pi \cos x + \cos \pi \sin x =$
$(0) \cos x + (-1) \sin x =$
$-\sin x$

29. $\sin(\pi - x) =$
$\sin \pi \cos x - \cos \pi \sin x =$
$(0) \cos x - (-1) \sin x =$
$\sin x$

30. $\sin\left(\dfrac{3\pi}{2} + x\right) =$

$\sin \dfrac{3\pi}{2} \cos x + \cos \dfrac{3\pi}{2} \sin x$

$= (-1) \cos x + (0) \sin x$
$= -\cos x$

31. $\sin\left(\dfrac{3\pi}{2} - x\right) =$

$\sin \dfrac{3\pi}{2} \cos x - \cos \dfrac{3\pi}{2} \sin x =$

$(-1) \cos x - (0) \sin x =$
$-\cos x$

32. $\sin\left(x - \dfrac{\pi}{2}\right) =$

$\sin x \cos \dfrac{\pi}{2} - \cos x \sin \dfrac{\pi}{2} =$

$\sin x(0) - \cos x(1) =$
$-\cos x$

33. $\sin(x - \pi) =$
$\sin x \cos \pi - \cos x \sin \pi =$
$\sin x(-1) - \cos x(0) =$
$-\sin x$

(continued)

34. $\sin\left(x - \dfrac{3\pi}{2}\right) =$

$\sin x \cos\dfrac{3\pi}{2} - \cos x \sin\dfrac{3\pi}{2} =$

$\sin x(0) - \cos x(-1) =$
$\cos x$

35. $\tan\left(\dfrac{\pi}{2} - x\right) =$

$\dfrac{\sin\left(\dfrac{\pi}{2} - x\right)}{\cos\left(\dfrac{\pi}{2} - x\right)} =$

$\dfrac{\sin\dfrac{\pi}{2}\cos x - \cos\dfrac{\pi}{2}\sin x}{\cos\dfrac{\pi}{2}\cos x + \sin\dfrac{\pi}{2}\sin x} =$

$\dfrac{(1)\cos x - (0)\sin x}{(0)\cos x + (1)\sin x} =$

$\dfrac{\cos x}{\sin x} = \cot x$

36. $\tan\left(\dfrac{3\pi}{2} + x\right) =$

$\dfrac{\sin\left(\dfrac{3\pi}{2} + x\right)}{\cos\left(\dfrac{3\pi}{2} + x\right)} =$

$\dfrac{\sin\dfrac{3\pi}{2}\cos x + \cos\dfrac{3\pi}{2}\sin x}{\cos\dfrac{3\pi}{2}\cos x - \sin\dfrac{3\pi}{2}\sin x}$

$= \dfrac{-\cos x}{\sin x} = -\cot x$

37. $\tan\left(\dfrac{3\pi}{2} - x\right) =$

$\dfrac{\sin\left(\dfrac{3\pi}{2} - x\right)}{\cos\left(\dfrac{3\pi}{2} - x\right)} =$

$\dfrac{\sin\dfrac{3\pi}{2}\cos x - \cos\dfrac{3\pi}{2}\sin x}{\cos\dfrac{3\pi}{2}\cos x + \sin\dfrac{3\pi}{2}\sin x}$

$= \dfrac{-\cos x}{-\sin x} = \cot x$

Use the formulas for the sine and cosine of the sum or difference to prove each identity. (*Hint*: In Exercises 35–38 rewrite tan *x* in terms of sin *x* and cos *x*.)

33. $\sin(x - \pi) = -\sin x$

34. $\sin\left(x - \dfrac{3\pi}{2}\right) = \cos x$

35. $\tan\left(\dfrac{\pi}{2} - x\right) = \cot x$

36. $\tan\left(\dfrac{3\pi}{2} + x\right) = -\cot x$

37. $\tan\left(\dfrac{3\pi}{2} - x\right) = \cot x$

38. $\tan\left(x - \dfrac{\pi}{2}\right) = -\cot x$

Use the definition of the function and the formulas for a sum or difference to find the given function value.

EXAMPLE $\quad \sec 15° = \dfrac{1}{\cos 15°}$

$= \dfrac{1}{\cos(45° - 30°)}$

$= \dfrac{1}{\cos 45° \cos 30° + \sin 45° \sin 30°}$

$= \dfrac{1}{\dfrac{1}{\sqrt{2}}\cdot\dfrac{\sqrt{3}}{2} + \dfrac{1}{\sqrt{2}}\cdot\dfrac{1}{2}}$

$= \dfrac{2\sqrt{2}}{\sqrt{3} + 1}\cdot\dfrac{\sqrt{3} - 1}{\sqrt{3} - 1}$

$= \sqrt{6} - \sqrt{2}$

39. $\sec 105°$ $-\sqrt{2} - \sqrt{6}$ **40.** $\csc 195°$ $-\sqrt{2} - \sqrt{6}$ **41.** $\cot 75°$ $2 - \sqrt{3}$ **42.** $\csc 285°$ $\sqrt{2} - \sqrt{6}$

43. $\cot\dfrac{11\pi}{12}$ $-2 - \sqrt{3}$ **44.** $\sec\dfrac{17\pi}{12}$ $-\sqrt{2} - \sqrt{6}$ **45.** $\csc\dfrac{7\pi}{12}$ $\sqrt{6} - \sqrt{2}$ **46.** $\cot\dfrac{19\pi}{12}$ $-2 + \sqrt{3}$

Find sin (*x* + *y*), where *x* and *y* satisfy the given conditions.

47. $\sin x = \dfrac{4}{5}; \quad 0 < x < \dfrac{\pi}{2}; \quad \sin y = \dfrac{12}{13}; \quad 0 < y < \dfrac{\pi}{2}$ $\dfrac{56}{65}$

48. $\sin x = \dfrac{5}{13}; \quad \dfrac{\pi}{2} < x < \pi; \quad \cos y = -\dfrac{3}{5}; \quad \pi < y < \dfrac{3\pi}{2}$ $\dfrac{33}{65}$

49. $\cos x = -\dfrac{15}{17}; \quad \dfrac{\pi}{2} < x < \pi; \quad \sin y = -\dfrac{4}{5}; \quad \dfrac{3\pi}{2} < y < 2\pi$ $\dfrac{84}{85}$

50. $\sin x = -\dfrac{1}{2}; \quad \pi < x < \dfrac{3\pi}{2}; \quad \cos y = -\dfrac{2\sqrt{2}}{3}; \quad \dfrac{\pi}{2} < y < \pi$ $\dfrac{2\sqrt{2} - \sqrt{3}}{6}$

Find sin (*x* − *y*), where *x* and *y* satisfy the given conditions.

51. $\sin x = \dfrac{24}{25}; 0 < x < \dfrac{\pi}{2}; \cos y = \dfrac{4}{5}; 0 < y < \dfrac{\pi}{2}$ $\dfrac{3}{5}$

52. $\sin x = \dfrac{2}{3}; \dfrac{\pi}{2} < x < \pi; \sin y = \dfrac{\sqrt{7}}{4}; 0 < y < \dfrac{\pi}{2}$ $\dfrac{6 + \sqrt{35}}{12}$

636 *Chapter 15*

53. $\cos x = -\dfrac{1}{4}$; $\pi < x < \dfrac{3\pi}{2}$; $\cos y = -\dfrac{3}{5}$; $\dfrac{\pi}{2} < y < \pi$ $\quad \dfrac{3\sqrt{15}+4}{20}$

54. $\cos x = \dfrac{\sqrt{5}}{3}$; $\dfrac{3\pi}{2} < x < 2\pi$; $\sin y = -\dfrac{1}{3}$; $\pi < y < \dfrac{3\pi}{2}$ $\quad \dfrac{4\sqrt{2}+\sqrt{5}}{9}$

Prove each of the following identities. Note any restrictions.

C **55.** $\cot (x + y) = \dfrac{\cot x \cot y - 1}{\cot x + \cot y}$

56. $\sec (x + y) = \dfrac{\sec x \sec y}{1 - \tan x \tan y}$

57. $\csc (x + y) = \dfrac{\csc x \csc y}{\cot x + \cot y}$

58. $\dfrac{\sin (x - y)}{\sin y} + \dfrac{\cos (x - y)}{\cos y} = \sin x \sec y \csc y$

59. $\dfrac{\cos (x + y)}{\sin y} + \dfrac{\sin (x + y)}{\cos y} = \cos x \sec y \csc y$

15–5 Double-Angle and Half-Angle Formulas

All the identities derived in the preceding sections of this chapter result from the basic sum and difference identities:

$$\sin (x_1 + x_2) = \sin x_1 \cos x_2 + \cos x_1 \sin x_2$$
$$\sin (x_1 - x_2) = \sin x_1 \cos x_2 - \cos x_1 \sin x_2$$
$$\cos (x_1 + x_2) = \cos x_1 \cos x_2 - \sin x_1 \sin x_2$$
$$\cos (x_1 - x_2) = \cos x_1 \cos x_2 + \sin x_1 \sin x_2$$

We will now derive still more identities.

If $x_1 = x_2 = x$, then

$$\sin (x + x) = \sin x \cos x + \sin x \cos x,$$

or

$$\sin 2x = 2 \sin x \cos x.$$

Similarly,

$$\cos (x + x) = \cos x \cos x - \sin x \sin x,$$

or

$$\cos 2x = \cos^2 x - \sin^2 x.$$

Trigonometric Identities **637**

38. $\tan\left(x - \dfrac{\pi}{2}\right) = \dfrac{\sin\left(x - \dfrac{\pi}{2}\right)}{\cos\left(x - \dfrac{\pi}{2}\right)} =$

$$\dfrac{\sin x \cos \dfrac{\pi}{2} - \cos x \sin \dfrac{\pi}{2}}{\cos x \cos \dfrac{\pi}{2} + \sin x \sin \dfrac{\pi}{2}}$$

$$= \dfrac{-\cos x}{\sin x} = -\cot x$$

(continued on p. 658)

Teaching Suggestions
p. T122

Related Activities p. T122

Key Ideas
Develop and use double- and half-angle formulas for the sine, cosine, and tangent functions.

1. Find $\tan \dfrac{\pi}{8}$.

Using $\tan \dfrac{x}{2} = \dfrac{\sin x}{1 + \cos x}$

with $x = \dfrac{\pi}{4}$, you have:

$$\tan \dfrac{\left(\dfrac{\pi}{4}\right)}{2} = \dfrac{\sin \dfrac{\pi}{4}}{1 + \cos \dfrac{\pi}{4}} =$$

$$\dfrac{\dfrac{1}{\sqrt{2}}}{1 + \dfrac{1}{\sqrt{2}}} = \dfrac{1}{\sqrt{2} + 1}$$

$$\therefore \tan \dfrac{\pi}{8} = \dfrac{1(\sqrt{2} - 1)}{(\sqrt{2} + 1)(\sqrt{2} - 1)} =$$

$$\sqrt{2} - 1$$

2. Find $\sin \dfrac{5\pi}{12}$.

Using $\sin \dfrac{x}{2} = \pm\sqrt{\dfrac{1 - \cos x}{2}}$

with $x = \dfrac{5\pi}{6}$, you have:

$$\sin \dfrac{\dfrac{5\pi}{6}}{2} =$$

$$\pm\sqrt{\dfrac{1 - \left(-\dfrac{\sqrt{3}}{2}\right)}{2}} =$$

$$\pm\sqrt{\dfrac{2 + \sqrt{3}}{4}}$$

$$\therefore \sin \dfrac{5\pi}{12} = \dfrac{\sqrt{2 + \sqrt{3}}}{2}$$

You can easily obtain two additional expressions for $\cos 2x$. Since

$$\sin^2 x + \cos^2 x = 1,$$

you have

$$\sin^2 x = 1 - \cos^2 x$$

and

$$\cos^2 x = 1 - \sin^2 x.$$

Therefore, on replacing $\sin^2 x$ with $(1 - \cos^2 x)$ in the identity for $\cos 2x$, you have

$$\cos 2x = \cos^2 x - (1 - \cos^2 x),$$

or

$$\cos 2x = 2 \cos^2 x - 1.$$

On the other hand, if you replace $\cos^2 x$ with $(1 - \sin^2 x)$, you have

$$\cos 2x = (1 - \sin^2 x) - \sin^2 x,$$

or

$$\cos 2x = 1 - 2 \sin^2 x.$$

Also, for all x for which $\tan x$ is defined and $|\tan x| \neq 1$,

$$\tan (x + x) = \dfrac{\tan x + \tan x}{1 - \tan x \tan x},$$

or

$$\tan 2x = \dfrac{2 \tan x}{1 - \tan^2 x}.$$

Corresponding identities can, of course, be written in terms of angles, and so these identities are often referred to as the **double-angle formulas** (or identities) for sine, cosine, and tangent.

On replacing x with $\dfrac{x}{2}$ in $\cos 2x = 2 \cos^2 x - 1$, you have

$$\cos x = 2 \cos^2 \dfrac{x}{2} - 1,$$

or

$$\cos^2 \dfrac{x}{2} = \dfrac{1 + \cos x}{2}.$$

Therefore

$$\cos \dfrac{x}{2} = \pm\sqrt{\dfrac{1 + \cos x}{2}}.$$

638   *Chapter 15*

If you replace x with $\frac{x}{2}$ in $\cos 2x = 1 - 2 \sin^2 x$, you can use a similar process to find that

$$\sin \frac{x}{2} = \pm \sqrt{\frac{1 - \cos x}{2}}.$$

In applying the formulas for $\sin \frac{x}{2}$ $\left(\text{or } \sin \frac{\alpha}{2}\right)$ and $\cos \frac{x}{2}$ $\left(\text{or } \cos \frac{\alpha}{2}\right)$, the choice of the positive or negative root depends on the quadrant that contains one endpoint of the unit-circle arc corresponding to $\frac{x}{2}$ $\left(\text{or the}\right.$ terminal side of $\left.\frac{\alpha}{2}\right)$. For example, if

$$0 < \frac{x}{2} < \frac{\pi}{2} \quad \left(\text{or } 0° < m\left(\frac{\alpha}{2}\right) < 90°\right),$$

you would select the positive root for each, but if $\frac{\pi}{2} < \frac{x}{2} < \pi$ $\left(\text{or}\right.$ $90° < m\left(\frac{\alpha}{2}\right) < 180°\Big)$, you would select the positive root for sine and the negative root for cosine, and so on.

For tangent, if x is a value for which $\tan \frac{x}{2}$ is defined, you have

$$\tan^2 \frac{x}{2} = \frac{\sin^2 \frac{x}{2}}{\cos^2 \frac{x}{2}} = \frac{\frac{1 - \cos x}{2}}{\frac{1 + \cos x}{2}} = \frac{1 - \cos x}{1 + \cos x},$$

from which

$$\tan \frac{x}{2} = \pm \sqrt{\frac{1 - \cos x}{1 + \cos x}}.$$

Again, the choice of root depends on the quadrant determined by $\frac{x}{2}$ $\left(\text{or } \frac{\alpha}{2}\right)$.

A simpler formula for $\tan \frac{x}{2}$ $\left(\text{or } \tan \frac{\alpha}{2}\right)$ can be found by using the double-angle formulas as follows:

$$2 \sin \frac{x}{2} \cos \frac{x}{2} = \sin x \qquad 2 \cos^2 \frac{x}{2} = 1 + \cos x$$

Then

$$\tan \frac{x}{2} = \frac{\sin \frac{x}{2}}{\cos \frac{x}{2}} = \frac{\sin \frac{x}{2} \cdot 2 \cos \frac{x}{2}}{\cos \frac{x}{2} \cdot 2 \cos \frac{x}{2}} = \frac{2 \sin \frac{x}{2} \cos \frac{x}{2}}{2 \cos^2 \frac{x}{2}},$$

and so

$$\tan \frac{x}{2} = \frac{\sin x}{1 + \cos x}.$$

Corresponding identities can be written in terms of angles, and so these identities are often called the **half-angle formulas** (or identities).

Trigonometric Identities **639**

Common Errors

Students sometimes have trouble using the half-angle formulas because they have to choose the positive or negative root. Remind students that the sign of the root depends upon the quadrant in which the *half-angle* lies, not the quadrant in which the whole angle (used in the formula) lies. Also remind them that the tangent function is positive if the angle lies in the first or third quadrant, and negative if the angle lies in the second or fourth quadrant.

Suggested Assignments

Extended Alg. with Trig.
  640/1–37 odd
S 623/10–20 even
Enriched Alg. with Trig.
  640/5, 11, 13–33 odd, 35–38
S 623/10–20 even

Additional A Exercises

Find $\sin 2\alpha$ for the angle α in the given quadrant satisfying the given condition.

1. II; $\sin \alpha = \dfrac{\sqrt{3}}{2}$ $-\dfrac{\sqrt{3}}{2}$

2. IV; $\sin \alpha = -\dfrac{8}{17}$ $-\dfrac{240}{289}$

Find $\cos 2\alpha$ for the angle α in the given quadrant satisfying the given condition.

3. III; $\cos \alpha = -\dfrac{3}{5}$ $\dfrac{-7}{25}$

4. II; $\sin \alpha = \dfrac{1}{17}$ $\dfrac{287}{289}$

Use the half-angle formulas to find each function value.

5. $\sin 67.5°$ $\dfrac{\sqrt{2 + \sqrt{2}}}{2}$

6. $\cos 157.5°$ $-\dfrac{\sqrt{2 + \sqrt{2}}}{2}$

7. $\tan \dfrac{7\pi}{12}$ $-\sqrt{7 + 4\sqrt{3}}$

Mixed Review

Solve.

1. $2x + 4y = 2$
   $x - 2y = 5$ $\{(3, -1)\}$

2. $3x + 2y = -3$
   $x + 3y = 6$ $\{(-3, 3)\}$

3. $4x - 6y + 2z = 6$
   $2y - 5z = 8$
   $x + 2y + 3z = -7$
   $\{(1, -1, -2)\}$

4. A rectangle with a perimeter of 56 m is three times as long as it is wide. Find the dimensions of the rectangle.   7 m × 21 m

640

EXAMPLE   Find $\tan \dfrac{\pi}{12}$.

SOLUTION   Using $\tan \dfrac{x}{2} = \dfrac{\sin x}{1 + \cos x}$ with $x = \dfrac{\pi}{6}$, you have:

$$\tan \dfrac{\left(\dfrac{\pi}{6}\right)}{2} = \dfrac{\sin \dfrac{\pi}{6}}{1 + \cos \dfrac{\pi}{6}} = \dfrac{\dfrac{1}{2}}{1 + \dfrac{\sqrt{3}}{2}} = \dfrac{1}{2 + \sqrt{3}}$$

Rationalizing the denominator yields:

$$\tan \dfrac{\pi}{12} = \dfrac{1(2 - \sqrt{3})}{(2 + \sqrt{3})(2 - \sqrt{3})} = 2 - \sqrt{3}$$

Oral Exercises

If x satisfies the given condition, state (a) the quadrant containing an endpoint of the unit-circle arc corresponding to $\dfrac{x}{2}$, (b) the sign of $\sin \dfrac{x}{2}$, (c) the sign of $\cos \dfrac{x}{2}$, and (d) the sign of $\tan \dfrac{x}{2}$.

1. $\dfrac{\pi}{2} < x < \pi$ **2.** $\pi < x < \dfrac{3\pi}{2}$ **3.** $3\pi < x < \dfrac{7\pi}{2}$ **4.** $2\pi < x < \dfrac{5\pi}{2}$

a. I   b. +   a. II   b. +   a. IV   b. −   a. III   b. −
c. +   d. +   c. −   d. −   c. +   d. −   c. −   d. +

Written Exercises

Find $\sin 2\alpha$ for the angle α in the given quadrant satisfying the given condition.

A   **1.** I; $\sin \alpha = \dfrac{4}{5}$ $\dfrac{24}{25}$ **2.** III; $\sin \alpha = -\dfrac{3}{5}$ $\dfrac{24}{25}$ **3.** II; $\cos \alpha = -\dfrac{5}{13}$ $-\dfrac{120}{169}$

4. IV; $\cos \alpha = \dfrac{2}{3}$ $\dfrac{-4\sqrt{5}}{9}$ **5.** IV; $\sin \alpha = -\dfrac{2}{\sqrt{13}}$ $-\dfrac{12}{13}$ **6.** I; $\tan \alpha = \dfrac{2\sqrt{10}}{3}$ $\dfrac{12\sqrt{10}}{49}$

Find $\cos 2\alpha$ for the angle α in the given quadrant satisfying the given condition.

7. I; $\cos \alpha = \dfrac{5}{13}$ $-\dfrac{119}{169}$ **8.** II; $\sin \alpha = \dfrac{2}{\sqrt{5}}$ $-\dfrac{3}{5}$ **9.** III; $\sin \alpha = -\dfrac{\sqrt{7}}{4}$ $\dfrac{1}{8}$

10. II; $\cos \alpha = -\dfrac{3}{\sqrt{10}}$ $\dfrac{4}{5}$ **11.** IV; $\cos \alpha = \dfrac{\sqrt{5}}{3}$ $\dfrac{1}{9}$ **12.** III; $\tan \alpha = \dfrac{2\sqrt{6}}{5}$ $\dfrac{1}{49}$

13. If $\sin x = \dfrac{2}{3}$ and $\dfrac{\pi}{2} < x < \pi$, find $\tan 2x$. $-4\sqrt{5}$

14. If $\tan x = \dfrac{1}{5}$, find **(a)** $\tan 2x$ and **(b)** $\tan 4x$. a. $\dfrac{5}{12}$ b. $\dfrac{120}{119}$

640   *Chapter 15*

Use the half-angle formulas to find each function value.

15. $\cos 75°$ $\frac{\sqrt{2-\sqrt{3}}}{2}$

16. $\sin 22.5°$ $\frac{\sqrt{2-\sqrt{2}}}{2}$

17. $\sin 105°$

18. $\tan 165°$ $-2+\sqrt{3}$

19. $\tan 112.5°$ $-\sqrt{2}-1$

20. $\cos 15°$

21. $\sin \frac{3\pi}{8}$ $\frac{\sqrt{2+\sqrt{2}}}{2}$

22. $\cos \frac{\pi}{8}$ $\frac{\sqrt{2+\sqrt{2}}}{2}$

23. $\tan \frac{\pi}{8}$

24. $\sin \frac{5\pi}{8}$ $\frac{\sqrt{2+\sqrt{2}}}{2}$

25. $\cos \frac{7\pi}{12}$ $-\frac{\sqrt{2-\sqrt{3}}}{2}$

26. $\tan \frac{11\pi}{12}$

Find $\sin \frac{\alpha}{2}$ and $\cos \frac{\alpha}{2}$ for the given angle α. Make a sketch of the angles

α and $\frac{\alpha}{2}$ to decide the sign of each answer.

27. $\cos \alpha = \frac{7}{9}, 0° < m(\alpha) < 90°$ $\frac{1}{3}; \frac{2\sqrt{2}}{3}$

28. $\cos \alpha = -\frac{7}{9}, 90° < m(\alpha) < 180°$ $\frac{2\sqrt{2}}{3}; \frac{1}{3}$

29. $\sin \alpha = \frac{24}{25}, 0° < m(\alpha) < 90°$

30. $\sin \alpha = \frac{24}{25}, 90° < m(\alpha) < 180°$

31. $\cos \alpha = -\frac{7}{8}, 180° < m(\alpha) < 270°$

32. $\sin \alpha = -\frac{7}{8}, 270° < m(\alpha) < 360°$

29. $\frac{3}{5}; \frac{4}{5}$ 30. $\frac{4}{5}; \frac{3}{5}$ 31. $\frac{\sqrt{15}}{4}; -\frac{1}{4}$ 32. $\frac{\sqrt{8-\sqrt{15}}}{4}; -\frac{\sqrt{8+\sqrt{15}}}{4}$

Use the half-angle and double-angle formulas to verify each of the following.

B 33. $\sin \left[2\left(\frac{x}{2}\right)\right] = \sin x$

34. $\cos \left[2\left(\frac{x}{2}\right)\right] = \cos x$

35. $\sin^2 \frac{x}{2} + \cos^2 \frac{x}{2} = 1$

36. $\sin^2 2x + \cos^2 2x = 1$

37. $1 + \tan^2 \frac{x}{2} = \sec^2 \frac{x}{2}$ $\left(Note: \sec \frac{x}{2} = \frac{1}{\cos \frac{x}{2}}\right)$

C 38. Show that if $\sin \alpha = \cos \beta$, where both α and β are in Quadrant I, then $\sin 2\alpha = \sin 2\beta$. Make a sketch of the angles α, β, 2α, and 2β showing how this is possible.

Computer Exercises   For students with computer experience

1. Write a program that will find $\sin \frac{\alpha}{2}$, given the measure of α as a positive integral number of degrees, possibly greater than 360°. The program should use the formula on page 639. Note that the program must determine from the size of $\frac{\alpha}{2}$ whether $\sin \frac{\alpha}{2}$ is positive or negative.

Run the program in Exercise 1 for each measure of α.

2. 300 0.5   3. 420 −0.5   4. 660 −0.5   5. 600 −0.8660

Trigonometric Identities   **641**

Supplementary Material

Test 48

Key Ideas

Summarize all the trigonometric identities and use
them to prove further identities.

Chalkboard Examples

Prove the identity. Note any
restrictions.

1. $\tan x - \sin 2x =$
$-\tan x \cos 2x$
Work on the left side.
$\tan x - \sin 2x$

$= \dfrac{\sin x}{\cos x} - 2 \sin x \cos x$

$= \dfrac{\sin x}{\cos x} - \dfrac{2 \sin x \cos^2 x}{\cos x}$

$= \dfrac{\sin x - 2 \sin x \cos^2 x}{\cos x}$

$= \dfrac{\sin x(1 - 2 \cos^2 x)}{\cos x}$

$= \dfrac{-\sin x(2 \cos^2 x - 1)}{\cos x}$

$= -\tan x \cos 2x, \cos x \neq 0$

Additional A Exercises

Prove the identity. Note any
restrictions.

1. $\cos^2 \alpha + \cos 2\alpha$
$= \cot \alpha \sin 2\alpha - \sin^2 \alpha$
Work on right side.

$= \dfrac{\cos \alpha}{\sin \alpha} \cdot 2 \sin \alpha \cos \alpha -$

$\sin^2 \alpha$
$= 2 \cos^2 \alpha - \sin^2 \alpha$
$= \cos^2 \alpha + (\cos^2 \alpha - \sin^2 \alpha)$
$= \cos^2 \alpha + \cos 2\alpha$

15–6 More on Identities

The following is a list of useful identities to which you can refer as
needed. These identities hold for all values of the variables for which the
function values are defined. Corresponding identities hold in terms of
angles under similar restrictions.

Summary of Useful Identities

Quotient Identities

$$\tan x = \dfrac{\sin x}{\cos x}$$

$$\cot x = \dfrac{\cos x}{\sin x}$$

Reciprocal Identities

$$\sin x \csc x = 1$$
$$\cos x \sec x = 1$$
$$\tan x \cot x = 1$$

Pythagorean Identities

$$\sin^2 x + \cos^2 x = 1$$
$$\tan^2 x + 1 = \sec^2 x$$
$$1 + \cot^2 x = \csc^2 x$$

Opposites

$$\sin (-x) = -\sin x$$
$$\cos (-x) = \cos x$$
$$\tan (-x) = -\tan x$$

Sum and Difference Identities

$$\sin (x_1 + x_2) = \sin x_1 \cos x_2 + \cos x_1 \sin x_2$$
$$\sin (x_1 - x_2) = \sin x_1 \cos x_2 - \cos x_1 \sin x_2$$
$$\cos (x_1 + x_2) = \cos x_1 \cos x_2 - \sin x_1 \sin x_2$$
$$\cos (x_1 - x_2) = \cos x_1 \cos x_2 + \sin x_1 \sin x_2$$
$$\tan (x_1 + x_2) = \dfrac{\tan x_1 + \tan x_2}{1 - \tan x_1 \tan x_2}$$
$$\tan (x_1 - x_2) = \dfrac{\tan x_1 - \tan x_2}{1 + \tan x_1 \tan x_2}$$

Double-Angle Identities

$$\sin 2x = 2 \sin x \cos x$$
$$\cos 2x = \cos^2 x - \sin^2 x$$
$$= 1 - 2 \sin^2 x$$
$$= 2 \cos^2 x - 1$$
$$\tan 2x = \dfrac{2 \tan x}{1 - \tan^2 x}$$

Half-Angle Identities

$$\sin \dfrac{x}{2} = \pm \sqrt{\dfrac{1 - \cos x}{2}}$$
$$\cos \dfrac{x}{2} = \pm \sqrt{\dfrac{1 + \cos x}{2}}$$
$$\tan \dfrac{x}{2} = \pm \sqrt{\dfrac{1 - \cos x}{1 + \cos x}}$$
$$= \dfrac{\sin x}{1 + \cos x}$$

---

642   Chapter 15

Written Exercises

Prove each identity. Note any restrictions.

A **1.** $\dfrac{\cos 2\alpha}{\sin^2 \alpha} = \cot^2 \alpha - 1$

2. $\tan \alpha (\cos 2\alpha + 1) = \sin 2\alpha$

3. $(\sin \alpha - \cos \alpha)^2 = 1 - \sin 2\alpha$

4. $\cos 2\alpha + 2 \sin^2 \alpha = 1$

5. $1 - \sin 2\alpha \tan \alpha = \cos 2\alpha$

6. $\cot \alpha \cos \alpha - \sin \alpha = \cos 2\alpha \csc \alpha$

7. $\dfrac{\cos 2\alpha}{\cos \alpha + \sin \alpha} = \cos \alpha - \sin \alpha$

8. $\dfrac{\sin 2\alpha}{1 - \cos 2\alpha} = \cot \alpha$

9. $(1 - \tan x)^2 = \dfrac{1 - \sin 2x}{\cos^2 x}$

10. $\dfrac{1 - \cos 2\alpha}{1 + \cos 2\alpha} = \tan^2 \alpha$

11. $\dfrac{1 - \tan^2 \alpha}{1 + \tan^2 \alpha} = \cos 2\alpha$

12. $\cot 2\alpha = \dfrac{1}{2}(\cot \alpha - \tan \alpha)$

B **13.** $\cos^4 \alpha - \sin^4 \alpha = \cos 2\alpha$

14. $1 - \tan^4 x = \dfrac{\cos 2x}{\cos^4 x}$

15. $\dfrac{1 - \cos x}{\sin x} = \tan \dfrac{x}{2}$ $\left(\textit{Hint:} \text{ Multiply the left side by } \dfrac{1 + \cos x}{1 + \cos x}.\right)$

16. $\tan \dfrac{\alpha}{2}(\cos 2\alpha - \cos^2 \alpha) = \dfrac{\sin 2\alpha}{2} - \sin \alpha$

17. $\sin 3\alpha = 3 \sin \alpha - 4 \sin^3 \alpha$

18. $\cos 3\alpha = 4 \cos^3 \alpha - 3 \cos \alpha$

19. $\cos 4\alpha = \cos^4 \alpha + \sin^4 \alpha - 6 \sin^2 \alpha \cos^2 \alpha$

20. $\sin 4\alpha = 4 \sin \alpha \cos \alpha - 8 \sin^3 \alpha \cos \alpha$

21. $\dfrac{\cos \alpha}{\cos \alpha - \sin \alpha} - \dfrac{\cos \alpha}{\cos \alpha + \sin \alpha} = \tan 2\alpha$

22. $\dfrac{\cos \alpha}{\cos \alpha + \sin \alpha} - \dfrac{\sin \alpha}{\cos \alpha - \sin \alpha} = 1 - \tan 2\alpha$

23. $\sin (\alpha + \beta) \sin (\alpha - \beta) = \sin^2 \alpha - \sin^2 \beta$

24. $\cos (\alpha + \beta) \cos (\alpha - \beta) = \cos^2 \alpha - \sin^2 \beta$

The following are sometimes called the *sum and product identities*.

25. $\sin (\alpha + \beta) + \sin (\alpha - \beta) = 2 \sin \alpha \cos \beta$

26. $\sin (\alpha + \beta) - \sin (\alpha - \beta) = 2 \cos \alpha \sin \beta$

27. $\cos (\alpha + \beta) + \cos (\alpha - \beta) = 2 \cos \alpha \cos \beta$

28. $\cos (\alpha + \beta) - \cos (\alpha - \beta) = -2 \sin \alpha \sin \beta$

C **29.** $\sin A + \sin B = 2 \sin \dfrac{A + B}{2} \cos \dfrac{A - B}{2}$

   (*Hint:* Let $\alpha = \dfrac{A + B}{2}$ and $\beta = \dfrac{A - B}{2}$ and use Exercise 25.)

30. $\sin A - \sin B = 2 \cos \dfrac{A + B}{2} \sin \dfrac{A - B}{2}$

31. $\cos A + \cos B = 2 \cos \dfrac{A + B}{2} \cos \dfrac{A - B}{2}$

32. $\cos A - \cos B = -2 \sin \dfrac{A + B}{2} \sin \dfrac{A - B}{2}$

Trigonometric Identities **643**

Suggested Assignments
Extended Alg. with Trig.
  643/1–23 odd, 25–29
R 644/*Self-Test 2*
Enriched Alg. with Trig.
  643/7–25 odd, 27–32
R 644/*Self-Test 2*

*Additional Answers
Written Exercises*

1. $\dfrac{\cos 2\alpha}{\sin^2 \alpha} = \dfrac{\cos^2 \alpha - \sin^2 \alpha}{\sin^2 \alpha}$

 $= \dfrac{\cos^2 \alpha}{\sin^2 \alpha} - \dfrac{\sin^2 \alpha}{\sin^2 \alpha} =$

 $\cot^2 \alpha - 1$; $\sin \alpha \neq 0$

2. $\tan \alpha (\cos 2\alpha + 1) =$

 $\dfrac{\sin \alpha}{\cos \alpha}(2 \cos^2 \alpha - 1 + 1)$

 $= \dfrac{\sin \alpha}{\cos \alpha}(2 \cos^2 \alpha)$

 $= 2 \sin \alpha \cos \alpha = \sin 2\alpha$;

 $\cos \alpha \neq 0$

3. $(\sin \alpha - \cos \alpha)^2 = \sin^2 \alpha - $

 $2 \sin \alpha \cos \alpha + \cos^2 \alpha = $

 $\sin^2 \alpha + \cos^2 \alpha - 2 \sin \alpha \cdot$

 $\cos \alpha = 1 - \sin 2\alpha$

4. $\cos 2\alpha + 2 \sin^2 \alpha = $

 $\cos^2 \alpha - \sin^2 \alpha + 2 \sin^2 \alpha$

 $= \cos^2 \alpha + \sin^2 \alpha = 1$

5. $1 - \sin 2\alpha \tan \alpha = 1 - $

 $(2 \sin \alpha \cos \alpha)\left(\dfrac{\sin \alpha}{\cos \alpha}\right) =$

 $1 - 2 \sin^2 \alpha = \cos 2\alpha$;

 $\cos \alpha \neq 0$

6. $\cot \alpha \cos \alpha - \sin \alpha = $

 $\dfrac{\cos \alpha}{\sin \alpha} \cdot \cos \alpha - \dfrac{\sin^2 \alpha}{\sin \alpha} =$

 $\dfrac{\cos^2 \alpha - \sin^2 \alpha}{\sin \alpha} = \dfrac{\cos 2\alpha}{\sin \alpha}$

 $= \cos 2\alpha \cdot \dfrac{1}{\sin \alpha} =$

 $\cos 2\alpha \csc \alpha$; $\sin \alpha \neq 0$

(continued on p. 659)

1. Express sin 50° cos 20° − cos 50° sin 20° as a function of a single angle and then evaluate the function.

$$\sin 30° = \frac{1}{2}$$

Evaluate.

2. cos 240° $-\dfrac{1}{2}$

3. sin $\dfrac{\pi}{12}$ $\dfrac{\sqrt{6} - \sqrt{2}}{4}$

4. Find sin 2α if sin α = $\dfrac{3}{5}$ and α is in Quadrant II.

$$\sin 2\alpha = -\frac{24}{25}$$

5. Use a half-angle formula to find

$$\sin \frac{11\pi}{12}. \quad \frac{\sqrt{2 - \sqrt{3}}}{2}$$

6. Prove that cot 2x = $\dfrac{1 - \tan^2 x}{2 \tan x}$, noting any restrictions.

$$\cot 2x = \frac{1}{\tan 2x}$$

$$= \frac{1}{\dfrac{2 \tan x}{1 - \tan^2 x}} = \frac{1 - \tan^2 x}{2 \tan x},$$

$$\sin x \neq 0, \cos x \neq 0$$

Teaching Suggestions
p. T123

Related Activities p. T123

Key Ideas

Develop and use the law of cosines.

Self-Test 2

VOCABULARY sum of α_1 and α_2 (p. 625)   reduction formula (p. 628)
difference of α_1 and α_2 (p. 625)   double-angle formula (p. 638)
opposite of α (p. 625)   half-angle formula (p. 639)

1. Express $\dfrac{\tan 165° - \tan 15°}{1 + \tan 165° \tan 15°}$ as a function of a single angle and   *Obj. 1, p. 625*
then evaluate the function. tan 150° = $-\dfrac{1}{\sqrt{3}}$

Use the formulas for the sine and cosine of a sum or difference, and the table on page 590, to evaluate each function.

2. cos $\dfrac{\pi}{12}$ **3.** sin 345°   2. $\dfrac{\sqrt{2} + \sqrt{6}}{4}$ 3. $\dfrac{-\sqrt{6} + \sqrt{2}}{4}$ 4. $\dfrac{1}{9}$

4. Find cos 2α if sin α = $\tfrac{2}{3}$ and α is in Quadrant II.   *Obj. 2, p. 625*
5. Use a half-angle formula to find sin 112.5°. $\dfrac{\sqrt{2 + \sqrt{2}}}{2}$

6. Prove that csc 2α = $\dfrac{\csc \alpha}{2 \cos \alpha}$ is an identity, noting any restrictions.   *Obj. 3, p. 625*

Check your answers with those at the back of the book.

Solving General Triangles

OBJECTIVES *for Sections 15-7 and 15-8:*
1. To apply the law of cosines and the law of sines to solve general triangles.
2. To apply the law of cosines and the law of sines to solve simple problems.

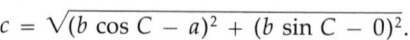

15–7 The Law of Cosines

The Pythagorean Theorem enables you to find the length of one side of a *right* triangle when you know the lengths of the other two sides. To find a more general relationship between the lengths of the three sides of *any* triangle, look at Figures 4, 5, and 6. In Figure 4, ∠BCA is in standard position, and so the coordinates of A are (b cos C, b sin C). By the distance formula then,

$$c = \sqrt{(b \cos C - a)^2 + (b \sin C - 0)^2}.$$

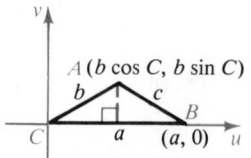

Figure 4

644 *Chapter 15*

Squaring and simplifying the right side, you have

$$c^2 = b^2 \cos^2 C - 2ab \cos C + a^2 + b^2 \sin^2 C$$
$$= a^2 + b^2 (\cos^2 C + \sin^2 C) - 2ab \cos C,$$

or, since $\cos^2 C + \sin^2 C = 1$,

$$c^2 = a^2 + b^2 - 2ab \cos C. \qquad (1)$$

By reorienting the axes so that angles A and B are, in turn, in standard position (see Figures 5 and 6), you can obtain the following analogous relationships:

$$b^2 = a^2 + c^2 - 2ac \cos B \qquad (2)$$
$$a^2 = b^2 + c^2 - 2bc \cos A \qquad (3)$$

These three relationships together are the law of cosines.

Notice that if one of the angles of the triangle, say C, is a right angle, then $\cos C = 0$ and you have $c^2 = a^2 + b^2$, which is the Pythagorean Theorem. Thus, the Pythagorean Theorem is a "special case" of the law of cosines.

EXAMPLE 1 Find, to three significant digits, the length of c in the triangle at the right.

SOLUTION Use the law of cosines in the form

$$c^2 = a^2 + b^2 - 2ab \cos C,$$

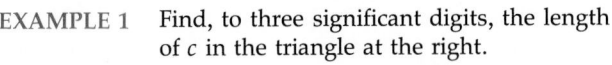

with $a = 18$, $b = 15$, and $m(C) = 112°$. You can find $\cos 112°$ directly with the aid of a calculator, or you can use Table 6:

$$\cos 112° = -\cos 68° \approx -0.3746 \quad \text{(from Table 6)}$$
$$c^2 \approx 18^2 + 15^2 - 2(18)(15)(-0.3746)$$
$$\approx 751.3$$
$$c \approx \sqrt{751.3} \approx 27.4$$

If you solve the relationship (3) for $\cos A$, you obtain

$$\cos A = \frac{b^2 + c^2 - a^2}{2bc}.$$

This relationship can be used to find $m(A)$ in $\triangle ABC$, given a, b, and c. Similarly, the relationships (1) and (2) yield formulas that can be used to find $m(C)$ and $m(B)$, respectively.

EXAMPLE 2 Find $m(A)$, to the nearest tenth of a degree, for $\triangle ABC$, in which $a = 5$, $b = 8$, and $c = 12$.

SOLUTION $\cos A = \dfrac{b^2 + c^2 - a^2}{2bc} = \dfrac{8^2 + 12^2 - 5^2}{2(8)(12)} \approx 0.9531$

From a calculator or Table 6, $m(A) \approx 17.6°$.

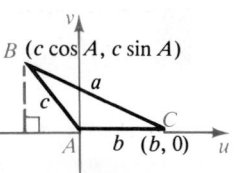

Figure 5

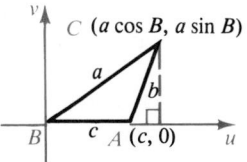

Figure 6

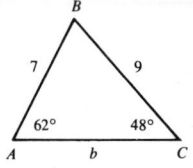

Suggested Assignments

Extended Alg. with Trig.
  First day
    646/1–17 odd
  Second day
    647/P:1–9
Enriched Alg. with Trig.
  First day
    646/5–17 odd, 14, 16
  Second day
    647/P:3–10

___

Additional A Exercises

For △*ABC*, find the indicated length to the nearest tenth.

1. $a = 7$, $b = 12$, $m(C) = 85°$, $c = $ _?_ 13.4

2. $a = 11$, $c = 13.5$, $m(B) = 68°$, $b = $ _?_ 13.9

For △*ABC*, find the measure of the desired angle to the nearest tenth of a degree.

3. $a = 6$, $b = 9.5$, $c = 11$, $m(B) = $ _?_ 59.6°

4. $a = 3$, $b = 5.2$, $c = 7$, $m(A) = $ _?_ 22.9°

5. A marathon race is run on a triangular course with a total length of 42 km. If the ratio of the three legs of the course is $2 : 3 : 2$, respectively, find the measure of the angle, to the nearest tenth of a degree, between the second and third legs of the course. 41.4°

6. After walking for 210 m, Bob makes an 80° turn and walks for another 174 m before stopping. How far is he, to the nearest meter, from his starting point? 248 m

Oral Exercises

State an equation you could use to find the value of *x*.

1.

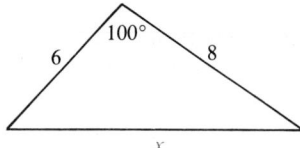

$x^2 = 6^2 + 8^2 - 2 \cdot 6 \cdot 8 \cos 100°$

2.

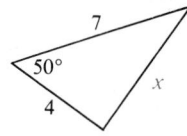

$x^2 = 4^2 + 7^2 - 2 \cdot 4 \cdot 7 \cos 50°$

3.

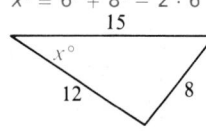

$\cos x = \dfrac{12^2 + 15^2 - 8^2}{2 \cdot 12 \cdot 15}$

4.

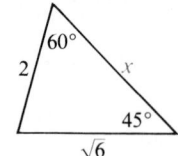

$x^2 = 2^2 + (\sqrt{6})^2 - 2 \cdot 2\sqrt{6} \cos 75°$

Written Exercises

For △*ABC* find the indicated length to the nearest tenth.

A **1.** $a = 5$, $b = 8$, $m(C) = 60°$, $c = $ _?_ 7.0

2. $a = 14$, $c = 16$, $m(B) = 120°$, $b = $ _?_ 26.0

3. $b = 6.5$, $c = 4\sqrt{3}$, $m(A) = 30°$, $a = $ _?_ 3.5

4. $a = 4$, $b = 7.5$, $m(C) = 60°$, $c = $ _?_ 6.5

5. $a = 15$, $c = 11$, $m(B) = 26°$, $b = $ _?_ 7.0

6. $b = 35$, $c = 9$, $m(A) = 75°$, $a = $ _?_ 33.8

For △*ABC* find the measure of the desired angle to the nearest tenth of a degree.

7. $a = 7$, $b = 15$, $c = 13$, $m(C) = $ _?_ 60.0°

8. $a = 7$, $b = 10$, $c = 14$, $m(B) = $ _?_ 42.3°

9. $a = 19$, $b = 22$, $c = 30$, $m(A) = $ _?_ 39.2°

10. $a = 8$, $b = 9$, $c = 13$, $m(C) = $ _?_ 99.6°

Solve each triangle. Give measurements to the nearest unit of length and the nearest degree. Answers may vary.

B **11.** $a = 4$, $b = 10$, $m(C) = 59°$

12. $a = 25$, $c = 5$, $m(B) = 38°$

13. $b = 8$, $c = 9$, $m(A) = 60°$

14. $a = 15$, $b = 8$, $m(C) = 63°$

15. $a = 4$, $b = 7$, $c = 8$

16. $a = 12$, $b = 7$, $c = 9$

17. Show that one angle of a triangle with sides of length 5, 7, and 8 has the same measure as one angle of a triangle whose sides have length 3, 5, and 7.

646 *Chapter 15*

Problems

In each problem, give lengths to three significant digits and angles to the nearest tenth of a degree.

A 1. How far apart are the outer ends of the hour and minute hands of a clock at two o'clock if the hour hand is 8 cm long and the minute hand is 15 cm long? 13.0 cm

2. Two ships leave from the same point at the same time on courses 33° apart at speeds of 12 km/h and 17 km/h, respectively. How far apart are the ships after 1 h? 9.53 km

3. Cities A, B, and C are located as in the diagram at the right. If two airplanes take off from city A flying toward cities B and C, respectively, what is the measure of the angle by which their headings differ? 48.2°

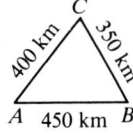

4. Two adjacent sides of a parallelogram form an acute angle and have lengths of 25 cm and 32 cm, respectively. If the shorter diagonal of the parallelogram has length 10 cm, what is the measure of the angle between the sides? 14.5°

5. A rear windshield wiper of length 31 cm (distance from the pivot to the outer end of the blade) sweeps out an angle of 120°. How far apart are the two outermost corners of the area swept out? 53.7 cm

6. A tunnel is to be dug from point A to point B. The distances from A and B to a third point C are 300 m and 280 m, respectively, and $m(\angle ACB) = 26°$. What is the distance between A and B? 132 m

B 7. Points A and B and points C and D are the vertices of a trapezoid in which the length of $\overline{AD}$ is 10, the length of $\overline{CD}$ is 42, and the length of $\overline{AB}$ is 10. If $m(\angle ADC) = 60°$ and $m(\angle DCA) = 14°$, use the law of cosines to find the length of $\overline{BC}$. 28.4

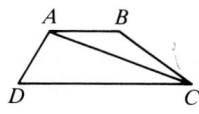

8. Each of the two diagonals from one vertex of a regular pentagon has length 50 cm. What is the length of a side of the pentagon? 30.9 cm

C 9. At sunset the lines of sight to the sun and the planet Venus from an observation point on Earth form an angle of measure 39°. If the sun is 1.5×10^8 km from Earth and Venus is 5.4×10^7 km from Earth at this moment, how far is Venus from the sun? 1.13×10^8 km

10. Show that, in the diagram at the right, if S is the midpoint of $\overline{PR}$, then $x^2 + y^2 = 2(u^2 + v^2)$.

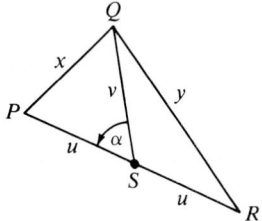

Trigonometric Identities **647**

Right $\triangle ABC$ has $m(C) = 90°$. Find the missing length for each, to the nearest tenth.

1. $a = 7$, $b = 5$, $c = $ _?_ 8.6
2. $a = $ _?_ , $b = 10$, $c = 12$ 6.6
3. $a = 4$, $b = $ _?_ , $c = 14$ 13.4
4. If the ratio of the angles in $\triangle QRS$ are $2 : 4 : 3$, respectively, what are their measures?
   $m(A) = 40°$, $m(B) = 80°$, $m(C) = 60°$

Additional Answers
Written Exercises

11. $c \approx 9$, $m(A) \approx 23°$, $m(B) \approx 98°$
12. $b \approx 21$, $m(A) \approx 134°$, $m(C) \approx 8°$
13. $a \approx 9$, $m(B) \approx 54°$, $m(C) \approx 66°$
14. $c \approx 13$, $m(A) \approx 85°$, $m(B) \approx 32°$
15. $m(A) \approx 30°$, $m(B) \approx 61°$, $m(C) \approx 89°$
16. $m(A) \approx 96°$, $m(B) \approx 35°$, $m(C) \approx 49°$

Additional Answers
Problems

10. $x^2 = u^2 + v^2 - 2uv \cos \alpha$; $y^2 = u^2 + v^2 - 2uv \cos (180° - \alpha)$; $x^2 + y^2 = 2(u^2 + v^2) - 2uv(\cos \alpha + \cos(180° - \alpha))$; $x^2 + y^2 = 2(u^2 + v^2) - 2uv(0)$; $x^2 + y^2 = 2(u^2 + v^2)$

Supplementary Material

Test 49

Key Ideas

Develop and use the area formula for triangles. Develop the law of sines and use it to solve general triangles.

15–8 The Law of Sines

As noted in Section 15-7, by choosing the coordinate system appropriately, you can position any $\triangle ABC$ so that any one of the angles A, B, or C is in standard position. Because the area of a triangle is equal to half the product of the length of a base and the corresponding altitude, you can express the area of $\triangle ABC$ in Figure 7 by

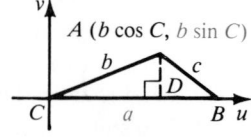

Figure 7

$$\text{Area} = \tfrac{1}{2}ab \sin C.$$

Similarly, with angles A and B in standard position,

$$\text{Area} = \tfrac{1}{2}bc \sin A \quad \text{and} \quad \text{Area} = \tfrac{1}{2}ca \sin B.$$

Thus, in general, you have the following.

Theorem. The area of a triangle is equal to half the product of the lengths of two sides and the sine of their included angle.

EXAMPLE 1 Find the area of a triangle in which $a = 7$, $b = 18$, and $m(C) = 26.7°$.

SOLUTION From Table 6 or a calculator, $\sin 26.7° \approx 0.4493$ so that

$$\text{Area} = \tfrac{1}{2}ab \sin C \approx \tfrac{1}{2}(7)(18)(0.4493) \approx 28.3.$$

Since each of the three expressions for the area of $\triangle ABC$ represents the same number, you know that

$$\tfrac{1}{2}bc \sin A = \tfrac{1}{2}ac \sin B = \tfrac{1}{2}ab \sin C.$$

Dividing each part of this compound sentence by $\tfrac{1}{2}abc$ produces

$$\frac{\sin A}{a} = \frac{\sin B}{b} = \frac{\sin C}{c}.$$

This relationship is formalized in the following theorem.

Law of Sines

The sines of the angles of a triangle are proportional to the lengths of the opposite sides.

Notice that if one of the angles of the triangle, say C, is a right angle, then $\sin C = 1$, and the law of sines yields the familiar right-triangle relationships

$$\sin A = \frac{a}{c} \quad \text{and} \quad \sin B = \frac{b}{c}.$$

648 *Chapter 15*

EXAMPLE 2   Solve the triangle pictured.

SOLUTION   Since $m(A) + m(B) + m(C) = 180°$, you
have

$$28.5° + 44.5° + m(C) = 180°,$$

or

$$m(C) = 107°.$$

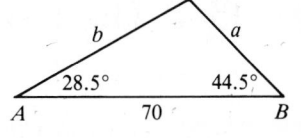

The law of sines gives:

$$\frac{\sin 28.5°}{a} = \frac{\sin 44.5°}{b} = \frac{\sin 107°}{70}$$

Solving separately for a and b, you find:

$$\frac{\sin 28.5°}{a} = \frac{\sin 107°}{70} \qquad \qquad \frac{\sin 44.5°}{b} = \frac{\sin 107°}{70}$$

$$a = \frac{70 \sin 28.5°}{\sin 107°} \qquad \qquad b = \frac{70 \sin 44.5°}{\sin 107°}$$

$$\approx \frac{70(0.4772)}{0.9563} \qquad \qquad \approx \frac{70(0.7009)}{0.9563}$$

$$\approx 34.9 \qquad \qquad \approx 51.3$$

$\therefore m(C) = 107°$, $a \approx 34.9$, and $b \approx 51.3$

The law of sines gives you a way to solve any triangle if you know

(a) the measures of two angles and the length of any one side, as illustrated in Example 2, or

(b) the lengths of two sides and the measure of an angle opposite one of them—insofar as there is a solution.

In the latter case, the data may be ambiguous. Figures 8 and 9 illustrate the possibilities for a triangle when you are given a, b, and $m(A)$.

Figure 8

$m(A) < 90°$

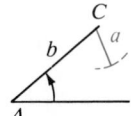

$a < b \sin A$
No solution

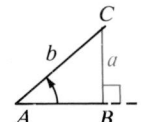

$a = b \sin A$
One solution

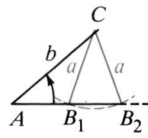

$b \sin A < a < b$
Two solutions

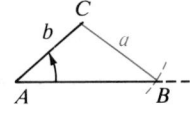

$a > b$
One solution

Figure 9

$90° \le m(A) < 180°$

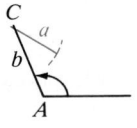

$a \le b$
No solution

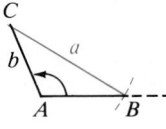

$a > b$
One solution

Trigonometric Identities   **649**

EXAMPLE 3 Solve the triangle for which $a = 60$, $b = 65$, and $m(A) = 35°$.

SOLUTION A sketch of the triangle, roughly to scale, suggests that there are two solutions. This can be verified by noting that

$$b \sin A ≈ 65(0.5736) ≈ 37.3$$

and

$$37.3 < 60 < 65$$

(see Figure 8, part 3).

To find $m(B)$, we use

$$\frac{\sin A}{a} = \frac{\sin B}{b},$$

or

$$\sin B = \frac{b \sin A}{a} ≈ \frac{65(0.5736)}{60} ≈ 0.6214.$$

∴ either $m(B) ≈ 38.4°$ or $m(B) ≈ 180° - 38.4° = 141.6°$.

If $m(B) = 38.4°$, then $m(C) = 180° - 35° - 38.4° = 106.6°$.
If $m(B) = 141.6°$, then $m(C) = 180° - 35° - 141.6° = 3.4°$.

From

$$\frac{\sin A}{a} = \frac{\sin C}{c}, \quad \text{or} \quad c = \frac{a \sin C}{\sin A},$$

taking $m(C) = 106.6°$, you find that $c ≈ \frac{60(0.9583)}{0.5736} ≈ 100.2$ and

taking $m(C) = 3.4°$, you find that $c ≈ \frac{60(0.0593)}{0.5736} ≈ 6.2$.

∴ the two possible triangles have the following parts:

$$m(B) ≈ 38.4°, \ m(C) ≈ 106.6°, \ c ≈ 100.2$$

or

$$m(B) ≈ 141.6°, \ m(C) ≈ 3.4°, \ c ≈ 6.2$$

Oral Exercises

For △ABC, use the law of sines to state an expression equivalent to the given one.

1. $\dfrac{\sin A}{\sin B}$ $\dfrac{a}{b}$

2. $\dfrac{b}{c}$ $\dfrac{\sin B}{\sin C}$

3. $\dfrac{a}{\sin A}$ $\dfrac{b}{\sin B}$ or $\dfrac{c}{\sin C}$

4. $\dfrac{c \sin B}{\sin C}$ b

How many solutions exist for $\triangle ABC$ with the given parts?

5. $m(A) = 30°$, $a = 5$, $b = 10$ one

7. $m(A) = 30°$, $a = 4$, $b = 10$ none

9. $m(A) = 150°$, $a = 10$, $b = 5$ one

6. $m(A) = 30°$, $a = 10$, $b = 5$ one

8. $m(A) = 30°$, $a = 4$, $b = 6$ two

10. $m(A) = 150°$, $a = 4$, $b = 6$ none

Written Exercises

Solve the $\triangle ABC$ with the given parts. Give lengths to the nearest tenth of a unit and angles to the nearest tenth of a degree. If no solution exists, so state. If two solutions exist, give both.

A

1. $m(A) = 30°$, $a = 12$, $m(B) = 54°$

3. $a = 40$, $b = 20$, $m(A) = 120°$

5. $m(A) = 14°$, $m(B) = 128°$, $a = 75$

7. $a = 48$, $m(B) = 33.3°$, $m(C) = 137°$

9. $m(B) = 17°$, $a = 50$, $b = 35$

2. $m(B) = 30°$, $m(C) = 67°$, $b = 20$

4. $a = 25$, $c = 8$, $m(C) = 30°$

6. $m(A) = 30°$, $a = 20$, $c = 34$

8. $a = 16$, $b = 35$, $m(A) = 61.7°$

10. $m(C) = 25°$, $a = 27$, $c = 55$

For each $\triangle ABC$ with the given parts, find the area to the nearest square unit. If no triangle exists, so state. If two triangles exist, give the area of each.

11. $a = 8$, $b = 9$, $m(C) = 30°$ 18

13. $m(C) = 63°$, $b = 16$, $a = 15$ 107

15. $a = 50$, $b = 40$, $m(B) = 30°$ 932, 151

12. $b = 24$, $c = 17$, $m(A) = 150°$ 102

14. $m(B) = 73°$, $a = 28$, $c = 18$ 241

16. $a = 25$, $b = 17$, $m(B) = 21°$

169, 40

(*Hint for Exercises 15 and 16*: Use the law of sines to find $m(C)$.)

In Exercises 17–19 use the diagram at the right.

B

17. Find an expression for the length of $\overline{CD}$ in terms of b and a trigonometric function of $\angle A$. $b \sin A$

18. Find an expression for the length of $\overline{CD}$ in terms of a and a trigonometric function of $\angle B$. $a \sin B$

19. Use the results of Exercises 17 and 18 above to prove the first part of the law of sines:

$$\frac{\sin A}{a} = \frac{\sin B}{b}$$

20. Use a diagram similar to the one above to prove $\dfrac{\sin B}{b} = \dfrac{\sin C}{c}$.

21. Use the fact that the area of a triangle can be given by Area $= \frac{1}{2}bc \sin A$ to prove the following formula: Area $= \dfrac{b^2 \sin A \sin C}{2 \sin B}$.

Substitute $\dfrac{b \sin C}{\sin B}$ for c

22. Show that if a, $m(B)$, and $m(C)$ are given in $\triangle ABC$, then b can be found from the equation $\dfrac{\sin (B + C)}{a} = \dfrac{\sin B}{b}$. $m(A) = 180° - m(B + C)$, $\sin A = \sin (B + C)$

Trigonometric Identities **651**

---

5. $m(C) = 38°$, $b \approx 244.3$, $c \approx 190.9$

6. $m(C) \approx 58.2°$, $m(B) \approx 91.8°$, $b \approx 40.0$ or $m(C) \approx 121.8°$, $m(B) \approx 28.2°$, $b \approx 18.9$

7. $m(A) = 9.7°$, $b \approx 156.4$, $c \approx 194.3$

8. no triangle

9. $m(A) \approx 24.7°$, $m(C) \approx 138.3°$, $c \approx 79.6$ or $m(A) \approx 155.3°$, $m(C) \approx 7.7°$, $c \approx 16.0$

10. $m(A) \approx 12.0°$, $m(B) \approx 143.0°$, $b \approx 78.3$

19. 1. $CD = b \sin A$, $CD = a \sin B$

    Result of Exs. 17, 18

    2. $a \sin B = b \sin A$

    Subs. prin.

    3. $\dfrac{\sin B}{b} = \dfrac{\sin A}{a}$

    Div. prop. =

20.

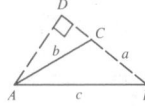

1. $\sin B = \dfrac{AD}{c}$

    Def. of sine

2. $c \sin B = AD$

    Mult. prop. =

3. $\sin(180° - C) = \dfrac{AD}{b}$

    Def. of sine

4. $\sin(180° - C) = \sin C$

    Supplements have = sine values

5. $\sin C = \dfrac{AD}{b}$

    Subs. prin.

6. $b \sin C = AD$

    Mult. prop. =

7. $c \sin B = b \sin C$

    Trans. prop. =

8. $\dfrac{\sin B}{b} = \dfrac{\sin C}{c}$

    Div. prop. =

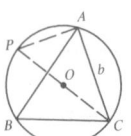 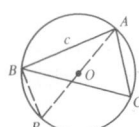
C **23.** Use the diagram at the right (in which circle O is the circumscribed circle about $\triangle ABC$) to prove the law of sines by means of the following steps:

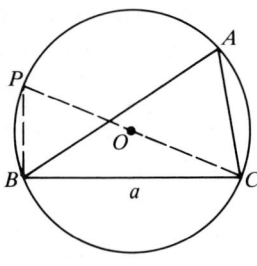

   a. Show that $m(P) = m(A)$.
   b. Show that the length of diameter $\overline{PC}$ of circle O equals $\dfrac{a}{\sin P}$ and $\dfrac{a}{\sin A}$.
   c. Draw similar diagrams to illustrate that the diameter of circle O also equals $\dfrac{b}{\sin B}$ and $\dfrac{c}{\sin C}$.

Problems

Give lengths to the nearest tenth of a unit and angles to the nearest tenth of a degree.

A **1.** Find the area of a parallelogram with adjacent sides of length 20 cm and 15 cm and with an angle of measure 117° included between the sides. 267.3 cm²

   2. Find the area of a regular 20-sided polygon inscribed in a circle of radius 12 cm. 445.0 cm²

   3. Each of two observers on the ground 15 km apart spots a plane as it passes over the line joining the two observers. The angles of elevation of the observers' lines of sight are 49.8° and 60.5°, respectively. How far is the plane from the second observer at this moment? 12.2 km

   4. Otis Washington starts out hiking from a camp at point A on a course 25.5° from the line through A and a town at point B 12 km away. He heads for a ridge at point C intending to change direction there and head for the town, but when he arrives at C he finds that the trail marker saying "10 km to town B" has fallen down. By what angle should he now turn in order to head straight for B? 31.1° or 148.9°

   5. The pitches of the front and back roofs of a saltbox house are 45° and 30°, respectively. If the slant height of the back roof is 10.2 m, what is the slant height, x, of the front roof? 7.2 m

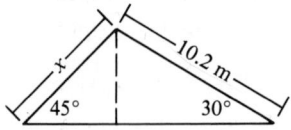

   6. A clock pendulum swings through an angle of 35°. If the distance between the two extreme positions of the bob at the end of the shaft is 9 cm, how long is the shaft?
   15.0 cm

B **7.** One side of a parallelogram has length 40 cm. An adjacent side of length 22 cm makes an angle of 46° with one of the diagonals of the parallelogram. Find the length of this diagonal. 52.0 cm

8. Sue Philips wants to know the distance between two points A and B on the opposite side of a river from her position. She finds that the length of $\overline{CD}$ is 75 m, $m(\angle ADC) = 60°$, $m(\angle ACD) = 54°$, and $m(\angle ACB) = 48°$. What is the distance from A to B if A, B, C, and D are the vertices of a trapezoid? 54.0 m

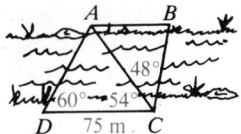

D   75 m   C

Computer Exercises   For students with computer experience

1. Write a program that will use the law of sines to find $m(B)$, in decimal degrees, for $\triangle ABC$, given a, b, and $m(A)$. The program should give both solutions for $m(B)$ if two are possible and should print a message that no triangle with the given parts exists if such is the case.

Run the program in Exercise 1 for $\triangle ABC$ with the given parts.

2. $a = 18$, $b = 25$, $m(A) = 30°$ 43.98°, 136.02°
3. $a = 7$, $b = 16$, $m(A) = 40°$ no triangle
4. $a = 21$, $b = 13$, $m(A) = 57°$ 31.28°

▪ Self-Test 3

VOCABULARY   law of cosines (p. 645)   law of sines (p. 648)

In $\triangle ABC$, find a to the nearest tenth of a unit.

1. $b = 5$, $c = 9$, $m(A) = 49°$ 6.9   *Obj. 1, p. 644*
2. $m(A) = 31°$, $m(B) = 57°$, $b = 95$ 58.3
3. Find the area of $\triangle ABC$ to the nearest square unit if $a = 12$, $b = 20$, and $m(C) = 49°$. 91 square units

Find each length to the nearest tenth of a unit.

4. Find the length of a side of a regular 9-sided polygon inscribed in a   *Obj. 2, p. 644*
circle of radius 50 cm. 34.2 cm
5. A ship is sighted from two radar stations 125 km apart. The angle between the line segment joining the two stations and the radar beam of the first station is 43°, and between this segment and the beam from the second station is 107°. How far is the ship from the second station? 170.5 km

Check your answers with those at the back of the book.

Trigonometric Identities   **653**

Additional Answers
Written Exercises
(continued from p. 623)

4. $\cos x + \sin x \tan x =$

$\cos x \left(\dfrac{\cos x}{\cos x} \right) + \sin x \cdot \dfrac{\sin x}{\cos x}$

$= \dfrac{\cos^2 x}{\cos x} + \dfrac{\sin^2 x}{\cos x} =$

$\dfrac{\cos^2 x + \sin^2 x}{\cos x} = \dfrac{1}{\cos x} =$

$\sec x; \cos x \neq 0$

5. $\cos x(\sec x + \cos x \csc^2 x)$

$= \cos x \left(\dfrac{1}{\cos x} + \cos x \cdot \right.$

$\left. \dfrac{1}{\sin^2 x} \right) = 1 + \dfrac{\cos^2 x}{\sin^2 x} = 1 +$

$\cot^2 x = \csc^2 x; \cos x \neq 0,$
$\sin x \neq 0$

6. $\dfrac{1 + \tan^2 x}{1 + \cot^2 x} = \dfrac{\sec^2 x}{\csc^2 x} =$

$\dfrac{\dfrac{1}{\cos^2 x}}{\dfrac{1}{\sin^2 x}} = \dfrac{\sin^2 x}{\cos^2 x} = \tan^2 x;$

$\cos x \neq 0, \sin x \neq 0$

7. $\dfrac{(\cos \alpha - \sin \alpha)^2}{\sin \alpha} =$

$\dfrac{\cos^2 \alpha - 2 \cos \alpha \sin \alpha + \sin^2 \alpha}{\sin \alpha}$

$= \dfrac{1 - 2 \cos \alpha \sin \alpha}{\sin \alpha}$

$= \dfrac{1}{\sin \alpha} - 2 \cos \alpha$

$= \csc \alpha - 2 \cos \alpha;$
$\sin \alpha \neq 0$

8. $\dfrac{\cos x}{1 - \cos x} - \dfrac{\cos x}{1 + \cos x} =$

$\dfrac{\cos x(1 + \cos x)}{(1 - \cos x)(1 + \cos x)} -$

$\dfrac{\cos x(1 - \cos x)}{(1 + \cos x)(1 - \cos x)} =$

$\dfrac{2 \cos x \cos x}{1 - \cos^2 x} = \dfrac{2 \cos^2 x}{\sin^2 x} =$

$\dfrac{2 \cos^2 x}{\sin^2 x} = 2 \cot^2 x;$

$\sin x \neq 0$

Chapter Summary

1. You use the *fundamental trigonometric identities* together with properties of real numbers to establish additional identities.

2. The formula $\cos (x_1 - x_2) = \cos x_1 \cos x_2 + \sin x_1 \sin x_2$ is the *formula for the cosine of a difference*. You can use this formula to derive *reduction formulas* and formulas for values of functions of sums and then for *double-* and *half-angle formulas*. (See page 642.)

3. If you know the measures of three parts of a triangle, including the length of at least one side, you can find the measures of the remaining parts by using the *law of cosines*, $c^2 = a^2 + b^2 - 2ab \cos C$, or the *law of sines*, $\dfrac{\sin A}{a} = \dfrac{\sin B}{b} = \dfrac{\sin C}{c}$.

   When given data for a triangle consist of the lengths of two sides and the measure of an angle opposite one of them, there may exist no, one, or two triangle(s) with the given measurements.

4. The area of a triangle is equal to half the product of the lengths of two sides and the sine of their included angle.

Chapter Review

Write the letter of the correct answer.

1. Rewrite $\dfrac{1}{\csc^2 \alpha}$ in terms of $\sin \alpha$.

   a. $\dfrac{1}{\sin^2 \alpha}$ **b.** $\dfrac{1}{1 - \sin^2 \alpha}$ **c.** $\sin^2 \alpha$ **d.** $-\sin^2 \alpha$

   15-1

2. Simplify $\dfrac{1}{\sec \alpha} (\tan \alpha + \cot \alpha)$ assuming $\sin \alpha \neq 0, \cos \alpha \neq 0$.

   a. $\csc \alpha$ **b.** 1 **c.** $\cos \alpha$ **d.** $\sec \alpha$

3. Which one of the following is equal to $\dfrac{(\sin x - \cos x)^2}{\cos x}$, assuming that $\cos x \neq 0$?

   a. $\sec x$ **b.** $\sec x - 2 \sin x$ **c.** $\sec x - 2 \tan x$ **d.** $-2 \sin x$

   15-2

4. Which one of the following is *not* equal to $\dfrac{1}{\sec \alpha - 1} + \dfrac{1}{\sec \alpha + 1}$, with suitable restrictions?

   a. $\dfrac{2 \sec \alpha}{\tan^2 \alpha}$ **b.** $\dfrac{2 \cos \alpha}{\sin^2 \alpha}$ **c.** $2 \cot \alpha \csc \alpha$ **d.** $2 \cos \alpha \sec \alpha$

5. Evaluate $\cos \dfrac{7\pi}{8} \cos \dfrac{\pi}{8} - \sin \dfrac{7\pi}{8} \sin \dfrac{\pi}{8}$. (First express it as the cosine of a single real number.)

   a. -1 **b.** 1 **c.** 0 **d.** $-\dfrac{1}{\sqrt{2}}$

   15-3

6. Given that $\sin x = \frac{4}{5}$, $0 < x < \frac{\pi}{2}$, and $\cos y = -\frac{12}{13}$, $\frac{\pi}{2} < y < \pi$, find $\sin (x - y)$.

   (a.) $-\frac{63}{65}$ b. $\frac{63}{65}$ c. $\frac{33}{65}$ d. $-\frac{33}{65}$

15-4

7. Which expression equals $\tan (x_1 + x_2)$?

   a. $\dfrac{\tan x_1 - \tan x_2}{1 - \tan x_1 \tan x_2}$ b. $\dfrac{\tan x_1 - \tan x_2}{1 + \tan x_1 \tan x_2}$

   (c.) $\dfrac{\tan x_1 + \tan x_2}{1 - \tan x_1 \tan x_2}$ d. $\dfrac{\tan x_1 + \tan x_2}{1 + \tan x_1 x_2}$

8. Which expression equals $\sin \frac{x}{2}$?

15-5

   a. $\pm\sqrt{\dfrac{1 + \cos x}{2}}$ (b.) $\pm\sqrt{\dfrac{1 - \cos x}{2}}$

   c. $\pm\sqrt{\dfrac{1 - \sin x}{2}}$ d. $\pm\sqrt{\dfrac{1 + \sin x}{2}}$

9. Which expression equals $\tan 2x$?

   a. $\dfrac{2 \tan x}{1 + \tan^2 x}$ b. $\dfrac{2 \sin x}{1 - \tan^2 x}$ c. $\dfrac{2 \sin x}{1 + \tan^2 x}$ (d.) $\dfrac{2 \tan x}{1 - \tan^2 x}$

10. Which one of the following is *not* equal to $\dfrac{\cos 2\alpha}{\cos \alpha}$, assuming that $\cos \alpha \neq 0$?

15-6

   a. $\cos \alpha - \sin \alpha \tan \alpha$ b. $\sec \alpha - \dfrac{2 \sin \alpha}{\cot \alpha}$

   c. $2 \cos \alpha - \sec \alpha$ (d.) $\sec \alpha - 2 \sin^2 \alpha \tan \alpha$

11. Which is the correct expression of the law of cosines for $\triangle ABC$?

15-7

   a. $b^2 = a^2 + c^2 - 2ab \cos C$ (b.) $b^2 = a^2 + c^2 - 2ac \cos B$

   c. $b^2 = a^2 + c^2 - 2ab \cos A$ d. $b^2 = a^2 + b^2 - 2ac \cos B$

12. In $\triangle ABC$, if $\sin A = 0.8$, $\sin B = 0.2$, and $a = 4$, find b.

15-8

   a. 0.8 b. 1.6 (c.) 1 d. 16

9.
$$\frac{\sin \alpha}{\sec \alpha - 1} + \frac{\sin \alpha}{\sec \alpha + 1} =$$
$$\frac{\sin \alpha(\sec \alpha + 1)}{(\sec \alpha - 1)(\sec \alpha + 1)} +$$
$$\frac{\sin \alpha(\sec \alpha - 1)}{(\sec \alpha + 1)(\sec \alpha - 1)} =$$
$$\frac{2 \sin \alpha \sec \alpha}{\sec^2 \alpha - 1} =$$
$$\frac{2 \sin \alpha \cdot \dfrac{1}{\cos \alpha}}{\tan^2 \alpha} =$$
$$\frac{2 \dfrac{\sin \alpha}{\cos \alpha}}{\tan^2 \alpha} = \frac{2 \tan \alpha}{\tan^2 \alpha} =$$
$$\frac{2}{\tan \alpha} = 2 \cot \alpha;$$
$$\cos \alpha \neq 0; \sin \alpha \neq 0$$

10. $\sec \alpha(\csc \alpha - \cot \alpha \cos \alpha)$
$$= \frac{1}{\cos \alpha} \cdot \frac{1}{\sin \alpha} - \frac{1}{\cos \alpha} \cdot$$
$$\cot \alpha \cos \alpha =$$
$$\frac{1}{\cos \alpha \sin \alpha} - \cot \alpha =$$
$$\frac{1}{\cos \alpha \sin \alpha} - \frac{\cos \alpha(\cos \alpha)}{\sin \alpha(\cos \alpha)} =$$
$$\frac{1 - \cos^2 \alpha}{\cos \alpha \sin \alpha} =$$
$$\frac{\sin^2 \alpha}{\cos \alpha \sin \alpha} = \frac{\sin \alpha}{\cos \alpha} = \tan \alpha;$$
$$\cos \alpha \neq 0, \sin \alpha \neq 0$$

(continued on p. 656)

(continued on p. 656)

Chapter Test

1. Express $\cot^2 \alpha$ in terms of $\tan \alpha$. $\dfrac{1}{\tan^2 \alpha}$; $\tan \alpha \neq 0$, $\cot \alpha \neq 0$

15-1

2. Prove that $(1 - \sin \alpha) \sec \alpha = \dfrac{\cos \alpha}{1 + \sin \alpha}$ and state any necessary restrictions.

15-2

3. Find $\cos (x - y)$, given that $\cos x = \frac{4}{5}$, $\frac{3\pi}{2} < x < 2\pi$, and $\cos y = -\frac{24}{25}$, $\pi < y < \frac{3\pi}{2}$. $-\frac{3}{5}$

15-3

Trigonometric Identities   **655**

Additional Answers

11. $\sec^2 x + \csc^2 x =$

$$\frac{1}{\cos^2 x} + \frac{1}{\sin^2 x} =$$

$$\frac{\sin^2 x + \cos^2 x}{\cos^2 x \cdot \sin^2 x} =$$

$$\frac{1}{\cos^2 x \sin^2 x} =$$

$$\frac{1}{\cos^2 x} \cdot \frac{1}{\sin^2 x} =$$

$\sec^2 x \csc^2 x$; $\sin x \neq 0$, $\cos x \neq 0$

12. $\sec x(\csc x - \sin x) =$

$$\frac{1}{\cos x}\left(\frac{1}{\sin x} - \sin x\right) =$$

$$\frac{1}{\cos x}\left(\frac{1 - \sin^2 x}{\sin x}\right) =$$

$$\frac{1}{\cos x}\left(\frac{\cos^2 x}{\sin x}\right) = \frac{\cos x}{\sin x} =$$

$\cot x$; $\sin x \neq 0$, $\cos x \neq 0$

13. $\dfrac{1 + \cot \alpha}{\sin \alpha + \cos \alpha} =$

$$\frac{1 + \dfrac{\cos \alpha}{\sin \alpha}}{\sin \alpha + \cos \alpha} =$$

$$\frac{\dfrac{\sin \alpha + \cos \alpha}{\sin \alpha}}{\sin \alpha + \cos \alpha} = \frac{1}{\sin \alpha} =$$

$\csc \alpha$; $\sin \alpha \neq 0$,
$\sin \alpha \neq -\cos \alpha$

14. $\dfrac{\sec \alpha + \csc \alpha}{1 + \cot \alpha} =$

$$\frac{\dfrac{1}{\cos \alpha}\left(\dfrac{\sin \alpha}{\sin \alpha}\right) + \dfrac{1}{\sin \alpha}\left(\dfrac{\cos \alpha}{\cos \alpha}\right)}{\dfrac{\sin \alpha}{\sin \alpha} + \dfrac{\cos \alpha}{\sin \alpha}}$$

$$\frac{\dfrac{\sin \alpha + \cos \alpha}{\cos \alpha \sin \alpha}}{\dfrac{\cos \alpha + \sin \alpha}{\sin \alpha}}$$

$$= \frac{\sin \alpha + \cos \alpha}{\cos \alpha \sin \alpha} \cdot$$

$$\frac{\sin \alpha}{\cos \alpha + \sin \alpha} = \frac{1}{\cos \alpha} =$$

$\sec \alpha$; $\sin \alpha \neq 0$, $\cos \alpha \neq 0$

4. Use a formula for the sine of a sum or difference to evaluate $\sin \dfrac{7\pi}{12}$. $\dfrac{\sqrt 6 + \sqrt 2}{4}$ *15-4*

5. Find $\tan 2x$ if $\tan x = 0.6$. 1.875   *15-5*

6. Given that $\sin x = -0.6$ and $\pi < x < \dfrac{3\pi}{2}$, find $\sin \dfrac{x}{2}$. $\dfrac{3}{\sqrt{10}}$

7. Prove that $\dfrac{2}{\sin 2\alpha} = \sec \alpha \csc \alpha$ and state any necessary restrictions.   *15-6*

8. In $\triangle ABC$, if $a = 15$, $b = 30$, and $m(C) = 60°$, find c to the nearest tenth. 26.0   *15-7*

9. In $\triangle ABC$, if $m(A) = 45°$, $m(B) = 60°$, and $a = 30$, find b to the nearest tenth. 36.7   *15-8*

APPLICATION

Hours of Daylight

Sine waves and other periodic functions are used to describe and analyze alternating-current (AC) electricity, electromagnetic radiation, and musical sounds. They are also used to represent fluctuations in temperature, water depth due to tides, and the number of hours of daylight at a given location over a period of time.

In general, the number of hours of daylight on a certain day at a given place depends on the latitude of the location and on the time of the year. In the Northern Hemisphere, the maximum amount of daylight occurs on June 21 and the minimum amount of daylight occurs on December 21. The graph at the top of the facing page shows the variation in daylight for one year in Chicago, Illinois, where x represents the number

656   *Chapter 15*

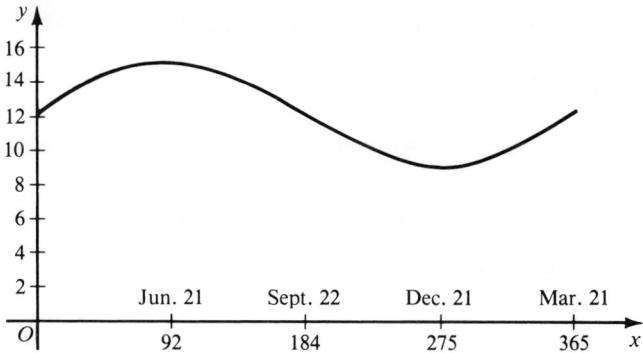

of days after March 21 and y is the number of hours of daylight (hours between sunrise and sunset). From this graph you can estimate the number of hours of daylight in Chicago on any day of the year. The general equation of the graph shown is $y = A \sin Bx + C$. The amplitude, A, of the sine wave shown is

$$A = \frac{15.2 - 9.2}{2} = 3.0.$$

The fundamental period of this sine wave is 365 days, since the graph represents one year. Thus $B = \frac{2\pi}{365}$. On March 21 and September 21, the amount of daylight is approximately equal to the amount of darkness. Thus $C \approx 12$. The actual value of C to three significant digits is 12.2. The change in this value is obtained by considering the following factors.

1. We see the sun when it is still below the horizon, due to the refraction, or bending, of sunlight in Earth's atmosphere.
2. We record the time of sunrise or sunset when the edge, rather than the center, of the solar disk appears or disappears.

Thus the equation for the number of hours of daylight in Chicago is

$$y = 3.0 \sin \frac{2\pi}{365}x + 12.2.$$

Exercises

1. Estimate the number of hours of daylight in Chicago on April 12.
   a. Use the graph shown above. 13 h
   b. Use the equation for the number of hours of daylight in Chicago. 13.3 h
2. What is the value of A in Vancouver, British Columbia, where the maximum number of hours of daylight is approximately 16 and the minimum number of hours of daylight is approximately 8? 4.0
3. What is the value of A in Barrow, Alaska, where the maximum number of hours of daylight is approximately 24 and the minimum number of hours of daylight is approximately 0? 12.0

15. $(\sec x + 1)(\csc x - \cot x) =$

$$\left(\frac{1}{\cos x} + 1\right)\left(\frac{1}{\sin x} - \frac{\cos x}{\sin x}\right)$$

$$= \left(\frac{1 + \cos x}{\cos x}\right)\left(\frac{1 - \cos x}{\sin x}\right)$$

$$= \frac{1 - \cos^2 x}{\cos x \sin x} = \frac{\sin^2 x}{\cos x \sin x}$$

$$= \frac{\sin x}{\cos x} = \tan x; \sin x \neq 0,$$

$$\cos x \neq 0$$

16. $(\tan x + \cos x)(\cot x + \tan x)$

$$= \left(\frac{\sin x}{\cos x} + \cos x\right)\left(\frac{\cos x}{\sin x} + \frac{\sin x}{\cos x}\right) = \left(\frac{\sin x}{\cos x} + \cos x\right) \cdot$$

$$\left(\frac{\cos^2 x + \sin^2 x}{\sin x \cos x}\right) =$$

$$\left(\frac{\sin x}{\cos x} + \cos x\right)\left(\frac{1}{\sin x \cos x}\right) =$$

$$\frac{\sin x}{\sin x \cos^2 x} + \frac{\cos x}{\sin x \cos x} =$$

$$\frac{1}{\cos^2 x} + \frac{1}{\sin x} = \sec^2 x +$$

$$\csc x; \sin x \neq 0, \cos x \neq 0$$

Trigonometric Identities **657**

55. $\cot(x + y) = \dfrac{\cos(x + y)}{\sin(x + y)} =$

$\dfrac{\cos x \cos y - \sin x \sin y}{\sin x \cos y + \cos x \sin y} =$

$\dfrac{\dfrac{\cos x \cos y}{\sin x \sin y} - \dfrac{\sin x \sin y}{\sin x \sin y}}{\dfrac{\sin x \cos y}{\sin x \sin y} + \dfrac{\cos x \sin y}{\sin x \sin y}} =$

$\dfrac{\cot x \cot y - 1}{\cot y + \cot x} =$

$\dfrac{\cot x \cot y - 1}{\cot x + \cot y}$;

$\sin(x + y) \neq 0, \sin x \neq 0,$
$\sin y \neq 0$

56. $\sec(x + y) = \dfrac{1}{\cos(x + y)} =$

$\dfrac{1}{\cos x \cos y - \sin x \sin y} =$

$\dfrac{\dfrac{1}{\cos x \cos y}}{\dfrac{\cos x \cos y}{\cos x \cos y} - \dfrac{\sin x \sin y}{\cos x \cos y}} =$

$\dfrac{\sec x \sec y}{1 - \tan x \tan y}$;

$\cos(x + y) \neq 0, \cos x \neq 0,$
$\cos y \neq 0$

57. $\csc(x + y) = \dfrac{1}{\sin(x + y)}$

$= \dfrac{1}{\sin x \cos y + \cos x \sin y}$

$= \dfrac{\dfrac{1}{\sin x \sin y}}{\dfrac{\sin x \cos y}{\sin x \sin y} + \dfrac{\cos x \sin y}{\sin x \sin y}}$

$= \dfrac{\csc x \csc y}{\cot y + \cot x} =$

$\dfrac{\csc x \csc y}{\cot x + \cot y}$; $\sin(x + y) \neq$

$0, \sin x \neq 0, \sin y \neq 0$

PREPARING FOR
COLLEGE ENTRANCE EXAMS

Strategy for Success: When you solve problems involving trigonometry, area, or distance it may be helpful to draw a sketch that shows the given information. Be careful to make no assumptions in drawing the figure. Use any available space in your test booklet or scrap paper, but avoid making any extra marks on your answer sheet.

Decide which is the best of the choices given and write the correct letter on your answer sheet.

1. Express $\dfrac{13\pi}{6}^R$ as a degree measure. C

(**A**) 150° (**B**) −390° (**C**) 390° (**D**) −150° (**E**) 315°

2. What is the radius of a circle in which an arc of length $\dfrac{8\pi}{5}$ is intercepted by a central angle with measure $\dfrac{3\pi}{2}^R$? B

(**A**) $\dfrac{15}{16}$ (**B**) $\dfrac{16}{15}$ (**C**) $\dfrac{5}{8}$ (**D**) $\dfrac{15}{5}$ (**E**) 4

3. Which of the following must be true for the function $y = -4 \sin x + 1$? E

I. The minimum value of the function is −3.
II. The maximum value of the function is 5.
III. The amplitude of the function is 4.

(**A**) I only (**B**) II only (**C**) III only (**D**) I and III only (**E**) I, II, and III

4. What is the common ratio of the geometric sequence $x_1, x_1 r,$ $x_1 r^2, \ldots$? E

(**A**) $\dfrac{1}{r}$ (**B**) r^2 (**C**) xr (**D**) x (**E**) r

5. For what value of n does $3(5^n) = 27$? D

(**A**) $\dfrac{\log 27}{\log 15}$ (**B**) $\dfrac{\log 5}{\log 9}$ (**C**) $\dfrac{\log 15}{\log 27}$ (**D**) $\dfrac{\log 9}{\log 5}$ (**E**) None of these

6. Simplify $\dfrac{1}{\csc \alpha}(\tan \alpha + \cot \alpha)$. (Assume no denominator is zero.) B

(**A**) $\sin \alpha$ (**B**) $\sec \alpha$ (**C**) $\tan \alpha$ (**D**) 1 (**E**) $\cos \alpha$

7. What is the value of $\sin 2\alpha$ if $\cos \alpha = \dfrac{-12}{13}$ and α lies in Quadrant II? A

(**A**) $-\dfrac{120}{169}$ (**B**) $-\dfrac{169}{120}$ (**C**) $-\dfrac{144}{169}$ (**D**) $-\dfrac{169}{144}$ (**E**) $\dfrac{120}{169}$

8. Simplify $2a^{-2}(30a^5 b - a^2)$. B

(**A**) $60a^3 b - 1$ (**B**) $60a^3 b - 2$ (**C**) $60a^3 - 2$ (**D**) $60a^3 - 1$ (**E**) $60a^7 b - 2$

658 *Chapter 15*

PROGRAMMING IN PASCAL

Exercises

1. **a.** Explain how one could compute all six circular functions in terms of sin or cos alone. Explain why it would therefore suffice for an implementation of Pascal to include only sin or cos to compute all the circular functions.
   b. Write Pascal functions to compute tan(x), cot(x), sec(x), csc(x); you may assume that the input x will always belong to the domain of the function.

2. Use the identities and formulas you learned in this chapter to write a Pascal program that solves triangle *ABC* (with angles measured in degrees) when the user enters two sides and an included angle, two angles and an included side, or three sides. The program should prompt the user to enter which set of data will be used, prompt the user for the appropriate data for the chosen option, then solve the triangle and display the side lengths and angle measures of triangle *ABC*. Use your program to check your solutions to several of the problems already solved by the law of cosines or the law of sines.

3. Research Problem
   Review the analysis in Section 15-8 of the ambiguous triangle case where the lengths of two sides and the measure of an angle opposite one of those sides is given. Write a program that accepts the lengths of two sides and the measure of an angle opposite one of them, and determines if there is a unique solution. If there is a unique solution, the data for the solved triangle should be displayed; otherwise, the user should be notified that there is no unique triangle that satisfies the conditions specified.

Additional Answers
Written Exercises
(continued from p. 643)

7. $\dfrac{\cos 2\alpha}{\cos \alpha + \sin \alpha} =$

$\dfrac{\cos^2 \alpha - \sin^2 \alpha}{\cos \alpha + \sin \alpha} =$

$\dfrac{(\cos \alpha - \sin \alpha)(\cos \alpha + \sin \alpha)}{(\cos \alpha + \sin \alpha)}$

$= \cos \alpha - \sin \alpha$;
$\cos \alpha \neq - \sin \alpha$

8. $\dfrac{\sin 2\alpha}{1 - \cos 2\alpha} =$

$\dfrac{2 \sin \alpha \cos \alpha}{1 - (1 - 2 \sin^2 \alpha)} =$

$\dfrac{2 \sin \alpha \cos \alpha}{1 - 1 + 2 \sin^2 \alpha} =$

$\dfrac{2 \sin \alpha \cos \alpha}{2 \sin^2 \alpha} = \dfrac{\cos \alpha}{\sin \alpha} =$

$\cot \alpha$; $\sin \alpha \neq 0$

10. $\dfrac{1 - \cos 2\alpha}{1 + \cos 2\alpha} =$

$\dfrac{1 - (1 - 2 \sin^2 \alpha)}{1 + (2 \cos^2 \alpha - 1)} =$

$\dfrac{1 - 1 + 2 \sin^2 \alpha}{1 + 2 \cos^2 \alpha - 1} =$

$\dfrac{2 \sin^2 \alpha}{2 \cos^2 \alpha} = \tan^2 \alpha$;

$\cos \alpha \neq 0$

11. $\dfrac{1 - \tan^2 \alpha}{1 + \tan^2 \alpha} = \dfrac{1 - \dfrac{\sin^2 \alpha}{\cos^2 \alpha}}{\sec^2 \alpha}$

$= \dfrac{\dfrac{\cos^2 \alpha - \sin^2 \alpha}{\cos^2 \alpha}}{\dfrac{1}{\cos^2 \alpha}} =$

$\cos^2 \alpha - \sin^2 \alpha =$
$\cos 2\alpha$; $\cos \alpha \neq 0$

12. $\cot 2\alpha = \dfrac{\cos 2\alpha}{\sin 2\alpha} =$

$\dfrac{\cos^2 \alpha - \sin^2 \alpha}{2 \sin \alpha \cos \alpha} =$

$\dfrac{\cos^2 \alpha}{2 \sin \alpha \cos \alpha} -$

$\dfrac{\sin^2 \alpha}{2 \sin \alpha \cos \alpha} = \dfrac{1}{2}\left(\dfrac{\cos \alpha}{\sin \alpha} - \right.$

$\left.\dfrac{\sin \alpha}{\cos \alpha}\right) = \dfrac{1}{2}(\cot \alpha - \tan \alpha)$;

$\sin \alpha, \cos \alpha \neq 0$

659

Spiral shapes, such as the spider web shown in the photo, are abundant in nature. These shapes can be easily drawn using polar coordinates. Section 16-3 describes polar equations and their graphs.

Chapter 16

Inverses; Polar Coordinates; Vectors

Problem Solving Strategies

Drawing a Diagram
Throughout the chapter diagrams are used to illustrate concepts and display information for problem solving.
Looking for a Pattern
Computation of z^2, z^3, and z^4, in Section 16-5 suggests DeMoivre's Theorem which is stated on page 680.

Inverse Functions

OBJECTIVES for Sections 16-1 and 16-2:
1. *To find values of inverse trigonometric and circular functions.*
2. *To solve trigonometric equations.*

16–1 Inverse Functions; Principal Values

You find the inverse of the function defined by

$$\{(x, y): y = \sin x\}$$

by interchanging the x and the y in the equation, thus obtaining the relation defined by

$$\{(x, y): x = \sin y\}.$$

Both relations are graphed in Figure 1. Notice that the graphs of $x = \sin y$ and $y = \sin x$ are reflections of each other in the line $y = x$. The inverse relation is denoted either by $\{(x, y): y = \sin^{-1} x\}$ (read "y = the inverse sine of x") or by $\{(x, y): y = \arcsin x\}$ (read "y = the arcsine of x"). Notice that the inverse relation $\{(x, y): y = \sin^{-1} x\}$ is *not* a function.

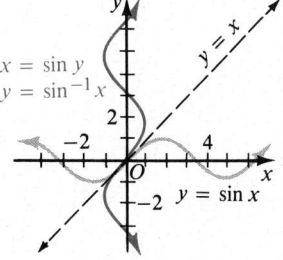

Figure 1

Inverses; Polar Coordinates; Vectors **661**

Teaching Suggestions
p. T124

Related Activities
p. T124

Key Ideas

Define inverses and principal value inverses for the circular functions. Use the definitions to find all the real values in the range of an inverse function.

Students should be aware of the difference between Sin^{-1} a, sin a^{-1}, and (sin a)$^{-1}$. Sin^{-1} a is read "the principal inverse sine of a," sin a^{-1} is read "the sine of the reciprocal of a," and (sin a)$^{-1}$ is read "one over the sine of a." Point out that the -1 is not an exponent in Sin^{-1} a, while it is an exponent in sin a^{-1} and in (sin a)$^{-1}$.

The equation $y = \sin^{-1} a$ is read "y is an angle whose sine is a."

Inverses of the other circular functions that you have studied are graphed in Figures 2 through 6 and are defined below.

$$\{(x, y): y = \cos^{-1} x\} \qquad \{(x, y): y = \tan^{-1} x\}$$
$$\{(x, y): y = \cot^{-1} x\} \qquad \{(x, y): y = \sec^{-1} x\}$$
$$\{(x, y): y = \csc^{-1} x\}$$

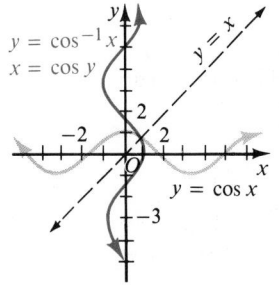

Figure 2

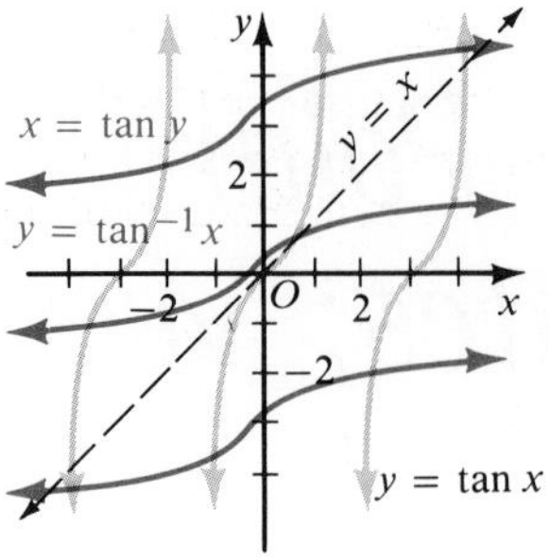

Figure 3

Figure 4

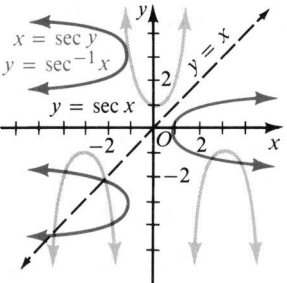

Figure 5

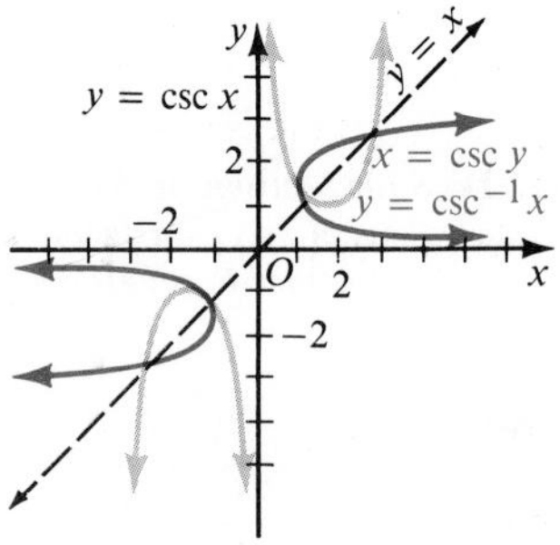

Figure 6

You can see from the graphs in Figures 1 through 6 that none of these inverse relations is a function.

By suitably restricting the ranges of the inverses of the circular functions, however, you can define for each such function any number of inverse functions. For example, as suggested by Figure 7, if the range is restricted to a suitable interval of length π, the resulting relation will then be a function.

We define the **principal-value inverse function for sine** to be the following:

$$\left\{(x, y): x = \sin y \quad \text{and} \quad -\frac{\pi}{2} \le y \le \frac{\pi}{2}\right\}$$

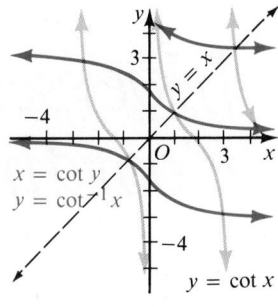

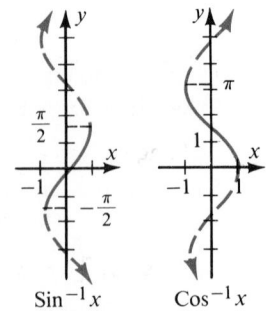

Sin^{-1}x Cos^{-1}x

Figure 7

This function is denoted by Arcsin x or Sin^{-1} x, with capital letters, and is read "principal arcsine of x" and "principal inverse sine of x" respectively. We can think of the expression Sin^{-1} x or Arcsin x as the measure of the angle between $-\frac{\pi}{2}$ and $\frac{\pi}{2}$, inclusive, whose sine is x. The domain of Sin^{-1} x is $\{x: |x| \leq 1\}$. Similarly, we define the **principal-value inverse function for cosine** to be

$$\{(x, y): x = \cos y \quad \text{and} \quad 0 \leq y \leq \pi\}$$

denoted Arcos x or Cos^{-1} x. The domain of Cos^{-1} x is also $\{x: |x| \leq 1\}$.

The graphs and definitions of the remaining principal-value inverses for the circular functions are shown in Figures 8 through 11.

$\left\{(x, y): x = \tan y \text{ and } -\frac{\pi}{2} < y < \frac{\pi}{2}\right\}$

Domain = $\mathcal{R}$

Range = $\left\{y: -\frac{\pi}{2} < y < \frac{\pi}{2}\right\}$

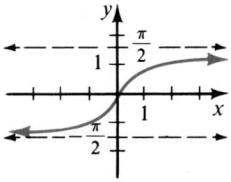

Figure 8

$\{(x, y): x = \cot y \text{ and } 0 < y < \pi\}$

Domain = $\mathcal{R}$

Range = $\{y: 0 < y < \pi\}$

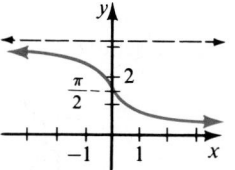

Figure 9

$\left\{(x, y): x = \sec y \text{ and } 0 \leq y \leq \pi, y \neq \frac{\pi}{2}\right\}$

Domain = $\{x: |x| \geq 1\}$

Range = $\left\{y: 0 \leq y \leq \pi, y \neq \frac{\pi}{2}\right\}$

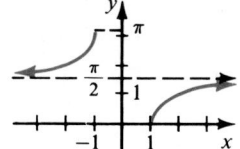

Figure 10

$\left\{(x, y): x = \csc y \text{ and } -\frac{\pi}{2} \leq y \leq \frac{\pi}{2}, y \neq 0\right\}$

Domain = $\{x: |x| \geq 1\}$

Range = $\left\{y: -\frac{\pi}{2} \leq y \leq \frac{\pi}{2}, y \neq 0\right\}$

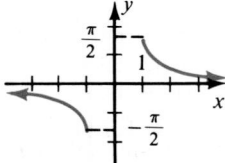

Figure 11

When 0, $\frac{\pi}{2}$, π, etc., are replaced with $0°$, $90°$, $180°$, etc., and y with $m(\alpha)$, in the preceding definitions, the results will define the principal-value inverse functions for the corresponding trigonometric functions.

Inverses; Polar Coordinates; Vectors **663**

1. **a.** Find all real values of

   y such that $\cos y = \dfrac{1}{\sqrt{2}}$.

   For $0 \leq y \leq 2\pi$, $\cos \dfrac{\pi}{4} =$

   $\dfrac{1}{\sqrt{2}}$ and $\cos \dfrac{7\pi}{4} = \dfrac{1}{\sqrt{2}}$.

   $\therefore y = \dfrac{\pi}{4}$ and $y = \dfrac{7\pi}{4}$

   and all the solutions
   are in the form

   $y = \dfrac{\pi}{4} + 2k\pi$ and

   $y = \dfrac{7\pi}{4} + 2k\pi, \ k \in J.$

   b. Find the real value of

   $\text{Cos}^{-1} \dfrac{1}{\sqrt{2}}$. For $\text{Cos}^{-1} x$

   the range is $0 \leq y \leq \pi$.

   $\therefore \text{Cos}^{-1} \dfrac{1}{\sqrt{2}} = \dfrac{\pi}{4}.$

2. **a.** Find all the real values
   of y to the nearest hundredth such that
   $\sin y = 0.3$.

   For $0 \leq y \leq \dfrac{\pi}{2}, y \approx 0.30$

   In Quadrant II, $\sin y$ is
   positive. $y \approx \pi - 0.30$
   $\approx 3.14 - 0.30 = 2.84.$
   $\therefore y \approx 0.30$ and
   $y \approx 2.84$ and all the solutions are in the form
   $y \approx 0.30 + 2k\pi$ and
   $y \approx 2.84 + 2k\pi.$

   b. Find the value of
   $\text{Sin}^{-1} \ 0.3$ for both the
   inverse circular and in-
   verse trigonometric
   function, correct to
   three significant digits.
   For $\text{Sin}^{-1} x$ the range is

   $-\dfrac{\pi}{2} \leq y \leq \dfrac{\pi}{2}$, or

   $-90° \leq m(\alpha) \leq 90°.$
   $\therefore \text{Sin}^{-1} \ 0.3 \approx 0.305$ or
   $\text{Sin}^{-1} \ 0.3 \approx 17.5°.$

EXAMPLE 1   **a.** Find all the real values of y such that $\cos y = -\dfrac{\sqrt{3}}{2}$.

   b. Find the real value of $\text{Cos}^{-1}\left(-\dfrac{\sqrt{3}}{2}\right)$.

SOLUTION   **a.** First, find the values of y such that $\cos y = -\dfrac{\sqrt{3}}{2}$ and $0 \leq y \leq 2\pi$. For

   this interval, $\cos \dfrac{5\pi}{6} = -\dfrac{\sqrt{3}}{2}$ and $\cos \dfrac{7\pi}{6} = -\dfrac{\sqrt{3}}{2}$. Thus, $y = \dfrac{5\pi}{6}$ and

   $y = \dfrac{7\pi}{6}$ and *all* the solutions are in the form

   $$y = \dfrac{5\pi}{6} + 2k\pi \text{ and } y = \dfrac{7\pi}{6} + 2k\pi, \ k \in J.$$

   b. For $\text{Cos}^{-1} x$ the range is $0 \leq y \leq \pi$.

   $$\therefore \text{Cos}^{-1}\left(-\dfrac{\sqrt{3}}{2}\right) = \dfrac{5\pi}{6}.$$

   Throughout the remainder of this chapter, the variable k *will always have {the integers} as its replacement set.*
   Sometimes you must use a table of function values or a calculator to approximate elements in the range of an inverse relation.

EXAMPLE 2   **a.** Find all the real values of y to the nearest hundredth such that $\sin y = 0.9$.
   b. Find the value of $\text{Sin}^{-1} \ 0.9$ for both the inverse circular and inverse trigonometric functions, correct to three significant digits.

SOLUTION   **a.** From Table 8, or a calculator, for the real values $0 \leq y \leq \dfrac{\pi}{2}, y \approx 1.12$.

   Since, in Quadrant II, $\sin y$ is positive,

   $$y \approx \pi - 1.12$$
   $$\approx 3.14 - 1.12$$
   $$\approx 2.02$$

   Thus, $y \approx 1.12$ and $y \approx 2.02$, and *all* solutions are in the form
   $$y \approx 1.12 + 2k\pi \quad \text{and} \quad y \approx 2.02 + 2k\pi.$$

   b. For $\text{Sin}^{-1} x$, the range is $-\dfrac{\pi}{2} \leq y \leq \dfrac{\pi}{2}$, or $-90° \leq m(\alpha) \leq 90°.$
   $$\therefore \text{Sin}^{-1} \ 0.9 \approx 1.12, \text{ or } \text{Sin}^{-1} \ 0.9 \approx 64.1°.$$

Thinking of principal-value inverses in terms of angles instead of real numbers gives the easiest way to obtain values for such compositions of functions as

$$\sin (\text{Cos}^{-1} x) \quad \text{and} \quad \text{Tan}^{-1}(\sin x).$$

For example, Figure 12 shows the angle function value $\alpha = \text{Cos}^{-1}\frac{2}{3}$. After using the Pythagorean Theorem to compute the length of side $\overline{AB}$ of $\triangle AOB$ to be $\sqrt{5}$, you can read from the diagram such values as $\sin\left(\text{Cos}^{-1}\frac{2}{3}\right) = \frac{\sqrt{5}}{3}$, $\tan\left(\text{Cos}^{-1}\frac{2}{3}\right) = \frac{\sqrt{5}}{2}$, and $\csc\left(\text{Cos}^{-1}\frac{2}{3}\right) = \frac{3}{\sqrt{5}}$.

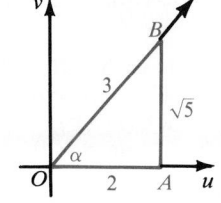

Figure 12

EXAMPLE 3   Simplify $\cos(\text{Sin}^{-1} a + \text{Cos}^{-1} b)$.

SOLUTION   To simplify write the expression as a variable expression in a and b. Using the sum formula for $\cos(\alpha_1 + \alpha_2)$ yields the following:

$$\cos(\text{Sin}^{-1} a + \text{Cos}^{-1} b) =$$
$$[\cos(\text{Sin}^{-1} a)][\cos(\text{Cos}^{-1} b)] - [\sin(\text{Sin}^{-1} a)][\sin(\text{Cos}^{-1} b)]$$

Now $\text{Sin}^{-1} a$ lies in Quadrant I or Quadrant IV, so that $\cos(\text{Sin}^{-1} a)$ is positive. Thus, $\text{Sin}^{-1} a$ need only be depicted as an angle in a right triangle as shown at the right and

$$\cos(\text{Sin}^{-1} a) = \sqrt{1 - a^2}.$$

Similarly, $\text{Cos}^{-1} b$ lies in Quadrant I or Quadrant II so that $\sin(\text{Cos}^{-1} b)$ is positive. Thus $\text{Cos}^{-1} b$ need only be depicted as an angle in a right triangle as shown at the right and

$$\sin(\text{Cos}^{-1} b) = \sqrt{1 - b^2}.$$

For every a in the domain of $\text{Sin}^{-1} a$, $\sin(\text{Sin}^{-1} a) = a$ and for every b in the domain of $\text{Cos}^{-1} b$, $\cos(\text{Cos}^{-1} b) = b$.

$$\therefore \cos(\text{Sin}^{-1} a + \text{Cos}^{-1} b) = \sqrt{1 - a^2} \cdot b - a \cdot \sqrt{1 - b^2}.$$

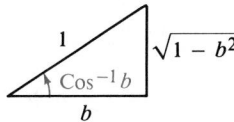

Oral Exercises

State the value of each of the following for both the inverse circular function and the inverse trigonometric function.

1. $\text{Cos}^{-1} 0$ $\frac{\pi}{2}$, $90°$

2. $\text{Cos}^{-1}\frac{1}{2}$ $\frac{\pi}{3}$, $60°$

3. $\text{Cos}^{-1}(-1)$ π, $180°$

4. $\text{Tan}^{-1} 1$ $\frac{\pi}{4}$, $45°$

5. $\text{Sin}^{-1}\frac{1}{\sqrt{2}}$ $\frac{\pi}{4}$, $45°$

6. $\text{Cos}^{-1}\left(-\frac{\sqrt{3}}{2}\right)$ $\frac{5\pi}{6}$, $150°$

7. $\text{Sin}^{-1}\left(-\frac{1}{\sqrt{2}}\right)$ $-\frac{\pi}{4}$, $-45°$

8. $\text{Tan}^{-1}(-1)$ $-\frac{\pi}{4}$, $-45°$

9. $\text{Sin}^{-1}\frac{\sqrt{3}}{2}$ $\frac{\pi}{3}$, $60°$

Inverses; Polar Coordinates; Vectors   **665**

3. Simplify:
$\sin(\text{Cos}^{-1} a + \text{Sin}^{-1} b)$
$\sin(\text{Cos}^{-1} a + \text{Sin}^{-1} b) =$
$[\sin(\text{Cos}^{-1} a)][\cos(\text{Sin}^{-1} b)] +$
$[\cos(\text{Cos}^{-1} a)][\sin(\text{Sin}^{-1} b)]$
$= \sqrt{1 - a^2}\sqrt{1 - b^2} + ab$
$= \sqrt{1 - b^2 - a^2 + a^2 b^2} + ab$

Mixed Review

1. Find $f^{-1}(x)$ when $y = 3x + 12$.

   $f^{-1}(x) = \frac{1}{3}x - 4$

2. What is the range for
   a. $\sin x$? $-1 \le \sin x \le 1$
   b. $\cos x$? $-1 \le \cos x \le 1$

3. As x increases from $0°$ to $90°$, what happens to the values of each trigonometric function?
   a. $\tan x$
      increases infinitely
   b. $\cos x$
      decreases from 1 to 0

Find the function values.

4. $\cos\frac{5\pi}{6}$ $-\frac{\sqrt{3}}{2}$

5. $\cot 45°$ 1

6. $\sin\frac{\pi}{4}$ $\frac{1}{\sqrt{2}}$

Suggested Assignments

Extended Alg. with Trig.
  666/1–47 odd
S 643/10–20 even
Enriched Alg. with Trig.
  666/5–51 odd
S 643/10–20 even

Additional A Exercises

Let α denote the degree measure of an angle and x denote a real number. Find all values of α or of x using an expression in which $k \in J$.

1. $\cos \alpha = \dfrac{\sqrt{3}}{2}$

  $\alpha = 30° + k \cdot 360°$ and
  $\alpha = 330° + k \cdot 360°$

2. $\sin \alpha = -\dfrac{1}{\sqrt{2}}$

  $\alpha = 225° + k \cdot 360°$ and
  $\alpha = 315° + k \cdot 360°$

3. $\sin x = -1 \quad x = \dfrac{3\pi}{2} + 2k\pi$

4. $\tan x = -1$

  $x = \dfrac{3\pi}{4} + 2k\pi$ and

  $x = \dfrac{7\pi}{4} + 2k\pi$

Find the value for both the inverse circular and inverse trigonometric functions.

5. $\text{Sin}^{-1}\dfrac{1}{2} \quad \dfrac{\pi}{6}; 30°$

6. $\text{Arctan } 1 \quad \dfrac{\pi}{4}; 45°$

7. $\text{Cos}^{-1} 0.5878 \quad 0.942; 54°$

Additional Answers
Written Exercises

3. $60° + k \cdot 360°, 300° + k \cdot 360°$

6. $30° + k \cdot 180°$

9. $\dfrac{4\pi}{3} + 2k\pi, \dfrac{5\pi}{3} + 2k\pi$

Written Exercises

In each of the following, α denotes the degree measure of an angle. Find all values of α using an expression in which $k \in J$.

A

1. $\sin \alpha = 0 \quad 0° + k \cdot 180°$

2. $\cos \alpha = 1 \quad 0° + k \cdot 360°$

3. $\cos \alpha = \dfrac{1}{2}$

4. $\sin \alpha = -\dfrac{1}{2}$

  $210° + k \cdot 360°, 330° + k \cdot 360°$

5. $\tan \alpha = 0 \quad 0° + k \cdot 180°$

6. $\tan \alpha = \dfrac{1}{\sqrt{3}}$

In each of the following, x denotes a real number. Find all values of x using an expression in which $k \in J$.

7. $\cos x = \dfrac{1}{\sqrt{2}} \quad \dfrac{\pi}{4} + 2k\pi; \dfrac{7\pi}{4} + 2k\pi$

8. $\sin x = 1 \quad \dfrac{\pi}{2} + 2k\pi$

9. $\sin x = -\dfrac{\sqrt{3}}{2}$

10. $\cos x = -\dfrac{1}{2}$

  $\dfrac{2\pi}{3} + 2k\pi, \dfrac{4\pi}{3} + 2k\pi$

11. $\tan x = -\dfrac{1}{\sqrt{3}} \quad \dfrac{5\pi}{6} + k\pi$

12. $\tan x = \sqrt{3} \quad \dfrac{\pi}{3} + k\pi$

Find the value of each of the following for both the inverse circular and the inverse trigonometric functions. Use Tables 6 and 8 or a calculator as needed.

13. $\text{Arcsin }\dfrac{\sqrt{3}}{2} \quad \dfrac{\pi}{3}, 60°$

14. $\text{Cos}^{-1}\left(-\dfrac{1}{\sqrt{2}}\right) \quad \dfrac{3\pi}{4}, 135°$

15. $\text{Arcsin }(-1) \quad -\dfrac{\pi}{2}, -90°$

16. $\text{Arccos }\dfrac{\sqrt{3}}{2} \quad \dfrac{\pi}{6}, 30°$

17. $\text{Tan}^{-1}(-\sqrt{3}) \quad -\dfrac{\pi}{3}, -60°$

18. $\text{Tan}^{-1}\dfrac{1}{\sqrt{3}} \quad \dfrac{\pi}{6}, 30°$

19. $\text{Sin}^{-1} 0.4067 \quad 0.42, 24.0°$

20. $\text{Cos}^{-1} 0.8090 \quad 0.63, 36.0°$

21. $\text{Arcsin }(-0.9898)$

22. $\text{Cos}^{-1}(-0.8910) \quad 2.7, 153°$

23. $\text{Tan}^{-1} 0.1584 \quad 0.16, 9°$

24. $\text{Arctan }(-3.078)$

21. $-1.43, -81.8°$
24. $-1.26, -72.0°$

Evaluate.

B

25. $\text{Sin}^{-1}\left(\cos \dfrac{\pi}{6}\right) \quad \dfrac{\pi}{3}$

26. $\text{Sin}^{-1}(\cos 180°) \quad -90°$

27. $\text{Cos}^{-1}(\sin 270°) \quad 180°$

28. $\text{Cos}^{-1}\left(\cos \dfrac{5\pi}{4}\right) \quad \dfrac{3\pi}{4}$

29. $\text{Sin}^{-1}\left(\sin \dfrac{5\pi}{4}\right) \quad -\dfrac{\pi}{4}$

30. $\text{Sin}^{-1}\left(\cos \dfrac{5\pi}{3}\right) \quad \dfrac{\pi}{6}$

31. $\text{Cos}^{-1}\left[\sin\left(-\dfrac{\pi}{6}\right)\right] \quad \dfrac{2\pi}{3}$

32. $\text{Tan}^{-1}(\tan 135°) \quad -45°$

33. $\text{Tan}^{-1}\left(\tan \dfrac{5\pi}{6}\right) \quad -\dfrac{\pi}{6}$

Use a diagram similar to Figure 12 on page 665 to find each of the following values.

34. $\sin\left(\text{Cos}^{-1}\dfrac{4}{5}\right) \quad \dfrac{3}{5}$

35. $\cos\left(\text{Sin}^{-1}\dfrac{12}{13}\right) \quad \dfrac{5}{13}$

36. $\tan\left(\text{Sin}^{-1}\dfrac{7}{25}\right) \quad \dfrac{7}{24}$

37. $\sin\left(\text{Tan}^{-1}\dfrac{15}{8}\right) \quad \dfrac{15}{17}$

38. $\cos\left(\text{Tan}^{-1}\dfrac{\sqrt{7}}{3}\right) \quad \dfrac{3}{4}$

39. $\tan\left(\text{Sin}^{-1}\dfrac{\sqrt{15}}{4}\right) \quad \sqrt{15}$

40. $\sin\left(\text{Cos}^{-1}\dfrac{2\sqrt{2}}{3}\right) \quad \dfrac{1}{3}$

41. $\cos\left[\text{Sin}^{-1}\left(-\dfrac{2\sqrt{6}}{5}\right)\right] \quad \dfrac{1}{5}$

Simplify.

42. $\cos \left(\text{Sin}^{-1} \dfrac{3}{5} + \text{Cos}^{-1} \dfrac{24}{25} \right)$ $\frac{3}{5}$

43. $\sin \left(\text{Sin}^{-1} \dfrac{4}{5} + \text{Cos}^{-1} \dfrac{1}{2} \right)$ $\dfrac{4 + 3\sqrt{3}}{10}$

44. $\sin \left(\text{Sin}^{-1} \dfrac{1}{\sqrt{2}} + \text{Tan}^{-1} \dfrac{3}{4} \right)$ $\frac{7}{5\sqrt{2}}$

45. $\cos \left(\text{Sin}^{-1} \dfrac{5}{13} - \text{Cos}^{-1} \dfrac{4}{5} \right)$ $\frac{63}{65}$

46. $\cos \left(2 \, \text{Sin}^{-1} \dfrac{7}{25} \right)$ $\frac{527}{625}$

47. $\sin \left(2 \, \text{Cos}^{-1} \dfrac{\sqrt{2}}{2} \right)$ 1

C 48. **a.** Is it true that if $0 \le x \le 1$, $\text{Sin}^{-1}(-x) = -\text{Sin}^{-1} x$? Yes
   b. Use part (a) above to show that if $0 \le x \le 1$, then $\cos (\text{Sin}^{-1}(-x)) = \cos (\text{Sin}^{-1} x)$.

49. **a.** Is it true for inverse circular functions that if $0 \le x \le 1$, $\text{Cos}^{-1}(-x) = \pi - \text{Cos}^{-1} x$? Yes
   b. Use part (a) above to show that if $0 \le x \le 1$, then $\sin (\text{Cos}^{-1}(-x)) = \sin (\text{Cos}^{-1} x)$.

Prove the following over the indicated domains.

50. $\text{Sin}^{-1} x + \text{Cos}^{-1} x = \dfrac{\pi}{2}$; $0 \le x \le 1$

51. $\text{Tan}^{-1} x + \text{Cot}^{-1} x = \dfrac{\pi}{2}$; $x \ge 0$

16–2 Equations Involving Circular and Trigonometric Functions

The equation $\cos x = c$, where $c \in \mathcal{R}$ is a constant, either has an empty solution set (if $|c| > 1$), or a solution set containing infinitely many members. For example, the solution set over $\mathcal{R}$ of $\cos x = -\dfrac{\sqrt{3}}{2}$ is

$$\left\{ x : x = \dfrac{5\pi}{6} + 2k\pi \right\} \cup \left\{ x : x = \dfrac{7\pi}{6} + 2k\pi \right\} \text{ where } k \in J.$$

The solution set of $\cos \alpha = -\dfrac{\sqrt{3}}{2}$ where α is an angle measured in degrees is $\{ \alpha : m(\alpha) = 150° + k \cdot 360° \} \cup \{ \alpha : m(\alpha) = 210° + k \cdot 360° \}$ where $k \in J$.

In either case, such a solution set is referred to as the general solution of the equation. The set of solutions in a specified interval is called the particular solution. The solutions of the equation $\cos x = -\dfrac{\sqrt{3}}{2}$ in the interval $0 \le x < 2\pi$ are $\dfrac{5\pi}{6}$ and $\dfrac{7\pi}{6}$ and the solution set is $\left\{ \dfrac{5\pi}{6}, \dfrac{7\pi}{6} \right\}$. In the interval $0° \le m(\alpha) < 360°$, the solutions of $\cos \alpha = -\dfrac{\sqrt{3}}{2}$ are $150°$ and $210°$ and the solution set is $\{ \alpha : m(\alpha) = 150° \} \cup \{ \alpha : m(\alpha) = 210° \}$. In this case we shall use the abbreviated notation $\{150°, 210°\}$. (Compare this with the abbreviated notation introduced in Section 14-4.)

Inverses; Polar Coordinates; Vectors   **667**

48. **b.** $\cos (\text{Sin}^{-1}(-x)) = \cos (-\text{Sin}^{-1}(x)) = \cos (\text{Sin}^{-1}(x))$

49. **b.** $\sin (\text{Cos}^{-1}(-x)) = \sin (\pi - \text{Cos}^{-1} x) = \sin \pi \cos (\text{Cos}^{-1} x) - \cos \pi \sin (\text{Cos}^{-1} x) = \sin (\text{Cos}^{-1} x)$

Teaching Suggestions
p. T124

Supplementary Materials
Test 51

Key Ideas
Solve trigonometric equations over $\mathcal{R}$ for both the general solution and the particular solution.

Common Errors
When solving equations, students frequently simplify the equation by dividing both sides by a variable expression. Remind them that some solutions may be lost when an expression containing a variable is cancelled. Two alternate methods of simplifying are transforming the equation so that one side is 0 and using trigonometric identities to substitute for an expression.

1. a. Find the general solution over $\mathcal{R}$ of
$3 \sin x \tan x = \sqrt{3} \sin x$.
$3 \sin x \tan x - \sqrt{3} \sin x = 0$
$\sin x(3 \tan x - \sqrt{3}) = 0$
$\sin x = 0$ or $\tan x = \dfrac{\sqrt{3}}{3}$;
the general solution is
$\left\{ 0 + 2k\pi, \pi + 2k\pi, \dfrac{\pi}{6} + 2k\pi, \dfrac{7\pi}{6} + 2k\pi, k \in J \right\}.$

b. Find the particular solution of $3 \sin x \tan x = \sqrt{3} \sin x$ in the interval $0 \le x < 2\pi$.
Let $k = 0$. ∴ the particular solution in the given interval is
$\left\{ 0, \pi, \dfrac{\pi}{6}, \dfrac{7\pi}{6} \right\}.$

2. Find the particular solution of $2 \sin^2 \alpha = \cos \alpha(\sec \alpha - \tan \alpha)$ in the interval $0° \le m(\alpha) < 360°$.
$2 \sin^2 \alpha = \cos \alpha (\sec \alpha - \tan \alpha)$
$2 \sin^2 \alpha = \cos \alpha \left(\dfrac{1}{\cos \alpha} - \dfrac{\sin \alpha}{\cos \alpha} \right)$
$2 \sin^2 \alpha = 1 - \sin \alpha$
$2 \sin^2 \alpha + \sin \alpha - 1 = 0$
$(2 \sin \alpha - 1)(\sin \alpha + 1) = 0$
$\sin \alpha = \dfrac{1}{2}$ or $\sin \alpha = -1$;
The solution set is
$\{30°, 150°, 270°\}$.

3. Solve $\sin 2\alpha = -2 \sin \alpha$ in the interval $0° \le m(\alpha) < 360°$.
$\sin 2\alpha = -2 \sin \alpha$
$2 \sin \alpha \cos \alpha = -2 \sin \alpha$
$2 \sin \alpha \cos \alpha + 2 \sin \alpha = 0$
$2 \sin \alpha(\cos \alpha + 1) = 0$
$\sin \alpha = 0$ or $\cos \alpha = -1$;
The solution set is
$\{0°, 180°\}$.

EXAMPLE 1   **a.** Find the general solution over $\mathcal{R}$ of $2 \sin x \cos x = \cos x$.
b. Find the particular solution of $2 \sin x \cos x = \cos x$ in the interval $0 \le x < 2\pi$.

SOLUTION   **a.** Do not divide each side by $\cos x$, since the solution resulting from $\cos x = 0$ will be lost. Rather, transform the equation so that one side is 0, and solve.

$$2 \sin x \cos x = \cos x$$
$$2 \sin x \cos x - \cos x = 0$$
$$(\cos x)(2 \sin x - 1) = 0$$

$\cos x = 0$	$2 \sin x - 1 = 0$
	$\sin x = \dfrac{1}{2}$

If $\cos x = 0$, then $x = \dfrac{\pi}{2} + 2k\pi$ and $x = \dfrac{3\pi}{2} + 2k\pi$. If $\sin x = \dfrac{1}{2}$, then $x = \dfrac{\pi}{6} + 2k\pi$ and $x = \dfrac{5\pi}{6} + 2k\pi$.

∴ the general solution is $\left\{ \dfrac{\pi}{2} + 2k\pi, \dfrac{3\pi}{2} + 2k\pi, \dfrac{\pi}{6} + 2k\pi, \dfrac{5\pi}{6} + 2k\pi, k \in J \right\}.$

b. The particular solution can be obtained by substituting $k = 0$.

∴ the solution set in the interval $0 \le x < 2\pi$ is $\left\{ \dfrac{\pi}{2}, \dfrac{3\pi}{2}, \dfrac{\pi}{6}, \dfrac{5\pi}{6} \right\}.$

Sometimes it is necessary to transform an equation into an equation involving a single function, as the following example illustrates.

EXAMPLE 2   Find the particular solution of $3 \sin^2 \alpha - 8 \cos \alpha = 0$ in the interval $0° \le m(\alpha) < 360°$.

SOLUTION   Begin by substituting $1 - \cos^2 \alpha$ for $\sin^2 \alpha$ in the original equation to obtain $3(1 - \cos^2 \alpha) - 8 \cos \alpha = 0$. Simplifying and factoring the left side yields

$$(3 \cos \alpha - 1)(\cos \alpha + 3) = 0$$

which is equivalent to

$$\cos \alpha = \dfrac{1}{3} \quad \text{or} \quad \cos \alpha = -3.$$

Since $|\cos \alpha| \le 1$ for all values of α, reject $\cos \alpha = -3$. If $\cos \alpha = \dfrac{1}{3}$, use Table 6 or a calculator to find that $\alpha = 70.5°$. In Quadrant IV $\cos \alpha > 0$. Thus $m(\alpha) = 360° - 70.5° = 289.5°$.

∴ the solution set is $\{70.5°, 289.5°\}$.

EXAMPLE 3 Solve $\tan^2 2\alpha - \tan 2\alpha = 0$ in the interval $0° \le m(\alpha) < 360°$.

SOLUTION The equation has the form of an equation in the single variable, $\tan 2\alpha$. Solve directly for the quantity 2α in the interval $0° \le m(2\alpha) < 720°$. Then solve for α in the interval $0° \le m(\alpha) < 360°$.

$$\tan^2 2\alpha - \tan 2\alpha = 0$$
$$\tan 2\alpha\,(\tan 2\alpha - 1) = 0$$
$$\tan 2\alpha = 0 \quad \text{or} \quad \tan 2\alpha = 1$$

If $\tan 2\alpha = 0$, then

$$m(2\alpha) = 0°,\ 180°,\ 360°,\ \text{or } 540°,\ \text{and}$$
$$m(\alpha) = 0°,\ 90°,\ 180°,\ \text{or } 270°.$$

If $\tan 2\alpha = 1$, then

$$m(2\alpha) = 45°,\ 225°,\ 405°,\ \text{or } 585°,\ \text{and}$$
$$m(\alpha) = 22.5°,\ 112.5°,\ 202.5°,\ \text{or } 292.5°.$$

Note that all the solutions are in the interval $0° \le m(\alpha) < 360°$.

∴ the solution set is $\{0°, 22.5°, 90°, 112.5°, 180°, 202.5°, 270°, 292.5°\}$.

1.a. $\{60° + k \cdot 360°, 120° + k \cdot 360°\}$ **b.** $\{60°, 120°\}$
2.a. $\{120° + k \cdot 360°, 240° + k \cdot 360°\}$ **b.** $\{120°, 240°\}$
3.a. $\{135° + k \cdot 360°, 225° + k \cdot 360°\}$ **b.** $\{135°, 225°\}$

Written Exercises
4.a. $\{45° + k \cdot 180°\}$ **b.** $\{45°, 225°\}$
5.a. $\{120° + k \cdot 360°, 240° + k \cdot 360°\}$ **b.** $\{120°, 240°\}$

a. Find the general solution of each equation for $m(\alpha)$ in degrees.
b. Find the particular solution in the interval $0° \le m(\alpha) < 360°$.

A
1. $2 \sin \alpha - \sqrt{3} = 0$
2. $2 \cos \alpha + 1 = 0$
3. $\sqrt{2} \cos \alpha + 1 = 0$
4. $\sin \alpha = \cos \alpha$
5. $\sec \alpha + 2 = 0$
6. $2 \sin^2 \alpha - 1 = 0$
7. $4 \cos^2 \alpha - 3 = 0$ **a.** $\{30° + k \cdot 180°, 150° + k \cdot 180°\}$ **8.** $1 - 3 \tan^2 \alpha = 0$
   b. $\{30°, 150°, 210°, 330°\}$ **a.** $\{30° + k \cdot 180°, 150° + k \cdot 180°\}$

a. Find the general solution of each equation for $x \in \mathcal{R}$. **b.** $\{30°, 150°, 210°, 330°\}$
b. Find the particular solution in the interval $0 \le x < 2\pi$.

9. $\sin x + \cos x = 0$ **a.** $\{\frac{3\pi}{4} + k\pi\}$ **b.** $\{\frac{3\pi}{4}, \frac{7\pi}{4}\}$ **10.** $3 \sin^2 x = \cos^2 x$
11. $\sin x = \sin 2x$ **12.** $\cos x = -\sin 2x$
13. $\cos 2x + 1 = 0$ **a.** $\{\frac{\pi}{2} + k\pi\}$ **b.** $\{\frac{\pi}{2}, \frac{3\pi}{2}\}$ **14.** $\cos 2x = 2 \sin^2 x - 1$
   a. $\{\frac{\pi}{4} + \frac{k\pi}{2}\}$ **b.** $\{\frac{\pi}{4}, \frac{3\pi}{4}, \frac{5\pi}{4}, \frac{7\pi}{4}\}$

Find the particular solution set for $0° \le m(\alpha) < 360°$. Use Table 6 or 7 or a calculator as needed. Give your answer to the nearest tenth of a degree or nearest 10'.

15. $2 \sin^2 \alpha + \sin \alpha = 1$ $\{30°, 150°, 270°\}$ **16.** $-\tan \alpha = \tan^2 \alpha$
17. $3 \cos \alpha = 2 \sin^2 \alpha$ $\{60°, 300°\}$ **18.** $\cos 2\alpha = -\cos \alpha$
19. $\sin 2\alpha = \sin^2 2\alpha$ $\{0°, 45°, 90°, 180°, 225°, 270°\}$ **20.** $2 \cos^2 2\alpha = 1$

Inverses; Polar Coordinates; Vectors **669**

Suggested Assignments
Extended Alg. with Trig.
  669/1–33 odd
R 670/*Self-Test 1*
Enriched Alg. with Trig.
  669/9–33 odd, 35–38
R 670/*Self-Test 1*

Additional A Exercises

a. Find the general solution for $m(\alpha)$ in degrees or for $x \in \mathcal{R}$.

b. Find the particular solution for $0° \le m(\alpha) < 360°$ or for $0 \le x < 2\pi$.

1. $2 \cos^2 \alpha - 1 = 0$
   a. $\{45° + k \cdot 90°\}$
   b. $\{45°, 135°, 225°, 315°\}$

2. $3 \csc^2 \alpha = 4$
   a. $\{60° + k \cdot 180°, 120° + k \cdot 180°\}$
   b. $\{60°, 120°, 240°, 300°\}$

3. $\cot^2 2\alpha - 3 = 0$
   a. $\{15° + k \cdot 90°, 75° + k \cdot 90°\}$
   b. $\{15°, 75°, 105°, 165°, 195°, 255°, 285°, 345°\}$

4. $\sec x - 2 \cos x = 1$
   a. $\left\{\pi + 2k\pi, \frac{\pi}{3} + 2k\pi, \frac{5\pi}{3} + 2k\pi\right\}$
   b. $\left\{\pi, \frac{\pi}{3}, \frac{5\pi}{3}\right\}$

5. $4 \tan^2 x = \sec^2 x$
   a. $\left\{\frac{\pi}{6} + k\pi, \frac{5\pi}{6} + k\pi\right\}$
   b. $\left\{\frac{\pi}{6}, \frac{5\pi}{6}, \frac{7\pi}{6}, \frac{11\pi}{6}\right\}$

Additional Answers
Written Exercises

6. a. $\{45° + k \cdot 90°\}$
   b. $\{45°, 135°, 225°, 315°\}$

10. a. $\left\{\frac{\pi}{6} + k\pi, \frac{5\pi}{6} + k\pi\right\}$

(continued)

669

Quick Quiz

1. Find all the real values of x such that $\cos x = \frac{1}{2}$.

$\frac{\pi}{3} + 2k\pi, \frac{5\pi}{3} + 2k\pi$

2. Find the real value of $\text{Sin}^{-1}\, 0.6816$.   0.75

3. Find the value for the inverse trigonometric function $\text{Cos}^{-1} \frac{\sqrt{3}}{2}$. $\frac{\pi}{6}$, or 30°

4. Solve $\sin 2x = -\sin x$ over $\mathcal{R}$.

$\left\{0 + 2k\pi, \frac{2\pi}{3} + 2k\pi, \right.$ $\left. \pi + 2k\pi, \frac{4\pi}{3} + 2k\pi\right\}$

5. Solve $3 \cot^2 \alpha = 1$ in the interval $90° \leq m(\alpha) \leq 270°$.   240°

Find the particular solution set for $0° \leq m(\alpha) < 360°$ or for the interval $0 \leq x < 2\pi$. Use a calculator or table as needed.

21. $8 \sin^2 x = 6 \sin x - 1$ $\{0.25, \frac{\pi}{6}, \frac{5\pi}{6}, 2.89\}$

22. $\sin x + \csc x = 2$ $\{\frac{\pi}{2}\}$

23. $3 \cos 2x = 2 \cos^2 x$

24. $2 \sin x \sin 2x = 3 \cos x$ $\{\frac{\pi}{3}, \frac{\pi}{2}, \frac{2\pi}{3}, \frac{4\pi}{3}, \frac{3\pi}{2}, \frac{5\pi}{3}\}$

25. $\tan^2 x = 3 - \sec^2 x$ $\{\frac{\pi}{4}, \frac{3\pi}{4}, \frac{5\pi}{4}, \frac{7\pi}{4}\}$

26. $2 \cos^2 \alpha - \sin \alpha = 1$ $\{30°, 150°, 270°\}$

B 27. $(\sin x - \cos x)^2 = \frac{1}{2}$ $\{\frac{\pi}{12}, \frac{5\pi}{12}, \frac{13\pi}{12}, \frac{17\pi}{12}\}$

28. $\tan^2 x - 1 = \sec x$

29. $3 \sin 2x = 2 \cos^2 2x$ $\{\frac{\pi}{12}, \frac{5\pi}{12}, \frac{13\pi}{12}, \frac{17\pi}{12}\}$

30. $2 \cos^2 2x = 5 \sin 2x - 1$

31. $(\sin \alpha + \cos \alpha)^2 = 2 \sin 2\alpha$ $\{45°, 225°\}$

32. $\cos 2x - 3 \sin x - 2 = 0$

33. $\cos 2x = \sqrt{3} \cos x - 1$ $\{\frac{\pi}{6}, \frac{\pi}{2}, \frac{3\pi}{2}, \frac{11\pi}{6}\}$

34. $8 \sin^4 \alpha - 10 \sin^2 \alpha + 3 = 0$ $\{45°, 60°, 120°, 135°, 225°, 240°, 300°, 315°\}$

C 35. $\sqrt{2 - 2 \cos \alpha} = 2 \cos \frac{\alpha}{2}$ $\{90°\}$

36. $\cot \alpha - \tan \alpha = 2$ $\{22.5°, 112.5°, 202.5°, 292.5°\}$

37. $\sqrt{3} \cos^2 \alpha - 2 \sin \alpha \cos \alpha = \sqrt{3} \sin^2 \alpha$ $\{30°, 120°, 210°, 300°\}$

38. $\frac{\cos 3x}{\sin x} + \frac{\sin 3x}{\cos x} = 2\sqrt{3}$ $\{\frac{\pi}{12}, \frac{7\pi}{12}, \frac{13\pi}{12}, \frac{19\pi}{12}\}$

23. $\{\frac{\pi}{6}, \frac{5\pi}{6}, \frac{7\pi}{6}, \frac{11\pi}{6}\}$ **28.** $\{\frac{\pi}{3}, \pi, \frac{5\pi}{3}\}$ **30.** $\{\frac{\pi}{12}, \frac{5\pi}{12}, \frac{13\pi}{12}, \frac{17\pi}{12}\}$ **32.** $\{\frac{7\pi}{6}, \frac{3\pi}{2}, \frac{11\pi}{6}\}$

Self-Test 1

VOCABULARY $\sin^{-1} x$ (p. 661)
arcsin x (p. 661)
$\cos^{-1} x$ (p. 662)
$\tan^{-1} x$ (p. 662)
$\cot^{-1} x$ (p. 662)
$\sec^{-1} x$ (p. 662)
$\csc^{-1} x$ (p. 662)
principal-value inverse function for sine (p. 662)
principal-value inverse function for cosine (p. 663)
general solution (p. 667)
particular solution (p. 667)

1. Find all the real values of y such that   *Obj. 1, p. 661*

$$\sin y = -\frac{1}{\sqrt{2}} \cdot \frac{5\pi}{4} + 2k\pi, \frac{7\pi}{4} + 2k\pi$$

2. Find the real value of $\text{Cos}^{-1}\, 0.8678$ 0.52

3. Find the value for the inverse trigonometric function of $\text{Tan}^{-1}(-\sqrt{3})$. −60°

4. Solve $\cos 2x + 1 = \cos x$ over $\mathcal{R}$.   *Obj. 2, p. 661*

5. Solve $4 \sin^2 \alpha - 3 = 0$ in the interval $0° \leq m(\alpha) \leq 180°$. $\{60°, 120°\}$

4. $\{\frac{\pi}{2} + 2k\pi, \frac{\pi}{3} + 2k\pi, \frac{3\pi}{2} + 2k\pi, \frac{5\pi}{3} + 2k\pi\}$

Check your answers with those at the back of the book.

Careers

Mining Engineering

Minerals are an important source of raw materials. They provide products used in various branches of industry. Mining engineers are responsible for locating, extracting, and processing minerals and rocks from the earth.

Before removing the ore from the earth, mining engineers must decide what method of mining will be the most economical for the company, the most effective for the mineral, and the least destructive for the environment. Some engineers supervise the maintenance and repair of equipment during the extraction of minerals, while others design improved mining machinery and develop new methods of removing minerals from the earth.

Mixed Review

1. Solve $2x^2 - x = 6$ over $\mathcal{R}$.

$$\left\{-\frac{3}{2}, 2\right\}$$

2. If α denotes the degree measure of an angle, find all the values of α such that $2 \sin \alpha = \sqrt{3}$. Use an expression in which $k \in J$.
$\alpha = 60° + k \cdot 360°$ and
$\alpha = 120° + k \cdot 360°, k \in J$

Complete the identity.

3. $\sin 2\alpha = \underline{?}$ $2 \sin \alpha \cos \alpha$
4. $\cos^2 \alpha = \underline{?}$ $1 - \sin^2 \alpha$
5. Factor $3n^2 - 1$ over $\mathcal{R}$.
$(\sqrt{3}n - 1)(\sqrt{3}n + 1)$

EXAMPLE   If a vein of copper or silver ore is near enough to the surface of the earth so that oxygen comes in contact with the ore, a chemical reaction occurs. During this chemical reaction any copper or silver in the ore dissolves and settles down to the level of the water table. Here the dissolved solution combines with existing sulfides to form more concentrated deposits of copper and silver. Engineers find a horizontal vein of ore on the face of a cliff and find a second vein at a 35° angle from the first. The water table is 30 m below the opening of the veins. How far along the top of the cliff should the miners sink the shaft to reach the richest deposits of copper and silver?

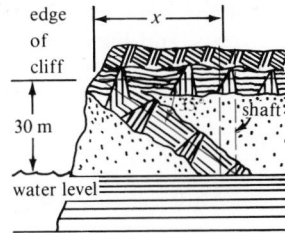

SOLUTION   The proposed shaft should be sunk to where the water level and the second vein meet. Let x be the distance from the edge of the cliff to the proposed shaft. Then:

$$\tan 35° = \frac{30}{x}$$

$$0.7002 \approx \frac{30}{x}$$

$$x \approx 42.8$$

∴ the miners should sink the shaft about 43 m from the edge of the cliff.

Inverses; Polar Coordinates; Vectors   **671**

Polar Coordinates

OBJECTIVES for Sections 16-3 through 16-5:
1. *To express the coordinates of a point in Cartesian or polar form when it is given in the other form.*
2. *To use De Moivre's theorem to find powers and roots of complex numbers.*

Teaching Suggestions
p. T125

Related Activities
p. T125

Key Ideas
Define and use the polar coordinate system. Convert polar coordinates to Cartesian coordinates. Convert Cartesian coordinates to polar coordinates. Graph polar equations.

Reading Algebra
In using the formulas that enable students to convert a pair of Cartesian coordinates into polar coordinates, students should realize that the values for sine and cosine can be either positive or negative depending on the value of *r* that is chosen. Thus, for the same point, four different pairs of polar coordinates can result.

16–3 Polar Coordinates; Polar Graphs

In Section 3-2 you learned that each point in a plane can be paired uniquely with an ordered pair of real numbers (x, y) in a given Cartesian coordinate system for the plane. Another coordinate system can be developed as follows.

Each point in the plane lies on a ray with initial point at the origin. If r represents the distance from the origin to a point P, and if an angle having the nonnegative x-axis as its initial side and the ray $\overrightarrow{OP}$ as its terminal side is denoted by θ, then the ordered pair

$$(r, m(\theta))$$

specifies the location of P in the plane (Figure 13). The components of the ordered pair $(r, m(\theta))$, usually written as simply (r, θ), are called a pair of **polar coordinates** of P. The nonnegative x-axis is then called the **polar axis**, and the origin is called the **pole**.

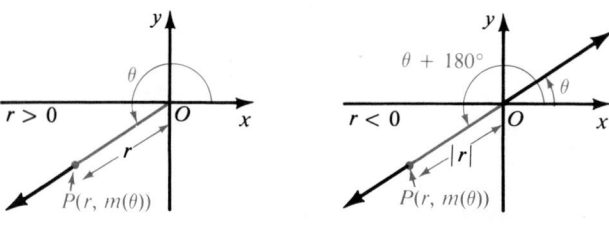

Figure 13   **Figure 14**

If r is negative, then we measure the distance $|r|$ along the extension of the terminal side of θ through the origin (Figure 14). For example, $(-3, 30°)$ is a pair of polar coordinates for the point P which also has coordinates $(3, 210°)$.

In this system, since θ is coterminal with infinitely many angles having the polar axis as initial side, the location of P can also be given, for example, by any of the ordered pairs

$$(r, \theta + 2k\pi) \quad \text{or} \quad (r, \theta + k \cdot 360°).$$

If $r = 0$, then *any* value might be assigned to $m(\theta)$. Thus the pole might, for example, be assigned polar coordinates $(0, 0°)$, $(0, 45°)$, or $(0, \frac{1}{2}\pi)$.

672 *Chapter 16*

In general:

> In a system of polar coordinates, each ordered pair (r, θ) can be associated with one and only one point of the plane, but each point of the plane may be associated with any number of ordered pairs (r, θ).

EXAMPLE 1 List all other polar coordinates of $P(2, 150°)$ for which $-360° \le m(\theta) \le 360°$.

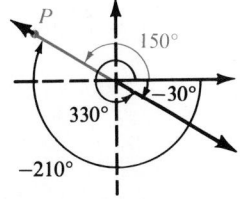

SOLUTION Sketch the graph of $P(2, 150°)$. By inspection, P has the additional coordinates

$$(-2, 330°), \quad (-2, -30°), \quad (2, -210°).$$

Since we have defined the polar axis of a polar coordinate system to coincide with the nonnegative x-axis of a Cartesian coordinate system, the polar and Cartesian coordinates are related as shown in Figure 15.

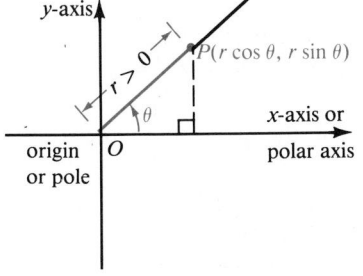

a   **Figure 15**   b

If $r > 0$ (Figure 15a), then $(x, y) = (r \cos \theta, r \sin \theta)$.
If $r = 0$, $(x, y) = (r \cos \theta, r \sin \theta) = (0, 0)$.
If $r < 0$ (Figure 15b), then

$$\begin{aligned}(x, y) &= (|r| \cos (\theta + 180°), |r| \sin (\theta + 180°)) \\ &= (-|r| \cos \theta, -|r| \sin \theta) \\ &= (r \cos \theta, r \sin \theta).\end{aligned}$$

In general:

> The polar and Cartesian coordinates of any point P are related by:
> $$\left.\begin{aligned} x &= r \cos \theta \\ y &= r \sin \theta \end{aligned}\right\} \tag{1}$$

Inverses; Polar Coordinates; Vectors **673**

These equations can be used to find Cartesian coordinates for a point whose polar coordinates are given.

EXAMPLE 2 Find the Cartesian coordinates of the point P with polar coordinates $(8, -60°)$.

SOLUTION Use Equations (1) to obtain the following:

$$x = r \cos \theta = 8 \cos (-60°) = 8\left(\frac{1}{2}\right) = 4$$

$$y = r \sin \theta = 8 \sin (-60°) = 8\left(-\frac{\sqrt{3}}{2}\right) = -4\sqrt{3}.$$

$\therefore$ the required Cartesian coordinates are $(4, -4\sqrt{3})$.

Given the Cartesian coordinates (x, y) of a point other than the origin, a pair of polar coordinates of the point can be found from the equations:

$$\sin \theta = \frac{y}{\pm\sqrt{x^2 + y^2}}, \qquad \cos \theta = \frac{x}{\pm\sqrt{x^2 + y^2}}, \qquad (2)$$

$$r = \pm\sqrt{x^2 + y^2}$$

EXAMPLE 3 Find a pair of polar coordinates of the point P whose Cartesian coordinates are $\left(-\frac{5\sqrt{3}}{2}, \frac{5}{2}\right)$.

SOLUTION From Equations (2), use $r = \sqrt{x^2 + y^2}$, to obtain

$$r = \sqrt{\left(-\frac{5\sqrt{3}}{2}\right)^2 + \left(\frac{5}{2}\right)^2} = \sqrt{\frac{75}{4} + \frac{25}{4}} = \sqrt{\frac{100}{4}} = \sqrt{25} = 5.$$

Since $\cos \theta = \dfrac{-\frac{5\sqrt{3}}{2}}{5} = -\dfrac{\sqrt{3}}{2}$ and $\sin \theta = \dfrac{\frac{5}{2}}{5} = \dfrac{1}{2}$, by inspection it can be seen that θ can be an angle of measure $150°$.

By choosing $r = -\sqrt{x^2 + y^2}$ another pair of polar coordinates can be obtained, namely $(-5, 330°)$.

$\therefore$ a pair of polar coordinates of P is $(5, 150°)$, or $(-5, 330°)$.

When an equation is written in terms of r and θ, it is called a *polar equation*. The set of points that satisfy polar equations such as

$$r = 4 \cos \theta, \quad r = 2, \quad r \sin \theta = 2, \quad \text{and} \quad \theta = 30°$$

are ordered pairs of the form (r, θ). The graph of the set of all solutions of such an equation is called the graph of the equation.

674 *Chapter 16*

Mixed Review

Use the distance formula to find the length of the segment from the origin to the point with the given coordinates.

1. $(8, 15)$ 17
2. $(2\sqrt{2}, -2\sqrt{7})$ 6
3. $\left(-\frac{8}{5}, -\frac{6}{5}\right)$ 2

4. Find all the values of α $(0° \le m(\alpha) \le 360°)$ for

$\cos \alpha = -\dfrac{1}{\sqrt{2}}$.   135°, 225°

5. Complete the table.

θ	45°	150°	270°
$\sin \theta$	$\dfrac{1}{\sqrt{2}}$	$\dfrac{1}{2}$	-1

Suggested Assignments

Extended Alg. with Trig.
  675/1–33 odd
S 600/4–12 even
Enriched Alg. with Trig.
  675/5–33 odd, 35–37
S 600/4–12 even

EXAMPLE 4 Sketch the graph of $r = 6 \cos \theta$.

SOLUTION The table shows selected solutions of the equation for $0° \leq m(\theta) \leq 360°$.

θ	0°	30°	45°	60°	90°	120°	135°	150°	180°
r	6	$3\sqrt{3}$	$3\sqrt{2}$	3	0	-3	$-3\sqrt{2}$	$-3\sqrt{3}$	-6
θ	210°		225°	240°	270°	300°	315°	330°	360°
r	$-3\sqrt{3}$		$-3\sqrt{2}$	-3	0	3	$3\sqrt{2}$	$3\sqrt{3}$	6

If you graph all these solutions in succession, you will twice graph the points in the figure at the left (below). That is, the points associated with

$$0° \leq m(\theta) \leq 180°$$

are the same as those associated with $180° \leq m(\theta) \leq 360°$.

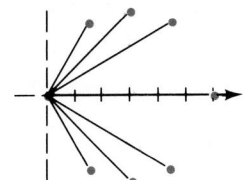 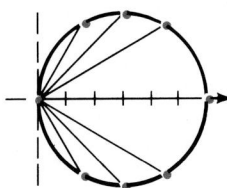

Connecting the points with a smooth curve yields the graph shown at the right above. You can verify (Exercise 27, page 676) that the graph is a circle.

Answers may vary
1. $(6\sqrt{2}, 45°)$, $(-6\sqrt{2}, -135°)$ 2. $(2, 120°)$, $(-2, -60°)$ 3. $(-8, 45°)$, $(8, -135°)$
4. $(4, -60°)$, $(-4, 120°)$ 5. $(5, 135°)$, $(-5, -45°)$ 6. $(2\sqrt{2}, -30°)$, $(-2\sqrt{2}, 150°)$
7. $(-5, 36.9°)$, $(5, -143.1°)$ 8. $(13, 112.6°)$, $(-13, -67.4°)$

Oral Exercises

For each point state two pairs of polar coordinates different from those given and such that (a) the first pair has $r > 0$, and (b) the second pair has $r < 0$.

1. $(4, -30°)$
   a. $(4, 330°)$
   b. $(-4, 150°)$

2. $(5, 135°)$
   a. $(5, -225°)$
   b. $(-5, -45°)$

3. $(-7, -45°)$
   a. $(7, 135°)$
   b. $(-7, 315°)$

4. $(9, 210°)$
   a. $(9, -150°)$
   b. $(-9, 30°)$

Written Exercises

Find a pair of polar coordinates with $-180° < m(\theta) \leq 180°$ for the point whose Cartesian coordinates are given. Graph the point in the plane.

A 1. $(6, 6)$
   5. $\left(\dfrac{-5\sqrt{2}}{2}, \dfrac{5\sqrt{2}}{2}\right)$

2. $(-1, \sqrt{3})$
6. $(\sqrt{6}, -\sqrt{2})$

3. $(-4\sqrt{2}, -4\sqrt{2})$
7. $(-4, -3)$

4. $(2, -2\sqrt{3})$
8. $(-5, 12)$

Inverses; Polar Coordinates; Vectors **675**

Additional A Exercises
Find a pair of polar coordinates, with $-180° < m(\theta) \leq 180°$, for the point whose Cartesian coordinates are given.
1. $(2, -2)$ $(2\sqrt{2}, -45°)$
2. $(-5\sqrt{3}, 5)$ $(10, 150°)$
3. $(0, 9)$ $(9, 90°)$
Find the Cartesian coordinates of the given point.
4. $(4\sqrt{2}, 135°)$ $(-4, 4)$
5. $(6, 240°)$ $(-3, -3\sqrt{3})$
6. $(2\sqrt{3}, 60°)$ $(\sqrt{3}, 3)$

Additional Answers
Written Exercises

24. a.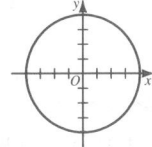

    b. $x^2 + y^2 = 16$

26. a.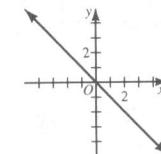

    b. $y = -x$

28. a.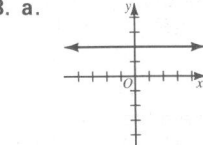

    b. $y = 2$

30. a.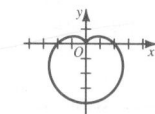

    b. $(x^2 + y^2 + 2y)^2 = 4(x^2 + y^2)$

(continued)

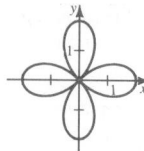
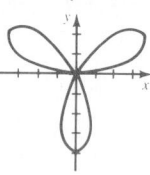

Find the Cartesian coordinates of the given point. Graph the point in the plane.

9. $(3\sqrt{2}, 315°)$ $(3, -3)$ **10.** $(5, 60°)$ $(\frac{5}{2}, \frac{5\sqrt{3}}{2})$ **11.** $(-2, 30°)$ $(-\sqrt{3}, -1)$

12. $(4, -150°)$ $(-2\sqrt{3}, -2)$ **13.** $\left(-1, \frac{\pi}{6}\right)$ $(-\frac{\sqrt{3}}{2}, -\frac{1}{2})$ **14.** $\left(-3, \frac{7\pi}{4}\right)$ $\left(\frac{-3}{\sqrt{2}}, \frac{3}{\sqrt{2}}\right)$

15. $\left(\sqrt{3}, \frac{2\pi}{3}\right)$ $(-\frac{\sqrt{3}}{2}, \frac{3}{2})$ **16.** $\left(-\sqrt{6}, -\frac{5\pi}{6}\right)$ $(\frac{3\sqrt{2}}{2}, \frac{\sqrt{6}}{2})$ **17.** $(-2, 180°)$ $(2, 0)$

Use Equations (1) on page 673 to rewrite each equation as a polar equation.

18. $x = 3$ $r = 3 \sec \theta$ **19.** $y = -2$ $r = -2 \csc \theta$ **20.** $x + y - 3 = 0$ $r = \frac{3}{\cos \theta + \sin \theta}$

21. $x^2 + y^2 = 9$ $r^2 = 9$ **22.** $x^2 - y^2 = 25$ $r^2 = \frac{25}{\cos 2\theta}$ **23.** $x^2 + y^2 - 8x = 0$ $r^2 - 8r \cos \theta = 0$

In Exercises 24–38, (a) sketch the graph of the given polar equation, and (b) transform the equation into an equation in Cartesian coordinates.

B   **24.** $r = 4$ **25.** $r = -3$ **26.** $\theta = -45°$

     27. $r = 6 \cos \theta$ **28.** $r = 2 \csc \theta$ **29.** $r = -4 \sec \theta$

     30. $r = 2 - 2 \sin \theta$ **31.** $r = 1 + \cos \theta$ **32.** $r = 2 \cos 2\theta$

     33. $r = \sin 2\theta$ **34.** $r = 4 \sin 3\theta$ **35.** $r = 1 - 2 \sin \theta$

C   **36.** $r = \theta$ (θ in radians)   **37.** $r^2 = 2 \sin \theta$ **38.** $r^2 = \cos 2\theta$
                                           $(x^2 + y^2)^3 = 4y^2$ $(x^2 + y^2)^2 = x^2 - y^2$

16–4 Graphs of Complex Numbers

Complex numbers can be graphed on a rectangular coordinate system similar to the Cartesian coordinate system. Each complex number $x + yi$, where $x, y \in \mathcal{R}$, is associated with the point (x, y). We represent the complex number by a point or by an arrow from the origin to the associated point. Such diagrams are also called **Argand diagrams**.

In Figure 16, $-2 + i$ is represented by a point and $2 + 3i$ is represented by an arrow. The graphs of all the real numbers

$$x + 0i$$

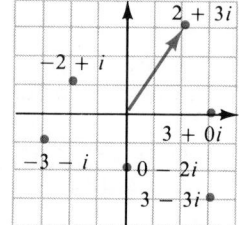

Figure 16

appear on the horizontal axis, called the real axis, and the graphs of the pure imaginary numbers

$$0 + yi$$

appear on the vertical axis, called the imaginary axis.

Complex numbers are frequently represented using polar coordinates. If $r \geq 0$, then $x + yi$ can be expressed as $r \cos \theta + (r \sin \theta)i$, or $r(\cos \theta + i \sin \theta)$, where

$$r = \sqrt{x^2 + y^2}, \qquad \cos \theta = \frac{x}{\sqrt{x^2 + y^2}}, \qquad \text{and} \qquad \sin \theta = \frac{y}{\sqrt{x^2 + y^2}}.$$

676   *Chapter 16*

The expression

$$r(\cos \theta + i \sin \theta)$$

is called the **polar form** (or the **trigonometric form**) for denoting the complex number $x + yi$. If

$$z = x + yi,$$

then the **modulus** or the **absolute value** of z is defined by $|z| = r = \sqrt{x^2 + y^2}$. An angle θ determined by the equations above is called an **amplitude** or an **argument** of z. Thus, $x + yi$ may be graphed as the point with rectangular coordinates (x, y) or polar coordinates (r, θ), $r \geq 0$, as shown in Figure 17.

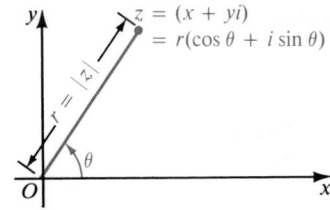

Figure 17

EXAMPLE 1 Express $-2 - 2i\sqrt{3}$ in polar form with $0° \leq m(\theta) < 360°$ and graph.

SOLUTION Determine the modulus of z.

$$|z| = \sqrt{(-2)^2 + (-2\sqrt{3})^2}$$
$$= \sqrt{16} = 4$$

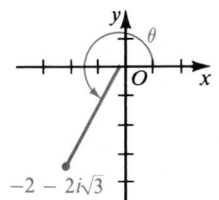

Then $\cos \theta = \dfrac{-2}{4} = -\dfrac{1}{2}$ and $\sin \theta =$

$\dfrac{-2\sqrt{3}}{4} = -\dfrac{\sqrt{3}}{2}$, so that $m(\theta) = 240°$.

$\therefore -2 - 2i\sqrt{3} = 4(\cos 240° + i \sin 240°)$.

By expressing complex numbers in polar form, you can compute their products and quotients by means of the following theorem.

Theorem.   If

$$z_1 = r_1(\cos \theta_1 + i \sin \theta_1) \text{ and } z_2 = r_2(\cos \theta_2 + i \sin \theta_2),$$

then: **1.** $z_1 z_2 = r_1 r_2 [\cos (\theta_1 + \theta_2) + i \sin (\theta_1 + \theta_2)]$,

2. $\dfrac{z_1}{z_2} = \dfrac{r_1}{r_2}[\cos (\theta_1 - \theta_2) + i \sin (\theta_1 - \theta_2)]$ $(z_2 \neq 0)$

PROOF OF PART 1

$$z_1 \cdot z_2 = r_1(\cos \theta_1 + i \sin \theta_1) \cdot r_2(\cos \theta_2 + i \sin \theta_2)$$
$$= r_1 r_2 [\cos \theta_1 \cos \theta_2 + i \sin \theta_1 \cos \theta_2 + i \cos \theta_1 \sin \theta_2 + i^2 \sin \theta_1 \sin \theta_2]$$
$$= r_1 r_2 [(\cos \theta_1 \cos \theta_2 - \sin \theta_1 \sin \theta_2) + i(\sin \theta_1 \cos \theta_2 + \cos \theta_1 \sin \theta_2)]$$
$$= r_1 r_2 [\cos (\theta_1 + \theta_2) + i \sin (\theta_1 + \theta_2)]$$

Inverses; Polar Coordinates; Vectors   **677**

PROOF OF PART 2

$$\frac{z_1}{z_2} = \frac{r_1(\cos\theta_1 + i\sin\theta_1)}{r_2(\cos\theta_2 + i\sin\theta_2)} = \frac{r_1(\cos\theta_1 + i\sin\theta_1)(\cos\theta_2 - i\sin\theta_2)}{r_2(\cos\theta_2 + i\sin\theta_2)(\cos\theta_2 - i\sin\theta_2)}$$

$$= \frac{r_1[\cos\theta_1\cos\theta_2 + i\sin\theta_1\cos\theta_2 - i\cos\theta_1\sin\theta_2 - i^2\sin\theta_1\sin\theta_2]}{r_2[\cos^2\theta - i^2\sin^2\theta_2]}$$

$$= \frac{r_1[(\cos\theta_1\cos\theta_2 + \sin\theta_1\sin\theta_2) + i(\sin\theta_1\cos\theta_2 - \cos\theta_1\sin\theta_2)]}{r_2[\cos^2\theta + \sin^2\theta]}$$

$$= \frac{r_1}{r_2}[\cos(\theta_1 - \theta_2) + i\sin(\theta_1 - \theta_2)]$$

EXAMPLE 2 If $z_1 = 6(\cos 135° + i\sin 135°)$ and $z_2 = 3(\cos 15° + i\sin 15°)$, express $z_1 z_2$ and $\dfrac{z_1}{z_2}$ in the form $a + bi$.

SOLUTION **a.** By Part 1 of the theorem:

$$z_1 z_2 = 6 \cdot 3[\cos(135° + 15°) + i\sin(135° + 15°)]$$
$$= 18(\cos 150° + i\sin 150°)$$
$$= 18\left(-\frac{\sqrt{3}}{2} + \frac{1}{2}i\right) = -9\sqrt{3} + 9i$$

b. By Part 2 of the theorem:

$$\frac{z_1}{z_2} = \frac{6}{3}[\cos(135° - 15°) + i\sin(135° - 15°)]$$
$$= 2(\cos 120° + i\sin 120°)$$
$$= 2\left(-\frac{1}{2} + i\frac{\sqrt{3}}{2}\right) = -1 + i\sqrt{3}$$

Oral Exercises

State the complex number represented by the given point.

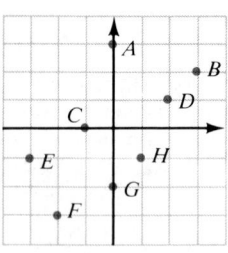

1. A $3i$
2. B $3 + 2i$
3. C -1
4. D $2 + i$
5. E $-3 - i$
6. F $-2 - 3i$
7. G $-2i$
8. H $1 - i$

Written Exercises

In Exercises 1–8, express the given complex number in polar form with $0° \le m(\theta) < 360°$. Graph each complex number.

A

1. $3 + 3i$
2. $-4i$
3. -7
4. $-1 + i\sqrt{3}$
5. $2\sqrt{2} - 2i\sqrt{2}$
6. $-\frac{1}{2} - \frac{\sqrt{3}}{2}i$
7. $-6 + 6i$
8. $\sqrt{2} - i\sqrt{6}$

Express each complex number in the form $a + bi$.

9. $4\sqrt{2}(\cos 135° + i \sin 135°)$ $-4 + 4i$

10. $10(\cos 30° + i \sin 30°)$ $5\sqrt{3} + 5i$

11. $2(\cos 120° + i \sin 120°)$ $-1 + i\sqrt{3}$

12. $12(\cos 330° + i \sin 330°)$ $6\sqrt{3} - 6i$

13. $6\left(\cos \frac{4\pi}{3} + i \sin \frac{4\pi}{3}\right)$ $-3 - 3i\sqrt{3}$

14. $\frac{1}{3}\left(\cos \frac{5\pi}{6} + i \sin \frac{5\pi}{6}\right)$ $-\frac{\sqrt{3}}{6} + \frac{1}{6}i$

15. $\sqrt{6}\left(\cos \frac{7\pi}{4} + i \sin \frac{7\pi}{4}\right)$ $\sqrt{3} - i\sqrt{3}$

16. $8\left(\cos \frac{11\pi}{6} + i \sin \frac{11\pi}{6}\right)$ $4\sqrt{3} - 4i$

In Exercises 17–22 find (a) $z_1 z_2$ and (b) $\dfrac{z_1}{z_2}$. Express each answer in the form $a + bi$.

19. a. $-12 - 12i\sqrt{3}$ b. $\frac{3\sqrt{3}}{4} + \frac{3}{4}i$

17. $z_1 = 2(\cos 300° + i \sin 300°)$, $z_2 = 5(\cos 60° + i \sin 60°)$ a. 10 b. $-\frac{1}{5} - \frac{\sqrt{3}}{5}i$

18. $z_1 = 2\sqrt{2}(\cos 45° + i \sin 45°)$, $z_2 = \sqrt{2}(\cos 225° + i \sin 225°)$ a. $-4i$ b. -2

19. $z_1 = 6(\cos 135° + i \sin 135°)$, $z_2 = 4(\cos 105° + i \sin 105°)$

20. $z_1 = 8(\cos 255° + i \sin 255°)$, $z_2 = \frac{1}{2}(\cos 75° + i \sin 75°)$ a. $2\sqrt{3} - 2i$ b. -16

21. $z_1 = \frac{3}{2}(\cos 105° + i \sin 105°)$, $z_2 = 4(\cos 15° + i \sin 15°)$ a. $-3 + 3i\sqrt{3}$ b. $\frac{3}{8}i$

22. $z_1 = \sqrt{2}(\cos 225° + i \sin 225°)$, $z_2 = 3\sqrt{2}(\cos 165° + i \sin 165°)$

a. $3\sqrt{3} + 3i$ b. $\frac{1}{6} + \frac{1}{6}i\sqrt{3}$

B 23. Let $z_1 = \frac{1}{2} + \frac{\sqrt{3}}{2}i$ and $z_2 = \frac{\sqrt{2}}{2} + \frac{\sqrt{2}}{2}i$.

   a. Find $z_1 z_2$ in the form $a + bi$ by using multiplication of complex numbers.

   b. Find $z_1 z_2$ in polar form by converting z_1 and z_2 to polar form and using the theorem on page 677.

   c. Use the answers to parts (a) and (b) and a calculator or table to find the value of $\sin 105°$ and $\cos 105°$. (*Hint:* Use the definition of equality of complex numbers on page 279.)

   d. Check your answer by using appropriate formulas for sine and cosine chosen from those on page 642.

C 24. Let $z = a + bi = r(\cos \theta + i \sin \theta)$ and $\bar{z} = a - bi$ denote the complex conjugate of z. Write $\bar{z}$ in polar form and use the theorem on page 677 to find each of the following: $\bar{z} = r(\cos(-\theta) + i \sin(-\theta))$

   a. $z\bar{z}$ in polar form $r^2(\cos 0° + i \sin 0°)$

   b. $\frac{z}{\bar{z}}$ in polar form $\cos 2\theta + i \sin 2\theta$

23. a. $\frac{\sqrt{2} - \sqrt{6}}{4} + \left(\frac{\sqrt{2} + \sqrt{6}}{4}\right)i$ b. $1(\cos 105° + i \sin 105°)$

   c. $0.9659; -0.2588$

Computer Exercises For students with computer experience

1. Let $z = x + yi$ be a complex number. Write a program that will compute the absolute value of z and an argument of z when x and y are input.

Additional A Exercises

Express the given complex number in polar form with $0° \le m(\theta) < 360°$.

1. $4 + 4i$

   $4\sqrt{2}(\cos 45° + i \sin 45°)$

2. $-i$

   $1(\cos 270° + i \sin 270°)$

3. $5 - 5i$

   $5\sqrt{2}(\cos 315° + i \sin 315°)$

Express each complex number in the form $a + bi$.

4. $4(\cos 60° + i \sin 60°)$

   $2 + 2i\sqrt{3}$

5. $6(\cos 210° + i \sin 210°)$

   $-3\sqrt{3} - 3i$

6. $2\sqrt{3}\left(\cos \frac{5\pi}{3} + i \sin \frac{5\pi}{3}\right)$

   $\sqrt{3} - 3i$

Mixed Review

Find the values of θ for $0° \le m(\theta) < 360°$.

1. $\tan \theta = \frac{\sqrt{3}}{3}$ $30°, 210°$

2. $\sin \theta = -\frac{1}{2}$, $\cos \theta = \frac{\sqrt{3}}{2}$

   $330°$

3. $\sin \theta = -\cos \theta$ $135°, 315°$

Multiply.

4. $(2 + i)(3 - 2i)$ $8 - i$

5. $(4 + i)(4 - i)$ 17

6. $(-3 + i)(5 + 2i)$ $-17 - i$

Inverses; Polar Coordinates; Vectors **679**

Run the program in Exercise 1 for each complex number.

5, 233°

2. $3 - 3i$ 4.243, 315°   **3.** $-2 + 2i$ 2.828, 135°   **4.** $-3 - 4i$

5. $-7i$ 7, 270°   **6.** $5 + 8i$ 9.43, 58°   **7.** 6

6, 0°

8. Extend the program in Exercise 1 so that it will use the theorem on page 677 to multiply two complex numbers. Have the output give the answer in $a + bi$ form.

Run the program in Exercise 8 for each pair of complex numbers.

9. $-1 + 3i, 4 - 5i$ **10.** $2 - i, -3i$ **11.** $2 + 2i, -3 - 4i$

11 + 17i $-3 - 6i$ $2 - 14i$

Teaching Suggestions
p. T126

Related Activities
p. T126

Key Ideas

Define DeMoivre's Theorem. Use the theorem to find powers and roots of complex numbers.

16–5 De Moivre's Theorem

If $z = r(\cos \theta + i \sin \theta)$, you can see that successive applications of Part 1 of the theorem on page 677 yield

$$z^2 = z \cdot z = r(\cos \theta + i \sin \theta) \cdot r(\cos \theta + i \sin \theta)$$
$$= r^2(\cos 2\theta + i \sin 2\theta),$$

$$z^3 = z^2 \cdot z = r^2(\cos 2\theta + i \sin 2\theta) \cdot r(\cos \theta + i \sin \theta)$$
$$= r^3(\cos 3\theta + i \sin 3\theta),$$

and

$$z^4 = z^3 \cdot z = r^3(\cos 3\theta + i \sin 3\theta) \cdot r(\cos \theta + i \sin \theta)$$
$$= r^4(\cos 4\theta + i \sin 4\theta).$$

Continuing this process suggests the general statement

$$z^n = z^{n-1}z = r^{n-1}[\cos (n - 1)\theta + i \sin (n - 1)\theta]r(\cos \theta + i \sin \theta)$$
$$= r^n(\cos n\theta + i \sin n\theta)$$

for all natural numbers n. A formal proof of this statement requires mathematical induction, which is discussed in more advanced courses.

By defining $z^0 = 1$ and $z^{-n} = \dfrac{1}{z^n}$, it is possible to extend this statement to include all integral powers of nonzero complex numbers. Thus we have the following theorem, named after the French mathematician Abraham De Moivre.

De Moivre's Theorem

If $z = r(\cos \theta + i \sin \theta)$ and $n \in \{$the integers$\}$, then

$$z^n = r^n(\cos n\theta + i \sin n\theta).$$

---

680   *Chapter 16*

EXAMPLE 1   Express $(1 + i)^6$ in the form $a + bi$.

SOLUTION   Express $1 + i$ in polar form:

$$r = \sqrt{(1)^2 + (1)^2} = \sqrt{2},$$

$$\cos\theta = \frac{1}{\sqrt{2}}, \text{ and } \sin\theta = \frac{1}{\sqrt{2}}$$

Thus $1 + i = \sqrt{2}(\cos 45° + i\sin 45°)$.

Then apply De Moivre's theorem.

$$\begin{aligned}
(1 + i)^6 &= [\sqrt{2}(\cos 45° + i\sin 45°)]^6 \\
&= (\sqrt{2})^6(\cos 6 \cdot 45° + i\sin 6 \cdot 45°) \\
&= 8(\cos 270° + i\sin 270°) \\
&= 8(0 - i) \\
&= -8i
\end{aligned}$$

$$\therefore (1 + i)^6 = -8i$$

EXAMPLE 2   Express $(\sqrt{3} + i)^{-4}$ in the form $a + bi$.

SOLUTION   First, express $\sqrt{3} + i$ in polar form:

$$\begin{aligned}
r &= |z| \\
&= \sqrt{(\sqrt{3})^2 + 1^2} \\
&= 2
\end{aligned}$$

Then $\cos\theta = \dfrac{\sqrt{3}}{2}$ and $\sin\theta = \dfrac{1}{2}$.

$$\therefore \sqrt{3} + i = 2(\cos 30° + i\sin 30°).$$

Then by De Moivre's theorem,

$$\begin{aligned}
[2(\cos 30° + i\sin 30°)]^{-4} &= 2^{-4}[\cos(-4 \cdot 30°) + i\sin(-4 \cdot 30°)] \\
&= \frac{1}{16}[\cos(-120°) + i\sin(-120°)] \\
&= \frac{1}{16}\left(-\frac{1}{2} - \frac{\sqrt{3}}{2}i\right) \\
&= -\frac{1}{32} - \frac{\sqrt{3}}{32}i.
\end{aligned}$$

$$\therefore (\sqrt{3} + i)^{-4} = -\frac{1}{32} - \frac{\sqrt{3}}{32}i.$$

Every complex number, including every real number, has n nth roots. To find the roots of complex numbers you can use De Moivre's theorem and the fact that two complex numbers in polar form are equal if and only if their moduli are equal, and their arguments differ in degree measure by a multiple of $360°$.

Chalkboard Examples

1. Express $(2 + 2i\sqrt{3})^5$ in the form $a + bi$.

$2 + 2i\sqrt{3} = 4(\cos 60° + i\sin 60°)$; $(2 + 2i\sqrt{3})^5 = 4^5(\cos 5 \cdot 60° + i\sin 5 \cdot 60°) = 1024\left(\dfrac{1}{2} - \dfrac{\sqrt{3}}{2}i\right) = 512 - 512i\sqrt{3}$

$\therefore (2 + 2i\sqrt{3})^5 = 512 - 512\sqrt{3}i$

2. Express $(-1 + i)^{-2}$ in the form $a + bi$.

$-1 + i = \sqrt{2}(\cos 135° + i\sin 135°)$; $(-1 + i)^{-2} = (\sqrt{2})^{-2}[\cos(-2 \cdot 135°) + i\sin(-2 \cdot 135°)] = \dfrac{1}{2}(0 + i) = \dfrac{1}{2}i$

$\therefore (-1 + i)^{-2} = 0 + \dfrac{1}{2}i.$

3. Find all cube roots of $27i$ and express them in the form $a + bi$.

$27i = 27(\cos 90° + i\sin 90°)$. Let $w = $ a cube root of $27i$. Then $w^3 = r^3(\cos 3\theta + i\sin 3\theta) = 27(\cos 90° + i\sin 90°)$ and $w = 3[\cos(30° + k \cdot 120°) + i\sin(30° + k \cdot 120°)]$.

If $k = 0$, $w = \dfrac{3\sqrt{3}}{2} + \dfrac{3}{2}i$.

If $k = 1$, $w = -\dfrac{3\sqrt{3}}{2} + \dfrac{3}{2}i$.

If $k = 2$, $w = -3i$.

If $k = 3$, $w = \dfrac{3\sqrt{3}}{2} + \dfrac{3}{2}i$, which is the same as the value when $k = 0$.

The three cube roots of $27i$ are $\dfrac{3\sqrt{3}}{2} + \dfrac{3}{2}i, -\dfrac{3\sqrt{3}}{2} + \dfrac{3}{2}i,$ and $-3i$.

EXAMPLE 3 Find all cube roots of $8i$ and express them in the form $a + bi$.

SOLUTION Express $8i = 0 + 8i$ in polar form:

$$r = 8, \cos \theta = \frac{0}{8} = 0, \text{ and } \sin \theta = \frac{8}{8} = 1$$

Thus $8i = 8(\cos 90° + i \sin 90°)$.

Now let $w = r(\cos \theta + i \sin \theta)$ represent a cube root of $8i$. Then

$$w^3 = r^3(\cos 3\theta + i \sin 3\theta) = 8(\cos 90° + i \sin 90°).$$

Thus, $r^3 = 8$ and $3\theta = 90° + k \cdot 360°$ where $k \in J$.

Therefore $r = \sqrt[3]{8} = 2$, $\theta = \dfrac{90° + k \cdot 360°}{3} = 30° + k \cdot 120°$,

and $w = 2[\cos (30° + k \cdot 120°) + i(\sin 30° + k \cdot 120°)]$.

Let $k = 0, 1, 2, 3$, and so on to find that:
If $k = 0$, then $w = 2(\cos 30° + i \sin 30°)$

$$= 2\left(\frac{\sqrt{3}}{2} + \frac{1}{2}i\right)$$

$$= \sqrt{3} + i.$$

If $k = 1$, then $w = 2[\cos (30° + 120°) + i \sin (30° + 120°)]$

$$= 2(\cos 150° + i \sin 150°)$$

$$= 2\left(-\frac{\sqrt{3}}{2} + \frac{1}{2}i\right)$$

$$= -\sqrt{3} + i.$$

If $k = 2$, then $w = 2[\cos (30° + 240°) + i \sin (30° + 240°)]$

$$= 2(\cos 270° + i \sin 270°)$$

$$= 2(0 - i) = -2i.$$

If $k = 3$, then $w = 2[\cos (30° + 360°) + i \sin (30° + 360°)]$

$$= 2(\cos 390° + i \sin 390°)$$

$$= 2(\cos 30° + i \sin 30°) = \sqrt{3} + i$$

which is the same as the value when $k = 0$.
Additional replacements for k will simply duplicate the three values of w already obtained.

$\therefore$ the three cube roots of $8i$ are $\sqrt{3} + i$, $-\sqrt{3} + i$, and $-2i$.

The three cube roots of $8i$ obtained in Example 3 can be graphed as shown in Figure 18. The points are equally spaced around a circle with center at the origin and radius 2.

The process used in Example 3 can be generalized as shown in the theorem on the next page.

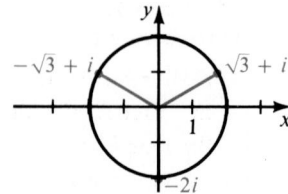

Figure 18

682 Chapter 16

> **Theorem.** The equation $z^n = r(\cos\theta + i\sin\theta)$, where $n \in$ {the natural numbers}, $r \in \mathcal{R}$, $r > 0$, has n roots:
>
> $$z = \sqrt[n]{r}\left(\cos\frac{\theta + k(360°)}{n} + i\sin\frac{\theta + k(360°)}{n}\right)$$
>
> for $k = 0, 1, 2, \ldots, n - 1$.

The nth roots of the complex number 1 are called the *nth roots of unity*. Following the reasoning in Example 3, the nth roots of unity can be found. Since $1 = 1(\cos 0° + i\sin 0°)$ the nth roots of unity are $\cos\frac{k \cdot 360°}{n} + i\sin\frac{k \cdot 360°}{n}$, $k = 0, 1, 2, \ldots, n - 1$. An nth root of unity, w, is said to be *primitive* if every nth root of unity is an integral power of w.

Oral Exercises

1. $4(\cos 60° + i\sin 60°)$
2. $27(\cos 135° + i\sin 135°)$
3. $16(\cos 120° + i\sin 120°)$
4. $32(\cos 150° + i\sin 150°)$

Let $z = 2(\cos 30° + i\sin 30°)$ and $w = 3(\cos 45° + i\sin 45°)$. State each of the following in polar form.

1. z^2 2. w^3 3. z^4 4. z^5 5. w^{-1} 6. z^{-3} 7. w^0 8. z^{-4}
5. $\frac{1}{3}(\cos(-45°) + i\sin(-45°))$
6. $\frac{1}{8}(\cos(-90°) + i\sin(-90°))$
7. $\cos 0° + i\sin 0°$
8. $\frac{1}{16}(\cos(-120°) + i\sin(-120°))$

Written Exercises

Use De Moivre's Theorem to express each of the following in the form $a + bi$.

7. $\frac{1}{2} + \frac{\sqrt3}{2}i$
11. $-\frac{1}{32} + \frac{\sqrt3}{32}i$

A 1. $\left(\frac{\sqrt3}{2} + \frac{1}{2}i\right)^3$ i
2. $(-1 + i\sqrt3)^3$ 8
3. $(-1 + i)^4$ -4
4. $(\sqrt2 - i\sqrt2)^4$ -16
5. $(\sqrt3 - i)^5$ $-16\sqrt3 - 16i$
6. $(-\sqrt3 + i)^4$ $-8 - 8i\sqrt3$
7. $\left(\frac{\sqrt3}{2} - \frac{1}{2}i\right)^{10}$ $\frac{64}{}$
8. $(-2 - 2i\sqrt3)^3$ 64
9. $(1 - i)^{-3}$ $-\frac{1}{4} + \frac{1}{4}i$
10. $(\sqrt2 + i\sqrt2)^{-4}$ $-\frac{1}{16}$
11. $(1 + i\sqrt3)^{-4}$
12. $\left(\frac{3\sqrt3}{2} - \frac{3}{2}i\right)^{-3}$ $\frac{1}{27}i$
13. $(0.2588 + 0.9659i)^2$ $-\frac{\sqrt3}{2} + \frac{1}{2}i$
14. $[2(-0.2588 + 0.9659i)]^3$ $\frac{8}{\sqrt2} - \frac{8}{\sqrt2}i$
   (*Hint:* $75° = \text{Cos}^{-1}\,0.2588 = \text{Sin}^{-1}\,0.9659$.)

In Exercises 15–19 find the required roots in the form $a + bi$ and graph the roots on a circle in the plane.

B 15. the three cube roots of unity $1, -\frac{1}{2} + \frac{\sqrt3}{2}i, -\frac{1}{2} - \frac{\sqrt3}{2}i$
16. the three cube roots of -8 $1 + \sqrt3 i, -2, 1 - \sqrt3 i$
17. the four fourth roots of -16 $\sqrt2 + i\sqrt2, -\sqrt2 + i\sqrt2, -\sqrt2 - i\sqrt2, \sqrt2 - i\sqrt2$
18. the four fourth roots of $-8 + 8i\sqrt3$ $\sqrt3 + i, -1 + \sqrt3 i, -\sqrt3 - i, 1 - i\sqrt3$
19. the four fourth roots of $-128 - 128i\sqrt3$ $2 + 2i\sqrt3, -2\sqrt3 + 2i, -2 - 2i\sqrt3, 2\sqrt3 - 2i$

Inverses; Polar Coordinates; Vectors **683**

Use DeMoivre's Theorem to express in the form $a + bi$.
1. $(\sqrt3 + i)^2$ $2 + 2i\sqrt3$
2. $(\sqrt2 + i\sqrt2)^2$ $4i$
3. $(1 - i\sqrt3)^3$ -8
4. $\left(\frac{\sqrt3}{2} - \frac{1}{2}i\right)^{-3}$ i
5. $(-\sqrt2 + i\sqrt2)^{-5}$
   $\frac{\sqrt2}{64} + \frac{\sqrt2}{64}i$

Additional Answers
Written Exercises

20. $2(\cos 18° + i\sin 18°)$,
   $2(\cos 90° + i\sin 90°)$,
   $2(\cos 162° + i\sin 162°)$,
   $2(\cos 234° + i\sin 234°)$,
   $2(\cos 306° + i\sin 306°)$

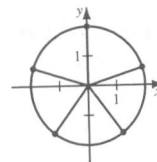

21. $z = a + bi =$
   $r(\cos\theta + i\sin\theta); \frac{1}{z} =$
   $\frac{1}{r(\cos\theta + i\sin\theta)} =$
   $\frac{1}{r} \cdot \frac{1}{\cos\theta + i\sin\theta} \cdot$
   $\frac{(\cos\theta - i\sin\theta)}{(\cos\theta - i\sin\theta)} =$
   $\frac{1}{r} \cdot \frac{\cos\theta - i\sin\theta}{\cos^2\theta + \sin^2\theta} =$
   $\frac{1}{r}(\cos\theta - i\sin\theta) =$
   $\frac{1}{r}(\cos(-\theta) + i\sin(-\theta))$.

(continued)

683

22. $z = a + bi = r(\cos \theta + i \sin \theta)$; $z^{-n} = (z^{-1})^n =$

$$\left[\frac{1}{r}(\cos(-\theta) + i \sin(-\theta))\right]^n =$$

$$\left(\frac{1}{r}\right)^n (\cos(-n\theta) + i \sin(-n\theta)) = r^{-n}(\cos(-n\theta) + i \sin(-n\theta))$$

28. $\cos 72° + i \sin 72°$,
$\cos 144° + i \sin 144°$,
$\cos 216° + i \sin 216°$,
$\cos 288° + i \sin 288°$

Quick Quiz

1. Find two pairs of polar co-ordinates for $P(1, -1)$ for which $-180° < m(\theta) \le 180°$.
$(\sqrt{2}, -45°), (-\sqrt{2}, 135°)$

2. Find the Cartesian coordinates for $P(-4, -120°)$.
$(2, 2\sqrt{3})$

3. Let $z_1 = 2(\cos 225° + i \sin 225°)$ and $z_2 = 4(\cos 15° + i \sin 15°)$.
Express in the form $a + bi$.
 a. $z_1 z_2$ $-4 - 4i\sqrt{3}$
 b. $\dfrac{z_1}{z_2}$ $-\dfrac{\sqrt{3}}{4} - \dfrac{1}{4}i$
 c. $(z_1)^2$ $4i$

4. Find the three cube roots of -1 in the form $a + bi$.
$\dfrac{1}{2} + \dfrac{\sqrt{3}}{2}i, -1, \dfrac{1}{2} - \dfrac{\sqrt{3}}{2}i$

---

20. Find and graph the five fifth roots of $32i$. Leave your answers in polar form.

21. Let $z = a + bi = r(\cos \theta + i \sin \theta)$. Show that $\dfrac{1}{z} = \dfrac{1}{r}(\cos(-\theta) + i \sin(-\theta))$.

22. Write z^{-n} as $(z^{-1})^n$ and show that De Moivre's Theorem holds if n is a negative integer. (*Hint*: Use the result of Exercise 21.)

23. Show that the product of the three cube roots of $1 - i$ equals $1 - i$.

24. Show that the product of the four fourth roots of -1 equals 1.

C 25. Show that if $z = r(\cos \theta + i \sin \theta)$, then the product of the two square roots of z equals $-z$.

26. Show that if $z = r(\cos \theta + i \sin \theta)$, then the product of the three cube roots of z equals z.

27. If $\bar{z}$ is the complex conjugate of z, show that $(\bar{z})^n$ equals the complex conjugate of z^n for all natural numbers n.

28. Find the primitive fifth roots of unity.

Self-Test 2

VOCABULARY
polar coordinates (p. 672)
polar axis (p. 672)
pole (p. 672)
graph of a polar equation (p. 674)
Argand diagrams (p. 676)
polar form, or trigonometric form, of a complex number (p. 677)
modulus, or absolute value, of a complex number (p. 677)

amplitude, or argument, of a complex number (p. 677)
De Moivre's Theorem (p. 680)
nth roots of unity (p. 683)
primitive root of unity (p. 683)

1. Find two pairs of polar coordinates for $P(-2, 2\sqrt{3})$ in the interval $-180° < m(\alpha) \le 180°$.   (4, 120°), (−4, −60°)   *Obj. 1, p. 672*

2. Find the Cartesian coordinates of the point with polar coordinates $(2, -45°)$. $(\sqrt{2}, -\sqrt{2})$

3. Let $z_1 = 3(\cos 330° + i \sin 330°)$ and $z_2 = 2(\cos 30° + i \sin 30°)$.   *Obj. 2, p. 672*
Express each of the following in the form $a + bi$.
 a. $z_1 z_2$ 6   b. $\dfrac{z_1}{z_2}$ $\dfrac{3}{4} - \dfrac{i 3\sqrt{3}}{4}$ c. $(z_1)^3$ $0 - 27i$

4. Find the three cube roots of $1 + i$ in polar form. $\sqrt[6]{2}(\cos 15° + i \sin 15°)$, $\sqrt[6]{2}(\cos 135° + i \sin 135°)$, $\sqrt[6]{2}(\cos 255° + i \sin 255°)$

Check your answers with those at the back of the book.

READING ALGEBRA Homonyms and Synonyms

In everyday life, we are accustomed to homonyms, that is words that sound the same but have different meanings. When a golfer cries "Fore!" we watch out *for* one, not *four* golf balls. Some mathematical words are also homonyms, such as sign and sine. Careful listening and the context in which the words are used, help to make the meaning clear.

Synonyms are different words that have the same meaning. Big, huge, great, large, massive, tremendous, and enormous are all words that describe size. Some have slightly different meanings than others. Usage determines which word we might ordinarily choose to describe a particular person or object. Algebra also has symbols that mean the same thing. The symbol we choose is dictated by convention, usage, and sometimes personal preference. For example,

$$3x, \qquad 3 \cdot x, \qquad \text{and } 3(x)$$

are all ways of indicating multiplication. Usually, we would use $3x$.

Hearing algebraic expressions read correctly can help to make the meaning clear. Do not look at an algebraic expression as you would look at an illustration. Algebra is a language and should be put into words. When you read $\sin x$ as "the sine of x" you are not confusing it with multiplication. The expression $(3x - 6)^2$ should be read "the quantity three times x minus six, all squared."

Functions such as the logarithmic and trigonometric functions should not be confused with multiplication. The symbols

$$\sin x, \qquad \log x, \qquad \text{and } g(x)$$

are all ways of describing functions, not the operation of multiplication.

Exercises

Translate each pair of expressions into words. Explain why each expression is read differently.

1. $9a - 2$; $9(a - 2)$

2. $\dfrac{2x + 3}{7y}$; $2x + \dfrac{3}{7y}$

3. $\cos 3x + 1$; $\cos (3x + 1)$

4. $\tan \frac{1}{2}x$; $\frac{1}{2} \tan x$

5. $\sqrt{1 - x^2}$; $1 - x$

6. 9^{2x-1}; $9^{2x} - 1$

7. $\log_5(y - 4)$; $\log_5 y - 4$

8. $8b - 3^2$; $(8b - 3)^2$

9. $\text{Sin}^{-1} 1$; $\sin^{-1} 1$

10. $\sec^2 x$; $\sec x^2$

11. $4x^2$; $(4x)^2$

12. $2y - 3x + 4$; $2y - (3y + 4)$

Vectors

OBJECTIVES for Sections 16-6 and 16-7:
1. *To find the norm and direction of a vector sum.*
2. *To apply the dot product of two vectors.*
3. *To solve problems using vectors.*

Teaching Suggestions
p. T126

Related Activities p. T126

Key Ideas

Define a vector in terms of
the norm and direction, the
horizontal and vertical components, or unit vectors. Find
the norm and direction of a
vector sum. Find the dot
product of two vectors.

16–6 Vectors in the Plane

In order to describe a physical quantity such as velocity or
force, you must specify both the quantity's magnitude and
its direction. Such a quantity can be represented by a
directed line segment in the plane and is called a **vector**. A
vector may be identified by naming its endpoints. The
symbol $\overrightarrow{AB}$ denotes a vector whose **initial point** is A and
whose **terminal point** is B. A single letter in boldface type,
such as **v,** can also be used to denote a vector. (In writing,
you can denote a vector by using a single letter with an
arrow over it, such as $\vec{w}$.) Figure 19 illustrates the vector
$\overrightarrow{OP}$, or **v,** with magnitude r and direction θ. The length or
magnitude of a vector is called its **norm**. The norm of $\overrightarrow{OP}$,
or **v,** in Figure 19, is denoted $\|\overrightarrow{OP}\|$ or $\|\mathbf{v}\|$. A point can be
thought of as a vector of length 0 and is called the **zero
vector**.

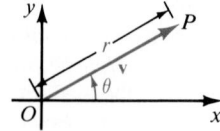

Figure 19

The angle θ ($-180° < \theta \leq 180°$) that the vector makes
with a ray directed parallel to, and in the direction of, the
positive x-axis is called the **direction angle**, or simply the
direction of the vector. In Figure 20, $\overrightarrow{OQ}$ and $\overrightarrow{AB}$ have
the same norm and the same direction. Such vectors are
called **equivalent**. A vector, such as $\overrightarrow{OQ}$, whose initial
point is at the origin of a coordinate system is said to be in
standard position. Every vector in the plane is equivalent
to a vector in standard position.

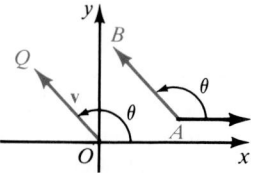

Figure 20

In Figure 21, if $\overrightarrow{AC}$ denotes the motion of a boat in still
water and $\overrightarrow{CB}$ denotes the motion of a current, then the
boat will move along $\overrightarrow{AB}$ shown by the red arrow. This
suggests that it is useful to define the **sum,**or **resultant** of
two vectors geometrically. That is, if $\overrightarrow{AC}$ and $\overrightarrow{CB}$ are
vectors, then

$$\overrightarrow{AC} + \overrightarrow{CB} = \overrightarrow{AB}.$$

Note that the initial and terminal points of the resultant $\overrightarrow{AB}$
are the initial point of $\overrightarrow{AC}$ and the terminal point of $\overrightarrow{CB}$,
respectively. The vectors $\overrightarrow{AC}$ and $\overrightarrow{CB}$ are called **components** of $\overrightarrow{AB}$.

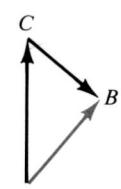

Figure 21

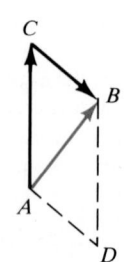

Figure 22

686   *Chapter 16*

Figure 22 shows that the resultant $\overrightarrow{AB}$ can be shown as the diagonal of a parallelogram with adjacent sides having the same lengths as $\overrightarrow{AC}$ and $\overrightarrow{CB}$ and parallel to them. You can use this fact to find the norm and direction of a vector sum.

EXAMPLE 1   Given that $\|\mathbf{u}\| = 4$ and the direction angle of $\mathbf{u}$ measures 75°, $\|\mathbf{v}\| = 5$, and the direction angle of $\mathbf{v}$ measures $-10°$, find an approximation to the nearest tenth of a unit for $\|\mathbf{u} + \mathbf{v}\|$, and the measure of the direction angle of $\mathbf{u} + \mathbf{v}$, to the nearest degree.

SOLUTION   Draw a sketch of the given vectors. By inspection $m(\angle DOB) = 85°$ so that $m(\angle ODC) = 95°$.

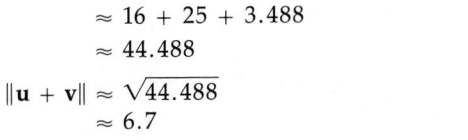

Then, to find $\|\mathbf{u} + \mathbf{v}\|$, use the law of cosines:

$$\|\mathbf{u} + \mathbf{v}\|^2 = \|\mathbf{u}\|^2 + \|\mathbf{v}\|^2 - 2\|\mathbf{u}\| \cdot \|\mathbf{v}\| \cos 95°$$
$$\approx 4^2 + 5^2 - 2(4)(5)(-0.0872)$$
$$\approx 16 + 25 + 3.488$$
$$\approx 44.488$$

$$\|\mathbf{u} + \mathbf{v}\| \approx \sqrt{44.488}$$
$$\approx 6.7$$

To find the measure of the direction angle θ of $\mathbf{u} + \mathbf{v}$, use the law of sines to first find $m(\angle COD)$:

$$\frac{\sin \angle COD}{4} \approx \frac{\sin 95°}{6.7}$$

$$\sin \angle COD \approx \frac{4(0.9962)}{6.7}$$

$$\approx \frac{3.9848}{6.7}$$

$$\approx 0.5947$$

$$m(\angle COD) \approx 36°$$

$$\therefore m(\theta) = 36° - 10° = 26° \text{ and } \|\mathbf{u} + \mathbf{v}\| = 6.7$$

An alternate way to determine the norm and direction angle of the sum $\mathbf{u} + \mathbf{v}$ is to resolve $\mathbf{u}$ and $\mathbf{v}$ into horizontal and vertical components that are parallel to the x-axis and y-axis respectively. Figure 23 shows $\mathbf{v}$ and its horizontal and vertical components denoted $\mathbf{v}_x$ and $\mathbf{v}_y$. Recalling that the norm of a vector is nonnegative and the definition of $\cos \theta$ on page 585, you can see that $|\cos \theta| = \dfrac{\|\mathbf{v}_x\|}{\|\mathbf{v}\|}$ and thus $\|\mathbf{v}_x\| = \|\mathbf{v}\| \, |\cos \theta|$. Similarly $\|\mathbf{v}_y\| = \|\mathbf{v}\| \, |\sin \theta|$.

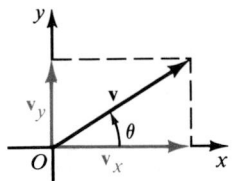

Figure 23

1. Given that $\|\mathbf{u}\| = 5$ and the direction angle of $\mathbf{u}$ is 87°, $\|\mathbf{v}\| = 7$ and the direction angle of $\mathbf{v}$ is 12°, find $\|\mathbf{u} + \mathbf{v}\|$ to the nearest tenth and the measure of the direction angle of $\mathbf{u} + \mathbf{v}$ to the nearest degree.
$m(\angle DOB) = 75°$ so $m(\angle ODC) = 105°$.
$\|\mathbf{u} + \mathbf{v}\|^2 \approx 5^2 + 7^2 - 2(5) \cdot (7)(-0.2588) \approx 92.116$;
$\|\mathbf{u} + \mathbf{v}\| \approx \sqrt{92.116} \approx 9.6$
$\dfrac{\sin \angle COD}{5} \approx \dfrac{\sin 105°}{9.6}$
$\sin \angle COD \approx 0.5031$
$m(\angle COD) \approx 30°$
$m(\theta) = 30° + 12° = 42°$

2. Find the angle between the vectors $\mathbf{u}$ and $\mathbf{v}$ defined by:
$\mathbf{u} = 5\mathbf{i} + 12\mathbf{j}$, $\mathbf{v} = 8\mathbf{i} + 15\mathbf{j}$
$a = 5$, $b = 12$, $c = 8$,
$d = 15$;
$\|\mathbf{u}\| = 13$, $\|\mathbf{v}\| = 17$.
$\therefore \mathbf{u} \cdot \mathbf{v} = 5(8) + 12(15)$
$= 220$
Then $(13)(17) \cos \theta = 220$,
$\cos \theta \approx 0.9955$.
$\therefore m(\theta) = 5.44°$.

3. Find a unit vector orthogonal to $4\mathbf{i} - 3\mathbf{j}$.
Let $\mathbf{v} = x\mathbf{i} + y\mathbf{j}$
$\mathbf{u} \cdot \mathbf{v} = 4x - 3y = 0$
$\therefore x = 3$, $y = 4$.
$\dfrac{\mathbf{v}}{\|\mathbf{v}\|} = \dfrac{3\mathbf{i} + 4\mathbf{j}}{5} = \dfrac{3}{5}\mathbf{i} + \dfrac{4}{5}\mathbf{j}$

Now, examining Figure 24, you can see that the horizontal and vertical components of the sum $\mathbf{u} + \mathbf{v}$ of two vectors are the sums of the corresponding components of $\mathbf{u}$ and $\mathbf{v}$:

$$(\mathbf{u} + \mathbf{v})_x = \mathbf{u}_x + \mathbf{v}_x \qquad \text{and} \qquad (\mathbf{u} + \mathbf{v})_y = \mathbf{u}_y + \mathbf{v}_y$$

so that

$$\|\mathbf{u} + \mathbf{v}\| = \sqrt{\|\mathbf{u}_x + \mathbf{v}_x\|^2 + \|\mathbf{u}_y + \mathbf{v}_y\|^2}.$$

Notice also that, for the direction angle θ of $\mathbf{u} + \mathbf{v}$, you have

$$|\cos \theta| = \frac{\|\mathbf{u}_x + \mathbf{v}_x\|}{\|\mathbf{u} + \mathbf{v}\|} \qquad \text{and} \qquad |\sin \theta| = \frac{\|\mathbf{u}_y + \mathbf{v}_y\|}{\|\mathbf{u} + \mathbf{v}\|}.$$

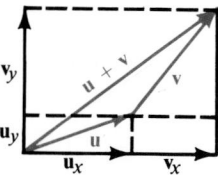

Figure 24

or

$$|\tan \theta| = \frac{\|\mathbf{u}_y + \mathbf{v}_y\|}{\|\mathbf{u}_x + \mathbf{v}_x\|}.$$

If we define the unit vectors $\mathbf{i}$ and $\mathbf{j}$ to be the vectors of magnitude 1 in the direction of the positive x-axis and the positive y-axis, respectively, then for any nonzero vector $\mathbf{v}$,

$$\mathbf{v} = a\mathbf{i} + b\mathbf{j}$$

where a is the magnitude of the horizontal component of $\mathbf{v}$ and b is the magnitude of the vertical component of $\mathbf{v}$.

Many physical applications of the concept of a vector suggest that we define the product of two vectors in a certain way. The **dot product** of two nonzero vectors, denoted $\mathbf{u} \cdot \mathbf{v}$, is defined to be

$$\mathbf{u} \cdot \mathbf{v} = \|\mathbf{u}\| \, \|\mathbf{v}\| \cos \theta$$

where θ is the angle between $\mathbf{u}$ and $\mathbf{v}$. If either $\mathbf{u}$ or $\mathbf{v}$ is the zero vector, then $\mathbf{u} \cdot \mathbf{v}$ is defined to be 0. Note that the dot product of two vectors is a real number and *not* a vector.

The following theorem can be used to determine the angle between two vectors whose horizontal and vertical components are given. The theorem is proved in Exercise 35 on page 691.

Theorem. If $\mathbf{v} = a\mathbf{i} + b\mathbf{j}$ and $\mathbf{u} = c\mathbf{i} + d\mathbf{j}$, then

$$\mathbf{u} \cdot \mathbf{v} = ac + bd.$$

EXAMPLE 2   Find the angle between the vectors $\mathbf{u}$ and $\mathbf{v}$ defined by the following vectors.

$$\mathbf{u} = 3\mathbf{i} - 4\mathbf{j}$$
$$\mathbf{v} = 5\mathbf{i} + 10\mathbf{j}$$

SOLUTION   By inspection $a = 3$, $b = -4$, $c = 5$, and $d = 10$. $\|\mathbf{u}\| = 5$, and $\|\mathbf{v}\| = 5\sqrt{5}$. Therefore, by the theorem on the previous page:

$$\mathbf{u} \cdot \mathbf{v} = (3)(5) + (-4)(10)$$
$$= -25$$

Substituting in the definition of the dot product yields the following:

$$(5)(5\sqrt{5}) \cos \theta = -25$$
$$\cos \theta = \frac{-25}{(5)(5\sqrt{5})}$$
$$= \frac{-1}{\sqrt{5}}$$
$$\approx -0.4472$$

$\therefore m(\theta) = 116.6°$.

Since $\cos 90° = 0$, one immediate consequence of the definition of the dot product is the following theorem.

> **Theorem.**   If $\mathbf{u}$ and $\mathbf{v}$ are two nonzero vectors, then $\mathbf{u} \cdot \mathbf{v} = 0$ if and only if the angle between $\mathbf{u}$ and $\mathbf{v}$ measures 90°.

If $\mathbf{u} \cdot \mathbf{v} = 0$, then $\mathbf{u}$ and $\mathbf{v}$ are said to be orthogonal.

EXAMPLE 3   Find a unit vector orthogonal to $2\mathbf{i} + 4\mathbf{j}$.

SOLUTION   Let $\mathbf{u} = 2\mathbf{i} + 4\mathbf{j}$ and $\mathbf{v} = x\mathbf{i} + y\mathbf{j}$. The vectors $\mathbf{u}$ and $\mathbf{v}$ are orthogonal if and only if $\mathbf{u} \cdot \mathbf{v} = 2x + 4y = 0$. One solution is $x = -2$ and $y = 1$. Thus, $\mathbf{v} = -2\mathbf{i} + \mathbf{j}$ is a vector that is orthogonal to $\mathbf{u}$. A nonzero vector divided by its norm has length 1. Thus:

$$\frac{\mathbf{v}}{\|\mathbf{v}\|} = \frac{-2\mathbf{i} + \mathbf{j}}{\sqrt{5}} = \frac{-2}{\sqrt{5}}\mathbf{i} + \frac{1}{\sqrt{5}}\mathbf{j}$$

is a unit vector orthogonal to $2\mathbf{i} + 4\mathbf{j}$.

To multiply a vector $\mathbf{v}$ by a real number or *scalar*, r, multiply the length of $\mathbf{v}$ by $|r|$ and reverse the direction if $r < 0$. Thus, $-\mathbf{v}$ is defined to be the vector with the same magnitude as $\mathbf{v}$ and opposite direction of $\mathbf{v}$ and vector subtraction is defined by

$$\mathbf{u} - \mathbf{v} = \mathbf{u} + (-\mathbf{v}).$$

Geometrically $\mathbf{u} - \mathbf{v}$ can be represented as the vector from the terminal point of $\mathbf{v}$ to the terminal point of $\mathbf{u}$ as Figure 25 illustrates.

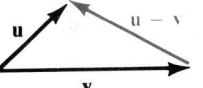

Figure 25

Mixed Review

1. Use the law of cosines to find the length, to the nearest tenth, of the third side of $\triangle ABC$ if $a = 12$, $b = 8$, and $m(C) = 53°$.   9.6

2. Use the law of sines to find the measure, to the nearest degree, of $\angle A$ in $\triangle ABC$ if $a = 7$, $c = 9$, and $m(C) = 114°$. $m(A) = 45°$

Additional A Exercises

For the following vectors $\mathbf{u}$ and $\mathbf{v}$, with the given norms and direction angles, draw a sketch showing $\mathbf{u}$, $\mathbf{v}$, and $\mathbf{u} + \mathbf{v}$, and find $\|\mathbf{u} + \mathbf{v}\|$.

1. $\mathbf{u}$: 6, 110°; $\mathbf{v}$: 9, 40°   12.4

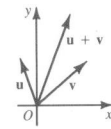

2. $\mathbf{u}$: 4, 80°; $\mathbf{v}$: 2, −40°   3.5

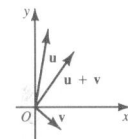

3. Find the measure of the angle between $\mathbf{u} = 2\mathbf{i} + 4\mathbf{j}$ and $\mathbf{v} = 5\mathbf{i} - 5\mathbf{j}$.   108.4°

4. Find a unit vector orthogonal to $3\mathbf{i} + \mathbf{j}$.

$$\frac{1}{\sqrt{10}}\mathbf{i} - \frac{3}{\sqrt{10}}\mathbf{j}$$

Oral Exercises

In Exercises 1–6 name the given vector by specifying its initial and terminal points.

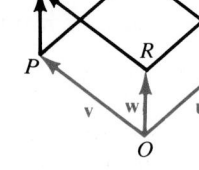

EXAMPLE   **u** + **w**   SOLUTION $\overrightarrow{OS}$

1. **u** + **v** $\overrightarrow{OM}$ 2. **v** + **w** $\overrightarrow{OQ}$ 3. **u** − **w** $\overrightarrow{RT}$
4. **w** − **v** $\overrightarrow{PR}$ 5. **w** − **u** $\overrightarrow{TR}$ 6. **u** − **v** $\overrightarrow{PT}$

In Exercises 7–9 state the magnitudes of the horizontal and vertical components of each vector.

7. $3\mathbf{i} - 2\mathbf{j}$ 3, 2   8. $5\mathbf{i} + 7\mathbf{j}$ 5, 7   9. $-2\mathbf{i} + 4\mathbf{j}$ 2, 4

Written Exercises

In Exercises 1–8, for the vectors **u** and **v** with the given norms and direction angles, draw a sketch showing **u**, **v**, and **u** + **v**, and find ‖**u** + **v**‖. If necessary leave answers in radical form.

A 1. **u**: 5, 135°; **v**: 8, 15° 7   2. **u**: 10, 100°; **v**: 6, 40° 14
3. **u**: 21, 150°; **v**: 5, 30° 19   4. **u**: 16, 120°; **v**: 5, 60° 19
5. **u**: 6, 65°; **v**: $\sqrt{2}$, 20° $5\sqrt{2}$ 6. **u**: 2, 30°; **v**: $3\sqrt{3}$, 0° 7
7. **u**: 12, 110°; **v**: $8\sqrt{2}$, −25° $4\sqrt{5}$ 8. **u**: $5\sqrt{3}$, 120°; **v**: 2, −30° 7

Find the measure of the angle θ between each pair of vectors, **u** and **v**.

9. $\mathbf{u} = -3\mathbf{i} + 4\mathbf{j}$; $\mathbf{v} = 7\mathbf{i} + 24\mathbf{j}$ 53.1°   10. $\mathbf{u} = 5\mathbf{i} - 12\mathbf{j}$; $\mathbf{v} = -2\mathbf{i} + \mathbf{j}$ 139.2°
11. $\mathbf{u} = 15\mathbf{i} + 8\mathbf{j}$; $\mathbf{v} = -\mathbf{i} - 13\mathbf{j}$ 122.5°   12. $\mathbf{u} = -3\mathbf{i} + 3\mathbf{j}$; $\mathbf{v} = 6\mathbf{i} - 9\mathbf{j}$ 168.7°

Find a unit vector orthogonal to **u**. Answers may vary.

13. $\mathbf{u} = 3\mathbf{i} + 4\mathbf{j}$ $\frac{4}{5}i - \frac{3}{5}j$ 14. $\mathbf{u} = 5\mathbf{i} - 12\mathbf{j}$ $\frac{12}{13}i + \frac{5}{13}j$
15. $\mathbf{u} = \mathbf{i} - \mathbf{j}$ $\frac{\sqrt{2}}{2}i + \frac{\sqrt{2}}{2}j$ 16. $\mathbf{u} = 5\mathbf{i} - 7\mathbf{j}$ $\frac{7}{\sqrt{74}}i + \frac{5}{\sqrt{74}}j$

For the vectors **u** and **v** with the given norms and direction angles, draw a sketch showing **u**, **v**, and **u** + **v**, and find ‖**u** + **v**‖ to the nearest tenth and the direction angle of **u** + **v** to the nearest degree. Answers may vary.

B 17. **u**: 32, 90°; **v**: 10, 30° 38, 77°   18. **u**: 8, 150°; **v**: 6, 60° 10, 113°
19. **u**: 20, 120°; **v**: 10, 45° 24.6, 97°   20. **u**: 10, 30°; **v**: 20, 105° 20. 24.6, 82°
21. **u**: 12, 45°; **v**: 8, −30° 16.1, 16°   22. **u**: 16, −90°; **v**: 20, 30° 22. 18.3, −19°
23. **u**: 10, 80°; **v**: 24, 50° 33.0, 59°   24. **u**: 8, 40°; **v**: 10, −40° 13.8, −5°

25-32. Use the vectors **u** and **v** in Exercises 17-24 to sketch **u** −**v**. Find ‖**u** − **v**‖ to the nearest tenth and the direction angle of **u** − **v** to the nearest degree.

Draw a diagram that verifies the following for vectors u, v, and w.

C **33.** **u** + **v** = **v** + **u**

34. **u** + (**v** + **w**) = (**u** + **v**) + **w**

35. Use the triangle shown to prove that if **u** = a**i** + b**j** and **v** = c**i** + d**j**, then **u** · **v** = ac + bd. (*Hint*: Find PQ using the distance formula and the law of cosines.)

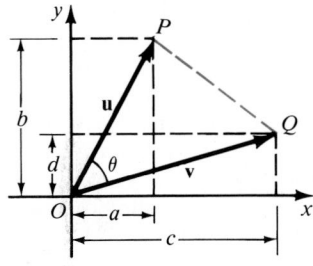

Thābit ibn Qurra
836–901

Arab scholars provided the bridge between the science of the Greco-Roman era and the rebirth of learning in Western Europe. A major contributor to this process was Thābit ibn Qurra.

Thābit translated many works of the ancient Greeks, including Archimedes' commentaries on Euclid's *Elements* from Greek into Arabic. Thābit's original mathematical works prepared the way for later discoveries in real numbers, integral calculus, and analytic geometry. In "Discourse on the Establishment of the Correctness of Algebraic Problems," he used Euclid's *Elements* to provide rules for solving the quadratic equations $x^2 + ax = b$, $x^2 + b = ax$, and $x^2 = ax + b$.

Thābit's translations, commentaries, and original works ranged over a wide variety of topics. He produced books and articles in medicine, psychology, politics, logic, and ethics.

25. 28.4, 108°

26. 10, −173°

27. 19.9, 149°

28. 19.9, −46°

29. 12.6, 83°

30. 31.2, −124°

31. 16.1, −148°

32. 11.7, 97°

33.

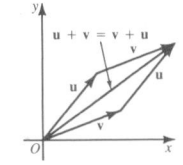

34.

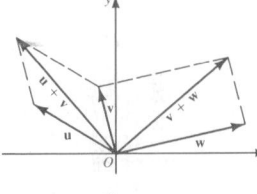

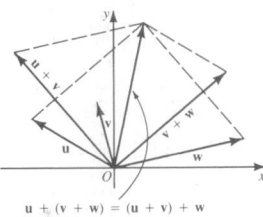

$u + (v + w) = (u + v) + w$

35. $PQ^2 = ‖\mathbf{u}‖^2 + ‖\mathbf{v}‖^2 − 2‖\mathbf{u}‖ · ‖\mathbf{v}‖ \cos \theta$. Since **u** · **v** = ‖**u**‖ ‖**v**‖ cos θ, we can substitute into the first equation to get $PQ^2 = ‖\mathbf{u}‖^2 + ‖\mathbf{v}‖^2 − 2\mathbf{u} · \mathbf{v}$. Using the distance formula, $PQ^2 = (a − c)^2 + (b − d)^2 = (a^2 + b^2) + (c^2 + d^2) − 2\mathbf{u} · \mathbf{v}$. Expanding and simplifying gives $ac + bd = \mathbf{u} · \mathbf{v}$.

Key Ideas

Use vectors to solve problems.

Chalkboard Example

A boat is traveling at 50 km/h with a heading of 70°. If the current is 30 km/h with a bearing of 310°, what is the resultant speed and the true course of the boat?

Let **u** = velocity of boat in still water, **w** = velocity of current, and **v** = resultant speed of boat.
$\|\mathbf{v}\|^2 = \|\mathbf{u}\|^2 + \|\mathbf{w}\|^2 - 2\|\mathbf{u}\|\|\mathbf{w}\| \cdot \cos 60° = 50^2 +$

$30^2 - 2(50)(30)\left(\frac{1}{2}\right) = 1900$

$\therefore \|\mathbf{v}\| = \sqrt{1900} \approx 44.$

$\cos \theta = \dfrac{50^2 + 44^2 - 30^2}{2(50)(44)} \approx$

0.8036, $m(\theta) \approx 37°$ ∴ the resultant speed of the boat is 44 km/h and its true course is $70° - 37° = 33°$.

16–7 Applications of Vectors

We can denote a vector by giving the magnitude of the vector and its direction angle. In navigation, the direction angle is called the **bearing**, and is defined as the angle θ, $0° \leq \theta < 360°$, measured clockwise from the vector pointing due north. For example, **v** and **u** in Figure 26 have bearings 120° and 315°, respectively.

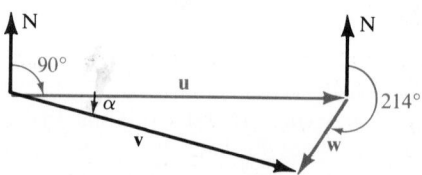

Figure 26

For a moving craft, such as a ship or airplane, the direction in which the front of the craft points (and thus the direction in which it is being propelled) is called the craft's **heading**. When a wind is blowing, an airplane's heading usually differs from its **true course**, or the bearing of the airplane's path relative to the ground. The **velocity** of an airplane is given by the direction and speed of the airplane and is thus a vector. The airplane's velocity relative to the ground is the resultant of its velocity in still air and the velocity of the wind. We refer to the magnitude of any velocity vector as its **speed**. (Recall the definitions of wind speed, air speed, and ground speed on page 139.)

EXAMPLE 1   The air speed of an airplane is 200 km/h and its heading is 90°. A 30 km/h wind is blowing with a bearing of 214°. Find the ground speed and the true course of the airplane.

SOLUTION   Let **u** = the velocity of the airplane in still air.
Let **w** = the wind velocity.
Let **v** = the velocity of the airplane relative to the ground.
Make a sketch showing the three vectors. Then by the law of cosines:

$$\|\mathbf{v}\|^2 = \|\mathbf{u}\|^2 + \|\mathbf{w}\|^2 - 2\|\mathbf{u}\|\|\mathbf{w}\| \cos 56°$$
$$\approx 200^2 + 30^2 - 2(200)(30)(0.5592) \approx 34{,}190$$

$\therefore \|\mathbf{v}\| = \sqrt{34{,}190} \approx 185$ (to the nearest kilometer per hour).
To find $m(\alpha)$ substitute in the following:

$$\cos \alpha = \frac{\|\mathbf{u}\|^2 + \|\mathbf{v}\|^2 - \|\mathbf{w}\|^2}{2\|\mathbf{u}\|\|\mathbf{v}\|} = \frac{200^2 + 185^2 - 30^2}{2(200)(185)} \approx 0.9909$$

$\therefore m(\alpha) \approx 8°$ (to the nearest degree).

$\therefore$ the ground speed of the airplane is 185 km/h, and its true course is $90° + 8° = 98°$.

692   Chapter 16

When a force **F** is exerted on an object to move it from A to B, work is done and energy is expended. Let $\overrightarrow{AB} = \mathbf{d}$. Then if **F** has the same direction as **d**, the work done by **F** is

$$W = \|\mathbf{F}\|\,\|\mathbf{d}\|.$$

This is also the energy expended by **F**.

The basic unit of work and energy is the *joule* (J). A joule is the work done by a force of one newton (N) in moving an object one meter in the direction of the force. Near the surface of Earth, the force exerted by gravity on an object whose mass is 1 kg is 9.8 N. Another unit of work is the *kilowatt-hour* (kW·h). One kilowatt-hour equals 3.6×10^6 J.

EXAMPLE 2   The mass of a loaded helicopter is 1500 kg. How much energy in joules does the engine expend in ascending vertically a distance of 80 m? How much energy in kW·h is expended for the same vertical move?

SOLUTION   The motor of the helicopter must overcome the force of gravity and exert an upward force, **F**, of magnitude

$$1500 \times 9.8 = 1.47 \times 10^4 \text{ N}.$$

The direction of the force is upward and has a magnitude of 80. Thus the work done is

$$\begin{aligned} W &= \|\mathbf{F}\|\,\|\mathbf{d}\| \\ &= (1.47 \times 10^4)(80) \\ &= 1.18 \times 10^6 \text{ J} \end{aligned}$$

Converting joules to kilowatt-hours yields the following:

$$\frac{1.18 \times 10^6}{3.6 \times 10^6} = 0.328 \text{ kW·h}$$

Problems

Solve. Find magnitudes to three significant digits and bearings to the nearest tenth of a degree.

A

1. An airplane flies due south at 120 km/h with a 40 km/h wind blowing from the east. What is the ground speed and the true course of the airplane? 126 km/h; bearing 198.4°

2. An airplane flies due south at 400 mi/h with a 25 mi/h wind with a bearing of 210°. What is the ground speed and the true course of the airplane? 422 mi/h; bearing 181.7°

3. How much energy does a 55 kg climber expend to climb a vertical formation that is 60 m high? Give your answer in joules and kilowatt-hours to three significant digits. 3.23×10^4 J; 0.00897 kW·h

4. How much energy is needed to lift a 2500 kg elevator 50 m? Give your answer in joules to two significant digits. 1.2×10^6 J

Inverses; Polar Coordinates; Vectors   **693**

Suggested Assignments

Extended Alg. with Trig.
  693/1–9
R 694/*Self-Test 3*
Enriched Alg. with Trig.
  693/4–10
R 694/*Self-Test 3*

Additional A Exercises

1. An airplane flies due east at 300 km/h with a 50 km/h tail wind. What is the ground speed and true course of the airplane? 350 km/h, 90°

2. A boat is traveling at 40 km/h with a heading of 75°. If the current is 25 km/h with a bearing of 315°, what is the resultant speed and the true course of the boat?   35 km/h, 37°

Mixed Review

Solve each triangle. Give lengths to the nearest tenth and angles to the nearest degree.

1. $a = 8$, $b = 12$, $m(C) = 83°$
   $c = 13.6$, $m(A) = 36°$,
   $m(B) = 61°$

2. $a = 4$, $c = 9$, $m(B) = 60°$
   $b = 7.8$, $m(C) = 88°$,
   $m(A) = 32°$

Simplify. Give the answer in standard notation. Let $a = 2.84 \times 10^5$ and $b = 7.19 \times 10^{-3}$.

3. $50a$ 1.42×10^7

4. $25b$ 1.80×10^{-1}

5. $\dfrac{b}{a}$ 2.53×10^{-8}

Quick Quiz

1. Given that $\|u\| = 5$ and the direction angle of **u** is 80°, and $\|v\| = 9$ and the direction angle of **v** is 15°, find $\|u + v\|$ to the nearest tenth and the direction angle of $u + v$ to the nearest degree.
   $\|u + v\| \approx 12.0$, $m(\theta) = 37°$

2. Find the angle between $5i + 7j$ and $2i - 3j$ to the nearest degree.
   $m(\theta) = 111°$

5. A ship sails 500 mi due east and then 200 mi on a heading of 120°. How far is the ship from its starting point and what is the ship's bearing from the starting point?   681 mi; bearing 98.4°

6. A power boat heads due south directly across a straight river at a speed of 15 km/h, while the current in the river flows due east at 6 km/h. What is the boat's speed and its true course?

7. A pilot wants to steer a true course of 90° with a ground speed of 500 km/h. If a 50 km/h wind with a bearing of 240° is blowing, what should be the pilot's air speed and heading?   544 km/h; bearing 87.4°

8. A ferryboat operator has found that by heading 25° upstream (measured from a line directly across the river), the boat will land at a point directly opposite its starting point. If the ferryboat travels at 12 km/h in still water, what is the speed of the current?   5.07 km/h

9. How much energy does a crane expend to vertically lift 30 containers 40 m above the deck of a ship? The mass of each container is 1800 kg. Give your answer in joules and kilowatt-hours to three significant digits. 2.12×10^7 J; 5.89 kW·h

B **10.** A boat heads directly east across a 45 m wide river that has a current flowing directly south at 5 km/h. If the speed of the boat in still water is 15 km/h, how far downstream will the boat land? What heading should the boat operator use in order to land directly opposite the starting point?   15 m; 70.5°

11. A plane flies at 300 km/h with a heading of 20°, while a wind with a bearing of 260° blows at 40 km/h. What is the true course of the plane and what is its ground speed?   282 km/h; bearing 13.0°

12. Suppose all the conditions of Problem 11 hold except that the plane's heading is 140°. Find the ground speed and true course of the plane.

C **13.** A search ship leaves Port Evans and sails due north for 200 mi and then due east for 50 mi. The ship then sails on a heading of 225° for 100 mi. What is the ship's distance and bearing from Port Evans?
   131 mi; bearing 350.9°

Self-Test 3

VOCABULARY   vector (p. 686)
             initial point (p. 686)
             terminal point (p. 686)
             zero vector (p. 686)
             direction angle (p. 686)
             equivalent vectors (p. 686)
             standard position (p. 686)
             sum, or resultant (p. 686)

components of a vector (p. 687)
dot product (p. 688)
orthogonal vectors (p. 689)
bearing (p. 692)
heading (p. 692)
true course (p. 692)
velocity (p. 692)
speed (p. 692)

1. Given the vectors **u** and **v** such that $\|\mathbf{u}\| = 6$ and the direction angle of **u** is 45° and $\|\mathbf{v}\| = 8$ and the direction angle of **v** is 105°, find an approximation of $\|\mathbf{u} + \mathbf{v}\|$ to the nearest tenth and the direction angle of $\mathbf{u} + \mathbf{v}$ to the nearest degree. 12.2; 80° *Obj. 1, p. 686*
2. Find the angle between $3\mathbf{i} - 4\mathbf{j}$ and $5\mathbf{i} + 12\mathbf{j}$. 120.5° *Obj. 2, p. 686*
3. A ship sails 75 km with a heading of 220° and then directly west for 50 km. How far is the ship from its starting point? What is the ship's bearing from its starting point? 114 km; 239.5° *Obj. 3, p. 686*

Check your answers with those at the back of the book.

3. A man walks 20 km due east, and then directly south for 17 km. How far is he from his starting point and what is his bearing from the starting point? $\sqrt{689} \approx 26$, km from the starting point at a bearing of approximately 130°

Chapter Summary

1. The inverse of each of the circular or trigonometric functions discussed in this chapter is a relation that is not a function unless the range is restricted. For the *principal-value inverse functions* the ranges are:

$$\text{Sin}^{-1}: -\frac{\pi}{2} \le y \le \frac{\pi}{2} \qquad \text{Cos}^{-1}: 0 \le y \le \pi \qquad \text{Cot}^{-1}: 0 < y < \pi$$

$$\text{Csc}^{-1}: -\frac{\pi}{2} \le y \le \frac{\pi}{2}, y \ne 0 \qquad \text{Sec}^{-1}: 0 \le y \le \pi, y \ne \frac{\pi}{2} \qquad \text{Tan}^{-1}: -\frac{\pi}{2} < y < \frac{\pi}{2}$$

2. Using trigonometric identities and the usual algebraic transformations, you can solve equations involving the circular or trigonometric functions.

3. Coordinates of points may be given in the Cartesian system as (x, y) or in the polar system as (r, θ), related as follows:

$$x = r \cos \theta \qquad y = r \sin \theta$$

$$r = \pm\sqrt{x^2 + y^2} \qquad \cos \theta = \frac{x}{\pm\sqrt{x^2 + y^2}} \qquad \sin \theta = \frac{y}{\pm\sqrt{x^2 + y^2}}$$

4. If $z_1 = r_1(\cos \theta_1 + i \sin \theta_1)$ and $z_2 = r_2(\cos \theta_2 + i \sin \theta_2)$, then:

$$z_1 z_2 = r_1 r_2[\cos (\theta_1 + \theta_2) + i \sin (\theta_1 + \theta_2)]$$

$$\frac{z_1}{z_2} = \frac{r_1}{r_2}[\cos (\theta_1 - \theta_2) + i \sin (\theta_1 - \theta_2)]$$

5. De Moivre's theorem states that if $z = r(\cos \theta + i \sin \theta)$ and $n \in \{\text{the integers}\}$, then $z^n = r^n(\cos n\theta + i \sin n\theta)$.

6. The equation $z^n = r(\cos \theta + i \sin \theta)$, where $n \in \{\text{the natural numbers}\}$, $r \in \mathcal{R}, r > 0$, has n solutions:

$$z = \sqrt[n]{r}\left(\cos \frac{\theta + k(360°)}{n} + i \sin \frac{\theta + k(360°)}{n}\right)$$

for $k = 0, 1, 2, \ldots, n - 1$.

Inverses; Polar Coordinates; Vectors   **695**

7. A *vector* may be written in the form $\overrightarrow{AB}$ or **u**; its norm, or length, is denoted by $\|\overrightarrow{AB}\|$ or by $\|\mathbf{u}\|$.

8. If $\overrightarrow{AB}$ and $\overrightarrow{CD}$ are any two vectors in the plane, and if $\overrightarrow{BE}$ is equivalent to $\overrightarrow{CD}$, then $\overrightarrow{AB} + \overrightarrow{CD} = \overrightarrow{AE}$. Vectors $\overrightarrow{AB}$ and $\overrightarrow{CD}$ are called *components* of $\overrightarrow{AE}$.

9. The *dot product* of two vectors **u** and **v** is defined as

$$\mathbf{u} \cdot \mathbf{v} = \|\mathbf{u}\| \, \|\mathbf{v}\| \cos \theta$$

where θ is the measure of the angle between the two vectors. Two vectors are *orthogonal* if their dot product is 0.

10. Problems in navigation and force can be solved using vectors. Vector directions used in navigation are called *bearings* and are measured clockwise from the vector pointing due north. When a force **F** moves an object a distance **d**, in the same direction as **d**, the work done by **F** is given by the formula $W = \|\mathbf{F}\| \, \|\mathbf{d}\|$.

Chapter Review

Write the letter of the correct answer.

1. Find the real value of $\text{Cos}^{-1} \dfrac{1}{\sqrt{2}}$.   *16-1*

   a. $\dfrac{7\pi}{4}$ **b.** $\dfrac{5\pi}{4}$ **(c.)** $\dfrac{\pi}{4}$ **d.** $\dfrac{\pi}{2}$

2. Evaluate $\text{Cos}^{-1}\left(\cos \dfrac{5\pi}{3}\right)$.

   a. $\dfrac{4\pi}{3}$ **b.** $\dfrac{5\pi}{3}$ **(c.)** $\dfrac{\pi}{3}$ **d.** $\dfrac{\pi}{2}$

3. Solve $4 \cos^2 x - 8 \cos x + 7 = 0$ for $0 \le x < 2\pi$.   *16-2*

   a. $\left\{\dfrac{\pi}{3}, \dfrac{5\pi}{3}\right\}$ **b.** $\left\{0, \dfrac{\pi}{3}\right\}$ **(c.)** $\varnothing$ **d.** $\left\{\dfrac{2\pi}{3}, \dfrac{4\pi}{3}\right\}$

4. Find the Cartesian coordinates of the point with polar coordinates $(5, 330°)$.   *16-3*

   a. $\left(-\dfrac{5\sqrt{3}}{2}, -\dfrac{5}{2}\right)$ **(b.)** $\left(\dfrac{5\sqrt{3}}{2}, -\dfrac{5}{2}\right)$

   c. $\left(\dfrac{5\sqrt{3}}{2}, \dfrac{5}{2}\right)$ **d.** $\left(\dfrac{5}{2}, -\dfrac{5\sqrt{3}}{2}\right)$

5. Find a pair of polar coordinates of the point whose Cartesian coordinates are $(3\sqrt{2}, -3\sqrt{2})$.

   (a.) $(6, 315°)$ **b.** $(6, 45°)$ **c.** $(-3, 135°)$ **d.** $(-6, 45°)$

6. Express $\sqrt{3} + i$ in polar form.

    a. $4(\cos 60° + i \sin 60°)$ **(b.)** $2(\cos 30° + i \sin 30°)$

    c. $4(\cos 210° + i \sin 210°)$ **d.** $2(\cos 60° + i \sin 60°)$

16-4

Let $z_1 = 9(\cos 180° + i \sin 180°)$ and $z_2 = 3(\cos 60° + i \sin 60°)$.

7. Express $\dfrac{z_1}{z_2}$ in the form $a + bi$.

    a. $\dfrac{3}{2} + \dfrac{3\sqrt{3}}{2}i$ **b.** $3 + 3i$ **(c.)** $-\dfrac{3}{2} + \dfrac{3\sqrt{3}}{2}i$ **d.** $\dfrac{3}{2} + \dfrac{\sqrt{3}}{2}i$

8. Express $(z_2)^3$ in the form $a + bi$.

    a. 27 **(b.)** -27 **c.** $1 - 27i$ **d.** $1 + 27i$

16-5

9. For the vectors **u** and **v** such that $\|u\| = 5$ with direction angle $120°$ and $\|v\| = 7$ with direction angle of $60°$, find $\|u + v\|$.

    a. 35 **(b.)** $\sqrt{109}$ **c.** $\sqrt{39}$ **d.** 7

16-6

10. Find the angle θ between $4i + 3j$ and $i + 2j$.

    a. $47°$ **b.** $101°$ **(c.)** $27°$ **d.** $11°$

11. An airplane flies 400 mi with a heading of $70°$ and then due east for 300 mi. What is the distance of the airplane (to the nearest mile) from its starting point?

    (a.) 690 mi   **b.** 576 mi   **c.** 500 mi   **d.** 161 mi

16-7

Chapter Test

1. Find the value of $\text{Cos}^{-1}\left(\sin \dfrac{5\pi}{6}\right)$. $\dfrac{\pi}{3}$

16-1

2. Solve $\tan x \sec x = \tan x$ over $\mathcal{R}$. $\{k\pi\}$

16-2

3. Transform $x^2 + y^2 = 4x$ into an equation in polar coordinates.

    $r^2 - 4r \cos \theta = 0$

16-3

Let $z_1 = 5(\cos 135° + i \sin 135°)$ and $z_2 = 2(\cos 15° + i \sin 15°)$.

4. Express $z_1 z_2$ in the form $a + bi$. $-5\sqrt{3} + 5i$

16-4

5. Express $\dfrac{z_1}{z_2}$ in the form $a + bi$. $-\dfrac{5}{4} + \dfrac{5\sqrt{3}}{4}i$

6. Express $(z_1)^3$ in the form $a + bi$. $\dfrac{125}{\sqrt{2}} + \dfrac{125}{\sqrt{2}}i$

16-5

7. Find the three cube roots of $1 - i$ in polar form.

8. For the vectors **u** and **v** such that $\|u\| = 3$ with direction angle $150°$ and $\|v\| = 8$ with direction angle $30°$, find $\|u + v\|$ to the nearest tenth and the direction angle of $u + v$ to the nearest degree.

16-6

9. An airplane flies due north at 300 km/h with a 45 km/h wind with a bearing of $120°$. What is the ground speed (to the nearest unit) and the true course (to the nearest tenth of a degree) of the airplane?

16-7

Inverses; Polar Coordinates; Vectors   **697**

Additional Answers
Chapter Test

7. $\sqrt[6]{2}(\cos 105° + i \sin 105°)$,
$\sqrt[6]{2}(\cos 225° + i \sin 225°)$,
$\sqrt[6]{2}(\cos 345° + i \sin 345°)$

8. 7.0; $52°$

9. 280 km/h; bearing $8°$

APPLICATION

Communication Among Bees

Honeybees communicate information about food sources using angles and distances. Hence the study of this behavior involves polar coordinates.

When a worker bee discovers a source of food, she flies back to the hive to give the nature and location of this source. The nature of the source is communicated by the odor of the food that the bee has collected. Its location is communicated by one of two dances.

If the food source is less than 80 m from the hive, the bee does a round dance in which she alternately circles in one direction and then in the other as shown in Figure 1. This information causes other bees to fly out and search for the food in all directions from the hive.

If the food is farther than 80 m from the hive, the bee does a "waggle dance." On the vertical side of the hive, the bee first moves through a half circle, then moves in a straight line, and then through another half circle in a direction opposite to that of the first half circle. Then she repeats the straight line portion and continues the dance. During the straight line portion of the dance, the bee moves her abdomen from side to side, hence the name "waggle dance." The straight line, or waggle portion of the dance communicates the direction of the food source relative to the sun. If the food source is in the same direction as the sun, the waggle path is straight up the vertical side of the hive (Figure 2).

Figure 1

Figure 2

If the direction is opposite to that of the sun, the waggle path is 180° from the vertical, or straight down the side of the hive (Figure 3). If the bees must fly at a heading of 60° to the right of the sun, the waggle path is 60° to the right of the vertically upward path (Figure 4). Thus the direction in

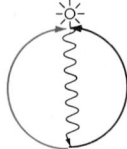

Figure 3

which bees must fly to find a food source is given in terms of an angle.

The distance to the food source is given by the number of waggle dances per unit of time. Karl von Frisch, a pioneer in the study of communications among bees, discovered that the number of waggle paths per second increases significantly as the distance of the food source decreases. The waggle dance merges into the round dance when the distance is about 80 m or less.

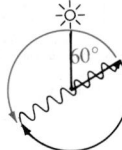

Figure 4

Exercises

1. Suppose that a beehive is at the center of a rectangular coordinate system. The system is in a plane parallel to the ground, and the positive y-axis points in the direction of the sun. What are the rectangular coordinates of a plant for which a bee indicates an angle of 60° clockwise from the sun and a distance of 600 m? $(300\sqrt{3}, 300)$

2. A certain food source has rectangular coordinates $(-156, -90)$ on the system described in Exercise 1. What distance and angle would a bee signal for this food source? 180; 210°

Cumulative Review

1. $\begin{bmatrix} 8 & 0 \\ 0 & -3 \end{bmatrix}$ 2. $\begin{bmatrix} -2 & -11 \\ 7 & -10 \end{bmatrix}$

3. $\begin{bmatrix} 1 & 0 \\ 0 & 1 \end{bmatrix}$ 4. $\begin{bmatrix} 1 & -1 \\ 1 & 0 \end{bmatrix}$

Chapter 13

Solve for the matrix X.

1. $X - \begin{bmatrix} 6 & -5 \\ 2 & 4 \end{bmatrix} = \begin{bmatrix} 2 & 5 \\ -2 & -7 \end{bmatrix}$

2. $X + \begin{bmatrix} 3 & 8 \\ -2 & 4 \end{bmatrix} = \begin{bmatrix} 1 & -3 \\ 5 & -6 \end{bmatrix}$

3. $3\begin{bmatrix} 1 & 0 \\ 0 & 1 \end{bmatrix} - 6X = \begin{bmatrix} -3 & 0 \\ 0 & -3 \end{bmatrix}$

4. $\begin{bmatrix} 1 & 2 \\ -1 & 3 \end{bmatrix} X = \begin{bmatrix} 3 & -1 \\ 2 & 1 \end{bmatrix}$

Perform the indicated operation.

5. $\begin{bmatrix} 3 & 1 \\ -2 & 0 \end{bmatrix} + \begin{bmatrix} -1 & 7 \\ 6 & 1 \end{bmatrix} \begin{bmatrix} 2 & 8 \\ 4 & 1 \end{bmatrix}$

6. $\begin{bmatrix} 3 & 2 & 0 \\ 4 & -1 & 5 \end{bmatrix} \begin{bmatrix} 1 \\ 2 \\ -3 \end{bmatrix} \begin{bmatrix} 7 \\ -13 \end{bmatrix}$

7. Find A^2 if $A = \begin{bmatrix} 1 & 3 \\ 0 & 0 \end{bmatrix} \cdot \begin{bmatrix} 1 & 3 \\ 0 & 0 \end{bmatrix}$

8. Find the inverse of $\begin{bmatrix} -2 & \frac{7}{2} \\ 1 & 2 \end{bmatrix}$. $\begin{bmatrix} -\frac{4}{15} & \frac{7}{15} \\ \frac{2}{15} & \frac{4}{15} \end{bmatrix}$

9. Solve $\begin{bmatrix} 1 & 2 \\ 3 & 2 \end{bmatrix} \begin{bmatrix} x \\ y \end{bmatrix} = \begin{bmatrix} -1 \\ 1 \end{bmatrix}$ for x and y. $x = 1, y = -1$

10. Determine the image of $P(1, -1)$ under the following transformation:
$\begin{bmatrix} x' \\ y' \end{bmatrix} = \begin{bmatrix} 3 & 7 \\ 9 & 5 \end{bmatrix} \begin{bmatrix} x \\ y \end{bmatrix}$ $(-4, 4)$

Inverses; Polar Coordinates; Vectors **699**

11. Determine the preimage of $P'(6, -4)$ under the transformation $\begin{bmatrix} x' \\ y' \end{bmatrix} = \begin{bmatrix} 0 & 2 \\ 2 & 0 \end{bmatrix}\begin{bmatrix} x \\ y \end{bmatrix}$. $(-2, 3)$

Chapter 14

12. Express $270°$ as a radian measure using π. $\frac{3\pi^R}{2}$

13. Express $-\frac{4\pi}{5}^R$ as a degree measure. $-144°$

14. Find $\cos \alpha$ when α is in Quadrant IV and $\sin \alpha = -\frac{8}{17}$. $\frac{15}{17}$

15. Find all $m(\alpha)$ for which the terminal side of α in standard position passes through the point $(12, 5)$. $22.6° + k \cdot 360°$

16. Find $\sin \left(-\frac{5\pi}{6}^R\right)$. $-\frac{1}{2}$

17. Find $\cos 1014°$. 0.4067

18. What is the amplitude and fundamental period of the function $y = -2 \cos 4x$? $2; \frac{\pi}{2}$

19. If $\sec \alpha = -3$ in Quadrant III, find the value of the five other trigonometric functions.

20. In right triangle ABC, if $m(C) = 90°$, $a = 19.7$, and $b = 19.6$, find $m(B)$. $44.9°$

21. In right triangle ABC, if $m(C) = 90°$, $m(A) = 36.8°$, and $b = 40$, find a. 29.9

19. $\sin \alpha = -\frac{2\sqrt{2}}{3}$, $\cos \alpha = -\frac{1}{3}$, $\tan \alpha = 2\sqrt{2}$, $\csc \alpha = -\frac{3}{2\sqrt{2}}$, $\cot \alpha = \frac{1}{2\sqrt{2}}$

Chapter 15

In Exercises 22 and 23 write an equivalent expression in simplest form using only the sine and/or cosine function. Note any restrictions.

22. $\frac{\sec^2 x - 1}{\cos^2 x - 1}$ $-\frac{1}{\cos^2 x}$, $\cos x \neq 0, \pm 1$

23. $(\tan x + \sin x)(1 - \cos x)$ $\frac{\sin^3 x}{\cos x}$, $\cos x \neq 0$

24. Express $\cos \frac{17\pi}{12}$ as the cosine of a sum, and evaluate the function. Leave your answer in simple radical form. $\cos \left(\frac{5\pi}{4} + \frac{\pi}{6}\right) = \frac{-\sqrt{6} + \sqrt{2}}{4}$

25. Express $\cos 175° \cos 25° + \sin 175° \sin 25°$ as a function of a single angle and evaluate the function. $\cos 150°$, $-\frac{\sqrt{3}}{2}$

26. Express $\sin 225° \cos 165° - \cos 225° \sin 165°$ as a function of a single angle and evaluate the function. Leave your answer in simple radical form. $\sin 60°$, $\frac{\sqrt{3}}{2}$

27. Use the half-angle formulas to find $\sin \frac{5\pi}{8}$. $\frac{1}{2}\sqrt{2 + \sqrt{2}}$

28. In $\triangle ABC$, $a = 9$, $b = 10$, and $c = 5$. Find $m(B)$ to the nearest degree. $86°$

29. In $\triangle ABC$, $m(A) = 30°$, $m(B) = 45°$ and $a = 20$. Find b to the nearest tenth. 28.3

700 *Chapter 16*

30. Find the value of arcsine (-1) for the inverse trigonometric and the inverse circular functions. $\frac{3\pi}{2} + 2k\pi$, $270° + k \cdot 360°$

31. Evaluate $\operatorname{Sin}^{-1}\left(\cos\frac{3\pi}{4}\right)$. $-\frac{\pi}{4}$

32. Solve $\sec^2 x - 2 = 0$ over $\mathcal{R}$. $\{\frac{\pi}{4} + k\frac{\pi}{2}\}$

33. Solve $\sin^2 x = 3\cos^2 x$ in the interval $0 \leq x \leq 2\pi$. $\{\frac{\pi}{3}, \frac{2\pi}{3}, \frac{4\pi}{3}, \frac{5\pi}{3}\}$

34. Find a pair of polar coordinates for the point $(-2\sqrt{3}, 2)$ with $-180° < m(\alpha) \leq 180°$. $(4, 150°), (-4, -30°)$

35. Find the Cartesian coordinates of $(-3, 150°)$. $(\frac{3\sqrt{3}}{2}, -\frac{3}{2})$

36. Express $4(\cos 240° + i \sin 240°)$ in the form $a + bi$. $-2 - 2i\sqrt{3}$

37. Express $(1 + i)^5$ in the form $a + bi$. $-4 - 4i$

38. Given vectors **u** and **v** such that $\|\mathbf{u}\| = 10$ and the direction angle of **u** is 45°, and $\|\mathbf{v}\| = 20$ and the direction angle of **v** is 120°, find $\|\mathbf{u} + \mathbf{v}\|$ to the nearest tenth. 24.6

PROGRAMMING IN PASCAL

Exercises

1. **a.** Write a Pascal procedure to convert polar coordinates into equivalent Cartesian coordinates. Assume θ is measured in degrees.

   b. Write a Pascal procedure to convert Cartesian coordinates into equivalent polar coordinates with θ measured in degrees. You may want to assume that $r \geq 0$ and $0° \leq \theta < 360°$. (The inverse trigonometric function must result with x and y in the correct quadrant.)

2. A complex number is written in standard form, $a + bi$.

   a. Write a Pascal procedure that is passed the values of a and b, and finds the reciprocal of $a + bi$ in standard form.

   b. Write a Pascal program that uses the function in part (a) to find and display in standard form $(a + bi)^n$, where n is an integer.

3. Write a Pascal procedure to compute the horizontal and vertical components, u_x and u_y, of a vector u when the norm of u and the direction angle (measured in degrees) is given. Test your program on some problems you have already solved.

4. Write a Pascal program to compute the norm and direction angle of the sum of two vectors, when the norm and direction angle for each vector are known.

The photo shows a continuous heat-treating line for the production of high strength steel. Some of the statistical methods used in this chapter can be used to measure the probability that the steel from this facility meets certain specifications.

702 *Chapter 17*

Chapter 17

Statistics

Problem Solving Strategies

Making a Table
Throughout the chapter tables are used to illustrate concepts and display information.
Drawing a Diagram
Frequency distributions and normal distributions are presented by making tables and by graphing the data given in those tables in Sections 17-1 (page 703) and 17-3 (page 717).

Organizing and Measuring Statistical Data

OBJECTIVES for Sections 17-1 and 17-2:
1. To represent data graphically, using histograms and frequency polygons.
2. To find the mean, median, and mode of a given set of data.
3. To find the variance and the standard deviation of a given set of data.
4. To use z-scores to compare data points from different sets of data.

17–1 Representing Data; Measures of Central Tendency

Consider the following list, which gives the EPA fuel economy ratings (in miles per gallon) of all the makes of cars sold in the United States in a particular year. The ratings have all been rounded to the nearest whole number.

20	19	20	26	19	25	18	16	16	21	18	18	19	20	19
21	21	20	19	26	21	19	15	19	19	18	18	17	21	36
26	32	27	23	25	17	21	28	27	18	18	17	28	24	22
26	30	20	19	16	20	16	27	35	14	30	16	20	16	30
16	20	21	19	19	17	23	17	37	16	25	25	21	19	19
21	21	21	20	16	24	28	25	21	27	25	22	39	29	42
						25	25	25						

Teaching Suggestions
p. T127

Related Activities
p. T127

Key Ideas

Prepare a frequency table, histogram, and frequency polygon for a given set of data. Define and explain the statistical measures of central tendency. Determine the mean, median, and mode for a given set of data.

Statistics **703**

The first thing you probably notice about the table is that it contains a great deal of information, or **data.** However, since the data lack organization, it is difficult to draw any general conclusions about them. One way to organize data is in a **frequency table.** This is a table that lists each number in the data, or **data point,** followed by its **frequency**—the number of times it occurs. It may also include the **relative frequency** of each data point. This is the fraction of the total number of data points that are represented by the frequency of a particular data point. For example, the list on page 703 contains 93 data points, of which 13 have the value 19. Therefore

$$\text{relative frequency of } 19 = \frac{13}{93}$$

$$\approx 0.140$$

Here are the first ten lines of the frequency table for the data on page 703.

Fuel economy rating	Frequency	Relative frequency
14	1	0.011
15	1	0.011
16	9	0.097
17	5	0.054
18	7	0.075
19	13	0.140
20	9	0.097
21	12	0.129
22	2	0.022
23	2	0.022

This kind of table is better than the unorganized array in which the data were originally presented, but, since the data contain 23 *distinct* numerical values, the complete table would have 23 lines. For some purposes, especially graphical display, this is inconvenient. A better way of organizing the data is to employ **classes** of fuel economy ratings rather than single values. For example, the economy ratings 18, 19, and 20 can be grouped as follows:

Fuel economy rating	Frequency	Relative frequency
18–20	29	0.312

This means that 29 of the fuel economy ratings fall within the **class interval** 18–20 miles per gallon, which includes both 18 and 20. In compiling a frequency table using classes, it is customary to make all the class intervals of equal size, except possibly the first and the last. You can estimate this size by subtracting the smallest data point from the largest and dividing by the number of intervals you want. A choice of about ten intervals is usually convenient.

704   *Chapter 17*

EXAMPLE 1 Prepare the complete frequency table for the fuel economy data given on page 703.

SOLUTION Ten intervals are chosen. Nine of the intervals cover three fuel economy ratings each. The last interval covers the four ratings from 39 to 42.

Fuel economy rating	Frequency	Relative frequency
12–14	1	0.011
15–17	15	0.161
18–20	29	0.312
21–23	16	0.172
24–26	15	0.161
27–29	8	0.086
30–32	4	0.043
33–35	1	0.011
36–38	2	0.022
Over 38	2	0.022

In order to present the data in an even more accessible form, you can draw a bar graph, or **histogram,** from a table like the one above. A histogram shows the **distribution** of the data. In drawing a histogram you should follow these guidelines:

1. Make the vertical bars of equal width. Leave no space between the bars.
2. List class intervals of the data in increasing order along the horizontal axis. List frequency along the vertical axis. For some purposes you may prefer to list the relative frequency instead of the frequency. You may even list both the frequency and relative frequency, as is shown in Example 2 below.
3. Make the units of equal size along each axis. However, the unit lengths of the two axes need not be the same.

EXAMPLE 2 Prepare a histogram based on the frequency table of Example 1.

SOLUTION The histogram is shown below.

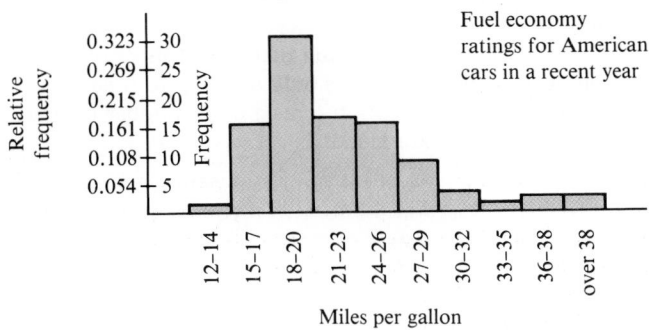

Fuel economy ratings for American cars in a recent year

Miles per gallon

The team scores for a high school soccer team during a six-week period are shown below.

3, 4, 0, 5, 4, 2,
0, 1, 3, 1, 1, 2

1. Prepare a frequency table.

Score	Freq.	Rel. Freq.
0	2	0.167
1	3	0.250
2	2	0.167
3	2	0.167
4	2	0.167
5	1	0.083

2. Prepare a histogram.

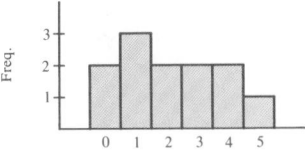

Points Scored

3. Prepare a frequency polygon.

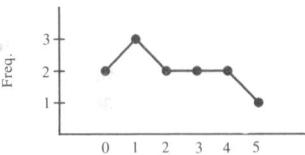

Points Scored

4. Find the mean score for the team.

$$\bar{x} = \frac{26}{12} \approx 2.2$$

5. Find the median score for the team.
Arrange data in increasing order;
median is 2

6. Find the mode for the team score. 1

705

Another way of presenting data is as a **frequency polygon.** In this kind of graph, the class intervals are represented by equally spaced points along the horizontal axis. Directly above each such point, another point is located whose height above the horizontal axis is the frequency of the class interval. These latter points are then joined in order by line segments. You can construct a frequency polygon by joining the midpoints of the tops of the bars of a histogram illustrating the same data.

EXAMPLE 3   Prepare a frequency polygon based on the frequency table of Example 1.

SOLUTION   The frequency polygon is shown below.

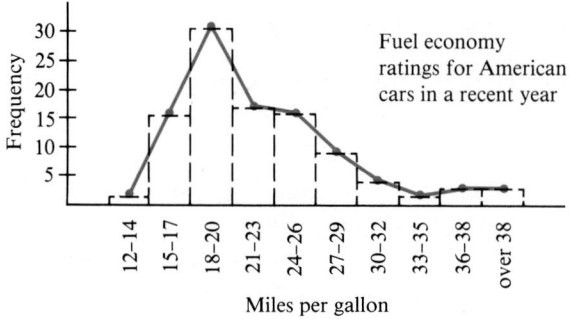

Fuel economy ratings for American cars in a recent year

(As a rule, you do *not* actually draw the corresponding histogram when constructing a frequency polygon.)

Often we would like to characterize a set of data by a single number, or **statistic,** that is derived from the data. Three such numbers indicate **central tendency,** that is, the numerical value around which the data points tend to cluster. They are described below.

1. The average, or **mean,** is the sum of all the data points divided by the number of data points (the sum of all the frequencies).
2. The **median** is the value that falls in the middle when all the data points are arranged in numerical order. If there is an *even* number of data points, the median is defined to be the mean of the two middle values.
3. The **mode** is the numerical value that occurs most often as a data point. If each value occurs only once, there is *no mode.* If more than one value occurs the maximum number of times, then *all* those values are modes.

If $\{x_i: i = 1, 2, \ldots, n\}$ is a set of n data points, we can use the summation notation of Section 8-3 to give a compact formula for the mean, denoted $\bar{x}$, of these data:

$$\bar{x} = \frac{\displaystyle\sum_{i=1}^{n} x_i}{n}$$

For the special case $n = 2$, this formula simplifies to the arithmetic mean of two numbers that was defined on page 297.

If a particular data point x_i has frequency f_i, then, since

$$\underbrace{x_i + x_i + \cdots + x_i}_{f_i \text{ times}} = x_i f_i,$$

we can rewrite the formula on the previous page as

$$\overline{x} = \frac{\sum_{i=1}^{r} x_i f_i}{\sum_{i=1}^{r} f_i},$$

where r is the number of *distinct* data points.

EXAMPLE 4   The number of children per family for each of 30 students in a math class at Eastville High School are shown below.

$$1, 2, 2, 4, 3, 1, 1, 1, 1, 2, 2, 6, 3, 1, 3,$$
$$1, 2, 2, 2, 4, 3, 3, 4, 2, 2, 1, 1, 1, 1, 2$$

Find each of the following:
a. the mean number of children per family
b. the median number of children per family
c. the mode for the number of children per family

SOLUTION   a. A frequency table for the given data is as follows:

Number of children	1	2	3	4	5	6
Frequency	11	10	5	3	0	1

Therefore the mean is given by

$$\overline{x} = \frac{1 \cdot 11 + 2 \cdot 10 + 3 \cdot 5 + 4 \cdot 3 + 5 \cdot 0 + 6 \cdot 1}{11 + 10 + 5 + 3 + 0 + 1} = \frac{64}{30} \approx 2.13$$

The result of this example demonstrates that the mean may not be an actual (or even a possible) data point.

b. Arrange the given data in increasing order.

$$1, 1, 1, 1, 1, 1, 1, 1, 1, 1, 1, 2, 2, 2, 2,$$
$$2, 2, 2, 2, 2, 3, 3, 3, 3, 3, 4, 4, 4, 6$$

Since there is an even number of data points, the median is the mean of the two middle values, 2 and 2. Therefore, the median is

$$\frac{2 + 2}{2} = 2.$$

c. The value 1 occurs 11 times and no other value occurs as frequently. Therefore, 1 is the only mode.

Statistics   **707**

1. A marathon runner completed a 26-km race in 4 h. What was the rate per hour?   6.5 km/h

2. A man drove his car at 60 km/h for 2 h and at 80 km/h for 3 h. How far did he drive and what was the car's average rate?   360 km; 72 km/h

Simplify and state the answer to the nearest tenth.

3. $\dfrac{15 \cdot 3 + 12 \cdot 5 + 11 \cdot 6}{14}$

   12.2

4. $(27 \cdot 2 + 20 \cdot 8 + 15 \cdot 7 + 10 \cdot 5) \div 22$ 16.8

5. $(15 \cdot 20 + 12 \cdot 15 + 10 \cdot 10 + 7 \cdot 23) \div 68$ 10.9

Additional A Exercises

In Exercises 1–4 find the mean, the median, and the mode of the given set of data.

1. Bowling scores of a team: 152, 110, 126, 191, 137, 160, 126, 130   141.5; 133.5; 126

2. The number of hours spent doing homework in a week: 6, 3, 10, 5, 4, 6, 8, 2, 9, 6, 7, 7, 6, 4, 2, 3, 9, 8, 10, 4, 5   5.9; 6; 6

3. In a science class the number of miles that each student lives from school: 4, 1.5, 2.5, 3, 1.25, 0.25, 0.5, 1, 2, 1, 1.25, 1.5, 0.5, 1.25, 1.75   1.55; 1.25; 1.25

4. The amount of money spent for lunch in a week: $10, $4, $6, $8, $10, $15, $11, $7, $9 $8.89; $9; $10

5. Construct a frequency table with 5 intervals for the data given in Exercise 2.

No. of hours	Freq.
1–2	2
3–4	5
5–6	6
7–8	4
9–10	4

Oral Exercises

Exercises 1–6 refer to the following list of numbers of passengers on a certain bus route on each day of a two-week period.

31, 30, 25, 34, 40, 86, 91, 95, 32, 30, 29, 30, 36, 34

1. Find the mean. 44.5   2. Find the mode. 30   3. Find the median. 31.5

4. Which of the three statistics from Exercises 1–3 seems to give the best measure of central tendency for this particular set of data? median

5. As a measure of the central tendency of individual annual income in the United States, economists generally use the median rather than the mean. Can you explain why on the basis of Exercises 1–4?

6. If each value in the list above were increased by 5, how would the mean, the median, and the mode be affected? Each is increased by 5.

5. The median is more representative of a greater number of incomes.

Written Exercises

In Exercises 1–6, find the mean, the median, and the mode of the given set of data.

A 1. Scores on a math test: 79, 82, 88, 93, 92, 82, 76, 70, 92, 82   83.6; 82; 82

2. The number of tomatoes on each of 11 tomato plants after 8 weeks of growth: 5, 6, 8, 7, 2, 3, 4, 4, 9, 4, 3   5; 4; 4

3. The number of lawnmowers sold by a salesperson during a nine-week period: 1, 5, 3, 2, 8, 10, 9, 12, 13   7; 8; no mode

4. The number of runs scored by a baseball team in each of its first 10 games: 5, 0, 1, 3, 8, 12, 2, 4, 0, 6   4.1; 3.5; 0

5. The normal Fahrenheit temperature in Phoenix, Arizona for a twelve-month period: 51, 55, 60, 68, 75, 85, 91, 88, 83, 72, 60, 53   70; 70; 60

6. The sizes of pants sold during one business day in a clothing store: 34, 40, 36, 44, 38, 40, 36, 42, 46, 34, 36, 36, 30, 44, 32   38; 36; 36
Of the three statistics you found, which is probably of most interest to the manager of the clothing store? mode

7. Construct a frequency table for the data given in Exercise 1.

8. Construct a frequency table for the data given in Exercise 6.

9. To improve its delivery system, a canned soup manufacturer has made up a frequency table that lists the numbers of cartons ordered weekly at stores it services and the frequency with which each number of cartons is ordered. Draw a histogram based on the data in the table below:

Number of cartons	1–3	4–6	7–9	10–12	13–15	16–18	19–21	over 21
Frequency	4	2	5	9	12	15	8	3

10. A teacher has made up a frequency table that shows the ages of students using the computer room. Draw a histogram based on the data shown in the table below.

Age	14	15	16	17	18	over 18
Frequency	2	12	27	38	20	1

11. The following is a list of the ages of the Presidents of the United States at the time of their first inauguration:

57	61	57	57	58	57	61	54	68	51
49	64	50	48	65	52	56	46	54	49
50	47	55	55	54	42	51	56	55	51
54	51	60	62	43	55	56	61	52	69

Construct a frequency table beginning with the class "35–45" and ending with the class "over 66," and including 9 classes altogether. Draw a histogram from your table.

12. A *degree-day* is a measure of fuel requirements for buildings. (See page 304.) The numbers of degree-days recorded for November in Boston, Massachusetts, were:

12	16	16	18	17	18	21	23	25	22
21	18	22	18	20	24	25	25	26	23
25	27	29	28	27	31	29	28	34	32

Construct a frequency table with 9 classes and draw a histogram from these data.

13–16. Draw a frequency polygon based on the data in Exercise 9–12.

17–20. Find the mean of the data given in Exercises 9–12.

17. 13.9 **18.** 16.7 **19.** 54.8 **20.** 23.3

B 21. A teacher compiled the following data based on the same test given to three different classes. What was the mean score of all the students? 79.3

	Number of students	Class average
Class *P*	38	76.5
Class *Q*	22	85.0
Class *R*	35	78.8

22. Three makers of television sets claim 5-year "average" life spans for their sets, based on the following life-span data (in years):

Brand X: 2, 2, 2, 3, 3, 4, 4, 9, 9, 12
Brand Y: 1, 2, 2, 3, 3, 4, 4, 5, 5, 5
Brand Z: 3, 3, 3, 4, 5, 5, 6, 6, 7, 7

a. Which indicator of central tendency did each maker use to compute its "average"? Brand *X*: mean; Brand *Y*: mode; Brand *Z*: median

b. Which brand seems to be the most likely to last 5 years or more? Brand *Z*

Statistics **709**

7.
Score	Freq.
70	1
76	1
79	1
82	3
88	1
92	2
93	1

8.
Size	Freq.
30	1
32	1
34	2
36	4
38	1
40	2
42	1
44	2
46	1

9.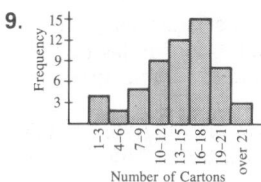

10.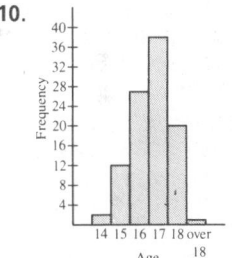

11.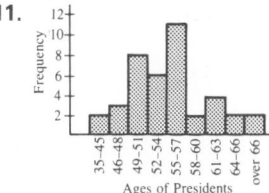

(continued)

12.

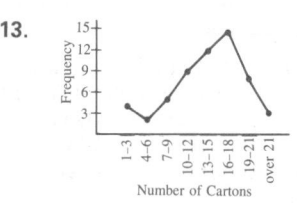

13.

14.

15.

16.

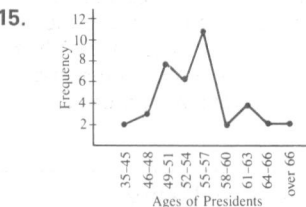

The *percentile rank* of a data point x = (percent of values less than x) + $\frac{1}{2}$(percent of values equal to x). Using the fuel economy ratings on page 703, find the percentile rank of each of the following. Round your answer to the nearest whole-number percentile.

C **23.** 32 94   **24.** 28 88   **25.** 21 55

Computer Exercises   For students with computer experience

1. Write a program that will sort the data on page 703 into the intervals indicated on page 705 and produce the table shown in Example 1.
2. Modify the program in Exercise 1 to find the actual mean of the data.
3. Modify the program in Exercise 2 to estimate the mean from the grouped data in the table. First, estimate the total by multiplying the *midpoint* of each interval by the frequency of that interval and finding the sum of these products. Then divide this sum by the number of data (the sum of the frequencies).
4. Modify the program in Exercise 3 to find the midpoint of the interval in which the median lies.

2. 22.172043   **3.** 22.225806   **4.** 22

17–2 Measures of Variation

If you were considering a hiking trip into a region of peaks and valleys, the mean or median altitude above sea level of these peaks and valleys would be of little interest to you. Instead you would want to know the amount of *variation* of these altitudes. In statistics, the question of variation also plays an important role.

    One measure of the variation (also called *variability* or *dispersion*) of data is the range. This is defined simply by

range = (greatest data point) − (least data point).

The range of the fuel economy data on page 703 is thus 42 − 14 = 28. Unfortunately, the range is of limited usefulness in statistics. To return to the proposed hiking trip, simply knowing the highest and lowest points of your journey would tell you nothing about the number of ups and downs that you would encounter along the way. The situation is similar with the range as a measure of the variation of data. Instead, statisticians often use a more sophisticated measure of variation, the variance, denoted σ^2 (read "sigma squared") and defined as follows for a set of data $\{x_i: i = 1, 2, \ldots, n\}$:

$$\sigma^2 = \frac{\sum_{i=1}^{n} (x_i - \bar{x})^2}{n}, \qquad (1)$$

710   *Chapter 17*

where $\bar{x}$ is the mean of the given data. If the data points are widely scattered about the mean, the variance will be relatively large, whereas if they are narrowly distributed about the mean, the variance will be relatively small. Note also that the squaring of the terms has the effect of avoiding the cancellation that would otherwise result from data points falling both above and below the mean.

If there are r distinct data points with frequencies $\{f_i: i = 1, 2, \ldots, r\}$, we can rewrite Formula (1) on the previous page as

$$\sigma^2 = \frac{\sum_{i=1}^{r} (x_i - \bar{x})^2 \cdot f_i}{\sum_{i=1}^{r} f_i}. \tag{1a}$$

The most important measure of the variation of a set of data is the **standard deviation**. The standard deviation, denoted σ, is defined as the nonnegative square root of the variance:

$$\sigma = \sqrt{\frac{\sum_{i=1}^{n} (x_i - \bar{x})^2}{n}} \tag{2}$$

As with the variance, if the data are given in a frequency table, we can write

$$\sigma = \sqrt{\frac{\sum_{i=1}^{r} (x_i - \bar{x})^2 \cdot f_i}{\sum_{i=1}^{r} f_i}}. \tag{2a}$$

EXAMPLE 1 The applicants for an accounting job made scores on a qualifying test that are summarized in the following frequency table:

Score	65	70	75	80	85	90	95
Frequency	1	6	14	12	8	4	3

Find the range, the variance, and the standard deviation of these data.

SOLUTION The range is $95 - 65 = 30$.
To find the variance, first compute the mean, $\bar{x}$:
$$\bar{x} = \frac{65 \cdot 1 + 70 \cdot 6 + 75 \cdot 14 + 80 \cdot 12 + 85 \cdot 8 + 90 \cdot 4 + 95 \cdot 3}{1 + 6 + 14 + 12 + 8 + 4 + 3}$$
$$= \frac{3820}{48} \approx 79.58$$
$$\sigma^2 \approx \frac{(65 - 79.58)^2 \cdot 1 + (70 - 79.58)^2 \cdot 6 + \ldots + (95 - 79.58)^2 \cdot 3}{48}$$
$$= \frac{2441.67}{48} \approx 50.87$$

The standard deviation, σ, is given by $\sigma = \sqrt{\sigma^2} = \sqrt{50.87} \approx 7.13$.

Statistics **711**

Supplementary Materials
Test 54

Key Ideas
Find the range, the variance, and the standard deviation of a given set of data. Use z-scores to compare data points from different sets of data.

Common Errors
In using the formula for variance (1b on page 712), students often forget to multiply the sum of the squares of the data values by the number of values. Point out that then the variance will be a negative number and the standard deviation, which is the square root of the variance, cannot be found.

Reading Algebra
Students should be able to read the symbol Σ and to state the statistical formulas in the chapter. Formula 1b on page 712 is read "the number of values multiplied by the sum of the squares of the values minus the square of the sum of the values, all divided by the square of the number of values." Formula 2b on page 712 is read "the square root of [the above statement for formula 1b]." Encouraging students to read the formulas slowly and correctly will help them understand and feel less overwhelmed by the notation.

1. The ages of the persons who applied for a job opening are summarized in the frequency table below.

Age	Frequency
17	7
18	12
19	19
20	4
21	13
22	6
24	2
25	1

Find the range, the variance, and the standard deviation of these data.

Range is $25 - 17 = 8$.

$\bar{x} = \dfrac{1254}{64} \approx 19.59$

$\sigma^2 \approx \dfrac{213.44}{64} \approx 3.34$

$\sigma = \sqrt{3.34} \approx 1.83$

2. The mean of a set of data is 45 and the standard deviation is 12.

a. Find the z-score of the data point 25.

$z\text{-score} = \dfrac{25 - 45}{12} \approx$

-1.67

b. Find the data point corresponding to the z-score 0.48.

$0.48 = \dfrac{x_i - 45}{12}, x_i =$

50.76, or 51

3. In a beanbag toss contest at a fair, each entrant threw 20 beanbags. The number of bull's-eyes for two age groups is shown below.

Group 1: 7, 9, 9, 10, 11, 13, 17;

Group 2: 8, 8, 8, 9, 10, 11, 12.

Even with the aid of a calculator, the computations in Example 1 on the previous page can be tedious. The following formula, which can be derived from Formula (1) (see Exercise 26), greatly simplifies the computation of the variance:

$$\sigma^2 = \frac{n\left(\sum\limits_{i=1}^{n} x_i^2\right) - \left(\sum\limits_{i=1}^{n} x_i\right)^2}{n^2} \qquad (1\text{b})$$

Likewise, the standard deviation can be given by

$$\sigma = \sqrt{\frac{n\left(\sum\limits_{i=1}^{n} x_i^2\right) - \left(\sum\limits_{i=1}^{n} x_i\right)^2}{n^2}}. \qquad (2\text{b})$$

These formulas can also be given in terms of the distinct data points and their frequencies (see Oral Exercise 6).

The concept of the standard deviation is especially valuable because it enables us to compare data points from *different sets* of data. For example, suppose two classes achieved the following grades on a mathematics test:

Class 1: 64, 70, 73, 77, 85, 90, 94

Class 2: 74, 75, 75, 76, 79, 80, 94

Although the mean grade is 79 in both classes, a grade of 94 is clearly more impressive in Class 2 because of the way the grades are distributed in that class. Statisticians express such information quantitatively by assigning to each value in a set of data a standard score, or *z-score*, defined by

$$z = \frac{x_i - \bar{x}}{\sigma},$$

where $\bar{x}$ is the mean and σ is the standard deviation of the set of data. Thus the z-score tells *how many standard deviations* the data point lies *above or below the mean*. (The z-score of a data point will obviously be negative when it is less than the mean.)

By using z-scores, we can now measure the relative importance of a grade of 94 in the mathematics test results of Classes 1 and 2. For Class 1, $\sigma = 10.2$, and for Class 2, $\sigma = 6.46$. Therefore the z-score of a grade of 94 is as follows:

$$\text{Class 1: } z = \frac{94 - 79}{10.2} \approx 1.47$$

$$\text{Class 2: } z = \frac{94 - 79}{6.46} \approx 2.32$$

When the two grades of 94 are measured in this way, as 1.47 standard deviations and 2.32 standard deviations, respectively, it is easy to see that 94 is higher above the mean in Class 2 than in Class 1. This indicates that 94 should be regarded as a better grade in Class 2 than in Class 1.

EXAMPLE 2   The mean of a set of data is 86 and the standard deviation is 34.
  a. Find the z-score of the data point 98.
  b. Find the data point corresponding to the z-score -1.6.

SOLUTION   **a.** The data point x_i is 98 and $\sigma = 34$.

The z-score is $\dfrac{98 - 86}{34} \approx 0.35$.

b. The z-score is -1.6. Therefore $-1.6 = \dfrac{x_i - 86}{34}$. When this equation is solved for x_i, the result is $x_i = (-1.6)34 + 86 = 31.6$, or about 32.

EXAMPLE 3   On an achievement test John scored 90 in language arts and 88 in mathematics. The mean score for language arts was 84 and the standard deviation was 6. The mean score for mathematics was 80 and the standard deviation was 3. On the basis of this test, in which of the two areas does John exhibit greater achievement?

SOLUTION   The z-scores of John's grades are as follows:

language arts: $z = \dfrac{x_i - \overline{x}}{\sigma} = \dfrac{90 - 84}{6}$

$= 1.00$

mathematics: $z = \dfrac{x_i - \overline{x}}{\sigma} = \dfrac{88 - 80}{3}$

≈ 2.67

A language arts grade of 90 is 1.00 standard deviations above the mean. A mathematics grade of 88 is 2.67 standard deviations above the mean. Therefore, John exhibits greater achievement in mathematics than in language arts, based on this test.

Oral Exercises

Exercises 1–5 refer to the following sets of data.

  Set A: 30, 52, 68
  Set B: 48, 49, 53

6. $\sigma^2 = \dfrac{\left(\sum\limits_{i=1}^{r} f_i\right)\left(\sum\limits_{i=1}^{r} f_i x_i^2\right) - \left(\sum\limits_{i=1}^{r} f_i x_i\right)^2}{\left(\sum\limits_{i=1}^{r} f_i\right)^2}$;

$\sigma = \sqrt{\sigma^2}$

1. State the range of each set. A: 38; B: 5
2. The mean of each set is 50. Which set do you think has a greater variance? a greater standard deviation? A; A
3. Find the variance and the standard deviation of Set B.   4.67; 2.16
4. If the number 70 were introduced into each set, in which set do you think it would have a greater z-score? Why? B
5. If the number 25 were introduced into each set, in which set would its z-score be greater? Why? A
6. State Formulas (1b) and (2b) in terms of r distinct data points $\{x_i\}$ and their frequencies $\{f_i\}$.

a. Find the mean, the variance, and the standard deviation for each group.

Group 1: $\overline{x} = \dfrac{76}{7} \approx$

10.86; $\sigma^2 \approx \dfrac{64.86}{7} \approx$

9.27; $\sigma = \sqrt{9.27} \approx$

3.04.

Group 2: $\overline{x} = \dfrac{66}{7} \approx$

9.43; $\sigma^2 \approx \dfrac{15.71}{7} \approx$

2.24; $\sigma = \sqrt{2.24} \approx 1.5$

b. In which group is a score of 11 a more impressive score?
z-score for Group 1:

$\dfrac{11 - 10.86}{3.04} \approx 0.046$; z-

score for Group 2:

$\dfrac{11 - 9.43}{1.5} \approx 1.047$;

Group 2

Mixed Review

Find the mean, the median, and the mode for each set of data.

1. 2, 3, 5, 7, 7, 8, 10
   6; 7; 7
2. 88, 90, 80, 96, 91, 82, 88, 89   88; 88.5; 88

Evaluate.

3. $\sum\limits_{i=1}^{4} (i - 0.3)^2$ 24.36

4. $\sum\limits_{i=1}^{5} (i - 1.2)^2$ 26.2

5. Simplify $\sqrt[5]{3n^4} \cdot \sqrt[5]{8n^3} \cdot \sqrt[5]{4n^2}$. $2n\sqrt[5]{3n^4}$

6. Solve $-2.4 = \dfrac{x - 19}{3.1}$. $\{11.56\}$

Suggested Assignments

Enriched Alg.
  714/1–13 odd, 14–17
R 716/Self-Test 1
S 708/1, 2
Enriched Alg. with Trig.
  714/1–13 odd, 14–17
R 716/Self-Test 1
S 708/1, 2

Additional A Exercises

Find the variance and the standard deviation of the given data.

1. Number of tests given per term by math teachers: 4, 4, 6, 6, 7, 7, 8 2; 1.41

2. Number of cars per hour driven through a given intersection: 20, 27, 25, 23, 18, 10 30.92; 5.56

3. Number of long-distance calls per day made by a company: 4, 6, 10, 12, 15, 18, 18, 19 28.69; 5.36

The mean of a set of data is 47 and the standard deviation is 16. Find the z-score of each data point.

4. 58 0.69

5. 83 2.25

6. 92 2.81

Additional Answers
Written Exercises

25. $\bar{y} = \dfrac{1}{n}\sum_{i=1}^{n} y_i = \dfrac{1}{n}\sum_{i=1}^{n} mx_i =$

$\dfrac{m}{n}\sum_{i=1}^{n} x_i = m\left(\dfrac{1}{n}\sum_{i=1}^{n} x_i\right) =$

$m\bar{x}.$

$\sigma_y^2 = \dfrac{\displaystyle\sum_{i=1}^{n}(y_i - \bar{y})^2}{n} =$

Written Exercises

Find the variance and the standard deviation of the given data.

A 1. Number of mushrooms in a supermarket container of a given size: 12, 17, 25, 25, 25, 30 35.89; 5.99

2. Masses of 5 U.S. coins (in grams): 2.6, 2.7, 2.9, 3.1, 3.2 0.052; 0.23

3. Average wind speed at a local airport on 7 successive days (in km/h): 4, 8, 12, 10, 7, 9, 6 6; 2.45

4. Numbers of visitors to the 7 most popular national parks in 1981 (in millions): 20.3, 17.0, 16.8, 12.5, 12.2, 12.2, 12.0 9.38; 3.06

5. Averages of the number of days per month with measurable precipitation in Buffalo, New York: 20, 17, 16, 14, 12, 10, 10, 11, 11, 11, 16, 20 12.67; 3.56

6. Average daily domestic crude oil production for the years 1973–1982 (in millions of barrels): 9.2, 8.8, 8.4, 8.1, 8.3, 8.7, 8.6, 8.6, 8.6, 8.7 0.08; 0.28

In Exercises 7–10, the mean of a set of data is 52 and the standard deviation is 20. Find the z-score of each data point.

7. 56 0.2   8. 82 1.5   9. 32 −1   10. 40 −0.6

In Exercises 11–14, the mean of a set of data is 48 and the standard deviation is 15. Find the data point corresponding to the given z-score.

11. $z = 1$ 63   12. $z = -2$ 18   13. $z = 0.8$ 60   14. $z = -1.5$ 25.5

15. Why do you think the following expression is never used as an indicator of variation in a set of data $\{x_i: i = 1, 2, \ldots, n\}$?

$$\dfrac{\displaystyle\sum_{i=1}^{n}(x_i - \bar{x})}{n}$$

(*Hint:* Compute the value of this expression for the two sets of data A and B in the Oral Exercises.) This expression is always equal to zero.

On an achievement test four students scored as follows:

					All students taking the test	
student	A	B	C	D	mean	standard deviation
law	80	90	84	85	71	6
medicine	76	90	92	108	90	8
engineering	77	66	74	70	72	2

On the basis of this test, which career, or careers, should each student consider pursuing? Explain your answer.

16. Student A engineering

17. Student B law

18. Student C law

19. Student D law

B 20. Herb and Maria scored 86 and 90, respectively, on a qualifying test for a job. Their z-scores, were 1.6 and 2.4 respectively. What were the mean and the standard deviation for this qualifying test?
$$\bar{x} = 78; \ \sigma = 5$$

In Exercises 21–24, tell how the mean and the standard deviation of a set of data are affected if the data are changed as described.

21. Every data point is increased by 5. $\bar{x}$ increased by 5; σ: no change
22. Every data point is multiplied by 2. $\bar{x}$ is doubled; σ is doubled
23. The *frequency* of each data point is doubled. $\bar{x}$: no change; σ: no change
24. If the mean of the original data $\{x_i\}$ is $\bar{x}$, the new data are $\{x_i'\}$, where $x_i' = \dfrac{x_i + \bar{x}}{2}$. $\bar{x}$: no change; σ is halved

C 25. Show that if $y_i = mx_i$, then $\bar{y} = m \cdot \bar{x}$ and $\sigma_y = |m| \cdot \sigma_x$.

26. Show that for data $\{x_1, x_2\}$, $\sigma = \dfrac{|x_2 - x_1|}{2}$.

27. Derive Formula (1b) on page 712 from Formula (1) on page 710. (*Hint*: For each term of the sum, you have
$$(x_i - \bar{x})^2 = x_i^2 - 2\bar{x}x_i + \bar{x}^2.$$
Note also that since $\bar{x}$ is a constant, $\displaystyle\sum_{i=1}^{n} \bar{x}x_i = \bar{x}\sum_{i=1}^{n} x_i .)$

Computer Exercises   For students with computer experience

1. Write a program that will find the mean, the variance, and the standard deviation using formula (1a) on page 711 for the variance.

Run the program for each of the following.

2. Example 1 on page 711
3. Exercises 1–6 on page 714
4. The variance for a frequency table may be found using the following formula
$$\sigma^2 = \frac{n\displaystyle\sum_{i=1}^{r} (x_i^2 f_i) - \left(\displaystyle\sum_{i=1}^{r} x_i f_i\right)^2}{n^2}$$
where $n = \displaystyle\sum_{i=1}^{r} f_i$. Change the program you wrote in Exercise 1 to use this formula for the variance.

Run the program for each of the following.

5. Example 1 on page 711
6. Exercise 10 on page 709

$$\frac{\displaystyle\sum_{i=1}^{n} (mx_i - m\bar{x})^2}{n} =$$

$$\frac{\displaystyle\sum_{i=1}^{n} (m(x_i - \bar{x}))^2}{n} =$$

$$m^2 \frac{\displaystyle\sum_{i=1}^{n} (x_i - \bar{x})^2}{n} = m^2 \sigma_x^{\,2}.$$

$$\therefore \sigma_y = \sqrt{m^2 \sigma_x^{\,2}} = |m|\sigma_x.$$

26. $\bar{x} = \dfrac{x_1 + x_2}{2}; \ \sigma^2 =$

$$\frac{\displaystyle\sum_{i=1}^{2} (x_i - \bar{x})^2}{2} =$$

$$\frac{(x_1 - \bar{x})^2 + (x_2 - \bar{x})^2}{2} =$$

$$\frac{1}{2}\left(\left(\frac{x_1}{2} - \frac{x_2}{2}\right)^2 + \left(\frac{x_2}{2} - \frac{x_1}{2}\right)^2\right) =$$

$$\left(\frac{x_1}{2} - \frac{x_2}{2}\right)^2 = \frac{1}{4}(x_1 - x_2)^2$$

$$\therefore \sigma = \sqrt{\frac{(x_1 - x_2)^2}{4}} =$$

$$\frac{1}{2}|x_1 - x_2|.$$

27. $\sigma^2 = \dfrac{\displaystyle\sum_{i=1}^{n} (x_i - \bar{x})^2}{n} =$

$$\frac{1}{n}\sum_{i=1}^{n} (x_i^2 - 2\bar{x}x_i + \bar{x}^2) =$$

$$\frac{1}{n}\left(\sum_{i=1}^{n} x_i^2 - 2\bar{x}\sum_{i=1}^{n} x_i + \sum_{i=1}^{n} \bar{x}^2\right). \text{ Since } \sum_{i=1}^{n} x_i =$$

$$n\bar{x} \text{ and } \sum_{i=1}^{n} \bar{x}^2 = n\bar{x}^2, \ \sigma^2 =$$

$$\frac{1}{n}\left(\sum_{i=1}^{n} x_i^2 - 2\bar{x}n\bar{x} + n\bar{x}^2\right) =$$

$$\frac{1}{n}\left(\sum_{i=1}^{n} x_i^2 - n\bar{x}^2\right). \text{ Since } \bar{x} =$$

$$\frac{1}{n}\sum_{i=1}^{n} x_i, \ \sigma^2 =$$

$$\frac{1}{n}\left(\sum_{i=1}^{n} x_i^2 - n\left(\frac{1}{n}\right)^2\left(\sum_{i=1}^{n} x_i\right)^2\right) =$$

$$\frac{1}{n^2}\left(n\sum_{i=1}^{n} x_i^2 - \left(\sum_{i=1}^{n} x_i\right)^2\right).$$

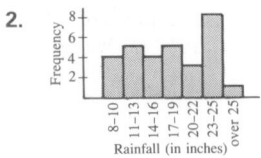
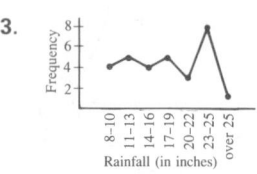
Self-Test 1

VOCABULARY   data (p. 703)
frequency table (p. 704)
data point (p. 704)
frequency (p. 704)
relative frequency (p. 704)
class (p. 704)
class interval (p. 704)
histogram (p. 705)
distribution of data (p. 705)
frequency polygon (p. 706)
statistic (p. 706)
central tendency (p. 706)
mean (p. 706)
median (p. 706)
mode (p. 706)
range (p. 710)
variance (p. 710)
standard deviation (p. 711)
z-score (p. 712)

The annual rainfall (inches) for a certain city for 30 consecutive years is listed below.

11	25	14	23	20	19	8	14	12	18
14	17	17	23	23	24	12	23	10	24
10	16	17	10	23	13	11	21	26	22

1. Construct a frequency table from these data using 7 classes.   *Obj. 1, p. 703*
2. Draw a histogram from the data above.
3. Draw a frequency polygon from the data above.
4. Find the mean, the median, and the mode of the following data. The   *Obj. 2, p. 703*
   number of appointment diaries sold each month between January
   and December: 98, 74, 37, 26, 10, 0, 4, 1, 18, 37, 89, 107 41.75; 31.5; 37
5. Find the variance and the standard deviation using the data in   *Obj. 3, p. 703*
   Exercise 4. 1449.02; 38.07
6. Two biology classes achieved the following grades on a test.   *Obj. 4, p. 703*

   School 1: 59, 62, 62, 83, 84
   School 2: 67, 48, 53, 83, 99

   Find the z-score for 83 in each class. In which class should 83 be
   regarded as the better grade? Class 1: 1.17; Class 2: 0.69; Class 1

Check your answers with those at the back of the book.

Applying the Normal Distribution

OBJECTIVES for Sections 17-3 and 17-4:
1. *To use the standard normal curve to find the percent of data that lie within a specified interval.*
2. *To use the standard normal curve to find the probability of a specified event.*
3. *To use the 95% confidence interval to estimate the fraction of a population that satisfies a specified condition.*

17–3 The Normal Distribution

In everyday life, sets of data arise out of widely varied situations. Nevertheless, certain similarities can sometimes be observed in the manner of the data's distribution. In particular, the data often seem to be distributed so that classes near the mean have the greatest frequencies, while those farther from the mean (either above or below it) have progressively smaller frequencies, approaching 0 for classes very far from the mean. A typical shape of such a distribution is shown in Figure 1 for the heights of pea plants.

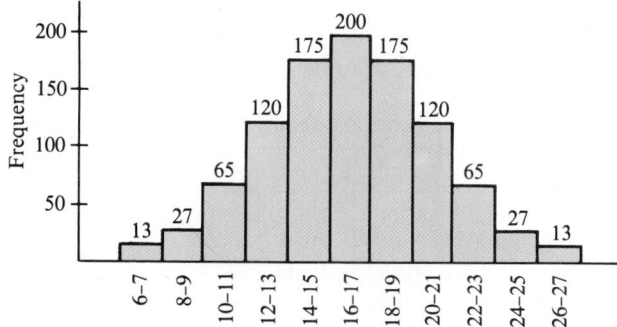

Heights of pea plants (cm)

Figure 1

Data exhibiting this pattern are said to be **normally distributed**. Some other examples of sets of data that represent normal distributions are: actual sizes of mass-produced parts, heights of adult females, life spans of organisms of a given species, and standardized test scores.

To understand why such a distribution occurs and how it can be defined mathematically, consider the pea-plant example above. To

Statistics **717**

2. Draw a histogram from the data on the previous page.

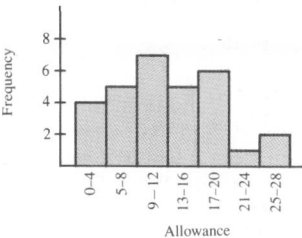

3. Draw a frequency polygon from the data on the previous page.

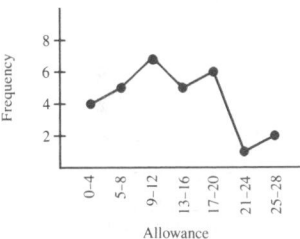

4. Find the mean, the median, and the mode of the following test scores: 70, 95, 90, 80, 90, 85, 75, 65, 90, 80.   82; 82.5; 90

5. Find the variance and the standard deviation using the data from Exercise 4. 86; 9.3

6. Two age groups achieved these scores in an archery competition. 15 yr olds: 68, 80, 52, 72, 60; 18 yr olds: 57, 72, 83, 86, 67. Find the *z*-score for 72 in each group. In which group should 72 be regarded as the better score?
15 yr: 0.579; 18 yr: −0.094; 15 yr olds

Key Ideas

Use the standard normal curve to find the percent of data that lie within a specified interval and to find the probability of a specified event.

greatly simplify the problem, suppose that only four factors—sunlight, water, soil quality, and fertilizer—contribute to a pea plant's growth and that each of these either is present and contributes 1 unit to the height of the plant, or is absent and contributes 0 units. We might then imagine that each plant, as it grows, passes through a "probability machine" like the one shown schematically below. Suppose that at each "fork" the plant accumulates 1 unit of height if it moves to the *right*, 0 units if it moves to the *left*. Suppose also that there is an equal probability ($\frac{1}{2}$) of moving either way at each fork. A typical "path" is shown in Figure 2.

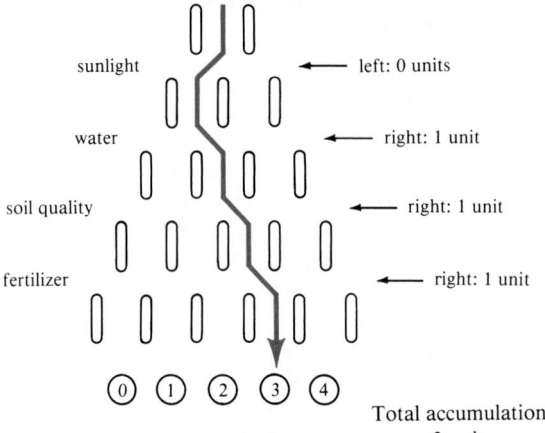

Figure 2

The table below shows the number of paths by which a plant can "arrive at" one of the values 0–4 that represent plant heights.

Ways of accumulating units

	0 units	1 unit	2 units	3 units	4 units
Paths	LLLL	RLLL LRLL LLRL LLLR	RRLL RLRL RLLR LRRL LRLR LLRR	RRRL RRLR RLRR LRRR	RRRR
Number of paths	1	4	6	4	1

Thus in a group of $1 + 4 + 6 + 4 + 1 = 16$ plants, we would ideally expect the following distribution (Figure 3).

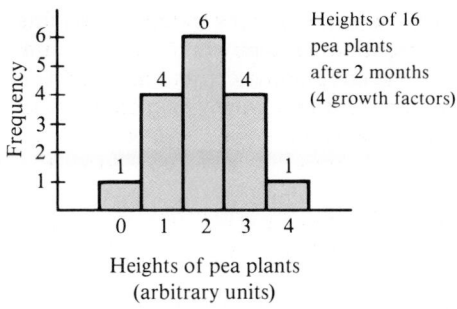

Heights of 16 pea plants after 2 months (4 growth factors)

Heights of pea plants (arbitrary units)

Figure 3

If 10 factors, instead of only 4, affected the plant's growth, we would expect the following distribution among 1024 plants (Figure 4).

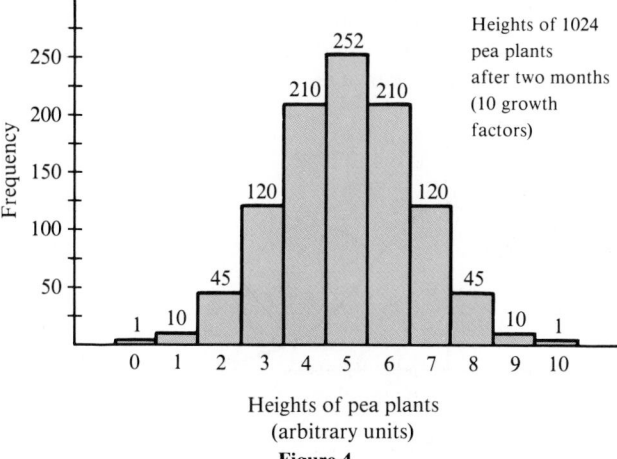

Heights of 1024 pea plants after two months (10 growth factors)

Heights of pea plants (arbitrary units)

Figure 4

You can see that as more and more factors are included, the contour of the tops of the bars approximate the bell-shaped curve, or **normal curve,** shown in Figure 5.

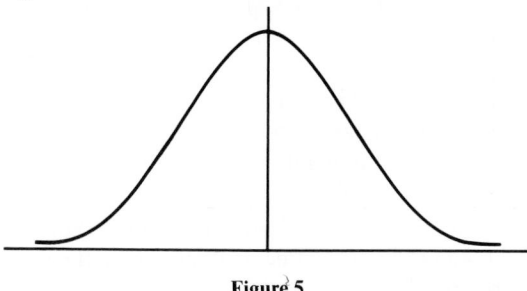

Figure 5

Statistics **719**

Since the scales used on the horizontal and vertical axes of histograms may vary from one normal distribution to another, there are many different normal curves. It is necessary, however, to have a single normal curve that can be used to describe *any* normally distributed set of data. The curve we use, called the **standard normal curve** has the following properties:

1. Its highest point occurs on the vertical axis, where it has a height of about 0.40.
2. The curve is symmetric about the vertical axis.
3. The height of the graph is arbitrarily close to 0 for values that are far enough to the right or left on the horizontal axis.
4. The total area under the graph is 1. (This can be verified by methods of calculus.)

For convenience, different units of measure are used on the vertical and horizontal axes. This results in some distortion of the standard normal curve from its true shape but will not cause any other practical difficulty (Figure 6). The equation of the normal curve is

$$y = \frac{1}{\sqrt{2\pi}} e^{\frac{-x^2}{2}}.$$

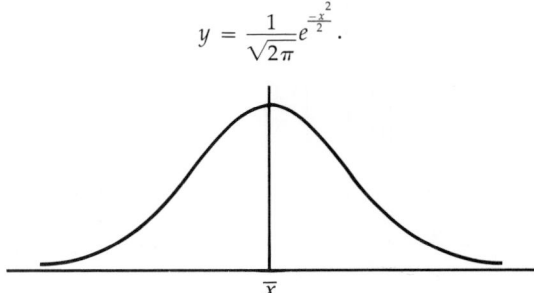

Figure 6

In order to use the standard normal curve to describe a normally distributed set of data, we associate *each data point with its z-score* along the horizontal axis. (In particular, 0 is associated with the mean of the data, and 1 with a data point 1 standard deviation greater than the mean.) The *area* under the graph between $z = a$ and $z = b$ then represents:

1. the expected *relative frequency* or *percent* of the data that fall between the data points corresponding to $z = a$ and $z = b$, or
2. the *probability* that a given data point will lie between the values corresponding to $z = a$ and $z = b$.

For example, the shaded region under the standard normal curve shown on page 721 (Figure 7) is known to have an area of about 0.68. Thus in any normal distribution about 68% of the data points fall within 1

720   *Chapter 17*

standard deviation of the mean and about 95% of the data points fall
between 2 standard deviations of the mean.

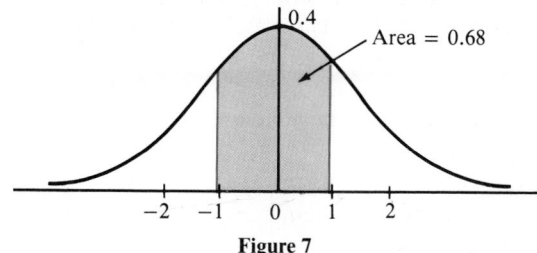

Figure 7

The area under the standard normal curve from 0 to x, denoted $A(x)$, is shown as the shaded area in Figure 8. A table of such areas is given below.

x	Area, $A(x)$	x	Area, $A(x)$
0.0	0.0000	2.1	0.4821
0.1	0.0398	2.2	0.4861
0.2	0.0793	2.3	0.4893
0.3	0.1179	2.4	0.4918
0.4	0.1554	2.5	0.4938
0.5	0.1915	2.6	0.4953
0.6	0.2257	2.7	0.4965
0.7	0.2580	2.8	0.4974
0.8	0.2881	2.9	0.4981
0.9	0.3159	3.0	0.4987
1.0	0.3413	3.1	0.4990
1.1	0.3643	3.2	0.4993
1.2	0.3849	3.3	0.4995
1.3	0.4032	3.4	0.4997
1.4	0.4192	3.5	0.4998
1.5	0.4332	3.6	0.4998
1.6	0.4452	3.7	0.4999
1.7	0.4554	3.8	0.4999
1.8	0.4641	3.9	0.5000
1.9	0.4713	4.0	0.5000
2.0	0.4772		

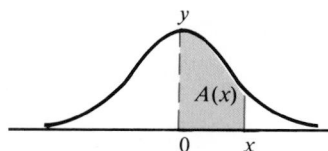

Figure 8

Using this table, we can make predictions about normally distributed data. In the examples that follow, the notation

$$P(a < z < b)$$

is used to represent the area under the standard normal curve between $z = a$ and $z = b$.

Statistics **721**

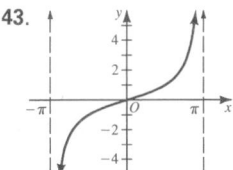
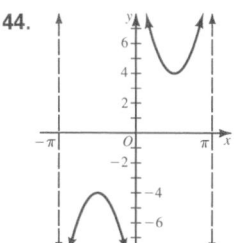

45.

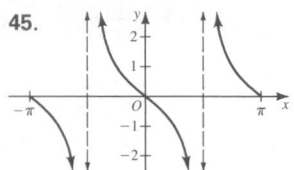

46.

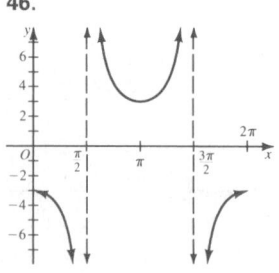

47.

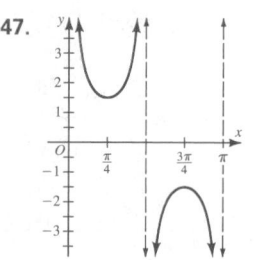

48.

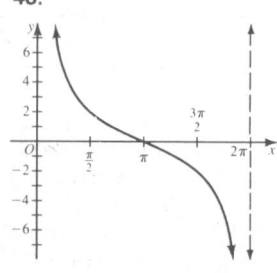

EXAMPLE 1   Find the percent of the area under the standard normal curve that is:

a. between $z = 1.2$ and $z = 2.5$

b. between $z = -1$ and $z = 1.6$

c. less than $z = 0.8$

SOLUTION   **a.** If $A(x)$ stands for the area under the standard normal curve between $z = 0$ and $z = x$ (as in the table on page 721), then

$P(1.2 < z < 2.5)$

$= A(2.5) - A(1.2)$

$= 0.4938 - 0.3849$

(from the table)

$= 0.1089$

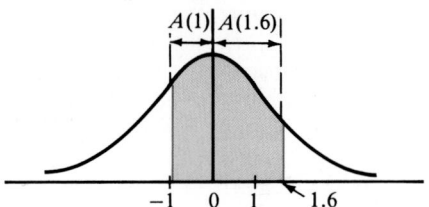

∴ the percent of the area between $z = 1.2$ and $z = 2.5$ is 10.89%.

b. As the diagram indicates,

$P(-1 < z < 1.6) = P(-1 < z < 0) + A(1.6).$

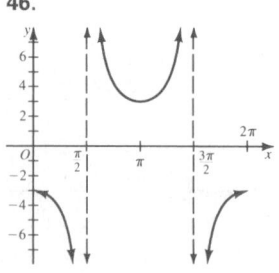

By the symmetry of the standard normal curve,

$$P(-1 < z < 0) = A(1).$$

Thus from the table the area is

$$A(1) + A(1.6) = 0.3413 + 0.4452$$
$$= 0.7865.$$

∴ the percent of the area between $z = -1$ and $z = 1.6$ is 78.65%.

c. As the diagram shows, $P(z < 0.8) = P(z < 0) + A(0.8)$.

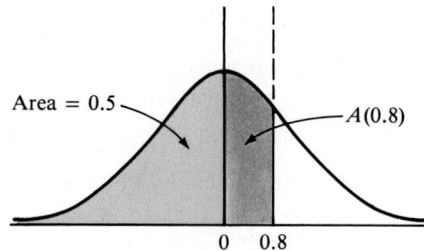

By the symmetry of the standard normal curve,

$$P(z < 0) = P(z > 0) = 0.5.$$

Thus

$$P(z < 0.8) = 0.5 + 0.2881 = 0.7881.$$

∴ the percent of the area less than $z = 0.8$ is 78.81%.

EXAMPLE 2 The mean score on a reading comprehension test is 72 and the standard deviation is 6. Assume that the scores are normally distributed.
a. What percent of the scores are above 81?
b. If a graded reading-test booklet is chosen at random, what is the probability that the score is between 63 and 81?

SOLUTION **a.** The z-score corresponding to 81 is

$$z = \frac{81 - 72}{6} = \frac{9}{6}$$
$$= 1.5.$$

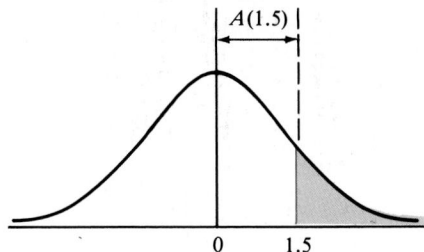

According to the diagram,
$$P(z > 1.5) = P(z > 0) - A(1.5)$$
$$= 0.5 - 0.4332 \text{ (from the table on page 721)}$$
$$= 0.0668.$$

∴ Thus 6.68%, or about 7% of the scores are above 81.

b. From part (a), the z-score corresponding to 81 is 1.5. The z-score corresponding to 63 is

$$z = \frac{63 - 72}{6} = -\frac{9}{6} = -1.5.$$

The probability that a test score lies between $z = -1.5$ and $z = 1.5$ is $P(-1.5 < z < 1.5)$.
Thus, from the table and from the symmetry of the standard normal curve, the specified probability is

$$2A(1.5) = 2(0.4332)$$
$$= 0.8664.$$

∴ the probability that a randomly chosen score lies between 63 and 81 is 0.8664, or about 0.87.

For Exs. 49–52, $u = \cos \alpha$ and $v = \sin \alpha$.

49. $\triangle ABO \sim \triangle PEO$

$$\frac{AB}{v} = \frac{OB}{u}$$

$$\frac{AB}{\sin \alpha} = \frac{1}{\cos \alpha}$$

$$AB = \frac{\sin \alpha}{\cos \alpha} = \tan \alpha$$

50. $\triangle ABO \sim \triangle PEO$

$$\frac{AO}{PO} = \frac{OB}{u}$$

$$\frac{AO}{1} = \frac{1}{\cos \alpha}$$

$$AO = \sec \alpha$$

51. $\triangle CDO \sim \triangle EOP$

$$\frac{CD}{u} = \frac{CO}{v}$$

$$\frac{CD}{\cos \alpha} = \frac{1}{\sin \alpha}$$

$$CD = \frac{\cos \alpha}{\sin \alpha} = \cot \alpha$$

52. $\triangle CDO \sim \triangle EOP$

$$\frac{OD}{PO} = \frac{CO}{v}$$

$$\frac{OD}{1} = \frac{1}{\sin \alpha}$$

$$OD = \csc \alpha$$

Additional A Exercises

Assume that the data are normally distributed.

1. A manufacturer offers a two-year warranty on his product based on a mean life expectancy of $2\frac{1}{2}$ years with a standard deviation of 3 months. What percent of the products will have to be repaired under warranty?   2.28%

2. The mean score on an aptitude test was 99 with a standard deviation of 11. What percent of the scores were over 109?   18.41%

3. The mean speed that cars travel on a highway is 80 km/h with a standard deviation of 10 km/h. What percent of the cars travel between 70 and 100 km/h?   81.85%

Oral Exercises

Use the table on page 721 to find each value.

1. $P(1 < z < 2)$ 0.1359
2. $P(z < 1)$ 0.8413
3. $P(z < -0.7)$ 0.242
4. $P(z > 1)$ 0.1587
5. $P(-2 < z < -1)$ 0.1359
6. $P(z < -1)$ 0.1587
7. $P(-1 < z < 1)$ 0.6826
8. $P(z < -1 \text{ or } z > 1)$ 0.3174

The mean number of defective bearings produced per day by a factory is 20 with a standard deviation of 5. Assuming that the numbers of defective bearings per day are normally distributed, find the probability that on a given day the number of defective bearings is:

9. between 20 and 25 0.3413
10. between 10 and 20 0.4772
11. greater than 25 0.1587
12. less than 30 0.9772

The mean score on a certain language test is 500 and the standard deviation is 100. Assume that the test scores are normally distributed. Approximately what percentage of scores are:

13. between 400 and 600? 0.6826
14. between 300 and 700? 0.9544
15. greater than 700? 0.0228
16. greater then 650? 0.0668
17. About what percent of normally distributed data falls within 2 standard deviations of the mean? 0.9544

Written Exercises

In Exercises 1–10, assume that the data in each situation are normally distributed.

A 1. The mean length of time that it takes a test animal to find its way through a certain maze is 42 seconds with a standard deviation of 15 seconds. What percent of the test animals might be expected to negotiate the maze in:
 a. between 30 and 60 seconds? 67.3%
 b. under 36 seconds? 34.46%
 c. over 54 seconds? 21.19%
 d. under 30 seconds or over 60 seconds? 32.7%

2. At a certain bank, mortgages are held by customers for a mean duration of 9.6 years, with a standard deviation of 4.0 years.
 a. The bank has a penalty for those who pay back their mortgages in under 2 years. What percent of their customers do this? 2.87%
 b. The bank wants to write letters to their customers whose mortgages are over 12 years old, offering favorable repayment terms. To what percent will they write? 27.43%

724   Chapter 17

3. The mean length of time for a telephone survey is 12 min with a standard deviation of 2 min. What is the probability that the telephone survey will last:
   a. more than 15 min? 0.0668
   b. less than 10 min? 0.1587
   c. between 8 and 17 min? 0.9710

4. The mean life of a certain brand of sparkplug is 16,000 km with a standard deviation of 2000 km. What is the probability that a given sparkplug of this brand will last: 0.3830
   a. between 15,000 and 17,000 km?
   b. between 18,000 and 19,000 km?
   c. less than 10,000 km? 0.0013
   d. more than 14,000 km? 0.8413

5. The mean flying time of a certain daily flight aboard Junket Airlines   4. b. 0.0919
is 84 minutes with a standard deviation of 5 minutes. To advertise its on-time record, the airline is considering offering to refund the fares of passengers on a flight that is over 11 minutes late. What percent of fares can it expect to refund (to the nearest tenth of a percent)? 1.4%

6. A certain bakery claims that each loaf of its whole wheat raisin bread contains an average of 48 raisins. If the standard deviation of the raisin count is 5, what percent of the loaves of bread could be expected to contain:
   a. fewer than 40 raisins? 5.48% b. between 40 and 50 raisins? 60.06%

B 7. The average grade on a placement test in English was 76 with a standard deviation of 5. The English department will waive the required course for the top 8% of the test takers. What is the cutoff grade for this group? 83

8. The sponsors of a magazine contest want to award prizes to the top 40 scorers on a word puzzle. They find that, out of 40,000 entrants, the mean score is 72 with a standard deviation of 10. What is the minimum score that wins a prize? 103

C 9. The maker of a certain kind of TV set would like to guarantee its set for a period of time that makes it likely that less than 1% of the sets sold will have to be replaced under the terms of the guarantee. If the sets have a mean life of 5 years with a standard deviation of 1.5 years, what is the longest guarantee the company can give? 1.55 yr

10. A company that makes an adjustable rowing machine has found that the mean height of potential buyers of their product is 170 cm with a standard deviation of 20 cm. If the company wants to minimize the adjustment it builds into the machine, what range of heights must it allow for in order to include at least 80% of potential buyers?
    144 cm to 196 cm

Statistics   **725**

1. Find the mean and the standard deviation of the set of data: 19, 21, 22, 22, 23, 26.   22.17; 2.11

The mean of a set of data is 65 with a standard deviation of 6. Find the z-score of each data point.

2. 50 −2.5

3. 77 2

4. 68 0.5

Simplify.

5. $\dfrac{27x^{\frac{5}{9}}y^{-\frac{1}{2}}}{9x^{\frac{4}{9}}y^{-\frac{3}{2}}}$ $3x^{\frac{1}{9}}y$

6. $\dfrac{x^{-2}y^{-1} - x^{-1}y^{-2}}{x^{-2} - y^{-2}}$ $\dfrac{1}{y+x}$

Key Ideas

Use a 95% confidence interval to test a hypothesis and to estimate the fraction of a population that has a specified characteristic.

17–4 Statistical Inference

Suppose a pollster questions 100 people who own cars (chosen by some method agreed to be random) and finds out that 70 of the 100 are satisfied with the performance of the car, while 30 are dissatisfied. What, if any, quantitative conclusions about the feelings of the entire population who own cars can be drawn from this poll?

Such a question asks for a **statistical inference**; that is, it asks us to draw general conclusions based on a relatively small amount of data. In studying such questions we use the term **population** to refer to the large set about which we want to draw general conclusions. (The set need not consist of human beings.) The term **sample** refers to the subset that provides the data. In a given situation, when more than one sample is considered, we almost always consider samples of a fixed size, say n individuals.

The poll results mentioned earlier can be examined by using an important branch of statistical inference called **hypothesis testing**. A *hypothesis* is a tentative assumption that is made in order to test its logical or empirical consequences. For example, a useful hypothesis concerning the people who own cars might be the following:

Of the population of people who own cars, approximately 50% are satisfied with the performance of their car.

This hypothesis appears to conflict with the poll results discussed earlier, which maintained that 70% of the sample approved of their car's performance. In view of these considerations, how credible is the hypothesis concerning the entire car-owning population?

Here it is important to mention that the techniques of hypothesis testing cannot answer this question with complete certainty, unless the sample size happens to equal the size of the population. However, we can specify the amount of *uncertainty* we are willing to accept. As will become clear shortly, this is more easily done if the task at hand is to *reject* a hypothesis (because we think it likely to be false) rather than to accept it (because we think it is likely to be true). For this reason, the appropriate question to ask in connection with any statistically tested hypothesis is the following:

Can the hypothesis be confidently rejected on the basis of the sample?

To answer this question, we make several assumptions, which can be proven in advanced statistics. Suppose that *all* samples of a given size of a certain population are polled to find out what fraction, $\overline{p}$, of each sample possesses a given characteristic. (In the example cited above, $\overline{p} = \frac{70}{100} = \frac{7}{10}$, which is the fraction of the sample that approves of the car's performance.) The assumptions are listed at the top of the following page.

726   *Chapter 17*

1. The mean of all the different values of $\bar{p}$ that we obtain equals the actual fraction, p, of the entire population that possesses the given characteristic.
2. The values of $\bar{p}$ are normally distributed.

In order to draw conclusions from a *single sample*, however, it will be important to know how widely varied is the distribution of sample fractions, $\bar{p}$. That is, we would like to know the variance, denoted $\sigma_{\bar{p}}^2$, and the standard deviation, denoted $\sigma_{\bar{p}}$, of the sample fractions. Naturally, if we have only one sample (or even several) we cannot compute $\sigma_{\bar{p}}^2$ or $\sigma_{\bar{p}}$ directly. Therefore, we make use of the following theorem, proved in advanced statistics courses.

Theorem. If p is the fraction of a large population that possesses a given characteristic, then for the fractions, $\bar{p}$, of samples of size n that possess the characteristic, we have

1. variance of the sample fractions: $\sigma_{\bar{p}}^2 = \dfrac{p(1 - p)}{n}$

2. standard deviation of the sample fractions:

$$\sigma_{\bar{p}} = \sqrt{\frac{p(1 - p)}{n}}.$$

To return to the hypothesis that only 50% of the car-owning population is satisfied with their car's performance, we find by means of the theorem above that, *if the hypothesis is true*, then $p = 0.5$, and the standard deviation of the sample fractions will be

$$\sigma_{\bar{p}} = \sqrt{\frac{p(1 - p)}{n}} = \sqrt{\frac{(0.5)(0.5)}{100}} = \frac{0.5}{10} = 0.05.$$

By assumption, however, the sample fractions are normally distributed, and thus according to the table on page 721, slightly over 95% of them fall within 2 standard deviations of the mean. In other words, for 95% of the sample fractions, $\bar{p}$, we would have

$$p - 2\sigma_{\bar{p}} < \bar{p} < p + 2\sigma_{\bar{p}} \qquad (1)$$
$$0.5 - 2(0.05) < \bar{p} < 0.5 + 2(0.05)$$
$$0.40 < \bar{p} < 0.60.$$

But since the given sample has a fraction of 0.7, it would have to fall in the remaining 5% of the sample fractions. While this does not rule out the hypothesis with certainty, it should convince us that the hypothesis is at least very unlikely, and we can reject it.

Notice that the method we used to test the hypothesis worked only

Chalkboard Examples

1. Of 300 people surveyed, 60 said they did not like yogurt. Suppose an advertisement states that 12% of the public does not like yogurt. Use a 95% confidence interval to show that the advertisement's assertion is unlikely.

$$\sigma_{\bar{p}} = \sqrt{\frac{0.12(1 - 0.12)}{300}} \approx$$

0.019; $0.2 - 2(0.019) < p < 0.2 + 2(0.019)$, $0.162 < p < 0.238$; $p = 0.12 < 0.162$

2. Of 200 voters polled, 60 support candidate B. Find a 95% confidence interval for the fraction of the population supporting that candidate.

$$\sigma_{\bar{p}} \approx \sqrt{\frac{\frac{60}{200}\left(1 - \frac{60}{200}\right)}{200}} =$$

$$\sqrt{\frac{(0.3)(0.7)}{200}} \approx 0.032$$

$0.3 - 2(0.032) < p < 0.3 + 2(0.032)$, $0.236 < p < 0.364$

Common Errors

In using the formula for the 95% confidence interval, students often forget to multiply the standard deviation, $\sigma_{\bar{p}}$, by 2.

Another error that students often make is to mix up the formulas for $\sigma_{\bar{p}}$. The first formula on page 729, which contains p, is used when there is a hypothesis about the entire population. The second formula, which contains only $\bar{p}$, is used when an estimation of the entire population is desired (i.e. no hypothesis is given).

because the given sample fell *outside* the interval containing 95% of the sample fractions. Had it fallen inside this interval, we would *not* have verified the hypothesis. We would have been able to draw *no conclusion* in that case. Therefore, in order to test a given hypothesis, we may have to test the *negation* of the hypothesis, using the theorem on page 727, in the hope that the sample fraction will fall *outside* the 95% interval for the negation.

If p is the actual fraction of a population that has a given characteristic, the reasoning in the example on page 727 shows that 95% of the sample fractions will be within 2 standard deviations of p. Since p is usually unknown in practical situations, however, a more useful statement equivalent to the Inequalities (1) is the following: If $\bar{p}$ is the fraction of a sample that has a given characteristic, there is a 95% probability that p, the actual fraction of the population that has the characteristic, will fall within 2 standard deviations of $\bar{p}$; that is, that

$$\bar{p} - 2\sigma_{\bar{p}} < p < \bar{p} + 2\sigma_{\bar{p}}. \tag{2}$$

The interval between $\bar{p} - 2\sigma_{\bar{p}}$ and $\bar{p} + 2\sigma_{\bar{p}}$ is therefore often called a **95% confidence interval** for p. In the example just discussed, the confidence interval for the fraction p of the population that approves of their car's performance is:

$$0.7 - 2(0.05) < p < 0.7 + 2(0.05)$$
$$0.6 < p < 0.8$$

(In Exercise 13 on page 731, you are asked to show that the Inequalities (1) and (2) are equivalent.)

EXAMPLE 1 A second poll of car performance is conducted. Of 200 individuals polled, 140 are found to be satisfied with their car's performance. Suppose that a magazine maintains that the true level of the public's approval for a car's performance is only 60%. Use a 95% confidence interval to show that the magazine's assertion is unlikely.

SOLUTION If the magazine's hypothesis is true, then $p = 0.6$, and therefore the standard deviation $\sigma_{\bar{p}}$ of the sample fractions is

$$\sigma_{\bar{p}} = \sqrt{\frac{p(1 - p)}{n}} = \sqrt{\frac{0.6(1 - 0.6)}{200}} = 0.035.$$

For the 95% confidence interval there is a 95% probability that the actual value of p falls within 2 standard deviations of $\bar{p}$. Therefore

$$\bar{p} - 2(\sigma_{\bar{p}}) < p < \bar{p} + 2(\sigma_{\bar{p}})$$
$$0.7 - 2(0.035) < p < 0.7 + 2(0.035)$$
$$0.63 < p < 0.77$$

Since the magazine's value for $p = 0.6$ does not fall within the confidence interval, this value is rejected as unlikely.

In **estimation,** the other branch of the study of statistical inference, we use 95% confidence intervals to give an **interval estimate** of the fraction p of a population that has a given characteristic. However, since we have no hypothesis for p, we cannot use the formula

$$\sigma_{\bar{p}} = \sqrt{\frac{p(1-p)}{n}}$$

to find the standard deviation of the sample fractions. However, by using the fraction $\bar{p}$ of the particular sample polled, we can *approximate* $\sigma_{\bar{p}}$ by

$$\sigma_{\bar{p}} \approx \sqrt{\frac{\bar{p}(1-\bar{p})}{n}}.$$

EXAMPLE 2   Out of a sample of 400 phonograph records produced by a certain machine, 20 were found to be defective. Find a 95% confidence interval for the defective fraction of all records produced.

SOLUTION   From the formula above, it follows that

$$\sigma_{\bar{p}} \approx \sqrt{\frac{\bar{p}(1-\bar{p})}{n}} = \sqrt{\frac{\frac{20}{400}\left(1-\frac{20}{400}\right)}{400}} = \sqrt{\frac{(0.05)(0.95)}{400}} \approx 0.011.$$

Since it can be stated with 95% confidence that

$$\bar{p} - 2\sigma_{\bar{p}} < p < \bar{p} + 2\sigma_{\bar{p}},$$

it follows that $0.05 - 2(0.011) < p < 0.05 + 2(0.011)$. That is, $0.028 < p < 0.072$.

∴ there is a 95% probability that the fraction of defective records is between 0.028 and 0.072.

Oral Exercises

At a certain bank, half of the customers are believed to have checking accounts. Samples of customers are polled as to whether they have checking accounts. Find the standard deviation of the sample fractions if the samples are all of size:

1. $4 \frac{1}{4}$ 2. $9 \frac{1}{6}$ 3. $16 \frac{1}{8}$ 4. $100 \frac{1}{20}$

5–8. For each of the sample sizes in Exercises 1–4, state the upper and lower boundaries of a 95% confidence interval for the sample fractions. **5.** 0, 1 **6.** $\frac{1}{6}, \frac{5}{6}$ **7.** $\frac{1}{4}, \frac{3}{4}$ **8.** $\frac{2}{5}, \frac{3}{5}$

9–10. For each of the sample sizes in Exercises 1 and 2, suppose that one person claims to have a checking account. On the basis of a 95% confidence interval, should the hypothesis that half of the customers have checking accounts be rejected? **9.** no   **10.** yes

Suggested Assignments

Enriched Alg.
730/1–13
R 731/*Self-Test 2*
Enriched Alg. with Trig.
730/1–13
R 731/*Self-Test 2*

Additional Answers
Written Exercises

1. $p = 0.8$, $\overline{p} = 0.7$, $n = 100$ ∴ $\sigma_{\overline{p}} = 0.04$, and the 95% confidence interval is $0.62 < p < 0.78$. Since p does not fall in this interval, the hypothesis is rejected.

2. $p = 0.1$, $\overline{p} = 0.24$, $n = 25$ ∴ $\sigma_{\overline{p}} = 0.06$, and the 95% confidence interval is $0.12 < p < 0.36$. Since p does not fall in this interval, the hypothesis is rejected.

10. a.

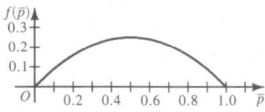

11. Since $\sigma_{\overline{p}} \le \sqrt{\dfrac{1}{4n}} = \dfrac{1}{2}\sqrt{\dfrac{1}{n}}$,

$2\sigma_{\overline{p}} \le \sqrt{\dfrac{1}{n}}$, so $\overline{p} - \sqrt{\dfrac{1}{n}} \le$

$\overline{p} - 2\sigma_{\overline{p}} < p < \overline{p} + 2\sigma_{\overline{p}}$

$\le \overline{p} + \sqrt{\dfrac{1}{n}}$.

13. If $p - 2\sigma_{\overline{p}} < \overline{p} < p + 2\sigma_{\overline{p}}$, then $p - 2\sigma_{\overline{p}} < \overline{p}$ and $\overline{p} < p + 2\sigma_{\overline{p}}$. ∴ $p < \overline{p} + 2\sigma_{\overline{p}}$ and $\overline{p} - 2\sigma_{\overline{p}} < p$. Combining the two shows $\overline{p} - 2\sigma_{\overline{p}} < p < \overline{p} + 2\sigma_{\overline{p}}$. The converse is proved similarly.

11–12. For each of the sample sizes in Exercises 3 and 4, suppose that about 37% of the people claim to have a checking account. On the basis of a 95% confidence interval, should the hypothesis that half of the customers have checking accounts be rejected? no; yes

13. In the formula for the standard deviation given in the theorem on page 727, how is the standard deviation of the sample fractions affected by an increase in the sample size? Explain your answer.
$\sigma_{\overline{p}}$ is decreased.

Written Exercises

In Exercises 1 and 2, use the 95% confidence interval as a criterion to show that each hypothesis is *unlikely*.

A 1. A restaurant that holds 100 diners claims that $\frac{4}{5}$ of all its customers order salad. On a particular night, 70 customers ordered salad.

2. A wild-life magazine claims that only 10% of the ponds and lakes in a certain state have no fish. A check of 25 ponds and lakes in the state reveals that 6 have no fish.
3. $0.388 < p < 0.592$

In Exercises 3–7, determine a 95% confidence interval for *p*, the proportion of the entire population having the indicated characteristic. 4. $0.254 < p < 0.346$

3. A poll of 96 voters shows that 47 support a candidate.

4. One hundred twenty out of 400 students polled have part-time jobs.

5. Out of 12 April days randomly selected from weather service records for a certain area, 3 were rainy. $0 < p < 0.5$

6. In a small sample of pond water, 150 microorganisms were found, of which 90 were plants. $0.52 < p < 0.68$

7. In a telephone survey, 90 out of 240 households called had their television sets tuned to Channel 2. $0.313 < p < 0.437$

8. In one batch of 72 machine parts, 8 were found to be defective. Find a 95% confidence interval for the number of defective parts in a daily output of 270. 10 to 50

9. Over the past 20 years 2% of the graduating class at Hilbert High School has passed the qualifying test for a state scholarship. This year, out of a senior class of 400, 16 qualified for scholarships. Are these results consistent with previous testing? If not, give some other possible explanations for the improvement in performance. No; the test may have been unusually easy. (Answers will vary.)

730   *Chapter 17*

B 10. **a.** Sketch a graph of the function $f(\bar{p}) = \bar{p}(1 - \bar{p})$.

    b. Determine the maximum value of this function. 0.25

    c. Explain why the standard deviation of $\bar{p}$ is less than or equal to $\sqrt{\dfrac{1}{4n}}$. Since $\bar{p}\,(1 - \bar{p}) \le \frac{1}{4}$, $\sigma_{\bar{p}} = \sqrt{\dfrac{\bar{p}(1-\bar{p})}{n}} \le \sqrt{\dfrac{1}{4n}}$.

11. Use Exercise 10(c) to verify that the widest possible 95% confidence interval is

$$\bar{p} - \sqrt{\frac{1}{n}} < p < \bar{p} + \sqrt{\frac{1}{n}}.$$

12. A pollster is seeking a confidence interval in which $\bar{p}$ is within 0.2 of p. How large a sample is needed for a 95% confidence interval? 25 or more

13. Show that the Inequalities (1) on page 727 and the Inequalities (2) on page 728 are equivalent inequalities.

C 14. The pollster for a Senate candidate has found that the candidate is favored by 36% of a sample. What is the minimum size of the sample if the pollster was able to estimate the candidate's popularity, with 95% confidence, to within ±4% of its actual value? 576

Self-Test 2

VOCABULARY   normal distribution (p. 717)   sample (p. 726)
                         normal curve (p. 719)   hypothesis testing (p. 726)
                         standard normal curve (p. 720)   95% confidence interval (p. 728)
                         statistical inference (p. 726)   estimation (p. 729)
                         population (p. 726)   interval estimate (p. 729)

In Exercises 1 and 2, assume that the data are normally distributed.

1. In a test of reaction times, the mean length of time required for a subject to react to a stimulus is 4.2 seconds with a standard deviation of 0.8 seconds. What percent of the subjects require between 3 and 5 seconds to react to the stimulus? 77.45%   *Obj. 1, p. 717*

2. At a certain insurance company, the mean processing time for a claim is 15 working days, with a standard deviation of 3.5 working days. What is the probability that a claim will be paid in less than 10 working days? 0.0808   *Obj. 2, p. 717*

3. Out of a sample of 100 electric switches made in a certain factory, 4 were found to be defective. Find a 95% confidence interval for the defective fraction of all electric switches produced by this factory.   *Obj. 3, p. 717*

$$0.00 < p < 0.08$$

Check your answers with those at the back of the book.

Statistics   **731**

Find the mean, the variance, and the standard deviation of the given data.

1. Bowling scores: 158, 210, 163, 178, 185, 137
   171.83; 525.14; 22.92

2. Test scores: 96, 98, 70, 70, 83, 85, 85, 87, 92, 65   83.1; 116.09; 10.77

Solve.

3. $x^4 - 10x^2 + 9 = 0$
   $\{-3, -1, 1, 3\}$

4. $4x^2 + 12x + 13 = 0$
   $\left\{ -\dfrac{3}{2} + i, -\dfrac{3}{2} - i \right\}$

5. $x^2 + 4x - 15 = 0$
   $\{-2 + \sqrt{19}, -2 - \sqrt{19}\}$

Quick Quiz

For Exercises 1 and 2, assume that the data are normally distributed.

1. The mean score on a test was 78 with a standard deviation of 14. What percent of those taking the test scored between 71 and 85? 28.3%

2. During the rush hour, in a given month, the mean number of cars that pass through a particular intersection is 48 with a standard deviation of 11.5. What is the probability that more than 71 cars pass through the intersection? 0.0228

3. Of 100 cars sold by a used car dealer, 22 needed major repairs within 90 days of purchase. Find a 95% confidence interval for the fraction of cars sold that will need repairs within the first 90 days.
   $0.138 < p < 0.302$

Leonardo Torres Quevedo
1852–1936

Leonardo Torres Quevedo was born into a family of technicians in Santander, Spain, studied engineering, and for a short time drew plans for railway lines in southern Spain. His scientific contributions included many inventions, which won him wide recognition.

In 1901 Torres Quevedo combined mechanical and electromechanical means to construct a machine that could solve algebraic equations of any degree. In 1913 he established that a machine could proceed by trial and error and in 1920 he produced a typewriter-operated electromechanical machine that could perform calculations.

His inventions, however, were not limited to calculating and computing machines. In 1914 he designed a new type of cable for the cable cars at Niagara Falls, Ontario. He also developed a lighter-than-air craft and built a robot that could play with and defeat a human opponent in chess.

Examining Relationships between Sets of Data

OBJECTIVES for Section 17-5:
1. To draw a scatter diagram for a set of paired data points.
2. To find the correlation coefficient of two variables.

Teaching Suggestions
p. T128

Related Activities
p. T128

Supplementary Materials
Test 55

17–5 Correlation

Suppose you conducted a survey of 8 people, asking each person how many hours each exercises per week and how many hours of sleep per week each person needs, with results as tabulated below.

Hours of exercise	2	4	5	8	8	10	12	15
Hours of sleep	50	48	49	52	54	56	60	63

It appears from this table that there is a numerical relationship, or correlation, between the amount of exercise each person engages in and

732 *Chapter 17*

the amount of sleep each needs. By graphing the pairs, such as (2, 50), that occur in the table, you can obtain a visual summary of the information it contains, called a scatter plot, or **scatter diagram**.

Key Ideas

Draw a scatter diagram for a set of paired data points. Find the correlation coefficient of two variables.

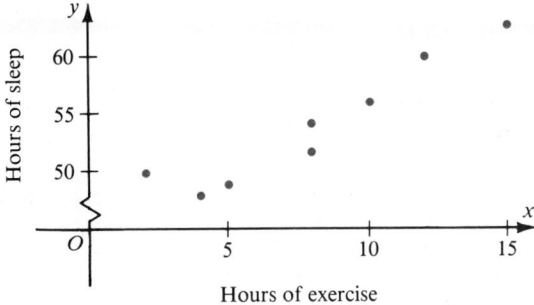

Figure 9

As the scatter diagram in Figure 9 shows, in this example an increase in the value of x is accompanied by an overall increase in the value of y. We say that in such a case there is a *positive correlation between the two variables*. If an increase in x is accompanied by an overall *decrease* in y, we say that there is a *negative correlation between the variables*. Of course, it is possible for a given pairing like the one above to show little or no correlation. The scatter diagrams in Figure 10 illustrate some of the possibilities.

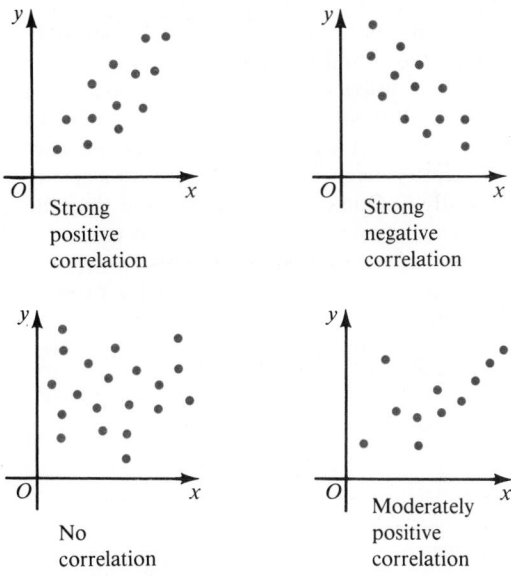

Figure 10

There are times, however, when we need a more precise indicator of the degree of correlation that exists between two variables than a scatter diagram. In order to define such an indicator, first consider a scatter diagram of the z-scores of the values in the table on page 732.

Hours of exercise	2	4	5	8	8	10	12	15
z-score	−1.5	−1.0	−0.7	0	0	0.5	1.0	1.7
Hours of sleep	50	48	49	52	54	56	60	63
z-score	−0.8	−1.2	−1.0	−0.4	0	0.4	1.2	1.8

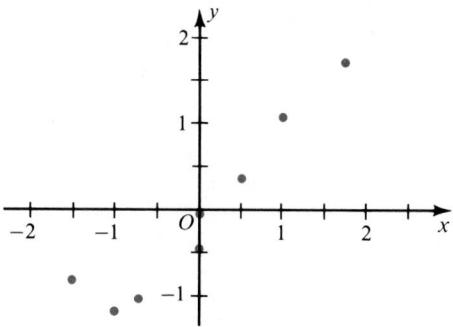

Figure 11

Note that the scatter diagram of the z-scores (Figure 11) strongly resembles the scatter diagram of the x- and y-values themselves (Figure 9), the major difference being the *placement* of the coordinate *axes* in relation to the graphed points. But since these points now lie in Quadrants I and III, their coordinates have the *same sign*. As you may have guessed, the points in a scatter diagram of the z-scores of a strong *negative* correlation will lie mostly in Quadrants II and IV, and thus the coordinates of these points will have *opposite signs*. In the case of a positive correlation, the products of corresponding z-scores $z_{x_i} \cdot z_{y_i}$ will be mostly positive, while in the case of a negative correlation the product of corresponding z-scores will be mostly negative.

This fact suggests the following definition of the **correlation coefficient**, r, between two quantities related by a set of ordered pairs $\{(x_i, y_i): i = 1, 2, \ldots, n\}$:

$$\text{correlation coefficient} = r = \frac{\sum_{i=1}^{n} z_{x_i} \cdot z_{y_i}}{n} \qquad (1)$$

In words, the correlation coefficient, r, is the mean of the products of corresponding z-scores of x and y.

It can be shown that

$$-1 \leq r \leq 1$$

for any values of x_i and y_i. It follows that a value of r that is close to 1 indicates a strong positive correlation, while a value of r close to -1 indicates a strong negative correlation. A value of r close to 0 indicates little or no correlation. We are now able to use correlation coefficients to supplement the information in the scatter diagrams on page 733 as we see in Figure 12.

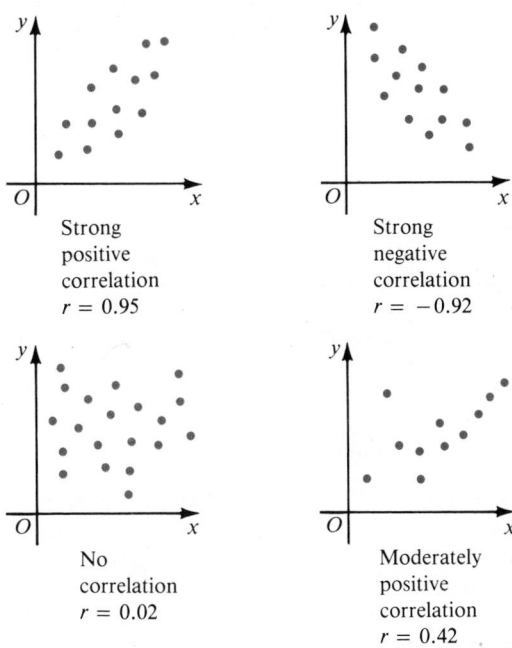

Strong
positive
correlation
$r = 0.95$

Strong
negative
correlation
$r = -0.92$

No
correlation
$r = 0.02$

Moderately
positive
correlation
$r = 0.42$

Figure 12

Although the formula for the correlation coefficient, r, given on page 734 is the easiest to understand, the formula below is more convenient for purposes of computation. You are asked to derive the formula in Exercise 17.

For the relation $\{(x_i, y_i): i = 1, 2, \ldots , n\}$,

$$r = \frac{\dfrac{\displaystyle\sum_{i=1}^{n} x_i \cdot y_i}{n} - \overline{x} \cdot \overline{y}}{\sigma_x \cdot \sigma_y}, \qquad (2)$$

where $\overline{x}$ and $\overline{y}$ are the means of the x- and y-values, respectively, and σ_x and σ_y are their respective standard deviations.

Chalkboard Example

A survey of cars and their owners produced the following results. Draw a scatter diagram displaying these data points and find the correlation coefficient.

Age of car (yr)	2	4	6	8	10
Number of owners	2	3	1	4	3

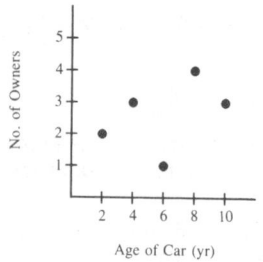

$$\bar{x} = \frac{30}{5} = 6; \ y = \frac{13}{5} = 2.6; \ \sigma_{\bar{x}}$$

$$\approx 2.83; \ \sigma_{\bar{y}} \approx 1.02; \ r \approx 0.42$$

EXAMPLE Applications of a certain type of weed killer on sample patches of ground produced the following results.

kg of weed killer used	1	3	4	5	7
number of weeds 1 month later	14	11	7	8	5

Draw a scatter diagram displaying these paired data points and find the correlation coefficient of the variables.

SOLUTION A scatter diagram of the given data is shown below.

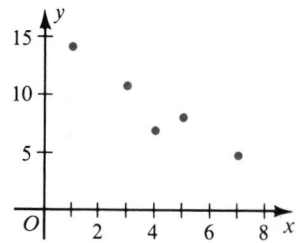

To find the correlation coefficient, r, first compute $\bar{x}$ and $\bar{y}$:

$$\bar{x} = \frac{1 + 3 + 4 + 5 + 7}{5} = 4 \qquad \bar{y} = \frac{14 + 11 + 7 + 8 + 5}{5} = 9$$

Compute the standard deviations σ_x and σ_y:

$$\sigma_x = \sqrt{\frac{(1-4)^2 + (3-4)^2 + (4-4)^2 + (5-4)^2 + (7-4)^2}{5}} = 2$$

$$\sigma_y = \sqrt{\frac{(14-9)^2 + (11-9)^2 + (7-9)^2 + (8-9)^2 + (5-9)^2}{5}} \approx 3.16$$

From Formula (2):

$$r = \frac{\left(\dfrac{\sum\limits_{i=1}^{n} x_i \cdot y_i}{n} \right) - \bar{x} \cdot \bar{y}}{\sigma_x \cdot \sigma_y}$$

$$= \frac{\dfrac{1 \cdot 14 + 3 \cdot 11 + 4 \cdot 7 + 5 \cdot 8 + 7 \cdot 5}{5} - 4 \cdot 9}{2 \cdot 3.16} \approx -0.95$$

∴ the correlation coefficient is approximately -0.95.

It is important to note that a strong correlation between two variables does not necessarily imply a cause-and-effect relationship. For example, in a survey of large cities, you might find that the cost of living and the air pollution index were highly correlated, but you would not be justified in concluding that air pollution increases the cost of living.

Oral Exercises

State whether each scatter diagram illustrates a positive, a negative, or no correlation.

1. 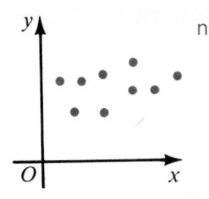 no

2. 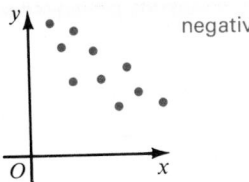 negative

3. 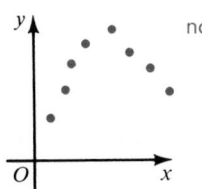 no

4. 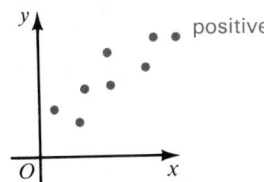 positive

Match each correlation coefficient to one of the scatter diagrams below.

5. $r = -0.5$ C   6. $r = 0.1$ B   7. $r = 0.8$ D   8. $r = -0.9$ A

A

B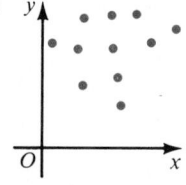

C

D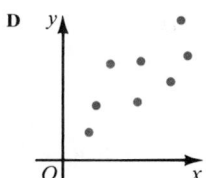

Use the following table to find each value.

x	1	7
y	10	20

9. $\overline{x}$ 4   10. $\overline{y}$ 15   11. σ_x 3   12. σ_y 5   13. $\dfrac{\sum\limits_{i=1}^{n} x_i y_i}{n}$ 75   14. r 1

Draw a scatter diagram and find the correlation coefficient for each set of data.

1.

Education after high school (yr)	Income ($1000)
0	12
1	15
2	16
3	18
4	24
5	30

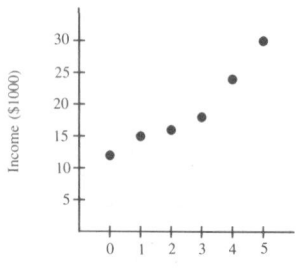

Education beyond H.S. (yr)

$r \approx 0.957$

2.

Number of assignments completed	Score on test
5	60
8	70
10	75
15	85
20	90
30	100

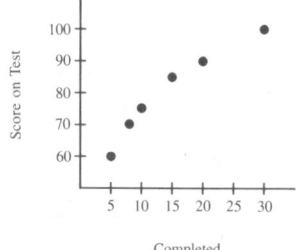

Completed Assignments

$r \approx 0.967$

Written Exercises

Draw a scatter diagram for each set of paired data points.

A 1.

Math achievement test score	60	75	80	85	90	95
Math grade	2.5	2	3	3.5	3	4

2.

Rainy days in March	5	8	9	11	12	15
March umbrella sales	20	25	22	30	36	35

3.

Height (in.)	64	65	67	69	70
Shoe size	$7\frac{1}{2}$	$8\frac{1}{2}$	8	9	12

4.

Inches of attic insulation		3	4	6	7	10	12
Average monthly heating cost ($)		150	140	125	120	120	119

5.

Age of a car (yr)	1	2	3	4	5	6	7
Value of car ($1000)	5	4	2.6	2.5	2.5	2.3	1.2

6.

Shoe size	8	$8\frac{1}{2}$	$9\frac{1}{2}$	10	11	13
Golf score	86	96	110	90	92	78

7–12. Find the correlation coefficient, r, for Exercises 1–6.

7. 0.74　　**8.** 0.90　　**9.** 0.81　　**10.** −0.85　　**11.** −0.92　　**12.** −0.46

B 13. Suppose the data for two variables x and y are as follows

x	a	b
y	c	d

where $a < b$ and $c < d$.

a. Find $\bar{x}$, $\bar{y}$, σ_x, and σ_y. (*Hint:* To find σ_x and σ_y, use the result of Exercise 26 on page 715.) $\bar{x} = \frac{a+b}{2}$; $\bar{y} = \frac{c+d}{2}$; $\sigma_x = \frac{b-a}{2}$; $\sigma_y = \frac{d-c}{2}$

b. Use the results of part (a) to show that $r = 1$.

14. Repeat Exercise 13, assuming $a < b$, $c > d$, and show that $r = -1$.

15. Suppose that for all pairs (x_i, y_i) in a relation, you have

$$y_i = mx_i$$

for some number $m > 0$. Use Formula (2) for r on page 735 to show that $r = 1$.

16. Repeat Exercise 15 with $m < 0$, and show that $r = -1$.

C **17.** In this exercise you will derive Formula (2) from Formula (1).

a. Show that $z_{x_i} \cdot z_{y_i} = \dfrac{x_i \cdot y_i - \overline{x} \cdot y_i - x_i \cdot \overline{y} + \overline{x} \cdot \overline{y}}{\sigma_x \sigma_y}$

b. Simplify $\displaystyle\sum_{i=1}^{n} (x_i y_i - \overline{x} y_i - x_i \overline{y} + \overline{x} \cdot \overline{y})$ using the facts that

$\displaystyle\sum_{i=1}^{n} \overline{x} y_i = \overline{x} \sum_{i=1}^{n} y_i = \overline{x}\, \dfrac{\left(n \displaystyle\sum_{i=1}^{n} y_i \right)}{n}$ and $\displaystyle\sum_{i=1}^{n} k = nk$, if k is a constant.

c. Derive Formula (2) from parts (a) and (b) above.

Computer Exercises For students with computer experience

1. Write a program that will find the correlation coefficient for the data in the Example on page 736.

2–7. By changing the data, run the program for Exercises 1-6 on page 738.

▊ Self-Test 3

VOCABULARY   correlation (p. 732)
scatter diagram (p. 733)
correlation coefficient (p. 734)

Draw a scatter diagram for each set of paired data points.

1.

Weight (kg)	56	58	61	64	67
Height (cm)	152	157	162	167	172

Obj. 1, p. 732

2.

Monthly average temperature (F)	0°	10°	20°	30°	40°	50°
Monthly earmuff sales		18	10	12	7	3	0

3. Find the correlation coefficient for Exercise 1. 0.998   *Obj. 2, p. 732*
4. Find the correlation coefficient for Exercise 2. −0.96

Check your answers with those at the back of the book.

Statistics   **739**

Additional Answers
Written Exercises
(See p. 743.)

Additional Answers
Self-Test 3

1.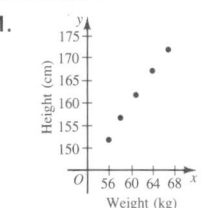

2.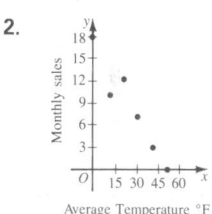

Quick Quiz

Draw a scatter diagram for each set of paired data points.

1.

Age (yr)	No. of children
10	0
20	1
30	2
40	3
50	3

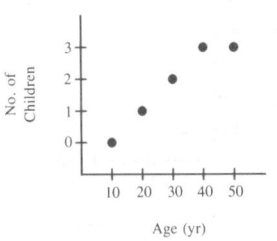

(continued)

2.

Months on a diet	Number of pounds lost
2	20
4	30
6	35
8	45
10	40
12	50

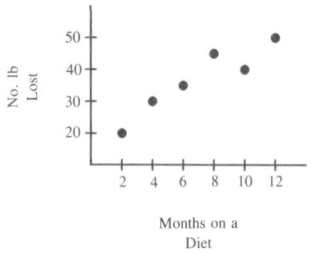

Months on a Diet

Find the correlation coefficient for:

3. Exercise 1. 0.97

4. Exercise 2. 0.94

Chapter Summary

1. Numerical data can be organized in a *frequency table* and represented graphically in a *histogram* or *frequency polygon*.

2. Three measures of central tendency are the *mean*, the *median*, and the *mode*. The mean is also referred to as the *average* and is defined by the 'formula

$$\bar{x} = \frac{\sum\limits_{i=1}^{n} x_i}{n}.$$

This mode is the measurement that occurs most often as a data point and the median is the value that falls in the middle when all the data points are arranged in numerical order.

3. The variability or dispersion of a set of data is measured by the *range*, the *variance*, and the *standard deviation*. These are defined on pages 710 and 711. The standard deviation makes it possible to compare data points from different sets of data and is given by the formula

$$\sigma = \sqrt{\frac{\sum\limits_{i=1}^{n} (x_i - \bar{x})^2}{n}}.$$

4. The *z-score* of a data point tells how many standard deviations a data point lies above or below the mean. Thus, a z-score can be used to *standardize* a test score or other data point.

5. The *normal distribution* often describes the way in which a set of data is distributed. About 68% of the distribution is within 1 standard deviation of the mean and about 95% of the distribution is within 2 standard deviations of the mean. Examples of normally distributed data are the heights of pea plants, the life spans of organisms, and standardized test scores.

6. Any normal curve can be transformed into the *standard normal curve* by using z-scores. Therefore, the standard normal curve can be used to describe any normally distributed set of data.

7. *Statistical inference* is the drawing of general conclusions about a population based on a relatively small *sample* of that population. The two branches of statistical inference are *hypothesis testing* and *estimation*. The 95% *confidence interval* is widely used as a criterion for statistical inference.

8. Two sets of numerical data may be plotted in a *scatter diagram*. According to the shape of the scatter diagram, there may be a *positive correlation*, a *negative correlation*, or *no correlation* between the two data

740   *Chapter 17*

sets. The correlation coefficient is found from the formula

$$r = \frac{\sum_{i=1}^{n} z_{x_i} \cdot z_{y_i}}{n},$$

and ranges in value from -1 to 1. A value of r equal to 0 indicates no correlation.

Chapter Review

Exercises 1–4 refer to the following cholesterol levels (mg/dL) for a group of American males of middle age.

147, 154, 172, 195, 195, 209, 218, 241, 283, 336

1. Find the mean of the given data.
  a. 195.5   **b.** 215   **c.** 195   **d.** 202

                                         17-1

2. Find the median of the given data.
  a. 202   **b.** 209   **c.** 215   **d.** 195

3. Find the standard deviation of the given data.
  a. 55.7   **b.** 59.1   **c.** 57.2   **d.** 3267

                                         17-2

4. Find the z-score of the data point 195.
  a. 0   **b.** -0.12 **c.** -0.36 **d.** 0.36

5. Use the table on page 721 to find the value of $P(z > 2)$.
  a. 0.4772   **b.** 0.0228   **c.** 0   **d.** 0.9772

                                         17-3

6. In an automobile assembly plant, the mean number of cars per week with faulty headlights is 16 with a standard deviation of 4. Assuming that the number of faulty headlights per week is normally distributed, find the probability that in a given week the number of faulty headlights is between 14 and 18.
  a. 0.24   **b.** 0.19   **c.** 0.12   **d.** 0.38

Exercises 7 and 8 refer to the following problem.
At a certain supermarket, 60% of the customers are believed to prefer low-fat milk to regular milk. Of 25 customers who are polled on their preferences, 13 say that they prefer low-fat milk.

7. If the belief that 60% of the customers prefer low-fat milk is true, find the standard deviation of all sample fractions of size 25.
  a. 0.136   **b.** 0.0096   **c.** 0.098   **d.** 0.0999

                                         17-4

8. Find the upper and lower boundaries of a 95% confidence interval for the sample fractions.
  a. $0.50 < \bar{p} < 0.70$ **b.** $0.40 < \bar{p} < 0.80$
  c. $0.32 < \bar{p} < 0.72$ **d.** $0.42 < \bar{p} < 0.62$

In Exercises 9 and 10, find the correlation coefficient, *r*.

9.

Year of Olympic game	1956	1960	1964	1968	1972	1976	1980	17-5
Winner's time (seconds)	62	61	60	60	59	56	55	

    (a.) -0.96 **b.** 0.96 **c.** 0.83 **d.** -0.83

10.

Number of job openings	13	47	48	63	77	108
Average weekly wage (dollars)	267	271	404	419	250	325

    a. 0 (b.) 0.13 **c.** -0.35 **d.** 0.35

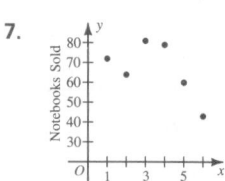
Chapter Test

1. Find the mean, median, and mode of the following data. The number of video cassette recorders sold during the first 10 days of the month: mean = 5.5; median = 6.5; mode = 7   17-1

$$7, 1, 10, 3, 0, 6, 7, 4, 9, 8$$

2. Find the variance and the standard deviation of the following data. Heights (in inches) of the starting players on a high school basketball team: $\sigma^2 = 10.16$; $\sigma = 3.19$ 17-2

$$70, 71, 75, 77, 78$$

3. Use the table on page 721 to find the value of $P(-1 < z < 2)$. 0.8185   17-3

4. The mean useful life of a certain model of automobile is 12 years, with a standard deviation of 2 years. What percent of the automobiles of this particular model will have a useful life of at least 7 years? 99.38%

5. A flower shop claims that 80% of its cut roses will survive for at least 4 days. When 16 customers were polled, only 7 customers claimed that cut roses they had purchased lasted as long as 4 days. Use the 95% confidence interval to show that the flower shop's claim is unlikely.   17-4

6. In a batch of 100 light bulbs, 8 were found to be defective. Find a 95% confidence interval for the number of defective light bulbs. 3 to 13

7. Draw a scatter diagram for the following set of paired data points: 17-5

Number of days before school opens	6	5	4	3	2	1
Number of notebooks sold	43	60	79	81	64	72

8. For the set of data in Exercise 7, find the correlation coefficient, r. -0.60

PREPARING FOR
COLLEGE ENTRANCE EXAMS

Strategy for Success: Some of the questions on the exam may be quantitative comparison questions, like questions 1–9 below. Read the directions carefully and practice questions like these before you take the exam.

Questions 1–9 each consist of two quantities, one in Column A and one in Column B. Mark A on your answer sheet if the quantity in Column A is greater, mark B if the quantity in Column B is greater, mark C if the quantities are equal, or mark D if the relationship cannot be determined from the information given.

Column A	Column B
1. $\sin \dfrac{10\pi}{3}$ B	$\cot \dfrac{3}{10\pi}$
2. $\tan(-x)$ D	$\tan x$
3. $\sin x + \cos x$ D	$\cot x + \tan x$
4. 1^n C	1^{2n}
5. $\sqrt{2} + \sqrt{5}$ A	$\sqrt{7}$
6. $a + b + c$ D	abc
7. Area of a square with side 2 B	Area of a circle with radius 2
8. $\dfrac{1}{y^2}$ C	$\left(\dfrac{1}{y}\right)^2$
9. $1 - \dfrac{a}{c}$ D	$\dfrac{-1}{\frac{c}{c-a}}$

Decide which is the best of the choices given and write the correct letter on your answer sheet.

10. Evaluate $\text{Cos}^{-1}(\sin 180°)$. B
  (A) 45°   **(B)** 90°   **(C)** −45°   **(D)** −90°   **(E)** 270°

11. Express $1 + i\sqrt{3}$ in polar form. E
  (A) $\sqrt{3}(\cos 60° + i \sin 60°)$ **(B)** $4(\cos 30° + i \sin 30°)$
  (C) $2(\cos 150° + i \sin 150°)$ **(D)** $2(\cos 30° + i \sin 30°)$ **(E)** $2(\cos 60° + i \sin 60°)$

12. Find $\|\mathbf{v}\|$ if $\mathbf{v} = 6\mathbf{i} + 8\mathbf{j}$. C
  (A) 6   **(B)** 8   **(C)** 10   **(D)** 12   **(E)** None of these

13. If the first two terms of a sequence are 2 and 3 and each successive term is found by adding all the preceding terms, what is the sixth term in the sequence? B
  (A) 20   **(B)** 40   **(C)** 21   **(D)** 42   **(E)** 11

Statistics **743**

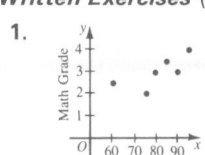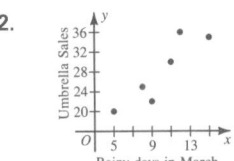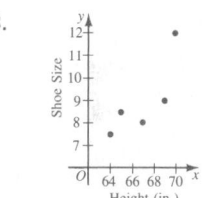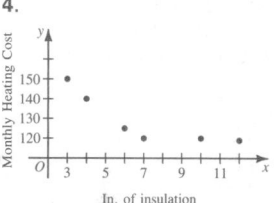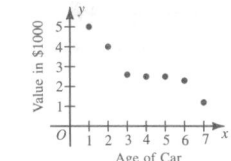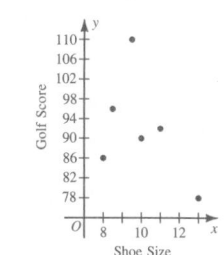

13. **b.** $r = \dfrac{\dfrac{\sum\limits_{i=1}^{2} x_i y_i}{n} - \overline{xy}}{\sigma_x \sigma_y} =$

$$\dfrac{\dfrac{ac + bd}{2} - \dfrac{(a + b)(c + d)}{4}}{\dfrac{(b - a)(d - c)}{4}}$$

$$= \dfrac{ac + bd - bc - ad}{bd - bc - ad + ac} = 1.$$

14. If $a < b, c < d$, then $\sigma_x = \dfrac{b - a}{2}$ and $\sigma_y = \dfrac{c - d}{2}$.

$$\therefore r = \dfrac{\dfrac{ac + bd}{2} - \dfrac{(a + b)(c + d)}{4}}{\dfrac{(b - a)(c - d)}{4}} =$$

$$\dfrac{ac + bd - ad - bc}{bc - bd - ac + ad} = -1.$$

15. By Exercise 25 on page 715, $\sigma_y = |m| \cdot \sigma_x = m\sigma_x$, and $\overline{y} = m\overline{x}$.

$$\therefore r = \dfrac{\dfrac{\sum\limits_{i=1}^{n} x_i y_i}{n} - \overline{x}(m\overline{x})}{\sigma_x(m\sigma_x)} =$$

$$\dfrac{\dfrac{\sum\limits_{i=1}^{n} mx_i^2}{n} - m\overline{x}^2}{m\sigma_x^2} =$$

$$\dfrac{m\left(\dfrac{\sum\limits_{i=1}^{n} x_i^2}{n} - \overline{x}^2\right)}{m\sigma_x^2} =$$

$$\dfrac{\dfrac{1}{n}\sum\limits_{i=1}^{n} x_i^2 - \left(\dfrac{1}{n}\sum\limits_{i=1}^{n} x_i\right)^2}{\sigma_x^2} =$$

$$\dfrac{\dfrac{1}{n^2}\left(n\sum\limits_{i=1}^{n} x_i^2 - \left(\sum\limits_{i=1}^{n} x_i\right)^2\right)}{\sigma_x^2} =$$

$$\dfrac{\sigma_x^2}{\sigma_x^2} = 1$$

Contest Problems

1. A cable consisting of 10 separate straight wires has been run under a river. You have access only to the 10 ends of the wires on each side of the river. Your equipment consists of as many labels and wire connectors as you need, and one continuity tester to test whether the circuit is open or closed. Starting on one side of the river, you may connect, test, and label as many wires as you want, then go to the other side of the river *once* and do the same. Finally, you may return to the first side, and after possibly retesting and relabeling on this side, the wires must be completely identified. That is, one wire must be labeled "#1" on both ends, one wire must be labeled "#2" on both ends, and so forth.

PROGRAMMING IN PASCAL

The following Pascal program fragment may be used to acquire data for statistical analysis of positive real numbers.

```
PROGRAM statistics (INPUT, OUTPUT);
CONST
   max = 20;   (* maximum number of items *)
TYPE
   values = ARRAY [1..max] OF real;
VAR
   data : values;
   counter : integer;

( ***********************************************************)
PROCEDURE obtain_data;

BEGIN
   counter := 0;
   writeln('Enter -1 to stop. ');   (* impossible data value as flag *)
   REPEAT
      counter := counter + 1;
      write ('Enter number: ');
      readln (data[counter]);
   UNTIL (data[counter] = -1) OR (counter = max);
   IF data [counter] = -1
      THEN counter : = counter -1;
END;   (* obtain_data *)
```

Exercises

1. Finish *statistics* by writing a Pascal procedure or procedures to find and display the following:
   a. the maximum, minimum, and range of the data
   b. the mean, $\bar{x}$, of the data
   c. the standard deviation, s, of the data
   Incorporate these procedures into a final version of the program, then test your program by entering several sets of data for which you have already computed the results.

2. The standardized variable or standard score is often used to record results on standardized tests. It is defined by $z = \dfrac{x - \bar{x}}{s} = \dfrac{1}{s}(x - \bar{x})$. We say that a set of scores has been standardized (and refer to it as a set of standard scores) when each item x has been replaced by
$$\frac{1}{s}(x - \bar{x}).$$
   a. Show that the mean of a set of standard scores is 0.
   b. Show that the standard deviation of a set of standard scores is 1.
   c. Enhance *statistics* to display the standardized scores of the items.

3. Suppose that the same constant k is added to each item of a data set.
   a. Show that the mean of the new set of data equals the original mean plus k.
   b. Show that both data sets have the same standard deviation.
   c. Explain how your teacher could adjust any set of your class test grades to have mean 75 (or any other desired average) and the same standard deviation as the original grades. Add (or subtract) the same quantity to all of the test grades.

4. Enhance *statistics* to offer the user the option of adjusting the data to have any desired mean and then to display the adjusted data.

16. Since $m < 0$, $\sigma_y = |m|\,\sigma_x = -m\sigma_x$. Proceed as in Ex. 15.
$$r = \frac{\dfrac{1}{n}\displaystyle\sum_{i=1}^{n} mx_i^2 - m\bar{x}^2}{-m\sigma_x^2} =$$
$$\frac{\sigma_x^2}{-\sigma_x^2} = -1.$$

17. a. $z_{x_i} \cdot z_{y_i} =$
$$\frac{(x_i - \bar{x})(y_i - \bar{y})}{\sigma_x \cdot \sigma_y} =$$
$$\frac{x_i y_i - \bar{x} y_i - x_i \bar{y} + \bar{x}\bar{y}}{\sigma_x \sigma_y}$$

   b. Since $\dfrac{1}{n}\displaystyle\sum_{i=1}^{n} x_i = \bar{x}$,
$$\sum_{i=1}^{n} x_i = n\bar{x}.$$
   Similarly, $\displaystyle\sum_{i=1}^{n} y_i = n\bar{y}$.
$$\therefore \sum_{i=1}^{n}(x_i y_i - \bar{x} y_i - x_i \bar{y} +$$
$$\bar{x}\bar{y}) = \sum_{i=1}^{n} x_i y_i -$$
$$\bar{x}\sum_{i=1}^{n} y_i - \bar{y}\sum_{i=1}^{n} x_i +$$
$$\sum_{i=1}^{n} \bar{x}\bar{y} = \sum_{i=1}^{n} x_i y_i -$$
$$n\bar{x}\bar{y} - n\bar{x}\bar{y} + n\bar{x}\bar{y} =$$
$$\sum_{i=1}^{n} x_i y_i - n\bar{x}\bar{y}.$$

   c. $r = \dfrac{\displaystyle\sum_{i=1}^{n} z_{x_i} \cdot z_{y_i}}{n} =$
$$\frac{\displaystyle\sum_{i=1}^{n}\left(\dfrac{x_i y_i - \bar{x} y_i - x_i \bar{y} + \bar{x}\bar{y}}{\sigma_x \sigma_y}\right)}{n} =$$
$$\frac{\dfrac{1}{n\sigma_x \sigma_y}\left(\displaystyle\sum_{i=1}^{n} x_i y_i - n\bar{x}\bar{y}\right)}{} =$$
$$\frac{\dfrac{\displaystyle\sum_{i=1}^{n} x_i y_i}{n} - \bar{x}\bar{y}}{\sigma_x \sigma_y}.$$

Tables

Table 1   Formulas

Circle	$A = \pi r^2$, $C = 2\pi r$	Cube	$V = s^3$
Parallelogram	$A = bh$	Rectangular Box	$V = lwh$
Right Triangle	$A = \frac{1}{2}bh$, $c^2 = a^2 + b^2$	Cylinder	$V = \pi r^2 h$
Square	$A = s^2$	Pyramid	$V = \frac{1}{3}Bh$
Trapezoid	$A = \frac{1}{2}h(b + b')$	Cone	$V = \frac{1}{3}\pi r^2 h$
Triangle	$A = \frac{1}{2}bh$	Sphere	$V = \frac{4}{3}\pi r^3$
Sphere	$A = 4\pi r^2$		

Table 2   Metric Units of Measure

Base Units

Length: meter (m)
Mass: kilogram (kg)*

Time

second (s), minute (min), hour (h)
day (d), month (mo), year (yr)

Temperature

degree Celsius (°C)
degree Kelvin (K)

Capacity

liter (L)
1 L = 1000 cm³

Force

Newton (N)

Pressure

Pascal (Pa)

Prefixes

Factor	Prefix	Symbol	Factor	Prefix	Symbol
10^{18}	exa	E	10^{-1}	deci	d
10^{15}	peta	P	10^{-2}	centi	c
10^{12}	tera	T	10^{-3}	milli	m
10^{9}	giga	G	10^{-6}	micro	μ
10^{6}	mega	M	10^{-9}	nano	n
10^{3}	kilo	k	10^{-12}	pico	p
10^{2}	hecto	h	10^{-15}	femto	f
10	deka	da	10^{-18}	atto	a

A prefix multiplies a unit by the factor given in the table.

Examples   gigameter: 1 Gm = 10^9 m = 1,000,000,000 m
milligram: 1 mg = 10^{-3} g = 0.001 g*

Compound units may be formed by division or multiplication.

Examples   kilometers per hour: km/h   square centimeters: cm²   cubic meters: m³

*Although the kilogram is defined as the base unit, the gram (g) is used with the prefixes to name other units of mass.

Table 3 Squares and Square Roots

N	N^2	$\sqrt{N}$	$\sqrt{10N}$	N	N^2	$\sqrt{N}$	$\sqrt{10N}$
1.0	1.00	1.000	3.162	5.5	30.25	2.345	7.416
1.1	1.21	1.049	3.317	5.6	31.36	2.366	7.483
1.2	1.44	1.095	3.464	5.7	32.49	2.387	7.550
1.3	1.69	1.140	3.606	5.8	33.64	2.408	7.616
1.4	1.96	1.183	3.742	5.9	34.81	2.429	7.681
1.5	2.25	1.225	3.873	6.0	36.00	2.449	7.746
1.6	2.56	1.265	4.000	6.1	37.21	2.470	7.810
1.7	2.89	1.304	4.123	6.2	38.44	2.490	7.874
1.8	3.24	1.342	4.243	6.3	39.69	2.510	7.937
1.9	3.61	1.378	4.359	6.4	40.96	2.530	8.000
2.0	4.00	1.414	4.472	6.5	42.25	2.550	8.062
2.1	4.41	1.449	4.583	6.6	43.56	2.569	8.124
2.2	4.84	1.483	4.690	6.7	44.89	2.588	8.185
2.3	5.29	1.517	4.796	6.8	46.24	2.608	8.246
2.4	5.76	1.549	4.899	6.9	47.61	2.627	8.307
2.5	6.25	1.581	5.000	7.0	49.00	2.646	8.367
2.6	6.76	1.612	5.099	7.1	50.41	2.665	8.426
2.7	7.29	1.643	5.196	7.2	51.84	2.683	8.485
2.8	7.84	1.673	5.292	7.3	53.29	2.702	8.544
2.9	8.41	1.703	5.385	7.4	54.76	2.720	8.602
3.0	9.00	1.732	5.477	7.5	56.25	2.739	8.660
3.1	9.61	1.761	5.568	7.6	57.76	2.757	8.718
3.2	10.24	1.789	5.657	7.7	59.29	2.775	8.775
3.3	10.89	1.817	5.745	7.8	60.84	2.793	8.832
3.4	11.56	1.844	5.831	7.9	62.41	2.811	8.888
3.5	12.25	1.871	5.916	8.0	64.00	2.828	8.944
3.6	12.96	1.897	6.000	8.1	65.61	2.846	9.000
3.7	13.69	1.924	6.083	8.2	67.24	2.864	9.055
3.8	14.44	1.949	6.164	8.3	68.89	2.881	9.110
3.9	15.21	1.975	6.245	8.4	70.56	2.898	9.165
4.0	16.00	2.000	6.325	8.5	72.25	2.915	9.220
4.1	16.81	2.025	6.403	8.6	73.96	2.933	9.274
4.2	17.64	2.049	6.481	8.7	75.69	2.950	9.327
4.3	18.49	2.074	6.557	8.8	77.44	2.966	9.381
4.4	19.36	2.098	6.633	8.9	79.21	2.983	9.434
4.5	20.25	2.121	6.708	9.0	81.00	3.000	9.487
4.6	21.16	2.145	6.782	9.1	82.81	3.017	9.539
4.7	22.09	2.168	6.856	9.2	84.64	3.033	9.592
4.8	23.04	2.191	6.928	9.3	86.49	3.050	9.644
4.9	24.01	2.214	7.000	9.4	88.36	3.066	9.695
5.0	25.00	2.236	7.071	9.5	90.25	3.082	9.747
5.1	26.01	2.258	7.141	9.6	92.16	3.098	9.798
5.2	27.04	2.280	7.211	9.7	94.09	3.114	9.849
5.3	28.09	2.302	7.280	9.8	96.04	3.130	9.899
5.4	29.16	2.324	7.348	9.9	98.01	3.146	9.950
5.5	30.25	2.345	7.416	10	100.00	3.162	10.000

Table 4 Cubes and Cube Roots

N	N^3	$\sqrt[3]{N}$	$\sqrt[3]{10N}$	$\sqrt[3]{100N}$	N	N^3	$\sqrt[3]{N}$	$\sqrt[3]{10N}$	$\sqrt[3]{100N}$
1.0	1.000	1.000	2.154	4.642	5.5	166.375	1.765	3.803	8.193
1.1	1.331	1.032	2.224	4.791	5.6	175.616	1.776	3.826	8.243
1.2	1.728	1.063	2.289	4.932	5.7	185.193	1.786	3.849	8.291
1.3	2.197	1.091	2.351	5.066	5.8	195.112	1.797	3.871	8.340
1.4	2.744	1.119	2.410	5.192	5.9	205.379	1.807	3.893	8.387
1.5	3.375	1.145	2.466	5.313	6.0	216.000	1.817	3.915	8.434
1.6	4.096	1.170	2.520	5.429	6.1	226.981	1.827	3.936	8.481
1.7	4.913	1.193	2.571	5.540	6.2	238.328	1.837	3.958	8.527
1.8	5.832	1.216	2.621	5.646	6.3	250.047	1.847	3.979	8.573
1.9	6.859	1.239	2.668	5.749	6.4	262.144	1.857	4.000	8.618
2.0	8.000	1.260	2.714	5.848	6.5	274.625	1.866	4.021	8.662
2.1	9.261	1.281	2.759	5.944	6.6	287.496	1.876	4.041	8.707
2.2	10.648	1.301	2.802	6.037	6.7	300.763	1.885	4.062	8.750
2.3	12.167	1.320	2.844	6.127	6.8	314.432	1.895	4.082	8.794
2.4	13.824	1.339	2.884	6.214	6.9	328.509	1.904	4.102	8.837
2.5	15.625	1.357	2.924	6.300	7.0	343.000	1.913	4.121	8.879
2.6	17.576	1.375	2.962	6.383	7.1	357.911	1.922	4.141	8.921
2.7	19.683	1.392	3.000	6.463	7.2	373.248	1.931	4.160	8.963
2.8	21.952	1.409	3.037	6.542	7.3	389.017	1.940	4.179	9.004
2.9	24.389	1.426	3.072	6.619	7.4	405.224	1.949	4.198	9.045
3.0	27.000	1.442	3.107	6.694	7.5	421.875	1.957	4.217	9.086
3.1	29.791	1.458	3.141	6.768	7.6	438.976	1.966	4.236	9.126
3.2	32.768	1.474	3.175	6.840	7.7	456.533	1.975	4.254	9.166
3.3	35.937	1.489	3.208	6.910	7.8	474.552	1.983	4.273	9.205
3.4	39.304	1.504	3.240	6.980	7.9	493.039	1.992	4.291	9.244
3.5	42.875	1.518	3.271	7.047	8.0	512.000	2.000	4.309	9.283
3.6	46.656	1.533	3.302	7.114	8.1	531.441	2.008	4.327	9.322
3.7	50.653	1.547	3.332	7.179	8.2	551.368	2.017	4.344	9.360
3.8	54.872	1.560	3.362	7.243	8.3	571.787	2.025	4.362	9.398
3.9	59.319	1.574	3.391	7.306	8.4	592.704	2.033	4.380	9.435
4.0	64.000	1.587	3.420	7.368	8.5	614.125	2.041	4.397	9.473
4.1	68.921	1.601	3.448	7.429	8.6	636.056	2.049	4.414	9.510
4.2	74.088	1.613	3.476	7.489	8.7	658.503	2.057	4.431	9.546
4.3	79.507	1.626	3.503	7.548	8.8	681.472	2.065	4.448	9.583
4.4	85.184	1.639	3.530	7.606	8.9	704.969	2.072	4.465	9.619
4.5	91.125	1.651	3.557	7.663	9.0	729.000	2.080	4.481	9.655
4.6	97.336	1.663	3.583	7.719	9.1	753.571	2.088	4.498	9.691
4.7	103.823	1.675	3.609	7.775	9.2	778.688	2.095	4.514	9.726
4.8	110.592	1.687	3.634	7.830	9.3	804.357	2.103	4.531	9.761
4.9	117.649	1.698	3.659	7.884	9.4	830.584	2.110	4.547	9.796
5.0	125.000	1.710	3.684	7.937	9.5	857.375	2.118	4.563	9.830
5.1	132.651	1.721	3.708	7.990	9.6	884.736	2.125	4.579	9.865
5.2	140.608	1.732	3.733	8.041	9.7	912.673	2.133	4.595	9.899
5.3	148.877	1.744	3.756	8.093	9.8	941.192	2.140	4.610	9.933
5.4	157.464	1.754	3.780	8.143	9.9	970.299	2.147	4.626	9.967
5.5	166.375	1.765	3.803	8.193	10	1000.000	2.154	4.642	10.000

Table 5 Common Logarithms of Numbers*

x	0	1	2	3	4	5	6	7	8	9
10	0000	0043	0086	0128	0170	0212	0253	0294	0334	0374
11	0414	0453	0492	0531	0569	0607	0645	0682	0719	0755
12	0792	0828	0864	0899	0934	0969	1004	1038	1072	1106
13	1139	1173	1206	1239	1271	1303	1335	1367	1399	1430
14	1461	1492	1523	1553	1584	1614	1644	1673	1703	1732
15	1761	1790	1818	1847	1875	1903	1931	1959	1987	2014
16	2041	2068	2095	2122	2148	2175	2201	2227	2253	2279
17	2304	2330	2355	2380	2405	2430	2455	2480	2504	2529
18	2553	2577	2601	2625	2648	2672	2695	2718	2742	2765
19	2788	2810	2833	2856	2878	2900	2923	2945	2967	2989
20	3010	3032	3054	3075	3096	3118	3139	3160	3181	3201
21	3222	3243	3263	3284	3304	3324	3345	3365	3385	3404
22	3424	3444	3464	3483	3502	3522	3541	3560	3579	3598
23	3617	3636	3655	3674	3692	3711	3729	3747	3766	3784
24	3802	3820	3838	3856	3874	3892	3909	3927	3945	3962
25	3979	3997	4014	4031	4048	4065	4082	4099	4116	4133
26	4150	4166	4183	4200	4216	4232	4249	4265	4281	4298
27	4314	4330	4346	4362	4378	4393	4409	4425	4440	4456
28	4472	4487	4502	4518	4533	4548	4564	4579	4594	4609
29	4624	4639	4654	4669	4683	4698	4713	4728	4742	4757
30	4771	4786	4800	4814	4829	4843	4857	4871	4886	4900
31	4914	4928	4942	4955	4969	4983	4997	5011	5024	5038
32	5051	5065	5079	5092	5105	5119	5132	5145	5159	5172
33	5185	5198	5211	5224	5237	5250	5263	5276	5289	5302
34	5315	5328	5340	5353	5366	5378	5391	5403	5416	5428
35	5441	5453	5465	5478	5490	5502	5514	5527	5539	5551
36	5563	5575	5587	5599	5611	5623	5635	5647	5658	5670
37	5682	5694	5705	5717	5729	5740	5752	5763	5775	5786
38	5798	5809	5821	5832	5843	5855	5866	5877	5888	5899
39	5911	5922	5933	5944	5955	5966	5977	5988	5999	6010
40	6021	6031	6042	6053	6064	6075	6085	6096	6107	6117
41	6128	6138	6149	6160	6170	6180	6191	6201	6212	6222
42	6232	6243	6253	6263	6274	6284	6294	6304	6314	6325
43	6335	6345	6355	6365	6375	6385	6395	6405	6415	6425
44	6435	6444	6454	6464	6474	6484	6493	6503	6513	6522
45	6532	6542	6551	6561	6571	6580	6590	6599	6609	6618
46	6628	6637	6646	6656	6665	6675	6684	6693	6702	6712
47	6721	6730	6739	6749	6758	6767	6776	6785	6794	6803
48	6812	6821	6830	6839	6848	6857	6866	6875	6884	6893
49	6902	6911	6920	6928	6937	6946	6955	6964	6972	6981
50	6990	6998	7007	7016	7024	7033	7042	7050	7059	7067
51	7076	7084	7093	7101	7110	7118	7126	7135	7143	7152
52	7160	7168	7177	7185	7193	7202	7210	7218	7226	7235
53	7243	7251	7259	7267	7275	7284	7292	7300	7308	7316
54	7324	7332	7340	7348	7356	7364	7372	7380	7388	7396

*Mantissas, decimal points omitted. Characteristics are found by inspection.

Table 5　Common Logarithms of Numbers

x	0	1	2	3	4	5	6	7	8	9
55	7404	7412	7419	7427	7435	7443	7451	7459	7466	7474
56	7482	7490	7497	7505	7513	7520	7528	7536	7543	7551
57	7559	7566	7574	7582	7589	7597	7604	7612	7619	7627
58	7634	7642	7649	7657	7664	7672	7679	7686	7694	7701
59	7709	7716	7723	7731	7738	7745	7752	7760	7767	7774
60	7782	7789	7796	7803	7810	7818	7825	7832	7839	7846
61	7853	7860	7868	7875	7882	7889	7896	7903	7910	7917
62	7924	7931	7938	7945	7952	7959	7966	7973	7980	7987
63	7993	8000	8007	8014	8021	8028	8035	8041	8048	8055
64	8062	8069	8075	8082	8089	8096	8102	8109	8116	8122
65	8129	8136	8142	8149	8156	8162	8169	8176	8182	8189
66	8195	8202	8209	8215	8222	8228	8235	8241	8248	8254
67	8261	8267	8274	8280	8287	8293	8299	8306	8312	8319
68	8325	8331	8338	8344	8351	8357	8363	8370	8376	8382
69	8388	8395	8401	8407	8414	8420	8426	8432	8439	8445
70	8451	8457	8463	8470	8476	8482	8488	8494	8500	8506
71	8513	8519	8525	8531	8537	8543	8549	8555	8561	8567
72	8573	8579	8585	8591	8597	8603	8609	8615	8621	8627
73	8633	8639	8645	8651	8657	8663	8669	8675	8681	8686
74	8692	8698	8704	8710	8716	8722	8727	8733	8739	8745
75	8751	8756	8762	8768	8774	8779	8785	8791	8797	8802
76	8808	8814	8820	8825	8831	8837	8842	8848	8854	8859
77	8865	8871	8876	8882	8887	8893	8899	8904	8910	8915
78	8921	8927	8932	8938	8943	8949	8954	8960	8965	8971
79	8976	8982	8987	8993	8998	9004	9009	9015	9020	9025
80	9031	9036	9042	9047	9053	9058	9063	9069	9074	9079
81	9085	9090	9096	9101	9106	9112	9117	9122	9128	9133
82	9138	9143	9149	9154	9159	9165	9170	9175	9180	9186
83	9191	9196	9201	9206	9212	9217	9222	9227	9232	9238
84	9243	9248	9253	9258	9263	9269	9274	9279	9284	9289
85	9294	9299	9304	9309	9315	9320	9325	9330	9335	9340
86	9345	9350	9355	9360	9365	9370	9375	9380	9385	9390
87	9395	9400	9405	9410	9415	9420	9425	9430	9435	9440
88	9445	9450	9455	9460	9465	9469	9474	9479	9484	9489
89	9494	9499	9504	9509	9513	9518	9523	9528	9533	9538
90	9542	9547	9552	9557	9562	9566	9571	9576	9581	9586
91	9590	9595	9600	9605	9609	9614	9619	9624	9628	9633
92	9638	9643	9647	9652	9657	9661	9666	9671	9675	9680
93	9685	9689	9694	9699	9703	9708	9713	9717	9722	9727
94	9731	9736	9741	9745	9750	9754	9759	9763	9768	9773
95	9777	9782	9786	9791	9795	9800	9805	9809	9814	9818
96	9823	9827	9832	9836	9841	9845	9850	9854	9859	9863
97	9868	9872	9877	9881	9886	9890	9894	9899	9903	9908
98	9912	9917	9921	9926	9930	9934	9939	9943	9948	9952
99	9956	9961	9965	9969	9974	9978	9983	9987	9991	9996

Table 6 Trigonometric Functions of θ (θ in decimal degrees)

θ Degrees	θ Radians	$\sin \theta$	$\cos \theta$	$\tan \theta$	$\cot \theta$	$\sec \theta$	$\csc \theta$		
0.0	.0000	.0000	1.0000	.0000	undefined	1.000	undefined	1.5708	**90.0**
0.1	.0017	.0017	1.0000	.0017	573.0	1.000	573.0	1.5691	89.9
0.2	.0035	.0035	1.0000	.0035	286.5	1.000	286.5	1.5673	89.8
0.3	.0052	.0052	1.0000	.0052	191.0	1.000	191.0	1.5656	89.7
0.4	.0070	.0070	1.0000	.0070	143.2	1.000	143.2	1.5638	89.6
0.5	.0087	.0087	1.0000	.0087	114.6	1.000	114.6	1.5621	89.5
0.6	.0105	.0105	.9999	.0105	95.49	1.000	95.49	1.5603	89.4
0.7	.0122	.0122	.9999	.0122	81.85	1.000	81.85	1.5586	89.3
0.8	.0140	.0140	.9999	.0140	71.62	1.000	71.62	1.5568	89.2
0.9	.0157	.0157	.9999	.0157	63.66	1.000	63.66	1.5551	89.1
1.0	.0175	.0175	.9998	.0175	57.29	1.000	57.30	1.5533	**89.0**
1.1	.0192	.0192	.9998	.0192	52.08	1.000	52.09	1.5516	88.9
1.2	.0209	.0209	.9998	.0209	47.74	1.000	47.75	1.5499	88.8
1.3	.0227	.0227	.9997	.0227	44.07	1.000	44.08	1.5481	88.7
1.4	.0244	.0244	.9997	.0244	40.92	1.000	40.93	1.5464	88.6
1.5	.0262	.0262	.9997	.0262	38.19	1.000	38.20	1.5446	88.5
1.6	.0279	.0279	.9996	.0279	35.80	1.000	35.81	1.5429	88.4
1.7	.0297	.0297	.9996	.0297	33.69	1.000	33.71	1.5411	88.3
1.8	.0314	.0314	.9995	.0314	31.82	1.000	31.84	1.5394	88.2
1.9	.0332	.0332	.9995	.0332	30.14	1.001	30.16	1.5376	88.1
2.0	.0349	.0349	.9994	.0349	28.64	1.001	28.65	1.5359	**88.0**
2.1	.0367	.0366	.9993	.0367	27.27	1.001	27.29	1.5341	87.9
2.2	.0384	.0384	.9993	.0384	26.03	1.001	26.05	1.5324	87.8
2.3	.0401	.0401	.9992	.0402	24.90	1.001	24.92	1.5307	87.7
2.4	.0419	.0419	.9991	.0419	23.86	1.001	23.88	1.5289	87.6
2.5	.0436	.0436	.9990	.0437	22.90	1.001	22.93	1.5272	87.5
2.6	.0454	.0454	.9990	.0454	22.02	1.001	22.04	1.5254	87.4
2.7	.0471	.0471	.9989	.0472	21.20	1.001	21.23	1.5237	87.3
2.8	.0489	.0488	.9988	.0489	20.45	1.001	20.47	1.5219	87.2
2.9	.0506	.0506	.9987	.0507	19.74	1.001	19.77	1.5202	87.1
3.0	.0524	.0523	.9986	.0524	19.08	1.001	19.11	1.5184	**87.0**
3.1	.0541	.0541	.9985	.0542	18.46	1.001	18.49	1.5167	86.9
3.2	.0559	.0558	.9984	.0559	17.89	1.002	17.91	1.5149	86.8
3.3	.0576	.0576	.9983	.0577	17.34	1.002	17.37	1.5132	86.7
3.4	.0593	.0593	.9982	.0594	16.83	1.002	16.86	1.5115	86.6
3.5	.0611	.0610	.9981	.0612	16.35	1.002	16.38	1.5097	86.5
3.6	.0628	.0628	.9980	.0629	15.89	1.002	15.93	1.5080	86.4
3.7	.0646	.0645	.9979	.0647	15.46	1.002	15.50	1.5062	86.3
3.8	.0663	.0663	.9978	.0664	15.06	1.002	15.09	1.5045	86.2
3.9	.0681	.0680	.9977	.0682	14.67	1.002	14.70	1.5027	86.1
4.0	.0698	.0698	.9976	.0699	14.30	1.002	14.34	1.5010	**86.0**
4.1	.0716	.0715	.9974	.0717	13.95	1.003	13.99	1.4992	85.9
4.2	.0733	.0732	.9973	.0734	13.62	1.003	13.65	1.4975	85.8
4.3	.0750	.0750	.9972	.0752	13.30	1.003	13.34	1.4957	85.7
4.4	.0768	.0767	.9971	.0769	13.00	1.003	13.03	1.4940	85.6
4.5	.0785	.0785	.9969	.0787	12.71	1.003	12.75	1.4923	85.5
4.6	.0803	.0802	.9968	.0805	12.43	1.003	12.47	1.4905	85.4
4.7	.0820	.0819	.9966	.0822	12.16	1.003	12.20	1.4888	85.3
4.8	.0838	.0837	.9965	.0840	11.91	1.004	11.95	1.4870	85.2
4.9	.0855	.0854	.9963	.0857	11.66	1.004	11.71	1.4853	85.1
5.0	.0873	.0872	.9962	.0875	11.43	1.004	11.47	1.4835	**85.0**
5.1	.0890	.0889	.9960	.0892	11.20	1.004	11.25	1.4818	84.9
5.2	.0908	.0906	.9959	.0910	10.99	1.004	11.03	1.4800	84.8
5.3	.0925	.0924	.9957	.0928	10.78	1.004	10.83	1.4783	84.7
5.4	.0942	.0941	.9956	.0945	10.58	1.004	10.63	1.4765	84.6
5.5	.0960	.0958	.9954	.0963	10.39	1.005	10.43	1.4748	84.5
5.6	.0977	.0976	.9952	.0981	10.20	1.005	10.25	1.4731	84.4
5.7	.0995	.0993	.9951	.0998	10.02	1.005	10.07	1.4713	84.3
5.8	.1012	.1011	.9949	.1016	9.845	1.005	9.895	1.4696	84.2
5.9	.1030	.1028	.9947	.1033	9.677	1.005	9.728	1.4678	84.1
6.0	.1047	.1045	.9945	.1051	9.514	1.006	9.567	1.4661	**84.0**
		$\cos \theta$	$\sin \theta$	$\cot \theta$	$\tan \theta$	$\csc \theta$	$\sec \theta$	θ Radians	θ Degrees

Table 6 Trigonometric Functions of θ (θ in decimal degrees)

θ Degrees	θ Radians	$\sin\theta$	$\cos\theta$	$\tan\theta$	$\cot\theta$	$\sec\theta$	$\csc\theta$		
6.0	.1047	.1045	.9945	.1051	9.514	1.006	9.567	1.4661	**84.0**
6.1	.1065	.1063	.9943	.1069	9.357	1.006	9.411	1.4643	83.9
6.2	.1082	.1080	.9942	.1086	9.205	1.006	9.259	1.4626	83.8
6.3	.1100	.1097	.9940	.1104	9.058	1.006	9.113	1.4608	83.7
6.4	.1117	.1115	.9938	.1122	8.915	1.006	8.971	1.4591	83.6
6.5	.1134	.1132	.9936	.1139	8.777	1.006	8.834	1.4574	83.5
6.6	.1152	.1149	.9934	.1157	8.643	1.007	8.700	1.4556	83.4
6.7	.1169	.1167	.9932	.1175	8.513	1.007	8.571	1.4539	83.3
6.8	.1187	.1184	.9930	.1192	8.386	1.007	8.446	1.4521	83.2
6.9	.1204	.1201	.9928	.1210	8.264	1.007	8.324	1.4504	83.1
7.0	.1222	.1219	.9925	.1228	8.144	1.008	8.206	1.4486	**83.0**
7.1	.1239	.1236	.9923	.1246	8.028	1.008	8.091	1.4469	82.9
7.2	.1257	.1253	.9921	.1263	7.916	1.008	7.979	1.4451	82.8
7.3	.1274	.1271	.9919	.1281	7.806	1.008	7.870	1.4434	82.7
7.4	.1292	.1288	.9917	.1299	7.700	1.008	7.764	1.4416	82.6
7.5	.1309	.1305	.9914	.1317	7.596	1.009	7.661	1.4399	82.5
7.6	.1326	.1323	.9912	.1334	7.495	1.009	7.561	1.4382	82.4
7.7	.1344	.1340	.9910	.1352	7.396	1.009	7.463	1.4364	82.3
7.8	.1361	.1357	.9907	.1370	7.300	1.009	7.368	1.4347	82.2
7.9	.1379	.1374	.9905	.1388	7.207	1.010	7.276	1.4329	82.1
8.0	.1396	.1392	.9903	.1405	7.115	1.010	7.185	1.4312	**82.0**
8.1	.1414	.1409	.9900	.1423	7.026	1.010	7.097	1.4294	81.9
8.2	.1431	.1426	.9898	.1441	6.940	1.010	7.011	1.4277	81.8
8.3	.1449	.1444	.9895	.1459	6.855	1.011	6.927	1.4259	81.7
8.4	.1466	.1461	.9893	.1477	6.772	1.011	6.845	1.4242	81.6
8.5	.1484	.1478	.9890	.1495	6.691	1.011	6.765	1.4224	81.5
8.6	.1501	.1495	.9888	.1512	6.612	1.011	6.687	1.4207	81.4
8.7	.1518	.1513	.9885	.1530	6.535	1.012	6.611	1.4190	81.3
8.8	.1536	.1530	.9882	.1548	6.460	1.012	6.537	1.4172	81.2
8.9	.1553	.1547	.9880	.1566	6.386	1.012	6.464	1.4155	81.1
9.0	.1571	.1564	.9877	.1584	6.314	1.012	6.392	1.4137	**81.0**
9.1	.1588	.1582	.9874	.1602	6.243	1.013	6.323	1.4120	80.9
9.2	.1606	.1599	.9871	.1620	6.174	1.013	6.255	1.4102	80.8
9.3	.1623	.1616	.9869	.1638	6.107	1.013	6.188	1.4085	80.7
9.4	.1641	.1633	.9866	.1655	6.041	1.014	6.123	1.4067	80.6
9.5	.1658	.1650	.9863	.1673	5.976	1.014	6.059	1.4050	80.5
9.6	.1676	.1668	.9860	.1691	5.912	1.014	5.996	1.4032	80.4
9.7	.1693	.1685	.9857	.1709	5.850	1.015	5.935	1.4015	80.3
9.8	.1710	.1702	.9854	.1727	5.789	1.015	5.875	1.3998	80.2
9.9	.1728	.1719	.9851	.1745	5.730	1.015	5.816	1.3980	80.1
10.0	.1745	.1736	.9848	.1763	5.671	1.015	5.759	1.3963	**80.0**
10.1	.1763	.1754	.9845	.1781	5.614	1.016	5.702	1.3945	79.9
10.2	.1780	.1771	.9842	.1799	5.558	1.016	5.647	1.3928	79.8
10.3	.1798	.1788	.9839	.1817	5.503	1.016	5.593	1.3910	79.7
10.4	.1815	.1805	.9836	.1835	5.449	1.017	5.540	1.3893	79.6
10.5	.1833	.1822	.9833	.1853	5.396	1.017	5.487	1.3875	79.5
10.6	.1850	.1840	.9829	.1871	5.343	1.017	5.436	1.3858	79.4
10.7	.1868	.1857	.9826	.1890	5.292	1.018	5.386	1.3840	79.3
10.8	.1885	.1874	.9823	.1908	5.242	1.018	5.337	1.3823	79.2
10.9	.1902	.1891	.9820	.1926	5.193	1.018	5.288	1.3806	79.1
11.0	.1920	.1908	.9816	.1944	5.145	1.019	5.241	1.3788	**79.0**
11.1	.1937	.1925	.9813	.1962	5.097	1.019	5.194	1.3771	78.9
11.2	.1955	.1942	.9810	.1980	5.050	1.019	5.148	1.3753	78.8
11.3	.1972	.1959	.9806	.1998	5.005	1.020	5.103	1.3736	78.7
11.4	.1990	.1977	.9803	.2016	4.959	1.020	5.059	1.3718	78.6
11.5	.2007	.1994	.9799	.2035	4.915	1.020	5.016	1.3701	78.5
11.6	.2025	.2011	.9796	.2053	4.872	1.021	4.973	1.3683	78.4
11.7	.2042	.2028	.9792	.2071	4.829	1.021	4.931	1.3666	78.3
11.8	.2059	.2045	.9789	.2089	4.787	1.022	4.890	1.3648	78.2
11.9	.2077	.2062	.9785	.2107	4.745	1.022	4.850	1.3631	78.1
12.0	.2094	.2079	.9781	.2126	4.705	1.022	4.810	1.3614	**78.0**
		$\cos\theta$	$\sin\theta$	$\cot\theta$	$\tan\theta$	$\csc\theta$	$\sec\theta$	θ Radians	θ Degrees

Table 6   Trigonometric Functions of θ (θ in decimal degrees)

θ Degrees	θ Radians	$\sin\theta$	$\cos\theta$	$\tan\theta$	$\cot\theta$	$\sec\theta$	$\csc\theta$		
12.0	.2094	.2079	.9781	.2126	4.705	1.022	4.810	1.3614	**78.0**
12.1	.2112	.2096	.9778	.2144	4.665	1.023	4.771	1.3596	77.9
12.2	.2129	.2113	.9774	.2162	4.625	1.023	4.732	1.3579	77.8
12.3	.2147	.2130	.9770	.2180	4.586	1.023	4.694	1.3561	77.7
12.4	.2164	.2147	.9767	.2199	4.548	1.024	4.657	1.3544	77.6
12.5	.2182	.2164	.9763	.2217	4.511	1.024	4.620	1.3526	77.5
12.6	.2199	.2181	.9759	.2235	4.474	1.025	4.584	1.3509	77.4
12.7	.2217	.2198	.9755	.2254	4.437	1.025	4.549	1.3491	77.3
12.8	.2234	.2215	.9751	.2272	4.402	1.025	4.514	1.3474	77.2
12.9	.2251	.2233	.9748	.2290	4.366	1.026	4.479	1.3456	77.1
13.0	.2269	.2250	.9744	.2309	4.331	1.026	4.445	1.3439	**77.0**
13.1	.2286	.2267	.9740	.2327	4.297	1.027	4.412	1.3422	76.9
13.2	.2304	.2284	.9736	.2345	4.264	1.027	4.379	1.3404	76.8
13.3	.2321	.2300	.9732	.2364	4.230	1.028	4.347	1.3387	76.7
13.4	.2339	.2317	.9728	.2382	4.198	1.028	4.315	1.3369	76.6
13.5	.2356	.2334	.9724	.2401	4.165	1.028	4.284	1.3352	76.5
13.6	.2374	.2351	.9720	.2419	4.134	1.029	4.253	1.3334	76.4
13.7	.2391	.2368	.9715	.2438	4.102	1.029	4.222	1.3317	76.3
13.8	.2409	.2385	.9711	.2456	4.071	1.030	4.192	1.3299	76.2
13.9	.2426	.2402	.9707	.2475	4.041	1.030	4.163	1.3282	76.1
14.0	.2443	.2419	.9703	.2493	4.011	1.031	4.134	1.3265	**76.0**
14.1	.2461	.2436	.9699	.2512	3.981	1.031	4.105	1.3247	75.9
14.2	.2478	.2453	.9694	.2530	3.952	1.032	4.077	1.3230	75.8
14.3	.2496	.2470	.9690	.2549	3.923	1.032	4.049	1.3212	75.7
14.4	.2513	.2487	.9686	.2568	3.895	1.032	4.021	1.3195	75.6
14.5	.2531	.2504	.9681	.2586	3.867	1.033	3.994	1.3177	75.5
14.6	.2548	.2521	.9677	.2605	3.839	1.033	3.967	1.3160	75.4
14.7	.2566	.2538	.9673	.2623	3.812	1.034	3.941	1.3142	75.3
14.8	.2583	.2554	.9668	.2642	3.785	1.034	3.915	1.3125	75.2
14.9	.2601	.2571	.9664	.2661	3.758	1.035	3.889	1.3107	75.1
15.0	.2618	.2588	.9659	.2679	3.732	1.035	3.864	1.3090	**75.0**
15.1	.2635	.2605	.9655	.2698	3.706	1.036	3.839	1.3073	74.9
15.2	.2653	.2622	.9650	.2717	3.681	1.036	3.814	1.3055	74.8
15.3	.2670	.2639	.9646	.2736	3.655	1.037	3.790	1.3038	74.7
15.4	.2688	.2656	.9641	.2754	3.630	1.037	3.766	1.3020	74.6
15.5	.2705	.2672	.9636	.2773	3.606	1.038	3.742	1.3003	74.5
15.6	.2723	.2689	.9632	.2792	3.582	1.038	3.719	1.2985	74.4
15.7	.2740	.2706	.9627	.2811	3.558	1.039	3.695	1.2968	74.3
15.8	.2758	.2723	.9622	.2830	3.534	1.039	3.673	1.2950	74.2
15.9	.2775	.2740	.9617	.2849	3.511	1.040	3.650	1.2933	74.1
16.0	.2793	.2756	.9613	.2867	3.487	1.040	3.628	1.2915	**74.0**
16.1	.2810	.2773	.9608	.2886	3.465	1.041	3.606	1.2898	73.9
16.2	.2827	.2790	.9603	.2905	3.442	1.041	3.584	1.2881	73.8
16.3	.2845	.2807	.9598	.2924	3.420	1.042	3.563	1.2863	73.7
16.4	.2862	.2823	.9593	.2943	3.398	1.042	3.542	1.2846	73.6
16.5	.2880	.2840	.9588	.2962	3.376	1.043	3.521	1.2828	73.5
16.6	.2897	.2857	.9583	.2981	3.354	1.043	3.500	1.2811	73.4
16.7	.2915	.2874	.9578	.3000	3.333	1.044	3.480	1.2793	73.3
16.8	.2932	.2890	.9573	.3019	3.312	1.045	3.460	1.2776	73.2
16.9	.2950	.2907	.9568	.3038	3.291	1.045	3.440	1.2758	73.1
17.0	.2967	.2924	.9563	.3057	3.271	1.046	3.420	1.2741	**73.0**
17.1	.2985	.2940	.9558	.3076	3.251	1.046	3.401	1.2723	72.9
17.2	.3002	.2957	.9553	.3096	3.230	1.047	3.382	1.2706	72.8
17.3	.3019	.2974	.9548	.3115	3.211	1.047	3.363	1.2689	72.7
17.4	.3037	.2990	.9542	.3134	3.191	1.048	3.344	1.2671	72.6
17.5	.3054	.3007	.9537	.3153	3.172	1.049	3.326	1.2654	72.5
17.6	.3072	.3024	.9532	.3172	3.152	1.049	3.307	1.2636	72.4
17.7	.3089	.3040	.9527	.3191	3.133	1.050	3.289	1.2619	72.3
17.8	.3107	.3057	.9521	.3211	3.115	1.050	3.271	1.2601	72.2
17.9	.3124	.3074	.9516	.3230	3.096	1.051	3.254	1.2584	72.1
18.0	.3142	.3090	.9511	.3249	3.078	1.051	3.236	1.2566	**72.0**
		$\cos\theta$	$\sin\theta$	$\cot\theta$	$\tan\theta$	$\csc\theta$	$\sec\theta$	θ Radians	θ Degrees

Table 6 Trigonometric Functions of θ (θ in decimal degrees)

θ Degrees	θ Radians	$\sin\theta$	$\cos\theta$	$\tan\theta$	$\cot\theta$	$\sec\theta$	$\csc\theta$		
18.0	.3142	.3090	.9511	.3249	3.078	1.051	3.236	1.2566	**72.0**
18.1	.3159	.3107	.9505	.3269	3.060	1.052	3.219	1.2549	71.9
18.2	.3177	.3123	.9500	.3288	3.042	1.053	3.202	1.2531	71.8
18.3	.3194	.3140	.9494	.3307	3.024	1.053	3.185	1.2514	71.7
18.4	.3211	.3156	.9489	.3327	3.006	1.054	3.168	1.2497	71.6
18.5	.3229	.3173	.9483	.3346	2.989	1.054	3.152	1.2479	71.5
18.6	.3246	.3190	.9478	.3365	2.971	1.055	3.135	1.2462	71.4
18.7	.3264	.3206	.9472	.3385	2.954	1.056	3.119	1.2444	71.3
18.8	.3281	.3223	.9466	.3404	2.937	1.056	3.103	1.2427	71.2
18.9	.3299	.3239	.9461	.3424	2.921	1.057	3.087	1.2409	71.1
19.0	.3316	.3256	.9455	.3443	2.904	1.058	3.072	1.2392	**71.0**
19.1	.3334	.3272	.9449	.3463	2.888	1.058	3.056	1.2374	70.9
19.2	.3351	.3289	.9444	.3482	2.872	1.059	3.041	1.2357	70.8
19.3	.3368	.3305	.9438	.3502	2.856	1.060	3.026	1.2339	70.7
19.4	.3386	.3322	.9432	.3522	2.840	1.060	3.011	1.2322	70.6
19.5	.3403	.3338	.9426	.3541	2.824	1.061	2.996	1.2305	70.5
19.6	.3421	.3355	.9421	.3561	2.808	1.062	2.981	1.2287	70.4
19.7	.3438	.3371	.9415	.3581	2.793	1.062	2.967	1.2270	70.3
19.8	.3456	.3387	.9409	.3600	2.778	1.063	2.952	1.2252	70.2
19.9	.3473	.3404	.9403	.3620	2.762	1.064	2.938	1.2235	70.1
20.0	.3491	.3420	.9397	.3640	2.747	1.064	2.924	1.2217	**70.0**
20.1	.3508	.3437	.9391	.3659	2.733	1.065	2.910	1.2200	69.9
20.2	.3526	.3453	.9385	.3679	2.718	1.066	2.896	1.2182	69.8
20.3	.3543	.3469	.9379	.3699	2.703	1.066	2.882	1.2165	69.7
20.4	.3560	.3486	.9373	.3719	2.689	1.067	2.869	1.2147	69.6
20.5	.3578	.3502	.9367	.3739	2.675	1.068	2.855	1.2130	69.5
20.6	.3595	.3518	.9361	.3759	2.660	1.068	2.842	1.2113	69.4
20.7	.3613	.3535	.9354	.3779	2.646	1.069	2.829	1.2095	69.3
20.8	.3630	.3551	.9348	.3799	2.633	1.070	2.816	1.2078	69.2
20.9	.3648	.3567	.9342	.3819	2.619	1.070	2.803	1.2060	69.1
21.0	.3665	.3584	.9336	.3839	2.605	1.071	2.790	1.2043	**69.0**
21.1	.3683	.3600	.9330	.3859	2.592	1.072	2.778	1.2025	68.9
21.2	.3700	.3616	.9323	.3879	2.578	1.073	2.765	1.2008	68.8
21.3	.3718	.3633	.9317	.3899	2.565	1.073	2.753	1.1991	68.7
21.4	.3735	.3649	.9311	.3919	2.552	1.074	2.741	1.1973	68.6
21.5	.3752	.3665	.9304	.3939	2.539	1.075	2.729	1.1956	68.5
21.6	.3770	.3681	.9298	.3959	2.526	1.076	2.716	1.1938	68.4
21.7	.3787	.3697	.9291	.3979	2.513	1.076	2.705	1.1921	68.3
21.8	.3805	.3714	.9285	.4000	2.500	1.077	2.693	1.1903	68.2
21.9	.3822	.3730	.9278	.4020	2.488	1.078	2.681	1.1886	68.1
22.0	.3840	.3746	.9272	.4040	2.475	1.079	2.669	1.1868	**68.0**
22.1	.3857	.3762	.9265	.4061	2.463	1.079	2.658	1.1851	67.9
22.2	.3875	.3778	.9259	.4081	2.450	1.080	2.647	1.1833	67.8
22.3	.3892	.3795	.9252	.4101	2.438	1.081	2.635	1.1816	67.7
22.4	.3910	.3811	.9245	.4122	2.426	1.082	2.624	1.1798	67.6
22.5	.3927	.3827	.9239	.4142	2.414	1.082	2.613	1.1781	67.5
22.6	.3944	.3843	.9232	.4163	2.402	1.083	2.602	1.1764	67.4
22.7	.3962	.3859	.9225	.4183	2.391	1.084	2.591	1.1746	67.3
22.8	.3979	.3875	.9219	.4204	2.379	1.085	2.581	1.1729	67.2
22.9	.3997	.3891	.9212	.4224	2.367	1.086	2.570	1.1711	67.1
23.0	.4014	.3907	.9205	.4245	2.356	1.086	2.559	1.1694	**67.0**
23.1	.4032	.3923	.9198	.4265	2.344	1.087	2.549	1.1676	66.9
23.2	.4049	.3939	.9191	.4286	2.333	1.088	2.538	1.1659	66.8
23.3	.4067	.3955	.9184	.4307	2.322	1.089	2.528	1.1641	66.7
23.4	.4084	.3971	.9178	.4327	2.311	1.090	2.518	1.1624	66.6
23.5	.4102	.3987	.9171	.4348	2.300	1.090	2.508	1.1606	66.5
23.6	.4119	.4003	.9164	.4369	2.289	1.091	2.498	1.1589	66.4
23.7	.4136	.4019	.9157	.4390	2.278	1.092	2.488	1.1572	66.3
23.8	.4154	.4035	.9150	.4411	2.267	1.093	2.478	1.1554	66.2
23.9	.4171	.4051	.9143	.4431	2.257	1.094	2.468	1.1537	66.1
24.0	.4189	.4067	.9135	.4452	2.246	1.095	2.459	1.1519	**66.0**
		$\cos\theta$	$\sin\theta$	$\cot\theta$	$\tan\theta$	$\csc\theta$	$\sec\theta$	θ Radians	θ Degrees

Table 6 Trigonometric Functions of θ (θ in decimal degrees)

θ Degrees	θ Radians	$\sin\theta$	$\cos\theta$	$\tan\theta$	$\cot\theta$	$\sec\theta$	$\csc\theta$		
24.0	.4189	.4067	.9135	.4452	2.246	1.095	2.459	1.1519	**66.0**
24.1	.4206	.4083	.9128	.4473	2.236	1.095	2.449	1.1502	65.9
24.2	.4224	.4099	.9121	.4494	2.225	1.096	2.439	1.1484	65.8
24.3	.4241	.4115	.9114	.4515	2.215	1.097	2.430	1.1467	65.7
24.4	.4259	.4131	.9107	.4536	2.204	1.098	2.421	1.1449	65.6
24.5	.4276	.4147	.9100	.4557	2.194	1.099	2.411	1.1432	65.5
24.6	.4294	.4163	.9092	.4578	2.184	1.100	2.402	1.1414	65.4
24.7	.4311	.4179	.9085	.4599	2.174	1.101	2.393	1.1397	65.3
24.8	.4328	.4195	.9078	.4621	2.164	1.102	2.384	1.1380	65.2
24.9	.4346	.4210	.9070	.4642	2.154	1.102	2.375	1.1362	65.1
25.0	.4363	.4226	.9063	.4663	2.145	1.103	2.366	1.1345	**65.0**
25.1	.4381	.4242	.9056	.4684	2.135	1.104	2.357	1.1327	64.9
25.2	.4398	.4258	.9048	.4706	2.125	1.105	2.349	1.1310	64.8
25.3	.4416	.4274	.9041	.4727	2.116	1.106	2.340	1.1292	64.7
25.4	.4433	.4289	.9033	.4748	2.106	1.107	2.331	1.1275	64.6
25.5	.4451	.4305	.9026	.4770	2.097	1.108	2.323	1.1257	64.5
25.6	.4468	.4321	.9018	.4791	2.087	1.109	2.314	1.1240	64.4
25.7	.4485	.4337	.9011	.4813	2.078	1.110	2.306	1.1222	64.3
25.8	.4503	.4352	.9003	.4834	2.069	1.111	2.298	1.1205	64.2
25.9	.4520	.4368	.8996	.4856	2.059	1.112	2.289	1.1188	64.1
26.0	.4538	.4384	.8988	.4877	2.050	1.113	2.281	1.1170	**64.0**
26.1	.4555	.4399	.8980	.4899	2.041	1.114	2.273	1.1153	63.9
26.2	.4573	.4415	.8973	.4921	2.032	1.115	2.265	1.1135	63.8
26.3	.4590	.4431	.8965	.4942	2.023	1.115	2.257	1.1118	63.7
26.4	.4608	.4446	.8957	.4964	2.014	1.116	2.249	1.1100	63.6
26.5	.4625	.4462	.8949	.4986	2.006	1.117	2.241	1.1083	63.5
26.6	.4643	.4478	.8942	.5008	1.997	1.118	2.233	1.1065	63.4
26.7	.4660	.4493	.8934	.5029	1.988	1.119	2.226	1.1048	63.3
26.8	.4677	.4509	.8926	.5051	1.980	1.120	2.218	1.1030	63.2
26.9	.4695	.4524	.8918	.5073	1.971	1.121	2.210	1.1013	63.1
27.0	.4712	.4540	.8910	.5095	1.963	1.122	2.203	1.0996	**63.0**
27.1	.4730	.4555	.8902	.5117	1.954	1.123	2.195	1.0978	62.9
27.2	.4747	.4571	.8894	.5139	1.946	1.124	2.188	1.0961	62.8
27.3	.4765	.4586	.8886	.5161	1.937	1.125	2.180	1.0943	62.7
27.4	.4782	.4602	.8878	.5184	1.929	1.126	2.173	1.0926	62.6
27.5	.4800	.4617	.8870	.5206	1.921	1.127	2.166	1.0908	62.5
27.6	.4817	.4633	.8862	.5228	1.913	1.128	2.158	1.0891	62.4
27.7	.4835	.4648	.8854	.5250	1.905	1.129	2.151	1.0873	62.3
27.8	.4852	.4664	.8846	.5272	1.897	1.130	2.144	1.0856	62.2
27.9	.4869	.4679	.8838	.5295	1.889	1.132	2.137	1.0838	62.1
28.0	.4887	.4695	.8829	.5317	1.881	1.133	2.130	1.0821	**62.0**
28.1	.4904	.4710	.8821	.5340	1.873	1.134	2.123	1.0804	61.9
28.2	.4922	.4726	.8813	.5362	1.865	1.135	2.116	1.0786	61.8
28.3	.4939	.4741	.8805	.5384	1.857	1.136	2.109	1.0769	61.7
28.4	.4957	.4756	.8796	.5407	1.849	1.137	2.103	1.0751	61.6
28.5	.4974	.4772	.8788	.5430	1.842	1.138	2.096	1.0734	61.5
28.6	.4992	.4787	.8780	.5452	1.834	1.139	2.089	1.0716	61.4
28.7	.5009	.4802	.8771	.5475	1.827	1.140	2.082	1.0699	61.3
28.8	.5027	.4818	.8763	.5498	1.819	1.141	2.076	1.0681	61.2
28.9	.5044	.4833	.8755	.5520	1.811	1.142	2.069	1.0664	61.1
29.0	.5061	.4848	.8746	.5543	1.804	1.143	2.063	1.0647	**61.0**
29.1	.5079	.4863	.8738	.5566	1.797	1.144	2.056	1.0629	60.9
29.2	.5096	.4879	.8729	.5589	1.789	1.146	2.050	1.0612	60.8
29.3	.5114	.4894	.8721	.5612	1.782	1.147	2.043	1.0594	60.7
29.4	.5131	.4909	.8712	.5635	1.775	1.148	2.037	1.0577	60.6
29.5	.5149	.4924	.8704	.5658	1.767	1.149	2.031	1.0559	60.5
29.6	.5166	.4939	.8695	.5681	1.760	1.150	2.025	1.0542	60.4
29.7	.5184	.4955	.8686	.5704	1.753	1.151	2.018	1.0524	60.3
29.8	.5201	.4970	.8678	.5727	1.746	1.152	2.012	1.0507	60.2
29.9	.5219	.4985	.8669	.5750	1.739	1.154	2.006	1.0489	60.1
30.0	.5236	.5000	.8660	.5774	1.732	1.155	2.000	1.0472	**60.0**
		$\cos\theta$	$\sin\theta$	$\cot\theta$	$\tan\theta$	$\csc\theta$	$\sec\theta$	θ Radians	θ Degrees

Table 6   Trigonometric Functions of θ (θ in decimal degrees)

θ Degrees	θ Radians	$\sin \theta$	$\cos \theta$	$\tan \theta$	$\cot \theta$	$\sec \theta$	$\csc \theta$		
30.0	.5236	.5000	.8660	.5774	1.732	1.155	2.000	1.0472	**60.0**
30.1	.5253	.5015	.8652	.5797	1.725	1.156	1.994	1.0455	59.9
30.2	.5271	.5030	.8643	.5820	1.718	1.157	1.988	1.0437	59.8
30.3	.5288	.5045	.8634	.5844	1.711	1.158	1.982	1.0420	59.7
30.4	.5306	.5060	.8625	.5867	1.704	1.159	1.976	1.0402	59.6
30.5	.5323	.5075	.8616	.5890	1.698	1.161	1.970	1.0385	59.5
30.6	.5341	.5090	.8607	.5914	1.691	1.162	1.964	1.0367	59.4
30.7	.5358	.5105	.8599	.5938	1.684	1.163	1.959	1.0350	59.3
30.8	.5376	.5120	.8590	.5961	1.678	1.164	1.953	1.0332	59.2
30.9	.5393	.5135	.8581	.5985	1.671	1.165	1.947	1.0315	59.1
31.0	.5411	.5150	.8572	.6009	1.664	1.167	1.942	1.0297	**59.0**
31.1	.5428	.5165	.8563	.6032	1.658	1.168	1.936	1.0280	58.9
31.2	.5445	.5180	.8554	.6056	1.651	1.169	1.930	1.0263	58.8
31.3	.5463	.5195	.8545	.6080	1.645	1.170	1.925	1.0245	58.7
31.4	.5480	.5210	.8535	.6104	1.638	1.172	1.919	1.0228	58.6
31.5	.5498	.5225	.8526	.6128	1.632	1.173	1.914	1.0210	58.5
31.6	.5515	.5240	.8517	.6152	1.625	1.174	1.908	1.0193	58.4
31.7	.5533	.5255	.8508	.6176	1.619	1.175	1.903	1.0175	58.3
31.8	.5550	.5270	.8499	.6200	1.613	1.177	1.898	1.0158	58.2
31.9	.5568	.5284	.8490	.6224	1.607	1.178	1.892	1.0140	58.1
32.0	.5585	.5299	.8480	.6249	1.600	1.179	1.887	1.0123	**58.0**
32.1	.5603	.5314	.8471	.6273	1.594	1.180	1.882	1.0105	57.9
32.2	.5620	.5329	.8462	.6297	1.588	1.182	1.877	1.0088	57.8
32.3	.5637	.5344	.8453	.6322	1.582	1.183	1.871	1.0071	57.7
32.4	.5655	.5358	.8443	.6346	1.576	1.184	1.866	1.0053	57.6
32.5	.5672	.5373	.8434	.6371	1.570	1.186	1.861	1.0036	57.5
32.6	.5690	.5388	.8425	.6395	1.564	1.187	1.856	1.0018	57.4
32.7	.5707	.5402	.8415	.6420	1.558	1.188	1.851	1.0001	57.3
32.8	.5725	.5417	.8406	.6445	1.552	1.190	1.846	.9983	57.2
32.9	.5742	.5432	.8396	.6469	1.546	1.191	1.841	.9966	57.1
33.0	.5760	.5446	.8387	.6494	1.540	1.192	1.836	.9948	**57.0**
33.1	.5777	.5461	.8377	.6519	1.534	1.194	1.831	.9931	56.9
33.2	.5794	.5476	.8368	.6544	1.528	1.195	1.826	.9913	56.8
33.3	.5812	.5490	.8358	.6569	1.522	1.196	1.821	.9896	56.7
33.4	.5829	.5505	.8348	.6594	1.517	1.198	1.817	.9879	56.6
33.5	.5847	.5519	.8339	.6619	1.511	1.199	1.812	.9861	56.5
33.6	.5864	.5534	.8329	.6644	1.505	1.201	1.807	.9844	56.4
33.7	.5882	.5548	.8320	.6669	1.499	1.202	1.802	.9826	56.3
33.8	.5899	.5563	.8310	.6694	1.494	1.203	1.798	.9809	56.2
33.9	.5917	.5577	.8300	.6720	1.488	1.205	1.793	.9791	56.1
34.0	.5934	.5592	.8290	.6745	1.483	1.206	1.788	.9774	**56.0**
34.1	.5952	.5606	.8281	.6771	1.477	1.208	1.784	.9756	55.9
34.2	.5969	.5621	.8271	.6796	1.471	1.209	1.779	.9739	55.8
34.3	.5986	.5635	.8261	.6822	1.466	1.211	1.775	.9721	55.7
34.4	.6004	.5650	.8251	.6847	1.460	1.212	1.770	.9704	55.6
34.5	.6021	.5664	.8241	.6873	1.455	1.213	1.766	.9687	55.5
34.6	.6039	.5678	.8231	.6899	1.450	1.215	1.761	.9669	55.4
34.7	.6056	.5693	.8221	.6924	1.444	1.216	1.757	.9652	55.3
34.8	.6074	.5707	.8211	.6950	1.439	1.218	1.752	.9634	55.2
34.9	.6091	.5721	.8202	.6976	1.433	1.219	1.748	.9617	55.1
35.0	.6109	.5736	.8192	.7002	1.428	1.221	1.743	.9599	**55.0**
35.1	.6126	.5750	.8181	.7028	1.423	1.222	1.739	.9582	54.9
35.2	.6144	.5764	.8171	.7054	1.418	1.224	1.735	.9564	54.8
35.3	.6161	.5779	.8161	.7080	1.412	1.225	1.731	.9547	54.7
35.4	.6178	.5793	.8151	.7107	1.407	1.227	1.726	.9529	54.6
35.5	.6196	.5807	.8141	.7133	1.402	1.228	1.722	.9512	54.5
35.6	.6213	.5821	.8131	.7159	1.397	1.230	1.718	.9495	54.4
35.7	.6231	.5835	.8121	.7186	1.392	1.231	1.714	.9477	54.3
35.8	.6248	.5850	.8111	.7212	1.387	1.233	1.710	.9460	54.2
35.9	.6266	.5864	.8100	.7239	1.381	1.235	1.705	.9442	54.1
36.0	.6283	.5878	.8090	.7265	1.376	1.236	1.701	.9425	**54.0**
		$\cos \theta$	$\sin \theta$	$\cot \theta$	$\tan \theta$	$\csc \theta$	$\sec \theta$	θ Radians	θ Degrees

Table 6   Trigonometric Functions of θ (θ in decimal degrees)

θ Degrees	θ Radians	$\sin\theta$	$\cos\theta$	$\tan\theta$	$\cot\theta$	$\sec\theta$	$\csc\theta$		
36.0	.6283	.5878	.8090	.7265	1.376	1.236	1.701	.9425	**54.0**
36.1	.6301	.5892	.8080	.7292	1.371	1.238	1.697	.9407	53.9
36.2	.6318	.5906	.8070	.7319	1.366	1.239	1.693	.9390	53.8
36.3	.6336	.5920	.8059	.7346	1.361	1.241	1.689	.9372	53.7
36.4	.6353	.5934	.8049	.7373	1.356	1.242	1.685	.9355	53.6
36.5	.6370	.5948	.8039	.7400	1.351	1.244	1.681	.9338	53.5
36.6	.6388	.5962	.8028	.7427	1.347	1.246	1.677	.9320	53.4
36.7	.6405	.5976	.8018	.7454	1.342	1.247	1.673	.9303	53.3
36.8	.6423	.5990	.8007	.7481	1.337	1.249	1.669	.9285	53.2
36.9	.6440	.6004	.7997	.7508	1.332	1.250	1.666	.9268	53.1
37.0	.6458	.6018	.7986	.7536	1.327	1.252	1.662	.9250	**53.0**
37.1	.6475	.6032	.7976	.7563	1.322	1.254	1.658	.9233	52.9
37.2	.6493	.6046	.7965	.7590	1.317	1.255	1.654	.9215	52.8
37.3	.6510	.6060	.7955	.7618	1.313	1.257	1.650	.9198	52.7
37.4	.6528	.6074	.7944	.7646	1.308	1.259	1.646	.9180	52.6
37.5	.6545	.6088	.7934	.7673	1.303	1.260	1.643	.9163	52.5
37.6	.6562	.6101	.7923	.7701	1.299	1.262	1.639	.9146	52.4
37.7	.6580	.6115	.7912	.7729	1.294	1.264	1.635	.9128	52.3
37.8	.6597	.6129	.7902	.7757	1.289	1.266	1.632	.9111	52.2
37.9	.6615	.6143	.7891	.7785	1.285	1.267	1.628	.9093	52.1
38.0	.6632	.6157	.7880	.7813	1.280	1.269	1.624	.9076	**52.0**
38.1	.6650	.6170	.7869	.7841	1.275	1.271	1.621	.9058	51.9
38.2	.6667	.6184	.7859	.7869	1.271	1.272	1.617	.9041	51.8
38.3	.6685	.6198	.7848	.7898	1.266	1.274	1.613	.9023	51.7
38.4	.6702	.6211	.7837	.7926	1.262	1.276	1.610	.9006	51.6
38.5	.6720	.6225	.7826	.7954	1.257	1.278	1.606	.8988	51.5
38.6	.6737	.6239	.7815	.7983	1.253	1.280	1.603	.8971	51.4
38.7	.6754	.6252	.7804	.8012	1.248	1.281	1.599	.8954	51.3
38.8	.6772	.6266	.7793	.8040	1.244	1.283	1.596	.8936	51.2
38.9	.6789	.6280	.7782	.8069	1.239	1.285	1.592	.8919	51.1
39.0	.6807	.6293	.7771	.8098	1.235	1.287	1.589	.8901	**51.0**
39.1	.6824	.6307	.7760	.8127	1.230	1.289	1.586	.8884	50.9
39.2	.6842	.6320	.7749	.8156	1.226	1.290	1.582	.8866	50.8
39.3	.6859	.6334	.7738	.8185	1.222	1.292	1.579	.8849	50.7
39.4	.6877	.6347	.7727	.8214	1.217	1.294	1.575	.8831	50.6
39.5	.6894	.6361	.7716	.8243	1.213	1.296	1.572	.8814	50.5
39.6	.6912	.6374	.7705	.8273	1.209	1.298	1.569	.8796	50.4
39.7	.6929	.6388	.7694	.8302	1.205	1.300	1.566	.8779	50.3
39.8	.6946	.6401	.7683	.8332	1.200	1.302	1.562	.8762	50.2
39.9	.6964	.6414	.7672	.8361	1.196	1.304	1.559	.8744	50.1
40.0	.6981	.6428	.7660	.8391	1.192	1.305	1.556	.8727	**50.0**
40.1	.6999	.6441	.7649	.8421	1.188	1.307	1.552	.8709	49.9
40.2	.7016	.6455	.7638	.8451	1.183	1.309	1.549	.8692	49.8
40.3	.7034	.6468	.7627	.8481	1.179	1.311	1.546	.8674	49.7
40.4	.7051	.6481	.7615	.8511	1.175	1.313	1.543	.8657	49.6
40.5	.7069	.6494	.7604	.8541	1.171	1.315	1.540	.8639	49.5
40.6	.7086	.6508	.7593	.8571	1.167	1.317	1.537	.8622	49.4
40.7	.7103	.6521	.7581	.8601	1.163	1.319	1.534	.8604	49.3
40.8	.7121	.6534	.7570	.8632	1.159	1.321	1.530	.8587	49.2
40.9	.7138	.6547	.7559	.8662	1.154	1.323	1.527	.8570	49.1
41.0	.7156	.6561	.7547	.8693	1.150	1.325	1.524	.8552	**49.0**
41.1	.7173	.6574	.7536	.8724	1.146	1.327	1.521	.8535	48.9
41.2	.7191	.6587	.7524	.8754	1.142	1.329	1.518	.8517	48.8
41.3	.7208	.6600	.7513	.8785	1.138	1.331	1.515	.8500	48.7
41.4	.7226	.6613	.7501	.8816	1.134	1.333	1.512	.8482	48.6
41.5	.7243	.6626	.7490	.8847	1.130	1.335	1.509	.8465	48.5
41.6	.7261	.6639	.7478	.8878	1.126	1.337	1.506	.8447	48.4
41.7	.7278	.6652	.7466	.8910	1.122	1.339	1.503	.8430	48.3
41.8	.7295	.6665	.7455	.8941	1.118	1.341	1.500	.8412	48.2
41.9	.7313	.6678	.7443	.8972	1.115	1.344	1.497	.8395	48.1
42.0	.7330	.6691	.7431	.9004	1.111	1.346	1.494	.8378	**48.0**
		$\cos\theta$	$\sin\theta$	$\cot\theta$	$\tan\theta$	$\csc\theta$	$\sec\theta$	θ Radians	θ Degrees

Table 6　Trigonometric Functions of θ (θ in decimal degrees)

θ Degrees	θ Radians	$\sin\theta$	$\cos\theta$	$\tan\theta$	$\cot\theta$	$\sec\theta$	$\csc\theta$		
42.0	.7330	.6691	.7431	.9004	1.111	1.346	1.494	.8378	**48.0**
42.1	.7348	.6704	.7420	.9036	1.107	1.348	1.492	.8360	47.9
42.2	.7365	.6717	.7408	.9067	1.103	1.350	1.489	.8343	47.8
42.3	.7383	.6730	.7396	.9099	1.099	1.352	1.486	.8325	47.7
42.4	.7400	.6743	.7385	.9131	1.095	1.354	1.483	.8308	47.6
42.5	.7418	.6756	.7373	.9163	1.091	1.356	1.480	.8290	47.5
42.6	.7435	.6769	.7361	.9195	1.087	1.359	1.477	.8273	47.4
42.7	.7453	.6782	.7349	.9228	1.084	1.361	1.475	.8255	47.3
42.8	.7470	.6794	.7337	.9260	1.080	1.363	1.472	.8238	47.2
42.9	.7487	.6807	.7325	.9293	1.076	1.365	1.469	.8221	47.1
43.0	.7505	.6820	.7314	.9325	1.072	1.367	1.466	.8203	**47.0**
43.1	.7522	.6833	.7302	.9358	1.069	1.370	1.464	.8186	46.9
43.2	.7540	.6845	.7290	.9391	1.065	1.372	1.461	.8168	46.8
43.3	.7557	.6858	.7278	.9424	1.061	1.374	1.458	.8151	46.7
43.4	.7575	.6871	.7266	.9457	1.057	1.376	1.455	.8133	46.6
43.5	.7592	.6884	.7254	.9490	1.054	1.379	1.453	.8116	46.5
43.6	.7610	.6896	.7242	.9523	1.050	1.381	1.450	.8098	46.4
43.7	.7627	.6909	.7230	.9556	1.046	1.383	1.447	.8081	46.3
43.8	.7645	.6921	.7218	.9590	1.043	1.386	1.445	.8063	46.2
43.9	.7662	.6934	.7206	.9623	1.039	1.388	1.442	.8046	46.1
44.0	.7679	.6947	.7193	.9657	1.036	1.390	1.440	.8029	**46.0**
44.1	.7697	.6959	.7181	.9691	1.032	1.393	1.437	.8011	45.9
44.2	.7714	.6972	.7169	.9725	1.028	1.395	1.434	.7994	45.8
44.3	.7732	.6984	.7157	.9759	1.025	1.397	1.432	.7976	45.7
44.4	.7749	.6997	.7145	.9793	1.021	1.400	1.429	.7959	45.6
44.5	.7767	.7009	.7133	.9827	1.018	1.402	1.427	.7941	45.5
44.6	.7784	.7022	.7120	.9861	1.014	1.404	1.424	.7924	45.4
44.7	.7802	.7034	.7108	.9896	1.011	1.407	1.422	.7906	45.3
44.8	.7819	.7046	.7096	.9930	1.007	1.409	1.419	.7889	45.2
44.9	.7837	.7059	.7083	.9965	1.003	1.412	1.417	.7871	45.1
45.0	.7854	.7071	.7071	1.0000	1.000	1.414	1.414	.7854	**45.0**
		$\cos\theta$	$\sin\theta$	$\cot\theta$	$\tan\theta$	$\csc\theta$	$\sec\theta$	θ Radians	θ Degrees

Table 7 Values of Trigonometric Functions for Angles in Degrees

m(α) Degrees	m(α) Radians	sin α	csc α	tan α	cot α	sec α	cos α		
0° 00′	.0000	.0000	Undefined	.0000	Undefined	1.000	1.0000	1.5708	90° 00′
10′	.0029	.0029	343.8	.0029	343.8	1.000	1.0000	1.5679	50′
20′	.0058	.0058	171.9	.0058	171.9	1.000	1.0000	1.5650	40′
30′	.0087	.0087	114.6	.0087	114.6	1.000	1.0000	1.5621	30′
40′	.0116	.0116	85.95	.0116	85.94	1.000	.9999	1.5592	20′
50′	.0145	.0145	68.76	.0145	68.75	1.000	.9999	1.5563	10′
1° 00′	.0175	.0175	57.30	.0175	57.29	1.000	.9998	1.5533	89° 00′
10′	.0204	.0204	49.11	.0204	49.10	1.000	.9998	1.5504	50′
20′	.0233	.0233	42.98	.0233	42.96	1.000	.9997	1.5475	40′
30′	.0262	.0262	38.20	.0262	38.19	1.000	.9997	1.5446	30′
40′	.0291	.0291	34.38	.0291	34.37	1.000	.9996	1.5417	20′
50′	.0320	.0320	31.26	.0320	31.24	1.001	.9995	1.5388	10′
2° 00′	.0349	.0349	28.65	.0349	28.64	1.001	.9994	1.5359	88° 00′
10′	.0378	.0378	26.45	.0378	26.43	1.001	.9993	1.5330	50′
20′	.0407	.0407	24.56	.0407	24.54	1.001	.9992	1.5301	40′
30′	.0436	.0436	22.93	.0437	22.90	1.001	.9990	1.5272	30′
40′	.0465	.0465	21.49	.0466	21.47	1.001	.9989	1.5243	20′
50′	.0495	.0494	20.23	.0495	20.21	1.001	.9988	1.5213	10′
3° 00′	.0524	.0523	19.11	.0524	19.08	1.001	.9986	1.5184	87° 00′
10′	.0553	.0552	18.10	.0553	18.07	1.002	.9985	1.5155	50′
20′	.0582	.0581	17.20	.0582	17.17	1.002	.9983	1.5126	40′
30′	.0611	.0610	16.38	.0612	16.35	1.002	.9981	1.5097	30′
40′	.0640	.0640	15.64	.0641	15.60	1.002	.9980	1.5068	20′
50′	.0669	.0669	14.96	.0670	14.92	1.002	.9978	1.5039	10′
4° 00′	.0698	.0698	14.34	.0699	14.30	1.002	.9976	1.5010	86° 00′
10′	.0727	.0727	13.76	.0729	13.73	1.003	.9974	1.4981	50′
20′	.0756	.0756	13.23	.0758	13.20	1.003	.9971	1.4952	40′
30′	.0785	.0785	12.75	.0787	12.71	1.003	.9969	1.4923	30′
40′	.0814	.0814	12.29	.0816	12.25	1.003	.9967	1.4893	20′
50′	.0844	.0843	11.87	.0846	11.83	1.004	.9964	1.4864	10′
5° 00′	.0873	.0872	11.47	.0875	11.43	1.004	.9962	1.4835	85° 00′
10′	.0902	.0901	11.10	.0904	11.06	1.004	.9959	1.4806	50′
20′	.0931	.0929	10.76	.0934	10.71	1.004	.9957	1.4777	40′
30′	.0960	.0958	10.43	.0963	10.39	1.005	.9954	1.4748	30′
40′	.0989	.0987	10.13	.0992	10.08	1.005	.9951	1.4719	20′
50′	.1018	.1016	9.839	.1022	9.788	1.005	.9948	1.4690	10′
6° 00′	.1047	.1045	9.567	.1051	9.514	1.006	.9945	1.4661	84° 00′
10′	.1076	.1074	9.309	.1080	9.255	1.006	.9942	1.4632	50′
20′	.1105	.1103	9.065	.1110	9.010	1.006	.9939	1.4603	40′
30′	.1134	.1132	8.834	.1139	8.777	1.006	.9936	1.4573	30′
40′	.1164	.1161	8.614	.1169	8.556	1.007	.9932	1.4544	20′
50′	.1193	.1190	8.405	.1198	8.345	1.007	.9929	1.4515	10′
7° 00′	.1222	.1219	8.206	.1228	8.144	1.008	.9925	1.4486	83° 00′
10′	.1251	.1248	8.016	.1257	7.953	1.008	.9922	1.4457	50′
20′	.1280	.1276	7.834	.1287	7.770	1.008	.9918	1.4428	40′
30′	.1309	.1305	7.661	.1317	7.596	1.009	.9914	1.4399	30′
40′	.1338	.1334	7.496	.1346	7.429	1.009	.9911	1.4370	20′
50′	.1367	.1363	7.337	.1376	7.269	1.009	.9907	1.4341	10′
8° 00′	.1396	.1392	7.185	.1405	7.115	1.010	.9903	1.4312	82° 00′
10′	.1425	.1421	7.040	.1435	6.968	1.010	.9899	1.4283	50′
20′	.1454	.1449	6.900	.1465	6.827	1.011	.9894	1.4254	40′
30′	.1484	.1478	6.765	.1495	6.691	1.011	.9890	1.4224	30′
40′	.1513	.1507	6.636	.1524	6.561	1.012	.9886	1.4195	20′
50′	.1542	.1536	6.512	.1554	6.435	1.012	.9881	1.4166	10′
9° 00′	.1571	.1564	6.392	.1584	6.314	1.012	.9877	1.4137	81° 00′
		cos α	sec α	cot α	tan α	csc α	sin α	Radians	Degrees
								m(α)	

Table 7 Values of Trigonometric Functions for Angles in Degrees

$m(\alpha)$ Degrees	$m(\alpha)$ Radians	$\sin \alpha$	$\csc \alpha$	$\tan \alpha$	$\cot \alpha$	$\sec \alpha$	$\cos \alpha$		
9° 00′	.1571	.1564	6.392	.1584	6.314	1.012	.9877	1.4137	81° 00′
10′	.1600	.1593	6.277	.1614	6.197	1.013	.9872	1.4108	50′
20′	.1629	.1622	6.166	.1644	6.084	1.013	.9868	1.4079	40′
30′	.1658	.1650	6.059	.1673	5.976	1.014	.9863	1.4050	30′
40′	.1687	.1679	5.955	.1703	5.871	1.014	.9858	1.4021	20′
50′	.1716	.1708	5.855	.1733	5.769	1.015	.9853	1.3992	10′
10° 00′	.1745	.1736	5.759	.1763	5.671	1.015	.9848	1.3963	80° 00′
10′	.1774	.1765	5.665	.1793	5.576	1.016	.9843	1.3934	50′
20′	.1804	.1794	5.575	.1823	5.485	1.016	.9838	1.3904	40′
30′	.1833	.1822	5.487	.1853	5.396	1.017	.9833	1.3875	30′
40′	.1862	.1851	5.403	.1883	5.309	1.018	.9827	1.3846	20′
50′	.1891	.1880	5.320	.1914	5.226	1.018	.9822	1.3817	10′
11° 00′	.1920	.1908	5.241	.1944	5.145	1.019	.9816	1.3788	79° 00′
10′	.1949	.1937	5.164	.1974	5.066	1.019	.9811	1.3759	50′
20′	.1978	.1965	5.089	.2004	4.989	1.020	.9805	1.3730	40′
30′	.2007	.1994	5.016	.2035	4.915	1.020	.9799	1.3701	30′
40′	.2036	.2022	4.945	.2065	4.843	1.021	.9793	1.3672	20′
50′	.2065	.2051	4.876	.2095	4.773	1.022	.9787	1.3643	10′
12° 00′	.2094	.2079	4.810	.2126	4.705	1.022	.9781	1.3614	78° 00′
10′	.2123	.2108	4.745	.2156	4.638	1.023	.9775	1.3584	50′
20′	.2153	.2136	4.682	.2186	4.574	1.024	.9769	1.3555	40′
30′	.2182	.2164	4.620	.2217	4.511	1.024	.9763	1.3526	30′
40′	.2211	.2193	4.560	.2247	4.449	1.025	.9757	1.3497	20′
50′	.2240	.2221	4.502	.2278	4.390	1.026	.9750	1.3468	10′
13° 00′	.2269	.2250	4.445	.2309	4.331	1.026	.9744	1.3439	77° 00′
10′	.2298	.2278	4.390	.2339	4.275	1.027	.9737	1.3410	50′
20′	.2327	.2306	4.336	.2370	4.219	1.028	.9730	1.3381	40′
30′	.2356	.2334	4.284	.2401	4.165	1.028	.9724	1.3352	30′
40′	.2385	.2363	4.232	.2432	4.113	1.029	.9717	1.3323	20′
50′	.2414	.2391	4.182	.2462	4.061	1.030	.9710	1.3294	10′
14° 00′	.2443	.2419	4.134	.2493	4.011	1.031	.9703	1.3265	76° 00′
10′	.2473	.2447	4.086	.2524	3.962	1.031	.9696	1.3235	50′
20′	.2502	.2476	4.039	.2555	3.914	1.032	.9689	1.3206	40′
30′	.2531	.2504	3.994	.2586	3.867	1.033	.9681	1.3177	30′
40′	.2560	.2532	3.950	.2617	3.821	1.034	.9674	1.3148	20′
50′	.2589	.2560	3.906	.2648	3.776	1.034	.9667	1.3119	10′
15° 00′	.2618	.2588	3.864	.2679	3.732	1.035	.9659	1.3090	75° 00′
10′	.2647	.2616	3.822	.2711	3.689	1.036	.9652	1.3061	50′
20′	.2676	.2644	3.782	.2742	3.647	1.037	.9644	1.3032	40′
30′	.2705	.2672	3.742	.2773	3.606	1.038	.9636	1.3003	30′
40′	.2734	.2700	3.703	.2805	3.566	1.039	.9628	1.2974	20′
50′	.2763	.2728	3.665	.2836	3.526	1.039	.9621	1.2945	10′
16° 00′	.2793	.2756	3.628	.2867	3.487	1.040	.9613	1.2915	74° 00′
10′	.2822	.2784	3.592	.2899	3.450	1.041	.9605	1.2886	50′
20′	.2851	.2812	3.556	.2931	3.412	1.042	.9596	1.2857	40′
30′	.2880	.2840	3.521	.2962	3.376	1.043	.9588	1.2828	30′
40′	.2909	.2868	3.487	.2994	3.340	1.044	.9580	1.2799	20′
50′	.2938	.2896	3.453	.3026	3.305	1.045	.9572	1.2770	10′
17° 00′	.2967	.2924	3.420	.3057	3.271	1.046	.9563	1.2741	73° 00′
10′	.2996	.2952	3.388	.3089	3.237	1.047	.9555	1.2712	50′
20′	.3025	.2979	3.357	.3121	3.204	1.048	.9546	1.2683	40′
30′	.3054	.3007	3.326	.3153	3.172	1.049	.9537	1.2654	30′
40′	.3083	.3035	3.295	.3185	3.140	1.049	.9528	1.2625	20′
50′	.3113	.3062	3.265	.3217	3.108	1.050	.9520	1.2595	10′
18° 00′	.3142	.3090	3.236	.3249	3.078	1.051	.9511	1.2566	72° 00′
		$\cos \alpha$	$\sec \alpha$	$\cot \alpha$	$\tan \alpha$	$\csc \alpha$	$\sin \alpha$	Radians	Degrees
								$m(\alpha)$	

Table 7   Values of Trigonometric Functions for Angles in Degrees

m(α) Degrees	Radians	sin α	csc α	tan α	cot α	sec α	cos α		
18° 00′	.3142	.3090	3.236	.3249	3.078	1.051	.9511	1.2566	72° 00′
10′	.3171	.3118	3.207	.3281	3.047	1.052	.9502	1.2537	50′
20′	.3200	.3145	3.179	.3314	3.018	1.053	.9492	1.2508	40′
30′	.3229	.3173	3.152	.3346	2.989	1.054	.9483	1.2479	30′
40′	.3258	.3201	3.124	.3378	2.960	1.056	.9474	1.2450	20′
50′	.3287	.3228	3.098	.3411	2.932	1.057	.9465	1.2421	10′
19° 00′	.3316	.3256	3.072	.3443	2.904	1.058	.9455	1.2392	71° 00′
10′	.3345	.3283	3.046	.3476	2.877	1.059	.9445	1.2363	50′
20′	.3374	.3311	3.021	.3508	2.850	1.060	.9436	1.2334	40′
30′	.3403	.3338	2.996	.3541	2.824	1.061	.9426	1.2305	30′
40′	.3432	.3365	2.971	.3574	2.798	1.062	.9417	1.2275	20′
50′	.3462	.3393	2.947	.3607	2.773	1.063	.9407	1.2246	10′
20° 00′	.3491	.3420	2.924	.3640	2.747	1.064	.9397	1.2217	70° 00′
10′	.3520	.3448	2.901	.3673	2.723	1.065	.9387	1.2188	50′
20′	.3549	.3475	2.878	.3706	2.699	1.066	.9377	1.2159	40′
30′	.3578	.3502	2.855	.3739	2.675	1.068	.9367	1.2130	30′
40′	.3607	.3529	2.833	.3772	2.651	1.069	.9356	1.2101	20′
50′	.3636	.3557	2.812	.3805	2.628	1.070	.9346	1.2072	10′
21° 00′	.3665	.3584	2.790	.3839	2.605	1.071	.9336	1.2043	69° 00′
10′	.3694	.3611	2.769	.3872	2.583	1.072	.9325	1.2014	50′
20′	.3723	.3638	2.749	.3906	2.560	1.074	.9315	1.1985	40′
30′	.3752	.3665	2.729	.3939	2.539	1.075	.9304	1.1956	30′
40′	.3782	.3692	2.709	.3973	2.517	1.076	.9293	1.1926	20′
50′	.3811	.3719	2.689	.4006	2.496	1.077	.9283	1.1897	10′
22° 00′	.3840	.3746	2.669	.4040	2.475	1.079	.9272	1.1868	68° 00′
10′	.3869	.3773	2.650	.4074	2.455	1.080	.9261	1.1839	50′
20′	.3898	.3800	2.632	.4108	2.434	1.081	.9250	1.1810	40′
30′	.3927	.3827	2.613	.4142	2.414	1.082	.9239	1.1781	30′
40′	.3956	.3854	2.595	.4176	2.394	1.084	.9228	1.1752	20′
50′	.3985	.3881	2.577	.4210	2.375	1.085	.9216	1.1723	10′
23° 00′	.4014	.3907	2.559	.4245	2.356	1.086	.9205	1.1694	67° 00′
10′	.4043	.3934	2.542	.4279	2.337	1.088	.9194	1.1665	50′
20′	.4072	.3961	2.525	.4314	2.318	1.089	.9182	1.1636	40′
30′	.4102	.3987	2.508	.4348	2.300	1.090	.9171	1.1606	30′
40′	.4131	.4014	2.491	.4383	2.282	1.092	.9159	1.1577	20′
50′	.4160	.4041	2.475	.4417	2.264	1.093	.9147	1.1548	10′
24° 00′	.4189	.4067	2.459	.4452	2.246	1.095	.9135	1.1519	66° 00′
10′	.4218	.4094	2.443	.4487	2.229	1.096	.9124	1.1490	50′
20′	.4247	.4120	2.427	.4522	2.211	1.097	.9112	1.1461	40′
30′	.4276	.4147	2.411	.4557	2.194	1.099	.9100	1.1432	30′
40′	.4305	.4173	2.396	.4592	2.177	1.100	.9088	1.1403	20′
50′	.4334	.4200	2.381	.4628	2.161	1.102	.9075	1.1374	10′
25° 00′	.4363	.4226	2.366	.4663	2.145	1.103	.9063	1.1345	65° 00′
10′	.4392	.4253	2.352	.4699	2.128	1.105	.9051	1.1316	50′
20′	.4422	.4279	2.337	.4734	2.112	1.106	.9038	1.1286	40′
30′	.4451	.4305	2.323	.4770	2.097	1.108	.9026	1.1257	30′
40′	.4480	.4331	2.309	.4806	2.081	1.109	.9013	1.1228	20′
50′	.4509	.4358	2.295	.4841	2.066	1.111	.9001	1.1199	10′
26° 00′	.4538	.4384	2.281	.4877	2.050	1.113	.8988	1.1170	64° 00′
10′	.4567	.4410	2.268	.4913	2.035	1.114	.8975	1.1141	50′
20′	.4596	.4436	2.254	.4950	2.020	1.116	.8962	1.1112	40′
30′	.4625	.4462	2.241	.4986	2.006	1.117	.8949	1.1083	30′
40′	.4654	.4488	2.228	.5022	1.991	1.119	.8936	1.1054	20′
50′	.4683	.4514	2.215	.5059	1.977	1.121	.8923	1.1025	10′
27° 00′	.4712	.4540	2.203	.5095	1.963	1.122	.8910	1.0996	63° 00′
		cos α	sec α	cot α	tan α	csc α	sin α	Radians	Degrees
									m(α)

Table 7 Values of Trigonometric Functions for Angles in Degrees

m(α) Degrees	m(α) Radians	sin α	csc α	tan α	cot α	sec α	cos α		
27° 00′	.4712	.4540	2.203	.5095	1.963	1.122	.8910	1.0996	63° 00′
10′	.4741	.4566	2.190	.5132	1.949	1.124	.8897	1.0966	50′
20′	.4771	.4592	2.178	.5169	1.935	1.126	.8884	1.0937	40′
30′	.4800	.4617	2.166	.5206	1.921	1.127	.8870	1.0908	30′
40′	.4829	.4643	2.154	.5243	1.907	1.129	.8857	1.0879	20′
50′	.4858	.4669	2.142	.5280	1.894	1.131	.8843	1.0850	10′
28° 00′	.4887	.4695	2.130	.5317	1.881	1.133	.8829	1.0821	62° 00′
10′	.4916	.4720	2.118	.5354	1.868	1.134	.8816	1.0792	50′
20′	.4945	.4746	2.107	.5392	1.855	1.136	.8802	1.0763	40′
30′	.4974	.4772	2.096	.5430	1.842	1.138	.8788	1.0734	30′
40′	.5003	.4797	2.085	.5467	1.829	1.140	.8774	1.0705	20′
50′	.5032	.4823	2.074	.5505	1.816	1.142	.8760	1.0676	10′
29° 00′	.5061	.4848	2.063	.5543	1.804	1.143	.8746	1.0647	61° 00′
10′	.5091	.4874	2.052	.5581	1.792	1.145	.8732	1.0617	50′
20′	.5120	.4899	2.041	.5619	1.780	1.147	.8718	1.0588	40′
30′	.5149	.4924	2.031	.5658	1.767	1.149	.8704	1.0559	30′
40′	.5178	.4950	2.020	.5696	1.756	1.151	.8689	1.0530	20′
50′	.5207	.4975	2.010	.5735	1.744	1.153	.8675	1.0501	10′
30° 00′	.5236	.5000	2.000	.5774	1.732	1.155	.8660	1.0472	60° 00′
10′	.5265	.5025	1.990	.5812	1.720	1.157	.8646	1.0443	50′
20′	.5294	.5050	1.980	.5851	1.709	1.159	.8631	1.0414	40′
30′	.5323	.5075	1.970	.5890	1.698	1.161	.8616	1.0385	30′
40′	.5352	.5100	1.961	.5930	1.686	1.163	.8601	1.0356	20′
50′	.5381	.5125	1.951	.5969	1.675	1.165	.8587	1.0327	10′
31° 00′	.5411	.5150	1.942	.6009	1.664	1.167	.8572	1.0297	59° 00′
10′	.5440	.5175	1.932	.6048	1.653	1.169	.8557	1.0268	50′
20′	.5469	.5200	1.923	.6088	1.643	1.171	.8542	1.0239	40′
30′	.5498	.5225	1.914	.6128	1.632	1.173	.8526	1.0210	30′
40′	.5527	.5250	1.905	.6168	1.621	1.175	.8511	1.0181	20′
50′	.5556	.5275	1.896	.6208	1.611	1.177	.8496	1.0152	10′
32° 00′	.5585	.5299	1.887	.6249	1.600	1.179	.8480	1.0123	58° 00′
10′	.5614	.5324	1.878	.6289	1.590	1.181	.8465	1.0094	50′
20′	.5643	.5348	1.870	.6330	1.580	1.184	.8450	1.0065	40′
30′	.5672	.5373	1.861	.6371	1.570	1.186	.8434	1.0036	30′
40′	.5701	.5398	1.853	.6412	1.560	1.188	.8418	1.0007	20′
50′	.5730	.5422	1.844	.6453	1.550	1.190	.8403	.9977	10′
33° 00′	.5760	.5446	1.836	.6494	1.540	1.192	.8387	.9948	57° 00′
10′	.5789	.5471	1.828	.6536	1.530	1.195	.8371	.9919	50′
20′	.5818	.5495	1.820	.6577	1.520	1.197	.8355	.9890	40′
30′	.5847	.5519	1.812	.6619	1.511	1.199	.8339	.9861	30′
40′	.5876	.5544	1.804	.6661	1.501	1.202	.8323	.9832	20′
50′	.5905	.5568	1.796	.6703	1.492	1.204	.8307	.9803	10′
34° 00′	.5934	.5592	1.788	.6745	1.483	1.206	.8290	.9774	56° 00′
10′	.5963	.5616	1.781	.6787	1.473	1.209	.8274	.9745	50′
20′	.5992	.5640	1.773	.6830	1.464	1.211	.8258	.9716	40′
30′	.6021	.5664	1.766	.6873	1.455	1.213	.8241	.9687	30′
40′	.6050	.5688	1.758	.6916	1.446	1.216	.8225	.9657	20′
50′	.6080	.5712	1.751	.6959	1.437	1.218	.8208	.9628	10′
35° 00′	.6109	.5736	1.743	.7002	1.428	1.221	.8192	.9599	55° 00′
10′	.6138	.5760	1.736	.7046	1.419	1.223	.8175	.9570	50′
20′	.6167	.5783	1.729	.7089	1.411	1.226	.8158	.9541	40′
30′	.6196	.5807	1.722	.7133	1.402	1.228	.8141	.9512	30′
40′	.6225	.5831	1.715	.7177	1.393	1.231	.8124	.9483	20′
50′	.6254	.5854	1.708	.7221	1.385	1.233	.8107	.9454	10′
36° 00′	.6283	.5878	1.701	.7265	1.376	1.236	.8090	.9425	54° 00′
		cos α	sec α	cot α	tan α	csc α	sin α	Radians	Degrees m(α)

Table 7 Values of Trigonometric Functions for Angles in Degrees

Degrees	Radians	sin α	csc α	tan α	cot α	sec α	cos α		
36° 00′	.6283	.5878	1.701	.7265	1.376	1.236	.8090	.9425	**54° 00′**
10′	.6312	.5901	1.695	.7310	1.368	1.239	.8073	.9396	50′
20′	.6341	.5925	1.688	.7355	1.360	1.241	.8056	.9367	40′
30′	.6370	.5948	1.681	.7400	1.351	1.244	.8039	.9338	30′
40′	.6400	.5972	1.675	.7445	1.343	1.247	.8021	.9308	20′
50′	.6429	.5995	1.668	.7490	1.335	1.249	.8004	.9279	10′
37° 00′	.6458	.6018	1.662	.7536	1.327	1.252	.7986	.9250	**53° 00′**
10′	.6487	.6041	1.655	.7581	1.319	1.255	.7969	.9221	50′
20′	.6516	.6065	1.649	.7627	1.311	1.258	.7951	.9192	40′
30′	.6545	.6088	1.643	.7673	1.303	1.260	.7934	.9163	30′
40′	.6574	.6111	1.636	.7720	1.295	1.263	.7916	.9134	20′
50′	.6603	.6134	1.630	.7766	1.288	1.266	.7898	.9105	10′
38° 00′	.6632	.6157	1.624	.7813	1.280	1.269	.7880	.9076	**52° 00′**
10′	.6661	.6180	1.618	:7860	1.272	1.272	.7862	.9047	50′
20′	.6690	.6202	1.612	.7907	1.265	1.275	.7844	.9018	40′
30′	.6720	.6225	1.606	.7954	1.257	1.278	.7826	.8988	30′
40′	.6749	.6248	1.601	.8002	1.250	1.281	.7808	.8959	20′
50′	.6778	.6271	1.595	.8050	1.242	1.284	.7790	.8930	10′
39° 00′	.6807	.6293	1.589	.8098	1.235	1.287	.7771	.8901	**51° 00′**
10′	.6836	.6316	1.583	.8146	1.228	1.290	.7753	.8872	50′
20′	.6865	.6338	1.578	.8195	1.220	1.293	.7735	.8843	40′
30′	.6894	.6361	1.572	.8243	1.213	1.296	.7716	.8814	30′
40′	.6923	.6383	1.567	.8292	1.206	1.299	.7698	.8785	20′
50′	.6952	.6406	1.561	.8342	1.199	1.302	.7679	.8756	10′
40° 00′	.6981	.6428	1.556	.8391	1.192	1.305	.7660	.8727	**50° 00′**
10′	.7010	.6450	1.550	.8441	1.185	1.309	.7642	.8698	50′
20′	.7039	.6472	1.545	.8491	1.178	1.312	.7623	.8668	40′
30′	.7069	.6494	1.540	.8541	1.171	1.315	.7604	.8639	30′
40′	.7098	.6517	1.535	.8591	1.164	1.318	.7585	.8610	20′
50′	.7127	.6539	1.529	.8642	1.157	1.322	.7566	.8581	10′
41° 00′	.7156	.6561	1.524	.8693	1.150	1.325	.7547	.8552	**49° 00′**
10′	.7185	.6583	1.519	.8744	1.144	1.328	.7528	.8523	50′
20′	.7214	.6604	1.514	.8796	1.137	1.332	.7509	.8494	40′
30′	.7243	.6626	1.509	.8847	1.130	1.335	.7490	.8465	30′
40′	.7272	.6648	1.504	.8899	1.124	1.339	.7470	.8436	20′
50′	.7301	.6670	1.499	.8952	1.117	1.342	.7451	.8407	10′
42° 00′	.7330	.6691	1.494	.9004	1.111	1.346	.7431	.8378	**48° 00′**
10′	.7359	.6713	1.490	.9057	1.104	1.349	.7412	.8348	50′
20′	.7389	.6734	1.485	.9110	1.098	1.353	.7392	.8319	40′
30′	.7418	.6756	1.480	.9163	1.091	1.356	.7373	.8290	30′
40′	.7447	.6777	1.476	.9217	1.085	1.360	.7353	.8261	20′
50′	.7476	.6799	1.471	.9271	1.079	1.364	.7333	.8232	10′
43° 00′	.7505	.6820	1.466	.9325	1.072	1.367	.7314	.8203	**47° 00′**
10′	.7534	.6841	1.462	.9380	1.066	1.371	.7294	.8174	50′
20′	.7563	.6862	1.457	.9435	1.060	1.375	.7274	.8145	40′
30′	.7592	.6884	1.453	.9490	1.054	1.379	.7254	.8116	30′
40′	.7621	.6905	1.448	.9545	1.048	1.382	.7234	.8087	20′
50′	.7650	.6926	1.444	.9601	1.042	1.386	.7214	.8058	10′
44° 00′	.7679	.6947	1.440	.9657	1.036	1.390	.7193	.8029	**46° 00′**
10′	.7709	.6967	1.435	.9713	1.030	1.394	.7173	.7999	50′
20′	.7738	.6988	1.431	.9770	1.024	1.398	.7153	.7970	40′
30′	.7767	.7009	1.427	.9827	1.018	1.402	.7133	.7941	30′
40′	.7796	.7030	1.423	.9884	1.012	1.406	.7112	.7912	20′
50′	.7825	.7050	1.418	.9942	1.006	1.410	.7092	.7883	10′
45° 00′	.7854	.7071	1.414	1.000	1.000	1.414	.7071	.7854	**45° 00′**
		cos α	sec α	cot α	tan α	csc α	sin α	Radians	Degrees
								m(α)	

Table 8　Values of Circular Functions and Trigonometric Functions for Angles in Radians

Real Number x or m(α)	m(α) in degrees	sin x or sin α	csc x or csc α	tan x or tan α	cot x or cot α	sec x or sec α	cos x or cos α
0	0°	0	Undefined	0	Undefined	1	1
0.01	0° 34′	0.0100	100.0	0.0100	100.0	1.000	1.000
.02	1° 09′	.0200	50.00	.0200	49.99	1.000	0.9998
.03	1° 43′	.0300	33.34	.0300	33.32	1.000	0.9996
.04	2° 18′	.0400	25.01	.0400	24.99	1.001	0.9992
0.05	2° 52′	0.0500	20.01	0.0500	19.98	1.001	0.9988
.06	3° 26′	.0600	16.68	.0601	16.65	1.002	.9982
.07	4° 01′	.0699	14.30	.0701	14.26	1.002	.9976
.08	4° 35′	.0799	12.51	.0802	12.47	1.003	.9968
.09	5° 09′	.0899	11.13	.0902	11.08	1.004	.9960
0.10	5° 44′	0.0998	10.02	0.1003	9.967	1.005	0.9950
.11	6° 18′	.1098	9.109	.1104	9.054	1.006	.9940
.12	6° 53′	.1197	8.353	.1206	8.293	1.007	.9928
.13	7° 27′	.1296	7.714	.1307	7.649	1.009	.9916
.14	8° 01′	.1395	7.166	.1409	7.096	1.010	.9902
0.15	8° 36′	0.1494	6.692	0.1511	6.617	1.011	0.9888
.16	9° 10′	.1593	6.277	.1614	6.197	1.013	.9872
.17	9° 44′	.1692	5.911	.1717	5.826	1.015	.9856
.18	10° 19′	.1790	5.586	.1820	5.495	1.016	.9838
.19	10° 53′	.1889	5.295	.1923	5.200	1.018	.9820
0.20	11° 28′	0.1987	5.033	0.2027	4.933	1.020	0.9801
.21	12° 02′	.2085	4.797	.2131	4.692	1.022	.9780
.22	12° 36′	.2182	4.582	.2236	4.472	1.025	.9759
.23	13° 11′	.2280	4.386	.2341	4.271	1.027	.9737
.24	13° 45′	.2377	4.207	.2447	4.086	1.030	.9713
0.25	14° 19′	0.2474	4.042	0.2553	3.916	1.032	0.9689
.26	14° 54′	.2571	3.890	.2660	3.759	1.035	.9664
.27	15° 28′	.2667	3.749	.2768	3.613	1.038	.9638
.28	16° 03′	.2764	3.619	.2876	3.478	1.041	.9611
.29	16° 37′	.2860	3.497	.2984	3.351	1.044	.9582
0.30	17° 11′	0.2955	3.384	0.3093	3.233	1.047	0.9553
.31	17° 46′	.3051	3.278	.3203	3.122	1.050	.9523
.32	18° 20′	.3146	3.179	.3314	3.018	1.053	.9492
.33	18° 55′	.3240	3.086	.3425	2.920	1.057	.9460
.34	19° 29′	.3335	2.999	.3537	2.827	1.061	.9428
0.35	20° 03′	0.3429	2.916	0.3650	2.740	1.065	0.9394
.36	20° 38′	.3523	2.839	.3764	2.657	1.068	.9359
.37	21° 12′	.3616	2.765	.3879	2.578	1.073	.9323
.38	21° 46′	.3709	2.696	.3994	2.504	1.077	.9287
.39	22° 21′	.3802	2.630	.4111	2.433	1.081	.9249
0.40	22° 55′	0.3894	2.568	0.4228	2.365	1.086	0.9211
.41	23° 30′	.3986	2.509	.4346	2.301	1.090	.9171
.42	24° 04′	.4078	2.452	.4466	2.239	1.095	.9131
.43	24° 38′	.4169	2.399	.4586	2.180	1.100	.9090
.44	25° 13′	.4259	2.348	.4708	2.124	1.105	.9048
0.45	25° 47′	0.4350	2.299	0.4831	2.070	1.111	0.9004
.46	26° 21′	.4439	2.253	.4954	2.018	1.116	.8961
.47	26° 56′	.4529	2.208	.5080	1.969	1.122	.8916
.48	27° 30′	.4618	2.166	.5206	1.921	1.127	.8870
.49	28° 05′	.4706	2.125	.5334	1.875	1.133	.8823

Table 8 Values of Circular Functions and Trigonometric Functions for Angles in Radians

Real Number x or m(α)	m(α) in degrees	sin x or sin α	csc x or csc α	tan x or tan α	cot x or cot α	sec x or sec α	cos x or cos α
0.50	28° 39′	0.4794	2.086	0.5463	1.830	1.139	0.8776
.51	29° 13′	.4882	2.048	.5594	1.788	1.146	.8727
.52	29° 48′	.4969	2.013	.5726	1.747	1.152	.8678
.53	30° 22′	.5055	1.978	.5859	1.707	1.159	.8628
.54	30° 56′	.5141	1.945	.5994	1.668	1.166	.8577
0.55	31° 31′	0.5227	1.913	0.6131	1.631	1.173	0.8525
.56	32° 05′	.5312	1.883	.6269	1.595	1.180	.8473
.57	32° 40′	.5396	1.853	.6410	1.560	1.188	.8419
.58	33° 14′	.5480	1.825	.6552	1.526	1.196	.8365
.59	33° 48′	.5564	1.797	.6696	1.494	1.203	.8309
0.60	34° 23′	0.5646	1.771	0.6841	1.462	1.212	0.8253
.61	34° 57′	.5729	1.746	.6989	1.431	1.220	.8196
.62	35° 31′	.5810	1.721	.7139	1.401	1.229	.8139
.63	36° 06′	.5891	1.697	.7291	1.372	1.238	.8080
.64	36° 40′	.5972	1.674	.7445	1.343	1.247	.8021
0.65	37° 15′	0.6052	1.652	0.7602	1.315	1.256	0.7961
.66	37° 49′	.6131	1.631	.7761	1.288	1.266	.7900
.67	38° 23′	.6210	1.610	.7923	1.262	1.276	.7838
.68	38° 58′	.6288	1.590	.8087	1.237	1.286	.7776
.69	39° 32′	.6365	1.571	.8253	1.212	1.297	.7712
0.70	40° 06′	0.6442	1.552	0.8423	1.187	1.307	0.7648
.71	40° 41′	.6518	1.534	.8595	1.163	1.319	.7584
.72	41° 15′	.6594	1.517	.8771	1.140	1.330	.7518
.73	41° 50′	.6669	1.500	.8949	1.117	1.342	.7452
.74	42° 24′	.6743	1.483	.9131	1.095	1.354	.7385
0.75	42° 58′	0.6816	1.467	0.9316	1.073	1.367	0.7317
.76	43° 33′	.6889	1.452	.9505	1.052	1.380	.7248
.77	44° 07′	.6961	1.437	.9697	1.031	1.393	.7179
.78	44° 41′	.7033	1.422	.9893	1.011	1.407	.7109
.79	45° 16′	.7104	1.408	1.009	.9908	1.421	.7038
0.80	45° 50′	0.7174	1.394	1.030	0.9712	1.435	0.6967
.81	46° 25′	.7243	1.381	1.050	.9520	1.450	.6895
.82	46° 59′	.7311	1.368	1.072	.9331	1.466	.6822
.83	47° 33′	.7379	1.355	1.093	.9146	1.482	.6749
.84	48° 08′	.7446	1.343	1.116	.8964	1.498	.6675
0.85	48° 42′	0.7513	1.331	1.138	0.8785	1.515	0.6600
.86	49° 17′	.7578	1.320	1.162	.8609	1.533	.6524
.87	49° 51′	.7643	1.308	1.185	.8437	1.551	.6448
.88	50° 25′	.7707	1.297	1.210	.8267	1.569	.6372
.89	51° 00′	.7771	1.287	1.235	.8100	1.589	.6294
0.90	51° 34′	0.7833	1.277	1.260	0.7936	1.609	0.6216
.91	52° 08′	.7895	1.267	1.286	.7774	1.629	.6137
.92	52° 43′	.7956	1.257	1.313	.7615	1.651	.6058
.93	53° 17′	.8016	1.247	1.341	.7458	1.673	.5978
.94	53° 52′	.8076	1.238	1.369	.7303	1.696	.5898
0.95	54° 26′	0.8134	1.229	1.398	0.7151	1.719	0.5817
.96	55° 00′	.8192	1.221	1.428	.7001	1.744	.5735
.97	55° 35′	.8249	1.212	1.459	.6853	1.769	.5653
.98	56° 09′	.8305	1.204	1.491	.6707	1.795	.5570
.99	56° 43′	.8360	1.196	1.524	.6563	1.823	.5487

Table 8 Values of Circular Functions and Trigonometric Functions for Angles in Radians

Real Number x or m(α)	m(α) in degrees	sin x or sin α	csc x or csc α	tan x or tan α	cot x or cot α	sec x or sec α	cos x or cos α
1.00	57° 18′	0.8415	1.188	1.557	0.6421	1.851	0.5403
1.01	57° 52′	.8468	1.181	1.592	.6281	1.880	.5319
1.02	58° 27′	.8521	1.174	1.628	.6142	1.911	.5234
1.03	59° 01′	.8573	1.166	1.665	.6005	1.942	.5148
1.04	59° 35′	.8624	1.160	1.704	.5870	1.975	.5062
1.05	60° 10′	0.8674	1.153	1.743	0.5736	2.010	0.4976
1.06	60° 44′	.8724	1.146	1.784	.5604	2.046	.4889
1.07	61° 18′	.8772	1.140	1.827	.5473	2.083	.4801
1.08	61° 53′	.8820	1.134	1.871	.5344	2.122	.4713
1.09	62° 27′	.8866	1.128	1.917	.5216	2.162	.4625
1.10	63° 02′	0.8912	1.122	1.965	0.5090	2.205	0.4536
1.11	63° 36′	.8957	1.116	2.014	.4964	2.249	.4447
1.12	64° 10′	.9001	1.111	2.066	.4840	2.295	.4357
1.13	64° 45′	.9044	1.106	2.120	.4718	2.344	.4267
1.14	65° 19′	.9086	1.101	2.176	.4596	2.395	.4176
1.15	65° 53′	0.9128	1.096	2.234	0.4475	2.448	0.4085
1.16	66° 28′	.9168	1.091	2.296	.4356	2.504	.3993
1.17	67° 02′	.9208	1.086	2.360	.4237	2.563	.3902
1.18	67° 37′	.9246	1.082	2.428	.4120	2.625	.3809
1.19	68° 11′	.9284	1.077	2.498	.4003	2.691	.3717
1.20	68° 45′	0.9320	1.073	2.572	0.3888	2.760	0.3624
1.21	69° 20′	.9356	1.069	2.650	.3773	2.833	.3530
1.22	69° 54′	.9391	1.065	2.733	.3659	2.910	.3436
1.23	70° 28′	.9425	1.061	2.820	.3546	2.992	.3342
1.24	71° 03′	.9458	1.057	2.912	.3434	3.079	.3248
1.25	71° 37′	0.9490	1.054	3.010	0.3323	3.171	0.3153
1.26	72° 12′	.9521	1.050	3.113	.3212	3.270	.3058
1.27	72° 46′	.9551	1.047	3.224	.3102	3.375	.2963
1.28	73° 20′	.9580	1.044	3.341	.2993	3.488	.2867
1.29	73° 55′	.9608	1.041	3.467	.2884	3.609	.2771
1.30	74° 29′	0.9636	1.038	3.602	0.2776	3.738	0.2675
1.31	75° 03′	.9662	1.035	3.747	.2669	3.878	.2579
1.32	75° 38′	.9687	1.032	3.903	.2562	4.029	.2482
1.33	76° 12′	.9711	1.030	4.072	.2456	4.193	.2385
1.34	76° 47′	,9735	1.027	4.256	.2350	4.372	.2288
1.35	77° 21′	0.9757	1.025	4.455	0.2245	4.566	0.2190
1.36	77° 55′	.9779	1.023	4.673	.2140	4.779	.2092
1.37	78° 30′	.9799	1.021	4.913	.2035	5.014	.1994
1.38	79° 04′	.9819	1.018	5.177	.1931	5.273	.1896
1.39	79° 39′	.9837	1.017	5.471	.1828	5.561	.1798
1.40	80° 13′	0.9854	1.015	5.798	0.1725	5.883	0.1700
1.41	80° 47′	.9871	1.013	6.165	.1622	6.246	.1601
1.42	81° 22′	.9887	1.011	6.581	.1519	6.657	.1502
1.43	81° 56′	.9901	1.010	7.055	.1417	7.126	.1403
1.44	82° 30′	.9915	1.009	7.602	.1315	7.667	.1304
1.45	83° 05′	0.9927	1.007	8.238	0.1214	8.299	0.1205
1.46	83° 39′	.9939	1.006	8.989	.1113	9.044	.1106
1.47	84° 14′	.9949	1.005	9.887	.1011	9.938	.1006
1.48	84° 48′	.9959	1.004	10.98	.0911	11.03	.0907
1.49	85° 22′	.9967	1.003	12.35	.0810	12.39	.0807

Table 8 Values of Circular Functions and Trigonometric Functions for Angles in Radians

Real Number x or m(α)	m(α) in degrees	sin x or sin α	csc x or csc α	tan x or tan α	cot x or cot α	sec x or sec α	cos x or cos α
1.50	85° 57′	0.9975	1.003	14.10	0.0709	14.14	0.0707
1.51	86° 31′	.9982	1.002	16.43	.0609	16.46	.0608
1.52	87° 05′	.9987	1.001	19.67	.0508	19.70	.0508
1.53	87° 40′	.9992	1.001	24.50	.0408	24.52	.0408
1.54	88° 14′	.9995	1.000	32.46	.0308	32.48	.0308
1.55	88° 49′	0.9998	1.000	48.08	0.0208	48.09	0.0208
1.56	89° 23′	.9999	1.000	92.62	.0108	92.63	.0108
1.57	89° 57′	1.000	†.000	1256	.0008	1256	.0008
$\frac{\pi}{2}$	90°	1	1	Undefined	0	Undefined	0

Appendix A

Computing with Logarithms

Every common (base 10) logarithm has two parts: a *characteristic* and a *mantissa*. In order to use a table of common logarithms, such as Table 5 on pages 749 and 750 of this book, to find an antilogarithm, the mantissa of the logarithm must be positive. For example, on page 458 you found that

$$\log 0.0124 = -1.9066.$$

Given the equation

$$\log x = -1.9066,$$

however, you cannot use the table directly to find x, since in this form the mantissa of $\log x$ is negative. On the other hand, given the relationship

$$\log x = 0.0934 + (-2),$$

You can find x using the table as follows:
1. Find the entry 0934 in the table. (The decimal points of the table entries are omitted but always precede the digits.) This entry is found in the row labeled 12 and in the column headed 4. Thus, since 12 denotes 1.2 and 4 extends this to 1.24, you find that

$$\text{antilog } 0.0934 = 1.24,$$

or

$$\log 1.24 = 0.0934.$$

2.
$$\log (1.24 \times 10^{-2}) = \log 1.24 + \log 10^{-2}$$
$$= \log 1.24 + (-2)$$
$$= 0.0934 + (-2)$$

Thus:
$$\text{antilog } [(0.0934) + (-2)] = 1.24 \times 10^{-2}$$
$$= 0.0124$$

Table 5 gives logarithms of numbers with three significant digits. Using *interpolation*, you can find approximations of the logarithms of numbers with four significant digits and also antilogarithms of numbers whose mantissas fall between two table entries.

EXAMPLE 1   Find: **a.** $\log 137.4$ **b.** antilog 1.3176

SOLUTION   **a.** Since $\log 137.4 = \log 1.374 + 2$, interpolate using Table 5 to find log 1.374, as shown on the next page.

	x	$\log x$	

$$0.010 \begin{bmatrix} & 1.380 & & 0.1399 & \\ 0.004 \begin{bmatrix} 1.374 \\ 1.370 \end{bmatrix} & \begin{bmatrix} \log 1.374 \\ 0.1367 \end{bmatrix} d & \end{bmatrix} 0.0032$$

$$\frac{d}{0.0032} = \frac{0.004}{0.010}$$

This yields

$$d \approx 0.0013.$$

Thus

$$\log 1.374 \approx 0.1367 + 0.0013 = 0.1380,$$

and

$$\log 137.4 \approx 2.1380.$$

b. First find two entries in Table 5 between which the mantissa of the given logarithm lies, and arrange the relevant facts as shown below.

	x	$\log x$	

$$0.010 \begin{bmatrix} & 2.080 & & 0.3181 & \\ d' \begin{bmatrix} \text{antilog } 0.3176 \\ 2.070 \end{bmatrix} & \begin{bmatrix} 0.3176 \\ 0.3160 \end{bmatrix} 0.0016 & \end{bmatrix} 0.0021$$

Solve for d':

$$\frac{d'}{0.010} = \frac{0.0016}{0.0021} = \frac{16}{21}$$

$$d' = 0.010\left(\frac{16}{21}\right) \approx 0.008$$

Therefore antilog $0.3176 \approx 2.070 + 0.008 = 2.078.$

Since the characteristic of the given logarithm is 1,

$$\text{antilog } 1.3176 \approx 20.78.$$

You can use logarithms read from a table to convert a tedious multiplication or division problem into a problem that involves only sums and differences of logarithms.

EXAMPLE 2   Compute using logarithms: $\dfrac{4.38 \times 0.410}{232}$

SOLUTION   Let $N = \dfrac{4.38 \times 0.410}{232}$.

1. Estimate N: $N \approx \dfrac{4 \times 0.4}{200} \approx 0.008$

2. Write an equation showing the plan of work:

$$\log N = \log 4.38 + \log 0.41 - \log 232$$

3. Substitute the values of the logarithms, using mantissas read from the table and characteristics arrived at by inspection:

$$\log N = 0.6415 + (0.6128 - 1) - 2.3655$$
$$= 0.6415 + 0.6128 - 0.3655 - 1 - 2 = 0.8888 - 3$$
$$N = \text{antilog } (0.8888 - 3) \approx 0.00774 \text{ (to three significant digits)}$$

(Note that this is in reasonable agreement with your estimate.)

Logarithm tables can also be used to find roots and powers of numbers.

EXAMPLE 3   Find $\sqrt[3]{0.83}$.

SOLUTION   Let $N = \sqrt[3]{0.83}$.

1. Estimate N: $N \approx \sqrt[3]{0.729} = 0.9$
2. Use the equation: $\log N = \log (0.83)^{\frac{1}{3}} = \frac{1}{3} \log 0.83$
3. Using the mantissa obtained from Table 5, you have

$$\log N = \tfrac{1}{3}(0.9191 - 1).$$

Since you will have to divide the value $0.9191 - 1$ by 3, a more convenient form of this value may be $2.9191 - 3$. Using this form, you get

$$\log N = \tfrac{1}{3}(2.9191 - 3) = 0.9730 - 1.$$

Thus:

$$N = \text{antilog } (0.9730 - 1) \approx 0.9398 \text{ or } 0.940$$

This agrees with the estimate.

$$\therefore \sqrt[3]{0.83} \approx 0.940$$

By writing an equation showing the plan of work, as in Example 2, you can use a table of logarithms to solve more complicated computational problems.

EXAMPLE 4   Find $\sqrt[5]{\dfrac{14,000}{(0.531)^3(725)}}$

SOLUTION   1. Let $N = \sqrt[5]{\dfrac{14,000}{(0.531)^3(725)}}$.

2. Estimate: $N \approx \sqrt[5]{\dfrac{14,000}{(0.5)^3(700)}} \approx \sqrt[5]{\dfrac{14,000}{(0.1)(700)}}$

$$\approx \sqrt[5]{\dfrac{14,000}{70}} = \sqrt[5]{200} \approx \sqrt[5]{243} = 3$$

3. Use the following equation:

$$\log N = \tfrac{1}{5}(\log 14,000 - [3 \log 0.531 + \log 725])$$
$$= \tfrac{1}{5}(4.1461 - [3(0.7251 - 1) + 2.8603]) = 0.4221$$
$$\therefore N = \text{antilog } (0.4221) = 2.64. \text{ (This agrees with the estimate.)}$$

Appendix B

Programming in Pascal

An Introduction to Pascal

Pascal was designed by Nicklaus Wirth (c. 1970) as a teaching language. It is an excellent medium in which to express many of the important ideas of computer science and is one of the leading languages in use at colleges and universities for introductory computer science courses.

This appendix is intended to cover those topics needed to work the exercises which follow each chapter of the text. Students using this appendix are assumed to be familiar with programming in BASIC and comfortable with many ideas related to programming. Before exploring some of the main structures of Pascal, you need to be made aware of some aspects of Pascal that differ from BASIC.

1. **Pascal is strongly typed.** The domain (type) of each variable must be explicitly declared at the beginning of the program. For example,

   VAR i : integer;

   declares that the variable identified by *i* can be assigned only integer values, the range of which depends upon the hardware and the implementation of Pascal being used.

   Pascal does *not* automatically assign the value zero to an otherwise unassigned numerical variable! Each variable, of whatever type, must be explicitly assigned a value (initialized) in a program before it is referenced.

2. **Pascal does not include the use of DATA lines.** This feature in BASIC allows the user to store values to be read into variables.

3. **Pascal nearly forces top-down design of the program structure.** In Pascal programming, the problem to be solved is analyzed into successively smaller subproblems until each subproblem can be solved relatively easily. The resulting procedures and functions can be coherently assembled into an over-all solution. All the analysis of the problem should be done first, then the code should be written. Typing the code into the computer and debugging are done last.

4. **Pascal is structured.** With sufficiently powerful commands governing the flow of program control, the GOTO command is almost never needed. Procedures, functions, and loops have precisely one entry point and one exit point. One should never attempt to "jump out" of a loop before it has terminated.

5. **Pascal programs ordinarily contain no line numbers.** The control structures of the language eliminate the need for line labels. If line numbers appear in a program listing, they are there merely to facilitate easy reference to parts of the program.

Keep these aspects of Pascal in mind as you begin to study the essentials of a Pascal program, and later to complete the exercises at the end of each chapter.

B–1 Program Essentials

Pascal variable names and type names, called *identifiers*, can be as long as the programmer wants them to be. This allows the programmer to use names that have meaning so that the program is understandable when it is read. For example, if a programmer is writing a program that keeps track of baseball statistics, he or she may choose the variable name *batters_average* to represent the batter's average. Thus it is clear throughout the program what the variable represents.

In choosing variable names a programmer must keep in mind, however, that most systems actually use only the *first eight characters* to identify a variable. The variable *batters_average* would be interpreted by the computer to be the same variable as *batters_age*. Below are some other rules that must be followed when choosing identifiers.

Rules for Variable Names

1. Reserved words may not be used.
2. The first character must be a letter.
3. Special characters may not be used.
4. Blanks may not appear in the middle of the name.

Reserved words are words that already have a special meaning in Pascal. An attempt to use one as an identifier would confuse the computer. A complete list of reserved words appears below.

Reserved Words

AND	FORWARD	PROGRAM
ARRAY	FUNCTION	RECORD
BEGIN	GOTO	REPEAT
CASE	IF	SET
CONST	IN	THEN
DIV	LABEL	TO
DO	MOD	TYPE
DOWNTO	NIL	UNTIL
ELSE	NOT	VAR
END	OF	WHILE
FILE	PACKED	WITH
FOR	PROCEDURE	

The programs you write will generally follow the outline on the top of the next page. Of course, not every program will contain every part indicated in the outline.

program heading
definition part
    constant definitions
    type definitions
declaration part
    variable declarations
    procedure and function declarations
statement part

The first line of every program must be the *program name* in the form:

PROGRAM program_name;

The word "program" must be followed by the program name which must be followed by a semi-colon. This line signals to the computer that a program is beginning. The semi-colon is important here and in all Pascal programs because it separates statements. Note that the same rules that apply to choosing variable names and type names apply also to choosing program, procedure, and function names.

Following the program name are the *definition sections*. In these sections constants, special types, variables, procedures, and functions must be defined. The *constant definition* section comes first. A constant is an identifier whose value remains the same throughout the program. An example of a constant definition section is:

CONST

price = 3.99;

tax = 0.05;

The reserved word CONST begins the section. Next, statements which set each constant identifier equal to its value are separated by semi-colons. In this example, the identifier *price* will maintain the value 3.99 throughout the program. Make no attempt to alter a constant during a program. Also be careful when defining constants *not* to use any identifiers whose values will vary (such as a variable or a function) or whose values must be computed. The following are *illegal* constant definitions.

CONST
    frac = 1/4;   }   *illegal constant*
    distance = abs(−6); } *definitions*

The *type definition* section follows the constant definition section. Here, types other than the pre-defined types must be defined by the user.

The *variable declaration* section comes after the type definitions. In this section, any variables that must be available throughout the program must be declared. Each variable is identified and its type specified. The variable declaration section must take the form:

VAR

    identifiers : type1;

    identifiers : type2;

The key word VAR heads the variable declaration section. Then each identifier is listed along with its type. If there are several identifiers of a particular type, they can be listed together, separated by commas. A colon must come directly before the type, and each declaration must be followed by a semi-colon.

The following is an example of a variable declaration section:

```
VAR
    battersave : real;
    hits, timesup, x : integer;
    firstinitial, lastinitial : char;
    newplayer : boolean;
```

Appearing in the above declaration section are the pre-defined data types: *integer*, *real*, *char*, and *boolean*.

Integer variables hold integers—the positive and negative counting numbers and zero.

Real variables can hold integers and other positive and negative numbers including numbers with decimal points. The computer, in actuality, stores only *rational* values.

Char variables hold any character on the keyboard including punctuation marks, numerals, and special characters.

Boolean variables take on one of the two values true and false.

Throughout the program, each variable can assume only values of its specified type. The following chart illustrates certain rules that must be kept in mind when assigning numbers to variables.

Variable Type	Valid	Invalid
integer	2	2.0
integer	2600	2,600
integer	25	25.
real	34.0	34.
real	0.67	.67
real	2476.2	2,476.2

After the variable declaration section come the *procedure* and *function declarations*, which will be discussed later. Lastly appears the *main body* of the program where the program statements appear. The main body of a program is initiated by the reserved word BEGIN and concluded by the reserved word END which must be followed by a period. The smallest possible legal program is:

```
PROGRAM does_nothing;
BEGIN
END.
```

Exercises

Classify each variable name as valid or invalid. If it is invalid, state why.

1. num2   **2.** 9players   **3.** a-okay   **4.** streetnum
   valid   invalid   invalid   valid

Classify each program declaration line as valid or invalid. If it is invalid, state why.

5. program batstats; valid   **6.** progname; invalid

7. program account-for   **8.** &check program invalid
                     invalid

Classify each statement as true or false.

9. A program terminates with the key word END followed by a semi-colon. false

10. Semi-colons are used in Pascal to separate complete statements. true

11. The definition section immediately follows the program heading. true

12. Special characters are allowed in variable names but not in program names. false

B–2 Assignment Statements

A programmer often wants to assign new values to variables in a program. In Pascal, assignment statements are made using a colon and an equals sign. For example,

$$x := 2;$$

means that the variable x is assigned the value 2. The equation

$$x = 2;$$

is *not* the same thing. (This is not an instruction to the computer to give the variable x a value, but rather a boolean expression that is either true or false. It does not make sense as a Pascal statement by itself.)

To assign a character to an appropriate variable, a single quote on each side of the variable must be used. For example, to assign the char variable *first_initial* the letter 'L', the form is as follows:

$$first\_initial := 'L';$$

To make an assignment statement with a boolean value, quotes are not needed. For example,

$$new\_player := true;$$

assigns the value *true* to the boolean variable *new_player*.

Recall that a variable used in an assignment statement must first be given a value (initialized). Attempts to use variables that have not been initialized will produce errors in the program. For example, the execution of the initialization statement $i := 3$; assigns the value 3 to i. Then the statement $i := i + 1$; adds 1 to the current value of i and then assigns the result 4, to *i*. If the initialization statement is absent, there is no value of *i* to which 1 can be added!

The following table illustrates the symbols used in Pascal for mathematical operations.

Real Operations	Symbol
Addition	+
Subtraction	−
Multiplication	*
Division	/

Integer Operations	Symbol
Addition	+
Subtraction	−
Multiplication	*
Whole Number Division	DIV
Remainder Division	MOD

The addition, subtraction, and multiplication operators can be applied to real or integer operands. The division operator represented by /, (a "slash"), can only be applied to real or integer operands, and its result assigned only to variables of type real.

Whole number division, represented by DIV in Pascal can only be used with integer operands. The DIV function divides the two operands, drops the remainder part of the quotient, and leaves only the *whole number part* as the result. For example, 21 DIV 5 = 4.

Remainder division, represented by MOD in Pascal, is also used only with integer operands. The MOD function divides the two operands, drops the whole number part of the quotient, and leaves only the *remainder part* as the result. For example, 21 MOD 5 = 1. A variable that is to take on a value obtained from a DIV or MOD operation can be of integer or real type.

Exercises

Given the following variable declaration section, state whether each assignment statement is valid or invalid. If it is invalid, state why.

```
VAR
    initial : char;
    num, x, y : real;
    i : integer;
    response : boolean;
```

3. valid
6. invalid
15. valid

1. x := 1.1; valid

2. i := 1.1/1; invalid

3. initial := 'k';

4. x := x + y; valid

5. num := true; invalid

6. response + x := y ;

7. num = x * y; valid

8. response := false; valid

9. i := 1; valid

10. y + x := y; invalid

11. num := 1,349.0; invalid

12. y := 0.3; valid

13. i := i + 3; valid

14. i := x DIV y; invalid

15. y := 3 * i/2;

B–3 Input and Output Procedures

The built-in procedure *writeln* is used in Pascal to print output. The text to be printed must be enclosed in parentheses and single quotes, and like all Pascal statements, the command must be follows by a semi-colon. The statement

writeln('Hello. This is the output.');

would print the following:

Hello. This is the output.

Another procedure *write* is used to collect the text to be printed. For example, the three *write* commands and one *writeln* command below

write('THIS');
write('WILL');
write('BE ON');
writeln('ONE LINE.');

would print the output in this form:

THIS WILL BE ON ONE LINE.

The text on the first three lines is collected on the same line until the *writeln* command finally outputs all the text. On the other hand if *writeln* were used each time,

writeln('THIS');
writeln('WILL NOT');
writeln('BE ON');
writeln('ONE LINE.');

the text would be printed line by line.

THIS
WILL NOT
BE ON
ONE LINE.

The text to be printed can *not* contain a carriage return. If the text to be printed is long, use as many writeln statements as necessary. For example, in the program make the instructions

writeln('THIS TEXT IS TOO LONG TO FIT');
writeln('ON ONE LINE ');

instead of

writeln('THIS TEXT IS TOO LONG TO FIT
ON ONE LINE');

If, in addition to text, the value of a variable is to be printed, the following form is taken:

writeln('TEXT TO BE PRINTED', num);

The text to be printed is enclosed in single quotes as before, but the variable identifier *num* whose value is to be printed is *not* enclosed in quotes. Below is another example of an output statement for both text and values.

writeln('TICKET NUMBER: ', tnum, 'SEAT NUMBER: ', seatnum);

Notice that commas separate the text in quotes from the variable identifiers. If the value of *tnum* at that point in the program is 1226, and the value of *seat_num is 15,* the output would be the following:

TICKET NUMBER: 1226   SEAT NUMBER: 15

Notice also that spaces must be included inside the quotes in order to be included in the output.

You have already seen how assignment statements are used to assign values to variables within a program. Values can also be read from an external source such as the computer keyboard. This allows the program user to supply information necessary for the program's execution. The built-in procedure *readln*, in Pascal is used to read input. For example, the statement

readln(num);

in a program tells the computer to read the value from the input device and assign that value to the identifier *num*. If the input device is a keyboard, the return key must be pressed in order for the input to be entered.

More than one value can be read using a single readln statement. The identifiers must be listed in the same order as the corresponding input will be read. The statement

readln(num1, num2);

instructs the computer to pause and read the first input value and assign it to *num1*, and then read the second input value and assign it to *num2*. For this command to work from the keyboard, each entry must be separated by a space or a comma. After all the numbers have been typed, the return key must be pressed for the values to be entered. If more values appear on the line than there are variables within the parentheses, the extra values will be ignored. If there are less values entered than there are variables, the system will "hang up" waiting for the next value.

Another procedure used to input information is called *read*. From the keyboard, *read* is similar to *readln* with an exception in regard to the cursor. The *read* procedure reads the input value, but does not send the input cursor down to the next line. The cursor remains on the same line, ready to read the next input value on that line. The statements

read(num1, num2);

.
. (* program fragment *)
.

readln(num3);
readln(num4);

expect to read the values *num1, num2,* and *num3* from the same line, and the value *num4* from the next line.

Examining the following sample program will be helpful in understanding input/output procedures in Pascal. Remember, the line numbers are *not* part of the actual program.

```
1   PROGRAM echo(input,output);
2   VAR   number : integer;
3   BEGIN   (* Main Program *)
4      write('Enter a number; press ⟨return⟩ : ');
5      readln(number);
6      write('Your number is : ');
7      writeln(number)
8   END.
```

Line 1 is the heading which names the program *echo* and indicates that *echo* may use the predefined input/output files. This indication is optional in some implementations but is required in standard Pascal for programs involving input and output.

Line 2 is the variable declaration. It names a global variable *number* and specifies its type as integer. *Global* variables are defined and accessible from *anywhere* in a program. The use of global variables is held to a minimum to avoid unintentional alteration of the value of a global variable.

Line 3 indicates where *echo* begins. The comments inside the (* *) symbols are there to make the program easy to understand and easy to maintain. Anything inside the (* *) symbols is ignored by the compiler itself, (no program commands should appear there.)

Line 4 displays on the output device the message within the single quotes. Since the *write* command is used, any following output will appear on the same line. A *writeln* statement would print the text and then move the cursor down to the next line.

Line 5 reads the value in the input device and assigns it to the variable *number*. If the input device is the keyboard, the user must press the carriage return in order to have the input read. Line 6 displays the indicated message.   Line 7 displays the value of *number* and moves the cursor down to the next line. Note that lines 6 and 7 could be combined into the single statement:

writeln('Your number is : ', number);

Line 8 indicates where *echo* ends. One and only one "END." must appear in each Pascal program.

Lines 3 through 8 form an *executable block* (in this case, the main executable block). Procedures and functions, being subprograms, also contain executable blocks. As we shall see, block structure is an important concept in Pascal.

Exercises

Write at least two different ways to produce each output in a Pascal program.

1. PLEASE ENTER YOUR CODE

2. CLIENT NUMBER
   10
   11
   12
   13

3. TODAY'S DATE _____
   TIME _____

State line for line the output from the instructions in each exercise.

4. writeln('The even numbers between 1 and 11 are');
   write('2, 4, 6, ');   The even numbers between 1 and 11 are
   writeln('8, 10 ');   2, 4, 6, 8, 10

5. writeln(' The batter''s average is');   The batter's average is
   writeln(' 0.365');   0.365

In Exercises 6 and 7 you are given the declaration

            VAR
                char1, char2, char3 : char;

and the following lines of input:

            Can you see the difference
            between the two commands?

Show what each fragment of code produces.

6. read(char1);   Cab
   readln(char2);
   readln(char3);
   writeln(char1, char2, char3);

7. read (char1, char2, char3);   Can
   write(char1);
   writeln(char2, char3);

8. **a.** Enter and execute *echo*. Study the program and its results carefully.

   b. Delete " ; press 〈return〉 " from line 4 and change the *readln* in line 5 to *read*. Execute the resulting program. After you enter a number, press the spacebar. Describe how the display has changed. Suggest an explanation for the change.

   c. In the program from part (b), replace each instance of *write* by *writeln*. Execute the resulting program. Describe how the result has changed. Suggest an explanation for the change.

   d. Compare and contrast the results of the commands in the following two pairs:
   (i.) *read* and *readln*   (ii.) *write* and *writeln*

9. Using the ideas of *echo*, write a Pascal program, *echo_and_square*, which produces a display similar to the following as it executes. Assume the user enters 10.

> Enter a number; press 〈return〉 10
>
> Your number is   10
>
> Its square is 100

B–4 Boolean Expressions; IF-THEN Statements

Recall that a boolean variable can be assigned one of the two values true and false. Similarly, *boolean expressions* are expressions that are evaluated as true or false. Boolean expressions are useful in programming because the programmer often wants statements to execute only if certain conditions hold. Before studying the Pascal programming structures that provide for this decision-making process, boolean expressions must be further explored.

In mathematics, you can evaluate certain relations as either true or false. We know that the expression

$$3 = 2 + 1$$

is true. We can also evaluate the expression

$$24 \div 3 > 8$$

to be false. In Pascal, *relational operators* are used to create similar expressions that can be evaluated as true or false. The table below demonstrates the relational symbols used in Pascal.

Mathematics	Pascal	English
$=$	$=$	equal to
$<$	$<$	less than
$\leq$	$<=$	less than or equal to
$>$	$>$	greater than
$\geq$	$>=$	greater than or equal to
$\neq$	$<>$	not equal to

The Pascal expression below, for example

$$\text{balance} > 850.00;$$

makes a claim that is either true or false.

Comparisons can also be made between non-numerical values. The statement

$$\text{more\_data} = \text{'Y'};$$

for example, asserts that the variable *more_data* has the character value 'Y'; this statement is either true or false. Comparisons can be made between boolean variables and expressions, too. Note that comparisons can only be made between values of the same ordinal type. An *ordinal type* is a type whose values can be ordered. (All the pre-defined types except real are ordinal.) Attempts to make comparisons between operands of different types create errors in Pascal programs.

The reserved words AND, OR, and NOT are *boolean operators*. The operator AND yields a true expression only if *both* the operands are true. The expression

$$(0 < 3) \text{ AND } (3 < 10)$$

is true. The operator OR yields a true expression if *either or both* of its operands are true. The expression

$$(0 < 3) \text{ OR } (10 <= 9)$$

is true since the operand $(0 < 3)$ is true. The operator NOT negates the value of its operand. For example, the expression

$$\text{NOT finished}$$

has the value true if the boolean variable *finished* has the value false. The expression has the value false if the variable *finished* has the value true.

Assignment statements using boolean expressions are just like other assignment statements. Be sure that boolean expressions are assigned only to boolean variables. Given the variable declarations

```
VAR
    withdraw, overdrawn : boolean;
    amount, balance : real;
    response : char;
    finished : boolean;
```

the following assignment statements are valid.

```
withdraw := true;
finished := (response = 'Y');
overdrawn := (balance − amount < 0);
```

The Pascal IF-THEN and IF-THEN-ELSE structures depend upon boolean expressions. The IF-THEN structure takes the form:

IF *boolean expression*
THEN *statement*;

If the boolean expression is true, the statement following the key-word THEN is executed. If the expression is false, the THEN statement is ignored and the program goes on to the next instruction. Note that a semi-colon does not appear until the end of the entire statement.

The IF-THEN-ELSE structure takes the form

IF *boolean expression*
THEN *statement(1)*
ELSE *statement(2)*;

If the boolean expression is true, then *statement(1)* is executed, otherwise (if the expression is false) *statement(1)* is ignored and *statement(2)* is executed. There should **not** be a semi-colon preceding the key word ELSE since the ELSE part of the statement is invalid if it is separated from the IF-THEN part.

Suppose you have to write a program to convert an integral score on a test to a letter grade according to the following scale:

A:90–100   D:60–69
B: 80–89   F: 0–59
C:70–79

One way to do this would be to use nested IF-THEN-ELSE statements.

```
1    PROGRAM award_grade(input,output);

2    VAR   grade : 'A' . . 'F';
3          score : integer;

4    BEGIN
5    writeln('Enter integer score between 0 and 100 inclusive');
6    readln(score);
7       IF score < 60
8         THEN grade := 'F'
9           ELSE IF score < 70
10              THEN grade := 'D'
11            ELSE IF score < 80
12                  THEN grade := 'C'
13                  ELSE IF score < 90
14                        THEN grade := 'B'
15                        ELSE grade := 'A'
16   writeln(grade);
17   END.
```

Line 1 is the program heading. Lines 2 and 3 make up the variable declaration section. The variable *grade* assumes any character value from *A* through *F*. Note the single quotes used to denote character values.

Line 4 begins the program, line 5 prompts the user for input and line 6 reads the input. Lines 7 through 15 make up one nested IF-THEN-ELSE statement that assigns the appropriate value to *grade*.

The nested IF-THEN-ELSE statement is followed by line 16, which outputs the grade. The program then ends on line 17.

Study *award_grade* carefully until you understand how IF-THEN-ELSE structures work. Keep in mind that ELSE statements are executed only when the IF expression preceding it is false. If you want several conditions to be checked, separate IF statements are necessary.

When the programmer wants more than one statement to be executed when a condition is true, the statements must be enclosed within the key words BEGIN and END. For example, the single IF-THEN-ELSE statement below contains compound action statements.

```
IF date <= 15 THEN
    BEGIN
        percent := 10;
        balance := balance + percent/100 * balance
    END
ELSE
    BEGIN
        percent := 12;
        balance := balance + percent/100 * balance;
    END;
```

When the value of *date* is less than or equal to 15, all of the statements within the following BEGIN . . . END block are executed, or in the case where *date* is greater than 15 the statements within the BEGIN . . . END block after the ELSE are executed.

Another way to make statement execution dependent upon certain conditions is to use a CASE structure. For example, the CASE statement below executes one and only one statement, depending upon the value of the variable *reply*.

```
CASE reply OF
        1: writeln('Enter cash deposit amount ');
    2, 3: writeln('Enter amount of check');
        4: writeln('Enter withdrawal amount');
        5: writeln('Enter transfer amount');
END;
```

During the program's execution, the user is prompted to enter an appropriate number *reply*. The value of *reply* controls which statement is executed. If *reply* = 4, then the output is

Enter withdrawal amount

The general form for CASE statements is the following:

CASE *case selector expression* OF

    value1: action;

    value2: action;

    value3, value4: action;

    value5: action;

    value6, value7: action;

END; (* CASE statement *)

The case selector (*reply* above) must be a simple ordinal type (char, integer, etc.). The value of the case selector determines which statement is executed. Note that each selector value must correspond to one execution statement only. For example, the case statement below is *illegal* because the selector value 2 corresponds to two different execution statements.

CASE num OF

    1 : writeln('Too small');

    2 : writeln('Okay');

    2 : writeln('Too big');

END;

Several different case values, however, may correspond to a single execution statement. The CASE statement below is *legal*.

CASE num OF

    1: writeln('Too small');

    2, 3: writeln('Okay');

    4: writeln('Too big');

END;

If *num* had the value 2 the statement writeln('Okay'); would execute. The same would occur if *num* had the value 3.

Exercises

In Exercises 1–4, evaluate each expression as true or false given the following assignments.

    continue := true;   num := 6;   answer := 'N';

    limit := 100;   val := 65;

                                                              false

1. num > limit false   **2.** (val > num) AND (NOT continue)

3. (answer = 'P') OR continue; true   4. val <> 100; true

5. Suppose you are grading a course in which the only grades are P (for Pass) and F (for Fail). One passes if and only if ones percent average for the course is at least 60. Write a program *pass_or_fail*, to handle the awarding of grades in this situation.

6. The postage on a first-class letter is 22¢ for the first ounce or fraction thereof and 17¢ for each additional ounce or fraction thereof. Assume that no first class letter will weigh more than 10 ounces. Use a nested IF-THEN-ELSE statement to write a Pascal program to display the postage for a first-class letter for which the weight is specified.

7. Suppose you wish to display an appropriate comment corresponding to the letter grade computed in program *award_grade*. The comments are to be as follows:

Grade	Comment
A	Excellent
B	Superior
C	Average
D	Needs Improvement
F	Unsatisfactory

Write a nested IF-THEN-ELSE statement which assigns the comments as specified above.

8. Complete Exercise 6 using a CASE statement instead of an IF-THEN-ELSE statement.

B–5 Procedures and Functions

The following example will illustrate the structure and use of procedures and functions.

```
1   PROGRAM echo_and_square(input, output)

2   VAR   number : real;

    (**********************************************************)
3   PROCEDURE get_num;
4   BEGIN
5     write('Enter a number; press (return) : ');
6     readln(number);
7   END;   (* get_num *)
```

(Program continues on next page.)

```
       (*****************************************************)
8    FUNCTION square(x : real) : real;

9    BEGIN
10     square := x * x;
11   END;   (* square *)


       (*****************************************************)
12   PROCEDURE display_num_and_sqr;

13   BEGIN
14     writeln('Your number is : ', number);
15     writeln('Its square is : ', square(number) );
16   END;   (* display_num_and_sqr *)


       (*****************************************************)
17   BEGIN   (* Main Program *)
18     get_num;
19     display_num_and_sqr
20   END.
```

Procedures are miniature programs that focus on accomplishing a task. A procedure or function can be *called* (instructed to execute) at any time during the course of the main program or during other procedures and functions. Stating the procedure name instructs the computer to send program control to the procedure. (See line 18 in program *echo_and_square* where procedure *get_num* is called.)

The outline of procedures is identical to that of programs except for the heading PROCEDURE instead of PROGRAM, and the END statement. Procedures have semi-colons after the END because the period is reserved for ending main programs. There is no need for the input, output indications in procedure headings since such an indication appears in the program heading when necessary.

Line 3 is the heading that declares the procedure *get_num*.

Lines 5 and 6 form the procedure's executable block of instructions—the body of *get_num*.

Refer to the program *echo_and_square* where line 18 calls the procedure *get_num*. This sends program control to the procedure *get_num* which then executes. When the execution of the procedure is complete, program control returns to the main body exactly where it had left (directly after calling *get_num*). The main body then calls *display_num_and_sqr* (line 19) and the sequence of control follows as with the other procedure.

Functions are similar to procedures except that when a function is called, *arguments* are passed to the functions and are used in the function to compute a value, which is assigned to the function identifier and then passed back to the main program. Line 8 is the function heading for *square*. The heading declares that the function will take in an argument of type real and assign it to the placeholder *x*, and that the result of the function will be of type real. In other words, the domain and range of the function are explicitly declared in the function heading.

FUNCTION square (x : real) : real;

domain range

The identifier x is called a *formal parameter*. This is the name for the placeholder for the argument passed to the function. The formal parameter x is defined only during the execution of *square*.

If a function has more than one argument, each must be listed with its type. For example, the heading

FUNCTION sale_price(price : real; percent_off : integer) : real;

declares the function *sale_price* which has two arguments, the first of type real and the second of type integer. To invoke this function, a call from the program must be made. To call a function, state the function name, followed by the actual parameters (the actual values) to be used in the function, listed in parentheses. If the following line appears in a program,

amount_due := sale_price(19.99, 10);

the actual parameter value 19.99 is assigned to the identifier *price* and 10 would be assigned to the identifier *percent_off*. The identifiers *price* and *percent_off* are also called the formal parameters of the function.

Refer back to line 10 of *echo_and_square*. This is the executable part of the function where the value of the square of x is computed and assigned to the function identifier *square*:

square := x * x;

There is no need to declare *square* within the function since *it is declared in the heading as the function identifier*. Be sure when assigning a value to a function identifier that the type of the value corresponds to the type of the function as declared in the heading. Here, the product of two real numbers is a real number and the specified function type is real, so the assignment is legal.

The function square is called in line 15 of *echo_and_square* within the procedure *display_num_and_sqr*.

writeln('Its square is : ', square(number));

function   actual parameter
call

Note again that the function is called by using its name with an actual parameter. At this point in execution, program control goes directly to the function *square* with the value of *number* placed in the parameter x. The square of x is then computed and assigned to *square*. Once that value is obtained, program control goes back to where the function was called (line 15).

The following is a procedure that uses *variable parameters*.

```
PROCEDURE increment(VAR count : integer);
BEGIN
   count := count + 1;
END;
```

The procedure will take in the parameter, execute, and then *return the new value of the parameter to the program* that called the procedure. For example if the following call is made,

```
increment(num2);
```

then the value of *num2* is passed to the procedure and assigned to *count*. The procedure then adds 1 to *count*. When the procedure is finished executing, in addition to program control returning to the main program, the value of *count* is assigned to the variable *num2* in the main program. This is because the key word VAR occurs before the *count* parameter in the procedure heading.

If *count* was not a variable parameter the procedure would receive its argument *num2*, increment it, but *not* return the new value (*num2* + 1) to the program. The value of *num2* would remain unchanged. Formal parameters not declared as variables are called *value parameters*.

Since a variable parameter will have a value assigned to it at the completion of the procedure, *the argument must be a variable*. If the argument is a constant or a function, there will be an attempt to assign a value to a constant or to a function, and this attempt will cause an error.

Study the following program:

```
PROGRAM see_difference(input,output);

VAR   global2, global3 : integer;

( ****************************************************************)
PROCEDURE change(num2 : integer; VAR num3 : integer);

VAR   temp : integer;

BEGIN
   temp := num2;
   num2 := num3;
   num3 := temp;
END;   (* change *)

( ****************************************************************)
BEGIN   (* main *)
   global2 := 5;
   global 3 := 6;
   writeln(global2, global3);
   change(global2, global3);
   writeln(global2, global3);
END.
```

Step through the program until you understand the process. You should see that the output is the following:

$$\begin{matrix} 5 & 6 \\ 5 & 5 \end{matrix}$$

Although the values of both *num2* and *num3* are changed *within* the procedure *change*, only the value of *global3* is changed in the main program. The new value of *num3* is passed back to *global3* since *num3* is a variable parameter. The new value of *num2* is *not* passed back to *global2* because the formal parameter *num2* is *not* a variable parameter, but rather a value parameter.

Recall the outline of a program from page 774. Note that all procedures and functions must be declared before they are called by other procedures or functions. Adjust the order of declarations as necessary.

Procedure-oriented Pascal helps to simplify the main body of programs. The main body of *echo_and_square* consists entirely of procedure calls. This makes the program more clear and easy to read.

In the next few sections, procedures and functions will be used to illustrate various program structures in Pascal. This will both introduce useful concepts to you, and give you practice in using functions and procedures.

Exercises

1. **a.** Enter and execute this version of *echo_and_square*.
   b. Write a Pascal function to compute the cube of a real number.
   c. Enhance the program in part (a) to display not only the number and its square but also its cube.
2. For each of the following functions, write a Pascal program that requests a real number value for x and then displays the values of both x and $f(x)$:
   a. $f(x) = 2x^2 - 3x - 5$ **b.** $f(x) = x^2 + x$ **c.** $f(x) = x^2 - 1$
3. **a.** Enter and execute *see_difference* as it appears.
   b. Change the heading of change so that the output becomes:

$$\begin{matrix} 5 & 6 \\ 6 & 5 \end{matrix}$$

B–6 FOR loops

Suppose you want a program to repeat an action a certain number of times. In Pascal a FOR loop is used to control the execution of the statement or block of statements. The general structure of a FOR loop is shown on the next page.

FOR *counter variable* := *initial value* TO *final value* DO
       action;

The counter variable is assigned an initial value, and the action is carried out for each different value of the counter variable.

The program below illustrates how the FOR . . . DO loop works.

```
PROGRAM show_for_loop(output);

  VAR   count : integer;

  BEGIN
    writeln('This program demonstrates a FOR loop.');
    FOR count := 1 TO 4 DO
      BEGIN
        Write('Hello.');
        Writeln(' This is loop number ', count :1);
      END;   (* loop *)
    writeln(' We are outside the loop now.');
  END.
```

The output from *show_for_loop* is as follows:

        This program demonstrates a FOR loop.
        Hello. This is loop number 1
        Hello. This is loop number 2
        Hello. This is loop number 3
        Hello. This is loop number 4
        We are outside the loop now.

When a FOR loop is executed, the counter variable is incremented (increased to the next higher value) each time the loop action is carried out. The loop in *show_for_loop* is executed exactly four times. During the first time in the loop, the value of *count* is 1. *Count* is incremented to the value of 2 as the loop is entered for the second time, incremented to the value of 3 for the third time through the loop and so on. After the fourth time through, the loop is exited.

Below is a list of things to remember when using the FOR structure.

1. The counter variable, initial value, and final value must all be of the same ordinal type.
2. There should be no attempts to alter the values of the counter variable or initial and final values within the loop.
3. The counter variables must be declared locally.

It is possible to use a *count down* form to control the number of loop executions. For example, the FOR . . . DO loop on the next page has initial value 10 and final value 1. The counter is incremented *down* to the next lowest value each time the loop is executed. Note the use of the reserved word DOWNTO instead of TO in the control structure.

```
FOR index := 10 DOWNTO 1 DO
    writeln(index);
```

The output would be the list of numbers 10, 9, 8, 7, 6, 5, 4, 3, 2, 1 printed in a column with 10 at the top and 1 at the bottom. Be sure, when using the DOWNTO form, that the initial value is greater than the final value.

FOR loops are often used in conjunction with arrays. An *array* is a structure that holds a specified number of elements all of the same type. For example, a variable declared as

```
VAR   nums : ARRAY[1 . . 100] OF integer;
```

is an array of 100 elements of type integer.

You may want to picture an array as a box with numbered cells that each hold the same type of element. The array nums might be pictured like this:

Each cell of *nums* holds an integer value.

To refer to the elements of an array in a program, the name of the array and the index of the particular element, enclosed in brackets, must appear. The elements of the array *nums*, for example, are referred to as *nums*[1], *nums*[2], . . . , *nums*[100]. The assignment statement

```
nums[9] := 27;
```

assigns the integer value 27 to cell 9 in the array *nums*.

Suppose you need a procedure to read 10 real numbers entered by a user. If the array *numbers* is declared in the program as

```
VAR   numbers : ARRAY[1 . . 10] OF real;
```

then the following procedure *get_nums* works to read the 10 values into the array *numbers*.

```
PROCEDURE get_nums;

VAR index : integer;

BEGIN
    FOR index := 1 TO 10 DO
        BEGIN
            write('Enter a number; press ⟨return⟩ : ');
            readln(numbers[index])
        END
END;   (* get_nums *)
```

Each time through the loop, a real number is entered by the user, and read from the input device into the correct cell of the array *numbers*. The first time through the loop, *index* has the value 1. If the user enters 2.35, then the *readln* statement reads 2.35 and assigns it to *numbers[index]*. That is, *numbers[1]* := 2.35 since *index* has the value 1 at that point. Next, *index* is incremented to 2, and the procedure will read a newly entered real number into *numbers[2]*, and so on, until all 10 compartments of array *numbers* are filled.

Exercises

In each of Exercises 1–6 the first line of a FOR loop is shown. Tell whether or not the line contains errors. If there are no errors, tell exactly how many times the loop executes. Assume that these assignments have been made.

lowbound := 10;   lastinitial := 'M';   upperbound := 20;

1. FOR i := 1 TO upperbound DO no; 20
2. FOR counter := 20 TO 10 DO yes
3. FOR pointer := 2 DOWNTO −2 DO no; 5
4. FOR index := 'A' TO lastinitial DO no; 13
5. FOR num := lowbound TO upperbound DO no; 11
6. FOR j := 1 DOWNTO 6 DO yes
7. Carefully study the program fragment below. Describe the output.

   ```
   num := 6;
   product := 1;
   FOR i := num DOWNTO 1 DO
     BEGIN
        product := product * i;
        writeln('This is loop number ', i, ', and product = ' , product);
     END;   (* FOR LOOP *)
   writeln('The final product is ,' product);
   ```

8. Write a function *square* that takes a real number as its argument and has the square of the number as its result.
9. Write a procedure *display_nums_and_sqrs* that uses a FOR loop to display ten numbers entered by the user, and their squares. (Use the procedure *get_nums* to input the numbers.)
10. Use *square* along with *get_nums* and *display_nums_and_sqrs* to write a complete program *echo_and_square* which takes 10 user-input real numbers, and prints out the numbers and their squares.

B–7 REPEAT loops

Suppose you have an array of integers called *list*,

VAR   list : ARRAY[1 . . 1000] of integer;

and you wish to determine whether or not the integer 12 is in the list. One way to find out would be to use a FOR loop to search the entire list.

FOR index := 1 TO 1000 DO
IF list[index] = 12
THEN writeln('Found 12. ');

This method of searching works by examining each and every element in the array. It would be more efficient, however, to carry on a search only until 12 is found. This prevents the entire list from being examined unnecessarily. The function below makes use of the REPEAT . . . UNTIL loop to search a list of 1000 elements.

```
1   FUNCTION in_list(num : integer) : boolean;
2   VAR found : boolean;
3        index : integer;
4   BEGIN
5     found := false;
6     index := 1;
7     REPEAT
8       found := (list[index] = num);
9       index := index + 1;
10    UNTIL (found) OR (index > 1000);
11   in_list := found;
12   END; (* in_list*)
```

The REPEAT loop structure is different from the FOR loop structure in that the number of loop executions is *not* defined in the loop structure itself. The REPEAT structure continues the loop execution until the boolean expression following the key word UNTIL is true. The general structure is

REPEAT
action
UNTIL boolean expression;

Be careful in choosing the conditions for these loops. If the boolean expression never becomes true, the loop will repeat infinitely.

The function *in_list* above works by comparing the elements of the array *list* to *num* (the number sought) until the element in the list equals *num* or the last element in *list* has been examined. The heading indicates that *in_list* accepts an integer as its argument and returns a *boolean* value *true* or *false*. The boolean variable *found* (declared in line 2) is used to indicate whether or not the number has been found. Line 5 initializes *found* by assigning it the value *false*. *Found* will retain the value *false* unless the element is in the list.

The local variable *index* is used as a counter during the search. Each time through the loop, *index* is incremented by 1 so that the next time through the loop the *next* item in the list (*list* [*index*]) is compared to *num*.

Lines 7 through 10 make up the REPEAT . . . UNTIL loop. The block of statements bracketed by the key words REPEAT and UNTIL are executed repeatedly until the boolean condition following the UNTIL is true. In this loop, the condition is true only if *found* is true (that is if *list*[*index*] = *num* in line 8), or *index* is greater than 1000 (which indicates that the entire array has been searched). Notice that a REPEAT . . . UNTIL loop, unlike the FOR loop, is always executed *at least once* since the boolean expression that controls termination of the loop is not evaluated until the *end* of the loop.

Line 11 gives *in_list* the final value of *found*. The value of *in_list* is false if *found* remained false throughout the function. The value of *in_list* is true if *found* became true during the execution of the function.

Exercises

1. a. Write a REPEAT . . . UNTIL loop that displays the even integers from 0 to 100 inclusive.
   b. Write a FOR loop which is equivalent to your solution to part (a).
2. a. Write a REPEAT . . . UNTIL loop that computes the sum of the odd integers from 1 to 99 inclusive.
   b. Write a FOR loop that is equivalent to your solution to part (a).
3. a. Use a REPEAT . . . UNTIL loop to write a fragment of Pascal code that computes the average of several numbers entered by the user. Assume that the user enters −1 to signal the end of input but do not include −1 in the averaging process.
   b. Is it possible to write a FOR loop that is equivalent to the loop in your solution to part (a)? Explain.

B–8 WHILE loops

Another loop structure in which the number of iterations is not defined by the structure itself is the WHILE . . . DO loop. The general form of this loop is shown below:

WHILE *boolean expression* DO
   action;

The loop action takes place *as long as the boolean expression is true*. Notice that because the boolean expression (which controls loop execution) is evaluated at the *top* of the structure, it is possible for a WHILE loop never to execute. That is, if the boolean expression is never true, the loop is never entered. Contrast this to the REPEAT loop, that always executes at least once, since its control boolean expression is evaluated at the *end* of the loop.

To demonstrate the use of WHILE loops, study the function *in_list* below which is equivalent to *in_list* in the previous section, but uses a WHILE loop to accomplish the search.

```
1    FUNCTION in_list(num : integer) : boolean;
2    VAR   done : boolean;
3            index : integer;
4    BEGIN
5      done := false;
6      index := 1;
7      WHILE (NOT done) AND (index <= 1000) DO
8        BEGIN
9          index := index + 1;
10         done := (list[index] = num)
11       END;   (* WHILE *)
12   in_list := done;
13   END;   (* in_list *)
```

Lines 7 through 11 make up the WHILE loop. The statement following the DO (lines 8 through 11) executes repeatedly as long as the boolean expression following the WHILE is true. Again, a compound expression (NOT *done*) AND (*index* <= 1000) controls the loop execution. Each time after completing the WHILE block of statements, program control goes back to line 7 and the conditions of the loop are evaluated. If *list[index]* does not equal *num*, (done is still false) and there are still elements of the array left to check, the loop is entered. If one of these conditions does not hold, the loop terminates. Whatever the value of *done* at the termination of the loop, it is assigned to *in_list*.

Be careful when choosing the condition of a WHILE loop because if the condition is never false, the loop will repeat infinitely. Note that, in addition to the fact that the REPEAT loop always executes at least once while the WHILE loop may never execute, a major difference between REPEAT and WHILE loops is that a WHILE loop terminates when a condition becomes *false* but a REPEAT loop terminates when a condition becomes *true*.

Exercises

1–3. Where possible, use WHILE loops to write solutions equivalent to those you wrote for part (a) of Exercises 1, 2, and 3 in the preceding section. Where not possible, explain why.

Appendix C

Preparing for College Entrance Exams

If you plan to attend college, you will most likely be required to take college entrance examinations. Some of these exams attempt to measure the extent to which your verbal and mathematical reasoning skills have been developed. Others test your knowledge of specific subject areas. Usually the best preparation for college entrance examinations is to follow a strong academic program in school, to study, and to read as extensively as possible.

The following test-taking strategies may prove useful:

- Familiarize yourself with the test you will be taking well in advance of the test date. Sample tests, with accompanying explanatory material, are available for many standardized tests. By working through this sample material, you become comfortable with the types of questions and directions that will appear on the test and you develop a feeling for the pace at which you must work in order to complete the test.

- Find out how the test is scored so that you know whether it is advantageous to guess.

- Skim sections of the test before starting to answer the questions, to get an overview of the questions. You may wish to answer the easiest questions first. In any case, do not waste time on questions you do not understand; go on to those that you do.

- Mark your answer sheet carefully, checking the numbering on the answer sheet about every five questions to avoid errors caused by misplaced answer marking.

- Write in the test booklet if it is helpful; for example, cross out incorrect alternatives and do mathematical calculations.

- Work carefully, but do not take time to double-check your answers unless you finish before the deadline and have extra time.

- Arrive at the test center early and come well prepared with any necessary supplies such as sharpened pencils and a watch.

College entrance examinations that test general reasoning abilities, such as the Scholastic Aptitude Test, usually include questions dealing with basic algebraic concepts and skills. The College Board Achievement Tests in mathematics (Level I and Level II) include many questions on algebra. The following second-year algebra topics often appear on these exams. For each of the topics listed, a page reference to the place in your textbook where this topic is discussed has been provided. As you prepare for college entrance exams, you may wish to review the topics on these pages.

Types of Numbers

Integers (p. 5)
Real, rational and irrational numbers
  (pp. 4, 221, 255)
Imaginary and complex numbers
  (pp. 276–277)
Polar form of complex numbers
  (pp. 677–678)

Properties of Real Numbers

Properties of equality (p. 8)
Properties of addition and multiplication
  (pp. 7, 14, 15, 17–19)
Properties of order (pp. 56–58, 65)
Axioms and methods of proof
  (pp. 7–8, 12–14, 66)

Sets and Logic

Set notation (p. 2)
Union and intersection (pp. 2, 61, 483)

Solving Equations

Linear equations (pp. 45–46)
Linear systems in two variables
  (pp. 128–130)
Linear systems in three variables
  (pp. 171–173)
Fractional equations (pp. 234–235)
Radical equations (pp. 273–274)
Quadratic equations, completing
  the square (pp. 341–342)
Quadratic equations, the quadratic
  formula (pp. 343–344)
Equations in quadratic form (pp. 341–344)
Polynomial equations (pp. 213–215,
  254–255)
Finding rational roots (pp. 254–255, 348)
Approximating irrational roots
  (pp. 376–378, 380–381)
Quadratic systems (pp. 423–426, 430–431)
Exponential equations (pp. 461–462)
Using matrices (pp. 547–551)
Trigonometric equations (pp. 667–669)

Solving Inequalities

Linear inequalities (pp. 57–58)
Combined inequalities (pp. 60–61)

Inequalities involving absolute value (p. 70)
Systems of inequalities (pp. 145–149)
Polynomial inequalities (pp. 218–219)

Factoring

Trinomial squares, difference of squares
  (p. 207)
Sum and difference of cubes (p. 209)
$ax^2 + bx + c$ (pp. 207–209)
Using the factor theorem (pp. 370–371)

Graphing

The number line (p. 4)
Inequalities (pp. 95–96)
Using a sign graph (pp. 218–219)
The coordinate plane (p. 88)
Linear equations (pp. 92–94)
Slope (pp. 98–101)
Parallel and perpendicular lines, intercepts
  (pp. 101, 104, 397–398)
Systems of linear equations (pp. 123–125)
Linear inequalities and systems of
  inequalities (pp. 95–96, 145–146)
Parabolas (pp. 354–355, 404–406)
Polynomial equations (pp. 357–359)
Circles (pp. 401–402)
Ellipses (pp. 408–410)
Hyperbolas (pp. 412–415)
Conics with center not at the origin
  (pp. 401–402, 406, 410, 415)
Quadratic systems (pp. 423–426)
Exponential functions (pp. 451–452)
Graphing the inverse of a function (p. 449)
Logarithmic functions (pp. 451–452)
Periodicity and symmetry (pp. 449, 591)
Sine and cosine, period and amplitude
  (pp. 595–598)
Tangent, cotangent, secant, cosecant
  (pp. 604–605)
Inverse trigonometric functions
  (pp. 662–663)
Polar equations (p. 675)

Functions

Functions and relations, domain and
  range (pp. 83–85)
Linear functions (p. 108)

Appendix D

Discrete Mathematics

An Introduction Discrete Mathematics

Besides the traditional association of mathematics with engineering and the physical sciences, mathematics is also used as a problem-solving tool in such fields as business management and the social sciences. As a result of this application of mathematics to other fields, mathematics itself has grown.

Since much applied mathematics is now done using computers, a branch of mathematics has evolved to deal with the discrete processes associated with computer science. We say that something is *discrete* if it consists of distinct parts. The integers, for example, form a discrete system. The real numbers, on the other hand, are not discrete; they have the property of continuity. You saw how discrete and continuous processes differ when you studied compound interest (pages 469–470).

You are already familiar with some of the topics from discrete mathematics: combinatorics (Chapter 12) and abstract algebraic systems (Chapters 1 and 2). You may also be familiar with symbolic logic from your study of geometry.

D–1 Mathematical Induction

You were told that the proof of De Moivre's Theorem (page 680) requires the use of mathematical induction. Mathematical·induction is often used to prove statements about positive integers.

If S is a statement in terms of a positive integer n, the proof of S by mathematical induction has two parts: First you show that S is true for $n = 1$, and second you show that *if* S is true for $n = k$, it is also true for $n = k + 1$, where k is some positive integer. Once both parts are proved, you can conclude that S is true for *all* positive integers.

EXAMPLE 1   Prove that $1 + 2 + 3 + \ldots + n = \dfrac{n(n + 1)}{2}$ for all positive integers n.

SOLUTION   The following is a proof using mathematical induction.

1. Show that the statement is true for $n = 1$.

$$1 \stackrel{?}{=} \frac{1(1 + 1)}{2}$$

$$1 = 1 \; \checkmark$$

2. Show that if the statement is true for $n = k$, it is also true for $n = k + 1$. In other words, assume that

$$1 + 2 + 3 + \ldots + k = \frac{k(k + 1)}{2}$$

is true, and prove that

$$1 + 2 + 3 + \ldots + k + (k + 1) = \frac{(k + 1)[(k + 1) + 1]}{2}$$

is also true.

Proof:

Assuming that

$$1 + 2 + 3 + \ldots + k = \frac{k(k + 1)}{2}$$

is true, add $k + 1$ to both sides and then simplify the right side:

$$
\begin{aligned}
1 + 2 + 3 + \ldots + k + (k + 1) &= \frac{k(k + 1)}{2} + (k + 1) \\
&= \frac{k(k + 1)}{2} + \frac{2(k + 1)}{2} \\
&= \frac{k(k + 1) + 2(k + 1)}{2} \\
&= \frac{(k + 1)(k + 2)}{2} \\
&= \frac{(k + 1)[(k + 1) + 1]}{2}
\end{aligned}
$$

$\therefore 1 + 2 + 3 + \ldots + n = \frac{n(n + 1)}{2}$ for all positive integers n.

Proving a statement by mathematical induction is comparable to knocking down an infinite row of dominoes. The first part of such a proof shows that the first domino falls. The second part shows that if any one domino falls, then the next in line also falls. Taken together, the two parts of the proof guarantee that all the dominoes fall.

Exercises

Use mathematical induction to prove each of the following statements for all positive integers n.

1. $1 + 3 + 5 + \ldots + (2n - 1) = n^2$

2. $2 + 4 + 6 + \ldots + 2n = n^2 + n$

3. $a + ar + ar^2 + \ldots + ar^{n-1} = \frac{a(1 - r^n)}{1 - r}$

4. $\frac{1}{1 \cdot 2} + \frac{1}{2 \cdot 3} + \frac{1}{3 \cdot 4} + \ldots + \frac{1}{n(n + 1)} = \frac{n}{n + 1}$

D–2 Algorithms and Recursion

Computers require precise sets of instructions in order to accomplish specific tasks. Sometimes the instructions are self–referencing (that is, a modified version of an instruction is embedded in itself). The dual need in computer science for precise sets of instructions and for self–referencing instructions is addressed by two topics from discrete mathematics: algorithms and recursion.

An *algorithm* is an orderly mathematical procedure completed in a finite number of steps. You are already familiar with the division algorithm for polynomials (page 224), the algorithm for converting a repeating decimal to a common fraction (page 262), and synthetic substitution (page 366). Another algorithm, called the Euclidean Algorithm, is an old but useful procedure for computing the greatest common factor (GCF) of two integers. It has the following steps:

Step 1: Divide the larger number by the smaller (noting the remainder).

Step 2: If the remainder is 0, stop; the divisor is the GCF. If the remainder is not 0, continue with Step 3.

Step 3: Make the remainder the new divisor, make the divisor the new dividend, and return to Step 1.

EXAMPLE 2 Use the Euclidean Algorithm to compute the GCF of 104 and 48.

SOLUTION
Step 1: $104 \div 48 = 2$ (remainder 8)
Step 2: Since $8 \neq 0$, continue with Step 3.
Step 3: New divisor $= 8$; new dividend $= 48$
Step 1: $48 \div 8 = 6$ (remainder 0)
Step 2: Since $0 = 0$, the GCF is 8.

Closely related to algorithms is the idea of recursion. You have already encountered recursion in your study of arithmetic and geometric sequences (pages 294, 306). We say that a statement is *recursive* if it makes reference to earlier, or simpler, versions of itself. For example, recall the definition of factorial given on page 486:

$$n! = n(n - 1)(n - 2) \ldots (3)(2)(1)$$

This definition is not recursive, but the one on page 524 is recursive:

$$n! = \begin{cases} n \times (n - 1)! & \text{if } n \geq 1 \\ 1 & \text{if } n = 0 \end{cases}$$

EXAMPLE 3 Write a recursive definition of nx, where n is a positive integer.

SOLUTION Because n is a positive integer, you can think of multiplication as repeated addition: $1x = x$, $2x = x + x$, $3x = x + 2x$, and so on.

$$\therefore nx = \begin{cases} x + (n - 1)x & \text{if } n \geq 2 \\ x & \text{if } n = 1 \end{cases}$$

Exercises

5. $x^n = \begin{cases} x \cdot x^{n-1} & \text{if } n \geq 2 \\ x & \text{if } n = 1 \end{cases}$

Use the Euclidean Algorithm to compute the GCF of each pair of numbers.

1. 42, 70 ₁₄

2. 456, 624 ₂₄

3. 1436, 465 ₁

4. 1122, 374 ³⁷⁴

5. Write a recursive definition of x^n, where n is a positive integer.

6. A function that describes the growth of a population that doubles every year is given by $P(t) = P_0 \cdot 2^t$, where P_0 is the initial population. Write a recursive definition of this function. $P(t) = \begin{cases} 2 \cdot P(t-1) & \text{if } t \geq 1 \\ P_0 & \text{if } t = 0 \end{cases}$

D–3 Graph Theory

In geometry you may have encountered the problem of the Seven Bridges of Königsberg. This problem is from graph theory, an important part of discrete mathematics. A *graph* consists of points and segments that connect related points. Graphs are used in counting, classifying, and expressing a variety of relationships.

EXAMPLE 4 An airline schedules flights between Springfield and Princeton, Rockland and Princeton, Cambridge and Rockland, and Cambridge and Princeton. Draw a graph to represent the connections that the airline makes between the cities.

SOLUTION

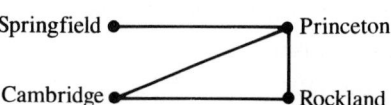

Notice that the graph in Example 4 makes certain facts about the airline apparent. For instance, someone traveling from Rockland to Springfield must take connecting flights through Princeton.

Exercises

1. a. $B{-}A{-}C{-}D{-}E{-}F$, $A{-}B{-}C{-}D{-}F{-}E$, $B{-}A{-}C{-}D{-}F{-}E$, $F{-}E{-}D{-}C{-}B{-}A$, $F{-}E{-}D{-}C{-}A{-}B$, $E{-}F{-}D{-}C{-}B{-}A$, $E{-}F{-}D{-}C{-}A{-}B$

1. The roads linking six towns are shown in the graph below. A tourist wants to start in one of these towns and visit each of the others without passing through any town more than once.

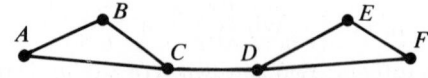

a. The path $A{-}B{-}C{-}D{-}E{-}F$ is one route that the tourist can take. Find all others.

b. From what town(s) is it impossible to begin such a trip? *C and D*

2. Tom, Sue, Joe, and Amy are four students. Tom and Joe have a class together, as do Amy, Joe, and Sue later in the day.
  a. Draw a graph to show which students have classes together.
  b. Which pair(s) of students never see each other in class?

  Tom and Sue, Tom and Amy

D–4 Theory of Formal Languages

When you read the sentence "The dog speaks French," you may be surprised at its meaning (semantics), but you would not question the fact that it has the structure (syntax) of an English sentence. The same cannot be said of "French dog the speaks." You have developed a feeling for correct English syntax based on years of speaking and writing the language. However, in the case of programming languages like BASIC and Pascal, "legal" statements must be specified in detail. As you might expect, there is a branch of mathematics that deals precisely with the theory of generating sentences.

Mathematically speaking, words are merely strings of symbols. We start with a finite (but not empty) set of symbols, A, which we call an *alphabet*. A *word* over A is a finite–length string of symbols from A. We consider words as being "generated" from A by the operation of concatenation (stringing together). After we form the set A^* of all words over A, any subset of A^* becomes a *language* over A.

EXAMPLE 5   If $A = \{a, b, c\}$, create three languages over A.

SOLUTION   Three languages over A are:
$$A_1^* = \{a, ac, baac, cabbc\}$$
$$A_2^* = \{aba, bab, abba, baab\}$$
$$A_3^* = \{aa, ba, ca\}$$

Once we have created a language, we must consider how the words of the language are to be arranged to form sentences. Acceptable arrangements of words are determined by a language's *grammar*. The problem of specifying a grammar for a language has received much attention both in computer science and in linguistics (the study of natural languages).

In the early 1950's the American linguist Noam Chomsky decided that a grammar for a language would be a set of symbols interrelated by an ordered set of rules having the form $X \longrightarrow Y$, which may be interpreted as an instruction to rewrite X as Y. The words used in the sentences of the language are called *terminal symbols*; they never appear to the left of an arrow. The other symbols are either *nonterminal symbols* (that is, intermediate symbols used to describe the structure of the sentences) or the *initial symbol* that begins the generation of any sentence in the language.

EXAMPLE 6   Generate three sentences from the language with the following grammar.

Let S be the initial symbol.

S → NP + VP
NP → ART + NOUN
ART → a, the
NOUN → student, teacher, principal
VP → VERB + NP
VERB → greeted, praised, helped

SOLUTION   The following sentences are all generated by this grammar:

The principal greeted a teacher.
The teacher praised the student.
A student helped the teacher.

One application of graph theory to linguistics is a *generation diagram*, which shows the "history" of a given sentence.

EXAMPLE 7   Draw a generation diagram for the sentence "A student helped the teacher."

SOLUTION

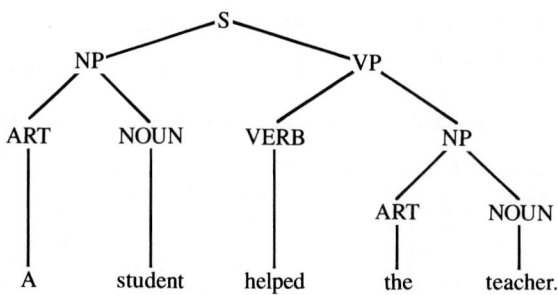

We have only touched on the theory of formal languages. There are many practical applications of this theory not only in computer science, where it is used in the construction of compilers for programming languages, but also in linguistics, where it forms the basis for research on machine translation of one human language to another (English to French, for example).

1. Answers may vary.

For example: $A_1^* = \{no, on\}$

$A_2^* = \{ono, noo, nono, noon\}$

$A_3^* = \{o, oo, ooo, oooo, ooooo\}$

Exercises

1. If $A = \{n, o\}$, create three languages over A.

2. Generate three more sentences using the grammar of Example 6.

3. Draw a generation diagram for the sentence "The teacher praised the student."

Glossary

abscissa (p. 88). The first coordinate of an ordered pair. Also, the x-coordinate.

absolute value (p. 69). The nonnegative (0 or positive) one of the pair a and the opposite of a where $a \in \mathcal{R}$.

accuracy of a measurement (p. 258). *See* relative error.

additive inverse (p. 8). For each $a \in \mathcal{R}$, there exists an additive inverse $-a \in \mathcal{R}$ such that $a + (-a) = 0$ and $(-a) + a = 0$.

algorithm (p. 50). A procedure for solving a problem.

amplitude of a periodic function (p. 597). When a periodic function attains a maximum value M and a minimum value m, the function has an amplitude of $\dfrac{M - m}{2}$.

angle (p. 576). The union of two noncollinear rays that have the same endpoint. *See also* directed angle.

angle of depression (p. 609). The angle between the line of sight to an object (below the observer) and a horizontal ray through the observer.

angle of elevation (p. 609). The angle between the horizontal ray through the observer and the line of sight to an object (above the observer).

antilogarithm (p. 459). If $\log x = a$, then x is called the antilogarithm of a.

Argand diagram (p. 676). A representation of a complex number on a rectangular coordinate system by a point or an arrow to the associated point.

arithmetic mean (or **average**) (p. 297). A single arithmetic mean inserted between two numbers is the average or *the* arithmetic mean of the two numbers.

arithmetic means (p. 297). The terms between two given terms of an arithmetic sequence.

arithmetic progression (p. 293). *See* arithmetic sequence.

arithmetic sequence (p. 293). Any sequence in which each term after the first is obtained by adding a fixed number, called the common difference, to the preceding term. Also called *arithmetic progression*.

arithmetic series (p. 301). A series whose terms are in arithmetic progression.

asymptote (p. 413). A line such that the distance between this line and a point P on the graph goes to 0 as the distance between P and the origin becomes greater and greater.

average (p. 297). *See* arithmetic mean.

binary operation (p. 6). An operation such as addition or multiplication that pairs two elements of a set to produce a third element of that set.

binomial expansions (p. 498). Powers of binomial expressions expanded for successive values of n often used to form combination patterns.

bounded sequence (p. 322). A sequence for which there exists a number that equals or exceeds the absolute value of every term of the sequence.

Cartesian coordinate system (p. 88). A rectangular system of picturing a relation in which each point corresponds to exactly one ordered pair of real numbers. Also called *plane rectangular coordinate system*.

Cartesian product (p. 484). If a finite set A contains r elements and a finite set B contains s elements, then the set of ordered pairs (a, b) with $a \in A$ and $b \in B$ is called the Cartesian product of A and B (denoted by $A \times B$) and contains rs elements.

central tendency (p. 706). The numerical value around which the data points tend to cluster.

characteristic (p. 459). The integral part of a logarithm to base 10 when the fractional part (the part between 0 and 1) is nonnegative.

circle (p. 401). In a plane, a circle is the set of all points at a given distance, called the radius, from a given point, called the center of the circle.

circular functions (p. 587). The trigonometric functions with domain pictured as lengths of arcs on the unit circle, rather than as the set of angles.

circular permutation (p. 487). An arrangement of objects in a circular pattern.

closed half-plane (p. 96). One of the two regions into which a line separates a coordinate plane, including the boundary.

column matrix (p. 527). A matrix with only one column.

combination (p. 493). An r-element subset of a set with n elements is called a combination of n elements taken r at a time, and is denoted by any of the following: $_nC_r$, $\binom{n}{r}$, $C(n, r)$ and C^n_r.

combined variation (p. 419). A relation defined by an equation where a given variable varies directly with another variable and inversely with a third.

common logarithm (p. 458). Logarithm to the base 10.

complement of E (p. 509). If an event E is in the sample space S, the complement of E (denoted $\bar{E}$) is the set of elements of S that are not members of E.

completing the square (p. 342). Transforming a quadratic expression into a square of a binomial.

complex conjugate (p. 281). For any real numbers a and b, the complex conjugate of $a + bi$ is $a - bi$; conversely, the complex conjugate of $a - bi$ is $a + bi$.

complex number (pp. 276–280). Any number of the form $a + bi$ where $a \in \mathcal{R}$ and bi is a pure imaginary number. If $b \neq 0$, $a + bi$ is also called an imaginary number.

composition of functions (p. 85). The composition of a function f with a function g is written $f \circ g$ and is defined to be $[f \circ g](x) = f(g(x))$.

compound sentence (p. 60). A sentence composed of two or more sentences joined together with *and* or *or*.

conditional probability (p. 514). The conditional probability of an event B occurring given that event A has already occurred, is given by $P(B|A) = \dfrac{P(A \cap B)}{P(A)}$.

conic sections (p. 415). The curves that are formed by

the intersection of a plane with a conical surface of two nappes (circle, ellipse, parabola, hyperbola).

conjunction (p. 60). A compound sentence formed by joining two sentences with the word *and*.

consistent system of equations (p. 124). A system of equations that has at least one solution.

constant function (p. 108). A linear function where $m = 0$ and, therefore, $y = b$ for all $x \in \mathcal{R}$.

constant of proportionality (p. 108). *See* constant of variation.

constant of variation (p. 248). In a direct variation $y = mk$ or in an inverse variation $xy = k$ ($k \neq 0$), k is the constant of variation. Also called *constant of proportionality*.

convergent infinite sequence (p. 321). An infinite sequence that has a limit.

convergent infinite series (p. 325). A series whose sequence of partial sums converges to a limit.

converse (p. 56). Any two "If . . . then" statements are converses of each other if each can be obtained from the other by interchanging the hypothesis and conclusion.

coordinate(s) of a point (p. 172). The number, or ordered pair, or ordered triple of numbers associated with a point.

correlation (p. 732). A numerical relationship between groups of data.

cosecant function (p. 602). A trigonometric function such that cosecant: $\alpha \rightarrow \dfrac{1}{\sin \alpha}$, $\sin \alpha \neq 0$.

cosine of an angle (p. 585). If α is an angle in standard position, with $P(u, v)$ any point other than the origin on the terminal side of α, and if $\sqrt{u^2 + v^2} = r$, then cosine: $\alpha \rightarrow \dfrac{u}{r}$.

cotangent function (p. 602). A trigonometric function such that cotangent: $\alpha \rightarrow \dfrac{\cos \alpha}{\sin \alpha}$, $\sin \alpha \neq 0$.

coterminal angles (p. 577). Angles that have the same initial side and the same terminal side.

data (p. 704). Numerical information.

data point (p. 704). A number in the data.

degree of rotation (p. 579). Each of the 360 equal parts into which a complete revolution is divided, in one system of measuring angles.

degree of a monomial (p. 42). In the monomial ax^n, $a \neq 0$, the number denoted by n.

degree of a polynomial (p. 43). The greatest of the degrees of its terms.

dependent equations (p. 125). A system of equations that has an infinite solution set.

dependent events (p. 515). Events that are not independent.

depressed equation (p. 370). Whenever r is a root of the polynomial equation $P(x) = 0$, $P(x) \div (x - r) = 0$ is called the depressed equation.

determinant (pp. 134, 176). A square array of numerals set off with vertical bars, which names a real number. The numerals in the array are called the entries (or elements) of the determinant. The order of the determinant is the number of rows (or columns).

dimensions of a matrix (p. 527). The number of (hori-zontal) rows and the number of (vertical) columns of entries in the matrix.

direct variation (p. 108). A linear function of the form $y = mx$, $m \neq 0$.

directed angle (p. 576). An ordered pair of rays with a common endpoint one ray called the initial side of the angle and the other called the terminal side of the angle, together with a rotation from the initial to the terminal side.

discriminant (p. 347). The number $b^2 - 4ac$, which is named under the radical sign in the quadratic formula, is called the discriminant of the quadratic equation $ax^2 + bx + c = 0$.

disjunction (p. 60). A compound sentence formed by joining two sentences by the word *or*.

distribution of data (p. 705). How frequently or infrequently data points occur.

divergent infinite sequence (p. 322). An infinite sequence that does not have a limit, or *diverges*.

divergent infinite series (p. 325). A series whose sequence of partial sums diverges.

domain (p. 83). The set of all the first components of a relation.

domain of a variable (p. 3). The set whose members may be used as a replacement for the variable. Also called *replacement set*.

double or multiple root (p. 214). A root that is obtained from a factor that appears two or more times in the factorization of the polynomial.

dot product (p. 688). The dot product of two nonzero vectors **u** and **v** is a real number defined to be
$$\mathbf{u} \cdot \mathbf{v} = \|\mathbf{u}\| \, \|\mathbf{v}\| \cos \theta$$
where θ is the angle between **u** and **v**.

element (p. 1). A member of a set.

ellipse (p. 408). In the plane, the set of points for each of which the sum of the distances from two fixed points, called foci (each a focus of the ellipse), is a constant. The distances from the foci to a point P on the curve are called the focal radii of P.

equivalent equations (p. 45). Equations that have the same solution set over a given set.

equivalent expressions (p. 43). Two expressions are equivalent if, when they are joined by the = symbol, the resulting equation is a true statement for every numerical replacement of the variable.

equivalent inequalities (p. 58). Inequalities that have the same solution set over a given set.

equivalent systems (p. 128). Systems with the same solution set.

equivalent vectors (p. 628). Vectors with the same norm and same direction.

even function (pp. 248, 262). A function such that whenever the function contains the ordered pair (a, b), it also contains $(-a, b)$.

event (p. 505). Any subset of possible outcomes for an experiment.

exponential equation (p. 461). An equation in which a variable appears in an exponent.

exponential function (p. 451). A function defined by an equation of the form $y = b^x$, ($b > 0$, $b \neq 1$).

exponential growth (p. 464). A situation where in a fixed time period growth occurs by a fixed percent.

expression (p. 3). A number, a variable, or a sum, difference, product, or quotient that contains one or more variables.

extremes of a proportion (p. 109). In the proportion of $\frac{y_1}{x_1} = \frac{y_2}{x_2}$, y_1 and x_2 are called the extremes.

feasibility region (p. 148). The graph of the system of constraints on a solution.

factor of a polynomial (pp. 6, 207). One of two or more polynomials whose product is the given polynomial.

factor set (p. 207). A designated set from which the factors of a number are chosen.

finite sequence (p. 213). A sequence that has a last term.

fractional equation (p. 234). An equation involving one or more rational expressions in which a variable appears in the denominator.

frequency (p. 704). The number of times a data point occurs in a group of data.

frequency polygon (p. 706). A representation of data made by joining the midpoints of the tops of the bars of a histogram illustrating the same data.

frequency table (p. 704). A table in which data is organized by listing each number that occurs in the data, and its frequency (the number of times that data point occurs).

function (p. 84). A set of ordered pairs in which no first component appears in more than one ordered pair. The set of first components is called the domain, and the set of second components is called the range of the function.

fundamental period (p. 591). If there is a least positive period p for a periodic function, p is called the fundamental period of the function.

geometric means (p. 312). The terms between two given terms of a geometric sequence.

geometric progression (p. 306). *See* geometric sequence.

geometric sequence (p. 306). Any sequence in which each term after the first is the product of the preceding term and a fixed number called the common ratio. Also called *geometric progression*.

geometric series (p. 316). A series whose terms are in geometric progression.

graph of an equation (p. 92). A line or curve consisting of all the points and only those points whose coordinates satisfy the given equation.

greatest common factor, or GCF (p. 207). The greatest integer that is a factor of all of the given integers.

greatest monomial factor of a polynomial (p. 207). The monomial with the greatest numerical coefficient and the greatest degree that is a factor of each term of the polynomial.

half-life (p. 465). The length of time that must elapse before only half of the original amount of substance remains unchanged.

histogram (p. 705). Organization of data distribution in a bar graph.

hypothesis (p. 12). The part of a theorem which states what is assumed to be true.

hyperbola (p. 412). A two-branched curve consisting of the set of points P in the plane such that for each point, the absolute value of the difference between the distances (called focal radii) from P to two fixed points (called foci) is a constant.

hypothesis testing (p. 728). The process of making a tentative assumption and testing its consequences.

identity (p. 586). An equation which is true for all real values of the variables.

imaginary number (p. 277). *See* complex number.

imaginary unit (p. 277). The number i having the property $i^2 + 1 = 0$, or $i^2 = -1$.

inconsistent system of equations (p. 124). A system of equations that has no solution set.

independent events (p. 515). Events such that the probability of one does not depend on the occurrence of the other. Two events A and B are independent if and only if $P(A \cap B) = P(A) \cdot P(B)$.

indirect proof (p. 65). A method of proof that first assumes that the conclusion of a theorem is false, and then reasons from this assumption to a contradiction of the hypothesis, or an axiom, or a previously proved theorem.

infinite sequence (p. 293). A sequence that has no last term.

inverse function (p. 447). Two functions f and g are inverses of each other if they are related such that $f(g(x)) = x$ for all x in the domain of g, and $g(f(x)) = x$ for all x in the domain of f.

inverse relation (p. 661). The relation (set of ordered pairs) which results when the components of each of the ordered pairs in a relation are interchanged.

inverse variation (p. 418). Any function defined by an equation of the form $xy = k$ where k, called the constant of variation, is a nonzero constant.

irrational numbers (p. 255). Real numbers that are not rational.

joint variation (p. 419). A relation defined by an equation of the form $y = kxz$ where y varies directly with x and also directly with z. We say y varies jointly with x and z.

leading coefficient (p. 254). The leading coefficient of a polynomial is the coefficient of the term of highest degree.

least common denominator of two rational algebraic expressions (p. 229). The polynomial of least degree and least positive constant factor that has each denominator as a factor.

limit of a sequence (p. 321). A number L such that the absolute value of the difference of a_n (the nth term of a sequence) and L can be made less than any given positive number, however small, by choosing n large enough.

linear combination (p. 128). When both sides of an equation are multiplied by the same nonzero constant and the resulting expressions are added to the corresponding sides of another equation, a linear combination of the two equations is obtained.

linear equation in three variables (p. 165). Any equation of the form $Ax + By + C = D$, where A, B, C, and D are real constants such that A, B, and C are not all 0.

linear equation in two variables (p. 92). An equation which can be transformed into the form $Ax + By = C$ where A, B, and $C \in \mathcal{R}$ and A and B are not both zero. The graph of such an equation is a straight line.

linear function (p. 108). A function f for which the rule for pairing is given by a linear equation of the form $y = mx + b$, where m, $b \in \mathcal{R}$.

linear interpolation (p. 381). The process of approximating a value of a polynomial function by using a line segment.

linear programming (p. 147). A means of finding maximum and minimum values of a linear expression over a region (the feasibility region) which satisfies a system of inequalities (constraints).

logarithm (p. 451). In the exponential function $x = b^y$ with base b where $b > 0$ and $b \neq 1$, the exponent y is called the logarithm of x to the base b.

mantissa (p. 459). The fractional part (the nonnegative part between 0 and 1) of the logarithm of a number when it is expressed as the sum of an integer and a nonnegative number less than 1.

matrix (plural, *matrices*) (p. 527). A rectangular array of numbers. Each number in the array is called an *entry* of the matrix. The number of rows and the number of columns of entries in the matrix are its *dimensions*.

maximum possible error (p. 257). Half the precision unit used in the measurement.

mean (p. 706). The sum of all the data points divided by the number of data points (the sum of all the frequencies).

means of a proportion (p. 109, 313). In the proportion $\dfrac{y_1}{x_1} = \dfrac{y_2}{x_2}$, x_1 and y_2 are called the means.

median (p. 706). The value(s) that falls in the middle when all the data points are arranged in order.

minor of an element in a determinant (p. 183). The determinant obtained by deleting the row and column containing the element.

mode (p. 706). The numerical value that occurs most often as a data point.

modulus, or absolute value, of a complex number (p. 677). Defined by $|z| = r = \sqrt{x^2 + y^2}$.

monomial (p. 41). A numeral, a variable, or an indicated product of a numeral and one or more variables.

multiplicative inverse (p. 8). For each nonzero $a \in \mathcal{R}$, there exists a multiplicative inverse $\dfrac{1}{a} \in \mathcal{R}$ such that $\dfrac{1}{a} \cdot a = 1$ and $a \cdot \dfrac{1}{a} = 1$. Also called *reciprocal*.

mutually exclusive events (p. 511). Events that have no outcome in common.

natural logarithm (p. 469). Logarithm to the base e. ($e = 2.7182818 \ldots$)

nondecreasing sequence (p. 322). Any sequence in which each term is less than or equal to the following term.

nonincreasing sequence (p. 322). Any sequence in which each term is greater than or equal to the following term.

nonsingular or, invertable, matrix (p. 549). A matrix A such that det $A \neq 0$.

norm of a vector (p. 686). The length of a vector.

normally distributed data (p. 717). A group of data that exhibits a pattern of having highest frequency in the classes of data near the mean, and progressively lower frequencies further from the mean, approaching 0 for classes very far from the mean.

***n*th root** (p. 251). Each solution of the equation $x^n = b$, where n is a positive integer, is called an nth root of b.

***n*th roots of unity** (p. 683). The nth roots of the complex number 1.

numerical coefficient (p. 41). *See* monomial.

octant (p. 162). Each of the eight regions into which space is divided by the three coordinate planes.

odd function (pp. 248, 626). A function with the property that whenever it contains (a, b), it also contains $(-a, -b)$.

odds (p. 509). The odds that the event E in sample space S will occur are given by $\dfrac{P(E)}{P(\overline{E})}$.

one-to-one function (p. 448). A function such that each element in the domain is paired with exactly one element in the range and each element in the range is paired with exactly one element in the domain.

open half-plane (p. 95.) One of the regions into which a line separates a coordinate graph, not including the boundary.

opposite of α (p. 625). If α has measure x^R, the opposite of α, denoted $-\alpha$, has measure $-x^R$.

ordinate (p. 88). The second component of an ordered pair. *Also* the y-coordinate.

orthogonal vectors (p. 689) Two vectors whose dot product is zero.

parabola (p. 404). A curve consisting of the set of points P whose perpendicular distance from a fixed line (called the directrix) is equal to its distance from a fixed point (called the focus) not on the line.

parallel lines (p. 124). Lines that lie in the same plane but do not intersect.

partial sum (p. 325) For any infinite series $a_1 + a_2 + \ldots + a_n + \ldots$, $S_n = \displaystyle\sum_{i=1}^{n} a_i$ is called a partial sum.

percentage (p. 231). The product obtained when a number called the base is multiplied by a percent.

periodic decimal (p. 262). *See* repeating decimal.

periodic function (p. 591). A function f is periodic if there is some nonzero constant p such that $f(x + p) = f(x - p) = f(x)$ for all x in the domain of f; p is called a *period* of the function.

permutation (p. 486). An arrangement of the elements of a set in a definite order.

perpendicular lines (p. 397). Two lines intersecting at right angles.

plane rectangular coordinate system (p. 88). *See* Cartesian coordinate system.

polar axis (p. 672). The nonnegative x-axis in the polar coordinate system.

polar coordinates (p. 672). The components of the ordered pair $(r, m(\theta))$, usually written (r, θ), that specify the location of point P in the plane in terms of r, the distance from the origin to P, and θ, an angle having the nonnegative x-axis as its initial side and the ray $\overrightarrow{OP}$ as its terminal side.

polar form (p. 681). The expression $r(\cos \theta + i \sin \theta)$ is called the polar form for denoting the complex number $x + yi$, where (r, θ) are the polar coordinates of the point (x, y).

pole (p. 672). The origin in the polar coordinate system.

polynomial (p. 42). A monomial or a sum of monomials.

polynomial function (p. 341). A function whose values are given by polynomials.

power function (pp. 42, 247). A function f defined by an equation of the form $f(x) = x^n$.

precision of a measurement (p. 257). The smallest unit on the scale of the measuring device used in making the measurement.

principal-value inverse function (p. 662). An inverse function defined by restricting (according to custom) the range of the inverse of a circular or trigonometric function.

probability (p. 171). Let S be a sample space of an experiment in which there are n possible outcomes, each equally likely. If an event E is a subset of S such that E contains h elements, then the probability of the event E is given by $\frac{h}{n}$.

proportion (p. 109). An equality of ratios.

pure imaginary number (p. 277). A number bi such that $(bi)^2 = -b^2$ when $b \neq 0$.

quadrantal angle (p. 577). An angle in standard position whose terminal side coincides with a coordinate axis.

quadratic function (pp. 354, 357). A function f with domain $\mathcal{R}$ and values given by a quadratic polynomial, that is, $f(x) = ax^2 + bx + c$, where a, b, and c are real numbers and $a \neq 0$.

quadratic polynomial (p. 43). A polynomial of degree 2.

radian (p. 580) A unit which can be used to measure any angle in standard position. If the length of an arc on the unit circle measured from point $(1, 0)$ is a units, then the measure of the angle in standard position interacting that arc is said to be a radians.

random experiment (p. 505). An experiment for which the outcome is unpredictable.

range of data (p. 710). A measure of the variation of data defined as the difference between the greatest data point and the least data point.

range of a relation (p. 83). The set of all the second components of a relation.

rational algebraic expression (p. 221). The quotient of two polynomials.

rational function (p. 221). A function defined by a quotient of two polynomials that are written in simplest form.

rational number (p. 221). Any number that is the quotient of two integers.

rationalizing the denominator (p. 268). The process of transforming an expression with a radical (or radicals) in its denominator into an equivalent expression with the denominator free of radicals.

real numbers (p. 4). The set of all the positive numbers, the negative numbers, and zero.

reciprocal (p. 8). *See* multiplicative inverse.

reducible polynomial (p. 207). A polynomial that can be expressed as a product of two or more polynomials of lower positive degree.

reduction formula (p. 628). A formula which can be used to reduce a given circular or trigonometric function value in any quadrant to a function value in Quadrant 1.

relation (p. 83). Any set of ordered pairs.

relative error (p. 258). The ratio of the maximum possible error in measurement to the measurement itself. Also called *accuracy of measurement*.

relative frequency (p. 704). The ratio of the frequency of a data point to the total number of data entries.

relatively prime integers (p. 207). Two integers whose greatest common factor is 1.

repeating decimal (p. 262). A decimal numeral for a rational number that contains an endlessly repeating block of digits (the repetend). Also called *periodic decimal*.

replacement set (p. 3). *See* domain of a variable.

right angle (p. 576). An angle having a rotation of $\frac{1}{4}$ of a revolution.

rounding error (p. 257). The difference between a number and its approximation.

sample (p. 726). The subset that provides the data from which statistical inference is made.

sample space (p. 505). The set of all possible outcomes of a random experiment.

scalars (p. 535). In dealing with matrices, we often refer to real numbers as scalars.

scatter diagram (p. 733). A diagram obtained by plotting pairs of data.

scientific notation (p. 257). *See* standard notation.

secant function (p. 602). A trigonometric function such that secant: $\alpha \to \dfrac{1}{\cos \alpha}$, $\cos \alpha \neq 0$.

sequence (pp. 293, 306). A set of numbers (some of which can be repeated) in a particular order.

series (pp. 300, 316). In general, given any sequence a_1, $a_2, \ldots$ with n or more terms, the associated series of n terms, S_n, is $S_n = a_1 + a_2 + \ldots + a_n$.

set (p. 1). A well-defined collection of objects.

significant digit (p. 258). In a numeral, each digit reporting the number of units of measure contained in a measurement.

similar monomials (p. 42). Monomials that are exactly the same or differ only in numerical coefficients. Also called *like monomials*.

simple form of a polynomial (p. 42). A polynomial in which no two terms are similar monomials.

simplifying a polynomial (p. 43). Replacing one polynomial with an equivalent polynomial in simple form.

sine of an angle (pp. 585, 604). If α is an angle in standard position, with $P(u, v)$ any point other than the origin on the terminal side of α, and if $\sqrt{u^2 + v^2} = r$, then sine: $\alpha \to \dfrac{v}{r}$.

sine wave (p. 596). The graph of the function $y = \sin x$.

singular matrix (p. 549). A square matrix A such that $\det A = 0$.

slope of a line (p. 98). Ratio of rise to run. Let (x_1, y_1) and (x_2, y_2) be the coordinates of any two different points of a nonvertical line. Then the slope m of the line is given by $m = \dfrac{y_2 - y_1}{x_2 - x_1}$.

solution set (p. 4). The set that consists of the values of the variable for which an open sentence is true. Also called *truth set*.

square matrix (p. 528). A matrix with the same number of rows as columns.

standard normal curve (p. 720). The curve that is the graph of $y = \dfrac{1}{\sqrt{2\pi}} e^{\frac{-x^2}{2}}$, the standard normal distribution.

standard notation (p. 257). An expression of a number as a product, $a \times 10^n$, where $1 \le |a| < 10$ and n is an integer. Also called *scientific notation*.

standard position (p. 577). The position of an angle placed on a rectangular coordinate system with the vertex of the angle at the origin and the initial side as the positive part of the horizontal axis.

straight angle (p. 577). An angle having a rotation of $\dfrac{1}{2}$ of a revolution.

statistic (p. 706). A single number that characterizes a data set.

statistical inference (p. 726). A general conclusion based on a relatively small amount of data.

sum of an infinite series (p. 325). If the sequence $S_1, S_2, \ldots, S_n, \ldots$ of partial sums converges and has limit S, then the sum of the infinite series is defined to be S.

synthetic substitution (p. 366). A sequence of substitution operations used to evaluate a polynomial function of any degree.

tangent function (p. 602). A trigonometric function such that tangent: $\alpha \to \dfrac{\sin \alpha}{\cos \alpha}$, $\cos \alpha \ne 0$.

terminating decimal (p. 261). A decimal numeral containing only a finite number of digits; that is, having 0 as a repetend.

terms of a polynomial (p. 42). The monomials in the expression for a polynomial.

trace of a plane (p. 167). The line in which a plane intersects a coordinate plane is called the trace of the given plane in that coordinate plane.

translation (p. 553). A transformation or mapping of the plane in which each point P of the plane is mapped onto a corresponding point P' of the plane. We say that P' of the plane is the *image* of P, and P is the *preimage* of P' under the mapping.

trigonometric form of a complex number (p. 681). *See* polar form.

trigonometric functions (p. 585). The set of functions (sine, cosine, tangent, cotangent, secant, cosecant) with the domain of each function a subset of the set of all angles.

unit circle (p. 579). The circle with center $(0, 0)$ and radius 1.

variable (p. 3). A symbol which may represent any one of the members of a specified set.

variation. *See* combined variation, direct variation, inverse variation, and joint variation.

vector (p. 686). A directed line segment in the plane or in space, from one endpoint called the initial point, to the other endpoint called the terminal point.

Venn diagram (p. 511). A diagram used to picture set relationships.

x-intercept (p. 104). The abscissa of the point where a graph intersects the x-axis.

y-intercept (p. 104). The ordinate of the point where a graph intersects the y-axis.

zero matrix (p. 529). A matrix each of whose entries is zero.

zero of a function (pp. 42, 221, 349). Any value of x in the domain of a function f which satisfies the equation $f(x) = 0$.

zero vector (p. 686). A vector of length 0 or, a point.

Index

graphing, 600
inverse, 661–665
reference angles and, 593–594
solving right triangles with, 607–609
Trigonometry, 575–669
Trinomial, 43, 205
factoring, 208
quadratic, 208
square, 207

Union of sets, 2, 483
Unit circle, 579, 587, 603, 626
Unity, nth roots of, 683
Upper bounds, 378

Values, 3
absolute, 69–70, 321
intermediate, property of, 380
maximum or minimum, 148–149, 359, 597
of polynomial functions, 365–366, 380–381
reciprocal, 633
and roots of open sentence, 4
of trigonometric functions, 593, 596–598, 599–600, 602–605

Variable(s), 3
Variance, 710
Variation
combined, 419
constant of, 418
direct, 108, 418
inverse, 418
joint, 419
of sign, 377
of statistical data, 710
Vector(s), 686–689
components, 686–687
equivalent, 686
norm of, 686
orthogonal, 689
Velocity, 692
Vertex, 355, 357, 394, 405, 414, 607
of an angle, 577
Vocabulary lists, 11, 29, 55, 72, 91, 98, 106, 113, 144, 152, 176, 189, 211, 220, 238, 256, 265, 275, 285, 305, 320, 330, 353, 364, 376, 383, 400, 422, 456, 491, 497, 518, 552, 563, 584, 601, 612, 644, 653, 670, 684, 694, 716, 731, 739

Whole numbers, 5

Wind speed, 139, 692

x-axis, 104, 162, 248, 393, 410
as polar axis, 672
x-coordinate, 88, 95, 162, 167, 413
x-intercept, 104, 125, 167, 409
xy-plane, 162
xz-plane, 162

y-axis, 104, 162, 248, 393, 410
y-coordinate, 88, 95, 162, 413
y-intercept, 104, 124, 125, 167, 409
yz-plane, 162

z-axis, 162
z-coordinate, 162
z-intercept, 167
Zero
defined as a real number, 4
of a function, 42, 221, 252, 349, 362, 370, 380
in imaginary units, 277
matrix, 529, 543, 544
and order of opposites, 65–66
as a remainder, 224, 261
and significant digits, 258
vector, 686, 688

Answers to Selected Exercises

Chapter 1   Review of Essentials

Written Exercises, pages 2–3

1. {1, 5, 9, 10, 11, 13, 15, 20, 25}   **3.** {15}   **5.** {5, 10}
7. $X \cap Y = X$; $X \cup Y = Y$; Example: $X = \{1, 3\}$,
$Y = \{1, 3, 5, 7\}$ **9.** 2^n

Computer Exercises, page 3

3. 3.63636, 4.27273, 4.90909, 5.54546, 6.18182, 6.81818,
7.45455, 8.09091, 8.72727, 9.36364

Written Exercises, page 6

1. {3}   **3.** ∅   **5.** {−2, −1, 1, 2, 3}   **7.** {−1, 1, 3}
9. {−2, −1, 1, 2, 3}

11.
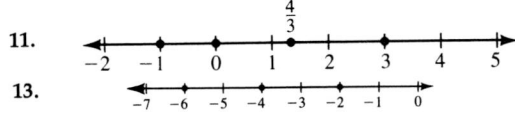

13.
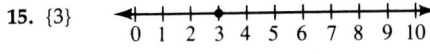

15. {3}

17. {1}

19. {1}

21. ∅

Written Exercises, pages 9–10

1. 3; comm. ax. for add.   **3.** y; Symm. prop. of =
5. −4; Ax. of add. inv.   **7.** (−2); Reflex. prop.
of =   **9.** $8 + c$; Comm. ax. for mult.   **11.** p, r;
Trans. prop. of =   **13.** 11   **15.** 120   **17.** yes; yes
19. 0; yes   **21.** $(1 \oplus 2) \oplus 3 = 3 = 3 = 0$; $1 \oplus$
$(2 \oplus 3) = 1 \oplus 1 = 0$; ∴ $(1 \oplus 2) \oplus 3 = 1 \oplus (2 \oplus 3)$
23. $2 * (1 \oplus 3) = 2 * 2 = 1$; $2 * 1 \oplus 2 * 3 = 2 \oplus 2 *$
$3 = 2 \oplus 3 = 1$; ∴ $2 * (1 \oplus 3) = 2 * 1 \oplus 2 * 3$
25. $(x + a)(x + b) = (x + a)x + (x + a)b$ (Dist. ax.)
$= x \cdot x + a \cdot x + x \cdot b + a \cdot b$
(Dist. ax.)
$= x^2 + ax + bx + ab$ (Comm. ax.
for mult.)
$= x^2 + (a + b)x + ab$ (Dist. ax.)
$(x + a)(x + b) = x^2 + (a + b)x + ab$ (Trans.
prop. of =)

Computer Exercises, page 10

3. −0.625, 1.6   **5.** −7, 0.142857   **7.** 3, −0.333333

Self-Test 1, page 11

1. $3 \in \{y: y < 5 \text{ and } y \in \mathcal{R}\}$ **2.** {0}   **3.** ∅

4.

5. {1}

6. Iden. ax. for add.   **7.** Assoc. ax. for add.
8. Dist. ax.   **9.** Ax. of mult. inv.   **10.** {−48}

Written Exercises, pages 15–17

1. −3   **3.** y **5.** 3   **7.** 4   **9.** z **11.** $7c + (−16)$
13. $3x + (−9)$ **15.** 2   **17.** $2w + (−3)$ **19.** {−23}
21. {−24}   **23.** {20}
27. 1. $c + a = c + b$ Hypothesis
2. $c + a = a + c$ Comm. ax. for add.
$c + b = b + c$
3. $a + c = b + c$ Subs. prin.
4. $a = b$ Theorem, page 13
29. 1. $b + a = a$ Hypothesis
2. $a = 0 + a$ Iden. ax. for add.
3. $b + a = 0 + a$ Trans. prop. of =
4. $b = 0$ Canc. prop. of add.
31. 1. $a, b, \in \mathcal{R}$ Hypothesis
2. $-a, -b \in \mathcal{R}$ Axiom of add. inv.
3. $-[(-a) + (-b)] =$ Prop. of opp. of a
$-(-a) + [-(-b)]$ sum
4. $-[(-a) + (-b)] = a + b$ Canc. prop. of
add. inv.
33. 1. $a + c = b + d$ Hypothesis
$c = d$
2. $a + c = b + c$ Subs. Prin.
3. $a = b$ Canc. prop. of add.
35. 1. $x + (-6) = -1$ Hypothesis
2. $= -1 + 0$ Iden. ax. for add.
3. $= -1 + [6 + (-6)]$ Ax. of add. inv.
4. $= (-1 + 6) + (-6)$ Assoc. ax. for add.
5. $= 5 + (-6)$ Subs. prin.
6. $x + (-6) = 5 + (-6)$ Trans. prop. of =
7. $x = 5$ Canc. prop. of add.
37. 1. $x + b = a$ Hypothesis
2. $= a + 0$ Iden. ax. for add.
3. $= a + [(-b) + b]$ Ax. of add. inv.
4. $= [a + (-b)] + b$ Assoc. ax. for add.
5. $x + b = [a + (-b)] + b$ Trans. prop. of =
6. $x = a + (-b)$ Canc. prop. of add.
39. 1. $2x + b = x$ Hypothesis
2. $(x + x) + b = x$ Subs. prin.
3. $x + (x + b) = x$ Assoc. ax. for add.
4. $x + (x + b) = x + 0$ Iden. ax. for add.
5. $x + b = 0$ Canc. prop. of add.
6. $x + b = (-b) + b$ Ax. of add. inv.
7. $x = -b$ Canc. prop. of add.

1. 420 **3.** $5xy$ **5.** -6 **7.** $30p + (-40pq)$
9. $4ac + 4bc$ **11.** $36h + (-132hk)$
13. $6cd + \left(-\dfrac{21}{2}c\right)$ **15.** $-4x + 13y + (-2xy)$
17. $2ab + 13a + (-40b)$ **19.** positive **21.** negative
23. zero **25.** 1. Hypothesis 2. Axiom of multiplicative inverses 3. Substitution principle
4. Associative axiom for multiplication 5. Axiom of multiplicative inverses 6. Identity axiom for multiplication

27. 1. $a \neq 0,\ b \neq 0$ — Hypothesis

2. $\dfrac{1}{ab}, \dfrac{1}{a}$, and $\dfrac{1}{b}$ are real numbers — Ax. of mult. inv.

3. $ab\left(\dfrac{1}{a} \cdot \dfrac{1}{b}\right) = \left[a\left(\dfrac{1}{a}\right)\right] \cdot \left[b\left(\dfrac{1}{b}\right)\right]$ — Comm. and assoc. ax. for mult.

4. $= 1 \cdot 1$ — Ax. of mult. inv.

5. $= 1$ — Iden. ax. for mult.

6. $ab\left(\dfrac{1}{a} \cdot \dfrac{1}{b}\right) = 1$ — Trans. prop. of =

7. $ab\left(\dfrac{1}{ab}\right) = 1$ — Ax. of mult. inv.

8. $ab\left(\dfrac{1}{ab}\right) = ab\left(\dfrac{1}{a} \cdot \dfrac{1}{b}\right)$ — Trans. prop. of =

9. $\dfrac{1}{ab} = \dfrac{1}{a} \cdot \dfrac{1}{b}$ — Canc. prop. of mult.

29. 1. $ab = 1$ — Hypothesis

2. $1 = a\left(\dfrac{1}{a}\right)$ — Ax. of mult. inv.

3. $ab = a\left(\dfrac{1}{a}\right)$ — Subs. prin.

4. $b = \dfrac{1}{a}$ — Canc. prop. of mult.

31. 1. $(-a)b = [(-1)a]b$ — Mult. prop. of -1
2. $= -1(ab)$ — Assoc. ax. for mult.
3. $= -ab$ — Mult. prop. of -1
4. $(-a)b = -ab$ — Trans. prop. of =

33. 1. $(-a)(-b) = [(-1)a][(-1)b]$ — Mult. prop. of -1
2. $= [(-1)(-1)](ab)$ — Comm. and assoc. ax. for mult.
3. $= [-(-1)](ab)$ — Mult. prop. of -1
4. $= 1 \cdot ab$ — Canc. prop. of add. inv.
5. $= ab$ — Iden. ax. for mult.
6. $(-a)(-b) = ab$ — Trans. prop. of =

35. 1. $ab = 0,\ b \neq 0$ — Hypothesis

2. $ab\left(\dfrac{1}{b}\right) = 0$ — Mult. prop. of zero

3. $ab\left(\dfrac{1}{b}\right) = a\left[b\left(\dfrac{1}{b}\right)\right]$ — Assoc. ax. for mult.

4. $= a \cdot 1$ — Ax. of mult. inv.

5. $= a$ — Iden. prop. for mult.

6. $ab\left(\dfrac{1}{b}\right) = a$ — Trans. prop. of =

7. $a = 0$ — Subs. prin. (in step 2)

37. 1. $x = a \cdot \dfrac{1}{b},\ b \neq 0$ — Hypothesis

2. $bx = b\left(a \cdot \dfrac{1}{b}\right)$ — Subs. prin.

3. $= a\left(b \cdot \dfrac{1}{b}\right)$ — Comm. and assoc. ax. for mult.

4. $= a \cdot 1$ — Ax. of mult. inv.

5. $= a$ — Iden. ax. for mult.

6. $bx = a$ — Trans. prop. of =

39. 1. $\dfrac{1}{x} = a\left(\dfrac{1}{b}\right)$ — Hypothesis

2. $b\left(\dfrac{1}{x}\right) = a$ — Exercise 37

3. $a = b\left(\dfrac{1}{x}\right)$ — Symm. prop. of =

4. $xa = b$ — Exercise 37

5. $x = b \cdot \dfrac{1}{a}$ — Exercise 36

41. 1. $a\left(\dfrac{1}{x}\right) = b$ — Hypothesis

2. $b = a\left(\dfrac{1}{x}\right)$ — Symm. prop. of =

3. $xb = a$ — Exercise 37

4. $x = a\left(\dfrac{1}{b}\right)$ — Exercise 36

Computer Exercises, page 22
3. zero **5.** positive **7.** zero **9.** positive
11. negative **13.** zero

Written Exercises, pages 23–24
1. $-2x + 2y$ **3.** $2mp - 2mq - 6m$ **5.** $60s - 12r - 48t$ **7.** $18a + 3ac - 3b$ **9.** $27 + 7w - 6z$
11. $2f + 14g - 15$ **13.** 2 **15.** -20 **17.** -19
19. 1. Hypothesis 2. Def. of subt. 3. Subs. prin.
4. Dist. ax. 5. Prop. of opp. in prod. 6. Def. of subt. 7. Trans. prop. of =
21. 1. $a, b \in \mathcal{R}$ — Hypothesis

2. $(a - b) + b = [a + (-b)] + b$ — Def. of subt.

3. $= a + [(-b) + b]$ — Assoc. ax. for add.

4. $= a + 0$ — Ax. of add. inv.

5. $= a$ — Iden. ax. for add.

6. $(a - b) + b = a$ — Trans. prop. of =

23. 1. $a, b, c, \in \mathcal{R}$ — Hypothesis
2. $-a(b - c) = -a[b + (-c)]$ — Def. of subt.
3. $= -ab - [a(-c)]$ — Exercise 22
4. $= -ab - (-ac)$ — Prop. of opp. in prod.
5. $= -ab + ac$ — Exercise 20
6. $= ac + (-ab)$ — Comm. ax. for add.
7. $= ac - ab$ — Def. of subt.
8. $-a(b - c) = ac - ab$ — Trans. prop. of =

25. 1. $c - a = c - b$ — Hypothesis
2. $c + (-a) = c + (-b)$ — Def. of subt.
3. $-a = -b$ — Canc. prop. of add.
4. $-1(a) = -a$ — Mult. prop. of -1
   $-1(b) = -b$
5. $-1(a) = -1(b)$ — Subs. prin.
6. $a = b$ — Canc. prop. of mult.

27. 1. $a = b$ — Hypothesis
2. $c - a = c - a$ — Refl. prop. of =
3. $c - a = c - b$ — Subs. prin.

29. 1. $x - b = a$ — Hypothesis
2. $x + (-b) = a$ — Def. of subt.
3. $= a + 0$ — Iden. ax. for add.
4. $= a + [b + (-b)]$ — Ax. of add. inv.
5. $= (a + b) + (-b)$ — Assoc. ax. for add.
6. $x + (-b) = (a + b) + (-b)$ — Trans. prop. of =
7. $x = a + b$ — Canc. prop. of add.

31. 1. $a, b, c, d \in \mathcal{R}$ — Hypothesis
2. $(a - b)(c + d)$ — Dist. ax.
   $= (a - b)c + (a - b)d$
3. $= [a + (-b)]c + [a + (-b)]d$ — Def. of subt.
4. $= ac + (-b)c + ad + (-b)d$ — Dist. ax.
5. $= ac + (-bc) + ab + (-bd)$ — Prop. of opp. in prod.
6. $= ac - bc + ad - bd$ — Def. of subt.
7. $(a - b)(c + d) = ac - bc + ad - bd$ — Trans. prop. of =

33. 1. $a, b, c, d \in \mathcal{R}$ — Hypothesis
2. $a[b - (c + d)] = ab - [a(c + d)]$ — Exercise 19
3. $= ab - a(c + d)$ — Prop. of opp. in prod.
4. $= ab - ac - ad$ — Exercise 22
5. $a[b - (c + d)] = ab - ac - ad$ — Trans. prop. of =

Written Exercises, pages 26–28
1. -3 3. $8x - 6$ 5. $-2k$ 7. $-62m$ 9. $29 - \dfrac{64}{n}$

11. $\dfrac{392r}{15}$ 13. a. $-2 = -2$ b. true 15. a. $18 = 18$ b. true 17. a. $14 \neq 11$ b. no 19. a. $-\dfrac{11}{3} \neq \dfrac{1}{3}$ b. no 21. 1. Hypothesis 2. Def. of div. 3. Dist. ax. 4. Def. of div. 5. Trans. prop. of =

23. 1. $a \in \mathcal{R}, a \neq 0$ — Hypothesis
2. $\dfrac{a}{a} = a \cdot \dfrac{1}{a}$ — Def. of div.
3. $= 1$ — Ax. of mult. inv.
4. $\dfrac{a}{a} = 1$ — Trans. prop. of =

25. 1. $a, b, \in \mathcal{R}, b \neq 0$ — Hypothesis
2. $(ab) \div b = (ab)\dfrac{1}{b}$ — Def. of div.
3. $= a\left(b \cdot \dfrac{1}{b}\right)$ — Assoc. ax. for mult.
4. $= a \cdot 1$ — Ax. of mult. inv.
5. $= a$ — Iden. ax. for mult.
6. $(ab) \div b = a$ — Trans. prop. of =

27. 1. $x = \dfrac{b}{a}, a \neq 0$ — Hypothesis
2. $1 \cdot x = 1\left(\dfrac{b}{a}\right)$ — Ident. ax. for mult.
3. $\left(\dfrac{1}{a} \cdot a\right)x = \left(\dfrac{1}{a} \cdot a\right) \cdot \dfrac{b}{a}$ — Ax. of mult. inv.
4. $\dfrac{1}{a}(ax) = \dfrac{1}{a}\left(a \cdot \dfrac{b}{a}\right)$ — Assoc. ax. for mult.
5. $ax = a \cdot \dfrac{b}{a}$ — Canc. prop. of mult.
6. $ax = \dfrac{b}{a} \cdot a$ — Comm. ax. for mult.
7. $= \left[b\left(\dfrac{1}{a}\right)\right] \cdot a$ — Def. of div.
8. $= b\left[\dfrac{1}{a} \cdot a\right]$ — Assoc. ax. for mult.
9. $= b \cdot 1$ — Ax. of mult. inv.
10. $= b$ — Iden. ax. for mult.
11. $ax = b$ — Trans. prop. of =

29. 1. $a, b, c, d \in \mathcal{R}$
   $b, d \neq 0$ — Hypothesis
2. $\dfrac{a}{b} \cdot \dfrac{c}{d} = a\left(\dfrac{1}{b}\right) \cdot c\left(\dfrac{1}{d}\right)$ — Def. of div.
3. $= (ac)\left(\dfrac{1}{b} \cdot \dfrac{1}{d}\right)$ — Comm. and assoc. ax. for mult.
4. $= ac\left(\dfrac{1}{bd}\right)$ — Prop. of recip. of a prod.
5. $= \dfrac{ac}{bd}$ — Def. of div.
6. $\dfrac{a}{b} \cdot \dfrac{c}{d} = \dfrac{ac}{bd}$ — Trans. prop. of =

31. 1. $a, b \in \mathcal{R}$
   $a, b \neq 0$ — Hypothesis
2. $\dfrac{1}{a} + \dfrac{1}{b} = \dfrac{1 \cdot b}{ab} + \dfrac{1 \cdot a}{ba}$ — Exercise 28
3. $= \dfrac{b}{ab} + \dfrac{a}{ba}$ — Iden. ax. for mult.

4. $\quad = \dfrac{b}{ab} + \dfrac{a}{ab}$ Comm. ax. for mult.

5. $\quad = \dfrac{b+a}{ab}$ Exercise 21

6. $\dfrac{1}{a} + \dfrac{1}{b} = \dfrac{b+a}{ab}$ Trans. prop. for $=$

33. 1. $a, b, \in \mathcal{R}$ Hypothesis
$\quad a, b, \neq 0$

2. $\dfrac{a}{b} \cdot \dfrac{b}{a} = \dfrac{ab}{ba}$ Exercise 29

3. $\quad = \dfrac{ab}{ab}$ Comm. ax. for mult.

4. $\quad = 1$ Exercise 23

5. $\dfrac{a}{b} \cdot \dfrac{b}{a} = 1$ Trans. prop. of $=$

35. 1. $a, b, c, d \in \mathcal{R}$ Hypothesis
$\quad b, c, d \neq 0$

2. $\dfrac{\frac{a}{b}}{\frac{c}{d}} = \dfrac{a}{b}\left(\dfrac{1}{\frac{c}{d}}\right)$ Def. of div.

3. $\quad = \dfrac{a}{b} \cdot \dfrac{d}{c}$ Exercise 34

4. $\quad = \dfrac{ad}{bc}$ Exercise 29

5. $\dfrac{\frac{a}{b}}{\frac{c}{d}} = \dfrac{ad}{bc}$ Trans. prop. of $=$

On the Calculator, page 28
1. 34.795 **3.** 18.23 **5.** 1810

Self-Test 2, page 29
1. 1. $a, b, \in \mathcal{R}$ Hypothesis
2. $-[(-a) + (-b)] = -[-(a+b)]$ Prop. of opp. of sum
3. $-[(-a) + (-b)] = a + b$ Canc. prop. of add. inv.

2. 1. $a, b \in \mathcal{R}$ Hypothesis
$\quad a \neq 0, b \neq 0$

2. $b \cdot \dfrac{1}{ab} = b\left(\dfrac{1}{a} \cdot \dfrac{1}{b}\right)$ Prop. of recip. of prod.

3. $\quad = \left(b \cdot \dfrac{1}{b}\right)\dfrac{1}{a}$ Comm. and assoc. ax. for mult.

4. $\quad = 1\left(\dfrac{1}{a}\right)$ Ax. of mult. inv.

5. $\quad = \dfrac{1}{a}$ Iden. ax. for mult.

6. $b \cdot \dfrac{1}{ab} = \dfrac{1}{a}$ Trans. prop. of $=$

3. -11 **4.** $20 - 24n$ **5.** $9 - b$ **6.** 1

Chapter Review, pages 30–31
1. b. **2.** a **3.** a **4.** b **5.** d **6.** d **7.** d **8.** d
9. c **10.** c **11.** a **12.** c **13.** b **14.** c **15.** b

16. a. **17.** a

Application, page 32
1. a. halved **b.** halved **3.** 5.4 V

Preparing for College Entrance Exams, page 33
1. D **3.** D **5.** C **7.** C

Chapter 2 Review of Essentials

Written Exercises, pages 44–45

1. $5x^3 + 3x^2 - 4x - 8$ **3.** $-7z^3 - 6z^2 + z - 11$
5. $2a^3 + 4a^2b - 2ab^2 - b^3$ **7.** $-3x^3 + 7x^2 - 10x +$
14 **9.** $-2z^4 - 7z^3 + 6z^2 + 5z + 9$ **11.** $8a^3 -$
$12a^2b + b^3$ **13.** $x - 3$ **15.** $-4a + 5b$ **17.** $-9p^2 - 3$
19. $5r^3 + 5r^2 - 9r - 3$ **21.** $-w^2 + 2w$ **23.** $7x^2 -$
$2xy$ **25.** $-4c^2 - 2c + 14d$ **27.** $16x - 96$
29. $-5x^4 + 3x^3 - 7x^2 + 2x - 5$

31. 1. $x^2 - (x^2 + 1)$ Def. of subt.
$\quad = x^2 + [-(x^2 + 1)]$
2. $\quad = x^2 + [(-x^2) + (-1)]$ Prop. of opp. of sum
3. $\quad = [x^2 + (-x^2)] + (-1)$ Assoc. ax. for add.
4. $\quad = 0 + (-1)$ Ax. of add. inv.
5. $\quad = -1$ Iden. ax. for add.
6. $x^2 - (x^2 + 1) = -1$ Trans. prop. of $=$

33. 1. $3x^2 - x - (3x^2 + x)$ Def. of subt.
$\quad = 3x^2 + (-x) +$
$\quad [-(3x^2 + x)]$
2. $\quad = 3x^2 + (-x) +$ Prop. of opp.
$\quad [(-3x^2) + (-x)]$ of sum
3. $\quad = [3x^2 + (-3x^2)] +$ Comm. and
$\quad [(-x) + (-x)]$ assoc. ax. for add.
4. $\quad = 0 + [(-x) +$ Ax. of add.
$\quad (-x)]$ inv.
5. $\quad = [(-x) + (-x)]$ Iden. ax. for add.
6. $\quad = [(-1 \cdot x) +$ Mult. prop.
$\quad (-1 \cdot x)]$ of -1
7. $\quad = [(-1) + (-1)]x$ Dist. ax.
8. $\quad = (-2)x$ Subs. prin.
9. $\quad = -2x$ Prop. of opp. in prod.
10. $3x^2 - x - (3x^2 + x) =$ Trans. prop.
$\quad -2x$ of $=$

Written Exercises, pages 47–48
1. Add 7 to each side; divide each side by 5.
3. Add 2 to each side; multiply each side by 3.
5. Simplify by the distributive axiom; simplify by the distributive axiom **7.** Subtract $\frac{1}{2}p$ from each side; add 5 to each side **9.** $\{-8\}$ **11.** $\{5\}$ **13.** $\{0\}$
15. $\{20\}$ **17.** $\{70\}$ **19.** $\{-27\}$ **21.** $\left\{-\dfrac{12}{5}\right\}$

23. $\{-14\}$ **25.** $\{6\}$ **27.** $\{-3\}$ **29.** $r = \dfrac{7b}{3a}$

31. $x = \dfrac{7h}{3k}$ **33.** $v = \dfrac{2r}{5r - s}$ **35.** $t = \dfrac{p - v}{g}$

37. $h = \dfrac{A - 2\pi r^2}{\pi r}$ **39.** $x = \dfrac{ac}{b}$

41. 1. $a = b;\ a, b, c \in \mathcal{R}$ Hypothesis
    2. $a + c = a + c$ Ref. prop. of $=$
    3. $a + c = b + c$ Subs. prin.

43. 1. $a = b;\ a, b, c \in \mathcal{R}$ Hypothesis
    2. $ac = ac$ Refl. prop. of $=$
    3. $ac = bc$ Subs. prin.

45. 1. $ax + b = c;\ a, b, c \in \mathcal{R}$ Hypothesis
      $a \neq 0$
    2. $ax + b - b = c - b$ Exercise 42
    3. $ax + b + (-b) = c - b$ Def. of subt.
    4. $ax + [b + (-b)] = c - b$ Assoc. ax. for add.
    5. $ax + 0 = c - b$ Ax. for add. inv.
    6. $ax = c - b$ Iden. ax. for add.
    7. $\dfrac{ax}{a} = \dfrac{c - b}{a}$ Exercise 44
    8. $ax\left(\dfrac{1}{a}\right) = \dfrac{c - b}{a}$ Def. of div.
    9. $x\left[a\left(\dfrac{1}{a}\right)\right] = \dfrac{c - b}{a}$ Comm. and assoc. ax. for mult.
    10. $x \cdot 1 = \dfrac{c - b}{a}$ Ax. of mult. inv.
    11. $x = \dfrac{c - b}{a}$ Iden. ax. for mult.

Computer Exercises, page 48
3. No solutions **5.** 0 **7.** 1

Problems, pages 52–54
1. 52, 53, 54 **3.** 67, 68 **5.** 15, 17, 19 **7.** 54°, 60°, 66° **9.** 14 cm **11.** 67.5 **13.** 1.5 lb **15.** 6 km **17.** 4 min **19.** 12 **21.** 22 shares at $7.50; 8 shares at $15.

Computer Exercises, page 54
3. $23.579 **7.** Impossible to achieve

Self-Test 1, page 55
1. $y^2 - 2y - 2$ **2.** $-6x^2 + 36$ **3.** $-a^3 - 5a^2 - 5a + 13$ **4.** $5xy + 2x - 2y - 16$ **5.** $\{1\}$ **6.** $\{16\}$
7. $a = \dfrac{y - 3}{4x}$ **8.** Each child receives $1700; the adult receives $4900 **9.** 12 cm and 20 cm **10.** 6, 8, 10

Written Exercises, pages 59–60
1. false; Example: $a = 1, b = 2$ **3.** true; multiply each side by -2 and reverse the direction of the inequality symbol **5.** false; Example: $a = 1, b = 2, c = -3, d = -2$ **7.** true; subtract b from each side

9. $\{y: y < -13\}$

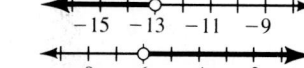

11. $\{a: a > -6\}$

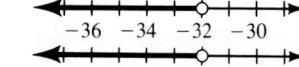

13. $\{c: c < -32\}$

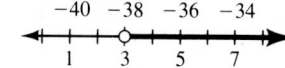

15. $\{r: r < -36\}$

17. $\{h: h > 3\}$

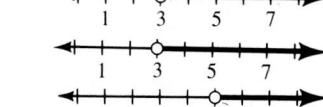

19. $\{d: d > 3\}$

21. $\{v: v > 2\}$

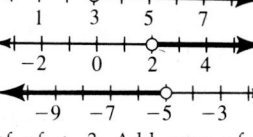

23. $\{x: x < -5\}$

25. 1. Hypothesis 2. Def. of $<$ 3. Add. prop. of $=$ 4. Comm. and assoc. ax. for add. 5. Def. of $<$

27. 1. $a > 0, a \in \mathcal{R}$ Hypothesis
    2. $a + (-a) > 0 + (-a)$ Add. prop. of order
    3. $0 > 0 + -a$ Ax. of add. inv.
    4. $0 > -a$ Iden. ax. for add.
    5. $-a < 0$ Inequality rewritten

29. 1. $a < b, c > 0;$ Hypothesis
      $a, b, c \in \mathcal{R}$
    2. For some positive   Def. of $<$
      number $p, a + p = b$
    3. $(a + p)c = bc$ Mult. prop. of $=$
    4. $ac + pc = bc$ Dist. ax.
    5. pc is a positive   Closure ax. for $\mathcal{R}_+$
      number
    6. $ac < bc$ Def. of $<$

31. 1. $a + c > b + c;$ Hypothesis
      $a, b, c \in \mathcal{R}$
    2. $(a + c) + (-c) > (b + c) + (-c)$ Add. prop. of order
    3. $a + [c + (-c)] > b + [c + (-c)]$ Assoc. ax. for add.
    4. $a + 0 > b + 0$ Ax. of add. inv.
    5. $a > b$ Iden. ax. for add.

33. 1. $a > b, c > d;$ Hypothesis
      $a, b, c, d \in \mathcal{R}$
    2. $a + c > b + c$ Add. prop. of order
      $b + c > b + d$
    3. $a + c > b + d$ Trans. prop. of order

Written Exercises, page 62
1. $\{x: x \leq -2\}$

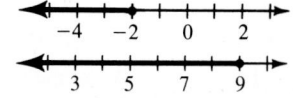

3. $\{a: a \leq 9\}$

5. $\left\{g: g \leq \dfrac{4}{3}\right\}$

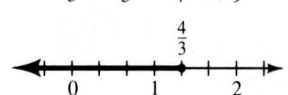

7. $\{k: 4 < k < 5\}$

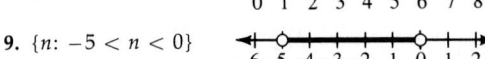

9. $\{n: -5 < n < 0\}$

11. $\{v: -5 \le v \le 10\}$

13. $\{u: u \ge 3\}$

15. $\{z: z \ge 3\}$

17. $\{x: x \le 4\}$

19. $\{b: b < -2\} \cup \{b: b > 4\}$

21. $\{t: -6 < t < -3\}$

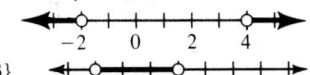

23. $\left\{h: h < \dfrac{3}{2}\right\} \cup \{h: h > 7\}$

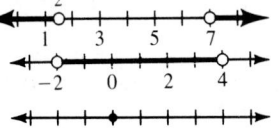

25. $\{v: -2 < v < 4\}$

27. $\{-1\}$

Problems, pages 63–64
1. 5 oz **3.** 11 dividers **5.** $135\,\pi\,\text{cm}^3$ **7.** 75 km
9. 60 min **11.** 2 tennis balls

Written Exercises, pages 67–69
1. 1. Mult. prop. of $=$ 2. Mult. prop. of 0 3. ax.
of mult. inv. 4. $1 > 0$ **3.** 1. Mult. prop. of $=$
2. Iden. ax. for mult. 3. Def. of div. 4. Assoc. ax.
for mult. 5. Ax. of mult. inv. 6. Iden. ax. for
mult. 7. Hypothesis: $a > b$

5. Case 1: Assume $\dfrac{1}{a} = 1$

1. $\dfrac{1}{a} = 1$ and $a > 1$ Hypothesis

2. $\dfrac{1}{a} \cdot a = 1 \cdot a$ Mult. prop. of $=$

3. $1 = 1 \cdot a$ Ax. of mult. inv.
4. $1 = a$ Iden. ax. for mult.
5. Contradiction   Hypothesis: $a > 1$

Case 2: Assume $\dfrac{1}{a} > 1$

1. $\dfrac{1}{a} > 1$ and $a > 1$ Hypothesis

2. $\dfrac{1}{a} \cdot a > 1 \cdot a$ Mult. prop. of order

3. $1 > 1 \cdot a$ Ax. of mult. inv.
4. $1 > a$ Iden. ax. for mult.
5. Contradiction   Hypothesis: $a > 1$

Therefore, $\dfrac{1}{a} < 1$

7. 1. $a > b$ Hypothesis
2. $a + a > a + b$ Add. prop. of order
3. $2a > a + b$ Dist. ax.
4. $2a\left(\dfrac{1}{2}\right) > (a + b)\dfrac{1}{2}$ Mult. prop. of order

5. $a\left(2 \cdot \dfrac{1}{2}\right) > (a + b)\dfrac{1}{2}$ Comm. and assoc. ax.
for mult.

6. $a \cdot 1 > (a + b)\dfrac{1}{2}$ Ax. of mult. inv.

7. $a > (a + b)\dfrac{1}{2}$ Iden. ax. for mult.

8. $a > \dfrac{a + b}{2}$ Def. of div.

9. $\dfrac{a + b}{2} < a$ Inequal. rewritten

9. 1. $0 < a < b$ Hypothesis
2. $\dfrac{1}{a} > 0, \dfrac{1}{b} > 0$ Theorem, page 65

3. $\dfrac{1}{a} \cdot \dfrac{1}{b} > 0$ Clos. ax. for $\mathcal{R}_+$

4. $a\left(\dfrac{1}{a} \cdot \dfrac{1}{b}\right) < b\left(\dfrac{1}{a} \cdot \dfrac{1}{b}\right)$ Mult. prop. of order

5. $\left(a \cdot \dfrac{1}{a}\right)\dfrac{1}{b} < \left(b \cdot \dfrac{1}{b}\right)\dfrac{1}{a}$ Comm. and assoc. ax.
for mult.

6. $1 \cdot \dfrac{1}{b} < 1 \cdot \dfrac{1}{a}$ Ax. of mult. inv.

7. $\dfrac{1}{b} < \dfrac{1}{a}$ Iden. ax. for mult.

8. $\dfrac{1}{a} > \dfrac{1}{b}$ Inequal. rewritten

11. Case 1: Assume $a = b$
1. $a = b, a^2 > b^2$ Hypothesis
2. $a \cdot a = a \cdot b; a \cdot b = b \cdot b$ Mult. prop. of $=$
3. $a^2 = ab; ab = b^2$ Subs. prin.
4. $a^2 = b^2$ Trans. prop. of $=$
5. Contradiction   Hypothesis $a^2 > b^2$
Case 2: Assume $a < b$
1. $a > 0, b > 0, a^2 > b^2$,   Hypothesis
   $a < b$
2. $a \cdot a < b \cdot a$ Mult. prop. of
   $b \cdot a < b \cdot b$ order
3. $a \cdot a < b \cdot b$ Trans. prop. of order
4. $a^2 < b^2$ Subs. Prin.
5. Contradiction   Hypothesis: $a^2 > b^2$
Therefore, $a > b$

13. Case 1: Assume $a = b$
1. $a = b, a^2 > b^2$ Hypothesis
2. $a \cdot a = b \cdot a$ Mult. prop. of $=$
   $b \cdot a = b \cdot b$
3. $a \cdot a = b \cdot b$ Trans. prop. of $=$
4. $a^2 = b^2$ Subs. Prin.
5. Contradiction   Hypothesis: $a^2 > b^2$
Case 2: Assume $a > b$.
1. $a > b, a < 0, b < 0$,   Hypothesis
   $a^2 > b^2$

2. $a \cdot a < b \cdot a$ Mult. prop. of order
$b \cdot a < b \cdot b$

3. $a \cdot a < b \cdot b$ Trans. prop. of order
4. $a^2 < b^2$ Subs. Prin.
5. Contradiction   Hypothesis: $a^2 > b^2$
Therefore, $a < b$

15. 1. $a > 0,\ b > 0,\ c > 0,$ Hypothesis
$d > 0,\ ab < cd,$
$b > d$

2. $ab > ad$ Mult. prop. of order
3. $ad < ab$ Inequal. rewritten
4. $ad < cd$ Trans. prop. of order
5. $\dfrac{1}{d} > 0$ Theorem, page 65

6. $ad\left(\dfrac{1}{d}\right) < cd\left(\dfrac{1}{d}\right)$ Mult. prop. of order

7. $a\left(d \cdot \dfrac{1}{d}\right) < c\left(d \cdot \dfrac{1}{d}\right)$ Assoc. prop. for mult.

8. $a \cdot 1 < c \cdot 1$.   Ax. of mult. inv.
9. $a < c$ Iden. ax. for mult.

17. 1. $a > b > 0$ Hypothesis
2. $a + a > a + b$ Add. prop. of order
3. $2a > a + b$ Dist. ax.
4. $2ab > (a + b)b$ Mult. prop. of order
5. $a + b > 0$ Clos. ax. for $\mathcal{R}_+$

6. $\dfrac{1}{a + b} > 0$ Theorem, page 65

7. $2ab\left(\dfrac{1}{a + b}\right) >$ Mult. prop. of order

$\quad [(a + b)b]\left(\dfrac{1}{a + b}\right)$

8. $\dfrac{2ab}{a + b} >$ Def. of div.

$\quad [(a + b)b]\left(\dfrac{1}{a + b}\right)$

9. $\dfrac{2ab}{a + b} >$ Comm. and assoc. ax.
     for mult.

$\quad b\left[(a + b)\left(\dfrac{1}{a + b}\right)\right]$

10. $\dfrac{2ab}{a + b} > b \cdot 1$ Ax. of mult. inv.

11. $\dfrac{2ab}{a + b} > b$ Iden. ax. for mult.

Written Exercises, pages 71–72
1. 5   **3.** 12   **5.** 7
7. $\{r: r \geq 4\} \cup \{r: r \leq -4\}$

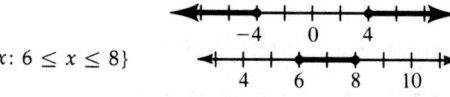

9. $\{x: 6 \leq x \leq 8\}$

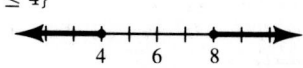

11. $\{c: c \geq 8\} \cup \{c: c \leq 4\}$

13. $\{1, 2\}$

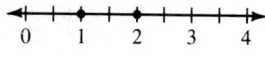

15. $\{n: n \geq -1\} \cup \{n: n \leq -8\}$

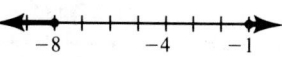

17. $\{q: -2 \leq q \leq 10\}$

19. $\left\{m: -1 \leq m \leq \dfrac{3}{2}\right\}$

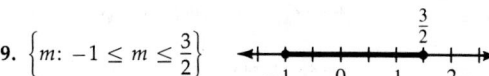

21. $\{u: u \geq 1\} \cup \{u: u \leq -8\}$

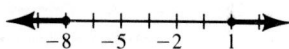

23. $\{a: a > 6\} \cup \{a: a < 3\}$

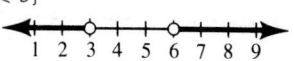

25. false; Example: $a = -1,\ b = 2$ **27.** true
29. true   **31.** true   **33.** false; Example: $a = -1,$
$b = 2$

35. 1. $a \in \mathcal{R}$ Hypothesis
2. $a \geq 0$ or $a < 0$ Comparison ax.
Case 1: $a \geq 0$
3. $|a| = a$ Def. of $|a|$ for $a > 0$
4. $|a|^2 = |a| \cdot |a|$ Def. of x^2
5. $\quad = a \cdot a$ Subs. prin.
6. $\quad = a^2$ Def. of x^2
7. $|a|^2 = a^2$ Trans. prop. of $=$
Case 2: $a < 0$
3. $|a| = -a$ Def. of $|a|$ for $a < 0$
4. $|a|^2 = (-a)(-a)$ Def. of x^2
5. $\quad = a \cdot a$ Prop. of opp. in prod.
6. $\quad = a^2$ Def. of x^2
7. $|a|^2 = a^2$ Trans. prop. of $=$

37. 1. $a, b \in \mathcal{R}$ Hypothesis
2. $|a + b|^2 = (a + b)^2$ Exercise 35
3. $|a + b|^2 =$ Dist. ax.
$\quad a^2 + 2ab + b^2$
4. $(|a| + |b|)^2 =$ Dist. ax.
$\quad |a|^2 + 2|a| \cdot |b| + |b|^2$
5. $|a|^2 = a^2,\ |b|^2 = b^2$ Exercise 35
6. $|a| \cdot |b| = |ab|$ Exercise 26
7. $(|a| + |b|)^2 =$ Subs. prin.
$\quad a^2 + 2|ab| + b^2$
8. $ab \leq |ab|$ Exercise 36
9. $2ab \leq 2|ab|$ Mult. prop. of
     order
10. $a^2 + 2ab + b^2 \leq$ Add. prop. of
$\quad a^2 + 2|ab| + b^2$ order
11. $|a + b|^2 \leq (|a| + |b|)^2$ Subs. prin.

Computer Exercises, page 72

3. $x < -7$ or $x > -1$ **5.** $19 < x < 37$

Self-Test 2, pages 72–73

1. $\{a: a > 4\}$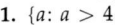

2. $\{x: x > -1\}$

3. $\left\{y: y \geq -\dfrac{23}{2}\right\}$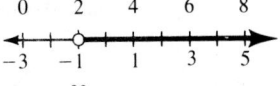

4. $\left\{b: \dfrac{1}{2} < b \leq \dfrac{3}{2}\right\}$

5. 8 min

6. Case 1: Assume $b = 0$

   1. $a < 0$, $ab > 0$, $b = 0$ Hypothesis
   2. $ab = ab$ Refl. prop. of =
   3. $ab = a \cdot 0$ Subs. prin.
   4. $ab = 0$ Mult. prop. of 0
   5. Contradiction   Hypothesis: $ab > 0$

   Case 2: Assume $b > 0$

   1. $a < 0$, $ab > 0$, $b > 0$ Hypothesis
   2. $a \cdot b < 0 \cdot b$ Mult. prop. of order
   3. $ab < 0$ Mult. prop. of 0
   4. Contradiction   Hypothesis: $ab > 0$
   Therefore, $b < 0$

7. $\{x: -2 \leq x \leq 12\}$

8. $\left\{y: y > \dfrac{2}{3}\right\} \cup \{y: y < -2\}$

Chapter Review, pages 74–75

1. c.   **2.** c.   **3.** b.   **4.** b   **5.** d.   **6.** b.   **7.** d.   **8.** b
9. b.   **10.** c.   **11.** d.   **12.** b.   **13.** d.   **14.** c.

Mixed Review, pages 76–77

1. 0   **3.** $50zw - 80z$ **5.** $\{-4\}$

7. $\{1\}$ **9.** $b = \dfrac{2a}{4a - 5}$ **11.** -16

13. $\{a: 2 < a \leq 7\}$

15. $\{t: -4 \leq t \leq 1\}$

17. Case 1: Assume $a - b = 0$

   1. $a > b$, $a - b = 0$ Hypothesis
   2. $a - b + b = 0 + b$ Add. prop. of =
   3. $a + (-b) + b = 0 + b$ Def. of subt.
   4. $a + [(-b) + b] = 0 + b$ Assoc. ax. for add.
   5. $a + 0 = 0 + b$ Ax. of add. inv.
   6. $a = b$ Iden. ax. of add.
   7. Contradiction   Hypothesis: $a > b$

   Case 2: Assume $a - b < 0$

   1. $a > b$, $a - b < 0$ Hypothesis

   2. $a - b + b < 0 + b$ Add. prop. of order
   3. $a + (-b) + b < 0 + b$ Def. of subt.
   4. $a + [(-b) + b] < 0 + b$ Assoc. ax. for add.
   5. $a + 0 < 0 + b$ Ax. of add. inv.
   6. $a < b$ Iden. ax. for add.
   7. Contradiction   Hypothesis: $a > b$
   Therefore, $a - b > 0$

Chapter 3   Linear Functions and Relations

Written Exercises, pages 86–87

1. 37   **3.** -13 **5.** 4   **7.** $\dfrac{9}{2}$ **9.** 37   **11.** 5   **13.** $y = \dfrac{x}{2}$;

a function   **15.** $x = \dfrac{1}{|y|}$; not a function   **17.** $y =$

$\dfrac{1}{x + 1}$; a function   **19.** $y = -x^2$; a function   **21.** $y =$

$\dfrac{1}{3}x + 1$; a function   **23.** $-4\dfrac{1}{2}$ **25.** $-4\dfrac{7}{9}$ **27.** $\dfrac{1}{49}$

29. $-4\dfrac{17}{25}$ **33. a.** 5, 10, 60   **b.** $2y = 5x$

Written Exercises, page 90

1. a function

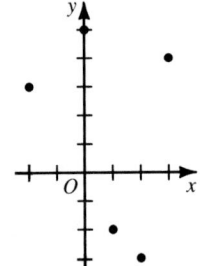

5. not a function

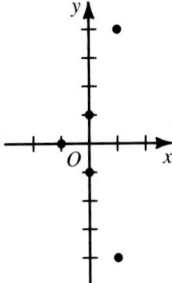

9.

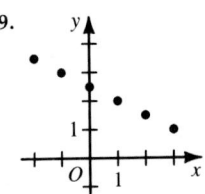

11.

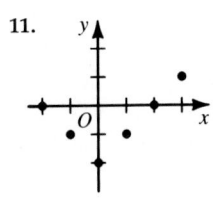

23.

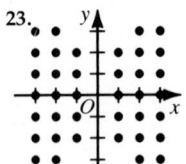

25.

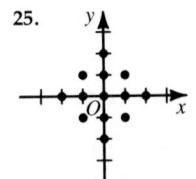

Computer Exercises, pages 90–91

3. $5.30   **5.** $26.00   **9.** $65.00   **11.** $6.20

Self-Test 1, page 91

1. $y = 2x$; a function   **2.** $y = 7$; a function   **3.** -6
4. 13   **5.** -118 **6.** 11

7. a function

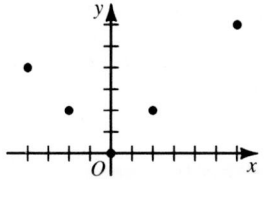

8. a function

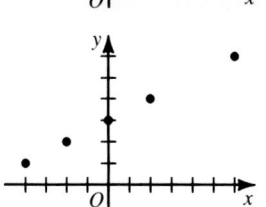

23.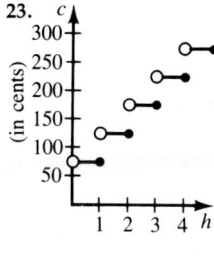

25.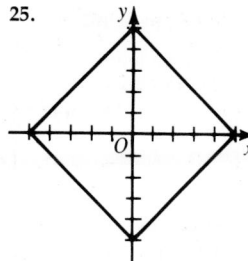

Written Exercises, pages 94–95

1.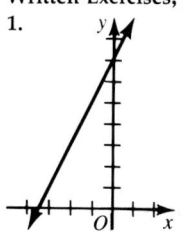

5.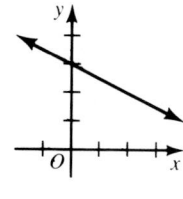

7.

11.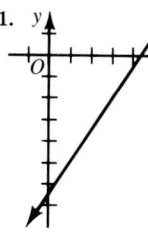

13.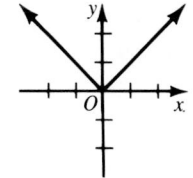

15.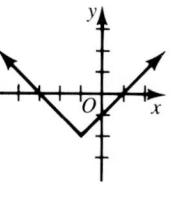

17.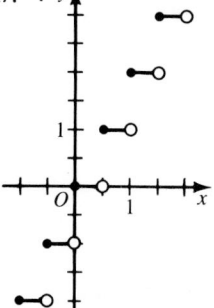

19. -8

21. -1

Written Exercises, page 97

1.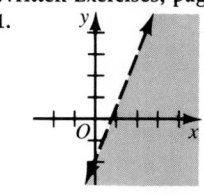

3.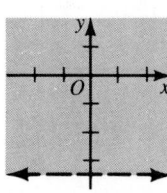

15.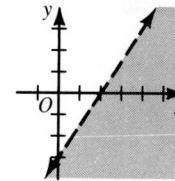

17.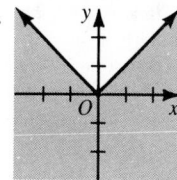

Self-Test 2, page 98

1.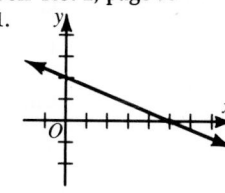

2.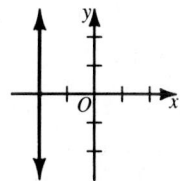

3.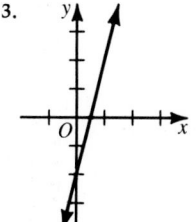

4.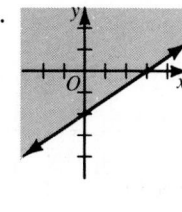

5.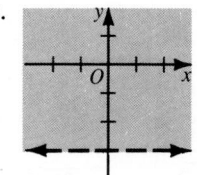

6.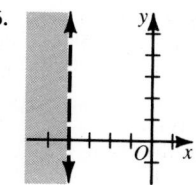

Written Exercises, page 102

1. 3 **3.** $-\dfrac{1}{2}$ **5.** undefined **7.** $\dfrac{4}{3}$ **9.** $\dfrac{3}{5}$ **11.** $-\dfrac{3}{2}$

15.
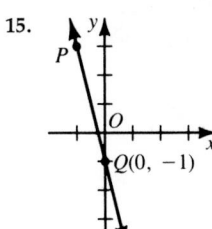

17.
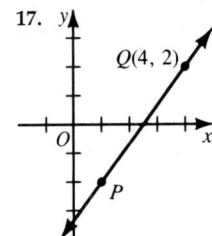

19. $-\dfrac{7}{2}$ **21.** $-\dfrac{15}{2}$ **23.** $-\dfrac{5}{3}$

25. $\dfrac{y_2 - y_1}{x_2 - x_1} = \dfrac{y_3 - y_2}{x_3 - x_2} = \dfrac{y_3 - y_1}{x_3 - x_1}$

Written Exercises, pages 105–106

1. $y = -x + 8$ **3.** $y = \dfrac{1}{2}x - 5$ **5.** $y = \dfrac{4}{3}x - 5$

7. $y = \dfrac{3}{2}x - \dfrac{7}{2}$ **9.** $y = \dfrac{3}{4}x$ **11.** $y = -\dfrac{2}{5}x + 10$

13. $y = 2x - 5$ **15.** $y = -x + 3$ **17.** $y = \dfrac{1}{2}x + 2$

19. $y = -4x + 1$ **21.** $y = \dfrac{1}{2}x - \dfrac{13}{6}$ **23.** $y = \dfrac{2}{3}x +$

1 **25.** $y = -2x - 5$ **27.** $y = -x + 7$ **29.** $y =$

$-\dfrac{1}{2}x - 10$ **31.** $-\dfrac{15}{2}$ **33.** $\dfrac{4}{7}$ **35.** 7 **37.** $k = \dfrac{2b}{b - 1}$

Computer Exercises, page 106

3. $y = -6.875x + 16.375$ **5.** $y = 0.8125x +$
47.6875 **7.** Vertical **9.** Vertical

Self-Test 3, page 106

1. $\dfrac{4}{5}$ **2.** $y = 2x - 4$ **3.** $y = \dfrac{3}{2}x + \dfrac{45}{8}$

4. $y = 4x - 2$ **5.** $y = -2x + 3$

Reading Algebra, page 107

1. Yes; No; No **5.** Sunday

Written Exercises, pages 110–111

1. $\dfrac{5}{2}$ **3.** $\dfrac{28}{3}$ **5.** $\dfrac{28}{5}$ **7.** $y = -3x + 1$

9. $y = 2x + 5$ **11.** $y = -\dfrac{1}{2}x + 4$ **13.** $y = 4x + 3$

15. -1 **17.** $\dfrac{7}{6}$ **19.** $\dfrac{5}{6}$ **21.** $\dfrac{45}{2}$ **23.** -15 **25.** $\dfrac{5}{3}$

27. 1. $\dfrac{y_1}{x_1} = \dfrac{y_2}{x_2}$ Hypothesis

  $x_1, y_1, x_2, y_2 \neq 0$
 2. $y_1 x_2 = y_2 x_1$ Prod. of means = prod. of
  extremes

 3. $\dfrac{y_1 x_2}{y_1 y_2} = \dfrac{y_2 x_1}{y_1 y_2}$ Div. prop. of =

 4. $\dfrac{x_2}{y_2} = \dfrac{x_1}{y_1}$ Exercise 28, page 27

5. $\dfrac{x_1}{y_1} = \dfrac{x_2}{y_2}$ Symm. prop. of =

29. 1. $\dfrac{y_1}{x_1} = \dfrac{y_2}{x_2}$ Hypothesis
  $(x_1, x_2, y_1, y_2 \neq 0, x_1 \neq x_2)$
 2. $y_1 = mx_1$ Def. of dir. variation

 3. $\dfrac{mx_1}{x_1} = \dfrac{y_1}{x_1}$ Div. prop. of =

 4. $\dfrac{mx_1}{x_1} = \dfrac{y_2}{x_2}$ Substitution

 5. $m = \dfrac{y_2}{x_2}$ Ex. 25, p. 27

 6. $m = \dfrac{y_1 - y_2}{x_1 - x_2}$ Def. of slope

 7. $\dfrac{y_1 - y_2}{x_1 - x_2} = \dfrac{y_2}{x_2}$ Substitution

31. 1. $g(x) = mx + b$ Hypothesis
  $g(r + s) = g(r) + g(s)$
 2. $g(r + s) = m(r + s) + b$ Subs. prin.
 3. $= mr + ms + b$ Dist. ax.
 4. $g(r) = mr + b$ Subs. prin.
  $g(s) = ms + b$
 5. $mr + ms + b$
  $= mr + b + ms + b$ Subs. prin.
 6. $0 = b$ Canc. prop. of add.
 7. $b = 0$ Symm. prop. of =

Problems, pages 112–113

1. \$625 **3.** 120 V **5.** $-9.5°C$ **7.** $\dfrac{1}{2}$ in./h **9.** 2000 s

Self-Test 4, page 113

1. $\dfrac{35}{4}$ **2.** 105 km

Chapter Review, pages 114–115

1. b **2.** d **3.** a **4.** a **5.** b **6.** d **7.** c **8.** c
9. d **10.** b **11.** d **12.** b **13.** d **14.** b **15.** c

Application, page 117

1. Substance A

Preparing for College Entrance Exams, page 118
1. D **3.** A **5.** E **7.** E

**Chapter 4 Systems of Linear Equations or
Inequalities**

Written Exercises, pages 126–127

1. $\{(3, 2)\}$; slopes: $-1, \dfrac{1}{2}$

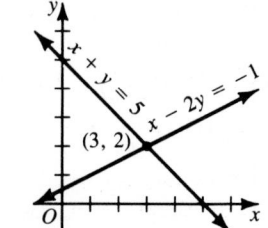

3. $\{(x, y): x + 3y = 6\}$;

slopes: $-\frac{1}{3}, -\frac{1}{3}$;

y-intercepts: 2, 2

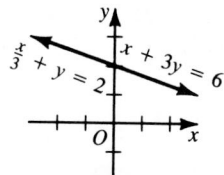

5. $\emptyset$; slopes: $-\frac{1}{2}, -\frac{1}{2}$;

y-intercepts: 1, 2

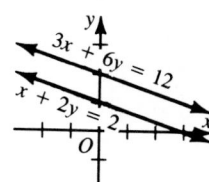

19. $\{(0, 4)\}$ **21.** $\emptyset$

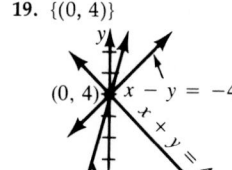 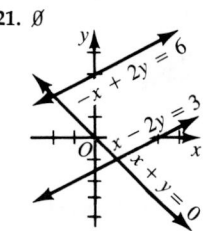

23. $\{(5, 2)\}$ **25.** $\{(0, 4)\}$

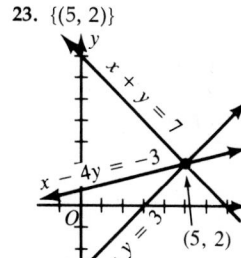 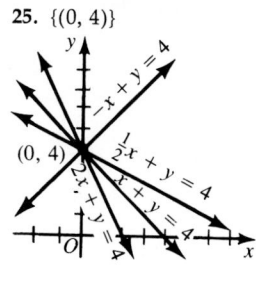

Computer Exercises, page 127

3. unique solution   **5.** unique solution

Written Exercises, pages 131–133

1. $\{(3, 5)\}$ **3.** $\{(-2, -6)\}$ **5.** $\{(7, 3)\}$

7. $\{(2, -5)\}$ **9.** $\left\{\left(4, \frac{1}{2}\right)\right\}$ **11.** $\{(6, 4)\}$

13. $\{(8, 7)\}$ **15.** $\left\{\left(\frac{3}{2}, 5\right)\right\}$ **17.** $\{(-2, 3)\}$

19. $\{(-6, -10)\}$ **21.** $\{(7, 2)\}$ **23.** $\left\{\left(\frac{1}{4}, -\frac{51}{8}\right)\right\}$

25. $\{(x, y): 4x - y = 2\}$ **27.** $\{(6, 0)\}$ **29.** $\{(-11, 7)\}$

31. $\frac{1}{x} = -1, \frac{1}{y} = 4; \left\{\left(-1, \frac{1}{4}\right)\right\}$ **33.** $a = 3, b = -2$

Written Exercises, pages 135–136

1. 22   **3.** 48   **5.** $63 - k^2$ **7.** 0   **9.** $\{(-38, 16)\}$

11. $\{(4, 6)\}$ **13.** $\left\{\left(\frac{3}{2}, \frac{1}{2}\right)\right\}$ **15.** $\{(2, 0)\}$

17. $\left\{\left(\frac{1}{3}, -\frac{2}{3}\right)\right\}$ **19.** $\left\{\left(-2, \frac{5}{3}\right)\right\}$ **21.** $\left\{\left(-\frac{1}{6}, \frac{2}{3}\right)\right\}$

Computer Exercises, page 136

3. $\{(1.79866, 6.95302)\}$ **5.** no solution

Problems, pages 142–144

1. 3 cucumbers and 9 heads of lettuce   **3.** congruent sides, 52 cm; base 36 cm   **5.** 57°, 88°   **7.** I_1, 1.5 A; I_2, 2.5 A   **9.** 175 tickets   **11.** $\overline{HG}$, 7 cm; $\overline{GF}$, 9 cm; $\overline{DE}$, 16 cm   **13.** 24 trucks   **15.** 0.5 km, 0.7 km

17. $A = \frac{12}{17}, B = -\frac{15}{17}$ **19.** $A = -2, B = 10$

21. $v_0 = 5$ m/s; $h_0 = 30$ m   **23.** 28 miles per gallon

25. 8 nickels, 5 dimes, 14 quarters

Self-Test 1, pages 144–145

1. $\{(1, -3)\}$

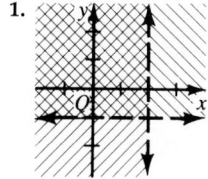

2. one solution

3. No solution; slopes: 1, 1; y-intercepts: -3, 4

4. Infinitely many solutions; slopes: $\frac{1}{2}, \frac{1}{2}$; y-intercepts: 2, 2   **5.** $\left\{\left(\frac{1}{5}, \frac{11}{5}\right)\right\}$

6. $\{(2, 3)\}$ **7.** $\{(3, 2)\}$ **8.** oranges: \$0.75; apples: \$0.60

Written Exercises, pages 146–147

1. 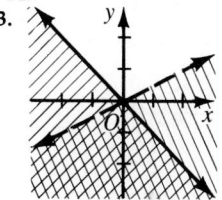   **3.**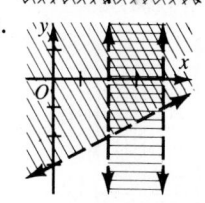

19. 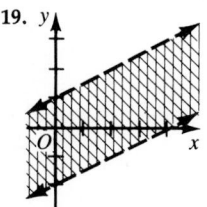   **21.**

33. 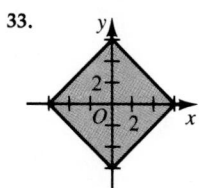   **35.**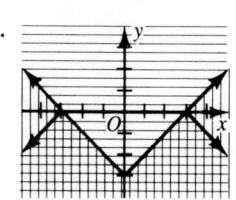

Written Exercises, pages 150–151

1. a.

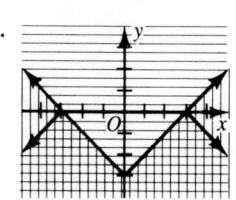

b. $(0, 0), (0, 6)$ $(4, 2), (4, 0)$

c. 0, 18, 10, 4

d. max.: 18; min: 0

3. a. 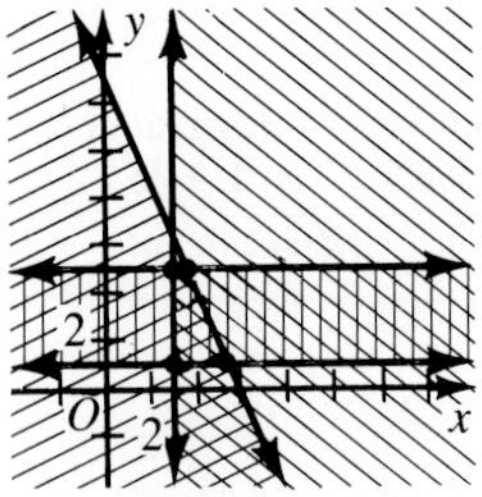

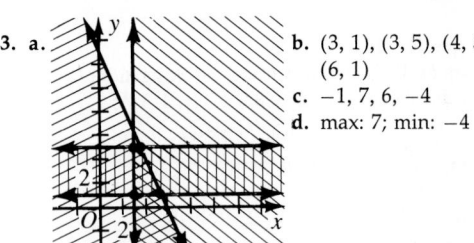

b. (3, 1), (3, 5), (4, 5), (6, 1) **c.** −1, 7, 6, −4 **d.** max: 7; min: −4

7. $x \geq 0$, $x \leq 12$, $y \geq 0$, $y \leq 8$, $x + 1.5y \leq 18$
9. (0, 0), (0, 8), (6, 8), (12, 4), (12, 0)

11. 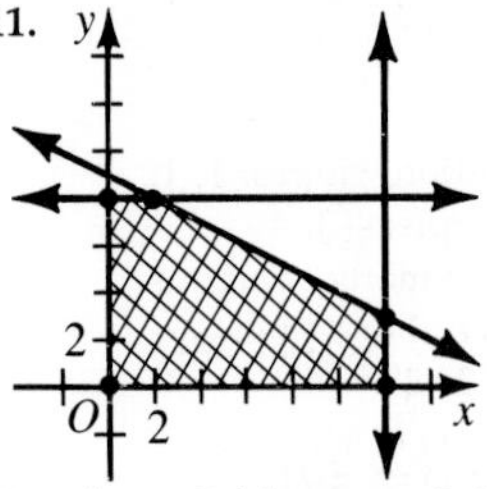

13. 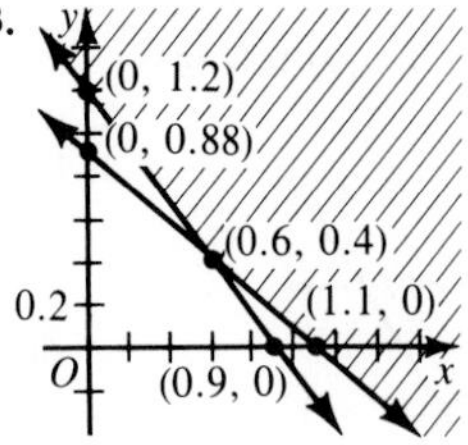

15. wheat: 0.6 kg; oats: 0.4 kg **17.** $a + bc$

Computer Exercises, page 151

3. max at (1, 26); min at (9, 16)

Self-Test 2, page 152

1. 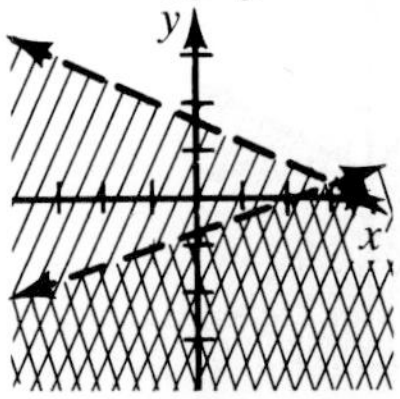

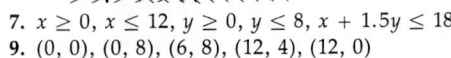

2. 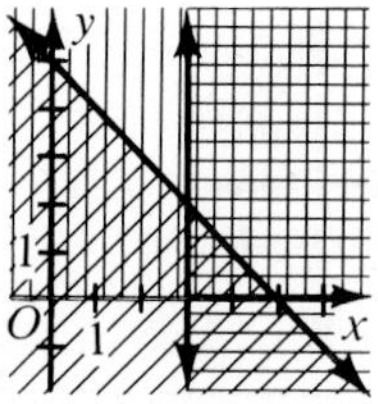

3. 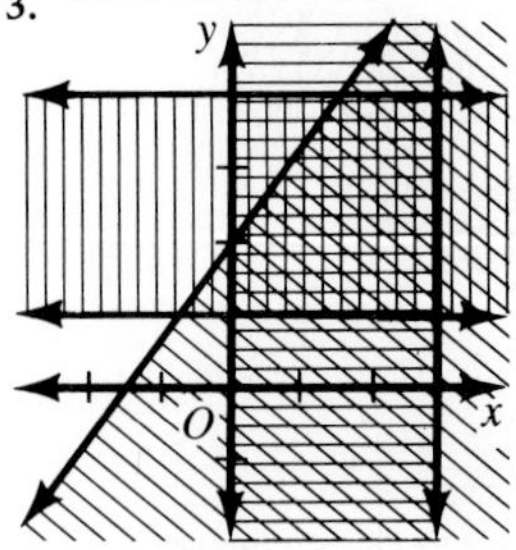

4. max: 15 at (5, 5); min: 4 at (2, 1) **5.** max: 21 at (6, 3); min: 7 at (2, 1) **6.** $845, from 200 shirts and 49 pairs of slacks

Chapter Review, pages 153–154

1. c **2.** b **3.** b **4.** c **5.** c **6.** c **7.** c **8.** a
9. c **10.** d **11.** b **12.** c

Cumulative Review, pages 155–157

1. 120 **3.** 5 **5.** $29z + 8$ **7.** {0} **9.** No solution
11. {−4, −2, 0, 2, 4}

13. 1. $\frac{a}{c} = \frac{b}{c}$, $c \neq 0$ Hypothesis

2. $a\left(\frac{1}{c}\right) = b\left(\frac{1}{c}\right)$ Def. of div.

3. $a = b$ Canc. prop. of mult.

15. 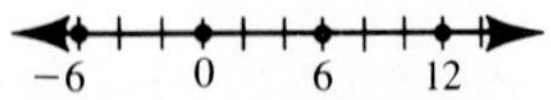

17. {−4} **19.** {12} **21.** {3} **23.** $b = 1$, $a \neq c$
29. 89 **31.** $\frac{16}{9}$ **33.** 25

37. 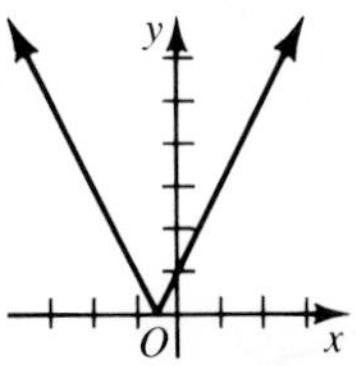

39. 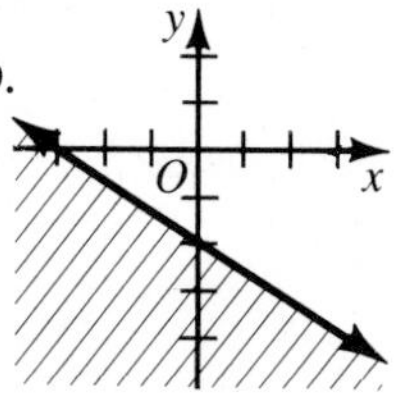

41. $k = 3$ **43.** $y = -3x + 1$ **45.** one solution: {(−2, 1)} **47.** no solution

49. 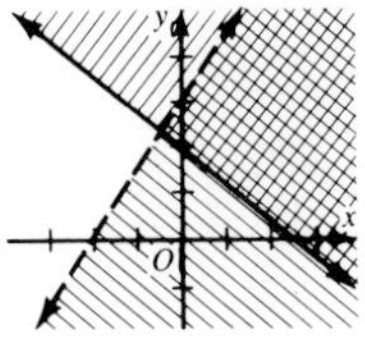

51. Adult tickets: 300; student tickets: 200

Chapter 5 Graphs in Space; Determinants

Written Exercises, page 164

1. 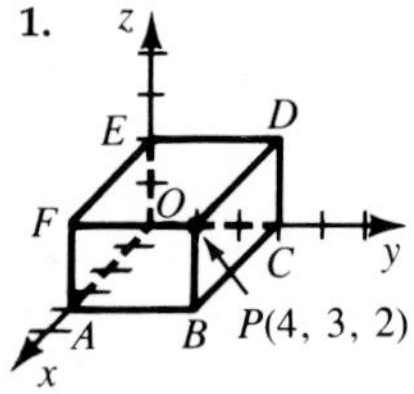

5. 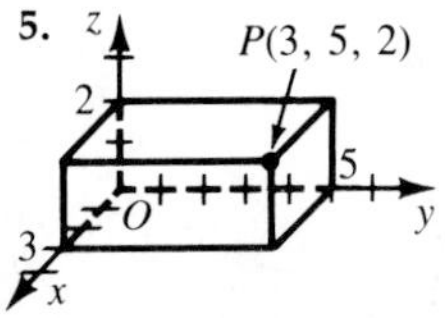

9. 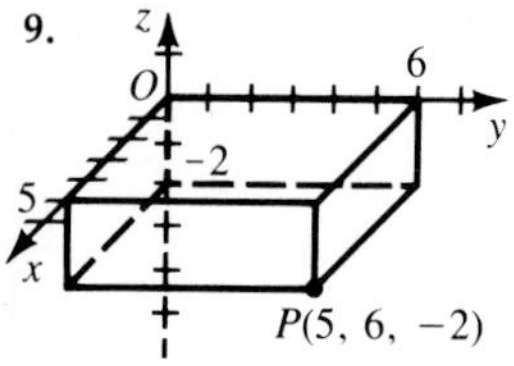

11. 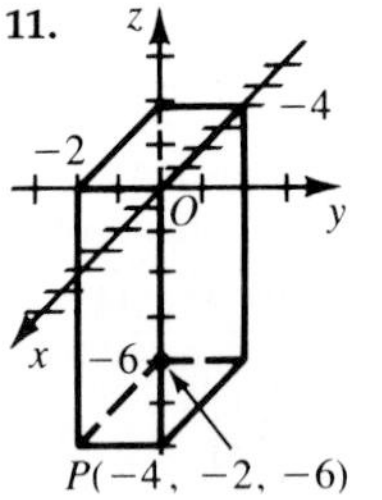

13. 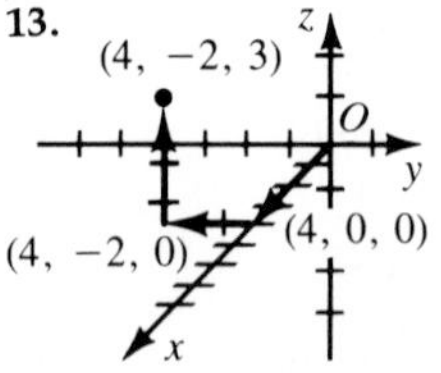

19. 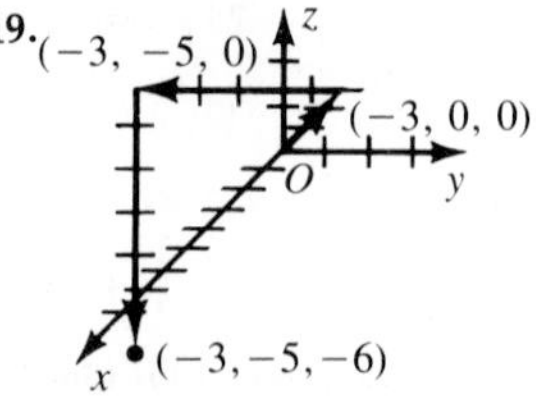

21. 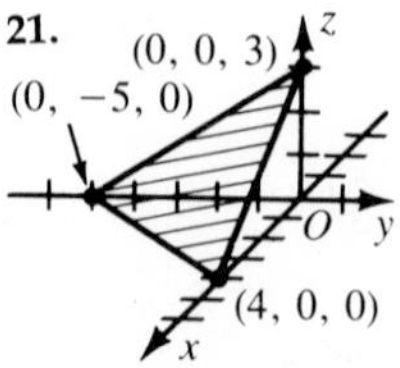

25. 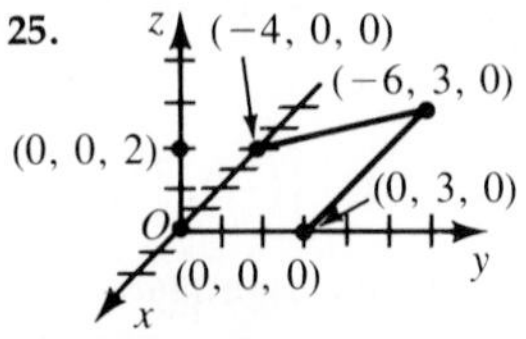

Written Exercises, page 170

1. a.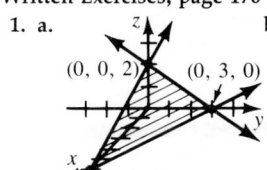

(0, 0, 2), (0, 3, 0), (6, 0, 0)

b. $\begin{cases} x = 0 \\ 2y + 3z = 6 \\ y = 0 \\ x + 3z = 6 \\ z = 0 \\ x + 2y = 6 \end{cases}$

5. a.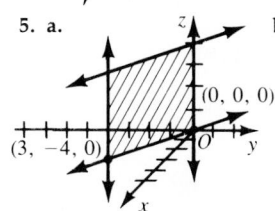

(3, −4, 0), (0, 0, 0)

b. $\begin{cases} z = 0 \\ 4x + 3y = 0 \\ x = 0 \\ y = 0 \end{cases}$

The graph contains the z-axis.

11. a.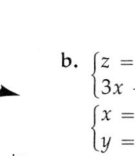

(0, 0, 0), (1, 3, 0)

b. $\begin{cases} z = 0 \\ 3x - y = 0 \\ x = 0 \\ y = 0 \end{cases}$

The plane contains the z-axis.

19.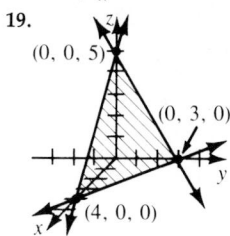

(0, 0, 5), (0, 3, 0), (4, 0, 0)

21.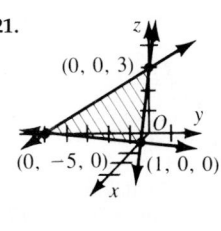

(0, 0, 3), (0, −5, 0), (1, 0, 0)

27. $4x + y = 4, z = 0$ **29.** $4x - 6y + 3z = 12$

Written Exercises, pages 174–175

1. $\{(3, -1, 5)\}$ **3.** $\{(0, 2, 5)\}$ **5.** $\{(3, 0, -2)\}$
7. $\{(-3, -2, 1)\}$ **9.** $\{(7, -6, -9)\}$ **11.** $\{(4, -7, 0)\}$
13. $\left\{\left(-\frac{68}{9}, \frac{47}{9}, -\frac{31}{9}\right)\right\}$ **15.** $\left\{\left(\frac{1}{2}, -\frac{3}{2}, \frac{5}{2}\right)\right\}$
17. $\{(8, 12, -9)\}$ **19.** consistent, $\{(0, 0, z): z \in \mathcal{R}\}$
21. consistent, $\{(x, y, 0): x + y = 3\}$
23. 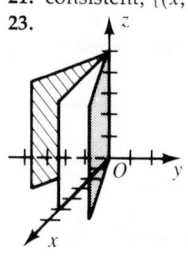 **25.**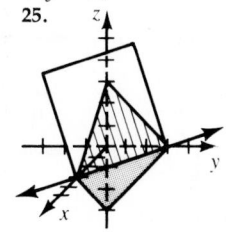

Computer Exercises, page 175
3. parallel to yz-plane; $(-3, 0, 0)$ **5.** parallel to
z-axis; $(20, 0, 0), (0, -35, 0)$ **7.** $(2.8571, 0, 0)$,

$(0, -2.5, 0), (0, 0, 4)$ **9.** xz-trace: $1.5x + 5z = 7.5$,
$y = 0$; $(5, 0, 0), (0, -2, 0), (0, 0, 1.5)$

Self-Test 1, page 176
1.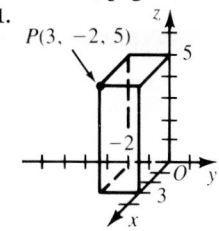

$P(3, -2, 5)$

2. 4, 3, 2

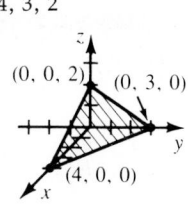

(0, 0, 2), (0, 3, 0), (4, 0, 0)

3. $\begin{cases} z = 0 \\ 4x + 13y = 24 \\ x = 0 \\ 13y - 7z = 24 \\ y = 0 \\ 4x - 7z = 24 \end{cases}$

4.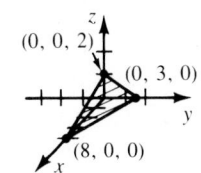

(0, 0, 2), (0, 3, 0), (8, 0, 0)

5. $\{(2, 1, 3)\}$

Written Exercises, page 179
1. -15 **3.** 33 **5.** $\{(2, 1, -2)\}$ **7.** $\{(-2, -4, 5)\}$
9. $\{(4, 1, -1)\}$ **11.** $D = 0$; infinite solution set
13. $D = -2; \{(3, 0, 4)\}$

Problems, pages 181–182
1. 20 nickels, 12 dimes, 6 quarters **3.** 8 cm ×
7 cm × 5 cm **5.** milk, 8 g; egg, 7 g; wheat, 2 g
7. $\overline{EF}$, 12; $\overline{EG}$, 34; $\overline{EH}$, 19 **9.** uphill, 16 km/h;
downhill, 24 km/h; level, 20 km/h **11.** 2 packs, $0.84;
4 packs, $1.44; 5 packs, $2.40 **13.** 30 km/hr; 80 km/hr

On the Calculator, page 183
1. 21.874 **3.** $\{(12, -5, 1.3)\}$

Written Exercises, pages 187–188
1. -138 **3.** -63 **5.** 33 **7.** $x = 1, y = -1$,
$z = -2, w = 3$

Computer Exercises, pages 188–189
3. $\{(2.5, 3, 7)\}$ **5.** Inconsistent **7.** 3718 **9.** -6

Self-Test 2, page 189
1. $\{(3, -1, 2)\}$ **2.** orange, $6; grapefruit, $7; lemon,
$4 **3.** -6

Chapter Review, pages 190–191
1. c **2.** d **3.** a **4.** a **5.** c **6.** b **7.** d **8.** a
9. c **10.** b

Application, page 193
1. 5×10^{-6} coulombs

Preparing for College Entrance Exams, page 194
1. E **3.** A **5.** A **7.** B

Chapter 6 Polynomials and Rational Expressions

Written Exercises, page 203
1. $-35u^{10}$ **3.** $\dfrac{6x^3}{y^2}$ **5.** $-\dfrac{a^4}{5b^5}$ **7.** $\dfrac{16e^{10}}{9d^6}$ **9.** $\dfrac{6x^2}{y^{18}}$

11. 1   13. $\frac{c}{d^6}$ 15. $\frac{8r^8}{s^5}$ 17. $\frac{4}{9u^2v^4}$ 19. $\frac{1}{x+y}$

21. $\frac{w^2 - 2wz + z^2}{wz}$ 23. a^2 25. 4^{k+1}

27. 1. $m < n,\ b \neq 0$ — Hypothesis
   2. $n - m > 0$ — Add. prop. of inequality
   3. $b^n = b^{n-m}b^m$ — Law 1
   4. $\dfrac{b^m}{b^n} = \dfrac{b^m}{b^{n-m}b^m}$ — Subs. prin.
   5. $= \dfrac{1}{b^{n-m}} \cdot \dfrac{b^m}{b^m}$ — Corollary, page 201
   6. $= \dfrac{1}{b^{n-m}} \cdot 1 \quad \dfrac{b^m}{b^m} = 1$
   7. $= \dfrac{1}{b^{n-m}}$ — Iden. ax. for $\times$
   8. $\dfrac{b^m}{b^n} = \dfrac{1}{b^{n-m}}$ — Trans. prop. of $=$

29. 1. $\dfrac{1}{b^{-n}} = \dfrac{1}{\frac{1}{b^n}}$ — Def. of b^{-n}
   2. $= 1\left(\dfrac{b^n}{1}\right)$ — Ex. 34, page 28
   3. $= b^n$ — Iden. ax. for $\times$
   4. $\dfrac{1}{b^{-n}} = b^n$ — Trans. prop. of $=$

31. 1. $b \in \mathcal{R}$, let $m = 0,\ n > 0$ — Hypothesis
   2. $b^0 b^n = b^{0+n}$ — Law 1
   3. $= b^n$ — Iden. ax. for $+$
   4. $= 1 \cdot b^n$ — Iden. ax. for $\times$
   5. $b^0 b^n = 1 \cdot b^n$ — Trans. prop. of $=$
   6. $b^0 = 1$ — Canc. prop. of $\times$

33. 1. $r, s, t, u \in \mathcal{R};\ t, u \neq 0$ — Hypothesis
   2. $\dfrac{rs}{tu} = rs \cdot \dfrac{1}{tu}$ — Def. of div.
   3. $= rs\left(\dfrac{1}{t} \cdot \dfrac{1}{u}\right)$ — Prop. of the reciprocal of a prod.
   4. $= \left(r \cdot \dfrac{1}{t}\right)\left(s \cdot \dfrac{1}{u}\right)$ — Assoc. and comm. ax. for mult.
   5. $= \dfrac{r}{t} \cdot \dfrac{s}{u}$ — Def. of div.
   6. $\dfrac{rs}{tu} = \dfrac{r}{t} \cdot \dfrac{s}{u}$ — Trans. prop. of $=$

Written Exercises, page 206

1. $x^2 + xy - 6y^2$ 3. $t^2 + 8t + 16$ 5. $25 - s^2$
7. $x^6 - 4y^2$ 9. $4u^4 - 20u^2v + 25v^2$ 11. $a^4 - 3a^2b + 2b^2$ 13. $6n^7 - 48n^5$ 15. $d^{10} + 2d^5e^3 + e^6$
17. $r^{2n} - 25$ 19. $4x^3 - 19x + 15$ 21. $u^3 - 64$
23. $z^3 + 9z^2 + 27z + 27$ 25. $x^4 - y^4$ 27. $p^6 - q^6$
29. $a^3 - b^3$ 31. $x^{3n} + 2x^{2n}y^n + 2x^n y^{2n} + y^{3n}$
33. $a^{2p} - a^2 b^{2q}$ 35. $a^5 - b^5$

Written Exercises, pages 210–211

1. $(5x + 3)^2$ 3. $(7 - 6k)(7 + 6k)$ 5. $(1 - 9r)^2$
7. $(3n - 4m)(9n^2 + 12mn + 16m^2)$ 9. $(3u - 5v) \cdot (u + 2v)$ 11. $(3w + 4z)(9w^2 - 12wz + 16z^2)$
13. irreducible   15. $(3ab - 8)(ab + 3)$

17. $x(8y - 3x)(8y + 3x)$ 19. $(3w - 2)(3w + 2) \cdot (9w^2 + 4)$ 21. $(z^3 - 7)^2$ 23. $2xy^2(5x + 2y)^2$
25. $bc(2b + 3c)(4b^2 - 6bc + 9c^2)$ 27. $(m + n) \cdot (m - n - 2)$ 29. $(p^n - q^n)(p^{2n} + p^n q^n + q^{2n})$
31. $(4v^{3n} - 7u^{2n})(4v^{3n} + 7u^{2n})$ 33. $(c^n + 3)(c^n - 3) \cdot (c^n + 2)(c^n - 2)$ 35. $(x^2 - 2xy + 2y^2) \cdot (x^2 + 2xy + 2y^2)$

Computer Exercises, page 211

3. $(11x^2 + 14y^5)(121x^4 - 154x^2y^5 + 196y^{10})$
5. $(x^7 + 12y^{13})(x^{14} - 12x^7y^{13} + 144y^{26})$

Self-Test 1, page 211

1. $\dfrac{4xy^8}{9z^2}$ 2. $\dfrac{c^4}{3ab^7}$ 3. $81x^4 - 216x^3 + 216x^2 - 96x + 16$ 4. $(4x - 3)(2x + 1)$ 5. $(2g - h)(4g^2 + 2gh + h^2)$ 6. $(x - 16)(x - 4)$ 7. irreducible

Written Exercises, page 216

1. $\{7, -4\}$ 3. $\{5, -5\}$ 5. $\left\{-\dfrac{3}{2}\right\}$ 7. $\{8, -8\}$
9. $\left\{0, \dfrac{9}{4}\right\}$ 11. $\{16, -3\}$ 13. $\left\{-\dfrac{4}{3}, \dfrac{3}{2}\right\}$ 15. $\left\{\dfrac{3}{4}, 2\right\}$
17. $\left\{\dfrac{9}{5}, \dfrac{4}{5}\right\}$ 19. $\{-4, 3\}$ 21. $\left\{0, \dfrac{2}{3}, 5\right\}$ 23. $\{5, -5\}$
25. $\left\{\dfrac{1}{3}, 1, -1\right\}$

Problems, pages 216–217

1. 16 cm × 9 cm   3. 1 m   5. 2.5 cm, 2 cm, 1.5 cm
7. 2.5 cm, 2.5 cm, 2.5 cm, 7.5 cm   9. 11:30 A.M.
11. 30 m × 8 m or 20 m × 12 m   13. 12 cm

Written Exercises, pages 219–220

1. $\{x: x \leq -3 \text{ or } x \geq 4\}$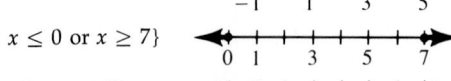

3. $\{a: -1 < a < 4\}$

5. $\{x: x \leq 0 \text{ or } x \geq 7\}$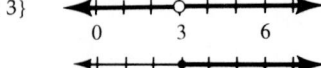

7. $\{c: -5 \leq c \leq 5\}$

9. $\{x: -1 < x < 6\}$

11. $\{k: k > 3 \text{ or } k < 3\}$

13. $\{y: y \geq 3\}$

15. $\left\{x: x < \dfrac{1}{4}, x \neq 0\right\}$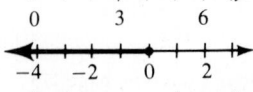

17. $\{z: z > 3\}$

19. $\{x: x \leq 0\}$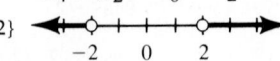

21. $\{x: x > 2 \text{ or } x < -2\}$

23. $\{a: a \le -3 \text{ or } a \ge 3, \text{ or } -1 \le a \le 1\}$

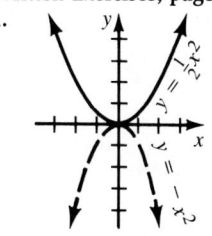
(number line) $-3 \quad -1 \quad 1 \quad 3$

Self-Test 2, page 220

1. $\{0, 5\}$ **2.** $\left\{-\dfrac{1}{3}, 4\right\}$ **3.** 26 cm

4. $\left\{x: -\dfrac{3}{5} < x < 1\right\}$

(number line) $-\dfrac{3}{5}$; $-1 \quad 0 \quad 1$

5. $\{x: x > 0 \text{ or } x < -2\}$

(number line) $-2 \quad -1 \quad 0 \quad 1$

Written Exercises, pages 222–223

1. $\dfrac{2x}{x-7}$ **3.** $\dfrac{3a-7}{8}$ **5.** $\dfrac{z^4}{3}$ **7.** $-\dfrac{3k^2}{2}$ **9.** $\dfrac{1}{7-p}$

11. $\dfrac{4}{3hk}$ **13.** $\dfrac{n^2+2n+4}{3(n+2)}$ **15.** $\dfrac{c-5}{c(c+5)}$

17. $\dfrac{16u^2-4u+1}{u+1}$ **19.** $\dfrac{(r+2)(r-2)}{r^2}$

21. $\dfrac{a^2-ab+b^2}{a^2+2ab+b^2}$ **23.** 1 **25.** $(k+5)(k+1)$

27. $\dfrac{r^6-8}{(4-r^2)(4+2r+r^2)}$ **29.** $x^2-4xy+8y^2$

Written Exercises, pages 225–226

1. $2x^3+5x-3$ **3.** $-b^2-ab+a^2$

5. $x-2+\dfrac{-5}{x-4}$ **7.** $2y-9$ **9.** r^2-2r+5

11. $-2v^2-4v+3+\dfrac{-2}{3v-1}$ **13.** a^2-3a+2

15. $9n^2-6n+4$ **17.** $3y^2+2y-4$

19. $16b^4-8b^3+4b^2-2b+1+\dfrac{8}{2b+1}$

21. $2w^2+4w-5$ **23.** $2v^2+v-5+\dfrac{v-30}{v^2-6}$

25. $a^{n-1}+a^{n-2}b+a^{n-3}b^2+\ldots+a^2b^{n-3}+ab^{n-2}+b^{n-1}$

Written Exercises, page 228

1. 5 **3.** $\dfrac{x}{4x+1}$ **5.** $\dfrac{7}{2(v+9)}$ **7.** $\dfrac{3}{a+b}$ **9.** $\dfrac{u+4}{3}$

11. $x(x+3)$ **13.** $\dfrac{2b-5a}{a^2+ab+b^2}$ **15.** $xy(x+y)$

17. $\dfrac{rs}{3(6s-r)}$ **19.** $\dfrac{k^2-1}{k}$ **21.** 1

Written Exercises, pages 230–231

1. 5 **3.** $\dfrac{3(d-3c+2cd)}{c^2d^2}$ **5.** 1 **7.** $\dfrac{a-5}{a-6}$

9. $\dfrac{3x}{x-2}$ **11.** $\dfrac{12xy}{(x+5y)(x-5y)^2}$

13. $-\dfrac{10x}{(x-2)(x+2)(x-3)}$ **15.** $\dfrac{12p}{3p-q}$ **17.** $a-b$

19. $\dfrac{(x-3)(x+1)}{x^2-4x-3}$ **21.** -1 **23.** $-\dfrac{1}{u^2v^2}$

Problems, pages 232–233

1. 18 min **3.** 6 min **5.** 7.5 km **7.** 100 mi **9.** 200 g **11.** 10 min **13.** 1.5 m

Written Exercises, pages 235–236

1. $\left\{\dfrac{11}{12}\right\}$ **3.** $\left\{-\dfrac{7}{12}\right\}$ **5.** $\left\{-\dfrac{2}{5}, 3\right\}$ **7.** $\{-6, 2\}$

9. $\{7, -2\}$ **11.** $\left\{\dfrac{1}{3}\right\}$ **13.** $\{-1, -2\}$ **15.** $\left\{\dfrac{7}{2}, 1\right\}$

17. $\{14, -1\}$ **19.** $\left\{1, \dfrac{3}{2}\right\}$ **21.** $\{3, -3\}$ **23.** $\left\{\dfrac{1}{3}, -\dfrac{1}{2}\right\}$

Problems, pages 236–237

1. 4 kPa **3.** $C_1 = 3\ \mu F$, $C_2 = 15\ \mu F$, $C_3 = 10\ \mu F$ **5.** 5 h **7.** 10.5 km/h **9.** 14 h **11.** 10 km/h **13.** 18 h

Self-Test 3, page 238

1. $\dfrac{3y^4+1}{y^5-y^3}$ **2.** $2x^2+2x-5+\dfrac{13}{3x+5}$

3. $\dfrac{(2y+3)^2}{y(y+1)(y-3)}$ **4.** 2 **5.** $\{-2, 6\}$

6. 10 students

Chapter Review, pages 239–240

1. b **2.** c **3.** d **4.** c **5.** a **6.** b **7.** d **8.** c **9.** a, d **10.** d **11.** b

Mixed Review, pages 241–242

1. 17 **3.** $20x-8y$ **5.** $\{2\}$ **7.** $\{m: m > -3\}$

9. $\{4, -1\}$ **11.** $b = \dfrac{3a^2}{4a-5}$ **13.** $x = -1, y = 3$

15. a. 5 **b.** 29 **17.** $y = -\dfrac{1}{2}x + 3$

19. 1 quarter, 38 dimes

Chapter 7 Rational and Irrational Numbers

Written Exercises, pages 249–250

1.

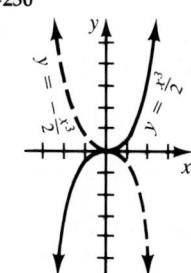

3.

5.

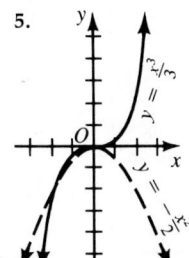

7. $\dfrac{2}{3}$ **9.** -2 **11.** 24

13. 147 **15.** 1, -1

17. The value of the function is multiplied by 8.

19. The value of the function is divided by 4.

21. The value of the function is multiplied by 16.

Problems, page 250
1. 0.3 m   **3.** 6.4 W   **5.** 67.5 W
7. a. 0.00016 J   **b.** 0.0000025   **9.** 14.70 h

Written Exercises, page 253
1. $\{12, -12\}$ **3.** $\{1, -1\}$ **5.** $\{-2\}$ **7.** $\emptyset$
9. $\left\{\frac{7}{4}, -\frac{7}{4}\right\}$ **11.** $\left\{\frac{1}{2}\right\}$ **13.** 3   **15.** 2   **17.** $\{4\}$
19. $\{-6\}$ **21.** $\{x: x \geq 0\}$ **23.** $\{8\}$ **25.** $\{y: y \geq -1\}$
27. $\{x: x \geq 2\}$
29. 1. $x \geq 0, y \geq 0$ .... Hypothesis
$\sqrt{x^2 + y^2} > x + y$
2. $\sqrt{x^2 + y^2} > 0$ .... Def. of principal root
3. $x + y > 0$ .... Closure ax. for $\mathcal{R}_+$
4. $(\sqrt{x^2 + y^2})^2 > (x + y)^2$ .... Ex. 34, page 60
5. $x^2 + y^2 > x^2 + 2xy + y^2$ .... Subs. prin.
6. $0 > 2xy$ .... Add. prop. of order
7. $x < 0$ or $y < 0$ .... Prop. of neg. in prod.
8. Contradiction .... Hypothesis
9. $\therefore \sqrt{x^2 + y^2} \leq x + y$ .... By indirect proof

Written Exercises, pages 255–256
1. $1, -1, 2$ **3.** $1, 2, -2$ **5.** $-1, -\frac{1}{2}, \frac{1}{3}$ **7.** $-2, \frac{1}{3}$
9. $1, -1, 5, -5$ **11.** $1, 4, -\frac{1}{2}$ **13.** $-2, \frac{1}{2}, \frac{3}{2}$
15. $-2, \frac{1}{2}$ **17.** $x^2 - 5 = 0$; possible rational roots
are $\pm 1, \pm 5$; $f(1) = -4, f(-1) = -4, f(5) = 20$,
$f(-5) = 20$ **19.** $x^2 - 7 = 0$; possible rational roots are
$\pm 1, \pm 7$; $f(1) = -6, f(-1) = -6, f(7) = 42$,
$f(-7) = 42$ **21.** $2x^3 - 1 = 0$; possible rational roots
are $\pm 1, \pm \frac{1}{2}$; $f(1) = 1, f(-1) = -3, f\left(\frac{1}{2}\right) = -\frac{3}{4}$,
$f\left(-\frac{1}{2}\right) = -\frac{5}{4}$
23. If $\sqrt{5} + 9$ is rational, $\sqrt{5} + 9 = \frac{a}{b}$, where a,
b are integers, $b \neq 0$. Then $\sqrt{5} = \frac{a - 9b}{b}$. $\frac{a - 9b}{b}$ is
a rational number, but Exercise 17 shows that $\sqrt{5}$
is irrational. This contradiction indicates that $\sqrt{5} + 9$
is irrational.
29. a. Since n is a root of $ax^3 + bx^2 + cx + d = 0$,
then n must satisfy the equation. Thus, $an^3 + bn^2 + cn + d = 0$, or $an^3 + bn^2 + cn = -d$. Factoring on
the left, we have $n(an^2 + bn + c) = -d$. Since $an^2 + bn + c$ and n are both integers, n must be a factor of
$-d$ and, hence, of d.   **b.** By similar reasoning, since
$\frac{1}{n}$ is a root of the equation, $\frac{a}{n^3} = -\frac{b}{n^2} - \frac{c}{n} - d$.

Multiplying both sides by n^3, we have $a = n(-b - cn - dn^2)$. Since n and $(-b - cn - dn^2)$
are both integers, n must be a factor of a.

Self-Test 1, page 256
1. 40   **2. a.** 12   **b.** -3 **3.** $\left\{-\frac{7}{2}\right\}$ **4.** -1

Written Exercises, page 260
1. 3.4×10^4 **3.** 5.1×10^{-4} **5.** 2.785×10^3
7. 2.023×10^1 **9.** 35,000   **11.** 0.0965   **13.** 0.303
15. 6,781,000   **17.** 8×10^5, or 800,000   **19.** 2×10^{15},
or 2,000,000,000,000,000   **21.** 2×10^4, or 20,000
23. 3×10^4, or 30,000   **25.** $a + b \approx 7$, accuracy,
7.1%; $a - b \approx 1$, accuracy, 50%; $ab \approx 10$, accuracy,
5%; $a \div b \approx 1$, accuracy, 50%; ab is most accurate.

On the Calculator, page 261
1. 1.3995×10^{14} **3.** 1.5665×10^{-2} **5.** 3.2088×10^{-8}

Written Exercises, pages 264–265
1. 0.4375   **3.** $0.1\overline{6}$ **5.** $0.74\overline{6}$ **7.** $0.\overline{216}$ **9.** Examples are
a. 1.35   **b.** 1.313113111 . . .   **11.** Examples are **a.** $\frac{71}{90}$
b. 0.7891011 . . .   **13.** $\frac{19}{40}$ **15.** $\frac{634}{125}$ **17.** $\frac{5}{11}$ **19.** $\frac{5}{37}$
21. $\frac{43}{18}$ **23.** $\frac{5}{12}$ **25.** $\frac{55}{27}$ **27.** $-\frac{2}{5}$
29. a. Since a, b, c, and d are positive integers, if
$ad < bc$, then $\frac{1}{bd}(ad) < \frac{1}{bd}(bc)$. Thus, $\frac{a}{b} < \frac{c}{d}$. Also, if
$\frac{a}{b} < \frac{c}{d}$, then $bd\left(\frac{a}{b}\right) < bd\left(\frac{c}{d}\right)$. Thus, $ad < bc$.
b. By Exercise 29(a), if $\frac{a}{b} < \frac{c}{d}$, then $ad < bc$. By
the addition property of order, $ab + ad < ab + bc$,
or $a(b + d) < b(a + c)$. Again by Exercise 29(a),
$\frac{a}{b} < \frac{a + c}{b + d}$.

Self-Test 2, page 265
1. 5.614×10^3 **2.** 8.37×10^{-3} **3.** 3.492×10^1
4. 7×10^6 **5.** $0.\overline{4}$ **6.** $\frac{31}{99}$ **7.** Examples are 0.451 and
0.454454445 . . .

Written Exercises, page 269
1. $8\sqrt{3}$ **3.** $\frac{6\sqrt{2}}{7}$ **5.** 25   **7.** $-\frac{1}{9}$ **9.** $\frac{\sqrt{3}}{2}$ **11.** $\frac{3}{5}$
13. $28\sqrt{3}$; 48.50   **15.** $\frac{4}{3}\sqrt{6}$; 3.27   **17.** $\frac{12\sqrt{3}}{5}$; 4.16
19. $\frac{5}{2}$; 2.50   **21.** $6x^8$ **23.** $2|y|\sqrt[5]{3x^3}$ **25.** $\frac{|a|^3\sqrt{b}}{b^3}$
27. $2c\sqrt[5]{c^2}$ **29.** $\frac{\sqrt{25d^2 - c^2}}{5|cd|}$ **31.** $\frac{\sqrt[3]{x^2y}}{x - y}$ **33.** $\frac{a}{a + b}$
35. It has been proved that $\sqrt[n]{a} \cdot \sqrt[n]{b}$ is one of the
nth roots of ab. Since exactly one of the numbers a, b

is negative, exactly one of $\sqrt[n]{a}$ and $\sqrt[n]{b}$ is negative, and $\sqrt[n]{a} \cdot \sqrt[n]{b}$ is nonpositive. Also $\sqrt[n]{ab}$ is nonpositive since exactly one of a, b is negative. $\therefore \sqrt[n]{ab}$, the single real root of ab (since n is odd), is equal to $\sqrt[n]{a} \cdot \sqrt[n]{b}$.

Written Exercises, page 271
1. $8\sqrt{5}$ 3. $\sqrt{3}$ 5. $20\sqrt[3]{3} - 6\sqrt[3]{2}$ 7. $18b\sqrt{b}$
9. $\sqrt{5x}(10x^2 + 8)$ 11. $-12\sqrt{5}$ 13. $\dfrac{\sqrt{5}}{3}$ 15. $9 -$
$2\sqrt{14}$ 17. 29 19. $3 - \sqrt[3]{9} - 6\sqrt[3]{3}$ 21. $4\sqrt{10} + 12$
23. $7\sqrt{3} - 13$ 25. 2 27. $a + b$ 29. $2 + 2\sqrt[3]{2} +$
$\sqrt[3]{4}$ 31. $\sqrt[3]{36} + 2\sqrt[3]{3} + 2\sqrt[3]{2}$
33. $(x\sqrt{3} - y)(x\sqrt{3} + y)$ 35. $(x + \sqrt{5})^2$
37. $(a - \sqrt[3]{10})(a^2 + a\sqrt[3]{10} + \sqrt[3]{100})$
39. $(x + 5\sqrt{2})^2$
43. $(x^2 - x\sqrt{2} + 1)(x^2 + x\sqrt{2} + 1)$

Problems, pages 272–273
1. 14 A   3. 62.2 Hz   5. 2.8 cm   7. $48°$ 9. 0.07

Written Exercises, page 275
1. $\{46\}$ 3. $\left\{\dfrac{27}{5}\right\}$ 5. $\left\{\dfrac{25}{2}\right\}$ 7. $\{128\}$ 9. $\{4\}$
11. $\{72\}$ 13. $\{4\}$ 15. $\{-1\}$ 17. $\left\{-\dfrac{13}{24}\right\}$
19. $\{16\}$ 21. $\varnothing$ 23. $\{5\}$ 25. $\{4, 12\}$ 27. $\{14, -2\}$
29. $\left\{\dfrac{1}{4}\right\}$ 31. $\{9\}$ 33. $\{0, 8\}$

Self-Test 3, page 275
1. $-19\sqrt{7}$ 2. $\sqrt[3]{4xy}(2y - 3x)$ 3. $5\sqrt{5} - 2$
4. $\dfrac{18 + 3\sqrt{7}}{29}$ 5. $\{40\}$ 6. $\varnothing$

Written Exercises, pages 278–279
1. -1 3. 1 5. $5i\sqrt{2}$ 7. $21i\sqrt{2}$ 9. $\dfrac{\sqrt{3}}{4}i$
11. $-\dfrac{\sqrt{10}}{4}i$ 13. $-15\sqrt{6}$ 15. 2 17. $\dfrac{-3\sqrt{10}}{5}i$
19. $-\dfrac{3}{2}i$ 21. -13 23. 8 25. $17i\sqrt{3}$ 27. $18i\sqrt{3}$
29. $-2i$ 31. $i\left(\dfrac{2}{5}\sqrt{5} - \dfrac{1}{10}\sqrt{30}\right)$ 33. $i\left(\dfrac{1}{4}\sqrt{3} - \dfrac{1}{3}\sqrt{15}\right)$
35. Let $r = -a$ and $s = -b$. Then $r, s > 0$ and
$-r = a, -s = b, (-1)r = -r$, and $(-1)s = -s$.
Then $\sqrt{a} = \sqrt{(-1)r} = i\sqrt{r}; \sqrt{b} = \sqrt{(-1)s} =$
$i\sqrt{s}$. Thus $\dfrac{\sqrt{a}}{\sqrt{b}} = \dfrac{i\sqrt{r}}{i\sqrt{s}} = \sqrt{\dfrac{r}{s}}$. Also $\dfrac{a}{b} = \dfrac{-r}{-s} = \dfrac{r}{s}$
$\therefore \sqrt{\dfrac{a}{b}} = \sqrt{\dfrac{r}{s}} = \dfrac{\sqrt{a}}{\sqrt{b}}$ 37. $\dfrac{i\sqrt{ab}}{a}$ and $-\dfrac{i\sqrt{ab}}{a}$

Written Exercises, page 282
1. $-2 + 2i$ 3. $-1 - 13i$ 5. $9 - 15i$ 7. $-\dfrac{1}{2} + 2i$

9. $\dfrac{7}{24} - \dfrac{3}{4}i$ 11. $1 + \dfrac{9}{4}i$ 13. 6 15. $3i$ 17. $-\dfrac{11}{24} + \dfrac{1}{2}i$
19. $x = 5, y = -3$ 21. $x = 3, y = -1$
23. $z = a + bi$ so $\bar{z} = a - bi$. But $z = \bar{z}$ so $a + bi =$
$a - bi$. Thus $a = a, bi = -bi$ or $2bi = 0$ so $b = 0$
$\therefore z = a + 0i = a$, a real number.   25. $z = a + bi$
so $z + 0 + 0i = (a + 0) + (b + 0)i = a + bi = z$.
Also, $0 + 0i + z = (0 + a) + (0 + b)i = a + bi = z$.
27. For all real numbers a and b, $(a + bi) + (a - bi) =$
$(a + a) + (b - b)i = 2a + 0 \cdot i = 2a$.
Also $(a + bi) - (a - bi) = (a - a) + (b + b)i = 2bi$.

Written Exercises, page 284
1. $19 - 13i$ 3. $21 - 20i$ 5. 1 7. $-\dfrac{1}{2} - \dfrac{\sqrt{3}}{2}i$
9. $-5i$ 11. $\dfrac{10}{37} + \dfrac{51}{37}i$ 13. $\dfrac{183}{221} + \dfrac{356}{221}i$ 15. -1
17. $(2y + 7i)(2y - 7i)$ 19. $(4a\sqrt{5} + bi\sqrt{15}) \cdot$
$(4a\sqrt{5} - bi\sqrt{15})$ 21. $(z\sqrt{2} + 3i\sqrt{3})(z\sqrt{2} - 3i\sqrt{3})$
23. $\dfrac{a + bi}{c + di} = \dfrac{a + bi}{c + di} \cdot \dfrac{c - di}{c - di} =$
$\dfrac{(ac + bd) + (bc - ad)i}{c^2 + d^2} = \dfrac{ac + bd}{c^2 + d^2} + \dfrac{bc - ad}{c^2 + d^2}i$
25. $\left(x - \dfrac{\sqrt{2}}{2} + \dfrac{\sqrt{2}}{2}i\right)\left(x + \dfrac{\sqrt{2}}{2} - \dfrac{\sqrt{2}}{2}i\right) \cdot$
$\left(x + \dfrac{\sqrt{2}}{2} + \dfrac{\sqrt{2}}{2}i\right)\left(x - \dfrac{\sqrt{2}}{2} - \dfrac{\sqrt{2}}{2}i\right)$

Computer Exercises, page 285
3. 1 5. $-i$ 7. $32i$ 9. $-278 + 29i$
13. $4 + i$ 15. $-1.423077 + 0.1153846i$

Self-Test 4, page 285
1. $10i$ 2. $8i\sqrt{3}$ 3. $3i\sqrt{6}$ 4. $i\sqrt{17}$ 5. $-5 + i$
6. $-6 + i$ 7. $26 - 7i$ 8. $\dfrac{8}{61} + \dfrac{27}{61}i$ 9. $-32 - 24i$
10. $\dfrac{1}{5} + \dfrac{1}{10}i$

Chapter Review, pages 287–288
1. b   2. a   3. c   4. c   5. d   6. b   7. a   8. d   9. b
10. a   11. b   12. d   13. a   14. d

Preparing for College Entrance Exams, page 289
1. C   3. D   5. D   7. E

Chapter 8   Sequences and Series

Written Exercises, pages 295–296
1. $2, -6, 18, -54$ 3. $\dfrac{5}{2}, 2, \dfrac{3}{2}, 1; d = -\dfrac{1}{2}$ 5. $-4,$
$-4 + k, -4 + 2k, -4 + 3k; d = k$ 7. $2, 5, 10, 17$
9. $5, 2, -1, -4; d = -3$ 11. $5, 10, 20, 40$
13. $a_1 = 2, a_{n+1} = a_n - 6$ 15. $a_1 = k, a_{n+1} = a_n k^2$
17. $a_1 = -\dfrac{1}{2}, a_{n+1} = a_n + 3$ 19. $a_n = n^2$
21. $a_n = 11\dfrac{1}{2} - \dfrac{7}{2}n$ 23. $a_n = 2 \cdot 5^{n-1}$

25. $a_n = \dfrac{n(n-1)}{2} + 1$; $a_1 = 1$, $a_{n+1} = a_n + n$

27. $a_n = \dfrac{n}{n+1}$; $a_1 = \dfrac{1}{2}$, $a_{n+1} = a_n + \dfrac{1}{(n+1)(n+2)}$

29. Since $d = a_2 - a_1$ and $d = a_3 - a_2$, $a_3 - a_2 = a_2 - a_1$. Thus, $a_1 + a_3 = 2a_2$ $\therefore \dfrac{a_1 + a_3}{2} = a_2$ **31.** $x = 2$

Written Exercises, pages 298–299

1. -28 **3.** 47 **5.** 77 **7.** $-\dfrac{65}{6}$ **9.** 404 **11.** -3

13. -5 **15.** -49 **17.** 19 **19.** 27 **21.** $28, 49, 70$ **23.** $7.5, 12, 16.5$ **25.** $-16.5, -15, -13.5, -12,$ $-10.5, -9, -7.5, -6, -4.5$ **27.** $-25, 3$ **29.** $-38,$ 7 **31.** $-16, 2.5$ **33.** -6 **35.** 0 **37.** 3 **39.** 10 **41.** 42nd

Problems, pages 299–300

1. 105 apples **3.** 11.2 cm **5.** 5:53 A.M.; March 16 **7.** 1992

Written Exercises, page 303

1. 84 **3.** -1600 **5.** 3354 **7.** 649 **9.** -154 **11.** 684 **13.** 20 **15.** 624 **17.** -329 **19.** 522

21. $\displaystyle\sum_{i=1}^{4} 2i + 7$ **23.** $\displaystyle\sum_{i=1}^{5} 4i - 9$ **25.** $\displaystyle\sum_{i=1}^{19} 2i + 2$

27. $-\dfrac{3}{2}$ **29.** 45 **31.** -12 **33.** 7455 **35.** 99

37. $a_1 = 1$, $d = 2$, $S_n = \dfrac{n}{2}[2 \cdot 1 + (n-1)2] = \dfrac{n}{2}(2 + 2n - 2) = n^2$

Problems, pages 303–305

1. \$4260 **3.** Kate with \$197,500; Kate with \$495,000 **5.** 8 s **7.** 325 segments **9.** 10 h

Self-Test 1, page 305

1. a. $a_1 = 4$; $a_{n+1} = a_n + 4$ **b.** $a_n = 4n$ **2. a.** $a_1 = 3$; $a_{n+1} = 3a_n$ **b.** $a_n = 3^n$ **3.** 98 **4.** 2 **5.** 4 **6.** $0, 2, 4, 6, 8$ **7.** -306 **8.** 310 **9.** March 17 **10.** \$13,500

Written Exercises, pages 309–310

1. $5, 25, 125, 625$ **3.** $40, 20, 10, 5$ **5.** $\dfrac{c^2}{d}, 2c, 4d, \dfrac{8d^2}{c}$ **7.** $\dfrac{5}{16}$ **9.** $\dfrac{2}{9}$ **11.** $\dfrac{c^3}{b^4}$ **13.** $\dfrac{1}{48}$ **15.** $\dfrac{2}{25}$

17. $\dfrac{1}{30,000}$ **19.** $\dfrac{3}{64}$ **21.** Arithmetic; 56

23. Geometric; $\dfrac{1}{450}$ **25.** Arithmetic; 2.72

29. $\dfrac{1}{r}$ **31.** r^2

Problems, pages 310–311

1. 243 neutrons **3.** 10,368 insects **5.** 96 great-great-grandparents **7.** 0.125 mg **9. a.** \$9600 **b.** \$9680 **c.** \$9724.05 **11.** \$1221.02

Written Exercises, pages 314–315

1. 3 **3.** $-\dfrac{1}{2}$ **5.** 4 or -4 **7.** $\dfrac{15}{4}, \dfrac{15}{2}$ **9.** $-\dfrac{5}{4}, -\dfrac{5}{2}$ **11.** $12,500; 2500$ **13.** $4, 20, 100, 500$ **15.** $-5, 15,$ $-45, 135$ **17.** $\dfrac{7}{4}, \dfrac{7}{2}, 7, 14, 28$; $\dfrac{7}{4}, -\dfrac{7}{2}, 7, -14, 28$

19. $4, 6, 9, \dfrac{27}{2}, \dfrac{81}{4}$; $4, -6, 9, -\dfrac{27}{2}, \dfrac{81}{4}$ **21.** 5

23. -32 **25.** $\dfrac{x^9}{y^5}$ **27.** $\dfrac{5}{4}$ **29.** 3 and -1

31. The common ratio of the given sequence squared, or $a_1{}^2$, $a_1{}^2r^2$, $a_1{}^2r^4$, $a_1{}^2r^6$, . . . is r^2 so the sequence is geometric.

Written Exercises, pages 317–318

1. $47\dfrac{5}{8}$ **3.** $485\dfrac{1}{3}$ **5.** -43 **7.** 341 **9.** 1275

11. 728 **13.** $-7\dfrac{15}{16}$ **15.** $26\dfrac{8}{9}$ **17.** $73\dfrac{8}{9}$ **19.** $\dfrac{131}{21}$

21. 135 **23.** 4 **25.** 1.75 **27.** $\dfrac{3}{2}$ or $-\dfrac{5}{2}$ **29.** $\dfrac{1}{3}$

Problems, pages 318–319

1. 126 **3.** 6.25 cm **5. a.** $\dfrac{243}{1024}$ **b.** $\dfrac{781}{1024}$ **7.** 4 s

Computer Exercises, page 320

3. 3071.25 **5.** -520.8336

Self-Test 2, page 320

1. 3072 **2.** $\dfrac{1}{3}$ **3.** 1000 or -1000 **4.** $30, 60, 120,$ 240 **5.** 4100 **6.** $\dfrac{182}{243}$ **7.** $29,282$ **8.** \$16,383

Written Exercises, pages 323–324

1. $4, 3\dfrac{1}{2}, 3\dfrac{1}{3}, 3\dfrac{1}{4}$; 3 **3.** $\dfrac{1}{20}, \dfrac{1}{5}, \dfrac{9}{20}, \dfrac{4}{5}$; not convergent

5. $-\dfrac{2}{3}, \dfrac{4}{5}, -\dfrac{8}{9}, \dfrac{16}{17}$; not convergent **7.** $1, \dfrac{1}{2}, \dfrac{1}{3}, \dfrac{1}{4}$; $\dfrac{1}{n}$ **9.** $\dfrac{1}{2}, \dfrac{1}{4}, \dfrac{1}{6}, \dfrac{1}{8}; \dfrac{1}{2n}$ **11.** $\dfrac{1}{10}, \dfrac{1}{18}, \dfrac{1}{26}, \dfrac{1}{34}; \dfrac{1}{8n+2}$

13. $n = 11$; $\dfrac{1}{11}$ **15.** $n = 6$; $\dfrac{1}{12}$ **17.** $n = 2$; $\dfrac{1}{18}$

Written Exercises, pages 327–328

1. 54 **3.** 36 **5.** $\dfrac{16}{21}$ **7.** Divergent **9.** $0.\overline{8}$ or $\dfrac{8}{9}$

11. 6 **13.** $\dfrac{10}{21}$ **15.** 10 **17.** 30 **19.** $-\dfrac{1}{6}$

21. $0.5 + 0.5(0.1) + 0.5(0.1)^2 + \cdots, \dfrac{5}{9}$

23. $0.21 + 0.21(0.01) + 0.21(0.01)^2 + \cdots, \dfrac{7}{33}$

25. $0.117 + 0.117(0.001) + 0.117(0.001)^2 + \cdots, \dfrac{13}{111}$

27. $0.03 + 0.03(0.1) + 0.03(0.1)^2 + \cdots, \dfrac{1}{30}$

29. $\frac{2}{5}$ **31.** $\frac{5}{6}$ **33.** $\frac{1}{3}$

Problems, pages 328–330

1. 160 cm **3.** 120 mm **5.** $\frac{3}{4}$ **7.** none **9.** $\frac{2\sqrt{3}}{5}$

Computer Exercises, page 330

3. $|r| \geq 1$

Self-Test 3, page 331

1. 3, 2, $\frac{5}{3}, \frac{3}{2}$; 1; $\frac{2}{n}$ **2.** $2\frac{2}{3}, 2\frac{8}{9}, 2\frac{26}{27}, 2\frac{80}{81}$; 3; $\frac{1}{3^n}$

3. $\frac{27}{4}$ **4.** $\frac{4}{5}$ **5.** $0.213 + 0.213(0.001) +$

$0.213(0.001)^2 + \cdots, \frac{71}{333}$

Chapter Review, pages 332–333

1. b **2.** d **3.** a **4.** c **5.** d **6.** b **7.** c **8.** a
9. c **10.** b **11.** c **12.** a **13.** d

Cumulative Review, pages 334–336

3. ∅ **5.** −68 **7.** $-r^8 s^5$ **9.** $-\frac{7}{2}x^3$

13. $\frac{33 - 5a}{(2 + a)(2 - a)}$ **15.** $\left\{\frac{-3}{2}, -1\right\}$ **17.** $\{-6, 2\}$

19. 11 cm, 17 cm **21.** 3 **23.** $\frac{25}{6}$ **25.** $-1 - \sqrt{3}$

27. $\left\{\frac{2}{3}, -\frac{2}{3}\right\}$ **29.** $\{12\}$ **31.** $8 - 10i$ **33.** $-40 + 20i$

35. −1, 2, −4, 8 **37.** 22 **39.** 15 **41.** 4 **43.** $27\frac{1}{9}$

Application, page 337

1. a. Exposure is doubled. **b.** Exposure is halved.

Chapter 9   Polynomial Functions

Written Exercises, pages 345–346

1. $\{5 + 3\sqrt{3}, 5 - 3\sqrt{3}\}$ **3.** $\left\{\frac{5 + i\sqrt{3}}{2}, \frac{5 - i\sqrt{3}}{2}\right\}$

5. $\left\{\frac{1 + \sqrt{6}}{5}, \frac{1 - \sqrt{6}}{5}\right\}$ **7.** $\left\{\frac{4 + i\sqrt{2}}{2}, \frac{4 - i\sqrt{2}}{2}\right\}$

9. $\left\{\frac{-8 + 2\sqrt{10}}{3}, \frac{-8 - 2\sqrt{10}}{3}\right\}$ **11.** $\{3 + \sqrt{5},$

$3 - \sqrt{5}\}$ **13.** $\left\{\frac{4 + i\sqrt{6}}{2}, \frac{4 - i\sqrt{6}}{2}\right\}$ **15.** $\left\{0, \frac{6}{7}\right\}$

17. $\left\{\frac{5 + \sqrt{15}}{2}, \frac{5 - \sqrt{15}}{2}\right\}$ **19.** $\{1 + 2i\sqrt{11},$

$1 - 2i\sqrt{11}\}$ **21.** $\left\{\frac{-6 + i\sqrt{14}}{2}, \frac{-6 - i\sqrt{14}}{2}\right\}$

23. $\left\{\frac{6 + 2\sqrt{2}}{7}, \frac{6 - 2\sqrt{2}}{7}\right\}$

25. $\left\{\frac{1 + i\sqrt{2}}{3}, \frac{1 - i\sqrt{2}}{3}\right\}$

27. $\left\{\frac{3 + \sqrt{2}}{2}, \frac{3 - \sqrt{2}}{2}\right\}$ **29.** $\{-\sqrt{3} + 2i,$

$-\sqrt{3} - 2i\}$ **31.** $\left\{1 + \frac{\sqrt{6}}{2}, 1 - \frac{\sqrt{6}}{2}\right\}$ **33.** $\{2, -2\}$

45. a. $(x^2 + 2x + 2)(x^2 - 2x + 2)$
b. $(x^2 + x\sqrt{2} + 1)(x^2 - x\sqrt{2} + 1)$

Problems, pages 346–347

1. a. $\frac{10 + \sqrt{30}}{7}$ s; $\frac{10 - \sqrt{30}}{7}$ s **b.** 2.2 s; 0.6 s

3. a. $17 + \sqrt{14}$ in.; $17 - \sqrt{14}$ in. **b.** 20.7 in.;
13.3 in. **5. a.** $6 + \sqrt{30}$ cm; $5 + \sqrt{30}$ cm

b. 11.5 cm; 10.5 cm **7. a.** $\frac{1 + 4\sqrt{3}}{2}$ cm **b.** 4.0 cm

9. a. $3 + 2\sqrt{11}$ m by $12 - 2\sqrt{11}$ m **b.** 9.6 m by 5.4 m

Written Exercises, pages 349–350

1. −3; 2 imaginary roots **3.** 1; 2 real rational roots
5. 89; 2 real irrational roots **7.** 0; one double root

9. 1728; 2 real irrational roots **11.** $2\frac{1}{2}$ **13.** 3, −2

15. $\{k: k < 3\}$ **17.** $\left\{k: k > \frac{17}{2}\right\}$

19. $c = 0$; $D = b^2 - 4ac = b^2 - 4 \cdot a \cdot 0 = b^2$; D is
the square of a rational number and the coefficients
of the equation are rational. Therefore the equation
has rational roots.

Written Exercises, page 352

1. $x^2 - 5x - 14 = 0$ **3.** $4x^2 + 12x + 9 = 0$
5. $x^2 - 72 = 0$ **7.** $x^2 - 6x + 14 = 0$
9. $2x^2 - 6x + 7 = 0$ **11.** $x^2 - 14x + 37 = 0$

13. −2 **15.** −11 **17.** $\frac{1}{5}$

Computer Exercises, page 353

3. $\{-2.5\}$ **5.** $\{-0.3333, -2\}$
7. $x^2 + 2x - 15 = 0$ **11.** $x^2 + 6x + 25 = 0$
15. $4x^2 - 12x + 13 = 0$ **17.** $100x^2 + 100x + 61 = 0$

Self-Test 1, pages 353–354

1. $\left\{\frac{3 + 3i}{2}, \frac{3 - 3i}{2}\right\}$ **2.** $\left\{\frac{-5 + \sqrt{35}}{5}, \frac{-5 - \sqrt{35}}{5}\right\}$
3. $\{3 + 2i\sqrt{5}, 3 - 2i\sqrt{5}\}$ **4.** 2 complex conjugate
roots **5.** 1 double rational root **6. a.** $\frac{5}{2}$ **b.** $\frac{3}{2}$

7. $x^2 - 4x + 29 = 0$

Written Exercises, pages 356–357

1. $y = x^2 + 3$ **5.**

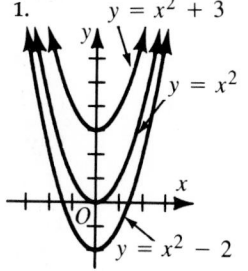
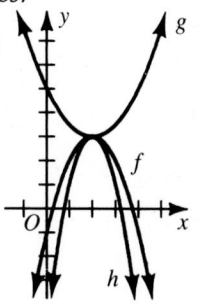

9. $y = 2(x - 1)^2 + 3$ **11.** $y = -\frac{1}{2}(x + 1)^2 + 4$

13. $y = 3(x + 3)^2 - 4$ **15.** $y = \frac{3}{2}x^2 - 5$

17. $y = (x - 5)^2 + 4$ or $y = (x + 1)^2 + 4$
19. $y = -2(x - 3)^2 + 7$ **21.** $y = 3x^2 - 16$
23. Assume $(h + r, s)$ is on the graph of
$y = a(x - h)^2 + k, a \neq 0$.
Then $y = a(h - r - h)^2 + k = ar^2 + k$. But
$s = a(h + r - h)^2 + k = ar^2 + k = y$.
So $y = s$ when $x = h - r$.

Written Exercises, pages 360–361
1. $x = 5$; $(5, -25)$ **5.** $x = 0$; $(0, -6)$

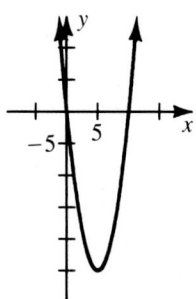
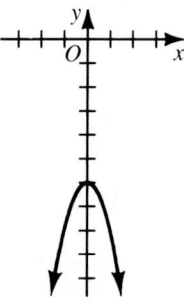

13. $y = 3x^2 - 4x + 2$ **15.** $y = \frac{5}{2}x^2 - \frac{3}{2}x - 4$
17. $y = rx^2 - rs$; $x = 0$; $(0, -rs)$ **19.** $y = r(x - s)^2 - (rs^2 + 2s^2)$; $x = s$; $(s, -rs^2 - 2s^2)$

Problems, pages 361–362
1. 12, 12 **3.** 18 m **5.** 74.1 m **7.** 6 m **9.** The wire should be cut at its midpoint. **11. a.** $y = \dfrac{24 - 2x}{3}$
b. $\dfrac{24x - 2x^2}{3}$ **c.** 6

Written Exercises, page 364
1. a. $x^2 - 16 > 0$ **3. a.** $5x - x^2 < 0$
  b. $\{4, -4\}$ **b.** $\{0, 5\}$
  c.   **c.**

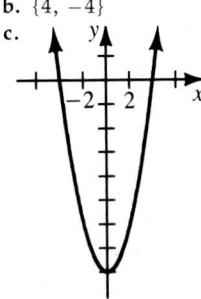
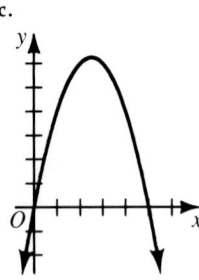

  d. $\{x: x > 4\} \cup$ **d.** $\{x: x < 0\} \cup$
  $\{x: x < -4\}$ $\{x: x > 5\}$
5. a. $x^2 - x - 12 \geq 0$ **b.** $\{-3, 4\}$
  d. $\{x: x \leq -3\} \cup \{x: x \geq 4\}$
7. a. $x^2 - 6x + 4 \leq 0$ **b.** $\{3 + \sqrt{5}, 3 - \sqrt{5}\}$
  d. $\{x: 3 - \sqrt{5} \leq x \leq 3 + \sqrt{5}\}$
9. a. $x^2 + 4x - 1 > 0$ **b.** $\{-2 + \sqrt{5}, -2 - \sqrt{5}\}$
  d. $\{x: x < -2 - \sqrt{5}\} \cup \{x: x > -2 + \sqrt{5}\}$
11. $x^2 - 5x \leq 0$ and $x^2 - 5x + 6 \geq 0$
  $\{x: 0 \leq x \leq 2\} \cup \{x: 3 \leq x \leq 5\}$

Self-Test 2, page 364
1. $x = 2$; $(2, 3)$ **2.** maximum   **3.**

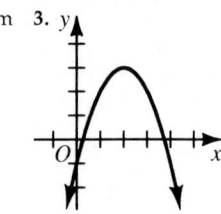

4. $y = 3(x + 1)^2 - 2$ **5.** 18 cm²
  6. $\{x: x < -1 - \sqrt{3}\} \cup$
  $\{x: x > -1 + \sqrt{3}\}$

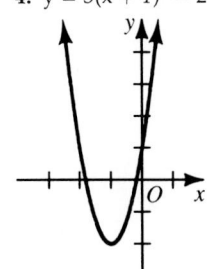
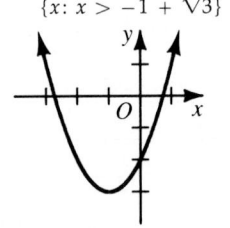

Written Exercises, pages 367–368
1. 0; a zero **3.** 0; a zero **5.** 20 **7.** 0; a zero
9. 60 **11.** -12 **13.** 0; a zero **15.** $30 + 114i$
17. 48 **19.** 0; a zero **21.** -1 **23.** -9
25. $\{-5, 1, 2\}$ **27.** $\left\{-2, 1, \dfrac{3}{2}\right\}$ **29.** $a = -2, b = -3$

Computer Exercises, page 368
3. -2 **7. a.** $-76 - 168i$ **b.** $-76 + 168i$

Written Exercises, pages 371–372
1. a. $3x^3 - 8x^2 + 5x + 6 = (x - 2) \cdot (3x^2 - 2x + 1) + 8$ **b.** $3x^3 - 8x^2 + 5x + 6 = (x + 1)(3x^2 - 11x + 16) - 10$ **3. a.** $4x^3 + 8x^2 - 7x + 5 = (x + 3)(4x^2 - 4x + 5) - 10$ **b.** $4x^3 + 8x^2 - 7x + 5 = \left(x - \dfrac{1}{2}\right)(4x^2 + 10x - 2) + 4$
5. a. $\dfrac{x^3 - 13x + 18}{x + 4} = (x^2 - 4x + 3) + \dfrac{6}{x + 4}$
  b. $\dfrac{x^3 - 13x + 18}{x - 2} = x^2 + 2x - 9$
7. a. $\dfrac{x^4 - 4x^3 - 3x - 9}{x - 5} = x^3 + x^2 + 5x + 22 + \dfrac{101}{x - 5}$ **b.** $\dfrac{x^4 - 4x^3 - 3x - 9}{x - 2i} = x^3 + (-4 + 2i)x^2 - (4 + 8i)x + 13 - 8i + \dfrac{7 + 26i}{x - 2i}$
9. yes **11.** no; $R = -10$ **13.** no; $R = -2$
15. $r_2 = -3$; $r_3 = 4$ **17.** $r_2 = -\dfrac{3}{2}$; $r_3 = -1$
19. $r_2 = 3i$; $r_3 = -3i$ **21.** $r_2 = -3$; $r_3 = 1$; $r_4 = 3$ **23.** $\{-2, -1, 2\}$
25. $\left\{2, \dfrac{3 + i\sqrt{7}}{2}, \dfrac{3 - i\sqrt{7}}{2}\right\}$ **27.** $\{-2, 3, i, -i\}$
29. 6 **31.** -3

Reading Algebra, page 373
1. the set **3.** equals approximately **5.** $|c|$

Written Exercises, page 375

1. $x^3 - 4x^2 + x + 6 = 0$ **3.** $x^3 - 4x^2 + 9x - 36 = 0$
5. $x^4 - x^2 - 12 = 0$ **7.** $2 - 3i$;
$(x + 1)(x - 2 - 3i)(x - 2 + 3i)$ **9.** $-i\sqrt{5}, 3$;
$(x + i\sqrt{5})(x - i\sqrt{5})(x - 3)$ **11.** $3 - i, -4$;
$(x - 3 - i)(x - 3 + i)(x + 4)$ **13.** $-2 - i, \pm i\sqrt{6}$;
$(x + 2 + i)(x + 2 - i)(x + i\sqrt{6})(x - i\sqrt{6})$

Self-Test 3, page 376

1. a. 75 **b.** $-3 + i$

2. $\dfrac{2x^3 - 5x^2 + 3x + 4}{x - 3} = 2x^2 + x + 6 + \dfrac{22}{x - 3}$

3. $\left\{-1, \dfrac{5}{2}, 2\right\}$ **4.** $\{2, 3 + 2i, 3 - 2i\}$

Written Exercises, page 379

1.

pos.	neg.	imag.
2	1	0
0	1	2

3.

pos.	neg.	imag.
3	1	0
1	1	2

7. $M = 1; L = -2$ **9.** $M = 3; L = -2$ **11.** $M = 3$;
$L = -2$ **13.** $M = 3; L = -2$

Written Exercises, page 382

1. $r_1 \approx -3; r_2 \approx 0; r_3 \approx 3$

3. $r_1 \approx -2; r_2 \approx \dfrac{1}{2}; r_3 \approx 3$

5. $-2 < r_1 < -\dfrac{3}{2}, 0 < r_2 < \dfrac{1}{2}, \dfrac{3}{2} < r_3 < 2$

9. $-2 < r_1 < -\dfrac{3}{2}, \dfrac{1}{2} < r_2 < 1, 2 < r_3 < \dfrac{5}{2}$

11. $2 < r < \dfrac{5}{2}$ **13.** $r \approx 2.22$

Self-Test 4, page 383

1.

pos.	neg.	imag.
3	0	2
1	2	2
1	0	4
3	2	0

2. $M = 3, L = -1$

3. $-\dfrac{3}{2} < r_1 < -1$,

$\dfrac{1}{2} < r_2 < 1, \dfrac{5}{2} < r_3 < 3$

Chapter Review, pages 384–385

1. c **2.** c **3.** d **4.** c **5.** d **6.** b **7.** c **8.** b
9. c **10.** a **11.** c **12.** d

Application, page 387

1. $f_2 = 2f_1$

Preparing for College Entrance Exams, page 388

1. E **3.** D **5.** B **7.** E

Chapter 10 Quadratic Relations and Systems

Written Exercises, pages 396–397

1. 5 **3.** 5 **5.** $\sqrt{5}$ **7.** $\dfrac{15}{2}$ **9.** $\sqrt{2}|a - b|$
11. $\sqrt{(0 + 2)^2 + (4 - 7)^2} = \sqrt{(0 - 2)^2 + (4 - 1)^2}$

13. a.

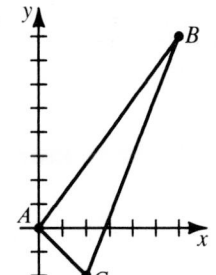

b. Not isosceles
c. Not right

15. b. Not isosceles **c.** Right: $AB = 2\sqrt{10}$,
$BC = \sqrt{10}, AC = 5\sqrt{2}; (AB)^2 + (BC)^2 = (AC)^2$
17. $(-3, 2)$ **19.** $(1, 5)$ **21.** 13 **23.** $P_1P_2 = $
$\sqrt{(b - a)^2 + (c - c)^2} = \sqrt{(b - a)^2} = |a - b|$
27. $4x - 3y = 1$
29. $AB = (s - r)\sqrt{1 + m^2}; BC = (t - s)\sqrt{1 + m^2}$;
$AC = (t - r)\sqrt{1 + m^2}; AC = AB + BC$

Written Exercises, page 399

1. $y = -3x + 1$ **3.** $3x + 4y = -20$
5. $y = -4x - 9$ **7.** $15x + 18y = -44$
9. $x + 6y = 8$ **11.** $5x - 3y = -10$

Computer Exercises, pages 399–400

3. Rectangle **5.** Quadrilateral

Self-Test 1, page 400

1. $\sqrt{101}$ **2.** $\left(\dfrac{9}{2}, 3\right)$ **3.** $y = \dfrac{1}{2}x$

Written Exercises, pages 402–403

1. $x^2 + y^2 - 16 = 0$ **3.** $x^2 + y^2 + 6x -$
$14y + 54 = 0$ **5.** $x^2 + y^2 - 3x + 5y + \dfrac{9}{2} = 0$
7. $x^2 + y^2 - 2ax - 2by + a^2 = 0$
9. $(x - 0)^2 + (y - 0)^2 = 6^2$
$C(0, 0)$;
$r = 6$

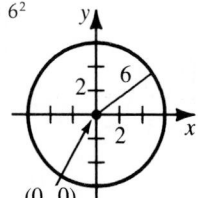

11. $(x - 0)^2 + (y - (-2))^2 = 2^2; C(0, -2); r = 2$
13. $(x - (-2))^2 + (y - 6)^2 = 7^2; C(-2, 6); r = 7$
15. $\left(x - \dfrac{5}{2}\right)^2 + \left(y - \left(-\dfrac{1}{2}\right)\right)^2 = 3^2; C\left(\dfrac{5}{2}, -\dfrac{1}{2}\right); r = 3$

17.

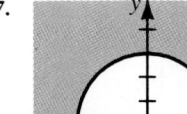

19.

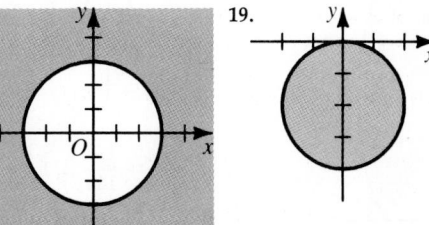

21. $(x - 5)^2 + (y - 6)^2 = 9$ **23.** $(x + 1)^2 +$ $(y - 5)^2 = 25$ **25.** $x^2 + y^2 - 6x + 4y - 12 = 0$
27. $x^2 + y^2 - 2kx = 0$
33. $(x - 4)^2 + (y + 1)^2 = 16$

29. $\dfrac{x^2}{36} + \dfrac{y^2}{49} = 1$ **31.** $\dfrac{(x - 4)^2}{25} + \dfrac{(y + 7)^2}{9} = 1$

33. $\dfrac{x^2}{36} + \dfrac{y^2}{27} = 1$ **35.** $\dfrac{x^2}{a^2} + \dfrac{y^2}{a^2 - c^2} = 1$

Computer Exercises, page 404
3. $C(5, 0)$; $r = 5$ **5.** $\varnothing$

Written Exercises, pages 407–408
1. $y^2 = 40x$ **3.** $x^2 = -2y$ **5.** $x^2 = 8y$

7. $y = \dfrac{1}{2}(x - 0)^2 - 3$ **9.** $x = (y - 1)^2 - 4$

11. $y = \dfrac{1}{4}(x + 6)^2 - 9$

13. $y = 2(x - 2)^2 - 3$

15. $x = \dfrac{1}{3}(y - 1)^2 + 3$

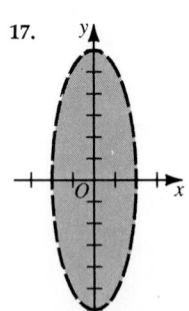

17. $y = (x - 3)^2 + 2$, $x \geq 3$

19. $y = \dfrac{1}{12}x^2 - 2$

21. $x = \dfrac{1}{8}y^2 - y + 3$

23. $y = \dfrac{1}{2}x^2 - 4x + 10$

25. $y = -\left(x + \dfrac{1}{2}\right)^2 + \dfrac{1}{4}$

27. $p = \dfrac{1}{4a}$

Written Exercises, pages 411–412
1. Foci: $(0, 3)$, $(0, -3)$ **3.** Foci: $(0, 2\sqrt{2})$, $(0, -2\sqrt{2})$
5. Foci: $(5, 0)$, $(-5, 0)$
7. Foci: $(-2\sqrt{5}, 0)$, $(2\sqrt{5}, 0)$
9. Foci: $(0, 10\sqrt{2})$, $(0, -10\sqrt{2})$
11. Foci: $\left(\dfrac{8}{3}, 0\right)$, $\left(-\dfrac{8}{3}, 0\right)$
13. Foci: $(3 + \sqrt{7}, 2)$, $(3 - \sqrt{7}, 2)$

17.

21.

25. $\dfrac{x^2}{169} + \dfrac{y^2}{144} = 1$

27. $\dfrac{x^2}{9} + \dfrac{y^2}{25} = 1$

Written Exercises, pages 416–417
1. Foci: $(5, 0)$, $(-5, 0)$ **3.** Foci: $(0, 5\sqrt{5})$, $(0, -5\sqrt{5})$
5. Foci: $(0, 7\sqrt{2})$, $(0, -7\sqrt{2})$
7. Foci: $(5\sqrt{2}, 0)$, $(-5\sqrt{2}, 0)$
9. Foci: $\left(0, \dfrac{13}{2}\right)$, $\left(0, -\dfrac{13}{2}\right)$
11. Foci: $(6, 0)$, $(-6, 0)$
13. Foci: $(2 + \sqrt{41}, -1)$, $(2 - \sqrt{41}, -1)$
15. Foci: $(-4, 3 + \sqrt{13})$, $(-4, 3 - \sqrt{13})$

17.

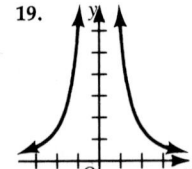
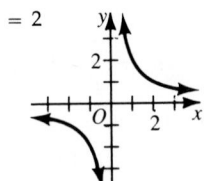

25. $\dfrac{y^2}{64} - \dfrac{x^2}{36} = 1$

27. $\dfrac{(x - 3)^2}{9} - \dfrac{(y - 1)^2}{16} = 1$

29. $\dfrac{x^2}{81} - \dfrac{y^2}{144} = 1$

31. $y^2 - \dfrac{x^2}{48} = 1$

33. $\dfrac{x^2}{a^2} - \dfrac{y^2}{c^2 - a^2} = 1$

Computer Exercises, page 417
3. Ellipse **5.** Point **7.** Hyperbola **9.** $\varnothing$

Written Exercises, page 420
1. $xy = 28$ **3.** $x^2y = 5$ **5.** $y\sqrt{x} = -\dfrac{15}{2}$ **7.** $\dfrac{1}{4}$
9. $\dfrac{3}{2}$ **11.** $\dfrac{3}{4}$ **13.** $\dfrac{14}{3}$ **15.** 20 **17.** $\dfrac{5}{3}$
19.   **21.** $xy = 2$

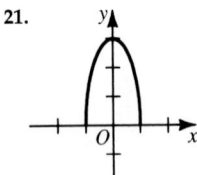

Problems, pages 421–422
1. 48 oscillations per minute **3.** 182 m³ **5.** 110 V
7. 5×10^{24} kg

Self-Test 2, page 422
1.

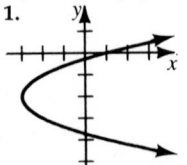

2.

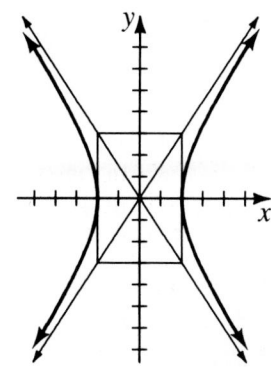

3. $x^2 + y^2 - 2x + 4y - 4 = 0$ **4.** $\dfrac{x^2}{144} + \dfrac{y^2}{169} = 1$

5. $a = 6$ **6.** 12 rev/min

Written Exercises, page 424

1. $\{(-1, -2), (3, 6)\}$ **3.** $\{(0, 1), (-2, 3)\}$
5. $\{(3, 2), (3, -2), (-3, 2), (-3, -2)\}$
7. $\{(3, 0), (-3, 0)\}$ **9.** $\{(-8, 6), (8, 6)\}$
11. $\{(3, 0), (-3, 0), (5, 4), (-5, 4)\}$
13. $\{(4, 3), (4, -3), (-4, 3), (-4, -3)\}$
15. $\{(0, -2), (4.5, 3), (-4.5, 3)\}$

Written Exercises, page 427

1. $\{(3, 12), (1, 4)\}$ **3.** $\{(-6, 2)\}$ **5.** $\{(i\sqrt{3}, -5i\sqrt{3}), (-i\sqrt{3}, 5i\sqrt{3})\}$ **7.** $\{(0, -1), (2, 0)\}$
11. $\{(3, -1), (-3, -19)\}$ **13.** $\left\{\left(\dfrac{1}{5}, \dfrac{4}{5}\right), \left(\dfrac{4}{5}, \dfrac{1}{5}\right)\right\}$

Problems, pages 427–428

1. 18 cm × 4 cm **3.** Door: 36 cm × 36 cm; panel:
12 cm × 12 cm **5.** $v_1 = 7$ m/s, $v_2 = 21$ m/s
7. $x = 5$ m, $y = 2$ m **9.** $x = 30$ m; $y = 10$ m
11. 105 cm² or 120 cm²

Written Exercises, pages 431–432

1. $\{(3, 1), (3, -1), (-3, 1), (-3, -1)\}$ **3.** $\{(4, 1), (4, -1), (-4, 1), (-4, -1)\}$ **5.** $\{(4, 2i\sqrt{3}), (4, -2i\sqrt{3}), (-4, 2i\sqrt{3}), (-4, -2i\sqrt{3})\}$
7. $\{(i\sqrt{3}, 3), (-i\sqrt{3}, 3), (i\sqrt{3}, -3), (-i\sqrt{3}, -3)\}$
9. $\left\{\left(\dfrac{1}{2}, \dfrac{3}{2}\right), \left(-\dfrac{1}{2}, \dfrac{3}{2}\right), \left(\dfrac{1}{2}, -\dfrac{3}{2}\right), \left(-\dfrac{1}{2}, -\dfrac{3}{2}\right)\right\}$
11. $\{(1, -3), (-1, -3), (\sqrt{6}, 2), (-\sqrt{6}, 2)\}$
13. $\{(4, 2i), (4, -2i), \left(-\dfrac{13}{3}, \dfrac{i\sqrt{61}}{3}\right),$
$\left(-\dfrac{13}{3}, -\dfrac{i\sqrt{61}}{3}\right)\}$ **15.** $\{(2\sqrt{3}, \sqrt{3}), (-2\sqrt{3}, -\sqrt{3}),$
$(\sqrt{3}, 2\sqrt{3}), (-\sqrt{3}, -2\sqrt{3})\}$

Problems, page 432

1. 5 m × 12 m **3.** (8, 6), (-8, 6) **5.** 8 cm, 15 cm,
17 cm **7.** Altitude: $\sqrt{10}$ cm; base edge: $2\sqrt{15}$ cm

Self-Test 3, page 433

1. $\left\{(0, 2), \left(\dfrac{2}{3}, 0\right)\right\}$

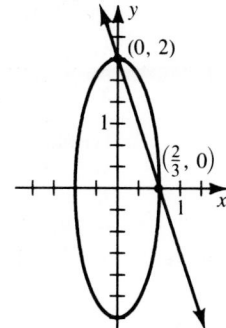

2. $\{(4, 2), (4, -2), (-4, 2), (-4, -2)\}$

3. $\left\{\left(\dfrac{15 + i\sqrt{71}}{2}, \dfrac{-5 + i\sqrt{71}}{2}\right),\right.$
$\left.\left(\dfrac{15 - i\sqrt{71}}{2}, \dfrac{-5 - i\sqrt{71}}{2}\right)\right\}$

Chapter Review, pages 434–435

1. b **2.** c **3.** d **4.** d **5.** a **6.** a **7.** a **8.** d
9. c **10.** a **11.** c

Mixed Review, page 436

1. 1 **3.** $-3v - 2$ **5.** $6 - 3i$ **7.** $\mathcal{R}$

9. $\{3 + \sqrt{15}, 3 - \sqrt{15}\}$ **11.** none

13.

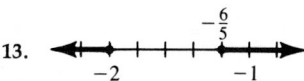

15.

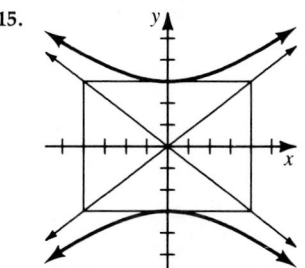

17. Max.: 8; min.: -3 **19.** -4815 **21.** 16 **23.** -5
25. $4608

Chapter 11 Exponents and Logarithms

Written Exercises, pages 443–444

1. 3 **3.** $\dfrac{1}{6}$ **5.** 8 **7.** $\dfrac{1}{9}$ **9.** 0.125 **11.** 13

13. $x\sqrt{x}$ **15.** $\sqrt{7n}$ **17.** $\sqrt{3}$ **19.** $\sqrt{5}$ **21.** $\dfrac{\sqrt{10}}{10}$

23. $\dfrac{16}{9}$ **25.** 6 **27.** $2\sqrt{2}$ **29.** $\sqrt[3]{3}$ **31.** $\dfrac{1}{5}$ **33.** $\sqrt[3]{3}$

35. $\sqrt[3]{100}$ **37.** $\sqrt[3]{7}$ **39.** $\left\{\dfrac{1}{32}\right\}$ **41.** $\{1, 9\}$ **43.** $\{1, 8\}$

45. $(b^{\frac{p}{r}} \cdot b^{\frac{q}{s}})^{rs} = b^{ps} \cdot b^{qr} = b^{ps+qr}; (b^{\frac{ps+rq}{rs}})^{rs} = b^{ps+rq};$
$\therefore b^{\frac{p}{r}} \cdot b^{\frac{q}{s}} = b^{\frac{ps+rq}{rs}}$ **47.** $(a^{\frac{p}{r}} \cdot b^{\frac{p}{r}})^r = a^p \cdot b^p; ((ab)^{\frac{p}{r}})^r = (ab)^p = a^p \cdot b^p; \therefore a^{\frac{p}{r}} \cdot b^{\frac{p}{r}} = (ab)^{\frac{p}{r}}$

Written Exercises, page 446

1. $3^{\sqrt{2}+\sqrt{3}}$ 3. $3^{3\sqrt{2}}$ 5. $3^{\sqrt{2}+2\sqrt{3}}$ 7. $3^{3\sqrt{2}-3}$ 9. $\{3\}$

11. $\{-2\}$ 13. $\{2\}$ 15. $\left\{\dfrac{1}{3}\right\}$

17. 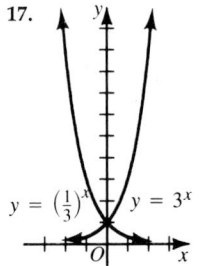   19.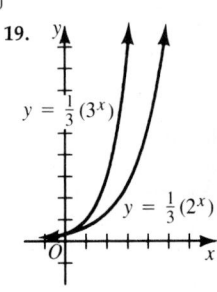

21. Answers may vary. For example, let $a = \sqrt{2}$ and $b = 2 - \sqrt{2}$, then $2^a \cdot 2^b = 2^2 = 4$.

Self-Test 1, page 447

1. 1000   2. $\dfrac{3}{5}$ 3. $\dfrac{1}{16}$ 4. $\sqrt[3]{5}$ 5. $\sqrt{7}$ 6. $\sqrt[3]{2}$

7. $\{3\}$ 8. $\{-2\}$

Written Exercises, page 450

1. $f^{-1}(x) = \dfrac{1}{4}x + 1$; $f^{-1}(f(x)) = \dfrac{1}{4}(4x - 4) + 1 = x$; $f(f^{-1}(x)) = 4\left(\dfrac{1}{4}x + 1\right) - 4 = x$ 3. $f^{-1}(x) = 21 - 3x$; $f^{-1}(f(x)) = 21 - 3\left(-\dfrac{1}{3}x + 7\right) = x$; $f(f^{-1}(x)) = -\dfrac{1}{3}(21 - 3x) + 7 = x$

7. $f^{-1}(x) = \dfrac{1}{3}x - \dfrac{2}{3}$ 9. $f^{-1}(x) = \sqrt[3]{x + 1}$

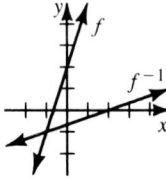   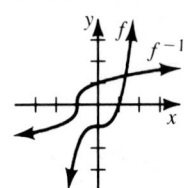

11. $f^{-1}(x) = -\sqrt{x^2 - 9}$, Domain: $\{x : x \geq 3\}$

13. $f^{-1}(x) = \dfrac{x^2}{4}$; Domain $\{x : x \geq 0\}$; $f^{-1}(f(x)) = \dfrac{(2\sqrt{x})^2}{4} = x$; $f(f^{-1}(x)) = 2\sqrt{\dfrac{x^2}{4}} = x$

15. $f^{-1}(1) = 0$, $f^{-1}(2) = 1$, $f^{-1}(4) = 2$, $f^{-1}(8) = 3$, $f^{-1}\left(\dfrac{1}{2}\right) = -1$, $f^{-1}\left(\dfrac{1}{4}\right) = -2$

17. A function that is not one-to-one; Domain $\{x : x \geq 1\}$; $f^{-1}(x) = 1 + \sqrt{x + 1}$, $x \geq -1$

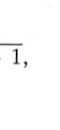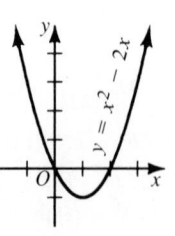

Written Exercises, page 453

1. 4   3. -3 5. $\dfrac{1}{2}$ 7. -3 9. 4   11. $\dfrac{2}{3}$ 13. $\{2\}$

15. $\{512\}$ 17. $\{27\}$ 19. $\{-4\}$ 21. $\{49\}$ 23. $\{216\}$

25. 10   27. 4   29. $\left\{\dfrac{3}{2}\right\}$ 31. $\{2\}$ 33. $\left\{\dfrac{4}{3}\right\}$ 35. $\{49\}$

37. $\{0\}$

Written Exercises, page 456

1. 3   3. -1 5. 2   7. $\{9\}$ 9. $\{16\}$ 11. $\{6\}$
13. $\{5\}$ 15. $\{3\}$ 17. 5   19. 49   21. 0.7   23. 5.2
25. 1.1   27. 3.5   29. $\{2\}$ 31. $\{2\}$ 33. $\{5, -2\}$

35. $x_1 = b^{\log_b x_1}$ and $x_2 = b^{\log_b x_2}$ $\dfrac{x_1}{x_2} = \dfrac{b^{\log_b x_1}}{b^{\log_b x_2}} = b^{\log_b x_1 - \log_b x_2}$ $\therefore \log_b \dfrac{x_1}{x_2} = \log_b x_1 - \log_b x_2$

Self-Test 2, pages 456–457

1. $f^{-1}(x) = 3x - 3$ 2. $f^{-1}(x) = \sqrt{x + 2}$, $x \geq -2$

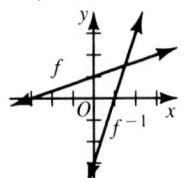   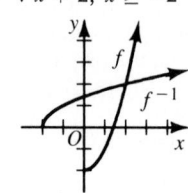

3. $\log_{\frac{1}{2}} 64 = -6$ 4. $\log_{36} 6 = \dfrac{1}{2}$ 5. $5^{-3} = \dfrac{1}{125}$

6. $10^3 = 1000$ 7. 4   8. $\dfrac{4}{3}$ 9. $\{4\}$ 10. $\{625\}$

11. $\{16\}$ 12. $\{9\}$

Written Exercises, pages 460–461

1. a. 1.3729   b. 4.3729   c. $0.3729 - 3$ or -2.6271
d. -0.3729 3. a. 2560   b. 256,000   c. 0.256
5. $\{36,200\}$ 7. $\{3\}$ 9. $\{3\}$ 11. 1.8751   13. 0.4771
15. $0.3010 - 1$ or -0.6990 17. 0.3010   19. $\{1.42\}$
21. $\{0.327\}$ 23. $\{14.7\}$ 25. a. $0.3010 - 1$ or -0.6990 b. 0.6990   c. -0.6020 d. 1.1761
27. 1.1132   29. a. 0 decibels   b. 140 decibels
c. 10^{12} 31. 73 decibels

Written Exercises, page 463

1. a. $\dfrac{\log 4.02}{\log 1.59}$ b. $\{3.00\}$ 3. a. $\dfrac{2 \log 145}{\log 19}$

b. $\{3.38\}$ 5. a. $\dfrac{\log 46}{4 \log 9.12}$ b. $\{0.433\}$

7. a. $\dfrac{\log 50.1 - \log 6}{\log 2}$ b. $\{3.06\}$

9. a. $\dfrac{\log 8.35 - \log 4780}{\log 3}$ b. $\{-5.78\}$

11. a. $\dfrac{\log 122 - \frac{1}{3} \log 25.3}{\log 6}$ b. $\{2.08\}$

13. a. $\dfrac{\frac{3}{4} \log 15 - \log 18}{\log 9}$ b. $\{-0.391\}$

15. a. $\dfrac{\frac{1}{3}\log 94}{\log 0.027} + 2$ **b.** $\{1.58\}$ **17.** 0.582

19. a. $\dfrac{\log 5}{\log 5 - \log 3}$ **b.** $\{3.15\}$

21. a. $\dfrac{\log 15}{2\log 6.7 - \log 15}$ **b.** $\{2.47\}$

23. a. $\dfrac{\log 39 + \log 8}{2\log 8 - \log 39}$ **b.** $\{11.6\}$ **25.** $\log_{25} 2 =$

$\dfrac{\log_5 2}{2\log_5 5} = \dfrac{\log 2}{2}$ **27.** $\log_b x = \dfrac{2\log x}{2\log b} =$

$\dfrac{2\log x}{\log b^2} = 2\log_{b^2} x$ **29.** $\log_{ab} x = \dfrac{\log_a x}{\log_a a + \log_a b} =$

$\dfrac{\log_a x}{1 + \log_a b}$ **31.** $\log_x y = \dfrac{\log_a y}{\log_a x} = \dfrac{\log_b y}{\log_b x}$,

$\therefore (\log_a x)(\log_b y) = (\log_a y)(\log_b x)$

Problems, pages 466–467
1. \$16,400 **3.** approx. 4.6 h **5.** approx. 25.5 yr
7. approx. 6 s **9.** 300 mo, or 25 yr

Computer Exercises, pages 467–468
3. 9 yr **5.** 4.6 yr **7.** 7.75 yr **9.** 7.75 yr

Written Exercises, page 471
1. $\ln 12.18 = 2.5$ **3.** $\ln 1.221 = \frac{1}{5}$ **5.** $e^{2.079} = 8$

7. $e^{-2.303} = 0.1$ **9.** $\ln 3$ **11.** $\ln\frac{5}{3}$ **13.** 42 **15.** $\frac{4}{3}$

17. \$7581 **19.** \$6264.50 **21.** \$6900 **23.** $\{1.12\}$
25. $\{1.05\}$ **27.**

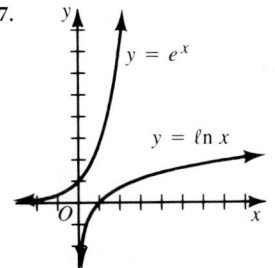

Computer Exercises, page 472
3. 2.718254 **5.** 2.7182818

On the Calculator, page 472
1. a. 7.2508% **b.** 8.3287% **c.** 11.6278%
d. 15.0274%

3.

compounded	9%	9.25%
quarterly	9.3083%	9.5758%
monthly	9.3807%	9.6524%
continuously	9.4174%	9.6913%
compounded	9.5%	9.75%
quarterly	9.8438%	10.1123%
monthly	9.9248%	10.1977%
continuously	9.9659%	10.2411%
compounded	10%	**5.** 12.2068%
quarterly	10.3813%	
monthly	10.4713%	
continuously	10.5171%	

Self-Test 3, page 473
1. $0.1818 - 3$ or -2.8182
2. $0.9031 - 1$ or -0.0969 **3.** $\{1.37\}$ **4.** $\{1030\}$

5. a. $\dfrac{\frac{1}{3}\log 34.6 - \log 4}{2\log 2.9}$ **b.** $\{-0.0963\}$

6. a. $\dfrac{\log 7}{\log 7 - 4\log 3}$ **b.** $\{-0.795\}$ **7.** approx. $22\frac{1}{2}$ yr

8. approx. $10\frac{1}{2}$ yr **9.** $\frac{1}{3}$ **10.** 0 **11.** \$6765

Chapter Review, pages 474–475
1. b **2.** d **3.** b **4.** c **5.** c **6.** d **7.** a **8.** a **9.** c
10. a **11.** d.

Application, page 478
1. $t = \dfrac{\ln N_0 - \ln N}{\lambda}$ **3.** 4220 yr

Preparing for College Entrance Exams, page 479
1. C **3.** E **5.** C **7.** A

Chapter 12 Permutations, Combinations, and Probability

Written Exercises, pages 485–486
1. 25 **3.** 360 **5.** 24 **7.** 72 **9.** 6,499,350 **11.** 147
13. 48

Written Exercises, page 489
1. a. 362, 880 **b.** 40,320 **3.** 5040 **5.** 151,200
7. 362,880 **9.** 5040 **11.** 360 **13.** 720 **15. a.** 1440
b. 3600

17.
$$(n - r)_n P_r \overset{?}{=} {}_n P_{r+1}$$
$$(7 - 4)_7 P_4 \overset{?}{=} {}_7 P_5$$
$$3 \cdot 7 \cdot 6 \cdot 5 \cdot 4 \overset{?}{=} 7 \cdot 6 \cdot 5 \cdot 4 \cdot 3$$
$$2520 = 2520$$
21. $(n - r)_n P_r = (n - r)(n)(n - 1)(n - 2) \cdots$
$(n - r + 1) = n(n - 1)(n - 2) \cdots (n - r + 1) \cdot$
$(n - r) = {}_n P_{r+1}$

Written Exercises, page 491
1. 180 **3.** 420 **5.** 453,600 **7.** 151,200 **9.** 360
11. 302,400 **13.** 12 **15. a.** 120 **b.** 240 **c.** 360

Self-Test 1, page 491
1. 75 **2.** 720 **3.** 360 **4.** 302,400

Written Exercises, pages 494–495
1. 1001 **3.** 36 **5.** 21 **7.** 45 **9.** 2002 **11.** 378
13. 31 **15.** 15

17. ${}_n C_r = \dfrac{n!}{r!(n - r)!}$; ${}_n C_{n-r} =$

$\dfrac{n!}{(n - r)![n - (n - r)]!} = \dfrac{n!}{(n - r)!r!}$; $\dfrac{n!}{r!(n - r)!} =$

$\dfrac{n!}{(n - r)!r!}$, so ${}_n C_r = {}_n C_{n-r}$

Computer Exercises, page 495
3. 3628800 **5.** 6.4023737×10^{15} **7.** 120 **9.** 2598960
13. 1663200

Written Exercises, pages 496–497
1. 5880 **3.** 8400 **5.** 45 **7.** 525 **9.** 24 **11.** 2970
13. 1024 **15.** 4512 **17.** 3744 **19.** 7200
21. 720; 48

Self-Test 2, page 497
1. 210 **2.** 462 **3.** 300 **4.** 420

Written Exercises, page 500
1. $b^5 + 10b^4 + 40b^3 + 80b^2 + 80b + 32$
3. $625 - 500r + 150r^2 - 20r^3 + r^4$
5. $243c^5 + 135c^4 + 30c^3 + \dfrac{10}{3}c^2 + \dfrac{5}{27}c + \dfrac{1}{243}$
7. $\dfrac{1}{64} - \dfrac{3}{8}b + \dfrac{15}{4}b^2 - 20b^3 + 60b^4 - 96b^5 + 64b^6$
9. $y^{32} - 8y^{28} + 28y^{24} - 56y^{20} + 70y^{16} - 56y^{12} + 28y^8 - 8y^4 + 1$
11. $2187 + 10{,}206b + 20{,}412b^2 + 22{,}680b^3 + 15{,}120b^4 + 6048b^5 + 1344b^6 + 128b^7$
13. $210a^6b^4$ **15.** $672p^6q^3$
17. $\dfrac{1287}{8}c^5d^8$ **19.** $x^{30} + 30x^{28} + 420x^{26} + 3640x^{24}$
23. $8 + \dfrac{1}{16} - \dfrac{1}{4096}$ **25.** $4 - \dfrac{1}{8}r - \dfrac{1}{512}r^2$

Computer Exercises, page 500
7. 2.01219647 **9.** 0.472161363

Written Exercises, page 504
1. $c^6 + 6c^5 + 15c^4 + 20c^3 + 15c^2 + 6c + 1$
3. $243x^5 + 405x^4y + 270x^3y^2 + 90x^2y^3 + 15xy^4 + y^5$ **5.** $1 - 7y^3 + 21y^6 - 35y^9 + 35y^{12} - 21y^{15} + 7y^{18} - y^{21}$ **7.** $r^{18} - 6r^{15}t^2 + 15r^{12}t^4 - 20r^9t^6 + 15r^6t^8 - 6r^3t^{10} + t^{12}$ **9.** $70m^4$ **11.** $15p^3q^7$
13. $7920x^4$ **15. a.** $6435x^7y^8$ **b.** $12{,}870x^8y^8$
17. $_{n-1}C_2 + {_{n-1}C_3} = \dfrac{(n-1)(n-2)}{2 \cdot 1} + \dfrac{(n-1)(n-2)(n-3)}{3 \cdot 2 \cdot 1} =$
$\dfrac{3(n-1)(n-2) + (n-1)(n-2)(n-3)}{3 \cdot 2 \cdot 1} =$
$\dfrac{(n-1)(n-2)[3 + (n-3)]}{3 \cdot 2 \cdot 1} =$
$\dfrac{(n-1)(n-2)n}{3 \cdot 2 \cdot 1} = \dfrac{n(n-1)(n-2)}{3 \cdot 2 \cdot 1} = {_nC_3}$
19. $f(n) = 2^{n-1}$; 128

Self-Test 3, page 504
1. $a^{10} - 10a^8 + 40a^6 - 80a^4 + 80a^2 - 32$
2. $3360p^4q^6$ **3.** $x^6 - 3x^5y + \dfrac{15}{4}x^4y^2 - \dfrac{5}{2}x^3y^3 + \dfrac{15}{16}x^2y^4 - \dfrac{3}{16}xy^5 + \dfrac{1}{64}y^6$

Written Exercises, pages 506–507
1. a. $\{1, 2, 3, \ldots, 14\}$ **b.** $\{4, 8, 12\}$
3. a. $\{(1, 1), (1, 2), (1, 3), (2, 1), (2, 2), (2, 3), (3, 1), (3, 2), (3, 3)\}$ **b.** $\{(1, 1), (2, 2), (3, 3)\}$ **5. a.** $\{(D, H), (D, C), (D, S), (C, H), (C, D), (C, S), (H, C), (H, D), (H, S), (S, H), (S, C), (S, D)\}$ **b.** $\{(D, H), (H, D)\}$

7. a. $\{(A, B), (A, C), (A, E), (B, A), (B, C), (B, E), (C, A), (C, B), (C, E), (E, A), (E, B), (E, C)\}$ **b.** $\{(A, B), (A, C), (A, E), (B, A), (B, E), (C, A), (C, E), (E, A), (E, B), (E, C)\}$
9. 52; 13 **11.** 1326; 325 **13.** 1326; 78 **15.** 2,598,960; 22,308

On the Calculator, page 507
1. 270,725; 14,950 **3.** 2,598,960; 42,504
5. 4.4814×10^{12}; 7,726,160

Written Exercises, pages 510–511
1. a. $\dfrac{1}{3}$ **b.** $\dfrac{2}{3}$ **c.** $\dfrac{4}{9}$ **d.** $\dfrac{5}{9}$ **3. a.** $\dfrac{1}{2}$ **b.** $\dfrac{2}{1}$
c. $\dfrac{4}{5}$ **d.** $\dfrac{5}{4}$ **5. a.** $\dfrac{1}{20}$ **b.** $\dfrac{1}{12}$ **c.** $\dfrac{7}{40}$ **d.** $\dfrac{11}{20}$
e. $\dfrac{11}{24}$ **f.** $\dfrac{3}{10}$ **7. a.** $\dfrac{1}{32}$ **b.** $\dfrac{5}{32}$ **c.** $\dfrac{5}{16}$ **d.** $\dfrac{5}{16}$
e. $\dfrac{31}{32}$ **9. a.** $\dfrac{1}{2}$ **b.** $\dfrac{1}{6}$ **c.** $\dfrac{1}{2}$ **d.** $\dfrac{19}{30}$ **11. a.** $\dfrac{2}{11}$
b. $\dfrac{3}{11}$ **c.** $\dfrac{3}{11}$ **d.** $\dfrac{8}{11}$ **13. a.** $\dfrac{1}{6}$ **b.** $\dfrac{5}{18}$ **15.** $\dfrac{1}{126}$

Written Exercises, page 513
1. a. $\dfrac{7}{64}$ **b.** $\dfrac{11}{32}$ **c.** $\dfrac{21}{32}$ **d.** $\dfrac{1}{64}$ **e.** 1 **3.** $\dfrac{7}{128}$
5. a. $\dfrac{3}{5}$ **b.** $\dfrac{4}{5}$ **7.** $\dfrac{2}{25}$ **9.** $\dfrac{24}{35}$

Written Exercises, pages 517–518
1. 0.22 **3. a.** $\dfrac{1}{24}$ **b.** $\dfrac{1}{6}$ **c.** $\dfrac{5}{8}$ **d.** $\dfrac{3}{8}$ **5. a.** $\dfrac{1}{338}$
b. $\dfrac{3}{338}$ **7.** $P(A) = \dfrac{1}{6}$; $P(B) = \dfrac{11}{36}$; $P(A|B) =$
$\dfrac{2}{11}$; $P(B|A) = \dfrac{2}{6} = \dfrac{1}{3}$ **9.** $P(A) = \dfrac{1}{3}$; $P(B) = \dfrac{1}{3}$;
$P(C) = \dfrac{1}{3}$; $P(D) = \dfrac{2}{3}$
11. No; given that event B occurs will change the probability of event C to $\dfrac{1}{2}$. **13.** Events A and C
15. a. $\dfrac{p}{2(p + q)}$ **b.** $\dfrac{2p + q}{2(p + q)}$

Self-Test 4, page 518
1. $\{(A, 1), (A, 2), (A, 3), (B, 1), (B, 2), (B, 3), (C, 1), (C, 2), (C, 3)\}$; $\{(A, 1), (A, 3), (B, 1), (B, 3)\}$ **2. a.** $\dfrac{1}{4}$
b. $\dfrac{3}{4}$ **3.** $\dfrac{1}{2}$ **4.** No; if the sum of the numbers equals 8, it is impossible for one die to show a 1.

Chapter Review, pages 519–520
1. d **2.** b **3.** a **4.** c **5.** d **6.** b **7.** a **8.** d
9. c **10.** a **11.** b **12.** c **13.** b **14.** d **15.** d

Cumulative Review, pages 521–523
1. $\{-2 + 4i, -2 - 4i\}$ **3.** $k = 3$ **5.** $x = \dfrac{3}{2}$
7. $y = 2x^2 - 8x + 5$ **9.** $-3, -1, 1$
11. $y = 2x - 7$ **13.** $y = -\dfrac{1}{12}x^2$ **15.** $\dfrac{y^2}{9} - \dfrac{x^2}{16} = 1$

17. $\{(i\sqrt{2}, -4i\sqrt{2}), (-i\sqrt{2}, 4i\sqrt{2})\}$

19. $\{(-\sqrt{3}, -\sqrt{3})\}$ **21.** 5 m, 12 m, 13 m

23. $\left\{\dfrac{5}{4}\right\}$ **25.** $\left\{\dfrac{9}{2}\right\}$ **27.** $\{2.67\}$ **29.** $\ln 1097 = 7$

31. 720 arrangements   **33.** 42   **35.** 21   **37.** 126

39. $16x^4 + 96x^3b + 216x^2b^2 + 216xb^3 + 81b^4$

Chapter 13   Matrices

Written Exercises, page 531

1. $\begin{bmatrix} -4 \\ -1 \end{bmatrix}$ **3.** $\begin{bmatrix} 4 & -1 \\ 12 & 1 \end{bmatrix}$ **5.** $\begin{bmatrix} 2 & 8 & -8 \\ -4 & 2 & 6 \end{bmatrix}$

7. $\begin{bmatrix} -9 & 5 \\ 1 & -7 \\ 7 & -6 \end{bmatrix}$

9. $w = -5$; $x = -3$; $y = 5$; $z = 5$ **11.** $x = -5$;
$y = -8$ **13.** $w = 3$; $x = -2$; $y = -1$;
15. $w = 6$; $x = -5$; $y = 0$ **17.** $|A| = 15$;
$|B| = 18$; $|A + B| = -6$

Written Exercises, page 534

1. $\begin{bmatrix} 5 & -4 \\ 5 & -6 \end{bmatrix}$ **3.** $\begin{bmatrix} 8 & -1 \\ 0 & -13 \end{bmatrix}$ **5.** $\begin{bmatrix} -1 & 3 \\ -8 & 8 \end{bmatrix}$

7. $\begin{bmatrix} -6 & 3 & -3 \\ 0 & -8 & 6 \end{bmatrix}$

9. $[a - c \quad b - d]$ **17.** $(-3, 5, -2), (-3, -2, -4)$,
$(-3, 5, -4), (-3, -2, -2)$

Written Exercises, pages 537–538

1. $\begin{bmatrix} 27 & -9 \\ -63 & 0 \end{bmatrix}$ **3.** $\begin{bmatrix} 29 & -13 \\ -41 & 10 \end{bmatrix}$ **5.** $\begin{bmatrix} 3 & -1 \\ -7 & 0 \end{bmatrix}$

7. $\begin{bmatrix} 13 & -11 \\ 23 & 20 \end{bmatrix}$ **9.** $\begin{bmatrix} 6 & -12 \\ 66 & 30 \end{bmatrix}$ **11.** $\begin{bmatrix} 1 & 1 \\ -13 & -4 \end{bmatrix}$

13. $\begin{bmatrix} -15 & 0 \\ 0 & 15 \end{bmatrix}$ **15.** $\begin{bmatrix} -36 & 12 \\ -24 & 60 \end{bmatrix}$ **17.** $\begin{bmatrix} -7 & -4 \\ 8 & -1 \end{bmatrix}$

Written Exercises, pages 541–542

1. $\begin{bmatrix} 11 \\ -11 \end{bmatrix}$ **3.** $\begin{bmatrix} 0 \\ -3 \end{bmatrix}$ **5.** $\begin{bmatrix} -2 & 7 \\ 6 & -4 \end{bmatrix}$ **7.** $\begin{bmatrix} 1 & -12 \\ 0 & 21 \end{bmatrix}$

9. Not defined   **11.** $\begin{bmatrix} -5 & 6 & 10 \\ 22 & -4 & 4 \\ -6 & 3 & 3 \end{bmatrix}$ **13.** $\begin{bmatrix} -7 & 9 \\ -9 & 14 \end{bmatrix}$

15. $\begin{bmatrix} 1 & 0 \\ 0 & 1 \end{bmatrix}$ **17.** $\begin{bmatrix} 9 & 0 \\ 18 & 9 \end{bmatrix}$

21. $x = 2$; $y = -1$ **23.** $x = 0$; $y = 4$
25. $x = 5$; $y = 3$; $z = 1$

27. $AB = \begin{bmatrix} ad - bc & 0 \\ 0 & ad - bc \end{bmatrix}$

29. $AB = \begin{bmatrix} 970 & 1095 & 1220 \\ 160 & 190 & 220 \end{bmatrix}$

Computer Exercises, page 542

3. $A^2 = \begin{bmatrix} -19 & -28 \\ 35 & 44 \end{bmatrix}$; $A^3 = \begin{bmatrix} -121 & -148 \\ 185 & 212 \end{bmatrix}$;

$A^4 = \begin{bmatrix} -619 & -700 \\ 875 & 956 \end{bmatrix}$

5. $A = A^2 = A^3 = A^4 = \begin{bmatrix} 1 & 0 & 0 \\ 0 & 1 & 0 \\ 0 & 0 & 1 \end{bmatrix}$

Written Exercises, pages 545–546

1. $AB = \begin{bmatrix} 0 & 0 \\ 0 & 0 \end{bmatrix}$; $BA = \begin{bmatrix} -28 & -14 \\ 56 & 28 \end{bmatrix}$; $AB \neq BA$

3. $AD = \begin{bmatrix} 8 & 4 \\ -16 & -8 \end{bmatrix}$; $DA = \begin{bmatrix} 20 & 10 \\ -40 & -20 \end{bmatrix}$; $AD \neq$

DA **5.** $CD = DC = \begin{bmatrix} 20 & 0 \\ 0 & 20 \end{bmatrix}$ **7.** $A(B + C) =$

$AB + AC = \begin{bmatrix} 40 & 20 \\ -80 & -40 \end{bmatrix}$ **9.** $B(AC) = (BA)C =$

$\begin{bmatrix} -280 & -140 \\ 560 & 280 \end{bmatrix}$ **11.** $A(BC) = \begin{bmatrix} 0 & 0 \\ 0 & 0 \end{bmatrix}$; $(BC)A =$

$\begin{bmatrix} -112 & -56 \\ 224 & 112 \end{bmatrix}$; $A(BC) \neq (BC)A$ **13.** $C(A + B) =$

$\begin{bmatrix} 12 & 20 \\ -24 & -40 \end{bmatrix}$; $(A + B)C = \begin{bmatrix} 44 & 36 \\ -88 & -72 \end{bmatrix}$;

$C(A + B) \neq (A + B)C$

15. a and c   **17.** a and b

Self-Test 1, pages 546–547

1. $\begin{bmatrix} 9 & 13 \\ 3 & 7 \end{bmatrix}$ **2.** $\begin{bmatrix} -7 & 7 & -4 \\ -6 & 6 & 11 \end{bmatrix}$ **3.** $\begin{bmatrix} -4 & -13 \\ 24 & -3 \end{bmatrix}$

4. $\begin{bmatrix} -8 & 1 \\ 12 & 12 \end{bmatrix}$ **5.** $\begin{bmatrix} 4 & -\dfrac{13}{2} \\ 2 & -10 \end{bmatrix}$ **6.** $\begin{bmatrix} 4 & -13 \\ -54 & -12 \end{bmatrix}$

7. $\begin{bmatrix} 1 \\ 33 \end{bmatrix}$ **8.** $AB = \begin{bmatrix} 12 & -15 \\ -12 & 51 \end{bmatrix}$; $BA = \begin{bmatrix} 32 & 14 \\ 40 & 31 \end{bmatrix}$;

$AB \neq BA$

Written Exercises, pages 551–552

1. $\begin{bmatrix} 3 & -2 \\ -7 & 5 \end{bmatrix}$ **3.** Singular   **5.** $\begin{bmatrix} \dfrac{1}{3} & 0 \\ 0 & \dfrac{1}{3} \end{bmatrix}$

7. $\begin{bmatrix} \dfrac{5}{2} & -\dfrac{3}{2} \\ -3 & 2 \end{bmatrix}$ **9.** $\begin{bmatrix} 1 & \dfrac{4}{3} \\ -2 & -3 \end{bmatrix}$ **11.** $\begin{bmatrix} -14 & -9 \\ 11 & 7 \end{bmatrix}$

13. $\begin{bmatrix} 7 & 9 \\ -11 & -14 \end{bmatrix}$ **15.** $\begin{bmatrix} \dfrac{5}{2} & 2 \\ -1 & -1 \end{bmatrix}$ **17.** $\begin{bmatrix} 2 & 4 \\ -2 & -5 \end{bmatrix}$

19. $\{(3, -1)\}$ **21.** $\{(-2, 1)\}$ **23.** $\left\{\left(\dfrac{1}{2}, 0\right)\right\}$

Computer Exercises, page 552

3. No inverse   **5.** $\begin{bmatrix} -1.5 & -2 \\ 2.5 & 3 \end{bmatrix}$

Self-Test 2, page 552

1. $\begin{bmatrix} \dfrac{-7}{19} & \dfrac{1}{19} \\ \dfrac{5}{19} & \dfrac{2}{19} \end{bmatrix}$ **2.** $\begin{bmatrix} -1 & 8 \\ 0 & -3 \end{bmatrix}$

3. $\begin{bmatrix} 2 & -1 \\ 7 & 8 \end{bmatrix}\begin{bmatrix} x \\ y \end{bmatrix} = \begin{bmatrix} -3 \\ 1 \end{bmatrix}$ **4.** $\left\{\left(\dfrac{1}{2}, -4\right)\right\}$

Written Exercises, page 556

1. $(-1, 3)$ **3.** $(-4, 3)$ **5.** $(2a, 0)$ **7.** $(-2, -2)$
9. $(-4, -5)$ **11.** $(-a, b)$
13. $\begin{bmatrix} x' \\ y' \end{bmatrix} = \begin{bmatrix} x \\ y \end{bmatrix} + \begin{bmatrix} 1 \\ -4 \end{bmatrix}$ **15.** $\begin{bmatrix} x' \\ y' \end{bmatrix} = \begin{bmatrix} x \\ y \end{bmatrix} + \begin{bmatrix} 5 \\ 3 \end{bmatrix}$
17. $\begin{bmatrix} x' \\ y' \end{bmatrix} = \begin{bmatrix} x \\ y \end{bmatrix} + \begin{bmatrix} c - a \\ d - b \end{bmatrix}$ **19.** $(13, -4)$
21. $(8, 14)$ **23.** $(1, 6)$ **25.** $(3, 5)$
27. $(a + h_1 + h_2, b + k_1 + k_2)$ **29.** Yes

Written Exercises, pages 562–563

1. $(10, 5)$ **3.** $(11, -4)$ **5.** $(-7, 3)$ **7.** $(-3, -5)$
9. $(-15, 11)$ **11.** $(-3, 4)$ **13.** $x' = 2x + 3y$;
$y' = 4x + 6y$; $m = 2$ **15.** $x' = -10x - 12y$;
$y' = -5x - 6y$; $m = \frac{1}{2}$ **17.** Square with vertices
$(0, 0)(3, 0)(3, 3)(0, 3)$ **19.** Square with vertices
$(0, 0)(0, -1)(-1, -1)(-1, 0)$
21. $(x', y') = (3x, 3y)$; an expansion by a factor of 3
23. $(x', y') = (-y, -x)$; a reflection in the line
$y = -x$ **25.** $(x', y') = (0, y)$; a projection onto the
y-axis **27.** $(x', y') = (0, 3x)$ a projection onto the
y-axis such that the ordinate of each image point is
three times that of the abscissa of the preimage.

Self-Test 3, page 563

1. $\begin{bmatrix} x' \\ y' \end{bmatrix} = \begin{bmatrix} x \\ y \end{bmatrix} + \begin{bmatrix} 1 \\ 7 \end{bmatrix}$ **2.** $(-5, 7)$ **3.** $(-3, 6)$
4. $(40, 18)$ **5.** $\left(\frac{37}{26}, -\frac{7}{13}\right)$

Chapter Review, pages 565–566

1. a **2.** a **3.** c **4.** d **5.** d **6.** a **7.** c **8.** c **9.** b
10. b **11.** d **12.** b **13.** a **14.** a **15.** d **16.** b

Application, page 569

1. $\begin{bmatrix} 0 & 1 & 0 \\ 1 & 0 & 1 \\ 1 & 0 & 0 \end{bmatrix}$ **3.** $\begin{bmatrix} 0 & 1 & 0 & 0 \\ 1 & 0 & 1 & 0 \\ 0 & 0 & 0 & 1 \\ 0 & 1 & 0 & 0 \end{bmatrix}$ **7.** Dot **9.** Dot

Preparing for College Entrance Exams, page 570

1. E **3.** C **5.** C **7.** A

Chapter 14 Trigonometric and Circular Functions

Written Exercises, page 578

1. 126 cm **3.** 260 m **5.** 37.5 cm **7.** 1.67 m
9. 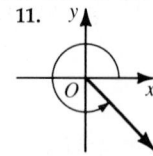 **11.**

15. $\frac{2}{3}$ revolution clockwise **17.** $\frac{1}{8}$ revolution

clockwise **19.** $\frac{1}{3}$ revolution counterclockwise

21. 22.0 cm **23.** 7120 km/h **25.** 3

Written Exercises, pages 582–583

1. $\frac{4\pi}{3}^R$ **3.** $\frac{5\pi}{6}^R$ **5.** $\frac{11\pi}{6}^R$ **7.** $\frac{7\pi}{4}^R$ **9.** $-\frac{3\pi}{5}^R$
11. $135°$ **13.** $-315°$ **15.** $150°$ **17.** $-330°$
19. $-160°$ **21.** 44 cm **23.** 275 cm **25.** 22 cm
27. 36 **29.** $\frac{3}{2}$ **31.** $\frac{3}{2}$

Computer Exercises, page 583

3. $65°19'2''$ **5.** $207°59'2''$

On the Calculator, page 583

1. 0.9774^R **3.** -0.4206^R **5.** 0.2342^R **7.** 0.4931^R
9. $25.9°$ **11.** $41.8°$ **13.** $138.7°$ **15.** $-259.2°$

Self-Test 1, page 584

1. 563.2 cm **2.** $\frac{8\pi}{3}^R$ **3.** $330°$

Written Exercises, page 588

1. $\sin \alpha = \frac{4}{5}$; $\cos \alpha = \frac{3}{5}$ **3.** $\sin \alpha = -\frac{12}{13}$; $\cos \alpha = \frac{5}{13}$

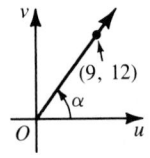
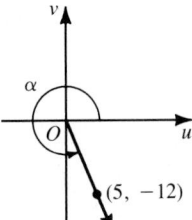

5. $\sin \alpha = 1$; $\cos \alpha = 0$ **7.** $\sin \alpha = \frac{2}{\sqrt{5}}$;
$\cos \alpha = \frac{1}{\sqrt{5}}$ **9.** $\sin \alpha = -\frac{3}{5}$, $\cos \alpha = -\frac{4}{5}$
11. $\cos \alpha = -\frac{5}{13}$ **13.** $\sin \alpha = \frac{8}{17}$ **15.** $\sin \alpha = -\frac{\sqrt{15}}{4}$
17. $\cos \alpha = -\frac{1}{5}$ **19.** $\sin \alpha = -\frac{2\sqrt{5}}{5}$; $\cos \alpha = -\frac{\sqrt{5}}{5}$
21. $\sin \alpha = \frac{1}{\sqrt{10}}$; $\cos \alpha = -\frac{3}{\sqrt{10}}$ **23.** $\sin \alpha = -\frac{1}{2}$;
$\cos \alpha = -\frac{\sqrt{3}}{2}$

Written Exercises, pages 591–592

1. $-\frac{1}{\sqrt{2}}$ **3.** $\frac{1}{\sqrt{2}}$ **5.** $\frac{1}{\sqrt{2}}$ **7.** $\frac{1}{2}$ **9.** $\frac{\sqrt{3}}{2}$ **11.** -1
13. a. $60° + k \cdot 360°$ **b.** $\frac{\pi}{3}^R + 2k\pi^R$
15. a. $210° + k \cdot 360°$ **b.** $\frac{7\pi}{6}^R + 2k\pi^R$
17. a. $225° + k \cdot 360°$ **b.** $\frac{5\pi}{4}^R + 2k\pi^R$
19. a. $225° + k \cdot 360°$ **b.** $\frac{5\pi}{4}^R + 2k\pi^R$ **23.** $\frac{2\pi}{n}$

Computer Exercises, page 592
3. 0.9949261

Written Exercises, pages 594–595
1. a. 75° **c.** 0.9659   **3. a.** 12° **c.** 0.9781
b. 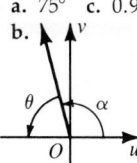   **b.**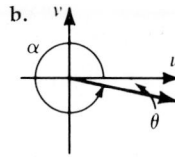

5. a. 15° **c.** −0.9659 **7. a.** 64.2° **c.** −0.9003
9. a. 1.09^R **c.** −0.4625 **11. a.** 1.28^R **c.** −0.2867
13. 0.6858 **15.** −0.6862 **17.** −0.7880 **19.** 0.9336
21. −0.0349 **23.** −0.8910 **25.** −0.9664 **27.** 0.6594
29. 192°0', or 192.0° **31.** 263°30', or 263.5°
33. 156°10', or 156.2° **35.** 117°40', or 117.7°
37. 133.6° **39.** 311.8° **41.** 306°50' **43.** 257°20'
45. 5.79^R **47.** 4.42^R

Computer Exercises, page 595
3. 37° **5.** 79°

Written Exercises, pages 598–599

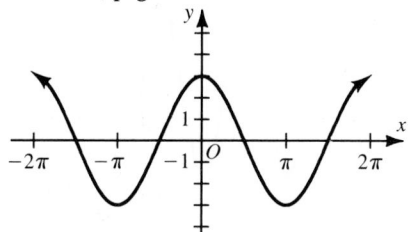

3.

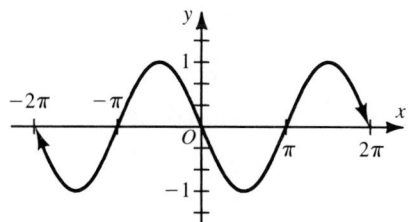

11. Maximum: 2; minimum: −4; $|A| = 3$
13. Maximum: 7; minimum: −1; $|A| = 4$

17.

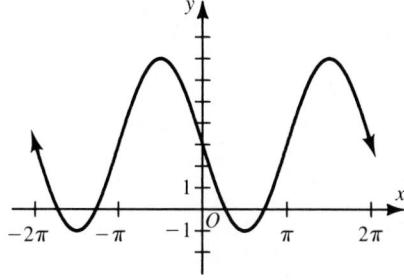

19.

— $y = \sin x$
-- $y = \sin(-x)$
and $y = -\sin x$

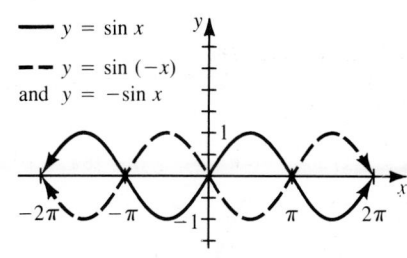

Written Exercises, page 600–601
1.

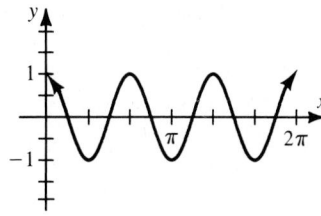

3.

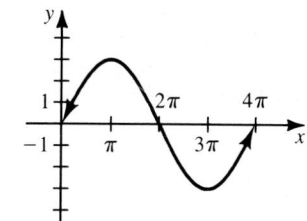

13.

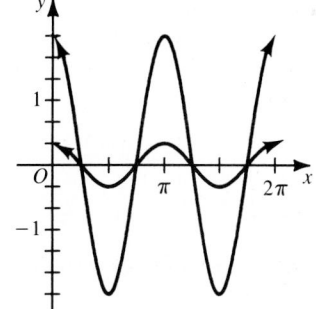

17.

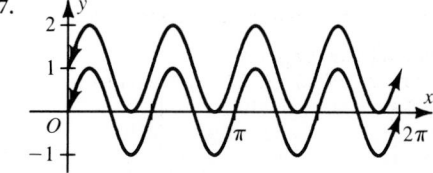

Self-Test 2, page 601
1. $\sin \alpha = \dfrac{\sqrt{3}}{2}$; $\cos \alpha = -\dfrac{1}{2}$ **2.** $\cos \alpha = -\dfrac{8}{17}$
3. 0.9940 **4.** −0.4695 **5.** 0.9636 **6.** 0.0508

7.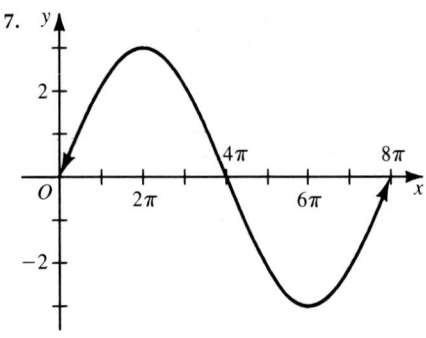

Written Exercises, pages 605–606
1. -1.111 **3.** -1.828 **5.** 0.4183 **7.** -1.286
9. -0.4346 **11.** 2.657 **13.** $62.3°$ **15.** $49.6°$
17. $38.1°$ **19.** $324°$ **21.** $205.5°$ **23.** $337.1°$

25. $\sin \alpha = -\dfrac{4}{5}$; $\cos \alpha = \dfrac{3}{5}$; $\tan \alpha = -\dfrac{4}{3}$;
$\cot \alpha = -\dfrac{3}{4}$; $\sec \alpha = \dfrac{5}{3}$; $\csc \alpha = -\dfrac{5}{4}$
27. $\sin \alpha = -1$; $\cos \alpha = 0$; $\tan \alpha$ not defined;
$\cot \alpha = 0$; $\sec \alpha$ not defined; $\csc \alpha = -1$
29. $\sin \alpha = 0$; $\cos \alpha = -1$; $\tan \alpha = 0$;
$\cot \alpha$ not defined; $\sec \alpha = -1$; $\csc \alpha$ not defined
31. $\sin \alpha = -\dfrac{5}{7}$; $\cos \alpha = \dfrac{2\sqrt{6}}{7}$;
$\tan \alpha = -\dfrac{5}{2\sqrt{6}}$; $\cot \alpha = -\dfrac{2\sqrt{6}}{5}$; $\sec \alpha = \dfrac{7}{2\sqrt{6}}$;
$\csc \alpha = -\dfrac{7}{5}$

33. $\cos \alpha = \dfrac{1}{2}$; $\tan \alpha = -\sqrt{3}$; $\cot \alpha = -\dfrac{1}{\sqrt{3}}$;
$\sec \alpha = 2$; $\csc \alpha = -\dfrac{2}{\sqrt{3}}$ **35.** $\cos \alpha = -\dfrac{2}{3}$;
$\tan \alpha = -\dfrac{\sqrt{5}}{2}$; $\cot \alpha = -\dfrac{2}{\sqrt{5}}$; $\sec \alpha = -\dfrac{3}{2}$;
$\csc \alpha = \dfrac{3}{\sqrt{5}}$ **37.** $\sin \alpha = \dfrac{\sqrt{15}}{7}$; $\cos \alpha = -\dfrac{\sqrt{34}}{7}$;
$\tan \alpha = -\dfrac{\sqrt{510}}{34}$; $\cot \alpha = -\dfrac{\sqrt{510}}{15}$; $\sec \alpha = -\dfrac{7}{\sqrt{34}}$
39. $\sin \alpha = -\dfrac{1}{\sqrt{26}}$; $\cos \alpha = -\dfrac{5}{\sqrt{26}}$; $\cot \alpha = 5$;
$\sec \alpha = -\dfrac{\sqrt{26}}{5}$; $\csc \alpha = -\sqrt{26}$

Written Exercises, page 610
1. $m(A) = 61.8°$; $a \approx 52.9$; $b \approx 28.4$
3. $m(B) = 71.7°$; $a \approx 23.4$; $b \approx 70.8$
5. $m(B) = 57.3°$; $a \approx 16.0$; $c \approx 29.7$
7. $m(B) = 33.9°$; $c \approx 57.8$; $b \approx 32.3$
9. $m(A) = 25.3°$; $c \approx 93.6$; $b \approx 84.6$
11. $m(B) \approx 34.8°$; $m(A) \approx 55.2°$; $c \approx 28.0$
13. $m(A) \approx 67.4°$; $m(B) \approx 22.6°$; $b \approx 25.0$
15. $m(A) \approx 32.2°$; $m(B) \approx 57.8°$; $c \approx 52.9$

17. $m(A) \approx 29.6°$; $m(B) \approx 60.4°$; $b \approx 21.8$
19. $m(A) \approx 53.3°$; $m(B) \approx 36.7°$; $a \approx 23.5$

Problems, pages 610–612
1. 30.9 ft **3.** 57.1° **5.** 11.5 **7.** 14.0 m **9.** 46.3 km
11. 89 m

Self-Test 3, page 612
1. $\sin \alpha = \dfrac{2\sqrt{10}}{7}$; $\cos \alpha = -\dfrac{3}{7}$; $\tan \alpha = -\dfrac{2\sqrt{10}}{3}$;
$\cot \alpha = -\dfrac{3}{2\sqrt{10}}$; $\sec \alpha = -\dfrac{7}{3}$; $\csc \alpha = \dfrac{7}{2\sqrt{10}}$
2. $c = 11.6$; $b = 9.22$; $m(B) = 52.8°$

Chapter Review, page 613
1. a **2.** c **3.** b **4.** d **5.** d **6.** c **7.** d **8.** b
9. a **10.** c **11.** c **12.** a **13.** b

Mixed Review, pages 615–616
1. $-9c$ **3.** $w^3 - 5w^2 + 12w - 18$ **5.** 16 **7.** $\{-20\}$
9. $\left\{y: -3 < y < \dfrac{1}{2}\right\}$ **11.** $\{(\sqrt{2}, 4), (-\sqrt{2}, 4), (0, -2)\}$
13.

15.
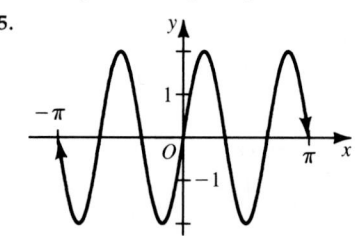

17. $x = -2$ **19.** 85 **21.** $720x^3y^2$ **23.** $c = 16$;
$b = 10.6$
25. $\begin{bmatrix} 15 & 20 \\ -30 & -40 \end{bmatrix}$ **27.** 13 years

Chapter 15 Trigonometric Identities

Written Exercises, pages 621–622
1. $\sin^2 \alpha$; $\cos \alpha \neq 0$ **3.** $\dfrac{1}{\cos \alpha}$; $\sin \alpha \neq 0$,
$\cos \alpha \neq 0$ **5.** $\dfrac{1}{\cos \alpha}$; $\cos \alpha \neq 0$ **7.** $\dfrac{1}{\cos^2 \alpha}$;
$\cos \alpha \neq 0$ **9.** $\dfrac{1}{\sin^2 \alpha}$; $\sin \alpha \neq 0$, $\cos \alpha \neq 0$
11. $\dfrac{1}{\cos^2 x}$; $\sin x \neq 0$, $\cos x \neq 0$
13. $\dfrac{1}{\sin x}$; $\sin x \neq 0$ **15.** $\cos x$; $\cos x \neq 0$, $\sin x \neq 0$
17. $\cos x$; $\cos x \neq 0$
19. $-2 \cot x$; $\sin x \neq 0$
21. $\dfrac{1}{\cos^2 \alpha}$; $\cos \alpha \neq 0$ **23.** $\tan x$; $\sin x \neq 0$, $\cos x \neq 0$
25. $\dfrac{2}{\sin x}$; $\cos x \neq 0$, $\sin x \neq 0$

27. $2 \sin \alpha$; $\cos \alpha \neq 0$ **29.** $\dfrac{1}{\sin \alpha}$; $\sin \alpha \neq 0$

31. $\dfrac{2}{\cos x}$; $\cos x \neq 0$

33. $\dfrac{1}{-\sqrt{\tan^2 \alpha + 1}}$ **35.** $\dfrac{1}{\sqrt{r^2 - 1}}$

Self-Test 1, page 624

1. $\sec \alpha (\cos \alpha + \sin \alpha \tan \alpha)$

$$= \frac{1}{\cos \alpha}\left(\cos \alpha + \sin \alpha \, \frac{\sin \alpha}{\cos \alpha}\right)$$

$$= \frac{\cos \alpha}{\cos \alpha} + \frac{\sin^2 \alpha}{\cos^2 \alpha}$$

$$= \frac{\cos^2 \alpha + \sin^2 \alpha}{\cos^2 \alpha}$$

$$= \frac{1}{\cos^2 \alpha}$$

$$= \sec^2 \alpha; \cos \alpha \neq 0$$

2. $\dfrac{\cos x}{\sec x} + \dfrac{\sin x}{\csc x}$

$$= \frac{\cos x}{\dfrac{1}{\cos x}} + \frac{\sin x}{\dfrac{1}{\sin x}}$$

$$= \cos^2 x + \sin^2 x$$

$$= 1; \cos x \neq 0, \sin x \neq 0$$

3. $\dfrac{\sin^2 \alpha}{1 - \cos \alpha} = \dfrac{1 - \cos^2 \alpha}{1 - \cos \alpha}$

$$= \frac{(1 - \cos \alpha)(1 + \cos \alpha)}{1 - \cos \alpha}$$

$$= 1 + \cos \alpha; \cos \alpha \neq 1$$

4. $\sec \alpha \csc \alpha - \tan \alpha$

$$= \frac{1}{\cos \alpha} \cdot \frac{1}{\sin \alpha} - \frac{\sin \alpha}{\cos \alpha}$$

$$= \frac{1 - \sin^2 \alpha}{\cos \alpha \sin \alpha}$$

$$= \frac{\cos^2 \alpha}{\cos \alpha \sin \alpha}$$

$$= \frac{\cos \alpha}{\sin \alpha}$$

$$= \cot \alpha; \cos \alpha \neq 0, \sin \alpha \neq 0$$

Written Exercises, pages 630–631

1. $\cos \dfrac{2\pi}{3} = -\dfrac{1}{2}$ **3.** $\cos \dfrac{\pi}{3} = \dfrac{1}{2}$ **5.** $\cos 1.5\pi = 0$

7. $\cos 150° = -\dfrac{\sqrt{3}}{2}$ **9.** $\cos (60° - 45°) =$

$\dfrac{\sqrt{6} + \sqrt{2}}{4}$ **11.** $\cos (120° + 45°) = \dfrac{-\sqrt{6} - \sqrt{2}}{4}$

13. $\cos (300° + 45°) = \dfrac{\sqrt{2} + \sqrt{6}}{4}$

15. $\cos (45° - 60°) = \dfrac{\sqrt{2} + \sqrt{6}}{4}$

17. $\cos \left(\dfrac{\pi}{3} - \dfrac{\pi}{4}\right) = \dfrac{\sqrt{2} + \sqrt{6}}{4}$

19. $\cos \left(\dfrac{\pi}{4} + \dfrac{7\pi}{6}\right) = \dfrac{-\sqrt{6} + \sqrt{2}}{4}$

29. $\sin x = \dfrac{3}{5}$; $\sin y = \dfrac{5}{13}$; $\cos (x - y) = \dfrac{63}{65}$

31. $\sin x = -\dfrac{24}{25}$; $\sin y = \dfrac{3}{5}$; $\cos (x - y) = -\dfrac{44}{125}$

33. $\sin x = \dfrac{\sqrt{5}}{3}$; $\sin y = -\dfrac{1}{2}$; $\cos (x + y) =$

$\dfrac{2\sqrt{3} + \sqrt{5}}{6}$ **35.** $\sin x = \dfrac{1}{4}$; $\sin y = -\dfrac{2}{3}$;

$\cos (x + y) = \dfrac{-5\sqrt{3} + 2}{12}$

Written Exercises, pages 635–637

1. $\sin 60° = \dfrac{\sqrt{3}}{2}$ **3.** $\sin 330° = -\dfrac{1}{2}$

5. $\tan 120° = -\sqrt{3}$ **7.** $\sin 225° = -\dfrac{1}{\sqrt{2}}$

9. $\tan 210° = \dfrac{1}{\sqrt{3}}$ **11.** $\dfrac{\sqrt{6} - \sqrt{2}}{4}$ **13.** $\dfrac{\sqrt{6} - \sqrt{2}}{4}$

15. $\dfrac{-\sqrt{2} - \sqrt{6}}{4}$ **17.** $\dfrac{-\sqrt{2} - \sqrt{6}}{4}$ **19.** $2 + \sqrt{3}$

21. $2 - \sqrt{3}$ **23.** $-2 - \sqrt{3}$ **25.** $2 - \sqrt{3}$

39. $-\sqrt{2} - \sqrt{6}$ **41.** $2 - \sqrt{3}$ **43.** $-2 - \sqrt{3}$

45. $\sqrt{6} - \sqrt{2}$ **47.** $\dfrac{56}{65}$ **49.** $\dfrac{84}{85}$ **51.** $\dfrac{3}{5}$ **53.** $\dfrac{3\sqrt{15} + 4}{20}$

Written Exercises, pages 640–641

1. $\dfrac{24}{25}$ **3.** $-\dfrac{120}{169}$ **5.** $-\dfrac{12}{13}$ **7.** $-\dfrac{119}{169}$ **9.** $\dfrac{1}{8}$ **11.** $\dfrac{1}{9}$

13. $-4\sqrt{5}$ **15.** $\dfrac{\sqrt{2} - \sqrt{3}}{2}$ **17.** $\dfrac{\sqrt{2} + \sqrt{3}}{2}$

19. $-\sqrt{2} - 1$ **21.** $\dfrac{\sqrt{2} + \sqrt{2}}{2}$ **23.** $\sqrt{2} - 1$

25. $-\dfrac{\sqrt{2} - \sqrt{3}}{2}$ **27.** $\dfrac{1}{3}, \dfrac{2\sqrt{2}}{3}$ **29.** $\dfrac{3}{5}, \dfrac{4}{5}$ **31.** $\dfrac{\sqrt{15}}{4}, -\dfrac{1}{4}$

Computer Exercises, page 641

3. -0.5 **5.** -0.8660

Self-Test 2, page 644

1. $\tan 150° = -\dfrac{1}{\sqrt{3}}$ **2.** $\dfrac{\sqrt{6} + \sqrt{2}}{4}$ **3.** $\dfrac{-\sqrt{6} + \sqrt{2}}{4}$

4. $\dfrac{1}{9}$ **5.** $\dfrac{\sqrt{2} + \sqrt{2}}{2}$ **6.** $\csc 2\alpha = \dfrac{1}{\sin 2\alpha} =$

$\dfrac{1}{2 \sin \alpha \cos \alpha} = \dfrac{\csc \alpha}{2 \cos \alpha}$; $\cos \alpha, \sin \alpha \neq 0$

Written Exercises, page 646

1. 7.0 **3.** 3.5 **5.** 7.0 **7.** $60.0°$ **9.** $39.2°$
11. $c = 9$, $m(A) = 23°$, $m(B) = 98°$ **13.** $a = 9$,
$m(B) = 54°$, $m(C) = 66°$ **15.** $m(A) = 30°$,
$m(B) = 61°$, $m(C) = 89°$
17. In the 5, 7, 8 triangle:
$5^2 = 8^2 + 7^2 - 2(8)(7) \cos \alpha$, and $\alpha = 38°$.
In the 3, 5, 7 triangle:
$5^2 = 3^2 + 7^2 - 2(3)(7) \cos \beta$, and $\beta = 38°$.

Problems, page 647

1. 13.0 cm **3.** $48.2°$ **5.** 53.7 cm **7.** 28.4
9. 1.13×10^8 km

Written Exercises, pages 651–652
1. $b = 19.4$, $c = 23.9$, $m(C) = 96°$
3. $m(B) = 25.7°$, $m(C) = 34.3°$, $c = 26.0$
5. $m(C) = 38°$, $b = 244.3$, $c = 190.9$
7. $m(A) = 9.7°$, $b = 156.4$, $c = 194.3$
9. $m(A) = 24.7°$, $m(C) = 138.3°$,
$c = 79.6$, or $m(A) = 155.3°$, $m(C) = 7.7°$, $c = 16.0$
11. 18 **13.** 107 **15.** 151 and 932 **17.** $b \sin A$
19. $b \sin A = a \sin B$ and $\dfrac{\sin A}{a} = \dfrac{\sin B}{b}$

Problems, pages 652–653
1. 267.3 cm² **3.** 12.2 km **5.** 7.2 m **7.** 52.0 cm

Computer Exercises, page 653
3. No triangle with the given parts exists.

Self-Test 3, page 653
1. 6.9 **2.** 58.3 **3.** 91 square units **4.** 34.2 cm
5. 170.5 km

Chapter Review, pages 654–655
1. c **2.** a **3.** b **4.** d **5.** a **6.** a **7.** c **8.** b
9. d **10.** d **11.** b **12.** c

Application, page 657
1. a. 13 h **b.** 13.3 h **3.** 12.0

Preparing for College Entrance Exams, page 658
1. C **3.** E **5.** D **7.** A

Chapter 16 Inverses; Polar Coordinates; Vectors

Written Exercises, pages 666–667
1. $0° + k \cdot 360°$, $180° + k \cdot 360°$ **3.** $60° + k \cdot 360°$,
$300° + k \cdot 360°$ **5.** $0° + k \cdot 360°$, $180° + k \cdot 360°$
7. $\dfrac{\pi}{4} + 2k\pi$, $\dfrac{7\pi}{4} + 2k\pi$ **9.** $\dfrac{4\pi}{3} + 2k\pi$, $\dfrac{5\pi}{3} + 2k\pi$
11. $\dfrac{5\pi}{6} + 2k\pi$, $\dfrac{11\pi}{6} + 2k\pi$ **13.** $\dfrac{\pi}{3}$ or 60° **15.** $-\dfrac{\pi}{2}$
or $-90°$ **17.** $-\dfrac{\pi}{3}$ or $-60°$ **19.** 0.42 or 24.0°
21. -1.43 or $-81.8°$ **23.** 0.16 or 9.0° **25.** $\dfrac{\pi}{3}$
27. 180° **29.** $-\dfrac{\pi}{4}$ **31.** $\dfrac{2\pi}{3}$ **33.** $-\dfrac{\pi}{6}$
35. $\dfrac{5}{13}$ **37.** $\dfrac{15}{17}$ **39.** $\sqrt{15}$ **41.** $\dfrac{1}{5}$
43. $\dfrac{4 + 3\sqrt{3}}{10}$ **45.** $\dfrac{63}{65}$ **47.** 1 **49. a.** Yes

Written Exercises, pages 669–670
1. a. $\{60° + k \cdot 360°, 120° + k \cdot 360°\}$
b. $\{60°, 120°\}$ **3. a.** $\{135° + k \cdot 360°,$
$225° + k \cdot 360°\}$ **b.** $\{135°, 225°\}$
5. a. $\{120° + k \cdot 360°, 240° + k \cdot 360°\}$
b. $\{120°, 240°\}$ **7. a.** $\{30° + k \cdot 180°,$
$150° + k \cdot 180°\}$ **b.** $\{30°, 150°, 210°, 330°\}$
9. a. $\left\{\dfrac{3\pi}{4} + k\pi\right\}$ **b.** $\left\{\dfrac{3\pi}{4}, \dfrac{7\pi}{4}\right\}$

11. a. $\left\{k\pi, \dfrac{\pi}{3} + 2k\pi, \dfrac{5\pi}{3} + 2k\pi\right\}$ **b.** $\left\{0, \dfrac{\pi}{3}, \pi, \dfrac{5\pi}{3}\right\}$
13. a. $\left\{\dfrac{\pi}{2} + k\pi\right\}$ **b.** $\left\{\dfrac{\pi}{2}, \dfrac{3\pi}{2}\right\}$
15. $\{30°, 150°, 270°\}$ **17.** $\{60°, 300°\}$
19. $\{0°, 45°, 90°, 180°, 225°, 270°\}$
23. $\left\{\dfrac{\pi}{6}, \dfrac{5\pi}{6}, \dfrac{7\pi}{6}, \dfrac{11\pi}{6}\right\}$ **25.** $\left\{\dfrac{\pi}{4}, \dfrac{3\pi}{4}, \dfrac{5\pi}{4}, \dfrac{7\pi}{4}\right\}$
31. $\{45°, 225°\}$ **33.** $\left\{\dfrac{\pi}{6}, \dfrac{\pi}{2}, \dfrac{3\pi}{2}, \dfrac{11\pi}{6}\right\}$
35. $\{90°\}$ **37.** $\{30°, 120°, 210°, 300°\}$

Self-Test 1, page 670
1. $\dfrac{5\pi}{4} + 2k\pi$, $\dfrac{7\pi}{4} + 2k\pi$ **2.** 0.52 **3.** $-60°$
4. $\left\{\dfrac{\pi}{2} + 2k\pi, \dfrac{\pi}{3} + 2k\pi, \dfrac{3\pi}{2} + 2k\pi, \dfrac{5\pi}{3} + 2k\pi\right\}$
5. $\{60°, 120°\}$

Written Exercises, pages 675–676
1. $(6\sqrt{2}, 45°)$ or $(-6\sqrt{2}, -135°)$
3. $(-8, 45°)$ or $(8, -135°)$ **5.** $(5, 135°)$ or $(-5, -45°)$
7. $(-5, 36.9°)$ or $(5, -143.1°)$ **9.** $(3, -3)$
11. $(-\sqrt{3}, -1)$ **13.** $\left(-\dfrac{\sqrt{3}}{2}, -\dfrac{1}{2}\right)$
15. $\left(-\dfrac{\sqrt{3}}{2}, \dfrac{3}{2}\right)$ **17.** $(2, 0)$ **19.** $r = -2 \csc \theta$
21. $r^2 = 9$ **23.** $r^2 - 8r \cos \theta = 0$
25. b. $x^2 + y^2 = 9$ **27. b.** $x^2 + y^2 = 6x$

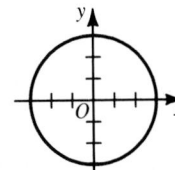 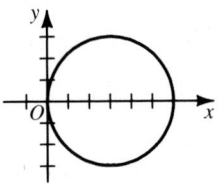

29. b. $x = -4$ **31. b.** $(x^2 + y^2 - x)^2 =$
$x^2 + y^2$

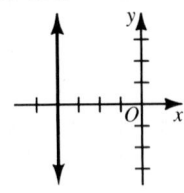 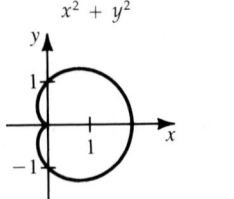

33. b. $(x^2 + y^2)^3 = 4x^2y^2$

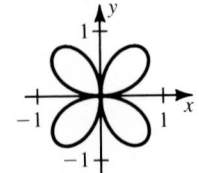

35. b. $(x^2 + y^2 + 2y)^2 = x^2 + y^2$ **37. b.** $(x^2 + y^2)^3 = 4y^2$

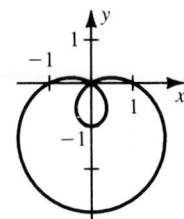
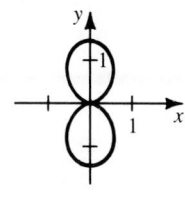

Written Exercises, pages 678–679

1. $3\sqrt{2}(\cos 45° + i \sin 45°)$ **3.** $7(\cos 180° + i \sin 180°)$

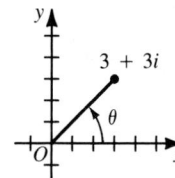
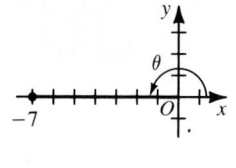

5. $4(\cos 315° + i \sin 315°)$
7. $6\sqrt{2}(\cos 135° + i \sin 135°)$
9. $-4 + 4i$ **11.** $-1 + i\sqrt{3}$
13. $-3 - 3i\sqrt{3}$

15. $\sqrt{3} - i\sqrt{3}$ **17. a.** $10 + 0i$ **b.** $-\dfrac{1}{5} - \dfrac{\sqrt{3}}{5}i$

19. a. $-12 - 12i\sqrt{3}$ **b.** $\dfrac{3\sqrt{3}}{4} + \dfrac{3}{4}i$

21. a. $-3 + 3i\sqrt{3}$ **b.** $0 + \dfrac{3}{8}i$

23. a. $\dfrac{\sqrt{2} - \sqrt{6}}{4} + \left(\dfrac{\sqrt{2} + \sqrt{6}}{4}\right)i$

b. $1(\cos 105° + i \sin 105°)$ **c.** $0.9659; -0.2588$

Computer Exercises, pages 679–680
3. $4\sqrt{2}; 135°$ **5.** $7; 270°$ **7.** $6; 0°$ **9.** $11 + 17i$
11. $2 - 14i$

Written Exercises, pages 683–684

1. $0 + i$ **3.** $-4 + 0i$ **5.** $-16\sqrt{3} - 16i$

7. $\dfrac{1}{2} + \dfrac{\sqrt{3}}{2}i$ **9.** $-\dfrac{1}{4} + \dfrac{1}{4}i$ **11.** $-\dfrac{1}{32} + \dfrac{\sqrt{3}}{32}i$

13. $-\dfrac{\sqrt{3}}{2} + \dfrac{1}{2}i$ **15.** $1, -\dfrac{1}{2} + \dfrac{\sqrt{3}}{2}i, -\dfrac{1}{2} - \dfrac{\sqrt{3}}{2}i$

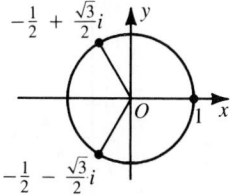

17. $\sqrt{2} + i\sqrt{2}, -\sqrt{2} + i\sqrt{2}, -\sqrt{2} - i\sqrt{2}, \sqrt{2} - i\sqrt{2}$

19. $2 + 2i\sqrt{3}, -2\sqrt{3} + 2i, -2 - 2i\sqrt{3}, 2\sqrt{3} - 2i$

21. $z = r(\cos \theta + i \sin \theta)$ and $\dfrac{1}{z} = z^{-1}$;
$z^{-1} = r^{-1}[\cos(-1 \cdot \theta) + i \sin(-1 \cdot \theta)] = \dfrac{1}{r}[\cos(-\theta) + i \sin(-\theta)]$.

Self-Test 2, page 684
1. $(4, 120°), (-4, -60°)$ **2.** $(\sqrt{2}, -\sqrt{2})$
3. a. $6 + 0i$ **b.** $\dfrac{3}{4} - i\dfrac{3\sqrt{3}}{4}$ **c.** $0 - 27i$
4. $\sqrt[6]{2}(\cos 15° + i \sin 15°)$,
$\sqrt[6]{2}(\cos 135° + i \sin 135°), \sqrt[6]{2}(\cos 255° + i \sin 255°)$

Reading Algebra, page 685
1. Nine times a, minus two; nine times the quantity a minus two.

Written Exercises, page 690
1. 7 **3.** 19 **5.** $5\sqrt{2}$ **7.** $4\sqrt{5}$ **9.** 53.1° **11.** 122.5°
13. Example: $\dfrac{4}{5}\mathbf{i} - \dfrac{3}{5}\mathbf{j}$ **15.** Example: $\dfrac{\sqrt{2}}{2}\mathbf{i} + \dfrac{\sqrt{2}}{2}\mathbf{j}$
17. 38.0, 77° **19.** 24.6, 97° **21.** 16.1, 16°
23. 33.0, 59° **25.** 28.4, 108° **27.** 19.9, 149°
29. 12.6; 83° **31.** 16.1, −148°

Problems, pages 693–694
1. 126 km/h; bearing 198.4° **3.** 3.23×10^4 J;
8.97×10^{-3} kW·h **5.** 681 mi; bearing 98.4°
7. 544 km/h; heading 87.4° **9.** 2.12×10^7 J;
5.89 kW·h **11.** 282 km/h; bearing 13.0°
13. 131 mi; bearing 350.9°

Self-Test 3, pages 694–695
1. 12.2; 80° **2.** 120.5° **3.** 114 km; 239.5°

Chapter Review, page 696
1. c **2.** c **3.** c **4.** b **5.** a **6.** b **7.** c **8.** b
9. b **10.** c **11.** a

Application, page 699
1. $(300\sqrt{3}, 300)$

Cumulative Review, pages 699–701
1. $\begin{bmatrix} 8 & 0 \\ 0 & -3 \end{bmatrix}$ **3.** $\begin{bmatrix} 1 & 0 \\ 0 & 1 \end{bmatrix}$ **5.** $\begin{bmatrix} 2 & 8 \\ 4 & 1 \end{bmatrix}$ **7.** $\begin{bmatrix} 1 & 3 \\ 0 & 0 \end{bmatrix}$
9. $x = 1, y = -1$ **11.** $(-2, 3)$ **13.** $-144°$
15. $22.6° + k \cdot 360°$ **17.** 0.4067 **19.** $\sin \alpha = -\dfrac{2\sqrt{2}}{3}, \cos \alpha = -\dfrac{1}{3}, \tan \alpha = 2\sqrt{2}, \csc \alpha = -\dfrac{3}{2\sqrt{2}},$
$\cot \alpha = \dfrac{1}{2\sqrt{2}}$ **21.** 29.9 **23.** $\dfrac{\sin^3 x}{\cos x}, \cos x \neq 0$
25. $\cos 150°, -\dfrac{\sqrt{3}}{2}$ **27.** $\dfrac{1}{2}\sqrt{2 + \sqrt{2}}$ **29.** 28.3
31. $-\dfrac{\pi}{4}$ **33.** $\left\{\dfrac{\pi}{3}, \dfrac{2\pi}{3}, \dfrac{4\pi}{3}, \dfrac{5\pi}{3}\right\}$ **35.** $\left(\dfrac{3\sqrt{3}}{2}, -\dfrac{3}{2}\right)$
37. $-4 - 4i$

Written Exercises, pages 708–710

1. 83.6; 82; 82 **3.** 7; 8; no mode **5.** 70; 70; 60

7.

Score	70	76	79	82	88	92	93
Frequency	1	1	1	3	1	2	1

9.
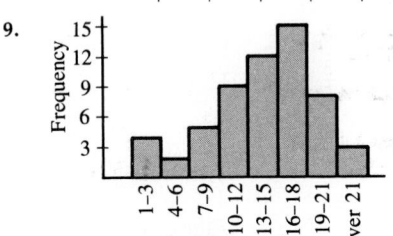

11.
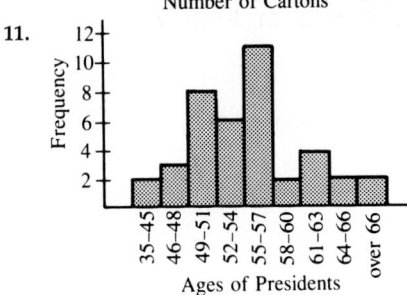

13.
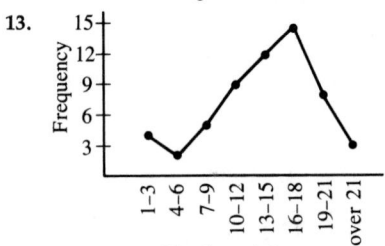

15.
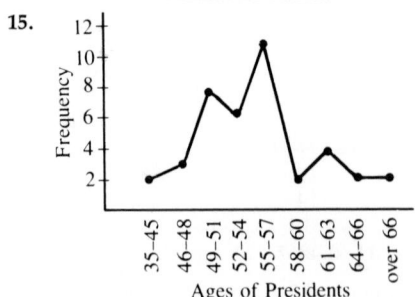

17. 13.9 **19.** 54.8 **21.** 79.3 **23.** 94 **25.** 55

Written Exercises, pages 714–715

1. 35.89; 5.99 **3.** 6; 2.45 **5.** 12.67; 3.56 **7.** 0.2
9. −1 **11.** 63 **13.** 60

15. Using this expression, the value computed for both data sets A and B is always zero. **17.** law
19. law **21.** mean: increased by 5; standard deviation: no change **23.** mean: no change; standard deviation: no change

Self-Test 1, page 716

1.

Inches	8–10	11–13	14–16	17–19	20–22	23–25	over 25
Frequency	4	5	4	5	3	8	1

2.
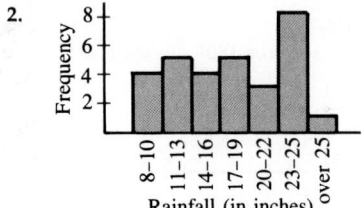

3.
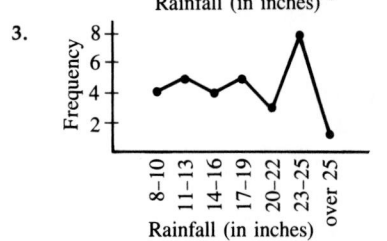

4. 41.75; 31.5; 37 **5.** 1449.02; 38.07
6. School 1: 1.17; School 2: 0.69; School 1

Written Exercises, pages 724–725

1. a. 67.3% **b.** 34.46% **c.** 21.19% **d.** 32.7%
3. a. 0.0668 **b.** 0.1587 **c.** 0.9710 **5.** 1.4% **7.** 83
9. 1.55 yr

Written Exercises, pages 730–731

1. $\sigma_{\bar{p}} = 0.04$; 95% confidence interval is 0.62 to
0.78; 0.8 > 0.78 **3.** $0.388 < p < 0.592$
5. $0 < p < 0.5$ **7.** $0.313 < p < 0.437$

Self-Test 2, page 731

1. 77.45% **2.** 0.0808 **3.** $0 < p < 0.08$

Written Exercises, pages 738–739

1.   **3.**
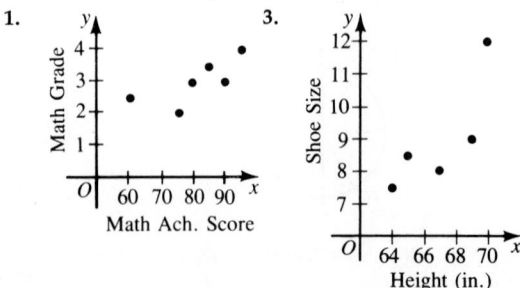

7. 0.74 **9.** 0.81 **11.** −0.92

13. a. $\overline{x} = \dfrac{a + b}{2}$; $\overline{y} = \dfrac{c + d}{2}$; $\sigma_x = \dfrac{|b - a|}{2}$;

$\sigma_y = \dfrac{|d - c|}{2}$

Self-Test 3, page 739

1.

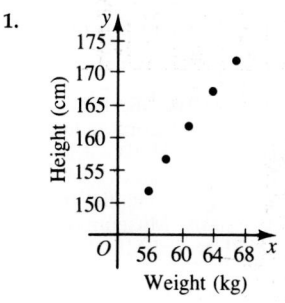

2.

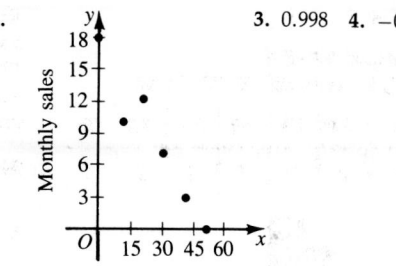

Average Temperature °F

3. 0.998 **4.** −0.96

Chapter Review, pages 741–742
1. b **2.** a **3.** a **4.** c **5.** b **6.** d **7.** c **8.** b **9.** a
10. b

Preparing for College Entrance Exams, page 743
1. B **3.** D **5.** A **7.** B **9.** D **11.** E **13.** B

Credits

Book designed by The Quarasan Group, Inc.
Cover concept and design by Lehman Millet Inc.
Mechanical art by ANCO
Portrait art created by Nathan Goldstein (pages 64, 171, 220, 315, 400, 457, 691, and 732)

Photos

page xiv: Bob Willis
page 32: © Melanie Wall / SOUTHERN LIGHT
page 35: © Harry Wilks / STOCK BOSTON
page 36: David S. Strickler / THE PICTURE CUBE
page 40: "Untitled" sculpture by Alexander Calder, 1978, East Wing, National Gallery of Art. © Herb Snitzer Photograph / STOCK BOSTON
page 49: © Michal Heron
page 53: © Terry E. Eiler 1983 / FOLIO, Inc.
page 78: Robert Schoen
page 82: H. Armstrong Roberts / E. P. JONES
page 112: Laimute E. Druskis / TAURUS PHOTOS
page 117: Barry L. Runk / GRANT HEILMAN PHOTOGRAPHY
page 119: © Larry Lawfer 1982 / THE PICTURE CUBE
page 122: © Peter Menzel MCMLXXXI
page 137: © Mike Mazzaschi / STOCK BOSTON
page 150: © Robert Perron / PHOTO RESEARCHERS, INC.
page 160: David L. Ryan / THE BOSTON GLOBE
page 182: © Robert Kalman / THE IMAGE WORKS, INC.
page 192: © Joel Gordon 1984
page 193: © Ellis Herwig / THE PICTURE CUBE
page 198: Ira Kirschenbaum, MD / STOCK BOSTON
page 204: © Bill Bachman 1983 / PHOTO RESEARCHERS, INC.
page 233: © Ellis Herwig / THE PICTURE CUBE
page 237: Hugh Rogers / Monkmeyer Press Photo Service, Inc.
page 246: Runk/Schoenberger / GRANT HEILMAN PHOTOGRAPHY
page 266: © Michal Heron
page 272: © Michael Hayman / PHOTO RESEARCHERS, INC.
page 292: David Wade
page 318: © Peter Menzel MCMXXXIV / STOCK BOSTON
page 336: © Ginger Chih 1978 / PETER ARNOLD, INC.
page 340: © Herb Snitzer Photography / STOCK BOSTON

page 386: © Ellis Herwig / THE PICTURE CUBE
page 389: © Robert Houser 1981 / PHOTO RESEARCHERS, INC.
page 392: GRANT HEILMAN PHOTOGRAPHY
page 421: Courtesy of Western Union Corporation
page 429: Steven Gerard / GERARD PHOTOGRAPHY
page 440: Section of a painted Egyptian coffin, from El Bersheh, XII Dynasty; Museum of Fine Arts, Boston; Harvard-MFA Expedition.
page 467: Omikron / TAURUS PHOTOS
page 476: The tomb of Pharoah Tutankhamun, Egyptian Dynasty XVIII; Interior of the Ante Chamber, the Hathor Couch. Photo by Harry Burton, The Metropolitan Museum of Art.
page 482: © Bohdan Hrynewych / SOUTHERN LIGHT
page 501: © The Photo Works / PHOTO RESEARCHERS, INC.
page 526: © Tim Carlson / STOCK BOSTON
page 557: © Russ Kinne 1981, National Science Foundation / PHOTO RESEARCHERS, INC.
page 571: © David M. Grossman / PHOTO RESEARCHERS, INC.
page 574: Courtesy of Hydro Quebec
page 610: © Ulrike Welsch
page 618: GRANT HEILMAN PHOTOGRAPHY
page 656: Read D. Brugger / THE PICTURE CUBE
page 660: © Frank Siteman MCMLXXXII / THE PICTURE CUBE
page 671: © Frank Fisher 1981 / LIAISON
page 698: Hans Pfletschinger / PETER ARNOLD, INC.
page 702: Courtesy of Inland Steel Company, Indiana Harbor Works / American Iron and Steel Institute
page 725: © Freda Leinwand / Monkmeyer Press Photo Service, Inc.
page 730: © Walter S. Silver / THE PICTURE CUBE